IRVING REINER

Selected Works

Irving Reiner

Selected Works

Edited by
Gerald J. Janusz

UNIVERSITY OF ILLINOIS PRESS
Urbana and Chicago

Publication of this work was supported in part by a grant
from the George A. Miller Endowment.

Manufactured in the United States of America
C 5 4 3 2 1

This book is printed on acid-free paper.

Library of Congress Cataloging-in-Publication Data
Reiner, Irving.
[Essays. Selections]
Selected works / Irving Reiner : edited by Gerald J. Janusz.
p. cm.
Bibliography: p.
ISBN 0-252-01611-4 (alk. paper)
1. Linear algebraic groups. 2. Integral representations.
3. Reiner, Irving—Bibliography. I. Janusz, Gerald J. II. Title.
QA171.R42 1989
512'.22—dc19 88-20894
CIP

Contents

---

*Numbers refer to bibliographic listing at the end of the biographical article.

Acknowledgments

Grateful acknowledgment is made to these publishers for permission to reprint the following articles:

Academic Press, Inc., publisher of **The Journal of Algebra**, for [45]* *Module extensions and blocks* (Copyright ©1967 by Academic Press, Inc.); [50] *Relative Grothendieck groups* (Copyright ©1969 by Academic Press, Inc.); [57] *A Mayer-Vietoris sequence for class groups* (Copyright ©1974 by Academic Press, Inc.); [68] *On Diederichsen's formula for extensions of lattices* (Copyright ©1979 by Academic Press, Inc.).

The American Mathematical Society, publisher of **Bulletin of the American Mathematical Society**, for [5S] *A survey of integral representation theory.*

The American Mathematical Society, publisher of **Proceedings of the American Mathematical Society**, for [10] *Real linear characters of the symplectic modular group*; [15] *A theorem on continued fractions*; [17] *Integral representations of cyclic groups of prime order*; [19] *Automorphisms of the two-dimensional general linear group over a euclidean ring* ; [24] *The nonuniqueness of irreducible constituents of integral group representations*; [48] *Finite generation of Grothendieck rings relative to cyclic subgroups.*

The American Mathematical Society, publisher of **The Transactions of the American Mathematical Society**, for [3] *A generalization of Meyer's theorem*; [4] *On the generators of the symplectic modular group*; [5] *Automorphisms of the unimodular group*; [6] *Automorphisms of the projective unimodular group*; [8] *Maximal sets of involutions*; [9] *Automorphisms of the symplectic modular group*; [21] *Inclusion theorems for congruence subgroups*; [41] *Integral representation algebras*; [49] *Reduction theorems for relative Grothendieck rings*; [56] *Class groups of integral group rings.*

Canadian Mathematical Society, publisher of **The Canadian Journal of Mathematics**, for [12] *Maschke modules over Dedekind rings*; [22] *On the class number of representations of an order*; [67] *Lifting isomorphisms of modules.*

Marcel Dekker, Inc., publisher of **Communications in Algebra**, for [70] *Zeta functions of integral representations.*

Oxford University Press, publisher of **Proceedings of the London Mathematical Society**, for [59] *Class groups and Picard groups of orders.*

Oxford University Press, publisher of **The Quarterly Journal of Mathematics**, for [76] *Left-vs-right zeta-functions.*

*Numbers refer to bibliographic listing at the end of the opening biographical article.

The Pacific Journal of Mathematics for [65] *Invariants of integral representations.*

Princeton University, publisher of **The Annals of Mathematics**, for [16] *A new type of automorphism of the general linear group over a ring*; [33] *Representations of cyclic groups in rings of integers, I*; [35] *Representations of cyclic groups in rings of integers, II.*

Springer–Verlag, Inc., publisher of **Mathematische Zeitschrift**, for [52] *An excision theorem for Grothendieck rings*; [69] *Zeta functions of arithmetic orders and Solomon's conjectures*; [75] *The prime ideal theorem in non-commutative arithmetic*; [80] *Zeta functions and composition factors for arithmetic orders.*

University College London, publisher of **Mathematika**, for [55] *Class groups for integral representations of metacyclic groups.*

University of Illinois, publisher of **The Illinois Journal of Mathematics**, for *Irving Reiner 1924–1986*; [23] *Behavior of integral group representations under ground ring extension*; [25] *Equivalence of representations under extension of local ground rings*; [26] *Indecomposable representations*; [30] *On the number of matrices with given characteristic polynomial*; [38] *Grothendieck groups of integral group rings.*

University of Michigan, publisher of **The Michigan Mathematical Journal**, for [31] *Failure of the Krull-Schmidt theorem for integral representations*; [39] *The integral representation ring of a finite group.*

Walter De Gruyter & Company, publisher of **Journal für die reine und angewandte Mathematik**, for [71] *L-functions of arithmetic orders, and asymptotic distribution of ideals*; [72] *Functional equations for L-functions of arithmetic orders*; [77] *Analytic continuation of partial zeta functions of arithmetic orders*; [78] *Functional equations for Hurwitz series and partial zeta functions of orders*; [79] *New asymptotic formulas for the distribution of left ideals of orders.*

Editor's Note

The papers of Irving Reiner reproduced here represent a part of the life's work of a dedicated and influential mathematician. His early papers were concerned with groups of invertible matrices over integral domains and their automorphisms. These studies led naturally to the study of integral representations of groups and orders with attempts to classify their representations using Grothendieck rings and their relative versions, Picard groups, and class groups. His later papers dealing with zeta-functions bring quantitative methods and new ideas to the representation theory. The influence of Reiner's work is most strongly felt by research mathematicians interested in problems related to integral representation theory. Recent applications of this theory to other areas of mathematics and to other branches of science have made it desirable to produce this volume as a convenient source for those wishing to read the original material.

A complete list of Reiner's publications can be found at the end of the biographical article. There are separate listings for his research papers, books, and survey articles and other miscellaneous works.

While space limitations did not permit the inclusion of all of Reiner's papers in this volume, forty-eight papers are reproduced in their entirety, along with abstracts of fourteen others. In preparing the abstracts, I have tried to make them sufficiently informative that interested readers can decide whether or not to seek out the original paper for their needs or interests. The research papers not reproduced here were, in some cases, early versions of work published later in more complete form and, in other cases, papers of an expository nature which contained alternate proofs or alternate approaches to known results.

The preparation of this volume was greatly aided by the cooperation of several people. The enthusiasm of Judith McCulloh, Executive Editor of the University of Illinois Press, for this project was greatly appreciated. I wish to thank Irma Reiner for her assistance and for giving me access to Irving's files. Thanks also go to the George A. Miller Endowment at the University of Illinois for its contribution which helped make publication of this volume possible.

IRVING REINER

1924–1986

Biography

Irving Reiner was born February 8, 1924 in Brooklyn, New York. He attended elementary and high school there and began his undergraduate work at Brooklyn College in 1940 with majors in mathematics and physics. His talent for mathematics was demonstrated early when he published his first mathematical paper [1], which appeared in his third year as an undergraduate. In the summer of 1943, he studied at the School of Advanced Instruction and Research in Mechanics at Brown University. It was there that he first met Alex Heller, a fellow student, with whom he later did joint work. He earned his B.A. magna cum laude from Brooklyn College in 1944 and entered the graduate school at Cornell University. His Master of Arts degree was awarded in February of 1945. His master's thesis, written under the direction of Burton W. Jones, was published in 1945 [2]; Jones also directed Reiner's doctoral work; his Ph.D. degree was awarded in June, 1947, and his thesis was published in 1949 [3].

While a graduate student at Cornell, Irving met Irma Moses, also a graduate student of Burton Jones. Irving and Irma were married in August of 1948 and they both came to the University of Illinois that year.

After completing his thesis, Irving Reiner went to the Institute for Advanced Study in Princeton where he met L. K. Hua. They worked together that year and agreed that they would seek employment at the same university in order to continue their collaboration. Both accepted offers from the University of Illinois. As it turned out, this was the only job offer Irv ever accepted. During his thirty-eight years at the University of Illinois, he wrote one hundred seven research papers, survey papers, books, and other scholarly works, many with co-authors. A discussion of his research is given below. He directed the thesis work for seventeen doctoral students. During his career, he received many awards and other recognition for his outstanding work, among these a Guggenheim Fellowship (1962), the Distinguished Alumnus Award from Brooklyn College (1963), and a NATO Senior Fellowship (1977). He received research support every year after 1953 either from the Office of Naval Research or the National Science Foundation. At various times he held visiting appointments at Queen Mary College and King's College of the University of London, the University of Paris, and the University of Warwick, Coventry. He made short term visits to and gave invited addresses at many universities around the world. He was a principal lecturer at mathematical conferences at Carleton College and the University of Sao Paulo, chairman of the organizing committee for the AMS Symposium on Representation Theory of Finite Groups and Related Topics held in 1970 in honor of Professor R. Brauer (resulting in [6E]), co-ordinator of the special year in algebra and algebraic number theory at the University of Illinois in 1981–82, and co-coordinator of the conference on Orders and their Applications held at Oberwolfach in 1984. From 1967 to 1972 he was National Counselor of the honorary mathematics fraternity, Pi Mu Epsilon. He served in various editorial capacities for the Proceedings of the American Mathematical Society, Contemporary Mathematics, and the Illinois Journal of Mathematics.

This rather impersonal recitation of activities and accomplishments indicates the high level of mathematical achievement reached by Irving Reiner. This level was not reached in a vacuum; it was the result of a life dedicated to mathematics. He gave encouragement to everyone in whom he saw some mathematical talent and, in return, he was stimulated by the success of other mathematicians with whom he had contact. Students, who might at first have felt intimidated by his reputation, were always quickly put at ease during their conversations with him about mathematics.

His Ph.D. students comment on his rather firm treatment of them; he was a strong advisor but not one who wrote the student's thesis. He met regularly with his students and required them to write up intermediate results. He would read these write-ups carefully and correct style as well as mathematics. He demanded the clear, precise writing from his students that could be found in all his own work. He treated his students with great respect, as, in fact, he treated everyone. He expected his doctoral students to attend seminars and participate in them on an equal basis with the faculty, to read the literature and to keep up on new developments. Many mathematicians, not only his former students, comment on the encouragement they received from him, especially as young scholars starting out. This was a reflection of his own enthusiasm for mathematical research, and his unselfish interest in having others share his excitement through their own succesess.

An inspection of the list of publications is enough to prove that Irving worked very hard. He would often work in streaks, putting in long hours on many consecutive days and nights. In order to refresh himself after such a period of intensive work, he would like to relax by attending a concert. He and Irma regularly attended the Krannert Center concert hall in Urbana or, for that matter, any concert hall where they happened to be (especially if there was a performance of a Beethoven quartet). During the summers, the vacation of choice would be a couple of weeks in the mountains. Irving loved the long hikes in the Canadian and American Rocky mountains. He always carried his camera on these walks; his wonderful pictures of the mountains and wild flowers are treasured by his family.

Visitors to the Illinois mathematics department were often entertained at the Reiner home. During the 1960s and '70s, when the Reiners' sons David and Peter were still living at home, there was a ping-pong table set up in the sun porch of their home. Irv enjoyed playing a game or two, especially with young visitors to the department who thought themselves good players. Irv was not an aggressive player; he played in a relaxed, gentle way, sometimes carrying on a conversation with his opponent during the game. But somehow he managed to return virtually every ball hit to his side of the table—much to the frustration of the opponent who was concentrating intensely on every shot. Lee Rubel tells of losing such a game to Irv and then attempting to excuse his loss with a remark to the effect that he had not been feeling well earlier that day and was not up to his usual game. Irv was understanding but remarked "I don't recall ever winning a game from a well man."

Mathematical work

It is awesome to consider the record of Irving Reiner's published research. He worked for nearly forty years on some of the most difficult problems in representation theory. I will comment very briefly upon his papers and books, in roughly chronological order.

His first paper [1] appeared while Reiner was still an undergraduate. It is a well-written paper in which he proves certain integer valued functions must take on non-prime values at some positive integer. He obtains a corollary which says that the function $f(x) = 2^{2^x} + k$, k a fixed positive integer, must assume a non-prime value for some positive integer x. His second paper [2] was his master's thesis. In August of 1947 he submitted his doctoral thesis for publication [3]. It was concerned with the values assumed by binary quadratic forms. The proofs made use of Dirichlet characters, L-series, and product representations of zeta-functions. After completing his thesis, he set these analytic methods aside for over thirty years, little knowing that he would return to them in thirteen of his last fourteen papers.

After leaving Cornell with his thesis complete, he went to the Institute for Advanced Study for the year 1947–48. He met L. K. Hua there and began the first of many successful collaborations. Their joint work dealt with classical subgroups of $GL(n, \mathbf{Z})$, the invertible $n \times n$ matrices over the ring of integers. They solved problems dealing with minimal sets of generators by direct matrix computations. Their joint papers [4], [5], [6] were followed by several individual works [7], [8], [9], [10] and the joint paper [11] with J. D. Swift. The results in these papers included a determination of generators for the automorphism groups of $GL(n, \mathbf{Z})$, $PGL(n, \mathbf{Z})$, and $Sp(2n, \mathbf{Z})$. Reiner spent two years 1954–56 at the Institute for Advanced Study in Princeton. During the first year, he participated with Charles W. Curtis, Peter Roquette and others, in a seminar directed toward an understanding of Richard Brauer's still relatively new theory of modular representations of finite groups. He and Curtis produced a set of notes from the seminar and subsequently decided to work together on the book [1B], which would appear some seven years later. In September of 1955, Irving submitted his first paper [12] in representation theory. The main result gave a test to determine if a module is relatively projective or relatively injective.

In late 1955, he wrote a review [1R] of Dieudonne's book on the classical groups. This book probably inspired him to return to some of the problems he had considered earlier. His papers [15], [16] and the joint papers [14], [18], [19] with Joe Landin were concerned with finding generators for $\mathrm{Aut}(GL(n, R))$ where R is a suitably restricted domain. Definitive results, (especially in [16]) were obtained when R is a principal ideal domain, or a skewfield (with the restriction $n \geq 3$), $Z[i]$ and $K[x]$, with K a field and $n = 2$.

In [17], a complete list of invariants of finitely generated ZG lattices was given for the case when G has prime order. Diederichsen (1938) had found the indecomposable ZG lattices but this is not sufficient to determine the decomposable lattices since the Krull-Schmidt theorem does not hold for ZG modules. This paper is of great importance in the development of integral representation theory, as it contains the first complete classification of all finitely generated modules over an integral group ring in a situation where the Krull-Schmidt theorem does not hold.

In [23] and [25] some positive and negative results are given to the following question:

Let R be a Dedekind domain with quotient field K, K' an extension field of K and R' the integral closure of R in K. Suppose that G is a finite group, M and N are RG modules which become isomorphic under the ground ring extension; that is $R'M \cong R'N$ as $R'G$ modules. Does it follow that $M \cong N$?

Questions of this sort reveal some of the subtleties of integral representation theory because in the classical case, $R = K$, $R' = K'$, a well-known theorem of Noether and Deuring says that $K'M \cong K'N$ implies $M \cong N$.

The next four years in the chronology, early 1960 to the middle of 1964, were a period of truly incredible activity. He supervised the thesis work of his first five students, collaborated with Alex Heller on five publications, completed eight individual papers, and produced his part of the classic book [1B]. (Bill Ferguson reminds me that at this time, the teaching load was nine hours per semester.) The papers with Heller made advances on several fronts. In [26], they considered characteristic p representations of abelian groups and showed that the classification problem, in general, was at least as difficult as a well-known unsolved problem in matrix theory. The problem was solvable in small cases and they used it to classify certain representations of the elementary abelian group of type (p, p).

Another problem which was receiving considerable attention at that time was concerned with the classification of those group rings RG which had only a finite number of indecomposable, finitely generated, R-torsion free modules (or RG lattices for short). The number of such indecomposable lattices is denoted by $n(RG)$. In [33] and [35] they proved if G is a p-group and $n(ZG)$ is finite, then G is cyclic of order p or p^2. They did not prove a converse. Alan Troy, in his 1961 thesis written under Reiner's supervision, proved that $n(ZG)$ is finite if G is the cyclic group of order four. This corrected a statement to the contrary in Diederichsen (1938). Then Heller and Reiner improved their result about p-groups to show for any finite group G, if $n(ZG)$ is finite, then for every prime p, the p-Sylow subgroup of G is cyclic of order p or p^2. Finally in his 1962 thesis written under Reiner's direction, Alfredo Jones proved $n(ZG)$ is finite if and only if every Sylow subgroup of G is cyclic of cube free order. Two of Reiner's other students solved related problems: Joseph Oppenheim obtained the classification of indecomposable ZG lattices when G is cyclic of square-free order; Myrna Lee classified the indecomposable ZG lattices for the case G is dihedral of order $2p$, p an odd prime.

Reiner's first student, Lawrence Levy, had already started on a problem when Troy and the others were working on the finiteness of $n(ZG)$. However the strong influence of this period was not entirely lost on Levy. In his 1983 paper (*Journal of Algebra*), Levy classified all finitely generated ZG modules, not just lattices, for the case G is cyclic of square free order.

The book [1B] was completed while Charlie Curtis was at the University of Wisconsin. Irving was fond of telling friends that this was probably the only mathematics book written in a museum and a hotel lobby. Since Chicago was approximately half way between Madison and Urbana, Curtis and Reiner would make day trips there, about once a month, to discuss the progress of the book, meeting first at the Art Institute (which was between their train stations) where the day's work was planned during a stroll through the galleries. When it was time to work on their manuscripts (and later the proofs), they usually found a congenial place in some out-of-the-way corner of the lobby of the Palmer House to spread out their papers. However unconventional that may have been, they succeeded in producing a book which influenced a generation of algebraists. The book contained material on non-semisimple algebras such as quasi-Frobenius algebras, a comprehensive treatment of integral representations, a new account of the theory of induced modules based upon the work of G. Mackey, and an introduction to Brauer's theory of blocks, which had previously not been available in book form.

With the solution of the finite type problem for ZG complete, Reiner turned to the study of Grothendieck groups and K-theory and a study of the papers of Swan on projective modules over group rings. Then he and Heller began an investigation of the representation ring, $a(RG)$, of a group ring RG. As a group, $a(RG)$ is generated by the symbols $[M]$, one for each isomorphism class of RG lattices, subject to the relations induced by direct sums: $[M \oplus N] = [M] + [N]$. The multiplication is induced by the tensor product over R. The coefficient ring R was usually a field of characteristic p or a discrete valuation ring of characteristic zero with finite residue field of characteristic p. This was related to the representation algebra, $A(RG)$, introduced by James Green sometime earlier, by the isomorphism $A(RG) = \mathbf{C} \otimes a(RG)$. Green, Conlon and others had found many cases where $A(RG)$ was a semisimple ring; this property was implied by the condition that $a(RG)$ had no nonzero nilpotent elements. Reiner and his students proved many results concerning existence or nonexistence of nonzero nilpotent elements in $a(RG)$ under various assumptions. The study of $a(RG)$ was viewed as weaker than a classification of the RG modules, but a strong step in that direction. Almost all of [36]–[47] deal with these ideas. In [36] and [38], Heller and Reiner studied Grothendieck groups and Whitehead groups of orders in semisimple algebras. They introduced a "relative G_0" group using torsion modules and established a 4-term exact sequence linking the Whitehead group to the Grothendieck group via a relative G_0-group. This has since been known as the Heller-Reiner exact sequence. These papers were written in the early 1960s when algebraic K-theory was in the formative stages and certainly not widely popular. Their use of this sequence was the among the first uses of long exact sequences, a technique which has since become extensively used.

The visit of T. Y. Lam to the University of Illinois during the summer of 1967 proved to be the beginning of another successful collaboration. In the joint papers [48]–[54], they introduced and studied relative Grothendieck rings, $a(G, H)$. The coefficient ring R is now suppressed from the notation. This relative ring is a homomorphic image of $a(RG)$ with kernel generated by all differences $[X] - [M] - [N]$, where there is an exact sequence

$$0 \to M \to X \to N \to 0$$

of RG modules which is split exact when restricted to RH, H a subgroup of G. They considered the question of when $a(G, H)$ is a free abelian group and obtained relations between $a(G, H)$ and $a(K, H)$ for $H \subset K \subset G$.

In 1968, Reiner gave a survey talk on integral representation theory [3S] at the University of Kentucky; a slightly modified version was the subject of an Invited Hour Address given at the AMS meeting in Chicago that year. The notes from this talk were expanded into a major survey article [5S] which listed results, methods, ideas and problems in integral representation theory. The bibliography of 265 items contained every important paper up to that time. His files contained the reprints of all these papers, and many more, along with his handwritten notes made while reading the papers, including the work of many Russian mathematicians, for whom he had high regard. He was pleased with his short paper [47] which was published in Russian.

During the same year, he wrote a small book [3B] intended as an introduction to matrix theory for students in the calculus sequence. This publication, which is written in the same clear and well-organized style as his other work, stands out from the advanced level texts and many research papers to show his interest in undergraduate education. He took an active interest in curriculum development and was one of the early advocates of the use of computers in undergraduate education. This may come as a surprise to those who know that Irv never used a computer in any of his research. He was fascinated by various results obtained by machine computations. "I must learn more about that someday," he said.

When Stephen Ullom joined the University of Illinois faculty, he brought a general interest in algebraic number theory and class groups in particular. He and Reiner began working together on the class group of an order. The papers [55], [56], [57] and [59] examined ways of computing class groups, $Cl(\Lambda)$, the kernel groups, $D(\Lambda)$, and locally free class groups, $LFP(\Lambda)$, of an order in a semisimple $\mathbf{Q}$-algebra. They refined a technique, namely the application of a Mayer-Vietoris sequence to a pull-back diagram, and used it to determine $Cl(ZG)$ for certain metacyclic groups G, and for $G = S_n$ or A_n, $n \leq 5$. Reiner continued the study of this problem in [62] and obtained further results on $D(ZS_n)$ and $D(ZA_n)$ for general n.

About this time, he published *Maximal orders* [4B]. This book came into being as a result of a sequence of graduate courses. It had been his custom for many years to teach a sequence of three one-semester courses: ordinary representation theory, modular representation theory, integral representation theory. The text for these was, of course, Curtis and Reiner [1B]. In the Fall of 1968, he began this sequence, and, because of an unusually large number of attendees, he offered a fourth semester entitled Algebras and Their Arithmetic. The topic was maximal orders. He prepared careful notes for this course, had them typed, and passed them out to the class. The resulting set of notes eventually became the book. After its publication, he said, "This book wrote itself." Here Irv was mistaken—the book did not write itself. It contained a comprehensive account of the application of local methods to noncommutative arithmetic. This theme was first developed in a book by Deuring in 1935, and, with many new methods and results due to Reiner and others, became one of the main topics in the books [1B], [5B] and [6B] as well as his *Maximal orders* [4B]. This book once again demonstrated Irv's instinct to know just what material to include so that the book would be useful to students and researchers alike for many years to come.

In January of 1977, Irving completed work on [65]. This is a remarkable paper in which he uses all the techniques acquired in his years of experience to give a complete classification of ZG lattices for G a cyclic group of order p^2, p a prime. This completed work begun some seventeen years earlier on group rings with only a finite number of indecomposable lattices. But the failure of the Krull-Schmidt theorem in this case means that more than the indecomposable lattices must be classified. To each lattice, he attaches a list of invariants: a genus invariant, a pair of ideal classes from the fields of p and p^2 roots of unity, a unit in a certain finite ring, and a quadratic residue character mod p. He then proves this list completely characterizes the module up to isomorphism. He also gives formulas for the number of indecomposable lattices; the formulas can be evaluated in terms of p only if p is a regular prime. In other cases the orders of certain unit groups appear which are not known explicitly. The methods used in this paper require the explicit computation of certain Ext groups. He applied these results again in [68] to compute certain Ext groups over ZG when G is any cyclic p-group. In a joint paper [66], he and Ed Green applied diagram methods, which had been so successful in the study of finite dimensional algebras of finite type, to give a new proof of a theorem of Jacobinski which gives necessary and sufficient conditions for a commutative order to have finite type.

The advances made in representation theory during the 60's and 70's prompted Curtis and Reiner to begin working on a second book on the subject. Instead of a new edition of [1B], they decided upon a completely new book. As the outline began to take shape, they recognized that the amount of research done since the publication of their first book made this project considerably larger than their first; they planned for two volumes. Eventually they produced two 1500-page manuscripts [5B] and [6B]. The influence of these books has yet to be determined, but it seems clear that these will be the standard reference works in the subject for many years.

After the two short papers [67], [68] were completed, Reiner began the study of a preliminary version of Louis Solomon's paper (Advances in Math., 1977) in which Solomon defines a zeta-function for any **Z**-order in a semisimple **Q**-algebra. For such an order Λ, it is defined as

$$\zeta_\Lambda(s) = \sum_M (\Lambda : M)^{-s}$$

where M runs through all left ideals of Λ having finite index $(\Lambda : M)$. Solomon computed $\zeta_{ZG}(s)$ for G of prime order p. This was a case Irving knew very well so naturally it interested him. He did the computation a different way and extended it to include the case where G was cyclic of order p^2 [70]. During the time he was working on this paper, Colin Bushnell (Kings's College, London) also was working on the conjectures in Solomon's paper. While Reiner used mainly algebraic methods, Bushnell was using the method of Tate and Weil's reworking of it to prove Solomon's conjecture in a special case. He sent a draft of his work to Reiner. This proved to be a fortunate coincidence; it was just the step Reiner needed since he knew how to reduce the general case to this special case. After an exchange of correspondence, they proved Solomon's conjectures in [69]. This proved to be the beginning of what was to become the most prolific and successful collaboration of Irving's career. They wrote twelve papers and five conference reports/announcements dealing with various partial zeta-functions (the M ranges over a restricted set of left ideals) in both the local and global cases, obtained product formulas, functional equations for the zeta-functions and the closely related L-functions associated with characters of the class group of the order, and analytic continuation of these functions. Much of the earlier work on integral representations, involving methods such as completions and localizations, was qualitative in nature, and the problem of actually computing integral representations remained a difficult one, even for special classes of finite groups. The new joint work with Bushnell added precise quantitative information about ways of counting isomorphism classes in the general case, and has added a new dimension to the theory.

Some of us in the department who, after hearing Irv lecture in seminars for years on modules and class groups, were surprised to hear his equally polished lectures, delivered in the usual organized, thorough manner, on Haar measure, p-adic harmonic analysis, and zeta-functions. These lectures were given as if he had spent his entire career working with these objects. The joint work with Bushnell is regarded by some as his best work. As he approached age sixty, an age when many would be content to take it easy and let the youngsters work on the hard problems, Irv mastered new ideas and methods which led to some of his deepest and most important contributions to the mathematical literature.

Conclusion

Irving Reiner died quietly in his sleep on October 28, 1986, after a long fight against cancer. His strong will to continue working was a source of inspiration to all his friends. After being confined to his home during the last year, he continued to receive visitors, took calls, saw some students, wrote letters, saw to the last details of the publication of [79] and [80], did some reviews for the problem section of the *American Mathematical Monthly*, and worked on [6B]. An indominable courage enabled him to give the impression that his illness was just an annoying problem that was temporarily delaying his return to mathematical research. His determination not to be changed by circumstance enabled him to project such an image of strength that one forgot that he was so terribly handicapped. We miss him very much but we have been enriched by knowing him and are inspired by his example.

Acknowledgments. I want to thank a number of people who helped in the preparation of this article. Carole Appel, Ken Appel, Charles W. Curtis, Colin Bushnell and T. Y. Lam provided information, corrected errors in the early draft and gave encouragement. And special thanks to Irma Reiner who helped in so many ways.

Gerald J. Janusz

Students of Irving Reiner

Lawrence S. Levy, 1961
Alan Troy, 1961
Alfredo Jones, 1962
Joseph H. Oppenheim, 1962
Myrna H.P. Pike, 1962
Lena Pu, 1964
Klaus Roggenkamp
Donald L. Stancl, 1966
Marshall M. Fraser, 1967
Thomas A. Hannula, 1967
Thomas G. Ralley, 1967
Paul R. Wilson, 1967
Janice Rose Zemanek, 1969
William H. Gustafson, 1970
How Ngee Ng, 1974
Gerald H. Cliff, 1975
Alberto G. Raggi-Cardenas, 1984
Hsin-Fong Chen, 1986
Anupam Srivastav

Reiner directed much of the work of K. Roggenkamp, but was not the official thesis advisor since Roggenkamp did not receive his degree from the University of Illinois.

The thesis advisor for M. Fraser was John Eagon who was then an instructor at Illinois; Reiner served as formal advisor and had only a minor role in the direction of the work.

Reiner had agreed to direct the thesis work of A. Srivastav and had introduced Srivastav to the subject matter of his thesis; Stephen Ullom directed Srivastav's thesis work.

Bibliography of Irving Reiner

Research Papers

1. *Functions not formulas for primes*, Amer. Math. Monthly, vol. 50 (1943), pp. 619–621.
2. *On genera of binary quadratic forms*, Bull. Amer. Math. Soc., vol. 51 (1945), pp. 909–912.
3. *A generalization of Meyer's theorem*, Trans. Amer. Math. Soc., vol. 65 (1949), pp. 170–186.
4. *On the generators of the symplectic modular group* (with L.K. Hua), Trans. Amer. Math. Soc., vol. 65 (1949), pp. 415–426.
5. *Automorphisms of the unimodular group* (with L.K. Hua), Trans. Amer. Math. Soc., vol. 71 (1951), pp. 331–348.
6. *Automorphisms of the projective unimodular group* (with L.K. Hua), Trans. Amer. Math. Soc., vol. 72 (1952), pp. 467–473.
7. *Symplectic modular complements*, Trans. Amer. Math. Soc. 77 (1954), pp. 498–505.
8. *Maximal sets of involutions*, Trans. Amer. Math. Soc., vol. 79 (1955), pp. 459–476.
9. *Automorphisms of the symplectic modular group*, Trans. Amer. Math. Soc., vol. 80 (1955), pp. 35–50.
10. *Real linear characters of the symplectic modular group*, Proc. Amer. Math. Soc., vol. 6 (1955), pp. 987–990.
11. *Congruence subgroups of matrix groups* (with J.D. Swift), Pacific J. Math., vol. 6 (1956), pp. 529–540.
12. *Maschke modules over Dedekind rings*, Canad. J. Math., vol. 8 (1956), pp. 329–334.
13. *Unimodular complements*, Amer. Math. Monthly, vol. 63 (1956), pp. 246–247.

14. *Automorphisms of the general linear group over a principal ideal domain* (with J. Landin), Ann. of Math., vol. 65 (1957), pp. 519–526.
15. *A theorem on continued fractions*, Proc. Amer. Math. Soc., vol. 8 (1957), pp. 1111–1113.
16. *A new type of automorphism of the general linear group over a ring*, Ann. of Math., vol. 66 (1957), pp. 461–466.
17. *Integral representations of cyclic groups of prime order*, Proc. Amer. Math. Soc., vol. 8 (1957), pp. 142–146.
18. *Automorphisms of the Gaussian unimodular group* (with J. Landin), Trans. Amer. Math. Soc., vol. 87 (1958), pp. 76–89.
19. *Automorphisms of the two-dimensional general linear group over a euclidean ring* (with J. Landin), Proc. Amer. Math. Soc., vol. 9 (1958), pp. 209–216.
20. *Normal subgroups of the unimodular group*, Illinois J. Math., vol. 2 (1958), pp. 142–144.
21. *Inclusion theorems for congruence subgroups* (with M. Newman), Trans. Amer. Math. Soc., vol. 91 (1959), pp. 369–379.
22. *On the class number of representations of an order*, Canad. J. Math., vol. 11 (1959), pp. 660–672.
23. *The behavior of integral group representations under ground ring extension*, Illinois J. Math., vol. 4 (1960), pp. 640–651.
24. *The non-uniqueness of irreducible constituents of integral group representations*, Proc. Amer. Math. Soc., vol. 11 (1960), pp. 655–658.
25. *Equivalence of representations under extension of local ground rings* (with H. Zassenhaus), Illinois J. Math., vol. 5 (1961), pp. 409–411.
26. *Indecomposable representations* (with A. Heller), Illinois J. Math., vol. 5 (1961), pp. 314–323.
27. *Subgroups of the unimodular group*, Proc. Amer. Math. Soc., vol. 12 (1961), pp. 173–174.
28. *The Krull-Schmidt theorem for integral group representations*, Bull. Amer. Math. Soc., vol. 67 (1961), pp. 365–367.
29. *The Schur index in the theory of group representations*, Michigan Math. J., vol. 8 (1961), pp. 39–47.
30. *The number of matrices with given characteristic polynomial*, Illinois J. Math., vol. 5 (1961), pp. 324–329.
31. *Failure of the Krull-Schmidt theorem for integral representations*, Michigan Math. J., vol. 9 (1962), pp. 225–232.
32. *Indecomposable representations of non-cyclic groups*, Michigan Math. J., vol. 9 (1962), pp. 187–191.
33. *Representations of cyclic groups in rings of integers, I* (with A. Heller), Ann. of Math., vol. 76 (1962), pp. 73–92.
34. *Extensions of irreducible modules*, Michigan Math. J., vol. 10 (1963), pp. 273–276.
35. *Representations of cyclic groups in rings of integers, II* (with A. Heller), Ann. of Math., vol. 77 (1963), pp. 318–328.
36. *Grothendieck groups of orders in semisimple algebras* (with A. Heller), Trans. Amer. Math. Soc., vol. 112 (1964), pp. 344–355.
37. *On the number of irreducible modular representations of a finite group*, Proc. Amer. Math. Soc., vol. 15 (1964), pp. 810–812.
38. *Grothendieck groups of integral group rings* (with A. Heller), Illinois J. Math., vol. 9 (1965), pp. 349–360.
39. *The integral representation ring of a finite group*, Michigan Math. J., vol. 12 (1965), pp. 11–22.
40. *Completion of primitive matrices*, Amer. Math. Monthly, vol. 73 (1966), pp. 380–381.
41. *Integral representation algebras*, Trans. Amer. Math. Soc., vol. 124 (1966), pp. 111–121.
42. *Nilpotent elements in rings of integral representations*, Proc. Amer. Math. Soc., vol. 17 (1966), pp. 270–274.
43. *Relations between integral and modular representations*, Michigan Math. J., vol. 13 (1966), pp. 357–372.

44. *Modular representation algebras* (with T. Hannula and T. Ralley), Bull. Amer. Math. Soc., vol. 73 (1967), pp. 100–101.
45. *Module extensions and blocks*, J. Algebra, vol. 5 (1967), pp. 157–163.
46. *Representation rings*, Michigan Math. J., vol. 14 (1967), pp. 385–391.
47. *An involution on $K^0(ZG)$*, Mat. Zametki, vol. 3 (1968), pp. 523–527.
48. *Finite generation of Grothendieck rings relative to cyclic subgroups* (with T.Y. Lam), Proc. Amer. Math. Soc., vol. 23 (1969), pp. 481–489.
49. *Reduction theorems for relative Grothendieck rings* (with T.Y. Lam), Trans. Amer. Math. Soc., vol. 142 (1969), pp. 421–435.
50. *Relative Grothendieck groups* (with T.Y. Lam), J. Algebra, vol. 11 (1969), pp. 213–242.
51. "Relative Grothendieck groups" (with T.Y. Lam) in *Theory of groups* (eds. R. Brauer and C.H. Sah), Symposium at Harvard University, Benjamin, New York, 1969, pp. 163–170.
52. *An excision theorem for Grothendieck rings* (with T.Y. Lam), Math. Z., vol. 115 (1970), pp. 153–164.
53. *Restriction maps on relative Grothendieck rings* (with T.Y. Lam), J. Algebra, vol. 14 (1970), pp. 260–298.
54. *Restriction of representations over field of characteristic p* (with T.Y. Lam and D. Wigner), Proc. Symp. Pure Math., vol. 21, Amer. Math. Soc., Providence, R.I., 1971, pp. 99–106.
55. *Class groups for integral representations of metacyclic groups* (with S. Galovich and S. Ullom), Mathematika, vol. 19 (1972), pp. 105–111.
56. *Class groups of integral group rings* (with S. Ullom), Trans. Amer. Math. Soc., vol. 170 (1972), pp. 1–30.
57. *A Meyer-Vietoris sequence for class groups* (with S. Ullom), J. Algebra, vol. 31 (1974), pp. 305–342.
58. *Hereditary orders*, Rend. Sem. Mat. Univ. Padova, vol. 52 (1974), pp. 219–225.
59. *Picard groups and class groups of orders* (with A. Frohlich and S. Ullom), Proc. London Math. Soc. (3), vol. 29 (1974), pp. 405–434.
60. *Locally free class groups of orders*, Carleton Lecture Notes (Section 21), vol. 9 (1974), pp. 1–29; also in Proc. International Conference on Representations of Algebras, Ottawa, Lecture Notes in Mathematics, vol. 488, Springer Verlag, New York, 1974, pp. 253–281.
61. *A proof of the Normal Basis Theorem* (with T.R. Berger), Amer. Math. Monthly, vol. 82 (1975), pp. 915–918.
62. *Projective class groups of symmetric and alternating groups*, Linear and Multilinear Algebra, vol. 3 (1975), pp. 115–121.
63. *Integral representations of cyclic groups of order p^2*, Proc. Amer. Math. Soc., vol. 58 (1976), pp. 8–12 (Erratum, Proc. Amer. Math. Soc., vol. 63 (1977), p. 374).
64. *Class groups and Picard groups of integral group rings and orders*, Regional Conference Math., vol. 26, Amer. Math. Soc., Providence, R.I., 1976.
65. *Invariants of integral representations*, Pacific J. Math., vol. 78 (1978), pp. 467–501.
66. *Integral representations and diagrams* (with E.L. Green), Michigan Math. J., vol. 25 (1978), pp. 53–84.
67. *Lifting isomorphisms of modules*, Canad. J. Math., vol. 31 (1979), pp. 808–811.
68. *On Diederichsen's formula for extensions of lattices*, J. Algebra, vol. 58 (1979), pp. 238–246.
69. *Zeta functions of arithmetic orders and Solomon's conjectures* (with C.J. Bushnell), Math. Z., vol. 173 (1980), pp. 135–161.
70. *Zeta functions of integral representations*, Comm. Algebra, vol. 8 (1980), 911–925.
71. *L-functions of arithmetic orders, and asymptotic distribution of left ideals* (with C.J. Bushnell), J. Reine Angew. Math., vol. 327 (1981), 156–183.
72. *Functional equations for L-functions of arithmetic orders* (with C.J. Bushnell), J. Reine Angew. Math., vol. 329 (1981), pp. 88–124.

73. *Matrix completions over Dedekind rings* (with W.H. Gustafson and M.E. Moore), Linear and Multilinear Algebra, vol. 10 (1981), pp. 141–143.
74. *Zeta functions of hereditary orders and integral group rings* (with C.J. Bushnell), Texas Tech. Univ. Math. Series, vol. 14 (1981), pp. 71–94.
75. *The prime ideal theorem in non-commutative arithmetic* (with C.J. Bushnell), Math. Z., vol. 181 (1982), pp. 143–170.
76. *Left-vs-right zeta-functions* (with C.J. Bushnell), Quart. J. Math. Oxford (2), vol. 35 (1984), pp. 1–19.
77. *Analytic continuation of partial zeta functions of arithmetic orders* (with C.J. Bushnell), J. Reine Angew. Math., vol. 349 (1984), 160–178.
78. *Functional equations for Hurwitz series and partial zeta functions of orders* (with C.J. Bushnell), J. Reine Angew. Math., vol. 364 (1986), pp. 130–148.
79. *New asymptotic formulas for the distribution of left ideals of orders* (with C.J. Bushnell), J. Reine Angew. Math., vol. 364 (1986), pp. 149–170.
80. *Zeta functions and composition factors for arithmetic orders*, (with C. Bushnell), Math. Z., vol. 194 (1987), pp. 415–428.

Books

1B. *Representation theory of finite groups and associative* algebras (with C.W. Curtis), John Wiley and Sons, New York, N. Y., 1962.
2B. *Representation theory of finite groups and associative algebras* (with C.W. Curtis), Russian translation, Moscow, 1969.
3B. *Introduction to matrix theory and linear algebra*, Holt, Rinehart and Winston, New York, N. Y., 1971.
4B. *Maximal orders* (London Math. Soc. Monograph), Academic Press, New York, 1975.
5B. *Methods of representation theory (with applications to finite groups and orders), Volume I* (with C.W. Curtis), Wiley-Interscience, New York, N. Y., 1981.
6B. *Methods of representation theory (with applications to finite groups and orders), volume II* (with C.W. Curtis), Wiley-Interscience, New York, N. Y., 1987.

Research Announcements, Survey Articles, Reviews and Editorships

1R. *Review of "La Geometrie des Groupes Classiques", by J. Dieudonne* (Springer Verlag, New York, 1956), Bull. Amer. Math. Soc., vol. 62 (1956), pp. 417–420.
2A. *Indecomposable representations of cyclic groups* (with A. Heller), Bull. Amer. Math. Soc., vol. 68 (1962), pp. 210–212.
3S. *A survey of integral representation theory* in Proc. Symp. Algebra, University of Kentucky, Lexington, Kentucky, 1968, pp. 8–14.
4A. *Relative Grothendieck rings* (with T.Y. Lam), Bull. Amer. Math. Soc., vol. 75 (1969), pp. 496–498.
5S. *A survey of integral representation theory*, Bull. Amer. Math. Soc., vol. 76 (1970), pp. 159–227.
6E. *Representation theory of finite groups and related topics* (Editor), Proc. Symp. Pure Math., vol. 21, Amer. Math. Soc., Providence, R.I., 1971.
7A. *Class groups of orders and a Mayer-Vietoris sequence* (with S. Ullom), Proc. Ohio State Univ. Conference on Orders and Group Rings, Lecture Notes in Mathematics, vol. 353 Springer Verlag, New York, 1973, pp. 139–151.
8A. *Remarks on class groups of integral group rings* (with S. Ullom), Symp. Math. Ist. Nazionale Alta Mat. (Rome), vol. 13 (1974), pp. 501–516.

9A. *Indecomposable integral representations of cyclic p-groups*, Proceedings of Philadelphia Conference, 1976, Dekker Lecture Notes, vol. 37, Marcel Dekker, New York, 1977, pp. 425–445.

10A. *Integral representations of cyclic p-groups*, Proc. Canberra Conference, 1978, Lecture Notes in Mathematics, vol. 697, Springer Verlag, New York, 1978, pp. 70–87.

11A. *Integral representations of finite groups*, Proc. Sem. Dubreil, 1977, Lecture Notes in Mathematics, vol. 641, Springer Verlag, New York, 1978, pp. 145–162.

12A. *Integral representations: genus, K-theory, and class groups*, Proc. Canberra Conference, 1978, Lecture Notes in Mathematics, vol. 697, Springer Verlag, New York, 1978, pp. 52–69.

13S. "Topics in integral representation theory" in *Integral Representations*, Lecture Notes in Mathematics, vol. 744, Springer Verlag, New York, 1979, pp. 1–143.

14A. *Solomon's conjectures and the local functional equation for zeta functions of orders* (with C.J. Bushnell), Bull. Amer. Math. Soc., vol. 2 (1980), pp. 306–310.

15S. "An overview of integral representation theory" 269–300 in *Ring Theory and Algebra III*, Proc. Third Oklahoma Conference, Marcel Dekker Lecture Notes, vol. 55, Marcel Dekker, New York, 1980, pp. 269–300.

16A. "Zeta-functions of orders" (with C.J. Bushnell) in *Orders and their applications*, K. Roggenkamp, ed., Lecture Notes in Mathematics, vol. 882, Springer Verlag, New York, 1980, pp. 159–173.

17A. *L-functions of arithmetic orders* (with C.J. Bushnell), C. R. Math. Rep. Acad. Sci. Canada, vol. 3 (1981), pp. 13–18.

18E. *Orders and their applications* (editor with K. Roggenkamp), June, 1984, Lecture Notes in Mathematics, No. 1142, Springer-Verlag, New York, 1985.

19A. *Analytic methods in noncommutative number theory* (with C.J. Bushnell), Proc. ICRA 1984 Conference, Carleton Univ., Ottawa, 1986, pp. 6.01–6.10.

20S. "A survey of analytic methods in noncommutative number theory, (with C. Bushnell) in *Orders and their application*, Lecture Notes in Mathematics, No. 1142, Springer Verlag, New York, 1985, pp. 50–87.

IRVING REINER

Selected Works

FUNCTIONS NOT FORMULAS FOR PRIMES

Irving Reiner, Brooklyn, New York

ABSTRACT

A function $F(x)$ is said to be a formula for primes if $F(x)$ is a prime for every positive integral value of x. We show that certain types of functions cannot be formulas for primes.

Theorem 1. *If $f_i(x)$, $g_i(x)$ $(i = 1, \cdots, n)$ are polynomials with integral coefficients and positive leading coefficients, the following is not a formula for primes:*

$$F(x) = \sum_{i=1}^{n} f_i(x)^{g_i(x)}.$$

Theorem 2. *Let $F(x)$ be defined as*

$$F(x) = \sum_{i=1}^{n} f_i(x)^{[h_i(x)^{g_i(x)}]}$$

where $f_i(x), g_i(x), h_i(x)$ $(i = 1, \cdots, n)$ are polynomials with integral coefficients and positive leading coefficients. Then $F(x)$ is not a formula for primes provided there exists an integer a such that:

1. All the $g_i(a)$ and $h_i(a)$ are positive;

2. If any $g_i(a)$ has a factor r_i in common with $q = F(a) - 1 = p - 1$, then any prime divisor of r_i occurs as a factor of q to a power less than or equal to $h_i(a)$.

Example 1. $2^{2^x} + k$ is not a formula for primes for integral k.

ON THE GENERA OF BINARY QUADRATIC FORMS

Irving Reiner

ABSTRACT

Let $\beta = ax^2 + 2bxy + cy^2$ be a properly primitive form with integral coefficients, and let the determinant $D = ac - b^2$ be written as $D = \pm 2^s \Delta$, where Δ is odd and positive, and the factorization of Δ into distinct primes is $\Delta = q_1^{\alpha_1} \cdots q_r^{\alpha_r}$. Suppose a is positive and prime to $2D$. Then Gauss showed that the genus of β is then determined by the Legendre symbols $(a \mid q_1), \cdots, (a \mid q_r)$, and $(-1 \mid a)$ if $D \equiv 0 \text{ or } 1 \pmod 4$, $(2 \mid a)$ if $D \equiv 0 \text{ or } 6 \pmod 8$, and $(-2 \mid a)$ if $D \equiv 0 \text{ or } 2 \pmod 8$. The characters are not independent since it can be shown that

$$(2|a)^s(-1|a)^{(\mp\Delta-1)/2}(a|q_1)^{\alpha_1} \cdots (a|q_r)^{\alpha_r} = 1. \tag{1}$$

In this paper, new proofs, using Dirichlet's theorem on primes in arithmetic progression, are given for the following two results known to Gauss.

Theorem 1. *For any prescribed set of characters satisfying (1), there exists a genus with the given characters.*

Theorem 2. *All existing properly primitive genera of binary quadratic forms of a given determinant contain the same number of classes.*

A GENERALIZATION OF MEYER'S THEOREM[1]

BY
IRVING REINER

1. **Introduction.** This paper is concerned with a generalization of the following theorem:

Every properly primitive binary quadratic form represents infinitely many primes in any preassigned arithmetic progression $Mx+N$ consistent with the generic character of the form, where M and N are relatively prime.

Dirichlet [1][2] conjectured this result in 1840, and sketched a method of proof. Weber [4] gave a complete proof for the special case where the words "in any preassigned arithmetic progression $\cdots$" are omitted, and the theorem as stated was finally proved by Meyer [3] in 1888.

The generalization consists of replacing the set of classes of properly primitive binary quadratic forms of given determinant D, which forms a group under composition, by an abelian group H whose elements we denote by θ_i. A correspondence is assumed between H and the set G of numbers $m>0$ which are prime to M and of which D is a quadratic residue. In §§2 and 3, the structures of G and H are examined, and in each case a basis is set up from which all elements may be generated. §4 then gives the specific details of the correspondence between G and H; these details parallel results about the representation of numbers by quadratic forms. Conclusions are then drawn from the correspondence which connects certain of the characters mod M set up in §2 with the group characters constructed in §3.

In §5, the basic Dirichlet series is first defined as

$$L(\chi, \tau; s) = \sum_{\theta_i \text{ in } H} \sum_{m \text{ in } G} a_m(\theta_i)\chi(\theta_i)\tau(m)m^{-s},$$

where $\tau(m)$ is a character mod M, $\chi(\theta_i)$ a group character, $a_m(\theta_i)$ is determined by the correspondence, and where the double summation extends over all elements of G and H. A few of its properties are discussed here, and in §6 sufficient conditions are found for the identity of two such series.

The series are then divided into three different types in §7. Those of the first type become infinite as $s \to 1^+$, while those of the second and third types are finite and different from 0 for $s=1$. The proof of the latter part of this statement is rather involved, and is given in §§7 and 8 after some additional conditions are imposed, without which the theorem fails to be true.

Presented to the Society, September 3, 1947; received by the editors August 6, 1947.

(1) The material of this paper comes from a thesis written under the direction of Professor Burton W. Jones.

(2) Numbers in brackets refer to the bibliography at the end of the paper.

In §9, the theorem is finally stated and its proof is completed. §10 is concerned with the additional conditions referred to in the preceding paragraph, and is devoted to proving a sufficient condition under which they hold. After a brief discussion of Meyer's results in §11, we conclude with an example showing that the new theorem is more general than the original theorem.

We shall use the following notation:

(a, b) denotes the greatest common divisor of a and b.

(a/b) denotes the Jacobi symbol $(\frac{a}{b})$.

$\phi(n)$ is the Euler ϕ-function.

$a|b$ means a divides b.

$a\nmid b$ means a does not divide b.

2. **The set G and characters mod M.** Let G be a set of positive integers which is closed under multiplication, with a basis consisting of the set of all primes f satisfying

(a) $(f, M)=1$, where $M=2^s\prod_{i=1}^r p_i^{\alpha_i}$ and the p_i are distinct odd primes and where $s>3$.

(b) $(D/f)=+1$, where $D=2^\sigma\prod_{i=1}^k p_i^{\beta_i}$ with $k\leqq r$, $\sigma\leqq s$ and $\beta_i\leqq\alpha_i$ for $i=1, \cdots, k$. Let g_i be a primitive root mod p_i^2 for $i=1, \cdots, r$; then, for every m in G, there exist integers $\alpha, \beta, \gamma_1, \cdots, \gamma_r$ which are unique mod 2, 2^{s-2}, $\phi(p_1^{\alpha_1}), \cdots, \phi(p_r^{\alpha_r})$ respectively, such that

$$(1) \qquad m \equiv (-1)^\alpha 5^\beta \pmod{2^s}, \qquad m \equiv g_i^{\gamma_i} \pmod{p_i^{\alpha_i}} \qquad (i = 1, \cdots, r).$$

Let $\vartheta, \eta, \rho_1, \cdots, \rho_r$ be defined by

$$(2) \qquad \vartheta^2 = 1, \qquad \eta^{2^{s-2}} = 1, \qquad \rho_i^{\phi(p_i^{\alpha_i})} = 1 \qquad (i = 1, \cdots, r),$$

and set

$$(3) \qquad \tau(\vartheta, \eta; \rho_1, \cdots, \rho_r; m) = \tau(m) = \vartheta^\alpha\eta^\beta\prod_{i=1}^r \rho_i^{\gamma_i}.$$

We shall call the number-theoretic function $\tau(m)$ a *character* mod M. Two characters τ and τ' are said to be the same only if their corresponding $\vartheta, \eta, \rho_1, \cdots, \rho_r$'s are the same. It may well happen that $\tau(m)=\tau'(m)$ for all m in G without τ and τ' being the same character. The set of values which $\tau(m)$ assumes for a fixed m as τ varies over the complete set of distinct characters is called the total character of m. Two numbers have the same total character if and only if they are congruent mod M. In virtue of (2), the number of distinct characters is

$$2\cdot 2^{s-2}\cdot\prod_{i=1}^r \phi(p_i^{\alpha_i}) = \phi(M).$$

We define the *principal character* τ_0 to be $\tau(1, 1; 1, \cdots, 1; m)$. By an *ambiguous character* τ_a we shall mean $\tau(\pm 1, \pm 1; \pm 1, \cdots, \pm 1; m)$, where any combination of signs is permitted. All other characters are *imaginary.* The complex conjugate of $\tau(\vartheta, \eta; \rho_1, \cdots, \rho_r; m)$ is clearly $\tau(\vartheta^{-1}, \eta^{-1}; \rho_1^{-1}, \cdots, \rho_r^{-1}; m)$.

3. **The group H and group characters.** Let H be a multiplicative abelian group with generators $\theta_1, \cdots, \theta_\nu$; let n_i be the degree of θ_i, that is, the smallest positive exponent r for which θ_i^r is the identity element θ_0. Every θ in H can then be written as

$$\theta = \prod_{i=1}^{\nu} \theta_i^{s_i}, \tag{4}$$

where s_i is uniquely determined mod n_i $(i=1, \cdots, \nu)$. The number of elements in H is $h = \prod_{i=1}^{\nu} n_i$. Let λ be defined by

$$n_1, \cdots, n_{\lambda-1} \quad \text{are even,} \quad n_\lambda, \cdots, n_\nu \quad \text{are odd.} \tag{5}$$

Define ω_i by

$$\omega_i^{n_i} = 1 \qquad (i = 1, \cdots, \nu), \tag{6}$$

and define

$$\chi(\omega_1, \cdots, \omega_\nu; \theta) = \chi(\theta) = \prod_{i=1}^{\nu} \omega_i^{s_i}. \tag{7}$$

Similar definitions hold for the $\chi(\theta)$ as held for the $\tau(m)$ above. The number of ambiguous characters χ_a is then $2^{\lambda-1}$.

Definition. An element θ in H is *ambiguous* if $\theta^2 = \theta_0$. The number of ambiguous elements is then $2^{\lambda-1}$, and these are representable as

$$\prod_{i=1}^{\lambda-1} (\theta_i^{n_i/2})^{\epsilon_i} \tag{8}$$

where each ϵ_i may be 0 or 1.

Definition. A *genus* is the aggregate of all elements

$$\prod_{i=1}^{\lambda-1} \theta_i^{2s_i+\epsilon_i} \cdot \prod_{i=\lambda}^{\nu} \theta_i^{s_i} \tag{9}$$

obtained by letting $s_1, \cdots, s_{\lambda-1}, s_\lambda, \cdots, s_\nu$ range over complete residue systems mod $n_1/2, \cdots, n_{\lambda-1}/2, n_\lambda, \cdots, n_\nu$ respectively, where $\epsilon_1, \cdots, \epsilon_{\lambda-1}$ are all fixed and are 0 or 1. The number of genera is seen to be $2^{\lambda-1}$.

4. **The correspondence between G and H.** We assume that there is a correspondence between G and H such that

(a) Every element of G corresponds to at least one element of H, and conversely.

(b) If m in G corresponds to θ in H, then θ corresponds to m, and conversely.

(c) If m_1, m_2 in G correspond to θ_1, θ_2 in H, then m_1m_2 corresponds to $\theta_1\theta_2$.

(d) If m_1, m_2 in G correspond to θ_1, θ_2 in H, and if $m_2 \equiv n^2m_1 \pmod{D}$ for some integer n, then for all χ_a, $\chi_a(\theta_1) = \chi_a(\theta_2)$.

(e) If θ_1, θ_2 in H correspond to m_1, m_2 in G, and if for all χ_a, $\chi_a(\theta_1) = \chi_a(\theta_2)$, then there exists an integer n for which $m_2 \equiv n^2m_1 \pmod{D}$.

(f) For every prime f in G, there exist two elements θ_f and θ_f^{-1} in H, whose product is θ_0, to which f corresponds. (The possibility $\theta_f = \theta_f^{-1}$ is not excluded; indeed, this relation will hold for any ambiguous θ_f.)

It is clear that the total ambiguous character of m depends only on the quadratic character of $m \pmod{M}$, where by the total ambiguous character of m we mean the set of values which $\tau_a(m)$ assumes as we keep m fixed and let τ_a range over all ambiguous characters. Among these τ_a, there will be some whose value depends only on the quadratic character of $m \pmod{D}$; call the set of these characters Σ. By the Σ-character of m we shall mean the set of values of $\tau_a(m)$, where m is kept fixed and τ_a ranges over all characters in Σ.

If we denote by χ_i that ambiguous character obtained by setting $\omega_i = -1$ and $\omega_j = 1$ if $j \neq i$, then these χ_i (for $i = 1, \cdots, \lambda - 1$) are $\lambda - 1$ independent characters in the sense that $\chi_1(\theta), \cdots, \chi_{\lambda-1}(\theta)$ may be any sequence of $+1$'s and -1's (depending, of course, on the choice of θ), and indeed every such sequence can be obtained by choosing θ of the proper genus. In fact, if the ϵ's are as in (9), the sequence $\chi_1(\theta), \cdots, \chi_{\lambda-1}(\theta)$ is simply $(-1)^{\epsilon_1}, \cdots, (-1)^{\epsilon_{\lambda-1}}$. Thus, to specify the values of these $\lambda - 1$ characters is equivalent to fixing a unique genus. This in turn means that the quadratic character of $m \pmod{D}$ is determined, where m corresponds to some θ of the genus. Conversely, if we specify the values of

$$(10)\qquad (-1/m) \text{ and/or } (2/m), \quad \text{or} \quad (-2/m), \text{ and } (m/p_i) \qquad (i = 1, \cdots, k),$$

where the p_i are given in §2, (b), then the quadratic character of $m \pmod{D}$ is uniquely determined, and so are the $\chi_a(\theta)$, where θ corresponds to m. Here, the characters in (10) other than the (m/p_i) are to be chosen according to the following scheme:

$$(11)\qquad \begin{aligned} D &\equiv 3 \pmod{4}: (-1/m), \\ D &\equiv 2 \pmod{8}: (2/m), \\ D &\equiv 6 \pmod{8}: (-2/m), \\ D &\equiv 4 \pmod{8}: (-1/m), \\ D &\equiv 0 \pmod{8}: (-1/m) \text{ and } (2/m). \end{aligned}$$

(We assume hereafter that $D \not\equiv 1 \pmod{4}$, and that D is not a perfect square.)

Suppose that there are λ_1 symbols in (10); since there is exactly one relation connecting them, namely $(D/m)=+1$, it follows that λ_1-1 of them are independent. In virtue of the relation between the symbols (10) and the $\lambda-1$ (multiplicatively) independent characters $\chi_1, \cdots, \chi_{\lambda-1}$ defined above, it follows that $\lambda_1-1=\lambda-1$, or $\lambda_1=\lambda$.

We further observe that each of the symbols (10) is an ambiguous character mod M; their $2^{\lambda-1}$ products are clearly then the same as the subset Σ. However, even more significant is the fact that these $\lambda-1$ independent symbols (10) are functions of the element θ to which m corresponds (in virtue of condition (e) in this section), and hence are also ambiguous characters χ_a. The $2^{\lambda-1}$ products of these symbols must therefore be the same as the set of ambiguous characters χ_a. Thus, every $\chi_a(\theta)$ may be written as

$$\chi_a(\theta) = \delta^{(m-1)/2}\cdot\epsilon^{(m^2-1)/8}\cdot(m/Q_1), \tag{12}$$

where m corresponds to θ, $\delta=\pm1$, $\epsilon=\pm1$, and Q_1 is an odd divisor of D. By the quadratic reciprocity law, this can be written as

$$\chi_a(\theta) = (Q/m), \tag{13}$$

where Q is a squarefree divisor of D, where m corresponds to θ, and where Q is even only if D is even; Q may, however, be negative. In any case, Q is determined by χ_a and does not depend on θ.

We may similarly write

$$\tau_a(m) = \vartheta^{(m-1)/2}\cdot\eta^{(m^2-1)/8}\cdot(m/P_1), \tag{14}$$

where P_1 is the product of all those odd primes p_i which divide M, for which $\rho_i=-1$. As above, we then have

$$\tau_a(m) = (P/m), \tag{15}$$

where P is a squarefree divisor of M, and where P is determined by the character τ_a and does not depend on m.

5. ***L*-series.** We define

$$L(\chi, \tau; s) = \prod_{f \text{ in } G} \{1 + [\chi(\theta_f) + \chi(\theta_f^{-1})]\tau(f)f^{-s} \tag{16}$$

$$+ [\chi(\theta_f^2) + \chi(\theta_f^{-2})]\tau(f^2)f^{-2s} + \cdots\}$$

$$= \prod_{f \text{ in } G} \frac{1-\tau(f^2)f^{-2s}}{[1-\chi(\theta_f)\tau(f)f^{-s}][1-\chi(\theta_f^{-1})\tau(f)f^{-s}]}. \tag{17}$$

Then if we set

$$K = \prod_{f \text{ in } G} [1-\tau(f^2)f^{-2s}], \tag{18}$$

we may write

$$(19)\qquad L = K\cdot \prod_{f \text{ in } G} \frac{1}{[1-\chi(\theta_f)\tau(f)f^{-s}][1-\chi(\theta_f^{-1})\tau(f)f^{-s}]}.$$

From (16) we have

$$(20)\qquad L = \sum_{\theta_i \text{ in } H} \sum_{m \text{ in } G} a_m(\theta_i)\chi(\theta_i)\tau(m)m^{-s},$$

where $a_m(\theta_i)$ is the number of products $\prod_{f^\alpha | m,\, f^{\alpha+1} \nmid m} (\theta_f^{\pm\alpha})$ which are the same as θ_i. Clearly

$$(21)\qquad \sum_{\theta_i \text{ in } H} a_m(\theta_i) = 2^{\nu(m)},$$

where $\nu(m)$ is the number of distinct prime factors of m. From (19) we obtain

$$(22)\qquad \log\frac{L}{K} = \sum_{f \text{ in } G} \sum_{r=1}^{\infty} \frac{[\chi(\theta_f^r) + \chi(\theta_f^{-r})]\tau(f^r)}{rf^{rs}},$$

for a properly chosen value of $\log L/K$. All of the results stated so far hold for $s>1$.

6. **Identity of L-series.** We shall write $L(\chi', \tau'; s) \equiv L(\chi'', \tau''; s)$, and call the two L-series identical, if corresponding terms m^{-s} in (20) have equal coefficients, that is, if

$$(23)\qquad \chi'(\theta)\tau'(m) = \chi''(\theta)\tau''(m)$$

for all m and their corresponding θ's. Define χ and τ by

$$(24)\qquad \chi'' = \chi\chi', \qquad \tau'' = \tau\tau'.$$

Then (23) holds if and only if for all m and their corresponding θ's,

$$(25)\qquad \chi(\theta)\tau(m) = 1.$$

In virtue of (3) and (7), (25) may be written as

$$(26)\qquad \prod_{i=1}^{\nu} \omega_i^{s_i}\cdot\vartheta^\alpha\eta^\beta \prod_{i=1}^{r} \rho_i^{\gamma_i} = 1,$$

where the correspondence between the s_i and α, β and the γ_i is determined by the correspondence between m and θ.

Let $\theta = \theta_0$. Then for all m corresponding to θ_0,

$$(27)\qquad \vartheta^\alpha\eta^\beta \prod_{i=1}^{r} \rho_i^{\gamma_i} = 1.$$

But for all primes f, conditions (c) and (f) of §4 imply that f^2 corresponds to $\theta_f\cdot\theta_f^{-1} = \theta_0$. Hence, if we choose one of the indices α, β, $\gamma_1, \cdots, \gamma_r$, say I, then we can find an m corresponding to θ_0 all of whose indices are 0, except for

I which has the value 2. Therefore

$$\vartheta^2 = 1, \qquad \eta^2 = 1, \qquad \rho_i^2 = 1 \qquad (i = 1, \cdots, r), \text{ or } \tau = \tau_a.$$

Next, set $\theta = \theta_i^2$ $(i=1, \cdots, \nu)$. If θ_i corresponds to m, then θ_i^2 corresponds to m^2, and all indices of m can be made even. Therefore $\omega_i^2 = 1$ for $i=1, \cdots, \nu$, or $\chi = \chi_a$. (23) thus holds if and only if

$$(28) \qquad (Q/m)(P/m) = 1 \qquad \text{for all } m \text{ in } G,$$

where P and Q are given by (15) and (13) respectively. Since m may be any integer prime to M for which $(D/m) = +1$, (28) implies either

$$(29a) \qquad P = Q$$

or

$$(29b) \qquad P = ((QD)),$$

where $((x))$ denotes the product of all prime divisors of x which occur to an odd exponent in the factorization of x.

Thus, we may choose for $\chi(\theta)$ any ambiguous χ_a, and then use either (29a) or (29b) to determine a character τ_a which satisfies (25); in other words, if we start with a fixed $L(\chi', \tau'; s)$, there will be 2^λ series identical with it: $L(\chi_a\chi', \tau_a\tau'; s)$, where χ_a and τ_a are any two ambiguous characters satisfying either (29a) or (29b). Thus the $h \cdot \phi(M)$ series $L(\chi, \tau; s)$ fall into sets of 2^λ identical ones.

Nonidentical series may (and, as we shall show, sometimes do) have the same value. However, exactly 2^λ series have the value

$$(30) \qquad L_1 = L_1(\chi_0, \tau_0; s) = \sum_{\theta_i \text{ in } H} \sum_{m \text{ in } G} a_m(\theta_i) m^{-s} = \sum_{m \text{ in } G} 2^{\nu(m)} m^{-s},$$

since if $L(\chi, \tau; s) = L_1$, we would have $\chi(\theta)\tau(m) = 1$ for all m and their corresponding θ's, and the previous discussion shows that this has exactly 2^λ solutions.

7. **L-series of the first and second kinds.** We divide the $L(\chi, \tau)$ into three kinds:

First kind (L_1): any L series identical with $L(\chi_0, \tau_0)$;

Second kind (L_2): any $L(\chi_a, \tau_a)$ not identical with $L(\chi_0, \tau_0)$;

Third kind (L_3): all other L series.

No $L_3 \equiv L_2$; for $L_2(\chi', \tau') \equiv L_3(\chi'', \tau'')$ would imply that χ'/χ'' and τ'/τ'' would both be ambiguous, whence so would both χ'' and τ'', and this is impossible.

We next show that $\lim_{s \to 1+} (s-1)L_1(s)$ exists and is not 0. We have

$$L_1 = \sum_m 2^{\nu(m)} m^{-s},$$

where $\sum_m$ extends over all $m>0$ for which $(m, M)=1$ and $(D/m)=1$. Set

$$M_1 = \sum_{m>0,\,(m,2D)=1,\,(D/m)=1} 2^{\nu(m)} m^{-s}.$$

Then

$$M_1 = L_1 \cdot \prod_{f|M,\,f|2D,\,(D/f)=1} [1 + 2f^{-s} + 2f^{-2s} + \cdots].$$

It therefore suffices to show that $\lim_{s\to 1+} (s-1)M_1$ exists and is not 0. This, however, follows from Dirichlet [2; paragraphs 89 and 96]. Thus

$$\lim_{s\to 1+} (s-1)L_1(s) = A \neq 0. \tag{31}$$

Next, we have

$$L_2 = \sum_{\theta_i \text{ in } H} \sum_{m \text{ in } G} a_m(\theta_i)(Q/m)(P/m)m^{-s} = \sum_{m \text{ in } G} 2^{\nu(m)}(PQ/m)m^{-s}, \tag{32}$$

where P and Q are given by (15) and (13) respectively, and neither (29a) nor (29b) holds. We shall prove that $L_2(s)$ exists and is not 0 for $s=1$, by use of the result of Dirichlet [2; paragraph 101] that $\sum_{n,(n,2D)=1} (D/n)n^{-s}$ exists and is not 0 for $s=1$, provided that D is not a perfect square.

Now, from (32) we obtain

$$\begin{aligned} L_2 &= \prod_{f \text{ in } G} \frac{1+(PQ/f)f^{-s}}{1-(PQ/f)f^{-s}} \\ &= \prod_{f \text{ in } G} \frac{1+(PQM^2/f)f^{-s}}{1-(PQM^2/f)f^{-s}}, \end{aligned} \tag{33}$$

where in each case $\prod_{f \text{ in } G}$ extends over all positive primes f such that $f\nmid M$ and $(D/f)=1$. Set $PQM^2 = D\cdot D_1$; then D_1 is not a square because neither (29a) nor (29b) holds. Then

$$L_2 = \prod_{f,\,(f,2DD_1)=1,\,(D/f)=1} \frac{1+(DD_1/f)f^{-s}}{1-(DD_1/f)f^{-s}} = \prod_p \frac{1+(DD_1/p)p^{-s}}{1-(D_1/p)p^{-s}},$$

where $\prod_p$ extends over all primes $p>0$ such that $p\nmid 2DD_1$. Therefore

$$\begin{aligned} L_2 &= \prod_p (1-p^{-2s}) \cdot \prod_p \frac{1}{1-(DD_1/p)p^{-s}} \cdot \prod_p \frac{1}{1-(D_1/p)p^{-s}} \\ &= \frac{\sum_n (DD_1/n)n^{-s} \cdot \sum_n (D_1/n)n^{-s}}{\sum_n n^{-2s}}, \end{aligned} \tag{34}$$

where in each case $\sum_n$ extends over all postive n for which $(n, 2DD_1)=1$.

The denominator of the right-hand side of (34) is positive for $s>1/2$; the first factor in the numerator of the right-hand side of (34), by Dirichlet's theorem quoted above, is finite and not 0 for $s=1$. We need only prove the same for the second factor $\sum_n (D_1/n)n^{-s}$. Now clearly

$$\sum_{n,q} (D_1/nq)(nq)^{-s} = \sum_{n'} (D_1/n')(n')^{-s}, \tag{35}$$

where q ranges over all divisors of D which are prime to $2D_1$ and n' ranges over all numbers prime to $2D_1$. However, the left-hand side of (35) is

$$\sum_n (D_1/n)n^{-s} \cdot \sum_q (D_1/q)q^{-s},$$

the second factor of which is merely

$$\prod_r \,[1 + (D_1/r)r^{-s} + (D_1/r^2)r^{-2s} + \cdots\,] = \prod_r \frac{1}{1-(D_1/r)r^{-s}},$$

where $\prod_r$ ranges over all primes $r>0$ for which $r \mid D$ and $r \nmid 2D_1$. Therefore

$$\sum_n (D_1/n)n^{-s} = \prod_r \,[1-(D_1/r)r^{-s}] \cdot \sum_{n'} (D_1/n')(n')^{-s}. \tag{36}$$

Since both factors on the right exist and are not 0 for $s=1$, the same is true for the left-hand side. This completes the proof that $L_2(1)$ exists and is not 0.

8. ***L*-series of the third kind.** To prove a similar result for $L_3(s)$, we begin by summing (22) over all $\phi(M)$ characters τ, and use the result

$$\sum_\tau \tau(a) = \begin{cases} \phi(M) & \text{if } a \equiv 1 \pmod{M}, \\ 0 & \text{otherwise.} \end{cases} \tag{37}$$

(We sketch a proof of this result:

If $a \equiv 1 \pmod{M}$, the result is clear; if $a \not\equiv 1 \pmod{M}$, then there exists a character τ' for which $\tau'(a) \neq 1$. But as τ ranges over all characters, so does $\tau\tau'$. Hence $\sum_\tau \tau(a) = \sum_\tau \tau\tau'(a) = \tau'(a) \cdot \sum_\tau \tau(a)$, whence $\sum_\tau \tau(a) = 0$.) We then obtain

$$\begin{aligned} \phi(M)\Bigg\{ &\sum_{f,\, f\equiv 1 \pmod{M}} [\chi(\theta_f) + \chi(\theta_f^{-1})]f^{-s} \\ &+ \frac{1}{2} \sum_{f,\, f^2\equiv 1 \pmod{M}} [\chi(\theta_f^2) + \chi(\theta_f^{-2})]f^{-2s} + \cdots \Bigg\} = \sum_\tau \log \frac{L}{K}. \end{aligned} \tag{38}$$

Now, from (18) it is clear that $K=K(\tau;\, s)$ exists and is not 0 for $s>1/2$, and is indeed continuous in that range. If $\tau=\tau_a$, K is real and positive; if τ is imaginary, K may also be imaginary. However, if $\bar{z}$ denotes the complex conjugate of z, we have $K(\tau;\, s) = \overline{K}(\tau;\, s)$. Similarly $L(\chi, \tau;\, s) = \overline{L}(\chi, \tau;\, s)$. The terms on the right-hand side of (38) may thus be grouped in pairs of conjugates

$$\log \frac{L}{K} + \log \frac{\bar{L}}{\bar{K}} = \log \frac{L\bar{L}}{K\bar{K}} \cdot$$

Suppose now that $\chi = \chi_a$. If $f \equiv 1 \pmod{M}$, then θ_f is in the same genus as θ_0, whence $\chi_a(\theta_f^{\pm 1}) = 1$. (38) then becomes

$$(39) \qquad 2\phi(M)\left\{\sum_{f, f\equiv 1 \pmod{M}} f^{-s} + \frac{1}{2}\sum_{f, f^2 \equiv 1 \pmod{M}} f^{-2s} + \cdots\right\} = \sum_{\tau} \log \frac{L}{K} \cdot$$

Since for each χ there are two characters τ for which two L series are identical, and in particular for each χ_a there are two τ_a for which $L = L_1$, it follows that on the right-hand side of (39) each L occurs in two terms corresponding to the same χ but having two different τ's. Keeping only one of these in each case, we have

$$(40) \qquad \phi(M)\left\{\sum_{f, f\equiv 1 \pmod{M}} f^{-s} + \cdots\right\} = \sum_{\tau}{}' \log \frac{L}{K} \cdot$$

Suppose now that we know that $L_2(s)$ and $L_3(s)$ are finite and continuous for $s \geqq 1$, with finite and continuous derivatives for $s \geqq 1$. If we set

$$(41) \qquad L(s) = f_1(s) + if_2(s),$$

where $f_1(s)$ and $f_2(s)$ are real functions of s, we have for L_2 and L_3

$$(42) \qquad \begin{aligned} L(s) = L(1) + (s-1)\left[f_1'(\vartheta_1(s-1)) + if_2'(\vartheta_2(s-1))\right], \\ 0 < \vartheta_1 < 1,\ 0 < \vartheta_2 < 1, \end{aligned}$$

where $f_i(s)$ and $f_i'(s)$ $(i = 1, 2)$, are continuous for $s \geqq 1$.

We may deduce from this that $L_3(1) \neq 0$. If this is not the case, (42) implies

$$(43) \qquad L_3(s) \cdot \bar{L}_3(s) = O((s-1)^2) \text{ as } s \to 1+.$$

We assume to begin with that $L_3(\chi_a, \tau; 1) = 0$, where χ_a is some particular ambiguous character. On the right side of (40) there occurs a corresponding term $\log (s-1)$, so that as $s \to 1^+$ the right side cannot become positively unbounded. By Dirichlet [2; paragraph 137], however, the left side of (40) becomes positively infinite as $s \to 1^+$. Therefore $L_3(\chi_a, \tau; 1) \neq 0$.

Next, suppose that $L_3(\chi, \tau; 1) = 0$ for some imaginary χ. Summing (38) over all characters χ and using the result

$$(44) \qquad \sum_{\chi} \chi(\theta) = \begin{cases} h & \text{if } \theta = \theta_0, \\ 0 & \text{otherwise,} \end{cases}$$

we obtain

$$2h\phi(M)\left\{\sum_{f,f\equiv 1(\mathrm{mod}\,M),f\leftrightarrow\theta_0} f^{-s}+\frac{1}{2}\sum_{f,f^2\equiv 1(\mathrm{mod}\,M),f^2\leftrightarrow\theta_0} f^{-2s}+\cdots\right\} \tag{45}$$
$$=\sum_{\chi,\tau}\log\frac{L}{K},$$

where $f\leftrightarrow\theta_0$ means that f corresponds to θ_0, and so on. On the right-hand side of (45), $\log L_1$ occurs exactly 2^λ times. However, for each $L_3(\chi, \tau; 1)$ which is 0 there are 2^λ such $L_3(\chi\chi_a, \tau\tau_a)$ (including the original $L_3(\chi, \tau)$), where χ_a and τ_a are related by (29a) or (29b). Now $L_3(\chi, \tau)=0$ implies $L_3(\bar{\chi}, \tau)=0$; if this $L_3(\bar{\chi}, \tau)$ is not included in the above 2^λ L_3's, there are at least $2^\lambda+1$ terms on the right-hand side of (45) each of which contributes the term $\log(s-1)$; this is enough to counterbalance the contribution of $-2^\lambda\log(s-1)$ from the $2^\lambda\log L_1$ terms. As a result, the right-hand side of (45) becomes negatively infinite as $s\to 1+$, while the left-hand side of (45) is non-negative. We thus have a contradiction in this case.

On the other hand, suppose that $L_3(\bar{\chi}, \tau)$ is included in the 2^λ L_3's previously mentioned; then for some χ_a we have $\bar{\chi}=\chi\chi_a$, whence $\chi^4=\chi_a^2=1$ for all θ. We shall now show that in this case there is another L_3 which is 0 and which is not included in any of the previous 2^λ L_3's. This will then establish the result that $L_3(\chi, \tau; 1)\neq 0$.

Since $\chi^4(\theta)=1$ for all θ, and since χ is imaginary, at least one of the roots $\omega_1, \cdots, \omega_{\lambda-1}$ must be $\pm i$, and also $\omega_\lambda=\cdots=\omega_\nu=i$. Suppose then that

$$\begin{array}{lll} \omega_1, \cdots, \omega_R & \text{are all } \pm i & (R\geqq 1), \\ \omega_{R+1}, \cdots, \omega_{\lambda-1} & \text{are all } \pm 1, & \\ \omega_\lambda, \cdots, \omega_\nu & \text{are all } \ \ 1. & \end{array}$$

This implies that $n_1, \cdots, n_R$ are all multiples of 4. In the series $L(\chi, \tau)$, the sum of those terms for which $\chi(\theta_i)=\pm i$ has the value 0, since the terms occur pairwise with opposite signs corresponding to inverse elements θ_i and θ_i^{-1} [using the fact that $a_m(\theta_i)=a_m(\theta_i^{-1})$]. Next, consider the remaining terms, that is, those for which $\chi(\theta)=\pm 1$. Since $\chi(\theta)=(\pm i)^{s_1}\cdots(\pm i)^{s_R}(\pm 1)^{s_{R+1}}\cdots(\pm 1)^{s_{\lambda-1}}=\pm 1$, we have $2\,|\,(s_1+\cdots+s_R)$. Let (P''/m) be that ambiguous character χ_a' obtained by letting $\omega_1=\cdots=\omega_R=-1$, $\omega_{R+1}=\cdots=\omega_\nu=1$. Then for all those θ's for which $\chi(\theta)=\pm 1$ we have $(P''/m)=1$, where θ corresponds to m, and where P'' is neither $((D))$ nor 1. Hence $L(\chi, \tau)$ has the same value as $L(\chi\chi_a', \tau)$, which is not one of the 2^λ series $L(\chi\chi_a, \tau\tau_a)$ mentioned above, since if it were $\chi_a=\chi_a'$, $\tau_a=\tau_0$, and then neither (29a) nor (29b) could hold. This completes the proof that $L_3(\chi, \tau; 1)\neq 0$.

9. **Completion of proof and statement of theorem.** In the proof that neither $L_2(1)$ nor $L_3(1)$ is 0, we needed the fact that $L_2(s)$, $L_2'(s)$, $L_3(s)$ and $L_3'(s)$ are all finite and continuous for $s\geqq 1$. We shall postpone the discussion of this to §10, and for the present follow the main line of the argument.

Let $(N, M)=1$. Multiplying (22) by $\tau^{-1}(N)$ and summing over all characters τ gives (from (37))

$$\begin{aligned}\phi(M)\Big\{ &\sum_{f, f\equiv N \pmod M} [\chi(\theta_f) + \chi(\bar{\theta}_f^{-1})] f^{-s} \\ &+ \frac{1}{2} \sum_{f, f^2\equiv N \pmod M} [\chi(\theta_f^2) + \chi(\theta_f^{-2})] f^{-2s} + \cdots \Big\} = \sum_\tau \tau^{-1}(N) \log \frac{L}{K}.\end{aligned} \tag{46}$$

If we now multiply (46) by $\chi(\theta_j)$ and sum over χ, we obtain

$$\begin{aligned} eh\Big\{ &\sum_{f, f\equiv N \pmod M), f\leftrightarrow\theta_j} f^{-s} + \frac{1}{2} \sum_{f, f^2\equiv N \pmod M), f^2\leftrightarrow\theta_j} f^{-2s} + \cdots \Big\} \\ &= \sum_\chi \sum_\tau \chi(\theta_j)\tau^{-1}(N) \log \frac{L}{K},\end{aligned} \tag{47}$$

where $e=2$ or 1 according as θ_j is ambiguous or not. On the right-hand side of (47), the term $\log L_1$ occurs exactly 2^λ times, namely twice for each

$$\chi_a(\theta) = (Q/m), \qquad m \leftrightarrow \theta, \tag{48}$$

in combination with τ_a given by

$$\tau_a(m) = (Q/m) \quad \text{or} \quad \tau_a(m) = (QD/m), \tag{49}$$

as can be seen from (29a) and (29b). The coefficient of $\log L_1$ on the right-hand side of (47) is therefore

$$\sum \chi_a(\theta_j)\tau_a(N), \tag{50}$$

since $\tau_a^{-1}(N) = \tau_a(N)$, and this in turn is

$$\sum_Q (Q/m_j)[(Q/N) + (QD/N)], \qquad m_j \leftrightarrow \theta_j, \tag{51}$$

and this is equal to

$$[1 + (D/N)] \sum_Q (Q/Nm_j), \tag{52}$$

where $\sum_Q$ is such as to make (Q/Nm_j) range over all ambiguous characters τ_a in Σ. This sum will be 0 except when N and m_j have the same Σ-character, in which case it will be positive. Thus, the right-hand side of (47) becomes positively infinite as $s\to 1+$ when the correspondent of N is in the genus of θ_j. This leads to the theorem:

If N is prime to M and corresponds to an element of the same genus as θ_j, then there are infinitely may primes f which correspond to θ_j for which $f\equiv N$ (mod M), *provided that conditions* (a)–(f) *of* §4 *hold, and that $L_i(s)$ and $L_i'(s)$* ($i=2, 3$) (*see* §7) *are continuous for $s\geqq 1$.*

We shall discuss this latter condition in the next section.

10. Sufficiency conditions. We shall now prove that $L_i(s)$ and $L_i'(s)$ ($i=2, 3$) are finite and continuous for $s \geqq 1$, provided that our correspondence established in §4 satisfies certain conditions. If we define

(53) $$R(m_0, \theta; s) = \sum_{m, m \equiv m_0 \pmod M} a_m(\theta) m^{-s},$$

we see that $R(m_0, \theta) = 0$ unless m_0 corresponds to an element in the genus of θ. Furthermore, by use of (20), we obtain

(54) $$L(\chi, \tau) = \sum_{m_0} \tau(m_0) \sum_{\theta_i} \chi(\theta_i) R(m_0, \theta_i),$$

where $\sum_{m_0}$ ranges over a reduced residue system mod M and $\sum_{\theta_i}$ ranges over all elements of H. We now impose the conditions:

(A) $$R(m_0, \theta_i; s) = A_1/(s-1) + B_1(m_0, \theta_i; s),$$

if m_0 corresponds to any element in the genus of θ_i, where A_1 is a constant ($\neq 0$) independent of m_0 and θ_i, and $B_1(m_0, \theta_i; s)$ is continuous for $s \geqq 1$.

(B) $$R'(m_0, \theta_i; s) = A_2/(s-1)^2 + B_2(m_0, \theta_i; s),$$

if m_0 corresponds to any element in the genus of θ_i, where A_2 is a constant ($\neq 0$) independent of m_0 and θ_i, and $B_2(m_0, \theta_i; s)$ is continuous for $s \geqq 1$.

To prove that $L_i(s)$ and $L_i'(s)$ ($i=2, 3$) are finite and continuous for $s \geqq 1$, in virtue of (54) and (A) and (B) it suffices to show that

(55) $$P(\chi, \tau) = \sum_{m_0, \theta_i} \tau(m_0) \chi(\theta_i)$$

is 0 provided that τ and χ do not satisfy (25) for all m and their corresponding θ's. We may write

(56) $$P(\chi, \tau) = \sum_{\theta} \chi(\theta) \left\{ \sum_{m_0} \tau(m_0) \right\},$$

where $\sum_{m_0}$ extends over all m_0 which correspond to elements in the genus of θ, and then $\sum_\theta$ extends over all θ. Further, the choice of the reduced residue system mod M to which the m_0 belong has no effect on the value of $P(\chi, \tau)$ because two numbers congruent mod M correspond to θ's of the same genus.

Let f be any prime in G; then, as m_0 ranges over those elements of a reduced residue system mod M which correspond to elements in the genus of θ, so does $f^2 m_0$, because $f^2 m_0 \equiv f^2 m_0'$ (mod M) holds if and only if $m_0 \equiv m_0'$ (mod M), and $f^2 m_0$ and m_0 correspond to elements in the same genus. Hence

(57) $$P(\chi, \tau) = \sum_{\theta} \chi(\theta) \left\{ \sum_{m_0} \tau(f^2 m_0) \right\} = \tau(f^2) P(\chi, \tau).$$

Hence $P=0$ unless $\tau(f^2)=1$ for each f in G; this in turn is true if and only if τ is an ambiguous character. But in virtue of (13) and (15), every ambiguous character τ_a is also an ambiguous character χ_a. Hence

$$P(\chi, \tau) = \sum_{\theta} \chi(\theta) \sum_{m_0} \chi_a(\theta), \tag{58}$$

where m_0 corresponds to elements in the same genus as θ. Now it is clear that the same number of elements in a reduced residue system correspond to each genus, since if m_0 corresponds to a genus containing θ, so do all of the numbers m_0n^2 obtained by letting n range over all numbers prime to M which give incongruent values of n^2, and conversely. The number of m_0 corresponding to elements in a given genus is thus $\phi(M)$ divided by $2^{\lambda-1}$. Call their quotient q. Then

$$P(\chi, \tau) = q\sum_{\theta} \chi(\theta)\chi_a(\theta) = q\sum_{\theta} \chi\chi_a(\theta). \tag{59}$$

Therefore $P(\chi, \tau)=0$ unless $\chi\chi_a=\chi_0$, that is, unless either (29a) or (29b) holds. This completes the proof that $L_i(s)$ and $L_i'(s)$ $(i=2, 3)$ are finite and continuous for $s\geqq 1$.

11. **Meyer's theorem.** A. Meyer [3] proved this theorem for the special case where H is the set of classes of properly primitive binary quadratic forms of determinant D, which forms a group under the operation of composition. In that case, it is known from the theory of quadratic forms that all of the conditions (a)–(f) of §4 hold, where the θ_i are the classes of forms of determinant D, and m corresponds to those classes θ which represent m. The crucial (and most difficult) part of the proof is to show the correctness of (A) and (B). Meyer proved this (or results equivalent to this) by using some theorems of H. Weber [4] who considered the simpler problem of representation of primes by quadratic forms (without considering the arithmetic progressions in which they lie). Weber's results were established by use of ϑ-function identities; simpler proofs for equivalent results are to be found (for suitably restricted values of D) in Landau, *Neuere Fortschritte der additiven Zahlentheorie*, pp. 87–90.

12. **An example.** We conclude with an example to show that this theorem is more general than the original theorem. We shall use the result of Dirichlet [2; p. 359] that $\sum_f f^{-s}-(1/\phi(M))\ln(1/(s-1))$ is analytic for $s>1/2$, where $\sum_f$ extends over all primes $f\equiv N \pmod{M}$, and where $(N, M)=1$.

Set $D=-20$, $M=-80$. G has a basis consisting of all primes $f\equiv 1$, 3, 7 or 9 (mod 20). Let H be the group $\{\theta, \theta_0\}$ with $\theta^2=\theta_0$. Since there are four reduced (nonequivalent) forms of determinant -20, this means that we are not working with quadratic forms in this case.

We establish the correspondence

$$f_1, f_9 \leftrightarrow \theta_0, \qquad f_3, f_7 \leftrightarrow \theta,$$

where $f_j\equiv j \pmod{20}$. Define

$$R_1 = R(1, \theta_0), \qquad R_9 = R(9, \theta_0), \qquad R_3 = R(3, \theta), \qquad R_7 = R(7, \theta). \tag{60}$$

We shall prove that (A) and (B) hold, simply by proving (A) and showing that the function $B_1(s)$ is analytic for $s > 1/2$. All other conditions of our hypothesis are automatically satisfied in this case.

If $m \leftrightarrow \theta_i$, then $a_m(\theta_i) = 2^{\nu(m)}$. Since we are working here with m (mod 20), we shall restrict ourselves (in this section) to those characters τ whose value does not depend on the exponent β in (1), that is, those for which η in (2) is 1. Then we have the following table:

m (mod 20)	α	γ
1	0	0
9	0	2
3	1	3
7	1	1

For any character for which $\eta = 1$, we have by setting $\chi = \chi_0$ in (17)

$$\sum_{m_0 \leftrightarrow \theta_i} \tau(m_0) R(m_0, \theta_i) = \prod_{f, (-20/f)=1} \frac{1 - \tau(f^2) f^{-2s}}{(1 - \tau(f) f^{-s})^2}, \tag{61}$$

or

$$\tau(1) R_1 + \tau(9) R_9 + \tau(3) R_3 + \tau(7) R_7 = \prod_{f, (-20/f)=1} \frac{1 + \tau(f) f^{-s}}{1 - \tau(f) f^{-s}}. \tag{62}$$

We shall choose four characters for τ, thus getting from (62) four equations from which we may find the R_j. Let $\tau = \tau_0, \tau_1, \tau_2, \tau_3$ successively, given by:

	ϑ	η	ρ
τ_0	1	1	1
τ_1	1	1	-1
τ_2	1	1	i
τ_3	1	1	$-i$

Let $P_i = \prod_{f_i} (1 + f_i^{-s})/(1 - f_i^{-s})$ $(i = 1, 9, 3, 7)$. We then obtain from (62):

$$
\begin{aligned}
&R_1 + R_9 + R_3 + R_7 = P_1P_9P_3P_7 = \xi_1,\\
&R_1 + R_9 - R_3 - R_7 = P_1P_9P_3^{-1}P_7^{-1} = \xi_2,\\
(63)\quad &R_1 - R_9 - iR_3 + iR_7 = P_1P_9^{-1}\prod_{f_3}\frac{1 - if_3^{-s}}{1 + if_3^{-s}}\prod_{f_7}\frac{1 + if_7^{-s}}{1 - if_7^{-s}} = \xi_3,\\
&R_1 - R_9 + iR_3 - iR_7 = P_1P_9^{-1}\prod_{f_3}\frac{1 + if_3^{-s}}{1 - if_3^{-s}}\prod_{f_7}\frac{1 - if_7^{-s}}{1 + if_7^{-s}} = \xi_4.
\end{aligned}
$$

Thence

$$
(64)\quad
\begin{aligned}
R_1 &= (\xi_1 + \xi_2 + \xi_3 + \xi_4)/4, & R_9 &= (\xi_1 + \xi_2 - \xi_3 - \xi_4)/4,\\
R_3 &= (\xi_1 - \xi_2 + i\xi_3 - i\xi_4)/4, & R_7 &= (\xi_1 - \xi_2 - i\xi_3 + i\xi_4)/4.
\end{aligned}
$$

Now

$$
\begin{aligned}
\ln P_j &= \sum_{f_j} \ln\left(1 + \frac{2f_j^{-s}}{1 - f_j^{-s}}\right)\\
(65)\quad &= \sum_{f_j}\left\{+\frac{2f_j^{-s}}{1 - f_j^{-s}} - \frac{1}{2}\left(\frac{2f_j^{-s}}{1 - f_j^{-s}}\right)^2 + \cdots\right\}\\
&= + \sum_{f_j}\{+ 2f_j^{-s}(1 + f_j^{-s} + \cdots) - \cdots\}\\
&= + 2\sum_{f_j} f_j^{-s} + \sum_{f_j} O(f_j^{-2s}).
\end{aligned}
$$

By Dirichlet's result, it then follows that

$$
(66)\quad \ln P_j - \frac{1}{4}\ln\left(\frac{1}{s - 1}\right)
$$

is analytic for $s>1/2$. Thus $\xi_1 = b/(s-1)$, where b is analytic for $s>1/2$; similarly, ξ_2 itself is analytic for $s>1/2$. In order to prove (A), it thus suffices to prove the analyticity of

$$
(67)\quad Z = \prod_{f_3}\frac{1 - if_3^{-s}}{1 + if_3^{-s}}\cdot\prod_{f_7}\frac{1 + if_7^{-s}}{1 - if_7^{-s}}
$$

for $s>1/2$. Now

$$
\begin{aligned}
(68)\quad Z &= \exp\left\{\sum_{f_3}\ln\left(1 - \frac{2if_3^{-s}}{1 + if_3^{-s}}\right) - \sum_{f_7}\ln\left(1 - \frac{2if_7^{-s}}{1 + if_7^{-s}}\right)\right\}\\
&= \exp\left\{2i\sum f_3^{-s} - 2i\sum f_7^{-s} + \text{remainder}\right\},
\end{aligned}
$$

where the remainder is analytic for $s>1/2$. The result then follows from Dirichlet's result quoted above.

Bibliography

1. G. L. Dirichlet, C. R. Acad. Sci. Paris vol. 10 (1840) p. 285.
2. ———, *Vorlesungen über Zahlentheorie*, 4th ed., 1894.
3. A. Meyer, *Ueber einen Satz von Dirichlet*, Journal für Mathematik vol. 103 (1888) p. 98.
4. H. Weber, *Quadratische Formen und Primzahlen*, Math. Ann. vol. 20 (1882) p. 301.

Cornell University,
Ithaca, N. Y.

ON THE GENERATORS OF THE SYMPLECTIC MODULAR GROUP

BY

L. K. HUA AND I. REINER

Introduction. Let n be a positive integer. Throughout this paper, unless the contrary is stated, we shall use capital Latin letters to denote n-rowed matrices and capital German letters to denote $2n$-rowed matrices. Furthermore, an r-rowed matrix R will be denoted by $R^{(r)}$. Let

$$\mathfrak{F} = \begin{pmatrix} 0 & I \\ -I & 0 \end{pmatrix},$$

where I and 0 denote the identity and zero matrices respectively. Let Γ be the group of all matrices $\mathfrak{M}$ with rational integral elements which satisfy

$$\mathfrak{M}\mathfrak{F}\mathfrak{M}' = \mathfrak{F}, \tag{1}$$

where $\mathfrak{M}'$ denotes the transpose of $\mathfrak{M}$. Let Γ_0 be the factor group of Γ over its centrum; Γ_0 is called the symplectic modular group. It can be thought of as being obtained from Γ by identifying the elements $\mathfrak{M}$ and $-\mathfrak{M}$. In applications to modular functions of the nth degree(1) and to the projective geometry of matrices(2) it is customary to identify $\mathfrak{M}$ and $-\mathfrak{M}$ as a single transformation. For this reason we have considered Γ_0 rather than Γ; it might be pointed out, however, that the generators of Γ_0 obtained in this paper happen to be a set of generators of Γ.

It is the aim of this paper to find the generators of the symplectic modular group. It will be proved here that this group is generated by two or four independent elements, according as $n=1$ or $n>1$. The method used here can be extended so as to give a set of generators for matrices with elements in any Euclidean ring. In particular, we give the details for the generalized Picard group at the end of this paper.

Problems of this type have been considered previously. Poincaré(3) stated without proof that every matrix $\mathfrak{M}$ with integral elements for which $\mathfrak{M}\mathfrak{G}\mathfrak{M}' = \mathfrak{G}$, where $\mathfrak{G}$ is the direct sum of n two-rowed skew-symmetric matrices, is expressible as a product of elementary matrices of two simple types. Brahana(4) proved this and extended the result to the case where $\mathfrak{G}$ is any skew-symmetric matrix by showing in this case that every such matrix $\mathfrak{M}$ is ex-

Presented to the Society, February 28, 1948; received by the editors March 6, 1948.

(1) C. L. Siegel, Math. Ann. vol. 116 (1939) pp. 617–657.

(2) L. K. Hua, Trans. Amer. Math. Soc. vol. 57 (1945) pp. 441–490.

(3) H. Poincaré, Rend. Circ. Mat. Palermo vol. 18 (1904) pp. 45–110.

(4) H. R. Brahana, Ann. of Math. (2) vol. 24 (1923) pp. 265–270.

pressible as a product of matrices taken from some finite set of matrices. From the results given in the present paper, a much stronger form of Brahana's result can be easily deduced.

1. If we set

$$\mathfrak{M} = \begin{pmatrix} A & B \\ C & D \end{pmatrix}, \tag{2}$$

(1) is equivalent to

$$AB' = BA', \qquad CD' = DC', \qquad AD' - BC' = I. \tag{3}$$

By taking inverses of both sides of (1) and using $\mathfrak{F}^{-1} = -\mathfrak{F}$, we can deduce that $\mathfrak{M}'\mathfrak{F}\mathfrak{M} = \mathfrak{F}$, so that

$$A'C = C'A, \qquad B'D = D'B, \qquad A'D - C'B = I. \tag{4}$$

We shall begin by showing in §3 that Γ_0 is generated by the following types of elements:

(I) *Translations*:

$$\mathfrak{T} = \begin{pmatrix} I & S \\ 0 & I \end{pmatrix},$$

where S is symmetric.

(II) *Rotations*:

$$\mathfrak{R} = \begin{pmatrix} U & 0 \\ 0 & U'^{-1} \end{pmatrix},$$

where U is unimodular, that is, abs $U = 1$ (where abs U denotes the absolute value of the determinant of U).

(III) *Semi-involutions*:

$$\mathfrak{S} = \begin{pmatrix} J & I - J \\ J - I & J \end{pmatrix}$$

where J is a diagonal matrix whose diagonal elements are 0's and 1's, so that $J^2 = J$ and $(I-J)^2 = I-J$.

It is easily verified that matrices of types I, II and III satisfy (1).

2. In this section we prove two lemmas on matrices.

LEMMA 1. *Let m be a nonzero integer, and let T be an n-rowed symmetric matrix at least one of whose elements is not divisible by m. There exists a symmetric matrix S with integral elements such that*

$$0 < \text{abs}\,(T - mS) < |m|^n. \tag{5}$$

Proof. The lemma is evident for $n = 1$. Consider next $n = 2$; let $T = (t_{ij})$, $S = (s_{ij})$. Then

$$\text{(6)} \qquad \operatorname{abs}(T - mS) = |(t_{11} - ms_{11})(t_{22} - ms_{22}) - (t_{12} - ms_{12})^2|.$$

If m divides both t_{11} and t_{22}, it cannot divide t_{12}; we can then choose S so that $t_{11}-ms_{11}=t_{22}-ms_{22}=0$ and $0<|t_{12}-ms_{12}|<|m|$. Suppose on the other hand that m does not divide one of the diagonal elements, say t_{11}. Fix s_{12} arbitrarily, and choose s_{11} so that $0<|t_{11}-ms_{11}|<|m|$. Since (6) can be written as

$$\operatorname{abs}(T - mS) = |-m(t_{11} - ms_{11})s_{22} + \cdots|,$$

where $\cdots$ represents terms not involving s_{22}, we can choose an integer s_{22} by the Euclidean algorithm so that

$$0 < \operatorname{abs}(T - mS) \leqq |m(t_{11} - ms_{11})| < |m|^2.$$

Suppose now that the result has been established for $n=r-1$ with $r\geqq 3$; we shall deduce it for $n=r$. Let $T=T^{(r)}$ and let some element t_{ij} of T be not divisible by m. Since $r\geqq 3$, there exists a diagonal element t_{kk} of T which is not in the same row or column as t_{ij}. Let T_1 be the symmetric $(r-1)$-rowed matrix obtained from T by omitting the kth row and kth column; let S_1 be similarly related to S. By the induction hypothesis, we may choose S_1 symmetric so that

$$\text{(7)} \qquad 0 < \operatorname{abs}(T_1 - mS_1) < |m|^{r-1}.$$

However, we have

$$\text{(8)} \qquad \operatorname{abs}(T - mS) = |(t_{kk} - ms_{kk}) \det(T_1 - mS_1) + \cdots|,$$

where $\cdots$ represents terms not involving s_{kk}. Choose s_{lk} arbitrarily for $l=1, 2, \cdots, k-1, k+1, \cdots, r$. Then by the Euclidean algorithm we can choose s_{kk} so that

$$0 < \operatorname{abs}(T - mS) \leqq |m| \operatorname{abs}(T_1 - mS_1) < |m|^r.$$

This completes the proof of the lemma.

LEMMA 2. *Let A and B satisfy $AB'=BA'$ and let* $\det A\neq 0$. *There exists a symmetric matrix S such that either*

$$\text{(9)} \qquad B - AS = 0$$

or

$$\text{(10)} \qquad 0 < \operatorname{abs}(B - AS) < \operatorname{abs} A.$$

Proof. From $AB'=BA'$ and $\det A\neq 0$, we may deduce that A^*B is symmetric, where A^* denotes the adjoint of A. We apply Lemma 1 with $T=A^*B$ and $m=\det A$. Either every element of A^*B is divisible by m, in which case there exists a symmetric matrix S with $A^*B=mS$, or else there exist symmetric matrices R and S such that $A^*B=mS+R$ with $0<\operatorname{abs} R<|m|^n$. In virtue of the relation $A^*A=mI$, these alternatives become: either $B=AS$ (in which case (9) holds), or $B-AS=AR/m$; however,

$$\operatorname{abs}\frac{AR}{m}=\frac{(\operatorname{abs} A)(\operatorname{abs} R)}{|m|^{n}}=\frac{\operatorname{abs} R}{|m|^{n-1}},$$

so that

$$0<\operatorname{abs}(B-AS)<|m|=\operatorname{abs} A.$$

3. We are now ready to show that Γ_0 is generated by matrices of types I, II and III. Let $\mathfrak{M}$ given by (2) be an arbitrary element of Γ_0. It suffices to prove that by multiplying $\mathfrak{M}$ by matrices of types I, II and III, one obtains a product of matrices of those types.

(3) implies that not both A and B are 0. Since

$$\begin{pmatrix} A & B \\ C & D \end{pmatrix}\begin{pmatrix} 0 & I \\ -I & 0 \end{pmatrix}=\begin{pmatrix} -B & A \\ * & * \end{pmatrix},$$

we may assume that A has rank $r>0$. Furthermore,

$$\begin{pmatrix} U & 0 \\ 0 & U'^{-1} \end{pmatrix}\begin{pmatrix} A & B \\ C & D \end{pmatrix}\begin{pmatrix} V & 0 \\ 0 & V'^{-1} \end{pmatrix}=\begin{pmatrix} UAV & * \\ * & * \end{pmatrix},$$

so that A may be taken to be of the form

$$A=\begin{pmatrix} A_1 & 0 \\ A_2 & 0 \end{pmatrix}, \tag{11}$$

where A_1 is an r-rowed nonsingular matrix. We similarly decompose B as

$$B=\begin{pmatrix} B_1^{(r)} & * \\ * & * \end{pmatrix}.$$

From (3) it is easily seen that $A_1B_1'=B_1A_1'$. By Lemma 2, there exists a symmetric matrix S_1 with either $A_1S_1+B_1=0$ or $0<\operatorname{abs} R_1<\operatorname{abs} A_1$, where $R_1=A_1S_1+B_1$. Define

$$S=\begin{pmatrix} S_1^{(r)} & 0 \\ 0 & 0 \end{pmatrix}.$$

Then

$$\begin{pmatrix} A & B \\ C & D \end{pmatrix}\begin{pmatrix} I & S \\ 0 & I \end{pmatrix}=\begin{pmatrix} A & AS+B \\ * & * \end{pmatrix}, \tag{12}$$

so that A remains unaltered while B_1 of B is replaced by 0 or R_1. If the second alternative occurs, we proceed as follows: let

$$J = \begin{pmatrix} 0 & 0 \\ 0 & I^{(n-r)} \end{pmatrix}. \tag{13}$$

Then

$$\begin{pmatrix} A & B \\ C & D \end{pmatrix} \begin{pmatrix} J & I - J \\ J - I & J \end{pmatrix} = \begin{pmatrix} \overline{A} & * \\ * & * \end{pmatrix}, \tag{14}$$

where

$$\overline{A} = AJ - B(I - J) = \begin{pmatrix} -R_1 & 0 \\ * & 0 \end{pmatrix}.$$

We now repeat the process as before, and so on. Since there are only finitely many positive integers less than abs A_1, this process eventually terminates. Thus, by multiplying $\mathfrak{M}$ by matrices of types I, II and III one arrives at a matrix

$$\begin{pmatrix} A_0 & B_0 \\ * & * \end{pmatrix}$$

with

$$A_0 = \begin{pmatrix} R & 0 \\ * & 0 \end{pmatrix}, \qquad B_0 = \begin{pmatrix} 0 & * \\ * & * \end{pmatrix}$$

and det $R \neq 0$. One readily deduces from $A_0 B_0' = B_0 A_0'$ that B_0 must be of the form

$$B_0 = \begin{pmatrix} 0 & * \\ 0 & * \end{pmatrix}.$$

But then

$$\begin{pmatrix} 0 & I \\ -I & 0 \end{pmatrix} \begin{pmatrix} A_0 & B_0 \\ C_0 & D_0 \end{pmatrix} \begin{pmatrix} J & I - J \\ J - I & J \end{pmatrix} = \begin{pmatrix} A^+ & B^+ \\ 0 & D^+ \end{pmatrix}$$

where J is given by (13). Finally we notice that for a matrix

$$\begin{pmatrix} A & B \\ 0 & D \end{pmatrix}$$

of Γ_0, we must have $A = U$ unimodular, $D = U'^{-1}$, and thence from (3), $B = SU'^{-1}$ with symmetric S. Therefore

$$\begin{pmatrix} A & B \\ 0 & D \end{pmatrix} = \begin{pmatrix} I & S \\ 0 & I \end{pmatrix} \begin{pmatrix} U & 0 \\ 0 & U'^{-1} \end{pmatrix}.$$

This completes the proof that Γ_0 is generated by the matrices of types I, II and III.

4. The set of matrices of types I, II and III which generate Γ_0 are certainly not independent generators. Let us reduce the number of generators as much as possible. Since

$$\begin{pmatrix} I & S_1 \\ 0 & I \end{pmatrix}\begin{pmatrix} I & S_2 \\ 0 & I \end{pmatrix} = \begin{pmatrix} I & S_1 + S_2 \\ 0 & I \end{pmatrix},$$

the subgroup formed by matrices of type I is generated by those type I matrices whose S's are given by

$$S_0 = \begin{bmatrix} 1 & 0 \cdots 0 \\ 0 & 0 \cdots 0 \\ \cdot & \cdot \quad \cdot \quad \cdot \\ 0 & 0 \cdots 0 \end{bmatrix}, \qquad S_1 = \begin{bmatrix} 0 & 1 & 0 \cdots 0 \\ 1 & 0 & 0 \cdots 0 \\ 0 & 0 & 0 \cdots 0 \\ \cdot & \cdot & \cdot \quad \cdot \quad \cdot \\ 0 & 0 & 0 \cdots 0 \end{bmatrix} \tag{15}$$

and all matrices obtained from these by interchanging any two rows and the corresponding columns. Next we note that

$$\begin{pmatrix} U & 0 \\ 0 & U'^{-1} \end{pmatrix}\begin{pmatrix} I & S \\ 0 & I \end{pmatrix}\begin{pmatrix} U^{-1} & 0 \\ 0 & U' \end{pmatrix} = \begin{pmatrix} I & USU' \\ 0 & I \end{pmatrix},$$

so that the group generated by matrices of types I and II is the same as that generated by all type II matrices and the two translations whose S's are given by (15). However, we have

$$\begin{pmatrix} 1 & -1 \\ 0 & 1 \end{pmatrix}\begin{pmatrix} 1 & 0 \\ 0 & -1 \end{pmatrix}\begin{pmatrix} 1 & 0 \\ -1 & 1 \end{pmatrix} = \begin{pmatrix} 0 & 1 \\ 1 & -1 \end{pmatrix} = \begin{pmatrix} 0 & 1 \\ 1 & 0 \end{pmatrix} - \begin{pmatrix} 0 & 0 \\ 0 & 1 \end{pmatrix}.$$

Hence the translation with S_1 is obtainable from that with S_0 and the matrices of type II. Therefore Γ_0 is generated by the matrix

$$\mathfrak{T}_0 = \begin{pmatrix} I & S_0 \\ 0 & I \end{pmatrix} \tag{16}$$

with S_0 given by (15), and all matrices of types II and III.

Since

$$\begin{pmatrix} U & 0 \\ 0 & U'^{-1} \end{pmatrix}\begin{pmatrix} V & 0 \\ 0 & V'^{-1} \end{pmatrix} = \begin{pmatrix} UV & 0 \\ 0 & (UV)'^{-1} \end{pmatrix},$$

in order to find the generators of the subgroup of rotations we have merely to find the generators of the group of unimodular matrices. These are given by the following theorem.

THEOREM 1. *Let $n \geqq 2$. Every unimodular matrix with rational integral elements is a product of the matrices U_1, U_2, U_3 and their inverses, where*

$$U_1 = \begin{pmatrix} 0 & 0 \cdots 0 & 1 \\ 1 & 0 \cdots 0 & 0 \\ \cdot & \cdot \quad \cdot \quad \cdot & \cdot \\ 0 & 0 \cdots 0 & 0 \\ 0 & 0 \cdots 1 & 0 \end{pmatrix}, \qquad U_2 = \begin{pmatrix} 1 & 1 \cdots 0 & 0 \\ 0 & 1 \cdots 0 & 0 \\ \cdot & \cdot \quad \cdot \quad \cdot & \cdot \\ 0 & 0 \cdots 1 & 0 \\ 0 & 0 \cdots 0 & 1 \end{pmatrix}, \tag{17}$$

$$U_3 = \begin{pmatrix} -1 & 0 \cdots 0 & 0 \\ 0 & 1 \cdots 0 & 0 \\ \cdot & \cdot \quad \cdot \quad \cdot & \cdot \\ 0 & 0 \cdots 1 & 0 \\ 0 & 0 \cdots 0 & 1 \end{pmatrix}.$$

Proof. It is known(5) that every unimodular matrix is a product of U_1, U_2, U_3, and

$$U_4 = \begin{pmatrix} 0 & 1 \cdots 0 & 0 \\ 1 & 0 \cdots 0 & 0 \\ \cdot & \cdot \quad \cdot \quad \cdot & \cdot \\ 0 & 0 \cdots 1 & 0 \\ 0 & 0 \cdots 0 & 1 \end{pmatrix},$$

and their inverses. It is sufficient to show that U_4 is expressible as a product of U_1, U_2, U_3 and their inverses. We define $T = U_2 U_1$ for the remainder of this proof, and let $\mathfrak{r} = (r_1, \cdots, r_n)'$ be a column vector. Then

$$T\mathfrak{r} = \begin{pmatrix} r_n + r_1 \\ r_1 \\ \cdot \\ \cdot \\ \cdot \\ r_{n-1} \end{pmatrix},$$

$$T^2\mathfrak{r} = \begin{pmatrix} r_{n-1} + r_n + r_1 \\ r_n + r_1 \\ r_1 \\ \cdot \\ \cdot \\ \cdot \\ r_{n-2} \end{pmatrix}, \cdots, \quad T^{n-1}\mathfrak{r} = \begin{pmatrix} r_2 + r_3 + \cdots + r_n + r_1 \\ r_3 + \cdots + r_n + r_1 \\ \cdot \\ \cdot \\ \cdot \\ r_n + r_1 \\ r_1 \end{pmatrix}.$$

(5) See for example, C. C. MacDuffee, *The theory of matrices*, Berlin, 1933, p. 34, Theorem 22.5.

Therefore

$$U_1^{-1}T^{n-1}\mathfrak{r} = \begin{pmatrix} r_3 + \cdots + r_n + r_1 \\ \vdots \\ r_n + r_1 \\ r_1 \\ r_2 + r_3 + \cdots + r_n + r_1 \end{pmatrix},$$

so that

$$(T^{-1})^{n-2}U_1^{-1}T^{n-1}\mathfrak{r} = \begin{pmatrix} r_1 \\ r_2 + \cdots + r_n + r_1 \\ r_3 \\ \vdots \\ r_n \end{pmatrix},$$

$$U_1(T^{-1})^{n-2}U_1^{-1}T^{n-1}\mathfrak{r} = \begin{pmatrix} r_n \\ r_1 \\ r_2 + \cdots + r_n + r_1 \\ r_3 \\ \vdots \\ r_{n-1} \end{pmatrix}.$$

From this we see that

$$T^{n-3}U_1(T^{-1})^{n-2}U_1^{-1}T^{n-1}\mathfrak{r} = \begin{pmatrix} r_3 + r_4 + \cdots + r_n \\ r_4 + \cdots + r_n \\ \vdots \\ r_n \\ r_1 \\ r_2 + \cdots + r_n + r_1 \end{pmatrix}$$

and

$$(T^{-1})^{n-2}U_1T^{n-3}U_1(T^{-1})^{n-2}U_1^{-1}T^{n-1}\mathfrak{r} = \begin{pmatrix} r_n \\ r_1 \\ r_2 + r_1 \\ \vdots \\ r_{n-1} \end{pmatrix}.$$

Define

$$U\dagger = U_1^{-1}(T^{-1})^{n-2}U_1T^{n-3}U_1(T^{-1})^{n-2}U_1^{-1}T^{n-1}.$$

Then

$$U\dagger = \begin{pmatrix} 1 & 0 \\ 1 & 1 \end{pmatrix} \dotplus I^{(n-2)},$$

where $\dotplus$ denotes the direct sum of matrices. But from

$$U_3 = \begin{pmatrix} -1 & 0 \\ 0 & 1 \end{pmatrix} \dotplus I^{(n-2)} \quad \text{and} \quad U_2^{-1} = \begin{pmatrix} 1 & -1 \\ 0 & 1 \end{pmatrix} \dotplus I^{(n-2)}$$

we deduce

$$\begin{aligned} U_3U\dagger U_2^{-1}U\dagger &= \begin{pmatrix} -1 & 0 \\ 0 & 1 \end{pmatrix}\begin{pmatrix} 1 & 0 \\ 1 & 1 \end{pmatrix}\begin{pmatrix} 1 & -1 \\ 0 & 1 \end{pmatrix}\begin{pmatrix} 1 & 0 \\ 1 & 1 \end{pmatrix} \dotplus I^{(n-2)} \\ &= \begin{pmatrix} 0 & 1 \\ 1 & 0 \end{pmatrix} \dotplus I^{(n-2)} = U_4. \end{aligned}$$

This completes the proof of the theorem.

COROLLARY. *Let $n \geqq 2$. Every unimodular matrix with rational integral elements of determinant $+1$ is a product of powers of U_2 and*

$$U_5 = \begin{bmatrix} 0 \cdots 0 & (-1)^{n-1} \\ 1 \cdots 0 & 0 \\ \cdots & \cdots \\ 0 \cdots 0 & 0 \\ 0 \cdots 1 & 0 \end{bmatrix}.$$

By Theorem 1 we see now that Γ_0 is generated by $\mathfrak{T}_0$ and the set of all semi-involutions and the three rotations defined by

$$\mathfrak{R}_i = \begin{pmatrix} U_i & 0 \\ 0 & U_i'^{-1} \end{pmatrix}, \qquad i = 1, 2, 3. \tag{18}$$

We finally consider type III matrices. Let J_r be the diagonal matrix obtained from the identity matrix by replacing the rth 1 by 0. In that case, if $r \neq s$, we have

$$\begin{pmatrix} J_r & I - J_r \\ J_r - I & J_r \end{pmatrix}\begin{pmatrix} J_s & I - J_s \\ J_s - I & J_s \end{pmatrix} = \begin{pmatrix} J_{rs} & I - J_{rs} \\ J_{rs} - I & J_{rs} \end{pmatrix},$$

where J_{rs} is obtained from the identity matrix by replacing the rth and sth ones by 0's. Therefore, in order to obtain all type III matrices, we need only those semi-involutions

$$\begin{pmatrix} J_r & I - J_r \\ J_r - I & J_r \end{pmatrix}, \qquad r = 1, 2, \cdots, n, \tag{19}$$

with J_r defined above. Now, let U be that unimodular matrix obtained from I by interchanging the 1st and rth rows; then we have

$$\begin{pmatrix} U & 0 \\ 0 & U'^{-1} \end{pmatrix} \begin{pmatrix} J_r & I - J_r \\ J_r - I & J_r \end{pmatrix} \begin{pmatrix} U^{-1} & 0 \\ 0 & U' \end{pmatrix} = \begin{pmatrix} J_1 & I - J_1 \\ J_1 - I & J_1 \end{pmatrix}.$$

Therefore Γ_0 is generated by the matrices $\mathfrak{T}_0$, $\mathfrak{R}_i$ $(i=1, 2, 3)$ and the matrix

$$\mathfrak{S}_0 = \begin{pmatrix} J_1 & I - J_1 \\ J_1 - I & J_1 \end{pmatrix}, \tag{20}$$

with J, previously defined. But

$$\mathfrak{S}_0^2 = \mathfrak{R}_3,$$

so that $\mathfrak{R}_3$ may be dropped from the list of generators. Therefore we have the following theorem.

THEOREM 2. *Γ_0 is generated by the four matrices $\mathfrak{T}_0$, $\mathfrak{R}_1$, $\mathfrak{R}_2$ and $\mathfrak{S}_0$ given by* (15), (18) *and* (20), *for $n>1$. For $n=1$, Γ_0 is generated by $\mathfrak{T}_0$ and $\mathfrak{S}_0$.*

5. In this section we shall prove the independence of the generators given in Theorem 2. For $n=1$, this is trivial because $\mathfrak{S}_0$ is of finite order while $\mathfrak{T}_0$ is not. Hereafter we suppose that $n>1$.

(1) *Independence of $\mathfrak{T}_0$.* We consider the transformation

$$(X_1, Y_1) = (X, Y)\mathfrak{M}; \tag{21}$$

if XY' is symmetric, it is easily verified that $X_1 Y_1'$ is also symmetric. We shall show that if the diagonal elements of XY' are even, those of $X_1 Y_1'$ will also be even if $\mathfrak{M}$ is $\mathfrak{R}_1$, $\mathfrak{R}_2$ or $\mathfrak{S}_0$, while if $\mathfrak{M} = \mathfrak{T}_0$, it is possible to choose X and Y so that some diagonal element of $X_1 Y_1'$ is odd. This will show that $\mathfrak{T}_0$ is not expressible as a product of $\mathfrak{R}_1$, $\mathfrak{R}_2$ and $\mathfrak{S}_0$ and their inverses.

Assume now that the diagonal elements of XY' are even. From (21) one readily deduces that if $\mathfrak{M}$ is a rotation, $X_1 Y_1' = XY'$, so that the diagonal elements of $X_1 Y_1'$ are also even. If secondly $\mathfrak{M}$ is a semi-involution, we have

$$X_1 = XJ + Y(I - J), \qquad Y_1 = -X(I - J) + YJ,$$

so that

$$X_1 Y_1' = XJY' - Y(I - J)X' = XJY' + YJX' - YX'.$$

Since XJY' is the transpose of YJX', it is again clear that the diagonal elements of X_1Y_1' are even. Finally, suppose $\mathfrak{M}=\mathfrak{T}_0$. Then we obtain

$$X_1Y_1' = X(XS_0+Y)' = XY' + XS_0X'$$

and for $X=I$ the first diagonal element of X_1Y_1' is odd. This completes the proof of the independence of $\mathfrak{T}_0$. We may remark in passing, however, that $\mathfrak{T}_0^2$ is expressible as a product of the powers of $\mathfrak{R}_1$, $\mathfrak{R}_2$ and $\mathfrak{S}_0$.

(2) *Independence of* $\mathfrak{R}_1$. Let $\mathfrak{r}=(r_1, \cdots, r_n, s_1, \cdots, s_n)'$ be a column vector with $2n$ components. It is clear that the second component r_2 is unaffected when $\mathfrak{r}$ is multiplied on the left by any of the matrices $\mathfrak{T}_0$, $\mathfrak{R}_2$, and $\mathfrak{S}_0$ and their inverses. Under multiplication on the left by $\mathfrak{R}_1$, however, r_2 is replaced by r_1. Hence $\mathfrak{R}_1$ cannot be expressed as a product of $\mathfrak{T}_0$, $\mathfrak{R}_2$ and $\mathfrak{S}_0$ and their inverses.

(3) *Independence of* $\mathfrak{R}_2$. Multiplying $\mathfrak{r}$ on the left by $\mathfrak{R}_1$ or $\mathfrak{S}_0$ or their inverses permutes components of $\mathfrak{r}$; under any such permutation, however, any r_i and its corresponding s_i remain n components apart. Since the effect of multiplying on the left by $\mathfrak{T}_0$ is to replace r_1 by r_1+s_1, it is clear that by multiplying $\mathfrak{r}$ on the left by a product of $\mathfrak{R}_1$, $\mathfrak{S}_0$ and $\mathfrak{T}_0$ and their inverses, r_1 may be replaced by a linear combination of r_1 and s_1 and its position may be changed. It is however impossible to replace r_1 by r_1+r_2 in this way, and this is exactly the effect of multiplication of $\mathfrak{r}$ on the left by $\mathfrak{R}_2$. This proves the independence of $\mathfrak{R}_2$.

(4) *Independence of* $\mathfrak{S}_0$. We note that

$$\begin{pmatrix} * & * \\ 0 & * \end{pmatrix}\begin{pmatrix} * & * \\ 0 & * \end{pmatrix} = \begin{pmatrix} * & * \\ 0 & * \end{pmatrix}.$$

Since $\mathfrak{T}_0$, $\mathfrak{R}_1$ and $\mathfrak{R}_2$ and their inverses are all of the form

$$\begin{pmatrix} * & * \\ 0 & * \end{pmatrix}$$

and $\mathfrak{S}_0$ is not of this form, it is clear that $\mathfrak{S}_0$ is not expressible as a product of $\mathfrak{T}_0$, $\mathfrak{R}_1$ and $\mathfrak{R}_2$ and their inverses.

6. Our previous method can be extended to any Euclidean ring; in particular, for the ring formed by the Gaussian integers, we have the following result:

THEOREM 3. *Let Γ' be the group of matrices $\mathfrak{M}$ with Gaussian integers as elements which satisfy* (1). *Let Γ_0' be obtained from Γ' by identifying the four elements $\pm\mathfrak{M}$ and $\pm i\mathfrak{M}$. Then for $n>1$, Γ_0' is generated by the matrices $\mathfrak{T}_0$, $\mathfrak{R}_1$, $\mathfrak{R}_2$ and $\mathfrak{S}_0$ defined previously, and the matrix*

$$\mathfrak{T}_1 = \begin{pmatrix} I & S_1 \\ 0 & I \end{pmatrix} \qquad \text{where } S_1 = iS_0. \tag{22}$$

For $n=1$, Γ_0' is generated by $\mathfrak{T}_0$, $\mathfrak{T}_1$ and $\mathfrak{S}_0$.

The independence of the generators is shown as follows (with suitable modifications when $n=1$):

(1) *Independence of* $\mathfrak{T}_0$. We use the method of §5, (1). Let XY' be a symmetric matrix with Gaussian integers as elements, such that the real part of each diagonal element is even. This property is preserved when (X, Y) is subjected to the transformations $\mathfrak{T}_1$, $\mathfrak{R}_1$, $\mathfrak{R}_2$ and $\mathfrak{S}_0$ according to (21), but not for the transformation $\mathfrak{T}_0$.

(2) *Independence of* $\mathfrak{T}_1$. This is clear since $\mathfrak{T}_1$ is the only generator which is not real.

(3) The independence of $\mathfrak{R}_1$, $\mathfrak{R}_2$ and $\mathfrak{S}_0$ follow exactly as before.

TSING HUA UNIVERSITY,
PEIPING, CHINA.
INSTITUTE FOR ADVANCED STUDY,
PRINCETON, N. J.

AUTOMORPHISMS OF THE UNIMODULAR GROUP

BY

L. K. HUA AND I. REINER

Notation. Let $\mathfrak{M}_n$ denote the group of $n\times n$ integral matrices of determinant ± 1 (the unimodular group). By $\mathfrak{M}_n^+$ we denote that subset of $\mathfrak{M}_n$ where the determinant is $+1$; $\mathfrak{M}_n^-$ is correspondingly defined. Let $I^{(n)}$ (or briefly I) be the identity matrix in $\mathfrak{M}_n$, and let X' represent the transpose of X. The direct sum of the matrices A and B will be represented by $A \dot{+} B$;

$$A \stackrel{\bullet}{=} B$$

will mean that A is similar to B. In this paper, we shall find explicitly the generators of the group $\mathfrak{A}_n$ of all automorphisms of $\mathfrak{M}_n$.

1. **The commutator subgroup of $\mathfrak{M}_n$.** The following result is useful, and is of independent interest.

THEOREM 1. *Let $\mathfrak{N}_n$ be the commutator subgroup of $\mathfrak{M}_n$. Then trivially $\mathfrak{N}_n \subset \mathfrak{M}_n^+$. For $n=2$, $\mathfrak{N}_n$ is of index 2 in $\mathfrak{M}_n^+$, while for $n>2$, $\mathfrak{N}_n=\mathfrak{M}_n^+$.*

Proof. Consider first the case where $n=2$. Define

$$S=\begin{pmatrix} 0 & 1 \\ -1 & 0\end{pmatrix}, \qquad T=\begin{pmatrix} 1 & 1 \\ 0 & 1\end{pmatrix}. \tag{1}$$

It is well known that S and T generate $\mathfrak{M}_2^+$. An element X of $\mathfrak{M}_2^+$ is called *even* if, when X is expressed as a product of powers of S and T, the sum of the exponents is even; otherwise, X is called *odd*. Since all relations satisfied by S and T are consequences of

$$S^2 = -I, \qquad (ST)^3 = I,$$

it follows that the parity of $X\in\mathfrak{M}_2^+$ depends only on X, and not on the manner in which X is expressed as a product of powers of S and T. Let $\mathfrak{E}$ be the subgroup of $\mathfrak{M}_2^+$ consisting of all even elements; then clearly $\mathfrak{E}$ is of index 2 in $\mathfrak{M}_2^+$. It suffices to prove that $\mathfrak{E}=\mathfrak{N}_2$.

We prove first that $\mathfrak{N}_2\subset\mathfrak{E}$. Since the commutator subgroup of a group is always generated by squares, it suffices to show that $A\in\mathfrak{M}_2$ implies $A^2\in\mathfrak{E}$. For $A\in\mathfrak{M}_2^+$, this is clear. If $A\in\mathfrak{M}_2^-$, set $A=XJ=JY$, where

$$J=\begin{pmatrix} 1 & 0 \\ 0 & -1\end{pmatrix}, \tag{2}$$

Presented to the Society, December 29, 1950; received by the editors January 8, 1951.

and X and $Y \in \mathfrak{M}_2^+$. Then $A^2 = XY = XJ^{-1}XJ$. Hence we need only prove that if $X \in \mathfrak{M}_2^+$, X and $J^{-1}XJ$ are of the same parity. This is easily verified for $X = S$ or T; since S and T generate $\mathfrak{M}_2^+$, and $J^{-1}X_1X_2J = J^{-1}X_1J \cdot J^{-1}X_2J$, the result follows.

On the other hand we can show that $\mathfrak{C} \subset \mathfrak{R}_2$. For, $\mathfrak{C}$ is generated by T^2 and ST, since $TS = (ST \cdot T^{-2})^2$. However, $T^2 = TJT^{-1}J^{-1} \in \mathfrak{R}_2$, and therefore also $(T')^{-2} \in \mathfrak{R}_2$. Furthermore, $ST = TST^{-1}S^{-1}(T')^{-2}T^2 \in \mathfrak{R}_2$. This completes the proof for $n = 2$.

Suppose now that $n > 2$, and define

$$(3) \qquad R = \begin{bmatrix} 0 \cdots 0 & (-1)^{n-1} \\ 1 \cdots 0 & 0 \\ \cdots\cdots & \cdot \\ 0 \cdots 1 & 0 \end{bmatrix} \in \mathfrak{M}_n^+, \qquad S = \begin{pmatrix} 0 & 1 \\ -1 & 0 \end{pmatrix} + I^{(n-2)},$$

$$T = \begin{pmatrix} 1 & 1 \\ 0 & 1 \end{pmatrix} + I^{(n-2)}.$$

(The symbols S and T defined here are the analogues in $\mathfrak{M}_n^+$ of those defined by (1). It will be clear from the context which are meant.) For $n > 2$ we have([1])

$$T' = [R^{-1}(TR)^{-(n-2)}R(TR)^{n-2}](TR)^{-1}[R(TR)^{-(n-2)}R^{-1}(TR)^{n-2}](TR) \in \mathfrak{R}_n.$$

Further $S = TST^{-1}S^{-1}(T')^{-2}T \in \mathfrak{R}_n$. Finally, for odd n there exists a permutation matrix P such that $R^2 = P^{-1}RP$, whence $R = R^{-1}P^{-1}RP \in \mathfrak{R}_n$. For even n, R represents the monomial transformation

$$\begin{pmatrix} x_1 & x_2 & \cdots & x_{n-1} & x_n \\ x_2 & x_3 & \cdots & x_n & -x_1 \end{pmatrix},$$

which is a product of

$$\begin{pmatrix} x_1 & x_2 & x_3 \cdots x_{n-1} & x_n \\ x_2 & -x_1 & x_3 \cdots x_{n-1} & x_n \end{pmatrix}, \quad \begin{pmatrix} x_1 & x_2 & x_3 & x_4 \cdots x_n \\ -x_3 & x_2 & x_1 & x_4 \cdots x_n \end{pmatrix}.$$

$$\begin{pmatrix} x_1 & x_2 & x_3 & x_4 \cdots x_n \\ x_4 & x_2 & x_3 & -x_1 \cdots x_n \end{pmatrix}, \ldots, \begin{pmatrix} x_1 & x_2 \cdots x_{n-1} & x_n \\ x_n & x_2 \cdots x_{n-1} & -x_1 \end{pmatrix}.$$

each factor of which is similar to S (and hence is in $\mathfrak{R}_n$). Since T and R generate $\mathfrak{M}_n^+$, the theorem is proved.

COROLLARY 1. *In any automorphism of $\mathfrak{M}_n$, always $\mathfrak{M}_n^+ \to \mathfrak{M}_n^+$.*

Proof. For $n > 2$ this is an immediate corollary, since the commutator subgroup goes into itself in any automorphism. For $n = 2$, let $S \to S_1$ and

(1) L. K. Hua and I. Reiner, Trans. Amer. Math. Soc. vol. 65 (1949) p. 423.

$T \to T_1$. Then $ST \in \mathfrak{N}_2$ implies $S_1 T_1 \in \mathfrak{N}_2$, so $\det(S_1 T_1) = 1$. Further, $S^2 = -I$ implies $S_1^2 = -I$, so $\det S_1 = 1$, since the minimum function of S_1 is x^2+1, and the characteristic function must therefore be a power of x^2+1. This completes the proof when $n=2$.

2. **Automorphisms of $\mathfrak{M}_2^+$.** We wish to determine the automorphisms of $\mathfrak{M}_2$. Since every automorphism of $\mathfrak{M}_2$ takes $\mathfrak{M}_2^+$ into itself, we shall first determine all automorphisms of $\mathfrak{M}_2^+$. For $X \in \mathfrak{M}_2^+$, define $\epsilon(X) = +1$ or -1, according as X is even or odd.

THEOREM 2. *Every automorphism of $\mathfrak{M}_2^+$ is of one of the forms*

$$\text{(I)} \qquad X \in \mathfrak{M}_2^+ \to AXA^{-1} \qquad A \in \mathfrak{M}_2$$

or

$$\text{(II)} \qquad X \in \mathfrak{M}_2^+ \to \epsilon(X) \cdot AXA^{-1}, \qquad A \in \mathfrak{M}_2.$$

That is, the automorphism group of $\mathfrak{M}_2^+$ is generated by the set of "inner" automorphisms $X \to AXA^{-1}$ $(A \in \mathfrak{M}_2)$ and the automorphism $X \to \epsilon(X) \cdot X$.

Proof. Let τ be an automorphism of $\mathfrak{M}_2^+$; it certainly leaves $I^{(2)}$ and $-I^{(2)}$ individually unaltered. Let S and T (as given by (1)) be mapped into S^τ and T^τ. Then $(S^\tau)^2 = -I$. Since all second order fixed points are equivalent, there exists a matrix $B \in \mathfrak{M}_2$ such that $BS^\tau B^{-1} = S$. Instead of τ, consider the automorphism τ': $X \to BX^\tau B^{-1}$, which leaves S unaltered. Assume hereafter that τ leaves S invariant. (It is this sort of replacement of τ by τ' which we shall mean when we refer to some property holding "after a suitable inner automorphism.") Set

$$T^\tau = \begin{pmatrix} a & b \\ c & d \end{pmatrix}.$$

From $(ST)^3 = I$ we obtain $(ST^\tau)^3 = I$, whence $b - c = 1$. Since $\det T^\tau = 1$, we get

$$ad = 1 + bc = c^2 + c + 1 > 0.$$

Set $N = |a+d|$. If $N \geqq 3$, consider the elements generated by S and T^τ (mod N). Since $a+d \equiv 0$ (mod N), we find that $(T^\tau)^2 \equiv I$ (mod N). Furthermore $(ST^\tau)^3 \equiv I$ (mod N); therefore S and T^τ generate (mod N) at most the 12 elements

$$\pm I, \ \pm S, \ \pm T^\tau, \ \pm ST^\tau, \ \pm T^\tau S, \ \pm ST^\tau S.$$

But if τ is an automorphism, S and T^τ generate $\mathfrak{M}_2^+$, which has more than 12 elements (mod N) for $N \geqq 3$.

Therefore $N \leqq 2$. Since $ad > 0$, either $a = d = 1$ or $a = d = -1$, and thence $b = 1$, $c = 0$ or $b = 0$, $c = -1$. There are 4 possibilities for T^τ:

$$T^\tau = \begin{cases} T_0 = \begin{pmatrix} 1 & 1 \\ 0 & 1 \end{pmatrix}, & T_2 = \begin{pmatrix} -1 & 1 \\ 0 & -1 \end{pmatrix}, \\ T_1 = \begin{pmatrix} 1 & 0 \\ -1 & 1 \end{pmatrix}, & T_3 = \begin{pmatrix} -1 & 0 \\ -1 & -1 \end{pmatrix}. \end{cases}$$

Since S and T generate $\mathfrak{M}_2^+$, to determine τ it is sufficient to specify S^τ and T^τ. Thus every automorphism of $\mathfrak{M}_2^+$ is of the form $S \to BSB^{-1}$, $T \to BT_iB^{-1}$ (for some i, $i=0, 1, 2, 3$), where $B \in \mathfrak{M}_2$. If J is given by (2), we have:

$$T_0 = T, \qquad T_1 = STS^{-1}, \qquad T_2 = -JTJ^{-1}, \qquad T_3 = -SJTJ^{-1}S^{-1},$$

and also $S = -JSJ^{-1}$. The possible automorphisms are:

$i = 0$: $\quad S \to BSB^{-1}, \qquad T \to BTB^{-1}.$

$i = 1$: $\quad S \to BS \cdot S \cdot S^{-1}B^{-1}, \qquad T \to BS \cdot T \cdot S^{-1}B^{-1}.$

$i = 2$: $\quad S \to -BJ \cdot S \cdot J^{-1}B^{-1}, \qquad T \to -BJ \cdot T \cdot J^{-1}B^{-1}.$

$i = 3$: $\quad S \to -BSJ \cdot S \cdot J^{-1}S^{-1}B^{-1}, \qquad T \to -BSJ \cdot T \cdot J^{-1}S^{-1}B^{-1}.$

These automorphisms are of two types: for $i=0$ and 1, $S \to ASA^{-1}$, $T \to ATA^{-1}$, which imply that $X \in \mathfrak{M}_2^+ \to AXA^{-1}$; for $i=2$ and 3, $S \to -ASA^{-1}$, $T \to -ATA^{-1}$, which imply that $X \in \mathfrak{M}_2^+ \to \epsilon(X) \cdot AXA^{-1}$. This completes the proof.

3. **Automorphisms of $\mathfrak{M}_n^+$ and $\mathfrak{M}_n$.** We are now faced with the problem of determining the automorphisms of $\mathfrak{M}_2$ from those of $\mathfrak{M}_2^+$. We shall have the same problem for $\mathfrak{M}_n$ and $\mathfrak{M}_n^+$. As we shall see, the passage from $\mathfrak{M}_n^+$ to $\mathfrak{M}_n$ is trivial, and most of the difficulty lies in determining the automorphisms of $\mathfrak{M}_n^+$. In this paper we shall prove the following results:

THEOREM 3. *For $n > 2$, the group of those automorphisms of $\mathfrak{M}_n^+$ which are induced by automorphisms of $\mathfrak{M}_n$ is generated by*

(i) *the set of all "inner" automorphisms*

$$X \in \mathfrak{M}_n^+ \to AXA^{-1} \qquad (A \in \mathfrak{M}_n),$$

and

(ii) *the automorphism*

$$X \in \mathfrak{M}_n^+ \to X'^{-1}.$$

REMARK. When $n=2$, the automorphism (ii) is the same as $X \to SXS^{-1}$, hence is included in (i). The automorphism $X \to \epsilon(X) \cdot X$ occurs only for $n=2$. Furthermore, for odd n all automorphisms of $\mathfrak{M}_n^+$ are induced by automorphisms of $\mathfrak{M}_n$.

THEOREM 4. *The generators of $\mathfrak{A}_n$ are*

(i) *the set of all inner automorphisms*

$$X \in \mathfrak{M}_n \to AXA^{-1} \qquad (A \in \mathfrak{M}_n),$$

(ii) *the automorphism* $X \in \mathfrak{M}_n \to X'^{-1}$,

(iii) *for even* n *only, the automorphism*

$$X \in \mathfrak{M}_n \to (\det X) \cdot X,$$

and

(iv) *for* $n=2$ *only, the automorphism*

$$X \in \mathfrak{M}_2^+ \to \epsilon(X) \cdot X, \qquad X \in \mathfrak{M}_2^- \to \epsilon(JX) \cdot X,$$

where J *is given by* (2).

Further, when $n=2$, *the automorphism* (ii) *may be omitted from this list.*

Let us show that Theorem 4 is a simple consequence of Theorem 3. Let τ be any automorphism of $\mathfrak{M}_n$. By Corollary 1, τ induces an automorphism on $\mathfrak{M}_n^+$ which, by Theorems 2 and 3, can be written as:

$$X \in \mathfrak{M}_n^+ \to \alpha(X) \cdot AX^*A^{-1},$$

where $A \in \mathfrak{M}_n$, $\alpha(X)=1$ for all X or $\alpha(X)=\epsilon(X)$ for all X (this can occur only when $n=2$), and where either $X^*=X$ for all X or $X^*=X'^{-1}$ for all X.

Let Y and $Z \in \mathfrak{M}_n^-$; then

$$Y^\tau Z^\tau = (YZ)^\tau = \alpha(YZ) \cdot A(YZ)^*A^{-1},$$

whence

$$Y^\tau = \alpha(YZ) \cdot AY^*Z^*A^{-1}(Z^\tau)^{-1}.$$

Let $Z \in \mathfrak{M}_n^-$ be fixed; then

$$Y^\tau = \alpha(YZ) \cdot AY^*B \qquad \text{for all } Y \in \mathfrak{M}_n^-,$$

where A and B are independent of Y. But then

$$AY^*B \cdot AY^*B = (Y^\tau)^2 = (Y^2)^\tau = \alpha(Y^2)A(Y^2)^*A^{-1},$$

so that

$$(BA)Y^*(BA) = \alpha(Y^2)Y^*.$$

Since this is valid for all $Y \in \mathfrak{M}_n^-$, we see that of necessity $\alpha(Y^2)=1$ for all Y, and $BA = \pm I$. This shows that either $Y^\tau = \alpha(YZ) \cdot AY^*A^{-1}$ for all $Y \in \mathfrak{M}_n^-$, or $Y^\tau = -\alpha(YZ) \cdot AY^*A^{-1}$ for all $Y \in \mathfrak{M}_n^-$. If $n=2$ and $\alpha(YZ)=\epsilon(YZ)$, it is trivial to verify that either $\epsilon(YZ)=\epsilon(JY)$ for all $Y \in \mathfrak{M}_2^-$ or $\epsilon(YZ)=-\epsilon(JY)$ for all $Y \in \mathfrak{M}_2^-$.

The remainder of the paper will be concerned with proving Theorem 3.

4. **Canonical forms for involutions.** In the proof of Theorem 3 we shall use certain canonical forms of involutions under similarity transformations.

LEMMA 1. *Under a similarity transformation, every involution* $X \in \mathfrak{M}_n$ *such that* $X^2 = I^{(n)}$ *can be brought into the form*

$$W(x, y, z) = \underbrace{L \dotplus \cdots \dotplus L}_{(x \text{ terms})} \dotplus (-I)^{(y)} \dotplus I^{(z)}, \tag{4}$$

where $2x+y+z=n$ *and*

$$L = \begin{pmatrix} 1 & 0 \\ 1 & -1 \end{pmatrix}.$$

Proof. We prove first, by induction on n, that every $X \in \mathfrak{M}_n$ satisfying $X^2 = I$ is similar to a matrix of the form

$$\begin{pmatrix} I^{(l)} & 0 \\ M & -I^{(n-l)} \end{pmatrix}. \tag{5}$$

For $n=1$ and 2, this is trivial. Let the theorem be proved for n, and assume that $X^2 = I^{(n+1)}$, where $n \geqq 2$. Then $X^2 - I = 0$, or $(X-I)(X+I)=0$. If $X-I$ is nonsingular, then $X = -I$ and the result is obvious. Hence, supposing that $X-I$ is singular (so that $\lambda = 1$ is a characteristic root of X), there exists a primitive column vector $\mathfrak{t} = (t_1, \cdots, t_{n+1})'$ with integral elements such that $\mathfrak{t}'X = \mathfrak{t}'$. Choose $P \in \mathfrak{M}_{n+1}$ with first row $\mathfrak{t}'$. Then

$$PXP^{-1} = \begin{pmatrix} 1 & \mathfrak{n}' \\ \mathfrak{x} & X_1 \end{pmatrix},$$

where $\mathfrak{n}$ denotes a vector whose components are 0; thus

$$X \overset{s}{=} \begin{pmatrix} 1 & \mathfrak{n}' \\ \mathfrak{x} & X_1 \end{pmatrix}.$$

But

$$I^{(n+1)} = X^2 \overset{s}{=} \begin{pmatrix} 1 & \mathfrak{n}' \\ (I + X_1)\mathfrak{x} & X_1^2 \end{pmatrix}$$

shows that $X_1^2 = I^{(n)}$ and $(I+X_1)\mathfrak{x} = \mathfrak{n}$. By the induction hypothesis,

$$X_1 \overset{s}{=} \begin{pmatrix} I^{(m)} & 0 \\ M & -I^{(n-m)} \end{pmatrix},$$

and, after making the similarity transformation, we have (as a consequence of $(I+X_1)\mathfrak{x} = \mathfrak{n}$)

$$\begin{pmatrix} 2I^{(m)} & 0 \\ M & 0 \end{pmatrix}\mathfrak{x} = \mathfrak{n}.$$

Therefore

$$\mathfrak{x} = (\underbrace{0, \cdots, 0}_{(m \text{ terms})}, \underbrace{*, \cdots, *}_{(n-m \text{ terms})})',$$

where * denotes an arbitrary element. Thus

$$X \stackrel{s}{=} \begin{pmatrix} 1 & & \mathfrak{n}' & \\ 0 & & & \\ \cdot & I^{(m)} & & 0 \\ \cdot & & & \\ 0 & & & \\ * & & & \\ \cdot & & & \\ \cdot & M & & -I^{(n-m)} \\ \cdot & & & \\ * & & & \end{pmatrix} = \begin{pmatrix} I^{(m+1)} & 0 \\ \overline{M} & -I^{(n-m)} \end{pmatrix}.$$

This completes the first part of the proof.

Suppose we now subject (5) to a further similarity transformation by

$$\begin{pmatrix} A^{(l)} & 0 \\ C & D^{(n-l)} \end{pmatrix} \in \mathfrak{M}_n.$$

A simple calculation shows that we obtain a matrix given by (5) with M replaced by $\overline{M}$, where $\overline{M} = 2CA^{-1} + DMA^{-1}$. Choosing firstly $C = 0$, A and D unimodular, we find that $\overline{M} = DMA^{-1}$, and by proper choice of A and D we can make $\overline{M}$ diagonal. Supposing this done, secondly put $A = I$, $D = I$; we find that $\overline{M} = M + 2C$. Since C is arbitrary, we can bring $\overline{M}$ into the form

$$\begin{pmatrix} I^{(k)} & 0 \\ 0 & 0 \end{pmatrix},$$

where k is the rank of M. Since we can interchange two rows and simultaneously interchange the corresponding columns by means of a similarity transformation, the lemma follows.

It is easily seen that

$$W(x, y, z) \stackrel{s}{=} W(\bar{x}, \bar{y}, \bar{z})$$

only when $x = \bar{x}$, $y = \bar{y}$, and $z = \bar{z}$. Furthermore, changing the order of terms in the direct summation does not alter the similarity class. The number A_n of nonsimilar involutions in $\mathfrak{M}_n$ is therefore equal to the number of solutions of $2x + y + z = n$, $x \geqq 0$, $y \geqq 0$, $z \geqq 0$. This gives

$$A_n = \begin{cases} \left(\dfrac{n+2}{2}\right)^2, & n \text{ even}, \\ \dfrac{(n+1)(n+3)}{4}, & n \text{ odd}. \end{cases} \tag{6}$$

Let B_n be the number of nonsimilar involutions in $\mathfrak{M}_n^+$, where the similarity factors are in $\mathfrak{M}_n$. One easily obtains

$$B_n = \begin{cases} (A_n - 1)/2, & \text{if } n \equiv 0 \pmod 4, \\ A_n/2, & \text{otherwise.} \end{cases} \tag{7}$$

5. **Automorphisms of $\mathfrak{M}_3^+$.** We shall now prove Theorem 3 for $n=3$. Let

$$I_1 = \begin{pmatrix} -1 & 0 & 0 \\ 0 & -1 & 0 \\ 0 & 0 & 1 \end{pmatrix}, \qquad I_2 = \begin{pmatrix} 1 & 0 & 0 \\ 1 & -1 & 0 \\ 0 & 0 & -1 \end{pmatrix} \in \mathfrak{M}_3^+.$$

Then $I_1^2 = I^{(3)}$. Let τ be any automorphism of $\mathfrak{M}_3^+$ and let $X = I_1^\tau$; then $X^2 = I^{(3)}$. By Lemma 1, the matrices I_1, I_2, and $I^{(3)}$ form a complete system of nonsimilar involutions in $\mathfrak{M}_3^+$. Therefore

$$X \stackrel{s}{=} I_1 \text{ or } I_2.$$

After a suitable inner automorphism, we may assume that either $I_1 \to I_1$ or $I_1 \to I_2$. We shall show that this latter case is impossible by considering the normalizer groups of I_1 and I_2. The normalizer group of I_1, that is, the group of matrices $\in \mathfrak{M}_3^+$ which commute with I_1, consists of all elements of $\mathfrak{M}_3^+$ of the form

$$\begin{pmatrix} a & b & 0 \\ c & d & 0 \\ 0 & 0 & e \end{pmatrix},$$

and is isomorphic to $\mathfrak{M}_2$. That of I_2 consists of all elements of $\mathfrak{M}_3^+$ of the form

$$\begin{pmatrix} a & 0 & 0 \\ (a-e)/2 & e & f \\ -h/2 & h & i \end{pmatrix},$$

and is isomorphic to that subgroup $\mathfrak{G}$ of $\mathfrak{M}_2$ consisting of the elements

$$\begin{pmatrix} e & f \\ h & i \end{pmatrix} \in \mathfrak{M}_2, \qquad \text{where } \left.\begin{matrix} e \equiv 1 \\ h \equiv 0 \\ i \equiv 1 \end{matrix}\right\} \pmod 2.$$

Since e and i are both odd, $\mathfrak{G}$ contains no element of order 3, and hence is not isomorphic to $\mathfrak{M}_2$. But then $I_1 \to I_2$ is impossible.

We may assume thus that after a suitable inner automorphism, I_1 is invariant. Thence elements of $\mathfrak{M}_3^+$ which commute with I_1 map into elements of the same kind, so that

$$\begin{pmatrix} X & \mathfrak{n}' \\ \mathfrak{n} & \pm 1 \end{pmatrix} \in \mathfrak{M}_3^+ \to \begin{pmatrix} X^\tau & \mathfrak{n}' \\ \mathfrak{n} & \pm 1 \end{pmatrix}.$$

Since this induces an automorphism $X \to X^\tau$ on $\mathfrak{M}_2$, we see that $\det X^\tau = \det X$, and hence the plus signs go together, and so do the minus signs. By Theorem 2 and that part of Theorem 4 which follows from Theorem 2, there exists a matrix $A \in \mathfrak{M}_2$ such that $X^\tau = \pm AXA^{-1}$; here, the plus sign certainly occurs when X is an even element of $\mathfrak{M}_2^+$, and if the minus sign occurs for one odd element of $\mathfrak{M}_2^+$, then it occurs for *every* odd element of $\mathfrak{M}_2^+$. By use of a further inner automorphism using the factor $A^{-1} \dotplus I^{(1)}$, we may assume that

$$\begin{pmatrix} X & \mathfrak{n}' \\ \mathfrak{n} & \pm 1 \end{pmatrix} \in \mathfrak{M}_3^+ \to \begin{pmatrix} \pm X & \mathfrak{n}' \\ \mathfrak{n} & \pm 1 \end{pmatrix}, \tag{8}$$

so that

$$M = \begin{pmatrix} 1 & 0 & 0 \\ 0 & -1 & 0 \\ 0 & 0 & -1 \end{pmatrix} \to M \quad \text{or} \quad M \to N = \begin{pmatrix} -1 & 0 & 0 \\ 0 & 1 & 0 \\ 0 & 0 & -1 \end{pmatrix}.$$

Since

$$N = \begin{pmatrix} 0 & -1 & 0 \\ 1 & 0 & 0 \\ 0 & 0 & 1 \end{pmatrix} \cdot M \cdot \begin{pmatrix} 0 & 1 & 0 \\ -1 & 0 & 0 \\ 0 & 0 & 1 \end{pmatrix},$$

we may assume (after a further inner automorphism, if necessary) that I_1, M, and N are all invariant under the automorphism (but (8) need not hold).

Thus, after a suitably chosen inner automorphism, we have I_1, M, and N invariant. Therefore there exist A, B, and $C \in \mathfrak{M}_2$ such that

$$\begin{gathered} \begin{pmatrix} X & \mathfrak{n} \\ \mathfrak{n}' & \pm 1 \end{pmatrix} \in \mathfrak{M}_3^+ \to \begin{pmatrix} \pm AXA^{-1} & \mathfrak{n} \\ \mathfrak{n}' & \pm 1 \end{pmatrix}, \\ \begin{pmatrix} \pm 1 & \mathfrak{n}' \\ \mathfrak{n} & X \end{pmatrix} \in \mathfrak{M}_3^+ \to \begin{pmatrix} \pm 1 & \mathfrak{n}' \\ \mathfrak{n} & \pm BXB^{-1} \end{pmatrix}, \\ \begin{pmatrix} a & 0 & b \\ 0 & \pm 1 & 0 \\ c & 0 & d \end{pmatrix} \in \mathfrak{M}_3^+ \to \begin{pmatrix} \alpha & 0 & \beta \\ 0 & \pm 1 & 0 \\ \gamma & 0 & \delta \end{pmatrix}, \end{gathered} \tag{9}$$

where

$$\begin{pmatrix} \alpha & \beta \\ \gamma & \delta \end{pmatrix} = \pm C \begin{pmatrix} a & b \\ c & d \end{pmatrix} C^{-1},$$

and $\mathfrak{n} = (0, 0)'$. Here, the $+1$ on the left goes with the $+1$ on the right always (and the -1's go together); further, when X is an even element of $\mathfrak{M}_2^+$, the plus sign occurs before AXA^{-1}, BXB^{-1}, and CXC^{-1}, while if the minus sign occurs before one of these for any odd $X \in \mathfrak{M}_2^+$, it occurs there for every odd $X \in \mathfrak{M}_2^+$.

Now we may assume that at most one of A, B, and C has determinant -1; for if both A and B (say) have determinant -1, apply a further inner automorphism (with factor N) which leaves I_1, M, and N invariant and changes the signs of det A and det B. Suppose hereafter, without loss of generality, that det $A =$ det $B = 1$.

Next, N is invariant, but by (9) goes into

$$\begin{pmatrix} \pm A\begin{pmatrix} -1 & 0 \\ 0 & 1 \end{pmatrix}A^{-1} & \mathfrak{n}' \\ \mathfrak{n} & -1 \end{pmatrix},$$

so that

$$\pm A\begin{pmatrix} -1 & 0 \\ 0 & 1 \end{pmatrix}A^{-1} = \begin{pmatrix} -1 & 0 \\ 0 & 1 \end{pmatrix}.$$

This gives two possibilities:

$$A = I^{(2)} \quad \text{or} \quad \begin{pmatrix} 0 & 1 \\ -1 & 0 \end{pmatrix}.$$

The same holds true for B (but not necessarily for C, since det $C = \pm 1$).

Suppose firstly that either A or B is $I^{(2)}$, say $A = I^{(2)}$. Then

$$T = \begin{pmatrix} 1 & 1 & 0 \\ 0 & 1 & 0 \\ 0 & 0 & 1 \end{pmatrix} \rightarrow \begin{pmatrix} \pm \begin{pmatrix} 1 & 1 \\ 0 & 1 \end{pmatrix} & \begin{matrix} 0 \\ 0 \end{matrix} \\ \begin{matrix} 0 & 0 \end{matrix} & 1 \end{pmatrix}.$$

Case 1. T invariant. Then

$$\begin{pmatrix} 0 & 1 & 0 \\ -1 & 0 & 0 \\ 0 & 0 & 1 \end{pmatrix} \quad \text{and} \quad \begin{pmatrix} 0 & 1 & 0 \\ 1 & 0 & 0 \\ 0 & 0 & -1 \end{pmatrix}$$

are both invariant. (The first matrix is invariant in virtue of the remarks after (9); the second is invariant because it is M times the first.) For either possible choice of B we find that

$$\begin{pmatrix} -1 & 0 & 0 \\ 0 & 0 & 1 \\ 0 & 1 & 0 \end{pmatrix} \rightarrow \begin{pmatrix} -1 & \begin{matrix} 0 & 0 \end{matrix} \\ \begin{matrix} 0 \\ 0 \end{matrix} & \pm \begin{pmatrix} 0 & 1 \\ 1 & 0 \end{pmatrix} \end{pmatrix}.$$

Therefore

$$U=\begin{pmatrix}0&1&0\\0&0&1\\1&0&0\end{pmatrix}=\begin{pmatrix}-1&0&0\\0&-1&0\\0&0&1\end{pmatrix}\begin{pmatrix}-1&0&0\\0&0&1\\0&1&0\end{pmatrix}\begin{pmatrix}0&1&0\\1&0&0\\0&0&-1\end{pmatrix}$$

is mapped into

$$\begin{pmatrix}-1&0&0\\0&-1&0\\0&0&1\end{pmatrix}\begin{pmatrix}-1&0\ \ 0\\0&\pm\begin{pmatrix}0&1\\1&0\end{pmatrix}\\0&\end{pmatrix}\begin{pmatrix}0&1&0\\1&0&0\\0&0&-1\end{pmatrix}=\begin{cases}U, & \text{if } + \text{ is used},\\ V, & \text{if } - \text{ is used},\end{cases}$$

where $V=I_1UI_1^{-1}$. Thus, in this case, $T\to T=I_1TI_1^{-1}$, and either $U\to U$ or $U\to I_1UI_1^{-1}$. Since T and U generate(²) $\mathfrak{M}_3^+$, the automorphism is inner.

Case 2.

$$T\to\begin{pmatrix}-1&-1&0\\0&-1&0\\0&0&1\end{pmatrix}.$$

Then

$$\begin{pmatrix}0&1&0\\1&0&0\\0&0&-1\end{pmatrix}\to\begin{pmatrix}0&-1&0\\-1&0&0\\0&0&-1\end{pmatrix},$$

and one finds in this case that

$$U\to\begin{pmatrix}0&-1&0\\0&0&1\\-1&0&0\end{pmatrix}\quad\text{or}\quad\begin{pmatrix}0&-1&0\\0&0&-1\\1&0&0\end{pmatrix}.$$

If we set $Z=TU^2$, then

$$\begin{pmatrix}1&0&0\\1&1&0\\0&0&1\end{pmatrix}=(UZ^{-1})^2UZ^2. \tag{10}$$

Now certainly the left side of (10) maps into

$$\begin{pmatrix}-1&0&0\\-1&-1&0\\0&0&1\end{pmatrix},$$

(²) L. K. Hua and I. Reiner, loc. cit.

whereas, knowing T^τ and U^τ, we can compute Z^τ and thence can find the image of the right side of (10). We readily find (for either value of U^τ) that the right side of (10) maps into

$$\begin{pmatrix} 1 & \cdot & \cdot \\ 3 & \cdot & \cdot \\ \cdot & \cdot & \cdot \end{pmatrix},$$

and hence we have a contradiction.

Therefore case 2 cannot occur, and so if either A or B equals $I^{(2)}$, the automorphism is inner. Suppose hereafter that

$$A = B = \begin{pmatrix} 0 & 1 \\ -1 & 0 \end{pmatrix}.$$

In this case we have

$$T \rightarrow \begin{pmatrix} \pm \begin{pmatrix} 1 & 0 \\ -1 & 1 \end{pmatrix} & \begin{matrix} 0 \\ 0 \end{matrix} \\ \begin{matrix} 0 & 0 \end{matrix} & 1 \end{pmatrix}.$$

Case 1*.

$$T \rightarrow \begin{pmatrix} 1 & 0 & 0 \\ -1 & 1 & 0 \\ 0 & 0 & 1 \end{pmatrix}.$$

Then as before

$$\begin{pmatrix} 0 & 1 & 0 \\ -1 & 0 & 0 \\ 0 & 0 & 1 \end{pmatrix} \text{ and } \begin{pmatrix} 0 & 1 & 0 \\ 1 & 0 & 0 \\ 0 & 0 & -1 \end{pmatrix}$$

are invariant, and again $U^\tau = U$ or V. After a further inner automorphism by a factor of I_1 (in the latter case) we also have $U \rightarrow U$. But then

$$T \rightarrow T'^{-1}, \qquad U \rightarrow U'^{-1}.$$

(This automorphism is easily shown to be a non-inner automorphism.)

Case 2*.

$$T \rightarrow \begin{pmatrix} -1 & 0 & 0 \\ 1 & -1 & 0 \\ 0 & 0 & 1 \end{pmatrix}.$$

Then

$$\begin{pmatrix} 0 & 1 & 0 \\ 1 & 0 & 0 \\ 0 & 0 & -1 \end{pmatrix} \rightarrow \begin{pmatrix} 0 & -1 & 0 \\ -1 & 0 & 0 \\ 0 & 0 & -1 \end{pmatrix},$$

and again we find that there are two possibilities for U^τ, each of which leads to a contradiction, just as in case 2. Therefore Theorem 3 holds when $n=3$.

6. **A fundamental lemma.** Theorem 3 will be proved by induction on n; the result has already been established for $n=2$ and 3. In going from $n-1$ to n, the following lemma is basic:

LEMMA 2. *Let* $n \geqq 4$, *and define* $J_1 = (-1) \dotplus I^{(n-1)}$. *In any automorphism* τ *of* $\mathfrak{M}_n$, $J_1^\tau = \pm A J_1 A^{-1}$ *for some* $A \in \mathfrak{M}_n$.

Proof. By Corollary 1, $J_1^\tau \in \mathfrak{M}_n^-$, and J_1^τ is an involution. After a suitable inner automorphism, we may assume that $J_1^\tau = W(x, y, z)$ (as defined by (4)), where $2x+y+z=n$ and $x+y$ is odd. Every element of $\mathfrak{M}_n$ which commutes with J_1 maps into an element of $\mathfrak{M}_n$ which commutes with W. Every matrix in $\mathfrak{M}_n^+$ maps into a matrix in $\mathfrak{M}_n^+$. Combining these facts, we see that the group $\mathfrak{G}_1$ consisting of those elements of $\mathfrak{M}_n^+$ which commute with J_1 is isomorphic to $\mathfrak{G}_2$, the corresponding group for W. If we prove that this can happen only for $x=0$, $y=1$, $z=n-1$ or $x=0$, $y=n-1$, $z=1$, the result will follow.

The group $\mathfrak{G}_1$ consists of the matrices in $\mathfrak{M}_n^+$ of the form $(\pm 1) \dotplus X_1$, $X_1 \in \mathfrak{M}_{n-1}$, and so clearly $\mathfrak{G}_1 \cong \mathfrak{M}_{n-1}$.

The group $\mathfrak{G}_2$ is easily found to consist of all matrices $C \in \mathfrak{M}_1^+$ of the form (we illustrate the case where $x=2$):

$$\begin{array}{cc}
C = \left[\begin{array}{cccccc}
a_1 & 0 & a_2 & 0 & 0 \cdots 0 & 2\beta_1 \cdots 2\beta_z \\
\dfrac{a_1-d_1}{2} & d_1 & \dfrac{a_2-d_2}{2} & d_2 & \alpha_1 \cdots \alpha_y & \beta_1 \cdots \beta_z \\
a_3 & 0 & a_4 & 0 & 0 \cdots 0 & 2\delta_1 \cdots 2\delta_z \\
\dfrac{a_3-d_3}{2} & d_3 & \dfrac{a_4-d_4}{2} & d_4 & \gamma_1 \cdots \gamma_y & \delta_1 \cdots \delta_z \\
\epsilon_1 & -2\epsilon_1 & \zeta_1 & -2\zeta_1 & & \\
\vdots & \vdots & \vdots & \vdots & U & 0 \\
\epsilon_y & -2\epsilon_y & \zeta_y & -2\zeta_y & & \\
\eta_1 & 0 & \theta_1 & 0 & & \\
\vdots & \vdots & \vdots & \vdots & 0 & V \\
\eta_z & 0 & \theta_z & 0 & &
\end{array}\right]
& \begin{array}{l}
2x \text{ rows} \\[4em] y \text{ rows} \\[4em] z \text{ rows}
\end{array} \\
\begin{array}{ccc}
2x \text{ columns} & \quad y \text{ columns} & \quad z \text{ columns}
\end{array} &
\end{array}$$

For the moment put

$$K = \begin{pmatrix} 1 & 0 \\ -1/2 & 1 \end{pmatrix} \underset{(x \text{ terms})}{+ \cdots +} \begin{pmatrix} 1 & 0 \\ -1/2 & 1 \end{pmatrix} + I^{(n-2x)}.$$

Then a simple calculation gives:

$$KCK^{-1} = \begin{bmatrix} a_1 & 0 & a_2 & 0 & 0 \cdots 0 & 2\beta_1 \cdots 2\beta_z \\ 0 & d_1 & 0 & d_2 & \alpha_1 \cdots \alpha_y & 0 \cdots 0 \\ a_3 & 0 & a_4 & 0 & 0 \cdots 0 & 2\delta_1 \cdots 2\delta_z \\ 0 & d_3 & 0 & d_4 & \gamma_1 \cdots \gamma_y & 0 \cdots 0 \\ 0 & -2\epsilon_1 & 0 & -2\zeta_1 & & \\ \vdots & \vdots & \vdots & \vdots & U & 0 \\ 0 & -2\epsilon_y & 0 & -2\zeta_y & & \\ \eta_1 & 0 & \theta_1 & 0 & & \\ \vdots & \vdots & \vdots & \vdots & 0 & V \\ \eta_z & 0 & \theta_z & 0 & & \end{bmatrix}$$

and so C is similar to

$$\begin{bmatrix} a_1 & a_2 & 2\beta_1 \cdots 2\beta_z \\ a_3 & a_4 & 2\delta_1 \cdots 2\delta_z \\ \eta_1 & \theta_1 & \\ \vdots & \vdots & V \\ \eta_z & \theta_z & \end{bmatrix} + \begin{bmatrix} d_1 & d_2 & \alpha_1 \cdots \alpha_y \\ d_3 & d_4 & \gamma_1 \cdots \gamma_y \\ -2\epsilon_1 & -2\zeta_1 & \\ \vdots & \vdots & U \\ -2\epsilon_y & -2\zeta_y & \end{bmatrix}$$

$$= \underset{x \qquad z}{\begin{bmatrix} S_1 & 2R_1 \\ Q_1 & T_1 \end{bmatrix}}\begin{matrix} x \\ z \end{matrix} + \underset{x \qquad y}{\begin{bmatrix} S_2 & Q_2 \\ 2R_2 & T_2 \end{bmatrix}}\begin{matrix} x \\ y \end{matrix},$$

with a fixed similarity factor depending only on W. Therefore $\mathfrak{G}_2 \cong \mathfrak{G}$, where $\mathfrak{G} = \mathfrak{G}(x, y, z)$ is the group of matrices in $\mathfrak{M}_n^+$ of the form

$$\underset{x \qquad z}{\begin{bmatrix} S_1 & 2R_1 \\ Q_1 & T_1 \end{bmatrix}}\begin{matrix} x \\ z \end{matrix} + \underset{x \qquad y}{\begin{bmatrix} S_2 & Q_2 \\ 2R_2 & T_2 \end{bmatrix}}\begin{matrix} x \\ y \end{matrix},$$

where $S_1 \equiv S_2 \pmod 2$. Here $2x+y+z=n$ and $x+y$ is odd.

We wish to prove that $\mathfrak{M}_{n-1} \cong \mathfrak{G}(x, y, z)$ only when $x=0$, $y=1$, $z=n-1$ or $x=0$, $y=n-1$, $z=1$. In order to establish this, we shall prove that in all other cases the number of involutions in $\mathfrak{G}$ which are nonsimilar in $\mathfrak{G}$ is greater than the number of involutions in $\mathfrak{M}_{n-1}$ which are nonsimilar in $\mathfrak{M}_{n-1}$;

this latter number is, of course, A_{n-1} (given by (6)).

We shall briefly denote the elements of $\mathfrak{G}$ by $A\dot{+}B$, where

$$A=\begin{pmatrix} S_1 & 2R_1 \\ Q_1 & T_1 \end{pmatrix} \quad \text{and} \quad B=\begin{pmatrix} S_2 & Q_2 \\ 2R_2 & T_2 \end{pmatrix}.$$

If $A_1\dot{+}B_1$ and $A_2\dot{+}B_2$ are two involutions in $\mathfrak{G}$, where either

$$A_1 \overset{s}{\neq} A_2$$

in $\mathfrak{M}_{x+z}$ or

$$B_1 \overset{s}{\neq} B_2$$

in $\mathfrak{M}_{x+y}$, then certainly

$$A_1\dot{+}B_1 \overset{s}{\neq} A_2\dot{+}B_2$$

in $\mathfrak{G}$ (these may be similar in $\mathfrak{M}_n$, however). Therefore, the matrices $A\dot{+}B$, where

$$A = I^{(a_1)} \dot{+} (-I)^{(b_1)} \dot{+} \underbrace{L \dot{+} \cdots \dot{+} L}_{(c_1 \text{ terms})},$$

$$B = I^{(a_2)} \dot{+} (-I)^{(b_2)} \dot{+} \underbrace{L \dot{+} \cdots \dot{+} L}_{(c_2 \text{ terms})},$$

obtained by taking different sets of values of $(a_1, b_1, c_1, a_2, b_2, c_2)$, if they lie in $\mathfrak{G}$, are certainly nonsimilar in $\mathfrak{G}$. Here we have

$$a_1+b_1+2c_1=x+z, \quad a_2+b_2+2c_2=x+y, \quad b_1+b_2+c_1+c_2 \text{ even}.$$

If $x\neq 0$, we impose the further restriction that $c_1 \leqq (z+1)/2$, $c_2 \leqq (y+1)/2$, and that in B instead of L we use L'. These conditions will insure that $A\dot{+}B\in\mathfrak{G}$. We certainly do not (in general) get all of the nonsimilar involutions of $\mathfrak{G}$ in this way, but instead we obtain only a subset thereof. Call the number of such matrices N.

For $x=0$, we have $N=B_yB_z+(A_y-B_y)(A_z-B_z)$. Since y is odd, $A_y=2B_y$, and therefore

$$N=B_yA_z=B_yA_{n-y}.$$

Case 1. n even. Then $N=(y+1)(y+3)(n-y+1)(n-y+3)/32$. If neither y nor $n-y$ is 1 (certainly neither can be zero), then

$$(y+1)(n-y+1) \geqq 4(n-2) \quad \text{and} \quad (y+3)(n-y+3) \geqq 6n,$$

so that

$$N \geqq (24/32)\, n(n-2).$$

For $n=4$, $x=0$, either $y=1$ or $z=1$. For $n\geqq 6$, we have $N>A_{n-1}$. Hence in this case $\mathfrak{G}$ is not isomorphic to $\mathfrak{M}_{n-1}$. (If either y or $n-y=1$, then $W(x, y, z) = \pm J_1$.)

Case 2. n odd. Then $N=(y+1)(y+3)(n-y+2)^2/32$. We find again that $N>A_{n-1}$ for $n\geqq 5$.

This settles the cases where $x=0$. Suppose that $x\neq 0$ hereafter. Then N is the number of solutions of

$$a_1+b_1+2c_1 = x+z, \quad a_2+b_2+2c_2 = x+y, \quad b_1+b_2+c_1+c_2 \text{ even},$$
$$0 \leqq c_1 \leqq \frac{z+1}{2}, \qquad 0 \leqq c_2 \leqq \frac{y+1}{2}.$$

Using $[r]$ to denote the greatest integer less than or equal to r, we readily find that N is given by

$$\frac{1}{2}\left[\frac{z+3}{2}\right]\left[\frac{y+3}{2}\right]\left(x+z+1-\left[\frac{z+1}{2}\right]\right)\left(x+y+1-\left[\frac{y+1}{2}\right]\right).$$

By considering separately the cases where y and z are both even, one even and one odd, and so on, it is easy to prove that $N\geqq A_{n-1}$ in all cases except when both y and z are zero. Leaving aside this case for the moment, consider the matrix $A_0\dotplus I^{(x+y)}\in\mathfrak{G}$, where $A_0\in\mathfrak{M}_{x+z}$ is given by

$$A_0 = \begin{pmatrix} 1 & 2 & 2\cdots & 2 \\ 0 & -1 & 0\cdots & 0 \\ 0 & 0 & -1\cdots & 0 \\ \cdot & \cdot & \cdots\cdot & \cdot \\ 0 & 0 & 0\cdots & -1 \end{pmatrix}.$$

The matrix $A_0\dotplus I^{(x+y)}$ is certainly an involution in $\mathfrak{G}$. Since, in $\mathfrak{M}_{x+z}$,

$$A_0 \overset{s}{=} \begin{pmatrix} 1 & 0\cdots & 0 \\ 0 & -1\cdots & 0 \\ \cdot & \cdots\cdot & \cdot \\ 0 & 0\cdots & -1 \end{pmatrix} = A_1,$$

$A_0\dotplus I^{(x+y)}$ can be similar (in $\mathfrak{G}$) only to that matrix (counted in the N matrices) of the form $A_1\dotplus I^{(x+y)}$. But from

$$A_1\cdot\begin{pmatrix} a_1 & a_2\cdots a_z & 2b_1\cdots 2b_z \\ \cdots & \cdots & \cdots \\ \cdots & \cdots & \cdots \\ \cdots & \cdots & \cdots \end{pmatrix} = \begin{pmatrix} a_1 & a_2\cdots a_z & 2b_1\cdots 2b_z \\ \cdots & \cdots & \cdots \\ \cdots & \cdots & \cdots \\ \cdots & \cdots & \cdots \end{pmatrix}\cdot A_0$$

we obtain

$$a_1 = a_2 = \cdots = a_z = 2b_1,$$

which is impossible. Hence $\mathfrak{G}$ contains at least $N+1$ nonsimilar involutions, and therefore $\mathfrak{G}$ is not isomorphic to $\mathfrak{M}_{n-1}$ in these cases.

We have left only the case $y=z=0$, $x=n/2$; then n is singly even. Here we may choose $A=W(c_1, b_1, a_1)$, $B=W(c_1, b_2, a_2)$, where

$$a_1 + b_1 + 2c_1 = x, \qquad a_2 + b_2 + 2c_1 = x, \qquad b_1 + b_2 \text{ even.}$$

Then $A \dotplus B \in \mathfrak{G}$, and the various matrices are nonsimilar. The number of such matrices is $(x+1)(x+2)(x+3)/12$, which is greater than A_{n-1} for $n \geqq 14$. For $n=6$, $\mathfrak{M}_{n-1}$ contains an element of order 5, while $\mathfrak{G}$ does not. For $n=10$, $\mathfrak{M}_{n-1}$ contains an element of order 7, while $\mathfrak{G}$ does not. This completes the proof of the lemma.

7. **Proof of Theorem 3.** We are now ready to give a proof of Theorem 3 by induction on n. Hereafter, let $n \geqq 4$ and suppose that Theorem 3 holds for $n-1$. If τ is any automorphism of $\mathfrak{M}_n$, by Corollary 1 and Lemma 2 we know that τ takes $\mathfrak{M}_n^+$ into itself, and $J_1^\tau = \pm AJ_1A^{-1}$. If we change τ by a suitable inner automorphism, then we may assume that $J_1 \to \pm J_1$. When n is odd, certainly $J_1 \to J_1$; when n is even, by multiplying τ by the automorphism $X \in \mathfrak{M}_n \to (\det X)\cdot X$ if necessary, we may again assume $J_1 \to J_1$.

Therefore, every $M \in \mathfrak{M}_n^+$ which commutes with J_1 goes into another such element, that is,

$$\begin{pmatrix} \pm 1 & \mathfrak{n}' \\ \mathfrak{n} & X \end{pmatrix}^\tau = \begin{pmatrix} \pm 1 & \mathfrak{n}' \\ \mathfrak{n} & X^\tau \end{pmatrix}.$$

Since this induces an automorphism on $\mathfrak{M}_{n-1}$, we have $\det X^\tau = \det X$, so that the plus signs go together, as do the minus signs. Furthermore, by our induction hypothesis,

$$X^\tau = \pm AX^*A^{-1},$$

where $A \in \mathfrak{M}_{n-1}$ and either $X^* = X$ for all $X \in \mathfrak{M}_{n-1}$ or $X^* = X'^{-1}$ for all $X \in \mathfrak{M}_{n-1}$; here the minus sign can occur only for $X \in \mathfrak{M}_{n-1}^-$, and if it occurs for one such X, it occurs for all $X \in \mathfrak{M}_{n-1}^-$. After changing our original automorphism by a factor of $I^{(1)} \dotplus A^{-1}$, we may assume that $X^\tau = \pm X^*$.

Let J_ν be obtained from $I^{(n)}$ by replacing the νth diagonal element by -1. Then

$$J_1J_n = \begin{pmatrix} -1 & 0\cdots 0 & 0 \\ 0 & 1\cdots 0 & 0 \\ \cdot & \cdots\cdots & \cdot \\ 0 & 0\cdots 1 & 0 \\ 0 & 0\cdots 0 & -1 \end{pmatrix} \to \begin{pmatrix} -1 & \mathfrak{n}' \\ \mathfrak{n} & \pm\begin{pmatrix} 1\cdots 0 & 0 \\ \cdots\cdot 0 & 0 \\ 0\cdots 1 & 0 \\ 0\cdots 0 & -1 \end{pmatrix}^* \end{pmatrix}.$$

The minus sign here is impossible by Lemma 2, since $n \geqq 4$. Hence $J_1 J_n$ is invariant, and therefore so is J_n. By the same reasoning all of the J_ν ($\nu = 1, \cdots, n$) are invariant.

From the above remarks we see that for $X \in \mathfrak{M}_{n-1}^+$,

$$\begin{pmatrix} 1 & \mathfrak{n}' \\ \mathfrak{n} & X \end{pmatrix}^\tau = \begin{pmatrix} 1 & \mathfrak{n}' \\ \mathfrak{n} & A_1 X^* A_1^{-1} \end{pmatrix}, \cdots, \begin{pmatrix} X & \mathfrak{n} \\ \mathfrak{n}' & 1 \end{pmatrix}^\tau = \begin{pmatrix} A_n X^* A_n^{-1} & \mathfrak{n} \\ \mathfrak{n}' & 1 \end{pmatrix},$$

where $A_\nu \in \mathfrak{M}_{n-1}$, and in fact $A_1 = I$. Now suppose that $Z \in \mathfrak{M}_{n-2}^+$, and form $I^{(2)} \dotplus Z$. Since it commutes with both J_1 and J_2, its image must do likewise. But then

$$A_1 \begin{pmatrix} 1 & \mathfrak{n}' \\ \mathfrak{n} & Z \end{pmatrix} A_1^{-1} = \begin{pmatrix} 1 & \mathfrak{n}' \\ \mathfrak{n} & \bar{Z} \end{pmatrix}$$

for every $Z \in \mathfrak{M}_{n-2}^+$. Setting

$$A_1 = \begin{pmatrix} a & \mathfrak{x}' \\ \mathfrak{y} & A \end{pmatrix}$$

we obtain $\mathfrak{x}'Z = \mathfrak{x}'$, $\mathfrak{y} = \bar{Z}\mathfrak{y}$. Since this holds for all $Z \in \mathfrak{M}_{n-2}^+$, we must have $\mathfrak{x} = \mathfrak{y} = \mathfrak{n}$, so that A_1 is itself decomposable. A similar argument (considering the matrices commuting with both J_1 and J_ν, for $\nu = 3, \cdots, n$) shows that A_1 is diagonal. Correspondingly, all of the A_ν are diagonal. It is further clear that all of the A_ν ($\nu = 1, \cdots, n$) are sections of a single diagonal matrix $D^{(n)}$. Using the further inner automorphism factor D^{-1}, we may henceforth assume that $X^\tau = X^*$ for every decomposable $X \in \mathfrak{M}_n^+$, where either $X^* = X$ always or $X^* = X'^{-1}$ always. Since $\mathfrak{M}_n^+$ is generated by the set of decomposable elements of $\mathfrak{M}_n^+$, the theorem is proved.

TSING HUA UNIVERSITY,
PEKING, CHINA.
UNIVERSITY OF ILLINOIS,
URBANA, ILL.

AUTOMORPHISMS OF THE PROJECTIVE UNIMODULAR GROUP

BY

L. K. HUA AND I. REINER

Notation. Let $\mathfrak{M}_n$ denote the group of $n\times n$ integral matrices of determinant ± 1 (the unimodular group). By $\mathfrak{M}_n^+$ we denote that subset of $\mathfrak{M}_n$ where the determinant is $+1$; $\mathfrak{M}_n^-$ is correspondingly defined. Let $\mathfrak{P}_{2n}$ be obtained from $\mathfrak{M}_{2n}$ by identifying $+X$ and $-X$, $X\in\mathfrak{M}_{2n}$. (This is the same as considering the factor group of $\mathfrak{M}_{2n}$ by its centrum.) We correspondingly obtain $\mathfrak{P}_{2n}^+$ and $\mathfrak{P}_{2n}^-$ from $\mathfrak{M}_{2n}^+$ and $\mathfrak{M}_{2n}^-$. Let $I^{(n)}$ (or briefly I) be the identity matrix in $\mathfrak{M}_n$, and let X' denote the transpose of X. The direct sum of A and B is represented by $A\dotplus B$, while

$$A\overset{s}{=}B$$

means that A is similar to B.

In this paper we shall find explicitly the generators of the group $\mathfrak{B}_{2n}$ of all automorphisms of $\mathfrak{P}_{2n}$, thereby obtaining a complete description of these automorphisms. This generalizes the result due to Schreier[1] for the case $n=1$.

We shall frequently refer to results of an earlier paper: *Automorphisms of the unimodular group*, L. K. Hua and I. Reiner, Trans. Amer. Math. Soc. vol. 71 (1951) pp. 331-348. We designate this paper by AUT.

1. **The commutator subgroup of $\mathfrak{P}_{2n}$.** The following useful result is an immediate consequence of the corresponding theorem for $\mathfrak{M}_{2n}$ (AUT, Theorem 1).

THEOREM 1. *Let $\mathfrak{S}_{2n}$ be the commutator subgroup of $\mathfrak{P}_{2n}$. Then clearly $\mathfrak{S}_{2n}\subset\mathfrak{P}_{2n}^+$. For $n=1$, $\mathfrak{S}_{2n}$ is of index 2 in $\mathfrak{P}_{2n}^+$, while for $n>1$, $\mathfrak{S}_{2n}=\mathfrak{P}_{2n}^+$.*

THEOREM 2. *In any automorphism of $\mathfrak{P}_{2n}$, always $\mathfrak{P}_{2n}^+$ goes into itself.*

Proof. This is a corollary to Theorem 1 when $n>1$, since the commutator subgroup goes into itself under any automorphism. For $n=1$, suppose that $\pm S\to\pm S_1$ and $\pm T\to\pm T_1$, where

$$(1)\qquad S=\begin{pmatrix}0&1\\-1&0\end{pmatrix},\qquad T=\begin{pmatrix}1&1\\0&1\end{pmatrix}.$$

Since S and T generate $\mathfrak{M}_2^+$, it follows that $\pm S$ and $\pm T$ generate $\mathfrak{P}_2^+$,

Received by the editors May 18, 1951.

(1) Abh. Math. Sem. Hamburgischen Univ. vol. 3 (1924) p. 167.

and hence so must $\pm S_1$ and $\pm T_1$. It is therefore sufficient to prove that det S_1=det $T_1=+1$. From $(ST)^3=I$ we deduce $S_1T_1=\pm T_1^{-1}S_1^{-1}T_1^{-1}S_1^{-1}$, so that det $S_1T_1=1$. Hence either S_1 and T_1 are both in $\mathfrak{P}_2^+$ or both in $\mathfrak{P}_2^-$; we shall show that the latter alternative is impossible.

Suppose that det S_1=det $T_1=-1$. From $S^2=I$ we deduce $S_1^2=\pm I$; if $S_1^2=-I$, then $S_1^2+I=0$ and the characteristic equation of S_1 is $\lambda^2+1=0$, from which it follows that det $S_1=1$; this contradicts our assumption that det $S_1=-1$, so of necessity $S_1^2=I$. But if this is the case, then it is easy to show that there exists a matrix $A\in\mathfrak{M}_2$ such that AS_1A^{-1} takes one of the two canonical forms

$$\begin{pmatrix}1 & 0\\ 0 & -1\end{pmatrix} \quad\text{and}\quad \begin{pmatrix}1 & 0\\ 1 & -1\end{pmatrix}.$$

By considering instead of the original automorphism τ, a new automorphism τ' defined by: $X^{\tau'}=AX^{\tau}A^{-1}$, we may hereafter assume that

$$S_1 = \pm\begin{pmatrix}1 & 0\\ 0 & -1\end{pmatrix} \quad\text{or}\quad \pm\begin{pmatrix}1 & 0\\ 1 & -1\end{pmatrix}.$$

Let

$$T_1 = \pm\begin{pmatrix}a & b\\ c & d\end{pmatrix};$$

then $ad-bc=-1$.

Now we observe that $J=(1)\dot{+}(-1)$ is distinct from $\pm I$ and $\pm S$, that it commutes with S, and that JT is an involution. Hence there exists a matrix $M\in\mathfrak{P}_2$ distinct from $\pm I$ and $\pm S_1$, such that M commutes with S_1, and MT_1 is an involution.

Case 1.

$$S_1 = \pm\begin{pmatrix}1 & 0\\ 0 & -1\end{pmatrix}.$$

Since $(S_1T_1)^3=\pm I$, we find that $a-d=\pm 1$. The only matrices commuting with S_1 which are distinct from $\pm I$ and $\pm S_1$ are

$$\pm\begin{pmatrix}0 & 1\\ 1 & 0\end{pmatrix} \quad\text{and}\quad \pm\begin{pmatrix}0 & 1\\ -1 & 0\end{pmatrix}.$$

If M is either of the first two matrices, then the condition that MT_1 be an involution yields $b+c=0$. Thus $a=d\pm 1$, $b=-c$, and $ad-bc=-1$. Combining these, we obtain $d(d\pm 1)+c^2=-1$, which is impossible. The other two choices for M imply $b=c$, and therefore $d(d\pm 1)-c^2=-1$. Hence $1-4(1-c^2)$ is a perfect square; but $4c^2-3=f^2$ implies $(2c+f)(2c-f)=1$, whence $c=\pm 1$.

But then $ad=0$; from $a-d=\pm 1$ we deduce that $a^2-d^2=\pm 1$, whence $(S_1T_1^2)^3 = \pm I$, which is impossible.

Case 2.

$$S_1 = \pm \begin{pmatrix} 1 & 0 \\ 1 & -1 \end{pmatrix}.$$

From $(S_1T_1)^3 = \pm I$ we obtain $a-d+b=\pm 1$. For M there are the four possibilities

$$\pm \begin{pmatrix} 1 & -2 \\ 0 & -1 \end{pmatrix} \quad \text{and} \quad \pm \begin{pmatrix} 1 & -2 \\ 1 & -1 \end{pmatrix}.$$

Since MT_1 is an involution, in the first two cases we have $a-2c-d=0$, whence

$$ad - bc = \{(a+d)^2 + (a-d\pm 1)^2 - 1\}/4 \neq -1.$$

In the second two cases we find that $a-2c+b-d=0$, so that $2c=a+b-d = \pm 1$, which is again a contradiction. This completes the proof of Theorem 2.

2. **Automorphisms of $\mathfrak{P}_2^+$.** Let us now determine all automorphisms of $\mathfrak{P}_2$. Since every such automorphism takes $\mathfrak{P}_2^+$ into itself, we begin by considering all automorphisms of $\mathfrak{P}_2^+$.

THEOREM 3. *Every automorphism of $\mathfrak{P}_2^+$ is of the form $X\in\mathfrak{P}_2^+\to AXA^{-1}$ for some $A\in\mathfrak{M}_2$; that is, all automorphisms of $\mathfrak{P}_2^+$ are "inner" (with $A\in\mathfrak{M}_2$ rather than $A\in\mathfrak{P}_2^+$.)*

Proof. Let τ be any automorphism of $\mathfrak{P}_2^+$, and define S and T as before; let $S_0\in\mathfrak{M}_2$ be a fixed representative of $\pm S^\tau$. By Theorem 2, $S_0\in\mathfrak{M}_2^+$, and therefore $S_0^2=-I$. Let T_0 be that representative of $\pm T^\tau$ for which $(S_0T_0)^3=I$ is valid. Then $S\to S_0$, $T\to T_0$ induces a mapping from $\mathfrak{M}_2^+$ onto itself. The mapping is one-to-one, for although an element of $\mathfrak{M}_2^+$ can be expressed in many different ways as a product of powers of S and T, these expressions can be gotten from one another by use of $S^2=-I$, $(ST)^3=I$; since S_0 and T_0 satisfy these same relations, the mapping is one-to-one. It is an automorphism because τ is one. Therefore (AUT, Theorem 2) there exists an $A\in\mathfrak{M}_2$ such that $S_0=\pm ASA^{-1}$, $T_0=\pm ATA^{-1}$. This proves the result.

COROLLARY. *Every automorphism of $\mathfrak{P}_2$ is of the form $X\in\mathfrak{P}_2\to AXA^{-1}$ for some $A\in\mathfrak{M}_2$.*

(This corollary is a simple consequence of Theorem 3, as is shown in AUT by the remarks following the statement of Theorem 4.)

3. **The generators of $\mathfrak{B}_{2n}$.** Our main result may be stated as follows:

THEOREM 4. *The generators of $\mathfrak{B}_{2n}$ are*

(i) *The set of all inner automorphisms*:

$$\pm X \in \mathfrak{P}_{2n} \to \pm AXA^{-1} \qquad (A \in \mathfrak{M}_{2n}),$$

and

(ii) *The automorphism* $\pm X \in \mathfrak{P}_{2n} \to \pm X'^{-1}$.

REMARK. For $n=1$, the automorphism (ii) is a special case of (i).

In the proof of Theorem 4 by induction on n, the following lemma (which has already been established for $n=1$) will be basic:

LEMMA 1. *Let* $J_1=(-1)\dot{+}I^{(2n-1)}$. *In any automorphism* τ *of* $\mathfrak{P}_{2n}$, $J_1^\tau = \pm AJ_1A^{-1}$ *for some* $A \in \mathfrak{M}_{2n}$.

Proof. The result is already known for $n=1$. Hereafter let $n \geqq 2$. Certainly $(J_1^\tau)^2 = \pm I$ and det $J_1^\tau = -1$. If $(J_1^\tau)^2 = -I$, then the minimum function of J_1^τ is λ^2+1, and its characteristic function must be some power of λ^2+1, whence det $J_1^\tau = 1$. Therefore $(J_1^\tau)^2 = I$ is valid in $\mathfrak{M}_{2n}$. After a suitable inner automorphism, we may assume that

$$J_1^\tau = W(x, y, z) = L \dot{+} \cdots \dot{+} L \dot{+} (-I)^{(y)} \dot{+} I^{(z)},$$

where

$$L = \begin{pmatrix} 1 & 0 \\ 1 & -1 \end{pmatrix}$$

occurs x times, $2x+y+z=2n$, and $x+y$ is odd. (This follows from AUT, Lemma 1.)

Let $\mathfrak{G}_1$ be the group consisting of all elements of $\mathfrak{P}_{2n}$ which commute with J_1, and $\mathfrak{G}_2$ the corresponding group for J_1^τ. The lemma will be proved if we can show that $\mathfrak{G}_1$ is not isomorphic to $\mathfrak{G}_2$ unless $J_1^\tau = \pm J_1$. The group $\mathfrak{G}_1$ consists of the matrices $\pm(1 \dot{+} X_1) \in \mathfrak{P}_{2n}$, so that $\mathfrak{G}_1 \cong \mathfrak{M}_{2n-1}$. The number of nonsimilar involutions in $\mathfrak{G}_1$ is therefore $n(n+1)$ (see AUT, §4). We shall prove that $\mathfrak{G}_2$ contains more than $n(n+1)$ involutions which are nonsimilar in $\mathfrak{G}_2$, except when $x=0$, $y=1$, $z=2n-1$ or $x=0$, $y=2n-1$, $z=1$.

Those elements $\pm C \in \mathfrak{P}_{2n}$ which commute with W must satisfy one of the two equations: $CW=WC$ or $CW=-WC$. The solutions of the first of these equations form a subgroup of $\mathfrak{G}_2$, and this subgroup is known (see AUT, proof of Lemma 2) to be isomorphic to $\mathfrak{G}_0 = \mathfrak{G}_0(x, y, z)$ consisting of all matrices in $\mathfrak{P}_{2n}$ of the form

$$\begin{pmatrix} S_1 & 2R_1 \\ Q_1 & T_1 \end{pmatrix} \dot{+} \begin{pmatrix} S_2 & Q_2 \\ 2R_2 & T_2 \end{pmatrix},$$

where S_1, S_2, T_1, and T_2 are square matrices of dimensions x, x, z, and y respectively, and where $S_1 \equiv S_2 \pmod 2$, $2x+y+z=2n$, and $x+y$ and $x+z$ are both odd.

Next we prove that $\overline{C}W = -W\overline{C}$ is solvable only when $y=z$. The space

$\mathfrak{U}$ of vectors $\mathfrak{u}$ such that $W\mathfrak{u}=\mathfrak{u}$ is of dimension $x+z$, while the space $\mathfrak{B}$ of vectors $\mathfrak{v}$ for which $W\mathfrak{v}=-\mathfrak{v}$ has dimension $x+y$. But if $\overline{C}W=-W\overline{C}$, then $W\overline{C}\mathfrak{u}=-\overline{C}\mathfrak{u}$ and $W\overline{C}^{-1}\mathfrak{v}=\overline{C}^{-1}\mathfrak{v}$, so the dimensions of $\mathfrak{U}$ and $\mathfrak{B}$ must be the same, whence $y=z$. Hence if $y\neq z$, there are no solutions of $\overline{C}W=-W\overline{C}$, $\overline{C}\in\mathfrak{M}_{2n}$.

We may now proceed to find a lower bound for the number of nonsimilar matrices in $\mathfrak{G}_0(x, y, z)$. We briefly denote the elements of $\mathfrak{G}_0$ by $A\dot{+}B$, where

$$A=\begin{pmatrix} S_1 & 2R_1 \\ Q_1 & T_1 \end{pmatrix} \quad\text{and}\quad B=\begin{pmatrix} S_2 & Q_2 \\ 2R_2 & T_2 \end{pmatrix}.$$

If $A_1\dot{+}B_1$ and $A_2\dot{+}B_2$ are two distinct involutions in $\mathfrak{G}_0$, where either

$$A_1 \overset{s}{\neq} A_2 \quad\text{in}\quad M_{x+z} \quad\text{or}\quad B_1 \overset{s}{\neq} B_2 \quad\text{in}\quad M_{x+y},$$

then certainly

$$A_1 \dot{+} B_1 \overset{s}{\neq} A_2 \dot{+} B_2 \quad\text{in}\quad \mathfrak{G}_0.$$

Now let

$$A = I^{(a_1)} \dot{+} (-I)^{(b_1)} \dot{+} L \dot{+} \cdots \dot{+} L,$$
$$B = I^{(a_2)} \dot{+} (-I)^{(b_2)} \dot{+} L \dot{+} \cdots \dot{+} L,$$

where L occurs c_1 times in A and c_2 times in B; the various elements $A\dot{+}B$ gotten by taking different sets of values of $(a_1, b_1, c_1, a_2, b_2, c_2)$, if they lie in $\mathfrak{G}_0$, are certainly nonsimilar in $\mathfrak{G}_0$, except that $A\dot{+}B$ and $(-A)\dot{+}(-B)$ are the same element of $\mathfrak{G}_0$. Hence the number N of nonsimilar involutions of $\mathfrak{G}_0$ is at least half of the number N_1 of solutions of

$$a_1 + b_1 + 2c_1 = x + z,$$
$$a_2 + b_2 + 2c_2 = x + y,$$

where if $x\neq 0$ we impose the restrictions that $c_1\leqq (z+1)/2$, $c_2\leqq (y+1)/2$, and that in B instead of L we use L'. (These conditions insure that $A\dot{+}B\in\mathfrak{G}_0$.) As in the previous paper, one readily shows that $N>n(n+1)$ unless $J_1^T=\pm J_1$. We omit the details.

This leaves only the case where $y=z$. If $\overline{C}W=-W\overline{C}$, then $\overline{C}^kW=(-1)^kW\overline{C}^k$; therefore no odd power of $\overline{C}$ can be $\pm I$. Let p be a prime such that $n<p<2n$. Since $x+y=n$, certainly n is odd, and $p\geqq n+2$. Now $\mathfrak{G}_1$ (being isomorphic to $\mathfrak{M}_{2n-1}$) contains infinitely many elements of order p. However, $\mathfrak{G}_2$ contains only two such elements, since $\overline{C}^p\neq\pm I$ by the above argument, while if $C\in\mathfrak{G}_0$ and $C^p=\pm I$, then setting $C=A^{(n)}\dot{+}B^{(n)}$ shows that $A^p=\pm I$ and $B^p=\pm I$. However, $A\in\mathfrak{M}_n$, and if $A^p=\pm I$, then the minimum function of A must divide $\lambda^p\mp 1$. But the degree of the minimum function is at most n, and therefore is less than $p-1$, whereas $\lambda^p\mp 1$ is the product of a linear factor $\lambda \mp 1$ and an irreducible factor of degree $p-1$; thence the minimum function of A is $\lambda \mp 1$, so $A = \pm I$. In the same way $B = \pm I$. Hence the only solutions are $C = I^{(n)} \dotplus I^{(n)}$ and $C = -I^{(n)} \dotplus I^{(n)}$. This completes the proof of the lemma. We remark that the use of the existence of the prime p could have been avoided, but the proof is much quicker this way.

4. **Proof of the main theorem.** We are now ready to prove Theorem 4 by induction on n. Hereafter, let $n \geqq 2$ and assume that Theorem 4 holds for $n-1$. Let τ be any automorphism of $\mathfrak{P}_{2n}$; then by Lemma 1, $J_1^\tau = \pm A J_1 A^{-1}$ for some $A \in \mathfrak{M}_{2n}$. If we change τ by a suitable inner automorphism, we may assume that $J_1^\tau = \pm J_1$.

Therefore, every $M \in \mathfrak{P}_{2n}$ which commutes with J_1 goes into another such element, that is,

$$\pm \begin{bmatrix} 1 & \mathfrak{n}' \\ \mathfrak{n} & X \end{bmatrix}^\tau = \pm \begin{bmatrix} 1 & \mathfrak{n}' \\ \mathfrak{n} & Y \end{bmatrix},$$

where $\mathfrak{n}$ denotes a column vector all of whose components are zero, and $X \in \mathfrak{M}_{2n-1}$. Thus, τ induces an automorphism on $\mathfrak{M}_{2n-1}$. Consequently (AUT, Theorem 4) there exists a matrix $A \in \mathfrak{M}_{2n-1}$ such that $Y = AX^*A^{-1}$ for all $X \in \mathfrak{M}_{2n-1}$, where either $X^* = X$ for all $X \in \mathfrak{M}_{2n-1}$ or $X^* = X'^{-1}$ for all $X \in \mathfrak{M}_{2n-1}$. After a further inner automorphism by a factor of $(1) \dotplus A^{-1}$, we may assume that $J_1^\tau = \pm J_1$ and also that $X^\tau = Y = X^*$ for all $X \in \mathfrak{M}_{2n-1}$.

Let J_ν be obtained from $I^{(2n)}$ by replacing the νth diagonal element by -1. Then

$$(J_1 J_{2n})^\tau = \pm \begin{bmatrix} 1 & 0 & \cdots & 0 & 0 \\ 0 & -1 & \cdots & 0 & 0 \\ \cdot & \cdot & \cdots & \cdot & \cdot \\ 0 & 0 & \cdots & -1 & 0 \\ 0 & 0 & \cdots & 0 & 1 \end{bmatrix}^\tau = \pm \begin{bmatrix} 1 & \mathfrak{n}' \\ \mathfrak{n} & \begin{bmatrix} -1 & \cdots & 0 & 0 \\ \cdot & \cdots & \cdot & \cdot \\ 0 & \cdots & -1 & 0 \\ 0 & \cdots & 0 & 1 \end{bmatrix}^* \end{bmatrix}$$

$$= \pm J_1 J_{2n},$$

so that $\pm J_{2n}$ is invariant. Similarly, all of the matrices $\pm J_\nu$ $(\nu = 1, \cdots, 2n)$ are invariant. Therefore for any $X \in \mathfrak{M}_{2n-1}$ we have

$$\pm \begin{pmatrix} 1 & \mathfrak{n}' \\ \mathfrak{n} & X \end{pmatrix}^\tau = \pm \begin{pmatrix} 1 & \mathfrak{n}' \\ \mathfrak{n} & A_1 X^* A_1^{-1} \end{pmatrix}, \cdots, \pm \begin{pmatrix} X & \mathfrak{n} \\ \mathfrak{n}' & 1 \end{pmatrix}^\tau = \pm \begin{pmatrix} A_{2n} X^* A_{2n}^{-1} & \mathfrak{n} \\ \mathfrak{n}' & 1 \end{pmatrix},$$

with $A_\nu \in \mathfrak{M}_{2n-1}$, and in fact $A_1 = I$.

Now suppose that $Z \in \mathfrak{M}_{2n-2}$, and consider $\pm(Z \dotplus I^{(2)})$; since it commutes with J_{2n-1} and J_{2n}, so does its image. But therefore

$$A_{2n}\begin{pmatrix} Z & \mathfrak{n} \\ \mathfrak{n}' & 1 \end{pmatrix} A_{2n}^{-1} = \begin{pmatrix} \bar{Z} & \mathfrak{n} \\ \mathfrak{n}' & 1 \end{pmatrix},$$

where $\bar{Z}$ denotes some matrix in $\mathfrak{M}_{2n-2}$. From this one easily deduces that A_{2n} must be of the form $B\dot{+}(1)$, with $B\in\mathfrak{M}_{2n-2}$. By considering the matrices commuting with J_ν and J_{2n} for $\nu=1, \cdots, 2n-2$ we see that A_{2n} must be diagonal. Furthermore, it is clear that all of the A_ν $(\nu=1, \cdots, 2n)$ must be diagonal, and all are sections of one diagonal matrix $D^{(2n)}$. Using the further inner automorphism factor D^{-1}, we find that $\pm X^\tau = \pm X^*$ for every decomposable matrix $\pm X\in\mathfrak{P}_{2n}$. Since $\mathfrak{P}_{2n}$ is generated by the set of its decomposable matrices, the theorem is proved.

TSING HUA UNIVERSITY,
PEKING, CHINA.
UNIVERSITY OF ILLINOIS,
URBANA, ILL.

SYMPLECTIC MODULAR COMPLEMENTS

Irving Reiner

ABSTRACT

Let Γ_{2n} denote the symplectic modular group; this is the subgroup of all $2n \times 2n$ integral matrices X such that

$$XEX^{tr} = E \qquad \text{where } E = \begin{pmatrix} 0 & I_n \\ -I_n & 0 \end{pmatrix},$$

where I_n is the $n \times n$ identity matrix and X^{tr} is the transpose of X.

Let $j, k \leq n$ be positive integers; let A, B be $j \times n$ integral matrices and C, D be $k \times n$ integral matrices.

Theorem. *The rectangular matrix*

$$\begin{pmatrix} A & B \\ C & D \end{pmatrix}$$

can be completed to an element of Γ_{2n} by placing $n-j$ rows below $(A\ B)$ and $n-k$ rows below $(C\ D)$ if and only if

$$AB^{tr} \text{ and } CD^{tr} \text{ are symmetric}$$

and

$$AD^{tr} - BC^{tr} = (I_j\ 0) \text{ or } \begin{pmatrix} I_k \\ 0 \end{pmatrix}$$

depending on whether $j \leq k$ or $k \leq j$.

The proof is a self-contained matrix theoretic computation. This generalizes results of C. L. Siegel (*Annals of Math.* v. 36(1935)) which consider the case $j = n$, $k = 0$.

The paper also gives a parametrization of all the possible completions of the given rectangular matrix.

MAXIMAL SETS OF INVOLUTIONS

BY

IRVING REINER

§1. Let U_n denote the unimodular group consisting of all $n \times n$ integral matrices of determinant ± 1. An element $W \in U_n$ is called an *involution* if $W^2 = I^{(n)}$ (the n-rowed identity matrix). We shall consider abelian sets of involutions in U_n of maximal size, and call such a set a *maximal set*. For a given involution $W \in U_n$, define $N(W)$ to be the maximum number of involutions, all conjugate to W in U_n, which can occur in a maximal set. Certainly $N(W)$ depends only on the class of W in U_n. We already know(1) that as x, y, and z range over all non-negative integers such that $2x+y+z=n$, the matrix

$$W(x, y, z) = \begin{pmatrix} 1 & 0 \\ 1 & -1 \end{pmatrix} \dotplus \cdots \dotplus \begin{pmatrix} 1 & 0 \\ 1 & -1 \end{pmatrix} \dotplus (-I)^{(y)} \dotplus I^{(z)} \tag{1}$$

(where x two-by-two blocks occur in the direct sum) gives a complete set of nonconjugate representatives of the classes of involutions in U_n. We shall obtain a new proof of this result during our investigation of $N(W)$.

§2. The analogous problem for the group R_n of rational nonsingular $n \times n$ matrices has previously been considered(2). We quote some of the known results:

Let V be the space of rational $n \times 1$ vectors. To each involution $W \in R_n$ we let correspond the plus-space $W^+ = \{x \in V: Wx = x\}$, and the minus-space $W^- = \{x \in V: Wx = -x\}$. Then V is the direct sum of W^+ and W^-. If p is the dimension of W^+, and q that of W^-, then $p+q=n$ and we call W a (p, q) involution. If W is a (p, q) involution, we may choose the coordinate system in such a way that the first p columns of $I^{(n)}$ span W^+, and the last q columns span W^-. This gives at once(3)

$$W = \begin{bmatrix} I^{(p)} & 0 \\ 0 & -I^{(q)} \end{bmatrix}.$$

Therefore, in R_n every (p, q) involution is a conjugate of the above matrix. On the other hand, any decomposition $V = A \dotplus B$ determines a unique involution $W \in R_n$ for which $W^+ = A$, $W^- = B$.

Presented to the Society, October 30, 1954; received by the editors September 25, 1954.

(1) L. K. Hua and I. Reiner, Trans. Amer. Math. Soc. vol. 71 (1951) pp. 331–348.

(2) J. Dieudonné, Memoirs of the American Mathematical Society, no. 2, 1951; also G. W. Mackey, Ann. of Math. vol. 43 (1942) pp. 244–260.

(3) Throughout this paper we shall use 0 to denote a null matrix of appropriate size.

Next we have for a pair of involutions W, $Z \in R_n$:
(i) $ZW = WZ$ if and only if $ZW^+ = W^+$, $ZW^- = W^-$.
(ii) $ZW = WZ$ implies

$$W^+ = (W^+ \dotplus Z^+) \dotplus (W^+ \cap Z^-),$$
$$W^- = (W^- \dotplus Z^+) \dotplus (W^- \cap Z^-).$$

From these it follows that a set of m mutually commutative involutions gives rise to a decomposition of V into a direct sum of subspaces $V_1, \cdots, V_h$, each of positive dimension. If we choose the direct sum of some of the V_i as W^+, and the direct sum of the remaining V_i as W^-, a unique involution W is determined. Since the h spaces $V_1, \cdots, V_h$ can be grouped into two disjoint sets in 2^h ways, we can obtain at most 2^h mutually commutative involutions from this decomposition of V; therefore $m \leqq 2^h$. If we now take the V_i all of dimension 1, an abelian set of 2^n involutions is generated, and this set is certainly maximal in size. Thus a maximal set in R_n has 2^n elements, and is gotten by starting with any n linearly independent vectors $v_1, \cdots, v_n$ in V, choosing as basis for W^+ any subset of these vectors, and letting the remaining vectors serve as basis for W^-. We shall say that the matrix with columns $v_1, \cdots, v_n$ *generates* the maximal set. Finally we observe that a maximal set contains exactly $C_{n,p}$ involutions of type (p, q).

§3. Let us call a maximal size abelian set of (p, q) involutions in R_n a *maximal* (p, q) *set*. The above reasoning shows that a maximal (p, q) set contains at least $C_{n,p}$ elements. We show now that there must be exactly $C_{n,p}$ elements in a maximal (p, q) set, and that such a set is embeddable in a uniquely determined maximal set. For, suppose that we have an abelian set of m involutions of type (p, q), and that they give rise to a decomposition $V = V_1 \dotplus \cdots \dotplus V_h$. Let $m_i > 0$ be the dimension of V_i. We obtain (p, q) involutions from this decomposition by choosing for W^+ those direct sums

$$V_{i_1} \dotplus \cdots \dotplus V_{i_s}$$

for which

$$m_{i_1} + \cdots + m_{i_s} = p.$$

Hence m cannot exceed the number of solutions of the above equation, so that

$$m \leqq \text{coefficient of } x^p \text{ in } (1 + x^{m_1}) \cdots (1 + x^{m_h}). \tag{2}$$

On the other hand, the right-hand side of (2) is $\leqq$coefficient of x^p in

$$(1 + x)^{m_1} \cdots (1 + x)^{m_h} = (1 + x)^n,$$

with equality if and only if each $m_i = 1$ (except when $p = 0$ or $q = 0$). Thus $m \leqq C_{n,p}$, and furthermore $m = C_{n,p}$ implies that each $m_i = 1$. Thus any maximal (p, q) set arises from a decomposition of V into n one-dimensional subspaces by choosing, in $C_{n,p}$ ways, any p of these subspaces to make up W^+, and using the remaining q of them for W^-. This decomposition of V generates a unique maximal set containing the given maximal (p, q) set.

§4. We now return to the unimodular group U_n, and consider an involution $W \in U_n$. Since also $W \in R_n$, we may associate with W the pair of spaces W^+, W^-. Let G denote the set of all $n \times 1$ vectors with integral elements, and define $W_+ = W^+ \cap G$, $W_- = W^- \cap G$. If W is a (p, q) involution, there exist vectors $v_1, \cdots, v_p \in W_+$ which form an integral basis for W_+, that is, every vector in W_+ is uniquely expressible as a linear combination of $v_1, \cdots, v_p$ with integral coefficients. Likewise there exists an integral basis for W_-.

In practice, these integral bases may be obtained as follows: let $x_1, \cdots, x_p \in V$ be a basis for W^+. Since $x \in W^+$ implies $ax \in W^+$ for rational a, we may take $x_1, \cdots, x_p \in G$, with each x_i primitive(4). As b ranges over all integers, let b_0 be such that the greatest common divisor of the elements in the column vector $x_2 + bx_1$ is maximal, say c_0. Then replace x_2 by $(x_2 + b_0 x_1)/c_0$, and repeat the procedure with $x_3 + b_1 x_1 + b_2 x_2$, etc. The integral $n \times p$ matrix whose columns are the p vectors finally obtained by this procedure will be primitive(4), and its columns will furnish an integral basis for W_+. In fact, a set of p vectors $y_1, \cdots, y_p \in W_+$ is an integral basis for W_+ if and only if the matrix $(y_1 \cdots y_p)$ is primitive(5).

§5. We have thus shown that to each involution $W \in U_n$ there correspond two primitive matrices $P^{n \times p}$ and $Q^{n \times q}$, whose columns give integral bases for W_+ and W_-, respectively. Set $T = (P\ Q)$; then we see that $T = (P\ Q)$ and $T_1 = (P_1\ Q_1)$ arise from the same involution in U_n if and only if there exist matrices $R \in U_p$, $S \in U_q$ such that $P_1 = PR$, $Q_1 = QS$. Furthermore, $T = (P\ Q)$ and $T_1 = (P_1\ Q_1)$ arise from conjugate involutions if and only if there exist matrices $A \in U_n$, $R \in U_p$, $S \in U_q$ such that

$$P_1 = APR, Q_1 = AQS, \text{ that is, } T_1 = AT\begin{pmatrix} R & 0 \\ 0 & S \end{pmatrix}.$$

THEOREM 1. *Using the above notation, let* $T = (P\ Q)$ *arise from an involution* $W \in U_n$. *Then the invariant factors of* T *are* $1, \cdots, 1, 2, \cdots, 2$, *and the number of* 2*'s is at most* $\min(p, q)$.

Proof. Let N be the module consisting of all integral linear combinations of the columns of T. The invariant factors $\epsilon_1, \cdots, \epsilon_n$ of T have the property that there exists an integral basis $u_1, \cdots, u_n$ of G for which $\epsilon_1 u_1, \cdots, \epsilon_n u_n$ is an integral basis of N(6). As we shall show in a moment, $u \in G$ implies $2u \in N$. From this we have at once that each ϵ_i is 1 or 2.

(4) An integral matrix is called *primitive* if the greatest common divisor of its maximal size minors is 1.

(5) See H. Weyl, Trans. Amer. Math. Soc. vol. 48 (1940) pp. 126–164; also C. L. Siegel, *Geometry of numbers*, New York University notes, 1946.

(6) van der Waerden, *Modern Algebra* II, 2d ed., Chap. XV.

For any $u \in G$ we write

$$2u = (I + W)u + (I - W)u.$$

Since $(I+W)u \in W_+$, and $(I-W)u \in W_-$, we have $2u \in N$.

Finally, P and Q are both primitive, so the pth and qth determinantal divisors of T are both 1. Therefore $\epsilon_1 = \cdots = \epsilon_p = 1$, $\epsilon_1 = \cdots = \epsilon_q = 1$, so the number of invariant factors which are 2 is at most min $(n-p, n-q) = \min(p, q)$.

DEFINITION. If there are x 2's occurring as invariant factors of T, we shall say that W is a $(p, q; x)$ *involution.*

We now assert that two involutions are conjugate in U_n if and only if they are of the same type. It is easy to see, using the criterion for conjugacy given at the beginning of this section, that conjugate involutions are of the same type. The converse will follow from:

THEOREM 2. *Let $W \in U_n$ be a $(p, q; x)$ involution. Then in U_n, W is conjugate to $W(x, q-x, p-x)$ (see Equation (1)).*

Proof. We shall give a proof which does not depend on the result stated in §1. Let W be a $(p, q; x)$ involution, and let $T = (P\ Q)$. The matrix T may be replaced by

$$T_1 = AT\begin{pmatrix} R & 0 \\ 0 & S \end{pmatrix}, \quad A \in U_n,\ R \in U_p,\ S \in U_q,$$

without changing the class of W. Since P is primitive, we may choose $A \in U_n$ so that T becomes

$$\begin{bmatrix} I^{(p)} & Q_1 \\ 0 & Q_2 \end{bmatrix}.$$

Now replace T by

$$\begin{bmatrix} I^{(p)} & 0 \\ 0 & A_1 \end{bmatrix}\begin{bmatrix} I^{(p)} & Q_1 \\ 0 & Q_2 \end{bmatrix}\begin{bmatrix} I^{(p)} & 0 \\ 0 & B \end{bmatrix}, \qquad A_1 \in U_q,\ B \in U_q.$$

This replaces Q_2 by A_1Q_2B, and by proper choice of A_1 and B we may diagonalize Q_2. Since none of these operations affects the invariant factors of T, it follows at once that Q_2 has $(q-x)$ 1's and x 2's along its main diagonal. Hence we may take

$$T = \begin{bmatrix} I^{(p)} & Q_3 & Q_4 \\ 0 & I^{(q-x)} & 0 \\ 0 & 0 & 2I^{(x)} \end{bmatrix}.$$

Replacing T by

$$\begin{bmatrix} I^{(p)} & -Q_3 & C \\ 0 & I^{(q-x)} & 0 \\ 0 & 0 & I^{(x)} \end{bmatrix} T,$$

we may make $Q_3=0$, and reduce the elements of Q_4 (mod 2). Next replace T by XTX^{-1}, where $X=A_2\dotplus I^{(q-x)}\dotplus B^{-1}$ with $A_2\in U_p$, $B\in U_x$. Then Q_4 is replaced by A_2Q_4B, and can be diagonalized. As above, reduce the elements of Q_4 (mod 2), so that Q_4 is now a diagonal matrix with diagonal elements 0's and 1's. None of them can be 0, since under all of these transformations each column of T remains primitive. Therefore T becomes

$$\begin{bmatrix} I^{(p)} & \begin{matrix} 0 \\ 0 \end{matrix} & \begin{matrix} I^{(x)} \\ 0 \end{matrix} \\ 0 & I^{(q-x)} & 0 \\ 0 & 0 & 2I^{(x)} \end{bmatrix},$$

and so W is conjugate to

$$\begin{bmatrix} I^{(p)} & \begin{matrix} 0 \\ 0 \end{matrix} & \begin{matrix} -I^{(x)} \\ 0 \end{matrix} \\ 0 & -I^{(q-x)} & 0 \\ 0 & 0 & -I^{(x)} \end{bmatrix}.$$

This latter matrix is clearly conjugate in U_n to $W(x, q-x, p-x)$.

§6. We shall now prove a result which is roughly the converse of Theorem 1, and which eliminates a great deal of computation in what follows.

THEOREM 3. *Let M be an integral nonsingular $n\times n$ matrix with invariant factors* $1, \cdots, 1, 2, \cdots, 2$, *where k 1's and $(n-k)$ 2's occur. Let $M_1^{n\times n_1}$ consist of any n_1 columns of M, and let $M_2^{n\times n_2}$ consist of the remaining n_2 columns of M. Then the involution W for which W^+ is spanned by the columns of M_1, and W^- by those of M_2, is an involution in U_n. Furthermore, if $m_i=$*rank (mod 2) *of M_i $(i=1, 2)$, then W is of type $(n_1, n_2; x)$ with $x=m_1+m_2-k$.*

Proof. 1. We show firstly that W is integral. Let e_j be the jth column of $I^{(n)}$. Since the invariant factors of M are 1's and 2's, $2e_j$ is an integral linear combination of the columns of M. Thus

$$2e_j = u_1 + u_2,$$

where u_i is an integral linear combination of the columns of M_i $(i=1, 2)$. But then

$$We_j = \frac{u_1 - u_2}{2} = e_j - u_2.$$

Therefore We_j, the jth column of W, is integral. This holds for each j; hence W itself is integral.

2. By the 2-rank of an integral array, we shall mean the rank (mod 2) of that array, that is, its rank over $GF(2)$. Since the kth determinantal divisor of M is 1, the $(k+1)$st 2, we see that the 2-rank of M is k. Consequently $m_1+m_2\geqq k$.

3. Suppose now that by elementary operations(7) on the columns of M, a new matrix N is obtained having a column $2u$, $u\in G$. The matrix N_1 gotten by replacing $2u$ by u then has invariant factors $1, \cdots, 1, 2, \cdots, 2$, where now $(k+1)$ 1's occur.

On the other hand, elementary operations on the columns of M cannot produce a matrix M_1 having a column $4u$, $u\in G$. For in that case, letting $x_1, \cdots, x_n$ be the columns of M, we have

$$4u = a_1x_1 + \cdots + a_nx_n,$$

where $a_1, \cdots, a_n$ are integers whose greatest common divisor is 1. But since M has invariant factors 1's and 2's, we have

$$2u = b_1x_1 + \cdots + b_nx_n$$

for some integers $b_1, \cdots, b_n$. From the linear independence of $x_1, \cdots, x_n$ we obtain $a_i=2b_i$ $(i=1, \cdots, n)$, contradiction.

4. Now let M be partitioned into M_1 and M_2 as in the hypothesis of the theorem. In general, M_1 will not be primitive; in fact, we may rearrange the columns of M_1 (thereby leaving unchanged the space spanned by its columns) so that the first m_1 of them are linearly independent (mod 2), and the remaining n_1-m_1 of the columns will then be linearly dependent (mod 2) on the first m_1 columns. By further elementary operations on the columns, we may take the last n_1-m_1 columns in the form $2v_i$, $v_i\in G$ $(i=m_1+1, \cdots, n_1)$. Upon replacing each column $2v_i$ by v_i $(i=m_1+1, \cdots, n_1)$, we obtain a "reduced" matrix $\overline{M}_1$ whose first m_1 columns, $u_1, \cdots, u_{m_1}$, coincide with those of M_1. We now verify that $\overline{M}_1$ is primitive, and for this it suffices to show $\overline{M}_1$ primitive (mod 2). If this were not the case, we would have a relation

$$\sum_{i=1}^{m_1} a_iu_i + \sum_{i=m_1+1}^{n_1} a_iv_i \equiv \text{null vector (mod 2)},$$

where each $a_i=0$ or 1, and at least one $a_i=1$. Not all of the last (n_1-m_1) a_i's vanish, since $u_1, \cdots, u_{m_1}$ are linearly independent (mod 2). Multiplying the above congruence by 2, we obtain

$$2\sum_{i=1}^{m_1} a_iu_i + \sum_{i=m_1+1}^{n_1} a_iu_i \equiv \text{null vector (mod 4)}.$$

(7) See C. C. MacDuffee, *The theory of matrices*, Springer, 1933, p. 32.

Hence, elementary operations on the columns of M_1 yield a column of the form $4u$, $u \in G$. This is impossible, since the columns of M_1 are also columns of M. Therefore $\overline{M}_1$ is primitive.

5. In the same manner we get a reduced matrix $\overline{M}_2$ from M_2; set $T = (\overline{M}_1 \overline{M}_2)$. The columns of $\overline{M}_1$ furnish an integral basis for W_+, those of $\overline{M}_2$ for W_-. We have made $(n_1 - m_1) + (n_2 - m_2)$ divisions by 2 in passing from $(M_1 M_2)$ to $(\overline{M}_1 \overline{M}_2)$, so T has invariant factors $1, \cdots, 1, 2, \cdots, 2$, where the number of 2's is

$$(n - k) - (n_1 - m_1) - (n_2 - m_2) = m_1 + m_2 - k.$$

This completes the proof of the theorem.

§7. We now consider maximal size abelian sets of involutions in U_n; every such set is also in R_n, hence cannot contain more than 2^n elements. On the other hand, there certainly exist maximal sets in U_n containing 2^n elements; for example, such a set is the set of diagonal matrices with ± 1's as diagonal elements. Thus, we shall consider abelian sets of 2^n involutions in U_n. From the discussion in §2, every maximal set in U_n arises from a generating matrix M consisting of n linearly independent primitive column vectors $u_1, \cdots, u_n$, by choosing in all possible ways a subset of the u_i's as basis for W^+, and the remaining u_i's as basis for W^-. We shall call M *permissible* if the 2^n involutions which it generates are all integral. Examples easily show that not every integral M is permissible.

Suppose now that the permissible generating matrices M and M_1 give rise to two maximal sets: $W_1, \cdots, W_{2^n}$, and $Z_1, \cdots, Z_{2^n}$, respectively. If there exists a matrix $Y \in U_n$ such that the W_i's are a rearrangement of the matrices YZ_iY^{-1}, we call the two maximal sets *conjugate*, and say that M and M_1 are *equivalent* (denoted by $M \sim M_1$).

THEOREM 4. *Let M and M_1 be permissible generating matrices. Then $M \sim M_1$ if and only if there exist matrices $A \in U_n$, $B \in U_n$ (where B is gotten from $I^{(n)}$ by permuting columns, possibly changing their signs), such that $M_1 = AMB$.*

Proof. The sufficiency of the condition is obvious. To prove necessity, let $W_1, \cdots, W_n$ be the $(n-1, 1)$ involutions in the first maximal set, and suppose them numbered so that u_i, the ith column of M, is basis for W_i^-. Renumber the Z's so that $Z_i = YW_iY^{-1}$; then $Z_1, \cdots, Z_n$ are also $(n-1, 1)$ involutions. Permute the columns of M_1 so that v_i, the ith column of M_1, is basis for Z_i^-. Then $W_iu_i = -u_i$ implies $Y^{-1}Z_iYu_i = -u_i$, so $Z_iYu_i = -Yu_i$. Therefore $v_i = \pm Yu_i$, and so, after changing the signs of some of the v_i if necessary, we have $M_1 = YM$. This completes the proof.

In R_n any two maximal sets are conjugate; simple examples show that this is no longer the case in U_n. We seek a complete system of nonequivalent permissible generating matrices. We may remark at once that $M \sim I^{(n)}$ if and only if M is unimodular. Further, $M \sim M_1$ implies that M and M_1 have the same invariant factors. This condition is not sufficient, however, as the example

$$M = \begin{bmatrix} 1 & 0 & 1 \\ 0 & 1 & 0 \\ 0 & 0 & 2 \end{bmatrix}, \qquad M_1 = \begin{bmatrix} 1 & 0 & 1 \\ 0 & 1 & 1 \\ 0 & 0 & 2 \end{bmatrix}$$

shows. The maximal set generated by M contains $\pm I$, two each $\pm(2, 1; 1)$ involutions, one each $\pm(2, 1; 0)$ involutions. That generated by M_1 contains $\pm I$ and three each $\pm(2, 1; 1)$ involutions. Hence the maximal sets are not conjugate.

§8. In order to simplify the statement and proof of the next theorem, we introduce here the following four operations on an integral $r \times s$ array R:

(i) Replace R by T, where $T \equiv R \pmod 2$.
(ii) Permute the rows of R.
(iii) Permute the columns of R.
(iv) If R is of the form

$$\begin{bmatrix} 1 & u \\ v & S \end{bmatrix}$$

replace R by

$$\begin{bmatrix} 1 & u \\ -v & S - vu \end{bmatrix}.$$

Or, more generally, if $R = (a_{ij})$ and some $a_{pq} = 1$, perform the corresponding replacement where now u represents the pth row from which a_{pq} has been deleted, v the qth column from which a_{pq} has been deleted, and S the $(r-1) \times (s-1)$ array obtained by deleting the pth row and qth column from R[8].

We call two $r \times s$ arrays *related* if one can be obtained from the other by a finite number of operations of the four types just described. It is easy to see that being related is an equivalence relation.

THEOREM 5. *Every permissible generating matrix is equivalent to one of the form*

$$\begin{bmatrix} I^{(r)} & R \\ 0 & 2I^{(n-r)} \end{bmatrix}, \tag{3}$$

where the elements of R are 0*'s and* 1*'s, and where R has no zero columns. Every matrix of this form is permissible. Furthermore, two such matrices*

---

(8) The case first illustrated is that where $p = q = 1$. The more general case could have been reduced to the case $p = q = 1$, by use of operations (ii) and (iii).

$$\begin{bmatrix} I^{(r)} & R \\ 0 & 2I^{(n-r)} \end{bmatrix} \quad \textit{and} \quad \begin{bmatrix} I^{(t)} & T \\ 0 & 2I^{(n-t)} \end{bmatrix}, \tag{4}$$

are equivalent if and only if $r=t$ and R is related to T.

Proof. 1. The generators equivalent to the permissible generator M are given by AMB, where $A \in U_n$ and where B permutes the columns of M, possibly changing some of their signs. Let $M_1 = AM$, $A \in U_n$; by suitable choice of A, we may take M_1 in Hermite canonical form:

$$M_1 = \begin{bmatrix} d_1 & d_{12} & \cdots & d_{1n} \\ 0 & d_2 & \cdots & d_{2n} \\ \cdot & \cdot & \cdots & \cdot \\ 0 & 0 & \cdots & d_n \end{bmatrix},$$

with each $d_i > 0$ ($i=1, \cdots, n$), and each d_{ij} reduced (mod d_j) ($i<j$; $i, j = 1, \cdots, n$). Since the first column of M is primitive, we have $d_1 = 1$. Suppose that exactly r of the diagonal elements $d_1, \cdots, d_n$ are 1; by permuting rows and permuting columns, we obtain from M_1 the equivalent generator

$$M_2 = \begin{bmatrix} I^{(r)} & & R & \\ & d_{r+1} & \cdots & d_{r+1,n} \\ 0 & \cdot & & \cdots \\ & 0 & \cdots & d_n \end{bmatrix}.$$

The matrix M_2 is equivalent to the permissible generator M, hence it too is permissible. In particular, if we choose the first r columns of M_2 as basis for W^+, the remaining $n-r$ columns as basis for W^-, and construct a reduced matrix $\overline{M}_2$ (as in the proof of Theorem 3) whose first r columns form an integral basis for W_+, and whose last $n-r$ columns form such a basis for W_-, then the invariant factors of $\overline{M}_2$ must be 1's and 2's. However, in this reduction of M_2 to $\overline{M}_2$, the first $r+1$ columns are unchanged. Upon subtracting from the $(r+1)$st column of $\overline{M}_2$ suitable multiples of each of the first r columns, a column vector is obtained all of whose elements are multiples of d_{r+1}. On the other hand, $\overline{M}_2$ has as invariant factors only 1's and 2's, and so (as in Part (3) of the proof of Theorem 3) we conclude that $d_{r+1}=2$. Next choose the first $r+1$ columns of M_2 as basis for W^+, the remaining columns as basis for W^-; then the same type of argument shows that $d_{r+2}=2$. Continuing in this manner, we obtain finally $d_{r+1}=d_{r+2}= \cdots =d_n=2$.

Next we show that $d_{ij}=0$ for $r<i<j$. For fixed $i>r$, let $j>i$ be minimal such that $d_{ij}=1$. (Certainly each such d_{ij} is 0 or 1, since it is reduced modulo d_j.) Upon interchanging the ith and jth columns of M_2, we obtain the equivalent generator

$$M_3 = \begin{bmatrix} & & i & & j & \\ I^{(r)} & \cdots & \vdots & & \cdots & \\ & & 1 & 0\cdots 0 & 2\cdots & \\ 0 & \cdots & \vdots & & \vdots & \\ & & 2 & \cdots & 0\cdots & \\ & & \vdots & & \vdots & \end{bmatrix} \begin{matrix} \\ \\ i \\ \\ j. \\ \\ \end{matrix}$$

By subtracting from the $(i+1)$st, $\cdots$, jth rows suitable multiples of the ith row, a new generator M_4 is found, where

$$M_4 = \begin{bmatrix} & & i & & j & \\ I^{(r)} & \cdots & \vdots & & \vdots & \\ 1 & & 1 & 0\cdots 0 & 2\cdots & \\ & & 0 & & \vdots & \\ 0 & \cdots & \vdots & & \vdots & \\ & & 0 & \cdots & -4\cdots & \\ & & \vdots & & \vdots & \end{bmatrix} \begin{matrix} \\ \\ i \\ \\ \\ j. \\ \\ \end{matrix}$$

But M_4 is a permissible generator, and is triangular. The above discussion has already shown that the diagonal elements in such a matrix are 1's and 2's. This gives a contradiction; therefore each $d_{ij}=0$, $r<i<j$. Thus we have

$$M_2 = \begin{bmatrix} I^{(r)} & R \\ 0 & 2I^{(n-r)} \end{bmatrix}$$

where each element of R is reduced (mod 2) and so must be 0 or 1. Furthermore, R has no zero column, since the column vectors of the original matrix M were all primitive. This completes the proof of the first statement in the theorem.

2. The matrix given by (3) has invariant factors 1's and 2's. Therefore, by Theorem 3, it is permissible.

3. Since equivalent generators have the same determinant, the two generators given by (4) cannot be equivalent unless $r=t$. Suppose hereafter that $r=t$. We show now that if R and T are related, the generators are equivalent.

(i) If $R\equiv T$ (mod 2), set $R=T+2S$. Then

$$\begin{bmatrix} I & R \\ 0 & 2I \end{bmatrix} = \begin{bmatrix} I & S \\ 0 & I \end{bmatrix} \begin{bmatrix} I & T \\ 0 & 2I \end{bmatrix}.$$

(ii) If $R=PTQ$, where P permutes the rows of T, Q the columns of T, then

$$\begin{bmatrix} I & R \\ 0 & 2I \end{bmatrix} = \begin{bmatrix} P & 0 \\ 0 & Q^{-1} \end{bmatrix} \begin{bmatrix} I & T \\ 0 & 2I \end{bmatrix} \begin{bmatrix} P^{-1} & 0 \\ 0 & Q \end{bmatrix}.$$

(iii) If T is obtained from R by an operation of type (iv), we may assume, after repeated use of operations (ii) and (iii), that

$$R = \begin{bmatrix} 1 & u \\ v & S \end{bmatrix} \quad \text{and} \quad T = \begin{bmatrix} 1 & u \\ -v & S - vu \end{bmatrix},$$

and we must show that

$$M = \begin{bmatrix} I & R \\ 0 & 2I \end{bmatrix} \sim \begin{bmatrix} I & T \\ 0 & 2I \end{bmatrix} = N.$$

Let B be the permutation matrix which interchanges the 1st and $(r+1)$st columns of M. Then

$$MB = \begin{bmatrix} 1 & 0 \cdots 0 & 1 & u \\ & & 0 & \\ v & I^{(r-1)} & \vdots & S \\ & & 0 & \\ 2 & 0 \cdots 0 & 0 & 0 \cdots 0 \\ 0 & 0 \cdots 0 & 0 & 2 \cdots 0 \\ \cdot & \cdots\cdots & \cdot & \cdots\cdots \\ 0 & 0 \cdots 0 & 0 & 0 \cdots 2 \end{bmatrix}.$$

Premultiplication of MB by

$$A_1 = \begin{bmatrix} 1 & 0 \cdots 0 \\ -v & \\ -2 & \\ 0 & I^{(n-1)} \\ \vdots & \\ 0 & \end{bmatrix}$$

gives

$$A_1MB = \begin{bmatrix} I^{(r)} & & T \\ & -2 & 0 \cdots 0 \\ & 0 & \\ 0 & \vdots & 2I \\ & 0 & \end{bmatrix},$$

so there exists a matrix $A \in U_n$ such that

$$AMB = N. \tag{5}$$

Therefore $M \sim N$. (We shall have occasion to use the above discussion again in the course of this proof.)

4. We must now prove, conversely, that if

$$\begin{bmatrix} I^{(r)} & R \\ 0 & 2I^{(n-r)} \end{bmatrix} = A \begin{bmatrix} I^{(r)} & T \\ 0 & 2I^{(n-r)} \end{bmatrix} B, \tag{6}$$

where $A \in U_n$, and where B permutes columns (possibly changing their signs), then R and T are related. We think of the columns of

$$N = \begin{bmatrix} I^{(r)} & T \\ 0 & 2I^{(n-r)} \end{bmatrix}$$

as partitioned into two sets, the first set consisting of the first r columns of N, the second of the last $(n-r)$ colunns. We say that B *displaces* a column of N if it moves the column out of its set. We now proceed by induction on the number of columns in the first set which B displaces.

If B does not displace any of the first r columns of N, we may write $B = C_1^{(r)} \dot{+} C_2^{(n-r)}$, where each C_i is a permutation (and possibly sign-changing) matrix. We then obtain

$$\begin{bmatrix} I & R \\ 0 & 2I \end{bmatrix} = A \begin{bmatrix} C_1 & 0 \\ 0 & C_2 \end{bmatrix} \cdot \begin{bmatrix} C_1^{-1} & 0 \\ 0 & C_2^{-1} \end{bmatrix} \begin{bmatrix} I & T \\ 0 & 2I \end{bmatrix} \begin{bmatrix} C_1 & 0 \\ 0 & C_2 \end{bmatrix}$$

$$= A_2 \begin{bmatrix} I & C_1^{-1} T C_2 \\ 0 & 2I \end{bmatrix} = A_2 \begin{bmatrix} I & T_1 \\ 0 & 2I \end{bmatrix},$$

where $T_1 = C_1^{-1} T C_2$ is related to T, and $A_2 \in U_n$. From the uniqueness of Hermite canonical form, we see at once that $R \equiv T_1 \pmod 2$, so that R is related to T_1. Hence R and T are related.

Suppose now that B displaces some of the first r columns of N, and (for simplicity in exposition) suppose that the first column of N is displaced by B. Then at least one of the last $(n-r)$ columns which B displaces must have 1 as its first component; for otherwise, the 2-rank of the first r columns of

$$\begin{bmatrix} I & T \\ 0 & 2I \end{bmatrix} B$$

would be less than r, while on the other hand these first r columns are also the first r columns of

$$A^{-1} \begin{bmatrix} I & R \\ 0 & 2I \end{bmatrix},$$

and hence have 2-rank equal to r. Let us suppose that the qth column of N $(q>r)$ has 1 as its first component, and is displaced by B. Let B_1 be the permutation matrix which interchanges the 1st and qth columns of a matrix; by the argument in Part 3, there exists a matrix $A_3 \in U_n$ such that

$$A_3 \begin{bmatrix} I & T \\ 0 & 2I \end{bmatrix} B_1 = \begin{bmatrix} I & T_1 \\ 0 & 2I \end{bmatrix},$$

where T_1 is related to T by an operation of type (iv). Hence (6) becomes

$$\begin{bmatrix} I & R \\ 0 & 2I \end{bmatrix} = AA_3^{-1} \begin{bmatrix} I & T_1 \\ 0 & 2I \end{bmatrix} B_1^{-1}B,$$

that is, we get a new equation in which T_1 is related to T, and where $B_1^{-1}B$ displaces one fewer of the first r columns than B does. This completes the proof of the theorem.

§9. While we are now in a position to find complete sets of nonequivalent generators, it will be more convenient to prove a type of duality theorem first. Let $W_1, \cdots, W_{2^n}$ be a maximal set in U_n. Then the set of their transposes $W_1', \cdots, W_{2^n}'$ is also a maximal set in U_n. Furthermore, if W is an involution of type $(p, q; x)$, so is W'. Thus the two maximal sets contain any given type of involution equally often, so that many times it will be enough to consider only one of the pair of sets.

THEOREM 6. *The permissible generating matrices*

$$M = \begin{bmatrix} I^{(r)} & R \\ 0 & 2I^{(n-r)} \end{bmatrix} \quad \text{and} \quad M_1 = \begin{bmatrix} 2I^{(r)} & 0 \\ -R' & I^{(n-r)} \end{bmatrix}$$

give rise to two maximal sets; the elements in one set are the transposes of the elements in the other set. Furthermore,

$$M_1 \sim M^* = \begin{bmatrix} I^{(n-r)} & R' \\ 0 & 2I^{(r)} \end{bmatrix}.$$

Proof. As D ranges over all 2^n diagonal matrices with diagonal elements ± 1, the involution W defined by

$$WM = MD \tag{7}$$

ranges over the 2^n elements in the maximal set which M generates. From (7) we obtain

$$W'^{-1}M'^{-1} = M'^{-1}D'^{-1}.$$

However, $W^{-1} = W$ and $D'^{-1} = D$. Thus

$$W'M'^{-1} = M'^{-1}D.$$

But if

$$M = \begin{bmatrix} I & R \\ 0 & 2I \end{bmatrix}, \quad \text{then} \quad M'^{-1} = \frac{1}{2}\begin{bmatrix} 2I & 0 \\ -R' & I \end{bmatrix},$$

so that

$$W'M_1 = M_1D.$$

Hence the elements in the set which M_1 generates are the transposes of those generated by M. The last statement, $M_1 \sim M^*$, is trivial.

We shall say that M and M^* are *dual* generators. It might be well to point out that although R has no zero columns, R' may very well have such. In this case, we merely make each column of M^* primitive. For example,

$$M = \begin{bmatrix} 1 & 0 & 1 & 1 \\ 0 & 1 & 0 & 1 \\ 0 & 0 & 2 & 0 \\ 0 & 0 & 0 & 2 \end{bmatrix}, \quad M^* = \begin{bmatrix} 1 & 0 & 1 & 0 \\ 0 & 1 & 1 & 1 \\ 0 & 0 & 2 & 0 \\ 0 & 0 & 0 & 2 \end{bmatrix};$$

$$M = \begin{bmatrix} 1 & 0 & 1 & 1 \\ 0 & 1 & 0 & 0 \\ 0 & 0 & 2 & 0 \\ 0 & 0 & 0 & 2 \end{bmatrix}, \quad M^* \sim \begin{bmatrix} 1 & 0 & 0 & 0 \\ 0 & 1 & 0 & 1 \\ 0 & 0 & 1 & 1 \\ 0 & 0 & 0 & 2 \end{bmatrix}.$$

Thus, in the first case M is self-dual (up to equivalence). A consequence of this result is that if the two generators given in (4) are equivalent, then R and T must have the same number of zero rows.

§10. In this section we shall list complete sets of nonequivalent generating matrices, and the types of involutions in the maximal sets they generate, for $n=2, 3, 4$. Since in a maximal set the elements may be paired as $\pm W$, and since the negative of a $(p, q; x)$ involution is of type $(q, p; x)$, we can list the elements in a maximal set according to type $\pm(p, q; x)$.

For $n=2$, there are 2 nonequivalent generators.

$\begin{pmatrix} 1 & 0 \\ 0 & 1 \end{pmatrix}$ generates 1 $\pm(2, 0; 0)$ involution, 2 $(1, 1; 0)$ involutions.

$\begin{pmatrix} 1 & 1 \\ 0 & 2 \end{pmatrix}$ generates 1 $\pm(2, 0; 0)$ involution, 2 $(1, 1; 1)$ involutions.

For $n=3$, there are 4 nonequivalent generators, given by

$$M_1 = \begin{bmatrix} 1 & 0 & 0 \\ 0 & 1 & 0 \\ 0 & 0 & 1 \end{bmatrix}, \quad M_2 = \begin{bmatrix} 1 & 0 & 1 \\ 0 & 1 & 0 \\ 0 & 0 & 2 \end{bmatrix}, \quad M_3 = \begin{bmatrix} 1 & 0 & 1 \\ 0 & 1 & 1 \\ 0 & 0 & 2 \end{bmatrix}, \quad M_4 = \begin{bmatrix} 1 & 1 & 1 \\ 0 & 2 & 0 \\ 0 & 0 & 2 \end{bmatrix}.$$

The types of involutions they generate are shown in the following table:

	M_1	M_2	M_3	M_4
$\pm(3, 0; 0)$	1	1	1	1
$\pm(2, 1; 0)$	3	1	0	0
$\pm(2, 1; 1)$	0	2	3	3

We may remark that M_1 and M_2 are each self-dual, while M_3 and M_4 are duals of one another.

For $n=4$, there are 8 nonequivalent generators, given by

$$M_1=\begin{bmatrix}1&0&0&0\\0&1&0&0\\0&0&1&0\\0&0&0&1\end{bmatrix},\quad M_2=\begin{bmatrix}1&0&0&0\\0&1&0&0\\0&0&1&1\\0&0&0&2\end{bmatrix},\quad M_3=\begin{bmatrix}1&0&0&0\\0&1&0&1\\0&0&1&1\\0&0&0&2\end{bmatrix},\quad M_4=\begin{bmatrix}1&0&0&1\\0&1&0&1\\0&0&1&1\\0&0&0&2\end{bmatrix},$$

$$M_5=\begin{bmatrix}1&0&1&1\\0&1&0&0\\0&0&2&0\\0&0&0&2\end{bmatrix},\quad M_6=\begin{bmatrix}1&0&1&0\\0&1&0&1\\0&0&2&0\\0&0&0&2\end{bmatrix},\quad M_7=\begin{bmatrix}1&0&1&1\\0&1&1&0\\0&0&2&0\\0&0&0&2\end{bmatrix},\quad M_8=\begin{bmatrix}1&1&1&1\\0&2&0&0\\0&0&2&0\\0&0&0&2\end{bmatrix}.$$

	M_1	M_2	M_3	M_4	M_5	M_6	M_7	M_8
$\pm(4, 0; 0)$	1	1	1	1	1	1	1	1
$\pm(3, 1; 0)$	4	2	1	0	1	0	0	0
$\pm(3, 1; 1)$	0	2	3	4	3	4	4	4
$(2, 2; 0)$	6	2	0	0	0	2	0	0
$(2, 2; 1)$	0	4	6	6	6	0	2	6
$(2, 2; 2)$	0	0	0	0	0	4	4	0

self-dual: M_1, M_2, M_6, M_7

duals: M_3 and M_5, M_4 and M_8.

In order to find a complete set of nonequivalent generators for a given n, we list for each r $(1\leqq r\leqq n)$ a complete set of $r\times(n-r)$ arrays of 0's and 1's, having no zero columns, such that no two arrays can be gotten from one another by row and column permutations. It is not too difficult to decide whether two listed arrays can be gotten one from the other by use of type

(iv) operations coupled with the other three types. By striking out all but one from each set of related arrays, we obtain a complete set of unrelated $r \times (n-r)$ arrays. As R ranges over all of these arrays, the matrix given by

$$M = \begin{bmatrix} I^{(r)} & R \\ 0 & 2I^{(n-r)} \end{bmatrix} \tag{8}$$

gives nonequivalent generators. The totality of these for all r $(1 \leqq r \leqq n)$ form a complete set of nonequivalent generators.

We let C_n denote the number of nonconjugate abelian sets of involutions in U_n containing 2^n elements; that is, C_n is the number of nonequivalent $n \times n$ generators. The method described above can be used to show that $C_5 = 16$ and $C_6 = 36$. (This last figure dashes one's expectations that $C_n = 2^{n-1}$.) However, to compute C_n by the above procedure is very tedious for large n. It would be of interest to have a simple method for evaluating C_n.

§11. Let $f(p, q; x)$ be the maximum number of involutions of type $(p, q; x)$ which occur in any maximal set; this maximum need be taken only over a complete set of nonconjugate maximal sets. Trivially $f(p, q; x) = f(q, p; x) \leqq C_{n,p}$. We now note some partial results on the evaluation of $f(p, q; x)$. By considering the maximal set generated by $I^{(n)}$, we see that $f(p, q; 0) = C_{n,p}$. Further, this is the only maximal set (up to conjugacy) all of whose involutions have $x = 0$.

Next set

$$M = \begin{bmatrix} 1 & 1 \cdots 1 \\ 0 & \\ \vdots & 2I^{(n-1)} \\ 0 & \end{bmatrix}.$$

In the maximal set which M generates, every involution (except $\pm I$) has $x = 1$. Therefore $f(p, q; 1) = C_{n,p}$ for $1 \leqq p \leqq n-1$. Further, the maximal sets generated by M and its dual M^* are the only sets (up to conjugacy) all of whose elements (except $\pm I$) have $x = 1$.

For $x > 1$, the problem of evaluating $f(p, q; x)$ becomes more difficult. For example, it may be shown that for fixed $x > 1$, and fixed $q > x$, we have $f(n-q, q; x) < C_{n,q}$ for all sufficiently large n. On the other hand, let $V^{(k)}$ denote a square matrix all of whose elements are $+1$, except for 0's along the main diagonal. Set

$$M = \begin{bmatrix} I^{(k)} & V \\ 0 & 2I^{(k)} \end{bmatrix} \quad \text{or} \quad \begin{bmatrix} I^{(k+1)} & \begin{matrix} 1 \cdots 1 \\ V \end{matrix} \\ 0 & 2I^{(k)} \end{bmatrix}$$

according as $n = 2k$ or $n = 2k+1$. Then the maximal set generated by M contains $C_{n,2}$ involutions of type $(n-2,\ 2;\ 2)$ for $n \geqq 6$. In general it is true that $f(n-x,\ x;\ x) = C_{n,x}$ for $x < [n/2]$.

§12. We shall now characterize the $\pm(n-1,\ 1;\ 0)$ involutions in U_n by inner properties; however, we shall not give any group-theoretic method for distinguishing between the $(n-1,\ 1;\ 0)$ and $(1,\ n-1;\ 0)$ involutions. For the moment, take $n>4$. Then we show that $f(p,\ q;\ x)>n$ except when $(p,\ q;\ x) =\pm(n-1,\ 1;\ 0)$, $\pm(n-1,\ 1;\ 1)$ or $\pm(n,\ 0;\ 0)$. Certainly $f(p,\ q;\ 0) = C_{n,p} > n$ when $\min(p,\ q) > 1$, and also $f(p,\ q;\ 1) = C_{n,p} > n$ when $\min(p,\ q)>1$. We must therefore prove that $f(p,\ q;\ x)>n$ for $1 < x \leqq q \leqq p < n-1$.

Let us write $n = ax+b$, $0 \leqq b < x$. Since $x \leqq n/2$, certainly $a \geqq 2$. Define

$$M = \begin{bmatrix} I^{(x)} & I^{(x)} & I^{(x)} \cdots I^{(x)} & \begin{matrix} I^{(b)} \\ 0 \end{matrix} \\ 0 & & 2I^{(n-x)} & \end{bmatrix}.$$

Case 1. $q \geqq x+b$. If we choose any q consecutive columns of M as basis for W^-, and the remaining p columns as basis for W^+, then W is of type $(p,\ q;\ x)$ because the 2-rank of each of the two submatrices is x. Thus, the maximal set generated by M contains at least n involutions of type $(p,\ q;\ x)$. We may obtain one extra $(p,\ q;\ x)$ involution by choosing the 1st column of M instead of the $(x+1)$st or $(2x+1)$st, etc., in one of the previously considered n partitions of M. (This extra involution does not arise when $n=4$ and $x=2$; indeed, $f(2,\ 2;\ 2)=4$.) Hence we have $f(p,\ q;\ x)>n$ when $q \geqq x+b$.

The same argument also works for $x=2$ and $x=3$ even when $q<x+b$, provided that we change M by replacing

$$\begin{pmatrix} I^{(b)} \\ 0 \end{pmatrix} \text{ by } \begin{pmatrix} 1 \\ 1 \end{pmatrix}, \quad \begin{pmatrix} 1 \\ 1 \\ 1 \end{pmatrix}, \quad \text{or} \quad \begin{pmatrix} 1 & 0 \\ 1 & 1 \\ 0 & 1 \end{pmatrix},$$

according as $x=2$, $b=1$, or $x=3$, $b=1$, or $x=3$, $b=2$, respectively.

Case 2. Now let $4 \leqq x \leqq q < x+b$. Choose any x linearly independent columns from the first ax columns of M, then pick $q-x$ other columns from the last b columns of M, and use these as basis for W^-; let the remaining p columns of M serve as basis for W^+. Every such W will be of type $(p,\ q;\ x)$, and there will be at least a^x of them. Thus

$$f(p,\ q;\ x) \geqq a^x.$$

Since $a \geqq 2$, certainly $f(p,\ q;\ x) \geqq 2^x > n$ when $x > \log_2 n$. When $4 \leqq x \leqq \log_2 n$, we have

$$a > \frac{n}{x} - 1 \geqq \frac{n}{\log_2 n} - 1,$$

so that

$$f(p, q; x) \geqq \left(\frac{n}{\log_2 n} - 1\right)^x \geqq \left(\frac{n}{\log_2 n} - 1\right)^4.$$

However, it is easy to verify that

$$\frac{n}{\log_2 n} - 1 > n^{1/4} \qquad \text{for } n \geqq 9.$$

Hence in all cases $f(p, q; x) > n$ for $n > 4$ except when $(p, q; x) = \pm(n-1, 1; 0)$, or $\pm(n-1, 1; 1)$, or $\pm(n, 0; 0)$. On the other hand, when $n=4$, the table in §10 permits us to characterize the $\pm(3, 1; 0)$ involutions by inner properties.

Now we may further distinguish the $\pm(n-1, 1; 0)$ involutions from the $\pm(n-1, 1; 1)$ involutions by observing that all maximal sets containing n involutions of type $(n-1, 1; 0)$ are conjugate, whereas this is false (for $n > 2$) for involutions of type $(n-1, 1; 1)$. For $n=2$, other arguments may be used to make this distinction[9].

§13. Define $g(p, q; x)$ to be the maximum number of elements in a maximal $(p, q; x)$ set. The previous discussion shows at once that $f(p, q; x) \leqq g(p, q; x) \leqq C_{n,p}$. We remark that neither equality sign can hold in general. For example, we have already shown that $f(2, 2; 2) = 4$; we prove now that $g(2, 2; 2) = 6$. Let

$$N = \begin{bmatrix} 1 & 1 & 1 & 1 \\ 0 & 2 & 0 & 2 \\ 0 & 0 & 2 & 2 \\ 0 & 0 & 0 & 4 \end{bmatrix}.$$

The maximal set which N generates contains 16 involutions; of these, the 8 involutions of types $\pm(3, 1)$ are not integral, while the remaining 8 are $\pm I$ and 6 integral involutions of type $(2, 2; 2)$. On the other hand, it is easy to verify that $g(3, 2; 2) < 10$. These remarks show that "maximal" sets in U_n in the sense of embeddability need not be maximal in size. It would be of interest to investigate the structure of abelian sets of involutions in U_n which could not be embedded in larger sets.

INSTITUTE FOR ADVANCED STUDY,
PRINCETON, N. J.
UNIVERSITY OF ILLINOIS,
URBANA, ILL.

(9) Hua and Reiner, op. cit.

AUTOMORPHISMS OF THE SYMPLECTIC MODULAR GROUP

BY
IRVING REINER

1. Introduction. Let Ω_n denote the unimodular group consisting of all $n\times n$ integral matrices of determinant ± 1, and let $I^{(n)}$ be the identity matrix in Ω_n. We shall use 0 to denote a null matrix whose size is determined by the context, X' for the tranpose of X, and $X\dotplus Y$ for the direct sum of X and Y. We call an integral matrix *primitive* if the greatest common divisor of its maximal size minors is 1.

Define

$$\mathfrak{F} = \begin{pmatrix} 0 & I^{(n)} \\ -I^{(n)} & 0 \end{pmatrix}, \tag{1}$$

and let the symplectic group Sp_{2n} consist of all rational $2n\times 2n$ matrices $\mathfrak{M}$ satisfying

$$\mathfrak{M}\mathfrak{F}\mathfrak{M}' = \mathfrak{F}. \tag{2}$$

We define the symplectic modular group Γ_{2n} to be the group of integral matrices in Sp_{2n}. Although we shall not do so in this paper, it is sometimes more convenient to work with the factor group of Γ_{2n} over its center $\pm\mathfrak{J}$; see [1; 2; 3][1]. We may also define an extended group Δ_{2n} consisting of all integral matrices $\mathfrak{M}$ for which $\mathfrak{M}\mathfrak{F}\mathfrak{M}' = \pm\mathfrak{F}$.

The automorphisms of Sp_{2n} (over any field) have previously been determined [5], as have the automorphisms of Γ_2 (see [4]). The object of this paper is to determine all automorphisms of Γ_{2n}. Let us call a homomorphism of Γ_{2n} into $\{\pm 1\}$ a *character*. Then we shall prove that every automorphism τ of Γ_{2n} is given by

$$\mathfrak{X}^\tau = \psi(\mathfrak{X})\mathfrak{A}\mathfrak{X}\mathfrak{A}^{-1} \qquad \text{for all } \mathfrak{X}\in\Gamma_{2n},$$

where ψ is a character, and $\mathfrak{A}\in\Delta_{2n}$. We may remark at this point that the mapping σ defined by

$$\mathfrak{X}^\sigma = \mathfrak{X}'^{-1} \qquad \text{for all } \mathfrak{X}\in\Gamma_{2n}$$

is obviously an automorphism. As we shall see, however, it is an inner automorphism.

Let us set

Presented to the Society, December 28, 1954; received by the editors November 26, 1954.

(1) Numbers in brackets refer to the bibliography at the end of this paper.

$$\mathfrak{M} = \begin{pmatrix} A & B \\ C & D \end{pmatrix}, \tag{3}$$

where A, B, C, D are integral $n \times n$ matrices. Then $\mathfrak{M} \in \Gamma_{2n}$ if and only if the following conditions are satisfied:

$$AB' \text{ symmetric, } CD' \text{ symmetric, } AD' - BC' = I. \tag{4}$$

We single out for future use certain types of elements of Γ_{2n}:

(1) Translations:

$$\mathfrak{T} = \begin{pmatrix} I & S \\ 0 & I \end{pmatrix} \text{ or } \begin{pmatrix} I & 0 \\ S & I \end{pmatrix}, \qquad S \text{ symmetric.}$$

(2) Rotations:

$$\mathfrak{R} = \begin{pmatrix} U & 0 \\ 0 & U'^{-1} \end{pmatrix}, \qquad U \in \Omega_n.$$

(3) Semi-involutions:

$$\mathfrak{S} = \begin{pmatrix} J & I - J \\ J - I & J \end{pmatrix}, \qquad J \text{ diagonal with diagonal elements 0's and 1's.}$$

Further, if $\mathfrak{M}$ given by (3) is in Γ_{2n}, then

$$\mathfrak{M}^{-1} = \begin{pmatrix} D' & -B' \\ -C' & A' \end{pmatrix}. \tag{5}$$

Finally, if

$$\mathfrak{M}_i = \begin{pmatrix} A_i & B_i \\ C_i & D_i \end{pmatrix} \in \Gamma_{2n_i} \qquad (i = 1, 2),$$

we define the *symplectic direct sum* $\mathfrak{M}_1 * \mathfrak{M}_2 \in \Gamma_{2(n_1+n_2)}$ by

$$\mathfrak{M}_1 * \mathfrak{M}_2 = \begin{bmatrix} A_1 & 0 & B_1 & 0 \\ 0 & A_2 & 0 & B_2 \\ C_1 & 0 & D_1 & 0 \\ 0 & C_2 & 0 & D_2 \end{bmatrix}.$$

We may remark that as $\mathfrak{M}$ ranges over all elements of Γ_{2n}, the matrix $[-I^{(n)} \dotplus I^{(n)}]\mathfrak{M}$ ranges over all elements in $\Delta'_{2n} = \{\mathfrak{X} \in \Delta_{2n} : \mathfrak{X} \notin \Gamma_{2n}\}$. Thence $\mathfrak{M}_i \in \Delta'_{2n_i}$ $(i=1,\ 2)$ implies $\mathfrak{M}_1 * \mathfrak{M}_2 \in \Delta'_{2(n_1+n_2)}$. However, $\mathfrak{M}_1 \in \Gamma_{2n_1}$ and $\mathfrak{M}_2 \in \Delta_{2n_2}$ implies $\mathfrak{M}_1 * \mathfrak{M}_2 \notin \Delta_{2(n_1+n_2)}$.

2. **Involutions in Γ_{2n}.** It is known [4] that as x, y, and z range over all non-negative integers such that $2x+y+z=n$, the matrix

$$(6) \qquad W(x, y, z) = \begin{pmatrix} 1 & 0 \\ 1 & -1 \end{pmatrix} \dotplus \cdots \dotplus \begin{pmatrix} 1 & 0 \\ 1 & -1 \end{pmatrix} \dotplus (-I)^{(y)} \dotplus I^{(z)}$$

(where x 2×2 blocks occur) gives a complete set of nonconjugate involutions in Ω_n. By an $[x, y, z]$ involution in Ω_n we mean any conjugate of $W(x, y, z)$ in Ω_n. Now define

$$\mathfrak{W}(x, y, z) = W(x, y, z) \dotplus W'(x, y, z) \in \Gamma_{2n}.$$

THEOREM 1. *The matrices* $\mathfrak{W}(x, y, z)$ *with* $2x+y+z=n$ *give a complete set of nonconjugate involutions in* Γ_{2n}.

Proof. We use induction on n. The result is trivial for $n=1$, so now let $\mathfrak{X}$ be an involution in Γ_{2n}, $n>1$. From $\mathfrak{X}^2=I^{(2n)}$ we conclude that the characteristic roots of $\mathfrak{X}$ are 1's and -1's. Let ϵ be a characteristic root of $\mathfrak{X}$; then there exists a primitive row vector $\mathfrak{x}$ such that $\mathfrak{x}\mathfrak{X}=\epsilon\mathfrak{x}$. We can then find [6] a matrix $\mathfrak{Y}\in\Gamma_{2n}$ whose first row is $\mathfrak{x}$. In that case the first row of $\mathfrak{X}_1=\mathfrak{Y}\mathfrak{X}\mathfrak{Y}^{-1}$ is $(\epsilon\ 0 \cdots 0)$. Since $\mathfrak{X}_1$ is an involution in Γ_{2n}, we obtain

$$\mathfrak{X}_1 = \begin{bmatrix} \epsilon & 0 \cdots 0 & 0 & 0 \cdots 0 \\ * & & 0 & \\ \vdots & A_1 & \vdots & B_1 \\ * & & 0 & \\ * & * \cdots * & \epsilon & 0 \cdots 0 \\ * & & * & \\ \vdots & C_1 & \vdots & D_1 \\ * & & * & \end{bmatrix},$$

where

$$\begin{pmatrix} A_1 & B_1 \\ C_1 & D_1 \end{pmatrix}$$

is itself an involution in $\Gamma_{2(n-1)}$. Continuing this procedure, we see that $\mathfrak{X}$ is conjugate in Γ_{2n} to a matrix of the form

$$\mathfrak{X}_2 = \begin{pmatrix} A & 0 \\ C & D \end{pmatrix}.$$

From the fact that $\mathfrak{X}_2$ is an involution in Γ_{2n}, we deduce at once that A is an involution in Ω_n, and $D=A'^{-1}$. However,

$$\begin{pmatrix} U & 0 \\ 0 & U'^{-1} \end{pmatrix}\begin{pmatrix} A & 0 \\ C & D \end{pmatrix}\begin{pmatrix} U^{-1} & 0 \\ 0 & U' \end{pmatrix} = \begin{pmatrix} UAU^{-1} & 0 \\ \overline{C} & U'^{-1}DU' \end{pmatrix},$$

and so by choosing $U \in \Omega_n$ properly, we find that $\mathfrak{X}$ is conjugate to $\mathfrak{X}_3$ given by

$$\mathfrak{X}_3 = \begin{pmatrix} W(x, y, z) & 0 \\ C & W'(x, y, z) \end{pmatrix}$$

with a new C. Since $\mathfrak{X}_3 \in \Gamma_{2n}$ is an involution, we have

$$(7) \qquad CW \text{ symmetric}, \qquad C \text{ skew-symmetric}.$$

The proof now splits into two cases:

CASE 1. If either $y \neq 0$ or $z \neq 0$, we may set $W(x, y, z) = W_1 \dotplus (\epsilon)$, $\epsilon = \pm 1$. From (7) we find that

$$\mathfrak{X}_3 = \begin{bmatrix} W_1 & 0 & 0 & 0 \\ 0 & \epsilon & 0 & 0 \\ C_1 & -\mathfrak{x}' & W_1' & 0 \\ \mathfrak{x} & 0 & 0 & \epsilon \end{bmatrix},$$

and that

$$\mathfrak{Z} = \begin{pmatrix} W_1 & 0 \\ C_1 & W_1' \end{pmatrix}$$

is an involution in $\Gamma_{2(n-1)}$. By the induction hypothesis there exist integers x_1, y_1, z_1 with $2x_1 + y_1 + z_1 = n-1$, such that $\mathfrak{Z}$ is conjugate to $\mathfrak{W}(x_1, y_1, z_1)$. For the moment set $P = W(x_1, y_1, z_1)$. Then in Γ_{2n}, $\mathfrak{X}_3$ is conjugate to $\mathfrak{X}_4$, where

$$\mathfrak{X}_4 = \begin{bmatrix} P & 0 & 0 & 0 \\ 0 & \epsilon & 0 & 0 \\ 0 & -\mathfrak{x}' & P' & 0 \\ \mathfrak{x} & 0 & 0 & \epsilon \end{bmatrix}$$

with a new $\mathfrak{x}$. But then

$$\mathfrak{X}_5 = \mathfrak{S}\mathfrak{X}_4\mathfrak{S}^{-1} = \begin{bmatrix} P & 0 & 0 & 0 \\ \mathfrak{x} & \epsilon & 0 & 0 \\ 0 & 0 & P' & \mathfrak{x}' \\ 0 & 0 & 0 & \epsilon \end{bmatrix} \text{ where } \mathfrak{S} = \begin{bmatrix} I^{(n-1)} & 0 & 0 & 0 \\ 0 & 0 & 0 & 1 \\ 0 & 0 & I^{(n-1)} & 0 \\ 0 & -1 & 0 & 0 \end{bmatrix}.$$

Since $\mathfrak{X}_5$ is now a direct sum $W \dotplus W'$, where W is an involution in Ω_n, the result follows upon transforming $\mathfrak{X}_5$ by a suitably chosen rotation in Γ_{2n}.

CASE 2. If both y and z are 0, we write $W(x, y, z) = L \dotplus W_1$, where

$$L = \begin{pmatrix} 1 & 0 \\ 1 & -1 \end{pmatrix}.$$

Then, as before, $\mathfrak{X}_3$ is conjugate to $\mathfrak{X}_4$ given by

$$\mathfrak{X}_4 = \begin{bmatrix} L & 0 & 0 & 0 \\ 0 & W_1 & 0 & 0 \\ \begin{matrix} 0 & b \\ -b & 0 \end{matrix} & B & L' & 0 \\ -B' & 0 & 0 & W_1' \end{bmatrix}.$$

However,

$$\mathfrak{M} = \begin{bmatrix} 0 & 0 & I^{(2)} & 0 \\ 0 & I^{(n-2)} & 0 & 0 \\ -I^{(2)} & 0 & 0 & 0 \\ 0 & 0 & 0 & I^{(n-2)} \end{bmatrix} \begin{bmatrix} I^{(n)} & 0 \\ 0 \dotplus b \dotplus 0^{(n-2)} & I^{(n)} \end{bmatrix} \in \Gamma_{2n},$$

and we have

$$\mathfrak{M}\mathfrak{X}_4\mathfrak{M}^{-1} = \begin{pmatrix} L' & B \\ 0 & W_1 \end{pmatrix} \dotplus \begin{pmatrix} L & 0 \\ B' & W_1' \end{pmatrix}.$$

The result then follows as in the previous case.

We have thus shown that any involution $\mathfrak{X} \in \Gamma_{2n}$ is conjugate to some $\mathfrak{W}(x, y, z)$. On the other hand, if $\mathfrak{W}(x, y, z)$ and $\mathfrak{W}(x_0, y_0, z_0)$ were conjugate in Γ_{2n}, they would certainly be conjugate in Ω_{2n}. This implies [4] that $x = x_0$, $y = y_0$, and $z = z_0$.

The conjugates of $\mathfrak{W}(x, y, z)$ in Γ_{2n} will be called (x, y, z) involutions.

3. **Characterization of the $\pm(0, 1, n-1)$ involutions.** In Sp_{2n}, every involution is conjugate to one of the form $I^{(2p)} * -I^{(2q)}$, with $p+q=n$. Any involution in the class of $I^{(2p)} * -I^{(2q)}$ is said to have *signature* $\{p, q\}$ (see [5]). One easily proves that any (x, y, z) involution in Γ_{2n} has signature $\{x+z, x+y\}$, and that the negative of an (x, y, z) involution is of type (x, z, y).

It is known that an abelian set of involutions of signature $\{p, q\}$ in Sp_{2n} cannot contain more than $C_{n,p}$ elements (see [5, Theorem 2; 7, §19]). We shall use this fact in proving the following basic result:

THEOREM 2. *Under any automorphism of* Γ_{2n}, *the image of a* $(0, 1, n-1)$ *involution is either a* $(0, 1, n-1)$ *involution or a* $(0, n-1, 1)$ *involution.*

Proof. (i) An abelian set of involutions in Γ_{2n}, each of type (x, y, z), we shall call an (x, y, z) set. Let $f(x, y, z)$ be the number of elements in an (x, y, z) set of largest size. The above-quoted result shows that

$$f(x, y, z) \leqq C_{n, x+z},$$

so for $(x, y, z) = \pm(0, 0, n)$, $\pm(0, 1, n-1)$, $\pm(1, 0, n-2)$ we have $f(x, y, z) \leqq n$.

We now show that $f(x, y, z) > n$ except for the 6 cases given above.

From an abelian set $\mathfrak{K}$ of $[x, y, z]$ involutions in Ω_n, one obtains an (x, y, z) set in Γ_{2n} by taking the set of matrices $U \dotplus U'^{-1}$, $U \in \mathfrak{K}$. We know, however, that there exist abelian sets of $[x, y, z]$ involutions in Ω_n containing more than n elements, except for the 6 cases listed above (see [8, §§12 and 13]).

(ii) The $\pm(0, 0, n)$ involutions in Γ_{2n} are $\pm I^{(2n)}$, so that certainly a $(0, 1, n-1)$ involution cannot be mapped onto a $\pm(0, 0, n)$ involution by an automorphism of Γ_{2n}. It remains to prove that the image cannot be of type $\pm(1, 0, n-2)$. To begin with, a simple calculation shows that two rotations $U \dotplus U'^{-1}$ and $V \dotplus V'^{-1}$ are conjugate in Γ_{2n} if and only if U and V are conjugate in Ω_n. For $n > 2$, there are at least two nonconjugate $[1, 0, n-2]$ sets in Ω_n, each containing n elements; on the other hand, there is a unique (up to conjugacy) abelian set of n $[0, 1, n-1]$ involutions in Ω_n (see [8, §12]). Hence for $n > 2$, the image of a $(0, 1, n-1)$ involution in Γ_{2n} must be of type $\pm(0, 1, n-1)$.

(iii) The case $n = 1$ is trivial, and so we are left with $n = 2$. Now we have

$$I^{(2)} * -I^{(2)} = \left(I^{(2)} * \begin{pmatrix} 0 & 1 \\ -1 & 0 \end{pmatrix}\right)^2,$$

so any $(0, 1, 1)$ involution in Γ_4 is the square of some element of Γ_4. We show that the $(1, 0, 0)$ involutions in Γ_4 are not squares. For suppose that

$$\begin{pmatrix} A & B \\ C & D \end{pmatrix}^2 = \begin{pmatrix} L & 0 \\ 0 & L' \end{pmatrix}, \text{ where } \begin{pmatrix} A & B \\ C & D \end{pmatrix} \in \Gamma_4 \text{ and } L = \begin{pmatrix} 1 & 0 \\ 1 & -1 \end{pmatrix}.$$

From (5) we then have

$$\begin{pmatrix} A & B \\ C & D \end{pmatrix} = \begin{pmatrix} L & 0 \\ 0 & L' \end{pmatrix} \begin{pmatrix} D' & -B' \\ -C' & A' \end{pmatrix}.$$

This implies that

$$\begin{pmatrix} A & B \\ C & D \end{pmatrix} = \begin{bmatrix} a & 0 & 0 & 0 \\ (a-d)/2 & d & 0 & b \\ c & -2c & a & (a+d)/2 \\ -2c & 4c & 0 & -d \end{bmatrix}.$$

Using $AD' - BC' = I$, we find that

$$-d^2 - 4bc = 1,$$

whence $d^2 \equiv -1 \pmod 4$, since a, b, c, d are integers. This is impossible, and so the theorem is proved.

4. Automorphisms of Γ_4. As is usually the case with determination of automorphisms of a group of matrices, the lower the dimension the more difficult the proof. We begin by stating in (i) some earlier results (see [4]) which will be needed.

(i) The group Δ_2 coincides with Ω_2, and Γ_2 is the subgroup Ω_2^+ consisting of all elements of Ω_2 with determinant $+1$. For the remainder of this paper we let

$$S = \begin{pmatrix} 0 & 1 \\ -1 & 0 \end{pmatrix}, \quad T = \begin{pmatrix} 1 & 1 \\ 0 & 1 \end{pmatrix}. \tag{8}$$

Then S and T generate Γ_2, and in any relation $S^{m_1}T^{n_1}S^{m_2}T^{n_2} \cdots = I$ the sum $m_1+n_1+m_2+n_2+ \cdots$ is always even. Hence the elements $X \in \Gamma_2$ can be classified as even or odd according to the parity of the sum of the exponents when X is expressed as a product of powers of S and T. The only nontrivial character of Γ_2 is defined by

$$\epsilon(X) = \begin{cases} 1, & X \text{ even}, \\ -1, & X \text{ odd}. \end{cases}$$

Then every automorphism τ of Γ_2 is given by

$$X^\tau = \lambda(X)AXA^{-1} \qquad \text{for all } X \in \Gamma_2,$$

where λ is a character, and $A \in \Omega_2$.

(ii) Now let τ be any automorphism of Γ_4. After changing τ by a suitable inner automorphism, we may assume that $\mathfrak{P}^\tau = \pm\mathfrak{P}$, where

$$\mathfrak{P} = I^{(2)} * -I^{(2)}.$$

Since $\mathfrak{P}$ and $-\mathfrak{P}$ are conjugate in Γ_4, assume in fact that $\mathfrak{P}^\tau = \mathfrak{P}$. Then any element of Γ_4 which commutes with $\mathfrak{P}$ maps into another such element, so that

$$(Y_1 * Z_1)^\tau = Y_2 * Z_2,$$

where Y_1, Y_2, Z_1, $Z_2 \in \Gamma_2$. Let us set

$$(Y * I)^\tau = Y^\alpha * Y^\beta \qquad \text{for } Y \in \Gamma_2,$$

$$(I * Z)^\tau = Z^\gamma * Z^\delta \qquad \text{for } Z \in \Gamma_2.$$

Then $\alpha, \beta, \gamma, \delta$ are all homomorphisms of Γ_2 into itself, since

$$(Y_1 * Z_1)(Y_2 * Z_2) = Y_1Y_2 * Z_1Z_2.$$

Further, since $Y * I$ and $I * Z$ commute, so do Y^α and Z^γ for all pairs of elements Y, $Z \in \Gamma_2$; also, every element of Γ_2 is a product $Y^\alpha Z^\gamma$ for some such pair. Since $S \in \Gamma_2$, there exists an element $X \in \Gamma_2$ such that $SX^{-1} \in \Gamma_2^\alpha$ and $X \in \Gamma_2^\gamma$. But then X commutes with SX^{-1}, whence $X = \pm I$ or $\pm S$. Therefore either $S \in \Gamma_2^\alpha$ or $S \in \Gamma_2^\gamma$.

Suppose now that $S \in \Gamma_2^\alpha$; since every element of Γ_2^γ commutes with S, we see that $\Gamma_2^\gamma \subset \{\pm I, \pm S\}$. However, $S \in \Gamma_2^\gamma$ would imply the finiteness of Γ_2^α, whence $\Gamma_2 = \Gamma_2^\alpha \Gamma_2^\gamma$ could not be true. Therefore $\Gamma_2^\gamma \subset \{\pm I\}$, and then certainly $\Gamma_2^\alpha = \Gamma_2$. Similarly, one of Γ_2^β, Γ_2^δ is Γ_2, and the other is included in $\{\pm I\}$.

Now we use the fact that $(-\mathfrak{P})^\tau = -\mathfrak{P}$, that is

$$(-I * I)^\tau = -I * I.$$

Therefore $(-I)^\alpha = -I$; but if $\Gamma_2^\alpha \subset \{\pm I\}$, the fact that $-I = S^2$ would imply $(-I)^\alpha = I$. Hence $\Gamma_2^\alpha = \Gamma_2$, $\Gamma_2^\gamma \subset \{\pm I\}$, and therefore $\Gamma_2^\beta \subset \{\pm I\}$, $\Gamma_2^\delta = \Gamma_2$.

Next we prove that α is an automorphism; we need merely prove that $Y^\alpha = I$ implies $Y = I$. But if $Y^\alpha = I$, then $(Y * I)^\tau = I * \pm I$. Since $(I * I)^\tau = I * I$ and $(I * -I)^\tau = I * -I$, this implies that $Y = I$. By the same reasoning, δ is also an automorphism.

(iii) Now define

$$Y_1 \circ Y_2 = \begin{pmatrix} a_1 & b_1 \\ c_1 & d_1 \end{pmatrix} \circ \begin{pmatrix} a_2 & b_2 \\ c_2 & d_2 \end{pmatrix} = \begin{bmatrix} 0 & a_1 & 0 & b_1 \\ a_2 & 0 & b_2 & 0 \\ 0 & c_1 & 0 & d_1 \\ c_2 & 0 & d_2 & 0 \end{bmatrix}.$$

Then $Y_1 \circ Y_2 \in \Gamma_4$ if and only if $Y_1, Y_2 \in \Gamma_2$. The elements of Γ_4 which anticommute with $\mathfrak{P}$ are of the form $Y_1 \circ Y_2$, and we have

$$(A * B)(C \circ D) = AC \circ BD,$$
$$(A \circ B)(C * D) = AD \circ BC,$$
$$(A \circ B)(C \circ D) = AD * BC.$$

Suppose now that $(I \circ I)^\tau = U \circ V$. Since $(I \circ I)^2 = I * I$, we have $(U \circ V)^2 = UV * VU = I * I$, so $V = U^{-1}$. But now let

$$\mathfrak{X}^\sigma = (U^{-1} * I)\mathfrak{X}^\tau(U * I).$$

Then $\mathfrak{P}^\sigma = \mathfrak{P}$, σ and τ differ by an inner automorphism, and $(I \circ I)^\sigma = I \circ I$. Changing notation, we henceforth assume $\mathfrak{P}^\tau = \mathfrak{P}$ and $(I \circ I)^\tau = I \circ I$. From

$$(I \circ I)(Y * Z)(I \circ I) = Z * Y$$

we deduce

$$(I \circ I)(Z^\gamma Y^\alpha * Y^\beta Z^\delta)(I \circ I) = Y^\gamma Z^\alpha * Z^\beta Y^\delta.$$

Therefore

$$Z^\gamma Y^\alpha = Z^\beta Y^\delta$$

for all $Y, Z \in \Gamma_2$. Hence $\beta = \gamma$, $\alpha = \delta$. We have thus shown that for any Y, $Z \in \Gamma_2$ we have

$$(Y * Z)^\tau = \lambda(Z)Y^\alpha * \lambda(Y)Z^\alpha,$$

where λ is a character, and α is an automorphism of Γ_2.

(iv) From the discussion in part (i) of this section, we know that there exists a character μ and an element $A \in \Delta_2$ such that $X^\alpha = \mu(X)AXA^{-1}$ for all $X \in \Gamma_2$. We remark next that if $\mathfrak{B} \in \Delta_{2n}$, the map ϕ defined by $\mathfrak{X}^\phi = \mathfrak{B}\mathfrak{X}\mathfrak{B}^{-1}$ for each $\mathfrak{X} \in \Gamma_{2n}$ is clearly an automorphism of Γ_{2n}. In particular, let us define an automorphism σ of Γ_4 by

$$\mathfrak{X}^\sigma = (A^{-1} * A^{-1})\mathfrak{X}^\tau(A * A) \qquad \text{for all } \mathfrak{X} \in \Gamma_4.$$

Calling this new automorphism τ instead of σ, we then know that

$$(Y * Z)^\tau = \lambda(Z)\mu(Y)Y * \lambda(Y)\mu(Z)Z$$

for each pair $Y, Z \in \Gamma_2$, and further that

$$(I \circ I)^\tau = (A^{-1} * A^{-1})(I \circ I)(A * A) = I \circ I.$$

Thence we have

$$(Y \circ Z)^\tau = (Y * Z)^\tau(I \circ I)^\tau = \lambda(Z)\mu(Y)Y \circ \lambda(Y)\mu(Z)Z.$$

(v) We apply the above results to the 4 generators of Γ_4, which are given by (see [3])

$$\mathfrak{R}_1 = I \circ I, \qquad \mathfrak{R}_2 = T \dotplus T'^{-1}, \qquad \mathfrak{S}_0 = S * I, \qquad \mathfrak{T}_0 = T * I$$

(where S and T are defined by (8)). We have at once

$$\mathfrak{R}_1^\tau = \mathfrak{R}_1, \qquad \mathfrak{S}_0^\tau = \pm S * \pm I, \qquad \mathfrak{T}_0^\tau = \pm T * \pm I, \qquad \mathfrak{S}_0^\tau\mathfrak{T}_0^\tau = \mathfrak{S}_0\mathfrak{T}_0,$$

(the last equation holding because $\mathfrak{S}_0\mathfrak{T}_0$ is a square).

We use now (and again later) an argument due to Hua [5] to find the possible images $\mathfrak{R}_2^\tau$. Observe that

$$\begin{bmatrix} I^{(2)} & \begin{matrix} 2n & 0 \\ 0 & 0 \end{matrix} \\ 0 & I^{(2)} \end{bmatrix} \quad \text{and} \quad \begin{bmatrix} I^{(2)} & \begin{matrix} 0 & 0 \\ 0 & 2m \end{matrix} \\ 0 & I^{(2)} \end{bmatrix}$$

are elements of Γ_4 which are invariant under τ; their product is also invariant. Hence the group of all elements of Γ_4 which commute element-wise with the set of matrices of the form

$$\begin{bmatrix} I^{(2)} & \begin{matrix} \lambda_1 & 0 \\ 0 & \lambda_2 \end{matrix} \\ 0 & I^{(2)} \end{bmatrix}, \qquad \lambda_1, \lambda_2 \text{ even integers,}$$

is mapped onto itself by τ. This group is readily found to consist of all elements of Γ_4 of the form

$$\begin{pmatrix} E & B \\ 0 & E \end{pmatrix}, \quad \text{where } E = \begin{pmatrix} \pm 1 & 0 \\ 0 & \pm 1 \end{pmatrix} \text{ and } EB' = BE.$$

The squares of these elements are the matrices of Γ_4 given by

$$\begin{pmatrix} I & M \\ 0 & I \end{pmatrix}, \tag{9}$$

where M is symmetric and all elements of M are even. Hence

$$\begin{pmatrix} I & M \\ 0 & I \end{pmatrix}^{\tau} = \begin{pmatrix} I & M_1 \\ 0 & I \end{pmatrix}$$

for even symmetric M, and M_1 is also even and symmetric.

Next observe that

$$\begin{pmatrix} 0 & I \\ -I & 0 \end{pmatrix}\begin{pmatrix} I & M \\ 0 & I \end{pmatrix}\begin{pmatrix} 0 & -I \\ I & 0 \end{pmatrix} = \begin{pmatrix} I & 0 \\ -M & I \end{pmatrix}.$$

Since

$$\begin{pmatrix} 0 & I \\ -I & 0 \end{pmatrix}^{\tau} = (S_* S)^{\tau} = \pm S_* \pm S,$$

we see that for even symmetric N we have

$$\begin{pmatrix} I & 0 \\ N & I \end{pmatrix}^{\tau} = \begin{pmatrix} I & 0 \\ N_1 & I \end{pmatrix}, \tag{10}$$

with N_1 even and symmetric.

Now let Σ be the group of matrices of the form (9) with M even and symmetric, and let Σ' be the group of matrices given by (10) with even symmetric N. Then τ maps both Σ and Σ' onto themselves, and so any element commuting with both Σ and Σ' maps into another such element. However, these elements are precisely the rotations in Γ_4. Hence for each $U \in \Omega_2$ we have

$$\begin{pmatrix} U & 0 \\ 0 & U'^{-1} \end{pmatrix}^{\tau} = \begin{pmatrix} U^{\sigma} & 0 \\ 0 & (U^{\sigma})'^{-1} \end{pmatrix}.$$

The map $U \to U^{\sigma}$ is an automorphism σ of Ω_2, and we already know from $\mathfrak{P}^{\tau} = \mathfrak{P}$ and $\mathfrak{R}_1^{\tau} = \mathfrak{R}_1$ that $S^{\sigma} = S$. Consequently (see [4]) there are only 4 possibilities for T^{σ}, given by

$$\begin{pmatrix} 1 & 1 \\ 0 & 1 \end{pmatrix}, \quad \begin{pmatrix} -1 & 1 \\ 0 & -1 \end{pmatrix}, \quad \begin{pmatrix} 1 & 0 \\ -1 & 1 \end{pmatrix}, \quad \begin{pmatrix} -1 & 0 \\ -1 & -1 \end{pmatrix}.$$

(vi) We next apply τ to both sides of the equation

$$(S*I)\begin{pmatrix} T^2 & 0 \\ 0 & T'^{-2} \end{pmatrix}(S*I)^{-1} = \begin{bmatrix} 1 & 0 & 0 & 0 \\ 0 & 1 & 0 & 0 \\ 0 & -2 & 1 & 0 \\ -2 & 0 & 0 & 1 \end{bmatrix}$$

and use equation (10). This shows that

$$(T^2)^\sigma = \begin{pmatrix} 1 & \pm 2 \\ 0 & 1 \end{pmatrix},$$

and so either

$$T^\sigma = T \text{ or } T^\sigma = \begin{pmatrix} -1 & 1 \\ 0 & -1 \end{pmatrix} = T_1 \text{ (say)}.$$

Now we show that $\mathfrak{S}_0^\tau = \pm \mathfrak{S}_0$, $\mathfrak{T}_0^\tau = \pm \mathfrak{T}_0$. For, $\mathfrak{R}_2$ and $\mathfrak{T}_0$ commute; hence so do $\mathfrak{R}_2^\tau$ and $\mathfrak{T}_0^\tau$. However, $\mathfrak{R}_2^\tau = \mathfrak{R}_2$ or $\mathfrak{R}_2^\tau = T_1 \dotplus T_1'^{-1}$, and it is easily verified that $\pm(T* -I)$ does not commute with either of these two possible images of $\mathfrak{R}_2$. Therefore $\mathfrak{T}_0^\tau = \pm(T*I)$, whence $\mathfrak{S}_0^\tau = \pm(S*I)$.

Next suppose that $\mathfrak{R}_2^\tau = T_1 \dotplus T_1'^{-1}$; then define τ_1 by $\mathfrak{X}^{\tau_1} = \mathfrak{P}\mathfrak{X}^\tau\mathfrak{P}^{-1}$. Then $\mathfrak{S}_0^{\tau_1} = \mathfrak{S}_0$, $\mathfrak{T}_0^{\tau_1} = \mathfrak{T}_0$, and $\mathfrak{R}_1^{\tau_1} = -\mathfrak{R}_1$, $\mathfrak{R}_2^{\tau_1} = -\mathfrak{R}_2$. We have therefore shown that apart from an "inner" automorphism by an element of Δ_4, every automorphism τ of Γ_4 can be described by

$$(\mathfrak{R}_1, \mathfrak{R}_2, \mathfrak{S}_0, \mathfrak{T}_0)^\tau = (\pm\mathfrak{R}_1, \pm\mathfrak{R}_2, \pm\mathfrak{S}_0, \pm\mathfrak{T}_0),$$

and the signs must satisfy

$$\mathfrak{R}_1^\tau\mathfrak{R}_2^\tau = \mathfrak{R}_1\mathfrak{R}_2, \qquad \mathfrak{S}_0^\tau\mathfrak{T}_0^\tau = \mathfrak{S}_0\mathfrak{T}_0.$$

Thus every automorphism τ is given by

$$\mathfrak{X}^\tau = \theta(\mathfrak{X})\mathfrak{A}\mathfrak{X}\mathfrak{A}^{-1} \qquad \text{for all } \mathfrak{X} \in \Gamma_4,$$

where $\mathfrak{A} \in \Delta_4$ and θ is a character of Γ_4.

(vii) It will be shown in a future note by the author [9] that Γ_4 has exactly one nontrivial character θ, where θ is the map of Γ_4 into $\{\pm 1\}$ induced by

$$\theta(\mathfrak{R}_1) = \theta(\mathfrak{R}_2) = \theta(\mathfrak{S}_0) = \theta(\mathfrak{T}_0) = -1.$$

This fact, together with the preceding discussion, settles the question of automorphisms of Γ_4. It will also be shown in the same note that Γ_{2n}, $n > 2$, has no nontrivial characters. This result will be needed in finding all automorphisms of Γ_{2n}.

5. **Automorphisms of Γ_{2n}, $n > 2$.** We are now ready to prove, by induction on n, the following result:

THEOREM 3. *For $n > 2$, every automorphism τ of Γ_{2n} is given by*

$$\mathfrak{X}^\tau = \mathfrak{A}\mathfrak{X}\mathfrak{A}^{-1},$$

where $\mathfrak{A}\in\Delta_{2n}$ *depends only on* τ.

Proof. (i) Let $n\geqq 3$; by the induction hypothesis and our previous results, we may assume that every automorphism σ of $\Gamma_{2(n-1)}$ is given by

$$X^\sigma = \theta(X)\cdot AXA^{-1},$$

where $A\in\Delta_{2(n-1)}$ and θ is a character of $\Gamma_{2(n-1)}$. Let τ be an automorphism of Γ_{2n}, and set

$$\mathfrak{P} = -I^{(2)} * I^{2(n-1)}.$$

We see from Theorem 2 that after changing τ by a suitable inner automorphism, we may take $\mathfrak{P}^\tau = \pm\mathfrak{P}$. The elements of Γ_{2n} which commute with $\mathfrak{P}$ are of the form $Y_1 * Z_1$, $Y_1\in\Gamma_2$, $Z_1\in\Gamma_{2(n-1)}$, so that we have

$$(Y_1 * Z_1)^\tau = Y_2 * Z_2.$$

Again we set

$$(Y * I)^\tau = Y^\alpha * Y^\beta \qquad \text{for } Y\in\Gamma_2,$$

$$(I * Z)^\tau = Z^\gamma * Z^\delta \qquad \text{for } Z\in\Gamma_{2(n-1)}.$$

Then Γ_2^α and $\Gamma_{2(n-1)}^\gamma$ commute elementwise, and Γ_2 is their product. As in §4, part (ii), we deduce that one of Γ_2^α, $\Gamma_{2(n-1)}^\gamma$ is Γ_2, and the other is contained in $\{\pm I\}$.

(ii) For the moment set $\mathcal{A}=\Gamma_2^\beta$, $\mathcal{B}=\Gamma_{2(n-1)}^\delta$. Then $\mathcal{A}$ and $\mathcal{B}$ commute elementwise, and their product is $\Gamma_{2(n-1)}$. This shows that $\mathcal{B}$ is a normal subgroup of $\Gamma_{2(n-1)}$. We shall show that $\mathcal{A}\subset\{\pm I\}$, $\mathcal{B}=\Gamma_{2(n-1)}$, and that δ is an automorphism.

For each involution $W\in\Gamma_{2(n-1)}$ we have $(W^\delta)^2=I^\delta=I$. Suppose that $W^\delta=\pm I$ for every involution $W\in\Gamma_{2(n-1)}$; since the involutions in $\Gamma_{2(n-1)}$ generate all of $\Gamma_{2(n-1)}$ (this follows readily from [3]), this would mean that $\mathcal{B}\subset\{\pm I\}$, and so β would map Γ_2 homomorphically *onto* $\Gamma_{2(n-1)}$. We may then show that β is an isomorphism; for, suppose that $Y^\beta=I$, $Y\neq I$. Then

$$(Y * I)^\tau = Y^\alpha * I.$$

Since $\mathcal{B}\subset\{\pm I\}$, certainly $\Gamma_{2(n-1)}^\gamma$ is not contained in $\{\pm I\}$, and so $\Gamma_2^\alpha \subset\{\pm I\}$, that is, α is a character. Therefore $Y^\alpha=\pm I$. But $Y^\alpha=I$ is impossible, since then $(Y * I)^\tau=I^{(2n)}$ and $Y=I$. On the other hand, $Y^\alpha=-I$ is impossible, since in that case $(Y * I)^\tau=\mathfrak{P}$, so $(Y * I)=\pm\mathfrak{P}$. Therefore we would have $Y=-I$, and this gives a contradiction because $-I=S^2$, and α a character, together imply $(-I)^\alpha=I$. Therefore β is an isomorphism. However, this is itself impossible because Γ_2 has no involutions other than $\pm I$, whereas $\Gamma_{2(n-1)}$ has such involutions for $n>2$.

We conclude from the above that there is at least one involution $W\in\Gamma_{2(n-1)}$

for which $W^\delta \neq \pm I$. However, $\mathcal{B}$ is a normal subgroup of $\Gamma_{2(n-1)}$, and $W^\delta \in \mathcal{B}$. Therefore $\mathcal{B}$ contains all of the conjugates of W^δ in $\Gamma_{2(n-1)}$. It is not difficult to see that if $W^\delta \neq \pm I$, the only elements of $\Gamma_{2(n-1)}$ which commute elementwise with all conjugates of W^δ are $\pm I$. Hence $\mathcal{A} \subset \{\pm I\}$, and $\mathcal{B} = \Gamma_{2(n-1)}$. Consequently

$$(Y * Z)^\tau = \theta(Z) Y^\alpha * \lambda(Y) Z^\delta,$$

where θ and λ are characters, α is a homomorphism of Γ_2 onto itself, and δ a homomorphism of $\Gamma_{2(n-1)}$ onto itself. We deduce readily that α and δ are automorphisms, whence incidentally $\mathfrak{P}^\tau = \mathfrak{P}$.

By the discussion at the beginning of the proof, we know that there exist matrices $C \in \Omega_2$, $D \notin \Delta_{2(n-1)}$, and characters μ, ν such that

$$Y^\alpha = \mu(Y) C Y C^{-1}, \qquad Z^\delta = \nu(Z) D Z D^{-1}.$$

If $C * D \in \Delta_{2n}$, define τ_1 by

$$\mathfrak{X}^{\tau_1} = (C * D)^{-1} \mathfrak{X}^\tau (C * D),$$

so that

$$(Y * Z)^{\tau_1} = \theta(Z)\mu(Y) Y * \lambda(Y)\nu(Z) Z.$$

However, possibly $C * D \in \Delta_{2n}$. In that case, if $K = (-1) \dotplus (1)$, then $CK * D \in \Delta_{2n}$, and we define τ_2 by

$$\mathfrak{X}^{\tau_2} = (CK * D)^{-1} \mathfrak{X}^\tau (CK * D).$$

Thus, changing notation, we may assume that

$$(Y * Z)^\tau = \theta(Z)\mu(Y) H Y H^{-1} * \lambda(Y)\nu(Z) Z, \tag{11}$$

for any $Y \in \Gamma_2$, $Z \in \Gamma_{2(n-1)}$, where θ, μ, λ, ν are characters, and where either $H = I^{(2)}$ or $H = K$.

(iii) Suppose now that $Y \in \Gamma_2$, $Z \in \Gamma_{2(n-1)}$ are given by

$$Y = \begin{pmatrix} a & b \\ c & d \end{pmatrix}, \qquad Z = \begin{pmatrix} A & B \\ C & D \end{pmatrix}.$$

Then define $Y *^i Z$ to be the $2n \times 2n$ matrix $\mathfrak{M}$ obtained by placing the elements of Y at the intersections of the ith and $(n+i)$th rows and columns, filling in the remaining places in those rows and columns with 0's, and letting the matrix obtained from $\mathfrak{M}$ by deleting the ith and $(n+i)$th rows and columns be identical with Z. Then $Y *^i Z$ is a generalization of the previously defined symplectic direct sum, and in fact $Y *^1 Z = Y * Z$.

Now set

$$\mathfrak{P}_i = -I^{(2)} *^i I^{2(n-1)} = I^{(2)} * Q_i, \text{ say.}$$

Then Q_i is a square in $\Gamma_{2(n-1)}$ (since $-I = S^2$), and so from (11) we have

$$\mathfrak{P}_i^\tau = I * Q_i = \mathfrak{P}_i.$$

As before it then follows for $Y \in \Gamma_2$, $Z \in \Gamma_{2(n-1)}$ that

$$(Y *^i Z)^\tau = (F_i(Z) f_i(Y) A_i Y A_i^{-1}) *^i (g_i(Y) G_i(Z) B_i Z B_i^{-1}), \tag{12}$$

where $A_i \in \Omega_2$, $B_i \in \Delta_{2(n-1)}$, and F_i, f_i, g_i, G_i are characters.

(iv) Next let X and $Y \in \Gamma_2$, $Z \in \Gamma_{2(n-2)}$. Applying τ to both sides of the equation

$$X * (Y * Z) = Y *^2 (X * Z)$$

and using (12), we obtain

$$\begin{aligned} &[F_1(Y*Z) f_1(X) A_1 X A_1^{-1}] * [g_1(X) G_1(Y*Z) B_1 (Y*Z) B_1^{-1}] \\ &\qquad = [F_2(X*Z) f_2(Y) A_2 Y A_2^{-1}] *^2 [g_2(Y) G_2(X*Z) B_2 (X*Z) B_2^{-1}]. \end{aligned} \tag{13}$$

In particular for $X = -I$, $Y = I$, $Z = I$ this yields

$$B_2(-I * I) B_2^{-1} = -I * I,$$

so that

$$B_2 = \pm A_1 * C_2$$

and further

$$B_1 = \pm A_2 * \pm C_2.$$

We use these expressions for B_1 and B_2 in (13) and obtain

$$\begin{aligned} F_1(Y*Z) f_1(X) &= g_2(Y) G_2(X*Z), \\ F_2(X*Z) f_2(Y) &= g_1(X) G_1(Y*Z), \\ g_1(X) G_1(Y*Z) &= g_2(Y) G_2(X*Z). \end{aligned}$$

These imply that $f_1 = g_1$ and $f_2 = g_2$.

Continuing in this way we see that each B_i decomposes completely, and in fact if

$$\mathfrak{D} = A_1 * A_2 * \cdots * A_n,$$

then B_i is obtained from $\mathfrak{D}$ by deleting A_i and possibly changing signs of some of the remaining A's. Furthermore, if any $A_i \in \Delta_2'$, then every $A_i \in \Delta_2'$, since each $B_i \in \Delta_{2(n-1)}$. Therefore $\mathfrak{D} \in \Delta_{2n}$. After a further inner automorphism of Γ_{2n} by a factor of $\mathfrak{D}^{-1}$, we may assume hereafter that

$$(Y *^i Z)^\tau = f_i(Y) [F_i(Z) Y *^i G_i(Z) B_i Z B_i^{-1}] \tag{14}$$

for $Y \in \Gamma_2$, $Z \in \Gamma_{2(n-1)}$, where f_i, F_i and G_i are characters and each B_i is of the form $(\pm I) * \cdots * (\pm I)$, and in fact we may take $B_1 = I$.

(v) Define

$$U_1 = \begin{bmatrix} 0 & 0 \cdots 0 & 1 \\ 1 & 0 \cdots 0 & 0 \\ 0 & 1 \cdots 0 & 0 \\ \cdot & \cdots\cdots & \cdot \\ \cdot & \cdots\cdots & \cdot \\ 0 & 0 \cdots 1 & 0 \end{bmatrix}, \qquad U_2 = T \dotplus I^{(n-2)},$$

where T is given by (8). Then the generators of Γ_{2n} are (see [3]):

$$\mathfrak{R}_1 = U_1 \dotplus U_1'^{-1}, \qquad \mathfrak{R}_2 = U_2 \dotplus U_2'^{-1}, \qquad \mathfrak{T}_0 = T * I, \qquad \mathfrak{S}_0 = S * I.$$

From (14) we find at once that

$$\mathfrak{T}_0^\tau = \pm \mathfrak{T}_0, \qquad \mathfrak{S}_0^\tau = \pm \mathfrak{S}_0, \quad \text{and} \quad \mathfrak{S}_0^\tau \mathfrak{T}_0^\tau = \mathfrak{S}_0 \mathfrak{T}_0.$$

Next, the rotations of Γ_{2n} map onto rotations under τ, since the rotations are generated by the elements $Y *^i Z$, $i = 1, \cdots, n$, where Y and Z have the forms

$$Y = \begin{pmatrix} a & 0 \\ 0 & d \end{pmatrix}, \qquad Z = \begin{pmatrix} A & 0 \\ 0 & D \end{pmatrix},$$

and the image of any such $Y *^i Z$ is of the same kind. Therefore τ induces an automorphism σ on the group Ω_n, where

$$\begin{pmatrix} V & 0 \\ 0 & V'^{-1} \end{pmatrix}^\tau = \begin{pmatrix} V^\sigma & 0 \\ 0 & (V^\sigma)'^{-1} \end{pmatrix}.$$

We then know [4] that there exists $H \in \Omega_n$ such that

$$V^\sigma = H V^\omega H^{-1} \qquad \text{for all } V \in \Omega_n,$$

where either $V^\omega = V$ for all V or $V^\omega = V'^{-1}$ for all V.

We know furthermore that τ maps every rotation $\mathfrak{P}_i$ onto itself, from which we see that H is diagonal, with diagonal elements ± 1's. Replace τ by τ_1 defined by

$$\mathfrak{X}^{\tau_1} = (H \dotplus H)\mathfrak{X}^\tau (H \dotplus H)$$

and change notation. We again have $\mathfrak{T}_0^\tau = \pm \mathfrak{T}_0$, $\mathfrak{S}_0^\tau = \pm \mathfrak{S}_0$, and $\mathfrak{S}_0^\tau \mathfrak{T}_0^\tau = \mathfrak{S}_0 \mathfrak{T}_0$, but now $V^\sigma = V^\omega$ for each $V \in \Omega_n$. The argument given in §4, parts (iii) and (iv) shows that $\mathfrak{R}_2^\tau = T'^{-1} \dotplus T$ is impossible, so $V^\sigma = V$ for all $V \in \Omega_n$. Therefore τ is given by

$$(\mathfrak{R}_1, \mathfrak{R}_2, \mathfrak{T}_0, \mathfrak{S}_0)^\tau = (\mathfrak{R}_1, \mathfrak{R}_2, \pm \mathfrak{T}_0, \pm \mathfrak{S}_0).$$

However, as we have already mentioned, Γ_{2n} has no nontrivial character for $n \geqq 3$. Hence $\mathfrak{T}_0^\tau = \mathfrak{T}_0$, $\mathfrak{S}_0^\tau = \mathfrak{S}_0$. This completes the proof of the theorem.

6. We remark finally that if $\mathfrak{M} \in \Gamma_{2n}$ is given by (3), then

$$\mathfrak{M}'^{-1} = \begin{pmatrix} D & -C \\ -B & A \end{pmatrix} = \begin{pmatrix} 0 & I \\ -I & 0 \end{pmatrix} \begin{pmatrix} A & B \\ C & D \end{pmatrix} \begin{pmatrix} 0 & -I \\ I & 0 \end{pmatrix},$$

so the automorphism σ: $\mathfrak{M}^{\sigma} = \mathfrak{M}'^{-1}$ is inner.

Furthermore, any element of Δ_{2n} can be written as the product of an element of Γ_{2n} and $-I^{(n)} \dotplus I^{(n)}$, so every automorphism of Γ_{2n} can be obtained by using inner automorphisms by elements in Γ_{2n}, coupled with the automorphism

$$\begin{pmatrix} A & B \\ C & D \end{pmatrix} \rightarrow \begin{pmatrix} -I & 0 \\ 0 & I \end{pmatrix} \begin{pmatrix} A & B \\ C & D \end{pmatrix} \begin{pmatrix} -I & 0 \\ 0 & I \end{pmatrix} = \begin{pmatrix} A & -B \\ -C & D \end{pmatrix}.$$

Bibliography

1. C. L. Siegel, Math. Ann. vol. 116 (1939) pp. 617–657.
2. L. K. Hua, Trans. Amer. Math. Soc. vol. 57 (1945) pp. 441–490.
3. L. K. Hua and I. Reiner, Trans. Amer. Math. Soc. vol. 65 (1949) pp. 415–426.
4. ———, Trans. Amer. Math. Soc. vol. 71 (1951) pp. 331–348.
5. L. K. Hua, Ann. of Math. vol. 49 (1948) pp. 739–759.
6. I. Reiner, Trans. Amer. Math. Soc. vol. 77 (1954) pp. 498–505.
7. J. Dieudonné, Memoirs of the American Mathematical Society, No. 2, 1951.
8. I. Reiner, Trans. Amer. Math. Soc. vol. 79 (1955) pp. 459–476.
9. ———, *Real linear characters of the symplectic modular group*, to appear in Proc. Amer. Math. Soc.

University of Illinois,
Urbana, Ill.
Institute for Advanced Study,
Princeton, N. J.

REAL LINEAR CHARACTERS OF THE SYMPLECTIC MODULAR GROUP

IRVING REINER

1. The symplectic modular group Γ_{2n} consists of all integral $2n\times 2n$ matrices $\mathfrak{M}$ for which $\mathfrak{M}\mathfrak{F}\mathfrak{M}'=\mathfrak{F}$, where

$$\mathfrak{F} = \begin{pmatrix} 0 & I^{(n)} \\ -I^{(n)} & 0 \end{pmatrix}.$$

In order to determine all possible automorphisms of Γ_{2n} [1],[1] it is necessary to find all real linear characters of Γ_{2n}, that is, all homomorphisms into $\{\pm 1\}$. In this note we prove that Γ_{2n} has no nontrivial real linear characters for $n>2$, while Γ_2 and Γ_4 each have exactly one nontrivial real linear character. We shall also determine Γ'_{2n}, the commutator subgroup of Γ_{2n}.

We define the symplectic direct sum $\mathfrak{M}_1 * \mathfrak{M}_2$ by

$$\mathfrak{M}_1 * \mathfrak{M}_2 = \begin{pmatrix} A_1 & B_1 \\ C_1 & D_1 \end{pmatrix} * \begin{pmatrix} A_2 & B_2 \\ C_2 & D_2 \end{pmatrix} = \begin{pmatrix} A_1 & 0 & B_1 & 0 \\ 0 & A_2 & 0 & B_2 \\ C_1 & 0 & D_1 & 0 \\ 0 & C_2 & 0 & D_2 \end{pmatrix}.$$

Set

$$S = \begin{pmatrix} 0 & 1 \\ -1 & 0 \end{pmatrix}, \qquad T = \begin{pmatrix} 1 & 1 \\ 0 & 1 \end{pmatrix}, \qquad V = \begin{pmatrix} 0 & 1 \\ 1 & 0 \end{pmatrix},$$

and define[2]

$$U_0 = S \dotplus I^{(n-2)}, \qquad U_1 = V \dotplus I^{(n-2)}, \qquad U_2 = T \dotplus I^{(n-2)}.$$

Then [2] Γ_{2n} is generated by $\mathfrak{R}_i = U_i \dotplus U_i'^{-1}$ $(i=0,\ 2)$, $\mathfrak{T}_0 = T * I^{2(n-1)}$, $\mathfrak{S}_0 = S * I^{2(n-1)}$, and their conjugates. When $n=1$, the $\mathfrak{R}_i$ are superfluous. Next we remark that

$$(1) \qquad \mathfrak{S}_0\mathfrak{T}_0 = (ST) * I = (ST)^{-2} * I,$$

$$(2) \qquad \mathfrak{R}_0\mathfrak{R}_2 = U_3 \dotplus U_3'^{-1}, \text{ where } U_3 = ST \dotplus I = (ST \dotplus I)^{-2},$$

$$(3) \qquad \mathfrak{T}_0 = \mathfrak{R}_1\mathfrak{R}_2\cdot\mathfrak{R}_0\mathfrak{T}_0\mathfrak{R}_0^{-1}\mathfrak{T}_0^{-1}\cdot(\mathfrak{R}_1\mathfrak{R}_2)^{-1}\cdot\mathfrak{S}_0\mathfrak{R}_1\cdot\mathfrak{R}_2\cdot(\mathfrak{S}_0\mathfrak{R}_1)^{-1}.$$

Presented to the Society, December 29, 1954 under the title *Characters of the symplectic modular group*; received by the editors November 13, 1954 and, in revised form, November 24, 1954.

[1] Numbers in brackets refer to the bibliography at the end of the paper.

[2] $A \dotplus B$ denotes the direct sum of the matrices A and B.

Therefore if θ is any real linear character of Γ_{2n}, we must have

$$\theta(\mathfrak{N}_0) = \theta(\mathfrak{N}_2) = \theta(\mathfrak{T}_0) = \theta(\mathfrak{S}_0) = \pm 1.$$

On the other hand, let Ω_n be the unimodular group consisting of all integral $n \times n$ matrices with determinant ± 1. Then for $n > 2$, Ω_n is its own commutator subgroup [**3**]. Hence for $n > 2$, U_0 is a product of commutators in Ω_n, and therefore $\mathfrak{N}_0$ is in the commutator subgroup of Γ_{2n}. Therefore $\theta(\mathfrak{N}_0) = +1$, so Γ_{2n} has no nontrivial real linear characters for $n > 2$.

Now we must prove that there exists a homomorphism of Γ_{2n} into $\{\pm 1\}$ which maps each generator $\mathfrak{N}_0$, $\mathfrak{N}_2$, $\mathfrak{T}_0$, $\mathfrak{S}_0$ into -1, for the cases $n=1$ and $n=2$. This is already known for $n=1$ [**3**], but we give an independent proof here. Let H be the normal subgroup of Γ_{2n} consisting of all matrices $\equiv I^{(2n)}$ (mod 2). Then it is known [**4**] that $\Gamma_{2n}/H \cong S_{3n}$ for $n=1, 2$, where S_k is the symmetric group on k symbols. Let π be the homomorphism mapping Γ_{2n} onto S_{3n}, and let A_{3n} be the alternating subgroup of S_{3n}. Then $\pi^{-1}(A_{3n})$ is a subgroup of index 2 of Γ_{2n}, $n=1, 2$. Therefore Γ_{2n} has a nontrivial real linear character for $n=1, 2$, and the previous discussion shows that it is unique, and maps each generator onto -1.

2. Now we consider Γ'_{2n}, and we begin with $n=\mathbf{1}$, the most difficult case. The commutator subgroup of $\Gamma_2/\{\pm I\}$ is known [**5**], but we shall not use this earlier result. According to [**6**], $\Gamma_2 = \{S, T\}$ has as defining relations

$$S^4 = TS^{-1}TS^{-1}TS = 1.$$

Then the sum of the exponents to which S (resp. T) occurs in any relation, must be of the form $4a-b$ (resp. $3b$), where a and b are integers. For $X \in \Gamma_2$ let α_X = sum of the exponents to which S occurs, and let β_X = sum of the exponents to which T occurs, when X is expressed as a power product of S and T. Then $X \in \Gamma'_2$ implies that α_X, β_X are of the form

$$\alpha_X = 4a - b, \qquad \beta_X = 3b, \tag{4}$$

for integral a, b. On the other hand,

$$S^{4a-b}T^{3b} = S^{-b}T^{3b} \equiv (S^{-1}T^3)^b \pmod{\Gamma'_2},$$

and

$$S^{-1}T^3 = S^{-1}T \cdot ST^{-1}ST^{-1}S^{-1} \cdot T \in \Gamma'_2.$$

Hence $X \in \Gamma'_2$ if and only if (4) holds. Consequently $T^{12} \in \Gamma'_2$, $T^m \notin \Gamma'_2$ for $m=1, \cdots, 11$, and we have

$$\Gamma_2 = \bigcup_{m=0}^{11} T^m \Gamma_2'.$$

Thus[3] $(\Gamma_2:\Gamma_2') = 12$.

Next we show how Γ_2' may be defined by means of congruences. Let

$$H_m = \{X \in \Gamma_2 : X \equiv I \pmod{m}\}.$$

Then H_m is a normal subgroup of Γ_2, and in particular Γ_2/H_3 is a group of order 24 consisting of all 2×2 matrices of determinant $+1$ with elements in $GF(3)$. This group contains a normal subgroup $\{C, D\}$ [4] of index 3, where

$$C = \begin{pmatrix} 0 & 1 \\ -1 & 0 \end{pmatrix}, \quad D = \begin{pmatrix} 1 & 1 \\ 1 & 2 \end{pmatrix}.$$

Hence the group K_3 consisting of all elements of Γ_2 congruent (mod 3) to a matrix in $\{C, D\}$ is a normal subgroup of Γ_2 of index 3. Therefore $\Gamma_2' \subset K_3$.

Next we remark that Γ_2/H_4 is of order 48, and contains the normal subgroup $\{A, E, F\}$ of order 12 generated by

$$A = \begin{pmatrix} -1 & 1 \\ -1 & 0 \end{pmatrix}, \quad E = \begin{pmatrix} -1 & 0 \\ 2 & -1 \end{pmatrix}, \quad F = \begin{pmatrix} -1 & 2 \\ 0 & -1 \end{pmatrix}$$

taken mod 4. If K_4 is the set of all elements of Γ_2 congruent mod 4 to a matrix in $\{A, E, F\}$, then K_4 is a normal subgroup of Γ_2 of index 4. Therefore $\Gamma_2' \subset K_4$. Since $K_4 K_3 = \Gamma_2$, it follows at once that

$$(\Gamma_2 : K_3 \cap K_4) = 12,$$

and so

$$\Gamma_2' = K_3 \cap K_4.$$[4]

3. We show next that $(\Gamma_4:\Gamma_4') = 2$, and also we determine Γ_4' by means of congruences. From [3] we find that $\mathfrak{R}_0\mathfrak{R}_2$ and $\mathfrak{R}_2^2 \in \Gamma_4'$. Hence $L = \Gamma_4' \cup \mathfrak{R}_2\Gamma_4'$ is a normal subgroup of Γ_4, and (using (3)) $\mathfrak{R}_0$, $\mathfrak{R}_2$, and $\mathfrak{T}_0$ are elements of L. Also we have

$$\mathfrak{S}_0\mathfrak{T}_0 = \mathfrak{T}_0\mathfrak{S}_0\mathfrak{T}_0^{-1}\mathfrak{S}_0^{-1}(\mathfrak{S}_0\mathfrak{T}_0\mathfrak{S}_0^{-1})^2\mathfrak{T}_0^2,$$

so $\mathfrak{S}_0 \in L$. Hence $L = \Gamma_4$, and therefore either $\Gamma_4' = \Gamma_4$ or $(\Gamma_4:\Gamma_4') = 2$.

[3] This result has been obtained independently by Professor J. L. Brenner.

[4] The author wishes to acknowledge with thanks some helpful conversations with Professor E. V. Schenkman on the material in §2.

However we have already seen that Γ_4 contains a subgroup K of index 2. Since $\Gamma_4' \subset K$, we then have $\Gamma_4' = K$.

Finally we remark that the previous discussion shows easily that $\Gamma_{2n}' = \Gamma_{2n}$ for $n > 2$.

Bibliography

1. I. Reiner, *Automorphisms of the symplectic modular group*, Trans. Amer. Math. Soc. vol. 80 (1955) pp. 35–50.

2. L. K. Hua and I. Reiner, Trans. Amer. Math. Soc. vol. 65 (1949) pp. 415–426.

3. ———, Trans. Amer. Math. Soc. vol. 71 (1951) pp. 331–348.

4. L. E. Dickson, *Linear groups*, Teubner, 1901.

5. H. Frasch, Math. Ann. vol. 108 (1933) pp. 229–252, especially p. 245, footnote.

6. J. Nielsen, Danske Videnskabernes Selskab. Matematisk-Fysiske Meddelelser vol. 5, no. 18 (1924) pp. 3–29.

Institute for Advanced Study and
University of Illinois

CONGRUENCE SUBGROUPS OF MATRIX GROUPS

Irving Reiner and J. D. Swift

ABSTRACT

Let M_r^+ denote the modular group consisting of all integral $r \times r$ matrices with determinant $+1$. Define the subgroup G_n, of M_2^+ by

$$G_n = \left\{ \begin{pmatrix} a & b \\ c & d \end{pmatrix} \in M_2^+ : c \equiv 0 \pmod{n} \right\}.$$

A theorem of M. Newman states that a subgroup H of G_n which contains G_{mn} must equal G_{an} for some divisor a of m. This result is extended in this paper in two directions: (i) Analogous results are obtained when the matrix entries lie in the ring of integers of an algebraic number field, and (ii) Results of this type are given for certain groups of matrices of arbitrary size.

Theorem 1. *Let $\mathcal{D}$ be the ring of integers in an algebraic number field and let $\mathcal{M}$ and $\mathcal{N}$ be ideals of $\mathcal{D}$. Let $G(\mathcal{N})$ be the group of 2×2 invertible matrices over $\mathcal{D}$ which have the lower left entry in $\mathcal{N}$. If H is a subgroup of $G(\mathcal{N})$ which contains $G(\mathcal{MN})$ where $\mathcal{M}$ is an ideal not containing (6), then $H = G(\mathcal{CN})$ for some ideal $\mathcal{C} \supseteq \mathcal{M}$.*

The proof is self–contained and uses a series of matrix computations. Examples are given to show the result fails if 2 or 3 is in $\mathcal{M}$.

In higher dimensions, consider the column groups and the row groups defined as:

$$C_m = \{(a_{ij}) \in M_r^+ : m|a_{21}, m|a_{31}, \cdots, m|a_{r1}\},$$
$$R_n = \{(a_{ij}) \in M_r^+ : n|a_{r1}, n|a_{r2}, \cdots, n|a_{r,r-1}\}.$$

Theorem 2. *Let H be a subgroup of M_r^+ satisfying*

$$(C_{am} \cap R_{bn}) \subseteq H \subseteq (C_m \cap R_n),$$

where $(am, bn) = 1$. Then $H = C_{\alpha m} \cap R_{\beta n}$, where $\alpha | a$ and $\beta | b$.

The proof of this result is also given by direct computation with matrices.

MASCHKE MODULES OVER DEDEKIND RINGS

IRVING REINER

1. Introduction. We use the following notation throughout:

$\mathfrak{o}$ = Dedekind ring (**8**; **12**, p. 83).
K = quotient field of $\mathfrak{o}$.
A = finite-dimensional separable algebra over K, with identity element e (**6**, p. 115).
G = $\mathfrak{o}$-order in A (**2**, p. 69).
$\mathfrak{p}$ = prime ideal in $\mathfrak{o}$.
$K_{\mathfrak{p}}$ = $\mathfrak{p}$-adic completion of K.
$\mathfrak{o}_{\mathfrak{p}}$ = $\mathfrak{p}$-adic integers in $K_{\mathfrak{p}}$.
$\mathfrak{p}^* = \pi\mathfrak{o}_{\mathfrak{p}}$ = unique prime ideal in $\mathfrak{o}_{\mathfrak{p}}$.
$\bar{K} = \mathfrak{o}/\mathfrak{p} = \mathfrak{o}_{\mathfrak{p}}/\mathfrak{p}^*$ = residue class field.

By a *G-module* we shall mean a left G-module R satisfying

1. R is a finitely generated torsion-free left $\mathfrak{o}$-module.
2. For $x, y \in G, r, s \in R$:

$$(xy)r = x(yr), \quad (x + y)r = xr + yr, \quad x(r + s) = xr + xs, \quad er = r.$$

Following Gaschütz and Ikeda (**3**; **5**; see also **7**; **10**) we call a G-module R an *M_u-G-module* (unterer Maschke Modul) if, whenever R is an $\mathfrak{o}$-direct summand of a G-module S, R is a G-direct summand of S. Likewise, R is an *M_0-G-module* (oberer Maschke Modul) if, whenever S/R_1 is G-isomorphic to R where the G-module S contains the G-module R_1 as $\mathfrak{o}$-direct summand, R_1 is a G-direct summand of S.

If all modules considered happen to have $\mathfrak{o}$-bases (for example, when $\mathfrak{o}$ is a principal ideal ring), then we may interpret these concepts in terms of matrix representations over $\mathfrak{o}$. Thus, a representation Γ of G in $\mathfrak{o}$ is an *M_0-representation* if for every reduced representation

$$\begin{pmatrix} \Gamma & \Lambda \\ 0 & \Delta \end{pmatrix}$$

of G in $\mathfrak{o}$, the binding system Λ is strongly-equivalent (**13**) to zero, that is, there exists a matrix T (over $\mathfrak{o}$) such that

$$\Lambda(x) = \Gamma(x)T - T\Delta(x) \qquad \text{for all } x \in G.$$

(Likewise we may define an M_u-representation of G in $\mathfrak{o}$.)

Received September 19, 1955; in revised form December 14, 1955. This work was supported in part by a contract with the National Science Foundation. The author wishes to thank Dr. P. Roquette for some helpful conversations during the preparation of this paper.

Starting with a prime ideal $\mathfrak{p}$ of $\mathfrak{o}$, we may form $\bar{G} = G/\mathfrak{p}G$, an algebra over $\bar{K}$. If R is a G-module, then $\bar{R} = R/\mathfrak{p}R$ can be made into a $\bar{G}$-module in obvious fashion, and $\bar{R}$ is then a vector space over $\bar{K}$. The main results of this note are as follows:

THEOREM 1. *If for each* $\mathfrak{p}$, $\bar{R}$ *is an* M_u-$\bar{G}$-*module* (*or* $M_{\mathfrak{o}}$-$\bar{G}$-*module*), *then* R *is an* M_u-G-*module* (*or* $M_{\mathfrak{o}}$-G-*module*).

THEOREM 2. *If* G *is a Frobenius algebra over* $\mathfrak{o}$, *and* R *is an* M_u-G-*module* (*or* $M_{\mathfrak{o}}$-G-*module*), *then for each* $\mathfrak{p}$, $\bar{R}$ *is an* M_u-$\bar{G}$-*module* (*or* $M_{\mathfrak{o}}$-$\bar{G}$-*module*).

The significance of Theorem 1 is that it reduces the problem of deciding whether an $\mathfrak{o}$-module R is an M_u-G-module to that of determining for each $\mathfrak{p}$ whether the vector space $\bar{R}$ over $\bar{K}$ is an M_u-$\bar{G}$-module. Thus, we pass from a *ring* problem to a *field* problem, which is in general much simpler.

In the important case where $G = \mathfrak{o}(H)$ is the group ring of a finite group H, then $\bar{G}$ is semi-simple whenever $\mathfrak{p}$ does not divide the order of H, and for such $\mathfrak{p}$ the module $\bar{R}$ is automatically an M-$\bar{G}$-module. More generally, we may form the ideal $I(G)$ of G defined by Higman **(4)**; his results show that $I(G) \neq 0$ in this case. From **(9)** we deduce at once that $\bar{G}$ is semi-simple whenever $\mathfrak{p}$ does not divide $I(G)$. Therefore:

COROLLARY 1. R *is an* M_u-G-*module* (*or* $M_{\mathfrak{o}}$-G-*module*) *if for each* $\mathfrak{p}$ *dividing* $I(G)$, $\bar{R}$ *is an* M_u-$\bar{G}$-*module* (*or* $M_{\mathfrak{o}}$-$\bar{G}$-*module*). (Note that only finitely many $\mathfrak{p}$'s are involved.)

Now let G be a Frobenius algebra over $\mathfrak{o}$, for example, $G = \mathfrak{o}(H)$. Then by **(5)** there is no distinction between $M_{\mathfrak{o}}$- and M_u-modules, and Theorems 1 and 2 tell us that R is an M-G-module if and only if for each $\mathfrak{p}$, $\bar{R}$ is an M-$\bar{G}$-module. Using the concept of *genus* introduced by Maranda in **(9)**, we have:

COROLLARY 2. *Let* G *be a Frobenius algebra over* $\mathfrak{o}$, *and let* R, S *be* G-*modules in the same genus. Then* R *is an* M-G-*module if and only if* S *is an* M-G-*module.*

2. $\mathfrak{p}$-adic completion. Theorem 1 will follow at once from two lemmas, of which we prove the more difficult first. Let R be a G-module, and define

$$G_{\mathfrak{p}} = G \otimes \mathfrak{o}_{\mathfrak{p}}, \quad R_{\mathfrak{p}} = \mathfrak{o}_{\mathfrak{p}} \otimes R,$$

both products being taken over $\mathfrak{o}$.

LEMMA 1. *If for each* $\mathfrak{p}$, $R_{\mathfrak{p}}$ *is an* M_u-$G_{\mathfrak{p}}$-*module* (*or* $M_{\mathfrak{o}}$-$G_{\mathfrak{p}}$-*module*), *then* R *is an* M_u-G-*module* (*or* $M_{\mathfrak{o}}$-G-*module*).

Proof. (We give the proof only for M_u-modules.) Let R be an $\mathfrak{o}$-direct summand of a G-module S. We wish to show that R is a G-direct summand of S, that is, that there exists $f \in \mathrm{Hom}_G(S, R)$ such that $f|R =$ identity. Using the Steinitz-Chevalley theory **(1; 11)** of the structure of finitely generated torsion-free modules over Dedekind rings, and taking into account the hypothesis that R is an $\mathfrak{o}$-direct summand of S, we may write

$$S = \mathfrak{A}_1 s_1 \oplus \ldots \oplus \mathfrak{A}_n s_n, \quad R = \mathfrak{A}_1 s_1 \oplus \ldots \oplus \mathfrak{A}_m s_m,$$

with $m \leqslant n$, where each $\mathfrak{A}_i$ is an $\mathfrak{o}$-ideal in K, and where $s_1, \ldots, s_n$ are linearly independent over K. For the remainder of this proof, let the index i range from 1 to n, and j from 1 to m.

To prove the lemma, it suffices to exhibit $f \in \mathrm{Hom}_A(KS, KR)$ such that $f|KR$ = identity, and f maps S into R. (We use KS to denote the K-module generated by S.) Let us set

$$f(s_i) = \sum a_{ij} s_j, \quad a_{ij} \in K, \tag{1}$$

thereby defining $f \in \mathrm{Hom}_K(KS, KR)$. Then f maps S into R if and only if for each $\alpha \in \mathfrak{A}_i$ we have $\alpha a_{ij} \in \mathfrak{A}_j$, that is, if and only if

$$a_{ij} \in (\mathfrak{A}_j : \mathfrak{A}_i) \qquad \text{for all } i, j. \tag{2}$$

On the other hand, the map f defined by (1) will be an A-homomorphism with $f|KR$ = identity, if and only if for all $x \in G$, $s \in S$, $r \in R$:

$$f(xs) = xf(s), \quad f(r) = r.$$

Let us set

$$G = \mathfrak{o}x_1 + \ldots + \mathfrak{o}x_t.$$

This is possible since **(2**, p. 70**)** G is a finitely generated $\mathfrak{o}$-module. Then f is an A-homomorphism with $f|KR$ = identity, if and only if

$$f(x_k s_i) = x_k f(s_i), \quad f(s_j) = s_j \qquad \text{for all } i, j, k, \tag{3}$$

where the index k ranges from 1 to t. Equations (3) are a set of linear equations with coefficients in K, to be solved for unknowns $\{a_{ij}\}$ satisfying (2).

From the hypotheses of the lemma we deduce that for each $\mathfrak{p}$, (3) has a solution $\{a_{ij}\}$ satisfying $a_{ij} \in (\mathfrak{A}_j : \mathfrak{A}_i)\mathfrak{o}_\mathfrak{p}$ for all i, j. Thus (3) is solvable over the extension field $K_\mathfrak{p}$ of K, and hence is also solvable over K. The general solution of (3) over K is given by

$$a_{ij} = e_{ij}/d_{ij}, \quad e_{ij} = e_{ij}(\mathfrak{t}) = b_{ij} + \sum_{\nu=1}^{N} c_{ij}^{(\nu)} t_\nu, \tag{4}$$

where the b_{ij}, $c_{ij}^{(\nu)}$, d_{ij} are fixed elements of $\mathfrak{o}$, $d_{ij} \neq 0$, and where $\mathfrak{t}$ ranges over all N-tuples in K^N. The general solution of (3) over $K_\mathfrak{p}$ is also given by (4) by letting $\mathfrak{t}$ range over $K_\mathfrak{p}{}^N$. Then for each $\mathfrak{p}$, we can find $\mathfrak{t}(\mathfrak{p})$ for which

$$e_{ij}(\mathfrak{t}(\mathfrak{p})) \in \mathfrak{B}_{ij}\mathfrak{o}_\mathfrak{p} \qquad \text{for all } i, j, \tag{5}$$

where $\mathfrak{B}_{ij} = (\mathfrak{A}_j : \mathfrak{A}_i) d_{ij}$.

For each $\mathfrak{p}$, let $b(\mathfrak{p})$ be the maximal exponent to which $\mathfrak{p}$ occurs in the prime ideal factorizations of the ideals $\mathfrak{B}_{ij}$. Then $b(\mathfrak{p}) = 0$ except for a finite set of primes. Set $P = \{\mathfrak{p}: b(\mathfrak{p}) > 0\}$, and choose an N-tuple $\mathfrak{t}$ with components in $\mathfrak{o}$ such that (componentwise)

$$\mathfrak{t} \equiv \mathfrak{t}(\mathfrak{p}) \pmod{\mathfrak{p}^{b(\mathfrak{p})}} \qquad \text{for each } \mathfrak{p} \in P.$$

In that case, $e_{ij}(\mathfrak{t}) \equiv e_{ij}(\mathfrak{t}(\mathfrak{p})) \pmod{\mathfrak{p}^{b(\mathfrak{p})}}$ for each $\mathfrak{p} \in P$, and all i, j, whence by (5) we have

$$\operatorname{ord}_{\mathfrak{p}} e_{ij}(\mathfrak{t}) \geqslant \operatorname{ord}_{\mathfrak{p}} \mathfrak{B}_{ij} \qquad \text{for all } i, j, \tag{6}$$

for all $\mathfrak{p} \in P$. But for $\mathfrak{p} \notin P$, equation (6) is certainly valid because $e_{ij}(\mathfrak{t}) \in \mathfrak{o}$, and $\operatorname{ord}_{\mathfrak{p}} \mathfrak{B}_{ij} \leqslant 0$. Hence we deduce that $e_{ij}(\mathfrak{t}) \in \mathfrak{B}_{ij} = (\mathfrak{A}_j : \mathfrak{A}_i) d_{ij}$ for all i, j, whence (4) gives a solution of (3) for which (2) holds.

We may remark that this lemma is almost trivial when $\mathfrak{o}$ is a principal ideal ring.

3. Modular representations. Now let $R_{\mathfrak{p}}$ be a $G_{\mathfrak{p}}$-module, and define $\bar{R}_{\mathfrak{p}} = R_{\mathfrak{p}}/\pi R_{\mathfrak{p}}$, $\bar{G}_{\mathfrak{p}} = G_{\mathfrak{p}}/\pi G_{\mathfrak{p}}$. To complete the proof of Theorem 1, we need only show:

LEMMA 2. *If $\bar{R}_{\mathfrak{p}}$ is an M_u-$\bar{G}_{\mathfrak{p}}$-module (or $M_{\mathfrak{o}}$-$\bar{G}_{\mathfrak{p}}$-module), then $R_{\mathfrak{p}}$ is an M_u-$G_{\mathfrak{p}}$-module (or $M_{\mathfrak{o}}$-$G_{\mathfrak{p}}$-module).*

Proof. Since $\mathfrak{o}_{\mathfrak{p}}$ is a principal ideal ring, we may express the proof (given here only for $M_{\mathfrak{o}}$-modules) in terms of matrix representations. We must show that if Γ is a representation of $G_{\mathfrak{p}}$ in $\mathfrak{o}_{\mathfrak{p}}$ for which $\bar{\Gamma}$ (the induced modular representation of $\bar{G}_{\mathfrak{p}}$ in $\bar{K}$) is an M_0-representation, then in any reduced representation

$$\begin{pmatrix} \Gamma & \Lambda \\ 0 & \Delta \end{pmatrix} \tag{7}$$

of $G_{\mathfrak{p}}$ in $\mathfrak{o}_{\mathfrak{p}}$, the binding system Λ is strongly-equivalent to zero.

We may write $G_{\mathfrak{p}} = \mathfrak{o}_{\mathfrak{p}} y_1 \oplus \ldots \oplus \mathfrak{o}_{\mathfrak{p}} y_n$, $\bar{G}_{\mathfrak{p}} = \bar{K} y_1 \oplus \ldots \oplus \bar{K} y_n$. We shall show the existence of a matrix T over $\mathfrak{o}_{\mathfrak{p}}$ such that

$$\Lambda(y_i) = \Gamma(y_i) T - T \Delta(y_i) \qquad \text{for each } i, \tag{8}$$

where in this proof the index i ranges from 1 to n. By taking residue classes mod $\mathfrak{p}^*$, the representation (7) gives a representation

$$\begin{pmatrix} \bar{\Gamma} & \bar{\Lambda} \\ 0 & \bar{\Delta} \end{pmatrix}$$

of $\bar{G}_{\mathfrak{p}}$ in $\bar{K}$. Since Γ is by hypothesis an $M_{\mathfrak{o}}$-representation, the binding system $\bar{\Lambda}$ is strongly-equivalent to zero over $\bar{K}$. Therefore there exists V_1 over $\mathfrak{o}_{\mathfrak{p}}$ such that

$$\Lambda(y_i) = \Gamma(y_i) V_1 - V_1 \Delta(y_i) + \pi \Lambda^{(1)}(y_i) \qquad \text{for each } i, \tag{9}$$

where $\Lambda^{(1)}$ is also over $\mathfrak{o}_\mathfrak{p}$. But then (7) with Λ replaced by $\Lambda^{(1)}$ gives another $\mathfrak{o}_\mathfrak{p}$-representation of $G_\mathfrak{p}$, whence the same argument shows

$$\Lambda^{(1)}(y_i) = \Gamma(y_i)\, V_2 - V_2\Delta(y_i) + \pi\,\Lambda^{(2)}(y_i) \qquad \text{for all } i,$$

where V_2 and $\Lambda^{(2)}$ are over $\mathfrak{o}_\mathfrak{p}$. Continuing in this way, we obtain a solution of (8) given by $T = V_1 + \pi V_2 + \pi^2 V_3 + \ldots$.

This proof could also have been stated in terms of cohomology groups.

4. Frobenius algebra. Suppose in this section that G is a Frobenius algebra over $\mathfrak{o}$, that is, there exist $\mathfrak{o}$-bases $\{u_i\}$, $\{v_i\}$ of G (called *dual* bases) such that the right regular representation of G with respect to $\{v_i\}$ coincides with the left regular representation with respect to $\{u_i\}$. Assume that G has an $\mathfrak{o}$-basis containing e. Ikeda showed **(5)** that $M_\mathfrak{o}$- and M_u-modules were the same, and that a G-module R is an M-G-module if and only if there exists an $\mathfrak{o}$-endomorphism ϕ of R such that

$$\sum u_i \phi v_i = \text{identity endomorphism of } R. \tag{10}$$

Gaschütz **(3)** had shown this for the case where $G = \mathfrak{o}(H)$, H = finite group, with (10) replaced by:

$$\sum_{h \in H} h\, \phi\, h^{-1} = \text{identity endomorphism of } R. \tag{11}$$

We may use Ikeda's result to obtain an immediate proof of Theorem 2. By hypothesis, R is an M-G-module, whence (10) holds for some $\mathfrak{o}$-endomorphism ϕ. But then clearly ϕ induces a $\bar{K}$-endomorphism $\bar{\phi}$ of $\bar{R}$, and $\sum u_i \bar{\phi}\, v_i$ = identity endomorphism of $\bar{R}$, so that $\bar{R}$ is an M-$\bar{G}$-module.

References

1. C. Chevalley, *L'arithmétique dans les algèbres de matrices,* Act. Sci. et Ind. *229* (1935).
2. M. Deuring, *Algebren* (Berlin, 1949).
3. W. Gaschütz, *Ueber den Fundamentalsatz von Maschke zur Darstellungstheorie der endlichen Gruppen,* Math. Z., *56* (1952), 376–387.
4. D. G. Higman, *On orders in separable algebras,* Can. J. Math., *7* (1955), 509–515.
5. M. Ikeda, *On a theorem of Gaschütz,* Osaka Math. J., *5* (1953), 53–58.
6. N. Jacobson, *The Theory of Rings* (New York, 1943).
7. F. Kasch, *Grundlagen einer Theorie der Frobeniuserweiterungen,* Math. Ann., *127* (1954), 453–474.
8. I. Kaplansky, *Modules over Dedekind rings and valuation rings,* Trans. Amer. Math. Soc., *72* (1952), 327–340.
9. J.-M. Maranda, *On the equivalence of representations of finite groups by groups of automorphisms of modules over Dedekind rings,* Can. J. Math., *7* (1955), 516–526.
10. H. Nagao and T. Nakayama, *On the structure of ($M_\mathfrak{o}$)- and (M_u)-modules,* Math. Z., *59* (1953), 164–170.

11. E. Steinitz, *Rechteckige Systeme und Moduln in algebraischen Zahlkörpern*, Math. Ann. I, *71* (1911), 328–354; II, *72* (1912), 297–345.
12. B. L. van der Waerden, *Modern Algebra*, II (New York, 1950).
13. H. Zassenhaus, *Neuer Beweis der Endlichkeit der Klassenzahl bei unimodularer Äquivalenz endlicher ganzzahliger Substitutionsgruppen*, Hamb. Abh., *12* (1938), 276–288.

Institute for Advanced Study and
University of Illinois

AUTOMORPHISMS OF THE GENERAL LINEAR GROUP OVER A PRINCIPAL IDEAL DOMAIN

Joseph Landin and Irving Reiner

ABSTRACT

Let R be a principapl ideal domain, not necessarily commutative, of characteristic $\neq 2$. Let n be an integer with $n \geq 3$, E a free right module over R on n generators $e_1, \ldots, e_n$, and $GL_n(R)$ the group of R-automorphisms of E. Let E^* denote the free left R-module $Hom_R(E, R)$ which has free basis $e_1^*, \ldots, e_n^*$ dual to the given basis of E. Let ΓL_n be the group of ***semi-linear*** automorphisms of E; that is automorphisms of E twisted by an automorphism of R. Finally, let χ be a homomorphism from $GL_n(R)$ into the group of units in the center of R which is identified with a subgroup of $GL_n(R)$. Assume χ satisfies the condition $\chi(w) = w^{-1}$ for a central unit w of R if and only if $w = 1$.

The main result of this paper determines all automorphisms of the group $GL_n(R)$.

Theorem. *An automorphism of $GL_n(R)$ has one of the two following forms:*

$$u \to \chi(u) \cdot gug^{-1} \qquad g \in \Gamma L_n, \quad \chi \text{ as above;}$$

$$u \to \chi(u) \cdot h^{-1}u^*h \qquad \chi \text{ as above}$$

where h is a one-to-one semi-linear map of E onto E^* relative to an antiautomorphism J of R, and where u^* is the contragriedient of u acting in E^* (i.e. the inverse transpose of u).

An example of such an h is the map h_0 from E to E^* defined by $h_0(e_i) = e_i^*$, and $h_0(x\alpha) = J(\alpha)h_0(x)$, for $x \in E, \alpha \in R$.

The proofs make use of a detailed analysis of the subspaces left invariant by elements of order two; the fundamental theorem of projective geometry is used.

A THEOREM ON CONTINUED FRACTIONS

IRVING REINER

1. **Introduction.** Let $[a_1, \cdots, a_n]$ denote the simple continued fraction whose successive quotients $a_1, \cdots, a_n$ are elements of a field K. Following the procedure of Milne-Thompson,[1] define the formal numerator P and formal denominator Q of this continued fraction by means of

$$\begin{pmatrix} a_1 & 1 \\ 1 & 0 \end{pmatrix}\begin{pmatrix} a_2 & 1 \\ 1 & 0 \end{pmatrix}\cdots\begin{pmatrix} a_n & 1 \\ 1 & 0 \end{pmatrix}\begin{pmatrix} 1 \\ 0 \end{pmatrix} = \begin{pmatrix} P \\ Q \end{pmatrix}. \tag{1}$$

For brevity, we write $[a_1, \cdots, a_n] \sim P/Q$.

Now let K be a skew-field, and let $R = K[x]$ be the ring of polynomials in an indeterminate x with coefficients in K, where we assume that x commutes with all elements of K. For $f_1, \cdots, f_n \in R$, define $[f_1, \cdots, f_n] \sim P/Q$ where (as above)

$$\begin{pmatrix} f_1 & 1 \\ 1 & 0 \end{pmatrix}\begin{pmatrix} f_2 & 1 \\ 1 & 0 \end{pmatrix}\cdots\begin{pmatrix} f_n & 1 \\ 1 & 0 \end{pmatrix}\begin{pmatrix} 1 \\ 0 \end{pmatrix} = \begin{pmatrix} P \\ Q \end{pmatrix}. \tag{2}$$

Next let $f \to f^*$ denote any homomorphism of $(R, +)$ into itself which leaves K elementwise fixed, and satisfies $(af)^* = af^*$ for all $a \in K, f \in R$. We shall prove:

THEOREM. *If $f_1, \cdots, f_k \in R$ are such that $[f_1, \cdots, f_k] \sim P/Q$ where $P, Q \in K$, then also $[f_1^*, \cdots, f_k^*] \sim P/Q$.*

Thus, any identity of the form $[f_1, \cdots, f_k] \sim P/Q$, $P, Q \in K$, depends only upon the additive structure of R, and not upon its multiplicative structure. This result will be basic in a future paper on the automorphisms of $GL_2(R)$.[2]

2. Several lemmas will be needed for the proof of the theorem.

LEMMA 1. Let $[f_1, \cdots, f_n] \sim P/Q$, *where $f_1, \cdots, f_n \in R$. Suppose that $f_r, f_{r+1}, \cdots, f_s$ $(s \geqq r)$ satisfy*

$$f_r, \cdots, f_s \in K, \quad \begin{pmatrix} f_r & 1 \\ 1 & 0 \end{pmatrix}\cdots\begin{pmatrix} f_s & 1 \\ 1 & 0 \end{pmatrix} = \begin{pmatrix} 0 & c \\ b & d \end{pmatrix}. \tag{3}$$

Set $e = dc^{-1}$, $f'_{s+1} = bf_{s+1}c^{-1}$, $f'_{s+2} = cf_{s+2}b^{-1}$, $\cdots$. Then

Presented to the Society April 6, 1957; received by the editors March 18, 1957.

[1] Proceedings of the Edinburgh Mathematical Society, 1933.

[2] I. Reiner, *A new type of automorphism of the general linear group over a ring*, to appear in Annals of Mathematics.

$$\begin{pmatrix} f_1 & 1 \\ 1 & 0 \end{pmatrix} \cdots \begin{pmatrix} f_{r-2} & 1 \\ 1 & 0 \end{pmatrix} \begin{pmatrix} f_{r-1} + e + f'_{s+1} & 1 \\ 1 & 0 \end{pmatrix} \begin{pmatrix} f'_{s+2} & 1 \\ 1 & 0 \end{pmatrix} \cdots \tag{4}$$
$$\begin{pmatrix} f'_n & 1 \\ 1 & 0 \end{pmatrix} \begin{pmatrix} t \\ 0 \end{pmatrix} = \begin{pmatrix} P \\ Q \end{pmatrix},$$

where t is either b or c, depending on the parity of $n-s$.

PROOF. From (2) we have

$$\begin{pmatrix} P \\ Q \end{pmatrix}$$
$$= \begin{pmatrix} f_1 & 1 \\ 1 & 0 \end{pmatrix} \cdots \begin{pmatrix} f_{r-1} & 1 \\ 1 & 0 \end{pmatrix} \begin{pmatrix} 0 & 1 \\ 1 & e \end{pmatrix} \begin{pmatrix} b & 0 \\ 0 & c \end{pmatrix} \begin{pmatrix} f_{s+1} & 1 \\ 1 & 0 \end{pmatrix} \cdots \begin{pmatrix} f_n & 1 \\ 1 & 0 \end{pmatrix} \begin{pmatrix} 1 \\ 0 \end{pmatrix}$$
$$= \begin{pmatrix} f_1 & 1 \\ 1 & 0 \end{pmatrix} \cdots \begin{pmatrix} f_{r-1} & 1 \\ 1 & 0 \end{pmatrix} \begin{pmatrix} 0 & 1 \\ 1 & e \end{pmatrix} \begin{pmatrix} f'_{s+1} & 1 \\ 1 & 0 \end{pmatrix} \cdots \begin{pmatrix} f'_n & 1 \\ 1 & 0 \end{pmatrix} \begin{pmatrix} t \\ 0 \end{pmatrix},$$

which readily implies (4).

LEMMA 2. *Let $f_1, \cdots, f_n \in R$, $f_1 \notin K$, $f_n \notin K$, $n > 1$. Define $[f_i, \cdots, f_n] \sim P_i/Q_i$, $1 \leqq i \leqq n$. If both P_1 and Q_1 lie in K, then there exists a run of consecutive elements $f_r, \cdots, f_s$ $(1 < r \leqq s < n)$ such that (3) holds.*

PROOF. Suppose that for every run of consecutive elements $f_r, \cdots, f_s$ which lie in K, we have $[f_r, \cdots, f_s] \sim a/b$ with $a \neq 0$. We shall obtain a contradiction. Using (2), we have

$$P_i = f_i P_{i+1} + Q_{i+1}, \qquad Q_i = P_{i+1}, \qquad 1 \leqq i \leqq n-1.$$

Consequently

(5) If $\deg P_i \leqq \deg Q_i$, and if $f_i \notin K$, then $\deg P_{i+1} < \deg Q_{i+1}$.

Since $\deg P_1 \leqq \deg Q_1$, and $f_1 \notin K$, (5) shows that $\deg P_2 < \deg Q_2$. If also $f_2 \notin K$, we have likewise: $\deg P_3 < \deg Q_3$. We may continue this process until we reach the first of the elements $f_1, f_2, \cdots$ which lies in K, say f_r; then $\deg P_r < \deg Q_r$. Let s be chosen so that $f_r, \cdots, f_s \in K$, $f_{s+1} \notin K$. By the supposition at the beginning of the proof, we have $[f_r, \cdots, f_s] \sim a/b$ with $a \neq 0$. Setting

$$\begin{pmatrix} f_r & 1 \\ 1 & 0 \end{pmatrix} \cdots \begin{pmatrix} f_s & 1 \\ 1 & 0 \end{pmatrix} = \begin{pmatrix} a & c \\ b & d \end{pmatrix},$$

we have at once

$$P_r = aP_{s+1} + cQ_{s+1}, \qquad Q_r = bP_{s+1} + dQ_{s+1}.$$

Since deg $P_r <$ deg Q_r, surely deg $P_{s+1} >$ deg Q_{s+1} is impossible. Hence deg $P_{s+1} \leqq$ deg Q_{s+1}.

We may therefore continue this process of descent, eventually obtaining deg $P_n \leqq$ deg Q_n. However, $P_n = f_n$, $Q_n = 1$, and $f_n \notin K$, so we have reached a contradiction. This completes the proof of the lemma.

We now prove the theorem stated in §1, by induction on k. The theorem obviously holds for $k=1$. Suppose it proved for $1 \leqq k \leqq n-1$, where $n \geqq 2$, and suppose that $[f_1, \cdots, f_n] \sim P/Q$, where both P and Q lie in K; we shall show that $[f_1^*, \cdots, f_n^*] \sim P/Q$.

To begin with, assume that $f_1 \in K$. Then (2) implies

$$\begin{pmatrix} f_2 & 1 \\ 1 & 0 \end{pmatrix} \cdots \begin{pmatrix} f_n & 1 \\ 1 & 0 \end{pmatrix} \begin{pmatrix} 1 \\ 0 \end{pmatrix} = \begin{pmatrix} 0 & 1 \\ 1 & -f_1 \end{pmatrix} \begin{pmatrix} P \\ Q \end{pmatrix}.$$

By the induction hypothesis this remains valid when each f_i $(1 \leqq i \leqq n)$ is replaced by f_i^*; hence (2) also remains valid under this replacement.

Secondly, assume that $f_n \in K$. Since the result follows trivially from the induction hypothesis when $f_n = 0$, we need only consider the case where $f_n \neq 0$. Equation (2) may then be written as

$$\begin{pmatrix} P \\ f_n^{-1} Q \end{pmatrix} = \begin{pmatrix} f_1 f_n & 1 \\ 1 & 0 \end{pmatrix} \begin{pmatrix} f_n^{-1} f_2 & 1 \\ 1 & 0 \end{pmatrix} \cdots \begin{pmatrix} f_n^{-1} f_{n-2} & 1 \\ 1 & 0 \end{pmatrix}$$
$$\times \begin{pmatrix} f_{n-1} f_n + 1 & 1 \\ 1 & 0 \end{pmatrix} \begin{pmatrix} 1 \\ 0 \end{pmatrix},$$

provided that n is even, with an analogous result for odd n. By the induction hypothesis, this remains valid under the substitution $f \to f^*$, which again implies the desired result.

We are thus left with the case where $n \geqq 2$, $f_1 \notin K$, $f_n \notin K$, and $[f_1, \cdots, f_n] \sim P/Q$ with both P and Q in K. By Lemma 2, there exists a run of consecutive elements $f_r, \cdots, f_s$ $(s \geqq r)$ satisfying (3). Keeping the notation of Lemma 1, equation (2) implies the validity of (4), with $t \neq 0$. Upon multiplying the successive quotients $f_n', \cdots, f_1$ in (4) by t or t^{-1} (according to the method used in the preceding paragraph), equation (4) reduces to an equation of type (2). This new equation of type (2) remains true under the substitution $f \to f^*$, as a consequence of the induction hypothesis. Therefore also (4) remains true under $f \to f^*$; as above, we deduce that the original equation (2) remains correct under the substitution $f \to f^*$. This completes the proof of the theorem.

UNIVERSITY OF ILLINOIS

A NEW TYPE OF AUTOMORPHISM OF THE GENERAL LINEAR GROUP OVER A RING

By Irving Reiner

(Received January 7, 1957)

1. Introduction

Let E be a free R-module of rank n, where R is a ring, and let $GL_n(R)$ denote the group of one-to-one R-linear maps of E onto itself. The structure of $A_n(R)$, the group of automorphisms of $GL_n(R)$, has previously been considered for the following cases:

(i) R = skew-field (1, 2, 3, 4, 10),
(i) $R = Z$ (ring of rational integers) (5),
(iii) R = ring of Gaussian integers (6),
(iv) R = non-commutative principal ideal domain ($n \geqq 3$) (7).

In all of these cases, it has been shown that $A_n(R)$ is generated by the following types of automorphisms:

$$(1) \qquad u \to tut^{-1}, \qquad t \in GL_n(R) \qquad \text{(inner)},$$

$$(2) \qquad u \to \chi(u)u,$$

where χ is a homomorphism of $GL_n(R)$ into the group of units of the center of R, satisfying the condition $\chi(\lambda I) = \lambda^{-1}$ if and only if $\lambda = 1$,

$$(3) \qquad u \to u^{\sigma}, \qquad \sigma = \text{automorphism of } R,$$

$$(4) \qquad u \to t^{-1}\check{u}t, \qquad \check{u} = \text{contragredient of } u,$$

where $t:E \to E^*$ is a correlation mapping E onto its dual E^* (see (7) for more details).

The purpose of this paper is to exhibit a new type of automorphism of $GL_2(R)$, which may appear even when R is a commutative integral domain with a euclidean algorithm. For the remainder of the paper, let $R = K[x]$ be the ring of polynomials in an indeterminate x over a field K. We shall construct certain automorphisms of $GL_2(R)$ which are clearly not expressible as a product of automorphisms of types (1)–(4). Using these new automorphisms, we shall obtain a set of generators of $A_2(R)$. The existence of such new automorphisms is especially surprising in view of the results on the structures of $A_2(Z)$ and $A_n(R)$ ($n \geqq 3$) given in (5, 7).

2. The new automorphisms

Since R is a commutative ring, $GL_2(R)$ may be identified with the group of 2×2 matrices over R with determinant in K^* (the set of non-zero elements of K). The group $GL_2(K)$ may then be embedded in $GL_2(R)$ in obvious fashion.

Define

$$S = \begin{pmatrix} 0 & 1 \\ -1 & 0 \end{pmatrix}, \quad D_a = \begin{pmatrix} a & 0 \\ 0 & 1 \end{pmatrix} (a \in K^*), \quad X_0 = \begin{pmatrix} 1 & 1 \\ 0 & 1 \end{pmatrix}, \tag{5}$$

and

$$X_m = \begin{pmatrix} 1 & x^m \\ 0 & 1 \end{pmatrix}, \quad m = 1, 2, \cdots. \tag{6}$$

The elements listed in (5) and (6) generate $GL_2(R)$, while those listed in (5) generate $GL_2(K)$. Hereafter, let the index $\underline{m}$ range over the set of positive integers, and let $\underline{a}$ range over the elements of K^*.

Let $\{y_m\}$ be any set of elements of R for which $1, y_1, y_2, \cdots$ form a K-basis of R. Putting

$$Y_m = \begin{pmatrix} 1 & y_m \\ 0 & 1 \end{pmatrix}, \tag{7}$$

we see that the elements $\{Y_m\}$ and those listed in (5) also form a set of generators of $GL_2(R)$.

Let $P = \Pi(S, \{D_a\}, X_0, \{X_m\})$ denote a formal power product (with negative exponents permitted) of the generators listed in (5) and (6). Let $\phi_y(P)$ be obtained from P by replacing each X_m by Y_m, thereby defining a map ϕ_y of $GL_2(R)$ onto itself. We shall show

$$\phi_y(P) = I \quad \text{if and only if } P = I, \tag{8}$$

both equalities being taken in $GL_2(R)$, where I denotes the identity element of $GL_2(R)$. This will imply that ϕ_y is a one-to-one map of $GL_2(R)$ onto itself, and so ϕ_y will be an automorphism of $GL_2(R)$.

Since P may be obtained from $\phi_y(P)$ by the process of replacing each Y_m by X_m, we see that (8) will be proved once we establish the following:

LEMMA. *Let* $1, y_1, y_2, \cdots$ *and* $1, z_1, z_2, \cdots$ *be two K-bases of R. Set*

$$Y_m = \begin{pmatrix} 1 & y_m \\ 0 & 1 \end{pmatrix}, \quad Z_m = \begin{pmatrix} 1 & z_m \\ 0 & 1 \end{pmatrix}, \quad m = 1, 2, \cdots.$$

Then $\Pi(S, \{D_a\}, X_0, \{Y_m\}) = I$ *implies* $\Pi(S, \{D_a\}, X_0, \{Z_m\}) = I$.

PROOF. Construct a K-linear map $f \to f^*$ of R onto itself by setting

$$f = a_0 + \sum a_m y_m \to f^* = a_0 + \sum a_m z_m, \quad a_0, a_1, \cdots \in K.$$

Thus we may write $Z_m = Y_m^*$. For convenience, let $Y_0 = Z_0 = X_0$, where X_0 is given in (5).

Next we note that any equation $\Pi(S, \{D_a\}, X_0, \{Y_m\}) = I$ can be written as

$$L_1 Y_{m_1}^{\varepsilon_1} L_2 Y_{m_2}^{\varepsilon_2} \cdots L_r Y_{m_r}^{\varepsilon_r} L_{r+1} = I, \tag{9}$$

where each L_i is a product $\Pi_i(S, \{D_a\})$, each $\varepsilon_i = \pm 1$, and each $m_i \geqq 0$. We must show that (9) implies

$$L_1 Z_{m_1}^{\varepsilon_1} L_2 Z_{m_2}^{\varepsilon_2} \cdots L_r Z_{m_r}^{\varepsilon_r} L_{r+1} = I. \tag{10}$$

We shall make use of the identities:

$$\begin{pmatrix}1 & t\\0 & 1\end{pmatrix} D_a = D_a \begin{pmatrix}1 & a^{-1}t\\0 & 1\end{pmatrix}, \qquad \begin{pmatrix}1 & 0\\t & 1\end{pmatrix} D_a = D_a \begin{pmatrix}1 & 0\\at & 1\end{pmatrix},$$

$$\begin{pmatrix}1 & t\\0 & 1\end{pmatrix} S = S \begin{pmatrix}1 & 0\\-t & 1\end{pmatrix}, \qquad \begin{pmatrix}1 & 0\\t & 1\end{pmatrix} S = S \begin{pmatrix}1 & -t\\0 & 1\end{pmatrix},$$

which are valid for $a \in K^*$, $t \in R$. Using these, the left-hand side of equation (9) can be written as $L_1 Y_{m_1}^{\varepsilon_1} \cdots L_{r-1} L_r \bar{Y} Y_{m_r}^{\varepsilon_r} L_{r+1}$, where $\bar{Y}$ is of the form

$$\bar{Y} = \begin{pmatrix}1 & by_{m_{r-1}}\\0 & 1\end{pmatrix}^{\theta}, \qquad b \in K^*,$$

and where θ denotes an operator which is either the identity or else maps a matrix onto its transpose. Using the same identities, the left-hand side of equation (10) can likewise be written as $L_1 Z_{m_1}^{\varepsilon_1} \cdots L_{r-1} L_r \bar{Z} Z_{m_r}^{\varepsilon_r} L_{r+1}$, where $\bar{Z} = \bar{Y}^*$.

Continuing this process, and working successively with $Y_{m_{r-1}}, Y_{m_{r-2}}, \cdots Y_{m_1}$, we find that the left-hand side of equation (9) equals

$$L_1 L_2 \cdots L_{r-1} L_r \bar{Y}_1 \bar{Y}_2 \cdots \bar{Y}_{r-1} Y_{m_r}^{\varepsilon_r} L_{r+1}, \tag{11}$$

where

$$\bar{Y}_i = \begin{pmatrix}1 & b_i y_{m_i}\\0 & 1\end{pmatrix}^{\theta_i} \qquad (i = 1, \cdots, r-1).$$

Correspondingly, the left-hand side of equation (10) equals

$$L_1 L_2 \cdots L_{r-1} L_r \bar{Z}_1 \bar{Z}_2 \cdots \bar{Z}_{r-1} Z_{m_r}^{\varepsilon_r} L_{r+1}, \tag{12}$$

where $\bar{Z}_i = \bar{Y}_i^*$. We must prove that if the expression in (11) equals I, then so does the expression in (12).

Now we remark that

$$\begin{pmatrix}1 & t_1\\0 & 1\end{pmatrix}^{\theta} \begin{pmatrix}1 & t_2\\0 & 1\end{pmatrix}^{\theta} = \begin{pmatrix}1 & t_1 + t_2\\0 & 1\end{pmatrix}^{\theta}.$$

Using this, and grouping those factors in (11) for which $\theta_i = \theta_{i+1}$, the expressions in (11) and (12) may be written as

$$A \begin{pmatrix}1 & 0\\f_1 & 1\end{pmatrix}\begin{pmatrix}1 & f_2\\0 & 1\end{pmatrix} \cdots \begin{pmatrix}1 & 0\\f_{k-1} & 1\end{pmatrix}\begin{pmatrix}1 & f_k\\0 & 1\end{pmatrix} B, \tag{13}$$

and

$$A \begin{pmatrix}1 & 0\\f_1^* & 1\end{pmatrix}\begin{pmatrix}1 & f_2^*\\0 & 1\end{pmatrix} \cdots \begin{pmatrix}1 & 0\\f_{k-1}^* & 1\end{pmatrix}\begin{pmatrix}1 & f_k^*\\0 & 1\end{pmatrix} B, \tag{14}$$

respectively, where $A, B \in GL_2(K)$ and where each $f_i \in R$.

If the product in (13) equals I, then the product

$$\begin{pmatrix}1 & 0\\f_1 & 1\end{pmatrix}\begin{pmatrix}1 & f_2\\0 & 1\end{pmatrix} \cdots \begin{pmatrix}1 & 0\\f_{k-1} & 1\end{pmatrix}\begin{pmatrix}1 & f_k\\0 & 1\end{pmatrix} \tag{15}$$

lies in $GL_2(K)$. To complete the proof of the lemma, it suffices to show that if the expression in (15) lies in $GL_2(K)$, then

$$\begin{pmatrix} 1 & 0 \\ f_1 & 1 \end{pmatrix} \cdots \begin{pmatrix} 1 & f_k \\ 0 & 1 \end{pmatrix} = \begin{pmatrix} 1 & 0 \\ f_1^* & 1 \end{pmatrix} \cdots \begin{pmatrix} 1 & f_k^* \\ 0 & 1 \end{pmatrix}. \tag{16}$$

This can be done most conveniently by the use of continued fractions. For $f_1, \cdots, f_n \in R$, let

$$[f_1, \cdots, f_n] = f_1 + 1/(f_2 + \cdots + 1/f_n).$$

WIth this continued fraction we associate a formal numerator P_n and a formal denominator Q_n (notation: $P_n/Q_n \sim [f_1, \cdots, f_n]$) which are defined by

$$\begin{matrix} P_0 = 1, & P_1 = f_1, & P_i = f_iP_{i-1} + P_{i-2} \\ Q_0 = 0, & Q_1 = 1, & Q_i = f_iQ_{i-1} + Q_{i-2} \end{matrix} \qquad (2 \leqq i \leqq n).$$

Let $[f_1, \cdots, f_i] \sim P_i/Q_i$ for $i = k - 1$ and $i = k$. A simple induction proof shows that

$$\begin{pmatrix} 1 & 0 \\ f_1 & 1 \end{pmatrix}\begin{pmatrix} 1 & f_2 \\ 0 & 1 \end{pmatrix} \cdots \begin{pmatrix} 1 & 0 \\ f_{k-1} & 1 \end{pmatrix}\begin{pmatrix} 1 & f_k \\ 0 & 1 \end{pmatrix} = \begin{pmatrix} Q_{k-1} & Q_k \\ P_{k-1} & P_k \end{pmatrix}.$$

If the product in (15) lies in $GL_2(K)$, then each of $Q_{k-1}, P_{k-1}, Q_k, P_k$ lies in K. We now use the following theorem (see (9)):

If $f_1, \cdots, f_n \in R$ are such that $[f_1, \cdots, f_n] \sim P_n/Q_n$, where both P_n and Q_n lie in K, then also $[f_1^*, \cdots, f_n^*] \sim P_n/Q_n$, where $f \to f^*$ is any K-linear map of R into itself which leaves K elementwise fixed.

This implies that (16) holds, and hence the lemma is established. Therefore the map ϕ_y defined above is an automorphism of $GL_2(R)$.

3. Generators of $A_2(R)$

In this section we shall use the methods developed in (8). Let $\tau \in A_2(R)$; after changing τ by an inner automorphism, we may assume that

$$\begin{pmatrix} 1 & t \\ 0 & 1 \end{pmatrix}^\tau = \alpha(t) \begin{pmatrix} 1 & \sigma(t) \\ 0 & 1 \end{pmatrix}, \qquad \alpha(t) \in K^*, \tag{17}$$

where $\alpha(t_1 + t_2) = \alpha(t_1)\alpha(t_2)$, and where σ is a one-to-one map of R onto itself satisfying $\sigma(t_1 + t_2) = \sigma(t_1) + \sigma(t_2)$.

From the relation

$$\begin{pmatrix} 1 & -t \\ 0 & 1 \end{pmatrix} = \begin{pmatrix} 1 & t \\ 0 & 1 \end{pmatrix}^{-1},$$

we find $\alpha(-t) = \alpha(t)$. Since also $\alpha(0) = 1$, we deduce that $\alpha(t) = \pm 1$ for all t. If the characteristic of K is different from 2, then $\alpha(t) = \alpha^2(t/2) = 1$ for all t; surely $\alpha(t) = 1$ for all t if the characteristic of K is 2. Thus in either case we have

$$\begin{pmatrix} 1 & t \\ 0 & 1 \end{pmatrix}^\tau = \begin{pmatrix} 1 & \sigma(t) \\ 0 & 1 \end{pmatrix}. \tag{18}$$

By making use of the relations $S^2 = -I$, $(SX_0)^3 = I$, it is not difficult to show that after changing τ by an inner automorphism, we may assume that $S^\tau = S$ and that (18) holds with $\sigma(1) = 1$. This implies

$$\begin{pmatrix} 1 & 0 \\ t & 1 \end{pmatrix}^\tau = \begin{pmatrix} 1 & 0 \\ \sigma(t) & 1 \end{pmatrix} \qquad \text{for all } t \in R.$$

Since the group of diagonal matrices is the intersection of the normalizers of the two groups

$$\left\{\begin{pmatrix} 1 & t \\ 0 & 1 \end{pmatrix}, \ t \in R\right\}, \qquad \left\{\begin{pmatrix} 1 & 0 \\ t & 1 \end{pmatrix}, \ t \in R\right\}$$

we see that

$$D_a^\tau = \lambda(a) D_{\rho(a)} \qquad \text{for } a \in K^*,$$

where both λ and ρ are multiplicative.

Now observe that $\sigma(at) = \rho(a)\sigma(t)$ for $a \in K^*$, $t \in R$; setting $t = 1$, we deduce that $\rho(a) = \sigma(a)$, whence ρ is additive. Upon defining $\rho(0) = 0$, we find readily that ρ is an automorphism of K. However, any automorphism η of K induces an automorphism η of $GL_2(R)$. For, the entries of any matrix $U \in GL_2(R)$ are elements of $K(x)$, and we may form U^η by applying η to the coefficients of the various powers of x. If we replace the automorphism τ by the automorphism $\tau\rho^{-1}$, and call this new automorphism τ, then in addition to (18) we shall have

$$S^\tau = S, \qquad D_a^\tau = \lambda_1(a) D_a, \qquad \sigma(at) = a\sigma(t), \qquad \sigma(a) = a$$

for $a \in K^*$, where λ_1 is also multiplicative.

The mapping σ is thus a K-linear map of R onto itself, which leaves K elementwise fixed. Set $y_m = \sigma(x^m)$, $m = 1, 2, \cdots$. Then $1, y_1, y_2, \cdots$ form a K-basis of R, and we may construct the automorphism ϕ_y of $GL_2(R)$ as in Section 2. In that case ϕ_y^{-1} is an automorphism of $GL_2(R)$ which leaves invariant each element of $GL_2(K)$, and furthermore

$$\begin{pmatrix} 1 & \sigma(x^m) \\ 0 & 1 \end{pmatrix}^{\phi_y^{-1}} = \begin{pmatrix} 1 & x^m \\ 0 & 1 \end{pmatrix}, \qquad m = 1, 2, \cdots.$$

Replacing τ by $\tau\phi_y^{-1}$, and calling this new automorphism τ, we now have

$$X_m^\tau = X_m, \qquad S^\tau = S, \qquad D_a^\tau = \lambda_1(a) D_a, \qquad X_0^\tau = X_0.$$

Consequently $U^\tau = \lambda(\det U)\cdot U$ for all $U \in GL_2(R)$.

We have thus proved:

THEOREM. *$A_2(R)$ is generated by the following types of automorphisms:*

(i) $u \to tut^{-1}$, $t \in GL_2(R)$ *(inner),*
(ii) $u \to u^\eta$, η = *automorphism of K,*
(iii) $u \to u^{\phi_y}$, ϕ_y *defined as in Section 2,*
(iv) $u \to \lambda\ (\det u)\ u$,

where λ is a multiplicative mapping of K^* into itself such that $\lambda(a^2) = a^{-1}$ if and only if $a = 1$.

It seems quite likely that analogous results are valid when $R = K[x]$, where K is a skew-field whose elements commute with the indeterminate x.

UNIVERSITY OF ILLINOIS

REFERENCES

1. J. DIEUDONNÉ, *On the automorphisms of the classical groups*, Mem. Amer. Math. Soc., 2 (1951), pp. 1–95.
2. ———, La Géométrie des Groupes Classiques, Springer, Berlin, 1955.
3. L. K. HUA, *On the automorphisms of the symplectic group over any field*, Ann. of Math., 49 (1948), pp. 739–759.
4. ———, Supplement to *On the automorphisms of the classical groups*, by J. DIEUDONNÉ, Mem. Amer. Math. Soc., 2 (1951), pp. 96–122.
5. L. K. HUA and I. REINER, *Automorphisms of the unimodular group*, Trans. Amer. Math. Soc., 71 (1951), pp. 331–348.
6. J. LANDIN and I. REINER, *Automorphisms of the Gaussian unimodular group*, to appear in Trans. Amer. Math. Soc.
7. ———, *Automorphisms of the general linear group over a principal ideal domain*, Ann. of Math., 65 (1957), pp. 519–526.
8. ———, *Automorphisms of the general linear group over a euclidean ring*, submitted to Trans. Amer. Math. Soc.
9. I. REINER, *A Theorem on continued fractions*, to appear in Proc. Amer. Math. Soc.
10. O. SCHREIER and B. L. VAN DER WAERDEN, *Über den Jordan-Hölderschen Satz*, Hamb. Abh., 6 (1928), pp. 303–322.

INTEGRAL REPRESENTATIONS OF CYCLIC GROUPS OF PRIME ORDER

IRVING REINER[1]

1. **Elementary facts.** In this paper we shall extend a result due to Diederichsen [2] on integral representations of cyclic groups of prime order, and shall simplify the proof thereof. Let Z denote the ring of rational integers, Q the rational field. If R is a ring, by a *regular* R-module we shall mean a finitely-generated torsion-free R-module.

LEMMA 1 (Zassenhaus [9]). *Let R be a regular Z-module contained in a field K, and suppose R contains a Q-basis of K. Then every irreducible regular R-module is R-isomorphic to an ideal in R. Two ideals in R are R-isomorphic (as R-modules) if and only if they lie in the same ideal class.*

REMARK. In terms of matrix representations, this lemma implies that there is a one-to-one correspondence between classes (under unimodular equivalence) of irreducible Z-representations of R and ideal classes of R. A full set of inequivalent irreducible matrix representations is obtained by restricting the regular representation of R to a full set of inequivalent ideals in R. In particular, let $f(x)\in Z[x]$ be irreducible, and set $R=Z[\theta]$ where θ is a zero of $f(x)$. Since every irreducible representation of R is described by $\theta\rightarrow X$, where X is an integral nonderogatory solution of $f(X)=0$, the number of unimodular classes of such matrix solutions coincides with the class number of $Z[\theta]$. (See [5; 8].)

Now let $\mathfrak{o}$ be a Dedekind ring (see [4]) which is assumed to be a regular Z-module. By Lemma 1, every irreducible regular $\mathfrak{o}$-module is $\mathfrak{o}$-isomorphic to an ideal in $\mathfrak{o}$.

LEMMA 2 (Steinitz [7], Chevalley [1]. This result can also be deduced from [6]). *Every regular $\mathfrak{o}$-module is $\mathfrak{o}$-isomorphic to a direct sum $\mathfrak{A}_1\oplus\cdots\oplus\mathfrak{A}_n$ of ideals in $\mathfrak{o}$. The $\mathfrak{o}$-rank n and the ideal class of $\mathfrak{A}_1\cdots\mathfrak{A}_n$ are the only invariants, and determine the module up to $\mathfrak{o}$-isomorphism.*

REMARK. Let $f(x)\in Z[x]$ be a monic irreducible polynomial, and let $f(\theta)=0$. Assume that $Z[\theta]$ coincides with the ring of all algebraic

Presented to the Society, September 2, 1955; received by the editors February 28, 1956.

[1] Part of this work was supported by a research contract with the National Science Foundation. The author wishes to thank the referee for his helpful suggestions.

integers in $Q(\theta)$. Then $Z[\theta]$ is a Dedekind ring, and the lemma implies that every integral matrix X for which $f(X)=0$, is integrally decomposable into a direct sum of irreducible matrices satisfying $f(X)=0$.

LEMMA 3. *Let $\mathfrak{e}$ and $\mathfrak{B}$ be ideals in $\mathfrak{o}$. Then there exists an $\mathfrak{o}$-automorphism of $\mathfrak{o}\oplus\mathfrak{B}$ which maps $\mathfrak{e}\oplus\mathfrak{B}$ isomorphically onto $\mathfrak{o}\oplus\mathfrak{e}\mathfrak{B}$.*

PROOF. Since only ideal classes are involved, we may assume $\mathfrak{e}+\mathfrak{B}=\mathfrak{o}$. Choose $e_0\in\mathfrak{e}$, $b_0\in\mathfrak{B}$ such that $e_0-b_0=1$. Then define an $\mathfrak{o}$-linear map ϕ: $\mathfrak{o}\oplus\mathfrak{B}\rightarrow\mathfrak{o}\oplus\mathfrak{B}$ by means of

$$\phi(a, b) = (a + b, ab_0 + e_0 b), \qquad a \in \mathfrak{o}, b \in \mathfrak{B}.$$

It is easily verified that ϕ is the desired o-automorphism of $\mathfrak{o}\oplus\mathfrak{B}$.

2. **Cyclic groups.** Let $G=\{g\}$ be a cyclic group of prime order p, and let $Z[g]$ be its group ring over the integers. We shall use the results of the previous section to classify all Z-regular $Z[g]$-modules. Define $s=1+g+\cdots+g^{p-1}\in Z[g]$. Let M be a Z-regular $Z[g]$-module, and define

$$(1) \qquad M_s = \{m \in M: sm = 0\}.$$

We may then view M_s as a $Z[g]/(s)$-module, where (s) is the principal ideal generated by s. However, $Z[g]/(s)\cong Z[\theta]$, where θ is a primitive pth root of 1. Further, $Z[\theta]$ is a Dedekind ring, hereafter denoted by $\mathfrak{o}$.

Now we observe that

$$(2) \qquad M_s \supset (g - 1)M \supset (\theta - 1)M_s,$$

all considered as $\mathfrak{o}$-modules. By Lemma 2, we may write

$$(3) \qquad M_s = \mathfrak{o} \oplus \cdots \oplus \mathfrak{o} \oplus \mathfrak{A},$$

where n (the number of summands) and the ideal class of the ideal $\mathfrak{A}$ in $\mathfrak{o}$ are uniquely determined. Using (2), we find that as $\mathfrak{o}$-module,

$$(4) \qquad (g - 1)M = \mathfrak{e}_1 \oplus \cdots \oplus \mathfrak{e}_{n-1} \oplus \mathfrak{e}_n\mathfrak{A},$$

with the $\mathfrak{e}_i$ ideals in $\mathfrak{o}$. From the second inclusion in (2), we see that each $\mathfrak{e}_i$ is either $\mathfrak{o}$ or the principal prime ideal $(\theta-1)$. By permuting the summands, and using Lemma 3 if necessary, we may then assume that

$$(5) \qquad \mathfrak{e}_1 = \cdots = \mathfrak{e}_r = 0, \qquad \mathfrak{e}_{r+1} = \cdots = \mathfrak{e}_n = (\theta - 1).$$

In that case, the quotient module

$$B = (g - 1)M/(\theta - 1)M_s \cong \mathfrak{o}/(\theta - 1) \oplus \cdots \oplus \mathfrak{o}/(\theta - 1),$$

where r summands occur. Since $(\theta-1)$ is an ideal of norm p, we see that B is an additive abelian group of type $(p, \cdots, p)$, and the integer r is thus uniquely determined as the rank of B. Let us fix β_k in the kth summand of (3) so that B is generated by the cosets $\beta_1+(\theta-1), \cdots, \beta_r+(\theta-1)$ (or $\beta_n+(\theta-1)\mathfrak{A}$ in case $r=n$). For example, we may choose β_k to be the unit element in $\mathfrak{o}$ for $k<n$, while if $r=n$, we choose $\beta_n \in \mathfrak{A}$ such that $\beta_n \notin (\theta-1)\mathfrak{A}$.

On the other hand, M/M_s is a regular Z-module, and therefore M_s is a Z-direct summand of M. Choose a regular Z-module X such that M is the direct sum of M_s and X. Then

$$(g-1)M = (\theta-1)M_s + (g-1)X,$$

so that the map $\phi: X \to B$ defined by

$$\phi(x) = (g-1)x + (\theta-1)M_s \qquad \text{for } x \in X$$

is a linear map of X onto B. With each $x \in X$ we may thus associate an r-tuple $(\alpha_1, \cdots, \alpha_r)$ (also denoted by $\phi(x)$) such that

$$(g-1)x \equiv \alpha_1\beta_1 + \cdots + \alpha_r\beta_r \pmod{(\theta-1)M_s},$$

with each $\alpha_i \in \bar{Z} = Z/pZ$. By choosing a suitable Z-basis $x_1, \cdots, x_m$ of X, we may assume that the vectors $\phi(x_1), \cdots, \phi(x_r)$ are linearly independent over $\bar{Z}$. Under a further change of Z-basis of X, we may then take

$$(g-1)x_i \equiv c_i\beta_i, \ (g-1)x_j \equiv 0 \ (\text{mod } (\theta-1)M_s, \qquad (1 \leqq i \leqq r, r < j \leqq m),$$

where each $c_i \in Z$, $c_i \not\equiv 0 \pmod{p}$. Set $(g-1)x_i = c_i\beta_i + (g-1)u_i$, $(g-1)x_j = (g-1)u_j$ $(1 \leqq i \leqq r, r < j \leqq m)$, with each $u_i \in M_s$, and define $y_i = x_i - u_i$ $(1 \leqq i \leqq m)$. Then we have

$$M = M_s \oplus Zy_1 \oplus \cdots \oplus Zy_m, \tag{6}$$

where

$$gy_i = y_i + c_i\beta_i, \qquad gy_j = y_j \qquad (1 \leqq i \leqq r, r < j \leqq m) \tag{7}$$

and where M_s defined by (3) is made into a $Z[g]$-module by

$$gm = \theta m \qquad \text{for } m \in M_s. \tag{8}$$

The structure of M is completely determined by the ideal class of $\mathfrak{A}$, the integers $r = Z$-rank of B, $m = Z$-rank of M/M_s, $n = \mathfrak{o}$-rank of M_s, and by the constants $c_1, \cdots, c_r$. We show now that we may in fact take each $c_i = 1$; this is a consequence of the following:

LEMMA 4. *Let* $\mathfrak{A}$ *be an ideal in* $\mathfrak{o}$, *let* $\beta \in \mathfrak{A}$ *be fixed, and let* $c \in Z$, $c \not\equiv 0 \pmod{p}$. *Let* $M_1 = \mathfrak{A} \oplus Zy_1$ *be made into a* $Z[g]$*-module by defining* $ga = \theta a$ *for* $a \in \mathfrak{A}$, $gy_1 = y_1 + \beta$. *Let* $M = \mathfrak{A} \oplus Zy_2$ *be made into a* $Z[g]$*-module by defining* $ga = \theta a$ *for* $a \in \mathfrak{A}$, $gy_2 = y_2 + c\beta$. *Then* M_1 *and* M *are* $Z[g]$*-isomorphic.*

PROOF. Set $u = 1 + \theta + \cdots + \theta^{c-1} =$ unit in $\mathfrak{o}$. Since $u - c = (\theta - 1) + (\theta^2 - 1) + \cdots + (\theta^{c-1} - 1)$, we may choose $t \in \mathfrak{A}$ so that $(\theta - 1)t = (u - c)\beta$. Now define a linear map ϕ: $M_1 \rightarrow M$ by

$$\phi(a) = ua, \quad a \in \mathfrak{A}, \quad \phi(y_1) = y_2 + t.$$

Then $g\phi(a) = \phi g(a)$ for all $a \in \mathfrak{A}$, and also

$$g\phi(y_1) = g(y_2 + t) = y_2 + c\beta + \theta t = y_2 + t + u\beta = \phi g(y_1).$$

Thus ϕ is a $Z[g]$-isomorphism of M_1 onto M.

To summarize, we have thus shown:

THEOREM. *Every Z-regular* $Z[g]$*-module is operator-isomorphic to a module defined by* (3), (6), (7), *and* (8), *with* $c_1 = \cdots = c_r = 1$. *The invariants which uniquely determine such a module (up to isomorphism) are: the ideal class of* $\mathfrak{A}$, $n = \mathfrak{o}$*-rank of* M_s, $m = Z$*-rank of* M/M_s, *and* $r = Z$*-rank of* $(g-1)M/(\theta-1)M_s$; *the only restrictions on these invariants are the conditions* $r \leqq m$, $r \leqq n$. *Conversely, for any such choice of invariants, equations* (3), (6), (7), *and* (8) *define a* $Z[g]$*-module with the given invariants.*

COROLLARY (See [**2**; **3**].) *The integrally-indecomposable regular* $Z[g]$*-modules are those for which either* $r = n = 0$, $m = 1$, or $r = m = 0$, $n = 1$, *or* $r = m = n = 1$. *The number of nonisomorphic modules of these types is* $2h + 1$, *where* h *is the class number of* $\mathfrak{o}$.

REFERENCES

1. C. Chevalley, *L'arithmétique dans les algèbres de matrices*, Actualités Scientifiques et Industrielles vol. 323 (1936) Paris.

2. F. E. Diederichsen, *Über die Ausreduktion ganzzahliger Gruppendarstellungen bei arithmetischer Äquivalenz*, Abhandlungen aus dem Mathematischen Seminar der Universität Hamburg vol. 14 (1938) pp. 357–412.

3. L. K. Hua and I. Reiner, *Automorphisms of the unimodular group*, Trans. Amer. Math. Soc. vol. 71 (1951) pp. 331–348.

4. I. Kaplansky, *Modules over Dedekind rings and valuation rings*, Trans. Amer. Math. Soc. vol. 72 (1952) pp. 327–340.

5. C. G. Latimer and C. C. MacDuffee, *A correspondence between classes of ideals and classes of matrices*, Ann. of Math. vol. 34 (1933) pp. 313–316.

6. I. Reiner, *Maschke modules over Dedekind rings*, in Canadian Journal of Mathematics vol. 8 (1956) pp. 329–334.

7. E. Steinitz, *Rechteckige Systeme und Moduln in algebraischen Zahlkörpern*, Math. Ann. vol. 71 (1911) pp. 328–354, vol. 72 (1912) pp. 297–345.

8. O. Taussky, *On a theorem of Latimer and MacDuffee*, Canadian Journal of Mathematics vol. 1 (1949) pp. 300–302.

9. H. Zassenhaus, *Neuer Beweis der Endlichkeit der Klassenzahl bei unimodularer Äquivalenz endlicher ganzzahliger Substitutionsgruppen*, Abhandlungen aus dem Mathematischen Seminar der Universität Hamburg vol. 12 (1938) pp. 276–288.

UNIVERSITY OF ILLINOIS

AUTOMORPHISMS OF THE GAUSSIAN UNIMODULAR GROUP

Joseph Landin and Irving Reiner

ABSTRACT

Let $G = Z[i]$ be the ring of Gaussian integers and G_n the group of $n \times n$ unimodular matrices with entries from G and $\mathcal{A}_n$ the automorphism group of G_n. Generators for the group $\mathcal{A}_n$ are determined. The case $n = 2$ requires special attention. Let

$$S_0 = \begin{pmatrix} 0 & 1 \\ -1 & 0 \end{pmatrix}, \quad T_0 = \begin{pmatrix} 1 & 1 \\ 0 & 1 \end{pmatrix}, \quad P_0 = \begin{pmatrix} i & 0 \\ 0 & 1 \end{pmatrix}$$

so that G_2 is the group generated by S_0, T_0, P_0.

Main Theorem. *The group $\mathcal{A}_n$ of automorphisms of G_n is generated by:*

1. $X \to AXA^{-1}$, $A \in G_n$;
2. $X \to X^*$, *the inverse transpose of* X, *(may be omitted if* $n = 2$*);*
3. $X \to \bar{X}$, *the complex conjugate matrix;*
4. $X \to (\det X)^k X$, *where* $k = 1$ *if* n *is even, and* $k = 2$ *if* n *is odd;*
5. *For the case* $n = 2$ *only,* $(P_0, S_0, T_0) \to (P_0, -S_0, -T_0)$.

The proof is not self-contained; it makes use of results from [5] and [8]. A canonical form for elements of order two is used. An alternate approach to this result can be found in [14] and [19].

AUTOMORPHISMS OF THE TWO-DIMENSIONAL GENERAI LINEAR GROUP OVER A EUCLIDEAN RING

JOSEPH LANDIN AND IRVING REINER

1. Introduction. Let E denote a free R-module of rank n over a ring R, and let $GL_n(R)$ be the group of one-to-one R-linear maps of E into itself. When R is (i) a skew-field, (ii) the ring Z of rational integers, (iii) the ring $Z[i]$ of Gaussian integers, or (iv) a noncommutative principal ideal domain ($n \geqq 3$ in this case), it has been proved that the group A_n of automorphisms of $GL_n(R)$ is generated by automorphisms of the following types:

(a) $u \rightarrow tut^{-1}$, $\quad t \in GL_n(R)$, (inner),

(b) $u \rightarrow \chi(u)u$,

where χ is a homorphism of $GL_n(R)$ into the group of units of the center of R satisfying $\chi(\lambda I) = \lambda^{-1}$ if and only if $\lambda = 1$.

(c) $u \rightarrow u^\sigma$, σ an automorphism of R,

(d) $u \rightarrow t^{-1}\check{u}t$, $\check{u}$ = contragredient of u, where $t: E \rightarrow E^*$ is a correlation mapping E onto its dual E^*. (For references concerning these results see [1].)

On the other hand, for the case where $R = K[x]$ is the ring of polynomials in an indeterminate x over a field K, it has been shown [1] that the above types of automorphisms do not generate all the automorphisms of $GL_2(R)$. It is thus clear that one cannot expect these types of automorphisms to generate A_2 unless fairly restrictive conditions are imposed on the ring R.

We shall assume henceforth:

(I) R is a commutative principal ideal domain, integrally closed in its quotient field.

(II) R is Euclidean.

(III) The group of units of R contains more than two elements.

(IV) There exist units α_λ, $\lambda \in \Lambda$, in R such that each $t \in R$ is expressible in the form

$$t = \sum_{i=1}^{m} n_i \alpha_i, \qquad n_i \in Z$$

where Z is the ring of rational integers and Λ is a set of indices. (If char $R = p \neq 0$, then the n_i are chosen from $GF(p)$.)

Integral domains satisfying these conditions certainly exist. For example, let R be the ring of all algebraic integers in a cyclotomic field over the rationals; if R is Euclidean it will satisfy (I)–(IV). As

Presented to the Society, June 15, 1957; received by the editors May 31, 1957.

another example, let R be the ring consisting of all expressions $x^k f(x)$ where $f(x) \in K[x]$ is a polynomial in an indeterminate x over a field K, and where k ranges over all rational integers.[1] Conditions (I)–(IV) are also valid for this ring.

We shall use the following notations:

K = quotient field of R; $(R, +)$ = additive group of R;

U = multiplicative group of units of R. We shall identify $GL_2(R)$ with the group of 2×2 matrices over R with determinant in U. Hereafter let

$$I = \begin{pmatrix} 1 & 0 \\ 0 & 1 \end{pmatrix}, \quad J = \begin{pmatrix} -1 & 0 \\ 0 & 1 \end{pmatrix}, \quad S = \begin{pmatrix} 0 & 1 \\ -1 & 0 \end{pmatrix}, \quad T = \begin{pmatrix} 1 & 1 \\ 0 & 1 \end{pmatrix},$$

$$X(t) = \begin{pmatrix} 1 & t \\ 0 & 1 \end{pmatrix}, \qquad t \in R.$$

Let ${}^tX$ denote the transpose of X and let $[\alpha, \beta]$ denote a diagonal matrix with diagonal entries α, β.

We shall find it convenient to introduce the subgroup V of $(R, +)$ generated by all differences of units:

$$V = \sum_{\alpha, \beta \in U} Z(\alpha - \beta),$$

where (as above) Z is replaced by $GF(p)$ if char $R = p \neq 0$. Since R has a unity element we see that $1 - (-1) = 2 \in V$. Assume that (IV) holds and let $t \in R$ be arbitrary, so that there are units $\{\alpha_i\}$ and integers $\{n_i\}$ such that

$$t = \sum_{i=1}^{m} n_i \alpha_i.$$

Since $\alpha_i - 1 \in V$ for each i, we find that

$$t \equiv \sum_{i=1}^{m} n_i \pmod{V}.$$

If $1 \notin V$, then since $2 \in V$ we see that

$$\sum_{i=1}^{m} n_i \equiv 0 \text{ or } 1 \pmod 2$$

according as $t \in V$ or $t \notin V$. Let $P(t)$ denote the residue of $\sum_{i=1}^{m} n_i$ (mod 2). Then $P(t)$ is a well-defined function of t whenever $1 \notin V$, even though the expression for t as a sum of units may not be unique.

On the other hand, if $1 \in V$ then there is an equation

[1] This example was given by Professor N. T. Hamilton.

$$(1) \qquad 1 = \sum_{i=1}^{m} n_i(\alpha_i - \beta_i), \qquad n_i \in Z, \qquad \alpha_i, \beta_i \in U.$$

We may remark that $1 \in V$ if and only if some sum of an odd number of units can be zero. Thus $1 \overline{\in} V$ for the cases $R=Z$ and $R=Z[i]$ (ring of Gaussian integers), while $1 \in V$ for the case where $R=K[x]$ is a polynomial domain over a field K of characteristic $\neq 2$.

Further we note that by virtue of (IV), the subgroup V is an ideal of R. For,

$$(\sum n_i\alpha_i)\cdot(\sum m_j(\beta_j - \gamma_j)) = \sum n_i m_j(\alpha_i\beta_j - \alpha_i\gamma_j) \in V,$$

where n_i, $m_j \in Z$, α_i, β_j, $\gamma_j \in U$.

2. **Transvections in** GL_2 (R). We begin by assuming that R satisfies (I) and (III). If char $R=0$ an element $u \in GL_2(R)$ will be called a *transvection* if there are more than two elements in $GL_2(R)$ conjugate to u and commuting with u. If char $R=p\neq 0$, an element $u \in GL_2(R)$, $u \neq I$, is called a transvection if $u^p = I$.

LEMMA 1. *An element $u \in GL_2(R)$ is a transvection if and only if u is conjugate in $GL_2(R)$ to an element of the form $\alpha X(t)$, $\alpha \in U$, $t\neq 0$. Furthermore, if* char $R=p\neq 0$, *then* $\alpha = 1$.

PROOF. (1) Char $R=0$. Consider u as an element of $GL_2(K)$. If u has distinct characteristic roots, then in some extension field of K, u is similar to $[a, b]$, $a\neq b$. On the one hand, only diagonal matrices commute with $[a, b]$; on the other, any matrix similar to $[a, b]$ must have the same characteristic roots. Hence, there are at most two elements in $GL_2(R)$ conjugate to u and commuting with it, contrary to the definition of transvection. Therefore u has a repeated characteristic root.

Since R is a principal ideal domain, then (as is well known) u is conjugate in $GL_2(R)$ to an element of the form $rX(t)$, $t \in R$. Then r^2 is a unit, whence so is r.

Conversely, let $u \in GL_2(R)$ be conjugate in $GL_2(R)$ to $\alpha X(t)$, $t\neq 0$, $\alpha \in U$. Let β_1, β_2, $\beta_3 \in U$ be distinct. Then the three matrices

$$[\beta_i, 1]\cdot \alpha X(t) \cdot [\beta_i^{-1}, 1] = \alpha X(\beta_i t), \qquad (i = 1, 2, 3)$$

commute with and are conjugate to $\alpha X(t)$, whence it is clear that U is a transvection.

(2) Char $R=p\neq 0$. If $u\neq I$ is a transvection it satisfies the equation $\lambda^p - 1 = (\lambda - 1)^p = 0$. Hence the characteristic polynomial of u is $(\lambda-1)^2$, so the characteristic roots are both 1. Therefore u is conjugate in $GL_2(R)$ to an element of the form $X(t)$.

Conversely, any element $u \in GL_2(R)$ conjugate to $X(t)$ clearly satisfies $u^p = I$. This completes the proof of the lemma.

Fix an element $t_0 \in R$, and let $\tau \in A_2$. It follows at once from Lemma 1 that to within inner automorphism

$$X(t_0)^\tau = \epsilon(t_0)X(\sigma(t_0)). \tag{2}$$

Since for each $t \in R$, $X(t)$ is a transvection commuting with $X(t_0)$ it follows (assuming (2)) that $X(t)^\tau$ is a transvection commuting with $X(\sigma(t_0))$. Consequently

$$X(t)^\tau = \epsilon(t)X(\sigma(t)), \qquad \sigma(t) \in R, \qquad \epsilon(t) \in U, \tag{3}$$

for all $t \in R$.

LEMMA 2. *The mapping* $t \to \epsilon(t)$ *is a homomorphism of* $(R, +)$ *into* U; *the mapping* $t \to \sigma(t)$ *is an automorphism of* $(R, +)$.

PROOF. It follows immediately from $X(s)X(t) = X(s+t)$ that ϵ and σ are both homomorphisms.

We now show that σ is an automorphism. If $\sigma(t) = 0$ then $X(t)$ is in the center of $GL_2(R)$, whence $t = 0$. Further, since

$$\{\alpha X(t) \colon \alpha \in U, t \in R, t \neq 0\}$$

is the set of all transvections commuting with $X(t_0)$ for fixed $t_0 \neq 0$, therefore $\{\alpha X(\sigma(t)) \colon \alpha \in U, t \in R, t \neq 0\}$ must be the entire set of transvections commuting with $X(\sigma(t_0))$. Hence σ is "onto," and therefore is an automorphism.

LEMMA 3. *For all* $t \in R$, $\epsilon(t) = \pm 1$.

PROOF. For $\tau \in A_2$ set

$$J^\tau = \begin{pmatrix} a & b \\ c & d \end{pmatrix},$$

where $J = [-1, 1]$. Then $a^2 + bc = d^2 + bc = 1$, $b(a+d) = c(a+d) = 0$. From $JX(t) = X(-t)J$ we deduce $c\sigma(t) + d = \alpha d$ and $c = \alpha c$, where $\alpha = \epsilon(t)^{-2}$. Consequently $c = 0$ or $= 1$. However, $c = 0$ implies $\alpha = 1$; therefore $\epsilon(t) = \pm 1$.

LEMMA 4. *Let* $\tau \in A_2$. *Changing* τ *by an inner automorphism we may assume* (3) *and* $S^\tau = S$.

PROOF. Set $Y = ST$; then $Y^3 = I$ implies $(Y^\tau)^3 = I$ for any $\tau \in A_2$. Therefore, the minimum and characteristic polynomials of Y^τ are equal and divide $\lambda^3 - 1$.

If char $R = 3$ then $\lambda^3 - 1 = (\lambda - 1)^3$ whence the characteristic polynomial of Y^τ is $\lambda^2 - 2\lambda + 1 = \lambda^2 + \lambda + 1$, and therefore

$$\text{Trace } Y^\tau = -1. \tag{4}$$

On the other hand, if char $R \neq 3$ and $\lambda^2+\lambda+1$ is irreducible over R equation (4) again holds. However, suppose $\lambda^2+\lambda+1$ is reducible over R; then the characteristic polynomial of Y is either

$$(\lambda-1)(\lambda-\omega),\ (\lambda-1)(\lambda-\omega^2) \text{ or } (\lambda-\omega)(\lambda-\omega^2) = \lambda^2+\lambda+1.$$

Now we have $T^\tau = \pm X(\sigma(1))$, whence det $T^\tau=1$. From $S^2=-1$ we deduce det $S^\tau=1$. Therefore det $Y^\tau=1$, whence the characteristic polynomial of Y^τ can only be $\lambda^2+\lambda+1$. Consequently (4) holds in all cases.

Set

$$S^\tau = \begin{pmatrix} a & b \\ c & d \end{pmatrix}.$$

Then $a^2+bc=d^2+bc=-1$, $b(a+d)=c(a+d)=0$. Suppose first $b=c=0$; then $a^2=d^2=-1$ implies $a=\pm i=d$. Now $a=d=\pm i$ is impossible since this would imply that S^τ is in the center of $GL_2(R)$. On the other hand, $a=-d=\pm i$ contradicts (4). Consequently $d=-a$.

For $t \in R$ we have

$$\begin{pmatrix} 1 & t \\ 0 & 1 \end{pmatrix}\begin{pmatrix} a & b \\ c & -a \end{pmatrix}\begin{pmatrix} 1 & t \\ 0 & 1 \end{pmatrix}^{-1} = \begin{pmatrix} a+ct & b-2at-ct^2 \\ c & -(a+ct) \end{pmatrix}.$$

Since

$$Y^\tau = \pm \begin{pmatrix} a & a\sigma(1)+b \\ c & c\sigma(1)-a \end{pmatrix}$$

and trace $Y^\tau=-1$, we have $c\sigma(1)=\pm 1$, whence $c \in U$. Hence there exists $t_0 \in R$ such that $a+ct_0=0$. Changing τ by an inner automorphism with factor $X(t_0)$, we now have

$$S^\tau = \begin{pmatrix} 0 & b \\ -b^{-1} & 0 \end{pmatrix}.$$

Finally, applying the inner automorphism with factor $[1, b]$ we obtain Lemma 4.

Lemma 5. *If τ is any automorphism of $GL_2(R)$ leaving S invariant and satisfying* (3) *then*

$${}^tX(t)^\tau = \epsilon(t)\,{}^tX(\sigma(t)).$$

This follows from ${}^tX(-t)=S^{-1}X(t)S$.

If τ is an automorphism of $GL_2(R)$ satisfying the hypotheses of Lemma 5 then $(T^\tau S)^3=I$ implies $\epsilon(1)\sigma(1)=1$. If $\sigma(1)=-1$, by introducing a further inner automorphism with factor J, we may obtain a new τ with $\sigma(1)=1$, but now $S^\tau = \pm S$. Then also $\epsilon(1) = \pm 1$.

The foregoing results may be summarized as

THEOREM 1. *If $\tau \in A_2$, then after changing τ by an inner automorphism if necessary, we have*

$$\begin{aligned} X(t)^\tau &= \epsilon(t) X(\sigma(t)), \qquad t \in R, \\ {}^tX(t)^\tau &= \epsilon(t)\, {}^tX(\sigma(t)), \qquad (5) \\ S^\tau &= \pm S,\ \epsilon(1) = \pm 1,\ \sigma(1) = 1, \end{aligned}$$

where τ induces the autmorphism σ: $(R, +) \to (R, +)$ and the homomorphism ϵ: $(R, +) \to U$, and where the plus signs go together as do the minus signs.

LEMMA 6. *If $\tau \in A_2$ satisfies* (5) *then*

$$[\alpha, 1]^\tau = \lambda(\alpha)[\rho(\alpha), 1]$$

where both λ and ρ are endomorphisms of U.

PROOF. Set

$$G = \{\alpha X(t) \colon \alpha \in U, t \in R\}, \qquad H = \{\alpha\, {}^tX(t) \colon \alpha \in U, t \in R\},$$

and let K denote the intersection of the normalizers of G and H. Then K consists of all diagonal matrices. Since $G^\tau = G$ and $H^\tau = H$ imply $K^\tau = K$, we see that $[\alpha, \beta]^\tau$ is also diagonal. In particular $[\alpha, 1]^\tau = \lambda(\alpha)[\rho(\alpha), 1]$.

LEMMA 7. *For all $\alpha \in U$, $t \in R$ we have*

$$\epsilon(\alpha t) = \epsilon(t), \qquad \rho(\alpha) = \sigma(\alpha), \qquad \sigma(\alpha t) = \sigma(\alpha)\sigma(t).$$

PROOF. The decomposition $X(\alpha t) = [\alpha, 1] \cdot X(t) \cdot [\alpha, 1]^{-1}$ yields $\epsilon(\alpha t) = \epsilon(t)$, $\sigma(\alpha t) = \rho(\alpha)\sigma(t)$, which implies the result.

Assuming next that R satisfies condition (IV) we prove

LEMMA 8. *Let $\tau \in A_2$ satisfy condition* (5). *Then the automorphism σ of $(R, +)$ induced by τ is a ring automorphism of R.*

PROOF. If $a \in Z$ (Char $R=0$) or if $a \in GF(p)$ (Char $R = p \neq 0$), then $\sigma(a) = a$. Hence, using (IV) it follows immediately that $\sigma(xy) = \sigma(x)\sigma(y)$ for all $x, y \in R$.

We henceforth assume that R satisfies condition (I)–(IV) of the introduction. We have seen that starting with an automorphism $\tau \in A_2$, after changing τ by an inner automorphism we obtain a new automorphism (again denoted by τ) satisfying

$$X(t)^\tau = \epsilon(t) X(\sigma(t)), \qquad S^\tau = \epsilon(1) S, \qquad [\alpha, 1]^\tau = \lambda(\alpha)[\sigma(\alpha), 1],$$

where $\epsilon\colon (R, +)\to U$ is a homomorphism satisfying $\epsilon(\alpha t)=\epsilon(t)$, $\alpha\in U$, where $\sigma\colon R\to R$ is a ring automorphism, and where λ is an endomorphism of U. Now replace τ by a new automorphism

$$U \to (U^{\tau})\sigma^{-1}$$

where σ^{-1} is the automorphism of $GL_2(R)$ induced by the ring automorphism σ^{-1} of R. Again calling this new automorphism τ, we now have an automorphism satisfying

$$X(t)^{\tau} = \epsilon(t)X(t), \qquad S^{\tau} = \epsilon(1)S, \qquad [\alpha, 1]^{\tau} = \lambda(\alpha)[\alpha, 1],$$

with possibly new maps ϵ and λ.

We find readily from the above that $[1, \alpha]^{\tau}=\lambda(\alpha)[1, \alpha]$, whence

$$[\alpha, \alpha]^{\tau} = \lambda^2(\alpha)[\alpha, \alpha].$$

From this equation we see that as α ranges over all elements of U so does $\alpha\lambda^2(\alpha)$. Thus $\alpha\to\alpha\lambda^2(\alpha)$ must be an automorphism of U, and from this it follows easily that

$$u \to \lambda(\det u)\cdot u$$

is an automorphism μ of $GL_2(R)$. Replacing τ by $\tau\mu^{-1}$, the new automorphism τ now satisfies

$$X(t)^{\tau} = \epsilon(t)X(t), \qquad S^{\tau} = \epsilon(1)S, \qquad [\alpha, 1]^{\tau} = [\alpha, 1].$$

Now let $t=\sum_{i=1}^{m} n_i\alpha_i$, $\alpha_i\in U$, $n_i\in Z$ (char $R=0$) or $n_i\in GF(p)$ (char $R=p\neq 0$). Then

$$\epsilon(t) = \prod_1^m \epsilon(n_i\alpha_i) = \prod_1^m \epsilon(n_i) = \prod_1^m (\epsilon(1))^{n_i} = \epsilon(1)^{\Sigma n_i}.$$

Set $\gamma=\epsilon(1)=\pm 1$. Then the automorphism τ satisfies

$$\text{(6)} \qquad X(t)^{\tau} = \gamma^{\Sigma n_i}X(t), \qquad S^{\tau} = \gamma S, \qquad [\alpha, 1]^{\tau} = [\alpha, 1].$$

We now show that if we define V (as before) to be the subgroup of $(R, +)$ generated by $\{\alpha-\beta;\ \alpha, \beta\in U\}$, then if $1\in V$ we must have $\gamma=1$, while if $1\notin V$ then equations (6) with $\gamma=-1$ define an automorphism η of $GL_2(R)$.

Indeed, if $1\in V$, then $1=\sum n_i(\alpha_i-\beta_i)$, $\alpha_i, \beta_i\in U$, so

$$\begin{aligned}\gamma = \epsilon(1) &= \prod \epsilon(n_i\alpha_i - n_i\beta_i) = \prod \epsilon(n_i\alpha_i)(\epsilon(n_i\beta_i))^{-1}\\ &= \prod \epsilon(n_i)(\epsilon(n_i))^{-1} = 1.\end{aligned}$$

On the other hand, if $1\notin V$, define $P(t)$ as in the introduction. Let $\eta\colon GL_2(R)\to GL_2(R)$ be defined by

$$(7) \qquad \eta: \begin{cases} X(t) \to (-1)^{P(t)} X(t), \\ S \to -S, \\ [\alpha, 1] \to [\alpha, 1]. \end{cases}$$

We shall prove that η induces an automorphism of $GL_2(R)$, and for this it suffices to show that η is well-defined. Thus, we need only prove that if a power product

$$\prod \{X(t_i), S, [\alpha_j, 1]\} = I$$

in $GL_2(R)$, then $n_s + \sum P(t_i) \equiv 0 \pmod 2$, where n_s is the number of factors equal to $S^{\pm 1}$.

For $t \in R$ we have $t = \sum n_i \alpha_i$ whence

$$X(t) = \prod X^{n_i}(\alpha_i) \equiv \prod X^{n_i}(1) \equiv T^{P(t)} \pmod V,$$

where $T = X(1)$. Also, $[\alpha, 1] \equiv I \pmod V$ for $\alpha \in U$. Hence, if

$$\prod \{X(t_i), S, [\alpha_j, 1]\} = I$$

then since the subgroup V of $(R\ +)$ is also an ideal in R we have

$$\prod \{T^{P(t_i)}, S, I\} \equiv I \pmod V.$$

However since $2 \in V$, the only power products of S and T which are distinct mod V are I, S, T, ST, TS and STS. Of these, only the first can be $\equiv I \pmod V$ because $1 \overline{\in} V$. But if a power product of S and T is $\equiv I \pmod 2$ then the total number of factors of S and T must be even. Hence $n_s + \sum P(t_i) \equiv 0 \pmod 2$. This completes the proof that $\eta \in A_2$ whenever $1 \overline{\in} V$.

To summarize our results we have:

THEOREM 2. *The group A_2 of automorphisms of $GL_2(R)$ is generated by*:

(1) *The inner automorphisms $u \to vuv^{-1}$, $v \in GL_2(R)$,*

(2) *The automorphisms induced by automorphisms of R,*

(3) *The scalar multiplications $U \to \lambda(\det u)u$, where λ is an endomorphism of U for which the map $\alpha \to \alpha\lambda^2(\alpha)$, $\alpha \in U$, is an automorphism of U,*

(4) *The automorphism η described in* (7), *provided that $1 \overline{\in} V$.*

REFERENCES

1. I. Reiner, *A new type of automorphism of the general linear group over a ring*, Ann. of Math. vol. 66 (1957) pp. 461–466.

2. J. Landin and I. Reiner, *Automorphisms of the Gaussian unimodular group*, Trans. Amer. Math. Soc. vol. 87 (1958) pp. 76–89.

UNIVERSITY OF ILLINOIS

INCLUSION THEOREMS FOR CONGRUENCE SUBGROUPS

BY

M. NEWMAN AND I. REINER[1]

1. **Introduction.** We shall use the following notation throughout: $A^{(r)}$ denotes an $r \times r$ matrix A; $I^{(r)}$ denotes the r-rowed identity matrix; 0 will be used for a zero matrix of appropriate size. Congruence of matrices will be interpreted as elementwise congruence. We write $a|b$ to indicate that a divides b. Lower case italic letters will always denote integers.

Let G_t be the proper unimodular group consisting of all $t \times t$ matrices with integral elements and determinant $+1$. For a fixed partition: $t=r+s$ of t into two positive integers r and s, and for a fixed positive integer n, define the subgroup

$$G_{r,s}(n) = \left\{ \begin{bmatrix} A^{(r)} & B \\ C & D^{(s)} \end{bmatrix} \in G_t: \quad C \equiv 0 \pmod{n} \right\}. \tag{1}$$

We shall prove:

THEOREM 1. *Let m, n be positive integers, and let H be a group such that*

$$G_{r,s}(mn) \subset H \subset G_{r,s}(n). \tag{2}$$

Then there exists a divisor d of m such that

$$H = G_{r,s}(dn). \tag{3}$$

Special cases of this have been proved in [1] and [3].

In the case where $t=2r$, define

$$G_r(m, n) = \left\{ \begin{bmatrix} A^{(r)} & B \\ C & D^{(r)} \end{bmatrix} \in G_{2r}: \begin{matrix} B \equiv 0 \pmod{m}, \\ C \equiv 0 \pmod{n} \end{matrix} \right\}. \tag{4}$$

Then we shall show:

THEOREM 2. *Let H be a group satisfying*

$$G_r(m, n) \subset H \subset G_{2r}. \tag{5}$$

If $(m, n)=1$, then there exist integers m_1, n_1 with $m_1|m$, $n_1|n$, and

$$H = G_r(m_1, n_1). \tag{6}$$

Presented to the Society, October 26, 1957; received by the editors October 14, 1957.

[1] The work of the first author was supported (in part) by the Office of Naval Research, and that of the second author was supported (in part) by a contract with the National Science Foundation.

A special case of this (with $r=1$) was proved in [2], where it was also shown that the hypothesis $(m, n)=1$ could not be dropped.

To generalize further, let $\mathfrak{n}=(n_1, \cdots, n_{t-1})$, and define

$$G_t(\mathfrak{n}) = G_{1,t-1}(n_1) \cap G_{2,t-2}(n_2) \cap \cdots \cap G_{t-1,1}(n_{t-1}). \tag{7}$$

Thus an element $M \in G_t$ lies in $G_t(\mathfrak{n})$ if and only if for every partition $t=r+s$ $(1 \leqq r \leqq t-1)$ we have

$$M = \begin{bmatrix} A^{(r)} & B \\ C & D^{(s)} \end{bmatrix}, \qquad C \equiv 0 \pmod{n_r}.$$

We shall prove:

THEOREM 3. *Let* $(m_i n_i, m_j n_j)=1$ *for* $1 \leqq i, j \leqq t-1$, $i \neq j$. *Let* H *be a group such that*

$$G_t(\mathfrak{m}\mathfrak{n}) \subset H \subset G_t(\mathfrak{n}), \tag{8}$$

where $\mathfrak{m}\mathfrak{n}$ *denotes* $(m_1 n_1, \cdots, m_{t-1} n_{t-1})$. *Then there exists a vector*

$$\mathfrak{d} = (d_1, \cdots, d_{t-1}),$$

with $d_1 | m_1, \cdots, d_{t-1} | m_{t-1}$, *such that*

$$H = G_t(\mathfrak{d}\mathfrak{n}). \tag{9}$$

Finally, we shall prove analogues of Theorems 1 and 2 for the symplectic modular group Γ_t of order t, which consists of all integral matrices

$$\begin{bmatrix} A^{(t)} & B \\ C & D^{(t)} \end{bmatrix}$$

satisfying

$$AB' = B'A, \qquad CD' = D'C, \qquad AD' - DC' = I.$$

2. We begin the proof of Theorem 1 with two lemmas.

LEMMA 1. *Let* $t=r+s$, *and let* n *be a fixed positive integer. For each*

$$M = \begin{bmatrix} A^{(r)} & B \\ C & D^{(s)} \end{bmatrix} \in G_t$$

there exists an integral $r \times s$ *matrix* X *such that* $(|A+XC|, n)=1$.

Proof. It is sufficient to show that for every prime p there exists an integral matrix X_p such that $p \nmid |A+X_pC|$. For we may then find an integral matrix X satisfying $X \equiv X_p \pmod{p}$ for each $p | n$. Since $|A+XC| \equiv |A+X_pC| \pmod{p}$, it then follows that $(|A+XC|, n)=1$.

Now let p be a fixed prime, and let $\alpha_1, \cdots, \alpha_r$ denote the rows of A, and

$\gamma_1, \cdots, \gamma_s$ those of C. Since the rows of X_pC are linear combinations of those of C, we need only show that there exist linear combinations

$$\beta_i = \sum_{j=1}^{s} x_{ij}\gamma_j \qquad (1 \leqq i \leqq r,\ x_{ij} \text{ integers})$$

such that $p \nmid \det(\alpha_i+\beta_i)$. Thus, we seek integers x_{ij} for which the vectors $\alpha_i+\beta_i$ $(1 \leqq i \leqq r)$ are linearly independent modulo p.

Since M is unimodular, the set $\{\alpha_1, \cdots, \alpha_r, \gamma_1, \cdots, \gamma_s\}$ contains exactly r linearly independent vectors modulo p. Suppose that r' of the α's are linearly independent modulo p $(r' \leqq r)$; for simplicity of notation, suppose that these are $\alpha_1, \cdots, \alpha_{r'}$. Then each α_k $(r' < k \leqq r)$ is a linear combination modulo p of $\alpha_1, \cdots, \alpha_{r'}$. Further, there exist $r-r'$ vectors $\gamma_1^*, \cdots, \gamma_{r-r'}^*$ among $\gamma_1, \cdots, \gamma_s$ such that the set $\{\alpha_1, \cdots, \alpha_{r'}, \gamma_1^*, \cdots, \gamma_{r-r'}^*\}$ is linearly independent modulo p. Then we need only choose $\beta_1 = \cdots = \beta_{r'} = 0$, $\beta_{r'+1} = \gamma_1^*, \cdots, \beta_r = \gamma_{r-r'}^*$ to achieve the desired result.

LEMMA 2. *Let $M \in G_{r,s}(n)$, and let m be a fixed positive integer. Then there exists an integral $r \times s$ matrix X and an integral $s \times r$ matrix Y such that*

$$W(nY)S(X)\,M \in G_{r,s}\,(m\,n), \tag{10}$$

where

$$W(nY) = \begin{bmatrix} I^{(r)} & 0 \\ nY & I^{(s)} \end{bmatrix}, \qquad S(X) = \begin{bmatrix} I^{(r)} & X \\ 0 & I^{(s)} \end{bmatrix}.$$

The entries of X and Y are integers determined only modulo m. Therefore the set of products $W(nY)S(X)$, as the entries of X and Y range over all residues modulo m, contains a full set of left coset representatives of $G_{r,s}(n)$ modulo $G_{r,s}(mn)$. Consequently $G_{r,s}(mn)$ is of finite index in $G_{r,s}(n)$.

Proof. Set

$$M = \begin{bmatrix} A^{(r)} & B \\ nC & D^{(s)} \end{bmatrix} \in G_{r,s}(n).$$

By Lemma 1, we can determine X modulo m such that $(|A+nXC|, m)=1$. Set $A_0 = A+nXC$. Then

$$S(X)M = \begin{bmatrix} A_0 & * \\ nC & * \end{bmatrix},$$

and

$$W(nY)S(X)M = \begin{bmatrix} * & * \\ n(YA_0+C) & * \end{bmatrix}.$$

In order for (10) to hold, we need only show that Y modulo m can be determined so that $YA_0+C\equiv 0 \pmod{m}$.

Now $(|A_0|, m)=1$, so that we may find an integer a with $a|A_0|\equiv 1 \pmod{m}$. Letting A_0^{adj} denote the adjoint of A_0, we set

$$Y \equiv -aCA_0^{\mathrm{adj}} \pmod{m}. \tag{11}$$

Using $A_0^{\mathrm{adj}}A_0=|A_0|I$, we obtain

$$YA_0 \equiv -C \pmod{m},$$

as desired.

The remainder of the lemma follows at once from (10).

We now proceed with the proof of Theorem 1. Let H be a group such that

$$G_{r,s}(mn) \subset H \subset G_{r,s}(n).$$

Using the argument in [1], we find by induction on the total number of prime factors of m that the conclusion of Theorem 1 is valid unless for every d dividing m, $d\neq 1$, we have

$$H \cap G_{r,s}(dn) = G_{r,s}(mn).$$

Suppose now that $H\neq G_{r,s}(mn)$. The above then shows that there exists a matrix

$$M = \begin{bmatrix} A^{(r)} & B \\ nC & D^{(s)} \end{bmatrix} \in H$$

such that $C\not\equiv 0 \pmod{d}$ for any divisor d of m, $d\neq 1$. Choose X, Y as in Lemma 2, and use the fact that $S(X)\in H$. Then we see that $W(nY)\in H$, where Y is chosen by use of (11). Hence also $Y\not\equiv 0 \pmod{d}$ for any divisor d of m, $d\neq 1$.

Call an $s\times r$ matrix T *permissible* if $W(nT)\in H$. We have shown the existence of a permissible matrix Y such that $Y\not\equiv 0 \pmod{d}$ for any divisor d of m, $d\neq 1$. We shall use this to deduce that every matrix is permissible. Since already $S(X)\in H$ for all X, it will then follow from Lemma 2 that $H=G_{r,s}(n)$, and the theorem will be proved.

Now we have

$$W(nT_1)\cdot W(nT_2) = W(n(T_1+T_2)),$$

and

$$\begin{bmatrix} V^{-1} & 0 \\ 0 & U \end{bmatrix} W(nT) \begin{bmatrix} V & 0 \\ 0 & U^{-1} \end{bmatrix} = W(nUTV), \quad U\in G_s,\ V\in G_r.$$

Therefore if T_1 and T_2 are permissible, so is T_1+T_2. If T is permissible, then so is $-T$; and if $U \in G_s$, $V \in G_r$, then UTV is also permissible.

Starting with the permissible Y above, set $Y_1 = UYV$, with $U \in G_s$, $V \in G_r$. Then Y_1 is also permissible, and with proper choice of U and V, we may take Y_1 in Smith normal form:

$$Y_1 = \begin{bmatrix} h_1 & & & & \\ & h_2 & & & \\ & & \cdot & & \\ & & & \cdot & \\ & & & & h_\mu \end{bmatrix}, \qquad \mu = \min(r, s),$$

where $h_1 | h_2 | \cdots | h_\mu$. If $(h_1, m) > 1$, then there is a prime $p | m$ such that $Y_1 \equiv 0 \pmod{p}$. Then also $Y \equiv 0 \pmod{p}$, which is impossible. Hence $(h_1, m) = 1$. Let us choose a so that $ah_1 \equiv 1 \pmod{m}$. Then $Y_2 = aY_1$ is also permissible. Since a permissible matrix remains permissible when multiples of m are added to its entries, we therefore have the permissible matrix

$$Y_3 = \begin{bmatrix} 1 & & & & \\ & k_2 & & & \\ & & \cdot & & \\ & & & \cdot & \\ & & & & k_\mu \end{bmatrix}.$$

Hence also

$$Y_4 = \begin{bmatrix} 0 & -k_2 & & & & \\ 1 & 0 & & & & \\ & & k_3 & & & \\ & & & \cdot & & \\ & & & & \cdot & \\ & & & & & k_\mu \end{bmatrix}$$

and

$$Y_5 = Y_3 - Y_4 = \begin{bmatrix} 1 & k_2 & & & & \\ -1 & k_2 & & & & \\ & & 0 & & & \\ & & & \cdot & & \\ & & & & \cdot & \\ & & & & & 0 \end{bmatrix}$$

are permissible. In Y_5 add the second row to the first row, and then subtract the matrix so obtained from Y_5, obtaining the permissible matrix which has 1 in the (1, 1) place, $-k_2$ in the (1, 2) place, and 0 elsewhere. In this matrix add k_2 times the first column to the second column, thereby obtaining the permissible matrix

$$Y_6 = \begin{bmatrix} 1 & & & \\ & 0 & & \\ & & \ddots & \\ & & & 0 \end{bmatrix}.$$

Since also UY_6V is permissible for all $U \in G_s$, $V \in G_r$, we find that every matrix whose entries are all zeros except for a single 1, must be permissible. Therefore all matrices are permissible, and Theorem 1 is proved.

3. We now prove Theorem 2. Let H be a group satisfying

$$G_r(m, n) \subset H \subset G_{2r},$$

where $G_r(m, n)$ is defined by (4), and where $(m, n) = 1$. Choose integers a, b satisfying $am - bn = 1$, and set

$$K = \begin{bmatrix} amI^{(r)} & I \\ bnI & I^{(r)} \end{bmatrix} \in G_{2r}.$$

Then as in [2] we find that $K^{-1}G_r(m, n)K = G_{r,r}(mn)$, and the remainder of the proof of Theorem 2 follows from Theorem 1 just as in [2].

Theorem 2 is false for $(m, n) > 1$, as is shown in [2].

4. To prove Theorem 3, we begin with several lemmas.

LEMMA 3. *Let $n_1, \cdots, n_{t-1}$ be pairwise coprime, and let $M \in G_t$. Then there exists an upper triangular matrix $S \in G_t$ such that for each r $(1 \leqq r \leqq t-1)$ we have*

$$M = \begin{bmatrix} A^{(r)} & B \\ C & D^{(t-r)} \end{bmatrix}, \qquad S \equiv \begin{bmatrix} I^{(r)} & X_r \\ 0 & I^{(t-r)} \end{bmatrix} \pmod{n_r}, \tag{12}$$

and

$$(|A^{(r)} + X_rC|, n_r) = 1. \tag{13}$$

Proof. Let M be fixed. For each r, write M in the form (12). By Lemma 1, we may then choose X_r such that (13) holds. We then use the Chinese remainder theorem to determine an upper triangular matrix S satisfying

$$S \equiv \begin{bmatrix} I^{(r)} & X_r \\ 0 & I^{(t-r)} \end{bmatrix} \pmod{n_r}, \qquad 1 \leqq r \leqq t-1.$$

This completes the proof of the lemma.

LEMMA 4. *Let S be an integral $t \times t$ matrix such that $|S| \equiv 1 \pmod{n}$. Then there exists a matrix $T \in G_t$ such that $T \equiv S \pmod{n}$.*

Proof. (Although this lemma is known, references are hard to come by, and so we insert a proof.)

Set $T=S+nY$; we need only choose Y so that $|S+nY|=1$. Let $U, V\in G_t$ be chosen so that $USV=D$ is diagonal, and set $X=UYV$. Then

$$|S+nY| = |D+nX|,$$

so it suffices to show that we can find X such that $|D+nX|=1$, where D is diagonal and $|D|\equiv 1 \pmod{n}$.

Let $D=\operatorname{diag}(d_1, \cdots, d_t)$, and set $|D|=1+nd$. Choose X so that

$$D+nX = \begin{bmatrix} d_1+nx & 0 & 0 & \cdots & 0 & ny \\ n & d_2 & 0 & \cdots & 0 & 0 \\ 0 & n & d_3 & \cdots & 0 & 0 \\ \cdot & \cdot & \cdot & \cdots & \cdot & \cdot \\ 0 & 0 & 0 & \cdots & n & d_t \end{bmatrix}.$$

Then

$$|D+nX| = 1+n(d+xd_2\cdots d_t \pm n^{t-1}y).$$

Since $(d_2\cdots d_t, n)=1$, we may choose integers x, y such that

$$d+xd_2\cdots d_t\pm n^{t-1}y=0,$$

which completes the proof.

LEMMA 5. *Let* $\mathfrak{m}=(m_1, \cdots, m_{t-1})$, $\mathfrak{n}=(n_1, \cdots, n_{t-1})$, *where* $(m_i, n_i)=1$ *for* $1\leqq i\leqq t-1$, $(m_in_i, m_jn_j)=1$ *for* $1\leqq i, j\leqq t-1$, $i\neq j$, *and let* $M\in G_r(\mathfrak{n})$. *Then there is an upper triangular matrix* $S\in G_t$ *and a lower triangular matrix* $W\in G_t$ *such that* $WSM\in G_r(\mathfrak{mn})$. *The entries of* W *and* S *are determined only modulo* $m_1\cdots m_{t-1}$, *and hence* $G(\mathfrak{mn})$ *is of finite index in* $G(\mathfrak{n})$.

Proof. This lemma follows readily from Lemma 3 in the same way that Lemma 2 follows from Lemma 1.

We now proceed with the proof of Theorem 3. Let $\mathfrak{m}$, $\mathfrak{n}$ be chosen as in the above lemma, and let H be a group such that

$$G_t(\mathfrak{mn})\subset H\subset G_t(\mathfrak{n}).$$

As in the proof of Theorem 1, by using induction on the total number of prime factors of $m_1m_2\cdots m_{t-1}$, we see that the theorem holds unless for every vector $\mathfrak{a}=(a_1, \cdots, a_{t-1})$ such that $a_1|m_1, \cdots, a_{t-1}|m_{t-1}$, except

$$\mathfrak{a}=(1, \cdots, 1),$$

we have

$$H\cap G_t(\mathfrak{an}) = G_t(\mathfrak{nm}). \tag{14}$$

Suppose that $H\neq G_t(\mathfrak{mn})$; then H must contain an element M such that for each r $(1\leqq r\leqq t-1)$ we have

$$M = \begin{bmatrix} A^{(r)} & B \\ n_r C & D^{(t-r)} \end{bmatrix}$$

with $C \not\equiv 0 \pmod{a_r}$ for each divisor a_r of m_r, $a_r \neq 1$.

Now choose an upper triangular matrix S and a lower triangular matrix W as in Lemma 5, such that $WSM \in G_r(\mathfrak{m}\mathfrak{n}) \subset H$. Since also $S \in H$, this shows that $W \in H$. Further, for each r we have

$$W \equiv \begin{bmatrix} I^{(r)} & 0 \\ n_r Y_r & I^{(t-r)} \end{bmatrix} \pmod{m_r}, \tag{15}$$

where $Y_r \not\equiv 0 \pmod{a_r}$ for any a_r dividing m_r, $a_r \neq 1$.

Call a lower triangular matrix in G_t *permissible* if it is an element of H. The above-constructed W is permissible. If we can show that all lower triangular matrices in $G_t(\mathfrak{n})$ are permissible, then using Lemma 5 we will deduce that $H = G_t(\mathfrak{n})$, and Theorem 3 will be established.

Define the non-negative integer k by $m_1 = \cdots = m_{k-1} = 1$, $m_k > 1$. (If $m_1 > 1$, then choose $k = 1$.) We shall show that also $m_{k+1} = \cdots = m_{t-1} = 1$. For let $m_0 = m_{k+1} \cdots m_{t-1}$; then $(m_0, m_k) = 1$.

Now we remark that the matrix Y_r was determined only modulo m_r, and hence since $(m_r, n_r) = 1$, we could have chosen the permissible matrix W so that instead of (15) we have (for each r)

$$W \equiv \begin{bmatrix} I^{(r)} & 0 \\ n_r Y_r & I^{(t-r)} \end{bmatrix} \pmod{m_r n_r}. \tag{16}$$

Then $W \in H$, so also $W^{m_0} \in H$. Now for each r $(1 \leqq r \leqq t-1)$ we have

$$W^{m_0} \equiv \begin{bmatrix} I^{(r)} & 0 \\ n_r m_0 Y_r & I^{(t-r)} \end{bmatrix} \pmod{m_r n_r},$$

whence

$$W^{m_0} \in G_t(n_1, \cdots, n_k, m_{k+1}n_{k+1}, \cdots, m_{t-1}n_{t-1}).$$

Unless $(1, \cdots, 1, m_{k+1}, \cdots, m_{t-1}) = (1, \cdots, 1)$, we deduce from (15) that $W^{m_0} \in G_t(\mathfrak{m}\mathfrak{n})$, which is impossible because $W^{m_0} \notin G_{k-1, t-k+1}(m_k n_k)$. We thus have shown that $\mathfrak{m} = (1, \cdots, 1, m_k, 1, \cdots, 1)$.

We are now supposing that

$$G_t(\mathfrak{m}\mathfrak{n}) \subset H \subset G_t(\mathfrak{n}),$$

where $\mathfrak{m} = (1, \cdots, 1, m_k, 1, \cdots, 1)$, $m_k > 1$, that (14) holds, and that $H \neq G_t(\mathfrak{m}\mathfrak{n})$. We have shown the existence of a lower triangular matrix $W \in H$ such that (16) holds, with $Y_k \not\equiv 0 \pmod{a_k}$ for any a_k dividing m_k, $a_k \neq 1$. We are trying to prove that every lower triangular matrix in $G_t(\mathfrak{n})$ is permissible (that is, lies in H), and consequently that $H = G_t(\mathfrak{n})$.

Let $U \in G_k$, $V \in G_{t-k}$ be arbitrary. By Lemma 4, there exists a matrix $R \in G_t$ such that

$$R \equiv I \quad (\text{mod } n_r), \qquad 1 \leqq r \leqq t-1,\ r \neq k,$$

$$R \equiv \begin{bmatrix} U & 0 \\ 0 & V \end{bmatrix} \quad (\text{mod } m_k n_k).$$

Then $R \in G_t(\mathfrak{mn}) \subset H$, and hence also $W_1 = RWR^{-1} \in H$. But we have

$$W_1 \equiv \begin{bmatrix} I^{(k)} & 0 \\ n_k V Y_k U^{-1} & I^{(t-k)} \end{bmatrix} \quad (\text{mod } m_k n_k),$$

and

$$W_1 \equiv \begin{bmatrix} I^{(r)} & 0 \\ n_r Y_r & I^{(t-r)} \end{bmatrix} \quad (\text{mod } n_r)$$

for $1 \leqq r \leqq t-1$, $r \neq k$. The same reasoning as in the proof of Theorem 1 then shows that all lower triangular matrices in $G_t(\mathfrak{n})$ lie in H, whence $H = G_t(\mathfrak{n})$ and Theorem 3 is proved.

5. We conclude with an examination of the symplectic modular group Γ_t of order t (see [4]). Let

$$\Gamma_t(m, n) = \left\{ \begin{bmatrix} A^{(t)} & B \\ C & D^{(t)} \end{bmatrix} \in \Gamma_t : \begin{matrix} B \equiv 0 \ (\text{mod } m), \\ C \equiv 0 \ (\text{mod } n) \end{matrix} \right\},$$

and set $\Gamma_t(n) = \Gamma_t(1, n)$. We shall prove analogues of Theorems 1 and 2. We begin with

LEMMA 6. *Let n be a fixed positive integer, and let*

$$M = \begin{bmatrix} A^{(t)} & B \\ C & D^{(t)} \end{bmatrix} \in \Gamma_t.$$

Then there exists a symmetric $t \times t$ matrix X such that $(|A + XC|, n) = 1$.

Proof. As in the proof of Lemma 1, it suffices to show for each prime p that there exists a symmetric matrix X_p for which $p \nmid |A + X_p C|$. For $U, V \in G_t$ we have

$$\begin{bmatrix} U & 0 \\ 0 & U'^{-1} \end{bmatrix} M \begin{bmatrix} V & 0 \\ 0 & V'^{-1} \end{bmatrix} = \begin{bmatrix} A_1^{(t)} & B_1 \\ C_1 & D_1^{(t)} \end{bmatrix} \in \Gamma_t,$$

with $A_1 = UAV$, $C_1 = U'^{-1}CV$. Set $Y_p = UX_pU'$; then

$$A_1 + Y_p C_1 = U(A + X_p C)V.$$

Hence we need only find a symmetric matrix Y_p such that $p \nmid |A_1 + Y_p C_1|$.

By proper choice of U, $V \in G_t$, we may assume that A_1 is diagonal. Let

$$A_1 \equiv \begin{bmatrix} E^{(k)} & 0 \\ 0 & 0 \end{bmatrix} \pmod{p},$$

where E is diagonal and nonsingular modulo p. (The case where $A \equiv 0 \pmod{p}$ is easily disposed of separately.) Setting

$$C_1 = \begin{bmatrix} C_{11}^{(k)} & C_{12} \\ C_{21} & C_{22}^{(t-k)} \end{bmatrix},$$

the symmetry of $A_1' C_1$ shows that $C_{12} \equiv 0 \pmod{p}$. Hence

$$\begin{bmatrix} A_1 \\ C_1 \end{bmatrix} \equiv \begin{bmatrix} E & 0 \\ 0 & 0 \\ C_{11} & 0 \\ C_{21} & C_{22} \end{bmatrix} \pmod{p},$$

whence $p \nmid |C_{22}|$. Then set

$$Y_p = \begin{bmatrix} 0 & 0 \\ 0 & I^{(t-k)} \end{bmatrix},$$

and obtain

$$A_1 + Y_p C_1 \equiv \begin{bmatrix} E & 0 \\ C_{21} & C_{22} \end{bmatrix} \pmod{p};$$

which shows that $p \nmid |A_1 + Y_p C_1|$. This completes the proof of the lemma.

LEMMA 7. *Let $M \in \Gamma_t(n)$, and let m be a fixed positive integer. Then there exist symmetric integral $t \times t$ matrices X, Y, whose entries are determined only modulo m, such that*

$$W(nY)S(X)M \in \Gamma_t(mn),$$

where

$$W(nY) = \begin{bmatrix} I^{(t)} & 0 \\ nY & I^{(t)} \end{bmatrix}, \quad S(X) = \begin{bmatrix} I^{(t)} & X \\ 0 & I^{(t)} \end{bmatrix}.$$

Proof. The proof follows that of Lemma 2. The only additional fact needed is that the matrix Y determined by Equation (11) can be chosen to be symmetric, since the symmetry of $A_0' C$ implies that of CA_0^{adj}.

We now have

THEOREM 4. *Let m, n be positive integers, and let H be a group such that*

$$\Gamma_t(mn) \subset H \subset \Gamma_t(n).$$

Then there exists a divisor d of m such that $H=\Gamma_t(dn)$.

Proof. This theorem follows from Lemmas 6 and 7 in the same manner that Theorem 1 follows from Lemmas 1 and 2. We omit the details.

THEOREM 5. *Let m, n be positive coprime integers, and let H be a group satisfying*

$$\Gamma_t(m, n) \subset H \subset \Gamma_t.$$

Then there exist integers m_1, n_1 *with* $m_1 | m$, $n_1 | n$, *and* $H=\Gamma_t(m_1, n_1)$.

Proof. The proof of Theorem 2 carries over to this case with minor modifications. We omit the details.

REFERENCES

1. Morris Newman, *Structure theorems for modular subgroups*, Duke Math. J. vol. 22 (1955) pp. 25–32.

2. ———, *An inclusion theorem for modular groups*, Proc. Amer. Math. Soc. vol. 8 (1957) pp. 125–127.

3. Irving Reiner and J. D. Swift, *Congruence subgroups of matrix groups*, Pacific J. Math. vol. 6 (1956) pp. 529–540.

4. L. K. Hua and Irving Reiner, *On the generators of the symplectic modular group*, Trans. Amer. Math. Soc. vol. 65 (1949) pp. 415–426.

NATIONAL BUREAU OF STANDARDS,
WASHINGTON, D. C.
UNIVERSITY OF ILLINOIS,
URBANA, ILL.

ON THE CLASS NUMBER OF REPRESENTATIONS OF AN ORDER

IRVING REINER

1. Introduction. We shall use the following notation throughout:

R = Dedekind ring **(5)**.
$\mathfrak{u}$ = multiplicative group of units in R.
h = class number of R.
K = quotient field of R.
p = prime ideal in R.
R_p = ring of p-adic integers in K.

We assume that h is finite, and that for each prime ideal p, the index $(R:p)$ is finite.

Let A be a finite-dimensional separable algebra over K, with an identity element e **(4**, p. 115**)**. Let G be an R-order in A, that is, G is a subring of A satisfying

(i) $e \in G$,
(ii) G contains a K-basis of A,
(iii) G is a finitely-generated R-module.

By a G-module we shall mean a left G-module which is a finitely-generated torsion-free R-module, on which e acts as identity operator. An A-module is defined analogously, replacing R by K. We shall assume, unless otherwise stated, that K is a splitting field for A; thus, the only possible A-endomorphisms of an irreducible A-module X are the scalar multiplications $x \to \alpha x$, $x \in X$, where $\alpha \in K$.

As in **(3)**, we may form the non-zero ideal $\mathfrak{g} \subset R$, defined as the intersection of the ideals which annihilate the one-dimensional cohomology groups $H(G, T)$, where T ranges over the set of two-sided G-modules. (In the special case where $G = R(\Pi)$ is the group ring of a finite group Π, the ideal $\mathfrak{g}$ is the principal ideal generated by the group order $(\Pi : 1)$.) Let $P = \{p_1, \ldots, p_t\}$ be the set of distinct prime divisors of $\mathfrak{g}$, and set

$$\mathfrak{g} = \prod_{p \in P} p^{\gamma(p)}. \tag{1}$$

For any G-module M, let KM be the A-module which consists of the K-linear combinations of the elements of M. If we set $A_p = R_pG$, we may likewise define the A_p-module $M_p = R_pM$. Two G-modules M and N are said to be in the same *genus* (notation: $M \vee N$) if and only if for each p, the modules

Received June 12, 1958. This research was supported in part by the Office of Naval Research.

M_p and N_p are A_p-isomorphic. As is shown in **(7)**, $M \vee N$ if and only if $KM \cong KN$ and $M_p \cong N_p$ for each $p \in P$.

For any A-module L', let $S(L')$ be the collection of G-modules L for which $KL \cong L'$. Suppose that $S(L')$ splits into r_g genera, and into r_G classes under G-isomorphism. As is shown in **(6**; **7**; and **9)**, both r_g and r_G are finite. The purpose of this paper is to consider the relation between r_g and r_G. For the special case where L' is irreducible, Maranda **(7)** has shown that $r_G = hr_g$. We shall restrict ourselves to the case where the irreducible constituents of L' are distinct from one another. If L' has k distinct irreducible constituents, we shall prove

$$r_G \geqslant h^k r_g. \tag{2}$$

Further, we shall show that equality holds provided that

(3) For each $\alpha \in R$ such that $(\alpha) + \mathfrak{g} = R$, there exists $\beta \in \mathfrak{u}$ for which $\beta \equiv \alpha \pmod{\mathfrak{g}^{k-1}}$.

Finally, we shall obtain formulas for r_g and r_G in the special case where $k = 2$. These formulas will show that if condition (3) fails, then r_G may exceed $h^2 r_g$ for this case.

2. Binding homomorphisms. In this section, we shall drop the hypothesis that K is a splitting field for the algebra A. Let L be a G-module which contains a submodule M, and assume that M is an R-direct summand of L. Define $N = L/M$ to be the factor G-module. Every element of L is then uniquely representable as an ordered pair (n, m), $n \in N$, $m \in M$, where the structure of L as R-module is given by

$$(n, m) + (n', m') = (n + n', m + m'), \qquad \alpha(n, m) = (\alpha n, \alpha m), \tag{4}$$

for $n, n' \in N$, $m, m' \in M, \alpha \in R$. Further, the action of G on L is given by

$$g(n, m) = (gn, \Lambda_g(n) + gm), \qquad g \in G, \text{ where } \Lambda_g \in \operatorname{Hom}_R(N, M). \tag{5}$$

Let $\Lambda : G \to \operatorname{Hom}_R(N, M)$ be the R-homomorphism defined by $g \to \Lambda_g$. The condition $(gh)(n, m) = g(h(n, m))$ is equivalent to

$$\Lambda_{gh}(n) = g\Lambda_h(n) + \Lambda_g(hn), \qquad g, h \in G, n \in N. \tag{6}$$

Call $\Lambda \in \operatorname{Hom}_R(G, \operatorname{Hom}_R(N, M))$ a *binding homomorphism* if **(6)** holds, and let $B(N, M)$ be the R-module consisting of all binding homomorphisms relative to N, M. The R-G-module L is then completely determined by equations (4) and (5), once an element $\Lambda \in B(N, M)$ is fixed. Let us denote this module L by $(N, M; \Lambda)$.

It is convenient to turn $\operatorname{Hom}_R(N, M)$ into a two-sided G-module T by defining

$$(gt)(n) = g(t(n)), \quad (tg)n = t(gn), \quad g \in G, n \in N, t \in \operatorname{Hom}_R(N, M).$$

We may then characterize $B(N, M)$ as the set of all $\Lambda \in \operatorname{Hom}_R(G, T)$ for which

$$\Lambda_{gh} = g\Lambda_h + \Lambda_g h, \qquad g, h \in G. \tag{7}$$

Now fix $t \in T$, and define $\Lambda \in \mathrm{Hom}_R(G, T)$ by

$$\Lambda_g = gt - tg, \qquad g \in G.$$

We find readily that $\Lambda \in B(N, M)$. Let $B'(N, M)$ be the R-module consisting of all the binding homomorphisms so obtained by letting t range over all elements of T. Define the R-module

$$C(N, M) = B(N, M)/B'(N, M).$$

From **(9)** we know that $C(N, M)$ contains only finitely many elements. Furthermore, from the definition of the ideal $\mathfrak{g}$, we have

$$\mathfrak{g} \cdot B(N, M) \subset B'(N, M)$$

for any N, M. Finally, if $[\Lambda]$ denotes the class $\Lambda + B'(N, M)$ of the element $\Lambda \in B(N, M)$, then we have:

$$[\Lambda] = [\Lambda'] \Rightarrow (N, M; \Lambda) \cong (N, M; \Lambda').$$

In fact, if $t \in T$ is such that $\Lambda_g' - \Lambda_g = gt - tg$, $g \in G$, then the map $(n, m) \to (n, m - tn)$ gives the desired isomorphism.

In the above discussion, replace R by R_p. If L^* is an A_p-module which contains a submodule M^* as R_p-direct summand, then $L^* = (N^*, M^*; \Lambda^*)$, where $N^* = L^*/M^*$, and where

$$\Lambda^* : A_p \to \mathrm{Hom}_{R_p}(N^*, M^*).$$

is an R_p-homomorphism satisfying $\Lambda^*_{xy} = x\Lambda^*_y + \Lambda^*_x y$, $x, y \in A_p$. Define $B(N^*, M^*)$, $B'(N^*, M^*)$ and $C(N^*, M^*)$ as above. For $\Lambda^* \in B(N^*, M^*)$, again let $[\Lambda^*] = \Lambda^* + B'(N^*, M^*)$. If $\gamma(p)$ is defined as in (1), we have

$$(8) \qquad \pi^{\gamma(p)} B(N^*, M^*) \subset B'(N^*, M^*)$$

where π is an element of p such that $\pi \notin p^2$.

Now let N, M be G-modules, and let N_p, M_p be the corresponding A_p-modules. There is a natural isomorphism of $B(N, M)$ into $B(N_p, M_p)$ which may be described as follows: for each $\Lambda \in B(N, M)$ and each $g \in G$, the map $\Lambda_g \in \mathrm{Hom}_R(N, M)$ may be extended in a unique manner to an element of $\mathrm{Hom}_{R_p}(N_p, M_p)$; we may then define Λ_x for each $x \in A_p$ by linearity. In this way, Λ is extended in a unique manner to an element $\Lambda^p \in B(N_p, M_p)$. The map $\Lambda \to \Lambda^p$ carries $B'(N, M)$ into $B'(N_p, M_p)$, and so induces an R-homomorphism of $C(N, M)$ into $C(N_p, M_p)$.

We may now define an R-homomorphism

$$\phi: \quad C(N, M) \to \sum_{p \in P} C(N_p, M_p)$$

by means of

$$\phi[\Lambda] = ([\Lambda^{p_1}], \ldots, [\Lambda^{p_t}]).$$

From **(8)**, we know that ϕ has kernel 0. We shall in fact show that ϕ is an isomorphism "onto."

THEOREM 1.

$$C(N, M) \cong \sum_{p \in P} C(N_p, M_p).$$

Remark. A slightly different version of this was first proved by deLeeuw **(1)**. We shall not use the results of **(8)**, but instead shall give a self-contained proof of the theorem.

Proof. We show firstly that the ϕ is an "onto" mapping. For each $p \in P$ suppose an element $\Omega^p \in B(N_p, M_p)$ chosen. We must prove the existence of an element $\Lambda \in B(N, M)$ such that $[\Lambda^p] = [\Omega^p], p \in P$. Let $T = \mathrm{Hom}_R(N, M)$, and let us set

$$T_p = \mathrm{Hom}_{R_p}(N_p, M_p) = R_p \,\mathrm{Hom}_R\,(N, M) = R_pT$$

for each prime ideal p.

For each $p \in P$, we may choose an element $\pi \in p$ such that $\pi \notin p^2$, and such that π does not lie in any other prime ideal in the set P. Set

$$a = \prod_{p \in P} \pi^{\gamma(p)}\,;$$

then $a \in R$, and for each $p \in P$ we may write

$$a = \pi^{\gamma(p)} d_p, \qquad d_p \in R, \qquad d_p = \text{unit in } R_p.$$

Define the integral ideal $\mathfrak{b}$ by

$$(a) = \mathfrak{b} \cdot \prod_{p \in P} p^{\gamma(p)}.$$

Then $\mathfrak{b}$ is not a multiple of any of the prime ideals in P.

We now make use of equation (8) to deduce that for each $p \in P$, there exists an element $u^p \in T_p$ such that

$$a \cdot \Omega_g^p = gu^p - u^p g, \qquad g \in G.$$

On the other hand, T is a finitely-generated R-module, so there exist elements $t_1, \ldots, t_r \in T$ such that

$$T = Rt_1 + \ldots + Rt_r,$$

whence

$$T_p = R_p t_1 + \ldots + R_p t_r.$$

We may therefore write (for $p \in P$)

$$u^p = \sum_{i=1}^{r} \beta_i^p t_i, \qquad \beta_i^p \in R_p.$$

Let us now choose $\alpha_1, \ldots, \alpha_r \in R$ such that

$$\alpha_i \equiv \beta_i^p \,(\mathrm{mod}\ \pi^{2\gamma(p)} R_p), \qquad p \in P, \qquad \alpha_i \equiv 0 \,(\mathrm{mod}\ \mathfrak{b}).$$

Set

$$t = a^{-1} \sum_{i=1}^{r} \alpha_i t_i \in KT,$$

and define $\Lambda \in \operatorname{Hom}_R(G, KT)$ by

$$\Lambda_g = gt - tg, \qquad g \in G.$$

We shall show that this is the desired Λ, that is, $\Lambda \in B(N, M)$, and $[\Lambda^p] = [\Omega^p]$ for $p \in P$. For $p \in P$ we have

$$a(\Omega_g^p - \Lambda_g) = gv^p - v^p g, \qquad g \in G,$$

where

$$v^p = u^p - at = \sum_{i=1}^{r} (\beta_i^p - \alpha_i)t_i.$$

From the way in which the α_i were chosen, we may therefore write

$$v^p = \pi^{2\gamma(p)} d_p w^p,$$

where $w^p \in T_p$, and thus

$$\Omega_g^p - \Lambda_g = \pi^{\gamma(p)}(gw^p - w^p g), g \in G.$$

This proves that for each $p \in P$,

$$\Omega^p - \Lambda^p \in \pi^{\gamma(p)} B(N_p, M_p) \subset B'(N_p, M_p),$$

and shows incidentally that

$$\text{(9)} \qquad \Lambda_g \in T_p, \qquad g \in G, \qquad p \in P.$$

On the other hand, we note that for each prime ideal $q \notin P$, the elements $a^{-1}\alpha_1, \ldots, a^{-1}\alpha_r$ all lie in R_q, and hence

$$\Lambda_g \in T_q, \qquad g \in G.$$

Coupled with (9), this implies that

$$\Lambda_g \in \bigcap_{q'} T_{q'}, \qquad g \in G,$$

where q' ranges over all prime ideals. The above intersection is precisely T, and so $\Lambda \in \operatorname{Hom}_R(G, T)$. That (7) holds follows at once from the definition of Λ; consequently, $\Lambda \in B(N, M)$. This completes the proof that ϕ is "onto."

In order to show that ϕ is an isomorphism, let $\Omega \in B(N, M)$ be such that $\Omega^p \in B'(N_p, M_p)$ for all $p \in P$; we must show that $\Omega \in B'(N, M)$. Since $\Omega^p \in B'(N_p, M_p)$, there exists for each $p \in P$ an element $u^p \in T_p$ such that

$$\Omega_g^p = gu^p - u^p g, \qquad g \in G.$$

By the preceding construction (with $a = 1$), there exists $\Lambda \in B'(N, M)$ (since now $t \in T$) such that

$$\Lambda_g^p \equiv \Omega_g^p \ (\text{mod } \pi^{\gamma(p)} T_p), \qquad g \in G.$$

Therefore

$$\Lambda - \Omega \in \mathfrak{g}B(N, M) \subset B'(N, M),$$

which shows that $\Omega \in B'(N, M)$.

COROLLARY. *If N, N^*, M, M^* are G-modules such that $N \vee N^*$ and $M \vee M^*$, then $C(N, M) \cong C(N^*, M^*)$ as R-modules.*

More generally, let

$$L_1 \supset L_2 \supset \ldots \supset L_k \supset (0)$$

be a set of G-modules such that each is an R-direct summand of its predecessor. Define $N_i = L_i/L_{i+1}$ to be the factor G-module. Then as above, every element of L_1 is uniquely representable as an ordered k-tuple $(n_1, \ldots, n_k)$ $n_i \in N_i$, where

$$(n_1, \ldots, n_k) + (n_1', \ldots, n_k') = (n_1 + n_1', \ldots, n_k + n_k'),$$
$$\alpha(n_1, \ldots, n_k) = (\alpha n_1, \ldots, \alpha n_k)$$

for $n_i, n_i' \in N_i$, $\alpha \in R$. The action of G on L_1 is given by

$$g(n_1, \ldots, n_k) = (gn_1, gn_2 + \Lambda_g^{12} n_1, \ldots, gn_k + \Lambda_g^{1k} n_1 + \ldots + \Lambda_g^{k-1,k} n_{k-1}),$$

where each $\Lambda_g^{ij} \in \mathrm{Hom}_R(N_i, N_j)$, and where the R-homomorphisms Λ^{ij} : $g \rightarrow \Lambda_g^{ij}$ satisfy conditions analogous to (7). Let $B(N_1, \ldots, N_k)$ denote the set of systems $\{\Lambda^{ij}\}$ satisfying these conditions. We denote the module L_1 by the symbol $(N_1, \ldots, N_k; \{\Lambda^{ij}\})$.

3. Isomorphisms of modules. Throughout this section, we fix an A-module L' with a composition series.

$$L = L_1' \supset L_2' \supset \ldots \supset L_k' \supset (0),$$

and let $N_i' = L_i'/L_{i+1}'$. We assume here that N_i' and N_j' are not isomorphic for $i \neq j$, and further that K is a splitting field for A. For any $L \in S(L')$, the A-module KL will have a composition series

$$KL = L_1'' \supset L_2'' \supset \ldots \supset L_k'' \supset (0)$$

in which $L_i''/L_{i+1}'' \cong N_i'$. Setting $L_i = L_i'' \cap L$, we see that L_i is a G-submodule of L for which $KL_i = L_i''$. Furthermore, L_{i+1} is a pure R-submodule of L_i, and therefore (by **5**) is an R-direct summand of L_i. Put $N_i = L_i/L_{i+1}$; then $KN_i \cong N_i'$, and

$$L = (N_1, \ldots, N_k; \{\Lambda^{ij}\})$$

for some $\{\Lambda^{ij}\} \in B(N_1, \ldots, N_k)$.

LEMMA 1. *Let $M_i, N_i, \in S(N_i')$, $1 \leqslant i \leqslant k$, and suppose that*

$$(M_1, \ldots, M_k; \{\Lambda^{ij}\}) \cong (N_1, \ldots, N_k; \{\Omega^{ij}\}).$$

Then $M_i \cong N_i$, $1 \leqslant i \leqslant k$.

Proof. (A modified version of this is given in **(2)**.) It suffices to prove that if $(N, M; \Lambda) \cong (\bar{N}, \bar{M}; \bar{\Lambda})$, where $KN \cong K\bar{N}$ and $KM \cong K\bar{M}$, and where KN and KM have no common irreducible constituent, then $M \cong \bar{M}$ and $N \cong \bar{N}$. Once this is established, a simple induction argument completes the proof.

Suppose that $\theta: (N, M; \Lambda) \cong (\bar{N}, \bar{M}; \bar{\Lambda})$ is given by

$$\theta(n, m) = \theta(n, 0) + \theta(0, m) = (\theta_1(n), \nu(n)) + (\mu(m), \theta_2(m)),$$

where

$$\theta_1 \in \mathrm{Hom}_R(N, \bar{N}),\ \nu \in \mathrm{Hom}_R(N, \bar{M}),\ \mu \in \mathrm{Hom}_R(M, \bar{N}),\ \theta_2 \in \mathrm{Hom}_R(M, \bar{M}).$$

From $\theta g(n, m) = g\theta(n, m)$ we obtain at once

$$(10.1, 10.2) \qquad \theta_1 g + \mu\Lambda_g = g\theta_1, \qquad \mu g = g\mu,$$

$$(10.3, 10.4) \qquad \bar{\Lambda}_g\theta_1 + g\nu = \nu g + \theta_2\Lambda_g, \qquad \theta_2 g = \bar{\Lambda}_g\mu + g\theta_2.$$

From (10.2) we have $\mu \in \mathrm{Hom}_G(M, \bar{N})$, and hence $\mu = 0$, since by hypothesis KM and $K\bar{N}$ have no common irreducible constituents. Equations (10.1) and (10.4) then imply that $\theta_1 \in \mathrm{Hom}_G(N, \bar{N})$ and $\theta_2 \in \mathrm{Hom}_G(M, \bar{M})$. Since θ is an isomorphism of $(N, M; \Lambda)$ onto $(\bar{N}, \bar{M}; \bar{\Lambda})$, we find readily that $\theta_1 : N \cong \bar{N}$ and $\theta_2 : M \cong \bar{M}$.

LEMMA 2. *Let $(N_1, \ldots, N_k; \{\Lambda^{ij}\})$ and $(N_1, \ldots, N_k; \{\Omega^{ij}\})$ be G-isomorphic modules in $S(L')$, where $N_i \in S(N_i')$. Then there exist units $\beta_1, \ldots, \beta_k \in \mathfrak{u}$, and homomorphisms $t_{ij} \in \mathrm{Hom}_R(N_i, N_j)$, such that the isomorphism between these G-modules is given by*

$$(n_1, \ldots, n_k) \to (\beta_1 n_1, \beta_2 n_2 + t_{12}n_1, \ldots, \beta_k n_k + t_{1k}n_1 + \ldots + t_{k-1,k}n_{k-1}).$$

Proof. From the proof of the preceding lemma, we find that the isomorphism must be given by

$$(n_1, \ldots, n_k) \to (\theta_1 n_1, \theta_2 n_2 + t_{12}n_1, \ldots, \theta_k n_k + t_{1k}n_1 + \ldots + t_{k-1,k}n_{k-1}),$$

with each $\theta_i: N_i \cong N_i$ and each $t_{ij} \in \mathrm{Hom}_R(N_i, N_j)$. Since KN_i is an absolutely irreducible A-module, θ_i must be given by scalar multiplication by a unit of R. This completes the proof.

If U, V are R-modules, and $f_1, f_2 \in \mathrm{Hom}_R(U, V)$, we shall often abbreviate the congruence $f_1 \equiv f_2 \pmod{\mathfrak{g}^a \mathrm{Hom}_R(U, V)}$ as $f_1 \equiv f_2 \pmod{\mathfrak{g}^a}$. A similar notation will be used for R_p-modules.

LEMMA 3. *Let $M_1, \ldots, M_k$ be G-modules, not necessarily irreducible, and let*

$$L = (M_1, \ldots, M_k; \{\Lambda^{ij}\}), \qquad \bar{L} = (M_1, \ldots, M_k; \{\Omega^{ij}\})$$

be G-modules for which

$$\Lambda^{ij} \equiv \Omega^{ij} \pmod{\mathfrak{g}^n}, \qquad 1 \leqslant i < j \leqslant k,$$

where n is a fixed integer $\geqslant k-1$. Then there exists a G-isomorphism $\theta: L \cong \bar{L}$ such that $\theta \equiv I \pmod{\mathfrak{g}^{n-k+1}}$, where $I: L \cong \bar{L}$ is the R-isomorphism given by $(m_1, \ldots, m_k) \rightarrow (m_1, \ldots, m_k)$.

Proof. The result is trivial for $k = 1$; let $k > 1$, and assume the result holds at $k-1$. Let us set

$$\Delta = (M_2, \ldots, M_k; \Lambda^{23}, \ldots, \Lambda^{k-1,k}),$$
$$\bar{\Delta} = (M_2, \ldots, M_k; \Omega^{23}, \ldots, \Omega^{k-1,k}).$$

From the induction hypothesis we deduce the existence of a G-isomorphism $\theta_1: \Delta \cong \bar{\Delta}$ such that

$$\theta_1 \equiv I \pmod{\mathfrak{g}^{n-k+2}}.$$

The map $(m_1, \delta) \rightarrow (m_1, \theta_1\delta)$, where $m_1 \in M_1$, $\delta \in \Delta$, then gives a G-isomorphism

$$\theta_2 : (M_1, \Delta; \Lambda^{12}, \ldots, \Lambda^{1k}) \cong (M_1, \bar{\Delta}; \overline{\Lambda}^{12}, \ldots, \overline{\Lambda}^{1k})$$

for some $(\overline{\Lambda}^{12}, \ldots, \overline{\Lambda}^{1k}) \in B(M_1, \bar{\Delta})$, and we have

$$\theta_2 \equiv I \pmod{\mathfrak{g}^{n-k+2}}.$$

Now set

$$\overline{\Lambda} = (\overline{\Lambda}^{12}, \ldots, \overline{\Lambda}^{1k}), \qquad \Omega = (\Omega^{12}, \ldots, \Omega^{1k}).$$

Then we see that both $\overline{\Lambda}$ and Ω are elements of $B(M_1, \bar{\Delta})$, and that $\overline{\Lambda} \equiv \Omega \pmod{\mathfrak{g}^{n-k+2}}$. By considering this congruence for the powers of the prime ideals dividing $\mathfrak{g}$, the method of proof of Theorem 1 shows the existence of an element $W \in \mathrm{Hom}_R(M_1, \bar{\Delta})$ such that

$$(\overline{\Lambda} - \Omega)_g = gW - Wg, \qquad g \in G,$$

and where, furthermore, $W \equiv 0 \pmod{\mathfrak{g}^{n-k+1}}$. The map $(m_1, \bar{\delta}) \rightarrow (m_1, \bar{\delta} - Wm_1)$ then yields a G-isomorphism $\theta_3 : (M_1, \bar{\Delta}; \Omega) \cong (M_1, \bar{\Delta}; \Lambda)$, where

$$\theta_3 \equiv I \pmod{\mathfrak{g}^{n-k+1}}.$$

Therefore

$$\theta_3^{-1}\theta_2 : (M_1, \Delta; \Lambda^{12}, \ldots, \Lambda^{1k}) \rightarrow (M_1, \bar{\Delta}; \Omega^{12}, \ldots, \Omega^{1k})$$

is a G-isomorphism of L onto $\bar{L}$ such that

$$\theta_3^{-1}\theta_2 \equiv I \pmod{\mathfrak{g}^{n-k+1}}.$$

4. Integral classes and genera for modules with two distinct constituents. Throughout this section, we suppose that L' is an A-module with two distinct irreducible constituents N' and M'; we assume again that K is a splitting field for A. Let $S(L')$ be partitioned into r_G classes under G-isomorphism, and into r_g genera. We shall obtain formulas for r_G and r_g.

LEMMA 4. *Let* $N \in S(N')$, $M \in S(M')$. *Then* $(N, M; \Lambda) \cong (N, M; \bar{\Lambda})$ *if and only if there exists* $\beta \in \mathfrak{u}$ *such that* $[\bar{\Lambda}] = \beta[\Lambda]$.

Proof. From Lemma 2 we deduce the existence of units $\beta_1, \beta_2 \in \mathfrak{u}$, and of $t \in \mathrm{Hom}_R(N, M)$, such that the isomorphism $(N, M; \Lambda) \cong (N, M; \bar{\Lambda})$ is given by $(n, m) \to (\beta_1 n, \beta_2 m + tn)$. This implies

$$\bar{\Lambda}_g = \beta_1^{-1}\beta_2\Lambda_g + g(\beta_1^{-1}t) - (\beta_1^{-1}t)g, \qquad g \in G.$$

Setting $\beta = \beta_1^{-1}\beta_2$, we have $[\bar{\Lambda}] = \beta[\Lambda]$. Conversely, starting from such a relation, we may reverse the steps to obtain an isomorphism of the modules.

LEMMA 5. *Let* $N \in S(N')$, $M \in S(M')$. *Then* $(N, M; \Lambda) \vee (N, M; \bar{\Lambda})$ *if and only if there exists an element* $\alpha \in R$ *such that* $(\alpha) + \mathfrak{g} = R$ *and* $[\bar{\Lambda}] = \alpha[\Lambda]$.

Proof. Let $(N, M; \Lambda) \vee (N, M; \bar{\Lambda})$. As in the preceding proof, we deduce that for each $p \in P$, there exists an element α_p which is a unit in R_p such that the classes $[\Lambda^p]$ and $[\bar{\Lambda}^p]$ in $C(N_p, M_p)$ are related by

$$[\bar{\Lambda}^p] = \alpha_p[\Lambda^p].$$

Choose $\alpha \in R$ such that $\alpha \equiv \alpha_p \pmod{p^{\gamma(p)}}$ for each $p \in P$; then $(\alpha) + \mathfrak{g} = R$. Furthermore, $(\alpha - \alpha_p)B(N_p, M_p) \subset B'(N_p, M_p)$, so that

$$\alpha[\Lambda^p] = \alpha_p[\Lambda^p], \qquad p \in P.$$

Therefore $[\bar{\Lambda}^p] = [\alpha\Lambda^p]$ for all $p \in P$, and so by Theorem 1 we have $[\bar{\Lambda}] = [\alpha\Lambda] = \alpha[\Lambda]$.

Suppose now that $S(N')$ splits into ν genera; according to **(7)**, each genus splits into h classes under G-isomorphism. Let us choose representatives of the $h\nu$ classes, say $\{N_j^i : 1 \leqslant i \leqslant \nu, 1 \leqslant j \leqslant h\}$, so that all the modules with the same subscript lie in the same genus. Likewise choose representatives $\{M_j^i : 1 \leqslant i \leqslant \mu, 1 \leqslant j \leqslant h\}$ of the $h\mu$ classes into which $S(M')$ splits. Let $(N, M; \Gamma) \in S(L')$, and suppose $N \vee N_1^i$, $M \vee M_1^j$. Then for each $p \in P$, there exists an element

$$\Omega^p \in B((N_1^i)_p, (M_1^j)_p)$$

such that

$$(N_p, M_p; \Gamma^p) \cong ((N_1^i)_p, (M_1^j)_p; \Omega^p)$$

as A_p-modules. By Theorem 1, there exists $\Lambda \in B(N_1^i, M_1^j)$ such that $[\Lambda^p] = [\Omega^p]$ for all $p \in P$. Therefore

$$(N, M; \Gamma)p \cong (N_1^i, M_1^j; \Lambda)_p, \qquad p \in P,$$

and so

$$(N, M; \Gamma) \vee (N_1^i, M_1^j; \Lambda).$$

Hence, every module in $S(L')$ is in the same genus as $(N_1{}^i, M_1{}^j; \Lambda)$ for some choice of i and j and some $\Lambda \in B(N_1{}^i, M_1{}^j)$. Further,

$$(N_1^i, M_1^j; \Lambda) \vee (N_1^{i'}, M_1^{j'}; \Lambda')$$

implies, by the method of proof of Lemma 1, that $i = i'$ and $j = j'$. Let us set

$$H_{ij} = \{(N_1^i, M_1^j; \Lambda) : \Lambda \in B(N_1^i, M_1^j)\}, 1 \leqslant i \leqslant \nu, 1 \leqslant j \leqslant \mu, \tag{11}$$

and suppose that H_{ij} splits into r_{ij} genera. Then we have at once

$$r_g = \sum_{i,j} r_{ij}. \tag{12}$$

On the other hand, any module in $S(L')$ is G-isomorphic to $(N_\rho{}^i, M_\sigma{}^j; \Lambda)$ for some i, j, ρ, σ and Λ. Further, by Lemma 1, two such modules cannot be isomorphic unless they have the same set of indices i, j, ρ, σ. Let us set

$$S(i, \rho; j, \sigma) = \{(N_\rho^i, M_\sigma^j; \Lambda) : \Lambda \in B(N_\rho^i, M_\sigma^j)\},$$

and suppose that $S(i, \rho; j, \sigma)$ splits into $s(i, \rho; j, \sigma)$ classes. Then

$$r_G = \sum_{i,j,\rho,\sigma} s(i, \rho; j, \sigma).$$

However, Lemma 4 states that $(N_\rho{}^i, M_\sigma{}^j; \Lambda) \cong (N_\rho{}^i, M_\sigma{}^j; \bar{\Lambda})$ if and only if there exists $\beta \in \mathfrak{u}$ such that $[\bar{\Lambda}] = \beta[\Lambda]$. Furthermore, the Corollary to Theorem 1 shows that $C(N_\rho{}^i, M_\sigma{}^j)$ is (as R-module) independent of ρ and σ. Therefore $s(i, \rho; j, \sigma) = s(i, 1; j, 1)$ for all ρ and σ, and we have

$$r_G = h^2 \sum_{i,j} s_{ij}, \tag{13}$$

where $s_{ij} = s(i, 1; j, 1)$ is the number of classes into which H_{ij} splits.

Before proceeding with the calculation of r_{ij} and s_{ij}, it will be convenient to introduce some notations. For a non-zero ideal $\mathfrak{a}$ in R, let $\phi(\mathfrak{a})$ denote the number of residue classes in $R/\mathfrak{a}$ which are relatively prime to $\mathfrak{a}$. If $\mathfrak{a} + \mathfrak{b} = R$, then $\phi(\mathfrak{a}\mathfrak{b}) = \phi(\mathfrak{a})\phi(\mathfrak{b})$. Next, let $u(\mathfrak{a})$ denote the number of distinct residue classes in $(\mathfrak{u} + \mathfrak{a})/\mathfrak{a}$; of course, $u(\mathfrak{a})$ is a divisor of $\phi(\mathfrak{a})$. However, $u(\mathfrak{a})$ is not a multiplicative function of $\mathfrak{a}$, as is seen from the example where K is the rational field.

LEMMA 6. *Let $N \in S(N')$, $M \in S(M')$, and $H = \{(N, M; \Lambda) : \Lambda \in B(N,M)\}$. Suppose H splits into r genera and s classes. Let $d(\mathfrak{a})$ be the number of elements in $C(N, M)$ with order ideal a. Then*

$$r = \sum_{\mathfrak{a}} d(\mathfrak{a})/\phi(\mathfrak{a}), \qquad s = \sum_{\mathfrak{a}} d(\mathfrak{a})/u(\mathfrak{a}),$$

both sums extending over all divisors of $\mathfrak{g}$.

(The order ideal of an element $c \in C(N, M)$ is $\{\alpha \in R : \alpha c = 0\}$.)

Proof. Let us use the symbol $(N, M; c)$ to denote the collection of mutually isomorphic modules $\{(N, M; \Lambda): \Lambda \in c\}$, where $c \in C(N, M)$. By Lemma 4, $(N, M; c)$ and $(N, M; c')$ cannot lie in the same genus unless c and c' have the same order ideal. Consider the set of $d(\mathfrak{a})$ elements of $C(N, M)$ with given order ideal $\mathfrak{a}$. For a fixed c in this set, all those c' of the form αc, where $\alpha \in R$ is such that $(\alpha) + \mathfrak{g} = R$, will yield modules in the same genus as those obtained from c. But as α ranges over all elements of R for which $(\alpha) + \mathfrak{g} = R$, αc gives exactly $\phi(\mathfrak{a})$ distinct elements of $C(N, M)$. Therefore

$$r = \sum_{\mathfrak{a}} d(\mathfrak{a})/\phi(\mathfrak{a}).$$

A similar argument gives the formula for s.

Let $d_p(p^n)$ denote the number of elements in $C(N_p, M_p)$ having order ideal p^n. Then

$$d_p(p^n) = \tau(p^n) - \tau(p^{n-1}),$$

where $\tau(p^n)$ denotes the number of elements of $C(N_p, M_p)$ which are annihilated by p^n. From Theorem 1,

$$d(\mathfrak{a}) = \prod_{p \in P} d_p(p^{a(p)}), \text{ where } \mathfrak{a} = \prod_{p \in P} p^{a(p)}.$$

We may therefore write

$$r = \prod_{p \in P} \left\{ \sum_{a=0}^{\gamma(p)} d_p(p^a)/\phi(p^a) \right\},$$

which confirms the result in (**7**) that the number of genera is the product over all $p \in P$ of the number of classes into which $S(L')$ splits under A_p-isomorphism. The corresponding multiplicative formula for s fails to hold, because $u(\mathfrak{a})$ is not multiplicative.

Applying Lemma 6 to our original problem, we may summarize our result as follows.

THEOREM 2. *Let $N^1, \ldots, N^\nu$ be representatives of the genera into which $S(N')$ splits, and $M^1, \ldots, M^\mu$ representatives of the genera of $S(M')$. For each divisor $\mathfrak{a}$ of $\mathfrak{g}$, let $d_{ij}(\mathfrak{a})$ denote the number of elements in $C(N^i, M^j)$ having order ideal $\mathfrak{a}$. Then $S(L')$ splits into r_g genera and r_G classes, where*

$$r_g = \sum_{\mathfrak{a}} \sum_{i,j} d_{ij}(\mathfrak{a})/\phi(\mathfrak{a}), \quad r_G = h^2 \sum_{\mathfrak{a}} \sum_{i,j} d_{ij}(\mathfrak{a})/u(\mathfrak{a}).$$

Here, $\phi(\mathfrak{a})$ is the number of residue classes in $R/\mathfrak{a}$ which are relatively prime to $\mathfrak{a}$, and $u(\mathfrak{a})$ is the number of distinct elements of $(\mathfrak{u} + \mathfrak{a})/\mathfrak{a}$.

COROLLARY. *We have $r_G \geqslant h^2 r_g$, with equality provided that $\phi(\mathfrak{g}) = u(\mathfrak{g})$. Furthermore, if any $C(N^i, M^j)$ contains an element of order ideal $\mathfrak{a}$, where $u(\mathfrak{a}) < \phi(\mathfrak{a})$, then $r_G > h^2 r_g$.*

5. Integral classes and genera in the general case. Now let L' be an A-module with k distinct irreducible constituents, and let K be a splitting field for A. We preserve the notation introduced at the beginning of § 3. In this section we shall generalize the results given in the Corollary to Theorem 2.

For each κ $(1 \leqslant \kappa \leqslant k)$, let $\{X_\kappa^{ij}: 1 \leqslant i \leqslant \nu(\kappa), 1 \leqslant j \leqslant h\}$ be a full set of representatives of the $h\nu(\kappa)$ classes into which the set $S(X_\kappa')$ splits; suppose these representative modules are so chosen that modules with the same indices i and κ lie in the same genus. Then every module in $S(L')$ is of the form

$$(X_1^{i_1 j_1}, \ldots, X_k^{i_k j_k}; \{\Lambda^{ij}\}).$$

Let $S(i_1, j_1; \ldots; i_k, j_k)$ be the set of all such modules obtained by letting $\{\Lambda^{ij}\}$ range over all systems in

$$B(X_1^{i_1 j_1}, \ldots, X_k^{i_k j_k}),$$

and let this set split into $r(i_1, j_1; \ldots; i_k, j_k)$ genera and $s(i_1, j_1; \ldots; i_k, j_k)$ classes. From the Corollary to Theorem 1, we see that $r(i_1, j_1; \ldots; i_k, j_k)$ is independent of $(j_1, \ldots, j_k)$, and therefore

$$r_g = h^{-k} \sum r(i_1, j_1, ; \ldots ; i_k, j_k), \quad r_G = \sum s(i_1, j_1,; \ldots ; i_k, j_k),$$

both summations extending over all possible values of the i's and j's. This implies the result that

$$r_G \geqslant h^k r_g.$$

Finally, we prove:

THEOREM 3. *If* $u(\mathfrak{g}^{k-1}) = \phi(\mathfrak{g}^{k-1})$, *then* $r_G = h^k r_g$.

Proof. We remark that the hypothesis of the Theorem is simply a restatement of condition (3) given in the introduction. To prove the theorem, we need only show that $r(i_1, j_1; \ldots; i_k, j_k) = s(i_1, j_1; \ldots; i_k, j_k)$. We simplify the notation by letting $M_\kappa \in S(X_\kappa')$, $1 \leqslant \kappa \leqslant k$. We shall prove that if

$$L = (M_1, \ldots, M_k; \{\Lambda^{ij}\}), \qquad \bar{L} = (M_1, \ldots, M_k; \{\bar{\Lambda}^{ij}\})$$

are such that $L \vee \bar{L}$, then also $L \cong \bar{L}$.

Since $L_p \cong \bar{L}_p$ for each $p \in P$, Lemma 2 shows the existence of units $\beta_1{}^p, \ldots, \beta_k{}^p$ in R_p, and homomorphisms

$$t_{ij}^p \in \operatorname{Hom}_p((M_i)_p, (M_j)_p)$$

such that the isomorphism $L_p \cong \bar{L}_p$ is given by

$$(m_1, \ldots, m_k) \to (\beta_1^p m_1, \beta_2^p m_2 + t_{12}^p m_1, \ldots, \beta_k^p m_k + t_{1k}^p m_1 + \ldots + t_{k-1,k}^p m_{k-1}).$$

By the hypothesis of the theorem, we may choose units $\beta_1, \ldots, \beta_k \in \mathfrak{u}$ such that

$$\beta_\kappa \equiv \beta_\kappa^p \qquad (\text{mod } p^{(k-1)\gamma(p)}), \qquad p \in P, \qquad 1 \leqslant \kappa \leqslant k.$$

As in the proof of Theorem 1, we may choose homomorphisms $w_{ij} \in \mathrm{Hom}_R(M_i, M_j)$ such that

$$w_{ij}^p \equiv t_{ij}^p \quad \bmod p^{(k-1)\gamma(p)}, \quad 1 \leqslant i < j \leqslant k, \quad p \in P.$$

Then the map

$$(m_1, \ldots, m_k) \to (\beta_1 m_1, \beta_2 m_2 + w_{12} m_1, \ldots, \beta_k m_k + w_{1k} m_1 + \ldots + w_{k-1,k} m_{k-1})$$

gives a G-isomorphism of L onto a module L^* where $L^* = (M_1, \ldots, M_k; \{\Omega^{ij}\})$ and $\Omega^{ij} \equiv \bar{\Lambda}^{ij} \pmod{\mathfrak{g}^{k-1}}$ for $1 \leqslant i < j \leqslant k$. By Lemma 3 we then have $L^* \cong \bar{L}$, which completes the proof of the theorem.

It would be of interest to obtain formulas for r_G and r_g which generalize those given in Theorem 2.

References

1. K. deLeeuw, *Some applications of cohomology to algebraic number theory and group representations*, unpublished.
2. F. E. Diederichsen, *Ueber die Ausreduktion ganzzahliger Gruppendarstellungen bei arithmetischer Äquivalenz*, Hamb. Abh., *14* (1938), 357–412.
3. D. G. Higman, *On orders in separable algebras*, Can. J. Math., *7* (1955), 509–515.
4. N. Jacobson, *The theory of rings* (New York, 1943).
5. I. Kaplansky, *Modules over Dedekind rings and valuation rings*, Trans. Amer. Math. Soc., *7* (1952), 327–40.
6. J.-M. Maranda, *On p-adic integral representations of finite groups*, Can. J. Math., *5* (1953), 344–355.
7. ——— *On the equivalence of representations of finite groups by groups of automorphisms of modules over Dedekind rings*, Can. J. Math., *7* (1955), 516–526.
8. I. Reiner, *Maschke modules over Dedekind rings*, Can. J. Math., *8* (1956), 329–334.
9. H. Zassenhaus, *Neuer Beweis der Endlichkeit der Klassenzahl bei unimodularer Äquivalenz endlicher ganzzahliger Substitutionsgruppen*, Hamb. Abh., *12* (1938), 276–288.

University of Illinois

BEHAVIOR OF INTEGRAL GROUP REPRESENTATIONS UNDER GROUND RING EXTENSION

BY

IRVING REINER[1]

1. Let K be an algebraic number field, and let R be a subring of K containing 1 and having quotient field K. Of primary interest will be the cases

(i) $R = K$,

(ii) $R =$ alg. int. $\{K\}$, the ring of all algebraic integers in K.

(iii) $R =$ valuation ring of a discrete valuation of K.

Given a finite group G, we denote by RG its group ring over R. By an *RG-module* we shall mean a left RG-module which as R-module is finitely generated and torsion-free, and upon which the identity element of G acts as identity operator. Each RG-module M is contained in a uniquely determined smallest KG-module

$$K \otimes_R M,$$

hereafter denoted by KM. For a pair M, N of RG-modules, we write

$$M \sim_R N$$

to denote the fact that $M \cong N$ as RG-modules. The notation

$$M \sim_K N$$

shall mean that $KM \cong KN$ as KG-modules.

Now let K' be an algebraic number field containing K, and let R' be a subring of K' which contains 1 and has quotient field K'. Suppose further that R' is a finitely generated R-module such that

$$R' \cap K = R.$$

Each RG-module M then determines an $R'G$-module denoted by $R'M$, given by

$$R'M = R' \otimes_R M.$$

If M, N are a pair of RG-modules, we write $M \sim_{R'} N$ if $R'M \cong R'N$ as $R'G$-modules. Surely

$$M \sim_R N \Rightarrow M \sim_{R'} N.$$

The reverse implication is false, as we shall see. We propose to investigate more closely the connection between R- and R'-equivalence.

As a first step we may quote without proof a well-known result [9, page 70] which is a consequence of the Krull-Schmidt theorem for KG-modules.

Received November 23, 1959.

[1] The research in this paper was supported in part by a contract with the Office of Naval Research.

THEOREM 1. *Let M, N be KG-modules, and let K' be an extension field of K. Then*

$$M \sim_{K'} N \quad \Rightarrow \quad M \sim_K N.$$

Remark. This result is valid for any pair of fields $K \subset K'$, even for those of nonzero characteristic.

COROLLARY. *If M, N are RG-modules, then*

$$M \sim_{R'} N \quad \Rightarrow \quad M \sim_K N.$$

2. An RG-module M is called *irreducible* if it contains no nonzero submodule of smaller R-rank. As is known [10], M is irreducible if and only if KM is irreducible as KG-module. Call M *absolutely irreducible* if for every field $K' \supset K$, the module $K'M$ is irreducible as $K'G$-module. Repeated use will be made of the following result [9, page 52]:

M is absolutely irreducible if and only if every KG-endomorphism of KM is given by a scalar multiplication

$$x \to ax, \qquad x \in KM,$$

for some $a \in K$.

As a first result, we prove

THEOREM 2. *Let R be a principal ideal ring, and let M, N be a pair of absolutely irreducible RG-modules. Then*

$$M \sim_{R'} N \quad \Rightarrow \quad M \sim_R N.$$

Proof. The preceding corollary shows that $M \sim_K N$. After replacing N by some new RG-module which is RG-isomorphic to N, we may in fact assume that $M \supset N$.

The isomorphism $R'M \cong R'N$ can be extended to an isomorphism $K'M \cong K'N$. As a consequence of the absolute irreducibility of M, and the fact that $K'M = K'N$, this latter isomorphism must be given by a scalar multiplication. Consequently there exists a scalar $\alpha \in K'$ such that

$$R'N = \alpha \cdot R'M. \tag{1}$$

Since R is a principal ideal ring, we may find an R-basis $\{m_1, \cdots, m_k\}$ of M, and nonzero elements $a_1, \cdots, a_k \in R$, such that

$$M = Rm_1 \oplus \cdots \oplus Rm_k, \tag{2}$$

$$N = Ra_1 m_1 \oplus \cdots \oplus Ra_k m_k. \tag{3}$$

Then

$$R'M = \sum R'm_i, \qquad R'N = \sum R'a_i m_i = \sum R'\alpha m_i. \tag{4}$$

Let $u(R')$ be the group of units of R', and $u(R)$ that of R. Then (4) implies the existence of $\beta_1, \cdots, \beta_k \in u(R')$ such that

$$a_i = \beta_i \alpha, \qquad 1 \leqq i \leqq k.$$

Therefore

$$a_i/a_1 = \beta_i/\beta_1 \in u(R'),$$

and so

$$b_i = a_i/a_1 \in u(R') \cap K = u(R).$$

Therefore

$$N = \sum Ra_i m_i = a_1 \sum Rb_i m_i = a_1 M,$$

which shows that N, M are R-equivalent, Q.E.D.

We next give an example to show that the result stated in Theorem 2 need not hold when R is not a principal ideal ring. Set

$$\mathfrak{o} = \text{alg. int. } \{K\}, \qquad \mathfrak{o}' = \text{alg. int. } \{K'\},$$

where $\mathfrak{o}$ is not a principal ideal ring. It is possible to choose K' so that for each ideal $\mathfrak{a}$ in $\mathfrak{o}$, the induced ideal $\mathfrak{o}'\mathfrak{a}$ in $\mathfrak{o}'$ is principal (see [4]). Now let M be any absolutely irreducible $\mathfrak{o}G$-module, $\mathfrak{a}$ any nonprincipal ideal in $\mathfrak{o}$, and set $N = \mathfrak{a}M$. Then M, N cannot be $\mathfrak{o}$-equivalent, since by the above remarks the isomorphism $M \cong N$ would imply that $N = aM$ for some $a \in K$. On the other hand,

$$\mathfrak{o}'N = \mathfrak{o}'\mathfrak{a}M = \alpha'\mathfrak{o}'M$$

for some $\alpha' \in K'$, and so M, N are $\mathfrak{o}'$-equivalent.

If M, N are $\mathfrak{o}G$-modules, we say that M, N are in the same *genus* (notation: $M \vee N$) if $RM \cong RN$ for each valuation ring R of a discrete valuation of K (see [5, 6]).

COROLLARY. *Let M, N be absolutely irreducible $\mathfrak{o}G$-modules. Then*

$$M \sim_{\mathfrak{o}'} N \Rightarrow M \vee N.$$

Proof. Let R be a valuation ring of a discrete valuation ϕ of K, and let ϕ' be an extension of ϕ to K', with valuation ring R'. Then R is a principal ideal ring, and so

$$M \sim_{\mathfrak{o}'} \mathrm{N} \Rightarrow \mathrm{M} \sim_{R'} \mathrm{N} \Rightarrow \mathrm{M} \sim_{R} \mathrm{N}$$

by Theorem 2, Q.E.D.

Maranda [5] showed that a pair of absolutely irreducible $\mathfrak{o}G$-modules M, N are in the same genus if and only if $M \cong \mathfrak{a}N$ for some $\mathfrak{o}$-ideal $\mathfrak{a}$ in K. But then $\mathfrak{o}'M \cong \mathfrak{o}'\mathfrak{a}N$, so M, N are $\mathfrak{o}'$-equivalent if and only if $\mathfrak{o}'\mathfrak{a}$ is a principal ideal in K'. Thus, the converse of the above corollary holds if and only if every ideal in $\mathfrak{o}$ induces a principal ideal in $\mathfrak{o}'$.

3. Throughout this section let R be the valuation ring of a discrete valuation ϕ of K, with unique maximal ideal P, and residue class field $\bar{K} = R/P$. Let ϕ' be an extension of ϕ to K', with valuation ring R', maximal ideal P', residue class field $\bar{K}' = R'/P'$. We shall give some *sufficient* conditions for the validity of the implication:

$$(5) \qquad M \sim_{R'} N \quad \Rightarrow \quad M \sim_R N,$$

where M, N denote RG-modules.

THEOREM 3. *If the group order* $(G:1)$ *is a unit in* R, *then* (5) *is valid.*

Proof. Use Theorem 1, together with the result [5] that if $(G:1)$ is a unit in R, then

$$M \sim_R N \quad \text{if and only if} \quad M \sim_K N.$$

THEOREM 4. *If* $\bar{K}' = \bar{K}$, *then* (5) *holds.*

Proof. Since R, R' are principal ideal rings, we may use matrix terminology. Let M, N be R-representations of G such that $M \sim_{R'} N$. Set

$$C = \{X \text{ over } R: M(g)X = XN(g),\ g \in G\},$$

$$C' = \{X \text{ over } R': M(g)X = XN(g),\ g \in G\}.$$

Since C is a finitely generated torsion-free R-module, we may choose an R-basis $\{X_1, \cdots, X_n\}$ of C. It is easily verified that this is also an R'-basis of C'.

The hypothesis $M \sim_{R'} N$ is equivalent to the statement that there exist elements $\alpha_1, \cdots, \alpha_n \in R'$ such that

$$\alpha_1 X_1 + \cdots + \alpha_n X_n$$

is unimodular over R', that is, has entries in R' and satisfies

$$| \alpha_1 X_1 + \cdots + \alpha_n X_n | \in u(R') \quad \text{(the group of units of } R').$$

Since $\bar{K}' = \bar{K}$, we may choose $a_1, \cdots, a_n \in R$ such that

$$a_i \equiv \alpha_i \pmod{P'}, \qquad 1 \leqq i \leqq n.$$

In that case,

$$a_1 X_1 + \cdots + a_n X_n \in C,$$

and is unimodular over R. Therefore $M \sim_R N$, Q.E.D.

In particular, suppose that K' is an *Eisenstein extension* of K relative to the valuation ϕ, that is, suppose that $K' = K(\alpha)$ where

$$\text{Irr}\ (\alpha, K) = x^m + b_1 x^{m-1} + \cdots + b_m$$

with $b_1, \cdots, b_m \in P$, $b_m \notin P^2$ (see [3]). In this case ϕ is uniquely extendable to K', and $\bar{K}' = \bar{K}$, so that (5) is true. We shall apply this later on.

Let us call a matrix of the form

$$\begin{bmatrix} 1 & & * \\ & \ddots & \\ & & 1 \end{bmatrix}$$

a *translation*; by such a notation, we mean to imply that the elements below the main diagonal are all zero. If M, N are R-representations of G, we write $M \approx N$ to indicate that M, N can be intertwined by a translation matrix.

On the other hand, suppose that

$$(6)\qquad M = \begin{bmatrix} M_1 & & * \\ & \ddots & \\ & & M_k \end{bmatrix}, \qquad N = \begin{bmatrix} M_1 & & * \\ & \ddots & \\ & & M_k \end{bmatrix}$$

are a pair of R-representations of G in which the $\{M_i\}$ are distinct (that is, not K-equivalent) and absolutely irreducible. If M, N can be intertwined by a matrix X over R of the form

$$(7)\qquad X = \begin{bmatrix} a_1 I & & * \\ & \ddots & \\ & & a_k I \end{bmatrix},$$

in which $a_i \in u(R)$, the group of units of R, then we shall say that M, N are *i-intertwinable*. Call M, N *everywhere intertwinable* if for each i, $1 \leqq i \leqq k$, M, N are i-intertwinable. Clearly if M, N are i-intertwinable, and if[2]

$$M \approx M', \qquad N \approx N',$$

then also M', N' are i-intertwinable.

LEMMA. *Let M, N be given by (6), and suppose the $\{M_i\}$ distinct and absolutely irreducible. Suppose that M, N are everywhere intertwinable, and further that they are intertwined by a matrix X given by (7) for which*

$$(8)\qquad a_1, \cdots, a_r \notin u(R), \qquad a_{r+1}, \cdots, a_k \in u(R).$$

Then

$$(9)\quad M \approx \left[\begin{array}{c|c} \begin{matrix} M_1 & & * \\ & \ddots & \\ & & M_r \end{matrix} & 0 \\ \hline & \begin{matrix} M_{r+1} & & * \\ & \ddots & \\ & & M_k \end{matrix} \end{array}\right], \qquad N \approx \left[\begin{array}{c|c} \begin{matrix} M_1 & & * \\ & \ddots & \\ & & M_r \end{matrix} & 0 \\ \hline & \begin{matrix} M_{r+1} & & * \\ & \ddots & \\ & & M_k \end{matrix} \end{array}\right].$$

Proof. Use induction on r. The result is trivial when $r = 0$, so assume $r \geqq 1$, and write

$$M = \begin{bmatrix} M_1 & * & * \\ & M' & \Lambda \\ & & M'' \end{bmatrix}, \qquad N = \begin{bmatrix} M_1 & * & * \\ & N' & \Delta \\ & & N'' \end{bmatrix},$$

[2] We use ${}^tM$ to denote the transpose of M; thus, M' is just another representation in this context.

where

$$M' = \begin{bmatrix} M_2 & & * \\ & \ddots & \\ & & M_r \end{bmatrix}, \quad M'' = \begin{bmatrix} M_{r+1} & & * \\ & \ddots & \\ & & M_k \end{bmatrix}, \quad \text{(submatrices of } M\text{)},$$

$$N' = \begin{bmatrix} M_2 & & * \\ & \ddots & \\ & & M_r \end{bmatrix}, \quad N'' = \begin{bmatrix} M_{r+1} & & * \\ & \ddots & \\ & & M_k \end{bmatrix}, \quad \text{(submatrices of } N\text{)}.$$

Then also

$$\begin{bmatrix} M' & \Lambda \\ & M'' \end{bmatrix}, \quad \begin{bmatrix} N' & \Delta \\ & N'' \end{bmatrix}$$

are everywhere intertwinable, and furthermore are intertwined by

$$\begin{bmatrix} a_2 I & & * \\ & \ddots & \\ & & a_k I \end{bmatrix},$$

a submatrix of X. It follows from the induction hypothesis that by transforming M, N by suitable translation matrices, we can make $\Lambda = \Delta = 0$. The new M, N will still be everywhere intertwinable, and also intertwinable by a new X for which (8) still holds.

Let us write

$$M = \left[\begin{array}{c|c|c} M_1 & * & \Lambda_{r+1} \cdots \Lambda_k \\ \hline & M' & 0 \\ \hline & & M'' \end{array}\right], \quad N = \left[\begin{array}{c|c|c} M_1 & * & \Delta_{r+1} \cdots \Delta_k \\ \hline & N' & 0 \\ \hline & & N'' \end{array}\right],$$

$$X = \left[\begin{array}{c|c|c} a_1 I & * & T_{r+1} \cdots T_k \\ \hline & X' & T \\ \hline & & X'' \end{array}\right], \quad X'' = \begin{bmatrix} a_{r+1} I & & * \\ & \ddots & \\ & & a_k I \end{bmatrix}.$$

Then

$$\begin{bmatrix} M' & 0 \\ & M'' \end{bmatrix}\begin{bmatrix} X' & T \\ & X'' \end{bmatrix} = \begin{bmatrix} X' & T \\ & X'' \end{bmatrix}\begin{bmatrix} N' & 0 \\ & N'' \end{bmatrix},$$

whence $M'T = TN''$. Since M', N'' have no common irreducible constituent, we conclude that $T = 0$.

It now follows that

$$\begin{bmatrix} M_1 & \Lambda_{r+1} \\ & M_{r+1} \end{bmatrix}, \quad \begin{bmatrix} M_1 & \Delta_{r+1} \\ & M_{r+1} \end{bmatrix} \tag{10}$$

are R-representations intertwined by

$$\begin{bmatrix} a_1 I & T_{r+1} \\ & a_{r+1} I \end{bmatrix}. \tag{11}$$

This implies that

$$M_1 T_{r+1} + a_{r+1} \Lambda_{r+1} = a_1 \Delta_{r+1} + T_{r+1} M_{r+1},$$

and hence (since $a_{r+1} \in u(R)$),

$$\Lambda_{r+1} = b\Delta_{r+1} + M_1 U - UM_{r+1}, \qquad b = a_{r+1}^{-1} a_1 \notin u(R), \tag{12}$$

for some U over R. On the other hand, the hypothesis that M, N are 1-intertwinable guarantees the existence of a matrix of the form (11) which intertwines the representations given in (10), but for which the element playing the role of a_1 is a unit in R. Therefore we also have

$$\Delta_{r+1} = c\Lambda_{r+1} + M_1 V - VM_{r+1} \tag{13}$$

for some $c \in R$ and some V over R. Combining (12) and (13), we obtain

$$(1 - bc)\,\Lambda_{r+1} = M_1 W - WM_{r+1}$$

for some W over R. Since $(1 - bc) \in u(R)$, we conclude that

$$\Lambda_{r+1} = M_1 Y - YM_{r+1}$$

for some Y over R. Hence by a translation transformation of M, we can make $\Lambda_{r+1} = 0$. From (13) it follows that we can also make $\Delta_{r+1} = 0$ by a translation transformation of N. For this new M, N we must have $T_{r+1} = 0$.

But now we observe that

$$\begin{bmatrix} M_1 & \Lambda_{r+2} \\ & M_{r+2} \end{bmatrix}, \qquad \begin{bmatrix} M_1 & \Delta_{r+2} \\ & M_{r+2} \end{bmatrix}$$

are representations intertwined by

$$\begin{bmatrix} a_1 I & T_{r+2} \\ & a_{r+2} I \end{bmatrix}.$$

The above type of argument shows that we can make $\Lambda_{r+2} = \Delta_{r+2} = 0$, and therefore also T_{r+2} must be 0. By continuing this process, we establish the validity of (9), Q.E.D.

We may now prove one of the main results of this paper.

THEOREM 5. *Let M, N be RG-modules which are R'-equivalent, and suppose that the irreducible constituents of KM (which coincide with those of KN) are distinct from one another and are absolutely irreducible. Then also M, N are R-equivalent.*

Proof. Again use matrix terminology, and proceed by induction on the number k of irreducible constituents of KM. The result for $k = 1$ follows from Theorem 2; suppose it known up to $k - 1$, and let KM have k distinct absolutely irreducible constituents. There will be no confusion from our using M to denote both the module and the R-representation it affords. The R-representations of G afforded by the RG-modules M, N may be taken to be of the form[3]

$$(14) \qquad M = \begin{bmatrix} M_1 & & * \\ & \ddots & \\ & & M_k \end{bmatrix}, \qquad N = \begin{bmatrix} N_1 & & * \\ & \ddots & \\ & & N_k \end{bmatrix},$$

where the $\{M_i\}$ and $\{N_i\}$ are absolutely irreducible, and where

$$(15) \qquad M_i \sim_K N_i, \qquad M_i \nsim_K M_j, \qquad j \neq i, \quad 1 \leqq i \leqq k.$$

Since M, N are R'-equivalent, they are intertwined by a matrix X' unimodular over R'. From (15) we find readily (see [6]) that X' has the form

$$(16) \qquad X' = \begin{bmatrix} X'_1 & & * \\ & \ddots & \\ & & X'_k \end{bmatrix},$$

and necessarily each X'_i is also unimodular over R'. But we have then

$$(17) \qquad M_i X'_i = X'_i N_i, \qquad 1 \leqq i \leqq k,$$

so that M_i, N_i are R'-equivalent for each i. By the induction hypothesis it follows that for each i, $1 \leqq i \leqq k$, M_i and N_i are R-equivalent. Consequently for each i there exists a matrix Y_i unimodular over R which intertwines M_i and N_i. Setting $Y = \operatorname{diag}(Y_1, \cdots, Y_k)$, we deduce that

$$N \sim_R YNY^{-1} = \begin{bmatrix} M_1 & & * \\ & \ddots & \\ & & M_k \end{bmatrix} \qquad \text{(say)}.$$

Replacing N by YNY^{-1}, we may henceforth assume that $N_1 = M_1, \cdots, N_k = M_k$, that is, that M, N are given by (6).

From the R'-equivalence of M, N it follows that they are intertwined by a unimodular matrix X' over R', given by (16). Since now $M_i = N_i$, and M_i is absolutely irreducible, (17) implies that each X'_i is a scalar matrix, so that we may write

$$(18) \qquad X' = \begin{bmatrix} \alpha_1 I & & * \\ & \ddots & \\ & & \alpha_k I \end{bmatrix}, \qquad \alpha_1, \cdots, \alpha_k \in u(R').$$

Let us now set

$$R' = R\beta_1 \oplus \cdots \oplus R\beta_n, \qquad \beta_1 = 1, \quad n = (K':K).$$

[3] This really follows from [10].

Then we may write

$$X' = \sum_{\nu=1}^{n} X^{(\nu)}\beta_\nu , \qquad X^{(\nu)} \text{ over } R;$$

we note that

$$X^{(\nu)} = \begin{bmatrix} a_1^{(\nu)}I & & * \\ & \ddots & \\ & & a_k^{(\nu)}I \end{bmatrix}, \qquad 1 \leqq \nu \leqq n,$$

where

$$(19) \qquad \alpha_i = \sum_\nu a_i^{(\nu)}\beta_\nu , \qquad a_i^{(\nu)} \in R.$$

Let us fix i, $1 \leqq i \leqq k$. Then $\alpha_i \in u(R')$, and so by (19) at least one of $a_i^{(1)}, \cdots, a_i^{(n)}$ is a unit in R. Since each $X^{(\nu)}$ intertwines M and N, and since $a_i^{(\nu)}$ occurs in the ith diagonal block of $X^{(\nu)}$, we may conclude that M, N are i-intertwinable. This shows then that if M, N given by (6) are R'-equivalent, they must be everywhere intertwinable.

Since M, N are 1-intertwinable, there exists an X (over R) given by (7) which intertwines M and N, and for which $a_1 \in u(R)$. If also $a_2, \cdots, a_k \in u(R)$, then X is unimodular over R, and so M, N are R-equivalent. For the remainder of the proof we may therefore suppose that not all of $a_2, \cdots, a_k$ are units in R. Let us write

$$a_1, \cdots, a_q \in u(R), \qquad a_{q+1}, \cdots, a_r \notin u(R), \qquad a_{r+1}, \cdots, a_s \in u(R), \cdots .$$

Partition X accordingly, say

$$X = \begin{bmatrix} Y_1 & & * \\ & \ddots & \\ & & Y_t \end{bmatrix}, \quad Y_1 = \begin{bmatrix} X_1 & & * \\ & \ddots & \\ & & X_q \end{bmatrix}, \quad Y_2 = \begin{bmatrix} X_{q+1} & & * \\ & \ddots & \\ & & X_r \end{bmatrix}, \cdots .$$

Correspondingly partition M, N, say

$$(20) \quad M = \begin{bmatrix} \bar{M}_1 & \Lambda_{12} & \Lambda_{13} & \\ & \bar{M}_2 & \Lambda_{23} & \\ & & \bar{M}_3 & * \\ & & & \ddots \\ & & & & \bar{M}_t \end{bmatrix}, \quad N = \begin{bmatrix} \bar{N}_1 & \Delta_{12} & \Delta_{13} & \\ & \bar{N}_2 & \Delta_{23} & \\ & & \bar{N}_3 & * \\ & & & \ddots \\ & & & & \bar{N}_t \end{bmatrix},$$

where

$$\bar{M}_1 = \begin{bmatrix} M_1 & & * \\ & \ddots & \\ & & M_q \end{bmatrix}, \qquad \bar{N}_1 = \begin{bmatrix} M_1 & & * \\ & \ddots & \\ & & M_q \end{bmatrix}, \cdots .$$

By repeated use of the lemma, we may transform M, N by translations so as to make successively

$$(21) \qquad \Lambda_{12} = \Delta_{12} = 0, \quad \Lambda_{23} = \Delta_{23} = 0, \quad \cdots, \quad \Lambda_{t-1,t} = \Delta_{t-1,t} = 0.$$

Such transformations do not affect the diagonal blocks of X, nor the R'-equivalence of M, N. We may therefore assume for the remainder of the proof that (21) holds. But in that case we see from (20) that

$$\begin{bmatrix} \bar{M}_1 & \Lambda_{14} \\ & \bar{M}_4 \end{bmatrix}, \quad \begin{bmatrix} \bar{N}_1 & \Delta_{14} \\ & \bar{N}_4 \end{bmatrix}$$

are R-representations of G, and again we may apply the lemma to conclude that M, N may be further transformed by translation matrices so as to make $\Lambda_{14} = \Delta_{14} = 0$, and so on. Continuing in this way, we find that

$$M \approx M' = \begin{bmatrix} \bar{M}_1 & & \Omega \\ & \ddots & \\ & & \bar{M}t \end{bmatrix}, \qquad N \approx N' = \begin{bmatrix} \bar{N}_1 & & \Sigma \\ & \ddots & \\ & & \bar{N}t \end{bmatrix},$$

where $\Omega_{ij} = \Sigma_{ij} = 0$ whenever the diagonal entries of X associated with $\bar{M}_i$ are units, those with $\bar{M}_j$ nonunits, or vice versa. But we may then find a permutation matrix F such that

$$FM'F^{-1} = \begin{bmatrix} M^* & 0 \\ & M^{**} \end{bmatrix}, \qquad FN'F^{-1} = \begin{bmatrix} N^* & 0 \\ & N^{**} \end{bmatrix},$$

where

$$M^* = \begin{bmatrix} \bar{M}_1 & & \\ & \bar{M}_3 & * \\ & & \ddots \end{bmatrix}, \qquad M^{**} = \begin{bmatrix} \bar{M}_2 & & \\ & \bar{M}_4 & * \\ & & \ddots \end{bmatrix},$$

$$N^* = \begin{bmatrix} \bar{N}_1 & & \\ & \bar{N}_3 & * \\ & & \ddots \end{bmatrix}, \qquad N^{**} = \begin{bmatrix} \bar{N}_2 & & \\ & \bar{N}_4 & * \\ & & \ddots \end{bmatrix}.$$

We now have

$$M \sim_R \begin{bmatrix} M^* & 0 \\ & M^{**} \end{bmatrix}, \qquad N \sim_R \begin{bmatrix} N^* & 0 \\ & N^{**} \end{bmatrix}, \tag{22}$$

and so (since $M \sim_{R'} N$),

$$\begin{bmatrix} M^* & 0 \\ & M^{**} \end{bmatrix} \sim_{R'} \begin{bmatrix} N^* & 0 \\ & N^{**} \end{bmatrix}.$$

Since M^*, M^{**} have no common irreducible constituents, this latter equivalence implies that

$$M^* \sim_{R'} N^*, \qquad M^{**} \sim_{R'} N^{**}.$$

We may (at last) use the induction hypothesis to conclude from this that

$$M^* \sim_R N^*, \qquad M^{**} \sim_R N^{**}.$$

This, together with (22), implies that M, N are R-equivalent. Thus the theorem is proved.

4. We shall apply the preceding result to the case of p-groups.

THEOREM 6. *Let G be a p-group, where p is an odd prime. Let R be the ring of p-integral elements of the rational field Q. Suppose that K' is an algebraic number field, and R' any valuation ring of K' such that $R' \supset R$. Then for any pair of irreducible RG-modules M, N we have*

$$M \sim_{R'} N \quad \Rightarrow \quad M \sim_R N. \tag{23}$$

Proof. Set $(G:1) = p^m$, $m > 1$, and let ζ be a primitive $(p^m)^{\text{th}}$ root of 1 over Q. Let M, N be R'-equivalent irreducible RG-modules. As a first step, let us set $K_1 = K'(\zeta)$, and let R_1 be a valuation ring of K_1 such that $R_1 \supset R'$. Then since

$$M \sim_{R'} N \quad \Rightarrow \quad M \sim_{R_1} N,$$

we may now restrict our attention to K_1, R_1 instead of K', R'.

Next we note that

$$f(x) = \text{Irr}\,(\zeta, Q) = x^{p^{m-1}(p-1)} + x^{p^{m-1}(p-2)} + \cdots + x^{p^{m-1}} + 1,$$

and that $f(x+1)$ is an Eisenstein polynomial at the prime p. If we set $K_0 = Q(\zeta)$, it follows that K_0 contains a uniquely determined valuation ring R_0 such that $R_0 \supset R$, and further that the residue class fields corresponding to R_0, R coincide. We may therefore conclude from Theorem 4 that

$$M \sim_{R_0} N \quad \Rightarrow \quad M \sim_R N. \tag{24}$$

The proof will be complete as soon as we establish

$$M \sim_{R_1} N \quad \Rightarrow \quad M \sim_{R_0} N. \tag{25}$$

This is a consequence of Theorem 5, however, as we now proceed to demonstrate. The modules $R_0 M$, $R_0 N$ are (in general) no longer irreducible. Since K_0 is an absolute splitting field for G (see [1]), the irreducible constituents of $K_0 M$ and $K_0 N$ are all absolutely irreducible. The multiplicity with which any absolutely irreducible constituent of $K_0 M$ occurs is precisely the Schur index of that constituent relative to the rational field (see [7]). On the other hand, for p-groups (p odd) it is known [2, 8] that this Schur index is 1. Hence the irreducible constituents of $R_0 M$ and $R_0 N$ are distinct and absolutely irreducible. We may therefore apply Theorem 5, and obtain

$$R_1 M \cong R_1 N \quad \Rightarrow \quad R_0 M \cong R_0 N,$$

so that (25) is proved, Q.E.D.

The referee has kindly pointed out that the preceding theorem is also valid for the more general case in which R is a valuation ring of an algebraic number field K such that R lies over the ring of p-integral elements of the rational field. Indeed, the above proof requires only a minor modification for the more general case.

5. We conclude by listing a number of open questions.

A. If $R \subset R'$ are valuation rings, does (5) hold without any restrictive hypotheses?

B. Using the notation of Section 2, under what conditions does $\mathfrak{o}'M \vee \mathfrak{o}'N$ imply $M \vee N$, where M and N are $\mathfrak{o}G$-modules?

C. If $\mathfrak{o}$ is a principal ideal ring, does $\mathfrak{o}'$-equivalence imply $\mathfrak{o}$-equivalence?

It may be of interest to mention yet one more special case in which additional information may be obtained. Suppose that M and N are projective RG-modules, where R is the valuation ring of a discrete valuation of K. (For example, M and N might be direct summands of RG.) Then it is known[4] that $M \sim_R N$ if and only if $M \sim_K N$. Using Theorem 1 and its corollary, we conclude that (5) holds in this case.

In particular, if M and N are projective $\mathfrak{o}G$-modules, then $\mathfrak{o}'M \vee \mathfrak{o}'N$ surely implies that M and N are K'-equivalent, and hence by the above discussion that $M \vee N$.

Added in proof. In a recently completed paper [11], Zassenhaus and the author have shown that (5) holds without any restrictive hypotheses, assuming still that R and R' are valuation rings as in Section 3. This settles questions A and B, but C is still open.

References

1. R. Brauer, *Applications of induced characters*, Amer. J. Math., vol. 69 (1947), pp. 709–716.
2. ———, *On the representations of groups of finite order*, Proceedings of the International Congress of Mathematicians 1950, vol. II, pp. 33–36.
3. H. Hasse, *Zahlentheorie*, Berlin, 1949.
4. E. Hecke, *Vorlesungen über die Theorie der algebraischen Zahlen*, Leipzig, 1923.
5. J.-M. Maranda, *On $\mathfrak{P}$-adic integral representations of finite groups*, Canadian J. Math., vol. 5 (1953), pp. 344–355.
6. I. Reiner, *On the class number of representations of an order*, Canadian J. Math., vol. 11 (1959), pp. 660–672.
7. ———, *The Schur index in the theory of group representations*, submitted to Michigan Math. J.
8. P. Roquette, *Realisierung von Darstellungen endlicher nilpotenter Gruppen*, Arch. Math., vol. 9 (1958), pp. 241–250.
9. B. L. van der Waerden, *Gruppen von linearen Transformationen*, Berlin, 1935.
10. H. Zassenhaus, *Neuer Beweis der Endlichkeit der Klassenzahl bei unimodularer Äquivalenz endlicher ganzzahliger Substitutionsgruppen*, Abh. Math. Sem. Univ. Hamburg, vol. 12 (1938), pp. 276–288.
11. H. Zassenhaus and I. Reiner, *Equivalence of representations under extensions of local ground rings*, to appear in Illinois J. Math.

University of Illinois
Urbana, Illinois

[4] R. G. Swan, *Induced representation and projective modules*, University of Chicago, mimeographed notes, 1959, Corollary 6.4.

THE NONUNIQUENESS OF IRREDUCIBLE CONSTITUENTS OF INTEGRAL GROUP REPRESENTATIONS[1]

IRVING REINER

Let Z denote the ring of rational integers, Q the rational field. As is well-known, the Z-representations of a finite group G can be classified either according to Q-equivalence or according to Z-equivalence. Thus, two Z-representations T, U of G are *Q-equivalent* ($T \sim_Q U$) if there exists a nonsingular rational matrix P such that

$$U(g) = P^{-1}T(g)P, \qquad g \in G. \tag{1}$$

On the other hand, we write $T \sim_Z U$ if (1) holds for some unimodular[2] matrix P.

If a representation T is equivalent to a "reduced" representation

$$g \to \begin{pmatrix} T_1(g) & * \\ 0 & T_2(g) \end{pmatrix}, \qquad g \in G$$

we say that T is *reducible.* Conceivably we must distinguish between Q- and Z-reducibility of a Z-representation. This difficulty does not in fact arise, because of the following theorem due to Zassenhaus [**3**].

(A) An integral representation is Q-reducible if and only if it is Z-reducible.

It is well-known that any given Q-representation T of G is Q-equivalent to a "completely reduced" representation

$$\begin{bmatrix} T_1 & & 0 \\ & \ddots & \\ 0 & & T_k \end{bmatrix}, \tag{2}$$

in which the T_i are irreducible. The Jordan-Hölder Theorem on modules asserts

(B) The irreducible representations $T_1, \cdots, T_k$ (often referred to as the *irreducible constituents* of T) are uniquely determined up to Q-equivalence and order of occurrence.

As an analogue of this, Zassenhaus [**3**] and Diederichsen [**1**] proved

(C) If the Z-representation T is Q-equivalent to a completely re-

Presented to the Society, September 4, 1959; received by the editors September 22, 1959.

[1] This work was supported in part by a contract with the Office of Naval Research.

[2] A square matrix with integral entries and determinant ± 1 is called *unimodular.*

duced representation (2), then also T is Z-equivalent to a reduced Z-representation

$$\begin{bmatrix} U_1 & & * \\ & \ddots & \\ 0 & & U_k \end{bmatrix}$$

in which

$$U_i \sim_Q T_i, \qquad 1 \leqq i \leqq k.$$

We may refer to $U_1, \cdots, U_k$ as a set of irreducible Z-constituents of T. From (B) it follows that they are unique up to Q-equivalence and order of occurrence. Diederichsen [1] (see also Maranda [2]) gave the following example to show that the irreducible Z-constituents were not necessarily unique up to Z-equivalence and order of occurrence.

Let

$$G = \{a, b: a^4 = b^2 = (ab)^2 = 1\}. \tag{3}$$

Then G is the group of symmetries of the square, and $(G: 1) = 8$. Set

$$A_1 = \begin{pmatrix} 0 & 1 \\ -1 & 0 \end{pmatrix}, \quad B_1 = \begin{pmatrix} 0 & 1 \\ 1 & 0 \end{pmatrix}; \quad A_2 = \begin{pmatrix} 0 & 1 \\ -1 & 0 \end{pmatrix}, \quad B_2 = \begin{pmatrix} 1 & 0 \\ 0 & -1 \end{pmatrix}; \tag{4}$$

$$T_i(a) = A_i, \qquad T_i(b) = B_i, \qquad i = 1, 2. \tag{5}$$

Then T_1 and T_2 are irreducible Z-representations of G which are Q-equivalent but not Z-equivalent. However, for suitable choice of integral Λ_1, Λ_2, Diederichsen showed that

$$\begin{pmatrix} T_1 & \Lambda_1 \\ 0 & T_1 \end{pmatrix} \sim_Z \begin{pmatrix} T_2 & \Lambda_2 \\ 0 & T_2 \end{pmatrix}.$$

Diederichsen attempted to show that this difficulty is due to the repetition of constituents, and asserted the following:

(D) Let T be an integral representation of some finite group, and suppose that

$$T \sim_Z \begin{bmatrix} U_1 & & * \\ & \ddots & \\ 0 & & U_k \end{bmatrix}$$

where for $i \neq j$, no irreducible constituent of U_i is Q-equivalent to any irreducible constituent of U_j. Then $U_1, \cdots, U_k$ are unique up to Z-equivalence and order of occurrence.

We shall show here that this statement is false.[3] If it were true, it would imply that if T is a Z-representation no two of whose irreducible constituents are Q-equivalent, then the irreducible Z-constituents of T are unique up to Z-equivalence and order of occurrence. We shall give a counterexample to show that this is not the case.

Keep the notation of equations (3)–(5). Let U be the Z-representation of G defined by

$$(6)\qquad U(a) = \left(\begin{array}{c|c} A_1 & \begin{matrix}1\\0\end{matrix} \\ \hline & 1 \end{array}\right), \qquad U(b) = \left(\begin{array}{c|c} B_1 & \begin{matrix}1\\-1\end{matrix} \\ \hline & 1 \end{array}\right),$$

so that U has irreducible Z-constituents T_1 and 1. Set

$$S = \begin{pmatrix} 1 & 0 & 1 \\ -1 & -1 & -1 \\ 2 & 1 & 1 \end{pmatrix}, \qquad S^{-1} = \begin{pmatrix} 0 & 1 & 1 \\ -1 & -1 & 0 \\ 1 & -1 & -1 \end{pmatrix}.$$

Then we find that $S^{-1}US = V$, where

$$(7)\qquad V(a) = \left(\begin{array}{c|c} 1 & 1\ 0 \\ \hline & A_2 \end{array}\right), \qquad V(b) = \left(\begin{array}{c|c} 1 & 0\ 1 \\ \hline & B_2 \end{array}\right).$$

Thus, V has irreducible Z-constituents 1, T_2. Since $U \sim_Z V$, we have our counterexample.

Keeping the notation and hypotheses of (D), the correct version (proved by Diederichsen) is

(D′) Once the order of occurrence of the U_i is fixed, then they are unique up to Z-equivalence. In other words, if

$$\begin{pmatrix} U_1 & & * \\ & \ddots & \\ 0 & & U_k \end{pmatrix}, \qquad \begin{pmatrix} V_1 & & * \\ & \ddots & \\ 0 & & V_k \end{pmatrix}$$

are a pair of Z-equivalent Z-representations of a finite group such that

$$U_i \sim_Q V_i, \qquad 1 \leqq i \leqq k,$$

and such that for $i \neq j$, no irreducible constituent of U_i is Q-equivalent to any irreducible constituent of U_j, then in fact $U_i \sim_Z V_i$, $1 \leqq i \leqq k$.

[3] The flaw in Diederichsen's argument is this: he shows that the order of occurrence of the irreducible constituents may be changed at will by unimodular transformation. Unfortunately, he overlooks the fact that such transformations may change the integral classes of the constituents which are involved.

References

1. F. E. Diederichsen, *Über die Ausreduktion ganzzahliger Gruppendarstellungen bei arithmetischer Äquivalenz*, Abh. Math. Sem. Hansischen Univ. vol. 13 (1940) pp. 357–412.

2. J.-M. Maranda, *On P-adic integral representations of finite groups*, Canad. J. Math. vol. 5 (1953) pp. 344–355.

3. H. Zassenhaus, *Neuer Beweis der Endlichkeit der Klassenzahl bei unimodularer Äquivalenz endlicher ganzzahliger Substitutionsgruppen*, Abh. Math. Sem. Hansischen Univ. vol. 12 (1938) pp. 276–288.

University of Illinois

EQUIVALENCE OF REPRESENTATIONS UNDER EXTENSIONS OF LOCAL GROUND RINGS[1]

BY

I. REINER AND H. ZASSENHAUS

We shall use the following notations: K = algebraic number field, R = valuation ring in K with maximal ideal P, K' = finite extension field over K, R' = valuation ring of K' containing R; A = finite-dimensional algebra over K, G = R-order in A (that is, G is a subring of A containing the unity element of A as well as a K-basis of A, and such that G has a finite R-basis). We define

$$A' = K' \otimes_K A, \qquad G' = R' \otimes_R G,$$

so that G' is an R'-order in the K'-algebra A'. By a *G-module* we shall mean a left unital G-module having a finite R-basis. To each G-module M there corresponds a G'-module M' defined by

$$M' = R' \otimes_R M.$$

Finally we assume that all G-modules have finite height at P (see Higman [2]). Thus for each pair M, N of G-modules there exists an integer $s \geqq 0$ such that

$$P^s \operatorname{Ext}^1(M, N) = 0. \tag{1}$$

The most interesting case is that in which $G = RH$ is the group ring of a finite group H; in this case we may choose for s any integer such that the group order $[H:1]$ lies in P^s. (In this connection see also Maranda [4].)

Our aim is to establish the following:

THEOREM. *Let M and N be G-modules. Then $M' \cong N'$ as G'-modules if and only if $M \cong N$ as G-modules.*

On the one hand we may regard this result as a generalization of the Noether-Deuring Theorem [1] which applies when $R = K$, and indeed the central idea of their proof is also used here. On the other hand the present theorem generalizes a result of the first author [5] in which the theorem was established under various restrictive hypotheses.

In order to prove this theorem it is sufficient to show that $M' \cong N'$ implies $M \cong N$, the reverse implication being obvious. Let s satisfy (1), set $t = s + 1$, and define

$$\bar{R} = R/P^t, \qquad \bar{R}' = R'/P^t R'.$$

Received September 3, 1960.

[1] This research was conducted at The Summer Institute on Finite Groups sponsored by the American Mathematical Society in August, 1960. The work of the first author was supported in part by the Office of Naval Research.

We may then view $\bar{R}$ as a subring of $\bar{R}'$. Furthermore R' is a free R-module with a finite basis, and so $\bar{R}'$ is a free $\bar{R}$-module with a finite basis. If we set

$$\bar{G} = G/P^tG, \qquad \bar{G}' = G'/P^tG',$$

we find readily that

$$\bar{G}' = \bar{R}' \otimes_{\bar{R}} \bar{G}.$$

Likewise for the G-module M we let

$$\bar{M} = M/P^tM, \qquad \bar{M}' = M'/P^tM',$$

and we have

$$\bar{M}' = \bar{R}' \otimes_{\bar{R}} \bar{M}. \tag{2}$$

Thus $\bar{M}$ is a $\bar{G}$-module, and by extension of the ground ring from $\bar{R}$ to $\bar{R}'$ we obtain the $\bar{G}'$-module $\bar{M}'$.

Suppose now that $M' \cong N'$; then $\bar{M}' \cong \bar{N}'$ as $\bar{G}'$-modules. If k is the number of elements in an $\bar{R}$-basis of $\bar{R}'$, it follows from (2) that as $\bar{G}$-module $\bar{M}'$ is isomorphic to a direct sum of k copies of $\bar{M}$, and likewise $\bar{N}'$ is isomorphic to a direct sum of k copies of $\bar{N}$. Thus

$$\bar{M} \oplus \cdots \oplus \bar{M} = \bar{N} \oplus \cdots \oplus \bar{N} \qquad \text{as } \bar{G}\text{-modules},$$

where k summands occur on each side. But now let

$$\bar{M} = M_1 \oplus \cdots \oplus M_a, \qquad \bar{N} = N_1 \oplus \cdots \oplus N_b$$

be the decompositions of $\bar{M}$ and $\bar{N}$ into indecomposable $\bar{G}$-submodules. Then we have

$$k(M_1 \oplus \cdots \oplus M_a) \cong k(N_1 \oplus \cdots \oplus N_b). \tag{3}$$

However $\bar{G}$ is a ring with minimum condition, and therefore (see Jacobson [3]) the Krull-Schmidt Theorem is valid for $\bar{G}$-modules. From (3) we conclude that the $\{M_i\}$ are up to isomorphism just a rearrangement of the $\{N_j\}$, and thus $\bar{M} \cong \bar{N}$.

To complete the proof we need only observe that $\bar{M} \cong \bar{N}$ implies $M \cong N$. a result due originally to Maranda [4] and generalized to the present context by Higman [2].

It is easy to see that the theorem is still valid under slightly more general hypotheses. For example K need not be an algebraic number field, so long as we know that R' has a finite R-basis and that $R' \cap K = R$. If furthermore K' is algebraic over K, the restriction that $(K':K)$ be finite can be dropped.

References

1. M. Deuring, *Galoissche Theorie und Darstellungstheorie*, Math. Ann., vol. 107 (1932), pp. 140–144.
2. D. G. Higman, *On representations of orders over Dedekind domains*, Canadian J. Math., vol. 12 (1960), pp. 107–125.

3. N. JACOBSON, *The theory of rings*, New York, 1943.
4. J.-M. MARANDA, *On 𝔓-adic integral representations of finite groups*, Canadian J. Math., vol. 5 (1953), pp. 344–355.
5. I. REINER, *Behavior of integral group representations under ground ring extension*, Illinois J. Math., vol. 4 (1960), pp. 640–651.

UNIVERSITY OF ILLINOIS
URBANA, ILLINOIS
UNIVERSITY OF NOTRE DAME
NOTRE DAME, INDIANA

INDECOMPOSABLE REPRESENTATIONS

BY

A. HELLER AND I. REINER[1]

1. Introduction

Let Λ be a finite-dimensional algebra over a field K. By a Λ-module we shall mean always a finitely generated left Λ-module on which the unity element of Λ acts as identity operator. It is well known that the Krull-Schmidt theorem holds for Λ-modules: each module is a direct sum of indecomposable Λ-modules, and these summands are uniquely determined up to order of occurrence and Λ-isomorphism. Thus the problem of classifying Λ-modules is reduced to that of finding the isomorphism classes of indecomposable Λ-modules. We denote the set of these by $M(\Lambda)$.

A central problem in the theory of group representations is that of determining a set of representatives of $M(\Lambda)$ for the special case where $\Lambda = KG$, the group algebra of a finite group G over the field K. A definitive answer can be given when the characteristic of K does not divide the group order $[G:1]$; in this case KG is semisimple, all indecomposable modules over KG are irreducible, and a full set of non-isomorphic minimal left ideals of KG constitute a set of representatives of $M(KG)$. For the case where the characteristic of K is p ($p \neq 0$), Higman [6] has proved the following remarkable result: *$M(KG)$ is finite if and only if the p-Sylow subgroups of G are cyclic.* If such is the case, Higman obtained an upper bound on the number of elements of $M(KG)$. A best possible upper bound was later obtained by Kasch, Kupisch, and Kneser [5].

We shall attempt to elucidate Higman's theorem by considering in detail the special case where G is an abelian p-group, and K a field of characteristic p. We shall exhibit some new classes of indecomposable modules. However we shall show that the problem of computing $M(KG)$, in case G is not cyclic, is at least as difficult as a classical unsolved problem in matrix theory

It should be pointed out that the question of determining all representations of a p-group in a field of characteristic p has been extensively treated by Brahana [1, 2, 3] from a somewhat different viewpoint. There is consequently a certain amount of overlapping between his results and ours, but we have thought it best to make this paper completely self-contained.

2. C-algebras

Inasmuch as we shall need to consider, together with modules over an algebra Λ, also modules over sub- and quotient-algebras of Λ, we cannot re-

Received June 10, 1960.

[1] Research of the first author was supported in part by the A. P. Sloan Foundation. Research of the second author was supported in part by a research contract with the Office of Naval Research.

strict our attention only to group algebras. Instead we shall work with a special type of commutative completely primary algebras.

DEFINITION. A *C-algebra* Λ over a field K of arbitrary characteristic is a finite-dimensional commutative algebra over K with a unity element, such that

$$\Lambda/R(\Lambda) \cong K,$$

where $R(\Lambda)$ denotes the radical of Λ. Any quotient algebra of a C-algebra is easily seen to be a C-algebra. Likewise any subalgebra Λ' of a C-algebra Λ, which contains the unity element of Λ, is a C-algebra.

We may describe a C-algebra Λ explicitly as follows. Let

$$u_1, \cdots, u_n \in R(\Lambda)$$

map onto a K-basis of $R(\Lambda)/R(\Lambda)^2$. From the nilpotency of $R(\Lambda)$ it follows readily that

$$\Lambda = K[u_1, \cdots, u_n], \tag{1}$$

though of course there are polynomial relations connecting the $\{u_i\}$. Let $x_1, \cdots, x_n$ be indeterminates over K, and define a K-homomorphism

$$\phi: K[x_1, \cdots, x_n] \to \Lambda \tag{2}$$

by means of

$$\phi(1) = 1, \quad \phi(x_1) = u_1, \quad \cdots, \quad \phi(x_n) = u_n. \tag{3}$$

Then ϕ is an algebra epimorphism; its kernel J has the property that

$$\sqrt{J} = (x_1, \cdots, x_n), \tag{4}$$

where as usual

$$\sqrt{J} = \{F \in K[x_1, \cdots, x_n]: F^r \in J \text{ for some } r\},$$

and where $(x_1, \cdots, x_n)$ denotes the ideal generated by the $\{x_i\}$. We have

$$K[x_1, \cdots, x_n]/J \cong \Lambda. \tag{5}$$

Conversely if J is an ideal in $K[x_1, \cdots, x_n]$ for which (4) holds, then equation (5) defines a C-algebra Λ. The integer n given by

$$n = \dim_K R(\Lambda)/R(\Lambda)^2$$

we shall call the *rank* of Λ.

In particular let G be an abelian p-group, and write

$$G = G_1 \times \cdots \times G_n,$$

where for each i, G_i is cyclic generated by an element g_i of order $r_i = p^{\alpha_i}$. Let K be any field of characteristic p. Then the K-homomorphism

$$\phi: K[x_1, \cdots, x_n] \to KG$$

defined by

$$\phi(1) = 1, \quad \phi(x_1) = g_1 - 1, \quad \cdots, \quad \phi(x_n) = g_n - 1,$$

is an algebra epimorphism with kernel

$$J = (x_1^{r_1}, \cdots, x_n^{r_n}).$$

Thus KG is a C-algebra of rank n.

3. Quotient algebras; the height of a module

Let Λ be a finite-dimensional K-algebra, and let $\Lambda' = \Lambda/W$ be a quotient algebra of Λ, where W is a two-sided ideal in Λ. Then each Λ'-module M may be made into a Λ-module by defining

$$\lambda \cdot m = (\lambda + W)m, \qquad \lambda \in \Lambda, \quad m \in M.$$

The Λ-modules obtained in this way are precisely those which are annihilated by W.

Moreover if a Λ-module is annihilated by W, then so is each sub- or quotient-module. In particular the direct sum of two Λ-modules is annihilated by W if and only if each summand is. Thus a Λ'-module is indecomposable if and only if it is indecomposable when considered as a Λ-module. This immediately implies the following result.

PROPOSITION 1. *If Λ' is a quotient algebra of Λ, then $M(\Lambda')$ may be canonically identified with a subset of $M(\Lambda)$.*

Now suppose that $R = R(\Lambda)$ is the radical of Λ; then for some integer m, $R^m = 0$. Thus for any Λ-module A there is a smallest integer h such that $R^h A = 0$. We call this h the *height* of A, and clearly $h \leqq m$.

Thus a module is of height $\leqq h$ if and only if it is annihilated by R^h, and so by Proposition 1 we may identify $M(\Lambda/R^h)$ with the subset of $M(\Lambda)$ consisting of the isomorphism classes of Λ-modules of height $\leqq h$.

If A is of height h, we have the upper Loewy series

$$A \supset RA \supset \cdots \supset R^{h-1}A \supset R^h A = 0,$$

and all inclusions are proper. On the other hand R annihilates each quotient of two successive terms, and so each quotient is semisimple. This establishes

PROPOSITION 2. *A Λ-module of height h is an $(h - 1)$-fold successive extension of semisimple modules. In particular a module of height 1 is semisimple, while a module of height 2 is an extension of one semisimple module by another.*

4. Height two modules over C-algebras

Let Λ be a C-algebra over K, and let R be its radical. Then $\Lambda/R \cong K$ shows that a semisimple Λ-module is just a vector space over K, so that $M(\Lambda/R)$ has just one element, namely, the class containing K.

As we have seen, the set of isomorphism classes of indecomposable Λ-modules of height $\leqq 2$ may be identified with $M(\Lambda/R^2)$. But Λ/R^2 depends only upon the rank of Λ, since we have

PROPOSITION 3. *Set* $\Delta_n = K[x_1, \cdots, x_n]/(x_1, \cdots, x_n)^2$, *where the* $\{x_i\}$ *are indeterminates over* K. *If* Λ *is any C-algebra over* K *of rank* n, *then*

$$\Lambda/R^2 \cong \Delta_n .$$

Proof. Let $u_1, \cdots, u_n \in R$ map onto a K-basis of R/R^2. For each $\lambda \in \Lambda$ let $\bar{\lambda}$ denote its image in Λ/R^2. Then we have at once

$$\Lambda/R^2 = K\bar{1} \oplus K\bar{u}_1 \oplus \cdots \oplus K\bar{u}_n .$$

On the other hand let $x \in K[x_1, \cdots, x_n]$ map onto $\tilde{x} \in \Delta_n$. Then

$$\Delta_n = K\tilde{1} \oplus K\tilde{x}_1 \oplus \cdots \oplus K\tilde{x}_n . \tag{6}$$

The map $\bar{1} \to \tilde{1}$, $\bar{u}_i \to \tilde{x}_i$ $(1 \leqq i \leqq n)$ thus gives the desired isomorphism.

COROLLARY. *The set of isomorphism classes of indecomposable* Λ-*modules of height* $\leqq 2$ *may be identified with* $M(\Delta_n)$, *where* n = *rank of* Λ.

We remark that (6) determines the structure of Δ_n, since $\tilde{1}$ is its unity element, and $\tilde{x}_i \tilde{x}_j = 0$ for all i, j. Set

$$S = K\tilde{x}_1 \oplus \cdots \oplus K\tilde{x}_n = \text{radical of } \Delta_n .$$

If A is any Δ_n-module, the sequence

$$0 \to SA \to A \to A/SA \to 0$$

is exact. Both SA and A/SA are annihilated by S, hence are semisimple Δ_n-modules, that is, they are vector spaces over K which are annihilated by S, and upon which K acts by scalar multiplication. For each i we define a K-homomorphism

$$\zeta_i : A/SA \to SA$$

by means of

$$a + SA \to \tilde{x}_i a, \qquad a \in A.$$

Then A is Δ_n-isomorphic to the space

$$A/SA \oplus SA,$$

the action on Δ_n on this space being given by

$$\tilde{x}_i(a + SA, b) = (0, \zeta_i a), \qquad a \in A,\ b \in SA,\ 1 \leqq i \leqq n. \tag{7}$$

We have thus shown that to each module A there corresponds a pair of vector spaces A/SA and SA, and an n-tuple of homomorphisms of the first space into the second. This pair of spaces, and the set of homomorphisms, completely determines A up to isomorphism.

Conversely let V, W be any pair of K-spaces, and let

$$\zeta_1, \cdots, \zeta_n \in \operatorname{Hom}_K(V, W)$$

be arbitrary. Define the action of Δ_n on $V \oplus W$ by letting K act by scalar multiplication, and using (7) to define the action of S. Then $V \oplus W$ becomes a Δ_n-module which we denote by

$$[V, W; \zeta_1, \cdots, \zeta_n],$$

and it is clear that the preceding construction associates with this module precisely the spaces V and W, and the homomorphisms $\zeta_1, \cdots, \zeta_n$.

Clearly $[V, W; \zeta_1, \cdots, \zeta_n] \cong [V', W'; \zeta_1', \cdots \zeta_n']$ if and only if there exist K-isomorphisms

$$\theta: V \cong V', \qquad \eta: W \cong W'$$

such that

$$\zeta_i' \theta = \eta \zeta_i, \qquad 1 \leqq i \leqq n.$$

We note further that the direct sum of the modules $[V, W; \zeta_1, \cdots, \zeta_n]$ and $[V', W'; \zeta_1', \cdots, \zeta_n']$ is just

$$[V \oplus V', W \oplus W'; \zeta_1 \oplus \zeta_1', \cdots, \zeta_n \oplus \zeta_n'].$$

We have thus introduced the concepts of isomorphism and decomposability for arrays $[V, W; \zeta_1, \cdots, \zeta_n]$, and have proved

PROPOSITION 4. *The elements of $M(\Delta_n)$ are in one-to-one correspondence with the set $S(n)$ of isomorphism classes of indecomposable arrays.*

(We have in fact constructed functors which connect the category of Δ_n-modules with that of arrays, and which provide a weak equivalence of these categories.)

The problem of determining a complete set of non-isomorphic indecomposable arrays is a classical problem of matrix theory, namely that of equivalence of matrix n-tuples. (In matrix terminology, we seek a complete set of non-equivalent indecomposable n-tuples of matrices, where "equivalence" is given by

$$(\mathbf{T}_1, \cdots, \mathbf{T}_n) \sim (\mathbf{P}\mathbf{T}_1 \mathbf{Q}, \cdots, \mathbf{P}\mathbf{T}_n \mathbf{Q}),$$

$\mathbf{P}$ and $\mathbf{Q}$ nonsingular.) The problem has been solved for $n \leqq 2$ (see [4], [7]), and is unsolved for $n > 2$. We shall use the solution for the case $n = 2$ to compute $M(\Delta_2)$, and hence to give a set of representatives for the isomorphism classes of indecomposable Λ-modules of height $\leqq 2$.

Since we are dealing with a C-algebra Λ of rank 2, we may write $\Lambda = K[u_1, u_2]$, where u_1 and $u_2 \in R(\Lambda)$ are such that their images form a K-basis for $R(\Lambda)/R(\Lambda)^2$. Then we have

PROPOSITION 5. *Up to isomorphism, there is only one indecomposable Λ-module of height* 1, *namely the space K on which K acts by scalar multiplication, and which is annihilated by u_1 and u_2. There are infinitely many non-isomorphic* *indecomposable Λ-modules of height 2, and a full set of these is given by the spaces $V \oplus W$, where*

$$V = Ka_1 \oplus \cdots \oplus Ka_r, \qquad W = Kb_1 \oplus \cdots \oplus Kb_s,$$

the action of K being scalar multiplication, and the action of u_1, u_2 given by

$$u_m \cdot a_i = \sum_{j=1}^{s} t_{ij}^{(m)} b_j, \qquad 1 \leqq i \leqq r, \quad m = 1, 2,$$

where

$$\mathbf{T}^{(1)} = (t_{ij}^{(1)}), \qquad \mathbf{T}^{(2)} = (t_{ij}^{(2)})$$

are $r \times s$ matrices over K given by the following choices:

(i)
$$\mathbf{T}^{(1)} = \mathbf{I}_{em}, \qquad \mathbf{T}^{(2)} = \mathbf{C}_e(p(x))$$

where $\mathbf{I}_{em}$ denotes the em-rowed identity matrix, e is an arbitrary positive integer, $p(x) = x^m - a_{m-1}x^{m-1} - \cdots - a_0$ is an arbitrary irreducible polynomial in $K[x]$, and $\mathbf{C}_e(p(x))$ is defined as

$$\mathbf{C}_e(p(x)) = \begin{bmatrix} \mathbf{B} & \mathbf{U} & \mathbf{0} & \cdots & \mathbf{0} \\ \mathbf{0} & \mathbf{B} & \mathbf{U} & \cdots & \mathbf{0} \\ \mathbf{0} & \mathbf{0} & \mathbf{B} & \cdots & \mathbf{0} \\ \cdot & \cdot & \cdot & \cdots & \mathbf{U} \\ \cdot & \cdot & \cdot & \cdots & \mathbf{B} \end{bmatrix}, \qquad e \ \mathbf{B}\text{'s occur,}$$

where

$$\mathbf{B} = \begin{bmatrix} 0 & 1 & 0 & \cdots & 0 \\ 0 & 0 & 1 & \cdots & 0 \\ \cdot & \cdot & \cdot & \cdots & \cdot \\ 0 & 0 & 0 & \cdots & 1 \\ a_0 & a_1 & a_2 & \cdots & a_{m-1} \end{bmatrix} = \text{companion matrix of } p(x),$$

$$\mathbf{U} = \begin{bmatrix} 0 & 0 & \cdots & 0 & 0 \\ 0 & 0 & \cdots & 0 & 0 \\ \cdot & \cdot & \cdots & \cdot & \cdot \\ 0 & 0 & \cdots & 0 & 0 \\ 1 & 0 & \cdots & 0 & 0 \end{bmatrix}.$$

(ii)

$$\mathbf{T}^{(1)} = \begin{bmatrix} 0 & \cdots & 0 & 0 \\ & & & 0 \\ & \mathbf{I}_m & & \vdots \\ & & & 0 \end{bmatrix}^{(m+1)\times(m+1)}, \qquad \mathbf{T}^{(2)} = \mathbf{I}_m,$$

(iii)

$$\mathbf{T}^{(1)} = \begin{bmatrix} 0 & \cdots & 0 \\ & \mathbf{I}_m & \end{bmatrix}^{(m+1)\times m}, \qquad \mathbf{T}^{(2)} = \begin{bmatrix} & \mathbf{I}_m & \\ 0 & \cdots & 0 \end{bmatrix}^{(m+1)\times m},$$

and

(iv)

$$\mathbf{T}^{(1)} = \begin{bmatrix} 0 & \\ \vdots & \mathbf{I}_m \\ 0 & \end{bmatrix}^{m\times(m+1)}, \qquad \mathbf{T}^{(2)} = \begin{bmatrix} & 0 \\ \mathbf{I}_m & \vdots \\ & 0 \end{bmatrix}^{m\times(m+1)},$$

where in (ii), (iii), *and* (iv) *m is an arbitrary positive integer.*

Remark. If K is algebraically closed, then $p(x) = x - \alpha$ for some $\alpha \in K$ and $\mathbf{C}_e(p(x))$ takes the simpler form

$$\mathbf{C}_e(p(x)) = \begin{bmatrix} \alpha & 1 & \cdots & 0 \\ 0 & \alpha & \cdots & 0 \\ \cdot & \cdot & \cdots & \cdot \\ 0 & 0 & \cdots & 1 \\ 0 & 0 & \cdots & \alpha \end{bmatrix}.$$

COROLLARY. *Let $G = G_1 \times G_2$, where for $i = 1, 2$, G_i is a cyclic group with generator g_i of order p^{α_i}, $\alpha_i > 0$. Let K be any field of characteristic p. Then there are infinitely many indecomposable KG-modules. A complete set of non-isomorphic indecomposable modules of height 2 is given by the above spaces $V \oplus W$, where the action of G is given as follows*:

$$(g_1 - 1)a_i = \sum t_{ij}^{(1)} b_j, \qquad (g_2 - 1)a_i = \sum t_{ij}^{(2)} b_j, \quad 1 \leqq i \leqq r,$$

and where

$$(g_1 - 1)W = (g_2 - 1)W = 0.$$

Finally we note that for $n \geqq 2$, Δ_2 is a quotient algebra of Δ_n, and hence by Proposition 1 we may conclude that $M(\Delta_n)$ is infinite. Thus $M(KG)$ is infinite whenever G is a non-cyclic abelian p-group, and K has characteristic p.

5. C-algebras of rank two

We have seen that if an abelian p-group G is a direct product of r cyclic groups, and K is a field of characteristic p, then KG is a C-algebra of rank r, and consequently $M(KG)$ contains a subset in one-to-one correspondence with $S(r)$, the set of isomorphism classes of indecomposable arrays $[V, W; \zeta_1, \cdots, \zeta_r]$. This shows that for $r > 2$ we cannot hope to find a complete system of non-isomorphic indecomposable KG-modules. We might expect, however, that this could be done for the special case where $r = 2$. The aim of this section is to show that even this special case leads to the problem of computing $S(p)$, and hence cannot be solved explicitly as soon as $p > 2$.

Let $G = G_1 \times G_2$, where for $i = 1, 2$, G_i is generated by an element g_i of order $r_i = p^{\alpha_i}$, $\alpha_i > 0$. Then we have seen that

$$KG \cong K[x_1, x_2]/(x_1^{r_1}, x_2^{r_2}),$$

and so surely

$$(x_1^{r_1}, x_2^{r_2}) \subset (x_1, x_2)^p.$$

We now prove generally

PROPOSITION 6. *Let $\Lambda = K[x, y]/J$ be a C-algebra of rank 2 such that for some $n > 2$,*

$$J \subset (x, y)^n.$$

Then $M(\Lambda)$ contains a subset in one-to-one correspondence with $S(n)$.

Proof. We begin by observing that

$$\Lambda_n = K[x, y]/(x, y)^n$$

is a quotient of Λ, so that $M(\Lambda_n) \subset M(\Lambda)$, and it suffices to prove the result for $M(\Lambda_n)$. Let x and y map onto X and Y, respectively, in Λ_n ; then

$$\Lambda_n = K[X, Y], \qquad (X, Y)^n = 0.$$

Using formula (6) for Δ_n, we embed Δ_n in Λ_n by the mapping

$$\psi(\bar{1}) = 1, \quad \psi(\bar{x}_1) = X^{n-1}, \quad \psi(\bar{x}_2) = X^{n-2}Y, \quad \cdots, \quad \psi(\bar{x}_n) = Y^{n-1},$$

which is easily seen to be an algebra isomorphism of Δ_n into Λ_n. By means of this embedding we may regard Δ_n as a subalgebra of Λ_n.

If A is a Δ_n-module, define

$$A^* = \Lambda_n \otimes_{\Delta_n} A, \tag{8}$$

which is a Λ_n-module. The correspondence $A \to A^*$ preserves isomorphisms and direct sums. In the other direction we proceed as follows. Let

$$R = (X, Y) = \text{radical of } \Lambda_n .$$

Then (as a subalgebra of Λ_n)

$$\Delta_n = K \cdot 1 \oplus R^{n-1}, \tag{9}$$

and $R^{n-1} = S$ is the radical of Δ_n. If B is a Λ_n-module, then for $1 \leqq i \leqq n$ we have

$$\bar{x}_i B = X^{n-i}Y^{i-1}B \subset R^{n-1}B,$$

$$\bar{x}_i \cdot RB \subset R^n B = 0,$$

and so there exists a K-homomorphism $\theta_i : B/RB \to R^{n-1}B$ given by

$$\theta_i(b + RB) = \bar{x}_i b, \qquad b \in B.$$

Setting

$$B' = B/RB \oplus R^{n-1}B,$$

we may therefore make B' into a Δ_n-module by defining for each i,

$$\bar{x}_i(\bar{b}, b_1) = (0, \theta_i \bar{b}), \qquad \bar{b} \in B/RB, \quad b_1 \in R^{n-1}B.$$

The correspondence $B \to B'$ maps Λ_n-modules onto Δ_n-modules and clearly preserves isomorphisms and direct sums.

We shall prove that for any Δ_n-module A, we have

$$(A^*)' \cong A, \tag{10}$$

so that each class in $M(\Delta_n)$ determines a class in $M(\Lambda_n)$, and the result follows from Proposition 4.

We have shown in Section 4 that

$$A \cong A/SA \oplus SA,$$

the action of Δ_n on the right-hand side being given by

$$\tilde{x}_i(a + SA, a_1) = (0, \tilde{x}_i a), \qquad a \in A, \quad a_1 \in SA.$$

On the other hand every element of A^* is expressible as a sum

$$\sum_{0 \leqq i+j \leqq n-1} X^i Y^j \otimes a_{ij}, \qquad a_{ij} \in A.$$

But we have

$$X^{n-i} Y^{i-1} \otimes a = 1 \otimes \tilde{x}_i a, \qquad a \in A,$$

and so every element of A^* is expressible as

$$a^* = 1 \otimes a_0 + \sum_{0 < i+j < n-1} X^i Y^j \otimes a_{ij}, \qquad a_0 \in A, \quad \{a_{ij}\} \in A.$$

To compute $(A^*)'$, we determine RA^*:

$$Xa^* = X \otimes a_0 \oplus \sum_{0 < i+j < n-2} X^{i+1} Y^j \otimes a_{ij} \oplus \sum_{i+j=n-2} 1 \otimes \tilde{x}_{n-i-1} a_{ij},$$

and likewise for Ya^*. Thus

$$A^*/RA^* \cong (1 \otimes A)/(1 \otimes SA) \cong A/SA.$$

Furthermore

$$R^{n-1}A^* = 1 \otimes SA \cong SA.$$

Thus

$$(A^*)' \cong A/SA \oplus SA,$$

where the action of each $\tilde{x}_i$ is given by

$$\begin{aligned} \tilde{x}_i(a + SA, a_1) &= \tilde{x}_i(1 \otimes a + 1 \otimes SA, 1 \otimes SA) \\ &= 1 \otimes \tilde{x}_i a = (0, \tilde{x}_i a), \qquad a \in A, \quad a_1 \in SA. \end{aligned}$$

This completes the proof of (10), and establishes the proposition.

REFERENCES

1. H. R. BRAHANA, *Metabelian groups of order p^{n+m} with commutator subgroups of order p^m*, Trans. Amer. Math. Soc., vol. 36 (1934), pp. 776–792.
2. ———, *Metabelian groups and trilinear forms*, Duke Math. J., vol. 1 (1935), pp. 185–197.
3. ———, *Metabelian groups and pencils of bilinear forms*, Amer. J. Math., vol. 57 (1935), pp. 645–667.
4. J. DIEUDONNÉ, *Sur la réduction canonique des couples de matrices*, Bull. Soc. Math. France, vol. 74 (1946), pp. 130–146.

5. F. KASCH, M. KNESER UND H. KUPISCH, *Unzerlegbare modulare Darstellungen endlicher Gruppen mit zyklischer p-Sylow-Gruppe*, Arch. Math., vol. 8 (1957), pp. 320–321.
6. D. G. HIGMAN, *Indecomposable representations at characteristic p*, Duke Math. J., vol. 21 (1954), pp. 377–381.
7. G. PICKERT, *Normalformen von Matrizen*, Enzyklopädie der mathematischen Wissenschaften (2. Aufl.), Bd. I, Nr. 7, Leipzig, 1953.

UNIVERSITY OF ILLINOIS
URBANA, ILLINOIS

ON THE NUMBER OF MATRICES WITH GIVEN CHARACTERISTIC POLYNOMIAL

BY
IRVING REINER[1]

1. Introduction

Let K be a finite field with q elements, and let K_n denote the ring of all $n \times n$ matrices with entries in K. Recently Fine and Herstein proved[2]

The number of nilpotent matrices in K_n is q^{n^2-n}.

We shall prove here the following generalizations.

THEOREM 1. *Let $f(x)$ be an irreducible polynomial in $K[x]$ of degree $d \geqq 1$. Then the number of matrices $X \in K_{nd}$ for which $f(X)$ is nilpotent is*

$$q^{n^2d^2-nd} \cdot \frac{(1-q^{-1})(1-q^{-2}) \cdots (1-q^{-nd})}{(1-q^{-d})(1-q^{-2d}) \cdots (1-q^{-nd})}. \tag{1}$$

Before stating the second result to be proved here, which includes the above theorem as a special case, we introduce some notation. Define

$$F(u, r) = (1-u^{-1})(1-u^{-2}) \cdots (1-u^{-r}), \tag{2}$$

where $F(u, 0) = 1$. Then we have[3]

THEOREM 2. *Let $g(x) \in K[x]$ be a polynomial of degree n, and let*

$$g(x) = f_1^{n_1}(x) \cdots f_k^{n_k}(x) \tag{3}$$

be its factorization in $K[x]$ into powers of distinct irreducible polynomials $f_1(x), \cdots, f_k(x)$. Set

$$d_i = \text{degree of } f_i(x), \qquad 1 \leqq i \leqq k.$$

Then the number of matrices $X \in K_n$ with characteristic polynomial $g(x)$ is

$$q^{n^2-n} \cdot \frac{F(q, n)}{\prod_{i=1}^{k} F(q^{d_i}, n_i)}. \tag{4}$$

The proofs of these theorems do not require a knowledge of the Fine-Herstein paper, except for the following combinatorial lemma which they establish and which we state without proof.

Received July 2, 1960.

[1] This research was supported by a contract with the Office of Naval Research.

[2] N. J. FINE AND I. N. HERSTEIN, *The probability that a matrix be nilpotent*, Illinois J. Math., vol. 2 (1958), pp. 499–504.

[3] Another proof of Theorems 1 and 2 is given by M. GERSTENHABER, *On the number of nilpotent matrices with coefficients in a finite field*, Illinois J. Math., vol. 5 (1961), pp. 330–333.

LEMMA 1 (Fine-Herstein). *Let u be any complex number which is not a root of unity. Let $\{r_1, \cdots, r_n\}$ range over all n-tuples of non-negative integers for which*

$$r_1 + 2r_2 + \cdots + nr_n = n,$$

and set

$$s_j = r_j + r_{j+1} + \cdots + r_n, \qquad 1 \leqq j \leqq n.$$

Then

$$(5) \qquad \sum_{\{r_1,\cdots,r_n\}} \frac{u^{s_1^2+s_2^2+\cdots+s_n^2}}{F(u^{-1}, r_1)F(u^{-1}, r_2) \cdots F(u^{-1}, r_n)} = \frac{u^n}{F(u^{-1}, n)}.$$

2. Automorphisms of modules over local rings

Throughout this section we let R be a commutative local ring with unity element, and let πR be its unique maximal ideal. Suppose further that π is nilpotent, say $\pi^n = 0$, and let

$$t = \text{number of elements in the field } R/\pi R.$$

Then

$$R \supset \pi R \supset \pi^2 R \supset \cdots \supset \pi^{n-1}R \supset \pi^n R = (0)$$

is a descending chain of ideals of R in which every ideal occurs, and each quotient is isomorphic (as R-module) to the field $R/\pi R$. Thus R contains t^n elements, and more generally $R/\pi^j R$ contains t^j elements.

We shall restrict our attention to R-modules which are finitely generated. Since R is a principal ideal domain, each such R-module V is a direct sum of cyclic R-modules. Moreover every nonzero cyclic R-module is a homomorphic image of R, hence is of the form $R/\pi^j R$ for some j, $1 \leqq j \leqq n$. Set

$$V_j = R/\pi^j R, \qquad 1 \leqq j \leqq n.$$

Then V_j contains t^j elements, and is indecomposable since it contains a unique minimal submodule $\pi^{j-1}V_j$. Thus every R-module V is expressible as

$$(6) \qquad \begin{aligned} V &= W_1 \oplus \cdots \oplus W_n, \\ W_j &= V_j \oplus \cdots \oplus V_j \qquad (r_j \text{ summands}), \end{aligned}$$

and such an expression is unique by the Krull-Schmidt Theorem.

LEMMA 2. *Let W_j be given in (6). The number of R-automorphisms of W_j is precisely*

$$t^{jr_j^2} F(t, r_j).$$

Proof. Since π^j annihilates W_j, we may regard W_j as an R'-module, where

$$R' = R/\pi^j R.$$

The number of R-automorphisms of W_j is then the same as the number of nonsingular $r_j \times r_j$ matrices X with entries in R'. Now a matrix X over R' is nonsingular if and only if $\bar{X}$ is nonsingular, where $\bar{X}$ is obtained from X by mapping each entry α of X onto its image $\bar{\alpha}$ in $R'/\pi R'$. Since $\bar{X}$ has its entries in the field $R'/\pi R' \cong R/\pi R$, there are

$$t^{r_j^2}F(t, r_j)$$

possible choices for $\bar{X}$. But for given $\bar{\alpha}$ there are t^{j-1} choices for $\alpha \in R'$, and thus the number of nonsingular matrices X over R' of size $r_j \times r_j$ is

$$(t^{j-1})^{r_j^2} \cdot t^{r_j^2}F(t, r_j).$$

This proves the lemma.

LEMMA 3. *Let V be given by* (6), *and set*

$$s_j = r_j + r_{j+1} + \cdots + r_n, \qquad 1 \leqq j \leqq n. \tag{7}$$

The number of R-automorphisms of V is then

$$N_V = \prod_{j=1}^n t^{s_j^2}F(t, r_j). \tag{8}$$

Proof. For convenience we rewrite (6) as

$$V = \sum_{j=1}^n \sum_{i=1}^{r_j} V_j e_{ji},$$

where e_{ji} is just an indexing mark, say

$$e_{ji} = (0, \cdots, 0, 1, 0, \cdots, 0)$$

with the 1 in an appropriate position. Any R-homomorphism is completely determined by its effect on the $\{e_{ji}\}$.

For $v \in V$, $v \neq 0$, define the *order* of v to be the smallest integer s for which $\pi^s v = 0$. Let us say that 0 has order zero. The elements of W_j have order $\leqq j$, clearly.

Now let θ be an R-automorphism of V. Then θ preserves order, so that for $1 \leqq j \leqq n$ we have

$$\theta(W_j) \subset W_1 + \cdots + W_{j-1} + W_j + \pi W_{j+1} + \cdots + \pi^{n-j}W_n.$$

Hence if we set

$$\theta(e_{ji}) = \sum_m a_{ji}^{(m)}, \qquad a_{ji}^{(m)} \in W_m, \tag{9}$$

then we see that for $m > j$ we have

$$a_{ji}^{(m)} \in \pi^{m-j}W_m. \tag{10}$$

Furthermore, for fixed j the mapping

$$e_{ji} \to a_{ji}^{(j)}, \qquad 1 \leqq i \leqq r_j, \tag{11}$$

must be an R-automorphism of W_j. It is easy to see that conversely if we define an R-homomorphism θ by means of (9) and (10), where for each j $(1 \leqq j \leqq n)$ equation (11) gives an R-automorphism of W_j, then θ is indeed an R-automorphism of V.

For fixed j, $1 \leqq j \leqq n$, the elements $\{a_{ji}^{(m)}: m < j\}$ may be chosen arbitrarily. Since there are r_j choices to be made, and $W_1 + \cdots + W_{j-1}$ contains

$$t^{1r_1+2r_2+\cdots+(j-1)r_{j-1}}$$

elements, this gives

$$t^{r_j(1r_1+2r_2+\cdots+(j-1)r_{j-1})} \tag{12}$$

possibilities for the $\{a_{ji}^{(m)}: m < j,\ 1 \leqq i \leqq r_j\}$.

Next the set of elements $\{a_{ji}^{(j)}: 1 \leqq i \leqq r_j\}$ may be chosen in

$$t^{jr_j^2}F(t, r_j) \tag{13}$$

ways, by Lemma 2. Finally, since for $m > j$ there are exactly t^{jr_m} elements in $\pi^{m-j}W_m$, there are

$$t^{jr_j(r_{j+1}+\cdots+r_n)} \tag{14}$$

choices for the elements $\{a_{ji}^{(m)}: m > j,\ 1 \leqq i \leqq r_j\}$. The number of R-automorphisms of V is therefore

$$N_V = \prod_{j=1}^{n} \{t^{u_j}F(t, r_j)\},$$

where for each j,

$$u_j = \sum_{m=1}^{j} mr_m r_j + jr_j \sum_{m=j+1}^{n} r_m .$$

If we define the symbols $\{s_j\}$ by (7), a routine calculation establishes (8).

(The above generalizes the formula for N_V obtained by Fine-Herstein in pp. 500–502, loc. cit., where N_V is referred to as μ in their paper.)

Now let V range over a full set of non-isomorphic R-modules having exactly t^n elements, so that $\{r_1, \cdots, r_n\}$ range over all n-tuples of non-negative integers for which

$$n = r_1 + 2r_2 + \cdots + nr_n .$$

LEMMA 4. *As V ranges over the above-mentioned R-modules, we have*

$$\sum_V 1/N_V = 1/t^n F(t, n). \tag{15}$$

Proof. Use the formula (8) for N_V, and then apply Lemma 1 with $u = t^{-1}$.

3. Nilpotent matrix polynomials

Let K be a field with q elements, $f(x) \in K[x]$ an irreducible polynomial of degree $d \geqq 1$, and let n be a fixed integer. We wish to determine the number of matrices $X \in K_{nd}$ for which $f(X)$ is nilpotent. We remark that $f(X)$ is nilpotent if and only if $f^n(X) = 0$, since $f(X)$ is nilpotent if and only if the characteristic polynomial of X is $f^n(x)$.

Define the ring R by

$$R = K[x]/(f^n(x)),$$

and for each polynomial $g(x) \in K[x]$ let $\overline{g(x)}$ denote its image in R. Then R is a commutative ring of the type discussed in the preceding section, with maximal ideal πR, where $\pi = \overline{f(x)}$. We have $\pi^n = 0$, and the number t of elements in the field $R/\pi R$ is given by

$$(16) \qquad t = q^d,$$

since $R/\pi R \cong K[x]/(f(x))$.

If V is any R-module of K-dimension nd, then V contains t^n elements. Furthermore V gives rise to a representation of R by matrices in K_{nd}, and the matrix X corresponding to $\bar{x}$ satisfies $f^n(X) = 0$. Conversely each such matrix X is obtainable in this way from some R-module with t^n elements.

For the rest of the proof we restrict ourselves to R-modules V with t^n elements. Each V gives rise to a set of equivalent matrix representations, and hence gives not only one matrix X corresponding to $\bar{x}$, but a system of matrices

$$\{P^{-1}XP : P \in K_{nd},\ P \text{ nonsingular}\}.$$

The number of distinct matrices in this system is just the number $q^{n^2d^2}F(q, nd)$ of nonsingular matrices in K_{nd}, divided by the number of nonsingular matrices $P \in K_{nd}$ satisfying

$$P^{-1}XP = X.$$

But since $\bar{x}$ generates the ring R, any such P yields an R-automorphism of V, and so there are N_V such nonsingular P's, where N_V is given by (8) with $t = q^d$.

On the other hand it is clear that non-isomorphic R-modules V, V^* give rise to matrices X, X^* which are not connected by any relation

$$X^* = P^{-1}XP, \qquad P \in K_{nd}, \quad P \text{ nonsingular}.$$

The above discussion shows therefore that the number of matrices $X \in K_{nd}$ for which $f(X)$ is nilpotent is precisely

$$\textstyle\sum_V q^{n^2d^2}F(q, nd)/N_V,$$

where V ranges over a full set of non-isomorphic R-modules having t^n elements. By using (15), the above is just

$$q^{n^2d^2}F(q, nd)/q^{nd}F(q^d, n),$$

that is,

$$q^{n^2d^2-nd}\cdot\frac{(1 - q^{-1})(1 - q^{-2})\cdots(1 - q^{-nd})}{(1 - q^{-d})(1 - q^{-2d})\cdots(1 - q^{-nd})}.$$

This completes the proof of Theorem 1.

4. Matrices with given characteristic polynomial

We are now ready to prove Theorem 2. Let $g(x)$ be given by (3), and let

$$S = K[x]/(g(x)) = R_1 \oplus \cdots \oplus R_k,$$

where

$$R_i = K[x]/(f_i^{n_i}(x)), \qquad 1 \leqq i \leqq k.$$

Any S-module V can be decomposed into a direct sum

$$V = V_1 \oplus \cdots \oplus V_k,$$

in which V_i is a left R_i-module, $1 \leqq i \leqq k$. We obtain all matrices $X \in K_n$ with characteristic polynomial $g(x)$ by letting V range over a full set of non-isomorphic S-modules of dimension n over K, chosen in such a way that

$$(V_1:K) = n_1 d_1, \quad \cdots, \quad (V_k:K) = n_k d_k,$$

and then for each such module V taking the set of matrices which correspond to $\bar{x} \in S$ (the image of $x \in K[x]$). Thus the number of matrices $X \in K_n$ with characteristic polynomial $g(x)$ is just

$$\textstyle\sum_V q^{n^2} F(q, n)/N_V.$$

It follows readily from the fact that the $\{f_i(x)\}$ are pairwise relatively prime that any S-automorphism of V maps each V_i onto itself, and thus is composed of a set of k automorphisms $\{\theta_i : 1 \leqq i \leqq k\}$, where $\theta_i : V_i \to V_i$. Therefore

$$N_V = N_{V_1} \cdots N_{V_k}.$$

Furthermore, a full set of non-isomorphic S-modules V of the type described above is obtained by letting each V_i range independently over a full set of non-isomorphic R_i-modules with $(V_i:K) = n_i d_i$, for $i = 1, \cdots, k$. Thus the number of matrices $X \in K_n$ with characteristic polynomial $g(x)$ is

$$\begin{aligned} q^{n^2} F(q, n) \textstyle\sum_V 1/N_{V_1} \cdots N_{V_k} &= q^{n^2} F(q, n) \textstyle\prod_{i=1}^k \{ \sum_{V_i} 1/N_{V_i} \} \\ &= q^{n^2} F(q, n) \cdot \{ \textstyle\prod_{i=1}^k q^{d_i n_i} F(q^{d_i}, n_i) \}^{-1}. \end{aligned}$$

Using the relation $n = \sum d_i n_i$, we obtain formula (4). This completes the proof of Theorem 2.

UNIVERSITY OF ILLINOIS
URBANA, ILLINOIS

FAILURE OF THE KRULL-SCHMIDT THEOREM FOR INTEGRAL REPRESENTATIONS

Irving Reiner

1. The following notation will be used throughout:

G is a finite group of order g;

K is an algebraic number field;

R is the ring of all algebraic integers in K;

P is a prime ideal in R;

R_P is the P-adic valuation ring in $K = \{\alpha/\beta: \alpha, \beta \in R, \beta \notin P\}$;

K_P^* is the P-adic completion of K, and R_P^* the ring of P-adic integers in K_P^*;

$\tilde{R} = \bigcap_{P|g} R_P = \{\alpha/\beta: \alpha, \beta \in R, R\beta + Rg = R\}$.

Let RG denote the group ring of G with coefficients from R. By an RG-*module* we shall always mean a finitely-generated left RG-module which is R-torsion-free, and upon which the identity element of G acts as identity operator. Analogous definitions hold for R_PG-modules, KG-modules, and so forth.

THEOREM 1.1 (Krull-Schmidt). *In any decomposition of a* KG-*module* M *into a direct sum of indecomposable submodules, the indecomposable summands are uniquely determined by* M, *up to* KG-*isomorphism and order of occurrence.*

The standard proof (see, for example, Curtis and Reiner [2, p. 83]) shows that K may be replaced by any commutative ring whose ideals satisfy the descending chain condition.

In the present paper we wish to consider the validity of the Krull-Schmidt theorem for RG-modules. Let us observe at once that the theorem already fails when $G = \{1\}$ if R contains non-principal ideals. Let $J_1, \cdots, J_n$ be ideals of R, and let $\dotplus$ denote the external direct sum operation. As is well known,

$$J_1 \dotplus \cdots \dotplus J_n \cong R \dotplus \cdots \dotplus R \dotplus J_1 \cdots J_n,$$

where $n - 1$ R's occur on the right-hand side.

Returning to an arbitrary finite group G, we might reasonably hope that the non-principal ideals of R are the only source of counterexamples. To avoid the difficulties arising from them, we may work with $\tilde{R}G$-modules instead of RG-modules, where $\tilde{R}$ is the principal ideal ring defined above.

To each RG-module M there corresponds an $\tilde{R}G$-module, denoted by $\tilde{R}M$ and defined by

$$\tilde{R}M = \tilde{R} \otimes_R M.$$

Received March 2, 1962.

This research was supported in part by a contract with the Office of Naval Research.

Clearly, $M \cong N$ implies $\tilde{R}M \cong \tilde{R}N$, but not conversely. On the other hand, $\tilde{R}M \cong \tilde{R}N$ if and only if for each P dividing g, $R_P M \cong R_P N$.

The usefulness of $\tilde{R}$ stems from the following result.

THEOREM 1.2. *An* RG-*module* M *is indecomposable if and only if the corresponding* $\tilde{R}G$-*module* $\tilde{R}M$ *is indecomposable.*

Proof. If M is decomposable, then obviously $\tilde{R}M$ is decomposable. Conversely, let X be an $\tilde{R}G$-direct summand of $\tilde{R}M$, and define $N = M \cap X$. It is easily verified that N is an RG-submodule of M for which $\tilde{R}N = X$, and such that M/N is R-torsion-free.

For each P dividing g, $\tilde{R}$ is a subring of R_P. Since X is a direct summand of $\tilde{R}M$, it follows at once that for each such P, $R_P N$ is an $R_P G$-direct summand of $R_P M$. This implies (see deLeeuw [3], Reiner [6]) that N is an RG-direct summand of M, and the theorem is proved.

The preceding result is quite useful in the determination of indecomposable RG-modules. Furthermore, if $\Sigma^{\oplus} M_i$ is a direct sum of indecomposable RG-modules, then $\Sigma^{\oplus} \tilde{R}M_i$ is a direct sum of indecomposable $\tilde{R}G$-modules. If one could establish a Krull-Schmidt theorem for $\tilde{R}G$-modules, then the $\{M_i\}$ would be unique up to order of occurrence and $\tilde{R}G$-isomorphism.

Our principal result, however, is that the Krull-Schmidt theorem does not hold either for $\tilde{R}G$- or for RG-modules, whenever G contains a normal subgroup of prime index and G is not a p-group. Indeed, in this case not even the R-ranks of the $\{M_i\}$ are uniquely determined.

To conclude this introduction, we recall that the Krull-Schmidt theorem holds for R_P^*G-modules (see Borevich and Fadeev [1], Reiner [7], Swan [8]). It also holds for $R_P G$-modules (see Heller [4]) whenever K is a splitting field for G, that is, whenever KG splits into a direct sum of full matrix algebras over K. Still unsettled is the question as to whether this latter hypothesis may be omitted.

2. In this section we shall show how to construct counterexamples to the Krull-Schmidt theorem for RG-modules, whenever there exist RG-modules with certain properties. We shall write Ext instead of Ext^1_{RG}, for convenience. If M and N are RG-modules, then to each $F \in \text{Ext}(N, M)$ there corresponds an RG-module which is an extension of N by M with extension class F. We denote this module by (M, N; F) or by

$$\begin{pmatrix} M & F \\ & N \end{pmatrix};$$

the latter notation is used to remind us of the matrix representation afforded by this module.

To each RG-module M there corresponds a KG-module denoted by KM, and defined as $K \otimes_R M$.

LEMMA 2.1. *Let* M *and* N *be indecomposable* RG-*modules such that*

$$\text{Hom}_{KG}(KM, KN) = 0 \quad \textit{and} \quad \text{Hom}_{KG}(KN, KM) = 0,$$

and let $F \in \text{Ext}(N, M)$. *Then the* RG-*module* (M, N; F) *is decomposable if and only if* $F = 0$.

Proof. See Heller and Reiner [5, II].

Suppose now that A, B, and C are RG-modules satisfying the following conditions:

(I) The modules KA, KB, and KC are irreducible, and no two of them are isomorphic.

(II) There exist non-zero elements $F \in \text{Ext}(B, A)$ and $F' \in \text{Ext}(C, A)$, such that the orders of F and F' are relatively prime integers.

THEOREM 2.2. *Let* A, B, *and* C *be* RG-*modules satisfying* (I) *and* (II). *Then the modules* A, (A, B; F), (A, C; F') *and* (A, B $\dotplus$ C; F + F') *are indecomposable, and*

$$A \dotplus (A, B \dotplus C; F + F') \cong (A, B; F) \dotplus (A, C; F').$$

Proof. Indecomposability of the above modules follows readily from the preceding lemma. Now let m be the order of F, let n be the order of F', and choose an integer k such that $kn \equiv 1 \pmod{m}$. In matrix notation, the module $A \dotplus (A, B \dotplus C; F + F')$ may be written as

$$M = \left[\begin{array}{c|ccc} A & & & \\ \hline & A & F & F' \\ & & B & 0 \\ & & & C \end{array}\right].$$

Let

$$X_1 = \left[\begin{array}{c|ccc} I & knI & & \\ \hline & I & & \\ & & I & \\ & & & I \end{array}\right],$$

the symbols I denoting identity matrices of appropriate sizes. Then

$$M_1 = X_1 M X_1^{-1} = \left[\begin{array}{c|ccc} A & 0 & knF & knF' \\ \hline & A & F & F' \\ & & B & 0 \\ & & & C \end{array}\right].$$

The entry knF' lies in Ext(C, A), and it lies in the zero class. Thus if we set

$$X_2 = \left[\begin{array}{c|ccc} I & & & T \\ \hline & I & & \\ & & I & \\ & & & I \end{array}\right],$$

then for a suitable choice of T we obtain the relation

$$M_2 = X_2 M_1 X_2^{-1} = \left[\begin{array}{c|ccc} A & 0 & knF & 0 \\ \hline & A & F & F' \\ & & B & 0 \\ & & & C \end{array}\right].$$

On the other hand, $knF = F$ in $\mathrm{Ext}(B, A)$. Set

$$X_3 = \begin{bmatrix} I & & & \\ -I & I & & \\ & & I & \\ & & & I \end{bmatrix}.$$

Then

$$M_3 = X_3 M_2 X_3^{-1} = \left[\begin{array}{c|ccc} A & 0 & F & 0 \\ \hline & A & 0 & F' \\ & & B & 0 \\ & & & C \end{array}\right].$$

Since $M_3 \cong (A, B; F) \dotplus (A, C; F')$, the theorem is established.

Thus, once we know the existence of RG-modules A, B, and C satisfying (I) and (II), the Krull-Schmidt theorem cannot possibly hold for RG-modules. Indeed, the R-ranks of the indecomposable summands in a direct sum are not uniquely determined by that direct sum.

3. We shall now show the existence of RG-modules for which (I) and (II) hold, provided that the group G satisfies certain hypotheses. One preliminary result will be needed.

LEMMA 3.1. *Let* p *be a prime divisor of the order of* G, *and let* A *be the* RG-*module* R *on which* G *acts trivially. Then there exists an* RG-*module* B *such that*

(i) KB *is irreducible,* $KB \ncong KA$, *and*

(ii) Ext (B, A) *contains a non-zero element of order* p.

Proof. Suppose the result false, and let P be a prime ideal of R which divides p. Then for each RG-module satisfying (i), the p-primary component of Ext (B, A) is zero, and thus

$$R_P \cdot \mathrm{Ext}(B, A) = 0.$$

This in turn implies that

$$\mathrm{Ext}(R_P B, R_P A) = 0.$$

Let M be the quotient module $R_P G/R_P A$. Then there is an exact sequence of $R_P G$-modules

$$0 \to R_P A \to R_P G \to M \to 0. \tag{3.2}$$

We shall show that $\operatorname{Ext}(M, R_P A) \neq 0$. For otherwise, the above sequence splits. If we write $\bar{R} = R/P$, $\overline{M} = M/PM$, and so on, then $\overline{R_P A} \cong \bar{R}$ as $\bar{R}G$-modules, where G acts trivially on $\bar{R}$. If the sequence (3.2) splits, then so does

$$(3.3) \qquad 0 \to \bar{R} \to \bar{R}G \to \overline{M} \to 0.$$

Let H be a p-Sylow subgroup of G; each $\bar{R}G$-module can be viewed as an $\bar{R}H$-module, and then (3.3) also splits as an exact sequence of $\bar{R}H$-modules. On the other hand, $\bar{R}H$ is an indecomposable $\bar{R}H$-module (see Curtis and Reiner [2, Section 54, Exercise 1], for example), and $\bar{R}G$ is (as $\bar{R}H$-module) a direct sum of $[G:H]$ copies of $\bar{R}H$. We have thus obtained a contradiction to the Krull-Schmidt theorem for $\bar{R}H$-modules. Therefore we have proved that $\operatorname{Ext}(M, R_P A) \neq 0$.

Next we observe that the irreducible module KA cannot occur as a composition factor of KM, since KA occurs with multiplicity 1 as a composition factor of the left regular module KG. Suppose for the moment that KM is itself irreducible. We may write $M = R_P M_0$ for some RG-module M_0, and then

$$\operatorname{Ext}(M, R_P A) = R_P \operatorname{Ext}(M_0, A).$$

This implies that $\operatorname{Ext}(M_0, A)$ has a non-zero p-primary component and so must contain a non-zero element of order p. Choosing $B = M_0$, we obtain the desired module.

On the other hand, if KM is reducible, we can find an R_P-pure $R_P G$-submodule N of M of lower R_P-rank, and then there exists an exact sequence

$$0 \to N \to M \to L \to 0,$$

say. From this we get an exact sequence

$$\operatorname{Ext}(L, R_P A) \to \operatorname{Ext}(M, R_P A) \to \operatorname{Ext}(N, R_P A),$$

and thus at least one of $\operatorname{Ext}(L, R_P A)$ and $\operatorname{Ext}(N, R_P A)$ is non-zero. Continuing in this manner, after a finite number of steps we arrive at an $R_P G$-module V such that KV is irreducible and is a composition factor of KM, and such that $\operatorname{Ext}(V, R_P A) \neq 0$. The rest of the argument is as in the preceding paragraph. This completes the proof.

We are now ready to prove

THEOREM 3.4. *Suppose that the order of* G *has at least two distinct prime divisors, and that* G *contains a normal subgroup of prime index. Then there exist* RG-*modules* A, B, *and* C *satisfying conditions* (I) *and* (II); *in fact,* A *may be chosen to be the* RG-*module* R *on which* G *acts trivially.*

Proof. Let G_0 be a normal subgroup of G, of prime index p, and let H be a cyclic group of order p. Then there is a homomorphism of G onto H with kernel G_0. Let $g \in G$ map onto $\bar{g} \in H$ under this homomorphism. Each RH-module M can be turned into an RG-module, again denoted by M, by defining

$$g \cdot m = \bar{g}m \qquad (g \in G,\ m \in M).$$

Indecomposable RH-modules become indecomposable RG-modules in this process, and irreducibility (as KH- or KG-modules) is also preserved. Furthermore, for a pair of RH-modules M and N,

$$\mathrm{Ext}_{RG}(M, N) = \mathrm{Ext}_{RH}(M, N),$$

where on the left-hand side M and N are viewed as RG-modules.

Choose A to be the RG-module on which G acts trivially. Then A is also an RH-module on which H acts trivially. By the preceding lemma, there exists an RH-module B satisfying conditions (i) and (ii) of that lemma. On viewing B as an RG-module, it is clear that KB is irreducible, $KB \not\cong KA$, and $\mathrm{Ext}_{RG}(B, A)$ contains a non-zero element of order p.

Now let q be a prime divisor of g distinct from p. By Lemma 3.1, there exists an RG-module C such that KC is irreducible, $KC \not\cong KA$, and Ext (C, A) contains a non-zero element of order q. Surely KC and KB are not KG-isomorphic. For if they were, then C could be viewed as an RH-module, and then the order of Ext (C, A) would be a power of p. This completes the proof of the theorem.

For any solvable group G which is not a p-group, the hypotheses of the preceding theorem hold automatically, and thus there exist RG-modules satisfying (I) and (II). We may conjecture that such modules exist for each finite group other than a p-group, but it is not clear how to prove their existence when G is a simple group, for example.

4. Since the Krull-Schmidt theorem fails for $\tilde{R}G$-modules, as well as for RG-modules, it is desirable to know under what conditions two direct sums of indecomposable $\tilde{R}G$-modules are isomorphic. This can be decided in a fairly trivial manner.

We have already remarked that if M and N are a pair of $\tilde{R}G$-modules, then $M \cong N$ if and only if $R_P M \cong R_P N$ for each prime ideal P dividing g. If M is an indecomposable $\tilde{R}G$-module, it may very well happen that $R_P M$ is decomposable as R_P G-module.

For convenience of notation, let b[M] denote the direct sum of b copies of the module M, where b is a positive integer. Now let $M_1, \cdots, M_r, N_1, \cdots, N_s$ be indecomposable $\tilde{R}G$-modules. We would like to know when there exists an isomorphism

$$a_1[M_1] \dotplus \cdots \dotplus a_r[M_r] \cong b_1[N_1] \dotplus \cdots \dotplus b_s[N_s], \tag{4.1}$$

where the $\{a_i\}$ and $\{b_i\}$ are positive integers.

For each P dividing g, let $V_1^P, V_2^P, \cdots$ denote a full set of non-isomorphic indecomposable $R_P G$-modules. Then we may write each $R_P M_i$ as a finite direct sum

$$R_P M_i = m_{i1}^P[V_1^P] \dotplus m_{i2}^P[V_2^P] \dotplus \cdots \qquad (1 \leq i \leq r),$$

where only finitely many non-zero coefficients occur. Likewise, let

$$R_P N_i = n_{i1}^P[V_1^P] \dotplus n_{i2}^P[V_2^P] \dotplus \cdots .$$

Obviously, if

$$\sum_{i=1}^{r} a_i m_{ij}^P = \sum_{i=1}^{s} b_i n_{ij}^P \text{ for each } P \text{ and each } j, \tag{4.2}$$

then (4.1) is valid.

Conversely, if the Krull-Schmidt theorem holds for $R_P G$-modules for each P dividing g, then (4.1) implies (4.2). In particular, (4.1) and (4.2) are equivalent statements whenever K is a splitting field for G.

In the special case where the set $\{V_j^P\}$ is finite for each P, the above considerations are especially useful in determining all relations of the form (4.1).

REFERENCES

1. Z. I. Borevich and D. K. Fadeev, *Theory of homology in groups, II,* Proc. Leningrad Univ. 7 (1959), 72-87.

2. C. W. Curtis and I. Reiner, *Representation theory of finite groups and associative algebras*, Wiley, New York, 1962.

3. K. deLeeuw, *Some applications of cohomology to algebraic number theory and group representations* (unpublished).

4. A. Heller, *On group representations over a valuation ring,* Proc. Nat. Acad. Sci. 47 (1961), 1194-1197.

5. A. Heller and I. Reiner, *Representations of cyclic groups in rings of integers, I, II,* Ann. of Math., (to appear).

6. I. Reiner, *On the class number of representations of an order*, Can. J. Math. 11 (1959), 660-672.

7. ———, *The Krull-Schmidt theorem for integral group representations*, Bull. Amer. Math. Soc. 67 (1961), 365-367.

8. R. G. Swan, *Induced representations and projective modules*, Ann. of Math. (2) 71 (1960), 552-578.

University of Illinois
Urbana, Illinois

REPRESENTATIONS OF CYCLIC GROUPS IN RINGS OF INTEGERS, I

By A. Heller and I. Reiner*
(Received September 6, 1961)

Introduction

Let G be a cyclic group of order p^2, where p is a fixed rational prime, and let Z be the ring of rational integers. By ZG we denote the group ring consisting of all formal linear combinations of the elements of G with coefficients from Z. We shall be concerned here with representations of G by matrices with entries in Z, or equivalently, with left ZG-modules having finite Z-bases.

The first systematic study of this problem occurred in a paper by Diederichsen [6] in 1938. At the end of that paper he gave an example purporting to show that there are infinitely many non-isomorphic indecomposable ZG-modules. As has been pointed out recently by Roiter [15] and Troy [16], however, this example is incorrect. Both Roiter and Troy proved that there are only nine (non-isomorphic) indecomposable ZG-modules for the case $p = 2$. In the present work we shall show that for arbitrary p, the number of non-isomorphic indecomposable ZG-modules is finite.

In § 1 we show how to reduce the problem to the analogous problem for representations of the group in the ring of p-adic integers. Section 2 is devoted to setting up the machinery needed in the rest of the paper, and to a discussion of the cyclic group of order p. The third section is the heart of the present work, and in it we prove the finiteness of the number of indecomposable representations of G in the ring of p-adic integers. In § 4 it is an easy task to list explicitly these indecomposable representations. The next section discusses briefly what occurs when we drop the restriction that the modules be torsion-free. Finally in § 6 we show that an arbitrary finite group will have infinitely many non-isomorphic indecomposable representations by matrices over Z if it contains a non-cyclic p-Sylow subgroup for some p. This extends a result due to Borevich and Faddeev [2].

In a sequel to this paper it will be shown that if G is a cyclic p-group of order greater than p^2, then there are *infinitely* many inde-

* The work of the first author was supported by a research contract with the National Science Foundation. The work of the second author was supported by a research contract with the Office of Naval Research.

composable ZG-modules.

1. Relations between global and local cases

Throughout this section let R be the ring of all algebraic integers in an algebraic number field K, and let P be a prime ideal in R which contains the rational prime p. Define

$$R_P = \{\alpha/\beta : \alpha, \beta \in R, \beta \notin P\} ,$$

so that R_P is the valuation ring in K corresponding to the P-adic valuation of K. Let K^* be the P-adic completion of K, R^* the ring of P-adic integers in K^*, and P^* the prime ideal in R^*.

To begin with let G be an arbitrary p-group, which later on will be assumed cyclic. By an RG-module we shall mean a left RG-module which is finitely generated and torsion-free as R-module. Analogous definitions hold for R_PG-modules, R^*G-modules, etc. For an RG-module M we denote $R_P \otimes_R M$ by R_PM, $R^* \otimes_R M$ by R^*M, etc., regarding M as imbedded in each of these.

Let N be a submodule of the R-module M; we call N an *R-pure* submodule of M if for each non-zero $\alpha \in R$,

$$\alpha m \in N,\ m \in M \Rightarrow m \in N .$$

Then (see Kaplansky [11]) N is R-pure if and only if N is an R-direct summand of M.

We shall need the following result, which we state without proof (see deLeeuw [5], Reiner [13]):

(1.1) LEMMA. *Let N be an R-pure RG-submodule of the RG-module M. Then N is an RG-direct summand of M if and only if for each prime ideal P of R which divides the order of G, R_PN is an R_PG-direct summand of R_PM.*

Using this we now establish

(1.2) THEOREM. *Let G be an arbitrary p-group, and assume that P is the only prime ideal of R which divides p. Then an RG-module M is decomposable if and only if the corresponding R_PG-module R_PM is decomposable.*

PROOF. If M is decomposable, obviously so is R_PM. Conversely let V be an R_PG-direct summand of R_PM, and let $N = M \cap V$. Clearly N is an RG-submodule of M, and $R_PN \subset V$. On the other hand for each $v \in V$ there exists a non-zero element $a \in R$ such that

$$av \in M\,, \qquad aR + P = R\,.$$

But then $av \in N$, and so $v \in a^{-1}N \subset R_P N$. This proves that $R_P N = V$.

To show that N is an R-pure submodule of M, suppose that $am \in N$, where $a \in R$, $a \neq 0$, $m \in M$. Then $am \in V$, so also $m \in V$, since V is an R_P-pure submodule of $R_P M$. Therefore $m \in M \cap V = N$, which shows that N is R-pure. The theorem is now an immediate consequence of the preceding lemma.

In (1.4), and also later in this paper, we shall need results on valuations and cyclotomic fields. As general references on these topics, we refer the reader to Artin [1] or Hasse [7].

We state without proof a theorem due to Heller [8] which will be used in the proof of Theorem 1.4.

(1.3) LEMMA. *Let H be an arbitrary finite group such that for each irreducible KH-module W, the K^*H-module $K^*W (= K^* \otimes_K W)$ is also irreducible. Then an $R_P H$-module M is decomposable if and only if the corresponding R^*H-module R^*M is decomposable.*

(1.4) THEOREM. *Let G be a cyclic group of order p^e, and let the algebraic number field L be obtained from K by adjoining the $p^{e\text{th}}$ roots of unity. Suppose that the p-adic valuation of the rational field Q has a unique extension to L. Then an $R_P G$-module M is decomposable if and only if the corresponding R^*G-module R^*M is decomposable.*

PROOF. In view of Lemma 1.3, it suffices to show that any irreducible KG-module W remains irreducible under extension of the ground field from K to K^*. Each such W is of the form

$$W = K[x]/(h(x))\,,$$

where x is an indeterminate over K, where $h(x)$ is a factor of

$$x^{p^e} - 1$$

irreducible in $K[x]$, and where the generator of G acts on W by left multiplication by x. Then

$$K^*W \cong K^*[x]/(h(x))\,,$$

and K^*W will be irreducible provided that $h(x)$ is irreducible in $K^*[x]$.

Let $\alpha \in L$ be a zero of $h(x)$. Then the number of distinct irreducible factors of $h(x)$ in $K^*[x]$ equals the number of extensions of the P-adic valuation from K to $K(\alpha)$. We have the inclusions

$$Q \subset K \subset K(\alpha) \subset L\,,$$

and by hypothesis there is only one extension of the p-adic valuation from Q to L. Hence there is only one extension of the P-adic valuation from K to $K(\alpha)$, and so $h(x)$ has only one type of irreducible factor in $K^*[x]$. But $h(x)$ has no repeated factors in any extension of the perfect field K, and so $h(x)$ is irreducible in $K^*[x]$. This completes the proof.

(1.5) COROLLARY. *Let G be a cyclic group of order p^e, and let R^* be the ring of p-adic integers in the p-adic completion of the rational field. Then a ZG-module M is indecomposable if and only if the corresponding R^*G-module R^*M is indecomposable.*

PROOF. We use Theorems 1.2 and 1.4 with $K = Q$, $R = Z$, $P = pZ$. The hypotheses of Theorem 1.2 are trivially satisfied, while those of Theorem 1.4 hold because of the well-known fact that the p-adic valuation of Q extends uniquely to the field $Q(\zeta)$, where ζ is a primitive $p^{e\text{th}}$ root of unity. We then have: M indecomposable $\Longleftrightarrow R_pM$ indecomposable $\Longleftrightarrow R^*M$ indecomposable, which completes the proof.

(1.6) COROLLARY. *Keeping the notation of the preceding corollary, suppose that there are only a finite number of non-isomorphic indecomposable R^*G-modules, say $M_1, \cdots, M_t$. Then there are only a finite number of non-isomorphic indecomposable ZG-modules.*

PROOF. Let M be an indecomposable ZG-module. Then by (1.5) R^*M is indecomposable, and so $R^*M \cong M_i$ for some i, $1 \leq i \leq t$. However for fixed i, the number of non-isomorphic ZG-modules M such that $R^*M \cong M_i$ is finite, as a consequence of the Jordan-Zassenhaus theorem (see Zassenhaus [17]). This proves the result.

For later use we now prove

(1.7) THEOREM. *Let G be an arbitrary p-group, R^* the ring of p-adic integers in the p-adic completion of the rational field. Then the group ring R^*G has a unique maximal two-sided ideal*

$$N = P^*G + \sum_{g \in G} R^*(g - 1) , \tag{1.8}$$

where P^* is the prime ideal in R^*. Any element of R^*G not in N is a unit of R^*G.

PROOF. Note that $\bar{R} = R^*/P^*$ is a finite field of characteristic p. Since G is a p-group, it is well-known (see Curtis and Reiner [4]) that the radical of $\bar{R}G$ is given by

$$\bar{N} = \sum_{g \in G} \bar{R}(g - 1) ,$$

and that

$$\bar{R}G/\bar{N} \cong \bar{R} .$$

We now define a ring homomorphism $\varphi: R^*G \to \bar{R}$ by means of

$$R^*G \longrightarrow R^*G/P^*G \cong \bar{R}G \longrightarrow \bar{R}G/\bar{N} \cong \bar{R} ,$$

and clearly the kernel of φ is the set N defined in (1.8). Thus N is a maximal two-sided ideal of R^*G.

In order to prove that N is the unique maximal two-sided ideal of R^*G, it suffices to show that if $x \in R^*G$ and $x \notin N$, then x is a unit in R^*G. The image of any such x in $\bar{R}G$ lies outside $\bar{N}$, hence is a unit in $\bar{R}G$. We may therefore choose $y \in R^*G$ such that $xy \equiv 1 \pmod{P^*G}$. Consequently $xy = 1 + z$ for some $z \in P^*G$. Since R^* is a complete local ring, the series $1 - z + z^2 - z^3 + \cdots$ converges to an element $w \in R^*G$, and we have $xyw = 1$. This shows that x has a right inverse, and a corresponding argument shows that x has a left inverse. Hence x has a two-sided inverse, and the theorem is proved.

2. Representations of cyclic groups of order p

Throughout this section let H denote a cyclic group generated by an element h of prime order p. Changing notation slightly, we shall let R denote the ring of p-adic integers in the p-adic completion Q^* of the rational field. From the results of Diederichsen [6] or Reiner [12], it follows easily that there are exactly three non-isomorphic indecomposable RH-modules. We shall re-prove this fact here, since the method used will be needed in § 3 in any case. For the results on algebraic number theory stated below, the reader is referred to Artin [1] or Hasse [7]; a discussion of the functor Ext and its properties may be found in Cartan and Eilenberg [3].

Let $\Phi_r(x)$ denote the cyclotomic polynomial of order r (and degree $\varphi(r)$, where φ is the Euler φ-function). Then the polynomial $\Phi_{p^e}(x)$ is irreducible in $R[x]$ for each e. Furthermore if θ_{p^e} denotes a zero of $\Phi_{p^e}(x)$, then there is a unique valuation ring of $Q^*(\theta_{p^e})$ which contains R, namely the ring $R[\theta_{p^e}]$.

For convenience we set

$$A = R , \qquad B = R[\theta_p] ,$$

both of which are principal ideal domains. We have ring isomorphisms

$$\frac{RH}{(h-1)RH} \cong A , \qquad \frac{RH}{\Phi_p(h)RH} \cong B , \tag{2.1}$$

given by $h \to 1$ and $h \to \theta_p$, respectively. Since A and B are quotient rings of RH, they are (canonically) left RH-modules. We shall show that the indecomposable RH-modules are A, B, and RH itself.

Let M be any RH-module, and set

$$N = \{m \in M: (h-1)m = 0\} .$$

Then N is an R-pure RH-submodule of M annihilated by $(h-1)$. We may therefore consider N as an A-module, and it is easily seen that N is A-torsion-free. Therefore $N \cong A^{(t)}$ for some t, where $A^{(t)}$ denotes the direct sum of t copies of A. This gives the structure of N both as A-module and (by(2.1)) as RH-module.

On the other hand $\Phi_p(h)$ annihilates M/N, so that the latter may be viewed as B-module. Since M/N is B-torsion-free, we have $M/N \cong B^{(u)}$ for some u; this exhibits M/N both as B-module and (by (2.1)) as RH-module. The problem of classifying all RH-modules is thus reduced to that of determining extensions of $B^{(u)}$ by $A^{(t)}$.

For convenience we write Ext instead of Ext^1 throughout this paper. Since RH is a commutative ring, it follows that $\text{Ext}(B, A)$ is itself an RH-module.

(2.2) LEMMA. *There is an RH-isomorphism*

$$\text{Ext}(B, A) \cong A/pA .$$

PROOF. The sequence

$$0 \longrightarrow \Phi_p(h) \cdot RH \xrightarrow{\tau} RH \longrightarrow B \longrightarrow 0 \tag{2.3}$$

is exact, hence so is

$$\begin{aligned} 0 \longrightarrow \text{Hom}_{RH}(B, A) \longrightarrow \text{Hom}_{RH}(RH, A) &\xrightarrow{\tau^*} \text{Hom}_{RH}(\Phi_p(h)RH, A) \\ \longrightarrow \text{Ext}(B, A) \longrightarrow \text{Ext}(RH, A) &\longrightarrow \cdots . \end{aligned}$$

The map τ^* is induced from τ as follows: for each $f \in \text{Hom}_{RH}(RH, A)$, we have

$$(\tau^* f)y = f(\tau y) , \qquad y \in \Phi_p(h)RH .$$

For convenience set $Y = \Phi_p(h)RH$. Since RH is free, we have $\text{Ext}(RH, A) = 0$, and so

$$\text{Ext}(B, A) \cong \text{Hom}_{RH}(Y, A)/\tau^*\text{Hom}_{RH}(RH, A) . \tag{2.4}$$

Now let $y = \Phi_p(h) \in Y$; then each $F \in \text{Hom}_{RH}(Y, A)$ is completely determined by the value $F(y) \in A$, and each $a \in A$ is of the form $F(y)$ for some such F. Hence

$$\text{Hom}_{RH}(Y, A) \cong A$$

as RH-modules. Let us determine which elements in A correspond to elements in the image of τ^*.

For $f \in \mathrm{Hom}_{RH}(RH, A)$, the value $f(1)$ may be arbitrary in A. Then we have

$$(\tau^* f)y = f(\tau y) = f(y) = \Phi_p(h)f(1) ,$$

using the fact that τ is just the inclusion map. Therefore the image of τ^* in A is precisely $\Phi_p(h)A$, and we have shown that

$$\mathrm{Ext}(B, A) \cong A/\Phi_p(h)A .$$

Since

$$\Phi_p(h)a = (h^{p-1} + \cdots + h + 1)a = pa , \qquad a \in A ,$$

the proof is complete.

Let us set $\bar{A} = A/pA$. Since $\mathrm{Ext}(B, A) \cong \bar{A}$, it is easy to see that $\mathrm{Ext}(B^{(u)}, A^{(t)})$ is isomorphic to the module of all $u \times t$ matrices with entries in $\bar{A}$. In order to calculate the effect of basis changes, it will be convenient to exhibit this isomorphism explicitly. Let $\sum_{i=1}^{u} RH\cdot x_i$ be a free RH-module with basis $x_1, \cdots, x_u$. Adding u copies of the exact sequence (2.3), we obtain the exact sequence

$$0 \longrightarrow \sum \Phi_p(h)\cdot RH\cdot x_i \xrightarrow{\tau} \sum RH\cdot x_i \longrightarrow \sum RH\cdot \bar{x}_i \longrightarrow 0 ,$$

where $\bar{x}_i$ is annihilated by $\Phi_p(h)$. Set $y_i = \Phi_p(h)x_i$. Then as above we obtain

$$\mathrm{Ext}(B^{(u)}, A^{(t)}) \cong \mathrm{Hom}_{RH}(\textstyle\sum RH\cdot y_i, A^{(t)})/\text{image of } \tau^* .$$

Let $A^{(t)} = Aa_1 \oplus \cdots \oplus Aa_t$. Then for each

$$F \in \mathrm{Hom}_{RH}(\textstyle\sum_{i=1}^{u} RH\cdot y_i, A^{(t)}) ,$$

we may write

$$\text{(2.5)} \qquad F(y_i) = \sum_{j=1}^{t} \alpha_{ij} a_j , \qquad \alpha_{ij} \in A,\ 1 \leqq i \leqq u .$$

The class $[F]$ which F determines in $\mathrm{Ext}(B^{(u)}, A^{(t)})$ then corresponds to the $u \times t$ matrix $\mathbf{F} = (\bar{\alpha}_{ij})$ with entries in $\bar{A}$.

Suppose we make a basis change in $B^{(u)}$ by leaving $\bar{x}_1, \bar{x}_3, \cdots, \bar{x}_u$ unchanged, but replacing $\bar{x}_2$ by $\bar{x}_2 - \lambda\bar{x}_1$ for some λ in RH. Then $y_1, y_3, \cdots, y_u$ are unchanged, but y_2 becomes $y_2 - \lambda y_1$, and α_{2j} is replaced by $\alpha_{2j} - \lambda\alpha_{1j}$, $1 \leqq j \leqq t$.

On the other hand, if $a_1 = a_1', a_2 = a_2' - \lambda a_1', a_3' = a_3, \cdots, a_t' = a_t$ is a basis change in $A^{(t)}$, then α_{i1} is replaced by $\alpha_{i1} - \lambda\alpha_{i2}$, $1 \leqq i \leqq u$.

We are now ready to prove

(2.6) THEOREM. *The only indecomposable RH-modules are (up to isomorphism) A, B and RH.*

PROOF. Let M be an indecomposable RH-module. From the discussion at the beginning of this section, we know that M must be an extension of $B^{(u)}$ by $A^{(t)}$ for some t and u. If $t = 0$, we must have $M \cong B$; while if $u = 0$, then $M \cong A$. For the rest of the proof, assume therefore that both t and u are positive.

Let $\mathbf{F} = (\bar{\alpha}_{ij})$ be the $u \times t$ matrix over $\bar{A}$ corresponding to the extension M of $B^{(u)}$ by $A^{(t)}$. If all entries of $\mathbf{F}$ are zero, then the extension splits, and M is decomposable. We may therefore assume that $\mathbf{F}$ has a non-zero entry, and in fact (after re-numbering basis elements) may assume that $\alpha_{11} \neq 0$. But then there exist elements $\lambda_2, \cdots, \lambda_u \in RH$ such that $\bar{\alpha}_{i1} - \lambda_i \bar{\alpha}_{11} = 0$, $2 \leqq i \leqq u$, so that by a basis change in $B^{(u)}$, we may make all of the elements in the first column of $\mathbf{F}$ below $\bar{\alpha}_{11}$ equal to zero. Analogously, a basis change in $A^{(t)}$ permits us to make the $(1, 2), \cdots, (1, t)$ entries of $\mathbf{F}$ equal to zero. In this case, however, the submodule $Aa_1 \oplus_R B\bar{x}_1$ is a direct summand of M. Since M is indecomposable, we must have $M \cong Aa_1 \oplus_R B\bar{x}_1$, that is, M must be an extension of B by A.

Consider now the extensions of B by A; each extension determines an extension class in $\mathrm{Ext}(B, A)$, which is represented by an element $\bar{\alpha}$ in $\bar{A}$. If $\bar{\alpha} = 0$, we have a split extension, which is obviously decomposable. On the other hand, a basis change in Aa_1, in which $a_1 \rightarrow \lambda a_1$ (where λ is a unit in A), has the effect of replacing $\bar{\alpha}$ by $\overline{\alpha\lambda}$. Since $\bar{A}$ is a field, if $\bar{\alpha} \neq 0$ we may choose λ so that $\overline{\alpha\lambda} = \bar{1}$. Thus there is only one isomorphism type of non-splitting extension E of B by A.

To complete the proof of the theorem, we need only show that $E \cong RH$, and that E is indecomposable. But it is easily seen that RH is an extension of B by A. Furthermore, RH is indecomposable since $\bar{R}H$ is indecomposable (by Theorem 1.7). Thus RH must be a non-splitting extension of B by A, and so $RH \cong E$.

3. Representations of cyclic groups of order p^2

Throughout this section let G be a cyclic group of order p^2, with generator g. We shall continue to use the notation of § 2. Our main result is

(3.1) THEOREM. *The number of non-isomorphic indecomposable RG-modules is finite.*

In the following section, we shall in fact prove that this number is precisely $4p + 1$. In any case, as a consequence of Theorem 3.1 and Corollary 1.6, we have

(3.2) COROLLARY. *The number of non-isomorphic indecomposable ZG-modules is finite.*

for some n, thereby exhibiting the structure of M/N both as C-module and (using (3.7)) as RG-module. To classify all RG-modules, we are thus led to the consideration of extensions of $C^{(n)}$ by modules N given in (3.5).

(3.8) LEMMA. *There is an RG-isomorphism*

$$\mathrm{Ext}(C, N) \cong N/pN .$$

PROOF. As in the proof of Lemma 2.2, the exact sequence

$$0 \longrightarrow \Phi_{p^2}(g)\cdot RG \longrightarrow RG \longrightarrow C \longrightarrow 0 \tag{3.9}$$

gives rise to an exact sequence

$$\longrightarrow \mathrm{Hom}_{RG}(RG, N) \longrightarrow \mathrm{Hom}_{RG}(\Phi_{p^2}(g)RG, N) \longrightarrow \mathrm{Ext}(C, N) \longrightarrow 0 .$$

We conclude, as in the proof of Lemma 2.2, that

$$\mathrm{Ext}(C, N) \cong N/\Phi_{p^2}(g)N .$$

Finally we have

$$\Phi_{p^2}(g)n = (g^{p(p-1)} + g^{p(p-2)} + \cdots + g^p + 1)n = pn , \qquad n \in N$$

which completes the proof.

Let us set $\bar{E} = E/pE$. If N is given by (3.5), then $\bar{N} \cong \bar{A}^{(k)} \oplus \bar{B}^{(l)} \oplus \bar{E}^{(m)}$. Each extension class in $\mathrm{Ext}(C^{(m)}, N)$ is therefore representable by a triple consisting of an $n \times k$ matrix over $\bar{A}$, an $n \times l$ matrix over $\bar{B}$, and an $n \times m$ matrix over $\bar{E}$. It will be convenient to obtain this correspondence explicitly. Let $\sum_{i=1}^{n} RG\cdot x_i$ be a free RG-module with basis $\{x_1, \cdots, x_n\}$. Adding n copies of (3.9) we obtain an exact sequence

$$0 \longrightarrow \sum \Phi_{p^2}(g)\cdot RG\cdot x_i \longrightarrow \sum RG\cdot x_i \longrightarrow \sum RG\cdot \bar{x}_i \longrightarrow 0$$

where $\bar{x}_i$ is annihilated by $\Phi_{p^2}(g)$. Set $y_i = \Phi_{p^2}(g)x_i$. Let

$$N = Aa_1 \oplus \cdots \oplus Aa_k \oplus Bb_1 \oplus \cdots \oplus Bb_l \oplus Ee_1 \oplus \cdots \oplus Ee_m . \tag{3.10}$$

(We could equally well write $RG\cdot a_1$, where a_1 is annihilated by $(g-1)$, and so on.) Then for each $F \in \mathrm{Hom}_{RG}(\sum RG\cdot y_i, N)$ we may write

$$F(y_i) = \sum_{\kappa=1}^{k} \alpha_{i\kappa} a_\kappa + \sum_{\lambda=1}^{l} \beta_{i\lambda} b_\lambda + \sum_{\mu=1}^{m} \gamma_{i\mu} e_\mu , \qquad 1 \leqq i \leqq n , \tag{3.11}$$

with $\alpha_{i\kappa} \in A$, $\beta_{i\lambda} \in B$, $\gamma_{i\mu} \in E$. The class $[F]$ which F determines in $\mathrm{Ext}(C^{(n)}, N)$ then corresponds to the triple of matrices

$$\mathbf{F}_A = (\bar{\alpha}_{i\kappa}), \qquad \mathbf{F}_B = (\bar{\beta}_{i\lambda}), \qquad \mathbf{F}_E = (\bar{\gamma}_{i\mu})$$

over $\bar{A}$, $\bar{B}$, and $\bar{E}$, respectively.

The ring $\bar{A}$ has already been discussed, and in fact $\bar{A} \cong \bar{R} = R/pR$. Next we have

$$\bar{B} = R[\theta_p]/pR[\theta_p] \cong \bar{R}G/(g-1)^{p-1}\bar{R}G ,$$

REMARK. It is not difficult to show that the number of non-isomorphic indecomposable ZG-modules is at least equal to

$$h_p \cdot h_{p^2} \cdot (4p - 4) + 2h_{p^2} + 2h_p + 1 , \tag{3.3}$$

where h_r is the number of ideal classes in the cyclotomic field of rth roots of 1 over Q. This formula follows readily from the discussion in § 4 and the results of Troy [16]. For $p = 2$ and $p = 3$, in fact, the number of non-isomorphic indecomposable ZG-modules is given precisely by (3.3), but we shall not attempt to prove this here. (For the case $p = 2$, see Roiter [15] or Troy [16].)

The proof of Theorem 3.1 is rather involved, and will occupy the rest of this section. Keeping the notation in the preceding section, we let H be a cyclic group with generator h of order p. There is a ring isomorphism

$$\frac{RG}{(g^p - 1)RG} \cong RH , \tag{3.4}$$

given by $g \to h$. By virtue of this, every RG-module which is annihilated by $(g^p - 1)$ can be viewed as RH-module, and conversely. In particular the RH-modules A, B, and $E(=RH)$ are indecomposable RG-modules.

Now let M be any RG-module, and define

$$N = \{m \in M \colon (g^p - 1)m = 0\} .$$

Then N is an R-pure RG-submodule of M, and N is annihilated by $(g^p - 1)$. By the above remarks it follows that

$$N \cong A^{(k)} \oplus B^{(l)} \oplus E^{(m)} , \tag{3.5}$$

where $A^{(k)}$ denotes a direct sum of k copies of A, etc., and where

$$A = R, \qquad B = R[\theta_p], \qquad E = RH , \tag{3.6}$$

the action of G being determined by formulas (3.4) and (2.1).

On the other hand we have a ring isomorphism

$$\frac{RG}{\Phi_{p^2}(g)RG} \cong R[\theta_{p^2}] = C \qquad \text{(say)} , \tag{3.7}$$

and C is the unique valuation ring in $Q^*(\theta_{p^2})$ which contains R. In particular, C is a principal ideal domain. By virtue of (3.7), every RG-module which is annihilated by $\Phi_{p^2}(g)$ may be viewed as C-module, and conversely.

Now the module M/N is annihilated by $\Phi_{p^2}(g)$, hence is a (finitely-generated) C-module. It is C-torsion-free because of the fact that to each non-zero element $c \in C$ there corresponds another element $c' \in C$ such that cc' is a non-zero element of R. We now have

$$M/N \cong C^{(n)}$$

since in $R[\theta_p]$ the element p can be expressed as a unit times $(\theta_p - 1)^{p-1}$. Finally we have

$$\bar{E} = RH/p\cdot RH \cong \bar{R}H \cong \bar{R}G/(g^p - 1) \cong \bar{R}G/(g-1)^p\bar{R}G\ .$$

Both $\bar{B}$ and $\bar{E}$ are commutative algebras over $\bar{R}$, with unique maximal ideals $(g-1)\bar{B}$ and $(g-1)\bar{E}$, respectively. Any non-zero element $\bar{\beta}$ of $\bar{B}$ is uniquely expressible as $(g-1)^e u$, where $0 \leqq e < p-1$, and u is a unit in $\bar{B}$. We call e the *exponent* of $\bar{\beta}$; the same terminology will be applied to elements of B, that is, for $\beta \in B$, the exponent of β is defined to be the exponent of $\bar{\beta} \in \bar{B}$. We may likewise speak of the exponent of a non-zero element of $\bar{E}$; here the exponents range from 0 to $p-1$, inclusive.

We have already seen that any RG-module M is an extension of $C^{(n)}$ by $A^{(k)} \oplus B^{(l)} \oplus E^{(m)}$ for some k, l, m, n. We shall determine here which of these extensions can possibly be indecomposable RG-modules, and then in § 4 we shall determine exactly which of these possibilities are indeed indecomposable. For $n = 0$ we know from § 2 that there are only three indecomposables, namely $A, B,$ and E. We may therefore restrict ourselves to the case $n > 0$.

Let M be an indecomposable extension of $C^{(n)}$ by N, and let its extension class determine the matrices $\mathbf{F}_A$, $\mathbf{F}_B$, and $\mathbf{F}_E$. Since $\mathbf{F}_A$ has entries in $\bar{A}$, we may proceed just as in the proof of Theorem 2.6, and after a basis change in $A^{(k)}$ and $C^{(n)}$, we may assume that

$$\mathbf{F}_A = \begin{pmatrix} 1 & & & 0 \\ & \ddots & & \\ & & 1 & \\ 0 & & & 0 \end{pmatrix},$$

where w 1's occur, each 1 denoting the unity element of $\bar{A}$. If w were less than k, then $Aa_{r+1} \oplus \cdots \oplus Aa_k$ would be a direct summand of M. Since M is indecomposable, this summand would be all of M, which contradicts the hypothesis that $n > 0$. Thus $w = k$, and we have

$$F(y_i) = a_i + \text{terms involving } b\text{'s and } e\text{'s}, \qquad 1 \leqq i \leqq k\ ,$$

$$F(y_i) = \text{terms involving } b\text{'s and } e\text{'s}, \qquad k+1 \leqq j \leqq n\ .$$

We now examine all entries of $\mathbf{F}_B$ and $\mathbf{F}_E$. If all entries are zero, then since M is indecomposable and we have assumed $n \geqq 1$, we must have $l = m = 0$, and either $M \cong C$ or else M is an extension of C by A. Excluding these cases hereafter, we may now assume that some entry of $\mathbf{F}_B$ or $\mathbf{F}_E$ is non-zero. The discussion now breaks into four cases, depending on where the smallest exponent occurs.

Case 1. Assume that the smallest exponent occurs somewhere in the last $n-k$ rows of $\mathbf{F}_E$. Re-numbering the basis of $E^{(m)}$ and re-numbering the last $n-k$ basis elements of $C^{(n)}$, we may hereafter assume that the smallest exponent occurs at $\gamma_{k+1,1}$. We may write

$$\gamma_{k+1,1} = (g-1)^r\{\bar{\alpha}_0 + \bar{\alpha}_1(g-1) + \cdots + \bar{\alpha}_{p-r-1}(g-1)^{p-r-1}\} ,$$

with each $\bar{\alpha}_i \in \bar{R}$, $\bar{\alpha}_0 \neq 0$. Set

$$z = \alpha_0 + \alpha_1(g-1) + \cdots + \alpha_{p-r-1}(g-1)^{p-r-1} .$$

Then by Theorem 1.7, z is a unit in RG. Thus $\gamma_{k+1,1}$ can be expressed as $(g-1)^r\bar{z}$, where z is a unit in RG.

Now we have

$$F(y_{k+1}) = \sum \beta_{k+1,\lambda}b_\lambda + \sum \gamma_{k+1,\mu}e_\mu = (g-1)^r \cdot e_1' ,$$

where $e_1' =$ (unit of RG)e_1 + linear combination of the $\{b_\lambda\}$ and the remaining $\{e_i\}$. (We have used the fact that the exponent of each β is at least r.) Now the annihilator of e_1' is (g^p-1), so $RG\cdot e_1' \cong RG/(g^p-1) \cong RG\cdot e_1$. Hence if we choose a new basis of $B^{(l)} \oplus E^{(m)}$ in which the only change is to replace e_1 by e_1', we may assume (changing notation) that

$$F(y_{k+1}) = (g-1)^r e_1 .$$

On the other hand, replacing $\bar{x}_i$ by $\bar{x}_i + \lambda_i\bar{x}_{k+1}$ $(i \neq k+1)$ has the effect of replacing γ_{i1} by $\gamma_{i1} + \lambda_i(g-1)^r$. Since e is minimal, we may choose the $\{\lambda_i\}$ in RG so as to make each new $\gamma_{i1} = 0$, $i \neq k+1$. But then $Ee_1 \oplus_R C\bar{x}_{k+1}$ is a direct summand of M, hence coincides with M. Thus in this case M is an extension of C by E, corresponding to an extension class $(g-1)^r \in \bar{E}$ for some r.

Case 2. Assume that the smallest exponent occurs somewhere in the first k rows of $\mathbf{F}_E$, but *not* in the last $n-k$ rows. After re-numbering basis elements of $C^{(n)}$, $E^{(m)}$ and $A^{(k)}$, we may assume that γ_{11} has smallest exponent. The method used in *Case* 1 shows that after further basis changes, we may in fact assume that

$$F(y_1) = a_1 + \gamma_{11}e_1, \qquad \gamma_{11} = (g-1)^r \qquad \text{for some } r .$$

For each i, $2 \leqq i \leqq k$, we let $\bar{x}_i' = \bar{x}_i - \rho_i\bar{x}_1$, $a_i' = a_i - \rho_i a_1$, $\rho_i \in RG$. Then

$$\begin{aligned} F(y_i') &= F(y_i) - \rho_i F(y_1) \\ &= a_i + \sum \beta_{i\lambda}b_\lambda + \sum \gamma_{i\mu}e_\mu - \rho_i(a_1 + \gamma_{11}e_1) \\ &= a_i' + \sum \beta_{i\lambda}b_\lambda + (\gamma_{i1} - \rho_i\gamma_{11})e_1 + \textstyle\sum_{\mu>1} \gamma_{i\mu}e_\mu . \end{aligned}$$

Choose ρ_i so that $\gamma_{i1} - \rho_i\gamma_{11} = 0$, $2 \leqq i \leqq k$. Then in the expansions for $F(y_i')$, $2 \leqq u \leqq k$, no e_1 term occurs.

On the other hand, for each j, $k+1 \leqq j \leqq n$, we let $\bar{x}'_j = \bar{x}_j - \rho_j\bar{x}_1$, with $\rho_j \in RG$ chosen so as to make $\gamma_{j1} - \rho_j\gamma_{11} = 0$. Then we have

$$F(y'_j) = -\rho_j a_1 + \sum \beta_{j\lambda} b_\lambda + \sum_{\mu>1} \gamma_{j\mu} e_\mu \,, \qquad k+1 \leqq j \leqq n \,.$$

However by hypothesis γ_{11} has smaller exponent than any γ_{j1} with $j \geqq k+1$, and hence each ρ_j above is a multiple of $g-1$. But $(g-1)a_1 = 0$, and so we have

$$F(y'_j) = \sum \beta_{j\lambda} b_\lambda + \sum_{\mu>1} \gamma_{j\mu} e_\mu \,, \qquad k+1 \leqq j \leqq n \,.$$

This shows that $Aa_1 \oplus_R Ee_1 \oplus_R C\bar{x}_1$ is a direct summand of M, hence coincides with M. Thus in this case M must be an extension of C by $A \oplus E$, with extension class $[F]$ where

$$F(y_1) = a_1 + (g-1)^r e_1$$

for some r.

Case 3. Now assume that the smallest exponent occurs in $\mathbf{F}_B$, but not in $\mathbf{F}_E$, and that it occurs in one of the last $n-k$ rows of $\mathbf{F}_B$. Re-numbering basis elements, we may hereafter assume that $\beta_{k+1,1}$ has smallest exponent. Now we have

$$F(y_{k+1}) = \sum \beta_{k+1,\lambda} b_\lambda + \sum \gamma_{k+1,\mu} e_\mu = (g-1)^r b'_1 \,,$$

where

$$b'_1 = \xi b_1 + \sum_{\lambda>1} \beta'_\lambda b_\lambda + \sum \varepsilon_\mu e_\mu$$

with ξ a unit in RG, and each ε_μ a multiple of $g-1$. But then $\Phi_p(g)$ is the annihilator of b'_1, and so

$$RG \cdot b'_1 \cong RG/\Phi_p(g)RG \cong RG \cdot b_1 \,.$$

Thus $\{b'_1, b_2, \cdots, b_l, e_1, \cdots, e_m\}$ form a new basis of $B^{(l)} \oplus E^{(m)}$, and relative to this new basis we have (dropping the primes)

$$F(y_{k+1}) = (g-1)^r b_1 \,.$$

Replacing $\bar{x}_i$ by $\bar{x}_i + \lambda_i \bar{x}_{k+1}$ as in *Case* 1, we may now make $\beta_{i1} = 0$ for $i \neq k+1$. In this case we see that $Bb_1 \oplus_R C\bar{x}_{k+1}$ is a direct summand of M, hence coincides with M. Thus M must be an extension of C by B, corresponding to an extension class $(g-1)^r \in \bar{B}$ for some r.

Case 4. Assume finally that the smallest exponent occurs in $\mathbf{F}_B$, but not in $\mathbf{F}_E$, and that it occurs in the first k rows of $\mathbf{F}_B$, and not in the last $n-k$ rows of $\mathbf{F}_B$. Re-numbering basis elements, we may assume that β_{11} has smallest exponent. After further basis change as in *Case* 3, we may in fact assume that

$$F(y_1) = a_1 + (g-1)^r b_1 \,.$$

For $2 \leqq i \leqq k$, choose ρ_i so that $\beta_{i1} - \rho_i(g-1)^r = 0$, and let

$$\bar{x}_i' = \bar{x}_i - \rho_i \bar{x}_1 , \qquad a_i' = a_i - \rho_i a_1 .$$

Then we have

$$F(y_i') = a_i' + \sum_{\lambda>1} \beta_{i\lambda} b_\lambda + \sum \gamma_{i\mu} e_\mu , \qquad 2 \leqq i \leqq k .$$

On the other hand for each j, $k+1 \leqq j \leqq n$, we let $\bar{x}_j' = \bar{x}_j - \rho_j \bar{x}_1$, with $\rho_j \in RG$ chosen so as to make $\beta_{j1} - \rho_j(g-1)^r = 0$. Then

$$F(y_j') = -\rho_j a_1 + \sum_{\lambda>1} \beta_{j\lambda} b_\lambda + \sum \gamma_{i\mu} e_\mu , \qquad k+1 \leqq j \leqq n ,$$

and since each ρ_j is a multiple of $g-1$, we have $\rho_j \cdot a_1 = 0$ for each j. This shows that $Aa_1 \oplus Bb_1 \oplus C\bar{x}_1$ is a direct summand of M, hence coincides with M. Thus in this case M must be an extension of C by $A \oplus B$, with extension class $[F]$, where

$$F(y_1) = a_1 + (g-1)^r b_1$$

for some r.

4. Indecomposable modules

In the notation of the preceding section, we have now shown that the only candidates for indecomposable RG-modules are the irreducible modules A, B, and C, the module $E = RH$, and extensions of C by N where N has one of the forms $A \oplus E$, E, $A \oplus B$, A, or B. Indecomposable modules of different forms are clearly non-isomorphic, so it remains to identify the indecomposables of each of the above types. Clearly A, B, C, and E are indecomposable, and nothing remains to be said about them.

Let us now determine when two extensions X and X' of C by N are isomorphic. We have two exact sequences

$$0 \longrightarrow N \longrightarrow X \longrightarrow C \longrightarrow 0 , \qquad 0 \longrightarrow N \longrightarrow X' \longrightarrow C \longrightarrow 0 .$$

Any isomorphism $\psi \colon X \cong X'$ takes N onto itself, and thus induces automorphisms on N and C. Any automorphism of C is given by multiplication by a unit of C, and hence (using Theorem 1.7) also by multiplication by a unit u of RG. But then $u^{-1}\psi$ is an isomorphism of X onto X' which induces the identity on C and an automorphism φ on N, so that the diagram

$$\begin{array}{ccccccccc} 0 & \longrightarrow & N & \longrightarrow & X & \longrightarrow & C & \longrightarrow & 0 \\ & & \downarrow{\scriptstyle \varphi} & & \downarrow{\scriptstyle u^{-1}\psi} & & \downarrow{\scriptstyle 1} & & \\ 0 & \longrightarrow & N & \longrightarrow & X' & \longrightarrow & C & \longrightarrow & 0 \end{array}$$

is commutative. It follows that φ takes the extension class of the first sequence into that of the second. Identifying $\operatorname{Ext}(C, N)$ with $\bar{N} = N/pN$

as in § 3, we have shown that two elements of $\bar{N}$ give rise to the isomorphic extensions of C by N if and only if there exists an automorphism φ of N whose induced automorphism $\bar{\varphi}$ on $\bar{N}$ carries one of these elements onto the other.

Let us apply this to the case where $N = Aa \oplus Ee$. We have already seen that, after basis changes, an indecomposable extension M of C by N must correspond to an extension class $[F]$ for which

$$F(y) = a + (g-1)^r e \; ,$$

and the element of $\bar{N}$ determined by this class is precisely $\bar{a} + (g-1)^r \bar{e}$. If $r = p$ then $(g-1)^r \bar{e} = 0$, and M is decomposable. Thus we have $0 \leqq r \leqq p-1$. Furthermore the exponent r is an invariant of M, since $(g-1)^{p-r}$ is the smallest power of $(g-1)$ which annihilates $F(y)$.

Next we note that M is decomposable when $r = p-1$. For in that case let

$$a' = a + \Phi_p(g)e \; , \qquad e' = e \; .$$

Then $\{a', e'\}$ form a new basis of N, since the annihilator of a' is $(g-1)$. Relative to this new basis we have $F(y) = a'$, and so M decomposes. Likewise M decomposes when $r = 0$, for we may set

$$a' = a \; , \qquad e' = e - a \; ,$$

getting $F(y) = -e'$.

Our possible indecomposable modules M which are extensions of C by $A \oplus E$ are thus the extensions for which

$$F(y) = a + (g-1)^r e \; ,$$

where $1 \leqq r \leqq p-2$. They are clearly non-isomorphic for different r, and we need only prove them indecomposable. Since r is an invariant of the extension, it follows from the preceding discussion that the only way in which M can decompose is for A to split off, that is, there must be an automorphism φ of N such that $\bar{\varphi}$ carries $(g-1)^r \bar{e}$ onto $\bar{a} + (g-1)^r \bar{e}$. But for such a map φ we must have

$$\varphi(e) = ue + ta \; ,$$

$u =$ unit of RG, $t \in R$. In that case

$$\bar{\varphi}((g-1)^r \bar{e}) = (g-1)^r(\bar{u}\bar{e} + \bar{t}\bar{a}) = (g-1)^r \bar{u}\bar{e} \neq \bar{a} + (g-1)^r \bar{e}$$

for any possible u. This shows that for each r, $1 \leqq r \leqq p-2$, we have indecomposable extensions. There are altogether $p-2$ non-isomorphic indecomposable extensions of C by $A \oplus E$.

Similar considerations apply to the other cases, although the arguments are somewhat simpler. We find readily that there are the following non-isomorphic indecomposable extensions:

(i) Extensions of C by $A \oplus E$; extension classes $\bar{a} + (g-1)^r \bar{e}$, $1 \leqq r \leqq p-2$.

(ii) Extensions of C by E; extension classes $(g-1)^r \bar{e}$, $0 \leqq r \leqq p-1$.

(iii) Extensions of C by $A \oplus B$; extension classes $\bar{a} + (g-1)^r \bar{b}$, $0 \leqq r \leqq p-2$.

(iv) Extension of C by A; extension class $\bar{a}$.

(v) Extensions of C by B; extension classes $(g-1)^r \bar{b}$, $0 \leqq r \leqq p-2$.

Counting the 4 modules A, B, C and E, the number of non-isomorphic indecomposable RG-modules is thus

$$4 + (p-2) + p + (p-1) + 1 + (p-1) = 4p + 1 .$$

5. Torsion modules

We keep the notation of the preceding section. Here we shall show that if modules with R-torsion are admitted, then there are infinitely many indecomposable RG-modules. In fact, we shall show that there are infinitely many indecomposable RH-modules.

Let $S = Z/p^2Z$, regarded as an R-module, and let

$$M_n = Sx_1 \oplus \cdots \oplus Sx_n ,$$

which we make into an RH-module by defining

$$hx_1 = x_1 + px_2, \quad hx_2 = x_2 + px_3, \cdots, \quad hx_{n-1} = x_{n-1} + px_n, \quad hx_n = x_n .$$

(5.1) THEOREM. *M_n is an indecomposable RH-module.*

PROOF. Any RH-direct summand N of M_n has the property that

$$(5.2) \qquad px \in N,\ x \in M_n \Rightarrow x \in N + pM_n .$$

Using this we shall show that each such N contains px_n. For by (5.2) we see that N must contain an element y which does not lie in pM_n. We may write

$$y = a_j x_j + \cdots + a_n x_n + pz , \qquad z \in M,\ a_i \in S ,$$

where j is minimal such that $p \nmid a_j$. If $j = n$ then

$$py = pa_n x_n \in N ,$$

whence also $px_n \in N$, since a_n is a unit in S. On the other hand if $j < n$, then since $(h-1)pz = 0$ for $z \in M_n$, we have

$$(h-1)y = p(a_j x_{j+1} + \cdots + a_{n-1} x_n) \in N .$$

From (5.2) we obtain

$$a_j x_{j+1} + \cdots + a_{n-1} x_n \in N + pM_n$$

and now repeat the argument $n - j - 1$ times. We have thus shown that any RH-direct summand of M_n must contain px_n, and therefore M_n must be indecomposable.

6. Indecomposable modules for non-cyclic groups

Throughout this section we let G be an arbitrary finite group, and let Z_p^* be the ring of p-adic integers in the p-adic completion of the rational field. Borevich and Faddeev [2] have shown that if the p-Sylow subgroup of G is non-cyclic, then there are infinitely many non-isomorphic indecomposable Z_p^*G-modules. We strengthen this result by proving

(6.1) THEOREM. *If for some p the group G has a non-cyclic p-Sylow subgroup, then there are infinitely many non-isomorphic indecomposable ZG-modules.*

PROOF. As in Borevich and Faddeev [2], we shall use the non-periodicity of the homology of G. The proof will split into two cases, depending on whether $p = 2$ or $p > 2$. To handle the case $p = 2$ in the easiest possible manner, we begin with a lemma which is of interest in itself.

(6.2) LEMMA. *Let H be an abelian group which is the direct product of the cyclic groups $[a]$ and $[b]$, each of order 2. For each n, let*

$$M_n = Zx_1 \oplus \cdots \oplus Zx_n \oplus Zy_0 \oplus Zy_1 \oplus \cdots \oplus Zy_n ,$$

and define the action of H on M_n by

$$(a + (-1)^i)x_i = y_{i-1} , \qquad (b + (-1)^i)x_i = y_i , \qquad 1 \leqq i \leqq n ,$$
$$(a - (-1)^i)y_i = (b - (-1)^i)y_i = 0 , \qquad 0 \leqq i \leqq n .$$

Then M_n is an indecomposable ZH-module.

PROOF. We must first verify that M_n is a ZH-module. It is easily seen that $(a^2 - 1)$ and $(b^2 - 1)$ both annihilate M_n, and also that

$$(a + (-1)^i)(b - (-1)^i)x_i = (b - (-1)^i)(a + (-1)^i)x_i , \qquad 1 \leqq i \leqq n ,$$

which implies that $abx_i = bax_i$ for each i. Trivially we have $aby_i = bay_i$ for each i, and thus ab and ba act in the same way on M_n. Thus M_n is a left ZH-module.

On the other hand we may form the $\bar{Z}H$-module $\bar{M}_n = M_n/2M_n$, where $\bar{Z} = Z/(2)$.

We have previously shown (Heller and Reiner [10]; see also Curtis and Reiner [4]) that $\bar{M}_n$ is indecomposable. Hence also M_n is indecomposable.

Suppose now that the finite group G contains a non-cyclic 2-Sylow subgroup S. Let us assume that there are finitely many non-isomorphic indecomposable ZG-modules, say $Y_1, \cdots, Y_t$, and let us obtain a contradiction.

To begin with we note that the group H defined in Lemma 6.2 is a homomorphic image of S, as is easily verified. Hence the indecomposable ZH-modules M_n constructed in Lemma 6.2 are also indecomposable ZS-modules on which the kernel of the homomorphism $S \to H$ acts trivially. For each n we form the induced ZG-module

$$M_n^G = ZG \otimes_{ZS} M_n .$$

Then we may write

$$M_n^G = \sum_{i=1}^{t\oplus} a_i Y_i , \qquad a_i \in Z,\ a_i \geqq 0 ,$$

where the above indicates that M_n^G is a direct sum of a_1 copies of Y_1, and so on. Therefore, passing over to the ring Z_2^* of 2-adic integers, we have

$$(6.3) \qquad (Z_2^* M_n)^G = \sum\nolimits^{\oplus} a_i (Z_2^* Y_i) .$$

Next we remark that for any Z_2^*S-module L, we may form $(L^G)_S$, the restriction of the induced module L^G to the operator domain Z_2^*S. From the formula

$$L^G = 1 \otimes L + \sum_{i=2}^{d} g_i \otimes L ,$$

where $\{1, g_2, \cdots, g_d\}$ is a set of left coset representatives of G modulo S, we find that L is a Z_2^*S-direct summand of $(L^G)_S$.

We may thus conclude that $Z_2^* M_n$ is a Z_2^*S-direct summand of the module $\sum a_i (Z_2^* Y_i)_S$. On the other hand we know from the proof of Lemma 6.2 that $Z_2^* M_n$ is indecomposable. Finally, the Krull-Schmidt theorem holds for Z_2^*S-modules (see Borevich and Faddeev [2], Reiner [14]). Consequently $Z_2^* M_n$ is a Z_2^*S-direct summand of $(Z_2^* Y_i)_S$ for some i, and hence the Z-rank of M_n cannot exceed the maximum of the Z-ranks of $Y_1, \cdots, Y_t$. This gives a contradiction, since the Z-rank of M_n is $2n+1$. We have thus proved Theorem 6.1 for the case where $p = 2$.

It is not clear what the analogue of Lemma 6.2 is for a direct product of two cyclic groups of order p, with $p > 2$. In order to prove Theorem 6.1 for odd p, we therefore use a different procedure. We assume for the rest of this section that $p > 2$, and that G has a non-cyclic p-Sylow subgroup. For simplicity we let R denote Z_p^*. Then (see Cartan and Eilenberg [3]) we know that $\hat{\mathrm{E}}\mathrm{xt}_{RG}^n(R,R)$ is just the p-primary component of $\hat{H}^n(G,Z)$, and therefore that $\hat{\mathrm{E}}\mathrm{xt}_{RG}^n(R, R)$ is not periodic in n.

Consider now the category of RG-modules which are finitely generated and R-torsion-free. In this category the classes of injectives and projectives coincide, and the Krull-Schmidt theorem holds. Any module M may be written in the form $M' \oplus M''$ with M'' projective and M' having no projective direct summands; both M' and M'' are uniquely determined (up to isomorphism) by M. The argument given by Heller [9] applies unchanged, and yields

(6.4) LEMMA. *If $0 \to N \to X \to M \to 0$ is exact, where X is projective, then N' is determined up to isomorphism by M'. Writing $N' = \Omega M'$, the map Ω carries indecomposables into indecomposables, thus determining a permutation on the set of isomorphism classes of non-projective indecomposable modules. Furthermore for any A, B, and n we have*

$$\hat{\text{Ext}}^n(\Omega A, B) \cong \hat{\text{Ext}}^{n+1}(A, B) .$$

Suppose now that there are only a finite number of non-isomorphic indecomposable ZG-modules, say $Y_1, \cdots, Y_t$. Then breaking each RY_i into a direct sum of indecomposable RG-modules, only a finite number of indecomposable summands occur (for $1 \leqq i \leqq t$), say $U_1, \cdots, U_q$.

Now let X be a projective resolution of Z over ZG. Then RX is a projective resolution of R over RG. By Lemma 6.4,

$$Z_n(RX) \cong \Omega^{n+1}R \oplus P_n , \qquad n \geqq 0 ,$$

with $\Omega^{n+1}R$ indecomposable and P_n projective. On the other hand, $Z_n(RX) = R \cdot Z_n(X)$ is a sum of U_i's. Thus $\Omega^{n+1}R = U_i$ for some i. But this implies that Ω is periodic on the orbit of R, and thus that $\hat{\text{Ext}}^n_{RG}(R, R)$ is periodic. We have obtained a contradiction, which completes the proof of Theorem 6.1.

UNIVERSITY OF ILLINOIS

REFERENCES

1. E. ARTIN, Theory of Algebraic Numbers, Göttingen, 1959.
2. Z. I. BOREVICH and D. K. FADDEEV, *Theory of homology in groups* II, Proc. Leningrad Univ., 7 (1959), 72–87.
3. H. CARTAN and S. EILENBERG, Homological Algebra, Princeton, 1956.
4. C. W. CURTIS and I. REINER, Representation Theory of Finite Groups and Associative Algebras, New York, 1962.
5. K. DELEEUW, Some applications of cohomology to algebraic number theory and group representations, unpublished.
6. F. E. DIEDERICHSEN, *Über die Ausreduktion ganzzahliger Gruppendarstellungen bei arithmetischer Äquivalenz,* Hamb. Abh., 13 (1940), 357–412.
7. H. HASSE, Zahlentheorie, Berlin, 1949.
8. A. HELLER, *On group representations over a valuation ring,* Proc. Nat. Acad. Sci. U.S.A., 47 (1961), 1194-1197.
9. ———, *Indecomposable modules and the loop space operation,* Proc. Amer. Math. Soc., 12 (1961), 640-643.
10. ——— and I. REINER, *Indecomposable representations,* Illinois J. Math., 5 (1961). 314-323.

11. I. KAPLANSKY, *Modules over Dedekind rings and valuation rings*, Trans. Amer. Math. Soc., 72 (1952), 327–340.
12. I. REINER, *Integral representations of cyclic groups of prime order*, Proc. Amer. Math. Soc., 8 (1957), 142–146.
13. ———, *On the class number of representations of an order*, Canad. J. Math., 11 (1959), 660–672.
14. ———, *The Krull-Schmidt theorem for integral group representations*, Bull. Amer. Math. Soc., 67 (1961), 365–367.
15. A. V. ROITER, *Integral representations of cyclic groups of the fourth order*, Proc. Leningrad Univ., 19 (1960), 65–74.
16. A. TROY, Integral representations of cyclic groups of order p^2, Thesis, Univ. of Illinois, 1961.
17. H. ZASSENHAUS, *Neuer Beweis der Endlichkeit der Klassenzahl bei unimodularer Äquivalenz endlicher ganzzahliger Substitutionsgruppen*, Hamb. Abh., 12 (1938), 276–288.

EXTENSIONS OF IRREDUCIBLE MODULES

Irving Reiner

ABSTRACT

Let G be a finite group of order g, R the ring of algebraic integers in an algebraic number field K, and RG the group ring of G over R. For two finitely generated RG-modules M, N write $Ext(M,N)$ for $Ext^1_{RG}(M,N)$. Then $Ext(M,N)$ is an R-module annihilated by g. Let $d(M)$ denote the set of elements of R which annihilate $Ext(M,N)$ for *all* N. Assume that M is R-torsion free; then M is contained in the K-vector space $KM = K \otimes_R M$. If KM is an absolutely irreducible KG-module with K-dimension $(KM : K)$, then the rational number

$$d_M = \frac{g}{(KM : K)}$$

is an integer.

Theorem. *Let M be an R-torsion free, RG-module such that KM is an absolutely irreducible KG-module. Then $d(M)$ is the principal ideal generated by d_M.*

The proof of the result is reduced to the local case; namely, consideration of the localization R_P for a maximal ideal P of R. A theorem of D. G. Higman (*Canadian Journ.* 9 (1957)19–34) is sketched and applied to give the proof.

REPRESENTATIONS OF CYCLIC GROUPS IN RINGS OF INTEGERS, II

By A. Heller and I. Reiner*

(Received January 19, 1962)

Introduction

Let G_k be a cyclic group of order k, and let ZG_k denote its group ring over the ring Z of rational integers. We denote by $n(ZG_k)$ the number of non-isomorphic indecomposable left ZG_k-modules having finite Z-bases. In 1938 Diederichsen [2] proved that $n(ZG_p)$ is finite for p a rational prime, and gave an incorrect proof that $n(ZG_4)$ is infinite. Recently Roiter [6] and Troy [8] independently showed that $n(ZG_4) = 9$. The present authors [4] have shown more generally that $n(ZG_{p^2})$ is finite for all primes p. The aim of this paper is to establish the somewhat surprising result that $n(ZG_{p^3})$ is infinite for each prime p.

Several interesting consequences of this result should be pointed out. To begin with, note that G_{p^3} is a homomorphic image of G_{p^r} for each $r \geqq 3$, and so each ZG_{p^3}-module may be viewed also as a ZG_{p^r}-module. Therefore $n(ZG_{p^r}) \geqq n(ZG_{p^3})$, which shows that $n(ZG_{p^r})$ is infinite for each $r \geqq 3$.

Next, let H_p denote (for the moment) a p-Sylow subgroup of an arbitrary finite group G. The authors have shown [4] that if $n(ZH_p)$ is infinite for some p, then also $n(ZG)$ is infinite. They proved furthermore that $n(ZH_p)$ is necessarily infinite if H_p is non-cyclic. Combining this result with that of the present paper, it follows that $n(ZH_p)$ is finite if and only if H_p is cyclic of order p or p^2. Thus, the possible groups G for which $n(ZG)$ is finite are small in number; in fact, if $n(ZG)$ is finite, then for each p dividing the order of G each p-Sylow subgroup of G must be cyclic of order p or p^2. Whether the converse is true is as yet unknown.†

Let Z^* denote the ring of p-adic integers in the p-adic completion of the rational field. Define $n(Z^*G_{p^r})$ to be the number of non-isomorphic indecomposable left $Z^*G_{p^r}$-modules having finite Z^*-bases. In their previous paper [4], the authors have shown that $n(ZG_{p^r})$ is finite if and only if $n(Z^*G_{p^r})$ is finite. It is therefore sufficient to prove that $n(Z^*G_{p^3})$ is infinite.

Section 1 contains some general remarks about extensions of one direct

* The research of the second author was supported in part by a research contract with the Office of Naval Research.

† Added in proof: The converse has recently been established by Dr. Alfredo Jones (Ph. D. thesis "Indecomposable integral representations", Univ. of Illinois, 1962). A sketch of Jones's proof is given in Curtis and Reiner, "Representation Theory of Finite Groups and Associative Algebras", Wiley, New York, 1962.

sum of modules by another. In §2, the ring of endomorphisms of a specific module is investigated. The third section gives the construction of an infinite set of non-isomorphic indecomposable $Z^*G_{p^3}$-modules.

1. Let R be a ring, later to be chosen as a group ring $Z^*G_{p^s}$. We begin with some generalities about R-modules, and write Hom instead of Hom_R, and Ext instead of Ext_R^1. Throughout this section, let M and N be a pair of R-modules for which

$$\mathrm{Hom}(M, N) = 0 , \quad \mathrm{Hom}(N, M) = 0 . \tag{1}$$

Let $M^{(k)}$ denote the sum of k copies of M, and define $N^{(k)}$ analogously. We shall assume that the Krull-Schmidt theorem holds for decompositions of $M^{(k)}$ and $N^{(k)}$. (This will certainly be the case when R is a group ring $Z^*G_{p^s}$; see [5] or [7]).

LEMMA 1. *Let X be an extension of N by M. Each $f \in \mathrm{Hom}(X, X)$ maps M into itself, and induces an endomorphism f^* of N. The mapping $f \to (f \mid M, f^*)$ of* $\mathrm{Hom}(X, X)$ *into the direct sum*

$$\mathrm{Hom}(M, M) + \mathrm{Hom}(N, N) ,$$

is a monomorphism.

PROOF. Let X be given by an exact sequence

$$0 \longrightarrow M \xrightarrow{i} X \xrightarrow{j} N \longrightarrow 0 ,$$

where we identify M and iM. Let $f \in \mathrm{Hom}(X, X)$. Then $jfi \in \mathrm{Hom}(M, N)$, so that $jfi = 0$ because of condition (1). Therefore $fi(M) \subset iM$, and so f carries M into itself. Thus $f \mid M \in \mathrm{Hom}(M, M)$.

The map $f^* \in \mathrm{Hom}(N, N)$ induced by f is given as follows. For $n \in N$, choose an element $x_n \in X$ such that $jx_n = n$, and define

$$f^*n = jfx_n , \qquad n \in N .$$

The map $f \to (f \mid M, f^*)$ is clearly a homomorphism.

Suppose now that both $f \mid M$ and f^* are zero maps; then $fi(M) = 0$ and $f^*N = 0$. The first of these shows that the mapping $n \to fx_n$ is a well-defined map of N into X. But $f^*N = 0$ then yields $jfx_n = 0$, so $fx_n \in iM$, and thus $x \to fx_n$ is a homomorphism of N into iM. This must be the zero map because of (1), which means that $fx_n = 0$ for all n. Hence $f = 0$, and the lemma is proved.

We are going to consider extensions of $N^{(u)}$ by $M^{(t)}$. Clearly $\mathrm{Ext}(N^{(u)}, M^{(t)})$ is isomorphic to the additive group A of all $t \times u$ matrices with entries in $\mathrm{Ext}(N, M)$. Each extension Y of $N^{(u)}$ by $M^{(t)}$ determines an element $a(Y) \in A$.

Now let

$$H = \mathrm{Hom}(M, M), \qquad H' = \mathrm{Hom}(N, N) .$$

Then $\mathrm{Ext}(N, M)$ is an (H, H')-bimodule, and to each $h \in H, h' \in H'$, $v \in \mathrm{Ext}(N, M)$ there corresponds an element $hvh' \in \mathrm{Ext}(N, M)$.

We shall say that a matrix $a(Y) \in A$ is (H, H')-*decomposable* if there exists an invertible $t \times t$ matrix T with entries in H, and an invertible $u \times u$ matrix U with entries in H', such that

$$T \cdot a(Y) \cdot U = \begin{pmatrix} B & 0 \\ 0 & D \end{pmatrix} \tag{2}$$

for some matrices B and D (not necessarily square matrices).

LEMMA 2. *Let M and N be indecomposable R-modules, and let Y be an extension of $N^{(u)}$ by $M^{(t)}$. Then Y is a decomposable R-module if and only if the corresponding matrix $a(Y)$ is (H, H')-decomposable.*

PROOF. From (1) we conclude that

$$\mathrm{Hom}(M^{(t)}, N^{(u)}) = 0 , \quad \mathrm{Hom}(N^{(u)}, M^{(t)}) = 0 ,$$

so by Lemma 1 any automorphism f of Y induces automorphisms f_1 (say) of $M^{(t)}$ and f_2 (say) of $N^{(u)}$. We may represent f_1 by an invertible $t \times t$ matrix T with entries in H, and f_2 by an invertible $u \times u$ matrix U with entries in H'. Applying f to Y has the effect of replacing $a(Y)$ by $T \cdot a(Y) \cdot U$.

The module Y is decomposable if and only if Y splits into a direct sum $Y = Y_1 \oplus Y_2$ in which Y_1 is an extension of N_1 by M_1, Y_2 of N_2 by M_2, where $M^{(t)} = M_1 \oplus M_2$ and $N^{(u)} = N_1 \oplus N_2$. By the Krull-Schmidt theorem, both M_1 and M_2 must be direct sums of copies of M, and N_1 and N_2 must be direct sums of copies of N, since we have assumed M and N indecomposable. But this shows that Y is decomposable if and only if there exist invertible matrices T and U such that (2) holds for some B and D. The lemma is thus established.

Now let $\mathrm{rad}\, H$ denote the Jacobson radical of the ring H, and $\mathrm{rad}\, H'$ that of H'. Let V be an (H, H')-submodule of $\mathrm{Ext}(N, M)$, and set

$$\tilde{V} = V/(\mathrm{rad}\, H) V , \quad \tilde{H} = H/\mathrm{rad}\, H , \quad \tilde{H}' = H'/\mathrm{rad}\, H' . \tag{3}$$

Assume that $(\mathrm{rad}\, H')\tilde{V} = 0$. Then we may view $\tilde{V}$ as an $(\tilde{H}, \tilde{H}')$-bimodule. Each matrix $a(Y)$ with entries in V determines a matrix $\tilde{a}(Y)$ with entries in $\tilde{V}$. Furthermore, if T is invertible over H, then the corresponding matrix $\tilde{T}$ is invertible over $\tilde{H}$. Analogously, if U is invertible over H', then $\tilde{U}$ is invertible over $\tilde{H}'$. We have at once

LEMMA 3. *Let Y be an extension of $N^{(u)}$ by $M^{(t)}$, and suppose that all entries of $a(Y)$ lie in V. Then Y is indecomposable if $\tilde{a}(Y)$ is $(\tilde{H}, \tilde{H}')$-indecomposable.*

In §§ 2 and 3 we shall construct a pair of indecomposable R-modules M and N for which (1) holds. For these modules it will turn out that $\tilde{H} \cong \bar{Z}$ and $\tilde{H}' \cong \bar{Z}$, where $\bar{Z} = Z^*/pZ^*$. We shall furthermore define a specific (H, H')-submodule V of $\operatorname{Ext}(N, M)$ for which $(\operatorname{rad} H')\tilde{V} = 0$, and such that

$$\tilde{V} = \bar{Z}v_1 \oplus \bar{Z}v_2 \,, \tag{4}$$

where

$$\tilde{H}v_i \subset \bar{Z}v_i \,, \quad \tilde{H}'v_i \subset \bar{Z}v_i \,, \qquad i = 1, 2 \,. \tag{5}$$

In terms of this notation, we have

LEMMA 4. *For each k, let Y_k be an extension of $N^{(k)}$ by $M^{(k)}$ such that*

$$\tilde{\alpha}(Y) = Iv_1 + Jv_2 \,,$$

where I is the k-rowed identity matrix, and J is the companion matrix of λ^k (λ = indeterminate). Then Y_k is an indecomposable R-module.

PROOF. By virtue of Lemma 3, it suffices to verify that $Iv_1 + Jv_2$ is $(\tilde{H}, \tilde{H}')$-indecomposable. Let K and L be invertible $k \times k$ matrices over $\bar{Z}$, and suppose that

$$K(Iv_1 + Jv_2)L = \begin{pmatrix} B & 0 \\ 0 & D \end{pmatrix} .$$

Write $B = B_1v_1 + B_2v_2$, $D = D_1v_1 + D_2v_2$. Then necessarily (by (5))

$$KL = \begin{pmatrix} B_1 & 0 \\ 0 & D_1 \end{pmatrix} , \qquad KJL = \begin{pmatrix} B_2 & 0 \\ 0 & D_2 \end{pmatrix} .$$

Comparing $\bar{Z}$-ranks on both sides of the first of these equations, we see that B_1 and D_1 must be non-singular square matrices. Replacing L by

$$L\begin{pmatrix} B_1 & 0 \\ 0 & D_1 \end{pmatrix}^{-1}$$

for the remainder of the proof, we have

$$KL = I \,, \quad KJL = \begin{pmatrix} B_2 & 0 \\ 0 & D_2 \end{pmatrix} ,$$

with B_2 and D_2 square matrices. But then $K = L^{-1}$, and so J must be decomposable under the transformation $J \to L^{-1}JL$, which is impossible. Hence $\tilde{\alpha}(Y)$ is $(\tilde{H}, \tilde{H}')$-indecomposable, and the result is proved.

2. In the preceding section we showed how to construct an infinite class of non-isomorphic indecomposable R-modules, where R is a ring such that the Krull-Schmidt theorem holds for R-modules. The construction depended on the existence of a pair of indecomposable R-modules M and N for which $\operatorname{Hom}(M, N) = 0$, $\operatorname{Hom}(N, M) = 0$, and for which $\operatorname{Hom}(M, M)$

and Hom(N, N) act in a suitable way on Ext(N, M). We are now going to define the module M to be used in this construction, postponing until § 3 the choice of N. It will turn out that for the module N chosen there, we will have Ext$(N, M) \cong \bar{M} = M/pM$. For this reason, we shall investigate in detail the action of Hom(M, M) on $\bar{M}$.

The following notation will be used throughout the remainder of this paper:

p = fixed rational prime, $p \geqq 2$.

Z^* = ring of p-adic integers in the p-adic completion of the rational field.

$\bar{Z}$ = Z^*/pZ^* = finite field with p elements.

G_{p^2} = cyclic group of order p^2, with generator g.

E_2 = $Z^*G_{p^2}$ = group ring of G_{p^2} with coefficients in Z^*.

φ_1 = $1 + g + g^2 + \cdots + g^{p-1}$, $\varphi_2 = 1 + g^p + g^{2p} + \cdots + g^{(p-1)p}$.

E_1 = $E_2/(g^p - 1)E_2 \cong Z^*G_p$, where G_p is cyclic of order p.

C = $Z^*[\zeta_2]$, where ζ_2 = primitive $p^{2\text{th}}$ root of 1.

a_L = left multiplication by the element a.

Since E_1 is a factor ring of E_2, it is canonically an E_2-module. Also C is an E_2-module, by means of $g \cdot c = \zeta_2 \cdot c$, $c \in C$. From $(g^p - 1)E_1 = 0$ and $\varphi_2 C = 0$ we find readily that

$$\text{Hom}(E_1, C) = 0 \,, \quad \text{Hom}(C, E_1) = 0 \,, \tag{6}$$

where Hom means Hom_{E_2}.

Next we note that

$$\varphi_2 E_2 = \varphi_2 \cdot \frac{Z^*[g]}{(g^{p^2} - 1)} \cong \frac{Z^*[g]}{(g^p - 1)} \cong E_1 \,. \tag{7}$$

Let us write $E_1 = E_2 e_1$, where $(g^p - 1)e_1 = 0$, and $E_2 = E_2 e_2$, where e_2 is the unity element of E_2. From (7) we see that the map $\mu \in \text{Hom}(E_1, E_2)$ defined by $\mu(e_1) = \varphi_2 e_2$ is an isomorphism of E_1 into E_2. The quotient $E_2/\varphi_2 E_2$ is isomorphic to C, and so we have an exact sequence

$$0 \longrightarrow E_1 \xrightarrow{\mu} E_2 \longrightarrow C \longrightarrow 0 \,.$$

This yields the exact sequence

$$\text{Hom}(E_2, E_1) \xrightarrow{\mu^*} \text{Hom}(E_1, E_1) \longrightarrow \text{Ext}(C, E_1) \longrightarrow 0 \,,$$

where Ext means $\text{Ext}^1_{E_2}$. Every element of Ext(C, E_1) is thus the image of some $f \in \text{Hom}(E_1, E_1)$. Identify Hom$(E_1, E_1)$ with E_1 by letting $f \rightarrow f(e_1) \in E_1$. For any $f' \in \text{Hom}(E_2, E_1)$ we have

$$(\mu^* f')e_1 = f'(\mu e_1) = f'(\varphi_2 e_2) = \varphi_2 f'(e_2) \in \varphi_2 E_1 \,.$$

This readily implies that

$$\operatorname{Ext}(C, E_1) \cong E_1/\varphi_2 E_1$$

as E_2-modules. But $\varphi_2 x = px$, $x \in E_1$, since $(g^p - 1)x = 0$. Therefore we have (compare [4])

$$\operatorname{Ext}(C, E_1) \cong E_1/pE_1 = \bar{E}_1 \tag{8}$$

(say) as E_2-modules.

We shall now take M to be an extension of C by E_1 which corresponds to the extension class $g - 1$ in $\bar{E}_1$, or equivalently, which corresponds to the map $(g - 1)_L \in \operatorname{Hom}(E_1, E_1)$. The E_2-module M is given explicitly by the exact sequence

$$0 \longrightarrow E_1 \xrightarrow{\theta} E_1 \dotplus E_2 \xrightarrow{\eta} M \longrightarrow 0 ,$$

where

$$\theta(e_1) = (g - 1)e_1 - \varphi_2 e_2 , \tag{9}$$

and η is the natural map onto $(E_1 \dotplus E_2)/\theta(E_1)$. Since $E_1 \dotplus E_2 = E_2 e_1 \oplus E_2 e_2$, we have $M = E_2\eta(e_1) + E_2\eta(e_2)$. It is easily seen that $E_2\eta(e_1)$ is a submodule of M which is isomorphic to E_1, and that $M/E_2\eta(e_1) \cong C$. This shows again that M is an extension of E_1 by C, and exhibits the submodule of M which is isomorphic to E_1. For future use we note that $\eta(e_1)$ generates this submodule, and the image of $\eta(e_2)$ generates the factor module.

We have already remarked in [4] that M is indecomposable. In fact, both E_1 and C are indecomposable E_2-modules, and (6) holds. Furthermore, $g - 1 \neq 0$ in $\bar{E}_1$, since $\bar{E}_1 \cong Z[g]/(g - 1)^p$. It follows then from Lemma 2 that M is indecomposable.

As in § 1, define $H = \operatorname{Hom}(M, M)$. By Lemma 1 there is a monomorphism

$$\rho\colon H \rightarrow \operatorname{Hom}(E_1, E_1) \dotplus \operatorname{Hom}(C, C)$$

given by $\rho(f) = (f|E_1, f^*)$, where $f \in H$ induces $f^* \in \operatorname{Hom}(C, C)$. Since E_2 is a commutative ring, ρ is indeed an E_2-monomorphism. Any E_2-endomorphism of E_1 is given by a left multiplication a_L for some $a \in E_2$, where a is determined only modulo $(g^p - 1)$, the annihilator of E_1. Likewise an element of $\operatorname{Hom}(C, C)$ is of the form b_L for some $b \in E_2$, where b is determined only modulo φ_2. We now characterize those pairs (a_L, b_L) which lie in $\rho(H)$.

Lemma 5. *The pair (a_L, b_L) lies in $\rho(H)$ if and only if*

$$(g - 1)(a - b) \in pE_2 + (g - 1)^p E_2 . \tag{10}$$

PROOF. (See Cartan-Eilenberg [1, p. 314], Heller [3]). Let

$$a_L \in \mathrm{Hom}(E_1, E_1) \quad \text{and} \quad b_L \in \mathrm{Hom}(C, C)$$

be given, and consider the diagram

$$\begin{array}{ccccccccc} 0 & \longrightarrow & E_1 & \longrightarrow & M & \longrightarrow & C & \longrightarrow & 0 \\ & & \downarrow a_L & & \downarrow h & & \downarrow b_L & & \\ 0 & \longrightarrow & E_1 & \longrightarrow & M & \longrightarrow & C & \longrightarrow & 0 \end{array} \tag{11}$$

with exact rows. We wish necessary and sufficient conditions for the existence of $h \in \mathrm{Hom}(M, M)$ making the diagram commute. We form the diagram

$$\begin{array}{ccc} \mathrm{Hom}(E_1, E_1) & \xrightarrow{\delta_2} & \mathrm{Ext}(C, E_1) \\ a_L^* \downarrow & & \downarrow b_L^* \\ \mathrm{Hom}(E_1, E_1) & \xrightarrow[\delta_1]{} & \mathrm{Ext}(C, E_1) \end{array}$$

in which δ_1 is the connecting homomorphism determined by the top row of (11), δ_2 by the bottom row. If $i: E_1 \to E_1$ is the identity map, then there exists a map h making (11) commutative if and only if

$$(\delta_1 a_L^*)i = (b_L^* \delta_2)i \; .$$

Now $(\delta_1 a_L^*)i = a_L(\delta_1 i)$, and $\delta_1 i$ is the extension class of the sequence which forms the top row of (11). Identifying $\mathrm{Ext}(C, E_1)$ with $\bar{E}_1$ by (8), we have

$$(\delta_1 a_L^*)i = a(g - 1) \qquad \text{in } \bar{E}_1 \; .$$

Likewise

$$(b_L^* \delta_2)i = b(g - 1) \qquad \text{in } \bar{E}_1 \; .$$

Hence h exists if and only if $(g - 1)(a - b) = 0$ in $\bar{E}_1$. However,

$$\bar{E}_1 \cong E_2/(pE_2 + (g^p - 1)E_2) = E_2/(pE_2 + (g - 1)^p E_2) \; ,$$

and so we obtain (10). This proves the lemma.

Now let $h \in H$ be such that $\rho(h) = (a_L, b_L)$. Then a_L gives the effect of h on the submodule $E_1\eta(e_1)$ of M described previously, and b_L gives the induced endomorphism of the factor module. Thus we have

$$h: \begin{cases} \eta(e_1) \longrightarrow a\eta(e_1) \\ \eta(e_2) \longrightarrow c\eta(e_1) + b\eta(e_2) \end{cases} \tag{12}$$

for some $c \in E_2$, where c is uniquely determined by a and b. We shall use this fact soon.

The ring H is commutative, since ρ is a monomorphism of H into the commutative ring $\mathrm{Hom}(E_1, E_1) \dotplus \mathrm{Hom}(C, C)$. An element $h \in H$ is invertible if and only if both a_L and b_L are invertible, where $\rho(h) = (a_L, b_L)$. However, the ring E_2 is a commutative local ring (see [4]) such that

$$\mathrm{rad}\, E_2 = pE_2 + (g-1)E_2 \,, \quad E_2/\mathrm{rad}\, E_2 \cong \bar{Z} \,.$$

Hence if a_L is not invertible then $a \in \mathrm{rad}\, E_2$, whence from (10) also $b \in \mathrm{rad}\, E_2$, and b_L is not invertible. The non-invertible elements of H therefore form an ideal, and so H is a local ring whose radical is the ideal of all non-invertible elements[1].

Furthermore, let $\rho(h) = (a_L, b_L)$ for some $h \in H$. We may choose an integer m, $0 \leqq m \leqq p-1$, such that $a - m \in \mathrm{rad}\, E_2$. Then $m_L \in H$, and $\rho(m_L) = (m_L, m_L)$. We have

$$\rho(h - m_L) = \big((a-m)_L \,, \quad (b-m)_L\big) \,,$$

so $h - m_L$ is not invertible in H. This shows that $H/\mathrm{rad}\, H \cong \bar{Z}$, and that to each $h \in H$ there corresponds an integer m, $0 \leqq m \leqq p-1$, such that $h - m_L \in \mathrm{rad}\, H$. We shall use this fact presently.

We are now ready to consider the module $\bar{M} = M/pM$. Since $M \cong E_1 \dotplus C$ as Z^*-modules, we have $\bar{M} \cong \bar{E}_1 \dotplus \bar{C}$ as $\bar{Z}$-modules, where $\bar{E}_1 = E_1/pE_1$, $\bar{C} = C/pC$. Therefore

$$(\bar{M}: \bar{Z}) = (\bar{E}_1: \bar{Z}) + (\bar{C}: \bar{Z}) = p + (p^2 - p) = p^2 \,. \tag{13}$$

On the other hand, we may determine $\bar{M}$ from the exact sequence

$$0 \longrightarrow \bar{E}_1 \xrightarrow{\bar{\theta}} \bar{E}_1 \dotplus \bar{E}_2 \longrightarrow \bar{M} \longrightarrow 0 \,,$$

where

$$\begin{aligned}\bar{\theta}(\bar{e}_1) &= (g-1)\bar{e}_1 - \varphi_2 \bar{e}_2 \\ &= (g-1)\bar{e}_1 - (g-1)^{p^2-p}\bar{e}_2 \,,\end{aligned}$$

since $\varphi_2 = (g-1)^{p^2-p}$ in $\bar{E}_2$. We set

$$u = \bar{e}_1 - (g-1)^{p^2-p-1}\bar{e}_2 + \bar{\theta}(\bar{E}_1) \,, \quad v = \bar{e}_2 + \bar{\theta}(\bar{E}_1) \,.$$

Then clearly $\bar{M} = E_2 u + E_2 v$. A simple calculation shows that

$$(g-1)u = 0 \,, \quad (g-1)^{p^2-1}v = 0 \,, \quad pu = pv = 0 \,, \tag{14}$$

and therefore

$$(\bar{M}: \bar{Z}) \leqq \big(\bar{E}_2/(g-1)\bar{E}_2: \bar{Z}\big) + \big(\bar{E}_2/(g-1)^{p^2-1}\bar{E}_2: \bar{Z}\big) = p^2 \,,$$

[1] The fact that H is a local ring could also be deduced from Swan [7], since H is the endomorphism ring of an indecomposable E_2-module, and E_2 is a Z^*-algebra.

since $\bar{E}_2 \cong \bar{Z}[g]/(g-1)^{p^2}$. Combining this with (13), we now have

$$\bar{M} = E_2 u \oplus E_2 v \ .$$

Now let us set

$$V = E_2 u \oplus E_2(g-1)v \subset \bar{M} \ . \tag{15}$$

Let us show that V is an H-submodule of $\bar{M}$. Using formulas (10) and (12), and the definitions of U and V, we obtain

$$h(u) = a\bar{e}_1 - (g-1)^{p^2-p-1}(c\bar{e}_1 + b\bar{e}_2) + \bar{\theta}(\bar{E}_1) = au \ , \tag{16}$$

$$\begin{aligned} h((g-1)v) &= (g-1)\{c\bar{e}_1 + b\bar{e}_2 + \bar{\theta}(\bar{E}_1)\} \\ &= \{b + c(g-1)^{p^2-p-1}\}(g-1)v \ . \end{aligned} \tag{17}$$

This shows that $hV \subset V$, as claimed.

We show next that $(\operatorname{rad} H)V = E_2(g-1)^2 v$. Clearly

$$(\operatorname{rad} H)V \supset (g-1)_L V = E_2(g-1)^2 v \ .$$

On the other hand, let $h \in \operatorname{rad} H$, and write $\rho(h) = (a_L, b_L)$, where both a and b are in $\operatorname{rad} E_2$. Then we have

$$a = pa_0 + (g-1)a_1 \ , \quad b = pb_0 + (g-1)b_1$$

for some $a_0, a_1, b_0 \in E_2$. Formulas (16) and (17) show at once that $h(V) \subset E_2(g-1)^2 v$, and thus $(\operatorname{rad} H)V \subset E_2(g-1)^2 v$.

Therefore

$$\tilde{V} = V/(\operatorname{rad} H)V = \bar{Z}v_1 \oplus \bar{Z}v_2 \ ,$$

where v_1 is the image of u, v_2 that of $(g-1)v$ in V. The first inclusion in (5) follows easily from the fact that each $h \in H$ is expressible as $h = m_L + h_1$, $0 \leqq m \leqq p-1$, $h_1 \in \operatorname{rad} H$.

3. We are now ready to prove that $n(Z^*G_{p^3})$ is infinite, where G_{p^3} is a cyclic group of order p^3 with generator g_3. As remarked in the introduction, this will imply that $n(ZG_{p^r})$ is infinite for $r \geqq 3$. We keep the notation of the preceding sections, and set $E_3 = Z^*G_{p^3} = Z^*[g_3]$. The map $g_3 \to g$ induces a ring homomorphism $E_3 \to E_2$, and so the indecomposable E_2-module M defined previously is also an indecomposable E_3-module. We shall apply the results of § 1 to this module M and an E_3-module N to be defined below. Note that the Krull-Schmidt theorem does hold for E_3-modules, (see [5] or [7]).

Now define

$$\varphi_3 = 1 + g_3^{p^2} + g_3^{2p^2} + \cdots + g_3^{(p-1)p^2} \in E_3 \ ,$$

and set

$$N = E_3/\varphi_3 E_3 = Z^*[\zeta_3] ,$$

where ζ_3 is a primitive $p^{3\text{th}}$ root of 1. Then N is indecomposable (in fact, irreducible), and we have

$$\varphi_3 N = 0, \quad (g_3^{p^2} - 1)M = 0 .$$

These relations at once imply the validity of (1).

To compute $\operatorname{Ext}(N, M)$, we use the exact sequence

$$0 \longrightarrow E_2 \longrightarrow E_3 \longrightarrow N \longrightarrow 0 ,$$

where the imbedding of E_2 in E_3 is given by $E_2 \cong \varphi_3 E_3 \subset E_3$.

From this we find (as in § 2) that

$$\operatorname{Ext}(N, M) \cong M/pM = \bar{M} .$$

Now define V by (15), set $H = \operatorname{Hom}(M, M)$, $H' = \operatorname{Hom}(N, N)$, and use the notation of (3). We have already discussed the way in which H acts on V, and must now determine the action of H' on V. This is not difficult, since every endomorphism of N is of the form x_L for some $x \in E_3$. Since V is an E_3-submodule of $\bar{M}$, we have $x_L V \subset V$. Thus V is an (H, H')-submodule of $\bar{M}$, as required.

It is easily verified that $\operatorname{rad} H' = (\operatorname{rad} E_2)_L$, and so

$$\tilde{H}' = H'/\operatorname{rad} H' \cong Z^*[\zeta_3]/(1 - \zeta_3) \cong \bar{Z} .$$

Furthermore,

$$(\operatorname{rad} H')\tilde{V} = (\operatorname{rad} E_3)_L \tilde{V} \subset (\operatorname{rad} H)\tilde{V} = 0 .$$

The second inclusion in (5) is a simple consequence of the above remarks.

We have thus obtained a pair of indecomposable E_3-modules M and N satisfying (1), and such that $\tilde{H} \cong \bar{Z}$, $\tilde{H}' \cong \bar{Z}$, and for which $\operatorname{Ext}(N, M)$ contains an (H, H')-submodule V satisfying (4) and (5).

THEOREM. *For each k, the module Y_k defined in Lemma 4 is an indecomposable E_3-module. Since*

$$(Y_k\colon Z^*) = k(M\colon Z^*) + k(N\colon Z^*) ,$$

the modules $Y_1, Y_2, \cdots$ are non-isomorphic, and thus $n(Z^* G_{p^3})$ is infinite.

We may remark that in the construction in Lemma 4, we could have used in place of J any matrix J' over $\bar{Z}$ which is indecomposable under the transformation $J' \to L^{-1}J'L$. Furthermore, if Y'_k is defined using J' instead of J, it is easily seen (using Lemmas 1 and 2) that $Y_k = Y'_k$ if and only if $J' = L^{-1}JL$ for some invertible L over $\bar{Z}$. Thus we may construct many families of non-isomorphic indecomposable E_3-modules. The classification of *all* indecomposable E_3-modules seems to us to be an impossible task.

It should also be pointed out that the construction given here cannot be used to obtain infinitely many indecomposable E_2-modules, since the Z^*G_p-module M_0 which plays the role of M remains indecomposable when taken modulo p, and there will not be any submodule V_0 of $\mathrm{Ext}(N_0, M_0)$ with the desired properties.

UNIVERSITY OF ILLINOIS

REFERENCES

1. H. CARTAN and S. EILENBERG, Homological Algebra, Princeton, 1956.
2. F. E. DIEDERICHSEN, *Über die Ausreduktion ganzzahliger Gruppendarstellungen bei arithmetischer Äquivalenz*, Abh. Math. Sem.Univ. Hamburg, 14 (1938), 357-412.
3. A. HELLER, *Homological algebra in abelian categories*, Ann. of Math., 68 (1958), 484-525.
4. ——— and I. REINER, *Representations of cyclic groups in rings of integers*, I, Ann. of Math., 76 (1962), 72-91.
5. I. REINER, *The Krull-Schmidt theorem for integral group representations*, Bull. Amer. Math. Soc., 67 (1961), 365-367.
6. A. V. ROITER, *Integral representations of cyclic groups of the fourth order*, Proc. Leningrad Univ., 19 (1960), 65-74.
7. R. G. SWAN, *Induced representations and projective modules*, Ann. of Math., 71 (1960), 552-578.
8. A. TROY, Integral representations of cyclic groups of order p^2, Ph. D. thesis, Univ. of Illinois, 1961.

GROTHENDIECK GROUPS OF ORDERS IN SEMISIMPLE ALGEBRAS

A. Heller and I. Reiner

ABSTRACT

Let A be a noetherian ring with unit; consider only finitely generated left modules. Let $\mathcal{A}$ be the free abelian group generated by (M), where M ranges over all A-modules. Define $\mathcal{A}_0$ to be the subgroup of $\mathcal{A}$ generated by all

$$(M) - (M') - (M''),$$

where there is an exact sequence

$$0 \to M' \to M \to M'' \to 0.$$

Then $K^0(A) = \mathcal{A}/\mathcal{A}_0$ is the *Grothendieck group* of A. The image of (M) in $K^0(A)$ is denoted by $[M]$.

The Whitehead group $K^1(A)$ is defined as follows: Consider the category whose objects are pairs (M, μ) with M a left A-module and μ an automorphism of M. A morphism $\phi : (M, \mu) \to (N, \nu)$ is an A-homomorphism $\phi : M \to N$ with the property $\phi\mu = \nu\phi$. A sequence is exact in this category if the corresponding sequence of A-modules is exact. Let $\mathcal{B}$ be the free abelian group generated by all pairs (M, μ); let $\mathcal{B}_0$ be the subgroup generated by all

$$(M, \mu) - (L, \lambda) - (N, \nu)$$

for which there is an exact sequence

$$0 \to (L, \lambda) \to (M, \mu) \to (N, \nu) \to 0$$

along with the elements

$$(M, \mu\mu') - (M, \mu) - (M, \mu').$$

Then $K^1(A) = \mathcal{B}/\mathcal{B}_0$ and $[M, \mu]$ denotes the image of (M, μ) in this group.

Now let R be a noetherian domain with quotient field F, A is an R-algebra and $A^* = F \otimes_R A$. Define $K_f^0(A)$ just as $K^0(A)$ except that only R-torsion free modules are used. The elements in this group are denoted by $[M]_f$. An application of Schanuel's lemma shows that the map $[M]_f \to [M]$ is an isomorphism of $K_f^0(A)$ with $K^0(A)$. The group $K_t^0(A)$ is defined as $K^0(A)$ except that only R-torsion modules are used; $[M]_t$ denotes an element of this group.

The main result of this paper is the following:

- If A^* is a semisimple ring then there is an exact sequence

$$K^1(A^*) \xrightarrow{\Delta} K_t^0(A) \xrightarrow{\eta} K_f^0(A) \xrightarrow{\theta} K^0(A^*) \longrightarrow 0$$

where the maps are defined as follows.

Let M^* be an A^*–module and μ^* an automorphism of M^*. Let M be an A–module for which $FM = M^*$. Let $U = M \cap \mu^*(M)$ so that both M/U and $\mu^*(M)/U$ are R–torsion modules. Define

$$\Delta : [M^*, \mu^*] \to [M/U] - [\mu^*(M)/U].$$

If M is an R–torsion A–module, there exists a projective A–module Y and a submodule $X \subset Y$ such that $M \cong Y/X$. Define η by the rule

$$\eta : [M]_t \to [Y]_f - [X]_f.$$

Finally θ is defined by $\theta[M]_f = [FM]$.

The proof that the maps are well–defined and the sequence is exact is given in detail. An application is made to the case $A = RG$ for a finite group G. By applying a result of Brauer concerning the rank of the decomposition matrix, information about the structure of $K_f^0(A)$ is obtained. The main application was also obtained using other methods in [38]; these methods did not require the use of Brauer's theorem and, in fact, they gave another proof of it.

GROTHENDIECK GROUPS OF INTEGRAL GROUP RINGS

BY

A. HELLER AND I. REINER

1. Introduction

Let A be a ring, and consider the category of A-modules. Unless otherwise stated, A-modules are assumed to be left modules which are finitely generated. Recall that the *Grothendieck group* $K^0(A)$ of this category is the abelian additive group defined by means of generators and relations, as follows: the generators are the symbols $[M]$, where M ranges over all A-modules; the relations are given by

$$[M] = [M'] + [M''],$$

corresponding to all short exact sequences of A-modules

$$0 \to M' \to M \to M'' \to 0.$$

In particular, let G be a finite group, and let R = alg. int. $\{F\}$, the ring of all algebraic integers in the algebraic number field F. Denote by FG the group algebra of G over F, and by RG the integral group ring of G over R. Swan [11] has already demonstrated the importance of the Grothendieck group $K^0(RG)$ for the study of RG-modules, and has recently obtained in [13] some new fundamental results on the structure of the group.

The present authors have given an explicit formula for $K^0(RG)$ under the restriction that F be a splitting field for G (see [9]). This formula involves the ideal theory of the Dedekind ring R, as well as the decomposition numbers of G relative to the set of those prime ideals of R which divide the order of G.

Here we shall generalize this formula to the case where F need not be a splitting field for G. Our results will involve the ideal theory of certain algebraic extension fields of R, as well as analogues of the decomposition matrices.

In our earlier paper, we made use of the following:

THEOREM 1 (Brauer [3], [4]). *If F is a splitting field for G, then the set of maximal size minors of the decomposition matrix of G (relative to any prime ideal of R) has greatest common divisor* 1.

As a by-product of the present approach, an independent proof of Brauer's theorem is obtained.

For the homological algebra used herein, we refer the reader to [5]. As a general reference for the representation theory needed here, we may cite [6].

2. Whitehead groups

This section is devoted to the introduction of notation, and the statement of one of the main results of our previous paper [9]. Let R be any Dedekind

Received January 17, 1964.

ring with quotient field F, and let A^* be a finite-dimensional semisimple F-algebra. By an R-order A of A^* is meant a subring A of A^* such that

(i) $1 \in A$,
(ii) A contains an F-basis of A^*, and
(iii) A is finitely generated as R-submodule of A^*.

We may then form the Grothendieck groups $K^0(A^*)$, $K^0(A)$, and $K_t^0(A)$, the last of which is obtained from the category of R-torsion A-modules.

For X an R-torsion-free A-module, we shall denote the A^*-module $F \otimes_R X$ by FX, for brevity, and shall regard X as embedded in FX. If X and Y are a pair of R-torsion-free A-modules for which $FX \cong FY$ as A^*-modules, we may identify FX and FY. Then we define

$$[X /\!\!/ Y] = [X/U] - [Y/U] \in K_t^0(A),$$

where U is any A-submodule of $X \cap Y$ such that $FX = FY = FU$.

Let us recall that the *Whitehead group* $K^1(A^*)$ is the abelian additive group defined by generators and relations as follows: the generators are the symbols $[M, \mu]$, where M ranges over all A-modules, and μ ranges over all automorphisms of M; the relations are, first, those of the form

$$[M, \mu\mu'] = [M, \mu] + [M, \mu']$$

for a pair of automorphisms μ, μ' of M; and second, those of the form

$$[M, \mu] = [M', \mu'] + [M'', \mu'']$$

for every short exact sequence of A^*-modules

$$0 \to M' \xrightarrow{\varphi} M \xrightarrow{\psi} M'' \to 0$$

such that $\mu\varphi = \varphi\mu'$, $\mu''\psi = \psi\mu$.

We quote without proof:

THEOREM 2 (Heller and Reiner [9]). *There is an exact sequence*

$$K^1(A^*) \xrightarrow{\delta} K_t^0(N) \xrightarrow{\eta} K^0(A) \xrightarrow{\theta} K^0(A^*) \to 0,$$

with the maps defined as follows:

(i) *Given* $[M, \mu] \in K^1(A^*)$, *let* M_0 *be any* A-*submodule of* M *such that* $FM_0 = M$, *and set*

$$\delta[M, \mu] = [\mu M_0 /\!\!/ M_0] \in K_t^0(A).$$

(ii) *The map* η *is induced by the inclusion of the category of* R-*torsion* A-*modules in the category of all* A-*modules.*

(iii) *For an* A-*module* L, *set* $\theta[L] = [F \otimes_R L]$.

For later use, we shall determine the Whitehead group of a simple algebra A^*. Suppose that A^* is a full matrix algebra over the division ring D, and let W be an irreducible A^*-module. Then we may write $D = \mathrm{Hom}_{A^*}(W, W)$, and view W as a right D-module. As is well known, we have

$$A^* = \mathrm{Hom}_D(W, W).$$

Now each A^*-module M is a direct sum of (say) t copies of W, and each automorphism μ of M is represented by an invertible $t \times t$ matrix $\bar{\mu}$ with entries in D. Let $\tilde{D}$ denote the multiplicative group of non-zero elements of D, and set

$$D^{\#} = \tilde{D}/[\tilde{D}, \tilde{D}],$$

the factor commutator group of $\tilde{D}$. We may then form the Dieudonné determinant $d(\bar{\mu}) \in D^{\#}$. It is easily seen that the relations which serve to define $K^1(A^*)$ are precisely those which characterize the Dieudonné determinant (see [8]). Thus we have

$$K^1(A^*) \cong D^{\#},$$

the isomorphism being given by $[M, \mu] \to d(\bar{\mu})$.

As a special case of the above, we have $K^1(D) \cong D^{\#}$. (In fact, Morita's theorem (see §3) implies that the categories of A-modules and D-modules are isomorphic. Consequently we may conclude that $K^1(A^*) \cong K^1(D)$.)

Suppose now that A is an R-order in the simple algebra A^*, and let W_0 be any A-submodule of W such that $FW_0 = W$. We may write

$$D^{\#} \cong K^1(A^*) \to K_t^0(A),$$

thereby obtaining a map $D^{\#} \to K_t^0(A)$ which we again denote by δ. For $\lambda \in \tilde{D}$, we have

$$\delta(\lambda) = [W_0 \lambda /\!\!/ W_0].$$

3. Maximal orders in central simple algebras

Let A^* be a central simple algebra over the algebraic number field F. Then A^* is isomorphic to a full matrix algebra over a division ring D whose center is F. Let W be an irreducible A^*-module, viewed as right D-module. Then we may write

$$D = \mathrm{Hom}_{A^*}(W, W), \qquad A^* = \mathrm{Hom}_D(W, W).$$

Now let R = alg. int. $\{F\}$, and let A be a maximal R-order in A^*. Such maximal orders always exist, but need not be unique. From the results of Auslander and Goldman [1], it follows that there exists a maximal R-order $\mathfrak{o}$ in D, and a right projective $\mathfrak{o}$-module M contained in W, such that $W = FM$ and

$$A = \mathrm{Hom}_{\mathfrak{o}}(M, M).$$

We shall use Morita's theorem to set up an isomorphism between the categories of left A-modules and left $\mathfrak{o}$-modules, following an approach due to Bass [2]. The right $\mathfrak{o}$-module M is called a *generator* (of the category of right $\mathfrak{o}$-modules) if given any pair of right $\mathfrak{o}$-modules X and Y, and any non-zero map f in $\mathrm{Hom}_{\mathfrak{o}}(X, Y)$, there exists a map $g \in \mathrm{Hom}_{\mathfrak{o}}(M, X)$ such that fg is not the zero map. It is convenient to rephrase this as follows: The map f induces a map

$$f^* : \mathrm{Hom}_{\mathfrak{o}}(M, X) \to \mathrm{Hom}_{\mathfrak{o}}(M, Y).$$

Then M is a generator if and only if for each $f \in \mathrm{Hom}_{\mathfrak{o}}(X, Y)$, $f \neq 0$ implies $f^* \neq 0$.

We now quote without proof.

THEOREM 3 (Morita [10]; see Bass [2]). *Let M be a right finitely generated projective $\mathfrak{o}$-module which is a generator for the category of right $\mathfrak{o}$-modules. Define $A = \mathrm{Hom}_{\mathfrak{o}}(M, M)$, viewed as a ring of left operators on M, and set $\tilde{M} = \mathrm{Hom}_{\mathfrak{o}}(M, \mathfrak{o})$, a left $\mathfrak{o}$-, right A-module. Then the categories of left A-modules and left $\mathfrak{o}$-modules are isomorphic, and the isomorphism is given as follows: a left $\mathfrak{o}$-module U corresponds to the left A-module $M \otimes_{\mathfrak{o}} U$, and inversely a left A-module V corresponds to the left $\mathfrak{o}$-module $\tilde{M} \otimes V$. Furthermore, $\mathfrak{o} = \mathrm{Hom}_A(M, M)$ as right operator domain on M.*

In order to apply the above, we must verify that in our case M is indeed a generator. Let X, Y be $\mathfrak{o}$-modules, and let $f \in \mathrm{Hom}_{\mathfrak{o}}(X, Y)$, $f \neq 0$. We need only show that $f^* \neq 0$. Let P denote a prime ideal of R, and let a subscript P indicate localization at P. Since $f \neq 0$, then also $f_P \neq 0$ for some P, where $f_P : X_P \to Y_P$. By the results of [1], the R_P-order $\mathfrak{o}_P$ is a hereditary principal ideal ring, so that M_P is a free $\mathfrak{o}_P$-module (see [13]). Consequently M_P is an $\mathfrak{o}_P$-generator, and therefore $(f_P)^* \neq 0$. But $(f_P)^* = (f^*)_P$, and therefore also $f^* \neq 0$, as desired.

Applying Morita's theorem, we have $\mathfrak{o} = \mathrm{Hom}_A(M, M)$, and the category of left A-modules is isomorphic to the category of left $\mathfrak{o}$-modules, under the isomorphisms given above. Therefore we have

$$K^0(A) \cong K^0(\mathfrak{o}).$$

Furthermore, the isomorphism of categories preserves R-torsion, so that also

$$K_t^0(A) \cong K_t^0(\mathfrak{o}), \qquad K^0(A/PA) \cong K^0(\mathfrak{o}/P\mathfrak{o})$$

for each prime ideal P of R. We note further that

$$K_t^0(A) = \textstyle\sum_P^{\oplus} K^0(A/PA) \cong \sum_P^{\oplus} K^0(\mathfrak{o}/P\mathfrak{o}) = K_t^0(\mathfrak{o}).$$

Let $I(R)$ denote the abelian multiplicative group of non-zero R-ideals in F. We shall show that $K_t^0(\mathfrak{o}) \cong I(R)$, and in fact shall give two descriptions of this isomorphism. For any R-torsion $\mathfrak{o}$-module X, define

$$\mathrm{ann}\ X = \{\alpha \in R : \alpha X = 0\} \in I(R).$$

Now let V be any R-torsion $\mathfrak{o}$-module, and let $V_1, \cdots, V_k$ be its $\mathfrak{o}$-composition factors. Define the *order ideal* of V to be

$$\mathrm{ord}\ V = \textstyle\prod_{i=1}^k (\mathrm{ann}\ V_i) \in I(R).$$

Then ord V is a well-defined ideal in R, and the map $[V] \to \text{ord } V$ defines a homomorphism of $K_t^0(\mathfrak{o})$ into $I(R)$, since the composition factors of an extension module are just those of the submodule together with those of the factor module.

Let us show that the above-defined map is an isomorphism. The additive group $K_t^0(\mathfrak{o})$ has as Z-basis the elements $[\mathfrak{o}/\mathfrak{m}]$, where $\mathfrak{m}$ ranges over all maximal left ideals of $\mathfrak{o}$. This is clear from the fact that every irreducible $\mathfrak{o}$-module is expressible as $\mathfrak{o}/\mathfrak{m}$, for some $\mathfrak{m}$. For fixed $\mathfrak{m}$, let $\mathfrak{p}$ be the unique maximal two-sided ideal of $\mathfrak{o}$ contained in $\mathfrak{m}$. Set $P = \mathfrak{p} \cap R$, a prime ideal in R. Then $\text{ord } (\mathfrak{o}/\mathfrak{m}) = P$.

On the other hand, there is a mapping $I(R) \to K_t^0(\mathfrak{o})$, defined as follows. Let P be any prime ideal of R. By [1] there is a unique maximal two-sided ideal $\mathfrak{p}$ of $\mathfrak{o}$ such that $\mathfrak{p} \cap R = P$. The ring $\mathfrak{o}/\mathfrak{p}$ is then a simple ring. If $\mathfrak{m}$ is any maximal left ideal of $\mathfrak{o}$ such that $\mathfrak{p} \subset \mathfrak{m}$, then $\mathfrak{o}/\mathfrak{m}$ is an irreducible $(\mathfrak{o}/\mathfrak{p})$-module, which is determined up to isomorphism by P, since $\mathfrak{o}/\mathfrak{p}$ is simple. Letting $P \to [\mathfrak{o}/\mathfrak{m}]$, we obtain a homomorphism of $I(R)$ onto $K_t^0(\mathfrak{o})$. It follows at once that $K_t^0(\mathfrak{o}) \cong I(R)$, the isomorphisms being given as above.

The referee has kindly pointed out a second description of the above isomorphism, which is more useful for later purposes. For $\mathfrak{a}$ any (non-zero) left ideal in $\mathfrak{o}$, let $N\mathfrak{a}$ be its reduced norm (see [7]). We recall the definition of reduced norm: take any R-composition series of the R-module $\mathfrak{o}/\mathfrak{a}$, and let $N'\mathfrak{a}$ be the product of the annihilators of the composition factors. Let n be the index of the division algebra D, so that $(D:F) = n^2$. The equation

$$(N\mathfrak{a})^n = N'\mathfrak{a}$$

then serves to define an ideal $N\mathfrak{a}$ in R, called the *reduced norm* of $\mathfrak{a}$.

We shall prove that $\text{ord } (\mathfrak{o}/\mathfrak{a}) = N\mathfrak{a}$, and for this it suffices to prove that $P = N\mathfrak{m}$, where $\mathfrak{m}$ and P are related as above. The simple ring $\mathfrak{o}/\mathfrak{p}$ is a full matrix algebra $(\bar{k})_r$ over some skewfield $\bar{k}$. Since $\bar{k}$ is a finite extension of R/P, it follows from Wedderburn's theorem that $\bar{k}$ is a field. The ring $\mathfrak{o}/\mathfrak{p}$ is a direct sum of r copies of the irreducible $(\mathfrak{o}/\mathfrak{p})$-module $\mathfrak{o}/\mathfrak{m}$, which implies that $N'\mathfrak{p} = (N'\mathfrak{m})^r$. On the other hand, $\mathfrak{o}/\mathfrak{p}$ is R-isomorphic to a direct sum of f copies of R/P, where $f = (\mathfrak{o}/\mathfrak{p}:R/P)$. Therefore

$$N'\mathfrak{p} = P^f, \qquad N\mathfrak{m} = P^{f/rn},$$

and we need only show that $f = rn$.

This may be accomplished by working over the P-adic completion $\hat{F}$ of the field F. Let $\hat{R}$ be the valuation ring of $\hat{F}$, and $\hat{P}$ its prime ideal. Set $\hat{D} = D \otimes_F \hat{F}$, $\hat{\mathfrak{o}} = \mathfrak{o} \otimes_R \hat{R}$. Then $\hat{D}$ is a simple ring with center $\hat{F}$, but is not necessarily a skewfield. Write $\hat{D} = (\hat{D}_1)_s$, a full matrix algebra over a skewfield $\hat{D}_1$. If we set $n_1^2 = (\hat{D}_1:\hat{F})$, then $n^2 = (\hat{D}:\hat{F}) = s^2n_1^2$, so $n = sn_1$. As in Theorem 3, we may write

$$\hat{\mathfrak{o}} = (\hat{\mathfrak{o}}_1)_s, \qquad \hat{\mathfrak{p}} = (\hat{\mathfrak{p}}_1)_s,$$

where $\hat{\mathfrak{o}}_1$ is a maximal order in $\hat{D}_1$, and $\hat{\mathfrak{p}}_1$ is a maximal two-sided ideal in $\hat{\mathfrak{o}}_1$. If $f_1 = (\hat{\mathfrak{o}}_1/\hat{\mathfrak{p}}_1 : \hat{R}/\hat{P})$, then

$$f = (\mathfrak{o}/\mathfrak{p} : R/P) = (\hat{\mathfrak{o}}/\hat{\mathfrak{p}} : \hat{R}/\hat{P}) = s^2 f_1 .$$

But also $\mathfrak{o}/\mathfrak{p} \cong (\hat{\mathfrak{o}}_1/\hat{\mathfrak{p}}_1)_s$, and since $\hat{F}$ is a complete P-adic field, it follows from [7] that $\hat{\mathfrak{o}}_1/\hat{\mathfrak{p}}_1$ is a field. Thus $r = s$, and so $f/rn = s^2 f_1/s^2 n_1 = f_1/n_1$. However, since $\hat{F}$ is complete, we have $f_1 = n_1$, which shows that $f = rn$, as claimed.

(Later on we shall need to know the number ν of composition factors of the $(\mathfrak{o}/\mathfrak{p})$-module $\mathfrak{o}/P\mathfrak{o}$. Let us compute this by comparing dimensions over R/P. The dimension of an irreducible $(\mathfrak{o}/\mathfrak{p})$-module is sf_1, while

$$\dim (\mathfrak{o}/P\mathfrak{o}) = s^2 \dim (\hat{\mathfrak{o}}_1/\hat{\mathfrak{p}}_1^{e_1}) = s^2 e_1 f_1 ,$$

where e_1 is the ramification index of P at $\hat{\mathfrak{p}}_1$. Since $\hat{F}$ is a P-adic field, we have $e_1 = f_1 = n_1$, and thus $\nu = s^2 e_1 f_1/sf_1 = se_1 = sn_1 = n$. This shows that $\mathfrak{o}/P\mathfrak{o}$ has n composition factors when viewed as $(\mathfrak{o}/\mathfrak{p})$-module.)

In §2 we had defined a map $\delta : D^\# \to K_t^0(A)$. Since

$$K_t^0(A) \cong K_t^0(\mathfrak{o}) \cong I(R),$$

δ gives a map of $D^\#$ into $I(R)$; denote by $J(R)$ the image of δ in $I(R)$. For $\lambda \in \tilde{D}$, we have $\delta(\lambda) = [W_0 \lambda /\!\!/ W_0] \in K_t^0(A)$, where W_0 is an A-submodule of A^* such that $FW_0 = W$; indeed, choose W_0 to be the module M in Theorem 3. Since M corresponds to $\mathfrak{o}$ itself in the correspondence given in Theorem 3, we see that

$$\delta(\lambda) = [\mathfrak{o}\lambda /\!\!/ \mathfrak{o}] \in K_t^0(\mathfrak{o}),$$

and hence (in $I(R)$)

$$\delta(\lambda) = \{N(\mathfrak{o}\lambda)\}^{-1}.$$

However, $N(\mathfrak{o}\lambda) = (N\lambda)R$, where $N\lambda$ is the reduced norm of the element λ. This shows that $J(R)$ is the subgroup of $I(R)$ generated by the principal ideals $(N\lambda)R$, where λ ranges over all non-zero elements of D.

As shown in [12], we may describe $J(R)$ explicitly. If P_0 is an infinite prime of R, and F_0 is the P_0-adic completion of F, we call D *ramified* at P_0 if $D \otimes_F F_0$ is a full matrix algebra over the real quaternions. Let U be the divisor of R consisting of all infinite primes P_0 at which D is ramified. Then $J(R)$ is precisely the ray mod U, that is,

$$J(R) = \{xR : x \in F, x > 0 \text{ at each } P_0 \in U\}.$$

We shall briefly discuss the projective class group $P(\mathfrak{o})$, and reduced projective class group $C(\mathfrak{o})$, of the ring $\mathfrak{o}$. The group $P(\mathfrak{o})$ is defined as the Grothendieck group of the category of projective $\mathfrak{o}$-modules, and there is an obvious map $P(\mathfrak{o}) \to K^0(\mathfrak{o})$. However, the ring $\mathfrak{o}$ is hereditary (by [1]), and as pointed out in [13], this easily implies that the above map is an isomorphism: $P(\mathfrak{o}) \cong K^0(\mathfrak{o})$. Since A is also hereditary, we have similarly $P(A) \cong K^0(A)$.

Swan [13] proved that

$$C(A) \cong C(\mathfrak{o}) \cong I(R)/J(R).$$

We may obtain this same result here by use of Theorems 2 and 3. Using the map θ defined in Theorem 2, we have

$$P(\mathfrak{o}) \cong K^0(\mathfrak{o}) \xrightarrow{\theta} K^0(D),$$

which defines a homomorphism (again denoted by θ) of $P(\mathfrak{o})$ into $K^0(D)$. In [13, Prop. 4.1], Swan showed that the kernel of θ (in $P(\mathfrak{o})$) is precisely $C(\mathfrak{o})$.

From Theorems 2 and 3, we obtain a pair of isomorphic exact sequences

$$\begin{array}{ccccccc} K^1(A) & \xrightarrow{\delta'} & K_t^0(A) & \xrightarrow{\eta'} & K^0(A) & \xrightarrow{\theta'} & K^0(A^*) \to 0 \\ \downarrow & & \downarrow & & \downarrow & & \downarrow \\ K^1(D) & \xrightarrow[\delta]{} & K_t^0(D) & \xrightarrow[\eta]{} & K^0(\mathfrak{o}) & \xrightarrow[\theta]{} & K^0(D) \to 0, \end{array}$$

in which each vertical arrow is an isomorphism. Therefore $\ker \theta' \cong \ker \theta$, that is, $C(A) \cong C(\mathfrak{o})$. Furthermore,

$$C(\mathfrak{o}) \cong \ker \theta = \operatorname{im} \eta \cong K_t^0(\mathfrak{o})/\operatorname{im} \delta \cong I(R)/J(R),$$

which gives the desired result.

As shown in [12] and [13], $C(\mathfrak{o})$ is always a finite group. The group $J(R)$ is an analogue of the group of principal ideals, and $I(R)/J(R)$ is an analogue of the ideal class group of R. Indeed, when $D = F$ then $\mathfrak{o} = R$, and in this case the quotient $I(R)/J(R)$ is precisely the ideal class group of R.

4. Grothendieck groups of group rings

Let G be a finite group, F an algebraic number field, and $R = \text{alg. int. }\{F\}$ Set $A = RG$, $A^* = FG$, and let $\mathfrak{O}$ be any maximal R-order of A^* which contains A. By restriction of operators, each $\mathfrak{O}$-module becomes an A-module, and R-torsion is preserved. Using Theorem 2, we obtain a commutative diagram with exact rows:

$$\begin{array}{ccccccc} K^1(A^*) & \xrightarrow{\delta'} & K_t^0(\mathfrak{O}) & \xrightarrow{\eta'} & K^0(\mathfrak{O}) & \xrightarrow{\theta'} & K^0(A^*) \to 0 \\ \downarrow 1 & & \downarrow \beta & & \downarrow \alpha & & \downarrow 1 \\ K^1(A^*) & \xrightarrow[\delta]{} & K_t^0(A) & \xrightarrow[\eta']{} & K^0(A) & \xrightarrow[\theta]{} & K^0(A^*) \to 0. \end{array}$$

In [13] Swan proved the difficult result that α is an epimorphism. Applying the "Five Lemma" to the above diagram, we conclude that also β is an epimorphism, and therefore $\ker \theta = \operatorname{im} \eta\beta$. Next, we note that $K^0(A^*)$ is a free Z-module, and therefore

$$K^0(A) = K^0(A^*) \oplus \ker \theta$$

as additive groups. Furthermore,

$$\ker \theta = \operatorname{im} \eta\beta \cong K_t^0(\mathfrak{O})/\ker \eta\beta.$$

Routine diagram-chasing yields

$$\ker \eta\beta = \ker \beta + \operatorname{im} \delta'$$

and consequently

$$K^0(A) \cong K^0(A^*) \quad \oplus \quad K_t^0(\mathfrak{O})/(\ker \beta + \operatorname{im} \delta').$$

Let $A^* = A_1^* \oplus \cdots \oplus A_n^*$ be the decomposition of A^* into simple rings A_i^*, and let M_i^* be an irreducible A_i^*-module. Set

$$D_i = \operatorname{Hom}_{A_i^*}(M_i^*, M_i^*),$$

so that D_i is a division algebra over F, and A_i^* is a full matrix algebra over D_i. Of course, $K^0(A^*)$ is the free Z-module with Z-basis $[M_1^*], \cdots, [M_n^*]$. Furthermore,

$$K^1(A^*) \cong \textstyle\sum_i K^1(A_i^*) \cong \prod_i D_i^\#,$$

the latter isomorphism determined as in §2.

Let F_i denote the center of D_i, and let $R_i = \text{alg. int.}\{F_i\}$. Each field F_i is then a finite extension of F, and each A_i^* is a central simple algebra over F_i.

Since $\mathfrak{O}$ is a maximal order, we may write

$$\mathfrak{O} = \mathfrak{O}_1 \oplus \cdots \oplus \mathfrak{O}_n,$$

where each $\mathfrak{O}_i$ is a maximal R-order in A_i^*. However R_i is finitely generated over R, and thus $\mathfrak{O}_i$ is also a maximal R_i-order in A_i^*. We may therefore apply the results of the preceding section.

To begin with, we deduce that for each i, there exists a maximal R_i-order $\mathfrak{o}_i$ in D_i, and a finitely generated projective right $\mathfrak{o}_i$-module M_i, such that $F_i M_i = M_i^*$, and

$$\mathfrak{O}_i = \operatorname{Hom}_{\mathfrak{o}_i}(M_i, M_i), \qquad \mathfrak{o}_i = \operatorname{Hom}_{\mathfrak{O}_i}(M_i, M_i).$$

Clearly $FM_i = F_i M_i = M_i^*$. The isomorphism between the categories of left $\mathfrak{o}_i$-modules and left $\mathfrak{O}_i$-modules is given by $X \to M_i \otimes_{\mathfrak{o}_i} X$, where X ranges over all left $\mathfrak{o}_i$-modules.

Next we have

$$K^0(\mathfrak{O}) \cong \textstyle\sum_i K^0(\mathfrak{O}_i) \cong \sum_i K^0(\mathfrak{o}_i),$$

and

$$K_t^0(\mathfrak{O}) \cong \textstyle\sum_i K_t^0(\mathfrak{O}_i) \cong \sum_i K_t^0(\mathfrak{o}_i).$$

Furthermore, R-torsion and R_i-torsion are equivalent concepts, and we need not distinguish between them. The results of §3 are thus directly applicable, and we deduce that

$$K_t^0(\mathfrak{O}_i) \cong K_t^0(\mathfrak{o}_i) \cong I(R_i),$$

with the isomorphisms given as in §3.

The map $\delta' : K^1(A^*) \to K_t^0(\mathfrak{O})$ induces maps $\delta_i' : K^1(A_i^*) \to I(R_i)$, and we have seen that the image of δ_i' is precisely $J(R_i)$.

Our next task is the consideration of the epimorphism $\beta : K_t^0(\mathfrak{O}) \to K_t^0(A)$. For each prime ideal P of R, β maps $K^0(\mathfrak{O}/P\mathfrak{O})$ onto $K^0(A/PA)$. Calling this map β_P, we have

$$\beta = \sum_P \beta_P, \qquad \ker \beta = \sum_P \ker \beta_P .$$

Let us show at once that β_P is an isomorphism whenever $P \nmid g$, where g is the order of G. For suppose that g is a unit in R_P, the localization of R at P. As shown in [13, Lemma 5.1], there are inclusions

$$A \subset \mathfrak{O} \subset g^{-1}A.$$

Therefore $A_P = \mathfrak{O}_P$, and so

$$A/PA \cong A_P/PA_P \cong \mathfrak{O}_P/P\mathfrak{O}_P \cong \mathfrak{O}/P\mathfrak{O}.$$

This implies that β_P is an isomorphism, as claimed. We have thus shown that

$$\ker \beta = \sum_{P|g} \ker \beta_P .$$

In order to investigate the map $\beta_P : K^0(\mathfrak{O}/P\mathfrak{O}) \to K^0(A/PA)$ for an arbitrary prime ideal P of R, we shall make use of the fact that

$$K^0(\mathfrak{O}/P\mathfrak{O}) \cong \sum_{i=1}^n K^0(\mathfrak{O}_i/P\mathfrak{O}_i).$$

Now we have seen that $I(R_i) \cong K_t^0(\mathfrak{O}_i)$, and in this isomorphism an element J of $I(R_i)$ maps onto an element of $K^0(\mathfrak{O}_i/P\mathfrak{O}_i)$ if and only if J is expressible as a product of powers of prime ideals of R_i which divide P. Let us denote by $I^{(P)}(R_i)$ the subgroup of $I(R_i)$ consisting of all such ideals J; then we have

$$I^{(P)}(R_i) \cong K^0(\mathfrak{O}_i/P\mathfrak{O}_i).$$

Let us specify this isomorphism explicitly. For a fixed prime ideal P of R, let P_{ij} range over the prime ideals of R_i which contain P. Then each P_{ij} is given by $P_{ij} = R_i \cap \mathfrak{p}_{ij}$ for some uniquely determined maximal two-sided ideal $\mathfrak{p}_{ij}$ of $\mathfrak{o}_i$. Let $V(\mathfrak{p}_{ij})$ denote an irreducible module over the simple ring $\mathfrak{o}_i/\mathfrak{p}_{ij}$. Then in the isomorphism $I(R_i) \cong K_t^0(\mathfrak{o}_i)$, the ideal P_{ij} maps onto $[V(\mathfrak{p}_{ij})]$. In the isomorphism $K_t^0(\mathfrak{o}_i) \cong K_t^0(\mathfrak{O}_i)$, the latter symbol $[V(\mathfrak{p}_{ij})]$ is mapped onto $[M_i \otimes_{\mathfrak{o}_i} V(\mathfrak{p}_{ij})]$. Summarizing our results, we have

$$\prod_{i=1}^n I^{(P)}(R_i) \cong K^0(\mathfrak{O}/P\mathfrak{O}), \tag{4.1}$$

with

$$P_{ij} \to [M_i \otimes_{\mathfrak{o}_i} V(\mathfrak{p}_{ij})], \qquad i \leq i \leq n,$$

where P_{ij} ranges over the prime ideals of R_i which divide P.

We have seen in §3 that the $(\mathfrak{o}_i/\mathfrak{p}_{ij})$-module $\mathfrak{o}_i/P_{ij}\,\mathfrak{o}_i$ has n_i composition factors $V(\mathfrak{p}_{ij})$, where $n_i^2 = (D_i:F_i)$. Hence

$$n_i[M_i \otimes_{\mathfrak{o}_i} V(\mathfrak{p}_{ij})] = [M_i \otimes_{\mathfrak{o}_i} (\mathfrak{o}_i/P_{ij}\,\mathfrak{o}_i)] = [M_i/P_{ij}\,M_i].$$

Thus, the isomorphism (4.1) is given by

$$\prod_{i,j} P_{ij}^{a_{ij}} \to \sum_{i,j} a_{ij} n_i^{-1}[M_i/P_{ij} M_i].$$

Now each M_i is an $\mathfrak{O}_i$-module, hence is an $\mathfrak{O}$-module annihilated by $\{\mathfrak{O}_l : 1 \leq l \leq n,\ l \neq i\}$. Then each $M_i/P_{ij} M_i$ is an $(\mathfrak{O}/P\mathfrak{O})$-module, hence by restriction of operators is also an (A/PA)-module. We may therefore conclude that the additive group $K^0(\mathfrak{O}/P\mathfrak{O})$ has Z-basis

$$\{n_i^{-1}[M_i/P_{ij} M_i] : P_{ij} \supset P,\ 1 \leq i \leq n\},$$

and the map β_P is obtained by viewing each $M_i/P_{ij} M_i$ as (A/PA)-module.

For fixed P, suppose that $\{Y_1, \cdots, Y_s\}$ is a full set of irreducible (A/PA)-modules. Then for each prime ideal P_{ij} of R_i which divides P, we may write

$$[M_i/P_{ij} M_i] = \sum_{k=1}^{s} d_{ij}^{(k)}[Y_k] \in K^0(A/PA),$$

where the $\{d_{ij}^{(k)}\}$ are non-negative integers. These integers may be regarded as a generalization of the decomposition numbers which occur in the theory of modular group representations. In terms of these $\{d_{ij}^{(k)}\}$, we have

$$\prod_{i=1}^{n} I^{(P)}(R_i) \cong K^0(\mathfrak{O}/P\mathfrak{O}) \xrightarrow{\beta_P} K^0(A/PA),$$

with

$$\prod_{i,j} P_{ij}^{a_{ij}} \to \sum_{i,j,k} a_{ij} n_i^{-1} d_{ij}^{(k)}[Y_k] \in K^0(A/PA).$$

Since β is an epimorphism, so is each map β_P.

In the special case where F is a splitting field for G, great simplifications occur. For each i, $1 \leq i \leq n$, the division algebra D_i coincides with F, and then also $F_i = F$. Furthermore, $\mathfrak{o}_i = R_i = R$, and each $n_i = 1$. Each $\mathfrak{O}_i$-module M_i is also an A-module, and $FM_i = M_i^*$, where $M_1^*, \cdots, M_n^*$ are a full set of irreducible A^*-modules. Then each P_{ij} coincides with P, and

$$[M_i/M_i P] = \sum_{k=1}^{s} d_i^{(k)}[Y_k] \in K^0(A/PA),$$

where the $\{d_i^{(k)}\}$ are now the ordinary decomposition numbers. The map β_P is then determined by

$$(P^{a_1}, \cdots, P^{a_n}) \to \sum_{i,k} a_i d_i^{(k)}[Y_k].$$

The statement that β_P is an epimorphism is easily seen to be equivalent to Brauer's Theorem 1.

Collecting our results in the general case, we have thus established the following theorem:

Let G be a finite group, F an algebraic number field, R = alg. int. $\{F\}$, and set $A = RG$, $A^* = FG$. Write $A^* = \sum_{i=1}^{n} A_i^*$, where A_i^* is isomorphic to a full matrix algebra over a division algebra D_i with center F_i; set $n_i^2 = (D_i{:}F_i)$. Define R_i = alg. int. $\{F_i\}$, and let $I(R_i)$ denote the multiplicative group of R_i-ideals in F_i. For each i let U_i be the divisor of R_i consisting of all infinite primes of R_i at which D_i is ramified. Set

$$J(R_i) = \{xR_i : x \in F_i, x > 0 \text{ at each prime in } U_i\}.$$

Choose any maximal R-order $\mathfrak{O}$ in A^* containing A, and write $\mathfrak{O} = \sum_{i=1}^n \mathfrak{O}_i$, with each $\mathfrak{O}_i$ a maximal R_i-order in A_i^*. For each i, there exists a maximal R_i-order $\mathfrak{o}_i$ in D_i, and a projective right $\mathfrak{o}_i$-module M_i, such that $\mathfrak{O}_i = \mathrm{Hom}_{\mathfrak{o}_i}(M_i, M_i)$. The modules $FM_1, \cdots, FM_n$ form a full set of irreducible A^*-modules.

For P a fixed prime ideal of R, define $I^{(P)}(R_i)$ as the subgroup of $I(R_i)$ generated by the prime ideals P_{ij} of R_i which contain P. Each $M_i/P_{ij}M_i$ may be viewed as an (A/PA)-module, and there is an epimorphism

$$\beta_P : \prod_{i=1}^n I^{(P)}(R_i) \to K^0(A/PA)$$

given by

$$\beta_P : \prod_{i,j} P_{ij}^{a_{ij}} \to \sum_{i,j} a_{ij} n_i^{-1}[M_i/P_{ij}M_i] \in K_0(A/PA).$$

The map β_P may be regarded as a generalization of the decomposition map, and is an isomorphism when P does not divide the order of G.

The additive structure of the Grothendieck group $K^0(A)$ is given by

$$K^0(A) \cong K^0(A^*) \oplus \frac{\prod_{i=1}^n I(R_i)}{\left\{\prod_{i=1}^n J(R_i)\right\}\left\{\prod_{P \mid [G:1]} \ker \beta_P\right\}}.$$

The Grothendieck group $K^0(A^*)$ is a free Z-module on the n generators $[M_1^*], \cdots, [M_n^*]$. The second summand on the right hand side is a finite abelian group, written multiplicatively, the determination of which depends on the ideal theory of each of the rings R_i, as well as the knowledge of the maps β_P.

REFERENCES

1. M. Auslander and O. Goldman, *Maximal orders*, Trans. Amer. Math. Soc., vol. 97 (1960), pp. 1–24.
2. H. Bass, *The Morita theorems*, mimeographed notes, Univ. of Oregon, 1962.
3. R. Brauer, *On the Cartan invariants of groups of finite order*, Ann. of Math. (2), vol. 42 (1941), pp. 53–61.
4. ———, *A characterization of the characters of groups of finite order*, Ann. of Math. (2), vol. 57 (1953), pp. 357–377.
5. H. Cartan and S. Eilenberg, *Homological algebra*, Princeton, Princeton Univ. Press, 1956.
6. C. W. Curtis and I. Reiner, *Representation theory of finite groups and associative algebras*, N.Y., Interscience, 1962.
7. M. Deuring, *Algebren*, Berlin, Springer, 1935.
8. J. Dieudonné, *Les determinants sur un corps non commutatif*, Bull. Soc. Math. France, vol. 71 (1943), pp. 27–45.
9. A. Heller and I. Reiner, *Grothendieck groups of orders in semisimple algebras*, Trans. Amer. Math. Soc., vol. 112 (1964), pp. 344–355.

10. K. MORITA, *Duality for modules*, Sci. Reports Tokyo Kyoiku Daigaku, Sect. A, vol. 6 (no. 150), pp. 53–61.
11. R. G. SWAN, *Induced representations and projective modules*, Ann. of Math. (2), vol. 71 (1960), pp. 552–578.
12. ———, *Projective modules over group rings and maximal orders*, Ann. of Math. (2), vol. 76 (1962), pp. 55–61.
13. ———, *The Grothendieck ring of a finite group*, Topology, vol. 2 (1963), pp. 85–110.

UNIVERSITY OF ILLINOIS
URBANA, ILLINOIS

THE INTEGRAL REPRESENTATION RING OF A FINITE GROUP

Irving Reiner

1. INTRODUCTION

We shall be concerned with matrix representations of a finite group G by nonsingular matrices with entries in a ring R, where R is a discrete valuation ring of characteristic zero. Let $\Gamma = RG$, the group ring of G with coefficients in R. We may then, equivalently, consider left Γ-modules having finite R-bases.

Let us assume that the Krull-Schmidt theorem is valid for such Γ-modules, that is, that every Γ-module is uniquely expressible as a direct sum of indecomposable modules. (Toward the end of this section we shall give some sufficient conditions for the validity of the Krull-Schmidt theorem.) Then we may define the *integral representation ring* $A(\Gamma)$, as follows. Denote by $\{M\}$ the isomorphism class of the Γ-module M. Form the additive abelian group generated by the symbols $\{M\}$ ranging over the distinct isomorphism classes of Γ-modules, with defining relations

$$\{M\} = \{M'\} + \{M''\} \qquad \text{whenever } M \cong M' \oplus M''.$$

On this additive group we impose a ring structure, by defining

$$\{M\}\{M'\} = \{M \otimes_R M'\},$$

where the action of G on $M \otimes_R M'$ is given (as is customary) by

$$g(m \otimes m') = gm \otimes gm' \qquad (g \in G,\ m \in M,\ m' \in M').$$

The ring thus obtained we denote by $A(\Gamma)$, and call it the integral representation ring of Γ. Clearly, $A(\Gamma)$ is a commutative associative ring. Its unity element is $\{R\}$; here, R denotes the *trivial* Γ-module, that is,

$$g\alpha = \alpha \qquad (g \in G,\ \alpha \in R).$$

The above representation ring is an analogue of the modular representation algebra recently introduced by J. A. Green [4]. Let k be a field of characteristic p (where $p \geq 0$), and let Ω denote the complex field. Form the Ω-algebra $A_\Omega(kG)$ consisting of the Ω-linear combinations of symbols corresponding to the isomorphism classes of kG-modules, with relations and multiplication defined in the manner above. If p does not divide the order of G, then the group algebra kG is semisimple, and Green showed easily that the representation algebra $A_\Omega(kG)$ is also semisimple. Indeed, if $p = 0$, then $A_\Omega(kG)$ is isomorphic to the (commutative) algebra of generalized characters (with coefficients from Ω). But this latter algebra has no nonzero nilpotent elements; for if η is a generalized character such that $\eta^m = 0$

Received September 14, 1964.

This research was supported by the Office of Naval Research and the National Science Foundation.

for some m, then obviously $\eta = 0$. If $p \neq 0$ and p does not divide the order of G, the same reasoning applies if we use Brauer characters in place of ordinary characters.

Green also established a much more difficult result: *For any cyclic group* G, *the modular representation algebra* $A_\Omega(kG)$ *is semisimple.* Since $A_\Omega(kG)$ is a commutative algebra, this is equivalent to the assertion that $A_\Omega(kG)$ does not contain any nonzero nilpotent elements.

Here we investigate the analogous question: Does $A(\Gamma)$ contain nonzero nilpotent elements? Our main result is as follows:

THEOREM. *Let* G *be a cyclic group of order* n. *Let* R *be a discrete valuation ring of characteristic zero, with maximal ideal* P. *Suppose that the Krull-Schmidt theorem holds for* RG-*modules. Assume that* $n \in P^2$, *and if* $2 \in P$, *assume further that* $n \in 2P$. *Then the integral representation ring* A(RG) *contains at least one nonzero nilpotent element.*

On the other hand, it is possible to choose G *and* R *such that* A(RG) *contains no nonzero nilpotent element.*

Throughout this paper, we let Z_p denote the p-adic valuation ring in the rational field Q, and Z_p^* the ring of p-adic integers in the p-adic completion of Q.

COROLLARY. *The representation ring* $A(Z_p G)$ *contains nonzero nilpotent elements if* G *is a cyclic group of order* p^e *with* $e > 1$.

A trivial observation should be made at this point. Let G be an arbitrary finite group whose order is a unit in the discrete valuation ring R. Let K be the quotient field of R. There is then a one-to-one correspondence between the isomorphism classes of RG-modules (having finite R-bases) and the isomorphism classes of KG-modules (see [3, Theorem 76.17]). This correspondence induces an isomorphism $A(RG) \cong A(KG)$. But A(KG) is a subring of the commutative semisimple algebra $A_\Omega(KG)$, hence contains no nonzero nilpotent elements. The same is therefore true for A(RG).

To conclude this section, we list several conditions, any one of which implies the Krull-Schmidt theorem for RG-modules.

i) The order of G is a unit in R.

ii) R is a complete discrete valuation ring.

iii) The quotient field of R is an algebraic number field which is a splitting field for G.

iv) G is an arbitrary p-group, where $R = Z_p$.

For the proofs that each of these imply the Krull-Schmidt theorem, we cite the following references: for i), see [3, Theorem 76.17]; for ii), see [2], [10], or [8]; for iii), see [5]. In order to show that iv) is a sufficient condition for the validity of the Krull-Schmidt theorem, one first uses the Witt-Berman theorem [3, Theorem 42.8] to show that an irreducible QG-module remains irreducible upon extension of the ground field from Q to its p-adic completion. The desired result now follows as in [3, Lemma 76.28 and Theorem 76.29].

2. TENSOR PRODUCTS OF MODULES

Let G be an arbitrary finite group of order n, and let R be a discrete valuation ring of characteristic zero. Let $P = \pi R$ be the maximal ideal of R, and set $\overline{R} = R/P$ ($\overline{R}$ is then a field of characteristic p). Set $\Gamma = RG$, $\overline{\Gamma} = \overline{R}G$, and consider finitely generated left Γ-modules. Since we shall need to work with Γ-modules that do not necessarily have R-bases, Γ-modules having R-bases will be called *R-free* Γ-*modules*.

Assume hereafter that the Krull-Schmidt theorem is valid for R-free Γ-modules. Let X be a fixed R-free Γ-module satisfying

$$P\Gamma \subset X \subset \Gamma. \tag{1}$$

If Y is an arbitrary R-free Γ-module, we shall show how the problem of calculating the tensor product module $Y \otimes_R X$ can be reduced to a calculation involving only $\overline{\Gamma}$-modules. Indeed, we shall see that (for fixed X) the isomorphism class of $Y \otimes_R X$ depends upon $\overline{Y} = Y/PY$ rather than upon Y.

Set $A = \Gamma/X$, so that there is an exact sequence of Γ-modules:

$$0 \to X \to \Gamma \to A \to 0. \tag{2}$$

From (1) we conclude that $PA = 0$, and thus that A may be viewed as $\overline{\Gamma}$-module.

Now let Y be any R-free Γ-module, and let $m = (Y: R)$ be the number of elements in an R-basis for Y. Set $\overline{Y} = Y/PY$, so that

$$0 \to PY \to Y \to \overline{Y} \to 0$$

is exact. Then

$$PY \otimes_R A \to Y \otimes_R A \to \overline{Y} \otimes_R A \to 0$$

is also exact. However, the image of $PY \otimes_R A$ in $Y \otimes_R A$ is zero, since $PA = 0$. This shows that $Y \otimes_R A \cong \overline{Y} \otimes_R A$. But both $\overline{Y}$ and A are $\overline{R}$-modules, which readily implies that $\overline{Y} \otimes_R A \cong \overline{Y} \otimes_{\overline{R}} A$. Thus

$$Y \otimes_R A \cong \overline{Y} \otimes_{\overline{R}} A \quad \text{as } \Gamma\text{-modules.}$$

Since Y is R-free, we obtain from (2) the exact sequence of Γ-modules

$$0 \to Y \otimes_R X \to Y \otimes_R \Gamma \to Y \otimes_R A \to 0. \tag{3}$$

Now $Y \otimes_R \Gamma \cong \Gamma^{(m)}$, where $\Gamma^{(m)}$ denotes the direct sum of m copies of Γ (see Swan [10, Lemma 5.1]). Thus (3) may be written as

$$0 \to Y \otimes_R X \to \Gamma^{(m)} \to \overline{Y} \otimes_{\overline{R}} A \to 0. \tag{4}$$

Let

$$\overline{Y} \otimes_{\overline{R}} A \cong B_1 \oplus \cdots \oplus B_t,$$

where the B_i are indecomposable $\overline{\Gamma}$-modules. Each B_i is expressible as a quotient of a free $\overline{\Gamma}$-module, hence also as a quotient of a free Γ-module. Thus for each i $(1 \leq i \leq t)$, there is an exact sequence

$$0 \to M_i \to \Gamma^{(n_i)} \to B_i \to 0,$$

for some n_i and some R-free Γ-module M_i. Therefore the sequence

$$(5) \qquad 0 \to M_1 \oplus \cdots \oplus M_t \to \Gamma^{(n_1+\cdots+n_t)} \to B_1 \oplus \cdots \oplus B_t \to 0$$

is exact.

SCHANUEL'S LEMMA (see Swan [11]). *If*

$$0 \to M \to L \to B \to 0 \quad \textit{and} \quad 0 \to M' \to L' \to B' \to 0$$

are two exact sequences of Γ*-modules, where both* L *and* L' *are projective, and* $B \cong B'$, *then*

$$M \oplus L' \cong M' \oplus L.$$

Applying this lemma to the sequences (4) and (5), we obtain

$$(6) \qquad Y \otimes_R X \oplus \Gamma^{(n_1+\cdots+n_t)} \cong M_1 \oplus \cdots \oplus M_t \oplus \Gamma^{(m)}.$$

Since the Krull-Schmidt theorem is assumed to hold for Γ-modules, we may use (6) to calculate $Y \otimes_R X$. Note that (for fixed X) the result depends only on m and $B_1, \cdots, B_t$, that is to say, only on the $\overline{\Gamma}$-module $\overline{Y}$.

Later (in Section 5) we shall systematically use this approach to calculate certain tensor product modules. For the moment, we shall content ourselves with the following simple consequence of formula (6).

If X and Y are a pair of R-free Γ-modules such that

$$(7) \qquad P\Gamma \subset X \subset \Gamma, \quad P\Gamma \subset Y \subset \Gamma, \quad \overline{X} \cong \overline{Y},$$

then the preceding discussion implies that

$$X \otimes_R X \cong X \otimes_R Y \cong Y \otimes_R X \cong Y \otimes_R Y.$$

Hence in $A(\Gamma)$ we have the relation

$$(\{X\} - \{Y\})^2 = \{X \otimes X\} - \{X \otimes Y\} - \{Y \otimes X\} + \{Y \otimes Y\} = 0.$$

Thus, if X and Y are a pair of nonisomorphic Γ-modules satisfying (7), then $\{X\} - \{Y\}$ is a nonzero nilpotent element of the integral representation ring $A(\Gamma)$.

3. CYCLIC GROUPS

We keep the notation of Section 2, and we assume throughout this section that G is a cyclic group generated by an element g of order n. Embed R in the group ring Γ by the usual map $\alpha \in R \to \alpha \cdot 1 \in \Gamma$, where 1 is the identity element of G.

Let us choose X to be the ideal generated by π and $g - 1$ in the commutative ring Γ. Then $P\Gamma \subset X \subset \Gamma$, and X has an R-basis $\{x_1, \cdots, x_n\}$, where

$$x_1 = g - 1,\ x_2 = g^2 - 1,\ \cdots,\ x_{n-1} = g^{n-1} - 1,\ x_n = \pi .$$

We find at once that

$$gx_i = x_{i+1} - x_1 \ (1 \leq i \leq n - 2),\quad gx_{n-1} = -x_1,\quad gx_n = x_n + \pi x_1. \tag{8}$$

Consequently $\overline{X} = X/PX = \overline{U} \oplus \overline{R}\overline{x}_n$, where $\overline{U} = \Sigma_{i=1}^{n-1} \overline{R}\overline{x}_i$, with the action of g on $\overline{U}$ given by the first $n - 1$ equations in (8), with the x_i replaced by $\bar{x}_i$. (Indeed, $\overline{U}$ is isomorphic to the augmentation ideal of $\overline{\Gamma}$.) In the matrix representation of G afforded by the module $\overline{U}$, the matrix corresponding to g is

$$C = \begin{bmatrix} -1 & -1 & -1 & \cdots & -1 & -1 \\ 1 & 0 & 0 & \cdots & 0 & 0 \\ 0 & 1 & 0 & \cdots & \cdot & \cdot \\ \cdot & \cdot & \cdot & \cdots & \cdot & \cdot \\ 0 & 0 & 0 & \cdots & 1 & 0 \end{bmatrix}.$$

Let us put $h(\lambda) = \Sigma_{k=0}^{n-1} \lambda^k$, where λ is an indeterminate over $\overline{R}$. Then C is similar to the companion matrix of $h(\lambda)$, which shows that

$$\overline{U} \cong \overline{R}[\lambda]/(h(\lambda)),$$

where g acts on the right-hand module as multiplication by λ.

We shall now show that if n satisfies the hypotheses of the theorem given in Section 1, then the trivial $\overline{\Gamma}$-module $\overline{R}$ cannot be a $\overline{\Gamma}$-direct summand of $\overline{U}$. Thus, we assume that $n \in P^2$, and if $2 \in P$, we assume in addition that $n \in 2P$. This guarantees that $n(n - 1)/2 \in P$ in all cases.

Suppose that $\overline{R}$ is a $\overline{\Gamma}$-direct summand of $\overline{U}$. Then there is a direct sum decomposition:

$$\frac{\overline{R}[\lambda]}{(h(\lambda))} \cong \frac{\overline{R}[\lambda]}{(\lambda - 1)} \oplus \frac{\overline{R}[\lambda]}{(k(\lambda))}. \tag{9}$$

It follows at once that $(\lambda - 1)k(\lambda) = h(\lambda)$, so that

$$k(\lambda) = \lambda^{n-2} + 2\lambda^{n-3} + \cdots + (n - 2)\lambda + (n - 1) \in \overline{R}[\lambda].$$

But then $k(1) = n(n - 1)/2 = 0$ in $\overline{R}$, and thus $\lambda - 1$ is a factor of $k(\lambda)$. From (9) we then conclude that $k(\lambda)$ annihilates $\overline{R}[\lambda]/(h(\lambda))$, which is obviously impossible. Thus, the assumptions about n imply that $\overline{R}$ is not a $\overline{\Gamma}$-direct summand of $\overline{U}$.

Next, define $s = 1 + g + \cdots + g^{n-1}$, and choose Y as the ideal generated by π and s in the ring Γ. Then $P\Gamma \subset Y \subset \Gamma$, and Y has an R-basis $\{y_1, \cdots, y_n\}$, where

$$y_1 = \pi, \quad y_2 = \pi g, \quad \cdots, \quad y_{n-1} = \pi g^{n-2}, \quad y_n = s.$$

Note that

$$y_1 + y_2 + \cdots + y_{n-1} + \pi g^{n-1} = \pi y_n.$$

We see at once that

$$gy_i = y_{i+1} \ (1 \leq i \leq n-2), \quad gy_{n-1} = \pi y_n - (y_1 + \cdots + y_{n-1}), \quad gy_n = y_n.$$

Therefore $\overline{Y} = Y/PY = \overline{W} \oplus \overline{R}\bar{y}_n$, where $g\bar{y}_n = \bar{y}_n$, and where $\overline{W} = \Sigma_{i=1}^{n-1} \overline{R}\bar{y}_i$. The action of G on $\overline{W}$ is given by

$$g\bar{y}_i = \bar{y}_{i+1} \quad (1 \leq i \leq n-2), \qquad g\bar{y}_{n-1} = -(\bar{y}_1 + \cdots + \bar{y}_{n-1}).$$

The matrix corresponding to g (acting on $\overline{W}$) is

$$D = \begin{bmatrix} 0 & 0 & 0 & \cdots & 0 & -1 \\ 1 & 0 & 0 & \cdots & 0 & -1 \\ 0 & 1 & 0 & \cdots & 0 & -1 \\ \cdot & \cdot & \cdot & \cdots & \cdot & \cdot \\ 0 & 0 & 0 & \cdots & 1 & -1 \end{bmatrix},$$

which is the transpose of the companion matrix of $h(\lambda)$. Hence C and D are similar over $\overline{R}$, which shows that $\overline{W} \cong \overline{U}$ as $\overline{\Gamma}$-modules.

We have thus shown that $\overline{X} \cong \overline{Y} \cong \overline{U} \oplus \overline{R}$, so that (7) holds. Our next task is to verify that X is not Γ-isomorphic to Y. We set $U = \Sigma_{i=1}^{n-1} Rx_i$ (this is a Γ-submodule of X; indeed, U is the augmentation ideal of Γ). Then U is an R-direct summand of X, and X/U is isomorphic to the trivial Γ-module R.

We shall determine a Γ-submodule V of Y such that

$$K \otimes_R V \cong K \otimes_R U, \qquad Y/V \cong R,$$

with V an R-direct summand of Y. The composition factors of the KG-module KG are the trivial module K, occurring with multiplicity 1, together with the composition factors of $K \otimes_R U$. Hence $\mathrm{Hom}_{KG}(K \otimes_R U, K) = 0$. Therefore any Γ-homomorphism of X into Y must induce a Γ-homomorphism of U into V. In particular, if $X \cong Y$ as Γ-modules, then $U \cong V$ as Γ-modules, and therefore $\overline{U} \cong \overline{V}$ as $\overline{\Gamma}$-modules. We shall show that this is impossible when n satisfies the hypotheses of the theorem in Section 1, since we shall verify that the trivial $\overline{\Gamma}$-module $\overline{R}$ is a $\overline{\Gamma}$-direct summand of $\overline{V}$. We may thus conclude that X is not isomorphic to Y, and hence that $\{X\} - \{Y\}$ is a nonzero nilpotent element of $A(\Gamma)$.

Since $n \in P^2$, we may write $n = \pi \cdot \pi'$ for some $\pi' \in P$. Now let V be the R-free P-module with basis $\{v_1, \cdots, v_{n-1}\}$, where

$$v_i = y_{i+1} - y_1 = \pi(g^i - 1) \qquad (1 \leq i \leq n-2),$$

$$v_{n-1} = \pi' y_1 - y_n = n - s = -\{(g-1) + (g^2-1) + \cdots + (g^{n-1}-1)\}.$$

Then V is an R-pure Γ-submodule of Y, hence an R-direct summand of Y. We have the relation

$$K \otimes_R V = \sum_{i=1}^{n-1}{}^{\oplus} K(g^i - 1),$$

so that $K \otimes_R V$ is the augmentation ideal of KG, and thus

$$K \otimes_R V \cong K \otimes_R U.$$

But then

$$K \otimes (Y/V) \cong (K \otimes Y)/(K \otimes V) \cong KG/(K \otimes_R U) \cong K,$$

which shows that $Y/V \cong R$. We see at once that

$$gv_i = v_{i+1} - v_1 \qquad (1 \leq i \leq n-3),$$

$$gv_{n-2} = -(2v_1 + v_2 + v_3 + \cdots + v_{n-2} + \pi v_{n-1}),$$

$$gv_{n-1} = \pi' v_1 + v_{n-1}.$$

Therefore $\overline{V} = V/PV = (\Sigma_{i=1}^{n-2} \overline{R}\overline{v}_i) \oplus \overline{R}\overline{v}_{n-1}$, $g\bar{v}_{n-1} = \bar{v}_{n-1}$, and so the trivial $\overline{\Gamma}$-module $\overline{R}$ is a $\overline{\Gamma}$-direct summand of $\overline{V}$, as claimed. This completes the proof of the first part of the theorem.

4. EXAMPLES

We shall now give several examples in which $A(RG)$ contains no nonzero nilpotent elements. We have already remarked in Section 1 that this is the case whenever the order of G is a unit in R.

We obtain a less trivial example by choosing G cyclic of order p, and taking $R = Z_p$. Here the Krull-Schmidt theorem is valid for RG-modules. Furthermore (see [7]), the mapping that assigns to each R-free RG-module M the $\overline{R}G$-module M/pM induces a monomorphism of $A(RG)$ into $A(\overline{R}G)$. Since Green has shown that $A(\overline{R}G)$ contains no nonzero nilpotent elements, the same is true for $A(RG)$.

For another example, take G cyclic of order pn', where $(n', p) = 1$, and let $R = Z_p^*$. Since R is a complete discrete valuation ring, again the Krull-Schmidt theorem holds in this case. Let G be a direct product $G_1 \times G_2$ of a cyclic group G_1 of order p and a cyclic group G_2 of order n'. From [1] or [6] it follows that each indecomposable R-free RG-module M is uniquely expressible in the form $M_1 \otimes_R M_2$, where M_1 is an indecomposable RG_1-module and M_2 is an irreducible RG_2-module, and where

$$(g_1, g_2)\cdot(m_1 \otimes m_2) = g_1 m_1 \otimes g_2 m_2 \qquad (g_i \in G_i,\ m_i \in M_i,\ i = 1, 2).$$

This shows that A(RG) is the direct product of the rings $A(RG_1)$ and $A(RG_2)$. Since neither of these rings contains a nonzero nilpotent element, A(RG) has the same property.

5. THE CYCLIC GROUP OF ORDER FOUR

Let G be a cyclic group of order 4, with generator g; let $R = Z_2$, $\overline{R} = R/2R$, $\Gamma = RG$, $\overline{\Gamma} = \overline{R}G$. The Krull-Schmidt theorem holds for Γ-modules, and we shall now compute the multiplication table of $A(\Gamma)$ by using equation (6).

Up to isomorphism, there are precisely 9 indecomposable R-free Γ-modules, as has been shown by Troy [12] and Roĭter [9]. We list these modules as

$$R,\ S,\ T,\ (R, S),\ (R, T),\ (S, T),\ X,\ Y,\ \Gamma .$$

Here, R is the trivial Γ-module, while S and T are given by

$$S = R\beta, \quad g\beta = -\beta; \quad T = Rt \oplus Ru, \quad gt = u, \quad gu = -t .$$

The module (R, S) is the unique nonsplit extension of the submodule R by the factor module S. The modules (R, T) and (S, T) are defined analogously. We have already defined the modules X and Y in Section 2.

For convenience, we write $\otimes$ instead of $\otimes_R$ or $\otimes_{\overline{R}}$, since it will be clear from the context which is intended. If M is any R-free Γ-module, and $m = (M : R)$, there are the obvious relations

$$(10) \qquad R \otimes M \cong M, \qquad \Gamma \otimes M \cong \Gamma^{(m)} \quad \text{(direct sum of } m \text{ copies of } \Gamma).$$

Straightforward simple calculations yield

$$(11) \quad \begin{cases} S \otimes S \cong R, \quad S \otimes T \cong T, \quad S \otimes (R, S) \cong (R, S), \quad S \otimes (R, T) = (S, T), \\ S \otimes (S, T) = (R, T), \quad T \otimes T \cong (R, S)^{(2)}, \quad T \otimes (R, S) = T^{(2)}, \\ (R, S) \otimes (R, S) \cong (R, S)^{(2)}. \end{cases}$$

Denote the elements $\{R\}, \{S\}, \cdots, \{\Gamma\}$ of $A(\Gamma)$ by $c_1, c_2, \cdots, c_9$, for convenience. We shall verify the following multiplication table for $A(\Gamma)$:

(12)

c_1	c_2	c_3	c_4	c_5	c_6	c_7	c_8	c_9
c_2	c_1							
c_3	c_3	$2c_4$						
c_4	c_4	$2c_3$	$2c_4$					
c_5	c_6	$c_4 + c_9$	$c_3 + c_9$	$c_1 + 2c_9$				
c_6	c_5	$c_4 + c_9$	$c_3 + c_9$	$c_2 + 2c_9$	$c_1 + 2c_9$			
c_7	c_7	$c_3 + c_4 + c_9$	$c_3 + c_4 + c_9$	$c_8 + 2c_9$	$c_8 + 2c_9$	$c_7 + c_8 + 2c_9$		
c_8	c_8	$c_3 + c_4 + c_9$	$c_3 + c_4 + c_9$	$c_7 + 2c_9$	$c_7 + 2c_9$	$c_7 + c_8 + 2c_9$	$c_7 + c_8 + 2c_9$	
c_9	c_9	$2c_9$	$2c_9$	$3c_9$	$3c_9$	$4c_9$	$4c_9$	$4c_9$

We already know that

$$c_i c_j = c_j c_i, \quad c_i c_1 = c_i \qquad (1 \leq i, j \leq 9),$$

and the products $c_i c_9$ are obtained immediately from the second relation in (10). Relations (11) give the products $c_2 c_j$ $(2 \leq j \leq 6)$, c_3^2, $c_3 c_4$, c_4^2.

We shall use equation (6) to compute the products that involve c_7 and c_8. We need some elementary facts about $\overline{\Gamma}$-modules. Set

$$A_j = \overline{\Gamma}/(g-1)^j\,\overline{\Gamma} \qquad (1 \leq j \leq 4).$$

Then A_j is an indecomposable $\overline{\Gamma}$-module of $\overline{R}$-dimension j, and $\{A_j\colon 1 \leq j \leq 4\}$ is a full set of indecomposable $\overline{\Gamma}$-modules. We see that $A_4 = \overline{\Gamma}$, while A_1 is the trivial $\overline{\Gamma}$-module.

Either by an easy direct calculation, or else by referring to Green [4], we obtain the isomorphisms

$$(13) \quad A_1 \otimes A_j \cong A_j, \quad A_4 \otimes A_j \cong A_4^{(j)}, \quad A_2 \otimes A_3 \cong A_2 \oplus A_4, \quad A_3 \otimes A_3 \cong A_1 \oplus A_4^{(2)}.$$

Furthermore, we see that

$$(14) \qquad \overline{R} \cong \overline{S} \cong A_1, \quad \overline{(R, S)} \cong \overline{T} \cong A_2, \quad \overline{(R, T)} \cong \overline{(S, T)} \cong A_3.$$

The discussion in Section 2 shows that

$$(15) \qquad \overline{X} \cong \overline{Y} \cong A_1 \oplus A_3.$$

Now X is the ideal of Γ generated by 2 and g - 1, and Y is the ideal generated by 2 and $1 + g + g^2 + g^3$. Let W be the ideal generated by 2 and $(g-1)^2$. Then X and Y are indecomposable, whereas

$$W \cong (R, S) \oplus T.$$

We may write four exact sequences

$$(16) \qquad \begin{cases} 0 \to X \to \Gamma \to A_1 \to 0, & 0 \to W \to \Gamma \to A_2 \to 0, \\ 0 \to Y \to \Gamma \to A_3 \to 0, & 0 \to \Gamma \to \Gamma \to A_4 \to 0, \end{cases}$$

where the embedding $\Gamma \to \Gamma$ in the last sequence is given by $\gamma \to 2\gamma$ $(\gamma \in \Gamma)$.

Now let M be any R-free Γ-module of R-rank m, and let

$$\overline{M} \otimes A_1 \cong \overline{M} \cong \sum_{i=1}^{4}{}^{\oplus} A_i^{(r_i)}.$$

Equation (6) then yields the relation

$$(17) \quad M \otimes X \oplus \Gamma^{(r_1+r_2+r_3+r_4)} \cong X^{(r_1)} \oplus W^{(r_2)} \oplus Y^{(r_3)} \oplus \Gamma^{(r_4)} \oplus \Gamma^{(m)}.$$

Let us compute $S \otimes X$; since $\overline{S} \cong A_1$, we get $r_1 = 1$, $r_2 = r_3 = r_4 = 0$, $m = 1$, and so

$$S \otimes X \oplus \Gamma \cong X \oplus \Gamma .$$

This shows that $c_2 c_7 = c_7$. Likewise, $\overline{T} \cong A_2$ gives

$$T \otimes X \oplus \Gamma \cong W \oplus \Gamma^{(2)},$$

so that $c_3 c_7 = c_3 + c_4 + c_9$. As a last illustration, the isomorphism $\overline{X} \cong A_1 \oplus A_3$ yields

$$X \otimes X \oplus \Gamma^{(2)} \cong X \oplus Y \oplus \Gamma^{(4)},$$

whence $c_7^2 = c_7 + c_8 + 2c_9$. In this manner, we evaluate all products involving c_7.

In order to compute $M \otimes Y$, we set

$$\overline{M} \otimes A_3 \cong \sum_{i=1}^{4}{}^{\oplus} A_i^{(s_i)} .$$

Equation (6) now becomes

$$M \otimes Y \oplus \Gamma^{(s_1+s_2+s_3+s_4)} \cong X^{(s_1)} \oplus' W^{(s_2)} \oplus Y^{(s_3)} \oplus \Gamma^{(s_4)} \oplus \Gamma^{(m)}.$$

Thus, $\overline{S} \cong A_1$ implies that $\overline{S} \otimes A_3 \cong A_3$, and so

$$S \otimes Y \oplus \Gamma \cong Y \oplus \Gamma,$$

that is, $c_2 c_8 = c_8$. Similarly,

$$\overline{T} \otimes A_3 \cong A_2 \otimes A_3 \cong A_2 \oplus A_4,$$

and this implies that

$$T \otimes Y \oplus \Gamma^{(2)} \cong W \oplus \Gamma \oplus \Gamma^{(2)}.$$

Therefore $c_3 c_8 = c_3 + c_4 + c_9$. Continuing in this manner, we easily evaluate all products involving c_8.

We are left with the problem of computing products such as $(S, T) \otimes (R, T)$. Direct calculation of these is a rather tedious process, and a better approach is to use the ideas of Section 2. There exist exact sequences

$$0 \to (R, T) \to \Gamma \to S \to 0, \tag{18}$$

$$0 \to R \to \Gamma \to (S, T) \to 0, \tag{19}$$

$$0 \to (R, S) \to \Gamma \to T \to 0, \tag{20}$$

(we omit the proof of this.) Tensoring the first of these sequences with (R, T), and using the isomorphism $(R, T) \otimes S \cong (S, T)$, we obtain an exact sequence

$$0 \to (R, T) \otimes (R, T) \to \Gamma^{(3)} \to (S, T) \to 0 .$$

The application of Schanuel's lemma to this last sequence and to the sequence in (19) shows that

$$(R, T) \otimes (R, T) \cong R \oplus \Gamma^{(2)}.$$

Thus $c_5^2 = c_1 + 2c_9$. Since $c_6 = c_2 c_5$, we deduce that

$$c_6^2 = c_5^2, \qquad c_5 c_6 = c_2 \cdot c_5^2 = c_2 (c_1 + 2c_9) = c_2 + 2c_9.$$

Similarly, tensoring (18) with T, we obtain an exact sequence

$$0 \to (R, T) \otimes T \to \Gamma^{(2)} \to T \to 0.$$

Comparing this with sequence (20), and using Schanuel's lemma once more, we have the isomorphism

$$(R, T) \otimes T \cong (R, S) \oplus \Gamma.$$

Therefore $c_3 c_5 = c_4 + c_9$. Multiply this last equation by c_2, thereby getting $c_3 c_6 = c_4 + c_9$.

Finally, we may compute $c_4 c_5$ as follows:

$$2c_4 c_5 = c_3^2 c_5 = c_3 (c_4 + c_9) = 2c_3 + 2c_9,$$

whence $c_4 c_5 = c_3 + c_9$. We multiply this last equation by c_2, getting

$$c_4 c_6 = c_3 + c_9.$$

This completes our verification of table (12).

REFERENCES

1. S. D. Berman, *A contribution to the theory of integral representations of finite groups*, Dokl. Akad. Nauk SSSR 157 (1964), 506-508.

2. Z. I. Borevič and D. K. Faddeev, *Theory of homology in groups, I*, Vestnik. Leningrad Univ. 11 (1956), No. 7, 3-39.

3. C. W. Curtis and I. Reiner, *Representation theory of finite groups and associative algebras*, Interscience, New York, 1962.

4. J. A. Green, *The modular representation algebra of a finite group*, Illinois J. Math. 6 (1962), 607-619.

5. A. Heller, *On group representations over a valuation ring*, Proc. Nat. Acad. Sci. U.S.A. 47 (1961), 1194-1197.

6. A. Jones, *Integral representations of abelian groups*, Illinois J. Math. (to appear).

7. I. Reiner, *Integral representations of cyclic groups of prime order*, Proc. Amer. Math. Soc. 8 (1957), 142-146.

8. ———, *The Krull-Schmidt theorem for integral group representations*, Bull. Amer. Math. Soc. 67 (1961), 365-367.

9. A. V. Roĭter, *On the representations of the cyclic group of fourth order by integral matrices*, Vestnik Leningrad Univ. 15 (1960), No. 19, 65-74.

10. R. G. Swan, *Induced representations and projective modules*, Ann. of Math. (2) 71 (1960), 552-578.

11. ———, *Periodic resolutions for finite groups*, Ann. of Math. (2) 72 (1960), 267-291.

12. A. Troy, *Integral representations of cyclic groups of order* p^2, Ph.D. thesis, Univ. of Illinois, 1961.

University of Illinois
Urbana, Illinois

INTEGRAL REPRESENTATION ALGEBRAS(1)

BY

IRVING REINER

1. **Introduction.** Let RG denote the group ring of a finite group G over a commutative ring R. By an RG-module we shall mean a left finitely generated module which is R-torsion free. The *representation ring* $a(RG)$ is an abelian additive group defined by generators and relations: the generators are the symbols $[M]$, where M ranges over a full set of representatives of the isomorphism classes of RG-modules, with relations

$$[M] = [M'] + [M'']$$

whenever $M \cong M' \oplus M''$. Multiplication in $a(RG)$ is defined by forming tensor products of modules:

$$[M][N] = [M \otimes_R N],$$

where as usual G acts on the tensor product by the formula

$$g(m \otimes n) = gm \otimes gn, \qquad g \in G, m \in M, n \in N.$$

Let C be the complex field, and define the *integral representation algebra* $A(RG)$ by the formula

$$A(RG) = C \otimes_Z a(RG).$$

Such representation algebras have recently been studied by Conlon [1], Green [3], [4], and O'Reilly [10], for the special case in which R is a field. They have shown that under suitable hypotheses, the algebra $A(RG)$ is semisimple.

The present author investigated $a(RG)$ when R is a ring of integers (see [11], [12]). Of particular interest are the following choices for R:

Z (the ring of rational integers),

Z_p (the p-adic valuation ring in the rational field Q),

Z_p^* (the ring of p-adic integers in the p-adic completion of Q),

$Z' = \bigcap_{p \mid [G:1]} Z_p$, a semilocal ring of integers in Q.

To give the reader the proper perspective, we quote two earlier results.

Received by the editors September 27, 1965.

(1) This research was supported in part by the National Science Foundation.

THEOREM 1. *If K is a field of characteristic p, and if G has a cyclic p-Sylow subgroup, then $A(KG)$ is a finite dimensional semisimple C-algebra (O'Reilly [10]).*

THEOREM 2. *Suppose that for some prime p, the group G contains an element of order p^2. Then both $a(Z_pG)$ and $a(Z_p^*G)$ contain at least one nonzero nilpotent element. The same is therefore true for $A(Z_pG)$ and $A(Z_p^*G)$ (Reiner [12]).*

The aim of the present paper is to present a partial converse to the latter theorem. We shall prove here:

THEOREM 3. *Let p be a fixed prime, and suppose that the p-Sylow subgroups of G are cyclic of order p. Then $A(Z_pG)$ and $A(Z_p^*G)$ contain no nonzero nilpotent elements.*

THEOREM 4. *If $[G:1]$ is squarefree, then $A(Z'G)$ contains no nonzero nilpotent element.*

In the course of the proof we shall establish the following fact, which is of independent interest, and is an immediate consequence of Theorem 5 below.

PROPOSITION. *Let p be an odd prime, and let G have a normal p-Sylow subgroup which is cyclic of order p. Let M and N be Z_pG-modules. Then $M \cong N$ if and only if $M/pM \cong N/pN$ as $(Z/pZ)G$-modules.*

2. **Preliminary remarks.** We collect here some definitions, remarks, and previously established results which will be needed in the paper.

(a) If R is a field, or if $R = Z_p^*$, the Krull-Schmidt theorem holds for RG-modules. (See [2, Theorem 14.5 and Theorem 76.26].)

(b) For R an arbitrary ring, every element of $a(RG)$ is expressible in the form $[M]-[N]$, where M and N are RG-modules, but is not uniquely so expressible. Furthermore, $[M]=[N]$ in $a(RG)$ if and only if there exists an RG-module X such that $M \oplus X \cong N \oplus X$. If the Krull-Schmidt Theorem holds for RG-modules, this last isomorphism implies that $M \cong N$. Furthermore, in this case $a(RG)$ has a Z-basis consisting of the symbols $[L]$, where L ranges over a full set of representatives of the isomorphism classes of indecomposable RG-modules.

(c) Let $A \to B$ be a monomorphism of abelian additive groups. Then also

$$C \otimes_Z A \to C \otimes_Z B$$

is a monomorphism. (See MacLane [7, p. 152, Theorem 6.2].)

(d) For M a Z_pG-module, let $M^* = Z_p^* \otimes_{Z_p} M$. The map $[M] \to [M^*]$ gives a ring homomorphism $a(Z_pG) \to a(Z_p^*G)$, which we claim is a monomorphism. For let M and N be Z_pG-modules such that $[M^*]=[N^*]$. Then $M^* \cong N^*$, which implies that $M \cong N$ (see Maranda [8], or [2, Theorem 76.9]). This proves that the above homomorphism is monic, so by the preceding remark, $A(Z_pG) \to A(Z_p^*G)$ is also monic.

Next, there is a ring homomorphism

$$a(Z'G) \to \prod_{p|[G:1]} a(Z_pG),$$

gotten by mapping $[M]$ onto the element whose pth component is $[Z_p \otimes_{Z'} M]$. This map is monic, since if M and N are $Z'G$-modules such that

$$Z_p \otimes_{Z'} M \cong Z_p \otimes_{Z'} N, \qquad p\,|\,[G:1],$$

then by Maranda [9] (see [2, Theorem 81.2]) it follows that $M \cong N$. The map

$$A(Z'G) \to \prod_{p|[G:1]} A(Z_pG)$$

is also monic, by (c) above.

(e) Suppose that R is either a field of characteristic p, or $R = Z_p^*$, and let H be a subgroup of G. An RG-module M is called (G,H)-*projective* if M is a direct summand of an induced module L^G for some RH-module L. If H is a p-Sylow subgroup of G, then every RG-module is (G,H)-projective (see [2, §63]). The $(G,\{1\})$-projective modules are just the ordinary projective RG-modules.

For D a subgroup of G, let $a_D(RG)$ be the ideal of $a(RG)$ generated by the set of all (G,D)-projective RG-modules. Denote by $a'_D(RG)$ the ideal generated by the (G,D')-projectives, where D' ranges over the proper subgroups of D. Define

$$w_D(RG) = a_D(RG)/a'_D(RG),$$

and set

$$W_D(RG) = C \otimes_Z w_D(RG), \qquad A_D = C \otimes a_D, \qquad A'_D = C \otimes a'_D.$$

Then we have:

TRANSFER THEOREM. *The algebra $A(RG)$ is semisimple if $W_D(R \cdot N_GD)$ is semisimple for each p-subgroup D of G. Here, N_GD is the normalizer of D in G (Green [4]).*

Using this, O'Reilly [10] was able to prove Theorem 1 by showing:

THEOREM. *If k is a field of characteristic p, and if G has a cyclic p-Sylow subgroup, then $W_D(k \cdot N_GD)$ is semisimple for each p-subgroup D of G.*

(f) Now let H be a normal subgroup of G, and suppose that the Krull-Schmidt Theorem holds for RG-modules. Let L be an RH-module, and let $x \in G$. We may form a new RH-module L^x, called a *conjugate* of L, by letting L^x have the same elements as L, but where each $h \in H$ acts on L^x as does xhx^{-1} on L.

For an RG-module M, denote by $\text{res}_H M$ the RH-module gotten from M by restriction of operators from G to H. From the Mackey Subgroup Theorem (see [2, Theorem 44.2]), it follows at once that $\operatorname{res}_H(L^G)$ is a direct sum of conjugates of L.

We note further that if L_1 and L_2 are RH-modules, then

$$L_1^G \otimes L_2^G \cong \sum^{\oplus} (L_1 \otimes L_2^y)^G,$$

the sum extending over certain elements $y \in G$ (see [2, Theorem 44.3]).

(g) Starting with a (left) RG-module M, we may form another (left) RG-module M^*, called the *contragredient* of M (see [2, §43]). As R-module, M^* is just $\operatorname{Hom}_R(M,R)$. Each $x \in G$ acts on M^* in the same way that x^{-1} acts on the right RG-module $\operatorname{Hom}_R(M,R)$. Then $(M_1 \oplus M_2)^* \cong M_1^* \oplus M_2^*$. Also, if

$$0 \to M_1 \to M_2 \to M_3 \to 0$$

is an exact sequence of RG-modules, then so is

$$0 \to M_3^* \to M_2^* \to M_1^* \to 0.$$

Finally, we have $(RG)^* \cong RG$, and therefore contragredients of projective modules are again projective.

(h) SCHANUEL'S LEMMA (SEE SWAN [13]). *Suppose we are given two exact sequences of RG-modules*:

$$0 \to U_i \to P_i \to V_i \to 0, \quad i = 1,2,$$

in which P_1 and P_2 are projective. If $V_1 \cong V_2$, then

$$U_1 \oplus P_2 \cong U_2 \oplus P_1.$$

3. **Main theorem.** Throughout this section we fix a prime p, and set $R = Z_p^*$, $\bar{R} = R/pR$. Let G contain a p-Sylow subgroup H which is cyclic of order p. In view of 2(d) above, Theorem 3 will be established as soon as we show that $A(RG)$ contains no nonzero nilpotent element.

Up to isomorphism, there are exactly three indecomposable RH-modules, namely R_H, I_H, and RH (see Heller and Reiner [6]). Here, R_H is the module R on which H acts trivially; I_H is the augmentation ideal of the group ring RH, and there is an exact sequence of RH-modules:

$$0 \to I_H \to RH \to R_H \to 0. \tag{1}$$

By 2(e) each indecomposable RG-module is a direct summand of one of the induced modules $(I_H)^G$, RG, $(R_H)^G$. (We have used the obvious isomorphism: $(RH)^G \cong RG$.) Thus the number of isomorphism classes of indecomposable RG-modules is finite, so $A(RG)$ is a finite dimensional commutative C-algebra. We must show that $A(RG)$ is semisimple, or equivalently, that $A(RG)$ contains no nonzero nilpotent element. To prove this, by 2(e) it is enough to show that $W_D(R \cdot N_G D))$. is semisimple for each p-subgroup D of G. But W_D is unchanged when D is replaced by one of its conjugates, and therefore we need only show that

$$W_{\{1\}}(RG) \text{ and } W_H(R \cdot N_G H)$$

are both semisimple.

The algebra $W_{\{1\}}(RG)$ is generated by the projective RG-modules. As is well known (see [2, §77]), there is a one-to-one isomorphism-preserving correspondence between the indecomposable direct summands of RG and those of $\bar{R}G$. In other words, we have

$$W_{\{1\}}(RG) \cong W_{\{1\}}(\bar{R}G),$$

the isomorphism being given by $[M] \to [M/pM]$. But $\bar{R}$ is a field of characteristic p, so by O'Reilly's Theorem of 2(e) it follows that $W_{\{1\}}(\bar{R}G)$ is semisimple. (This can also be proved easily by use of Brauer characters; see, for example, Conlon [1].)

It remains for us to show that $W_H(R \cdot N_G H)$ is semisimple. Changing notation, we may hereafter assume that G has a normal p-Sylow subgroup H which is cyclic of order p, and we must prove that $W_H(RG)$ is semisimple. For p odd, this is an immediate consequence of O'Reilly's Theorem together with the following result:

THEOREM. 5. *Let G have a normal p-Sylow subgroup H which is cyclic of order p, where p is an odd prime. Then the algebra homomorphism*

$$W_H(RG) \to W_H(\bar{R}G)$$

is monic.

(Before starting the proof, we may remark that the theorem fails to be true when $p = 2$. Nevertheless, most of the details of the proof are valid for $p = 2$, and will be used for that case in the following section.)

Proof. As was pointed out earlier in this section, the nonisomorphic indecomposable direct summands of $(R_H)^G$, $(I_H)^G$, and RG, give a full set of indecomposable RG-modules. The direct summands of RG are RG-projective, and generate the ideal $a_H(RG)$ of $a_H(RG)$. Thus $w_H(RG)$ is generated as Z-module by the indecomposable direct summands of $(R_H)^G$ and $(I_H)^G$. Let us write

$$(R_H)^G = N_1 \oplus \cdots \oplus N_k, \; N_i \text{ indecomposable,} \tag{2}$$

where the summands are numbered so that the first m of them are a full set of nonisomorphic modules from the set of summands. It will turn out that

$$(I_H)^G = L_1 \oplus \cdots \oplus L_k, \; L_i \text{ indecomposable,} \tag{3}$$

with the first m summands a full set of nonisomorphic modules from the set $\{L_1, \cdots, L_k\}$. Thus $\{[N_1], \cdots, [N_m], [L_1], \cdots, [L_m]\}$ form a Z-basis for $w_H(RG)$.

Let $\bar{N}_i = N_i/pN_i$, $\bar{L}_i = L_i/pL_i$, viewed as $\bar{R}G$-modules. In order to prove that $W_H(RG) \to W_H(\bar{R}G)$ is monic, it suffices by 2(c) to show that $w_H(RG) \to w_H(\bar{R}G)$ is monic; and for this, we need only show that $\{[\bar{N}_1], \cdots, [\bar{N}_m], [\bar{L}_1], \cdots, [\bar{L}_m]\}$ are linearly independent (over Z) in $w_H(\bar{R}G)$.

By Schur's Theorem [2, Theorem 7.5], there exists a subgroup F of G such that $G = HF$, $F \cong G/H$. It is easily seen that

$$(R_H)^G \cong RF \qquad \text{as } RG\text{-modules},$$

where H acts trivially on the module RF. Indeed, $(R_H)^G = RG \otimes_{RH} R_H$, and the isomorphism is given by

$$\sum_{x \in F; y \in H} \alpha_{x,y}(xy \otimes 1) \to \sum_{x,y} \alpha_{x,y} x, \qquad \alpha_{x,y} \in R.$$

Hence the $\{N_i\}$ occurring in (2) are gotten by decomposing RF into a direct sum of indecomposable left ideals. However, $p \nmid [F:1]$, so if K is the quotient field of R, we have

$$KF = KN_1 \oplus \cdots \oplus KN_k,$$

where the $\{KN_i\}$ are minimal left ideals of KF. Furthermore, it follows from [2, Theorem 76.17 and Theorem 76.23], that each $\bar{N}_i$ is indecomposable, and that for $1 \leqq i, j \leqq k$,

$$N_i \cong N_j \Leftrightarrow KN_i \cong KN_j \Leftrightarrow \bar{N}_i \cong \bar{N}_j.$$

Thus $\{\bar{N}_1, \cdots, \bar{N}_m\}$ are distinct indecomposable $\bar{R}G$-modules, on each of which H acts trivially.

Turning next to the consideration of $(I_H)^G$, we observe first that forming induced modules preserves exactness, and so from (1) we obtain an exact sequence of RG-modules

$$0 \to (I_H)^G \to RG \to (R_H)^G \to 0. \tag{4}$$

Each N_i is a quotient module of $(R_H)^G$, hence also of RG, and so there exist exact sequences

$$0 \to M_i \to RG \to N_i \to 0, \qquad 1 \leqq i \leqq k. \tag{5}$$

If $N_i \cong N_j$, then by 2(h) we have $M_i \cong M_j$. Conversely, if $M_i \cong M_j$, then taking contragredients (see 2(g)) and using 2(h) again, we obtain $N_i^* \cong N_j^*$, and $N_i \cong N_j$.

If $RG^{(k)}$ denotes a direct sum of k copies of RG, then from (5) we obtain an exact sequence

$$0 \to M_1 \oplus \cdots \oplus M_k \to RG^{(k)} \to N_1 \oplus \cdots \oplus N_k \to 0.$$

Comparing this with (4) and using 2(h), we find that

(6) $$M_1 \oplus \cdots \oplus M_k \cong (I_H)^G \oplus RG^{(k-1)}.$$

For each i, $1 \leqq i \leqq k$, let us write

$$M_i = L_i \oplus P_i,$$

where P_i is projective, and L_i has no projective direct summand. It follows from the Krull-Schmidt Theorem for RG-modules that M_i determines L_i and P_i uniquely, up to isomorphism. By (6), each L_i is a direct summand of $(I_H)^G$. On the other hand, $(I_H)^G$ has no projective direct summand, since $\text{res}_H (I_H)^G$ is a direct sum of conjugates of I_H, hence of copies of I_H, whereas for X a projective RG-module, $\text{res}_H X$ is free. Consequently

$$L_1 \oplus \cdots \oplus L_k \cong (I_H)^G,$$

and $\{L_1, \cdots, L_m\}$ are a full set of nonisomorphic modules from the set $\{L_1, \cdots, L_k\}$. To show that each L_i is indecomposable, we shall establish the stronger result that $\{\bar{L}_1, \cdots, \bar{L}_m\}$ are a set of distinct indecomposable $\bar{R}G$-modules.

From (5) we obtain exact sequences

$$0 \to \bar{N}_i^* \to \bar{R}G \to \bar{L}_i^* \oplus \bar{P}_i^* \to 0, \qquad 1 \leqq i \leqq m.$$

If $\bar{L}_i \cong \bar{L}_j$ for some i, j, where $1 \leqq i, j \leqq m$, the above implies (using 2(h)) that $\bar{N}_i^* \oplus \bar{P}_j^* \cong \bar{N}_j^* \oplus \bar{P}_i^*$. But $\bar{N}_i^*$ is indecomposable, and is not projective because H acts trivially on $\bar{N}_i^*$. Hence $\bar{N}_i^* \cong \bar{N}_j^*$, so $N_i \cong N_j$ and $i = j$.

Next, suppose $\bar{L}_i$ decomposable; then so is $\bar{L}_i^*$, and we may write $\bar{L}_i^* = U_1 \oplus U_2$, say. Each U_j is a homomorphic image of $\bar{R}G$, so there exist $\bar{R}G$-modules X_1, X_2 with

$$0 \to X_j \to \bar{R}G \to U_j \to 0, \qquad j = 1, 2,$$

exact. Thus

$$0 \to X_1 \oplus X_2 \to \bar{R}G^{(2)} \oplus \bar{P}_i^* \to U_1 \oplus U_2 \oplus \bar{P}_i^* \to 0$$

is exact. By 2(h) it follows that

$$\bar{N}_i^* \oplus \bar{R}G \oplus \bar{P}_i^* \cong X_1 \oplus X_2.$$

But $\bar{N}_i^*$ is indecomposable, and $\bar{P}_i^*$ is projective, so either X_1 or X_2 must be projective; say X_1 is projective. Then $\text{res}_H X_1$ is free. On the other hand, X_1 is a direct summand of $\bar{L}_i^*$, and $\text{res}_H \bar{L}_i^*$ is a direct sum of copies of $\bar{I}_H$ (since $\bar{I}_H^* \cong \bar{I}_H$). This gives a contradiction, and so indeed each $\bar{L}_i$ is indecomposable.

We may remark that in terms of the loop space functor Ω introduced by Heller [5], we have $L_i = \Omega(N_i)$.

Let us show at once that $\bar{N}_i \cong \bar{L}_j$ is impossible, and it is precisely for this purpose that the hypothesis $p > 2$ is needed. We know that $\text{res}_H \bar{N}_i$ is a direct sum of copies of $\bar{R}_H$, whereas $\text{res}_H \bar{L}_j$ is a direct sum of copies of $\bar{I}_H$. But $\bar{I}_H$ is indecomposable, and for $p > 2$, $\bar{I}_H$ is not isomorphic to $\bar{R}_H$. Thus $\bar{N}_i \not\cong \bar{L}_j$ for any i,j, and we have shown that $\{\bar{N}_1,\cdots,\bar{N}_m,\bar{L}_1,\cdots,\bar{L}_m\}$ are a set of nonisomorphic indecomposable $\bar{R}G$-modules. Obviously none of them lies in $a_H'(\bar{R}G)$, and so they are Z-linearly independent when considered as elements of $w_H(\bar{R}G)$. This completes the proof of Theorem 5.

Since we have already shown that $W_{\{1\}}(RG) \cong W_{\{1\}}(\bar{R}G)$, it follows that the map $A(RG) \to A(\bar{R}G)$ is monic when restricted to $A_H'(RG)$. Combining this with Theorem 5, we may conclude that the algebra homomorphism

$$A(RG) \to A(\bar{R}G)$$

is also monic, provided the hypotheses of Theorem 5 are satisfied. But this establishes the validity of the proposition stated at the end of §1.

4. **The case** $p = 2$. In this section we shall prove Theorem 3 for the case $p = 2$. We use the notation of the preceding section, and we are assuming now that G has a cyclic 2-Sylow subgroup H of order 2. As we have seen, we need only show that $W_H(RG)$ contains no nonzero nilpotent element, and it suffices to prove this for the case where H is normal in G. Furthermore, in order to prove that $W_H(RG)$ has no nilpotent elements except 0, it is enough to show that if $x \in W_H(RG)$ satisfies $x^2 = 0$, then necessarily $x = 0$.

As in §3, we let $\{N_1,\cdots,N_m\}$ be the nonisomorphic indecomposable summands of $(R_H)^G$, and $\{L_1,\cdots,L_m\}$ those of $(I_H)^G$. Now $R_H \not\cong I_H$, even for $p = 2$, so by considering restrictions to H it is clear that $N_i \not\cong L_j$ for any i,j. Hence $W_H(RG)$ has C-basis $\{[N_1],\cdots,[N_m], [L_1],\cdots,[L_m]\}$. Furthermore, we know from §3 that $\{\bar{N}_1,\cdots,\bar{N}_m\}$ are the nonisomorphic indecomposable summands of $(\bar{R}_H)^G$, while $\{\bar{L}_1,\cdots,\bar{L}_m\}$ are those of $(\bar{I}_H)^G$. However, since $p = 2$ we have $\bar{R}_H \cong \bar{I}_H$, and so the $\bar{L}$'s are a rearrangement of the $\bar{N}$'s. Thus the maps $W_H(RG) \to W_H(\bar{R}G)$, $A(RG) \to A(\bar{R}G)$, are no longer monomorphisms.

In this case we have $[G:F] = 2$, so F is normal in G, and $G/F \cong H$. If h is the generator of H, we may form the RH-module Y having the same elements as R, but where

$$h\alpha = -\alpha, \qquad \alpha \in Y.$$

Then use the homomorphism of G onto H to turn Y into an RG-module, that is, let F act trivially on Y. The RG-module thus obtained will also be denoted by Y. Obviously

$$\bar{Y} \cong \bar{R}_G, \qquad Y \otimes Y \cong R_G,$$

where R_G is the trivial RG-module.

Consider now the RG-modules $Y \otimes N_1,\cdots,Y \otimes N_m$. Each is indecomposable, since

$$\overline{Y \otimes N_i} \cong \bar{Y} \otimes \bar{N}_i \cong \bar{N}_i.$$

urthermore, it cannot happen that $Y \otimes N_i \cong N_j$, since h acts on $Y \otimes N_i$ as multiplication by -1, whereas h acts trivially on N_j. Thus, the modules $\{Y \otimes N_i : 1 \leqq i \leqq m\}$ coincide with the modules $\{L_i : 1 \leqq i \leqq m\}$ in some order.

Let us set $Q_i = Y \otimes N_i$, $1 \leqq i \leqq m$. The above discussion shows that $\{[N_1], \cdots, [N_m], [Q_1], \cdots, [Q_m]\}$ is a Z-basis for $w_H(RG)$, hence also a C-basis for $W_H(RG)$. Furthermore, the kernel of the algebra homomorphism $W_H(RG) \to W_H(\bar{R}G)$ has C-basis $\{[N_i] - [Q_i] : 1 \leqq i \leqq m\}$.

We shall now investigate $N_i \otimes N_j$. Since N_i and N_j are direct summands of $(R_H)^G$, their tensor product is a direct summand of $(R_H)^G \otimes (R_H)^G$. By 2(f) we see that this latter module is a direct sum of modules of the form $(R_H \otimes R_H^y)^G$, for some elements $y \in G$. However, $R_H^y \cong R_H$ and $R_H \otimes R_H \cong R_H$. Therefore $N_i \otimes N_j$ is a direct sum of copies of $N_1, \cdots, N_m$; suppose that N_s occurs with multiplicity α_{ijs} as a direct summand of $N_i \otimes N_j$. We have then

$$[N_i][N_j] = \sum_{s=1}^{m} \alpha_{ijs}[N_s] \qquad \text{in } W_H(RG).$$

Furthermore we obtain

$$Q_i \otimes Q_j = (Y \otimes N_i) \otimes (Y \otimes N_j) \cong (Y \otimes Y) \otimes (N_i \otimes N_j) \cong N_i \otimes N_j,$$

so

$$[Q_i][Q_j] = \sum_{s=1}^{m} \alpha_{ijs}[N_s] \qquad \text{in } W_H(RG).$$

Finally we note that

$$[Q_i][N_j] = [Y \otimes (N_i \otimes N_j)] = \sum_{s} \alpha_{ijs}[Y \otimes N_s] = \sum_{s} \alpha_{ijs}[Q_s] \qquad \text{in } W_H(RG).$$

Suppose now that $x \in W_H(RG)$ and $x^2 = 0$; we are trying to prove that x must be 0. Since the image of x in $W_H(\bar{R}G)$ is also nilpotent, and since $W_H(\bar{R}G)$ contains no nonzero nilpotent element, it follows that x lies in the kernel of the map $W_H(RG) \to W_H(\bar{R}G)$. Thus we may write

$$x = \sum_{i=1}^{m} c_i([N_i] - [Q_i]), \qquad c_i \in C.$$

Then

$$\begin{aligned} x^2 &= \sum_{i,j=1}^{m} \{c_i c_j [N_i][N_j] - 2c_i c_j [N_i][Q_j] + c_i c_j [Q_i][Q_j]\} \\ &= \sum_{i,j,s=1}^{m} 2c_i c_j \alpha_{ijs}([N_s] - [Q_s]). \end{aligned}$$

But $x^2 = 0$, and so

$$2 \cdot \sum_{i,j=1}^{m} c_i c_j \alpha_{ijs} = 0, \qquad 1 \leqq s \leqq m.$$

Therefore $\sum_{i,j} c_i c_j \alpha_{ijs} = 0$, which shows that $\sum_{i=1}^{m} c_i[N_i]$ has square 0. Thus $\sum_i c_i[\bar{N}_i] = 0$ in $W_H(\bar{R}G)$, and consequently each $c_i = 0$. This proves that $x = 0$, and completes the demonstration of Theorem 3 for the case $p = 2$.

5. **Concluding remarks.** Let us show that Theorem 4 is an easy consequence of Theorem 3. Suppose that $[G:1]$ is squarefree; then by using Theorem 3 for each prime p dividing $[G:1]$, we see that each $A(Z_p^* G)$ contains no nonzero nilpotent element. Hence also the product

$$\prod_{p \mid [G:1]} A(Z_p^* G)$$

contains no nonzero nilpotent element. But by 2(d) the algebra $A(Z'G)$ may be embedded in the above product, and hence also $A(Z'G)$ contains no nonzero nilpotent element.

It would be of interest to consider the corresponding question for $A(ZG)$. The difficulty seems to arise from the fact that the map $A(ZG) \to A(Z'G)$ need not be monic.

CONJECTURE 1. The kernel of the map $A(ZG) \to A(Z'G)$ is a torsion Z-module.

As remarked in Theorem 2, if G contains an element of order p^2, then $A(Z_p^* G)$ contains nonzero nilpotent elements. On the other hand, we have shown that if the p-Sylow subgroup of G is cyclic of order p, then $A(Z_p^* G)$ is semisimple. We are left with a large class of groups which fall into neither category, for example an elementary abelian (p, p) group.

CONJECTURE 2. If the p-Sylow subgroup of G is not cyclic of order p, then $A(Z_p^* G)$ contains nonzero nilpotent elements.

We may remark that Theorem 5 is best possible, in the following sense. Let $R = Z_p^*$, and let H be a p-Sylow subgroup of G. If H is not normal in G, or if H is not cyclic of order p, then the maps

$$W_H(RG) \to W_H(\bar{R}G), \quad A(RG) \to A(\bar{R}G),$$

are not monic. Indeed, even when G is cyclic of order p^2, the map $A(RG) \to A(\bar{R}G)$ is not monic.

Finally, the proof of Theorem 5 suggests that the proposition at the end of §1 may be a special case of a more general result. This will be investigated more fully in a future work (to appear in Michigan Math. J.)

REFERENCES

1. S. B. Conlon, *Certain representation algebras*, J. Austral. Math. Soc. **4** (1964), 152–173.

2. C. W. Curtis and I. Reiner, *Representation theory of finite groups and associative algebras*, Interscience, New York, 1962.

3. J. A. Green, *The modular representation algebra of a finite group*, Illinois J. Math. **6** (1962), 607–619.

4. ———, *A transfer theorem for modular representations*, J. of Algebra **1** (1964), 73–84.

5. A. Heller, *Indecomposable representations and the loop space operation*, Proc. Amer. Math. Soc. **12** (1961), 640–643.

6. A. Heller and I. Reiner, *Representations of cyclic groups in rings of integers*. I, Ann. of Math. (2) **76** (1962), 73–92.

7. S. MacLane, *Homology*, Academic Press, New York, 1963.

8. J. M. Maranda, *On p-adic integral representations of finite groups*, Canad. J. Math. **5** (1953), 344–355.

9. ———, *On the equivalence of representations of finite groups by groups of automorphisms of modules over Dedekind rings*, Canad. J. Math. **7** (1955), 516–526.

10. M. F. O'Reilly, *On the semisimplicity of the modular representation algebra of a finite group*, Illinois J. Math. **9** (1965), 261–276.

11. I. Reiner, *The integral representation ring of a finite group*, Michigan Math. J. **12** (1965), 11–22.

12. ———, *Nilpotent elements in rings of integral representations*, Proc. Amer. Math. Soc. **17** (1966), 270–274.

13. R. G. Swan, *Periodic resolutions for finite groups*, Ann. of Math. (2) **72** (1960), 267–291.

University of Illinois,
Urbana, Illinois

NILPOTENT ELEMENTS IN RINGS OF INTEGRAL REPRESENTATIONS

Irving Reiner

ABSTRACT

The result of a previous paper [39], *The integral representation ring of a finite group* is extended in the following way.

Theorem. *Suppose R is a discrete valuation ring of characteristic zero with maximal ideal P and with R/P of characteristic $p \neq 0$. Let G be a finite group containing a cyclic subgroup G^* of order n such that*

i. the Krull–Schmidt theorem holds for RG^*-modules,

ii. $n \in P^2$ and if $2 \in P$ then $n \in 2P$. Then the integral representation ring $A(RG)$ contains a nonzero nilpotent element.

The proof makes use of the proof of the analogous theorem (the case $G^* = G$) in the paper cited above. A corollary obtains the same conclusion if R is a valuation ring in an algebraic number field, even though the Krull–Schmidt theorem may fail in this case.

RELATIONS BETWEEN INTEGRAL AND MODULAR REPRESENTATIONS

Irving Reiner

ABSTRACT

Let R be a noetherian complete local integral domain, with meximal ideal P, residue field $\bar{R} = R/P$, and field of quotients K. Let A be an R algebra which is finitely generated as an R–module. Set $\bar{A} = A/PA$ so that $\bar{A}$ is a finite dimensional $\bar{R}$–algebra; if M is an A–modules, $\bar{M} = M/PM$ is an $\bar{A}$–module. Moreover the correspondence $M \to \bar{M}$ gives a one–to–one, isomorphism preserving correspondene between the set of *projective* A–modules M and the *projective* $\bar{A}$–modules $\bar{M}$. One of the main results of this paper is to generalize this correspondence to include non–projective modules in the case that A is the group ring RG of a finite group G.

For an A–module M let

$$E(M) = \mathrm{Hom}_A(M, M), \quad \widehat{E}(M) = E(M)/\mathrm{rad}\, E(M),$$

with the analogous meaning if M is an $\bar{A}$–module.

Theorem 1. *Let H be a normal subgroup of the finite group G and let M be an RH–module such that $\bar{M}$ is indecomposable, $\widehat{E}(M) \cong \widehat{E}(\bar{M})$, and the stabilizer S of M in G equals the stabilizer of $\bar{M}$. Then there is a ono–to–one, isomorphism preserving correspondence $X \to \bar{X}$ between the indecomposable summands $\{X\}$ of M^G and the indecomposable summands $\{\bar{X}\}$ of $\bar{X}^G$.*
Moreover for each decomposition into indecomposable summands,

$$M^S = \sum_{i=1}^{n} \oplus L_i,$$

there are decompositions into indecomposable summands given by

$$\bar{M}^S = \sum_{i=1}^{n} \oplus \bar{L}_i, \quad M^G = \sum_{i=1}^{n} \oplus L_i^G, \bar{M}^G = \sum_{i=1}^{n} \oplus \bar{L}_i^G.$$

Finally, for each i, the R–algebras $\widehat{E}(L_i)$, $\widehat{E}(\bar{L}_i)$, $\widehat{E}(L_i^G)$, and $\widehat{E}(\bar{L}_i^G)$ are all isomorphic.

The proof makes use of general ring theoretic results which follow mainly from idempotent lifting techniques. The paper contains slightly more information than is reprinted in the statement of the theorem. Application of this result is made to obtain isomorphisms between certain subrings of the representation ring $a(RG)$ and corresponding subrings of $a(\bar{R}G)$.

An application of results of P. Tucker and S. Conlon are combined with Theorem 1 to give a proof of a result of J. Green:

Proposition. *Let G be a p–group, where $\bar{R}$ has characteristic p. Let G_0 be any subgroup of G and N an RG_0–module for which $\widehat{E}(\bar{N}) \cong \bar{R}$. Then N^G is indecomposable, and $\widehat{E}(N^G) \cong \bar{R}$.*

Several more results along this line are given.

Module Extensions and Blocks*

IRVING REINER

Department of Mathematics, University of Illinois, Urbana, Illinois

Communicated by W. Feit

Received January 18, 1966

1. INTRODUCTION

Let R be a Dedekind ring of characteristic zero, with quotient field K, and let G be a finite group. All RG-modules are assumed finitely generated and R-torsionfree. For each prime ideal P of R, let R_P denote the P-adic completion of R, with quotient field K_P. For convenience, we shall write Ext in place of Ext^1_{RG} or $\mathrm{Ext}^1_{R_PG}$. If $\{\mathrm{Ext}\}_P$ denotes the P-primary component of Ext, then for any pair of RG-modules M, N we have

$$\mathrm{Ext}(M, N) \cong \sum_P \{\mathrm{Ext}(M, N)\}_P ,$$

where P ranges over the divisors of $[G:1]$. Furthermore, if we set $M_P = R_P \otimes_R M$, then

$$\{\mathrm{Ext}(M, N)\}_P \cong \mathrm{Ext}(M_P , N_P).$$

For each KG-module U, denote by $\sigma(U)$ the collection of all RG-modules M such that $K \otimes_R M \cong U$. We shall say that two KG-modules U, V are *linked at* P if there exist modules $M \in \sigma(U)$, $N \in \sigma(V)$, such that $\{\mathrm{Ext}(M, N)\}_P \neq 0$. Call U and V *linked* if they are linked at some P, that is, if $\mathrm{Ext}(M, N) \neq 0$ for some choice of $M \in \sigma(U)$, $N \in \sigma(V)$.

Let U_i be a KG_i-module, $i = 1, 2$, and let $G = G_1 \times G_2$. We may form the KG-module $U_1 \# U_2$, defined as the K-space $U_1 \otimes_K U_2$, with the action of G given by

$$(g_1 , g_2)(u_1 \otimes u_2) = g_1u_1 \otimes g_2u_2 .$$

If $[G_1 : 1]$ and $[G_2 : 1]$ are relatively prime, then as U_i ranges over all irreducible KG_i-modules ($i = 1, 2$), the product $U_1 \# U_2$ ranges over all irreducible KG-modules.

* This research was supported in part by the National Science Foundation.

Let Z denote the ring of rational integers, Q the rational field. Berman and Lichtman [1] recently proved

THEOREM 1. *Let G be a nilpotent group, and write $G = G_1 \times \cdots \times G_r$, where $G_1, \ldots, G_r$ are the Sylow subgroups of G belonging to the distinct prime divisors of $[G:1]$. For each i, $1 \leqslant i \leqslant r$, let U_i and V_i be a pair of irreducible KG_i-modules, and set*

$$U = U_1 \# \cdots \# U_r, \qquad V = V_1 \# \cdots \# V_r.$$

Assume that the $\{U_i\}$ differ from the $\{V_i\}$ for at least one subscript i. Then U and V are linked if and only if the $\{U_i\}$ differ from the $\{V_i\}$ for exactly one subscript i.

The object of the present note is to generalize this result, and at the same time to put it into a more natural setting involving block idempotents. As general reference for the techniques used herein, we may cite [2].

2. DISJOINT MODULES

Let $c(R_pG)$ denote the center of R_pG, and write $1 \in c(R_pG)$ as

$$1 = e_1 + \cdots + e_t,$$

a sum of orthogonal primitive idempotents in $c(R_pG)$. These $\{e_i\}$ are called the *block idempotents* of R_pG, and are not necessarily primitive in $c(K_pG)$. For each i, $1 \leqslant i \leqslant t$, let us write

$$e_i = \sum_j \epsilon_{ij},$$

where the $\{\epsilon_{ij}\}$ are orthogonal primitive idempotents in $c(K_pG)$. Then the decomposition of K_pG into simple components is given by

$$K_pG = \sum_{i,j}^{\oplus} K_pG \cdot \epsilon_{ij},$$

and the irreducible K_pG-modules are just the minimal left ideals in these simple components. Hence for each irreducible K_pG-module X there is a unique block idempotent e such that $ex = x$, $x \in X$. We say that X *belongs* to e; in this case, X is annihilated by all the other block idempotents. Let us call two K_pG-modules X, Y *disjoint* if the set of block idempotents to which the composition factors of X belong has no elements in common with the corresponding set for Y.

If in particular $P \nmid [G:1]$, then each ϵ_{ij} lies in $R_P G$, and so the $\{\epsilon_{ij}\}$ coincide with the $\{e_i\}$. Thus, in this case, if X and Y are $K_P G$-modules having no common composition factor, then X and Y are disjoint.

Returning to the general case, for any KG-module U, set $U_P = K_P \otimes_K U$. If M is an RG-module in $\sigma(U)$, then $K_P \otimes_{R_P} M_P \cong U_P$, that is, $M_P \in \sigma(U_P)$. Let us call two KG-modules U, V *disjoint at* P if U_P and V_P are disjoint.

THEOREM 2. *Let X and Y be disjoint $K_P G$-modules. Then*

$$\operatorname{Ext}(M, N) = 0 \quad \text{for all } M \in \sigma(X), \quad N \in \sigma(Y). \tag{1}$$

Hence if U and V are KG-modules which are disjoint at P, then U and V are not linked at P.

Proof. It suffices to prove the first assertion, which clearly implies the second. Let $M \in \sigma(X)$, $N \in \sigma(Y)$. Given any exact sequence

$$0 \rightarrow X_1 \rightarrow X \rightarrow X_2 \rightarrow 0$$

of $K_P G$-modules, we may find an exact sequence of $R_P G$-modules

$$0 \rightarrow M_1 \rightarrow M \rightarrow M_2 \rightarrow 0$$

in which $M_i \in \sigma(X_i)$, $i = 1, 2$. Then

$$\operatorname{Ext}(M_2, N) \rightarrow \operatorname{Ext}(M, N) \rightarrow \operatorname{Ext}(M_1, N)$$

is exact, and thus it is enough to show that $\operatorname{Ext}(M_i, N) = 0$, $i = 1, 2$. By repeated use of this argument, we are reduced to proving that if X and Y are irreducible $K_P G$-modules belonging to different block idempotents, then (1) holds.

Suppose that X belongs to the block idempotent e; then $em = m$, $m \in M$, whereas e annihilates N. Let us view $\operatorname{Ext}(M, N)$ as the group of binding functions from G to $\operatorname{Hom}_{R_P}(M, N)$, modulo the subgroup of inner binding functions. Recall that a binding function is a map $F : G \rightarrow \operatorname{Hom}(M, N)$, satisfying

$$F_{gh} = gF_h + F_g h, \qquad g, h \in G.$$

Call F *inner* if there exists $u \in \operatorname{Hom}(M, N)$ such that

$$F_g = gu - ug, \qquad g \in G.$$

If F is any binding function, then eF and Fe are also binding functions. We claim that their difference is an inner binding function. Indeed, since $\operatorname{Ext}(K_P M, K_P N) = 0$, there exists $t \in \operatorname{Hom}(K_P M, K_P N)$ such that

$$F_g = gt - tg, \qquad g \in G,$$

and for which

$$(gt - tg)M \subset N, \qquad g \in G. \tag{2}$$

Setting $u = et - te$, we have

$$eF_g - F_g e = gu - ug, \qquad g \in G.$$

Furthermore, since e is an R_P-linear combination of group elements, it follows from (2) that $u \in \text{Hom}(M, N)$. Therefore $eF - Fe$ is inner, as claimed.

We have now shown that eF and Fe determine the same element of $\text{Ext}(M, N)$. But $Fe = F$, since e acts as the identity on M; on the other hand, $eF = 0$ since e annihilates N. This shows that F is in the zero class, and completes the proof of the theorem.

We shall need two easy observations.

Lemma 1. *Let U, V be KG-modules having no common composition factor, and suppose that $P \nmid [G:1]$. Then U and V are disjoint at P.*

Proof. Let U' denote any composition factor of U, and V' of V. Then U' and V' are distinct irreducible KG-modules, whence $(U')_P$ and $(V')_P$ have no common composition factor. Since $P \nmid [G:1]$, this shows that $(U')_P$ and $(V')_P$ are disjoint, which implies that U and V are disjoint at P.

Lemma 2. *Let $G = G_1 \times G_2$, and let U_1, V_1 be KG_1-modules which are disjoint at P. Let U_2, V_2 be arbitrary KG_2-modules. Then $U_1 \# U_2$ and $V_1 \# V_2$ are disjoint at P.*

Proof. The set of composition factors of $(U_1 \# U_2)_P$ is the union of the sets of composition factors of $X \# (U_2)_P$, where X ranges over the composition factors of $(U_1)_P$. If X belongs to the block idempotent e of $R_P G_1$, then e is a central idempotent of $R_P G$ which acts trivially on each composition factor of $X \# (U_2)_P$, and annihilates each composition factor of $(V_1 \# V_2)_P$. This shows that $(U_1 \# U_2)_P$ and $(V_1 \# V_2)_P$ are disjoint, as claimed.

Corollary. *Using the notation of Theorem 1, suppose that the $\{U_i\}$ differ from the $\{V_i\}$ at two or more places. Then U and V are not linked.*

Proof. Suppose that U_i and V_i are nonisomorphic, $i = 1, 2$. We shall show that for each P, U_i and V_i are not linked at P. Since $[G_1 : 1]$ and $[G_2 : 1]$ are relatively prime, either $P \nmid [G_1 : 1]$ or $P \nmid [G_2 : 1]$. Without loss of generality, suppose $P \nmid [G_1 : 1]$. By Lemma 1, U_1 and V_1 are disjoint at P, whence so are U and V by Lemma 2. The result now follows from Theorem 2.

3. Primitive Block Idempotents

In order to obtain a converse to Theorem 2, we shall have to restrict the types of block idempotents under consideration. If e is a block idempotent of R_PG, then $R_PG \cdot e$ is a two-sided ideal of R_PG which cannot be decomposed into a direct sum of two-sided ideals. It may happen that in fact $R_PG \cdot e$ cannot be decomposed into a direct sum of *left* ideals; in this case, e is a primitive idempotent not only in $c(R_PG)$, but in the ring R_PG itself. This will occur if and only if the ring $R_PG \cdot e$ is completely primary, that is, $R_PG \cdot e/\text{rad}(R_PG \cdot e)$ is a skewfield.

Theorem 3. *Let e be a block idempotent of R_PG which is primitive in R_PG. Let U and V be distinct irreducible KG-modules such that e acts as the identity on both U_P and V_P. Then U and V are linked at P.*

Proof. Choose primitive idempotents ϵ_1, ϵ_2 in $c(KG)$ such that U is a minimal left ideal in $KG \cdot \epsilon_1$, V in $KG \cdot \epsilon_2$. Then the RG-module $RG \cdot \epsilon_1$ is a successive extension of modules in $\sigma(U)$, and likewise, $RG \cdot \epsilon_2$ is a successive extension of modules in $\sigma(V)$. Hence if U and V are not linked at P, the left R_PG-module $R_PG(\epsilon_1 + \epsilon_2)$ decomposes into a direct sum

$$R_PG(\epsilon_1 + \epsilon_2) \cong R_PG \cdot \epsilon_1 \dotplus R_PG \cdot \epsilon_2 . \tag{3}$$

We shall show that this is impossible.

Since e acts as identity on both U_P and V_P, we have $e\epsilon_1 = \epsilon_1$, $e\epsilon_2 = \epsilon_2$. Setting $\epsilon = \epsilon_1 + \epsilon_2$, we conclude that $e\epsilon = \epsilon$. The ring epimorphism

$$R_PG \cdot e \to R_PG \cdot \epsilon$$

given by $x \to x\epsilon$ induces an epimorphism

$$R_PG \cdot e/\text{rad}(R_PG \cdot e) \to R_PG \cdot \epsilon/\text{rad}(R_PG \cdot \epsilon),$$

and therefore $R_PG \cdot \epsilon$ is also a completely primary ring. This shows that $R_PG \cdot \epsilon$ is indecomposable as left R_PG-module, which contradicts (3), and establishes the theorem.

Corollary. *Let G be a p-group, and let U, V be distinct irreducible KG-modules. Then for each P dividing p, U and V are linked at P.*

Proof. The ring R_PG is completely primary, and so we may take $e = 1$ in the above theorem.

Remark. Theorem 3 may also be obtained from a result of Thompson ([3], Theorem 1). For if we set $W = K_P(U + V) \cap R_PG \cdot e$, and use his approach, we see that W is an R_P-pure R_PG-submodule of $R_PG \cdot e$. Let bars denote reduction mod P; then $\bar{W}$ is an $\bar{R}G$-submodule of $\bar{R}G \cdot \bar{e}$. Since this latter module has a unique minimal submodule, $\bar{W}$ must be indecomposable, whence so is W. But on the other hand, if U and V are not linked at P, then W is decomposable.

Next comes a lemma of independent interest. Suppose that $E = G \times H$ is a direct product, and let T be an RH-module. Given an exact sequence of RG-modules

$$0 \to N \to L \to M \to 0,$$

we obtain an exact sequence of RE-modules:

$$0 \to N \# T \to L \# T \to M \# T \to 0.$$

Thus there exists a map

$$\varphi_T : \mathrm{Ext}^1_{RG}(M, N) \to \mathrm{Ext}^1_{RE}(M \# T, N \# T).$$

LEMMA 3. *The map φ_T is monic.*

Proof. If $F : G \to \mathrm{Hom}_R(M, N)$ is a binding function representing an element of $\mathrm{Ext}(M, N)$, then its image F' under φ_T is given by

$$F'_{(g,h)}(m \otimes t) = F_g m \otimes t, \qquad g \in G, h \in H, m \in M, t \in T.$$

In order to show φ_T monic, it suffices to establish this fact when R_P is used in place of R. To avoid notational difficulties, we may instead restrict ourselves to the case where R is a principal ideal domain.

We must show that if F' is inner, then so is F. Now if F' is inner, there exists a map $w \in \mathrm{Hom}_R(M \# T, N \# T)$ such that

$$F'_{(g,h)}(m \otimes t) = (g, h)w(m \otimes t) - w(gm \otimes ht)$$

for all g, h, m, t. Setting $h = 1$, this becomes

$$F_g m \otimes t = (g, 1)w(m \otimes t) - w(gm \otimes t). \tag{4}$$

Suppose that $T = \sum^{\oplus} Rt_i$, and set

$$w(m \otimes t_i) = \sum_j \omega_{ij}(m) \otimes t_j, \qquad \omega_{ij} \in \mathrm{Hom}(M, N). \tag{5}$$

Set $t = t_i$ in (4) and use (5), thereby obtaining

$$F_g m \otimes t_i = \sum_j \{g\omega_{ij}(m) - \omega_{ij}(gm)\} \otimes t_j$$

for all g, m, i. But $N \# T = \sum^{\oplus} N \otimes t_j$, and therefore $F_g m = g\omega_{ii}(m) - \omega_{ii}(gm)$ for each g, m, i. Fixing i, we obtain $F_g = g\omega_{ii} - \omega_{ii}g$, proving that F is inner, and establishing the Lemma.

Corollary 1. *Let U and V be KG-modules, Y a KH-module, and set $E = G \times H$. If U and V are linked at P, then $U \# Y$ and $V \# Y$ are also linked at P.*

Proof. Since U and V are linked at P, there exist RG-modules $M \in \sigma(U)$, $N \in \sigma(V)$ such that $\{\text{Ext}(M, N)\}_P \neq 0$. Choose any RH-module $T \in \sigma(Y)$. Then by Lemma 3,

$$\{\text{Ext}(M \# T, N \# T)\}_P \neq 0.$$

Since $M \# T \in \sigma(U \# Y)$ and $N \# T \in \sigma(V \# Y)$, the Corollary is proved.

Corollary 2. *Using the notation of Theorem* 1, *suppose that the $\{U_i\}$ and $\{V_i\}$ differ at exactly one subscript i. Then U and V are linked.*

Proof. Let us suppose that U_1 and V_1 are distinct, and that $U_i = V_i$ for $i = 2, \ldots, r$. Choose P dividing $[G_1 : 1]$; then U_1 and V_1 are linked at P, by the Corollary to Theorem 3. Set $H = G_2 \times \cdots \times G_r$, $Y = U_2 \# \cdots \# U_r$, and use Corollary 1 above. Then U and V are also linked at P, and the proof is complete.

References

1. Berman, S. D. and Lichtman, A. I. Integral representations of finite nilpotent groups. *Usp. Mat. Nauk* **20** (1965), 186-188.
2. Curtis, C. W. and Reiner, I. "Representation Theory of Finite Groups and Associative Algebras." Wiley (Interscience), New York, 1962.
3. Thompson, J. G. "Vertices and Sources" (to be published).

REPRESENTATION RINGS

Irving Reiner

ABSTRACT

Let Λ be a ring with unit. The Grothendieck group $G(\Lambda)$ is the quotient of the free abelian group generated by the isomorphism classes $[M]$ of left Λ–modules by the subgroup generated by $[V] - [U] - [W]$, where there is an exact sequence $0 \to U \to V \to W \to 0$. For any category $\mathcal{C}$ of Λ–modules, one defines the Grothendieck group $G(\mathcal{C})$ by restricting the modules M to be in $\mathcal{C}$.

In the special case where $\Lambda = RG$, G a finite group and R an integral domain, take the category $\mathcal{C}$ to consist of all R–torsion free RG–modules. The Grothendieck group of this category becomes a ring with multiplication induced by $[M \otimes_R N] = [M][N]$. This ring is called the *representation ring* of RG and is denoted by $a(RG)$.

Let Z be the ring of rational integers, G a group of order n, and let

$$Z' = \{\frac{a}{b} : a, b \in Z, (b, n) = 1\}.$$

One of the results in this paper is the proof that, as additive groups

$$a(ZG) = b(ZG) \oplus a(Z'G)$$

where $b(ZG)$ is a finite group which is an ideal in $a(ZG)$. An explicit description of $b(ZG)$ is given; it is the subgroup generated by the differences $[ZG] - [P]$ where P is a ZG–module which is locally isomorphic to ZG at every prime of Z. A slightly more general result is proved about the representation ring $a(RG)$, for a Dedekind domain R, in which the Higman ideal plays the role of nZ in the above statement.

FINITE GENERATION OF GROTHENDIECK RINGS RELATIVE TO CYCLIC SUBGROUPS[1]

T. Y. LAM AND I. REINER

Let Ω be a field of characteristic p ($p \neq 0$), G a finite group, and consider finitely generated left ΩG-modules ("G-modules" for short). The isomorphism classes of G-modules, relative to direct sum, form a free Z-module $a(G)$ called the *representation ring* of G. Let $\boldsymbol{U}$ be some family of subgroups of G. An exact sequence of ΩG-modules

$$0 \to L \to M \to N \to 0$$

is called $\boldsymbol{U}$-split if its restriction to each $H \in \boldsymbol{U}$ is split. Let $i(G, \boldsymbol{U})$ be the ideal of $a(G)$ generated by all expressions $[M]-[L]-[N]$ arising from $\boldsymbol{U}$-split G-exact sequences. Then

$$a(G, \boldsymbol{U}) = a(G)/i(G, \boldsymbol{U})$$

is the *relative Grothendieck ring*, and is a commutative ring with unity. Finally, we use the notation $M \mid N$ to indicate that M is a direct summand of N; M_H denotes the restriction of M, and V^G an induced module.

Let us say that a G-module M is $(G, \boldsymbol{U})$*-projective* if

$$M \,\Big|\, \sum_{H \in \boldsymbol{U}}^{\oplus} (M_H)^G.$$

(This generalizes the usual notion when $\boldsymbol{U}$ consists of a single subgroup of G.) The $(G, \boldsymbol{U})$-projective modules generate an ideal $k(G, \boldsymbol{U})$ in $a(G)$, and the map

$$\kappa\colon k(G, \boldsymbol{U}) \to a(G, \boldsymbol{U})$$

given by $[M] \to [M]$ is the *Cartan map*. As shown by Dress [2], κ is monic and its cokernel is a p-torsion group.

The present authors have investigated $a(G, H)$ extensively, and have shown [4], [7] that:

(i) If H is a normal cyclic p-subgroup of G, then $a(G, H)$ has a finite free Z-basis.

(ii) If H is any cyclic subgroup of a p-group G, then $a(G, H)$ has a finite free Z-basis (and indeed, $a(G, H) \cong a(H)$).

We cannot expect that $a(G, H)$ will always have a *finite* free

Received by the editors May 26, 1969.

[1] This research was partially supported by the National Science Foundation.

Z-basis; in fact, when $H=G$ the relative ring $a(G, H)$ is just $a(G)$, and this is finitely generated if and only if G has a cyclic p-Sylow subgroup, by [**3**]. On the other hand, all of the available evidence [**4**]–[**7**] indicates that $a(G, H)$ is always Z-free, but this is a difficult open question.

The purpose of this article is to establish the comforting theorem:

If H is cyclic, then $a(G, H)$ is a finitely generated Z-module.

This will appear as a special case of the Main Theorem given below. We shall also sketch our calculations showing that $a(G, H)$ has a finite free Z-basis when $G=A_4$ or S_4, and $|H|=2$, where $p=2$.

MAIN THEOREM. *The relative Grothendieck ring $a(G, \boldsymbol{U})$ is finitely generated as Z-module if and only if each $H\in \boldsymbol{U}$ has a cyclic p-Sylow subgroup.*

PROOF. In one direction the result is straightforward: Assume $a(G, \boldsymbol{U})$ finitely generated, and let $H\in\boldsymbol{U}$. There is an epimorphism $a(G, \boldsymbol{U})\rightarrow a(G, H)$, whence $a(G, H)$ is finitely generated. Since κ embeds $k(G, H)$ into $a(G, H)$, also $k(G, H)$ is finitely generated. This implies at once (as in [**3**], say) that H has a cyclic p-Sylow subgroup.

Conversely, assume that each $H\in\boldsymbol{U}$ has a cyclic p-Sylow subgroup. Replacing each H by one of its p-Sylow subgroups does not affect $\boldsymbol{U}$-splitting or $(G, \boldsymbol{U})$-projectivity, so $a(G, \boldsymbol{U})$ and $k(G, \boldsymbol{U})$ are unchanged. We may thus assume hereafter that $\boldsymbol{U}=\{H_1, \cdots, H_d\}$ is some collection of cyclic p-subgroups of G, and we must show that $a(G, \boldsymbol{U})$ is finitely generated.

Let $F_1', \cdots, F_s'$ be a full set of irreducible ΩG-modules. They can all be realized in some *finite* subfield Ω_0 of Ω. In fact, let F_i' afford the (ordinary) character ϕ_i, and let Ω_0 be obtained from the prime field $GF(p)$ of Ω by adjoining all of the character values $\{\phi_i(g): g\in G, 1\leqq i\leqq s\}$. Then (see [**1**, Theorem 70.23]) each F_i' is realizable in Ω_0, that is, there exist $\Omega_0 G$-modules F_i with $F_i'=\Omega F_i$, $1\leqq i\leqq s$.

Now let $H\in\boldsymbol{U}$ be cyclic with generator a of order h, and set $\alpha=a-1\in\Omega H$, so $\alpha^h=0$. Let F be an $\Omega_0 H$-submodule of the ΩH-module M, and put $n=(F:\Omega_0)$. We now prove by induction on n that there exists an $\Omega_0 H$-submodule T of M such that

$$(T:\Omega_0)\leqq hn, \quad F\subset T\subset M, \quad \Omega T\,|\, M.$$

Let $n\geqq 1$ and assume the result known if $(F:\Omega_0)<n$. Since H is a p-group, there exists a nonzero H-trivial element $x\in F$. Let $x\in\alpha^m M$ with m maximal, and write $x=\alpha^m y$, $y\in M$. Note that $\alpha x=0$ since x

is H-trivial. Now set $Y=\Omega_0 H\cdot y$; we claim that $\Omega Y=\Omega H\cdot y$ is an H-direct summand of M, and for this it suffices to prove that

$$\alpha^r M \cap \Omega Y = \alpha^r \Omega Y, \qquad 0 \leqq r \leqq h-1.$$

Now each $u \in \Omega Y$ is expressible as $u=\sum_0^m a_i \cdot \alpha^i y$, $a_i \in \Omega$. Let $u \in \alpha^r M$, and let a_j be the first nonzero coefficient, so

$$u = a_j \cdot \alpha^j y + \cdots + a_m \cdot \alpha^m y \subset \alpha^r M.$$

Multiply by α^{m-j}, getting $a_j x \in \alpha^{r+m-j} M$, whence (since m was maximal) $j \geqq r$. Then $u \in \alpha^j \Omega Y \subset \alpha^r \Omega Y$, as desired, and we have shown that $\Omega Y | M$.

Next observe that $(F+Y)/Y$ is an $\Omega_0 H$-submodule of $M/\Omega Y$ of Ω_0-dimension $\leqq n-1$, so by the induction hypothesis there exists an $\Omega_0 H$-submodule T/Y of $M/\Omega Y$ such that

$$(T/Y:\Omega_0) \leqq h(n-1), \frac{F+Y}{Y} \subset \frac{T}{Y} \subset \frac{M}{\Omega Y}, \ \frac{\Omega T}{\Omega Y}\bigg| \frac{M}{\Omega Y}.$$

Here T is itself an $\Omega_0 H$-submodule of M, and $Y \subset F+Y \subset T \subset M$. Now write

$$M = \Omega T + L, \ L \cap \Omega T = \Omega Y, \quad L = \Omega H\text{-submodule of } M;$$

this is possible since $\Omega T/\Omega Y$ is a direct summand of $M/\Omega Y$. But now $\Omega Y | M$, $\Omega Y \subset L$, so $\Omega Y | L$; hence $L=\Omega Y \oplus W$, say. Then $M=\Omega T \oplus W$, proving that $\Omega T | M$. Finally,

$$(T:\Omega_0) = (T/Y:\Omega_0) + (Y:\Omega_0) \leqq h(n-1) + h = hn.$$

This completes the induction proof.

We apply the above procedure to the irreducible $\Omega_0 G$-modules $F_1, \cdots, F_s$ and an arbitrary ΩG-module M. Namely, if $F_r \subset M$ then for each $H \in \boldsymbol{U}$ we can find an $\Omega_0 H$-submodule T of M satisfying

$$F_r \subset T \subset M, \Omega T \,|\, M_H, (T:\Omega_0) \text{ bounded},$$

with the bound independent of M. Hence there exists a positive constant c, depending only on the group G, with the following property:

If F is any irreducible $\Omega_0 G$-module contained in an ΩG-module M, then for each i, $1 \leqq i \leqq d$, there exists an $\Omega_0 H_i$-submodule N_i of M satisfying the conditions

$$F \subset N_i \subset M, \quad (N_i:\Omega_0) \leqq c, \Omega N_i \,|\, M_{H_i}.$$

We shall refer to the system

$$[F; N_1, \cdots, N_d; \alpha_1, \cdots, \alpha_d],$$

where each α_i is the $\Omega_0 H_i$-embedding of F in N_i, as a *system* derived from M. Obviously each nonzero M gives rise to one or more such systems.

Let M' be another ΩG-module containing the same F, and let the system $[F; N_1', \cdots, N_d'; \alpha_1', \cdots, \alpha_d']$ be derived from M'. We call this system *equivalent* to the preceding one if there exist $\Omega_0 H_i$-isomorphisms ν_i: $N_i \cong N_i'$ $(1 \leqq i \leqq d)$ such that $\nu_i \alpha_i = \alpha_i'$, $1 \leqq i \leqq d$. Now each N_i is a vector space over the finite field Ω_0, and the dimensions of all $\{N_i\}$ are uniformly bounded. Hence the number of equivalence classes of systems which begin with any one of $F_1, \cdots, F_s$, must be finite.

For each σ, $1 \leqq \sigma \leqq s$, choose a finite set of ΩG-modules $\{M_\sigma^r: 1 \leqq r \leqq u_\sigma\}$ whose derived systems

$$[F_\sigma; \overset{r}{N}_{1,\sigma}, \cdots, \overset{r}{N}_{d,\sigma}; \overset{r}{\alpha}_{1,\sigma}, \cdots, \overset{r}{\alpha}_{d,\sigma}]$$

represent all equivalence classes of systems which begin with F_σ. We claim that the elements

$$(1) \qquad \{[\overset{r}{M}_\sigma] - [\overset{r}{M}_\sigma/\Omega F_\sigma]: 1 \leqq r \leqq u_\sigma, 1 \leqq \sigma \leqq s\}$$

generate the additive group $a(G, \boldsymbol{U})$. Indeed, let M be any nonzero ΩG-module, and let us show by induction on $(M:\Omega)$ that $[M]$ is expressible as a Z-linear combination of the above elements. Pick any irreducible ΩG-submodule $\Omega F_\sigma \subset M$, and form a derived system $[F_\sigma; N_1, \cdots, N_d; \alpha_1, \cdots, \alpha_d]$, so that

$$F_\sigma \subset N_i \subset M, \Omega N_i \mid M_{H_i}, \qquad 1 \leqq i \leqq d.$$

Choose r so that this system is equivalent to the one derived above from M_σ^r. Then for each i, $1 \leqq i \leqq d$, the diagram

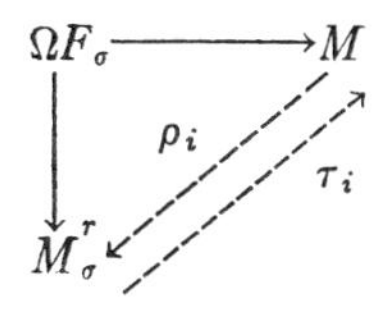

in which the horizontal and vertical arrows are ΩG-inclusions, can always be completed by ΩH_i-homomorphisms ρ_i, τ_i to a commutative diagram. Indeed, we have:

$$\begin{array}{ccc} \Omega F_\sigma & \xrightarrow{\alpha_i} & \Omega N_i \mid M_{H_i} \\ \alpha^r_{i,\sigma} \downarrow & & \\ \Omega N^r_{i,\sigma} \mid (M^r_\sigma)_{H_i} & & \end{array}$$

But there exists an ΩH_i-isomorphism ν_i: $\Omega N_i \cong \Omega N^r_{i,\sigma}$ with $\nu_i \alpha_i = \alpha^r_{i,\sigma}$. Hence it suffices to choose $\rho_i = \nu_i$ on ΩN_i, $\rho_i = 0$ on a complement of ΩN_i in M_{H_i}; likewise, let $\tau_i = \nu_i^{-1}$ on $\Omega N^r_{i,\sigma}$, and $\tau_i = 0$ on a complement of $\Omega N^r_{i,\sigma}$ in $(M^r_\sigma)_{H_i}$.

It now follows that if P denotes the ΩG-module defined by the pushout diagram

$$\begin{array}{ccc} \Omega F_\sigma & \rightarrow & M \\ \downarrow & & \downarrow \\ M^r_\sigma & \rightarrow & P, \end{array}$$

then the sequences

$$0 \rightarrow M^r_\sigma \rightarrow P \rightarrow M/\Omega F_\sigma \rightarrow 0, \quad 0 \rightarrow M \rightarrow P \rightarrow M^r_\sigma/\Omega F_\sigma \rightarrow 0$$

are H_i-split for each $H_i \in \boldsymbol{U}$. Thus in $a(G, \boldsymbol{U})$ we have the relation

$$[M] = [M^r_\sigma] - [M^r_\sigma/\Omega F_\sigma] + [M/\Omega F_\sigma].$$

The last term is a combination of the elements in (1), by the induction hypothesis. Hence so is $[M]$, and the theorem is proved.

Of course this procedure is rather wasteful, since we obtain far too many generators for $a(G, \boldsymbol{U})$. However, if it is true that $a(G, \boldsymbol{U})$ is Z-free, then since its Z-rank is that of $k(G, \boldsymbol{U})$, it can be generated by just as many elements as needed to generate $k(G, \boldsymbol{U})$, and we can presumably determine the latter quite explicitly.

We now show what happens when $p=2$, $\boldsymbol{U} = \{H\}$, $|H| = 2$, $G = A_4$ or S_4, to indicate the difficulties that one faces in trying to reduce the number of generators for $a(G, H)$.

Consider first $A_4 = \langle a, c \mid a = (12)(34), c = (123) \rangle$, $H = \langle a \rangle$, Ω any field of characteristic 2 containing ω, a primitive cube root of 1. We claim $a(A_4, H)$ is Z-free on 4 generators. There are 3 irreducible A_4-modules, given by

$$F_i\colon a \rightarrow 1,\ c \rightarrow \omega^i, \qquad i = 0, 1, 2.$$

For $i \neq j$ there is an A_4-module Y_{ij} with

$$0 \rightarrow F_i \rightarrow Y_{ij} \rightarrow F_j \rightarrow 0 \text{ exact}, \ (Y_{ij})_H \cong \Omega H, \ F_i = \alpha \cdot Y_{ij},$$

where $\alpha = a - 1$.

Let M be any A_4-module containing F_i. If $F_i \not\subset \alpha M$, then since $(F_i: \Omega) = 1$ we have $F_i \cap \alpha M = 0 = \alpha F_i$, and so F_i is an H-direct summand of M. Hence when $F_i \not\subset \alpha M$ we have

$$[M] = [F_i] + [M/F_i] \quad \text{in } a(A_4, H).$$

On the other hand if $F_i \subset \alpha M$, then there is a system derived from M of the form $[F_i;\ N;\ \alpha_1]$ in which $F_i \subset N \subset M$, $N | M_H$, $(N: \Omega) = 2$, $F_i = \alpha N$. But for $j \neq i$, Y_{ij} gives a system equivalent to this one. Hence we obtain in $a(A_4, H)$

$$[M] = [Y_{ij}] - [Y_{ij}/F_i] + [M/F_i], \tag{2}$$

and of course $[Y_{ij}/F_i] = [F_j]$. It follows at once that the 6 elements $[F_0]$, $[F_1]$, $[F_2]$, $[Y_{10}]$, $[Y_{21}]$, $[Y_{02}]$ generate $a(A_4, H)$. Further, (2) yields identities such as

$$[Y_{10}] - [F_0] = [Y_{12}] - [F_2],\ [Y_{20}] - [F_0] = [Y_{21}] - [F_1] \tag{3}$$

etc. Finally let $C = \langle c \rangle$, 1_C = trivial C-module of dimension 1. It turns out that there are a pair of H-split A_4-exact sequences:

$$0 \to Y_{01} \to (1_C)^{A_4} \to Y_{20} \to 0,$$

$$0 \to Y_{02} \to (1_C)^{A_4} \to Y_{10} \to 0,$$

so $[Y_{20}] + [Y_{01}] = [Y_{02}] + [Y_{10}]$. Using (3) we obtain

$$[Y_{10}] = [Y_{20}] - [F_2] + [F_1].$$

Multiplying by $[F_1]$, we get $[Y_{21}]$ in terms of $[Y_{01}]$ and the F's. This result, together with equations (3), show that $a(A_4, H)$ is spanned by the 4 elements $[F_0]$, $[F_1]$, $[F_2]$ and $[Y_{10}]$. Using the maps $a(A_4, H) \to a(H)$ and $a(A_4, H) \to a(A_4, 1)$, it is easily seen that $a(A_4, H)$ is Z-free on these 4 generators. In fact it follows that $\phi: a(A_4, H) \to a(H) \oplus a(A_4, 1)$ is a monomorphism. This means that the class of any A_4-module M in $a(A_4, H)$ depends only on the composition factors of M and on the H-isomorphism type of M_H. This remark enables us to resolve any $[M]$ in $a(A_4, H)$ as a combination of the four indicated generators. The ring monomorphism ϕ shows also that $a(A_4, H)$ is without nilpotent elements. Finally, routine calculation shows that coker κ is a (2, 2)-group.

Now let $G = S_4$, and keep all of the above notation. We wish to investigate $a(G, H)$ with H of order 2. If $H = \langle (12) \rangle$, then put K = Klein 4-group, so $K \triangle G$, $K \cap S_3 = 1$, $G = K \cdot S_3$. By the Excision Theorem [5, 3.2] we obtain $a(G, H) \cong a(S_3, H) = a(S_3)$, the latter since H is a 2-Sylow subgroup of S_3. Hence $a(G, H)$ is easily computed in this case.

Suppose hereafter that $H=\langle a\rangle$, with $a=(12)(34)$ as above, and set $c=(123)$, $d=(1324)$, $\delta=d-1$, so $\alpha=\delta^2$, and c, d generate G. There are two irreducible Ω-representations of G:

$$E_0\colon c\to 1,\ d\to 1, \qquad E_1\colon c\to\begin{pmatrix}1 & 1\\ 1 & 0\end{pmatrix},\ d\to\begin{pmatrix}0 & 1\\ 1 & 0\end{pmatrix},$$

and $E_1=F_1^G=F_2^G$, with F_1, F_2 the representations of A_4 defined previously. There exist G-modules X_{01}, X_{10}, X_{11} giving nonsplit extensions

$$0\to E_i\to X_{ij}\to E_j\to 0, \qquad (i,j) = (0,1),(1,0),(1,1),$$

and such that

$$E_0 = \alpha X_{01}, \qquad E_1 = \alpha X_{11}, \qquad (E_1\cap\alpha X_{10}\colon\Omega) = 1.$$

Indeed we have

$$X_{11} = Y_{12}^G, \qquad X_{01} = Y_{01}^G/E_0,$$

where E_0 is embedded in Y_{01}^G via $E_0\to F_0^G\subset Y_{01}^G$. The relation

$$[Y_{12}]+[F_0] = [Y_{01}]+[F_2] \quad \text{in } a(A_4,H)$$

gives

$$[X_{11}]+[F_0^G] = [Y_{01}^G]+[E_1] \quad \text{in } a(G,H). \tag{4}$$

Now F_0^G is H-trivial, so $[F_0^G]=2[E_0]$. Also, $E_0\subset\alpha\cdot Y_{01}^G$, so Y_{01}^G gives the system $[E_0;N]$, $N\cong\Omega H$. But X_{01} gives an equivalent system, so

$$[X_{01}]-[X_{01}/E_0] = [Y_{01}^G]-[Y_{01}^G/E_0]. \tag{5}$$

From (4) and (5) we get

$$[X_{11}] = 2[X_{01}]-2[E_0] \quad \text{in } a(G,H). \tag{6}$$

As a last preliminary step, we note the existence of a G-module L with both sequences exact:

$$0\to E_0\to L\to X_{11}\to 0, \qquad 0\to E_1\to L\to X_{01}\to 0, \tag{7}$$

and

$$E_0\not\subset\alpha L, \qquad (E_1\cap\alpha L\colon\Omega) = 1.$$

We are now ready to prove that $[E_0]$, $[E_1]$, $[X_{01}]$ are a free Z-basis of $a(G,H)$. Let M be any nonzero G-module. If $E_0\subset M$, we have

Case i. $E_0\not\subset\alpha M$. Then $E_0\cap\alpha M=0=\alpha E_0$, so $(E_0)_H\mid M_H$, and $[M]=[E_0]+[M/E_0]$ in $a(G,H)$.

Case ii. $E_0 \subset \alpha M$. Then M yields a system $[E_0; N]$ with $N = \Omega H$, $N | M_H$, $E_0 = \alpha N$. Since X_{01} yields an equivalent system, we obtain $[M] = [X_{01}] - [X_{01}/E_0] + [M/E_0]$, and here $[X_{01}/E_0] = [E_1]$.

On the other hand let $E_1 \subset M$.

Case iii. $E_1 \cap \alpha M = 0$. Then $E_1 \cap \alpha M = \alpha E_1$, so $[M] = [E_1] + [M/E_1]$.

Case iv. $E_1 \subset \alpha M$. Then M yields the system $[E_1; N]$, $E_1 = \alpha N$, $N \cong \Omega H \dot{+} \Omega H$, equivalent to that gotten from X_{11}. Hence $[M] = [X_{11}] - [X_{11}/E_1] \dot{+} [M/E_1]$.

Case v. $(E_1 \cap \alpha M : \Omega) = 1$. There is a unique one-dimensional subspace $\Omega e \subset E_1$ such that $\delta e = 0$, where $\delta = d - 1$. But $\delta(E_1 \cap \alpha M) \subset E_1 \cap \alpha M$, whence $\delta(E_1 \cap \alpha M) = 0$, and so $E_1 \cap \alpha M = \Omega e$. This shows that $E_1 \cap \alpha M$ is independent of M.

Now $E_1 \cap \alpha X_{10} = \Omega e$, and we may write $e = \alpha x$, $x \in X_{10}$; choose $z \in X_{10}$ with

$$E_1 = \Omega \cdot \alpha x \oplus \Omega z, \qquad X_{10} = \Omega x \oplus \Omega \cdot \alpha x \oplus \Omega z.$$

Also $E_1 \cap \alpha M = \Omega e$, so $e = \alpha m$ for some $m \in M$. The embedding $E_1 \to M$ carries αx onto αm, and takes z onto some element $n \in M$. Set $N = \Omega m \oplus \Omega \alpha m \oplus \Omega n$; then $N | M_H$ since $N \cap \alpha M = \alpha N$. There is an H-isomorphism $\tau: X_{10} \cong N$ given by $x \to m$, $\alpha x \to \alpha m$, $z \to n$, making the diagram

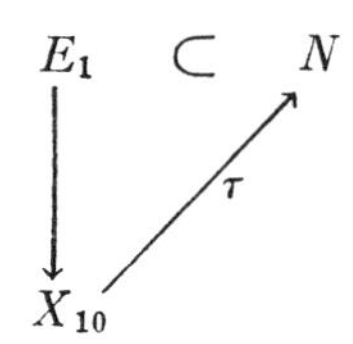

commute. Extend τ to a map of X_{10} into M, and define a map $M \to X_{10}$ by letting it coincide with τ^{-1} on N, and vanish on an H-complement of N. The Pushout Lemma [5, (3.1)] gives

$$[M] = [X_{10}] - [X_{10}/E_1] + [M/E_1] = [X_{10}] - [E_0] + [M/E_1].$$

These arguments show that $[E_0]$, $[E_1]$, $[X_{10}]$ and $[X_{01}]$ generate $a(G, H)$. By applying Cases (i) and (v) to the module L in (7), we obtain

$$[L] = [E_0] + [X_{11}] = [X_{10}] - [E_0] + [X_{01}].$$

By (6) this gives $[X_{10}] = [X_{01}]$, so indeed the generator $[X_{01}]$ may be omitted. Using the maps $a(G, H) \to a(G, 1)$ and $a(G, H) \to a(H)$, it is easily seen that $a(G, H)$ is Z-free on $[E_0]$, $[E_1]$, $[X_{10}]$. Also, as in the case of A_4, the class of $[M] \in a(G, H)$ is uniquely determined by M_H

and the S_4-composition factors of M; furthermore $a(S_4, H)$ is free of nonzero nilpotent elements.

Unfortunately we have not been able to generalize these methods to establish freeness of $a(G, H)$ in the case where H is *any* (nonnormal) cyclic p-group.

References

1. C. W. Curtis and I. Reiner, *Representation theory of finite groups and associative algebras*, Interscience, New York, 1962.

2. A. Dress, *On relative Grothendieck rings*, Bull. Amer. Math. Soc. **75** (1969), 955–958.

3. D. G. Higman, *Indecomposable representations at characteristic p*, Duke Math. J. **21** (1954), 377–381.

4. T. Y. Lam and I. Reiner, *Relative Grothendieck groups*, J. Algebra **11** (1969), 213–242.

5. ———, *Reduction theorems for relative Grothendieck rings*, Trans. Amer. Math. Soc. **142** (1969), 421–435.

6. ———, *Relative Grothendieck rings*, Bull. Amer. Math. Soc. **75** (1969), 496–498.

7. ———, *Restriction maps on relative Grothendieck rings*, J. Algebra (to appear).

University of California, Berkeley and
University of Illinois, Urbana

REDUCTION THEOREMS FOR RELATIVE GROTHENDIECK RINGS(1)

BY

T. Y. LAM AND I. REINER

1. **Introduction.** Relative Grothendieck rings arise naturally when one considers modular representations of a finite group G, and their restrictions to some fixed subgroup H of G. This article continues our earlier work on the subject [4], but can be read independently of that work.

Let G be a finite group, and let Ω be a field of characteristic p, where we assume $p \neq 0$ to avoid trivial cases. By a "G-module" we mean always a finitely generated left ΩG-module. Form the free abelian group $\mathscr{A}$ on the symbols $[M]$, where M ranges over the isomorphism classes of G-modules; let $\mathscr{B}$ be the subgroup of $\mathscr{A}$ generated by all expressions $[M_1]-[M_2]-[M_3]$, where $M_1 \cong M_2 \oplus M_3$. The factor group $\mathscr{A}/\mathscr{B}$ will be called the *Green ring* or *representation ring* of G. We shall denote it by $a(G, G)$, in order to conform with notation to be introduced later. Define multiplication in $a(G, G)$ by

$$[M][N] = [M \otimes_{\Omega} N],$$

where each $g \in G$ acts on $M \otimes N$ according to the usual formula: $g(m \otimes n) = gm \otimes gn$. Then $a(G, G)$ becomes a commutative ring whose identity element is $[1_G]$, where 1_G denotes the 1-representation of G

Now let

$$(1.1) \qquad E: 0 \to L \to M \to N \to 0$$

be an exact sequence of G-modules, and define

$$(1.2) \qquad \gamma_E = [M]-[L]-[N] \in a(G, G).$$

Denote by $i(G, H)$ the additive subgroup of $a(G, G)$ generated by the expressions γ_E, where E ranges over all exact sequences of G-modules which are split when restricted to H. Then $i(G, H)$ is an ideal in $a(G, G)$, and we define the *relative Grothendieck ring* $a(G, H)$ to be the factor ring $a(G, H) = a(G, G)/i(G, H)$.

When $H = G$, the ideal $i(G, G) = 0$, and in this case $a(G, H)$ coincides with the Green ring $a(G, G)$; this explains the notation $a(G, G)$ for the Green ring. Note that $a(G, G)$ is a free Z-module on the symbols $[M]$, M = indecomposable G-module.

Presented to the Society, August 21, 1968; received by the editors August 5, 1968.

(1) This research was supported by the National Science Foundation.

On the other hand, when $H=1$, the relative Grothendieck ring $a(G, 1)$ is the Z-free module spanned by the irreducible G-modules. Thus $a(G, 1)$ is isomorphic to the ring of generalized Brauer characters of G. (See [2, §82].)

There are surjective ring homomorphisms

$$a(G, G) \to a(G, H), \qquad a(G, H) \to a(G, 1).$$

Thus we expect that a relation $[M]=[N]$ in $a(G, H)$ will yield more information about the modules M and N than merely that they have the same Brauer characters (and the same composition factors). On the other hand, this relation may hold true even when M, N are not isomorphic, so that $a(G, H)$ is easier to investigate than the Green ring $a(G, G)$. We hope that eventually we can use relative Grothendieck rings to study modular representations of groups with noncyclic Sylow p-subgroups.

Our previous work [4] dealt mainly with the structure of $a(G, H)$ when H was a normal cyclic p-subgroup of G. In the present paper, we drop the restrictions on H, and obtain some "reduction formulas" which can be used in the calculation of $a(G, H)$. The principal results are Theorems 3.2 and 5.1.

2. Preliminary results. Throughout this paper, we assume that G is a finite group, with subgroups H, K satisfying $H \subset K \subset G$, where the inclusions need not be proper. In this section we shall give some miscellaneous propositions needed later.

The restriction map, taking G-modules into K-modules, induces a ring homomorphism

$$a(G, H) \to a(K, H).$$

On the other hand, each K-module M determines an induced G-module M^G, given by

$$M^G = \Omega G \otimes_{\Omega K} M.$$

If

$$E: 0 \longrightarrow L \xrightarrow{\lambda} M \xrightarrow{\mu} N \longrightarrow 0$$

is an exact sequence of K-modules, there is an exact sequence of G-modules

$$E^G: 0 \longrightarrow L^G \xrightarrow{\lambda'} M^G \xrightarrow{\mu'} N^G \longrightarrow 0,$$

where $\lambda' = 1 \otimes \lambda$, $\mu' = 1 \otimes \mu$. One can find examples where E is H-split but E^G is not. However, we prove(2)

(2.1) Proposition. *Suppose that G can be written as a disjoint union* $G = g_1K \cup \cdots \cup g_sK$, *where the elements* $\{g_i\}$ *are such that* $g_1H \cup \cdots \cup g_sH = Hg_1 \cup \cdots \cup Hg_s$. *(This is surely possible when* $H \triangle G$, *for example.) Then for each H-split K-exact*

(2) This generalizes Lemma 2.6 of [4].

sequence E, the G-exact sequence E^G is also H-split. In this case, therefore, the induction map $[M] \to [M^G]$ defines an additive homomorphism $a(K, H) \to a(G, H)$.

Proof. We may write

$$L^G = \sum_{i=1}^{s}{}^{\oplus} g_i \otimes L, \qquad M^G = \sum_{i=1}^{s}{}^{\oplus} g_i \otimes M,$$

and for each i,

$$\lambda'(g_i \otimes l) = g_i \otimes \lambda(l), \qquad l \in L.$$

Since the sequence E is H-split, there exists an H-retraction of λ, that is, there exists a map $\rho \in \operatorname{Hom}_H (M, L)$ such that $\rho\lambda = 1$ on L. Define $\rho' \in \operatorname{Hom}_\Omega (M^G, L^G)$ by

$$\rho'(g_i \otimes m) = g_i \otimes \rho(m), \qquad m \in M, 1 \leqq i \leqq s.$$

Then ρ' is well defined, and $\rho'\lambda' = 1$ on L^G. It remains to prove that ρ' is an H-homomorphism.

Fix the integer i, $1 \leqq i \leqq r$, and let $x \in H$, $m \in M$. Since $xg_i \in Hg_i$, the hypothesis implies that $xg_i = g_j y$ for some $y \in H$ and some j, $1 \leqq j \leqq r$. Therefore

$$\begin{aligned}\rho'(x \cdot (g_i \otimes m)) &= \rho'(xg_i \otimes m) = \rho'(g_j \otimes ym) = g_j \otimes \rho(ym) \\ &= g_j \otimes y\rho(m) = xg_i \otimes \rho(m) = x \cdot \rho'(g_i \otimes m).\end{aligned}$$

Hence ρ' is an H-homomorphism, and the proposition is established.

Given a G-module M, if there exists an H-split G-exact sequence (1.1) with $L \neq 0$, $N \neq 0$, then of course $[M] = [L] + [N]$ in $a(G, H)$. If no such sequence exists, call M an *H-simple* G-module. Thus, M is H-simple if and only if no nontrivial G-submodule of M is an H-direct summand of M. Each H-simple G-module must be G-indecomposable. By way of illustration, we note that the 1-simple G-modules are the irreducible modules, and the G-simple G-modules are the indecomposable modules.

We have defined $i(G, H)$ to be the additive subgroup of $a(G, G)$ generated by the expressions γ_E given in (1.2), where E ranges over all H-split G-exact sequences. Let $i_0(G, H)$ be the subgroup of $i(G, H)$ generated by those γ_E in which the sequence E given by (1.1) ranges over all those H-split G-exact sequences for which the module L is H-simple.

(2.2) Proposition. $i_0(G, H) = i(G, H)$.

Proof. It suffices to show that $\gamma_E \in i_0(G, H)$ for every H-split G-exact sequence E given by (1.1). We use induction on $\dim L$, and remark that the result follows from the definition of $i_0(G, H)$ when L is H-simple. Suppose then that L is not H-simple, so there exists a nontrivial H-split G-exact sequence

$$E_1 : 0 \to U \to L \to L/U \to 0.$$

By composing the H-retractions of M onto L, and L onto U, we get an H-retraction $M \to U$. Thus there is another H-split G-exact sequence

$$E_2 : 0 \to U \to M \to M/U \to 0.$$

Finally, using the H-projection $L \to L/U$, we obtain an H-homomorphism $M \to L/U$, and hence also there is an H-split G-exact sequence

$$E_3 : 0 \to L/U \to M/U \to M/L \to 0.$$

Since $M/L \cong N$, we find readily that in $a(G, G)$ we have

$$\gamma_E = \gamma_{E_2} + \gamma_{E_3} - \gamma_{E_1}.$$

But $\dim U < \dim L$ and $\dim L/U < \dim L$, so by the induction hypothesis each $\gamma_{E_j} \in i_0(G, H)$, $j = 1, 2, 3$. Therefore also $\gamma_E \in i_0(G, H)$, as desired.

The remaining propositions in this section are probably well known, but since they are not readily available in the literature, we have included their proofs for the convenience of the reader.

(2.3) Proposition. *Let E and F be finite-dimensional algebras over the field Ω, and let $\hat{E} = E/\text{rad}\, E$, where* rad E *is the radical of E. Define $\hat{F}$ analogously. If $\hat{E} \cong \Omega$, or more generally if $\hat{E}$ is a separable algebra over Ω, then*

$$\frac{E \otimes F}{\text{rad}\,(E \otimes F)} \cong \hat{E} \otimes \hat{F},$$

where $\otimes$ means $\otimes_\Omega$.

Proof. Since $\hat{E}$ is separable, the algebra $\hat{E} \otimes \hat{F}$ is semisimple (see [2, §71]). There is an algebra epimorphism $\phi : E \otimes F \to \hat{E} \otimes \hat{F}$, and $\ker \phi = (\text{rad}\, E) \otimes F + E \otimes (\text{rad}\, F)$. Clearly $\ker \phi$ is a nilpotent two-sided ideal in $E \otimes F$, and so $\ker \phi \subset \text{rad}\,(E \otimes F)$. On the other hand, ϕ carries $\text{rad}\,(E \otimes F)$ onto a nilpotent two-sided ideal in the semisimple algebra $\hat{E} \otimes \hat{F}$, and so $\text{rad}\,(E \otimes F) \subset \ker \phi$. This proves that $\text{rad}\,(E \otimes F) = \ker \phi$, and establishes the proposition.

We need next some properties of outer tensor products of modules (see [2, §43]). Let B and K be groups, M a B-module, N a K-module. Make $M \otimes_\Omega N$ into a $(B \times K)$-module, called the *outer tensor product* of M and N and denoted by $M \# N$, by setting

$$(x, y)(m \otimes n) = xm \otimes yn, \qquad x \in B,\ y \in K.$$

(2.4) Proposition. *Let $G = B \times K$, and let M_1, M_2 be B-modules, and N_1, N_2 K-modules. There is an Ω-isomorphism*

$$\psi : \text{Hom}_B\,(M_1, M_2) \otimes_\Omega \text{Hom}_K\,(N_1, N_2) \cong \text{Hom}_G\,(M_1 \# N_1, M_2 \# N_2),$$

given by $f \otimes g \to f \# g$.

Proof. If f and g are operator homomorphisms, so is $f \# g$. Thus ψ carries $\mathrm{Hom}_B \otimes \mathrm{Hom}_K$ into Hom_G. Furthermore, ψ is monic since there is an Ω-isomorphism

$$(2.5) \qquad \mathrm{Hom}_\Omega(M_1, M_2) \otimes_\Omega \mathrm{Hom}_\Omega(N_1, N_2) \cong \mathrm{Hom}_\Omega(M_1 \# N_1, M_2 \# N_2).$$

It remains to show that ψ is epic.

By (2.5), each $u \in \mathrm{Hom}_G(M_1 \# N_1, M_2 \# N_2)$ is expressible as

$$u = \sum_{i=1}^{k} f_i \# g_i, \qquad f_i \in \mathrm{Hom}_\Omega(M_1, M_2),\ g_i \in \mathrm{Hom}_\Omega(N_1, N_2),$$

and we may assume that the $\{f_i\}$ are linearly independent over Ω. Then

$$u(m \otimes n) = \sum f_i(m) \otimes g_i(n), \qquad m \in M_1,\ n \in N_1.$$

Since $u(m \otimes yn) = y \cdot u(m \otimes n)$, $y \in K$, we obtain

$$\sum f_i(m) \otimes \{g_i(yn) - yg_i(n)\} = 0, \qquad m \in M_1,\ n \in N_1,\ y \in K.$$

But the $\{f_i\}$ are linearly independent, so we may choose $m \in M_1$ such that $f_1(m) \neq 0$, $f_2(m) = 0, \ldots, f_k(m) = 0$. This yields $g_1(yn) = yg_1(n)$, $n \in N_1$, $y \in K$, so that $g_1 \in \mathrm{Hom}_K(N_1, N_2)$. The same holds for $g_2, \ldots, g_k$, and we may thus rewrite u as

$$u = \sum f_j' \# g_j', \qquad f_j' \in \mathrm{Hom}_\Omega(M_1, M_2),\ g_j' \in \mathrm{Hom}_K(N_1, N_2),$$

where now the $\{g_j'\}$ are linearly independent over Ω. The preceding argument can be used again, so that each $f_j' \in \mathrm{Hom}_B(M_1, M_2)$. This proves that $u \in$ image of ψ, as desired.

For a B-module M, we put $E(M) = \mathrm{Hom}_B(M, M)$, $\hat{E}(M) = E(M)/\mathrm{rad}\, E(M)$, and we use analogous notation for other modules.

(2.6) Proposition. *Let M be a B-module, N a K-module. Then $E(M \# N) \cong E(M) \otimes E(N)$, where $\otimes$ is $\otimes_\Omega$. If M and N are indecomposable, and if $\hat{E}(M) \cong \Omega$, then $M \# N$ is also indecomposable. In particular, $M \# N$ is indecomposable if M is irreducible, N indecomposable, and Ω is a splitting field for the group B.*

Proof. The first assertion follows at once from (2.4). Next, a module M is indecomposable if and only if $E(M)$ is completely primary (see [2, §54]), that is, if and only if $\hat{E}(M)$ is a skewfield. Suppose now that M and N are indecomposable, and that $\hat{E}(M) \cong \Omega$. By (2.3),

$$\hat{E}(M \# N) \cong \frac{E(M) \otimes E(N)}{\mathrm{rad}\,(E(M) \otimes E(N))} \cong \hat{E}(M) \otimes \hat{E}(N) \cong \hat{E}(N).$$

But $\hat{E}(N)$ is a skewfield, whence so is $\hat{E}(M \# N)$, and thus $M \# N$ is indecomposable.

If M is irreducible, and Ω is a splitting field for B, then $E(M) \cong \Omega$, and so also $\hat{E}(M) \cong \Omega$. This completes the proof.

3. **Excision of normal p-subgroups.** We begin with a simple result which shows how one may obtain relations in a relative Grothendieck ring. (A special case of this lemma was used in [4, Lemma 3.5].)

(3.1) PUSHOUT LEMMA. *Let M, U, V be G-modules, and let $\alpha: M \to U$, $\beta: M \to V$ be G-monomorphisms. Suppose that there exist H-homomorphisms $\theta: U \to V$, $\eta: V \to U$ making the following diagram commute:*

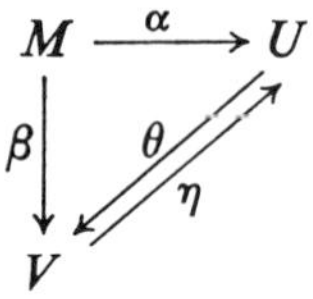

$$\begin{array}{ccc} M & \xrightarrow{\alpha} & U \\ {\scriptstyle\beta}\downarrow & \theta \swarrow \nearrow \eta & \\ V & & \end{array}$$

Then in $a(G, H)$ there is a relation $[V]-[V/\beta M]=[U]-[U/\alpha M]$.

Proof. Define the G-module X by the pushout diagram

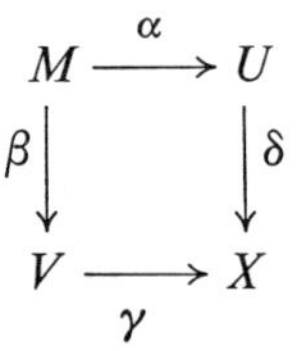

$$\begin{array}{ccc} M & \xrightarrow{\alpha} & U \\ {\scriptstyle\beta}\downarrow & & \downarrow{\scriptstyle\delta} \\ V & \xrightarrow{\gamma} & X \end{array}$$

so that

$$X = (U \oplus V)/\{(\alpha m, \ -\beta m) : m \in M\}.$$

Then γ and δ are G-monomorphisms, and $U/\alpha M \cong X/\gamma V$. Now the G-exact sequence $0 \to V \xrightarrow{\gamma} X \to X/\gamma V \to 0$ is H-split if and only if there exists a map $\theta \in \operatorname{Hom}_H(U, V)$ such that $\theta\alpha = \beta$. Since such a map θ exists by hypothesis, this sequence is H-split, and so in $a(G, H)$ we have the formula

$$[X] = [V]+[X/\gamma V] = [V]+[U/\alpha M].$$

Similarly,

$$[X] = [U]+[V/\beta M].$$

Eliminating $[X]$ between these two equations gives the desired formula in $a(G, H)$.

(3.2) THEOREM. (*Excision of normal p-subgroups.*) *Let G be a semidirect product $P \cdot K$, where $P \triangle G$, $P \cap K=1$, and P is a p-group. Let $H \subset K$. Then the restriction map φ: $a(G, H) \to a(K, H)$ is a ring isomorphism.*

Proof. Each K-module M determines a G-module M', consisting of the same elements as M, with the action of G defined by using the homomorphism $G \to G/P \cong K$. In other words, the K-module M becomes the G-module M' on which the elements of P act trivially. The map ψ: $a(K, H) \to a(G, H)$, given by $[M] \to [M']$, is thus a well-defined ring homomorphism. If the subscript K denotes restriction to K, then for each K-module M we have

$$\varphi\psi[M] = \varphi[M'] = [(M')_K] = [M].$$

Thus $\varphi\psi=1$, so in order to prove that φ is a ring isomorphism with inverse ψ, it suffices to show that ψ is an epimorphism.

Let L be a G-module, and consider the element $[L] \in a(G, H)$. We shall show by induction on $\dim L$ that $[L] \in$ image of ψ always. Let M be an irreducible G-submodule of L. Since P is a normal p-subgroup of G, P acts trivially on M; therefore the mapping $\alpha\colon M \to (L_K)'$ is a G-monomorphism. Let $\beta\colon M \to L$ be the inclusion map. In order to use the Pushout Lemma, we must find H-homomorphisms θ, η making the following diagram commute:

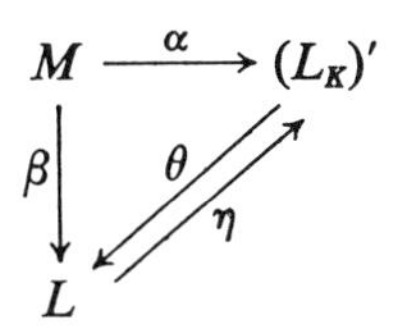

Since $H \subset K$, and $(L_K)'$ is the lifting of L_K to a G-module with trivial action of P, we may in fact choose θ and η to be the identity map on L.

It follows now from (3.1) that in $a(G, H)$ we have the equality

$$[L] = [(L_K)'] - [(L_K)'/\alpha M] + [L/\beta M].$$

The first two terms on the right lie in the image of ψ, from the definition of ψ. The last term also lies in the image of ψ by the induction hypothesis, since $\dim(L/\beta M) < \dim L$. This completes the proof.

Keeping the notation used in the above proof, we note that φ and ψ have been shown to be inverses of one another. Therefore also $\psi\varphi=1$, which means that for each G-module L, there is an equality in $a(G, H)$: $[L]=[(L_K)']$.

We list also the following simple consequence of the theorem.

(3.3) Corollary. *Let P be any p-group. Then the restriction map $a(P\times H, H) \to a(H, H)$ is a ring isomorphism of the relative Grothendieck ring $a(P\times H, H)$ onto the Green ring $a(H, H)$.*

Next we give an example to show how the Excision Theorem may be used to calculate $a(G, H)$. Take Ω any field of characteristic 2, and let $G=S_4$ (symmetric group on the symbols 1, 2, 3, 4), $K=$subgroup of G consisting of the permutations on the symbols 1, 2, 3, and $H=\{(1), (12)\}$. If we choose

$$P = \{(1), (12)(34), (13)(24), (14)(23)\},$$

then $P \triangle G$, and G is a semidirect product $P\cdot K$. The theorem then implies that $a(G, H) \cong a(K, H)$, the isomorphism being given by restriction from G to K.

However, H is a Sylow 2-subgroup of K, so that an exact sequence of K-modules is H-split if and only if it is K-split (see [2, §63]). Therefore $a(K, H) = a(K, K)$, and the latter is just the Green ring of S_3. Since the indecomposable S_3-modules are easily determined, the structure of $a(G, H)$ is completely known in this case.

We conclude the section with a somewhat deeper consequence of the Excision Theorem. Recall that a G-module M is called *(G, H)-projective* if M is a G-direct summand of an induced module X^G, for some H-module X. These (G, H)-projectives generate an additive subgroup $k(G, H)$ of the Green ring $a(G, G)$, and indeed $k(G, H)$ is an ideal in $a(G, G)$. Furthermore, it is easily found that

$$k(G, H) \cdot i(G, H) = 0. \tag{3.4}$$

There is a ring homomorphism

$$\kappa \colon k(G, H) \to a(G, H)$$

defined by composition of maps:

$$k(G, H) \to a(G, G) \to a(G, G)/i(G, H) = a(G, H),$$

where the first map is inclusion. We have called κ the *Cartan homomorphism* for the pair (G, H), since in the special case where $H = 1$, the matrix of κ is the Cartan matrix associated with the group algebra ΩG.

Making use of a result of Conlon [1], we now prove that κ is monic. This fact was established in [4] under the restrictive hypothesis that $H \triangle G$. As we shall see, this restriction is unnecessary. Let $\boldsymbol{C}$ be the complex field, and set

$$K(G, H) = \boldsymbol{C} \otimes_{\boldsymbol{Z}} k(G, H), \qquad A(G, H) = \boldsymbol{C} \otimes_{\boldsymbol{Z}} a(G, H),$$

and define $A(G, G)$, $I(G, H)$ analogously. There is a commutative diagram

$$\begin{array}{ccc} k(G, H) & \xrightarrow{\kappa} & a(G, H) \\ {\scriptstyle\alpha}\downarrow & & \downarrow{\scriptstyle\beta} \\ K(G, H) & \xrightarrow[\kappa']{} & A(G, H) \end{array}$$

where α, β arise from tensoring with $\boldsymbol{C}$, and where κ' is defined by composition of maps

$$K(G, H) \to A(G, G) \to A(G, G)/I(G, H) \cong A(G, H).$$

It follows at once that

$$\ker \kappa' = K(G, H) \cap I(G, H), \tag{3.5}$$

while from (3.4) we have

$$K(G, H) \cdot I(G, H) = 0. \tag{3.6}$$

Now $k(G, H)$ is Z-free, since it is a Z-submodule of the free Z-module $a(G, G)$. Therefore α is monic. To prove κ monic, it thus suffices to show κ' monic. By Conlon [1, Theorem 3.21(a)], $K(G, H)$ is an ideal direct summand of $A(G, G)$. Hence there exists an idempotent $e \in A(G, G)$ such that $K(G, H) = e \cdot A(G, G)$. By (3.6) we deduce that $e \cdot I(G, H) = 0$. However, e acts as the identity on $K(G, H)$, and therefore $K(G, H) \cap I(G, H) = 0$. This completes the proof that κ' is monic, and thus also that κ is monic.

We restate our conclusion as follows.

(3.7) PROPOSITION. *The Cartan homomorphism* $\kappa: k(G, H) \to a(G, H)$ *is monic. Thus, if* M *and* N *are* (G, H)*-projective* G*-modules, then* $M \cong N$ *if and only if* $[M] = [N]$ *in* $a(G, H)$.

From this we may deduce

(3.8) THEOREM. *Let* $G = P \cdot K$, $P \triangle G$, $P = p$*-group,* $P \cap K = 1$. *Let* M *and* N *be* (G, K)*-projective* G*-modules. Then* M *and* N *are* G*-isomorphic if and only if they are* K*-isomorphic.*

Proof. By (3.7), $M \cong N$ if and only if $[M] = [N]$ in $a(G, K)$. From the Excision Theorem 3.2 with $H = K$, it follows that the restriction map gives an isomorphism $a(G, K) \cong a(K, K)$. Thus $[M] = [N]$ in $a(G, K)$ if and only if $[M_K] = [N_K]$ in $a(K, K)$. The latter equality holds if and only if $M_K \cong N_K$, since $a(K, K)$ is the Green ring of K. This completes the proof.

It would be of interest to give a direct proof of the preceding theorem.

4. **Direct product with p-free factor.** We begin with an easy result.

(4.1) PROPOSITION. *Let* $G = B \times K$ *be a direct product of groups, and let* $H \subset K$. *Define*

$$\tau: a(B, 1) \otimes_Z a(K, H) \to a(G, H)$$

by letting $[X] \otimes [M] \to [X \# M]$, *where* $X \# M$ *is the outer tensor product of modules (see §2). Then* τ *is a well-defined ring homomorphism.*

Proof. In order to prove that τ is well defined, we must verify two things:

(i) If $0 \to X_1 \to X_2 \to X_3 \to 0$ is an exact sequence of B-modules, then for each K-module M, the sequence

$$0 \to X_1 \# M \to X_2 \# M \to X_3 \# M \to 0$$

is an H-split G-exact sequence. This is indeed the case; exactness is clear, and the sequence is H-split since $(X_i \# M)_H$ is a direct sum of $(X_i : \Omega)$ copies of M_H.

(ii) If $0 \to L \to M \to N \to 0$ is an H-split K-exact sequence, then for each B-module X, the sequence

$$0 \to X \# L \to X \# M \to X \# N \to 0$$

is H-split and G-exact. Again, exactness is clear; further, if $\mu: M \to L$ is an H-retraction splitting the original K-exact sequence, then $1 \# \mu: X \# M \to X \# L$ is an H-retraction splitting the G-exact sequence.

This proves that τ is a well-defined additive homomorphism, since we extend it by linearity. But τ is also a ring homomorphism, by virtue of the G-isomorphism

$$(X \# M) \otimes (X' \# M') \cong (X \otimes X') \# (M \otimes M'),$$

where $\otimes$ means $\otimes_\Omega$.

For the remainder of this section, let $G = A \times K$, where $H \subset K$, and where A is a p-free group, that is, p does not divide $[A:1]$. Suppose also that Ω is a splitting field for A. Since K contains a Sylow p-subgroup of G, it follows that every G-module is (G, K)-projective, and hence that every indecomposable G-module L is a G-direct summand of an induced module M^G for some indecomposable K-module M (see [2, §63]). However,

$$M^G = \Omega G \otimes_{\Omega K} M \cong \Omega A \# M \cong \sum\nolimits^{\oplus} (X_i \# M),$$

where the $\{X_i\}$ are irreducible A-modules such that $\Omega A = \sum^{\oplus} X_i$. By the last statement in (2.6), each $X_i \# M$ is indecomposable; hence L must be isomorphic to some $X_i \# M$, and therefore $[L]$ lies in the image of τ. Since every G-module is a direct sum of indecomposables, this proves that the map

$$\tau: a(A, 1) \otimes_Z a(K, H) \to a(G, H) \tag{4.2}$$

is a surjective ring homomorphism in this case.

Consider for a moment the special case where $H = K$. Since K contains a Sylow p-subgroup of G, a G-exact sequence is K-split if and only if it is G-split. Hence the relative Grothendieck ring $a(G, K)$ coincides with the Green ring $a(G, G)$ in this case, and as in (4.2) there is a surjective ring homomorphism

$$\tau_0: a(A, 1) \otimes_Z a(K, K) \to a(G, G). \tag{4.3}$$

Now let X range over the distinct irreducible A-modules, and M over the distinct indecomposable K-modules. Then the products $[X] \otimes [M]$ form a free Z-basis for $a(A, 1) \otimes a(K, K)$, and

$$\tau_0: [X] \otimes [M] \to [X \# M].$$

If $X \# M \cong X' \# M'$, with X, X' irreducible and M, M' indecomposable, then by (2.4) we may conclude that $\operatorname{Hom}_A(X, X') \neq 0$, and hence that $X \cong X'$. Furthermore, $(X \# M)_K$ is a direct sum of $(X : \Omega)$ copies of M, whence also $M \cong M'$. Thus the elements $[X \# M]$ form a free Z-basis for the Green ring $a(G, G)$, which proves that the map τ_0 in (4.3) is a ring isomorphism.

We shall use this fact in proving

(4.4) THEOREM. *Let $G = A \times K$, where $H \subset K$ and where A is a p-free group. Assume that Ω is a splitting field for A. Then the map τ given by* (4.2) *is a ring isomorphism.*

Proof. We have already shown that τ is a ring epimorphism, and it remains to prove that τ is monic. It follows from the definition of relative Grothendieck rings that there are two exact sequences of additive groups:

$$0 \to i(K, H) \to a(K, K) \to a(K, H) \to 0,$$
$$0 \to i(G, H) \to a(G, G) \to a(G, H) \to 0.$$

We apply $a(A, 1) \otimes_Z \cdot$ to the first sequence; this preserves exactness since $a(A, 1)$ is Z-free. Therefore we obtain a commutative diagram of Z-modules, with exact rows:

$$\begin{array}{ccccccccc} 0 \to & a(A, 1) \otimes i(K, H) & \to & a(A, 1) \otimes a(K, K) & \to & a(A, 1) \otimes a(K, H) & \to 0 \\ & \downarrow \tau_1 & & \downarrow \tau_0 & & \downarrow \tau & \\ 0 \longrightarrow & i(G, H) & \longrightarrow & a(G, G) & \longrightarrow & a(G, H) & \longrightarrow 0, \end{array}$$

where $\otimes$ is $\otimes_Z$, and each τ is given by $\#$. The image of τ_1 clearly lies in $i(G, H)$, and we have already shown that τ_0 is an isomorphism. To complete the proof, it thus suffices to show that τ_1 is epic.

Consider any H-split G-exact sequence

$$E: 0 \longrightarrow L \xrightarrow{\lambda} M \xrightarrow{\mu} N \longrightarrow 0$$

in which L is H-simple, and set $\gamma_E = [M] - [L] - [N] \in a(G, G)$. By Proposition 2.2, such γ_E's generate $i(G, H)$, and so we need only show that $\gamma_E \in$ image of τ_1. Since L is an H-simple G-module, it must be G-indecomposable, and so we may write $L = X \# T$ for some irreducible A-module X and some indecomposable K-module T. Let us express M as a direct sum of indecomposable G-modules, say $M = \sum^{\oplus} X_i \# T_i$, X_i irreducible, T_i indecomposable. Collecting terms with common first factor, we may then write $M = \sum^{\oplus} X_i \# U_i$, where now the X_i are distinct irreducible A-modules, and the K-modules U_i may be decomposable. However, there exists a nonzero $\lambda \in \mathrm{Hom}_G(L, M)$, defined in the sequence E, and so by Proposition 2.4 some one of the X_i's must be isomorphic to X. To fix the notation, suppose that $X_1 \cong X$; by (2.4), we then have $\lambda(L) \subset X_1 \# U_1$. Replacing L by a module isomorphic to it, we may assume that X coincides with X_1, so now we have

$$L = X_1 \# T, \qquad M = X_1 \# U_1 \oplus \sum_{i>1}^{\oplus} X_i \# U_i,$$

and $\lambda: X_1 \# T \to X_1 \# U_1$ is a G-monomorphism. However, $\mathrm{Hom}_A(X_1, X_1) \cong \Omega$ since Ω is a splitting field for A, and so by (2.4) we may write $\lambda = 1 \# g$ for some monomorphism $g \in \mathrm{Hom}_K(T, U_1)$. Thus there is a G-exact sequence

$$E_1: 0 \longrightarrow X_1 \# T_1 \xrightarrow{1 \# g} X_1 \# U_1 \longrightarrow X_1 \# \operatorname{coker} g \longrightarrow 0.$$

Now let $\rho \in \mathrm{Hom}_H(M, L)$ be an H-retraction which splits E, so that $\rho\lambda = 1$ on L; let ρ_1 be the restriction of ρ to the summand $X_1 \# U_1$ of M. Since $\lambda(L) \subset X_1 \# U_1$, we see that $\rho_1\lambda = 1$ on L, and thus the sequence E_1 is H-split. But then the sequence E_1 is also $(A \times H)$-split, since H contains a Sylow p-subgroup of $A \times H$. Hence there exists a map

$$\theta \in \mathrm{Hom}_{A \times H}(X_1 \# U_1, X_1 \# T)$$

such that $\theta(1 \# g) = 1$. By (2.4) we may write θ in the form $\theta = 1 \# f$ for some $f \in \mathrm{Hom}_H(U_1, T)$; in that case it follows that $fg = 1$ on T. Therefore the sequence

$$E_2 : 0 \longrightarrow T \xrightarrow{g} U_1 \longrightarrow \operatorname{coker} g \longrightarrow 0$$

is an H-split K-exact sequence. Consequently

$$\gamma_{E_1} = \tau_1\{[X_1] \otimes \gamma_{E_2}\} \in \text{image of } \tau_1.$$

We are trying to prove that γ_E lies in the image of τ_1. However, in $a(G, G)$ we have the equations

$$\begin{aligned} \gamma_E &= [M] - [L] - [M/L] \\ &= \Big\{[X_1 \# U_1] + \sum_{i>1} [X_i \# U_i]\Big\} - [X_1 \# T] - \Big\{[X_1 \# \operatorname{coker} g] + \sum_{i>1} [X_i \# U_i]\Big\} \\ &= \gamma_{E_1}. \end{aligned}$$

This completes the proof of the theorem.

5. **Direct product in general case.** The main result of this article is as follows:

(5.1) THEOREM. *Let G be a direct product $B \times K$ of arbitrary finite groups B and K, and let $H \subset K$. Assume that the field Ω is a splitting field for B and all of its subgroups. Then the map*

$$\tau : a(B, 1) \otimes_Z a(K, H) \to a(G, H)$$

defined by $[X] \otimes [M] \to [X \# M]$, *is a ring isomorphism.*

We have already shown in (4.1) that τ is a ring homomorphism. Furthermore, we have established the above theorem for the special case where the first factor B is p-free (see (4.4)). An easy argument, based on the material in §3, shows that the theorem is also valid when B is a p-group; however, we shall not use the results of §3 in this form, but will rely directly on Theorem 3.2.

The proof of Theorem 5.1 for the general case depends on reducing it to the two special cases mentioned above, the key device being the Brauer Induction Theorem (see [2, §40]). The quickest way of accomplishing this reduction is by means of Lam's theory of Frobenius functors [3]. In order to make this article self-contained, however, we have included a direct proof of the reduction step.

Let us begin the proof of Theorem 5.1 with a simple observation. Let B_1 be a subgroup of B, and set $G_1 = B_1 \times K$; remember that $G = B \times K$, and that $H \subset K$. There are restriction maps

$$r\colon a(B, 1) \to a(B_1, 1), \qquad r'\colon a(G, H) \to a(G_1, H).$$

On the other hand, if $B = \bigcup b_i B_1$ is a coset decomposition of B, then $G = \bigcup b_i G_1$ is a coset decomposition of G. Clearly $\bigcup b_i H = \bigcup H b_i$, so by (2.1) there exists an induction map $i'\colon a(G_1, H) \to a(G, H)$, where $i'[M] = [M^G]$, $M = G_1$-module. Finally, there is an induction map $i\colon a(B_1, 1) \to a(B, 1)$, given by $i\colon [U] \to [U^B]$, $U = B_1$-module. The induction and restriction maps are connected by the usual Frobenius relations (see [5], for example):

$$(5.2) \qquad u \cdot i(v) = i(r(u) \cdot v), \qquad u \in a(B, 1),\ v \in a(B_1, 1),$$

$$(5.3) \qquad x \cdot i'(y) = i'(r'(x) \cdot y), \qquad x \in a(G, H),\ y \in a(G_1, H).$$

Next we note that there is a diagram consisting of two commutative squares:

$$(5.4) \qquad \begin{array}{ccc} a(B, 1) \otimes_Z a(K, H) & \xrightarrow{\tau} & a(G, H) \\ r \otimes 1 \Big\downarrow \Big\uparrow i \otimes 1 & & r' \Big\downarrow \Big\uparrow i' \\ a(B_1, 1) \otimes_Z a(K, H) & \xrightarrow[\tau_1]{} & a(G_1, H) \end{array}$$

that is,

$$\tau_1(r \otimes 1) = r'\tau, \qquad \tau(i \otimes 1) = i'\tau_1.$$

We apply this first to the case where $B = A \times P$, $B_1 = A$, with A a p-free group and P a p-group. There is a commutative diagram

$$(5.5) \qquad \begin{array}{ccc} a(A \times P, 1) \otimes_Z a(K, H) & \xrightarrow{\tau} & a(A \times P \times K, H) \\ r \otimes 1 \Big\downarrow & & r' \Big\downarrow \\ a(A, 1) \otimes_Z a(K, H) & \xrightarrow[\tau_1]{} & a(A \times K, H) \end{array}$$

Now the map $r\colon a(A \times P, 1) \to a(A, 1)$ is an isomorphism, since P acts trivially on every irreducible $(A \times P)$-module. Thus $r \otimes 1$ is also an isomorphism. Theorem 3.2 tells us that r' is an isomorphism, while τ_1 is an isomorphism by Theorem 4.4. This implies that the map τ in (5.5) is an isomorphism, that is,

$$(5.6) \qquad a(A \times P, 1) \otimes_Z a(K, H) \cong a(A \times P \times K, H).$$

Let 1_B be the 1-representation of B. From the Brauer Induction Theorem [2, §40] it follows that there exist elementary subgroups B_α of B, and elements

$x_\alpha \in a(B_\alpha, 1)$, such that

$$[1_B] = \sum_\alpha i_\alpha(x_\alpha) \quad \text{in } a(B, 1), \tag{5.7}$$

where $i_\alpha: a(B_\alpha, 1) \to a(B, 1)$ is the induction map. (Actually, the theorem asserts such a result for representations over a splitting field of characteristic zero, but this easily implies the same result for modular representations.) Each elementary subgroup B_α can be written as a direct product $A_\alpha \times P_\alpha$, with A_α p-free and P_α a p-group. By (5.6) we know that

$$\tau_\alpha: a(B_\alpha, 1) \otimes_Z a(K, H) \to a(B_\alpha \times K, H) \tag{5.8}$$

is a ring isomorphism, where τ_α is given by outer tensor product.

For each α, there is a diagram consisting of two commutative squares:

$$\begin{array}{ccc} a(B, 1) \otimes_Z a(K, H) & \xrightarrow{\ \tau\ } & a(B \times K, H) \\ r_\alpha \otimes 1 \downarrow \uparrow i_\alpha \otimes 1 & & r'_\alpha \downarrow \uparrow i'_\alpha \\ a(B_\alpha, 1) \otimes_Z a(K, H) & \xrightarrow[\ \tau_\alpha\ ]{} & a(B_\alpha \times K, H) \end{array}$$

As remarked previously, the map τ defined in the statement of Theorem 5.1 is a ring homomorphism. Let us show now that τ is monic. Suppose that $\tau(v)=0$, where

$$v = \sum_j u_j \otimes t_j, \qquad u_j \in a(B, 1),\ t_j \in a(K, H).$$

Then for each α, $0 = r'_\alpha \tau(v) = \tau_\alpha(r_\alpha \otimes 1)v$. Since τ_α is monic, this gives $(r_\alpha \otimes 1)v = 0$, that is,

$$\sum_j r_\alpha(u_j) \otimes t_j = 0. \tag{5.9}$$

Using (5.7), (5.2) and (5.9), we obtain

$$\begin{aligned} v = v\{[1_B] \otimes [1_K]\} &= \Big(\sum_j u_j \otimes t_j\Big)\Big(\sum i_\alpha(x_\alpha) \otimes [1_K]\Big) \\ &= \sum_{j,\alpha} (u_j \cdot i_\alpha(x_\alpha)) \otimes t_j = \sum_{j,\alpha} i_\alpha(r_\alpha(u_j) \cdot x_\alpha) \otimes t_j \\ &= \sum_\alpha \Big((i_\alpha \otimes 1) \cdot \Big\{\Big(\sum_j r_\alpha(u_j) \otimes t_j\Big) \cdot (x_\alpha \otimes 1)\Big\}\Big) = 0. \end{aligned}$$

This completes the proof that τ is monic.

In order to show that τ is epic, we note first that by (5.7),

$$\begin{aligned} [1_G] = \tau\{[1_B] \otimes [1_K]\} &= \tau\Big\{\sum_\alpha i_\alpha(x_\alpha) \otimes [1_K]\Big\} \\ &= \tau \sum_\alpha (i_\alpha \otimes 1)(x_\alpha \otimes [1_K]) = \sum_\alpha i'_\alpha \tau_\alpha(y_\alpha), \end{aligned}$$

where $y_\alpha = x_\alpha \otimes [1_K]$. Therefore for $u \in a(B \times K, H)$ we have (by (5.3))

$$u = u \cdot [1_G] = u \cdot \sum_\alpha i'_\alpha \tau_\alpha(y_\alpha)$$
$$= \sum_\alpha i'_\alpha \{r'_\alpha(u) \cdot \tau_\alpha(y_\alpha)\}.$$

But for each α the map τ_α is epic, and so there exists an element

$$z_\alpha \in a(B_\alpha, 1) \otimes a(K, H)$$

such that $r'_\alpha(u) \cdot \tau_\alpha(y_\alpha) = \tau_\alpha(z_\alpha)$. Hence

$$u = \sum_\alpha i'_\alpha \tau_\alpha(z_\alpha) = \tau\left\{\sum_\alpha (i_\alpha \otimes 1) z_\alpha\right\},$$

that is, u lies in the image of τ. This completes the proof of the theorem.

References

1. S. B. Conlon, *Relative components of representations*, J. Algebra **8** (1968), 478–501.

2. C. W. Curtis and I. Reiner, *Representation theory of finite groups and associative algebras*, Pure and Applied Mathamatics, Vol. 11, Interscience, New York, 1962.

3. T. Y. Lam, *Induction theorems for Grothendieck groups and Whitehead groups of finite groups*, Ann. Sci. École Norm. Sup. (4) **1** (1968), 91–148.

4. T. Y. Lam and I. Reiner, *Relative Grothendieck groups*, J. Algebra **11** (1969), 213–242.

5. R. G. Swan, *Induced representations and projective modules*, Ann. of Math. (2) **71** (1960), 552–578.

University of California,
Berkeley, California
University of Illinois,
Urbana, Illinois

Relative Grothendieck Groups

T. Y. LAM

Mathematics Department, University of California, Berkeley, California 94720

AND

I. REINER*

Mathematics Department, University of Illinois, Urbana, Illinois 61801

Communicated by J. A. Green

Received May 15, 1968

SECTION 1. INTRODUCTION

A number of recent articles have dealt with various Grothendieck groups associated with certain categories of modules over group algebras and group rings (see [*1*]-[*5*], [*7*]-[*12*], [*14*]-[*22*], [*25*]-[*27*]). These Grothendieck groups provide a natural framework for the study of modules, and have proven useful in representation theory. Here we shall investigate relative Grothendieck groups arising from categories of modules over a group algebra when we consider the behavior of these modules upon restriction to some fixed subgroup of the original group.

Let Λ be a ring with unity element, C some category of Λ-modules, and S some collection of short exact sequences

$$0 \to L \to M \to N \to 0 \tag{1.1}$$

from C. We define the *Grothendieck group of C relative to S*, denoted for the moment by $a(C, S)$, as follows. Let $a(C)$ be the free abelian group on the symbols $[M]$, where M ranges over a full set of representatives of the isomorphism classes of modules in C. Further, let $a(S)$ be the subgroup of $a(C)$ generated by all expressions $[M]$-$[L]$-$[N]$ arising from sequences (1.1) in S. Then $a(C, S)$ is the factor group $a(C)/a(S)$, and is an abelian additive group. We shall also use the notation $[M]$ for elements of $a(C, S)$, and shall say that the modules $\{M_i\}$ *span* $a(C, S)$ if $a(C, S) = \sum_i Z[M_i]$.

Now let Ω be a field of characteristic p; the theory to be developed here becomes trivial when $p = 0$, so hereafter assume that $p \neq 0$. Let G be a finite group; the term "G-module" shall always mean a finitely generated

* This work was partially supported by a research contract with the National Science Foundation.

left ΩG-module. Let H be a subgroup of G, and let (1.1) be a sequence of G-modules. Call the sequence *H-split* if upon restriction to H, the sequence of modules is a split exact sequence. Form the relative Grothendieck group $a(C, S)$, where C is the category of all G-modules, and S the collection of H-split sequences from C. Changing notation slightly, we shall denote this hereafter by $a(G, H)$.

If there exists an H-split sequence (1.1) from C, then of course $[M] = [L] + [N]$ in $a(G, H)$. If no such sequence exists (with L, N nonzero), call M *H-simple*. Thus, a G-module M is H-simple if no nontrivial G-submodule is an H-direct summand. An obvious argument by induction on $(M : \Omega)$ shows that the H-simple G-modules span $a(G, H)$.

For a pair of G-modules X, Y, define their *product* to be $X \otimes_\Omega Y$, where as usual

$$g(x \otimes y) = gx \otimes gy, \; g \in G.$$

If (1.1) is any H-split sequence of G-modules, then so is

$$0 \to L \otimes X \to M \otimes X \to N \otimes X \to 0.$$

Setting $[M][X] = [M \otimes X]$, this shows that $a(G, H)$ is a commutative ring, whose identity element is $[1_G]$, where 1_G is the trivial G-module Ω.

When $H = 1$, a G-module is H-simple if and only if it is irreducible. The Jordan-Hölder Theorem implies that $a(G, 1)$ is Z-free on $[F_1],\dots, [F_s]$, where the $\{F_i\}$ are a full set of non-isomorphic irreducible G-modules. Thus $a(G, 1)$ is isomorphic to the ring $\sum_{i=1}^{s} Z\varphi_i$, where the $\{\varphi_i\}$ are the irreducible Brauer characters of G relative to Ω (see [*6*], Section 82). Of course in this case $a(G, 1)$ contains no nonzero nilpotent elements.

In the other extreme case, a G-module is G-simple if and only if it is indecomposable. The Krull-Schmidt Theorem implies that $a(G, G)$ is a free Z-module spanned by the indecomposable modules. This ring $a(G, G)$ is called the *representation ring* or *Green ring* of ΩG, since Green ([*8*]) initiated the study of the algebra $\mathbf{C} \otimes_Z a(G, G)$, where $\mathbf{C}$ is the complex field. To simplify notation, we hereafter write $a(G)$ in place of $a(G, G)$; this was Green's original notation. It is still an open question as to whether $a(G)$ can contain nonzero nilpotent elements, in general.

It is reasonable to conjecture that the H-simple G-modules form a free Z-basis for $a(G, H)$. As we shall see, however, this is false in general (see comments at end of Section 3). Indeed, one of our main results is the assertion that in certain cases $a(G, H)$ is Z-free. Even in these cases, moreover, the H-simple G-modules need not form a free Z-basis for $a(G,H)$. We shall also determine the ring structure of $a(G, H)$ in various special cases, in all of which H is assumed to be a normal cyclic p-subgroup of G. The main theorems are Theorems 2.1, 3.4, 4.3, 5.2, 6.1 and 6.7.

Section 2. The Cartan Homomorphism

Let H be any subgroup of G. Recall that a G-module N is (G, H)-*projective* if it has the property that every exact sequence of G-modules

$$0 \to X \to Y \to N \to 0$$

which is H-split is also G-split (see [6], Section 63). For V any H-module, define the induced G-module V^G by

$$V^G = \Omega G \otimes_{\Omega H} V,$$

where G acts on the left on the first factor. Then N is (G, H)-projective if and only if N is a G-direct summand of some induced module V^G. Further, if L is any G-module then $L \otimes N$ is also (G, H)-projective in this case.

We now form the relative Grothendieck group $a(C, S)$, where C is the category of all (G, H)-projective G-modules, and S the collection of all H-split (and hence G-split) sequences from C. Denote this group by $k(G, H)$ hereafter. Then $k(G, G)$ coincides with the Green ring $a(G)$, and $k(G, H)$ is an ideal in $a(G)$. Thus $k(G, H)$ is Z-free on the indecomposable (G, H)-projectives. In particular, $k(G, 1)$ is Z-free on the indecomposable projective G-modules (often referred to as the *principal indecomposable* G-modules).

Let $i(G, H)$ be the additive subgroup of $a(G)$ generated by all expressions $[M]$-$[L]$-$[N]$, where $0 \to L \to M \to N \to 0$ ranges over all H-split exact sequences of G-modules. Obviously $i(G, H)$ is an ideal of $a(G)$, and

$$a(G, H) \cong a(G)/i(G, H).$$

For convenience, we summarize our definitions thus:

$$\begin{aligned} a(G) &= a(G, G) = k(G, G) = \text{Green ring}, \\ a(G, H) &= G\text{-modules}/H\text{-split sequences}, \\ a(G, 1) &\cong \text{character ring}, \\ k(G, H) &= (G, H)\text{-projectives/direct sum}, \\ k(G, 1) &= G\text{-projectives/direct sum}. \end{aligned}$$

Now define the *Cartan homomorphism*

$$\kappa : k(G, H) \to a(G, H)$$

by composition of maps

$$k(G, H) \to a(G) \to a(G)/i(G, H) \cong a(G, H),$$

where the first arrow is inclusion. If $U_1, \ldots, U_s$ are the non-isomorphic principal indecomposable G-modules, and $F_1, \ldots, F_s$ the irreducible G-modules, then

$$\kappa : k(G, 1) \to a(G, 1)$$

is given by

$$\kappa[U_i] = \sum_{j=1}^{s} c_{ij}[F_j], \qquad 1 \leqslant i \leqslant s,$$

where (c_{ij}) is the Cartan matrix associated with ΩG. As shown by Brauer (see [*6*], Section 84, and [*23*]), $\det(c_{ij})$ is a power of p. Thus, when $H = 1$, κ is a monomorphism whose cokernel is a finite p-group.

The following is a slight modification of results of Conlon [*3*], and we include the proof for the convenience of the reader.

(2.1) THEOREM. *Let H be a normal subgroup of G. Then the Cartan homomorphism*

$$\kappa : k(G, H) \to a(G, H)$$

is monic, and the cokernel of κ is a p-torsion abelian group. Furthermore, any torsion element of $a(G, H)$ must be a p-torsion element.

Proof. Let $\bar{G} = G/H$, and view each $\bar{G}$-module as a G-module upon which H acts trivially. Then every exact sequence of $\bar{G}$-modules is an H-split sequence of G-modules, that is,

$$i(\bar{G}, 1) \subset i(G, H).$$

We now show that

$$i(G, H) \cdot k(G, H) = 0 \quad \text{in} \quad a(G).$$

For let

$$0 \to L \to M \to N \to 0 \tag{2.2}$$

be any H-split exact sequence of G-modules. Then for any G-module X, the sequence

$$0 \to L \otimes X \to M \otimes X \to N \otimes X \to 0 \tag{2.3}$$

is also H-split. However, if X is (G, H)-projective, then so is $N \otimes X$, and thus (2.3) is also G-split. Therefore

$$\{[M] - [L] - [N]\} \cdot [X] = 0 \quad \text{in} \quad a(G),$$

which establishes the assertion. It also shows that if X is some fixed (G, H)-projective module, and M is any $\bar{G}$-module, then the element $[X][M]$ of $a(G)$ depends only upon the $\bar{G}$-composition factors of M.

Now let Z' be the ring of quotients of Z relative to the multiplicative set $\{1, p, p^2, \ldots\}$. Then

$$Z' = \{a/b : a, b \in Z, \quad b = \text{power of } p\}. \tag{2.4}$$

Put $a'(G, H) = Z' \otimes_Z a(G, H)$, etc. The preceding discussion shows that

$$i'(G, H) \cdot k'(G, H) = 0 \quad \text{in} \quad a'(G), \quad \text{and} \quad i'(\bar{G}, 1) \subset i'(G, H).$$

We next construct an idempotent generator e for the ideal $k'(G, H)$ of the ring $a'(G)$. Since p is invertible in Z', the Cartan map $k(\bar{G}, 1) \to a(\bar{G}, 1)$ extends to an isomorphism

$$k'(\bar{G}, 1) \cong a'(\bar{G}, 1).$$

Hence there exists an element $e \in k'(\bar{G}, 1)$ mapping onto the identity elemen $[1_{\bar{G}}]$ of $a'(\bar{G}, 1)$. Write

$$e = \sum r_i[X_i], \qquad r_i \in Z', \qquad X_i = \bar{G}\text{-projective}.$$

The existence of a G-isomorphism

$$\Omega\bar{G} \cong (1_H)^G,$$

where 1_H is the trivial H-module Ω, shows that each $X_i \in k(G, H)$, and thus that $e \in k'(G, H)$. The earlier remarks show that for fixed $[M] \in k'(G, H)$, the element $e[M]$ depends only on the image of e in $a'(\bar{G}, 1)$. This proves that e is the identity element of $k'(G, H)$.

On the other hand, we claim that $[1_G] - e \in i'(G, H)$, and for this it suffices to show that

$$[1_{\bar{G}}] - e \in i'(\bar{G}, 1). \tag{2.5}$$

But in fact $a'(\bar{G})/i'(\bar{G}, 1) \cong a'(\bar{G}, 1)$, and since $[1_{\bar{G}}] - e$ has zero image in $a'(\bar{G}, 1)$, (2.5) is established.

(Therefore $k'(G, H)$ has idempotent generator e, and $i'(G, H)$ has idem-aotent generator $1 - e$. Consequently

$$a'(G) = k'(G, H) \oplus i'(G, H),$$

and so

$$k'(G, H) \cong a'(G)/i'(G, H) \cong a'(G, H).$$

ence there is a commutative diagram

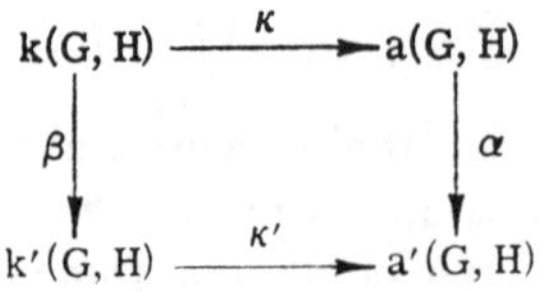

$$\begin{array}{ccc} k(G, H) & \xrightarrow{\kappa} & a(G, H) \\ \beta \downarrow & & \downarrow \alpha \\ k'(G, H) & \xrightarrow{\kappa'} & a'(G, H) \end{array}$$

where the vertical arrows are gotten by tensoring with Z'. Since $k(G, H)$ is Z-free, β is monic and $k'(G, H)$ is Z'-free. But κ' is an isomorphism, and therefore monic, whence κ is also monic.

Furthermore, $Z' \otimes \text{coker } \kappa = 0$, which shows that coker κ is a p-torsion abelian group. Finally, the torsion submodule of $a(G, H)$ lies in ker α, since $a'(G, H)$ is Z'-free. On the other hand, ker α is the p-torsion submodule of $a(G, H)$. This completes the proof of the theorem.

The following result is of interest:

(2.6) LEMMA. *Let $H \subset E \subset G$, where H is normal in G. To each E-module M corresponds an induced G-module M^G. This induction map carries $a(E, H)$ into $a(G, H)$.*

Proof. We must show that for each H-split exact sequence of E-modules

$$0 \longrightarrow L \xrightarrow{\lambda} M \xrightarrow{\mu} N \longrightarrow 0, \tag{2.7}$$

the sequence of G-modules

$$0 \longrightarrow L^G \xrightarrow{\lambda'} M^G \xrightarrow{\mu'} N^G \longrightarrow 0 \tag{2.8}$$

is also H-split. Here, $\lambda' = 1 \otimes \lambda$; the sequence (2.8) is obviously exact, since ΩG is ΩE-free.

Let $\rho : M \to L$ be an H-homomorphism splitting (2.7), that is, satisfying $\rho\lambda = 1$ on L. Define $\rho' : M^G \to L^G$ thus: put $G = \bigcup_{i=1}^{n} g_i E$, so $L^G = \sum_1^n g_i \otimes L$, $M^G = \sum_1^n g_i \otimes M$. Then set $\rho'(g_i \otimes m) = g_i \otimes \rho m$. In that case $\rho'\lambda' = 1$ on L^G, and it suffices to verify that ρ' is an H-homomorphism. Let $x \in H$, $g \in G$, $m \in M$, and set $g^{-1}xg = y \in H$. Then

$$\rho'\{x(g \otimes m)\} = \rho'(g \otimes ym) = g \otimes \rho(ym) = g \otimes y\rho(m),$$

whereas

$$x \cdot \rho'(g \otimes m) = x(g \otimes \rho m) = xg \otimes \rho m = g \otimes y\rho(m).$$

This completes the proof.

We conclude this section with a few simple remarks. Let H be any subgroup of G, and let P be a Sylow p-subgroup of H. By a result of Higman [*13*] (see also [*6*], Section 63), every H-module is (H, P)-projective. This immediately implies that

$$k(G, H) \cong k(G, P).$$

Furthermore, an exact sequence of G-modules is H-split if and only if it is P-split, and therefore

$$a(G, H) \cong a(G, P).$$

These formulas show that in calculating relative Grothendieck groups, it suffices to assume that the subgroup involved is a p-group. It is also clear that $a(G, H)$ depends only on the conjugacy class of H.

SECTION 3. FREENESS OF THE RELATIVE GROUP

For the remainder of the paper, we shall restrict ourselves to the case where H is a normal cyclic p-subgroup of G. Our methods do not seem to extend readily to more general cases. We fix the following notation once and for all:

G = finite group.
Ω = any field of characteristic p.
G-module = finitely generated left ΩG-module.
$H \triangle G$, H generated by element x of order h, $h = p^e$.
$\omega = x - 1 \in \Omega H \subset \Omega G$.
$U_1, \ldots, U_s$ = full set of non-isomorphic principal indecomposable G-modules.
F_j = unique minimal submodule of U_j, $1 \leqslant j \leqslant s$. (Hence $F_1, \ldots, F_s$ are a full set of non-isomorphic irreducible G-modules).

We shall repeatedly use the fact that each $\omega F_j = 0$. This is well-known, following either from Clifford's Theorem ([6], Section 49) or from the obvious fact that ω lies in the radical of ΩG. Now define

$$V_{nj} = \{u \in U_j : \omega^{n-1}u \in F_j\}, \quad 1 \leqslant n \leqslant h, \quad 1 \leqslant j \leqslant s. \tag{3.1}$$

We show at once that each V_{nj} is a G-submodule of U_j. Since $H \triangle G$, the group G acts by conjugation on ΩH and on each of its ideals $\omega^r \Omega H$. Now $\omega^r \Omega H$ has dimension $h - r$ over Ω, and hence the quotients

$$F^{(r)} = \frac{\omega^r \Omega H}{\omega^{r+1} \Omega H}, \quad 0 \leqslant r \leqslant h - 1, \tag{3.2}$$

afford one-dimensional representations $\lambda_r : G \to \Omega$. We may thus write

$$g\omega^r g^{-1} = \lambda_r(g)\, \omega^r + \xi \cdot \omega^{r+1}, \quad \lambda_r(g) \in \Omega, \tag{3.3}$$

for some $\xi = \xi(g, r) \in \Omega H$. For $u \in V_{nj}$ we obtain

$$\begin{aligned} \omega^{n-1} \cdot gu &= g(g^{-1}\omega^{n-1}g)\, u \\ &= g(\lambda_{n-1}(g^{-1})\, \omega^{n-1} + \xi \cdot \omega^n)\, u \in F_j\,. \end{aligned}$$

Therefore also $gu \in V_{nj}$, and so each V_{nj} is a G-submodule of U_j. Clearly $V_{1j} = F_j$.

Obviously $V_{nj} \supset \omega^{h-(n-1)} U_j$ since $\omega^h = 0$. Let us show that

$$V_{nj} = \omega^r V_{n+r,j}, \qquad 1 \leqslant n \leqslant n + r \leqslant h. \tag{3.3a}$$

On the one hand,

$$\omega^{n-1} \cdot \omega^r V_{n+r,j} = \omega^{n+r-1} V_{n+r,j} \subset F_j,$$

so $V_{nj} \supset \omega^r V_{n+r,j}$. On the other hand, let $v \in V_{nj}$; then $\omega^n v = 0$. However, the restriction $(U_j)_H$ is H-free, and so from $\omega^n v = 0$ it follows that $v = \omega^{h-n} z$ for some $z \in U_j$. Then $\omega^{h-1} z = \omega^{n-1} v \in F_j$, and so we have $v = \omega^r u$, where $u = \omega^{h-n-r} z \in V_{n+r,j}$. This proves (3.3a), and so there is a chain of G-modules:

$$U_j \supset V_{hj} \supset \omega U_j \supset \cdots \supset \omega^{h-n} U_j \supset V_{nj} \supset \omega^{h-n+1} U_j \supset \cdots \supset V_{1j} = F_j \supset 0.$$

We may now state one of the main theorems of this paper.

(3.4) THEOREM. *The set* $\{[V_{nj}] : 1 \leqslant n \leqslant h, 1 \leqslant j \leqslant s\}$ *is a Z-free basis for $a(G, H)$. The cokernel of the Cartan homomorphism*

$$\kappa : k(G, H) \to a(G, H)$$

is isomorphic to the direct sum of h copies of the cokernel of the ordinary Cartan homomorphism $k(G/H, 1) \to a(G/H, 1)$.

As we shall see in Section 4, the G-modules V_{nj} need not be H-simple. Indeed, under the assumption that H has a complement in G, it will be possible to find a new Z-basis for $a(G, H)$ consisting of H-simple G-modules.

The proof of Theorem 3.4 is based on the lemma which follows. The authors wish to thank Dr. Conlon for the opportunity of reading a preprint of his article [5]; his calculations contain a special case of one of the basic ideas in the lemma.

3.5 LEMMA. *Let $\phi : a(G) \to a(G, 1)$ be the natural projection of the Green ring $a(G)$ onto the ring $a(G, 1)$ of Brauer characters. Let M be any G-module such that $\omega^r M = 0$, and let*

$$\phi[\omega^{r-1} M] = \sum_{j=1}^{s} a_j [F_j] \qquad \text{in} \qquad a(G, 1).$$

Then the following equality holds in $a(G, H)$:

$$[M] = \sum_j a_j \left\{ [V_{rj}] - \left[\frac{V_{rj}}{F_j} \right] \right\} + \left[\frac{M}{\omega^{r-1}M} \right]. \tag{3.6}$$

Proof. Step 1. Some simple observations about "pushout" diagrams will be needed. Let R, S, T be G-modules, and let $\alpha : R \to S, \beta : R \to T$ be G-homomorphisms. Define

$$U = (S \dotplus T)/\{(\alpha r, -\beta r) : r \in R\},$$

the *pushout* of the pair of maps (α, β). There are obvious maps

$$S \to S \dotplus T \to U, \qquad T \to S \dotplus T \to U,$$

giving rise to a commutative diagram

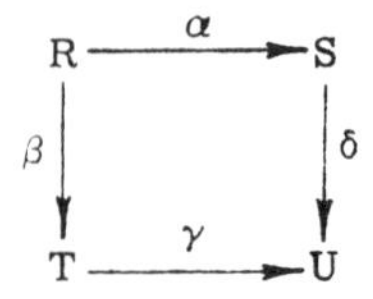

of G-modules and G-homomorphisms. If α and β are monic, then so are γ and δ, and we have

$$U/\gamma T \cong S/\alpha R, \qquad U/\delta S \cong T/\beta R. \tag{3.7}$$

Finally, the sequence

$$0 \longrightarrow T \xrightarrow[\gamma]{} U \longrightarrow U/\gamma T \longrightarrow 0$$

splits if and only if there exists a homomorphism $\rho : S \to T$ such that $\rho\alpha = \beta$. A similar remark holds for the sequence $0 \to S \to U \to U/\delta S \to 0$.

Step 2. We shall use induction on the number t of composition factors of $\omega^{r-1}M$. For $t = 0$ we have $\omega^{r-1}M = 0$, so each $a_j = 0$, and (3.6) is an obvious identity. Now let $t > 0$; choose any minimal submodule of $\omega^{r-1}M$ and identify it with F_k. Putting $N = M/F_k$, we have

$$\omega^r N = 0, \qquad \omega^{r-1}N = \omega^{r-1}M/F_k.$$

Thus N satisfies the same hypotheses as M, and $\omega^{r-1}N$ has $t - 1$ composition factors. Indeed

$$\begin{aligned} \phi[\omega^{r-1}N] &= \phi[\omega^{r-1}M] - \phi[F_k] \\ &= \left(\sum a_j[F_j]\right) - [F_k]. \end{aligned}$$

By the induction hypothesis, (3.6) holds for N, and gives

$$[N] = \sum a_j\{[V_{rj}] - [V_{rj}/F_j]\} - \{[V_{rk}] - [V_{rk}/F_k]\} + [N/\omega^{r-1}N]. \quad (3.8)$$

Step 3. Let us also view F_k as a submodule of V_{rk}, and form the pushout diagram

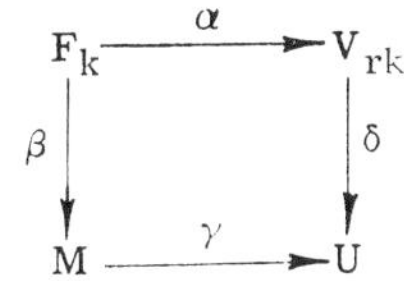

where α and β are inclusion maps. We shall show that the G-exact sequences

$$\begin{gathered} 0 \longrightarrow M \xrightarrow{\gamma} U \longrightarrow U/\gamma M \longrightarrow 0, \\ 0 \longrightarrow V_{rk} \longrightarrow U \longrightarrow U/\delta V_{rk} \longrightarrow 0 \end{gathered} \quad (3.9)$$

are both H-split. To see this, we first pick elements $m_1, \ldots, m_d \in M$ such that

$$\{\omega^{r-1}m_1, \ldots, \omega^{r-1}m_d\}$$

is an Ω-basis for F_k (this is possible because $F_k \subset \omega^{r-1}M$). Set $M_1 = \sum_{i=1}^d \Omega H \cdot m_i$. If we define the ring $\Lambda = \Omega H/\omega^r\Omega H$, then $\omega^r M = 0$ and $\omega^r V_{rk} = 0$, so each of the modules occurring in (3.9) can be considered as a Λ-module, and the sequences are H-split if and only if they are Λ-split.

Now Λ is a self-injective ring (see [*6*], Exercise 58.2 (c)), and hence the free Λ-module M_1 is a Λ-direct summand of M; thus $M = M_1 \oplus M_2$ for some Λ-submodule $M_2 \subset M$. On the other hand, we have

$$F_k = \omega^{r-1}V_{rk} \subset V_{rk},$$

so we can find elements $v_1, \ldots, v_d \in V_{rk}$ such that

$$\omega^{r-1}v_i = \omega^{r-1}m_i, \quad 1 \leqslant i \leqslant d.$$

Set $Y_1 = \sum_{i=1}^d \Omega H \cdot v_i$. Then Y_1 is a Λ-free submodule of V_{rk}, and so there is a decomposition $V_{rk} = Y_1 \oplus Y_2$, where Y_2 is some Λ-submodule of V_{rk}. There clearly exists a Λ-isomorphism $\mu : M_1 \cong Y_1$, given by $\mu(m_i) = v_i$. This μ restricts to the identity map on F_k.

We shall construct Λ-homomorphisms ρ and σ such that $\rho\alpha = \beta$, $\sigma\beta = \alpha$.

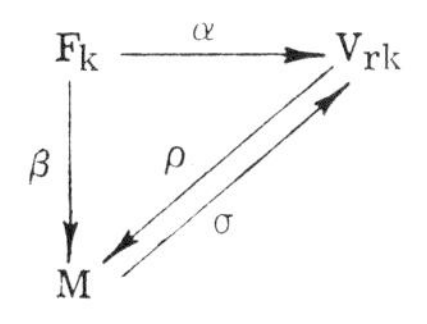

Once this is done, it will follow from the remarks at the end of step 1 that both sequences in (3.9) are Λ-split, and hence also H-split. To define $\rho : V_{rk} \to M$, we simply set

$$\rho \mid_{Y_1} = \mu^{-1}, \qquad \rho \mid_{Y_2} = 0.$$

Since $F_k \subset Y_1$, the equation $\rho\alpha = \beta$ follows from the fact that μ is the identity map on F_k. Similarly, define $\sigma : M \to V_{rk}$ by

$$\sigma \mid_{M_1} = \mu, \qquad \sigma \mid_{M_2} = 0.$$

Step 4. We have thus proved that the sequences (3.9) are H-split, and so we have the following equations in $a(G, H)$:

$$[U] = [M] + [U/\gamma M], \qquad [U] = [V_{rk}] + [U/\delta V_{rk}]. \tag{3.10}$$

From (3.7) we have

$$U/\gamma M \cong V_{rk}/F_k\,, \qquad U/\delta V_{rk} \cong M/F_k = N.$$

Substituting into (3.10) and eliminating $[U]$, we obtain

$$[M] = [N] + [V_{rk}] - [V_{rk}/F_k].$$

Using (3.8), this becomes

$$[M] = \sum a_j\{[V_{rj}] - [V_{rj}/F_j]\} + [N/\omega^{r-1}N].$$

The proof of the lemma is completed by observing that

$$N/\omega^{r-1}N \cong M/\omega^{r-1}M.$$

To begin the proof of Theorem 3.4, we show first that the *hs* modules V_{nj} span $a(G, H)$. Indeed, let S be the subgroup they generate. We shall prove that every $[M] \in S$ by using induction on the integer r such that $\omega^r M = 0$, $\omega^{r-1}M \neq 0$. When $r = 1$, Lemma 3.5 yields

$$[M] = \sum a_j[V_{1j}] \in S$$

since $V_{1j}/F_j = 0$. For $r > 1$, again use (3.6). Since ω^{r-1} annihilates both $M/\omega^{r-1}M$ and V_{rj}/F_j, the induction hypothesis shows that both $[M/\omega^{r-1}M]$ and $[V_{rj}/F_j]$ lie in S. From (3.6) it then follows that also $[M] \in S$, and this completes the proof that the V_{nj} span $a(G, H)$.

In order to prove that the $\{[V_{nj}]\}$ are a Z-free basis of $a(G, H)$, we may proceed in either of two ways. The first approach uses the fact the Cartan homomorphism in (2.1) is monic, and so

$$Z\text{-rank of } a(G, H) \geqslant Z\text{-rank of } k(G, H).$$

As shown in [18] or [24], there are precisely hs indecomposable (G, H)-projective modules, namely the modules

$$\{\omega^n U_j : 0 \leqslant n \leqslant h-1, \qquad 1 \leqslant j \leqslant s\}. \tag{3.11}$$

Since these are a Z-free basis for $k(G, H)$, it follows that $k(G, H)$ has Z-rank hs. Therefore the Z-rank of $a(G, H)$ is at least hs. But $a(G, H)$ is spanned by the hs elements $\{[V_{nj}]\}$, and therefore these elements must be a Z-free basis of $a(G, H)$, as claimed.

We shall give another proof of this fact in the course of some calculations to be used later. Note that the natural epimorphism $\phi : a(G, H) \to a(G, 1)$ has a Z-splitting given by $[F_k] \to [F_k]$. Therefore $a(G, 1)$ may be identified with the subring R of $a(G, H)$ defined by

$$R = \sum_{j=1}^{s} Z[F_j].$$

Then ϕ yields a contraction $x \in a(G, H) \to x^* \in R$, namely

$$[M]^* = \sum a_j[F_j] \in R \quad \text{whenever} \quad \phi[M] = \sum a_j[F_j] \quad \text{in } a(G, 1).$$

Let us number the $\{F_j\}$ so that F_1 is the 1-representation of G; then $[F_1]$ is the identity element of $a(G, H)$. Define

$$A_n = [V_{n1}] - [V_{n1}/F_1] \in a(G, H).$$

Then $A_1 = [F_1]$, and

$$[F_j]\, A_n = [V_{nj}] - [V_{nj}/F_j], \quad 1 \leqslant j \leqslant s.$$

(3.12) LEMMA. *The following identity holds in $a(G, H)$:*

$$[M] = [\omega^{h-1}M]^*\, A_h + \left[\frac{\omega^{h-2}M}{\omega^{h-1}M}\right]^* A_{h-1} + \cdots + \left[\frac{M}{\omega M}\right]^* A_1 .$$

Proof. Let $\omega^r M = 0$, $\omega^{r-1}M \neq 0$, and use induction on r. For $r = 1$, (3.6) yields

$$[M] = \sum a_j[V_{1j}] = \sum a_j[F_j] = [M]^*,$$

since $\phi[M] = \sum a_j[F_j]$. Thus the lemma holds for $r = 1$, since when $\omega M = 0$

the right-hand side of the desired equation becomes $[M]^* A_1$, which equals $[M]^*$.

For $r > 1$, let $[\omega^{r-1}M]^* = \sum a_j[F_j]$. By (3.6),

$$[M] = \left\{\sum a_j[F_j]\right\} A_r + [M'], \qquad M' = M/\omega^{r-1}M.$$

Then $\omega^{r-1}M' = 0$, so by induction

$$[M'] = [\omega^{r-2}M']^* A_{r-1} + \left[\frac{\omega^{r-3}M'}{\omega^{r-2}M'}\right]^* A_{r-2} + \cdots + \left[\frac{M'}{\omega M'}\right] A_1 .$$

The desired formula for $[M]$ now follows from the isomorphisms

$$\omega^{r-2}M' \cong \omega^{r-2}M/\omega^{r-1}M, \ldots, M'/\omega M' \cong M/\omega M.$$

This proves the lemma.

Remark. This lemma shows that the element $[M]$ in $a(G, H)$ depends only on the composition factors of the modules $M/\omega M$, $\omega M/\omega^2 M, \ldots$, This generalizes a result of O'Reilly ([*18*], Lemma 9).

Viewing $a(G, H)$ as a module over the ring $R \cong a(G, 1)$, we have shown that

$$a(G, H) = \sum_{r=1}^{h} R \cdot A_r .$$

Let us prove that $\{A_1, \ldots, A_h\}$ form a free R-basis for $a(G, H)$. This is in fact clear from the already-established result that the $\{[V_{nj}]\}$ are a Z-free basis for $a(G, H)$; however, we are giving here another proof of this result, and so we shall not use it in proving the above assertion about the $\{A_r\}$.

We define

$$\psi_r : a(G, H) \to R$$

by

$$\psi_r[M] = \left[\frac{\omega^{r-1}M}{\omega^r M}\right]^*, \qquad 1 \leqslant r \leqslant h.$$

To prove that ψ_r is well-defined, let $0 \to L \to M \to N \to 0$ be an H-split G-exact sequence. Then $M_H \cong L_H \oplus N_H$, and so there is an H-isomorphism

$$\frac{\omega^{r-1}M}{\omega^r M} \cong \frac{\omega^{r-1}L}{\omega^r L} \oplus \frac{\omega^{r-1}N}{\omega^r N} .$$

Hence the G-sequence

$$0 \to \frac{\omega^{r-1}L}{\omega^r L} \to \frac{\omega^{r-1}M}{\omega^r M} \to \frac{\omega^{r-1}N}{\omega^r N} \to 0$$

is exact (and even H-split), which shows that ψ_r annihilates $[M] - [L] - [N]$, as desired.

From the fact that $\omega F_j = 0$, it follows at once that for each G-module M,

$$\omega(a \otimes b) = a \otimes \omega b, \ a \in F_j\,, \ b \in M.$$

Hence

$$\omega^r(F_j \otimes M) = F_j \otimes \omega^r M.$$

Therefore each ψ_r is R-linear, since

$$\psi_r([F_j][M]) = \left[\frac{\omega^{r-1}(F_j \otimes M)}{\omega^r(F_j \otimes M)}\right]^* = \left[\frac{F_j \otimes \omega^{r-1}M}{F_j \otimes \omega^r M}\right]^*$$
$$= \left[F_j \otimes \frac{\omega^{r-1}M}{\omega^r M}\right]^* = [F_j]\,\psi_r[M].$$

Let us now compute $\psi_r(A_n)$; we have

$$\psi_r(A_n) = \left[\frac{\omega^{r-1}V_{n1}}{\omega^r V_{n1}}\right]^* - \left[\frac{\omega^{r-1}(V_{n1}/F_1)}{\omega^r(V_{n1}/F_1)}\right]^*.$$

For $n < r$, $\omega^{r-1}V_{n1} = 0$, and so $\psi_r(A_n) = 0$. When $n = r$, we have

$$\omega^{n-1}V_{n1} = F_1\,, \qquad \omega^n V_{n1} = 0, \qquad \psi_n(A_n) = [F_1] = 1.$$

Finally, for $n > r$ there is a G-isomorphism

$$\frac{\omega^{r-1}(V_{n1}/F_1)}{\omega^r(V_{n1}/F_1)} \cong \frac{\omega^{r-1}V_{n1}}{\omega^r V_{n1}}$$

and thus again $\psi_r(A_n) = 0$.

Suppose now that

$$\sum_{n=1}^{h} \lambda_n A_n = 0, \qquad \lambda_n \in R.$$

Applying ψ_r to both sides, we conclude that $\lambda_r = 0$. This holds for each r, so the $\{A_n\}$ are a free R-basis for $a(G, H)$. Consequently the set

$$\{[F_j]\,A_n : 1 \leqslant j \leqslant s, \ 1 \leqslant n \leqslant h\}$$

is a Z-free basis for $a(G, H)$. Therefore the generating set $\{[V_{nj}]\}$ is also Z-free, as claimed.

We may rewrite the statement of Lemma 3.12 as follows:

$$[M] = \psi_h[M]\,A_h + \psi_{h-1}[M]\,A_{h-1} + \cdots + \psi_1[M]\,A_1 \tag{3.13}$$

for each G-module M.

We shall now determine the cokernel of the Cartan homomorphism $\kappa : k(G, H) \to a(G, H)$, and we begin by evaluating ψ_r on each of the modules $\omega^n U_j$ listed in (3.11). For each G-module M, and for fixed t, let

$$M' = \{m \in M : \omega^t m \in \omega^{t+1} M\}.$$

Then there is an isomorphism

$$\frac{M}{M'} \cong \frac{\omega^t M}{\omega^{t+1} M} \otimes F^{(t)}, \tag{3.14}$$

where $F^{(t)}$ is defined by (3.2). In particular

$$\frac{\omega^{h-m} U_j}{\omega^{h-m+1} U_j} \cong \omega^{h-1} U_j \otimes F^{(m-1)}, \qquad 1 \leqslant m \leqslant h. \tag{3.15}$$

By definition

$$\psi_r(\omega^{h-n} U_j) = \left[\frac{\omega^{h-n+r-1} U_j}{\omega^{h-n+r} U_j}\right]^*,$$

and this is clearly zero for $r \geqslant n + 1$. On the other hand, for $r \leqslant n$ we use (3.15) to obtain

$$\psi_r(\omega^{h-n} U_j) = [\omega^{h-1} U_j \otimes F^{(n-r)}]^* = [\omega^{h-1} U_j]^* [F^{(n-r)}].$$

Setting $Y_j = [\omega^{h-1} U_j]^*$, $1 \leqslant j \leqslant s$, we may summarize our results thus

$$\psi_r(\omega^{h-n} U_j) = \begin{cases} Y_j[F^{(n-r)}], & n \geqslant r \\ 0\quad, & n < r. \end{cases}$$

Substituting into (3.13), we obtain for $1 \leqslant n \leqslant h$,

$$[\omega^{h-n} U_j] = Y_j([F^{(0)}]\,A_n + [F^{(1)}]\,A_{n-1} + \cdots + [F^{(n-1)}]\,A_1).$$

Let us write $Y_j = [\omega^{h-1} U_j]^* = \sum y_{kj}[F_k]$ in R. Since $[F^{(0)}] = 1$, we have

$$[\omega^{h-n} U_j] = \sum_k y_{kj}[F_k]\,A_n + \cdots,$$

where the suppressed terms involve terms $[F_k]\,A_m$ with $m < n$. Now the

$\{[\omega^{h-n}U_j]\}$ form a Z-basis for $k(G, H)$, while the $\{[F_k]\ A_n\}$ form a Z-basis for $a(G, H)$. Numbering the basis elements suitably, the matrix of the Cartan map κ is given by

$$C = \begin{bmatrix} (y_{kj}) & & & * \\ & (y_{kj}) & & \\ 0 & & \ddots & \\ & & & (y_{kj}) \end{bmatrix}$$

with each of the h diagonal blocks an $s \times s$ matrix (y_{kj}). We shall show that (y_{kj}) is in fact the Cartan matrix of the group algebra $\Omega\bar{G}$, where $\bar{G} = G/H$. This will at once imply that the cokernel of κ is the direct sum of h copies of the cokernel of the Cartan homomorphism $k(\bar{G}, 1) \to a(\bar{G}, 1)$, and will complete the proof of our theorem.

Now each F_j is also an irreducible $\bar{G}$-module, since $\omega F_j = 0$, and so $F_1, \ldots, F_s$ are a full set of irreducible $\bar{G}$-modules. We need only show that $\omega^{h-1}U_1, \ldots, \omega^{h-1}U_s$ are a full set of principal indecomposable $\bar{G}$-modules. Since $\omega \cdot \omega^{h-1}U_j = 0$, each $\omega^{h-1}U_j$ is a $\bar{G}$-module. Further, $\omega^{h-1}U_j$ contains the unique minimal submodule F_j, so the $\{\omega^{h-1}U_j\}$ are indecomposable and no two of them are isomorphic. It therefore suffices to show that each is $\bar{G}$-projective. But $\omega^{h-1}U_j$ is a direct summand of $\omega^{h-1}\Omega G$, since U_j is a direct summand of ΩG. Further, by (3.15),

$$\omega^{h-1}\Omega G \otimes F^{(h-1)} \cong \frac{\Omega G}{\omega \Omega G} \cong \Omega\bar{G},$$

so $\omega^{h-1}\Omega G$ is $\bar{G}$-projective. Therefore each $\omega^{h-1}U_j$ is $\bar{G}$-projective, and Theorem 3.4 is proved.

In Section 2 we observed that

$$a(G, H) \cong a(G)/i(G, H),$$

where $i(G, H)$ is the ideal of $a(G)$ generated by all expressions $[M] - [L] - [N]$, where $0 \to L \to M \to N \to 0$ ranges over all H-split G-exact sequences. As a corollary to the main theorem, we obtain an alternative description of $i(G, H)$.

(3.16) Corollary. *The ideal $i(G, H)$ consists of all elements of $a(G)$ of the form $[M] - [N]$, where M and N are G-modules such that for each r, $0 \leqslant r \leqslant h - 1$, $\omega^r M$ and $\omega^r N$ have the same Brauer characters.*

Proof. By Lemma 3.12, a necessary and sufficient condition for $[M] - [N]$ to belong to $i(G, H)$ is that $\psi_k[M] = \psi_k[N]$ for $1 \leqslant k \leqslant h$. This happens if and only if for each r, $0 \leqslant r \leqslant h - 1$, $\omega^r M$ and $\omega^r N$ have the same Brauer character.

Using now the fact that $i(G, H)$ is an ideal, we record

(3.17) COROLLARY. *Suppose that for each* r, $0 \leqslant r \leqslant h - 1$, $\omega^r M$ *and* $\omega^r N$ *have the same Brauer characters. Then for every* G-*module* T, *and for each* r, *the modules* $\omega^r(T \otimes M)$ *and* $\omega^r(T \otimes N)$ *also have the same Brauer characters.*

To conclude this section, we make some remarks about the case where G is a cyclic p-group, say of order p^f, with generator y. There are p^f indecomposable G-modules, given by

$$M_r = (y - 1)^{p^f - r} \cdot \Omega G, \qquad 1 \leqslant r \leqslant p^f,$$

and $(M_r : \Omega) = r$. In this case $s = 1$, and $U_1 = \Omega G$. We find readily that

$$V_{n1} = M_{(n-1)p^{f-e}+1}, \qquad 1 \leqslant n \leqslant h.$$

Thus for example if $|G| = 32$ and $|H| = 4$, then $a(G, H)$ has free Z-basis $\{[M_1], [M_9], [M_{17}], [M_{25}]\}$.

It happens that these basis elements are H-simple; but there are other H-simple G-modules as well. Indeed, it is easy to verify that the H-simple G-modules are precisely the modules

$$M_1, \quad \{M_t : 1 + p^{f-e} \leqslant t \leqslant p^f\}.$$

SECTION 4. DIRECT AND SEMIDIRECT PRODUCTS

Throughout this section we make the simplifying hypothesis that G is a semidirect product $H \cdot A$, where A is some subgroup of G, and $H \cap A = 1$. This will permit us to obtain a simpler Z-free basis for $a(G, H)$, and to prove the existence of a ring isomorphism

$$a(G, H) \cong a(H) \otimes_Z a(A, 1)$$

in case G is the direct product $H \times A$.

To begin with, we have seen that each irreducible G-module F_j is annihilated by ω, and hence can also be viewed as irreducible A-module by means of the isomorphism $G/H \cong A$. Let $W_1, \ldots, W_s$ be a full set of principal indecomposable A-modules, and F_j a minimal A-submodule of W_j.

Now ΩA is a direct sum of copies of the $\{W_j\}$, and so ΩG is a direct sum of copies of the induced modules $\{W_j^G\}$. Let us show that these $\{W_j^G\}$ are a full set of nonisomorphic principal indecomposable G-modules (and are thus a rearrangement of the $\{U_j\}$ of Section 3). We may write

$$W_j^G = \Omega G \otimes_{\Omega A} W_j = \sum_{n=0}^{h-1} \oplus\, \omega^n \otimes W_j\,. \tag{4.1}$$

The action of G on W_j^G is described completely by the formulas

$$\omega(\omega^n \otimes z) = \omega^{n+1} \otimes z, \quad a(\omega^n \otimes z) = \lambda_n(a)\, \omega^n \otimes az + \xi\omega^{n+1} \otimes z,$$

where (compare (3.3)) for $a \in A$,

$$a\omega^n a^{-1} = \lambda_n(a)\, \omega^n + \xi \cdot \omega^{n+1}, \qquad \xi = \xi(a, n) \in \Omega H.$$

Any minimal submodule of W_j^G must be annihilated by ω, hence lies in $\omega^{h-1} \otimes W_j$. Since $F^{(h-1)}$ affords the representation λ_{h-1}, we have

$$\omega^{h-1} \otimes W_j \cong F^{(h-1)} \otimes W_j$$

as A-modules. But $F^{(h-1)}$ is one-dimensional, and thus W_j^G has the unique minimal submodule $\omega^{h-1} \otimes F_j$. This shows that each W_j^G is indecomposable, and that no two of them are isomorphic. Thus $\{W_j^G : 1 \leqslant j \leqslant s\}$ are a full set of nonisomorphic principal indecomposable G-modules.

Let $j \to j'$ be the permutation on the set $\{1,..., s\}$ defined by the relation

$$F_{j'} \cong F^{(h-1)} \otimes F_j \tag{4.2}$$

as G-modules. Then in the notation of §3, $U_{j'} = W_j^G$, and $a(G, H)$ is Z-free on the $\{[V_{nj'}]\}$. Let us compute $V_{nj'}$ explicitly. By definition,

$$V_{nj'} = \{u \in W_j^G : \omega^{n-1}u \in \omega^{h-1} \otimes F_j\}.$$

It follows immediately that

$$V_{nj'} = \omega^{h-n} \otimes F_j + \omega^{h-n+1} \otimes W_j + \cdots + \omega^{h-1} \otimes W_j\,.$$

We are now ready to introduce a new collection of G-modules. Define

$$M_{nj} = \omega^{h-n}F_j^G, \qquad 1 \leqslant n \leqslant h, \quad 1 \leqslant j \leqslant s,$$

where in forming the induced module F_j^G, we view F_j as A-module only. Then M_{nj} is a G-submodule of $V_{nj'}$, and

$$M_{nj} = \omega^{h-n} \otimes F_j + \omega^{h-n+1} \otimes F_j + \cdots + \omega^{h-1} \otimes F_j\,.$$

Denote by ϕ the natural map from the relative Grothendieck ring onto the character ring. The main result of this section is as follows:

(4.3) THEOREM. *The symbols* $\{[M_{nj}] : 1 \leqslant n \leqslant h, 1 \leqslant j \leqslant s\}$ *form a Z-free basis for* $a(G, H)$. *Each module* M_{nj} *is H-simple. If* $C = (c_{jk})$ *is the Cartan matrix of the group algebra* ΩA, *then the matrix giving the Cartan homomorphism* $\kappa : k(G, H) \to a(G, H)$ *can be written as a diagonal matrix with h copies of C along the main diagonal. Specifically,*

$$[\omega^{h-n} W_j{}^G] = \sum_{k=1}^{s} c_{jk}[M_{nk}],$$

where

$$\phi[W_j] = \sum c_{jk}[F_k] \qquad \text{in} \quad a(A, 1).$$

Proof. Step 1. Let $\Lambda = \Omega H/\omega^n \Omega H$. Then both M_{nj} and $V_{nj'}$ are Λ-modules, and M_{nj} is Λ-free on $(F_j : \Omega)$ generators. As remarked in Section 3, the ring Λ is self-injective. Hence the G-exact sequence

$$0 \to M_{nj} \to V_{nj'} \to V_{nj'}/M_{nj} \to 0$$

is Λ-split, and thus also H-split. We may therefore write

$$[V_{nj'}] = [M_{nj}] + [V_{nj'}/M_{nj}] \text{ in } a(G, H).$$

However, ω^{n-1} annihilates $V_{nj'}/M_{nj}$. It follows from formula (3.6) that the $\{[M_{nj}]\}$ span $a(G, H)$. In fact, if M is any G-module and $\omega^r M = 0$, we may express $[M]$ as a sum of terms $[M_{rj}]$ for various j's, together with terms in which the modules are annihilated by ω^{r-1}.

We have now shown that the hs symbols $\{[M_{nj}]\}$ span $a(G, H)$. But by Theorem 3.4, $a(G, H)$ is Z-free of rank hs. Therefore the $\{[M_{nj}]\}$ are a free Z-basis for $a(G, H)$, as claimed.

[For later use, we make the following observation. If M is a G-module such that $\omega^n M = 0$, $\omega^{n-1} M \neq 0$, then we have

$$[M] = \sum a_j[V_{nj}] + [M_1], \qquad \omega^{n-1} M_1 = 0,$$

where

$$\phi[\omega^{n-1} M] = \sum a_j[F_j] \qquad \text{in} \quad a(G, 1).$$

Now $[V_{nj'}] = [F^{(h-1)}] \cdot [V_{nj}]$ by (4.2), so

$$\begin{aligned} [F^{(h-1)}][M] &= \sum a_j[V_{nj'}] + [M_2], \qquad \omega^{n-1} M_2 = 0 \\ &= \sum a_j[M_{nj}] + [M_3], \qquad \omega^{n-1} M_3 = 0.] \end{aligned}$$

Step 2. Let us show next that each M_{nj} is an H-simple G-module. If Y

is any maximal G-submodule of M_{nj}, then M_{nj}/Y is irreducible, hence annihilated by ω. Therefore $\omega M_{nj} \subset Y$. On the other hand, it is easily verified that

$$M_{nj}/\omega M_{nj} \cong F^{(h-n)} \otimes F_j ,$$

and the right-hand module is irreducible. Thus ωM_{nj} is the unique maximal submodule of M_{nj}. This shows that the proper G-submodules of M_{nj} are $\{\omega^l M_{nj} : l = 1,..., n-1\}$. Define Λ as in Step 1; then Λ is indecomposable, and M_{nj} is Λ-free. If $\omega^l M_{nj}$ were an H-direct summand of M_{nj}, it would follow that also $\omega^l M_{nj}$ is Λ-free. This is false if $l > 0$, and so no G-submodule of M_{nj} is an H-direct summand thereof. This completes the proof that M_{nj} is H-simple.

Step 3. Since the symbols $\{[\omega^{h-n} W_j^G]\}$ span $k(G, H)$, we need only compute their images in $a(G, H)$ in order to describe the Cartan homomorphism κ. Let

$$0 \to L \to M \to N \to 0$$

be any exact sequence of A-modules. Then

$$0 \to L^G \to M^G \to N^G \to 0$$

is an exact sequence of G-modules. Each induced module can be expressed as in (4.1), which shows that the sequence

$$0 \to \omega^{h-n} L^G \to \omega^{h-n} M^G \to \omega^{h-n} N^G \to 0$$

is also exact. It is obviously H-split, and so in $a(G, H)$ we have

$$[\omega^{h-n} M^G] = [\omega^{h-n} L^G] + [\omega^{h-n} N^G].$$

In other words, the element $[\omega^{h-n} M^G]$ in $a(G, H)$ depends only on the A-composition factors of the A-module M. Therefore, if

$$\phi[W_j] = \sum c_{jk}[F_k] \quad \text{in} \quad a(A, 1),$$

then

$$[\omega^{h-n} W_j^G] = \sum c_{jk}[\omega^{h-n} F_j^G],$$

as claimed. The theorem is thus established.

An obvious consequence is

(4.4) Corollary. *The map*

$$\theta : a(G, H) \to a(H) \otimes_Z a(A, 1)$$

defined by

$$\theta[M_{nj}] \to [\omega^{h-n}\Omega H] \otimes [F_j]$$

is an isomorphism of additive groups.

In general, the map θ need not be a ring homomorphism. However, if the elements of H commute with those of A, then each module M_{nj} is the outer tensor product

$$M_{nj} = \omega^{h-n}\Omega H \otimes_{\Omega} F_j$$

of an H-module with an A-module. Therefore we have

(4.5) THEOREM. *If $G = H \times A$, the map θ above is a ring isomorphism.*

Remark. In the special case where A is a p-group, we have $a(A, 1) \simeq Z$. The isomorphism θ in Theorem 4.5 is then precisely the restriction map which carries $[M] \in a(G, H)$ onto $[M_H] \in a(H)$, and we have $a(G, H) \simeq a(H)$. This generalizes Conlon's result [5] dealing with the case where G is a (2, 2)-group and H has order 2.

SECTION 5. BRAUER CHARACTERS AND p-COMPLEMENTS

Reverting to the notation of Section 3, we now make the additional hypothesis that G admits a p-complement, that is, G contains a subgroup B whose index $[G : B]$ is the p-part of $[G : 1]$. We remark that a p-complement automatically exists if any one of the following conditions holds:

i) G has a normal p-Sylow subgroup.

ii) G is solvable.

iii) G has a cyclic p-Sylow subgroup, and $H \neq 1$.

The existence of a p-complement in case i) follows from the Schur-Zassenhaus Theorem, in case ii) from Hall's Theorem, and in case iii) from Burnside's Theorem.

We shall show here that under suitable hypotheses on Brauer characters, the ring $a(G, H)$ is isomorphic to the Green ring $a(HB)$. To begin with, define the map $\rho : a(G) \to a(HB)$ by restriction, that is, $\rho[M] = [M_{HB}]$. This map carries H-split sequences into H-split sequences, thereby inducing a map $\theta : a(G, H) \to a(HB, H)$. However, since H is a p-Sylow subgroup of HB, a sequence of HB-modules is HB-split if and only if it is H-split. Therefore $a(HB, H) = a(HB)$, so that restriction of modules induces a ring homomorphism $\theta : a(G, H) \to a(HB)$. Furthermore, H acts trivially on each irreducible HB-module, so $a(HB, 1) \simeq a(B, 1)$. Hence there is a ring homomorphism $\theta_0 : a(G, 1) \to a(B, 1)$. There is clearly a commutative diagram

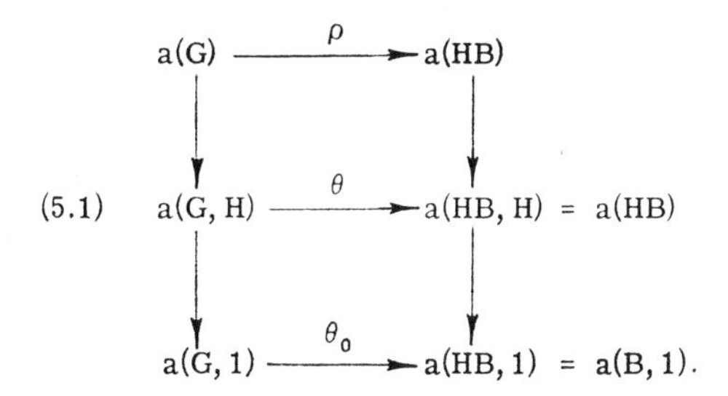

Our first main result in this section is as follows.

(5.2) THEOREM. *The map θ is epic if and only if θ_0 is epic. Further, θ is an isomorphism if and only if θ_0 is an isomorphism.*

Proof. Define the modules $\{V_{nj} : 1 \leqslant n \leqslant h, 1 \leqslant j \leqslant s\}$ as in Section 3. We can use the results of §4 for the semidirect product HB to obtain a Z-free basis for $a(HB, H)$. Let $\{K_1, \ldots, K_t\}$ be a full set of irreducible B-modules, and set

$$M'_{nq} = \omega^{h-n}(K_q)^{HB}, \qquad 1 \leqslant n \leqslant h, \qquad 1 \leqslant q \leqslant t.$$

Then the $\{[M'_{nq}]\}$ are a Z-free basis for $a(HB)$. (We are using M' instead of M to avoid confusion with the notation of Section 4.)

Now we note

$$s = Z\text{-rank of } a(G, 1), \qquad hs = Z\text{-rank of } a(G, H),$$
$$t = Z\text{-rank of } a(B, 1), \qquad ht = Z\text{-rank of } a(HB).$$

These show at once that if the first assertion in the theorem is true, then so is the second assertion.

To establish the first assertion, we remark first that by (5.1), if θ is epic so is θ_0. Suppose conversely that θ_0 is epic. We shall prove by induction on n that each $[M'_{nq}]$ lies in the image of θ. Let us compute $\theta[V_{nj}]$, which is by definition $[(V_{nj})]_{HB}]$ viewed as element of $a(HB)$. In order to express this element in terms of the symbols $\{[M'_{lq}]\}$, we use the remark at the end of Step 1 of the proof of Theorem 4.3. According to the procedure given there, we must first find the composition factors of the HB-module $(\omega^{n-1}V_{nj})_{HB}$, that is, of $(F_j)_{HB}$. If we set

$$\theta_0[F_j] = \sum_{q=1}^{t} a_{jq}[K_q] \quad \text{in} \quad a(B, 1),$$

then by the remark referred to above we have (in $a(HB)$)

$$[F^{(h-1)}][(V_{nj})_{HB}] = \sum_{q=1}^{t} a_{jq}[M'_{nq}] + [X],$$

where $\omega^{n-1}X = 0$. This $[X]$ is expressible as a sum of terms $[M'_{lq}]$ with $l < n$, and thus lies in the image of θ by the induction hypothesis. The left-hand expression is also in the image of θ, since it is $\theta[F^{(h-1)} \otimes V_{nj}]$. Therefore $\sum a_{jq}[M'_{nq}]$ is contained in the image of θ, for each j. But θ_0 is epic, and hence also each $[M'_{nq}]$ is in the image of θ, which completes the proof of the theorem.

(5.3) COROLLARY. *Suppose G has a normal p-Sylow subgroup, and let B be a p-complement. Then $\theta : a(G, H) \to a(HB)$ is an isomorphism. In particular, if G is a p-group then $a(G, H) \cong a(H)$.*

Proof. Since there is a homomorphism $G \to B$, the irreducible G-modules coincide with the irreducible B-modules. Thus θ_0 is an isomorphism, whence so is θ. When G is a p-group, then of course $B = 1$.

In Section 2 we defined $i(G, H)$ to be the additive subgroup of $a(G)$ generated by all expressions $[M] - [L] - [N]$, where

$$0 \to L \to M \to N \to 0 \tag{5.4}$$

is a G-exact sequence. Then $i(G, H)$ is an ideal in $a(G)$, and is the kernel of the homomorphism $a(G) \to a(G, H)$. We now prove a result reminiscent of the long exact cohomology sequence for a pair of spaces.

(5.5) THEOREM. *If θ_0 is an isomorphism, then for any subgroup H' of H, there is a short exact sequence*

$$0 \longrightarrow i(G, H) \xrightarrow{\epsilon} i(G, H') \xrightarrow{\lambda} i(HB, H') \longrightarrow 0.$$

Proof. Since every H-split sequence is also H'-split, the map ϵ is just the inclusion of one ideal of $a(G)$ in another. The map λ is induced by restriction of G-modules to HB.

Let us show first that $\lambda\epsilon = 0$ that is, λ annihilates $[M] - [L] - [N]$ for every H-split G-exact sequence (5.4). The sequence still is H-split when the modules are restricted to HB; but H is a p-Sylow subgroup of HB, and thus the restricted sequence is HB-split. But then

$$\lambda\{[M] - [L] - [N]\} = [M_{HB}] - [L_{HB}] - [N_{HB}] = 0$$

in $a(HB)$, as desired.

We claim next that the kernel of λ is contained in the image of ϵ. Let

$x = [M] - [N] \in i(G, H') \subset a(G)$, where M and N are G-modules, and suppose that $\lambda(x) = 0$ in $a(HB)$. The hypothesis implies that $a(HB) \cong a(G, H)$, which shows that the image of x in $a(G, H)$ is also zero. Hence $x \in i(G, H)$, as desired.

Finally, we show λ is epic. Let $y \in i(HB, H') \subset a(HB)$. Since θ is epic, we may write $y = \rho(z)$ for some $z \in a(G)$. Consider the commutative diagram

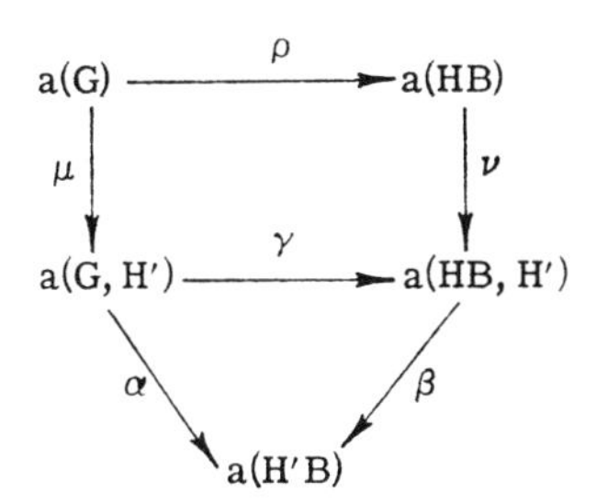

By Theorem 5.2, both maps α, β are isomorphisms. Hence γ is also an isomorphism. Now $\nu(y) = 0$ implies

$$0 = \nu\rho(z) = \gamma\mu(z), \quad \text{so} \quad \mu(z) = 0.$$

Therefore $z \in \ker \mu = i(G, H')$, and $y = \lambda(z)$, so λ is epic. This completes the proof of the theorem.

(5.6) Corollary. *Assume that θ_0 is an isomorphism. Let M, N be G-modules having the same Brauer characters, and such that M, N are HB-isomorphic. Then for any G-module T, and for $0 \leqslant k \leqslant h - 1$, the G-modules*

$$\omega^k(T \otimes M) \quad \text{and} \quad \omega^k(T \otimes N)$$

have the same Brauer characters.

Proof. Take $H' = 1$ in (5.5), and set $x = [M] - [N] \in a(G)$. Since M and N have the same Brauer characters, it follows that the image of x in $a(G, 1)$ is zero. Hence $x \in i(G, 1)$. Further, since M and N are HB-isomorphic, the map $\lambda : i(G, 1) \to i(HB, 1)$ annihilates x. By Theorem 5.5, we conclude that $x \in i(G, H)$. Thus $[M] = [N]$ in $a(G, H)$, and the desired result now follows from (3.16) and (3.17).

(5.7) Corollary. *Assume that θ_0 is an isomorphism, and let M, N be a pair of G-modules which are (G, H)-projective. If M and N are HB-isomorphic, then M and N are G-isomorphic.*

Proof. By Theorems 2.1 and 5.2, the composition of maps

$$k(G, H) \xrightarrow{\kappa} a(G, H) \longrightarrow a(HB)$$

is a monomorphism. The element $[M]-[N]$ of $k(G, H)$ has zero image in $a(HB)$, and thus is zero in $k(G, H)$. This shows that M and N are G-isomorphic, as claimed.

Section 6. The Ring Structure

We continue to use the notation of Section 3, but drop the special hypotheses of Sections 4, 5. Assuming as always that H is a normal cyclic p-subgroup of G, we shall show that the ring $a(G, H)$ has no nonzero nilpotent elements, and shall investigate its internal structure as an algebra over its subring $R \cong a(G, 1)$.

Let $\mathbf{C}$ be the complex field, and set

$$A(G, H) = \mathbf{C} \otimes_Z a(G, H), \qquad K(G, H) = \mathbf{C} \otimes_Z k(G, H).$$

Since $a(G, H)$ is Z-free, tensoring with $\mathbf{C}$ embeds it in the commutative finite dimensional $\mathbf{C}$-algebra $A(G, H)$. To show that $a(G, H)$ contains no nonzero nilpotent elements, we establish the equivalent assertion that $A(G, H)$ is semisimple.

By Theorem 2.1, the Cartan homomorphism κ induces an isomorphism of $\mathbf{C}$-algebras: $K(G, H) \cong A(G, H)$. But we already know from Conlon's extension of results of Green and O'Reilly that $K(G, H)$ is semisimple. Indeed, let $K'(G, H)$ be the ideal of $K(G, H)$ spanned by the (G, H')-projective G-modules, where H' ranges over all proper subgroups of H. Set

$$W(G, H) = K(G, H)/K'(G, H).$$

Conlon ([*3*], [*4*]) showed that there is a ring isomorphism

$$K(G, H) \cong \sum_{H'}^{\oplus} W(G, H'),$$

the sum extending over all subgroups H' of H. (Conlon's result reduces to the above because H is cyclic and normal). Each H' is also cyclic and normal, so by O'Reilly's theorem [*18*] it follows that $W(G, H')$ is semisimple. Hence we have proved

(6.1) Theorem. *The ring $a(G, H)$ contains no nonzero nilpotent elements.*

As in Section 3, define the subring R of $a(G, H)$ by

$$R = \sum_{j=1}^{s} Z[F_j] \simeq a(G, 1).$$

For convenience we set $v_n = [V_{n1}]$, where F_1 is the 1-representation of G. Since $[V_{nj}] = [F_j]\, v_n$, we have

$$a(G, H) = \sum_{n=1}^{h}{}^{\oplus}\ Rv_n\,.$$

For $u \in a(G, H)$, write $u = O(n)$ to indicate that u is expressible as

$$u = \alpha_1 v_1 + \alpha_2 v_2 + \cdots + \alpha_n v_n\,, \quad \alpha_i \in R.$$

We are going to show that the ring $a(G, H)$ may be obtained by successively adjoining to R the elements

$$\{v_{p^k+1}\,,\ k = 0, 1,\ldots, e-1\}.$$

Further, each of these e elements will satisfy a p-th degree equation over the preceding extension of R. We begin with a basic lemma.

(6.2) Lemma. *Let $0 \leqslant k \leqslant e-1$, $1 \leqslant r < p^{e-k}$, $n = rp^k + 1$. Then*

i) *For $m = p^k + 1$,*

$$v_m v_n = \begin{cases} v_{m+n-1} + O(m+n-2), & p \nmid (r+1) \\ O(m+n-2), & p \mid (r+1). \end{cases}$$

ii) *For $1 \leqslant m \leqslant p^k$, $v_m v_n = v_{m+n-1} + O(m+n-2)$.*

Proof. Let $a \in V_m$, $b \in V_n$. Then $\omega^m a = \omega^n b = 0$, and x acts trivially on $\omega^{n-1}b$. We may write

$$\omega(a \otimes b) = \omega a \otimes xb + a \otimes \omega b = (D_1 + D_2)(a \otimes b),$$

where

$$D_1(a \otimes b) = \omega a \otimes xb, \qquad D_2(a \otimes b) = a \otimes \omega b.$$

Since D_1 and D_2 commute, this gives

$$\omega^{m+n-2}(a \otimes b) = (D_1 + D_2)^{m+n-2}(a \otimes b).$$

When $m = p^k + 1$ we have $m + n - 2 = (r + 1) p^k$, and so

$$(D_1 + D_2)^{(r+1)p^k} (a \otimes b) = (D_1^{p^k} + D_2^{p^k})^{r+1} (a \otimes b)$$
$$= (r + 1) D_1^{m-1} D_2^{n-1}(a \otimes b).$$

Therefore

$$\omega^{m+n-2}(V_m \otimes V_n) = (r + 1) F_1 \otimes F_1 \cong (r + 1) F_1 ,$$

so part (i) follows from Lemma 3.5.
On the other hand, when $m \leqslant p^k$ we have $D_1^{p^k} a = 0$, and so

$$(D_1 + D_2)^{m+n-2} (a \otimes b) = (D_1 + D_2)^{m-1} (D_1^{p^k} + D_2^{p^k})^r (a \otimes b)$$
$$= (D_1 + D_2)^{m-1} D_2^{n-1}(a \otimes b)$$
$$= D_1^{m-1} D_2^{n-1}(a \otimes b).$$

Thus $\omega^{m+n-2}(V_m \otimes V_n) \cong F_1$, and we may again use (3.5). This proves the lemma.

Taking $k = 0$ in case i), we obtain

$$v_2 v_{r+1} = \begin{cases} v_{r+2} + O(r + 1), & p \nmid (r + 1) \\ O(r + 1), & p \mid (r + 1). \end{cases}$$

This shows that

$$v_2^{\,2} = v_3 + O(2), \qquad v_2^{\,3} = v_4 + O(3),\ldots,$$
$$v_2^{p-1} = v_p + O(p - 1), \qquad v_2^{\,p} = O(p).$$

Therefore v_2 satisfies a monic p-th degree equation over R, and no lower degree equation. Furthermore, we may conclude that

$$\sum_{j=1}^{n} Rv_j = R + Rv_2 + \cdots + Rv_2^{n-1}, \qquad 1 \leqslant n \leqslant p.$$

We shall use this to prove by induction on k and n that

$$\sum_{j=1}^{np^k} Rv_j = S_k + S_k v_{p^k+1} + \cdots + S_k(v_{p^k+1})^{n-1}, \qquad 1 \leqslant n \leqslant p, \quad (6.3)$$

where

$$S_k = R[v_2 , v_{p+1} , v_{p^2+1} ,\ldots, v_{p^{k-1}+1}], \qquad 1 \leqslant k \leqslant e. \qquad (6.4)$$

We have just established (6.3) for the case $k = 0$. Now let $k > 0$, and assume it true at $k - 1$ for $1 \leqslant n \leqslant p$. Since (6.3) holds at $(k - 1, p)$, it also holds at $(k, 1)$, since the two assertions are identical. It remains to show that if (6.3) holds, then also

$$\sum_{j=1}^{(n+1)p^k} Rv_j = S_k + S_k v_{p^k+1} + \cdots + S_k(v_{p^k+1})^n. \tag{6.5}$$

By Lemma 6.2 we have

$$(v_{p^k+1})^2 = v_{2p^k+1} + O(2p^k), \ldots, (v_{p^k+1})^n = v_{np^k+1} + O(np^k). \tag{6.6}$$

Hence

$$S_k(v_{p^k+1})^n = O(p^k) \cdot O(np^k + 1) = O((n + 1)\, p^k)$$

by (6.2ii). Thus the left-hand side of (6.5) contains the right-hand side. To prove the reverse inclusion, note that the last equation in (6.6) implies that v_{np^k+1} is contained in the right-hand side of (6.5). Further, by (6.2ii),

$$v_m \cdot v_{np^k+1} = v_{np^k+m} + O(np^k + m - 1), \qquad 1 \leqslant m \leqslant p^k,$$

and thus also each v_{np^k+m}, $1 \leqslant m \leqslant p^k$, lies in the righthand side of (6.5). This establishes the formula (6.5), and completes the proof of the identity (6.3). An immediate consequence is

(6.7) Theorem. *The ring $a(G, H)$ coincides with S_e defined in* (6.4). *Therefore $a(G, H)$ may be obtained from R by successively adjoining to R the elements*

$$v_2, v_{p+1}, v_{p^2+1}, \ldots, v_{p^{e-1}+1}.$$

Each of these elements satsfies a monic p-th degree equation over the preceding extension of R, and no lower degree equation.

It is possible to find the explicit pth degree equation which v_{p^k+1} satisfies over S_k, and from this to deduce anew that $a(G, H)$ does not contain any nonzero nilpotent elements. This calculation becomes somewhat simpler if one works instead with the elements $\{A_j\}$ introduced in Section 3, since it is now clear that

$$a(G, H) = R[A_2, A_{p+1}, \ldots, A_{p^{e-1}+1}].$$

Even so, the calculation is rather complicated, and is closely related to that given by O'Reilly in [*18*]. We merely record the result which gives the successive powers of A_2: let $\varphi \in R$ be the specific element $[F^{(1)}]$, where $F^{(k)}$ is defined in Section 3. Then

$$A_2 A_n = A_{n+1} - \varphi A_n + \varphi A_{n-1}, \quad 2 \leqslant n \leqslant p-1,$$
$$A_2 A_p = (1-\varphi) A_p + \varphi A_{p-1}.$$

In conclusion we make the following remarks. The methods used here depend strongly on the assumption that H is a normal cyclic p-subgroup of G. We have carried out some preliminary investigations in the case where H need not be normal, but is still assumed to be a cyclic p-group. These investigations suggest that $a(G, H)$ is always Z-free, but so far we have not been able to prove this in general.

There are also some scattered results for noncyclic H, but these are as yet too fragmentary to suggest a general theory. In all cases, a crucial role is played by the "pushout lemma" (3.5).

References

1. Conlon, S. B. Certain representation algebras. *J. Austral. Math. Soc.* **5** (1965), 83–99.
2. Conlon, S. B. The modular representation algebra of groups with Sylow 2-subgroup $Z_2 \times Z_2$. *J. Austral. Math. Soc.* **6** (1966), 76–88.
3. Conlon, S. B., Structure in representation algebras. *J. Algebra* **5** (1967), 274–279.
4. Conlon, S. B., Relative components of representations. *J. Algebra* **8** (1968), 478–501.
5. Conlon, S. B. $g^0(G, H)$ when Sylow 2-subgroup of G is V_4. *J. Austral. Math. Soc.* (to appear).
6. Curtis, C. W. and Reiner, I. "Representation Theory of Finite Groups and Associative Algebras." Wiley (Interscience), New York (1962).
7. Feit, W. Groups with a cyclic Sylow subgroup. *Nagoya Math. J.* **27** (1966), 571–584.
8. Green, J. A. The modular representation algebra of a finite group. *Ill. J. Math.* **6** (1962), 607–619.
9. Green, J. A. A transfer theory for modular representations. *J. Algebra* **1** (1964), 73–84.
10. Hannula, T. A., Ralley, T. G. and Reiner, I. Modular representation algebras. *Bull. Am. Math. Soc.* **73** (1967), 100–101.
11. Heller, A. and Reiner, I. Grothendieck groups of orders in semisimple algebras. *Trans. Am. Math. Soc.* **112** (1964), 344–355.
12. Heller, A. and Reiner, I. Grothendieck groups of integral group rings. *Ill. J. Math.* **9** (1965), 349–360.
13. Higman, D. G. Indecomposable representations at characteristic p. *Duke Math. J.* **21** (1954), 369–376.
14. Lam, T. Y. Induction theorems for Grothendieck groups and Whitehead groups of finite groups. *Ann. École Normale Sup.* 4e serie, t. **1** (1968), 91–148.
15. Lam, T. Y. A theorem on Green's modular representation ring. *J. Algebra* (to appear).

16. Lam, T. Y. Artin exponent of finite groups. *J. Algebra* (to appear).
17. O'Reilly, M. F. On the modular representation algebra of metacyclic groups. *J. London Math. Soc.* **39** (1964), 267–276.
18. O'Reilly, M. F. On the semisimplicity of the modular representation algebra of a finite group. *Ill. J. Math.* **9** (1965), 261–276.
19. Reiner, I. The integral representation ring of a finite group. *Michigan Math. J.* **12** (1965), 11–22.
20. Reiner, I. Integral representation algebras. *Trans. Am. Math. Soc.* **124** (1966), 111–121.
21. Reiner, I. Relations between integral and modular representations. *Michigan Math. J.* **13** (1966), 357–372.
22. Reiner, I. Representation rings. *Michigan Math. J.* **14** (1967), 385–391.
23. Serre, J.-P. Introduction à la théorie de Brauer. Séminaire I.H.E.S. (1965–1966), 22 March 1966.
24. Srinivasan, B. On the indecomposable representations of a certain class of groups. *Proc. London Math. Soc.* (3) **10** (1960), 497–513.
25. Stancl, D. L. Multiplication in Grothendieck rings of integral group rings. *J. Algebra* **7** (1967), 77–90.
26. Swan, R. G. Induced representations and projective modules. *Ann. Math.* (2) **71** (1960), 552–578.
27. Swan, R. G. The Grothendieck ring of a finite group. *Topology* **2** (1963), 85–110.

An Excision Theorem for Grothendieck Rings *

TSIT-YUEN LAM and IRVING REINER

§ 1. Introduction

In this paper we shall investigate excision properties of relative Grothendieck rings associated with finite groups. The setting is that of modular representation theory: we start with a finite group G, and a field Ω of nonzero characteristic p, and we consider ΩG-modules of finite dimension over Ω (called "G-modules" for short). Now let H be some subgroup of G, and define the *relative Grothendieck ring* $a(G, H)$ as the quotient A/A', where A is the free abelian group generated by the symbols $[M]$ where M ranges over representatives of all isomorphism classes of G-modules, and A' is the subgroup of A generated by all expressions $[M]-[M']-[M'']$ arising from H-split exact sequences of G-modules

$$0 \to M' \to M \to M'' \to 0.$$

The ring structure of $a(G, H)$ is obtained by setting $[M][N]=[M \otimes_\Omega N]$, with G acting diagonally on $M \otimes_\Omega N$. For the case $H=G$, note that $a(G, G)$ is just Green's representation ring; on the other hand when $H=1$, $a(G, 1)$ is the ring of generalized Brauer characters of G. The structure of $a(G, H)$ has been studied in [3-7].

In any relative theory dealing with functors on pairs of objects, it is reasonable to expect certain Excision Theorems to hold. This point was first made emphatically by Eilenberg-Steenrod [2] in their axiomatization of algebraic topology, in which the Excision Axiom played a crucial role. Not surprisingly, such a theorem also holds for homotopy theory and topological K-theory for pairs of spaces (see the books by Hu, Atiyah, respectively). Various algebraic versions of the Excision Theorem have also been obtained: for the cohomology theory of pairs of groups, see [9], Theorem 2.2; for algebraic K-theory for rings relative to two-sided ideals, see [1], Chapter 7, § 6 for K_1, and [8], Lemma 6.3 for K_2.

From now on we restrict our attention to relative Grothendieck rings. If K is any normal p-subgroup of G, then by Clifford's Theorem K acts trivially on every irreducible G-module. Consequently there is an isomorphism $a(G, 1) \cong a(G/K, 1)$ between the rings of Brauer characters of G and of G/K. This special instance of an Excision Theorem was considerably generalized in [4], Theorem 3.2. The aim of the present paper is to obtain an even stronger generalization, given in Theorem 3.1 below. Section 4 will be devoted to some applica-

* This research was partially supported by the National Science Foundation.

tions of the excision theorem, the first being a new product isomorphism theorem which refines our earlier result in [6], Theorem 5.1. The second application shows how an important conjecture about restriction maps can always be reduced to the split case.

For a G-module M, its restriction to H is denoted by M_H. On the other hand, if L is any H-module, the induced G-module $\Omega G \otimes_{\Omega H} L$ will be written as L^G. We use the notation $\dot{\bigcup}$ to indicate a disjoint union. Finally, $N_G(H)$ denotes the normalizer of H in G, and $C_G(H)$ the centralizer of H in G.

The present article can be read independently of our earlier work, except for the following two results which are needed.

(1.1) **Pushout Lemma.** *Let $H \subset G$, and let F, M, N be G-modules. Suppose that there exists a commutative diagram*

$$\begin{array}{ccc} F & \xrightarrow{\alpha} & M \\ {\scriptstyle\beta}\downarrow & \rho \swarrow\!\!\nearrow \sigma & \\ N & & \end{array} \qquad \rho\alpha = \beta, \quad \sigma\beta = \alpha,$$

where α, β are G-monomorphisms, and ρ, σ are H-homomorphisms. Then the following relation is valid in $a(G, H)$:

$$[M] - [M/\alpha F] = [N] - [N/\beta F].$$

Remark. For the proof, see [3], Lemma 3.5 or [4], Lemma 3.1. This result is a generalization of Schanuel's Lemma, and provides a basic method for obtaining relations in $a(G, H)$.

(1.2) **Theorem.** *Let $C \times E$ be a direct product of groups, and let $H \subset E$. Suppose that Ω is a splitting field for C and all of its subgroups. Then there is a ring isomorphism $a(C \times E, H) \cong a(C, 1) \otimes_Z a(E, H)$.*

Proof. See [4], Theorem 5.1; a generalization is given in [6], Theorem 5.1.

§ 2. Basic Lemmas

Throughout this section let T be a subgroup of the finite group S, and let M be a T-module. By a *quasi-S-structure* on M we mean a family $\{\psi_x \colon x \in S\}$ of Ω-automorphisms of M such that $\psi_1 = \text{identity}$, and

$$(2.1) \qquad t\psi_y = \psi_z t' \quad \text{whenever} \quad ty = zt', \qquad y, z \in S, \; t, t' \in T.$$

Taking $y = t' = 1$, this shows that $\psi_t = t$, $t \in T$. The concept of quasi-S-structure was used in [6] for the case where $T \triangle S$.

(2.2) **Lemma.** *Let $S = \dot{\bigcup}_{i=1}^{r} x_i T$, $x_1 = 1$. Let there be given a family $\{\psi_{x_i} \colon 1 \leqq i \leqq r\}$ of Ω-automorphisms of the T-module M, such that $\psi_1 = \text{identity}$, and*

$$(2.3) \qquad t\psi_{x_i} = \psi_{x_j} t' \quad \text{whenever} \quad tx_i = x_j t', \qquad t, t' \in T.$$

For arbitrary $x = x_i t \in S$, $t \in T$, set $\psi_x = \psi_{x_i} t$. Then $\{\psi_x \colon x \in S\}$ is a quasi-S-structure on M.

Proof. To verify (2.1), let $t\,y=z\,t'$, $t, t'\in T$. If we write $y=x_i\,t_1$, $z=x_j\,t_2$, with $t_1, t_2\in T$, then

$$\psi_y=\psi_{x_i}\,t_1, \quad \psi_z=\psi_{x_j}\,t_2, \quad t\,x_i=x_j\,t_2\,t'\,t_1^{-1},$$

so

$$t\,\psi_y=t\,\psi_{x_i}\,t_1=\psi_{x_j}\cdot t_2\,t'\,t_1^{-1}\cdot t_1=\psi_z\,t',$$

as desired, which proves the lemma.

Now let $\{\psi_x\}$ be a quasi-S-structure on the T-module M, and let M_0 be a T-submodule of M. We shall say that M_0 is *suitable* relative to $\{\psi_x\}$ if for all $x, y\in S$,

$$\psi_x(M_0)=M_0, \quad \psi_{xy}=\psi_x\,\psi_y \text{ on } M_0.$$

Such a suitable T-submodule M_0 can always be made into an S-module by defining $x\cdot m_0=\psi_x(m_0)$, $x\in S$, $m_0\in M_0$. Note that for $t\in T$, $\psi_t(m_0)=t\,m_0$, where $t\,m_0$ is calculated by using the fact that M_0 is a T-submodule of M.

In the above situation, then, we are given a T-module M and a T-submodule M_0 which can be made into an S-module in such a way as to preserve the original action of T. Whenever this occurs, we shall refer to M_0 as an *(S, T)-submodule* of the T-module M.

Suppose now that M_0 is any (S, T)-submodule of M. The following construction was introduced in [6], §3: let $M^S=\Omega S\otimes_{\Omega T}M$, and define $\lambda(M_0)$ to be the Ω-subspace of M^S spanned by the set of elements

$$\{x\otimes x^{-1}m_0-1\otimes m_0\colon x\in S,\ m_0\in M_0\}.$$

Writing $S=\dot{\bigcup}_{i=1}^{r} x_i\,T$, $x_1=1$, we have for $t\in T$

$$x\otimes x^{-1}m_0=(x_i\,t)\otimes(x_i\,t)^{-1}m_0=x_i\otimes x_i^{-1}m_0,$$

so $\lambda(M_0)$ is also spanned by the set

$$\{x_i\otimes x_i^{-1}m_0-1\otimes m_0\colon 2\leqq i\leqq r,\ m_0\in M_0\}.$$

Furthermore, $\lambda(M_0)$ is an S-submodule of M^S, since for $y\in S$,

$$\begin{aligned} y\{x\otimes x^{-1}m_0-1\otimes m_0\}&=\{y\,x\otimes(y\,x)^{-1}\,y\,m_0-1\otimes y\,m_0\}\\ &\quad-\{y\otimes y^{-1}(y\,m_0)-1\otimes y\,m_0\}\in\lambda(M_0). \end{aligned}$$

Let us put

$$(2.4)\qquad [M_0, M]^{S,T}=\left[\frac{M^S}{\lambda(M_0)}\right]-\left[\left(\frac{M}{M_0}\right)^{S}\right]\in a(S, T).$$

When there is no danger of confusion, we shall omit the superscripts S, T, and write just $[M_0, M]$. The basic properties of this symbol are given in the next three results.

(2.5) **Comparison Theorem.** *Let $M_0\subset M$ be T-modules, and suppose that M_0 is suitable relative to the quasi-S-structure $\{\psi_x\}$ on M, so that M_0 may be viewed* *as (S, T)-submodule of M. Let X be any S-submodule of* M_0, *and set* $\bar{M}=M/X$, $\bar{M}_0=M_0/X$. *Then* $\bar{M}_0$ *is a suitable submodule of* $\bar{M}$ *relative to the induced quasi-S-structure* $\{\bar{\psi}_x\}$, *and*

$$[M_0, M]=[X, M]+[\bar{M}_0, \bar{M}] \quad \text{in} \quad a(S, T).$$

Proof. Clearly X is itself suitable in M relative to $\{\psi_x\}$, whence also $\bar{M}_0$ is suitable in $\bar{M}$. Now $1\otimes X+\lambda(X)=X^S\subset M^S$, so there is an S-exact sequence

$$0\to X \xrightarrow{\varepsilon} \frac{M^S}{\lambda(X)} \to \bar{M}^S \to 0,$$

where $\varepsilon(u)=1\otimes u+\lambda(X)$, $u\in X$. (To verify that ε is an S-homomorphism, observe that for $s\in S$, $u\in X$,

$$\varepsilon(s\,u)=1\otimes s\,u\equiv s\otimes u=s(1\otimes u) \mod \lambda(X).)$$

In the same way, there is another S-exact sequence

$$0\to X \xrightarrow{\varepsilon'} \frac{M^S}{\lambda(M_0)} \to \frac{\bar{M}^S}{\lambda(\bar{M}_0)} \to 0,$$

where $\varepsilon'(u)=1\otimes u+\lambda(M_0)$, $u\in X$.

Now consider the diagram

$$\begin{array}{ccc} X & \xrightarrow{\varepsilon'} & \dfrac{M^S}{\lambda(M_0)} \\ {\scriptstyle\varepsilon}\downarrow & \sigma \nearrow \swarrow \rho & \\ \dfrac{M^S}{\lambda(X)} & & \end{array}$$

where σ is the canonical S-epimorphism. To define ρ, note that $M^S=\dot{\sum} x_i\otimes M$, where $S=\dot{\bigcup} x_i T$; now set

$$\rho(x_i\otimes m)=1\otimes\psi_{x_i} m, \quad 1\leqq i\leqq r, \quad m\in M.$$

Let us check that ρ is a T-homomorphism: if $t\in T$, set $t x_i=x_j t'$ with $t'\in T$. Then

$$\begin{aligned}\rho\{t(x_i\otimes m)\}&=\rho\{x_j\otimes t' m\}=1\otimes\psi_{x_j}(t' m)\\ &=1\otimes t\,\psi_{x_i}(m)=t\cdot\rho(x_i\otimes m),\end{aligned}$$

as desired.

Clearly $\sigma\varepsilon=\varepsilon'$, $\rho\varepsilon'=\varepsilon$, so by (1.1) we have in $a(S, T)$

$$\begin{aligned}\left[\frac{M^S}{\lambda(X)}\right]-[\bar{M}^S]&=\left[\frac{M^S}{\lambda(M_0)}\right]-\left[\frac{\bar{M}^S}{\lambda(\bar{M}_0)}\right]\\ &=[M_0, M]-[\bar{M}_0, \bar{M}],\end{aligned}$$

since $M/M_0\cong\bar{M}/\bar{M}_0$. The left-hand expression is $[X, M]$, and the theorem is established.

(2.6) **Corollary** (see [6], Theorem 3.1). *Let $X \subset M$ be a pair of S-modules. Then*

$$[M]=[X, M_T]+[M/X] \quad \text{in} \quad a(S, T).$$

Proof. Since M is an S-module, we may define ψ_x on M by $\psi_x(m)=xm$, $x \in S$, $m \in M$. Then X is suitable relative to this quasi-S-structure on M_T. Take $M_0=M$ in (2.5), and use the obvious facts that $[M, M_T]=[M]$, $[\bar{M}, \bar{M}_T]=[\bar{M}]$ in $a(S, T)$. Then (2.5) yields the desired result.

(2.7) **Theorem.** *Let X be any (S, T)-submodule of the T-module M, and let M' be any T-direct summand of M containing X. Then $[X, M]=[X, M']$ in $a(S, T)$.*

Proof. Write $M=M' \oplus N$, $N=T$-module. There are S-direct sum decompositions:

$$\frac{M^S}{\lambda(X)} \cong \frac{(M')^S}{\lambda(X)} \oplus N^S, \qquad \left(\frac{M}{X}\right)^S \cong \left(\frac{M'}{X}\right)^S \oplus N^S.$$

Eliminating N^S, we obtain

$$\left[\frac{M^S}{\lambda(X)}\right]-\left[\left(\frac{M}{X}\right)^S\right]=\left[\frac{(M')^S}{\lambda(X)}\right]-\left[\left(\frac{M'}{X}\right)^S\right]$$

in $a(S, T)$, that is, $[X, M]=[X, M']$.

(The proof actually shows that the equation holds in the Green ring $a(S)$.)

§3. Excision Theorem

Throughout this section we assume that K is a normal p-subgroup of G such that $H \cap K=1$, where H is some given subgroup of G. Put $\bar{G}=G/K$, $\bar{H}=HK/K \cong H$. Every $\bar{G}$-module N can be viewed as a G-module upon which K acts trivially. The map $[N] \mapsto [N]$ induces a ring homomorphism φ: $a(\bar{G}, \bar{H}) \to a(G, H)$ called the *inflation* map. Our main result is

(3.1) **Excision Theorem.** *Suppose that G is expressible as a disjoint union $\dot{\bigcup}_{i=1}^{r} x_i HK$, $x_1=1$, where the $\{x_i\}$ satisfy*

$$\text{Condition } C\colon \quad \dot{\bigcup_{i=1}^{r}} H x_i \subset \dot{\bigcup_{i=1}^{r}} x_i H.$$

Then the inflation map

$$\varphi\colon a(\bar{G}, \bar{H}) \to a(G, H)$$

is a ring isomorphism.

Before proving the theorem, we record

(3.2) **Corollary**[1]. *If either $G=K \cdot N_G(H)$ or if G is a semidirect product $K \cdot B$, where $K \cap B=1$ and $H \subset B$, then φ: $a(\bar{G}, \bar{H}) \cong a(G, H)$.*

[1] The second part of this corollary is our previous Theorem 3.2 in [4].

Proof of Corollary. If $G=K\cdot N_G(H)$, then the $\{x_i\}$ may be chosen as elements of $N_G(H)$. In this case, $x_iH=Hx_i$ for each i, so (C) is obviously true.

On the other hand, if $G=K\cdot B$ is a semidirect product, then the $\{x_i\}$ may be chosen as left coset representatives of H in B. In this case, $\bigcup x_iH=B$, so condition C is again valid.

Turning to the proof of Theorem 3.1, we suppose henceforth that its hypotheses are satisfied. We shall construct an additive homomorphism $\mu\colon a(G,H)\to a(\bar{G},\bar{H})$ which will turn out to be the inverse of φ. Since φ is in any case a ring homomorphism, this will imply that both φ and μ are ring isomorphisms. To begin with, for each isomorphism type of G-module M, we shall define an element $\mu(M)\in a(\bar{G},\bar{H})$. The definition is a recursive one, and proceeds by induction on $\dim M$. Once $\mu(M)$ is defined for each M, we shall prove that μ behaves properly on H-split G-exact sequences, and is therefore well-defined on $a(G,H)$.

Given any G-module M, let M_0 denote the K-socle of M. Since K is a normal p-subgroup of G, M_0 consists of all K-trivial elements of M, and is a G-submodule of M. If $M\neq 0$, then clearly $M_0\neq 0$. We may view M_0 canonically as a $\bar{G}$-module, and any G-submodule of M_0 as a $\bar{G}$-submodule of M_0.

On the other hand, the isomorphism $H\cong\bar{H}$ permits us to view M_H as an $\bar{H}$-module, which we denote by $M_{\bar{H}}$. It is then clear that M_0 is a $(\bar{G},\bar{H})$-submodule of $M_{\bar{H}}$ in the sense of §2, and thus the element $[M_0,M_{\bar{H}}]^{\bar{G},\bar{H}}$ in $a(\bar{G},\bar{H})$ is defined. Indeed, we have explicitly

$$[M_0,M_{\bar{H}}]^{\bar{G},\bar{H}}=[(M_{\bar{H}})^{\bar{G}}/\lambda(M_0)]-[\{(M/M_0)_{\bar{H}}\}^{\bar{G}}],$$

where $\lambda(M_0)$ is the $\bar{G}$-submodule of $(M_{\bar{H}})^{\bar{G}}$ spanned by the elements

$$\{\bar{x}\otimes\bar{x}^{-1}m_0-1\otimes m_0\colon \bar{x}\in\bar{G},\ m_0\in M_0\}.$$

We are now ready to define $\mu(M)$ by induction on $\dim M$. Set $\mu(0)=0$, and for $M\neq 0$, set

$$\mu(M)=[M_0,M_{\bar{H}}]+\mu(M/M_0)\in a(\bar{G},\bar{H}). \tag{3.3}$$

(We omit the superscripts $\bar{G}$, $\bar{H}$ for convenience.) Thus $\mu(M)$ is now defined for every G-module M.

The following lemma is vital.

(3.4) **Lemma.** *Under the hypotheses of* (3.1), *there exists a quasi-$\bar{G}$-structure on the $\bar{H}$-module $M_{\bar{H}}$, such that every $\bar{G}$-submodule X of M_0 is suitable relative to this structure.*

Proof. Note that $\bar{G}=\dot{\bigcup}_{i=1}^{r}\bar{x}_i\bar{H}$, $\bar{x}_1=1$. We must construct a family $\{\psi_{\bar{x}}\colon \bar{x}\in\bar{G}\}$ of Ω-automorphisms of $M_{\bar{H}}$ such that $\psi_1=$identity, and

$$t\psi_{\bar{x}_i}=\psi_{\bar{x}_j}t' \quad \text{whenever } t\bar{x}_i=\bar{x}_jt',\quad t,\quad t'\in\bar{H}. \tag{3.5}$$

Let us in fact choose $\psi_{\bar{x}_i}$ to be multiplication by x_i, $1 \leqq i \leqq r$; this is meaningful since M is to start with a G-module. In order to verify (3.5), choose any $h \in H$ mapping onto t; by Condition C, we may write $h x_i = x_j h'$ for some j and some $h' \in H$, in which case $t' = \bar{h}'$. Then t acts as h on $M_{\bar{H}}$, so

$$t \cdot \psi_{\bar{x}_i} = h x_i = x_j h' = \psi_{\bar{x}_j} t',$$

as desired. It is then clear that *every* $\bar{G}$-submodule X of M_0 is suitable for this quasi-$\bar{G}$-structure $\{\psi_{\bar{x}}\}$. This completes the proof of the lemma.

Now suppose that X is any G-submodule of M_0. Then both X and M_0 are $(\bar{G}, \bar{H})$-submodules of $M_{\bar{H}}$, and are suitable relative to the quasi-$\bar{G}$-structure on $M_{\bar{H}}$ defined in the preceding lemma. We may then apply the Comparison Theorem 2.5 to conclude that

$$[M_0, M_{\bar{H}}] = [X, M_{\bar{H}}] + [\bar{M}_0, \bar{M}_{\bar{H}}] \quad \text{in} \quad a(\bar{G}, \bar{H}), \tag{3.6}$$

where $\bar{M}_0 = M_0/X$, $\bar{M} = M/X$. We shall use this relation in proving that formula (3.3) remains valid when M_0 is replaced by any of its G-submodules. We claim

(3.7) **Lemma.** *If X is any G-submodule of M_0, then*

$$\mu(M) = [X, M_{\bar{H}}] + \mu(M/X) \quad \text{in} \quad a(\bar{G}, \bar{H}).$$

Proof. Use induction on $\dim M$, the result being clear if $\dim M = 0$ or if $X = 0$. Now let $M \neq 0$, let X be any nonzero G-submodule of M_0, and put $\bar{M} = M/X$, $\bar{M}_0 = M_0/X$. Then $\bar{M}_0$ is a K-trivial G-submodule of $\bar{M}$, thus lies in the K-socle of $\bar{M}$. By the induction hypothesis,

$$\mu(\bar{M}) = [\bar{M}_0, \bar{M}_{\bar{H}}] + \mu(M/M_0),$$

where we have taken account of the fact that $\bar{M}/\bar{M}_0 \cong M/M_0$. Using (3.6), we obtain

$$\begin{aligned} \mu(M) &= [M_0, M_{\bar{H}}] + \mu(M/M_0) = [M_0, M_{\bar{H}}] + \mu(\bar{M}) - [\bar{M}_0, \bar{M}_{\bar{H}}] \\ &= [X, M_{\bar{H}}] + \mu(\bar{M}), \end{aligned}$$

as desired. This proves the lemma.

We are now ready to show that μ induces an additive homomorphism from $a(G, H)$ into $a(\bar{G}, \bar{H})$. We must prove that if $0 \to M' \to M \to M'' \to 0$ is an H-split G-exact sequence, then

$$\mu(M) = \mu(M') + \mu(M'').$$

Let us prove this by induction on $\dim M'$, the result being clear when $M' = 0$. Now let $M' \neq 0$, let X be any nonzero K-trivial G-submodule of M', and set $\bar{M}' = M'/X$, $\bar{M} = M/X$. The sequence

$$0 \to \bar{M}' \to \bar{M} \to M'' \to 0$$

is also H-split G-exact, so by the induction hypothesis

$$\mu(\overline{M})=\mu(\overline{M}')+\mu(M'').$$

By (2.7) we have $[X, M_{\bar{H}}]=[X, M'_{\bar{H}}]$ in $a(\bar{G}, \bar{H})$, and thus using (3.7) we get

$$\begin{aligned}\mu(M)&=[X, M_{\bar{H}}]+\mu(\overline{M})=[X, M'_{\bar{H}}]+\mu(\overline{M}')+\mu(M'')\\&=\mu(M')+\mu(M''),\end{aligned}$$

as claimed. This completes the proof that μ: $a(G, H)\to a(\bar{G}, \bar{H})$ is well-defined.

To prove that $\mu\cdot\varphi=1$, let N be any $\bar{G}$-module. Its inflation $\varphi(N)$ is a K-trivial G-module, again denoted by N. Then N is its own K-socle, so

$$\mu\varphi[N]=[N, N_{\bar{H}}]^{\bar{G}, \bar{H}}+\mu(N/N)=[N],$$

using (2.6).

Instead of showing directly that $\varphi\cdot\mu=1$, it is easier to complete the proof of the Excision Theorem by showing that φ is epic. We use induction on dim M to prove that $[M]\in$ image of φ, for each G-module M. As usual, let M_0 be the K-socle of M, and consider the $\bar{G}$-exact sequence

$$0\to M_0\xrightarrow{i'}\frac{(M_{\bar{H}})^{\bar{G}}}{\lambda(M_0)}\to\left\{\left(\frac{M}{M_0}\right)_{\bar{H}}\right\}^{\bar{G}}\to 0,$$

where $i'(m_0)=1\otimes m_0+\lambda(M_0)$. View the above modules as K-trivial G-modules, and consider the diagram

$$\begin{array}{ccc} M_0 & \xrightarrow{i'} & \dfrac{(M_{\bar{H}})^{\bar{G}}}{\lambda(M_0)} \\ {\scriptstyle i}\downarrow & \overset{\sigma}{\nearrow}\ \underset{\rho}{\swarrow} & \\ M & & \end{array}$$

where both i and i' are G-inclusions. Define $\sigma(m)=1\otimes m+\lambda(M_0)$, so σ is an H-homomorphism for which $\sigma i=i'$. On the other hand, define ρ on $(M_{\bar{H}})^{\bar{G}}$ by setting

$$\rho(\bar{x}_i\otimes m)=x_i m,\qquad 1\leqq i\leqq r,\qquad m\in M.$$

Since $\lambda(M_0)$ is defined by viewing M_0 as $(\bar{G}, \bar{H})$-submodule of $M_{\bar{H}}$, it follows that $\lambda(M_0)$ is spanned by the set of elements

$$\{\bar{x}_i\otimes\bar{x}_i^{-1}m_0-1\otimes m_0\colon 2\leqq i\leqq r,\ m_0\in M_0\}.$$

Hence $\rho\{\lambda(M_0)\}=0$, so ρ is well-defined on the quotient $(M_{\bar{H}})^{\bar{G}}/\lambda(M_0)$. Clearly $\rho i'=i$, and it remains to check that ρ is an H-homomorphism. For $h\in H$, we may use condition C to write $hx_i=x_jh'$ for some j and some $h'\in H$. Then

$$\begin{aligned}\rho\{h(\bar{x}_i\otimes m)\}&=\rho\{\bar{h}\bar{x}_i\otimes m\}=\rho\{\bar{x}_j\otimes h'm\}\\&=x_j\cdot h'm=h\cdot\rho(\bar{x}_i\otimes m),\end{aligned}$$

as claimed. The Pushout Lemma 1.1 then yields the following relation in $a(\bar{G}, \bar{H})$:

$$[M]=\varphi\{[M_0, M_{\bar{H}}]^{\bar{G}, \bar{H}}\}+[M/M_0].$$

By the induction hypothesis, $[M/M_0] \in$ image of φ. Thus also $[M] \in$ image of φ, so φ is epic, and the proof of Theorem 3.1 is complete.

We suspect that the Excision Theorem is no longer valid if Condition C is waived, but we have no counterexample to support this suspicion. Of course, it is possible that some slightly weaker condition than C would be sufficient. However, it is easy to see that one cannot omit the hypotheses that K be a normal p-subgroup of G such that $H \cap K = 1$.

§4. Product Isomorphism Theorem and the Restriction Conjecture

We shall now give two applications of the Excision Theorem. The first refines our earlier Product Isomorphism Theorem 5.1 in [6]; not only is the present proof simpler, but a stronger result is obtained. The second application is concerned with the question as to when the restriction map

$$\text{res}: a(G, H) \to a(E, H)$$

is monic, where $H \subset E \vartriangle G$ and $[G:E]$ is a power of p.

In both applications we shall start with the situation $H \subset E \vartriangle G$, and suppose that $G = EA$ for some subgroup A such that $E \cap A$ is a p-group. For each $a \in A$, the map $e \to a^{-1} e a = e^a$, $e \in E$, is an automorphism of E. Form the semidirect product

$$\tilde{G} = \{(a, e): a \in A, e \in E\},$$

where

$$(a_1, e_1)(a, e) = (a_1 a, e_1^a e).$$

Then $\tilde{G} = E_0 \cdot A_0$, $E_0 \vartriangle \tilde{G}$, $E_0 \cap A_0 = 1$, where E_0 is the image of E in $\tilde{G}$, and A_0 that of A; let H_0 be the image of H in $\tilde{G}$, so $H_0 \subset E_0$.

There is a group epimorphism $f: \tilde{G} \to G$ given by $f(a, e) = a e$, with kernel $K = \{(a, a^{-1}): a \in E \cap A\}$. The map $a \to (a, a^{-1})$ yields an isomorphism $E \cap A \cong K$, so K is a normal p-subgroup of $\tilde{G}$ such that $\tilde{G}/K \cong G$, and $K \cap H_0 = 1$. We wish to apply the Excision Theorem, so we must verify that condition C holds for some set of left $H_0 K$-coset representatives in $\tilde{G}$. Write

$$G = \dot{\bigcup_{k=1}^{s}} a_k E, \quad E = \dot{\bigcup_{l=1}^{t}} e_l H, \quad a_k \in A, \quad a_1 = e_1 = 1.$$

Then $G = \dot{\bigcup} a_k e_l H$, and $[\tilde{G}:H_0 K] = [\tilde{G}/K:H_0 K/K] = [G:H] = st$. Setting $x_{kl} = (a_k, e_l) \in \tilde{G}$, $1 \leqq k \leqq s$, $1 \leqq l \leqq t$, it follows readily that the $\{x_{kl}\}$ are a full set of left coset representatives of $H_0 K$ in $\tilde{G}$.

Now take any $(1, h) \in H_0$. Then $(1, h) x_{kl} = (1, h)(a_k, e_l) = (a_k, b)$ for some $b \in E$. We may write $b = e_n h'$ for some n and some $h' \in H$, whence

$$(1, h) x_{kl} = (a_k, e_n)(1, h') \in x_{kn} H_0.$$

Therefore

$$\bigcup_{k,l} H_0 x_{kl} \subset \bigcup_{k,l} x_{kl} H_0, \tag{4.1}$$

as desired. We may therefore apply the Excision Theorem to get

(4.2) $$a(\tilde{G}, H_0) \cong a(\tilde{G}/K, H_0 K/K) \cong a(G, H).$$

Note further that if we set $\tilde{E} = E_0 K$, then the elements $\{x_{1l}\}$ are left $H_0 K$-coset representatives in $\tilde{E}$. Analogously to (4.1), we obtain $\bigcup H_0 x_{1l} \subset \bigcup x_{1l} H_0$, so again using the Excision Theorem, we have

(4.3) $$a(\tilde{E}, H_0) \cong a(E, H).$$

We are ready to apply these results. Refining [6], Theorem 5.1, we have

(4.4) **Product Isomorphism Theorem.** *Let $H \subset E \vartriangle G$, where $G = E \cdot C$ for some subgroup C of $C_G(E)$. Suppose that $C \cap E$ is a p-group, and that Ω is a splitting field for C and all of its subgroups. Then there is a ring isomorphism*

$$a(G, H) \cong a(G/E, 1) \otimes_Z a(E, H).$$

Proof. By (4.2), $a(G, H) \cong a(\tilde{G}, H_0)$ with $A = C$. Since C centralizes E, it is clear that $\tilde{G}$ is a direct product $C \times E_0$, with $H_0 \subset E_0$. Using (1.2) we obtain

$$a(\tilde{G}, H_0) = a(C \times E_0, H_0) \cong a(C, 1) \otimes_Z a(E, H).$$

Finally, $C \cap E$ is a normal p-subgroup of C, and $C/(C \cap E) \cong G/E$. By Clifford's Theorem, we have

$$a(G/E, 1) \cong a\big(C/(C \cap E), 1\big) \cong a(C, 1),$$

and the proof is complete.

(4.5) **Corollary.** *Let $H \subset E \vartriangle G$, where $G = E \cdot C_G(E)$ and where the greatest common divisor of $[G:E]$ and $[E:1]$ is a power of p, possibly equal to 1. Assume that Ω is a splitting field for $C_G(E)$ and all of its subgroups. Then there is a ring isomorphism*

$$a(G, H) \cong a(G/E, 1) \otimes_Z a(E, H).$$

Proof. It suffices to check that the hypotheses of (4.4) hold in this case. Let $C_0 = C_G(E)$, $T = C_0 \cap E$, so T is an abelian group central in C_0. Write $T = P \times Q$ with P the p-primary component of T, and $[Q:1]$ p-free. Since $[Q:1]$ divides $[E:1]$, and since $[C_0:Q] = [C_0:T_0][P:1] = [G:E][P:1]$, it follows from the hypotheses that $[Q:1]$ is relatively prime to $[C_0:Q]$. Thus Q is a central Hall subgroup of C_0. Let C be a complement of Q in C_0, so $C_0 = Q \times C$. We have $G = EC_0 = EC$, and $E \cap C$ is a p-group since in fact $E \cap C = P$. The corollary now follows from (4.4).

For our second application consider the following

(4.6) *Restriction Conjecture.* Let $H \subset E \vartriangle G$, where $[G:E]$ is a power of p. Then the restriction map

$$\text{res}\colon a(G, H) \to a(E, H)$$

is monic[2].

[2] In the case $H = 1$, we have $a(G, 1) \to a(E, 1)$, i.e. restriction of Brauer characters. Since the p-regular elements of G all lie in E, the restriction map is clearly a monomorphism in this case.

Call G a *split extension* of E if $G=EB$ for some subgroup B of G such that $E\cap B=1$. We now prove

(4.7) **Theorem.** *If the restriction conjecture holds whenever G is a split extension of E, then it holds true in general.*

Proof. Let $H\subset E \vartriangle G$, where G is an arbitrary extension of E, and $[G:E]$ is a power of p. If P is a Sylow p-subgroup of G, then $G=EP$, and $E\cap P$ is obviously a p-group. Hence both of the isomorphisms (4.2) and (4.3) are valid. (Using $A=P$.)

Consider

$$\begin{array}{ccccc} a(G,H) & \xrightarrow{\varphi} & a(\tilde{G},H_0) & & \\ \downarrow r_1 & & \downarrow r_2 & \searrow r_3 & \\ a(E,H) & \xrightarrow{\varphi'} & a(\tilde{E},H_0) & \xrightarrow{r_4} & a(E_0,H_0), \end{array}$$

where φ and φ' are inflation maps, and each r_i is a restriction map. Since the various maps do nothing to the modules involved, only changing their operator domains, the diagram is commutative. Both φ and φ' are isomorphisms, by (4.2) and (4.3). Further, $\tilde{G}$ is a *split* extension of E_0 by A_0, and $\tilde{G}/E_0$ is a p-group; hence r_3 is monic, since we have assumed the validity of the restriction conjecture for split extensions. But then r_1 must also be monic, as claimed, which proves the theorem.

Unfortunately, we are unable to prove the Restriction Conjecture in the split case. We shall merely conclude with two consequences of (4.6).

(4.8) **Corollary.** *Assume* (4.6) *true, and let G be nilpotent. If* $H\subset J\subset G$, *where* $[G:J]$ *is a power of p, then* res: $a(G,H)\to a(J,H)$ *is monic.*

Proof. Since G is nilpotent, every proper subgroup of G is properly contained in its normalizer. Hence there is a chain of subgroups $H\subset J=J_0 \vartriangle J_1 \vartriangle \cdots \vartriangle J_n=G$. By (4.6), each

$$\text{res: } a(J_{i+1},H)\to a(J_i,H), \qquad 0\leqq i\leqq n-1,$$

is monic. Hence so is res: $a(G,H)\to a(J,H)$.

(4.9) **Corollary.** *Assume* (4.6) *true, and G nilpotent. Then* $a(G,H)$ *is* $\mathbb{Z}$*-free for all* $H\subset G$.

Proof. Since $a(G,H)$ is unchanged when H is replaced by one of its Sylow p-subgroups, we may assume that H is a p-group. Next, G is the direct product of its Sylow subgroups, so we may write $G=P\times Y$, where P is the Sylow p-subgroup of G, and Y is the product of all the other Sylow subgroups of G. Set $E=H\times Y$, so $H\subset E\subset G$ and $[G:E]$ is a power of p. By (4.8), $a(G,H)\to a(E,H)$ is monic. However, $a(E,H)\cong a(E,E)$ since H is the Sylow p-subgroup of E. Thus $a(G,H)$ is embedded in $a(E,E)$, and the latter is $\mathbb{Z}$-free by virtue of the Krull-Schmidt Theorem.

References

1. Bass, H.: Algebraic K-theory. New York: Benjamin 1969.
2. Eilenberg, S., Steenrod, N.: Foundations of algebraic topology. Princeton University Press 1951.
3. Lam, T.-Y., Reiner, I.: Relative Grothendieck groups. J. Algebra **11**, 213 – 242 (1969).
4. – – Reduction theorems for relative Grothendieck rings. Trans. Amer. Math. Soc. **142**, 421 – 435 (1969).
5. – – Relative Grothendieck rings. Bull. Amer. Math. Soc. **75**, 496 – 498 (1969).
6. – – Restriction maps on relative Grothendieck rings. J. Algebra **14**, 260 – 298 (1970).
7. – – Finite generation of Grothendieck rings relative to cyclic subgroups. Proc. Amer. Math. Soc. **23**, 481 – 489 (1969).
8. Milnor, J.: Notes on algebraic K-theory. Mass. Inst. of Technology (1969).
9. Ribes, L.: On a cohomology theory for pairs of groups. Proc. Amer. Math. Soc. **21**, 230 – 234 (1968).

Prof. T.-Y. Lam
University of California
Berkeley, California 94720
USA

Prof. Irving Reiner
University of Illinois
Urbana, Illinois 61801
USA

(Received December 8, 1969)

RESTRICTION MAPS ON RELATIVE GROTHENDIECK RINGS

T. Y. Lam and I. Reiner

ABSTRACT

Let G be a finite group, H a subgroup of G, and Ω a field of characteristic p. The *relative Grothendieck ring* $a(G,H)$ is the free abelain group generated by the symbols $[M]$, where M ranges over all isomorphism types of ΩG modules, modulo the subgroup generated by all expressions $[M]-[L]-[N]$ where

$$0 \to L \to M \to N \to 0$$

is an exact sequence of ΩG modules which is split as a sequence of ΩH modules. The ring structure is induced by the tensor product of modules. This idea includes the *representation ring*, $a(G,G) \cong a(G)$, that has been studied elsewhere, and the *character ring* $a(G,1)$ which is isomorphic to the ring of Brauer characters.

The first result obtained in this paper deals with the case that G is a p-group and H is a cyclic subgroup. Let res:$a(G,H) \to a(H)$ denote the usual restriction of operators mapping. Let $k(G,H)$ denote the ideal of of the representation ring $a(G)$ generated by all relatively (G,H)-projective ΩH-modules, *i.e.* direct summands of induced modules V^G, V an ΩH-modules. The natural map, $\kappa : k(G,H) \to a(G,H)$ induced by the inclusion, is called the *Cartan map.*

Theorem. *Let H be a cyclic subgroup of the p-group G. Then the restriction map*

$$res{:}a(G,H) \to a(H)$$

is a ring isomorphism. The cokernel of the Cartan map $\kappa : k(G,H) \to a(G,H)$ is a finite abelian p-group, whose order is the product of the integers

$$[N_G(K) : H]^{\varphi([K:1])},$$

where K ranges over all subgroups of H, and φ is Euler's function.

The proof involves very explicit use of the module theory for cyclic groups and the Pushout Lemma to obtain relations in $a(G,H)$.

The two main results in the remainder of the paper deal with reduction formulas that identify $a(G,H)$ in terms of "smaller" groups for various special assumptions placed upon G and H. These results were given, with different proofs, in [52].

A SURVEY OF INTEGRAL REPRESENTATION THEORY[1]

BY IRVING REINER

TABLE OF CONTENTS

1. Introduction. Notation and definitions. First of all I wish to acknowledge with thanks the many helpful conversations I have had with Professors Olga Taussky, Peter Roquette and Hans Zassenhaus, when I first began studying the subject of integral representations.

Historically, the subject received its main impetus from two branches of algebra. One branch is algebraic number theory, especially that part concerned with ideal theory; and the other is matrix theory, mainly that portion dealing with matrix representations of associative algebras. Methods of homological algebra have played an increasingly important role in the subject in recent years.

An expanded version of an address delivered before the Chicago meeting of the Society by invitation of the Committee to Select Hour Speakers for Western Sectional Meetings, April 20, 1968, under the title *Recent progress in the theory of integral representations*; received by the editors September 17, 1969.

AMS Subject Classifications. Primary 1075, 1548, 2080, 1640; Secondary 1069, 1620.

Key Words and Phrases. Integral representations, orders, representation lattices, Grothendieck groups, noncommutative arithmetic, indecomposable representations, group representations, genus.

[1] This work was partially supported by a research contract with the National Science Foundation.

We wish to study representations of certain algebraic systems, called *orders*, by means of matrices with entries in an integral domain R. It is often more convenient to consider, instead of matrices, the underlying spaces on which these matrices act. Such spaces are special instances of *R-lattices*, by which we mean finitely generated torsionfree R-modules. Only those R-lattices having a free R-basis give rise to matrices. Of course, when R is a principal ideal domain, every R-lattice has a free R-basis.

Let Λ be a ring with unity element 1, containing R in its center (identify R with $R\cdot 1$). This makes Λ into an R-module. Call Λ an *R-order* if Λ is finitely generated and torsionfree as R-module. Let F be the quotient field of R, and set $A = F \otimes_R \Lambda$, so that A is a finite dimensional algebra over F. We always identify Λ with $1 \otimes \Lambda$, so Λ is embedded in A, and we may write $A = F\cdot\Lambda$ (the set of F-linear combinations of the elements of Λ). Likewise, every R-lattice M is embedded in the vector space over F given by $F \otimes_R M = FM$. The dimension $(FM : F)$ is called the *R-rank* of M.

Here are some examples of orders:

1. The ring of all algebraic integers in an algebraic number field F is a Z-order in F, and will be denoted hereafter by alg. int.$\{F\}$. We shall use this notation only when F is an algebraic number field.
2. If G is a finite group, the integral group ring RG is an R-order in the group algebra FG.
3. The ring $(R)_n$ of all $n \times n$ matrices with entries in R is an R-order in the matrix algebra $(F)_n$.
4. For α an algebraic integer, the ring $Z[\alpha]$ is a Z-order in the field $Q(\alpha)$, where Q is the rational field.

Let Λ be an R-order; a *Λ-lattice* is an R-lattice which is also a Λ-module. The fundamental problem in the theory of integral representations is as follows:

Given an R-order Λ, determine all Λ-lattices. Once we have obtained some answers to this question, we may then proceed to the next problem:

Apply the theory of integral representations to investigate properties of various orders.

The purpose of this survey is to describe the present state of knowledge about these problems, especially the first one. References which give a general introduction to the theory of orders are CR[2] Chapter 11, Roggenkamp-Dyson [**208b**], Faddeev [**80**].

---

[2] CR denotes the reference Curtis-Reiner [46] throughout.

It is clearly an impossible task to give a complete description of the contents of the 250 or so bibliographical references for this article. I have of course been guided by my own interests in making a selection from the wealth of material available, and I regret that many references are mentioned only briefly in passing. Certain important topics on the periphery of the subject have been omitted entirely, and would require their own survey articles; in particular, there is no discussion of cohomology of groups, nor of Galois theory of rings.

In addition to the earlier notation, we also fix the following once and for all:

char F : characteristic of the field F
P = maximal ideal of the integral domain R
R_P = localization of R at $P = \{\alpha/\beta : \alpha \in R, \beta \in R - P\}$
R_P^* = P-adic completion of R; F_P^* = quotient field of R_P^*
$M^{(k)}$ = direct sum of k copies of M
$(D)_n$ = ring of all $n \times n$ matrices with entries in D
G = finite group of order g
$\dot{+}$: external direct sum
$\sum^{\cdot}$, $\oplus$: internal direct sum
$a \mid b$: a divides b
$a \nmid b$: a does not divide b
rad Λ = Jacobson radical of the ring Λ

A *free* left Λ-module is a direct sum of copies of the left Λ-module Λ. A *projective* left Λ-module is a direct summand of a free module. The ring Λ is *left hereditary* if every left ideal of Λ is a projective Λ-module. (Equivalently, Λ is left hereditary if every submodule of a free Λ-module is isomorphic to an external direct sum of left ideals of Λ, each of which is projective. M. Auslander has shown that if Λ is both left and right noetherian, then Λ is left hereditary if and only if Λ is right hereditary. See Rotman [**218**], Cartan-Eilenberg [**40**].)

A *Dedekind domain* is an integral domain in which every nonzero ideal is uniquely expressible as a product of prime ideals. (Equivalently, a Dedekind domain is a hereditary integral domain. See Cartan-Eilenberg [**40**].) In particular alg. int.$\{F\}$ is always a Dedekind domain.

(1.1). Definition. Let R be a Dedekind domain. An *R-ideal* in F is a nonzero R-lattice in F. Products of R-ideals are again R-ideals; relative to this multiplication, the R-ideals in F form a multiplicative group denoted by $I\{R\}$.

The following theorem lists some standard properties of lattices over Dedekind domains.

(1.2). THEOREM. *Let R be a Dedekind domain.*

(i) *Every R-lattice is a projective R-module.*

(ii) *Let $N \subset M$ be R-lattices. Then N is an R-direct summand of M if and only if M/N is R-torsionfree.*

(iii) *Each R-ideal J in F can be generated (as R-module) by two elements. Further, each J is invertible, that is, there is an R-ideal J' such that $J \cdot J' = R$.*

(iv) *Given any R-lattice M, there exist elements $m_1, \cdots, m_k \in M$ and ideals $J_1, \cdots, J_k \in I\{R\}$ such that*

$$M = J_1 m_1 \oplus \cdots \oplus J_k m_k.$$

(v) *Given R-lattices $N \subset M$, there exist elements $m_1, \cdots, m_k \in M$, R-ideals $J_1, \cdots, J_k \in I\{R\}$, and integral ideals $E_1, \cdots, E_k$ in R (some of which may be zero), such that*

$$M = J_1 m_1 \oplus \cdots \oplus J_k m_k,$$
$$N = E_1 J_1 m_1 \oplus \cdots \oplus E_k J_k m_k,$$
$$E_1 \supset E_2 \supset \cdots \supset E_k.$$

References. Cartan-Eilenberg [**40**, Chapter 7], Chevalley [**41**], CR Chapter 3, Kaplansky [**138**], [**139**], Levy [**152**], Steinitz [**228**], Zariski-Samuel [**262**, Chapter 5].

Let $J, J' \in I\{R\}$; they are in the same *ideal class* if $J' = Jx$ for some $x \in F$. The number of ideal classes is called the (ideal) class number of R (or of F), and is finite when F is an algebraic number field. (CR Chapter 3).

Now let $N \subset M$ be R-lattices for which M/N is a torsion R-module. Then each of the ideals $\{E_i\}$ occurring in part (v) of Theorem 1.2 is nonzero, and we define the *order ideal* of M/N to be the product $E_1 \cdots E_k$. Equivalently, this order ideal is the product of the R-annihilators of the composition factors of the R-module M/N. Denote by $\operatorname{ord}(M/N)$ the order ideal of M/N. References for this concept: Fröhlich [**85**], CR §80.

2. **General remarks. Jordan-Zassenhaus Theorem.** To begin with, let R be an arbitrary noetherian domain, and Λ an R-order in the F-algebra A. Given a Λ-lattice M, form the A-module $F \otimes_R M$. Identifying M with $1 \otimes M$, we write $F \cdot M$ in place of $F \otimes M$. Such identifications will be made hereafter without specific comment.

Let $N \subset M$ be Λ-lattices, and call N *R-pure* in M if M/N is R-torsionfree. An easy result is

(2.1). THEOREM. *Given any Λ-lattice M, there is a one-to-one correspondence $W \leftrightarrow N$ between A-submodules W of FM and R-pure Λ-sublattices N of M. The correspondence is given by*

$$N = M \cap W, \qquad W = FN.$$

Further, each finitely generated A-module V is of the form FM for some Λ-lattice M in V.

References. Zassenhaus [263]; CR §73.

The latter part of the above theorem has important consequences for the theory of group representations. Let G be a finite group; an *F-representation* of G of *degree n* is a homomorphism of G into $(F)_n$, carrying the identity of G onto the identity matrix. Two F-representations $T:G\to(F)_n$, $U:G\to(F)_n$ are *F-equivalent* if there exists an invertible $C\in(F)_n$ such that

$$C^{-1}\cdot T(x)\cdot C = U(x) \quad \text{for all } x \in G.$$

In other words, T and U are equivalent if they are afforded by the same FG-module by using different F-bases of that module.

Taking $\Lambda = RG$ in the latter part of (2.1), we obtain

(2.2). THEOREM. *If R is a principal ideal domain, every F-representation of the finite group G is F-equivalent to an R-representation of G.*

Even when R is not a principal ideal domain, the above result is useful. In particular, suppose $R = \text{alg. int.}\{F\}$, and let $R' = \text{alg. int.}\{F'\}$ for some $F' \supset F$. It is known that F' may be chosen so that for each $J \in I\{R\}$, the R'-ideal JR' is principal. It is also known that every complex representation of a finite group G is equivalent (over the complex field) to an F-representation of G, for some algebraic number field F. This yields

(2.3). THEOREM. *Every complex representation of G is equivalent (over the complex field) to a representation by matrices whose entries are algebraic integers.*

Reference. CR (75.4).

In this direction, we quote a result due to Schur.

(2.4). THEOREM. *Let $R = \text{alg. int.}\{F\}$, and let h be the number of ideal classes in R. If $(h, n) = 1$, then every F-representation of G of degree n is F-equivalent to an R-representation of G.*

Reference. CR (75.5).

Returning now to the general case where Λ is an R-order in A, we start with any Λ-lattice M, and write a composition series for the A-module FM, say

$$FM = W_0 \supset W_1 \supset \cdots \supset W_k = 0.$$

Then there is a chain of Λ-lattices

$$(2.5) \qquad M = M_0 \supset M_1 \supset \cdots \supset M_k = 0, \quad M_i = M \cap W_i.$$

For each i, M_{i+1} is an R-pure Λ-sublattice of M_i such that

$$F \otimes_R (M_i/M_{i+1}) \cong W_i/W_{i+1}.$$

We may call (2.5) an "R-composition series" for M, with "R-composition factors" $\{M_i/M_{i+1}: 0 \leqq i < k\}$. The analogue of the Jordan-Hölder Theorem is *not* valid, however, since indeed the R-composition factors are (in general) not uniquely determined by M. (See Reiner [**184**]; CR, Example 2 at the end of §73.)

As is well known, the number of ideal classes in alg. int. $\{F\}$ is finite (see CR, §20, for example). A far-reaching generalization of this is the fundamental result

(2.6). JORDAN-ZASSENHAUS THEOREM. *Let R be a Dedekind domain whose quotient field F is an algebraic number field, and let Λ be an R-order in a semisimple F-algebra A. Let W be any finitely generated A-module, and set*

$$\sigma(W) = \{M : M = \Lambda\text{-lattice}, FM \cong W\}.$$

Then the number $s(W)$ of Λ-isomorphism classes in $\sigma(W)$ is finite.

References. Zassenhaus [**263**], CR §79. For corresponding results when $R = k[X]$, where X is an indeterminate over the finite field k, see Roggenkamp [**202**], [**203**]. For the more general case where R is the integral closure of $k[X]$ in a finite extension of $k(X)$, see Higman-McLaughlin [**122**]. The case where R is a P-adic ring is treated by Jenner [**134**].

The actual determination of $s(W)$ is a difficult problem. The most fruitful approach is to partition $\sigma(W)$ into certain collections of Λ-isomorphism classes, called *genera*. We place two Λ-lattices M and N in the same genus if and only if

(2.7). *$FM \cong FN$ and $R_P M \cong R_P N$ for each maximal ideal P of R.*

It is sometimes possible to determine explicitly the number of genera in $\sigma(W)$, and then to find the number of Λ-isomorphism classes in each genus. We shall discuss this in more detail in §§6, 8.

We give some examples where $s(W)$ can be calculated. For instance, suppose

$$\Lambda = Z[x]/(x^n - 1),$$

so $\Lambda \cong ZG$, G cyclic of order n. Let $W = Q(\theta_d)$, the cyclotomic field generated by a primitive dth root of unity θ_d, where $d \mid n$. Then W is an irreducible A-module, and each $M \in \sigma(W)$ is Λ-isomorphic to a $Z[\theta_d]$-ideal in the field $Q(\theta_d)$. In this case, $s(W)$ is precisely the number of ideal classes in the ring $Z[\theta_d]$. (See also Taussky-Todd **[250]**, **[251]**.)

In this direction, let us consider the case where $f(x)$ is a monic nth degree polynomial in $Z[x]$ with distinct zeros. Set

$$A = Q[x]/(f(x)), \qquad \Lambda = Z[x]/(f(x)),$$

so Λ is a Z-order in A. Any $X \in (Q)_n$ with minimal polynomial $f(x)$ gives rise to an algebra isomorphism $A \cong Q[X]$ via $x \to X$, and so this map is just a Q-representation of A afforded by A itself. Those matrices $M \in (Z)_n$ with minimal polynomial $f(x)$ are afforded by Λ-lattices in A. Isomorphism classes of lattices correspond to classes of integral matrices under unimodular equivalence, where the matrices M, N are *unimodularly equivalent* if there exists a matrix $U \in (Z)_n$ such that $U^{-1} \in (Z)_n$, and $M = UNU^{-1}$. Further, two Λ-lattices L, L' are isomorphic if and only if $L' = L\lambda$ for some unit $\lambda \in A$, so the number of isomorphism classes of lattices is the same as the number of left ideal classes in Λ.

These results, due to Latimer-MacDuffee **[148]**, were simplified by Taussky **[240]** in case $f(x)$ is irreducible. In that case, let α be an algebraic integer for which $f(\alpha) = 0$, so $A = Q(\alpha)$ is a field, and $\Lambda = Z[\alpha]$ is a Z-order in A. Each ideal $\mathfrak{m}$ of Λ gives rise to a matrix M (with $f(M) = 0$) by letting α act on a Z-basis for $\mathfrak{m}$. Assuming that $\Lambda = \text{alg. int.}\{A\}$, Taussky **[244]** proved that the ideal class of $\mathfrak{m}^{-1}$ corresponds to the transpose M^{T}. Further, let $X \in (Z)_n$ carry the Z-basis $\{1, \alpha, \cdots, \alpha^{n-1}\}$ of Λ onto a Z-basis of the ideal $\mathfrak{m}$; such a matrix X is an *ideal matrix*. Taussky studied such ideal matrices in detail in **[246]** and **[247]**; see also **[248]** for a description of matrices U such that $M^{\mathsf{T}} = UMU^{-1}$.

Other references for this section: Bender **[15]**, Dade **[49]**.

3. Extensions. As general references for this section we cite Cartan-Eilenberg **[40]**, CR, Rotman **[218]**. We assume throughout that R is a noetherian domain, and that Λ is any R-order. Given a pair of Λ-lattices N, L, we may form the finitely generated R-module

$\operatorname{Ext}^1_\Lambda(N, L)$, whose elements are in one-to-one correspondence with classes of exact sequences of Λ-lattices $0 \to L \to M \to N \to 0$. Two such sequences are placed in the same class if there is a commutative diagram

$$\begin{array}{ccccccccc} 0 & \to & L & \to & M & \to & N & \to & 0 \\ & & 1\downarrow & & \downarrow & & 1\downarrow & & \\ 0 & \to & L & \to & M' & \to & N & \to & 0. \end{array}$$

(Addition of extensions is gotten from the Baer sum, and the elements of R act on Ext by acting on either one of the lattices.)

Hence if we are given a pair of Λ-lattices L and N, and wish to determine all Λ-lattices M containing L such that $M/L \cong N$, we must first calculate $\operatorname{Ext}^1_\Lambda(N, L)$, and then decide when extensions M from different classes are Λ-isomorphic. (See Reiner [**183**], for example.)

Let R_P denote the localization of R at a maximal ideal P of R. For each R-module T, we may set $T_P = R_P \otimes_R T$. Then Λ_P is an R_P-order, and each Λ-lattice L determines a Λ_P-lattice L_P.

(3.1). THEOREM. *Let L, N be Λ-lattices, and let P range over the set of maximal ideals of R. Then*:

(i) *For each P,*

$$\{\operatorname{Ext}^1_\Lambda(N, L)\}_P \cong \operatorname{Ext}^1_{\Lambda_P}(N_P, L_P).$$

(ii) *There is a monomorphism*

$$\operatorname{Ext}^1_\Lambda(N, L) \to \prod_P \operatorname{Ext}^1_{\Lambda_P}(N_P, L_P).$$

(iii) *If R is a Dedekind domain, and if there exists a nonzero ideal J in R such that $J \cdot \operatorname{Ext}^1_\Lambda(N, L) = 0$, then*

$$\operatorname{Ext}^1_\Lambda(N, L) \cong \sum_{P \supset J}{}^{\bullet} \operatorname{Ext}^1_{\Lambda_P}(N_P, L_P).$$

(iv) *If R^*_P denotes the P-adic completion of R_P, then*

$$R^*_P \otimes_{R_P} \operatorname{Ext}^1_{\Lambda_P}(N_P, L_P) \cong \operatorname{Ext}^1_{\Lambda^*_P}(N^*_P, L^*_P).$$

References. CR (75.25), Auslander-Goldman [**1**], Cartan-Eilenberg [**40**], DeLeeuw [**54**], Nunke [**74**], Reiner [**183**]. We remark that all of the above isomorphisms are natural.

An exact sequence of Λ-lattices

$$0 \to L \xrightarrow{f} M \xrightarrow{g} N \to 0 \tag{3.2}$$

is said to be *split* if $f(L)$ is a Λ-direct summand of M, or equivalently, if there exists a map $h \in \mathrm{Hom}_\Lambda(N, M)$ such that gh = identity map on N. From the definition of Ext, it follows at once that the sequence (3.2) is split if and only if it represents the zero class in $\mathrm{Ext}^1_\Lambda(N, L)$.

For each maximal ideal P of R, tensoring (3.2) with R_P yields an exact sequence of Λ_P-lattices

$$(3.3) \qquad 0 \to L_P \xrightarrow[f_P]{} M_P \xrightarrow[g_P]{} N_P \to 0.$$

Then (ii) of (3.1) may be rephrased as follows: the sequence (3.2) splits if and only if (3.3) splits for each P.

Suppose now that the exact sequence (3.2) of Λ-lattices is split as a sequence of R-modules. Then every element of M can be represented uniquely as an ordered pair (l, n), $l \in L$, $n \in N$. The action of the elements of Λ is given by a formula

$$(3.4) \qquad x(l, n) = (xl + F_x n, xn), \qquad x \in \Lambda,$$

where $F_x \in \mathrm{Hom}_R(N, L)$. We have thus obtained an R-homomorphism

$$(3.5) \qquad F: \Lambda \to \mathrm{Hom}_R(N, L),$$

given by $x \to F_x$, and it is easily seen that F satisfies the condition

$$(3.6) \qquad F_{xy} = x \cdot F_y + F_x \cdot y, \qquad x, y \in \Lambda.$$

Conversely, any R-homomorphism F (as in (3.5)) for which (3.6) holds gives rise to an extension of N by L by means of formula (3.4). Zassenhaus [**263**] calls F a *binding system*. For each $t \in \mathrm{Hom}_R(N, L)$, the map $x \to xt - tx$, $x \in \Lambda$, is an *inner binding system*. Assuming that N is R-projective, we have

$$\mathrm{Ext}^1_\Lambda(N, L) \cong \text{binding systems/inner binding systems.}$$

This holds in particular whenever R is a Dedekind domain, and N is any R-lattice.

Assuming now that R is a Dedekind domain, let M be any left Λ-lattice, and set $M^* = \mathrm{Hom}_R(M, R)$. Then M^* becomes a right Λ-lattice by means of the formula

$$(m^* x)m = m^*(xm), \qquad m^* \in M^*, x \in \Lambda, m \in M.$$

There is a natural isomorphism $(M^*)^* \cong M$ as left Λ-lattices. Every exact sequence (3.2) of left Λ-lattices gives rise to an exact sequence $0 \to N^* \to M^* \to L^* \to 0$ of right Λ-lattices. Thus there is an R-isomorphism

$$\mathrm{Ext}^1_\Lambda(N, L) \cong \mathrm{Ext}^1_\Lambda(L^*, N^*).$$

Consider now the special case where $\Lambda = RG$, and let M be any left Λ-lattice. The above-defined M^* can then be made into a *left* RG-lattice M', called the *contragredient* of M, by defining

$$(xm^*)m = m^*(x^{-1}m), \quad m^* \in M', \, x \in G, \, m \in M.$$

(If M affords a matrix representation $x \to M(x)$, $x \in G$, then M' affords the representation $x \to (M(x^{-1}))^{\mathsf{T}}$ where T denotes transpose.)

It is easily seen that $(RG)' \cong RG$ as left RG-modules, and that there is an R-isomorphism

$$\mathrm{Ext}^1_{RG}(N, L) \cong \mathrm{Ext}^1_{RG}(L', N')$$

for any pair of left RG-lattices N, L.

Call a left RG-lattice L *weakly injective* if every exact sequence $0 \to L \to M \to N \to 0$ is RG-split, where M, N are arbitrary RG-lattices. Then L is weakly injective if and only if its contragredient L' is projective.

4. **Higman ideal.** Let Λ be an R-order, where R is any noetherian domain. A Λ-*bimodule* is a two-sided Λ-module T such that

$$x(ty) = (xt)y, \qquad \alpha t = t\alpha, \quad x, y \in \Lambda, \; t \in T, \; \alpha \in R.$$

A 1-*cocycle* is a map $F \in \mathrm{Hom}_R(\Lambda, T)$ such that

$$F_{xy} = xF_y + F_x y, \qquad x, y \in \Lambda.$$

For each $t_0 \in T$, the map F given by

$$F_x = xt_0 - t_0 x, \qquad x \in \Lambda$$

is a 1-*coboundary*. The 1-*cohomology group* $H^1(\Lambda, T)$ is defined by

$$H^1(\Lambda, T) = \text{1-cocycles/1-coboundaries.}$$

For example, if N and L are Λ-lattices, then $T = \mathrm{Hom}_R(N, L)$ is naturally a Λ-bimodule; and if N is R-projective, then

$$H^1(\Lambda, T) = \mathrm{Ext}^1_\Lambda(N, L).$$

Given a finite group G of order g, a version of Maschke's theorem (see Zassenhaus [**263**], CR 73.22) asserts that

$$g \cdot H^1(RG, T) = 0$$

for all RG-bimodules T. To generalize this result to an arbitrary R-order Λ, we begin by setting

$$i(\Lambda) = \bigcap_T \mathrm{ann}_R\, H^1(\Lambda, T),$$

where T ranges over all Λ-bimodules, and ann_R denotes annihilator in R. Then $i(\Lambda)$ is an ideal in R, called the *Higman ideal* of Λ. (See D. G. Higman [**118**], [**119**], CR §75, for the results in this section.)

Recall that the algebra A is called *separable* over F if and only if any one of the following equivalent conditions holds:

(i) For each field E containing F, the algebra $F \otimes_E A$ is semisimple.

(ii) The algebra A is semisimple, and the centers of its simple components are separable field extensions of F.

(iii) There exists a nondegenerate bilinear form $f: A \times A \to F$ satisfying $f(ab, c) = f(a, bc)$, $a, b, c \in A$, and such that if $\{a_1, \cdots, a_n\}$ and $\{b_1, \cdots, b_n\}$ are F-bases of A for which $f(a_i, b_j) = \delta_{ij}$, then for some $x \in A$ we have

$$1 = \sum_1^n b_i x a_i.$$

(4.1). THEOREM. *The ideal $i(\Lambda)$ is nonzero if and only if A is separable over F. For any maximal ideal P of R,*

$$i(\Lambda_P) = \{i(\Lambda)\}_P, \qquad i(\Lambda_P^*) = i(\Lambda_P)^*.$$

If G is a finite group of order g, then $i(RG) = gR$.

Suppose now that A is separable over F, and that Λ is R-projective (the latter surely holds when R is a Dedekind domain). Starting with the bilinear form f in (iii) above, we may compute $i(\Lambda)$ as follows. Define

$$\textit{inverse different} = I(\Lambda) = \{a \in A : f(\Lambda, a) \subset R\},$$
$$\textit{different} = D(\Lambda) = \{x \in A : I(\Lambda)\cdot x \subset \Lambda\}.$$

(4.2). THEOREM. *An element α of R lies in $i(\Lambda)$ if and only if α is expressible in the form*

$$\alpha = \sum_{i=1}^n b_i x a_i$$

for some $x \in D(\Lambda)$, where the $\{a_i\}$, $\{b_i\}$ are as in (iii) *above.*

(For generalizations of the concept of the different, see Fossum [**84**], Watanabe [**258**], [**259**].)

Suppose that Λ is a hereditary order in the semisimple algebra A; then submodules of free Λ-modules are projective (Cartan-Eilenberg [**40**, Chapter I, §5]). Since each Λ-lattice is embeddable in a free

Λ-module, each Λ-lattice is projective. Thus for such an order Λ, we have $\mathrm{Ext}^1_\Lambda(N, L)=0$ for every pair of Λ-lattices N, L. On the other hand, it may well happen that the Higman ideal $i(\Lambda)$ is a proper ideal of R, and that $H^1(\Lambda, T)\neq 0$ for some T. Since representation theory is more concerned with extensions of modules than with cohomology of bimodules, it is of interest to investigate the R-annihilator of Ext. In this direction we have

(4.3). THEOREM. *Let R be a Dedekind domain of characteristic zero, G a finite group of order g, M an RG-lattice of R-rank m. Suppose that FM is an absolutely irreducible FG-module (that is, $\mathrm{Hom}_{FG}(FM, FM)\cong F$). Let $\alpha\in R$. Then*

$$\alpha\cdot \mathrm{Ext}^1_{RG}(M, N) = 0 \tag{4.4}$$

for every RG-lattice N if and only if $\alpha\in(g/m)R$. Dually, if FN is absolutely irreducible, then (4.4) *holds for each RG-lattice M if and only if $\alpha\in(g/n)R$, where n is the R-rank of N.*

References. Reiner **[189]**; see also D. G. Higman **[120]**, **[121]**.

This result has been generalized by Jacobinski **[125]** and Roggenkamp **[208a]**. Let $R=$alg. int.$\{F\}$, and set

$$FG = \sum{}^{\bullet} A_i \text{ (simple components)},$$
$$F_i = \text{center of } A_i,\ R_i = \text{alg. int.}\{F_i\}, \qquad (A_i:F_i) = n_i^2.$$

Let D_i^{-1} be the inverse different of R_i with respect to R, relative to the ordinary trace form from F_i to F. Define the integral ideal d_i of R by setting $d_i^{-1}=D_i^{-1}\cap F$.

Now let Λ_0 be any maximal order of FG containing RG. Since Λ_0 is a hereditary ring (see §7), $\mathrm{Ext}^1_{\Lambda_0}(N_0, L_0)=0$ for each pair of Λ_0-lattices N_0, L_0. Using this fact, together with arithmetic properties of maximal orders, one obtains

(4.5). THEOREM. *Let e_i be a primitive central idempotent of FG.*

(i) *$(g/n_i)d_i^{-1}$ is an integral ideal of R.*

(ii) *If M, N are RG-lattices for which either $e_iM=M$ or $e_iN=N$, then*

$$(g/n_i)d_i^{-1}\cdot \mathrm{Ext}^1_{RG}(M, N) = 0.$$

(iii) *Let M be an RG-lattice such that $e_iM=M$, and let $\alpha\in R$ have the property that $\alpha\cdot\mathrm{Ext}^1_{RG}(M, N)=0$ for every RG-lattice N. Then $\alpha\in(g/n_i)d_i^{-1}$.*

References. Parts (i), (ii) are due to Jacobinski **[126]**, and (iii) was proved by Roggenkamp **[208a]**.

We quote some miscellaneous results on the vanishing of Ext.

(4.6). THEOREM (BERMAN [21]). *Suppose that G has a normal cyclic Sylow p-subgroup, and let M, N be Z_p^*G-lattices such that $Q_p^*M \cong Q_p^*N$. Then*

$$\mathrm{Ext}^1_{Z_p^*G}(M, N) = 0.$$

Gudivok [100] remarks that this no longer holds if in place of Z_p^* we use R_P^*, where $R =$ alg. int.$\{F\}$.

Berman-Lihtman [30], generalized by Reiner [194], discuss the question as to when $\mathrm{Ext}^1_{RG}(M, N) = 0$ for all RG-lattices M, N such that FM, FN are a fixed pair of FG-modules.

5. **Representations over local domains.** Throughout this section let R be a discrete valuation ring with quotient field F, maximal ideal P, residue class field $\bar{R} = R/P$. Let R^* be the P-adic completion of R, with quotient field F^*. For any R-lattice M, let $M^* = R^*M$, $\bar{M} = M/PM \cong M^*/P^*M^*$. Now let Λ^* be any R^*-order; then $\bar{\Lambda} = \Lambda^*/P^*\Lambda^*$ is a finite dimensional $\bar{R}$-algebra. If $e \in \Lambda^*$ is idempotent, so also is $\bar{e}$. A fundamental result is

(5.1). THEOREM ON LIFTING IDEMPOTENTS. *Every idempotent in $\bar{\Lambda}$ is of the form $\bar{e}$ for some idempotent $e \in \Lambda^*$.*

References. CR §77 for this specific case. A more general discussion of the problem is given in Jacobson [132, Chapter III, §8].

This theorem is a generalization of Hensel's Lemma; its validity means that P-adic methods will play as large a role in integral representation theory as they already do in algebraic number theory. The most important consequence of (5.1) is

(5.2). KRULL-SCHMIDT THEOREM. *Every Λ^*-lattice is a finite direct sum of indecomposable lattices, which are uniquely determined up to Λ^*-isomorphism and order of occurrence.*

References. The result, due to Azumaya [3], was rediscovered by various authors; see Borevič-Faddeev [32, II], Reiner [186], [193], Swan [232].

The key step in the proof is

(5.3). PROPOSITION. *Let M be a Λ^*-lattice, and set $E(M) = \mathrm{Hom}_{\Lambda^*}(M, M)$, $\hat{E}(M) = E(M)/\mathrm{rad}\, E(M)$. Then M is indecomposable if and only if $\hat{E}(M)$ is a skewfield.*

(5.4). COROLLARY (REINER [193]). *Let M be a Λ^*-lattice; call M*

absolutely indecomposable if M remains indecomposable under all ground ring extensions. Then M is absolutely indecomposable if and only if $\hat{E}(M)$ is a field which is purely inseparable over $\overline{R}$.

Suppose now that Λ is an R-order in the separable F-algebra A, and let $i(\Lambda)$ be the Higman ideal of Λ. Then $i(\Lambda)=P^{k_0}$ for some k_0, and by (4.1), $i(\Lambda^*)=(P^*)^{k_0}$. For a Λ-lattice M, the Λ-module M/P^kM will be denoted by $\langle M\rangle_k$, and a similar notation will be used for Λ^*-lattices.

(5.5). LEMMA. *Let M, N be Λ-lattices, and let $k>k_0$. Given any $f\in \mathrm{Hom}_\Lambda(\langle M\rangle_k, \langle N\rangle_k)$, there exists a map $F\in\mathrm{Hom}_\Lambda(M, N)$ such that f and F induce the same map from $\langle M\rangle_{k-k_0}$ into $\langle N\rangle_{k-k_0}$.*

References. Maranda [**154**], D. G. Higman [**120**], [**121**], Nazarova-Roĭter [**169**].

Using the preceding lemma, one obtains

(5.6). THEOREM. *Let M, N be Λ-lattices, X a Λ^*-lattice.*

(i) *If $M\cong N$ then $\langle M\rangle_k\cong\langle N\rangle_k$ for all $k\geqq 0$.*

(ii) *If $\langle M\rangle_k\cong\langle N\rangle_k$ for some $k>k_0$, then $M\cong N$.*

(iii) *$M\cong N$ if and only if $M^*\cong N^*$.*

(iv) *If $\langle X\rangle_k$ is decomposable for some $k>k_0$, then X is decomposable.*

(v) *Suppose that for some $k>2k_0$, $\langle X\rangle_k$ contains a Λ^*-submodule U as $\langle R\rangle_k$-direct summand. Then X contains a Λ^*-sublattice Y as R^*-direct summand, and Y coincides with U modulo $(P^*)^{k-k_0}$.*

(vi) *If $k_0=0$, then $M\cong N$ if and only if $FM\cong FN$.*

References. These results are originally due to Maranda [**154**], with a somewhat weaker version of (iv). Improvements and generalizations are given in Heller [**109**], D. G. Higman [**121**], CR §76.

The passage from Λ^*-lattices to Λ-lattices is facilitated by

(5.7). PROPOSITION. *Let L be a Λ^*-lattice, V an A-module such that $F^*L=F^*V$. Set $M=L\cap V$; then M is a Λ-lattice for which $M^*=L$, $FM=V$.*

References. Bourbaki [**36**, Chapter 3, §3, no. 5]; Heller [**109**]; Takahashi [**238**]; CR §76.

(5.8). COROLLARY. *If A is a direct sum of full matrix algebras over F, then every Λ^*-lattice is of the form R^*M for some Λ-lattice M. The Krull-Schmidt Theorem holds for Λ-lattices. Every idempotent in $\overline{\Lambda}$ is of the form $\bar{e}$ for some idempotent $e\in\Lambda$.*

REMARK. See Heller [**109**], CR §76. The same conclusions are valid if we merely assume that A is a direct sum of full matrix algebras over skewfields S_i such that for each i, $S_i \otimes_{F^*} F$ is also a skewfield.

(5.9). THEOREM (REINER [186]). *Let* L, M, N *be* Λ*-lattices. Then*
(i) $M^{(t)} \cong N^{(t)}$ *if and only if* $M \cong N$.
(ii) $M \oplus L \cong N \oplus L$ *if and only if* $M \cong N$.

(5.10). THEOREM. *Let* F' *be a finite separable extension of* F, *and let* R' *be a valuation ring in* F' *containing* R. *For* M *any* Λ*-lattice, define*

$$M' = R' \otimes_R M, \qquad \Lambda' = R' \otimes_R \Lambda,$$

so that M' *is a* Λ'*-lattice. Then for any pair of* Λ*-lattices* M, N *we have* $M \cong N$ *if and only if* $M' \cong N'$.

References. Reiner-Zassenhaus [199]. See Bialnicki-Birula [31a] for generalization.

6. **Genus.** Throughout this section let A be a semisimple algebra over the algebraic number field F, and let R be any Dedekind domain with quotient field F. Let Λ be an R-order in A. By (4.1) the Higman ideal $i(\Lambda)$ is a nonzero ideal in R, and $i(\Lambda)$ annihilates $H^1(\Lambda, T)$ for every Λ-bimodule T. Let M, N be any Λ-lattices; since R is a Dedekind domain, from (1.2) we see that M is R-projective. The discussion in §3 then yields

$$\mathrm{Ext}^1_\Lambda(M, N) = H^1(\Lambda, \mathrm{Hom}_R(M, N)),$$

and consequently

$$(6.1) \qquad i(\Lambda) \cdot \mathrm{Ext}^1_\Lambda(M, N) = 0 \qquad \text{for any } \Lambda\text{-lattices } M, N.$$

We have denoted by R_P the localization of R at its maximal ideal P, and by R_P^* the P-adic completion of R. Two Λ-lattices M, N are in the same *genus* (notation: $M \vee N$) if

$$(6.2) \qquad FM \cong FN, \qquad M_P \cong N_P \quad \text{for each } P.$$

Denote by $\Gamma(M)$ the set of all Λ-lattices in the same genus as M. Since $FM \cong FN$ implies that $M_P \cong N_P$ whenever $P \not\supset i(\Lambda)$ (see (5.6)), in condition (6.2) it suffices to let P range over the maximal ideals of R containing $i(\Lambda)$. The requirement $FM \cong FN$ is usually superfluous, and is inserted here for emphasis, since in fact if $M_P \cong N_P$ for even one P, then automatically $FM \cong FN$. We remark also that by (5.6), $M_P \cong N_P$ if and only if $M_P^* \cong N_P^*$.

(Instead of using the Higman ideal, it is sometimes more convenient to introduce the finite set $\mathfrak{J}$ of all maximal ideals P of R for which Λ_P is not a maximal R_P-order in A. In that case, $M \vee N$ if and only if $FM \cong FN$ and $M_P \cong N_P$, $P \in \mathfrak{J}$. See (7.6).)

Let us introduce the semilocal Dedekind domain $\tilde{R}$ defined by

$$\tilde{R} = \bigcap_{P \supset i(\Lambda)} R_P.$$

Of course F is also the quotient field of $\tilde{R}$. We set $\tilde{\Lambda} = \tilde{R}\Lambda$, $\tilde{M} = \tilde{R}M$, and so on, so $\tilde{M}$ is a $\tilde{\Lambda}$-lattice.

(6.3). THEOREM. *Let M, N be Λ-lattices.*

(i) *$M \vee N$ if and only if $\tilde{M} \cong \tilde{N}$.*

(ii) $\mathrm{Ext}^1_\Lambda(M, N) \cong \mathrm{Ext}^1_{\tilde{\Lambda}}(\tilde{M}, \tilde{N})$.

(iii) *M is decomposable if and only if $\tilde{M}$ is decomposable.*

(iv) *$M \vee N$ if and only if for every nonzero ideal J in R, there exists a Λ-monomorphism $\phi: M \to N$ such that $N/\phi(M)$ is an R-torsion module whose R-annihilator is relatively prime to J. (In particular, if ϕ can be chosen so that* $\mathrm{ann}_R(N/\phi(M))$ *is relatively prime to $i(\Lambda)$, then $M \vee N$.)*

References. CR §81, Maranda [**153**], Reiner [**188**], Takahashi [**238**].

The following generalization of the "strong approximation theorem" of algebraic number theory is often useful.

(6.4). THEOREM. *Let W be any A-module. Let $\mathcal{S}$ be some finite set of maximal ideals of R, and suppose that for each $P \in \mathcal{S}$ we are given a Λ^*_P-lattice $X^{(P)}$ such that*

$$F^*_P \otimes_{R^*_P} X^{(P)} \cong F^*_P \otimes_F W$$

as A^*_P-modules. Then there exists a Λ-lattice M for which $FM \cong W$ and

$$R^*_P \otimes_R M \cong X^{(P)}, \qquad P \in \mathcal{S}.$$

References. Bourbaki [**36**, Chapter 7], Takahashi [**238**].

For an A-module W, we have denoted by $\sigma(W)$ the collection of Λ-lattices M such that $FM \cong W$. If M, $N \in \sigma(W)$, we call M and N *R_P-equivalent* if $M_P \cong N_P$.

(6.5). THEOREM (MARANDA [**153**], TAKAHASHI [**238**]). *The number of genera in $\sigma(W)$ equals the number of $\tilde{\Lambda}$-isomorphism classes in $\sigma(W)$, and this number is given by*

$$\prod_{P \supset i(\Lambda)} h_P,$$

where h_P is the number of classes in $\sigma(W)$ relative to R_P-equivalence.

To determine the number of Λ-isomorphism classes in $\sigma(W)$, we may first use (6.5) to count the number of genera in $\sigma(W)$. Then we are left with the question of finding the number $|\Gamma(M)|$ of Λ-isomorphism classes in the genus $\Gamma(M)$ containing a Λ-lattice M. It may happen that $|\Gamma(M)|$ is not the same for each $M \in \sigma(W)$; see for example Nazarova-Roĭter [**167**] for the case where $\Lambda = ZG$, $G = S_3$.

The A-module W is called *absolutely irreducible* if W is irreducible, and remains irreducible under arbitrary extension of the ground field. Since A is semisimple, W is absolutely irreducible if and only if $\mathrm{Hom}_A(W, W) \cong F$.

(6.6). THEOREM (MARANDA [**153**], TAKAHASHI [**238**]). *Let M be a Λ-lattice such that FM is absolutely irreducible. Then each $N \in \Gamma(M)$ is isomorphic to a Λ-lattice JM for some nonzero R-ideal J in F. Further, $J_1M \cong J_2M$ if and only if $J_2 = J_1\alpha$ for some $\alpha \in F$. Hence $|\Gamma(M)|$ equals the number of ideal classes in R.*

In studying questions of decomposability, it is convenient to introduce the following concept: let M, N be Λ-lattices, and call N a *local summand* of M if for each maximal ideal P of R, N_P^* is isomorphic to a Λ_P^*-direct summand of M_P^*. (It suffices to assume this condition for those P which contain $i(\Lambda)$; if $i(\Lambda) = R$, assume instead that FN is isomorphic to a direct summand of FM.)

(6.7). THEOREM. *Let M, N be Λ-lattices such that N is a local summand of M. Then there exists a decomposition $M = X \oplus Y$ into Λ-lattices, with $X \in \Gamma(N)$.*

References. Jones [**136**], Jacobinski [**128**], Reiner [**188**].

A much deeper result of this nature is

(6.8). THEOREM (JACOBINSKI [**129**]). *Let M, N be Λ-lattices such that N is a local summand of M. Assume that every irreducible A-module appearing as a composition factor of FN appears with greater multiplicity in FM. Then there exists a decomposition $M \cong N \oplus Y$ for some Λ-lattice Y.*

(6.9). COROLLARY. *Let M, N be Λ-lattices in the same genus. Then N is isomorphic to a direct summand of $M \dotplus M$.*

Call a Λ-lattice T *faithful* if no nonzero element of Λ annihilates T.

The following elegant result was proved by Roĭter [212], [217], and is essentially equivalent to Theorem 6.8.

(6.10). THEOREM. *Let M, N be Λ-lattices in the same genus, and let T be any faithful Λ-lattice. Then there exists a Λ-lattice $T' \in \Gamma(T)$ such that $M \dotplus T \cong N \dotplus T'$.*

As a consequence of this result, it is possible to prove

(6.11). THEOREM (ROĬTER [212], JACOBINSKI [129]). *Let Λ be an R-order. There exists a positive integer b depending only on Λ, such that $|\Gamma(M)| \leq b$ for every Λ-lattice M.*

We shall return to the study of genera in §8. Further references on genera are Faddeev [79], [80], [82] (the first of these articles is semi-expository), Drozd-Turčin [75], Jacobinski [128], [130], Roĭter [210]–[214], Drozd [70a], [706], Roggenkamp [205].

7. **Maximal orders.** The importance of the theory of maximal orders in integral representations has become increasingly evident from the work of Heller-Reiner [115], Jacobinski [126], [128], [129], and Swan [234]. We devote this section to a summary of the main results of the theory, and refer the reader to the following references: Auslander-Goldman [1], Chevalley [41], Deuring [56], Roggenkamp-Dyson [208b], Jacobson [131], Reiner [198], Schilling [222], Weil [260].

Assume throughout that R is a Dedekind domain with quotient field F. (The more general case where R is a noetherian integrally closed domain is treated in Auslander-Goldman [1]; see also Fossum [84a].) Let A be a finite dimensional F-algebra; an element $x \in A$ is *integral* over R if x is a zero of a monic polynomial with coefficients in R. The set of all such elements x is the *integral closure* of R in A, and is a ring when A is commutative.

If A is a simple algebra with center F, we may choose an extension field E of F for which $E \otimes_F A \cong (E)_m$. Then each $x \in A$ maps onto $1 \otimes x \in E \otimes A$, and is therefore representable as a matrix $\boldsymbol{M}(x) \in (E)_m$. The characteristic polynomial

$$\det(\lambda I - \boldsymbol{M}(x)) = \lambda^m - T(x)\lambda^{m-1} + \cdots + (-1)^m N(x)$$

has coefficients in F, and is called the *reduced characteristic polynomial* of x. Call $T(x)$ the *reduced trace* of x, and $N(x)$ the *reduced norm* of x. Then $T: A \to F$ is an F-linear map, while $N: A \to F$ is multiplicative.

Returning to the general case of an arbitrary F-algebra A, we call J a *full* R-lattice in A if J is an R-lattice for which $F \cdot J = A$. For such a lattice, define

$$\text{left order of } J = O_l(J) = \{x \in A : xJ \subset J\},$$
$$\text{right order of } J = O_r(J) = \{x \in A : Jx \subset J\}.$$

These are R-orders in J, and J is a left $O_l(J)$-lattice and a right $O_r(J)$-lattice. Put

$$J^{-1} = \{x \in A : JxJ \subset J\} = \{x \in A : Jx \subset O_l(J)\},$$

again a full R-lattice in A, called the *inverse* of J.

An R-order in A is a *maximal* R-order if it is not properly contained in any bigger R-order in A.

(7.1). THEOREM. *If A is a separable F-algebra,*[3] *every R-order in A is contained in a maximal R-order.*

(7.2). THEOREM. *Let A be a separable F-algebra, and suppose that $A = A_1 \oplus \cdots \oplus A_r$ (simple components). Let F_i be the center of A_i, and R_i = integral closure of R in F_i $(1 \leqq i \leqq r)$. Then the maximal R-orders Λ in A are precisely those orders of the form $\Lambda_1 \oplus \cdots \oplus \Lambda_r$, where each Λ_i is a maximal R-order in A_i. Further, maximal R-orders in A_i coincide with maximal R_i-orders in A_i. Every element of an R-order in A is integral over R. When A is commutative, $R_1 \oplus \cdots \oplus R_r$ is the unique maximal R-order in A.*

For the remainder of this section, *assume that A is a separable F-algebra.* The preceding theorem shows that the study of maximal orders can always be reduced to the central simple case. However, for the time being, we shall not make such a reduction.

(7.3). THEOREM. *Let Λ be any R-order in A, and let P range over the maximal ideals of R. The following statements are equivalent:*

(i) *Λ is a maximal R-order in A.*
(ii) *For each P, Λ_P is a maximal R_P-order in A.*
(iii) *For each P, Λ_P^* is a maximal R_P^*-order in A_P^*.*

REMARK. For each fixed P, (ii) and (iii) are equivalent.

In the case of central simple algebras, we have

(7.4). THEOREM. *Let Λ_P be an R_P-order in the simple algebra A with center F. Then Λ_P is a maximal R_P-order if and only if Λ_P is a left hereditary ring whose Jacobson radical* rad Λ_P *is the unique maximal two-sided ideal of Λ_P.*

[3] See §4. When char $F=0$, this merely means that A is semisimple. Without some hypothesis on A, maximal orders need not exist (see Deuring [56], Faddeev [80]).

(7.5). THEOREM. *Let Λ_P be a maximal R_P-order in the central simple F-algebra A. Then:*

(i) rad $\Lambda_P \supset P \cdot \Lambda_P$, *and* $\Lambda_P/\text{rad } \Lambda_P \cong \Lambda_P^*/\text{rad } \Lambda_P^* =$ *simple* (R/P)-*algebra.*

(ii) *Every full two-sided ideal in Λ_P is a power of* rad Λ_P.

(iii) *Every one-sided ideal in Λ_P (full or not) is principal.*

(iv) *Two Λ_P-lattices M, N are isomorphic if and only if $FM \cong FN$.*

(v) *The maximal R_P-orders in A are precisely the orders of the form $t^{-1}\Lambda_P t$, $t =$ unit in A.*

As a consequence of (7.2) and (7.5), we obtain

(7.6). THEOREM. *Let A be any separable algebra over F. Then every maximal R-order Λ in A is left and right hereditary, and every Λ-lattice is projective. If M and N are Λ-lattices such that $FM \cong FN$, then $M_P \cong N_P$ for each maximal ideal P of R.*

REMARK. There exist hereditary orders which are not maximal; see Brumer [**39**], Harada [**104**], [**105**], and also §13 of this article.

We turn next to the multiplicative theory of ideals in maximal orders. The first basic result is

(7.7). THEOREM. *Let J be a full R-lattice in A, where A is any separable F-algebra. Then $O_l(J)$ is a maximal order if and only if $O_r(J)$ is a maximal order.*

A *normal ideal* in A is a full R-lattice J in A both of whose orders $O_l(J)$, $O_r(J)$ are maximal. Call J *integral* if $J \subset O_l(J)$, or equivalently, if $J \subset O_r(J)$. If J is a normal ideal, so is J^{-1}, and the following relations are valid:

$$O_r(J^{-1}) = O_l(J), \quad O_l(J^{-1}) = O_r(J), \quad J \cdot J^{-1} = O_l(J), \quad J^{-1} \cdot J = O_r(J).$$

Let Λ be a maximal R-order in the separable F-algebra A. By a Λ-*ideal* in A we shall mean a full R-lattice J in A for which $O_l(J) = O_r(J) = \Lambda$. A *prime ideal* of Λ is a maximal Λ-ideal contained in Λ. Equivalently, a prime ideal is an integral Λ-ideal $\mathfrak{P}$ such that whenever $\mathfrak{P}$ contains the product of two integral Λ-ideals, it must contain one of the factors. If $\mathfrak{P}$ is a prime ideal of Λ, then $\Lambda/\mathfrak{P}$ is a simple algebra over the field R/P, where $P = R \cap \mathfrak{P}$.

(7.8). THEOREM. *The set $I\{\Lambda\}$ of Λ-ideals in A is an abelian group relative to the multiplication $J \cdot J'$ of Λ-ideals. The identity element of the group is Λ itself, and the inverse of J is J^{-1}. Further, $I\{\Lambda\}$ is the free abelian group generated by the prime ideals of Λ.*

If A is given as in (7.2), and $\Lambda = \Lambda_1 \oplus \cdots \oplus \Lambda_r$, then the prime ideals of Λ are obtained by replacing exactly one summand Λ_i by one of its prime ideals $\mathfrak{P}$. Furthermore, for each Λ_i there is a one-to-one correspondence $\mathfrak{P} \leftrightarrow R_i \cap \mathfrak{P}$ between the prime ideals $\mathfrak{P}$ of Λ_i and the maximal ideals $R_i \cap \mathfrak{P}$ of R_i. We discuss this in more detail below (see (7.13)).

While the theory of (two-sided) Λ-ideals in A is not too hard, that of one-sided ideals in Λ is more complicated, since together with Λ one must also consider other maximal orders as well. A product $J_1 J_2 \cdots J_n$ of normal ideals in A is called *proper* if $O_r(J_i) = O_l(J_{i+1})$, $1 \leqq i \leqq n-1$. This product is again a normal ideal, with left order $O_l(J_1)$ and right order $O_r(J_n)$.

(7.9). THEOREM. *Let J, J' be normal ideals with $J \supset J'$. Then there exist normal integral ideals N_1, N_2 such that J' is a proper product $N_1 J N_2$. If J and J' have the same left order, the factor N_1 may be omitted.*

A *maximal integral ideal*[4] is a normal ideal $\mathfrak{M}$ which is a maximal left ideal in its left order Λ (or equivalently, a maximal right ideal in its right order Λ'). The largest Λ-ideal $\mathfrak{P}$ contained in $\mathfrak{M}$ is a prime ideal of Λ, and $\Lambda/\mathfrak{M}$ is an irreducible module for the simple algebra $\Lambda/\mathfrak{P}$. We may characterize $\mathfrak{P}$ as the Λ-annihilator of $\Lambda/\mathfrak{M}$.

If Λ and Λ' are maximal orders in A, we may choose a normal ideal J_0 with left order Λ, right order Λ'. The map $N \rightarrow J_0^{-1} N J_0$, $N \in I\{\Lambda\}$, gives an isomorphism $I\{\Lambda\} \cong I\{\Lambda'\}$. This isomorphism is independent of the choice of J_0. The normal ideals N and $J_0^{-1} N J_0$ are said to be *similar.*

(7.10). THEOREM. *Let J be any integral normal ideal properly contained in its left order Λ. Then J is expressible as a proper product of maximal integral ideals $\mathfrak{M}_1, \cdots, \mathfrak{M}_l$:*

$$J = \mathfrak{M}_1 \cdots \mathfrak{M}_l.$$

The number l equals the number of composition factors of the left Λ-module Λ/J. The prime ideals associated with the $\{\mathfrak{M}_i\}$ are similar to the Λ-annihilators of the composition factors of Λ/J, and hence are uniquely determined by J up to similarity and order of occurrence. Finally, the set of integral normal ideals in A forms a groupoid[5] *relative to proper products.*

---

[4] Some authors use the term "indecomposable ideal" instead; this leads to some confusion when one studies lattices, and so we propose the clearer terminology "maximal integral ideal."

[5] This is *Brandt's groupoid*; see Jacobson [131, p. 132].

For the remainder of this section we shall concentrate our attention on the study of maximal orders in central simple algebras. Let A be a simple algebra with center F, and say $A \cong (D)_n$ where D is a skewfield with center F. We call $((D:F))^{1/2}$ the *index* of A. Let W be an irreducible left A-module, viewed as right D-space of dimension n. Then

$$(7.11) \qquad A = \mathrm{Hom}_D(W, W), \qquad D = \mathrm{Hom}_A(W, W).$$

We begin with a pair of theorems relating the study of maximal orders in A with those in D.

(7.12). THEOREM. *Let R be a Dedekind domain with quotient field F, and let Δ be some fixed maximal R-order in D. Let M be any full right Δ-lattice in W, and put $\Lambda = \mathrm{Hom}_\Delta(M, M)$. Then Λ is a maximal R-order in A. Every other maximal R-order in A is of the form $\mathrm{Hom}_\Delta(N, N)$ for some full right Δ-lattice N in W.*

Furthermore there is a one-to-one inclusion-preserving correspondence (the Morita correspondence) $X \leftrightarrow L$ between the set of all left Λ-lattices X and the set of all left Δ-lattices L, with

$$X = M \otimes_\Delta L, \qquad L = \tilde{M} \otimes_\Lambda X.$$

Here, $\tilde{M} = \mathrm{Hom}_\Delta(M, \Delta)$ is viewed as left Δ-, right Λ-lattice.

We may embed D in A by identifying each element $d \in D$ with a diagonal matrix all of whose diagonal entries are equal to d. Keeping the notation of the preceding theorem, we have

(7.13). THEOREM. *To each maximal ideal P of R there corresponds a unique prime ideal $\mathfrak{p}$ of Δ containing P, and*

$$P = \mathfrak{p} \cap R, \qquad P\Delta = \mathfrak{p}^e \quad \text{for some } e.$$

Set $\mathfrak{P} = \mathfrak{p}\Lambda$; then $\mathfrak{P}$ is the unique prime ideal of Λ containing P, and

$$P = \mathfrak{P} \cap R, \qquad P\Lambda = \mathfrak{P}^e, \qquad \Lambda/\mathfrak{P} \cong (\Delta/\mathfrak{p})_n.$$

Both $\Lambda/\mathfrak{P}$ and $\Delta/\mathfrak{p}$ are simple (R/P)-algebras.

Further, the group $I\{\Lambda\}$ of Λ-ideals in A is free abelian on the set of generators $\mathfrak{P}$, one for each maximal ideal P of R. If $\mathfrak{M}$ is any maximal left ideal of Λ, then $\mathfrak{M}$ contains a unique prime ideal $\mathfrak{P}$ of Λ, and $\mathfrak{M}$ is the inverse image under the map $\Lambda \to \Lambda/\mathfrak{P}$ of some maximal left ideal of $\Lambda/\mathfrak{P}$.

The two preceding theorems show how to reduce questions about maximal orders in simple algebras to the case of maximal orders in skewfields. Concerning these we have a basic "local" result:

(7.14). THEOREM. *Let R be a complete discrete valuation ring, D a skewfield with center F. Let $N: D \to F$ be the reduced norm map. Then an element $a \in D$ is integral over R if and only if $N(a) \in R$. The integral closure Δ of R in D is a ring, and is the unique maximal order in D. In Δ there is a unique maximal two-sided ideal $\mathfrak{p}$, given by*

$$\mathfrak{p} = \{a \in D : N(\alpha) \in P\}.$$

In fact, $\mathfrak{p}$ is a maximal left ideal of Δ, and every nonzero one-sided ideal of Δ is a power of $\mathfrak{p}$. The ideal $\mathfrak{p}$ is a principal left ideal, and $\Delta/\mathfrak{p}$ is a skewfield of finite dimension over R/P.

Suppose in particular that R is the P-adic completion of a ring of algebraic integers, and let m be the index of D. In that case $\Delta/\mathfrak{p}$ is a field of dimension m over R/P, and furthermore $P\Delta = \mathfrak{p}^m$.

Turning now to the global case, let A be a central simple F-algebra, where F is an algebraic number field which is the quotient field of some Dedekind domain R. If P is any maximal ideal of R, we may form the central simple F_P^*-algebra A_P^*; the index m_P of A_P^* is called the *P-index* of A. Set $A_P^* \cong (S)_t$, where S is a skewfield of index m_P; let $\bar{S}$ denote the residue class field of the maximal R_P^*-order in S modulo its prime ideal.

(7.15). THEOREM. *Let Λ be a maximal R-order in the central simple F-algebra A, where F is an algebraic number field. Let P be a maximal ideal of R, and $\mathfrak{P}$ the corresponding prime ideal of Λ. Denote by m_P the P-index of A. Then (using the notation above)*

$$\Lambda/\mathfrak{P} \cong (\bar{S})_t, \qquad (\bar{S} : R/P) = m_P, \qquad P\Lambda = \mathfrak{P}^{m_P}.$$

Further, if $\mathfrak{d}$ is the different of Λ with respect to R relative to the reduced trace map from A to F, then $\mathfrak{P}$ appears with exponent $m_P - 1$ in $\mathfrak{d}$. (If $m_P > 1$, we say that A is ramified *at P.)*

Keeping the above notation, let $N: A \to F$ be the reduced norm map. For a full R-lattice J in A, define its *reduced norm* $N(J)$ to be the R-ideal of F generated by the set of elements $\{N(a) : a \in J\}$.

(7.16). THEOREM. *If $\mathfrak{M}$ is a maximal integral ideal, its reduced norm $N(\mathfrak{M})$ is the maximal ideal $\mathfrak{M} \cap R$ of R. If J is any normal integral ideal with left order Λ, then $N(J)$ is the product of the R-annihilators of the Λ-composition factors of Λ/J. Further, if JJ' is a proper product of normal ideals, then*

$$N(JJ') = N(J) \cdot N(J'), \qquad N(J^{-1}) = N(J)^{-1}.$$

We have denoted by $I\{R\}$ the group of R-ideals in F. Set

$$(7.17) \qquad I'\{R\} = \{N(a)R : a = \text{unit in } A\}.$$

Then it is clear that the reduced norm maps the set of normal ideals in A *onto* $I\{R\}$, carrying the set of principal ideals onto $I'\{R\}$. The factor group $I\{R\}/I'\{R\}$ is finite.

A *prime* of the algebraic number field F is an equivalence class of valuations of F. The *finite primes* are those which come from P-adic valuations of F, where P is a maximal ideal in alg. int.$\{F\}$. On the other hand, an *infinite prime* P is one whose class of valuations extends the ordinary absolute value on the rational field Q. Let F_P^* be the P-adic completion of F, where P is some infinite prime of F; then either $F_P^* = \mathbf{R}$ (real field) or $F_P^* = \mathbf{C}$ (complex field), and there is an embedding $\epsilon_P: F \rightarrow F_P^*$.

Now let A be any central simple F-algebra, and for P an infinite prime of F, put $A_P^* = F_P^* \otimes_F A$, a central simple F_P^*-algebra. Then $A_P^* \cong (S)_t$ for some skewfield S with center F_P^*, and there are only the following possibilities:

(i) $F_P^* = \mathbf{C}$, $\quad S = \mathbf{C}$.

(ii) $F_P^* = \mathbf{R}$, $\quad S = \mathbf{R}$.

(iii) $F_P^* = \mathbf{R}$, $\quad S = \mathbf{H}$ (=quaternion skewfield over $\mathbf{R}$).

In case (iii), we say that P *ramifies* in A.

Hence for any prime P (finite or infinite) of F, we say that P ramifies in A if and only if A_P^* is not a full matrix algebra over F_P^*. The following theorems are rather deep.

(7.18). THEOREM (HASSE). *If A is not a full matrix algebra over F, then at least two primes (finite or infinite) of F must ramify in A.*

(7.19). THEOREM (HASSE). *A nonzero element $\alpha \in F$ is the reduced norm $N(a)$ of some unit $a \in A$ if and only if $\epsilon_P(\alpha) > 0$ at every infinite prime P of F which ramifies in A. Hence the group $I'\{R\}$ defined in (7.17) is given by*

$I'\{R\} = \{\alpha R : \alpha \in F,\ \epsilon_P(\alpha) > 0$ at every infinite prime P of F ramified in $A\}$.

Call A a *totally definite quaternion algebra* if A is a skewfield with center F, such that for each infinite prime P of F, $A_P^* \cong \mathbf{H}$. This can occur only when $(A:F) = 4$ and $F_P^* = \mathbf{R}$ for every infinite prime P of F.

(7.20). THEOREM (EICHLER). *Suppose that A is not a totally definite quaternion algebra, and let J be a normal ideal in A. Then J is a principal ideal if and only if the reduced norm $N(J)$ lies in $I'\{R\}$.*

8. **Further results on genera.** Throughout this section let R be a Dedekind domain whose quotient field F is an algebraic number field,

and let A be a semisimple F-algebra:

$$\begin{aligned} (8.1)\qquad & A = A_1 \oplus \cdots \oplus A_r \quad \text{(simple components).} \\ & A_i \cong (D_i)_{n_i}, \qquad D_i = \text{skewfield with center } F_i. \\ & R_i = \text{integral closure of } R \text{ in } F_i. \end{aligned}$$

Denote by e_i the central idempotent of A generating A_i. As in §7, let $I\{R_k\}$ be the ideal group in F_k. Let $N_k: A_k \rightarrow F_k$ be the reduced norm map, and as in (7.17) and (7.19), let

$$I'\{R_k\} = \{N_k(a) \cdot R_k : a = \text{unit in } A_k\}.$$

Each central idempotent $e \in A$ is expressible as a sum $e_{i_1} + \cdots + e_{i_s}$, say. Define

$$(8.2)\qquad I\{e\} = \prod_{\nu=1}^{s} I(R_{i_\nu}), \qquad I'\{e\} = \prod_{\nu=1}^{s} I'\{R_{i_\nu}\}.$$

These will be used later.

Now let M be a left Λ-lattice, where Λ is any R-order in A, and set

$$E(M) = \mathrm{Hom}_\Lambda(M, M), \qquad E(FM) = \mathrm{Hom}_A(FM, FM).$$

(8.2a). LEMMA. *$E(M)$ is an R-order in $E(FM)$. If Λ is a maximal order in A, then $E(M)$ is a maximal R-order in $E(FM)$.*

We shall view M as left Λ-, right $E(M)$-lattice. Define

$$(8.3)\qquad e_M = \sum_{e_i M \neq 0} e_i = e_{i_1} + \cdots + e_{i_s} \text{ (say).}$$

Then e_M is a central idempotent of A which depends only on FM, and we have

$$(8.4)\qquad E(FM) \cong \sum_{\nu=1}^{s}{}^{\cdot} (D_{i_\nu})_{k_\nu},$$

where $A_{i_\nu} \cdot FM$ is a direct sum of k_ν irreducible A-modules. Now let $\mathcal{S}(\Lambda)$ be the family of all left Λ-lattices M such that no simple component of $E(FM)$ is a totally definite quaternion algebra (see end of §7). Thus $M \in \mathcal{S}(\Lambda)$ if and only if for each ν, $1 \leqq \nu \leqq s$, either $k_\nu > 1$ or else D_{i_ν} is not a totally definite quaternion algebra. We shall say that the Λ-lattice M *satisfies the Eichler condition* if and only if $M \in \mathcal{S}(\Lambda)$; of course this is really a condition on the A-module FM.

We have denoted by $\Gamma(M)$ the genus of M, and by $|\Gamma(M)|$ the number of Λ-isomorphism classes in $\Gamma(M)$. Suppose in particular that Λ_0 is a maximal R-order in A, and that M_0 is a left Λ_0-lattice. Since $E(M_0)$ is a maximal order in $E(FM_0)$, it can be written as

$$E(M_0) = \sum_{\nu=1}^{s}{}^{\boldsymbol{\cdot}}\,\Omega_{i_\nu}, \quad \text{where} \quad e_{M_0} = e_{i_1} + \cdots + e_{i_s},$$

and Ω_{i_ν} is a maximal R_{i_ν}-order in the i_νth simple component of $E(FM_0)$. Each full left $E(M_0)$-ideal J in $E(FM_0)$ likewise decomposes as $J = \sum^{\boldsymbol{\cdot}} J_{i_\nu}$. Define the *reduced norm*

$$N^*(J) = \prod_{i=1}^{s} N_{i_\nu}(J_{i_\nu}) \in I\{e_{M_0}\}.$$

From §7 we see that if J is principal then $N^*(J) \in I'\{e_{M_0}\}$; and if $M_0 \in \mathcal{S}(\Lambda_0)$, then the converse holds true by (7.20). Hence we obtain

(8.5). THEOREM. *Let Λ_0 be a maximal R-order in A, and let M_0, N_0 be left Λ_0-lattices.*

(i) *$N_0 \in \Gamma(M_0)$ if and only if $FN_0 \cong FM_0$.*

(ii) *Each $N_0 \in \Gamma(M_0)$ is Λ_0-isomorphic to $M_0 J$ for some full left $E(M_0)$-ideal J in $E(FM_0)$.*

-(iii) *For J_1, J_2 full left $E(M_0)$-ideals in $E(FM_0)$, $M_0 J_1 \cong M_0 J_2$ if and only if $J_1 = J_2 a$ for some unit $a \in E(FM_0)$.*

(iv) *The reduced norm map N^* maps the set of all full left $E(M_0)$-ideals onto $I\{e_{M_0}\}$, carrying principal ideals onto $I'\{e_{M_c}\}$.*

(v) *If M_0 satisfies the Eichler condition, then J is principal if and only if $N^*(J) \in I'\{e_{M_0}\}$. Therefore*

$$|\,\Gamma(M_0)\,| = [I\{e_{M_0}\} : I'\{e_{M_0}\}],$$

and this index is finite.

References. Chevalley [41], Jacobinski [128], [129].

Consider now an arbitrary R-order Λ in A, and choose some maximal order Λ_0 containing Λ. Let $\mathfrak{f}$ be a full two-sided Λ_0-ideal in Λ (such exist: take $\mathfrak{f} = \alpha\Lambda_0$, with suitable $\alpha \in R$). Given a Λ-lattice M, we may form $\Lambda_0 M$ in FM, and then $M_0 = \Lambda_0 M$ is a Λ_0-lattice in FM. Call a full left ideal J in $E(M_0)$ *relatively prime* to $\mathfrak{f}$ if for each maximal ideal P of R, either $\mathfrak{f}_P = (\Lambda_0)_P$ or else $J_P = \{E(M_0)\}_P$.

(8.6). THEOREM (JACOBINSKI [128]). *Every Λ-lattice N in the genus of M is Λ-isomorphic to $M \cap M_0 J$ for some full left ideal J in $E(M_0)$ relatively prime to $\mathfrak{f}$. Further, $M \cap M_0 J \cong M$ if and only if $J = E(M_0)a$ for some $a \in E(M)$.*

Let $\mathfrak{a} = \prod \mathfrak{a}_i$ be an ideal in the ideal group $I\{e_M\}$; here i ranges over all indices such that $e_i M \neq 0$. We call $\mathfrak{a}$ *relatively prime* to $\mathfrak{f}$ if for each maximal ideal P of R such that $\mathfrak{f}_P \neq (\Lambda_0)_P$, the numerator and denominator of each R_i-ideal $\mathfrak{a}_i$ in F_i is prime to P. The collection of all such ideals $\mathfrak{a}$ is a subgroup $I\{e_M, \mathfrak{f}\}$ of finite index in $I\{e_M\}$. We now let $I'\{e_M, \mathfrak{f}\}$ be the subgroup of $I\{e_M, \mathfrak{f}\}$ generated by all principal ideals of the form $N^*(\Lambda_0 a)$, $a \in E(M)$, which are prime to $\mathfrak{f}$. Define

(8.7) $$V(M, \mathfrak{f}) = I\{e_M, \mathfrak{f}\} \Big/ I'\{e_M, \mathfrak{f}\},$$

a finite multiplicative group.

(8.8). THEOREM (JACOBINSKI [128], [129]). *The mapping which assigns to $M \cap M_0 J$ the coset of $N^*(J)$ in $V(M, \mathfrak{f})$ carries the set of Λ-isomorphism classes in $\Gamma(M)$ onto $V(M, \mathfrak{f})$. When M satisfies the Eichler condition, the mapping is one-to-one, and*

$$|\Gamma(M)| = \text{order of } V(M, \mathfrak{f}).$$

This result has many interesting consequences:

(8.9). COROLLARY (JACOBINSKI [128], [129], ROĬTER [212]). *Given an R-order Λ, there exists a positive integer k depending only on Λ, such that for any pair of Λ-lattices M and N, we have $N \in \Gamma(M)$ if and only if $M^{(k)} \cong N^{(k)}$.*

(8.10). COROLLARY (JACOBINSKI [128]). *For each given R-order Λ, there exists a finite extension field F' of F such that for any pair of Λ-lattices M and N, $N \in \Gamma(M)$ if and only if*

$$R' \otimes_R M \cong R' \otimes_R N \qquad \text{as } R' \otimes_R \Lambda\text{-lattices.}$$

Here, R' denotes the integral closure of R in F'.

We remark that, in general, $R' \otimes M \cong R' \otimes N$ need not imply that $M \cong N$ (see Berman-Gudivok [27], for instance).

The *strict genus* of the Λ-lattice M, denoted by $\Gamma^s(M)$, consists of all $N \in \Gamma(M)$ such that $\Lambda_0 M \cong \Lambda_0 N$ for some maximal order Λ_0 containing Λ. Jacobinski [128] showed that when M satisfies the Eichler condition, then this isomorphism holds for every Λ_0 as soon as it is valid for one Λ_0. Hereafter let M satisfy the Eichler condition; up to Λ-isomorphism, the Λ-lattices in $\Gamma(M)$ are of the form $N = M \cap M_0 J$ as in (8.6), and it turns out that $N \in \Gamma^s(M)$ if and only if $J = E(M_0)b$ for some $b \in E(M_0)$. Using the fact that each class of full left ideals in $E(M_0)$ contains an ideal relatively prime to $\mathfrak{f}$, one obtains

(8.11). THEOREM (JACOBINSKI [128]). *If M satisfies the Eichler condition, the genus $\Gamma(M)$ splits into h strict genera, with h equal to the number of classes of full left ideals in $E(M_0)$. Each strict genus $\Gamma^s(M)$ contains the same number $|\Gamma^s(M)|$ of Λ-isomorphism classes. Further, $h = [I\{e_M\} : I'\{e_M\}]$.*

Takahashi [**238**] and Jacobinski [**128**] compute $|\Gamma^{\bullet}(M)|$ as the number of double cosets in some group.

(8.12). THEOREM (JACOBINSKI [**128**], [**129**]). *Let M and N be Λ-lattices satisfying the Eichler condition. Let T be a faithful*[6] *Λ_0-lattice satisfying the Eichler condition, where Λ_0 is a maximal order containing Λ. Then $N \in \Gamma^{\bullet}(M)$ if and only if $M \dotplus T \cong N \dotplus T$ as Λ-lattices.*

(8.13). THEOREM. *Let Λ be an order in a simple algebra, and let M, N be Λ-lattices in the same genus satisfying the Eichler condition. Then N is isomorphic to a maximal sublattice of M.*

References. Drozd [**70a**], Jacobinski [**130**]. This provides a partial answer to a question raised by Roĭter [**212**].

9. **Projective modules and relative projective modules.** Let R be a Dedekind ring with quotient field F, and let Λ be an R-order in the F-algebra A. A left Λ-lattice M is *projective* if M is a direct summand of a free module, or equivalently, if every exact sequence $0 \rightarrow U \rightarrow V \rightarrow M \rightarrow 0$ of Λ-modules splits.

(9.1). THEOREM. *Let R be a discrete valuation ring with maximal ideal P, and set $\bar{R} = R/P$, $\bar{\Lambda} = \Lambda/P\Lambda$, $\bar{M} = M/PM$.*

(i) *A left Λ-lattice M is projective if and only if $\bar{M}$ is a projective $\bar{\Lambda}$-module.*

(ii) *If M and N are projective Λ-modules, then $M \cong N$ if and only if $\bar{M} \cong \bar{N}$.*

References. Nakayama [**161**]–[**163**], Reiner [**181**], Swan [**232**], CR §77.

Much deeper results can be obtained when $\Lambda = RG$, as indicated below.

(9.2). THEOREM. *Let R be a discrete valuation ring in the algebraic number field F, and let G be a finite group. For a pair of projective RG-lattices M, N, we have $M \cong N$ if and only if $FM \cong FN$.*

References. This was first proved by Swan [**232**]. Other proofs are due to Rim (unpublished), Giorgiutti [**89**], and Bass [**8**]; these make use of the nonsingularity of the Cartan matrix of $\bar{R}G$ (see CR §77). A proof of an entirely different nature may be found in Hattori [**107**].

The preceding theorem is one of the key steps in proving the following striking result on projective RG-lattices in the global case.

(9.3). THEOREM (SWAN [**232**]). *Let R be any Dedekind domain of*

[6] This means that for $x \in \Lambda_0$, $xT = 0$ implies $x = 0$.

characteristic zero, with quotient field F, and let G be a finite group of order g. Suppose that every prime divisor of g is a nonunit in R. Let P range over the maximal ideals of R.

(i) *An RG-lattice M is projective if and only if for each P dividing gR, M_P is R_PG-projective, or equivalently, M/PM is $(R/P)G$-projective.*

(ii) *If M is a projective RG-lattice, then FM is FG-free, and for each P, M_P is R_PG-free and M/PM is $(R/P)G$-free. Hence an RG-lattice is projective if and only if it is in the same genus as some free RG-lattice.*

(iii) *For each projective RG-lattice M there exists a decomposition*

$$M \cong RG^{(r)} \dotplus L$$

for some $r \geqq 0$ and some projective left ideal L in RG. Here, $RG^{(r)}$ denotes the direct sum of r copies of RG. The ideal L is in the same genus as RG, and $F\cdot L = FG$.

REMARKS. 1. Part (i) is valid more generally: if Λ is an R-order in a separable algebra over F, then a Λ-lattice M is projective if and only if M_P is projective for each P dividing the Higman ideal $i(\Lambda)$.

2. The crucial result in the above is part (ii), and especially the fact that if M is projective then FM is free. This readily implies the remaining results (see CR §78). It also shows that the group ring RG contains no nontrivial idempotent elements; for a direct proof of this, see Coleman [**44**].

Now let M, N be any RG-lattices; then $M \otimes_R N$ is also an RG-lattice, with the action of G given by

$$x(m \otimes n) = xm \otimes xn, \qquad x \in G, m \in M, n \in N.$$

(9.4). THEOREM (SWAN [**232**]). *If M is any R-free RG-lattice, then the RG-lattice $RG \otimes_R M$ (with diagonal action of G, as above) is RG-free on $(FM:F)$ generators. Consequently if L is any RG-lattice and N is RG-projective, then $L \otimes_R N$ is also RG-projective.*

We turn now to a discussion of the relations between RG-lattices and RH-lattices, where H is a subgroup of G. Given an RH-lattice V, define the *induced* RG-lattice V^G as

$$V^G = RG \otimes_{RH} V,$$

with the elements of G acting on the left on the first factor only. If $G = \bigcup x_i H$, then $V^G = \sum^{\cdot} x_i \otimes V$ (as R-modules). On the other hand, each RG-lattice M can be restricted to H to give an RH-lattice M_H.

(9.5). THEOREM. *Let $H \subset K \subset G$ be finite groups, V an RH-lattice, M an RG-lattice, and let V' denote the contragredient of V (see end of §3).*

Then:

(i) *Induction is transitive*: $(V^K)^G \cong V^G$.

(ii) *Frobenius reciprocity theorem holds*: $V^G \otimes_R M \cong (V \otimes_R M_H)^G$.

(iii) $(V')^G \cong (V^G)'$.

(iv) *The Mackey subgroup theorem and tensor product theorem are valid.*

References. The proofs in CR §§38, 43, 44 carry over unchanged to the present situation. The Mackey theorems (Mackey [**153**]) are stated in CR(44.2), (44.3).

If M, N are lattices we write $M|N$ to indicate that M is isomorphic to a direct summand of N.

(9.6). THEOREM. *Let H be a subgroup of G, $V = RH$-lattice, $M = RG$-lattice.*

(i) $V|(V^G)_H$ *always.*

(ii) $M|(M_H)^G$ *if* $[G:H]$ *is a unit in* R.

References. See CR(63.6), (63.7); the second result is originally due to D. G. Higman [**116**].

We introduce next the concept of relative projective lattices. Let M be an RG-lattice, and H a subgroup of G. We call M *(G, H)-projective* if any one of the following equivalent conditions holds true:

(i) $M|(M_H)^G$.

(ii) $M|L^G$ for some RH-lattice L.

(iii) Every exact sequence of RG-lattices $0 \to X \to Y \to M \to 0$ which is RH-split is also RG-split.

(iv) Every exact sequence of RG-lattices $0 \to M \to Y \to W \to 0$ which is RH-split is also RG-split.

(v) Let $G = \bigcup x_i H$. Then there exists an element $\gamma \in \mathrm{Hom}_{RH}(M, M)$ such that $\sum_i x_i \, \gamma x_i^{-1} =$ identity map on M. (See CR §63.)

This concept plays a vital role in Green's theory of vertices and sources of indecomposable RG-lattices. While the theory originally dealt with FG-modules, the entire discussion carries over almost unchanged to RG-lattices, provided the domain R is sufficiently nice so that the Krull-Schmidt theorem holds for RG-lattices (see 5.2).

(9.7). THEOREM. *Let R be a complete discrete valuation ring, and let M be an indecomposable RG-lattice.*

(i) *There exists a subgroup H of G, called the vertex of M, such that M is (G, H)-projective, and such that for K a subgroup of G, M is (G, K)-projective if and only if K contains a conjugate of H. The subgroup H is uniquely determined by M up to conjugacy in G.*

(ii) *If $H =$ vertex of M, then there exists an indecomposable RH-lattice*

L (unique up to conjugacy) such that $M | L^G$. Call L the source of M.

(iii) *If the rational prime p is a nonunit in R, then the vertex of an indecomposable lattice is always a p-group.*

(iv) *Let K be any subgroup of G such that $[G:K]$ is a power of p, where p is a prime which is a nonunit in R. Let V be any absolutely indecomposable*[7] *RG-lattice. Then V^G is an absolutely indecomposable RG-lattice.*

References. Green [**90**], [**91**], [**93**], CR §65.

Other references for this section: Thompson [**253**], Dress [**59**], [**61**]–[**63**], Conlon [**44a**]–[**44d**].

10. **Grothendieck groups and Whitehead groups.** Let Λ be a ring with unity element, and let $\mathcal{C}$ be some category of Λ-modules (always left finitely generated modules). Let $\mathcal{S}$ be some collection of short exact sequences

$$(10.1) \qquad 0 \to L \to M \to N \to 0, \qquad L, M, N \in \mathcal{C}.$$

Form the free abelian group $\mathfrak{A}$ on the symbols (M), one for each isomorphism class of Λ-modules in $\mathcal{C}$. Let $\mathfrak{A}_0$ be the subgroup of $\mathfrak{A}$ generated by all expressions $(M)-(L)-(N)$, where in (10.1) we use all sequences in the given collection $\mathcal{S}$. We call the additive group $\mathfrak{A}/\mathfrak{A}_0$ the *Grothendieck group* of $\mathcal{C}$ relative to $\mathcal{S}$, and denote it by $K(\mathcal{C}, \mathcal{S})$. For each $M \in \mathcal{C}$ the image of (M) in $K(\mathcal{C}, \mathcal{S})$ will be written as $[M]$.

As general references for this topic, we cite Bass [**14**], Heller [**109**], Swan [**235**].

Of special interest for us are the following cases:

(10.2). Choose $\mathcal{C}$ the category of all Λ-modules, $\mathcal{S}$ the collection of all short exact sequences from $\mathcal{C}$. In this case $K(\mathcal{C}, \mathcal{S})$ is called the *Grothendieck group of* Λ, and will be denoted by $K(\Lambda)$ for brevity.

(10.3). Take $\mathcal{C}$ to be the category of all projective Λ-modules, and $\mathcal{S}$ the set of all short exact sequences from $\mathcal{C}$ (these are necessarily split). We call this $K(\mathcal{C}, \mathcal{S})$ the *projective class group* of Λ, and denote it by $P(\Lambda)$. (See Rim [**200**], [**201**].)

For the remainder of this section, we assume that Λ is an R-order in the F-algebra A, where R is any noetherian domain.

(10.4). Let $\mathcal{C}$ be the category of all Λ-lattices, $\mathcal{S}$ the set of all short exact sequences from $\mathcal{C}$. Denote by $K_f(\Lambda)$ the Grothendieck group $K(\mathcal{S}, \mathcal{C})$ thus obtained. An easy argument (see Swan [**232**]) shows that the obvious map $[M] \to [M]$ gives an isomorphism $K_f(\Lambda) \cong K(\Lambda)$.

[7] See (5.4).

The Grothendieck group $K(A)$ is easily described: if $X_1, \cdots, X_n$ are a full set of nonisomorphic irreducible A-modules, then $K(A)$ is the free abelian group with basis $[X_1], \cdots, [X_n]$. (The same holds true for any artinian ring A.) Thus when G is a finite group and char $F=0$, we may identify $K(FG)$ with the ring of generalized characters of FG-modules. On the other hand when char $F\neq 0$, we may identify $K(FG)$ with the ring of generalized Brauer characters of FG-modules (see CR §§38, 82).

In analyzing the structure of $K(\Lambda)$, one begins by comparing $K(\Lambda)$ with $K(\Lambda_S)$, where Λ_S is gotten from Λ by passing from R to a ring of quotients R_S. Specifically, we quote

(10.5). THEOREM (SWAN [232], [234]). *Let S be any multiplicative subset of R (with $1\in S$, $0\notin S$), and let R_S be the ring of quotients $\{\alpha/\beta:\alpha\in R, \beta\in S\}$. Set $\Lambda_S=R_S \otimes_R \Lambda$. Then there is an exact sequence of additive groups.*

$$\sum_P{}^{\cdot} K\left(\frac{\Lambda}{P\Lambda}\right) \xrightarrow{\eta} K(\Lambda) \xrightarrow{\theta} K(\Lambda_S) \rightarrow 0,$$

where in the first direct sum P ranges over all those prime ideals in R for which $P\cap S$ is nonempty. The map θ is given by $[M]\rightarrow[M_S]$, $M=\Lambda$-module. The map η is defined on each $K(\Lambda/P\Lambda)$ by viewing $(\Lambda/P\Lambda)$-modules as Λ-modules.

The result can be greatly strengthened when Λ is a group ring. Call a Dedekind domain R *semilocal* if R has only a finite number of prime ideals. The following difficult and important result is due to Swan [234]:

(10.6). THEOREM. *Let R be a semilocal Dedekind domain, and let G be a finite group. Then there is an isomorphism $\theta:K(RG)\cong K(FG)$, where θ is the map defined by $[M]\rightarrow[FM]$ (as in (10.5), with S chosen to be the set of nonzero elements of R).*

(10.7). COROLLARY. *Let G be a finite group, and let $\mathfrak{Z}$ be some finite set of prime ideals of the Dedekind domain R. There is an exact sequence of additive groups*

$$\sum_{P\notin\mathfrak{Z}}{}^{\cdot} K\left(\frac{R}{P}G\right) \xrightarrow{\eta} K(RG) \xrightarrow{\theta} K(FG) \rightarrow 0,$$

where in the direct sum P ranges over all nonzero prime ideals of R distinct from those in the preassigned set $\mathfrak{Z}$.

A projective Λ-module M is *special* if FM is free as A-module. Let $P_0(\Lambda)$ be the Grothendieck group of the category of all special projective Λ-modules, using for $\mathcal{S}$ the collection of all sequences of such modules. Equivalently, $P_0(\Lambda)$ is the subgroup of $P(\Lambda)$ generated by special projectives. Denote by $C_0(\Lambda)$ the subgroup of $P_0(\Lambda)$ generated by the set of all differences

$$\{[M]-[N], \quad M, N \text{ special projective } \Lambda\text{-modules}, FM \cong FN\}.$$

Terminology:

$P(\Lambda)$ = projective class group of Λ.
$P_0(\Lambda)$ = special projective class group of Λ.
$C_0(\Lambda)$ = reduced special projective class group of Λ.

(See Rim [**200**], [**201**], Strooker [**229**], for further discussion of these concepts.)

There is an obvious mapping

$$\alpha: P(\Lambda) \to K(\Lambda)$$

gotten by mapping $[M]$ onto $[M]$, for M any projective Λ-module. This induces a map $C_0(\Lambda) \to K(\Lambda)$.

An immediate consequence of (10.7) is

(10.8). THEOREM (SWAN [**234**]). *Let R be a Dedekind domain such that* char $F \nmid [G:1]$. *Then there is an exact sequence of additive groups*:

$$C_0(RG) \xrightarrow{\alpha} K(RG) \xrightarrow{\theta} K(FG) \to 0.$$

When R=alg. int.$\{F\}$, it follows at once from (9.3) that *every* projective RG-module is special, and is expressible as the direct sum of a free module and a projective left ideal L in the group ring RG. Each element of $C_0(RG)$ can thus be expressed as a difference $[RG]-[L]$, with L a projective left ideal in RG. The Jordan-Zassenhaus theorem (2.6) then implies that the group $C_0(RG)$ is finite, whence also ker θ is a finite abelian group.

Keeping the assumption that R=alg. int.$\{F\}$, let Λ_0 be any maximal R-order in FG containing RG; then $[G:1]\Lambda_0 \subset RG$. By (7.6) every left Λ_0-lattice is Λ_0-projective, whence

$$K(\Lambda_0) \cong K_f(\Lambda_0) \cong P(\Lambda_0).$$

Since each Λ_0-module may be viewed as an RG-module by restriction of the operator domain, there is a homomorphism

$$K(\Lambda_0) \xrightarrow{\text{res}} K(RG).$$

In the other direction, we may define a map $\beta: C_0(RG) \to C_0(\Lambda_0)$ by

$$\beta\{[M] - [N]\} = [\Lambda_0 \otimes_{RG} M] - [\Lambda_0 \otimes_{RG} N],$$

where M, N are special projective RG-modules with $FM \cong FN$. Using Theorem 10.6 and its corollaries, Swan [**234**] proved

(10.9). THEOREM. *Let R = alg. int.$\{F\}$, and let Λ_0 be a maximal R-order in FG containing RG. Then there is a commutative diagram of additive groups, with exact rows:*

$$\begin{array}{ccccccc} & C_0(RG) & \xrightarrow{\alpha} & K(RG) & \xrightarrow{\theta} & K(FG) & \to 0 \\ & \beta\downarrow & & \uparrow \text{res} & & \uparrow 1 & \\ 0 \to & C_0(\Lambda_0) & \xrightarrow{\alpha'} & K(\Lambda_0) & \xrightarrow{\theta'} & K(FG) & \to 0 \end{array}$$

Furthermore, both β and res *are epimorphisms, and* ker θ *is the image of $C_0(\Lambda_0)$ in $K(RG)$.*

REMARK. Swan shows by example that neither β nor $\text{res} \circ \alpha'$ need be monic.

Swan's Theorem 10.6 is the starting point for the explicit calculation of the additive structure of $K(RG)$, R=alg. int.$\{F\}$, due to Heller-Reiner [**114**], [**115**]. This calculation uses the concept of the *Whitehead group* $K^1(\Lambda)$ of a ring Λ, which we proceed to define. (General references: Bass [**12**], [**14**], Heller-Reiner [**114**], Heller [**110**], Swan [**235**].)

Let Λ be an arbitrary ring with unity, and consider only finitely generated Λ-modules. Now form the collection of all ordered pairs (M, μ) in which M ranges over all projective Λ-modules, and for each M, μ ranges over all automorphisms of M. Let $\mathfrak{B}$ be the free abelian group generated by all such pairs. On the other hand, consider commutative diagrams of projective Λ-modules, with exact rows and with λ, μ, ν automorphisms:

$$\begin{array}{ccccccccc} 0 \to & L & \xrightarrow{f} & M & \xrightarrow{g} & N & \to 0 \\ & \lambda\downarrow & & \mu\downarrow & & \nu\downarrow & \\ 0 \to & L & \xrightarrow{f} & M & \xrightarrow{g} & N & \to 0 \end{array} \tag{10.10}$$

Let $\mathfrak{B}_0$ be the subgroup of $\mathfrak{B}$ generated by all expressions

$$(M, \mu) - (L, \lambda) - (N, \nu)$$

arising from diagrams (10.10). Next, let $\mathfrak{B}_1$ be the subgroup of $\mathfrak{B}$ generated by all expressions

$$(M, \mu\mu') - (M, \mu) - (M, \mu')$$

where M is any projective Λ-module, and μ, μ' are any automorphisms of M. Then we set

$$K^1(\Lambda) = \mathfrak{B}/(\mathfrak{B}_0 + \mathfrak{B}_1),$$

an abelian additive group. The image of (M, μ) in $K^1(\Lambda)$ will be denoted by $[M, \mu]$.

(Alternatively, we may also define

$$K^1(\Lambda) = \text{inj lim} \frac{GL(n, \Lambda)}{[GL(n, \Lambda), GL(n, \Lambda)]},$$

with the obvious embedding of $GL(n, \Lambda)$ into $GL(n+1, \Lambda)$. See Bass [**12**], [**14**] for details.)

If $F^\#$ denotes the multiplicative group $F-\{0\}$, then $K^1(F) \cong F^\#$, the isomorphism being given by $[M, \mu] \to \det \mu$ for each vector space M over F and each automorphism μ of M. More generally, let D be a skewfield, and let $D^\#$ be the commutator factor group of the multiplicative group $D-\{0\}$. Then $K^1(D) \cong D^\#$, where the isomorphism is given by mapping $[M, \mu]$ onto the Dieudonné determinant of μ.

Now let R be any noetherian domain, and let Λ be an R-order in the F-algebra A. Denote by $K_t(\Lambda)$ the Grothendieck group of the category of R-torsion Λ-modules, using for $\mathcal{S}$ the set of all exact sequences of such modules. For a pair of Λ-lattices M, N satisfying $FM = FN$, define

$$[M /\!/ N] = [M/(M \cap N)] - [N/(M \cap N)] \in K_t(\Lambda).$$

Now let $[M^*, \mu^*] \in K^1(A)$, with μ^* any automorphism of the A-module M^*. Choose a Λ-lattice M such that $FM = M^*$, and set

$$\Delta[M^*, \mu^*] = [M /\!/ \mu^*(M)] \in K_t(\Lambda).$$

(10.11). THEOREM (HELLER-REINER [**114**]). *The map* $\Delta: K^1(A) \to K_t(\Lambda)$ *given above is well defined, and there is an exact sequence of additive groups*

$$K^1(A) \xrightarrow{\Delta} K_t(\Lambda) \xrightarrow{\eta} K(\Lambda) \xrightarrow{\theta} K(A) \to 0,$$

with η, θ *analogous to those in* (10.5).

Next take $R = \text{alg. int.}\{F\}$, and as in (10.9) let Λ_0 be a maximal

R-order in FG containing RG. Then there is a commutative diagram of additive groups with exact rows

$$\begin{array}{ccccccccc} K^1(FG) & \xrightarrow{\Delta'} & K_t(\Lambda_0) & \xrightarrow{\eta'} & K(\Lambda_0) & \xrightarrow{\theta'} & K(FG) & \to & 0 \\ 1\downarrow & & \gamma\downarrow & & \gamma'\downarrow & & 1\downarrow & & \\ K^1(FG) & \xrightarrow{\Delta} & K_t(RG) & \xrightarrow{\eta} & K(RG) & \xrightarrow{\theta} & K(FG) & \to & 0 \end{array}$$

in which γ, γ' are restriction maps. Since γ' is epic by (10.9), we obtain

$$(10.12) \quad K(RG) \cong K(FG) \dotplus \ker\theta, \quad \ker\theta \cong \frac{K_t(\Lambda_0)}{\text{image of } \Delta' + \ker\gamma}.$$

We now give an explicit description of the finite abelian group ker θ. From §7 it follows easily that Δ' is monic, and that

$$K_t(\Lambda_0) \cong I\{1\}, \qquad \text{image of } \Delta' \cong I'\{1\},$$

using the notation of (8.1) and (8.2) with $A = FG$. For each i, $1 \leqq i \leqq r$, choose a left Λ_0-lattice M_i such that FM_i is an irreducible A_i-module. We proceed to construct a set of "generalized decomposition numbers" corresponding to a fixed maximal ideal P of R dividing $[G:1]$, as follows: let $\{P_{i1}, \cdots, P_{i\lambda_i}\}$ be the maximal ideals of R_i dividing P, where $1 \leqq i \leqq r$. Let $\overline{R} = R/P$, and let $Y_1, \cdots, Y_w$ be a full set of irreducible $\overline{R}G$-modules. For each i and j, we may view $M_i/P_{ij}M_i$ as $\overline{R}G$-module; let $d_{ij}^{(k)}$ be the number of its composition factors which are isomorphic to Y_k, where $1 \leqq k \leqq w$. Now set

$$u(P) = \left\{ \prod_{i=1}^{r} \prod_{j=1}^{\lambda_i} P_{ij}^{a_{ij}} \in I\{1\} : \sum_{i,j} a_{ij} m_i^{-1} d_{ij}^{(k)} = 0, \ 1 \leq k \leq w \right\},$$

where m_i is the index of the skewfield D_i occurring in (8.1). (Thus, $D_i \cong \operatorname{Hom}_A(FM_i, FM_i)$.)

(10.13). THEOREM (HELLER-REINER [**114**], [**115**]). *The kernel of* γ *is precisely the subgroup* $\prod_P u(P)$ *of* $I\{1\}$, *where* P *ranges over all maximal ideals of* R *dividing* $[G:1]$. *Hence*

$$\ker\theta \cong I\{1\} \Big/ \left(I'\{1\} \cdot \prod_P u(P) \right),$$

$$K(FG) \cong \sum_{i=1}^{r}{}^{\cdot} Z[FM_i],$$

and $K(RG) \cong K(FG) \dotplus \ker\theta$ *as additive groups.*

It is also of interest to give an explicit formula for the reduced special projective class group $C_0(RG)$.

(10.14). THEOREM (JACOBINSKI [**129**]). *Let $R = alg.\ int.\{F\}$, and let $\mathfrak{f}$ be the product of all those maximal ideals P of R for which $R_P G$ is not a maximal order. Let M be the left RG-lattice defined as follows: $M = RG^{(2)}$ if some simple component of FG is a totally definite quaternion algebra, while $M = RG$ otherwise. Then*

$$C_0(RG) \cong V\{M, \mathfrak{f}\},$$

where $V\{M, \mathfrak{f}\}$ is the group defined in (8.7).

REMARK. In Theorems 10.9, 10.13, 10.14 and 10.15 (below), instead of assuming that $R = \text{alg. int.}\{F\}$, we need only assume that R is a Dedekind domain whose quotient field is an algebraic number field, and that no prime divisor of $[G:1]$ is a unit in R.

For R any Dedekind domain, we may define a ring structure on $K(RG)$ by setting, for each pair of RG-lattices M and N,

$$[M]\cdot[N] = [M \otimes_R N].$$

Here, the elements of G act diagonally on $M \otimes N$ (as in §9). If 1_G denotes the RG-lattice R on which G acts trivially, then $K(RG)$ is a commutative ring with unity element $[1_G]$. Despite Theorem 10.13 which gives such precise information about $K(RG)$ as additive group, very little is known about the multiplicative structure of $K(RG)$; see Obayashi [**175**], Stancl [227], Swan [**234**], Uchida [**255**]. The only general result which we quote is the striking theorem:

(10.15). THEOREM (SWAN [**234**]). $(\ker\,\theta)^2 = 0$, *where* $\theta: K(RG) \rightarrow K(FG)$ *is given as in* (10.9), *and* $R = alg.\ int.\{F\}$.

Let R be any Dedekind domain; as in §9, we may define a restriction map $K(RG) \rightarrow K(RH)$, $(x \rightarrow x_H)$, for H any subgroup of G. Likewise the induction map $K(RH) \rightarrow K(RG)$, $(y \rightarrow y^G)$, is well defined. The analogues of Theorem 9.5, (i) and (ii), remain valid. If $\mathcal{H}$ is some collection of subgroups of G, set

$$K_{\mathcal{H}}(RG) = \sum_{H \in \mathcal{H}} \{K(RH)\}^G \subset K(RG).$$

(10.16). THEOREM (SWAN [**233**]). *Let R be a Dedekind domain, and m a positive integer such that $m \cdot K(FG) \subset K_{\mathcal{H}}(FG)$. Then $m^2 \cdot K(RG) \subset K_{\mathcal{H}}(RG)$. In particular, if $\mathcal{H}$ is the collection of all hyperelementary subgroups of G (see* CR §42), *then $K(RG) = K_{\mathcal{H}}(RG)$.*

This circle of ideas involving the Frobenius reciprocity theorem has been systematically investigated by Lam [**146**].

Now let R be a Dedekind domain of characteristic zero. There is a map

$$\tau\colon P(RG) \to K(FG), \qquad [M] \to [FM],$$

and it is of interest to identify the image of τ. The elements of $K(FG)$ can be thought of as generalized characters (CR §38), that is, as differences $\phi_1 - \phi_2$ of characters ϕ_1, ϕ_2 afforded by FG-modules. Call an element $x \in G$ *R-singular* if the order of x is not a unit in R.

(10.17). THEOREM (SWAN [**234**]). *Let χ be a generalized character, viewed as an element of $K(FG)$. Then χ lies in the image of τ if and only if $\chi(x) = 0$ for every R-singular element $x \in G$.*

Call G *p-solvable* if each composition factor of G is either a p-group or else has order prime to p.

(10.18). THEOREM (SWAN [**234**]). *Let R be a Dedekind domain of characteristic zero, and let X be an FG-module whose character vanishes on all R-singular elements of G. Suppose that for each of those prime factors p of $[G:1]$ which are not units in R, the group G is p-solvable. Then there exists a projective RG-lattice M for which $X \cong FM$.*

The result is no longer true if the hypothesis on G is omitted.

The Whitehead group $K^1(RG)$ is considerably harder to deal with than $K^0(RG)$. Indeed, even the calculation of $K^1(R)$ is already difficult. Bass [**13**] has shown how to compute the Z-rank of $K^1(ZG)$. A complete discussion of these problems, as well as a wealth of material on related subjects, may be found in the reference by Bass [**14**].

Other references for this section: Krugljak [**145**], Roggenkamp [**204**], [**207**].

11. Commutative orders and related results. Throughout this section, R is a Dedekind domain and Λ is an R-order in the F-algebra A. The semiexpository paper by Faddeev [**80**] will serve as general reference for the material in this section.

For J a full R-lattice in A, define $O_l(J)$, $O_r(J)$ and J^{-1} as at the beginning of §7. Call J *right invertible* if $J \cdot J^{-1} = O_l(J)$, or equivalently, if $JN = O_l(J)$ for some full R-lattice N in A. If J is right invertible, then J^{-1} is left invertible, and $O_r(J^{-1}) = O_l(J)$.

Suppose M is any faithful left Λ-lattice not necessarily contained in A. (Call M *faithful* if for $\lambda \in \Lambda$, $\lambda M = 0$ implies that $\lambda = 0$.) Since A acts on FM, we may define

$$O_l(M) = \{x \in A : xM \subset M\}.$$

This is an R-order in A, called the *left multiplier ring* of M, and $O_l(M) \supset \Lambda$. We say that M is an *exact* Λ-lattice if $\Lambda = O_l(M)$.

(11.1). PROPOSITION. *For J a full left Λ-lattice in A, set*

$$J' = \{x \in A : Jx \subset \Lambda\} \cong \mathrm{Hom}_\Lambda(J, \Lambda).$$

Then J is Λ-projective if and only if $J' \cdot J = O_r(J)$.

(11.2). PROPOSITION. *Let A be commutative, and suppose that J is an exact full left Λ-lattice in A. Then the following statements are equivalent:*
(i) *J is Λ-projective.*
(ii) *J is invertible.*
(iii) *For each maximal ideal P of R, the localization J_P is a principal Λ_P-ideal in A.*

Reference. Faddeev [**80**, Propositions 18.2 and 27.1].

Following Borevič-Faddeev [**34**], we introduce the concept of orders of cyclic index. Let A be a commutative separable algebra over F, so A is a direct sum of fields F_i, each of which is a finite separable extension of F. Let R_i be the integral closure of R in F_i. Then $\Lambda_0 = \sum^{\cdot} R_i$ is the unique maximal R-order in A. An arbitrary R-order Λ is said to have *cyclic index* (in Λ_0) if $\Lambda_0 = \Lambda + \Lambda\omega$ for some $\omega \in \Lambda_0$. Thus, orders of cyclic index are "relatively close" to being maximal orders.

(11.3). THEOREM (BOREVIČ-FADDEEV [**34**]). *Let A be a commutative separable F-algebra, and Λ an R-order in A of cyclic index. Then every full left Λ-lattice in A is invertible. Further, each full left Λ-lattice M in a free A-module $A^{(s)}$ uniquely determines an ascending chain of R-orders in A:*

$$\Lambda \subset \Lambda_1 \subset \cdots \subset \Lambda_s,$$

such that for each i, $1 \leqq i \leqq s$, there exists an exact full left Λ_i-lattice J_i in A, and

$$M \cong J_1 \dotplus \cdots \dotplus J_s \qquad \text{as } \Lambda\text{-lattices.}$$

The ideal class of the product $J_1 \cdots J_s$ in Λ_s is also uniquely determined by M. This chain of R-orders, and this ideal class, are the only invariants of M, and we may in fact choose $J_i = \Lambda_i$, $1 \leqq i \leqq s-1$, in the above decomposition.

References. See also Brooks [**38**].

Using a completely different approach, Bass [**10**] obtained the following fruitful result.

(11.4). BASS' THEOREM. *Let Λ be a noetherian (commutative) integral domain whose integral closure $\tilde{\Lambda}$ in the quotient field of Λ is finitely generated as Λ-module. Then every finitely generated torsionfree Λ-module is isomorphic to an external direct sum of ideals of Λ if and only if every ideal of Λ can be generated by two or less elements.*

REMARKS. Let F be the quotient field of Λ; for M a Λ-lattice, we call $(FM:F)$ the *rank* of M. Suppose that Λ is a noetherian domain for which every Λ-lattice is a direct sum of Λ-lattices of rank at most k. Bass then shows that for each maximal ideal $\mathfrak{m}$ of Λ, every ideal of $\Lambda_{\mathfrak{m}}$ can be generated by at most $k+1$ elements. This in turn implies (by a result of I. S. Cohen) that each nonzero prime ideal of Λ is maximal, and that each ideal of Λ can be generated by $\mathrm{Max}(2, k+1)$ elements. Taking $k=1$, we obtain the theorem in one direction.

The proof in the other direction is more difficult. Assume that $\tilde{\Lambda}$ is finitely generated as Λ-module, and that each ideal of Λ can be generated by two elements. The key step is the proof that under these hypotheses, every ideal of Λ is Λ'-projective for some uniquely determined domain Λ' containing Λ.

A related reference is Swan [**237**].

(11.5). COROLLARY. *Let x be an indeterminate over the quotient field F of the Dedekind domain R. Given any finitely generated projective left $R[x]$-module M, there exists an R-lattice M_0 such that*

$$M \cong R[x] \otimes_R M_0.$$

In particular, when R is a principal ideal domain, every finitely generated projective $R[x]$-module is free.

References. The result for the case where R is a principal ideal domain was proved by Seshadri [**225**], and then extended by Bass [**10**] to Dedekind domains.

As pointed out by Borevič-Faddeev [**35**], Bass' Theorem yields a partial converse to (11.3).

(11.6). PROPOSITION. *Let A be a finite separable field extension of F, and let Λ be an R-order in A. If every left Λ-lattice is isomorphic to an external direct sum of ideals of Λ, then Λ is an order of cyclic index.*

Roĭter [**215**] obtains a partial generalization of (11.4) to the noncommutative case:

(11.7). THEOREM. *Let A be a separable F-algebra, not necessarily commutative, and let Λ be an R-order in A. If every left ideal of Λ can be* *generated by two elements, then every left Λ-lattice is isomorphic to an external direct sum of left ideals of Λ.*

In the proof of this theorem, Roĭter uses his concept of "divisibility" of modules (see §12). We shall return in §13 to the question as to other sufficient conditions which guarantee that every indecomposable Λ-lattice is isomorphic to an ideal in Λ.

We turn next to some other aspects of orders in commutative algebras. Let A be a commutative separable F-algebra, and let Λ_0 be the unique maximal R-order in A. Let Λ be any R-order in A, and M a full Λ-lattice in $A^{(n)}$. Define $M_0 = \Lambda_0 M$, and consider the order ideals $\text{ord}(M_0/M)$, $\text{ord}(\Lambda_0/\Lambda)$, defined as at the end of §1.

(11.8). THEOREM (FRÖHLICH [87]). *The relation*

$$\text{ord}(M_0/M) \supset \{\text{ord}(\Lambda_0/\Lambda)\}^n$$

is always true, and these ideals of R are equal if and only if M is Λ-projective. Further, M is Λ-projective if and only if M is an external direct sum $\sum^{\cdot} M_i$ with each M_i a Λ-lattice in the same genus as Λ.

We remark that this surprisingly simple criterion for projectivity does not extend readily to the noncommutative case (see Ballew [4], [4a]).

Next let A be a field, and let J be a full R-lattice in A. Then each power J^n is also a full R-lattice in A, and $O_l(J) \subset O_l(J^n)$, but these orders need not be equal. We have called J *invertible* if $J \cdot L = O_l(J)$ for some full R-lattice L in A; if J is invertible, so is each J^n. However, it may well happen that J^n is invertible though J itself is not. The most striking result of this type is as follows:

(11.9). THEOREM (DADE-TAUSSKY-ZASSENHAUS [52], [53]). *Let Λ be a Z-order in an algebraic number field A, where $(A:Q) = d > 1$. Then for every full Λ-lattice J in A, J^{d-1} and all higher powers of J are invertible.*

This is in fact a special case of their more general result:

(11.10). THEOREM. *Let Λ be a noetherian integral domain with quotient field A, such that every nonzero prime ideal of Λ is a maximal ideal of Λ.*

(i) *For each full left Λ-lattice J in A, some power of J is invertible.*

(ii) *If the integral closure of Λ in A is finitely generated as Λ-module, then there exists a positive integer n such that J^n is invertible for every J.*

(iii) *If Λ is an R-order (where R = Dedekind domain) and $(A:F) = d$, then J^{d-1} is invertible for every full Λ-lattice J in A.*

Further references on this topic: Dade [**47**], Dade-Robinson-Taussky-Ward [**50**], Singer [**226a**] Dade-Taussky [**51a**].

12. **Divisibility of modules.** This concept of Roĭter's [**210**], [**211**], [**213**]–[**216**] has already shown its importance in several investigations. To begin with, let Λ be any noetherian ring (with 1), and consider only finitely generated left Λ-modules. If M, N are Λ-modules, denote by $M\cdot\mathrm{Hom}(M, N)$ the additive subgroup of N generated by the set of elements

$$\{f(m) \colon f \in \mathrm{Hom}_\Lambda(M, N),\ m \in M\}.$$

We shall say that M *covers* N (notation: $M \succ N$) if $M\cdot\mathrm{Hom}(M, N) = N$. For example, M covers every direct summand of each of its homomorphic images. (Roĭter uses the term "divides" in place of "covers," but this conflicts with the standard notation $U | V$ indicating that U is a direct summand of V.)

Let M' be a submodule of M. Call M' *supercharacteristic* if $M'\cdot\mathrm{Hom}(M', M) = M'$. For example, if M and N are arbitrary modules, then $M\cdot\mathrm{Hom}(M, N)$ is a supercharacteristic submodule of N.

As usual let $M^{(k)}$ denote the direct sum of k copies of M. An easy argument shows that $M \succ N$ if and only if there is a Λ-exact sequence $M^{(k)} \to N \to 0$ for some k. In particular, $\Lambda \succ N$ for every N.

A *D-submodule* of a Λ-module M is a supercharacteristic proper submodule M' covered by M, and such that for M'' a supercharacteristic submodule of M, $M' + M'' = M$ implies $M'' = M$. The sum of all D-submodules of M is again a D-submodule, the largest such in M, and is denoted by $D(M)$. Its importance stems from

(12.1). THEOREM (ROĬTER [**213**, THEOREM 2]). *Let R be a complete discrete valuation ring, Λ an R-order in the separable F-algebra A, and let M be a left Λ-lattice. Then $D(M) = 0$ if and only if for each k, every exact sequence of Λ-lattices of the form*

$$0 \to M_1 \to M^{(k)} \to M_2 \to 0$$

is split. In particular, Λ is hereditary if and only if $D(\Lambda) = 0$.

We return briefly to the general situation for the following basic idea. Let Λ be any noetherian ring; a left Λ-module M is said to have a *normal* decomposition if M can be expressed as a direct sum of nonzero submodules: $M = M_1 \oplus \cdots \oplus M_r$, $r > 1$, such that $M_i \succ M_j$ for $1 \leqq i < j \leqq r$. If no such decomposition exists, call M *normally indecomposable.*

(12.2). THEOREM. *Let Λ be as in* (12.1), *and let M, N be left Λ-lattices* *such that* $N \succ M$. *If* N *is normally indecomposable, or if* $\mathrm{Hom}_\Lambda(N, N)$ *is commutative, then every exact sequence* $M \to N \to 0$ *is* Λ*-split.*

Let B be an arbitrary finite dimensional algebra over a field F, and assume B has a unity element. If in place of Λ-lattices we consider finitely generated B-modules, then the analogues of Theorems 12.1 and 12.2 remain valid. The latter theorem is a key step in Roĭter's recent proof [**216**] of the Brauer-Thrall conjecture:

If there exist infinitely many nonisomorphic indecomposable B*-modules, there cannot be a uniform upper bound on their dimensions over* F.

13. **Hereditary orders and related results.** Throughout this section let R denote a Dedekind domain, and let A be a separable F-algebra. An R-order Λ is *hereditary* if every left ideal of Λ is a projective Λ-module. In our case, this is equivalent to the condition that every left Λ-lattice be projective.

The material in this section comes from the work of Brumer [**39**], Drozd-Kiričenko [**71**], Drozd-Kiričenko-Roĭter[8] [**73**], and Harada [**104**]–[**106**]. One may also consult the lecture notes by Roggenkamp-Dyson [**208b**].

(13.1). Theorem. *Let* Λ *be an* R*-order in* A.

(i) *If* Λ *is hereditary, so is every bigger order.*

(ii) Λ *is hereditary if and only if for each maximal ideal* p *of* R, *the order* Λ_p *is hereditary.*

(iii) *A decomposition of* A *into simple components gives rise to an analogous decomposition of every hereditary order in* A.

(iv) *If* A *is a simple algebra with center* F', *then the center* R' *of a hereditary* R*-order* Λ *in* A *is a Dedekind domain with quotient field* F'.

(v) *If* Λ *is hereditary, every* Λ*-lattice is isomorphic to an external direct sum of full* Λ*-lattices in irreducible* A*-modules.*

Parts (ii)–(iv) of the preceding theorem enable us to reduce the study of hereditary orders to the case where A is a central simple algebra, and R_p is a discrete valuation ring. The main results in this case are as follows (Brumer [**39**]):

(13.2). Let D be a skewfield with center F, and let $A = \mathrm{Hom}_D(V, V) \cong (D)_n$, where V is a right vector space over D of dimension n. Let Δ be a maximal R_p-order in D; then its Jacobson radical rad Δ is the unique maximal two-sided ideal of Δ (see (7.13)), and we define the skewfield Ω (over R/p) by

$$\Delta/\mathrm{rad}\ \Delta \cong (\Omega)_l \quad \text{for some } l.$$

[8] Abbreviated as D-K-R [**73**] hereafter.

On the other hand, we have

$$F_p^* \otimes_F A \cong (B)_m \qquad \text{for some skewfield } B,$$

and indeed $m = ln$.

A Δ-*chain* $\mathcal{E}$ *of period* r is a strictly decreasing chain of full right Δ-lattices in V:

$$\mathcal{E}: \cdots E_{-1} \supset E_0 \supset E_1 \supset \cdots \supset E_{r-1} \supset E_r \supset \cdots$$

such that $E_{k+r} = E_k \cdot (\text{rad } \Delta)$ for each k. Set

$$\Lambda = \text{Hom}_\Delta(\mathcal{E}, \mathcal{E}) = \{f \in A : f(E_k) \subset E_k \text{ for all } k\}.$$

Then Λ is a hereditary R_p-order in A, and

$$\text{rad } \Lambda = \{f \in A : f(E_k) \subset E_{k+1} \text{ for all } k\}.$$

The quotient $\Lambda/\text{rad } \Lambda$ is a direct sum of r full matrix algebras over the skewfield Ω, and for $0 \leqq k \leqq r-1$, the kth simple component $(\Omega)_{n_k}$ is the ring of $(\Lambda/\text{rad } \Lambda)$-endomorphisms of E_k/E_{k+1}. The sequence $(n_0, n_1, \cdots, n_{r-1})$ is called the *invariant* of $\mathcal{E}$, and satisfies $n_0 + \cdots + n_{r-1} = m$; the invariant is defined up to cyclic permutation.

Now let us define

$$P_k = \{f \in \Lambda : f(E_k) \subset E_{k+1}\}, \qquad \Gamma_k = \text{Hom}_\Delta(E_k, E_k).$$

Both P_k and Γ_k depend only on k (mod r). There are precisely r distinct maximal two-sided ideals in Λ, namely $P_1, \cdots, P_r$. There are precisely r distinct maximal R_p-orders in A containing Λ, namely $\Gamma_1, \cdots, \Gamma_r$, and their intersection is Λ. Furthermore, $\mathcal{E}$ is the set of *all* left Λ-, right Δ-lattices in V, and $(\text{rad } \Lambda)E_k = E_{k+1}$ for all k.

Conversely, if Λ is a hereditary R_p-order in A, the set of all (Λ, Δ)-lattices in V forms a Δ-chain $\mathcal{E}$ of period r (for some $r \leqq n$), and $\Lambda = \text{Hom}_\Delta(\mathcal{E}, \mathcal{E})$. There is a one-to-one inclusion-reversing correspondence between R_p-orders containing Λ, and subchains of $\mathcal{E}$.

Let M be an indecomposable Λ-lattice, where Λ is a hereditary R_p-order in A; then M is isomorphic to a full Λ-lattice in V, and $\text{Hom}_\Lambda(M, M)$ is a hereditary R_p-order in D. If D has a unique maximal R_p-order Δ (which is certainly true when $D = F$, and also when R is a complete discrete valuation ring), then $\text{Hom}_\Lambda(M, M) = \Delta$; further, under this assumption, the invariant $(n_0, \cdots, n_{r-1})$ completely determines the structure of Λ. Specifically, if Λ' is another hereditary R_p-order in A, whose invariant is a cyclic permutation of $(n_0, \cdots, n_{r-1})$, then $\Lambda' = x\Lambda x^{-1}$ for some unit $x \in A$. Finally, when $\text{Hom}_\Lambda(M, M) = \Delta$, we may display Λ as the set of all those matrices in $(\Delta)_n$ of the form

$$\begin{bmatrix} \Delta & \mathfrak{M} & \mathfrak{M} \cdots \mathfrak{M} \\ \Delta & \Delta & \mathfrak{M} \cdots \mathfrak{M} \\ \Delta & \Delta & \Delta \cdots \mathfrak{M} \\ \cdot & \cdot & \cdot \quad \cdots \quad \cdot \\ \Delta & \Delta & \Delta \cdots \Delta \end{bmatrix},$$

partitioned into $n_i \times n_j$ *blocks* all of whose entries belong to the indicated symbol (either Δ or $\mathfrak{M}$, where $\mathfrak{M} = \text{rad } \Delta$).

Returning to the global case, we have

(13.3). THEOREM. *Let A be a central simple F-algebra. Suppose that Λ_0 is some fixed R-order in A. Let there be given for each maximal ideal p of R, a hereditary R_p-order $\Lambda(p)$ in A, satisfying $\Lambda(p) = (\Lambda_0)_p$ for all but a finite number of p's. Set*

$$\Lambda = \bigcap_p \Lambda(p).$$

Then Λ is a hereditary R-order in A, and is the unique R-order such that $\Lambda_p = \Lambda(p)$ for all p. If we set

$$M(p) = \Lambda \cap \text{rad } \Lambda(p)$$

then $M(p)$ is a full two-sided ideal in Λ, and satisfies

$$(M(p))_p = \text{rad } \Lambda(p), \qquad (M(p))_q = \Lambda_q \quad \text{if } q \neq p.$$

Let $\mathcal{I}(\Lambda)$ be the set of full two-sided Λ-ideals in A. An element $J \in \mathcal{I}(\Lambda)$ is *right Λ-invertible* if there exists a $J' \in \mathcal{I}(\Lambda)$ with $JJ' = \Lambda$.

(13.4). THEOREM. *Let Λ be a hereditary order in the central simple F-algebra A. Then $\mathcal{I}(\Lambda)$ is the free abelian group on the generators $\{M(p): p = \text{maximal ideal of } R\}$, and for $J \in \mathcal{I}(\Lambda)$, we have $J^{-1} = \{x \in A: xJ \subset \Lambda\}$.*

We may mention some other criteria for an order to be hereditary:

(13.5). THEOREM (D-K-R [73]). *Let Λ be an R-order in the separable algebra A. Then the following are equivalent:*

(i) *Λ is hereditary.*

(ii) *There are no nonzero D-submodules of Λ (see §12).*

(iii) *Given any pair of R-orders Γ_1, Γ_2 in A such that $\Lambda \subset \Gamma_1 \subset \Gamma_2$, there exists an R-order Γ_3 in A for which $\Lambda \subset \Gamma_3$ and $\Gamma_1 = \Gamma_2 \cap \Gamma_3$.*

(iv) *Every maximal R-order in A containing Λ is projective as left Λ-module.*

For a hereditary order Λ we know that every left Λ-lattice is isomorphic to an external direct sum of left ideals[9] in Λ. We have seen in §11 that this property may hold for a larger class of orders. Here we shall sketch some further results in this direction.

(13.6). Let Λ be an R-order in the separable F-algebra A. Call Λ a *Gorenstein order* if any one of the following equivalent conditions is true:

(i) $\mathrm{Ext}^2_\Lambda(Y, \Lambda)=0$ for every left Λ-module Y.

(ii) $\mathrm{Ext}^1_\Lambda(M, \Lambda)=0$ for every left Λ-lattice M.

(iii) Λ^* covers Λ (see §12), where Λ^* is the left Λ-module defined by $\Lambda^*=\mathrm{Hom}_R(\Lambda, R)$.

(iv) For every exact[10] faithful left Λ-lattice M, $M\cdot\mathrm{Hom}_\Lambda(M,\Lambda)=\Lambda$.

(v) Every exact full left Λ-lattice in A is right invertible.[11]

(Five more statements, each equivalent to (i), may be obtained by interchanging "left" and "right" in all of the above.)

References. D-K-R [**73**], Bass [**11**], Nazarova-Roĭter [**170**].

A Λ-lattice M is called *weakly injective* if every exact sequence $0\to M\to X\to Y\to 0$ of Λ-lattices is split; in other words, M is weakly injective if and only if M is a Λ-direct summand of every bigger Λ-lattice of which M is an R-direct summand.

(13.7). Proposition. *Let Λ be an R_p^*-order in a separable algebra. Then Λ is a Gorenstein order if and only if every projective left Λ-lattice is weakly injective, or equivalently, if and only if every weakly injective Λ-lattice is projective.*

(13.8). Definition. An R-order Λ in A is a *Bass order* if every R-order in A containing Λ is a Gorenstein order.

Every hereditary order is a Bass order. Roĭter [**215**] proved that if every bigger R-order can be generated by two elements as left Λ-module, then Λ is a Bass order. When A is commutative, the converse also holds (Bass [**11**]).

(13.9). Theorem (D-K-R [**73**]). *Let Λ be an R-order in A. The following are equivalent:*

(i) *Λ is a Bass order.*

(ii) *Λ_p^* is a Bass order for each maximal ideal p of R.*

(iii) *The set of all full two-sided Λ-lattices in A forms a groupoid[12] relative to proper[13] products.*

---

[9] Indeed, part (v) of Theorem 13.1 asserts an even stronger result.

[10] See definition immediately preceding (11.1).

[11] See beginning of §11.

[12] See Jacobson [**131**, p. 132]. Compare this theorem with the earlier Theorem 7.10.

[13] The product JJ' is proper if replacing either factor by a larger lattice increases the product.

(13.10). DEFINITION. Let $V_1, \cdots, V_t$ be a full set of nonisomorphic irreducible A-modules, and let M be a Λ-lattice. We shall say that M has *signature* $(m_1, \cdots, m_t)$ if

$$FM \cong V_1^{(m_1)} \dotplus \cdots \dotplus V_t^{(m_t)}.$$

(13.11). DEFINITION. Let Λ be an R_p^*-order in the separable algebra A, and suppose that Λ is a Bass order. We consider three types of orders:

Type I. The algebra A has two simple components, and every indecomposable Λ-lattice has signature (1, 0), (0, 1) or (1, 1) with the last type actually occurring.

Type II. The algebra A is simple, and every indecomposable Λ-lattice has signature (1) or (2), with the latter type actually occurring.

Type III. The algebra A is simple, and every indecomposable Λ-lattice has signature (1).

(13.12). THEOREM (D-K-R [**73**, THEOREM 9.7]). *Every Bass R_p^*-order in a separable algebra is a ring direct sum of Bass orders of Types* I, II *and* III.

REMARKS. (i) Hereditary orders have Type III.

(ii) If G is cyclic of order p, Z_p^*G is of Type I.

(iii) Those Z_p^*-orders in the matrix ring $(Q_p^*)_2$, having finitely many indecomposable lattices, are of Type II (see Drozd-Kiričenko [**71**]).

(iv) Quadratic Z_p^*-rings are nonhereditary Bass orders of Type III (see Borevič-Faddeev [**33**]).

Detailed theorems on the structure of Bass orders are given in D-K-R [**73**]. Other references for this subject: Drozd-Roĭter [**74**], Roggenkamp [**204**]. Also see Michler [**160a**], and references therein, for general results on hereditary orders.

14. Finiteness of the number of indecomposable representations. Let R be a Dedekind domain with quotient field F, and let Λ be an R-order in the separable F-algebra A. Denote by $n(\Lambda)$ the number of isomorphism classes of indecomposable left Λ-lattices. When is $n(\Lambda)$ finite?

(If the algebra A is not semisimple, it follows from Faddeev [**80**, Proposition 25.1] that $n(\Lambda)$ is infinite. For this reason we limit our attention to separable algebras hereafter.)

The basic tool for passing from the global to the local case is as follows:

(14.1). THEOREM. *Let F be an algebraic number field. Then $n(\Lambda)$ is finite if and only if for each maximal ideal P of R, $n(\Lambda_P^*)$ is finite.*

References. Jones [**135**], Kneser [**143**], CR §81A.

Let G be a finite group, p a prime, and denote by G_p a Sylow p-subgroup of G. The preceding theorem tells us that $n(RG)$ is finite if and only if $n(R_P^*G)$ is finite for each P.

(14.2). THEOREM. *Let the rational prime p be contained in the maximal ideal P of R. Then $n(R_P^*G)$ is finite if and only if $n(R_P^*G_p)$ is finite.*

REMARK. This follows readily from Theorem 9.6.

(14.3). THEOREM. *Keep the above notation. If G_p is not cyclic, or if G_p is cyclic of order greater than p^2, then $n(R_P^*G_p)$ is infinite.*

The above result was proved by Borevič-Faddeev [**32**, II] and Heller-Reiner [**112**, II]. It is also a special case of a more general result:

(14.4). THEOREM (DADE [**48**]). *Let R be any noetherian domain with quotient field F, and let A be any finite dimensional F-algebra which is a direct sum of at least four subalgebras. Let Λ be an R-order in A, and suppose there exists a maximal ideal P of R for which $\Lambda/P\Lambda$ is completely primary.*[14] *Then $n(\Lambda)$ is infinite.*

In particular, $n(\Lambda)$ is infinite whenever Λ is an indecomposable R_P^*-order in a semisimple algebra having four or more simple components.

Generalizations of this result may be found in Drozd-Roĭter [**74**], Gudivok [**95**]–[**101**], Jacobinski [**127**].

In the other direction, we have

(14.5). THEOREM. *Let G_p be cyclic of order p or p^2. Then $n(Z_p^*G_p)$ is finite.*

References. Berman [**25**], Berman-Gudivok [**29**], Heller-Reiner [**112**, I].

(14.6). COROLLARY. *The number of isomorphism classes of indecomposable ZG-lattices is finite if and only if for each prime p dividing $[G:1]$, G_p is cyclic of order p or p^2.*

In order to generalize (14.5) and (14.6), it is necessary to determine for which cyclic p-groups H the number $n(R_P^*H)$ is finite. Many partial results have been obtained (see Gudivok [**101**] for a detailed description, as well as for calculations which give a solution in most cases). The complete solution was obtained by Jacobinski [**127**], who proved the following:

[14] A ring Γ is *completely primary* if $\Gamma/\text{rad}\,\Gamma$ is a skewfield.

(14.7). THEOREM. *Let* $R=$ *alg. int.* $\{F\}$. *For each maximal ideal* P *of* R, *let* p *be the rational prime contained in* P, *and let* $e(P)$ *be the largest integer for which* $pR \subset P^{e(P)}$. *Let* G_p *be a Sylow* p-*subgroup of* G. *Then* $n(RG)$ *is finite if and only if for every maximal ideal* P *dividing* $[G:1]$, *one of the following conditions holds*:

(i) $e(P)=1$ *and* G_p *is cyclic of order* p^2.

(ii) $e(P) \leqq 2$, $p>3$, *and* G_p *is cyclic of order* p.

(iii) $e(P) \leqq 3$, $p=3$, *and* G_p *is cyclic of order* p.

As a matter of fact, Jacobinski [127] derived this as a consequence of his much more general theorem, which gives necessary and sufficient conditions for the finiteness of $n(\Lambda)$, where Λ is any commutative R_P^*-order in a separable algebra. This problem has also been solved by Drozd-Roĭter; their results are easier to state, though perhaps more difficult to apply.

(14.8). THEOREM (DROZD-ROĬTER [74]). *Let* $R=$ alg. int. $\{F\}$, *and let* Λ *be an* R-*order in the commutative semisimple algebra* A. *Let* Λ_0 *be the unique maximal* R-*order in* A, *and define* $\operatorname{rad}(\Lambda_0/\Lambda)$ *as the intersection of the maximal* Λ-*submodules of* Λ_0/Λ. *Then* $n(\Lambda)$ *is finite if and only if both of the following conditions are satisfied*:

(i) *As* Λ-*module*, Λ_0/Λ *can be generated by two elements.*

(ii) rad (Λ_0/Λ) *is a cyclic* Λ-*module.*

We mention one further general result.

(14.9). THEOREM (JACOBINSKI [129]). *Let* F *be an algebraic number field, and let* Λ *be any* R-*order in the semisimple* F-*algebra* A. *Let* M *be some fixed* Λ-*lattice, and let* B_M *be the collection of direct summands of the lattices* M, $M^{(2)}$, $M^{(3)}$, $\cdots$. *Then the number of isomorphism classes of indecomposable lattices in* B_M *is finite. In particular, there are finitely many isomorphism classes of indecomposable projective* Λ-*lattices.*

Other references: Roggenkamp [208].

15. **Representations of specific groups and orders.** The aim of this section is to present a guide to the many articles containing explicit calculations and results on the classification of Λ-lattices, where Λ is some order.

1. ZG-lattices, $G=$ cyclic group.

(a) $[G:1]=p$. References: Diederichsen [57], Reiner [182], CR §74.

(b) $[G:1]=4$. References: Diederichsen [57], Knee [142], Matuljauskas [156], Roĭter [209], Troy [254].

(c) $[G:1]=p^2$. References: Berman-Gudivok [27]–[29], Heller-Reiner [112], [113].

(d) $[G:1]$ squarefree. References: Knee [142], Oppenheim [176].

2. ZG-lattices, G arbitrary.

(a) G=dihedral group of order $2p$. References: Leahey [149], Lee [150], Matuljauskas [157], Nazarova-Roĭter [167].

(b) G=nonabelian group of order pq (p, q distinct primes). Reference: Pu [179].

3. In the following cases, there are infinitely many indecomposable ZG-lattices, and these are fully classified:

(a) G=abelian (2, 2)-group. Reference: Nazarova [164].

(b) $G=A_4$. Reference: Nazarova [165].

4. Further references on integral representations of groups: Barannik-Gudivok [6], Berman [21]–[25], Berman-Gudivok [27]–[29], Berman-Lihtman [30], Drobotenko [65], Gudivok [95]–[101], Gudivok-Rud'ko [102], Jones [136], [137], Kneser [143], Matuljauskas-Matuljauskene [159], Nazarova-Roĭter [168], [171], Reiner [187], [190], [191], Roggenkamp [202], [203].

5. Representations over residue class rings Z/mZ: Drobotenko-Drobotenko-Žilinskaja-Pogoriljak [66], Drobotenko-Gudivok-Lihtman [68], Drobotenko-Lihtman [67], Hannula [103], Nazarova-Roĭter [169].

6. Representations of orders in algebras: Bass [11], Borevič-Faddeev [34], [35], Drozd [70], Drozd-Kiričenko [71], Drozd-Kiričenko-Roĭter [73], Drozd-Roĭter [74], Drozd-Turčin [75], Faddeev [81], [83], Jacobinski [127], Kaplansky [140], Kiričenko [141], Nazarova [166], Nazarova-Roĭter [170], [171].

16. **Representation rings.** Let R be a Dedekind domain whose quotient field F is an algebraic number field, and let Λ be an R-order in the semisimple F-algebra A. By the *representation group* of Λ we shall mean the abelian additive group $a(\Lambda)$ generated by the symbols $[M]$, one for each isomorphism class of Λ-lattices, and with relations $[M]=[M']+[M'']$ whenever $M\cong M'\dotplus M''$. If M and N are Λ-lattices then $[M]=[N]$ in $a(\Lambda)$ if and only if $M\dotplus X\cong N\dotplus X$ for some Λ-lattice X.

(Note that $a(\Lambda)$ is just the Grothendieck group $K(\mathcal{C}, \mathcal{S})$ of §10 where $\mathcal{C}$ is the category of all left Λ-lattices and $\mathcal{S}$ the collection of all split short exact sequences from $\mathcal{C}$.)

Let us consider the additive homomorphism

$$\phi\colon\ a(\Lambda) \to \sum_P{}^{\cdot}\, a(\overset{*}{\Lambda}_P),$$

where P ranges over all maximal ideals of R containing the Higman ideal $i(\Lambda)$ defined in §4.

(16.1). THEOREM (REINER [**195**]). *The kernel of ϕ is precisely the torsion submodule of the additive group $a(\Lambda)$, and may be characterized as the finite group consisting of all elements*

$$\{[\Lambda] - [M]\colon\ M \in \textit{genus of } \Lambda\}.$$

As in (6.2), we introduce the semilocal ring $\tilde{R}$ defined by

$$\tilde{R} = \bigcap_{P \supset i(\Lambda)} R_P,$$

and set $\tilde{\Lambda} = \tilde{R}\Lambda$. Then the map ϕ factors through $a(\tilde{\Lambda})$, that is, we may write

$$a(\Lambda) \underset{\mu}{\to} a(\tilde{\Lambda}) \underset{\phi'}{\to} \sum_{P \supset i(\Lambda)}{}^{\cdot}\, a(\overset{*}{\Lambda}_P).$$

Here μ and ϕ' are homomorphisms with $\phi' \cdot \mu = \phi$, and μ is epic, ϕ' is monic. The additive group $a(\tilde{\Lambda})$ is generated by indecomposable genera; these need not form a free Z-basis of $a(\tilde{\Lambda})$, however, since the Krull-Schmidt Theorem is not generally true for $\tilde{\Lambda}$-lattices (see §19). Nevertheless, $a(\tilde{\Lambda})$ is Z-free, since ϕ' embeds it in a Z-free module.

More generally let $\mathcal{B}$ be some category of left Λ-lattices closed under formation of direct sums, and let $\mathcal{S}$ be the collection of all split short exact sequences from $\mathcal{B}$. The relative Grothendieck group $K(\mathcal{B}, \mathcal{S})$ will be denoted briefly by $a(\mathcal{B})$. Just as above we may consider the map

$$\phi\colon\ a(\mathcal{B}) \to \sum_P{}^{\cdot}\, a(\overset{*}{\mathcal{B}}_P),$$

where now P ranges over some nonempty finite set of maximal ideals of R, including all those for which Λ_P^* is not a maximal order. Jacobinski [**129**] showed that the kernel of ϕ is precisely the torsion subgroup of $a(\mathcal{B})$, and can be given explicitly as a subgroup of $V(M, \mathfrak{f})$ for some suitably chosen Λ-lattice M in the category $\mathcal{B}$. (Here, $V(M, \mathfrak{f})$ is as defined in (8.7).)

(16.2). THEOREM (JACOBINSKI [**129**]). *Let M be some fixed Λ-lattice,* *and choose* $\mathfrak{B}$ *to be the category of all* Λ*-lattices which are direct summands of* $M^{(n)}$ *for some* n. *Then (see* §8 *for notation):*

(i) $\ker \phi \cong V(M^{(t)}, \mathfrak{f})$, *where* $t=1$ *if* M *satisfies the Eichler condition, while* $t=2$ *otherwise.*

(ii) *The image of* ϕ *is a finitely generated free* Z*-module.*

REMARK. Theorem 14.9 follows readily from (ii). On the other hand, Theorem 10.14 is the special case of (i) in which $M=\Lambda$.

When $\Lambda=RG$, we can make $a(\Lambda)$ into a commutative ring, defining multiplication by $[M][N]=[M\otimes_R N]$, with the elements of G acting diagonally on $M\otimes N$. It follows at once from (16.1) that $(\ker \phi)^2 = 0$. On the other hand, while both $a(R_P G)$ and $a(R_P^* G)$ are free Z-modules, very little is known about their multiplicative structure. Up to this point, attention has centered on the question as to whether $a(R_P G)$ can contain any nonzero nilpotent elements; the corresponding question for $a(kG)$, where k is a modular field, has been only partially settled.†

(16.3). THEOREM (REINER **[190]**–**[192]**). *Let* R_P *be a valuation ring in* F, *and let* G *contain a cyclic subgroup of order* n, *where* $n\in P^2$; *and if* $2\in P$ *assume further that* $n\in 2P$. *Then* $a(R_P G)$ *and* $a(R_P^* G)$ *contain nonzero nilpotent elements.*

On the other hand, suppose that $[G:1]$ *is squarefree. Then* $a(Z_p G)$ *and* $a(Z_p^* G)$ *contain no nonzero nilpotent elements.*

Rud'ko **[219]** has determined the multiplication table for the ring $a(Z_p^* G)$, where G is cyclic of order p^2.

Complementing the preceding theorem, we quote

(16.4). THEOREM (ZEMANEK **[265]**). *If* G *has a noncyclic Sylow* p*-subgroup, then* $a(ZG)$, $a(Z_p G)$ *and* $a(Z_p^* G)$ *all contain nonzero nilpotent elements.*

17. Group rings. Let G, H be finite groups, and K a field. It is known that the group algebra KG need not determine the group G up to isomorphism. Thus, for example, Berman **[20]** proved that if p is an odd prime, and G, H are nonisomorphic noncommutative groups of order p^3, then $QG\cong QH$. On the other hand, Passman **[178]** showed that if K is a field of characteristic p, there exist many nonisomorphic groups G, H with $KG\cong KH$. For a survey of the subject of group algebras, we refer the reader to Holvoet **[124]**.

Now let $R=$ alg. int. $\{F\}$. The following problem is still unsettled:

† Using the results of **[265]**, Zemanek has recently shown that $a(kG)$ contains nonzero nilpotent elements whenever char $k=p>2$, and G is an abelian group of type (p, p).

If $RG \cong RH$, does it necessarily follow that $G \cong H$? Partial answers so far obtained are given below:

(17.1). THEOREM. *If G, H are abelian and $ZG \cong ZH$, then $G \cong H$.*

References: G. Higman [**123**], CR §37.

(17.2). THEOREM (MAY [**160**]). *If $ZG \cong ZH$, then G and H have isomorphic commutator factor groups.*

(17.3). THEOREM. *If G, H are metabelian groups, then $ZG \cong ZH$ implies that $G \cong H$.*

References: Jackson [**124b**], [**124c**], Whitcomb [**261**].

(17.4). THEOREM (PASSMAN [**178**]). *Let R = alg. int. $\{F\}$. The integral group ring RG uniquely determines the lattice of normal subgroups of G, and also the upper and lower central series of G. If $Z_1(G)$ is the center of G, and $Z_2(G)/Z_1(G)$ is the center of $G/Z_1(G)$, then $RG \cong RH$ implies that $Z_2(G) \cong Z_2(H)$. Hence if G is nilpotent of class 2, and $RG \cong RH$, then $G \cong H$.*

(17.5). COROLLARY. *If $G/Z_1(G)$ is a p-group of order at most p^3, then $RG \cong RH$ implies that $G \cong H$.*

The article by Passman [**178**] contains many other results of this nature.

Also of interest is the question of determining the units in the group ring RG.

(17.6). THEOREM. *Let R = alg. int. $\{F\}$. Any unit of finite order in RG is of the form ϵx, where $x \in G$ and ϵ is a root of unity in R.*

References. G. Higman [**123**], CR §37.

For further discussion of the units in RG, and their significance for algebraic K-theory, we refer the reader to Bass [**14**, Chapter 11, §7].

Let I denote the augmentation ideal of ZG, that is,

$$I = \sum{}' Z(x-1)$$

where x ranges over all elements of G except 1. Gruenberg [**94**] proved:

$$\bigcap_{n=1}^{\infty} I^n = 0 \quad \text{if and only if} \quad G \text{ is a } p\text{-group.}$$

For other results of this nature, as well as for a general algebraic study of group rings, see Connell [**45**], Passman [**177**], [**178**].

Further references on group rings: Banaschewski [**5**], Berman [**16**]–[**20**], Bovdi [**37**], Cohn-Livingstone [**42**], [**43**], Larson [**147**], Saksonov [**220**], [**221**], Schneider-Weissglass [**223**], Sehgal [**224**], Takahashi [**239**].

18. **Algebraic number theory.** Clearly the methods of algebraic number theory play a large role in the study of integral representations of orders. Recently, however, representation theory has begun to yield results in algebraic number theory. The chief application deals with the following problem:

Let R be a Dedekind domain with quotient field F, and let F' be a finite normal separable field extension of F, with Galois group G. Denote by R' the integral closure of R in F'. Then the elements of G act on R', and R' is an RG-lattice. What can be said about R' as RG-lattice?

More generally, an R'-ideal J in F' is *ambiguous* if $\sigma(J)=J$ for all $\sigma\in G$. Each ambiguous ideal J in F' is also an RG-lattice, and J is RG-free if and only if J has a normal integral basis over R (that is, $J=\sum_{\sigma\in G}R\cdot\sigma(x)$ for some $x\in J$). In particular, then, R' has a normal integral basis over R if and only if $R'\cong RG$ as RG-modules.

(i) Let $R=$ alg. int. $\{F\}$. Call F' *tamely ramified* over F if no maximal ideal of R divides the different of F' over F, or equivalently, if some element of R' has trace equal to 1. In this case, every ambiguous ideal is RG-projective; conversely, if R' is RG-projective, then F' is tamely ramified over F.

(ii) Hilbert proved that if F' is an abelian extension of the rational field Q, then $R'\cong ZG$ if and only if F' is tamely ramified over Q. (See Ullom [**256**], [**257**] for a partial generalization to the case of ambiguous ideals in R'.)

(iii) Fröhlich [**86**] gave a necessary and sufficient condition for the isomorphism $R'\cong RG$, assuming that F' is an abelian extension of F of a suitably restricted type (Kummer extension).

(iv) Leopoldt [**151**] showed that when F' is an abelian extension of Q, then

$$O_l(R') = \{x \in QG : x\cdot R' \subset R'\}$$

is isomorphic to R' as left ZG-module. Further, $O_l(R')$ is an order in QG which can be obtained from ZG by adjunction of certain idempotents arising from wild ramification. If F' is tamely ramified over Q, then indeed $O_l(R')=ZG$, and we recover Hilbert's result (ii). Jacobinski [**125**] investigated $O_l(R')$ in more general circumstances.

(v) Martinet [**155a**] recently proved that if F is a normal tamely ramified extension of Q whose Galois group G is dihedral of order $2p$ (p an odd prime), then alg. int $\{F\}\cong ZG$, so a normal integral basis exists in this case.

Other references for this section: Newman-Taussky [**173**].

19. **Krull-Schmidt and Cancellation Theorems.** Let R be a Dedekind domain, Λ any F-algebra. We shall say that the Krull-Schmidt Theorem is valid for Λ-lattices if every Λ-lattice is expressible as a direct sum of indecomposable lattices which are uniquely determined up to Λ-isomorphism and order of occurrence. As pointed out in Theorems 5.2 and 5.8, we have

(19.1). THEOREM. *The Krull-Schmidt theorem holds for Λ-lattices if*

(i) *R is a complete discrete valuation ring, or*

(ii) *R is a discrete valuation ring, and A is a direct sum of full matrix algebras over F, or*

(iii) *R is a discrete valuation ring and A is a direct sum of full matrix algebras over skewfields which remain skewfields upon passing from F to its completion.*

When R=alg. int. $\{F\}$, the Krull-Schmidt Theorem holds for R-lattices if and only if R is a principal ideal domain. Hence in order to insure that the Krull-Schmidt Theorem is valid for Λ-lattices, it is usually necessary to impose strong restrictions on R. Often one deals instead with the easier case of R_P-lattices.

Let Z_p denote the localization of Z at the prime ideal (p) of Z. Berman-Gudivok [**27**] observed that the Krull-Schmidt Theorem already fails for $Z_p\,G$-lattices when G is a suitably chosen cyclic group of squarefree order. In this direction we state

(19.2). THEOREM (JONES [**137**]). *Let G be an abelian group of exponent $k\cdot p^n$ where p is a prime, $p\nmid k$. The Krull-Schmidt Theorem holds for $Z_p\,G$-lattices if and only if one of the following holds true*:

(i) $k=1$.

(ii) $n=0$.

(iii) *The element p generates the multiplicative group of integers* (mod k) *which are relatively prime to k.*

Other references. Dress [**58**], Roggenkamp [**206**].

Now let A be a separable algebra over F, and let $i(\Lambda)$ be the Higman ideal of Λ (see §4) (so $i(\Lambda)=[G:1]R$ when $\Lambda=RG$). As in §6, define

$$\tilde{R} = \bigcap_P R_P,$$

where P ranges over all maximal ideals of R containing $i(\Lambda)$. Set

$\tilde{\Lambda} = \tilde{R}\Lambda$, an $\tilde{R}$-order in A useful in the study of genera. As shown by Reiner [**188**], the Krull-Schmidt Theorem need not hold for $\tilde{\Lambda}$-lattices; indeed, the $\tilde{R}$-ranks of the indecomposable direct summands of a $\tilde{\Lambda}$-lattice $\tilde{M}$ need not be uniquely determined by $\tilde{M}$. This question as to when the Krull-Schmidt Theorem holds for $\tilde{\Lambda}$-lattices is studied in detail in Jacobinski [**129**].

Generalizing Theorem 5.8, we have

(19.3). THEOREM (JACOBINSKI [**129**]). *Let F be an algebraic number field, and let Λ be an R-order in the semisimple F-algebra A. Suppose there exists a maximal ideal P of R such that*

(i) *for every maximal ideal P' of R distinct from P, the order $\Lambda_{P'}$ is a maximal $R_{P'}$-order in A, and*

(ii) *every A_P^*-module is of the form $F_P^* \otimes_F X$ for some A-module X.*

Then the Krull-Schmidt Theorem is valid for $\tilde{\Lambda}$-lattices (and for Λ_P-lattices as well).

(19.4). COROLLARY. *Let G be a finite p-group, where $p \neq 2$, p prime. Then the Krull-Schmidt Theorem holds for Z_pG-lattices.*

We shall say that the Cancellation Theorem holds for Λ-lattices if for each triple of Λ-lattices M, N, X, the isomorphism $M \dotplus X \cong N \dotplus X$ implies that $M \cong N$. Of course, if the Krull-Schmidt Theorem holds for Λ-lattices, then so does the Cancellation Theorem. When $R =$ alg. int. $\{F\}$, the Cancellation Theorem need not be true for RG-lattices; see for example Berman-Gudivok [**27**], as well as Theorem 8.12 above.

Indeed, an important example shows that the Cancellation Theorem is not necessarily true for the category of projective RG-lattices. We quote

(19.5). THEOREM (SWAN [**233**]). *Let G be a generalized quaternion group of order 32. There exists a projective left ideal M in the group ring ZG, such that*

$$ZG \dotplus ZG \cong M \dotplus ZG, \qquad M \ncong ZG.$$

In fact, if Γ is a maximal Z-order in QG containing ZG, a corresponding result holds for Γ-lattices.

Given Λ-lattices M and N, we have called N a *local summand* of M if $N_P^* | M_P^*$ for each maximal ideal P of R.

(19.6). THEOREM (JACOBINSKI [**129**]). *Let F be an algebraic number field, and let Λ be an R-order in the semisimple F-algebra A. Suppose that M and N are left Λ-lattices which satisfy the Eichler condition (see*

§8), *and let the* Λ*-lattice* X *be a local summand of* $M^{(n)}$ *for some* n. *Then* $M \dotplus X \cong N \dotplus X$ *implies that* $M \cong N$.

(19.7). COROLLARY (BASS [12, §9]). *Let* X *be a projective* Λ*-lattice, and* M *any* Λ*-lattice having* $\Lambda^{(2)}$ *as local summand. Then for any* Λ*-lattice* N, $M \dotplus X \cong N \dotplus X$ *implies that* $M \cong N$.

Finally we note

(19.8). THEOREM (JACOBINSKI [129]). *Let* F *be an algebraic number field, and let* G *be a finite group such that no prime divisor of* $[G:1]$ *is a unit in* R. *Suppose that no simple component of the group algebra* FG *is a totally definite quaternion algebra.*[15] *Then the Cancellation Theorem holds in the category of projective* RG*-lattices.*

REFERENCES

1. M. Auslander and O. Goldman, *Maximal orders*, Trans. Amer. Math. Soc. **97** (1960), 1–24. MR **22** #8034.

2. ———, *The Brauer group of a commutative ring*, Trans. Amer. Math. Soc. **97** (1960), 367–409. MR **22** #12130.

3. G. Azumaya, *Corrections and supplementaries to my paper concerning Krull-Remak-Schmidt's theorem*, Nagoya Math. J. **1** (1950), 117–124. MR **12**, 314.

4. D. Ballew, *The module index, projective modules and invertible ideals*, Ph.D. Thesis, University of Illinois, Urbana, Ill., 1969.

4a. ———, *The module index and invertible ideals*, Trans. Amer. Math. Soc. **148** (1970), (to appear).

5. B. Banaschewski, *Integral group rings of finite groups*, Canad. Math. Bull. **10** (1967), 635–642. MR **38** #1187.

6. L. F. Barannik and P. M. Gudivok, *Projective representations of finite groups over rings*, Dopovīdī Akad. Nauk Ukraīn. RSR Ser. A **1968**, 294–297. (Ukrainian) MR **37** #4177.

7. ———, *On indecomposable projective representations of finite groups*, Dopovīdī Akad. Nauk Ukraīn. RSR Ser. A **1969**, 391–393. (Ukrainian)

8. H. Bass, *Finitistic dimension and a homological generalization of semi-primary rings*, Trans. Amer. Math. Soc. **95** (1960), 466–488. MR **28** #1212.

9. ———, *Projective modules over algebras*, Ann. of Math. (2) **73** (1961), 532–542. MR **31** #1278.

10. ———, *Torsion free and projective modules*, Trans. Amer. Math. Soc. **102** (1962), 319–327. MR **25** #3960.

11. ———, *On the ubiquity of Gorenstein rings*, Math. Z. **82** (1963), 8–28. MR **27** #3669.

12. ———, *K-theory and stable algebra*, Inst. Hautes Études Sci. Publ. Math. No. 22 (1964), 5–60. MR **30** #4805.

13. ———, *The Dirichlet unit theorem, induced characters, and Whitehead groups of finite groups*, Topology **4** (1966), 391–410. MR **33** #1341.

14. ———, *Algebraic K-theory*, Math. Lecture Note Series, Benjamin, New York, 1968.

[15] See end of §7.

15. E. A. Bender, *Classes of matrices and quadratic fields*, Linear Algebra and Appl. **1** (1968), 195–201. MR **37** #6301.

16. S. D. Berman, *On certain properties of integral group rings*, Dokl. Akad. Nauk SSSR **91** (1953), 7–9. (Russian) MR **15**, 99.

17. ———, *On isomorphism of the centers of group rings of p-groups*, Dokl. Akad. Nauk SSSR **91** (1953), 185–187. (Russian) MR **15**, 99.

18. ———, *On a necessary condition for isomorphism of integral group rings*, Dopovīdī Akad. Nauk Ukraīn. RSR **1953**, 313–316. (Ukrainian) MR **15**, 599.

19. ———, *On the equation $x^m=1$ in an integral group ring*, Ukrain. Mat. Ž. **7** (1955), 253–261. (Russian) MR **17**, 1048.

20. ———, *On certain properties of group rings over the field of rational numbers*, Užgorod. Gos. Univ. Naučn. Zap. Him. Fiz. Mat. **12** (1955), 88–110. (Russian) MR **20** #3920.

21. ———, *On automorphisms of the center of an integral group ring*, Užgorod. Gos. Univ. Naučn. Zap. Him. Fiz. Mat. (1960), no. 3, 55. (Russian)

22. ———, *Integral representations of finite groups*, Dokl. Akad. Nauk SSSR **152** (1963), 1286–1287=Soviet Math. Dokl. **4** (1963), 1533–1535. MR **27** #4854.

23. ———, *On the theory of integral representations of finite groups*, Dokl. Akad. Nauk SSSR **157** (1964), 506–508=Soviet Math. Dokl. **5** (1964), 954–956. MR **29** #2308.

24. ———, *Integral representations of a cyclic group containing two irreducible rational components*, In Memoriam: N. G. Čebotarev, Izdat. Kazan Univ., Kazan, 1964, pp. 18–29. (Russian) MR **33** #4154.

25. ———, *On integral monomial representations of finite groups*, Uspehi Mat. Nauk **20** (1965), no. 4 (124), 133–134. (Russian) MR **33** #4155.

26. ———, *Representations of finite groups over an arbitrary field and over rings of integers*, Izv. Akad. Nauk SSSR Ser. Mat. **30** (1966), 69–132; English transl., Amer. Math. Soc. Transl. (2) **64** (1967), 147–215. MR **33** #5747.

27. S. D. Berman and P. M. Gudivok, *Integral representations of finite groups*, Dokl. Akad. Nauk SSSR **145** (1962), 1199–1201 =Soviet Math. Dokl. **3** (1962), 1172–1174. MR **25** #3095.

28. ———, *Integral representations of finite groups*, Užgorod. Gos. Univ. Naučn. Zap. Him. Fiz. Mat. (1962), no. 5, 74–76. (Russian)

29. ———, *Indecomposable representations of finite groups over the ring of p-adic integers*, Izv. Akad. Nauk SSSR Ser. Mat. **28** (1964), 875–910; English transl., Amer. Math. Soc. Transl. (2) **50** (1966), 77–113. MR **29** #3550.

30. S. D. Berman and A. I. Lihtman, *On integral representations of finite nilpotent groups*, Uspehi Mat. Nauk **20** (1965), no. 5 (125), 186–188. (Russian) MR **34** #7673.

31. S. D. Berman and A. R. Rossa, *On integral group-rings of finite and periodic groups*, Algebra and Math. Logic: Studies in Algebra, Izdat. Kiev Univ., Kiev, 1966, pp. 44–53. (Russian) MR **35** #265.

31a. A. Bialnicki-Birula, *On the equivalence of integral representations of finite groups*, Proc. Amer. Math. Soc. (to appear).

32. Z. I. Borevič and D. K. Faddeev, *Theory of homology in groups*. I, II, Vestnik Leningrad. Univ. **11** (1956), no. 7, 3–39; **14** (1959), no. 7, 72–87. (Russian) MR **18**, 188; MR **21** #4968.

33. ———, *Integral representations of quadratic rings*, Vestnik Leningrad. Univ. **15** (1960), no. 19, 52–64. (Russian) MR **27** #3668.

34. ———, *Representations of orders with cyclic index*, Trudy Mat. Inst. Steklov. **80** (1965), 51–65. Proc. Steklov Inst. Math. **80** (1965), 56–72. MR **34** #5805.

35. ———, *Remarks on orders with a cyclic index*, Dokl. Akad. Nauk SSSR **164** (1965), 727–728=Soviet Math. Dokl. **6** (1965), 1273–1274. MR **32** #7601.

36. N. Bourbaki, *Algèbre commutative*, Actualités Sci. Indust., no. 1293, Hermann, Paris, 1961. MR **30** #2027.

37. A. A. Bovdi, *Periodic normal divisors of the multiplicative group of a group ring*, Sibirsk Mat. Ž. 9 (1968), 495–498 = Siberian Math. J. 9 (1968), 374–376. MR **37** #2853.

38. J. O. Brooks, *Classification of representation modules over quadratic orders*, Ph.D. Thesis, University of Michigan, Ann Arbor, Mich., 1964.

39. A. Brumer, *Structure of hereditary orders*, Bull. Amer. Math. Soc. **69** (1963), 721–724; Addendum, ibid. **70** (1964), 185. MR **27** #2543.

40. H. Cartan and S. Eilenberg, *Homological algebra*, Princeton Univ. Press, Princeton, N. J., 1956. MR **17**, 1040.

41. C. Chevalley, *L'arithmétique dans les algèbres de matrices*, Actualités Sci. Indust., no. 323, Hermann, Paris, 1936.

42. J. A. Cohn and D. Livingstone, *On groups of order p^3*, Canad. J. Math. **15** (1963), 622–624. MR **27** #3700.

43. ———, *On the structure of group algebras*, Canad. J. Math. **17** (1965), 583–593. MR **31** #3514.

44. D. B. Coleman, *Idempotents in group rings*, Proc. Amer. Math. Soc. **17** (1966), 962. MR **33** #1379.

44a. S. B. Conlon, *Structure in representation algebras*, J. Algebra **5** (1967), 274–279. MR **34** #2719.

44b. ———, *Relative components of representations*, J. Algebra **8** (1968), 478–501.

44c. ———, *Decompositions induced from the Burnside algebra*, J. Algebra **10** (1968), 102–122. MR **38** #5945.

44d. ———, *Monomial representations under integral similarity*, J. Algebra **13** (1969), 496–508.

45. I. G. Connell, *On the group ring*, Canad. J. Math. **15** (1963), 650–685. MR **27** #3666.

46. C. W. Curtis and I. Reiner, *Representation theory of finite groups and associative algebras*, Pure and Appl. Math., vol. XI, Interscience, New York, 1962; 2nd ed., 1966. MR **26** #2519.

47. E. C. Dade, *Rings in which no fixed power of ideal classes becomes invertible*, Math. Ann. **148** (1962), 65–66. MR **25** #3963.

48. ———, *Some indecomposable group representations*, Ann. of Math. (2) **77** (1963), 406–412. MR **26** #2521.

49. ———, *The maximal finite groups of 4×4 integral matrices*, Illinois J. Math. **9** (1965), 99–122. MR **30** #1192.

50. E. C. Dade, D. W. Robinson, O. Taussky, and M. Ward, *Divisors of recurrent sequences*, J. Reine Angew. Math. **214/215** (1964), 180–183. MR **28** #5079.

51. E. C. Dade and O. Taussky, *Some new results connected with matrices of rational integers*, Proc. Sympos. Pure Math., vol. 8, Amer. Math. Soc., Providence, R. I., 1965, pp. 78–88. MR **32** #2395.

51a. ———, *On the different in orders in an algebraic number field and special units connected with it*, Acta Arith. 9 (1964), 47–51.

52. E. C. Dade, O. Taussky and H. Zassenhaus, *On the semigroup of ideal classes in an order of an algebraic number field*, Bull. Amer. Math. Soc. **67** (1961), 305–308. MR **25** #65.

53. ———, *On the theory of orders, in particular on the semigroup of ideal classes and genera of an order in an algebraic number field*, Math. Ann. **148** (1962), 31–64. MR **25** #3962.

54. K. deLeeuw, *Some applications of cohomology to algebraic number theory and group representations* (unpublished).

55. F. R. DeMeyer, *The trace map and separable algebras*, Osaka J. Math. **3** (1966), 7–11. MR **37** #4122.

56. M. Deuring, *Algebren*, Springer-Verlag, Berlin, 1935; rev. ed., Ergebnisse der Mathematik und ihrer Grenzgebiete, Band 41, 1968. MR **37** #4106.

57. F. E. Diederichsen, *Über die Ausreduktion ganzzahliger Gruppendarstellungen bei arithmetischer Äquivalenz*, Abh. Math. Sem. Univ. Hamburg **14** (1938), 357–412.

58. A. Dress, *A remark on the Krull-Schmidt theorem for integral group representations of rank* 1 (to appear).

59. ———, *An intertwining number theorem for integral representations and applications* (to appear).

60. ———, *On the decomposition of modules*, Bull. Amer. Math. Soc. **75** (1969), 984–986.

61. ———, *On integral representations*, Bull. Amer. Math. Soc. **75** (1969), 1031–1034.

62. ———, *The ring of monomial representations.* I: *Structure Theory* (to appear).

63. ———, *Vertices of integral representations*, Math. Z. (to appear).

64. ———, *On relative Grothendieck rings*, Bull. Amer. Math. Soc. **75** (1969), 955–958.

65. V. S. Drobotenko, *Integral representations of primary abelian groups*, Algebra and Math. Logic: Studies in Algebra, Izdat. Kiev. Univ., Kiev, 1966, pp. 111–121. (Russian) MR **34** #4375.

66. V. S. Drobotenko, E. S. Drobotenko, Z. P. Žilinskaja and E. Y. Pogoriljak, *Representations of the cyclic group of prime order p over residue classes* mod p^s, Ukrain. Mat. Ž. **17** (1965), no. 5, 28–42; English transl., Amer. Math. Soc. Transl. (2) **69** (1968), 241–256. MR **32** #5743.

67. V. S. Drobotenko and A. I. Lihtman, *Representations of finite groups over the ring of residue classes* mod p^s, Dokl. Užgorod Univ. **3** (1960), 63. (Russian)

68. V. S. Drobotenko, P. M. Gudivok and A. I. Lihtman, *On representations of finite groups over the ring of residue classes* mod m, Ukrain. Mat. Ž. **16** (1964), 82–89. (Russian) MR **29** #4810.

69. V. S. Drobotenko and V. P. Rud'ko, *Representations of a cyclic group by groups of automorphisms of a certain class of modules*, Dopovīdī Akad. Nauk Ukraīn. RSR Ser. A. **1968**, 302–304. (Ukrainian) MR **37** #2873.

70. Ju. A. Drozd, *Representations of cubic Z-rings*, Dokl. Akad. Nauk SSSR **174** (1967), 16–18 = Soviet Math. Dokl. **8** (1967), 572–574. MR **35** #6659.

70a. ———, *On the distribution of maximal sublattices*, Mat. Zametki **6** (1969), 19–24.

70b. ———, *Adèles and integral representations*, Izv. Akad. Nauk SSSR Ser. Mat. **33** (1969), 1080–1088.

71. Ju. A. Drozd, and V. V. Kiričenko, *Representation of rings in a second order matrix algebra*, Ukrain. Mat. Ž. **19** (1967), no. 3, 107–112. (Russian) MR **35** #1632.

72. ———, *Hereditary orders*, Ukrain. Mat. Ž. **20** (1967), 246–248. (Russian).

73. Ju. A. Drozd, V. V. Kiričenko and A. V. Roĭter, *On hereditary and Bass orders*, Izv. Akad. Nauk SSSR Ser. Mat. **31** (1967), 1415–1436 = Math. USSR Izv. **1** (1967), 1357–1376. MR **36** #2608.

74. Ju. A. Drozd and A. V. Roĭter, *Commutative rings with a finite number of indecomposable integral representations*, Izv. Akad. Nauk SSSR Ser. Mat. **31** (1967), 783–798 = Math. USSR Izv. **1** (1967), 757–772. MR **36** #3768.

75. Ju. A. Drozd and V. M. Turčin, *Number of representation modules in a genus* *for integral second order matrix rings*, Mat. Zametki **2** (1967), 133–138 = Math. Notes **2** (1967), 564–566. MR **37** #5253.

76. V. H. Dyson and K. W. Roggenkamp, *Modules over orders*, Springer Lecture Notes (to appear).

77. M. Eichler, *Über die Idealklassenzahl total definiter Quaternionenalgebren*, Math. Z. **43** (1938), 102–109.

78. ———, *Über die Idealklassenzahl hyperkomplexer Zahlen*, Math. Z. **43** (1938), 481–494.

79. D. K. Faddeev, *On the semigroup of genera in the theory of integer representations*, Izv. Akad. Nauk SSSR Ser. Mat. **28** (1964), 475–478; English transl., Amer. Math. Soc. Transl. (2) **64** (1967), 97–101. MR **28** #5089.

80. ———, *An introduction to multiplicative theory of modules of integral representations*, Trudy Mat. Inst. Steklov. **80** (1965), 145–182 = Proc. Steklov Inst. Math. **80** (1965), 164–210. MR **34** #5873.

81. ———, *On the theory of cubic Z-rings*, Trudy Mat. Inst. Steklov. **80** (1965), 183–187 = Proc. Steklov Inst. Math. **80** (1965), 211–215. MR **33** #4083.

82. ———, *Equivalence of systems of integer matrices*, Izv. Akad. Nauk SSSR Ser. Mat. **30** (1966), 449–454; English transl., Amer. Math. Soc. Transl. (2) **71** (1968), 43–48. MR **33** #2642.

83. ———, *The number of classes of exact ideals for Z-rings*, Mat. Zametki **1** (1967), 625–632 = Math. Notes **1** (1967), 415–419. MR **35** #5466.

84. R. Fossum, *The Noetherian different of projective orders*, J. Reine Angew. Math. **224** (1966), 207–218. MR **36** #5119.

84a. ———, *Maximal orders over Krull domains*, J. Algebra **10** (1968), 321–332. MR **38** #2130.

85. A. Fröhlich, *Ideals in an extension field as modules over the algebraic integers in a finite number field*, Math. Z. **74** (1960), 29–38. MR **22** #4708.

86. ———, *The module structure of Kummer extensions over Dedekind domains*, J. Reine Angew. Math. **209** (1962), 39–53. MR **28** #3988; p. 1247.

87. ———, *Invariants for modules over commutative separable orders*, Quart. J. Math. Oxford Ser. (2) **16** (1965), 193–232. MR **35** #1583.

88. ———, *Resolvents, discriminants, and trace invariants*, J. Algebra **4** (1966), 173–198. MR **34** #7499.

89. I. Giorgiutti, *Modules projectifs sur les algèbres de groupes finis*, C.R. Acad. Sci. Paris **250** (1960), 1419–1420. MR **23** #A1691.

90. J. A. Green, *On the indecomposable representations of a finite group*, Math. Z. **70** (1958/59), 430–445. MR **24** #A1304.

91. ———, *Blocks of modular representations*, Math. Z. **79** (1962), 100–115. MR **25** #5114.

92. ———, *The modular representation algebra of a finite group*, Illinois J. Math. **6** (1962), 607–619. MR **25** #5106.

93. ———, *A transfer theorem for modular representations*, J. Algebra **1** (1964), 73–84. MR **29** #147.

94. K. Gruenberg, *Residual properties of infinite soluble groups*, Proc. London Math. Soc. (3) **7** (1957), 29–62. MR **19**, 386.

95. P. M. Gudivok, *Integral representations of a finite group with a noncyclic Sylow p-subgroup*, Uspehi Mat. Nauk **16** (1961), 229–230.

96. ———, *Integral representations of groups of type (p, p)*, Dokl. Užgorod Univ. Ser. Phys.-Mat. Nauk (1962), no. 5, 73.

97. ———, *On p-adic integral representations of finite groups*, Dokl. Užgorod Univ. Ser. Phys.-Mat. Nauk (1962), no. 5, 81–82.

98. ———, *Representations of finite groups over certain local rings*, Dopovīd Akad. Nauk Ukraīn. RSR **1964,** 173–176. (Ukrainian) MR **29** #3551.

99. ———, *Representations of finite groups over quadratic rings*, Dokl. Akad. Nauk SSSR **159** (1964), 1210–1213 = Soviet Math. Dokl. **5** (1964), 1669–1672. MR **30** #174.

100. ———, *Representations of finite groups over local number rings*, Dopovīdī Akad. Nauk Ukraīn. RSR **1966,** 979–981. (Ukrainian) MR **34** #1407.

101. ———, *Representations of finite groups over number rings*, Izv. Akad. Nauk SSSR Ser. Mat. **31** (1967), 799–834 = Mat. USSR Izv. **1** (1967), 773–805. MR **36** #1554.

102. P. M. Gudivok and V. P. Rud′ko, *On p-adic integer-valued representations of a cyclic p-group*, Dopovīdī Akad. Nauk Ukraīn. RSR **1966,** 1111–1113. (Ukrainian) MR **34** #1409.

103. T. Hannula, *The integral representation ring* $a(R_kG)$, Trans. Amer. Math. Soc. **133** (1968), 553–559.

104. M. Harada, *Hereditary orders*, Trans. Amer. Math. Soc. **107** (1963), 273–290. MR **27** #1474.

105. ———, *Structure of hereditary orders over local rings*, J. Math. Osaka City Univ. **14** (1963), 1–22. MR **29** #5879.

106. ———, *Multiplicative ideal theory in hereditary orders*, J. Math. Osaka City Univ. **14** (1963), 83–106. MR **29** #5880b.

107. A. Hattori, *Rank element of a projective module*, Nagoya Math. J. **25** (1965), 113–120. MR **31** #226.

108. ———, *Semisimple algebras over a commutative ring*, J. Math. Soc. Japan **15** (1963), 404–419. MR **28** #2125.

109. A. Heller, *On group representations over a valuation ring*, Proc. Nat. Acad. Sci. U.S.A. **47** (1961), 1194–1197. MR **23** #A2468.

110. ———, *Some exact sequences in algebraic K-theory*, Topology **3** (1965), 389–408. MR **31** #3477.

111. A. Heller and I. Reiner, *Indecomposable representations*, Illinois J. Math. **5** (1961), 314–323. MR **23** #A222.

112. ———, *Representations of cyclic groups in rings of integers*. I, II, Ann. of Math. (2) **76** (1962), 73–92; (2) **77** (1963), 318–328. MR **25** #3993; MR **26** #2520.

113. ———, *On groups with finitely many indecomposable integral representations*, Bull. Amer. Math. Soc. **68** (1962), 210–212. MR **25** #1222.

114. ———, *Grothendieck groups of orders in semisimple algebras*, Trans. Amer. Math. Soc. **112** (1964), 344–355. MR **28** #5093.

115. ———, *Grothendieck groups of integral group rings*, Illinois J. Math. **9** (1965), 349–360. MR **31** #211.

116. D. G. Higman, *Indecomposable representations at characteristic p*, Duke Math. J. **21** (1954), 377–381. MR **16**, 794.

117. ———, *Induced and produced modules*, Canad. J. Math. **7** (1955), 490–508. MR **19,** 390.

118. ———, *On orders in separable algebras*, Canad. J. Math. **7** (1955), 509–515. MR **19,** 527.

119. ———, *Relative cohomology*, Canad. J. Math. **9** (1957), 19–34. MR **18,** 715.

120. ———, *On isomorphisms of orders*, Michigan Math. J. **6** (1959), 255–257. MR **22** #62.

121. ———, *On representations of orders over Dedekind domains*, Canad. J. Math. **12** (1960), 107–125. MR **22** #63.

122. D. G. Higman and J. E. MacLaughlin, *Finiteness of class numbers of representations of algebras over function fields*, Michigan Math. J. **6** (1959), 401–404. MR **22** #39.

123. G. Higman, *The units of group-rings*, Proc. London Math. Soc. (2) **46** (1940), 231–248. MR **2**, 5.

124. R. Holvoet, *Sur l'isomorphie d'algèbres de groupes*, Bull. Soc. Math. Belg. **20** (1968), 264–282.

124a. D. A. Jackson, *On a problem in the theory of integral group rings*, Ph.D. thesis, Oxford University, Oxford, 1967.

124b. ———, *The group of units of the integral group ring of finite metabelian and finite nilpotent groups*, Quart. J. Math. **20** (1969), 319–331.

125. H. Jacobinski, *Über die Hauptordnung eines Körpers als Gruppenmodul*, J. Reine Angew. Math. **213** (1963/64), 151–164. MR **29** #1200.

126. ———, *On extensions of lattices*, Michigan Math. J. **13** (1966), 471–475. MR **34** #4377.

127. ———, *Sur les ordres commutatifs avec un nombre fini de réseaux indécomposables*, Acta Math. **118** (1967), 1–31. MR **35** #2876.

128. ———, *Über die Geschlechter von Gittern über Ordnungen*, J. Reine Angew. Math. **230** (1968), 29–39. MR **37** #5250.

129. ———, *Genera and decompositions of lattices over orders*, Acta. Math. **121** (1968), 1–29.

130. ———, *On embedding of lattices belonging to the same genus*, Proc. Amer. Math. Soc. **24** (1970), 134–136.

131. N. Jacobson, *The theory of rings*, Math. Surveys, no. II, Amer. Math. Soc., Providence, R. I., 1943. MR **5**, 31.

132. ———, *Structure of rings*, Amer. Math. Soc. Colloq. Publ., vol. 37, Amer. Math. Soc., Providence, R. I., 1956. MR **18**, 373.

133. W. E. Jenner, *Block ideals and arithmetics of algebras*, Compositio Math. **11** (1953), 187–203. MR **16**, 7.

134. ———, *On the class number of non-maximal orders in $\mathcal{P}$-adic division algebras*, Math. Scand. **4** (1956), 125–128. MR **18**, 375.

135. A. Jones, *Groups with a finite number of indecomposable integral representations*, Michigan Math. J. **10** (1963), 257–261. MR **27** #3698.

136. ———, *Integral representations of the direct product of groups*, Canad. J. Math. **15** (1963), 625–630. MR **27** #4870.

137. ———, *On representations of finite groups over valuation rings*, Illinois J. Math. **9** (1965), 297–303. MR **31** #257.

138. I. Kaplansky, *Elementary divisors and modules*, Trans. Amer. Math. Soc. **66** (1949), 464–491. MR **11**, 155.

139. ———, *Modules over Dedekind rings and valuation rings*, Trans. Amer. Math. Soc. **72** (1952), 327–340. MR **13**, 719.

140. ———, *Submodules of quaternion algebras*, Proc. London Math. Soc. (3) **19** (1969), 219–232.

141. V. V. Kiričenko, *Orders whose representations are all completely reducible*, Mat. Zametki 2 (1967), 139–144 = Math. Notes 2 (1967), 567–570. MR **36** #2609.

142. D. I. Knee, *The indecomposable integral representations of finite cyclic groups*, Ph.D. Thesis, M.I.T., Cambridge, Mass., 1962.

143. M. Kneser, *Einige Bemerkungen über ganzzahlige Darstellungen endlicher Gruppen*, Arch. Math **17** (1966), 377–379. MR **34** #1408.

144. S. A. Krugljak, *Exact ideals in a second order integral matrix ring*, Ukrain. Mat. Ž. **18** (1966), no. 3, 58–64. (Russian) MR **33** #7378.

145. ———, *The Grothendieck group*, Ukrain. Mat. Ž. **18** (1966), no. 5, 100–105. (Russian) MR **34** #204.

146. T. Y. Lam, *Induction theorems for Grothendieck groups and Whitehead groups of finite groups*, Ann. Sci. École Norm. Sup. (4) **1** (1968), 91–148.

147. R. Larson, *Group rings over Dedekind domains.* I, II, J. Algebra **5** (1967), 358–361; **7** (1967), 278–279. MR **35** #266; MR **35** #5525.

148. C. G. Latimer and C. C. MacDuffee, *A correspondence between classes of ideals and classes of matrices*, Ann. of Math, (2) **34** (1933), 313–316.

149. W. J. Leahey, *The classification of the indecomposable integral representations of the dihedral group of order 2p*, Ph.D. Thesis, M.I.T., Cambridge, Mass., 1962.

150. M. P. Lee, *Integral representations of dihedral groups of order 2p*, Trans. Amer. Math. Soc. **110** (1964), 213–231. MR **28** #139.

151. H. Leopoldt, *Über die Hauptordnung der ganzen Elemente eines abelschen Zahlkörpers*, J. Reine Angew. Math. **201** (1959), 119–149. MR **21** #7195.

152. L. S. Levy, *Decomposing pairs of modules*, Trans. Amer. Math. Soc. **122** (1966), 64–80. MR **33** #2677.

153. G. W. Mackey, *On induced representations of groups*, Amer. J. Math. **73** (1951), 576–592. MR **13**, 106.

154. J.-M. Maranda, *On $\mathfrak{P}$-adic integral representations of finite groups*, Canad. J. Math. **5** (1953), 344–355. MR. **15**, 100.

155. ———, *On the equivalence of representations of finite groups by groups of automorphisms of modules over Dedekind rings*, Canad. J. Math. **7** (1955), 516–526. MR **19**, 529.

155a. J. Martinet, *Sur l'arithmétique des extensions galoisiennes à groupe de Galois diédral d'ordre 2p*, Ann. Inst. Fourier (Grenoble) **19** (1969) (to appear).

156. A. Matuljauskas, *Integral representations of a fourth-order cyclic group*, Litovsk. Mat. Sb. **2** (1962), no. 1, 75–82. (Russian) MR **26** #6274.

157. ———, *Integral representations of the cyclic group of order six*, Litovsk. Mat. Sb. **2** (1962), no. 2, 149–157. (Russian) MR **27** #5835.

158. ——, *On the number of indecomposable representations of the group Z_8*, Litovsk. Mat. Sb **3** (1963), no. 1, 181–188. (Russian) MR **29** #2309.

159. A. Matuljauskas and M. Matuljauskene, *On integral representations of a group of type* (3, 3), Litovsk. Mat. Sb. **4** (1964), 229–233. (Russian) MR **29** #4812.

160. Warren May, *Commutative group algebras*, Trans. Amer. Math. Soc. **136** (1969), 139–149. MR **38** #2224.

160a. G. O. Michler, *Structure of semi-perfect hereditary noetherian rings*, J. Algebra **13** (1969), 327–344.

161. T. Nakayama, *A theorem on modules of trivial cohomology over a finite group*, Proc. Japan Acad. **32** (1956), 373–376. MR **18**, 191.

162. ———, *On modules of trivial cohomology over a finite group*, Illinois J. Math. **1** (1957), 36–43. MR **18**, 793.

163. ———, *On modules of trivial cohomology over a finite group.* II: *Finitely generated modules*, Nagoya Math. J. **12** (1957), 171–176. MR **20** #4587.

164. L. A. Nazarova, *Unimodular representations of the four group*, Dokl. Akad. Nauk SSSR **140** (1961), 1011–1014 = Soviet Math. Dokl. **2** (1961), 1304–1307. MR **24** #A770.

165. ———, *Unimodular representations of the alternating group of degree four*, Ukrain. Mat. Ž. **15** (1963), 437–444. (Russian) MR **28** #2148.

166. ———, *Representations of a tetrad*, Izv. Akad. Nauk SSSR Ser. Mat. **31** (1967), 1361–1378 = Math. USSR Izv. **1** (1967), 1305–1322. MR **36** #6400.

167. L. A. Nazarova and A. V. Roĭter, *Integral representations of the symmetric group of third degree*, Ukrain. Mat. Ž. **14** (1962), 271–288. (Russian) MR **26** #6273.

168. ———, *On irreducible representations of p-groups over $Z_p(\epsilon)$*, Ukrain. Mat. Ž. **18** (1966), no 1, 119–124. (Russian) MR **34** #254.

169. ———, *On integral p-adic representations and representations over residue class rings*, Ukrain. Mat. Ž. **19** (1967), no. 2, 125–126. (Russian) MR **35** #267.

170. ———, *Refinement of a theorem of Bass*, Dokl. Akad. Nauk SSSR **176** (1967), 266–268 = Soviet Math. Dokl. **8** (1967), 1089–1092. MR **37** #1402.

171. ———, *Finitely generated modules over a dyad of a pair of local Dedekind rings, and finite groups having an abelian normal subgroup of index p*, Izv. Akad. Nauk SSSR Ser. Mat. **33** (1969), 65–89 = Math. USSR Izv. **3** (1969).

172. M. Newman and O. Taussky, *Classes of positive definite unimodular circulants*, Canad. J. Math. 9 (1957), 71–73. MR **18**, 634.

173. ———, *On a generalization of the normal basis in abelian algebraic number fields*, Comm. Pure Appl. Math. **9** (1956), 85–91. MR **17**, 829.

174. R. J. Nunke, *Modules of extensions over Dedekind rings*, Illinois J. Math. **3** (1959), 222–241. MR **21** #1329.

175. T. Obayashi, *On the Grothendieck ring of an abelian p-group*, Nagoya Math. J. **26** (1966), 101–113. MR **37** #1438.

176. J. Oppenheim, *Integral representations of cyclic groups of squarefree order*, Ph.D. Thesis, University of Illinois, Urbana, Ill., 1962.

177. D. S. Passman, *Nil ideals in group rings*, Michigan Math. J. **9** (1962), 375–384. MR **26** #2470.

178. ———, *Isomorphic groups and group rings*, Pacific J. Math. **15** (1965), 561–583. MR **33** #1381.

179. L. C. Pu, *Integral representations of non-abelian groups of order pq*, Michigan Math. J. **12** (1965), 231–246. MR **31** #2321.

180. T. Ralley, *Decomposition of products of modular representations*, J. London Math. Soc. **44** (1969), 480–484.

181. I. Reiner, *Maschke modules over Dedekind rings*, Canad. J. Math. **8** (1956), 329–334. MR **18**, 7.

182. ———, *Integral representations of cyclic groups of prime order*, Proc. Amer. Math. Soc. **8** (1957), 142–146, MR **18**, 717.

183. ———, *On the class number of representations of an order*, Canad. J. Math. **11** (1959), 660–672. MR **21** #7229.

184. ———, *The nonuniqueness of irreducible constituents of integral group representations*, Proc. Amer. Math. Soc. **11** (1960), 655–658. MR **23** #A223.

185. ———, *Behavior of integral group representations under ground ring extension*, Illinois J. Math. **4** (1960), 640–651. MR **22** #12145.

186. ———, *The Krull-Schmidt theorem for integral group representations*, Bull. Amer. Math. Soc. **67** (1961), 365–367. MR **25** #2132.

187. ———, *Indecomposable representations of non-cyclic groups*, Michigan Math. J. **9** (1962), 187–191. MR **25** #3994.

188. ———, *Failure of the Krull-Schmidt theorem for integral representations*, Michigan Math. J. **9** (1962), 225–231. MR **26** #2482.

189. ———, *Extensions of irreducible modules*, Michigan Math J. **10** (1963), 273–276. MR **27** #5807.

190. ———, *The integral representation ring of a finite group*, Michigan Math. J. **12** (1965), 11–22. MR **30** #3152.

191. ———, *Nilpotent elements in rings of integral representations*, Proc. Amer. Math. Soc. **17** (1966), 270–274. MR **32** #5745.

192. ———, *Integral representation algebras*, Trans. Amer. Math. Soc. **124** (1966), 111–121. MR **34** #2722.

193. ———, *Relations between integral and modular representations*, Michigan Math. J. **13** (1966), 357–372. MR **36** #5240.

194. ———, *Module extensions and blocks*, J. Algebra **5** (1967), 157–163. MR **35** #4316.

195. ———, *Representation rings*, Michigan Math. J. **14** (1967), 385–391. MR **36** #1555.

196. ———, *An involution on* $\tilde{K}^0(ZG)$, Mat. Zametki **3** (1968), 523–527. (Russian) MR **37** #5270.

197. ———, *A survey of integral representation theory*, Proc. Algebra Sympos., University of Kentucky (Lexington, 1968), pp. 8–14.

198. ———, *Maximal orders*, Mimeograph Notes, University of Illinois, Urbana, Ill., 1969.

199. I. Reiner and H. Zassenhaus, *Equivalence of representations under extensions of local ground rings*, Illinois J. Math. **5** (1961), 409–411. MR **23** #A3764.

200. D. S. Rim, *Modules over finite groups*, Ann. of Math. (2) **69** (1959), 700–712. MR **21** #3474.

201. ———, *On projective class groups*, Trans. Amer. Math. Soc. **98** (1961), 459–467. MR **23** #A1690.

202. K. W. Roggenkamp, *Gruppenringe von unendlichem Darstellungstyp*, Math. Z. **96** (1967), 393–398. MR **34** #5948.

203. ———, *Darstellungen endlicher Gruppen in Polynomringen*, Math. Z. **96** (1967), 399–407. MR **34** #5949.

204. ———, *Grothendieck groups of hereditary orders*, J. Reine Angew. Math. **235** (1969), 29–40.

205. ———, *On the irreducible lattices of orders*, Canad. J. Math. **21** (1969), 970–976.

206. ———, *Das Krull-Schmidt Theorem für projektive Gitter in Ordnungen über lokalen Ringen*, Math. Seminar (Giessen, 1969).

207. ———, *Projective modules over clean orders*, Compositio Math. **21** (1969), 185–194.

208. ———, *A necessary and sufficient condition for orders in direct sums of complete skewfields to have only finitely many nonisomorphic indecomposable integral representations*, Bull. Amer. Math. Soc. **76** (1969), 130–134.

208a. ———, *Projective homorphisms and extensions of lattices*, J. Reine Angew. Math. (to appear).

208b. K. W. Roggenkamp and V. H. Dyson, *Modules over orders*, Springer Lecture Notes (to appear).

209. A. V. Roĭter, *On the representations of the cyclic group of fourth order by integral matrices*, Vestnik Leningrad. Univ. **15** (1960), no. 19, 65–74. (Russian) MR **23** #A1730.

210. ———, *Categories with division and integral representations*, Dokl. Akad. Nauk SSSR **153** (1963), 46–48 = Soviet Math. Dokl. **4** (1963), 1621–1623. MR **33** #2704.

211. ———, *On a category of representations*, Ukrain. Mat. Ž. **15** (1963), 448–452. (Russian) MR **28** #3072.

212. ———, *On integral representations belonging to a genus*, Izv. Akad. Nauk SSSR Ser. Mat. **30** (1966), 1315–1324; English transl., Amer. Math. Soc. Transl. (2) **71** (1968), 49–59. MR **35** #4255.

213. ———, *Divisibility in the category of representations over a complete local Dedekind ring*, Ukrain. Mat. Ž. **17** (1965), no. 4, 124–129. (Russian) MR **33** #5699.

214. ———, *E-systems of representations*, Ukrain. Mat. Ž. **17** (1965), no. 2, 88–96. (Russian) MR **32** #7620.

215. ———, *An analog of Bass' theorem for representation modules of non-commutative orders*, Dokl. Akad. Nauk SSSR **168** (1966), 1261–1264 = Soviet Math. Dokl. **7** (1966), 830–833. MR **34** #2632.

216. ———, *Unboundedness of the dimensions of indecomposable representations of algebras having infinitely many indecomposable representations*, Izv. Akad. Nauk SSSR Ser. Mat. **32** (1968), 1275–1282 = Math. USSR Izv. **2** (1968) (to appear).

217. ———, *On the theory of integral representations of rings*, Mat. Zametki **3** (1968), 361–366. (Russian) MR **38** #187.

218. J. Rotman, *Homological algebra*, Van Nostrand, Princeton, N. J., 1970.

219. V. P. Rud'ko, *On the integral representation algebra of a cyclic group of order p^2*, Dopovīdī Akad. Nauk Ukrain. RSR Ser. A **1967**, 35–39. (Ukrainian) MR **35** #268.

220. A. I. Saksonov, *On group rings of finite p-groups over certain integral domains*, Dokl. Akad Nauk BSSR **11** (1967), 204–207. (Russian) MR **35** #270.

221. ———, *Group-algebras of finite groups over a number field*, Dokl. Akad. Nauk BSSR **11** (1967), 302–305. MR **35** #1681.

222. O. F. G. Schilling, *The theory of valuations*, Math. Surveys, no. 4, Amer. Math. Soc., Providence, R. I., 1950. MR **13**, 315.

223. H. Schneider and J. Weissglass, *Group rings, semigroup rings and their radicals*, J. Algebra **5** (1967), 1–15. MR **35** #4317.

224. S. K. Sehgal, *On the isomorphism of integral group rings*. I, II, Canad. J. Math. **21** (1969), 410–413, 1182–1188.

225. C. S. Seshadri, *Triviality of vector bundles over the affine space K^2*, Proc. Nat. Acad. Sci. U. S. A. **44** (1958), 456–458. MR **21** #1318.

226. ———, *Algebraic vector bundles over the product of an affine curve and the affine line*, Proc. Amer. Math. Soc. **10** (1959), 670–673. MR **29** #2263.

226a. M. Singer, *Invertible powers of ideals over orders in commutative separable algebras*, Proc. Cambridge Philos. Soc. (to appear).

227. D. L. Stancl, *Multiplication in Grothendieck rings of integral group rings*, J. Algebra **7** (1967), 77–90. MR **36** #6476.

228. E. Steinitz, *Rechteckige Systeme und Moduln in algebraischen Zahlenkörpern*. I, II, Math. Ann. **71** (1911), 328–354; **72** (1912), 297–345.

229. J. R. Strooker, *Faithfully projective modules and clean algebras*, Ph.D. Thesis, University of Utrecht, 1965.

230. R. G. Swan, *Projective modules over finite groups*, Bull. Amer. Math. Soc. **65** (1959), 365–367. MR **22** #5660.

231. ———, *The p-period of a finite group*, Illinois J. Math. **4** (1960), 341–346. MR **23** #A188.

232. ———, *Induced representations and projective modules*, Ann. of Math. (2) **71** (1960), 552–578. MR **25** 2131.

233. ———, *Projective modules over group rings and maximal orders*, Ann. of Math. (2) **76** (1962), 55–61. MR **25** #3066.

234. ———, *The Grothendieck ring of a finite group*, Topology **2** (1963), 85–110. MR **27** #3683.

235. ———, *Algebraic K-theory*, Springer Lecture Notes, Berlin, 1968.

236. ———, *Invariant rational functions and a problem of Steenrod*, Invent. Math. **7** (1969), 148–158.

237. ———, *The number of generators of a module*, Math. Z. **102** (1967), 318–322. MR **36** #1434.

238. S. Takahashi, *Arithmetic of group representations*, Tôhoku Math. J. (2) **11** (1959), 216–246. MR **22** #733.

239. ———, *A characterization of group rings as a special class of Hopf algebras*, Canad. Math. Bull. **8** (1965), 465–475. MR **32** #2459.

240. O. Taussky, *On a theorem of Latimer and MacDuffee*, Canad. J. Math. **1** (1949), 300–302. MR **11**, 3.

241. ———, *Classes of matrices and quadratic fields*, Pacific J. Math. **1** (1951), 127–132. MR **13**, 201.

242. ———, *Classes of matrices and quadratic fields*. II, J. London Math. Soc. **27** (1952), 237–239. MR **13**, 717.

243. ———, *Unimodular integral circulants*, Math. Z. **63** (1955), 286–289. MR **17**, 347.

244. ———, *On matrix classes corresponding to an ideal and its inverse*, Illinois J. Math. **1** (1957), 108–113. MR **20** #845.

245. ———, *Matrices of rational integers*, Bull. Amer. Math. Soc. **66** (1960), 327–345. MR **22** #10994.

246. ———, *Ideal matrices*. I, Arch. Math. **13** (1962), 275–282. MR **27** #168.

247. ———, *Ideal matrices*. II, Math. Ann. **150** (1963), 218–225. MR **28** #105.

248. ———, *On the similarity transformation between an integral matrix with irreducible characteristic polynomial and its transpose*, Math. Ann. **166** (1966), 60–63. MR **33** #7355.

249. ———, *The discriminant matrices of an algebraic number field*, J. London Math. Soc. **43** (1968), 152–154. MR **37** #4053.

250. O. Taussky and J. Todd, *Matrices with finite period*, Proc. Edinburgh Math. Soc. (2) **6** (1940), 128–134. MR **2**, 118.

251. ———, *Matrices of finite period*, Proc. Roy. Irish Acad. Sect. A **46** (1941), 113–121. MR **2**, 243.

252. O. Taussky and H. Zassenhaus, *On the similarity transformation between a matrix and its transpose*, Pacific J. Math. 9 (1959), 893–896. MR **21** #7216.

253. John G. Thompson, *Vertices and sources*, J. Algebra **6** (1967), 1–6. MR **34** #7677.

254. A. Troy, *Integral representations of cyclic groups of order p^2*, Ph.D. Thesis, University of Illinois, Urbana, Ill., 1961.

255. K. Uchida, *Remarks on Grothendieck groups*, Tôhoku Math. J. (2) **19** (1967), 341–348. MR **37** #2838.

256. S. Ullom, *Normal bases in Galois extensions of number fields*, Nagoya Math. J. **34** (1969), 153–167.

257. ———, *Galois cohomology of ambiguous ideals*, J. Number Theory **1** (1969), 11–15.

258. Y. Watanabe, *The Dedekind different and the homological different*, Osaka J. Math. **4** (1967), 227–231. MR **37** #2795.

259. ———, *The Dedekind different and the homological different of an algebra*, J. Math. Soc. Japan (to appear).

260. A. Weil, *Basic number theory*, Die Grundlehren der math. Wissenschaften, Band 114, Springer-Verlag, New York, 1967. MR **38** #3244.

261. A. R. Whitcomb, *The group ring problem*, Ph.D. thesis, University of Chicago, Chicago, Ill., 1968.

262. O. Zariski and P. Samuel, *Commutative algebra*. Vol. I, University Series in Higher Math., Van Nostrand, Princeton, N. J., 1958. MR **19,** 833.

263. H. Zassenhaus, *Neuer Beweis der Endlichkeit der Klassenzahl bei unimodularer Aquivalenz endlicher ganzzahliger Substitutionsgruppen*, Abh. Math. Sem. Univ. Hamburg **12** (1938), 276–288.

264. ———, *Über die Äquivalenz ganzzahliger Darstellungen*, Nachr. Akad. Wiss. Göttingen Math.-Phys. Kl. II **1967,** 167–193. MR **37** #6319.

265. J. Zemanek, *On the semisimplicity of integral representation rings*, Ph.D. Thesis, University of Illinois, Urbana, Ill., 1970.

UNIVERSITY OF ILLINOIS, URBANA, ILLINOIS 61801

CLASS GROUPS FOR INTEGRAL REPRESENTATIONS OF METACYCLIC GROUPS

S. GALOVICH, I. REINER AND S. ULLOM

1. *Introduction.* Let R be a Dedekind domain whose quotient field K is an algebraic number field, and let Λ be an R-order in a semisimple K-algebra A with 1. A Λ-*lattice* is a finitely generated R-torsionfree left Λ-module. We shall call a Λ-lattice M *locally free of rank n* if for each maximal ideal $\mathfrak{p}$ of R, $M_{\mathfrak{p}}$ is $\Lambda_{\mathfrak{p}}$-free on n generators. (The subscript $\mathfrak{p}$ denotes localization.) The (locally free) *class group* of Λ is the additive group $C(\Lambda)$ generated by symbols

$$x_M = [\Lambda]-[M],\quad M = \text{locally free rank 1 } \Lambda\text{-lattice},$$

where

$$x_{M_1}+x_{M_2} = x_{M_3} \text{ whenever } M_1 \dotplus M_2 \cong \Lambda \dotplus M_3,$$

and where $x_M = 0$ if and only if M is stably free (that is, $M \dotplus \Lambda^{(k)} \cong \Lambda \dotplus \Lambda^{(k)}$ for some k).

Let ZG be the integral group ring of a finite group G. A number of recent articles have been devoted to the calculation of the class group $C(ZG)$ for various groups G (see Fröhlich **[1]**, Kervaire and Murthy **[3]**, Martinet **[5]**, Reiner and Ullom **[7, 8]**, and Ullom **[11]**). For the most part, it is only in rare cases that the order $|C(ZG)|$ can be computed explicitly. In such cases, the formula for $|C(ZG)|$ usually involves the ideal class numbers of certain cyclotomic fields, and makes use of detailed information about units in integral group rings.

The purpose of this note is to compute $C(ZG)$ for the case where G is a metacyclic group of order pq. Let

$$G = \langle x, y : x^p = 1, y^q = 1, yxy^{-1} = x^r\rangle, \tag{1.1}$$

where p is an odd prime, q is any divisor of $p-1$, and r is a primitive q-th root of 1 mod p. Let ω be a primitive p-th root of 1 over Q, and set $K = Q(\omega)$, $R = Z[\omega]$; thus $R = \text{alg. int.}\{K\}$, the ring of all algebraic integers in K. Let L be the unique subfield of K such that $(K\colon L) = q$, and put $S = \text{alg. int.}\{L\}$. Denote by $C(S)$ the ideal class group of S. Let $H = \langle y\rangle$, a cyclic group of order q.

Our main result is

(1.2) THEOREM. *There is an epimorphism*

$$C(ZG) \to C(S) \dotplus C(ZH),$$

whose kernel $D_0(ZG)$ is a finite cyclic group of order q, q odd, and of order $q/2$, q even.

The case where $q = 2$ is already known (see Lee **[4]**, Reiner and Ullom **[7, 8]**). It is rather surprising that such an explicit formula can be obtained, especially since as yet there is no analogous result for the seemingly simpler case of a cyclic group of order pq, with p, q distinct primes. When q is prime, we know from Rim **[9]** or Reiner **[6]** that $C(ZH) \cong C(R')$, where $R' = \text{alg. int.}\{K'\}$, and $K' = Q(\sqrt[q]{1})$. For

---

The last two named authors were partially supported by a contract with the National Science Foundation.

composite q, there is no such simple expression for $C(ZH)$, except for certain special cases.

2. *Fibre products and the analogue of the Mayer–Vietoris sequence.* We introduce the following notation in addition to that listed in §1. Let σ be the automorphism of K defined by the map $\omega \to \omega^r$; then L is the subfield of K fixed by σ. Put

$$P = (1 - \omega) R, \; P_0 = P \cap S,$$

so P_0 is the unique maximal ideal of S containing the rational prime p, and we have

$$R/P \cong S/P_0 \cong Z/pZ = \bar{Z} \text{ (say)}.$$

We set $H = \langle y \rangle$, a cyclic group of order q, where q is any divisor of $p - 1$ (not necessarily prime). Then let $R \circ H$ be the twisted group ring of H over R, that is,

$$R \circ H = R \oplus Ry \oplus \ldots \oplus Ry^{q-1}; \; y^q = 1, \; y\alpha = \alpha^\sigma y, \; \alpha \in R.$$

The corresponding K-algebra $K \circ H$ is central simple over L.

We now choose $\Lambda = ZG$, and set

$$I = (x^{p-1} + x^{p-2} + \ldots + x + 1)\Lambda, \quad J = (x - 1)\Lambda.$$

Then I, J are two-sided ideals of Λ such that $I \cap J = 0$, and there are obvious isomorphisms

$$\Lambda/I \cong R \circ H, \; \Lambda/J \cong ZH, \; \Lambda/(I+J) \cong \bar{Z}H.$$

Hence Λ is given by the fibre product

$$\begin{array}{ccc} \Lambda & \longrightarrow & R \circ H \\ \downarrow & & \downarrow \phi_1 \\ ZH & \xrightarrow[\phi_2]{} & \bar{Z}H \end{array},$$

that is,

$$\Lambda \cong \{(\xi, \eta) : \xi \in R \circ H, \; \eta \in ZH, \; \phi_1(\xi) = \phi_2(\eta)\}.$$

Let $u(R \circ H)$ denote the group of units in $R \circ H$, and let $u^*(R \circ H)$ be its image in the unit group $u(\bar{Z}H)$. Since the algebra QG satisfies the Eichler condition, it follows from (1.9) of **[8]** that there is an exact sequence

$$(2.1) \qquad u^*(ZH) \dotplus u^*(R \circ H) \to u(\bar{Z}H) \to C(\Lambda) \to C(ZH) \dotplus C(R \circ H) \to 0.$$

Here, $u^*(ZH)$ denotes the image of $u(ZH)$ in $u(\bar{Z}H)$. Since $u(ZH) \subset u(R \circ H)$, it is clear that $u^*(ZH) \subset u^*(R \circ H)$, so the first factor in the first term in (2.1) may be omitted.

Now Rosen **[10]** has proved that $C(R \circ H) \cong C(S)$, by using the fact that the order $R \circ H$ is hereditary. Alternatively, since $R \circ H$ is hereditary and satisfies the Eichler conditon, the existence of an isomorphism $C(R \circ H) \cong C(S)$ is a consequence of the following result of Jacobinski **[2]**: A locally free $R \circ H$-lattice M is determined up to isomorphism by the lattice $\Lambda' M$, where Λ' is a maximal order containing $R \circ H$.

It follows from (2.1) and the preceding remarks that there is an exact sequence

$$u^*(R \circ H) \to u(\bar{Z}H) \to C(\Lambda) \xrightarrow{\mu} C(ZH) \dotplus C(S) \to 0.$$

Letting $D_0(\Lambda)$ denote the kernel of μ, we obtain

$$D_0(\Lambda) \cong u(\bar{Z}H)/u^*(R \circ H). \tag{2.2}$$

In order to prove Theorem 1.2, it therefore suffices to show that $u(\bar{Z}H)/u^*(R \circ H)$ is a cyclic group of the desired order.

Remark. Let Λ^* be a maximal order in QG containing Λ. In **[8]** we defined $D(\Lambda)$ as the kernel of the epimorphism $C(\Lambda) \to C(\Lambda^*)$. The groups $D_0(\Lambda)$ and $D(\Lambda)$ are related by an exact sequence

$$0 \to D_0(\Lambda) \to D(\Lambda) \to D(ZH) \to 0.$$

In particular, when q is prime, then $D(ZH) = 0$ and thus $D_0(\Lambda) = D(\Lambda)$.

3. *Some preliminary lemmas.* Let $u(S)$ denote the group of units of S, and $u^*(S)$ its image in $u(\bar{Z})$ under the mapping $S \to S/P_0 \cong \bar{Z}$. Write

$$p - 1 = qm,$$

and set

$$T_0 = \{x \in u(\bar{Z}) : x^{2m} = 1\},\ T_1 = \{x \in u(\bar{Z}) : x^m = 1\}.$$

Then T_0, T_1 are subgroups of the cyclic group $u(\bar{Z})$, and

$$|T_0| = (p - 1,\ 2m) = m.(q, 2),\ |T_1| = m. \tag{3.1}$$

(3.2) LEMMA. *The group $u(\bar{Z})/u^*(S)$ is cyclic of order $q/(q, 2)$.*

Proof. Case 1. Let q be even; we shall show that $u^*(S) = T_0$, whence $u(\bar{Z})/u^*(S)$ will necessarily be cyclic of order $(p - 1)/2m$. This will give the desired result, since $(p - 1)/2m = q/2$.

We have chosen $P_0 = S \cap (1 - \omega) R$. For each $u \in u(S)$ and each $\tau \in \mathrm{Gal}\,(L/Q)$, we have

$$u^\tau \equiv u \pmod{P_0}.$$

Let $N_{L/Q}$ denote the norm from L to Q; since $(L\colon Q) = m$, it follows from the preceding congruence that

$$N_{L/Q}(u) \equiv u^m \pmod{P_0}.$$

Of course $N_{L/Q}(u) = \pm 1$, since u is a unit in S. Let $\bar{u} \in u(\bar{Z})$ denote the image of u. Then the above yields

$$\bar{u}^{2m} = 1 \quad \text{in} \quad u(\bar{Z}),$$

which shows that $u^*(S) \subset T_0$. This argument is valid whether or not q is even.

Now put

$$K^+ = Q(\omega + \omega^{-1}),\ R^+ = \text{alg. int.}\{K^+\},$$

and for each $a \in Z$ relatively prime to p, set

$$\xi_a = (\omega^a - \omega^{-a})/(\omega - \omega^{-1}). \tag{3.3}$$

Then $\xi_a \in u(R^+)$. Now $(K : L) = q$, $(K : K^+) = 2$, so since q is even and $\mathrm{Gal}\,(K/Q)$ is cyclic, we may conclude that $L \subset K^+$, and $(K^+ : L) = q/2$. Set

$$u_a = N_{K^+/L}(\xi_a) \in u(S).$$

We have

$$\xi_a \equiv a \pmod{P},$$

and the same congruence holds true for any algebraic conjugate of ξ_a. Therefore

$$u_a \equiv a^{q/2} \pmod{P}.$$

But u_a and $a^{q/2}$ lie in S, whence

$$u_a \equiv a^{q/2} \pmod{P_0}.$$

Passing to $u(\bar{Z})$, we have

$$\bar{u}_a = \bar{a}^{q/2}, \ a \in Z, \ (a, p) = 1.$$

Since $\bar{u}_a \in u^*(S)$, we have shown that $u^*(S)$ contains the $(q/2)$-th powers of all elements in $u(\bar{Z})$. Hence $u^*(S)$ contains all elements of $u(\bar{Z})$ of order dividing $2m$, that is, $u^*(S) \supset T_0$. This proves that $u^*(S) = T_0$, and the lemma is established for even q.

Case 2. Let q be odd. It suffices to show that $u^*(S) = T_1$. Since q is odd, the field L is totally imaginary, and thus in the preceding discussion we have

$$N_{L/Q}(u) = +1.$$

Therefore $\bar{u}^m = 1$ for all $u \in u(S)$, which shows that $u^*(S) \subset T_1$.

Let ξ_a be given by (3.3), but now define

$$u_a = N_{K/L}(\xi_a) \in u(S).$$

Since $\xi_a \equiv a \pmod{P}$ and $(K : L) = q$, we obtain

$$u_a \equiv a^q \pmod{P_0}.$$

Therefore $\bar{u}_a = \bar{a}^q$ in $u(\bar{Z})$, for each $a \in Z$ prime to p. Hence $u^*(S)$ contains the q-th powers of all elements in $u(\bar{Z})$, and therefore $u^*(S) \supset T_1$. Consequently $u^*(S) = T_1$, and the lemma is proved.

In order to apply the results in Rosen **[10]**, we begin with some simple observations. First note that

$$R = Z[\omega] = S[\omega] = S \oplus S\omega \oplus \dots \oplus S\omega^{q-1},$$

so R is a free S-module. Next, R is tamely ramified over S, since P_0 is the only prime of S which can ramify in R. An R-ideal J in K is called *ambiguous* if $J^\sigma = J$; every ambiguous ideal is uniquely expressible as $P^e \cdot B$, with $0 \leqslant e \leqslant q-1$, and B an S-ideal in L. Ambiguous ideals may be viewed as $R \circ H$-lattices, with the generator y of H acting as σ. As a consequence of these remarks and Rosen's results, we obtain

(3.4) PROPOSITION. (i) *The ring $R \circ H$ is a hereditary order.*

(ii) *Every $R \circ H$-lattice M is isomorphic to an external direct sum*:

$$M \cong P^{e_1} B_1 \dotplus \dots \dotplus P^{e_s} B_s,$$

where for each i, $1 \leqslant i \leqslant s$, $0 \leqslant e_i \leqslant q-1$, and B_i = S-ideal of L. Here, s is the R-rank of M. The $R \circ H$-lattice M is determined up to $R \circ H$-isomorphism by the set of integers $\{e_1, \dots, e_s\}$, and by the ideal class of the S-ideal $B_1 \dots B_s$,

(iii) *There is an isomorphism of left $R \circ H$-lattices*:

$$R \circ H \cong R \dotplus P \dotplus P^2 \dotplus \dots \dotplus P^{q-1}.$$

We shall use this to calculate $u(R \circ H)$, the group of units of the twisted group ring $R \circ H$.

(3.5) LEMMA. *The group of units $u(R \circ H)$ is isomorphic to the group of all $q \times q$ matrices (α_{ij}) satisfying*

$$\begin{cases} \alpha_{ij} \in S, \ 1 \leqslant i \leqslant j \leqslant q, \ \alpha_{ij} \in P_0, \ 1 \leqslant j < i \leqslant q, \\ \det(\alpha_{ij}) \in u(S). \end{cases}$$

Proof. Let $\operatorname{End}(R \circ H)$ denote the ring of all $R \circ H$-endomorphisms of the left $R \circ H$-lattice $R \circ H$. Then $u(R \circ H)$ is anti-isomorphic to the group of units of $\operatorname{End}(R \circ H)$. We have

$$\operatorname{End}(R \circ H) \cong \sum_{e,\, d} \cdot \operatorname{Hom}_{R \circ H}(P^e, P^d),$$

where e and d range independently from 0 to $q - 1$. Now $K.P^e$ and $K.P^d$ are irreducible modules for the central simple L-algebra $K \circ H$, and therefore

$$\operatorname{Hom}_{K \circ H}(K \cdot P^e, K \cdot P^d) \cong L.$$

Consequently,

$$\operatorname{Hom}_{R \circ H}(P^e, P^d) = L \cap P^{d-e} = \begin{cases} S, & d \leqslant e, \\ P_0, & d > e, \end{cases}$$

where $P_0 = L \cap P$. Thus the units of $\operatorname{End}(R \circ H)$ may be represented by invertible $q \times q$ matrices (γ_{ij}), such that $\gamma_{ij} \in S$, $i \geqslant j$, $\gamma_{ij} \in P_0$, $i < j$. Now use the anti-isomorphism of $u(R \circ H)$ with the group of units of $\operatorname{End}(R \circ H)$, to obtain the desired description of $u(R \circ H)$.

We have defined a ring homomorphism $\phi_1 : R \circ H \to \bar{Z}H$ in §2, given by

$$R \circ H \to \frac{R \circ H}{P \circ H} \cong \bar{Z}H.$$

It is necessary for us to calculate the image $u^*(R \circ H)$ in $u(\bar{Z}H)$ under this mapping ϕ_1. We have

$$R \circ H \cong \sum_{e=0}^{q-1} \cdot P^e, \quad P \circ H \cong \sum_{e=0}^{q-1} \cdot P^{e+1},$$

so

$$\bar{Z}H \cong \frac{R \circ H}{P \circ H} \cong \sum_{e=0}^{q-1} \cdot \frac{P^e}{P^{e+1}}.$$

The above is a left $\bar{Z}H$-isomorphism, and hence gives the decomposition of the semi-simple algebra $\bar{Z}H$ into simple components P^e/P^{e+1}, each of which is a one-dimensional space over $\bar{Z}$. The units of $\bar{Z}H$ are thus representable by $q \times q$ diagonal matrices, with diagonal entries in $u(\bar{Z})$. Now

$$\operatorname{Hom}_{\bar{Z}H}(P^e/P^{e+1}, P^d/P^{d+1}) = 0, \ d \neq e, \ 0 \leqslant d, \ e \leqslant q - 1.$$

Hence if $u \in u(R \circ H)$ is represented by a matrix (α_{ij}) satisfying the conditions of (3.5), then $\phi_1(u)$ is represented by a diagonal matrix with entries $\bar{\alpha}_{11}, \ldots, \bar{\alpha}_{qq}$. Here, $\bar{\alpha}_{ii}$ represents the image of α_{ii} under the map $S \to S/P_0 \cong \bar{Z}$. Note that each $\bar{\alpha}_{ii} \in u(\bar{Z})$, since

$$\prod \alpha_{ii} \equiv \det(\alpha_{ij}) \qquad (\operatorname{mod} P_0),$$

and $\det(\alpha_{ij}) \in u(S)$.

4. *Conclusion.* We have shown in §2 that there is an epimorphism

$$C(ZG) \to C(ZH) \dotplus C(S),$$

whose kernel $D_0(ZG)$ is given by an isomorphism

$$D_0(ZG) \cong u(\bar{Z}H)/u^*(R \circ H).$$

We are going to establish that

$$u(\bar{Z}H)/u^*(R \circ H) \cong u(\bar{Z})/u^*(S). \tag{4.1}$$

We have already shown in (3.2) that $u(\bar{Z})/u^*(S)$ is cyclic of order $q/(q, 2)$. Hence once (4.1) is proved, we may conclude that $D_0(ZG)$ is also cyclic of order $q/(q, 2)$, as asserted by Theorem 1.2.

In proving (4.1), we shall represent the units in $R \circ H$ by invertible $q \times q$ matrices (α_{ij}) satisfying conditions (3.5), and the units in $\bar{Z}H$ by invertible $q \times q$ diagonal matrices over $\bar{Z}$. As shown in §3, we may identify $u^*(R \circ H)$ with the subgroup of $u(\bar{Z}H)$ represented by matrices

$$\begin{bmatrix} \bar{\alpha}_{11} & 0 & \dots & 0 \\ 0 & \bar{\alpha}_{22} & \dots & 0 \\ & \dots & \dots & \\ 0 & 0 & \dots & \bar{\alpha}_{qq} \end{bmatrix},$$

where (α_{ij}) ranges over all invertible $q \times q$ matrices such that

$$\alpha_{ij} \in S,\ i \leqslant j,\ \alpha_{ij} \in P_0,\ i > j,\ \det(\alpha_{ij}) \in u(S).$$

There is a commutative diagram

$$\begin{array}{ccc} u(R \circ H) & \xrightarrow{\phi_1} & u(\bar{Z}H) \\ \downarrow \det & & \downarrow \det \\ u(S) & \xrightarrow[\phi']{} & u(\bar{Z}) \end{array},$$

where the vertical arrows arise from the determinant map, and where ϕ' is induced by the map $S \to S/P_0 \cong \bar{Z}$. Since $u(\bar{Z}H) \to u(\bar{Z})$ is surjective, the map

$$\theta : \frac{u(\bar{Z}H)}{u^*(R \circ H)} \to \frac{u(\bar{Z})}{u^*(S)}$$

is also surjective. To prove (4.1), we need only show that θ is injective. Let

$$X = \begin{bmatrix} \beta_1 & 0 & \dots & 0 \\ 0 & \beta_2 & \dots & 0 \\ \cdot & \cdot & \dots & 0 \\ 0 & 0 & \dots & \beta_q \end{bmatrix}, \quad \beta_i \in u(\bar{Z}),$$

represent an element of $u(\bar{Z}H)$, and suppose that $\theta(X) = 1$, that is, $\prod \beta_i \in u^*(S)$. Then there exists an element $u \in u(S)$ such that $\prod \beta_i = \bar{u}$ in $\bar{Z}$. Choose $\alpha_1, \dots, \alpha_q \in S$ such that $\bar{\alpha}_i = \beta_i$, $1 \leqslant i \leqslant q$. Then

$$\alpha_1 \dots \alpha_q = u + \pi_0$$

for some $\pi_0 \in P_0$. Let Y be the $q \times q$ matrix given as follows: if $q = 1$, choose $Y = (u)$. If $q > 1$, choose

$$Y = \begin{bmatrix} \alpha_1 & 0 & 0 & \cdots & 0 & (-1)^q \pi_0 \\ 1 & \alpha_2 & 0 & \cdots & 0 & 0 \\ 0 & 1 & \alpha_3 & \cdots & & 0 \\ \cdot & \cdot & \cdot & \cdots & \alpha_{q-1} & 0 \\ 0 & 0 & 0 & \cdots & 1 & \alpha_q \end{bmatrix}.$$

Then for $q > 1$,

$$\det Y = \alpha_1 \dots \alpha_q - \pi_0 = u \in u(S),$$

so Y represents a unit in $R \circ H$. Thus for all $q \geqslant 1$, Y represents a unit in $R \circ H$, and $\phi_1(Y) = X$. It follows that $X = 1$ in $u(\bar{Z}H)/u^*(R \circ H)$, and hence that θ is injective. This completes the proof of (4.1) and of the Theorem.

References

1. A. Fröhlich, " On the classgroup of integral group rings of finite Abelian groups ", *Mathematika,* 16 (1969), 143–152. MR 41, 5512.
2. H. Jacobinski, " Two remarks about hereditary orders ", *Proc. Amer. Math. Soc.*, 28 (1971), 1–8.
3. M. A. Kervaire and M. P. Murthy, " On the projective class group of cyclic groups of prime power order " (to appear).
4. M. P. Lee, " Integral representations of dihedral groups of order $2p$ ", *Trans. Amer. Math. Soc.* 110 (1964), 213–231. MR 28, 139.
5. J. Martinet, " Modules sur l'algebre du groupe quaternionien ", *Ann. Sci. Ecole Norm. Sup.*, 4 (1971), 399–408.
6. I. Reiner, " Integral representations of cyclic groups of prime order ", *Proc. Amer. Math. Soc.* 8 (1957), 142–146. MR 18, 717.
7. ——— and S. Ullom, " Class groups of integral group rings, *Trans. Amer. Math. Soc.* (to appear)
8. ———, and ———, "A Mayer-Vietoris sequence for class groups " *J. Algebra* (to appear).
9. D. S. Rim, " Modules over finite groups ", *Ann. of Math.* (2), 69 (1959), 700–712. MR 21, 3474.
10. M. Rosen, " Representations of twisted group rings ", *Thesis, Princeton University* (Princeton, N.J., 1963).
11. S. Ullom, "A note on the classgroups of integral group rings of some cyclic groups ", *Mathematika,* 17 (1970), 79–81. MR 42, 4650.

Carleton College,
Northfield, Minnesota 55057
University of Illinois,
Urbana, Illinois 61801

12B25: *Algebraic number fields; local fields; Class field theory.*

(*Received on the* 28*th of February*, 1972.)

CLASS GROUPS OF INTEGRAL GROUP RINGS[1]

BY

I. REINER AND S. ULLOM

ABSTRACT. Let Λ be an R-order in a semisimple finite dimensional K-algebra, where K is an algebraic number field, and R is the ring of algebraic integers of K. Denote by $C(\Lambda)$ the reduced class group of the category of locally free left Λ-lattices. Choose $\Lambda = ZG$, the integral group ring of a finite group G, and let Λ' be a maximal Z-order in QG containing Λ. There is an epimorphism $C(\Lambda) \to C(\Lambda')$, given by $M \to \Lambda' \otimes_\Lambda M$, for M a locally free Λ-lattice. Let $D(\Lambda)$ be the kernel of this epimorphism; the groups $D(\Lambda)$, $C(\Lambda)$ and $C(\Lambda')$ are all finite. Our main theorem is that $D(ZG)$ is a p-group whenever G is a p-group. This generalizes Fröhlich's result for the case where G is an abelian p-group. Our proof uses some facts about the center F of QG, as well as information about reduced norms. We also calculate $D(ZG)$ explicitly for G cyclic of order $2p$, dihedral of order $2p$, or the quaternion group. In these cases, the ring ZG can be conveniently described by a pullback diagram.

1. Introduction. Let $R = \text{alg int}\{K\}$ be the ring of all algebraic integers in an algebraic number field K, and let Λ be an R-order in a semisimple finite dimensional K-algebra A. By a Λ-*lattice* we mean always a left Λ-module which is finitely generated and torsionfree as R-module. For each maximal ideal P of R, let R_P denote the localization of R at P, that is,

$$R_P = \{\alpha/\beta \colon \alpha, \beta \in R, \beta \notin P\}.$$

Set $\Lambda_P = R_P \otimes_R \Lambda$, an R_P-order in A.

Two Λ-lattices M, N are in the same *genus* if $M_P \cong N_P$ for each maximal ideal P of R. We shall call M *locally free* if M is in the same genus as some free Λ-lattice $\Lambda^{(r)}$; here, $\Lambda^{(r)}$ denotes the direct sum of r copies of Λ. In this case, we call r the Λ-rank of M, and write $r = \text{rank}_\Lambda M$. Of course, every locally free Λ-lattice is projective as left Λ-module.

We next define the (locally free) projective class group $P(\Lambda)$ as follows: let $\mathfrak{B}$ denote the free abelian group generated by symbols $[M]$, one for each isomorphism class of locally free Λ-lattices. Let $\mathfrak{B}_0$ be the subgroup of $\mathfrak{B}$ generated by all expressions $[L \oplus M] - [L] - [M]$ where L, M are locally free Λ-lattices. Then set $P(\Lambda) = \mathfrak{B}/\mathfrak{B}_0$, an abelian additive group. There is an additive epimorphism $P(\Lambda) \longrightarrow Z$, induced by the mapping $[M] \longrightarrow \text{rank}_\Lambda M$; this epimorphism is split by the mapping $n \longrightarrow \Lambda^{(n)}$, $n \in Z$, $n > 0$. Let $C(\Lambda)$ be the kernel of this

Received by the editors July 1, 1971.

AMS 1970 *subject classifications*. Primary 16A54; Secondary 20C05, 20C10.

(1) This research was partially supported by a grant from the National Science Foundation.

epimorphism, so $C(\Lambda)$ is the subgroup of $P(\Lambda)$ consisting of all expressions $[M]-[N]$, with M, N locally free of the same Λ-rank. Clearly,

$$P(\Lambda) \cong Z \dotplus C(\Lambda)$$

as additive groups. For brevity, we refer to $C(\Lambda)$ as the *class group* of Λ. It is easily seen that $[M]-[N]=0$ in $C(\Lambda)$ if and only if M and N are stably isomorphic, that is, $M \dotplus \Lambda^{(s)} \cong N \dotplus \Lambda^{(s)}$ for some nonnegative integer s.

The methods of proof in Swan [13] show at once that every element of the class group $C(\Lambda)$ is expressible in the form $[J]-[\Lambda]$, where J is a locally free left Λ-lattice in A. Since there are only a finite number of isomorphism classes of such J's (by the Jordan-Zassenhaus theorem), it follows that $C(\Lambda)$ is necessarily a finite group. Note further that $[J_1]-[\Lambda]=[J_2]-[\Lambda]$ in $C(\Lambda)$ if and only if J_1 and J_2 are stably isomorphic; that is, $J_1 \dotplus \Lambda^{(s)} \cong J_2 \dotplus \Lambda^{(s)}$ for some nonnegative integer s.

For the case where $\Lambda = RG$, the integral group ring of a finite group G, Swan [13] proved that *every* projective Λ-lattice is locally free. Thus our $C(\Lambda)$ is the usual "reduced projective class group."

On the other hand, let Λ' be a maximal R-order in a semisimple K-algebra A. A projective Λ'-lattice M' is called *special* if $K \otimes_R M'$ is A-free. From the theory of maximal orders (see [4], [12] or [15]), we know that *every* Λ'-lattice M' is projective, and that M' is locally free if and only if M' is special. Thus our $C(\Lambda')$ is the usual "reduced special projective class group." Since most of our calculations below depend on working with locally free lattices, it seems desirable to single out this class of projective lattices rather than the class of special projectives.

Suppose now that Λ' is a maximal R-order in A containing Λ. There is then an additive homomorphism $P(\Lambda) \longrightarrow P(\Lambda')$, induced by letting

$$[M] \longrightarrow [\Lambda' \otimes_\Lambda M], \quad M = \text{locally free } \Lambda\text{-lattice}.$$

This homomorphism induces a homomorphism η: $C(\Lambda) \longrightarrow C(\Lambda')$, and Swan [14] showed that η is in fact an epimorphism. Let $D(\Lambda)$ be the kernel of η, so there is an exact sequence of additive groups

$$0 \longrightarrow D(\Lambda) \longrightarrow C(\Lambda) \longrightarrow C(\Lambda') \longrightarrow 0,$$

and obviously

$$|C(\Lambda)| = |D(\Lambda)| \cdot |C(\Lambda')|.$$

The class group $C(\Lambda')$ of the maximal order Λ' is easily determined, and hence we can find the order of $C(\Lambda)$ by calculating that of $D(\Lambda)$. In this direction, Fröhlich [8] has shown that $|D(ZG)|$ is a power of p whenever G is an abelian p-group. One of the main results of the present article is the fact that $|D(ZG)|$ is a power of p, for an arbitrary finite p-group G, not necessarily abelian. We may remark that this conclusion need not hold, even for abelian p-groups, when Z is replaced by some larger ring of algebraic integers. This already follows from the work of Ullom [16].

We have also included explicit calculations of $C(ZG)$ and $D(ZG)$ for the following groups G: cyclic group of order $2p$, dihedral group of order $2p$, quaternion group of order 8. Here, p is any odd prime. Some of these calculations could be simplified slightly by using Milnor's Mayer-Vietoris sequence in K-theory (see Bass [1]). We have avoided using this sequence, however, in order to illustrate the computational aspects of our present approach. A later work [18] will be devoted to K-theory calculations.

The following notation will be used throughout:

rad = Jacobson radical.

$(R)^{n\times n}$ = ring of all $n \times n$ matrices over a ring R.

$M^{(k)}$ = direct sum of k copies of M.

$R_{\hat{P}}$ = P-adic completion of the integral domain R at a prime P of the quotient field of R.

$\Sigma^{\cdot}$, $\dot{+}$: direct sum.

$\operatorname{ann}_R X = \{\alpha \in R: \alpha X = 0\}$ = R-annihilator of the R-module X.

$u(\Lambda)$ = group of units(2) of the ring Λ.

General references for this paper are [11], [12] and [15]. The authors wish to thank Professor A. Fröhlich for some helpful conversations, and for pointing out Lemma 3.4 below.

2. **Addition in the class group.** The purpose of this section is to review some well-known results for the convenience of the reader; proofs may be found in the general references cited at the end of §1. As before, we let Λ be any R-order in the semisimple K-algebra A.

Suppose that M is a locally free Λ-lattice of Λ-rank r. It follows readily from the approximation theorem for algebraic numbers that, given any nonzero ideal $\mathfrak{q}$ of R, there exists a Λ-embedding $M \longrightarrow \Lambda^{(r)}$ such that

$$\operatorname{ann}_R(\Lambda^{(r)}/M) + \mathfrak{q} = R.$$

The methods of proof in Swan [13] then show that there exists a locally free left Λ-lattice J in A such that $M \cong \Lambda^{(r-1)} \dot{+} J$. Hence every element of the (locally free) class group $C(\Lambda)$ is expressible in the form

$$x_j = [J] - [\Lambda], \qquad J = \text{locally free } \Lambda\text{-free lattice in } A.$$

As pointed out in §1, we have $x_{J_1} = x_{J_2}$ if and only if J_1 and J_2 are stably isomorphic.

How can we calculate $x_{J_1} + x_{J_2}$ in $C(\Lambda)$? First of all, embed J_1 in Λ so as to get a Λ-exact sequence

(2) A *unit* of Λ is an element of Λ which has a two-sided inverse in Λ.

(2.1) $$0 \to J_1 \xrightarrow{i_1} \Lambda \to T_1 \to 0,$$

with T_1 an R-torsion Λ-module. Then choose a Λ-exact sequence

(2.2) $$0 \to J_2 \xrightarrow{i_2} \Lambda \to T_2 \to 0,$$

with T_2 an R-torsion Λ-module such that

(2.3) $$\operatorname{ann}_R T_1 + \operatorname{ann}_R T_2 = R.$$

It follows easily that the map $(i_1, i_2): J_1 \dotplus J_2 \to \Lambda$ is a Λ-epimorphism, and hence that

(2.4) $$J_1 \dotplus J_2 \cong \Lambda \dotplus J_3$$

for some Λ-lattice J_3 in A. We may conclude from this that for each maximal ideal P of R,

$$\Lambda_{\hat{P}} \dotplus \Lambda_{\hat{P}} \cong \Lambda_{\hat{P}} \dotplus (J_3)_{\hat{P}},$$

where the subscript $\hat{P}$ indicates passage to P-adic completions. Since the Krull-Schmidt theorem is valid for $\Lambda_{\hat{P}}$-modules, the above isomorphism implies that $\Lambda_{\hat{P}} \cong (J_3)_{\hat{P}}$. This in turn implies that $\Lambda_P \cong (J_3)_P$, so we have verified that J_3 is also locally free. The operation of addition in $C(\Lambda)$ is then given by the formula: $x_{J_1} + x_{J_2} = x_{J_3}$.

For the remainder of this section, suppose that A is commutative. Let us show that the Λ-lattice J_3 occurring in (2.4) is such that

$$J_3 \cong J_1 \cdot J_2 \quad (= \text{product of } \Lambda\text{-ideals in } A).$$

Indeed, from (2.1) we obtain an exact sequence of Λ-modules

$$0 \to J_1 \otimes J_2 \to J_2 \to T_1 \otimes J_2 \to 0,$$

where $\otimes$ means $\otimes_\Lambda$. On the other hand, from (2.2) we get an exact sequence

$$\operatorname{Tor}_1^\Lambda(T_1, T_2) \to T_1 \otimes J_2 \to T_1 \to T_1 \otimes T_2 \quad (\to 0).$$

By virtue of (2.3), the first and last terms of the above sequence vanish, and thus $T_1 \otimes J_2 \cong T_1$. Consequently there is an exact sequence

(2.5) $$0 \to J_1 \otimes J_2 \to J_2 \to T_1 \to 0.$$

Applying Schanuel's lemma to the pair of sequences (2.1) and (2.5), we obtain

$$J_1 \dotplus J_2 \cong \Lambda \dotplus (J_1 \otimes J_2).$$

Since the map $\xi \otimes \eta \to \xi\eta$ gives an isomorphism $J_1 \otimes J_2 \cong J_1 \cdot J_2$, we may conclude that

(2.6) $$J_1 \dotplus J_2 \cong \Lambda \dotplus J_1 J_2 \quad \text{and} \quad x_{J_1} + x_{J_2} = x_{J_1 J_2} \quad \text{in } C(\Lambda).$$

On the other hand, since A is commutative, we may define an ideal class group $\overline{C}(\Lambda)$ in a more natural manner, as follows: Relative to the usual multiplication of Λ-ideals in A, the locally free Λ-lattices[3] in A form a multiplicative group $\overline{I}(\Lambda)$. The set of all principal ideals

---

[3] Note that every locally free Λ-lattice in A is invertible.

$$\{\Lambda x\colon x \in A,\ x \text{ invertible in } A\}$$

forms a subgroup $\overline{I}_0(\Lambda)$ of $\overline{I}(\Lambda)$. We now set

$$\overline{C}(\Lambda) = \overline{I}(\Lambda)/\overline{I}_0(\Lambda),$$

the multiplicative group of classes of locally free Λ-ideals in A. (This group $\overline{C}(\Lambda)$ is the same as Pic (Λ); see Bass [1].)

Let us show that the multiplicative group $\overline{C}(\Lambda)$ is isomorphic to the additive group $C(\Lambda)$. Define a map $\mu\colon \overline{C}(\Lambda) \to C(\Lambda)$ by

$$\mu(\text{ideal class of } J) = [J] - [\Lambda],$$

where J is any locally free left Λ-lattice in A. The definition is meaningful, since if J_1 is in the same ideal class as J_2, then $J_1 \cong J_2$ as left Λ-modules, and thus

$$[J_1] - [\Lambda] = [J_2] - [\Lambda] \quad \text{in } C(\Lambda).$$

We have already remarked at the beginning of this section that μ is epic. We claim that μ is one-to-one. For if $[J] - [\Lambda] = 0$ in $C(\Lambda)$, then

$$J \dotplus \Lambda^{(r)} \cong \Lambda \dotplus \Lambda^{(r)} \tag{2.7}$$

for some nonnegative integer r. But the $(r+1)$st exterior power (over Λ) of $J \dotplus \Lambda^{(r)}$ is J itself. Hence (2.7) implies that $J \cong \Lambda$, that is, $J \in \overline{I}_0(\Lambda)$. This completes the proof that μ gives a one-to-one mapping of $\overline{C}(\Lambda)$ onto $C(\Lambda)$. Equations (2.6) show that μ is a group homomorphism, so we have established the isomorphism $C(\Lambda) \cong \overline{C}(\Lambda)$.

In a forthcoming article [17], Fröhlich has considered the group Pic(ZG) for non-abelian G. The authors wish to thank Professor Fröhlich for the opportunity of reading a preliminary version of this article. The relationship between Pic(ZG) and our $C(ZG)$ will be investigated in a future work [19].

3. **Explicit formulas for the class group.** Returning to the general case, we fix the following notation for the remainder of this article. Let Λ be any R-order in the semisimple K-algebra A, and let Λ' be some maximal R-order in A containing Λ. We may write

$$A = \sum_{i=1}^{m}{}^{\cdot} A_i \quad \text{(simple components)},$$

$$\Lambda' = \sum_{i=1}^{m}{}^{\cdot} \Lambda_i, \quad \Lambda_i = \text{maximal } R\text{-order in } A_i, \tag{3.1}$$

$$F = \sum_{i=1}^{m}{}^{\cdot} K_i, \quad K_i = \text{center of } A_i, \qquad C = \sum_{i=1}^{m}{}^{\cdot} R_i, \quad R_i = \text{alg int}\,\{K_i\} \subset \Lambda_i.$$

Each Λ_i is then a maximal R_i-order in the ith simple component A_i of A. Further, F is the center of A, and C is the unique maximal R-order in F.

(i) *Reduced norms.* Let $N_i: A_i \to K_i$ be the reduced norm map. If A_i happens to be a full matrix algebra $(K_i)^{m_i \times m_i}$ over the field K_i, and if we let the element $x \in A_i$ map onto the $m_i \times m_i$ matrix $\mathbf{x}$ over K_i, then $N_i(x) = \det \mathbf{x}$. In the general case (see [2]), the reduced norm N_i is obtained by first passing to a splitting field for A_i, and then using the preceding comment.

Now define the reduced norm map $N: A \to F$ by working in each simple component separately, that is, $N = \Sigma^{\cdot} N_i$.

In order to describe the image $N_i(A_i)$ in K_i, we proceed as follows. For P any prime of K_i (finite or infinite), let $(K_i)_{\hat{P}}$ denote the P-adic completion of K_i. Set

$$(A_i)_{\hat{P}} = (K_i)_{\hat{P}} \otimes_{K_i} A_i,$$

a simple algebra with center $(K_i)_{\hat{P}}$. We shall say that P *ramifies* in A_i if $(A_i)_{\hat{P}}$ is *not* a full matrix algebra over the field $(K_i)_{\hat{P}}$. A proof of the following result of Hasse may be found in Swan-Evans [15, Theorem 7.6]:

(3.2) **Theorem.** *A nonzero element $\alpha \in K_i$ lies in $N_i(A_i)$ if and only if $\alpha_{\hat{P}} > 0$ for every real infinite prime P of K_i which ramifies in A_i. Here, $\alpha_{\hat{P}}$ denotes the image of α in $(K_i)_{\hat{P}}$.*

(ii) *The Eichler condition.* Call A_i a *totally definite quaternion algebra* (over its center K_i) if every infinite prime P of K_i is real, and for every such P the algebra $(A_i)_{\hat{P}}$ is a quaternion skewfield over its center $(K_i)_{\hat{P}}$.

We shall say that the algebra A *satisfies the Eichler condition* if no simple component A_i of A is a totally definite quaternion algebra. Commutative algebras automatically satisfy the Eichler condition.

(iii) *Localization.* Let $\mathfrak{f}$ be a nonzero ideal of R (eventually to be chosen so that $\mathfrak{f} \cdot \Lambda' \subset \Lambda$). Call an element $\alpha \in F$ *prime* to $\mathfrak{f}$ if, for each prime ideal P of R dividing $\mathfrak{f}$, α is a unit in the localization C_P. Rather than use idèles, we introduce the semilocal ring

$$R_{\mathfrak{f}} = \bigcap_{P \text{ divides } \mathfrak{f}} R_P = \{\alpha/\beta : \alpha, \beta \in R, \ \beta \text{ prime to } \mathfrak{f}\}.$$

This ring is a principal ideal domain, with maximal ideals $\{P \cdot R_{\mathfrak{f}} : P \text{ divides } \mathfrak{f}\}$. Now set

$$\Lambda_{\mathfrak{f}} = R_{\mathfrak{f}} \otimes_R \Lambda, \qquad C_{\mathfrak{f}} = R_{\mathfrak{f}} \otimes_R C,$$

and so on. It is then clear that an element $\alpha \in F$ is prime to $\mathfrak{f}$ if and only if $\alpha \in u(C_{\mathfrak{f}})$, the group of units of $C_{\mathfrak{f}}$.

Likewise, an element $x \in \Lambda$ is said to be *prime* to $\mathfrak{f}$ if $x \in u(\Lambda_{\mathfrak{f}})$, that is, if $\Lambda_{\mathfrak{f}} = \Lambda_{\mathfrak{f}} \cdot x$. In this case, it follows that $\Lambda'_{\mathfrak{f}} = \Lambda'_{\mathfrak{f}} \cdot x$, so that x is prime to $\mathfrak{f}$ when we view x as an element of the larger order Λ'. The converse is also true, however:

(3.3) **Lemma.** *Let* $x \in \Lambda \subset \Lambda'$, *and suppose that* x *is a unit in* $\Lambda'_{\mathfrak{f}}$. *Then* x *is also a unit in* $\Lambda_{\mathfrak{f}}$. *Hence the property of being "prime to* $\mathfrak{f}$" *is independent of the choice of* R-*order.*

This lemma is an immediate consequence of the following more general fact:

(3.4) **Lemma.** *Let* $\Lambda \subset \Lambda'$ *be* R-*orders in* A. *Then* $\Lambda \cap u(\Lambda') = u(\Lambda)$.

Proof. Let $x \in \Lambda$; we must show that $x \in u(\Lambda)$ if and only if $x \in u(\Lambda')$. Now $x \in u(\Lambda)$ if and only if $\Lambda = \Lambda x$, so we need to prove that $\Lambda = \Lambda x$ if and only if $\Lambda' = \Lambda' x$. Let $[\Lambda : \Lambda x]$ denote the order ideal of the R-torsion Λ-module $\Lambda/\Lambda x$, that is, $[\Lambda : \Lambda x]$ is the product of the R-annihilators of the R-composition factors of $\Lambda/\Lambda x$ (see [3] or [7]). Relative to a K-basis of A, right multiplication by x acting on A gives rise to a matrix $\mathbf{M}(x)$, and we have

$$[\Lambda : \Lambda x] = R \cdot \det \mathbf{M}(x).$$

Since the same equation holds with Λ replaced by Λ', we obtain $[\Lambda : \Lambda x] = [\Lambda' : \Lambda' x]$. This implies the desired result.

Remark. Using the notation of the preceding proof, we may observe that every prime ideal of R which divides $C \cdot N(x)$ also divides $R \cdot \det \mathbf{M}(x)$, and conversely: Hence an element $x \in \Lambda$ is a unit in Λ if and only if $N(x)$ is a unit in R.

Now let $u(\Lambda_{\mathfrak{f}})$ be the group of units of the ring $\Lambda_{\mathfrak{f}}$. Obviously

$$\{x \in \Lambda: x \text{ prime to } \mathfrak{f}\} \subset u(\Lambda_{\mathfrak{f}}),$$

by definition of "prime to $\mathfrak{f}$". On the other hand, any element $y \in u(\Lambda_{\mathfrak{f}})$ is expressible as $y = x/r$, with $x \in \Lambda$, $r \in R$, where r is prime to $\mathfrak{f}$. Then both x and r are elements of Λ prime to $\mathfrak{f}$. This shows that $u(\Lambda_{\mathfrak{f}})$ is the multiplicative group generated by the set

$$\{x \in \Lambda: x \text{ prime to } \mathfrak{f}\}.$$

Furthermore, the preceding remark shows that an element $x \in \Lambda$ is prime to $\mathfrak{f}$ if and only if its reduced norm $N(x)$ is prime to $\mathfrak{f}$.

(iv) *Explicit formulas when Eichler condition holds.* Hereafter, let $\mathfrak{f}$ be a nonzero ideal of R such that $\mathfrak{f} \cdot \Lambda' \subset \Lambda$, where Λ' is a maximal R-order in A containing Λ. Such an ideal $\mathfrak{f}$ always exists. (For example, when Λ is the integral group ring RG of a finite group G, we may choose $\mathfrak{f} = |G| \cdot R$.) Denote by $I(C, \mathfrak{f})$ the multiplicative group of all C-ideals in F, nonzero at each component, which are prime to $\mathfrak{f}$.

Let $I(\Lambda)$ be the subgroup of $I(C, \mathfrak{f})$ generated by all ideals

$$\{C \cdot N(x): x \in \Lambda, x \text{ prime to } \mathfrak{f}\}.$$

Since $(C \cdot N(x)) \cdot (C \cdot N(y)) = C \cdot N(xy)$ for $x, y \in \Lambda$ prime to $\mathfrak{f}$, and since the set of elements $\{x \in \Lambda: x \text{ prime to } \mathfrak{f}\}$ generates the group $u(\Lambda_{\mathfrak{f}})$, we conclude at once that

$$I(\Lambda) = \{C \cdot N(x) \colon x \in u(\Lambda_{\mathfrak{f}})\}. \tag{3.5}$$

Likewise, we set

$$I(\Lambda') = \{C \cdot N(y) \colon y \in u(\Lambda'_{\mathfrak{f}})\}. \tag{3.6}$$

We now quote the following basic result due to Jacobinski [9]:

(3.7) **Theorem.** *If the algebra A satisfies the Eichler condition, then there is a commutative diagram*

$$\begin{array}{ccc} C(\Lambda) & \cong & I(C, \mathfrak{f})/I(\Lambda) \\ \eta\downarrow & & \eta'\downarrow \\ C(\Lambda') & \cong & I(C, \mathfrak{f})/I(\Lambda') \end{array}$$

where η is defined by applying $\Lambda' \otimes_\Lambda \cdot$ to Λ-lattices, and η' is the natural epimorphism. Consequently,

$$D(\Lambda) = \ker \eta \cong I(\Lambda')/I(\Lambda).$$

Remark. The groups $I(\Lambda')$ and $I(\Lambda)$ depend on $\mathfrak{f}$. However, Jacobinski showed that (up to isomorphism) the groups $C(\Lambda)$, $C(\Lambda')$ and $D(\Lambda)$ are independent of the choice of the maximal order Λ' containing Λ, and are also independent of the choice of the nonzero ideal $\mathfrak{f}$ of R such that $\mathfrak{f} \cdot \Lambda' \subset \Lambda$.

Let us now define a homomorphism

$$\theta \colon u(\Lambda'_{\mathfrak{f}}) \to I(\Lambda')$$

by setting $\theta(x) = C \cdot N(x)$, $x \in u(\Lambda'_{\mathfrak{f}})$. Then θ is an epimorphism with kernel

$$\{x \in u(\Lambda'_{\mathfrak{f}}) \colon N(x) = \text{unit in } C\}.$$

Denote by C^* the group of units of C, and set

$$N^{-1}(C^*) = \{x \in A \colon N(x) \in C^*\}.$$

Then

$$\ker \theta = u(\Lambda'_{\mathfrak{f}}) \cap N^{-1}(C^*),$$

and therefore

$$u(\Lambda'_{\mathfrak{f}})/\{u(\Lambda'_{\mathfrak{f}}) \cap N^{-1}(C^*)\} \cong I(\Lambda').$$

However, $N^{-1}(C^*) \lhd u(\Lambda'_{\mathfrak{f}}) \cdot N^{-1}(C^*)$, since $N^{-1}(C^*)$ contains the commutator subgroup of $u(\Lambda'_{\mathfrak{f}})$. We thus may deduce from the above isomorphism the fact that

$$(u(\Lambda'_{\mathfrak{f}}) \cdot N^{-1}(C^*))/N^{-1}(C^*) \cong I(\Lambda'),$$

where the isomorphism is induced by letting an element x in the numerator map onto $C \cdot N(x)$.

In the same manner, we obtain

$$(u(\Lambda_{\mathfrak{f}})\cdot N^{-1}(C^*))/N^{-1}(C^*)\cong I(\Lambda).$$

Consequently, we have

$$(3.8)\quad D(\Lambda)\cong\frac{I(\Lambda')}{I(\Lambda)}\cong\frac{u(\Lambda'_{\mathfrak{f}})\cdot N^{-1}(C^*)}{u(\Lambda_{\mathfrak{f}})\cdot N^{-1}(C^*)}\cong\frac{u(\Lambda'_{\mathfrak{f}})}{u(\Lambda'_{\mathfrak{f}})\cap\{u(\Lambda_{\mathfrak{f}})\cdot N^{-1}(C^*)\}}.$$

Note that $u(\Lambda_{\mathfrak{f}})\cdot N^{-1}(C^*)\lhd u(\Lambda'_{\mathfrak{f}})N^{-1}(C^*)$, since $N^{-1}(C^*)$ contains all commutators. Formula (3.8) avoids the use of reduced norms, except for the calculation of $N^{-1}(C^*)$.

We conclude this subsection by giving Jacobinski's construction for the element $\delta_x\in D(\Lambda)$ which corresponds to an element $x\in u(\Lambda'_{\mathfrak{f}})$ in the isomorphism (3.8). Namely, write $x=yz^{-1}$ with both y and z in Λ' and prime to $\mathfrak{f}$. Then $\delta_x=[\Lambda\cap\Lambda'y]-[\Lambda\cap\Lambda'z]$ is the desired element of $D(\Lambda)$. (It is easily checked that both $\Lambda\cap\Lambda'y$ and $\Lambda\cap\Lambda'z$ are locally free Λ-lattices, and that δ_x lies in the kernel of the mapping $C(\Lambda)\to C(\Lambda')$.)

(v) *Explicit formulas in the general case.* We now turn to the case where A need not satisfy the Eichler condition. Let us form

$$E(A\dotplus A)=\mathrm{Hom}_A(A\dotplus A,\,A\dotplus A).$$

Suppose that the ith simple component A_i of A is given by $A_i\cong(D_i)^{m_i\times m_i}$, $D_i=$ skewfield with center K_i. Then

$$E(A\dotplus A)\cong\sum_{i=1}^{m}{}^{\cdot}(D_i)^{2m_i\times 2m_i}.$$

Thus $E(A\dotplus A)$ has the same center F as A, and is a semisimple K-algebra which automatically satisfies the Eichler condition. Furthermore,

$$E(\Lambda'\dotplus\Lambda')=\mathrm{Hom}_{\Lambda'}(\Lambda'\dotplus\Lambda',\,\Lambda'\dotplus\Lambda')$$

is a maximal R-order in $E(A\dotplus A)$.

Let $N^*\colon E(A\dotplus A)\to F$ be the reduced norm map. We shall say that an element $x\in E(\Lambda\dotplus\Lambda)$ is prime to $\mathfrak{f}$ if x is a unit in $\{E(\Lambda\dotplus\Lambda)\}_{\mathfrak{f}}$, or equivalently, if $N^*(x)$ is prime to $\mathfrak{f}$. There is an obvious ring isomorphism

$$\{E(\Lambda\dotplus\Lambda)\}_{\mathfrak{f}}\cong E(\Lambda_{\mathfrak{f}}\dotplus\Lambda_{\mathfrak{f}}),$$

and we shall always identify these rings with one another. As before, let $I(C,\mathfrak{f})$ be the group of all C-ideals of F prime to $\mathfrak{f}$, and let $J(\Lambda)$ be the subgroup of $I(C,\mathfrak{f})$ generated by the set of all principal ideals

$$\{C\cdot N^*(y)\colon y\in E(\Lambda\dotplus\Lambda),\ y\text{ prime to }\mathfrak{f}\}.$$

As in the preceding section, we obtain

(3.9) $$J(\Lambda) = \{C \cdot N^*(y):\ y = \text{unit in } E(\Lambda_{\mathfrak{f}} \dotplus \Lambda_{\mathfrak{f}})\}.$$

Likewise, we set

(3.10) $$J(\Lambda') = \{C \cdot N^*(y):\ y = \text{unit in } E(\Lambda'_{\mathfrak{f}} \dotplus \Lambda'_{\mathfrak{f}})\}.$$

The following basic result is due to Jacobinski [9]:

(3.11) **Theorem.** *Whether or not A satisfies the Eichler condition, there is a commutative diagram*

$$\begin{array}{ccc} C(\Lambda) \cong I(C, \mathfrak{f})/J(\Lambda) \\ \eta\downarrow \qquad\qquad \eta''\downarrow \\ C(\Lambda') \cong I(C, \mathfrak{f})/J(\Lambda') \end{array}$$

where η is defined by applying $\Lambda' \otimes_{\Lambda}$ to Λ-lattices, and η'' is the natural epimorphism. Consequently,

$$D(\Lambda) = \ker \eta \cong J(\Lambda')/J(\Lambda).$$

Up to isomorphism, these formulas are independent of the choice of maximal order Λ' containing Λ, and of the ideal $\mathfrak{f}$ of R such that $\mathfrak{f} \cdot \Lambda' \subset \Lambda$. Furthermore, when A satisfies the Eichler condition, we have

$$J(\Lambda') = I(\Lambda'), \qquad J(\Lambda) = I(\Lambda).$$

There are obvious analogues of formula (3.8), which we shall not write down explicitly.

4. **Functorial property of class groups.** We intend to show that any group epimorphism $G \twoheadrightarrow \overline{G}$ induces epimorphisms (of additive groups)

(4.1) $$C(RG) \twoheadrightarrow C(R\overline{G}), \qquad D(RG) \twoheadrightarrow D(R\overline{G}).$$

This will be a consequence of some fairly general considerations, together with the formulas given in §3. Let Λ be any R-order in the semisimple algebra A, let Λ' be a maximal R-order in A containing Λ, and let $\mathfrak{f}$ be a nonzero ideal of R such that $\mathfrak{f} \cdot \Lambda' \subset \Lambda$. Let $A \to \overline{A}$ be an epimorphism of K-algebras, and let Λ map onto the R-order $\overline{\Lambda}$. We shall show that there are epimorphisms

(4.2) $$C(\Lambda) \twoheadrightarrow C(\overline{\Lambda}), \qquad D(\Lambda) \twoheadrightarrow D(\overline{\Lambda})$$

which of course imply the analogous results for the special case in (4.1). For convenience, we treat only the case where A satisfies the Eichler condition, since the proof in the general case proceeds in an entirely similar manner.

The kernel of the map $A \to \overline{A}$ is a two-sided ideal in A, hence is a direct sum of some of the simple components of A. This kernel is therefore expressible in the form $A(1 - e)$, where e is some central idempotent in A. Then e maps onto $\overline{e} \in \overline{A}$, where $\overline{e}$ is the identity element of $\overline{A}$, and we have $A = Ae \oplus A(1 - e)$,

$Ae \cong \bar{A}$, $\Lambda e \cong \bar{\Lambda}$. Letting C again denote the maximal R-order in the center F of A, it is clear that $e \in C$, and thus $C = Ce \oplus C(1-e)$.

Let N', N'' be reduced norm maps, where

$$N': Ae \longrightarrow Fe, \qquad N'': A(1-e) \longrightarrow F(1-e).$$

Then $N = N' \dotplus N''$; that is,

$$N(x) = N'(xe) + N''(x(1-e)), \qquad x \in A.$$

There is an obvious map from the group of C-ideals in F onto the group of Ce-ideals in Fe, given by $\mathfrak{a} \longrightarrow \mathfrak{a}e$. This map induces an epimorphism

$$\tau: I(C, \mathfrak{f}) \longrightarrow I(Ce, \mathfrak{f}).$$

If $x \in u(\Lambda_{\mathfrak{f}})$, then $xy = 1$ for some $y \in u(\Lambda_{\mathfrak{f}})$. Therefore, $xe \cdot ye = e$, so $xe \in u(\Lambda_{\mathfrak{f}} \cdot e)$. Furthermore, for $x \in u(\Lambda_{\mathfrak{f}})$, we obtain

$$\tau\{C \cdot N(x)\} = Ce \cdot N(x) = Ce \cdot N'(xe). \tag{4.3}$$

This implies at once that τ maps $I(\Lambda)$ into $I(\Lambda e)$, where

$$I(\Lambda) = \{C \cdot N(x): x \in u(\Lambda_{\mathfrak{f}})\}, \qquad I(\Lambda e) = \{Ce \cdot N'(z): z \in u(\Lambda_{\mathfrak{f}} \cdot e)\}.$$

Hence τ induces an epimorphism

$$I(C, \mathfrak{f})/I(\Lambda) \twoheadrightarrow I(Ce, \mathfrak{f})/I(\Lambda e).$$

Since $\Lambda e \cong \bar{\Lambda}$, we deduce from Theorem 3.7 that there is an epimorphism $C(\Lambda) \twoheadrightarrow C(\bar{\Lambda})$, as claimed.

Next we note that $e \in \Lambda'$, and therefore $\Lambda' = \Lambda' e \oplus \Lambda'(1-e)$. This shows that $C(\Lambda') = C(\Lambda' e) \oplus C(\Lambda'(1-e))$, so there is also an epimorphism $C(\Lambda') \twoheadrightarrow C(\Lambda' e)$.

It remains for us to prove the existence of an epimorphism $D(\Lambda) \twoheadrightarrow D(\bar{\Lambda})$. Now $\Lambda' e$ is a maximal order in Ae containing Λe, and

$$I(\Lambda' e) = \{Ce \cdot N'(z): z = \text{unit in } \Lambda'_{\mathfrak{f}} \cdot e\}.$$

Given any unit $z \in \Lambda'_{\mathfrak{f}} \cdot e$, we may find an element $x \in u(\Lambda'_{\mathfrak{f}})$ such that $z = xe$. But then formula (4.3) shows that $Ce \cdot N'(z)$ lies in the image of τ. This proves that τ maps $I(\Lambda')$ *onto* $I(\Lambda' e)$. Therefore τ induces an epimorphism

$$I(\Lambda')/I(\Lambda) \twoheadrightarrow I(\Lambda' e)/I(\Lambda e),$$

so we have shown the existence of an epimorphism $D(\Lambda) \twoheadrightarrow D(\bar{\Lambda})$, as desired.

5. **Odd p-groups.** Let G be a finite p-group, where p is any prime, and set $A = QG$, $\Lambda = ZG$. Keep the notation of §3, especially that listed in (3.1). We may choose the ideal $\mathfrak{f} = |G| \cdot Z$, and hence we may write "prime to p" in place of

"prime to $\mathfrak{f}$" throughout. The discussion splits into two cases, depending on whether or not p is odd. We shall consider first the easier case where p is odd.

We begin by recalling some preliminary results from representation theory and algebraic number theory.

(5.1) **Theorem.** *Let G be an odd p-group, and let K_i denote the center of the ith simple component of QG. Then for each i, the field K_i is a cyclotomic field $Q(\omega)$ for some p^nth root of unity ω. There is exactly one simple component, say A_1, for which $K_1 = Q$; in fact, $A_1 = K_1 = Q$. Furthermore, each A_i is a full matrix algebra over K_i.*

Reference. Feit [6, (14.5)]; the result is due to Witt and Roquette.

(5.2) **Theorem.** *Let $K_i = Q(\omega)$, where ω is a primitive p^nth root of unity, with $n \geq 1$. Let $R_i = \operatorname{alg\,int}\{K_i\}$. Then there is a unique prime ideal P_i of R_i containing p, and $R_i/P_i \cong Z/pZ$. Given any element $\alpha \in R_i$ prime to P_i, there exists an element $\beta \in R_i$ such that $\beta \equiv 1 \pmod{P_i}$, and $R_i \cdot \alpha = R_i \cdot \beta$.*

Proof. The uniqueness of P_i, and the fact that $R_i/P_i \cong Z/pZ$, both follow from the well-known result that p is completely ramified in K_i. Now let $\alpha \in R_i$ be prime to P_i; then we can choose $m \in Z$ such that $\alpha \equiv m \pmod{P_i}$, and clearly $p \nmid m$. Set $u = (\omega^m - 1)/(\omega - 1)$. Then u is a unit in R_i, and $u \equiv m \pmod{P_i}$ since $P_i = (1-\omega)R_i$. The element $\beta = u^{-1}\alpha$ satisfies the desired conditions.

It is clear from (5.1) that A satisfies the Eichler condition, and that for each i, no infinite prime of K_i ramifies in A_i. As in §3, let $I(C, p)$ be the group of all C-ideals in F which are prime to p, and set

$$I(\Lambda') = \{C \cdot N(x) \colon x \in u(\Lambda'_p)\}, \qquad I(\Lambda) = \{C \cdot N(x) \colon x \in u(\Lambda_p)\}.$$

We have seen that $C(\Lambda') \cong I(C, p)/I(\Lambda')$, $C(\Lambda) \cong I(C, p)/I(\Lambda)$, and that $D(\Lambda) \cong I(\Lambda')/I(\Lambda)$. Now

$$\Lambda'_p \cong \sum_{i=1}^{m}{}^{\cdot} (\Lambda_i)_p, \qquad (\Lambda_i)_p \cong \text{full matrix algebra over } (R_i)_p,$$

from which it is clear that $I(\Lambda')$ consists of all principal C-ideals in F which are prime to p. Therefore,

$$C(\Lambda') \cong \prod_{i=1}^{m} C(R_i),$$

where $C(R_i)$ is the ideal class group of R_i.

One of our main results is as follows:

(5.3) **Theorem.** *Let G be an odd p-group. Then $|D(ZG)|$ is a power of p (possibly equal to 1).*

This theorem was established by Fröhlich [8] for the case where G is an abelian p-group. We devote the remainder of this section to proving the result for an arbitrary p-group, p odd. In the next section, we shall prove the corresponding result for the case where $p = 2$.

(5.4) **Lemma.** *There is an epimorphism of multiplicative groups*

$$\theta\colon 1 + \operatorname{rad} \Lambda'_p \longrightarrow I(\Lambda')/I(\Lambda),$$

given by

$$\theta(x) = C \cdot N(x) \bmod I(\Lambda), \qquad x \in 1 + \operatorname{rad} \Lambda'_p.$$

Proof. First of all, it is clear that $1 + \operatorname{rad} \Lambda'_p$ is a multiplicative group. Next, each $x \in 1 + \operatorname{rad} \Lambda'_p$ is a unit in Λ'_p, and hence $C \cdot N(x) \in I(\Lambda')$. Therefore, θ maps $1 + \operatorname{rad} \Lambda'_p$ into $I(\Lambda')/I(\Lambda)$. Since the reduced norm map N is multiplicative, θ is a homomorphism, and we need to prove that θ is an epimorphism.

Let $C\alpha \in I(\Lambda')$, where $\alpha \in F$ is prime to p. Then $\alpha = \Sigma\, \alpha_i$, with $\alpha_i \in K_i$ prime to P_i (using the notation of (5.2)). Fix the notation so that $A_1 = K_1 = Q$, and $K_i \neq Q$ for $i > 1$. Choose $r \in Z$ prime to p so that $r\alpha \in C$ and $r\alpha_1 \equiv 1 \pmod p$. If we replace $C\alpha$ by $C \cdot \alpha N(r)$, the coset $\bmod\, I(\Lambda)$ is unchanged, but now each new α_i is an element of R_i, and $\alpha_1 \equiv 1 \pmod p$.

Changing notation, consider the element $C\alpha \in I(\Lambda')$, where $\alpha = \Sigma\, \alpha_i$, with each $\alpha_i \in R_i$ prime to P_i, and where $\alpha_1 \equiv 1 \pmod p$. By (5.2) we can choose $\beta_i \in R_i$ with

$$\beta_i \equiv 1 \pmod{P_i}, \qquad R_i \cdot \beta_i = R_i \cdot \alpha_i \quad \text{for all } i.$$

(For $i = 1$, take $\beta_1 = \alpha_1$.) Letting $\beta = \Sigma\, \beta_i$, we see that $C\alpha = C\beta$, and hence it suffices to show the existence of an element $x \in 1 + \operatorname{rad} \Lambda'_p$ such that $N(x) = \beta$.

We may write

$$\Lambda' = \sum\nolimits^{\cdot} \Lambda_i, \qquad (\Lambda')_p = \sum\nolimits^{\cdot} (\Lambda_i)_p = \sum\nolimits^{\cdot} (\Lambda_i)_{P_i},$$

and thus

$$\operatorname{rad} \Lambda'_p = \sum\nolimits^{\cdot} \operatorname{rad} (\Lambda_i)_{P_i}.$$

Now each A_i is a full matrix algebra over K_i, and so the theory of maximal orders tells us that each $(\Lambda_i)_{P_i}$ is a full matrix ring over $(R_i)_{P_i}$, and furthermore $\operatorname{rad} (\Lambda_i)_{P_i} = P_i \cdot (\Lambda_i)_{P_i}$. For each i, choose $x_i \in 1 + P_i \cdot (\Lambda_i)_{P_i} \subset A_i$ such that x_i is represented by a diagonal matrix with diagonal entries $\beta_i, 1, \cdots, 1$. Then $N_i(x_i) = \beta_i$, and setting $x = \Sigma\, x_i$, we have found an element $x \in 1 + \operatorname{rad} \Lambda'_p$ such that $N(x) = \beta$. This completes the proof of the lemma.

(5.5) **Lemma.** *Let* Γ *be a ring with unity, and let* L *be a left ideal of* Γ *such that* $L \subset \operatorname{rad} \Gamma$. *Define* $1 + L = \{1 + x\colon x \in L\}$. *Then* $1 + L$ *is a multiplicative group contained in* Γ.

Proof. Clearly $1 + L$ is closed under multiplication, and we need only check the existence of inverses. Each $x \in L$ lies in rad Γ, so $1 + x$ has a two-sided inverse w in Γ. Then $w(1 + x) = 1$, so $w = 1 - wx \in 1 + L$, as desired.

(5.6) **Lemma.** *Let us set* $X' = \operatorname{rad} \Lambda'_p$, $X = X' \cap \Lambda_p$. *Then* $1 + X$ *is a subgroup of the multiplicative group* $1 + X'$, *of index a finite power of* p.

Proof. We imitate Fröhlich's proof in [8]. For sufficiently large t, we have

$$(X')^t \subset p \cdot \Lambda'_p, \qquad p^t \cdot \Lambda'_p \subset \Lambda_p.$$

Thus for large r we obtain

$$(X')^r \subset p \cdot \Lambda_p, \qquad X^r \subset p \cdot \Lambda_p \subset \operatorname{rad} \Lambda_p. \tag{5.7}$$

Since X is a two-sided ideal of Λ_p, and $X^r \subset \operatorname{rad} \Lambda_p$, it follows that also $X \subset \operatorname{rad} \Lambda_p$. Hence $1 + X$ is a multiplicative group, by (5.5).

For $i \geq 1$, we observe that $(X')^i + X$ is a two-sided ideal of the ring $(X')^i + \Lambda_p$, and

$$\{(X')^i + X\}^{r+i} \subset (X')^{r+i} + X^{r+i} \subset p\{(X')^i + \Lambda_p\} \subset \operatorname{rad} \{(X')^i + \Lambda_p\}.$$

It follows from (5.5) that $1 + (X')^i + X$ is a multiplicative subgroup of $1 + X'$.

Now consider the epimorphism

$$\mu: 1 + (X')^i + X \longrightarrow ((X')^i + X)/((X')^{i+1} + X)$$

defined by

$$\mu(1 + x) = x + (X')^{i+1} + X, \qquad x \in (X')^i + X.$$

It is easily seen that μ is a homomorphism of a multiplicative group onto an additive group, and that

$$\text{kernel of } \mu = 1 + (X')^{i+1} + X.$$

Therefore

$$\frac{1 + (X')^i + X}{1 + (X')^{i+1} + X} \cong \frac{(X')^i + X}{(X')^{i+1} + X}. \tag{5.8}$$

Since $(X')^r \subset X$ by (5.7), we have

$$[1 + X' : 1 + X] = \prod_{i=1}^{r-1} [1 + (X')^i + X : 1 + (X')^{i+1} + X]$$

$$= \prod_{i=1}^{r-1} [(X')^i + X : (X')^{i+1} + X],$$

with the last equality a consequence of (5.8). Thus

$$[1 + X' : 1 + X] = [X' : X].$$

But $p^t X' \subset X$, and X' is finitely generated over R. Hence the index $[X' : X]$ is a finite power of p, and the lemma is established.

We are now ready to conclude the proof of our Theorem 5.3. We have just shown that $1 + X$ is a subgroup of $1 + X'$, of p-power index. Denote by $(1 + X)^*$ the normal closure of $1 + X$ in $1 + X'$. Then $(1 + X)^*$ is generated by elements of the form $y(1 + x)y^{-1}$, $y \in 1 + X'$, $x \in X$. Let θ: $1 + X' \to I(\Lambda')/I(\Lambda)$ be the epimorphism defined in (5.4). Then

$$\theta\{y(1 + x)y^{-1}\} = C \cdot N(1 + x) \bmod I(\Lambda),$$
$$= 1 \bmod I(\Lambda),$$

since $1 + x \in u(\Lambda_p)$. Hence θ induces an epimorphism

$$(1 + X')/(1 + X)^* \twoheadrightarrow I(\Lambda')/I(\Lambda) \cong D(\Lambda).$$

Since $[1 + X' : (1 + X)^*]$ is a divisor of $[1 + X' : 1 + X]$, it follows from (5.6) that $(1 + X')/(1 + X)^*$ is a finite p-group. Hence $D(\Lambda)$ is also a finite p-group, and the theorem is proved.

Professor L. McCulloh has suggested an alternative proof of (5.3). Denote by $I_0(C, \mathfrak{f})$ the subgroup consisting of all principal ideals of $I(C, \mathfrak{f})$, and by $S(C, \mathfrak{f})$ the subgroup of $I_0(C, \mathfrak{f})$ consisting of ideals which possess a generator α such that $\alpha \equiv 1 \pmod{^* \mathfrak{f}}$.(4) Those ideals $C \cdot \Sigma\, \alpha_i$, $\alpha_i \in K_i$, which satisfy the additional condition that $(\alpha_i)_{\hat{P}} > 0$ at those infinite primes P of K_i ramified in A_i, form a subgroup denoted by $S^+(C, \mathfrak{f})$. Of course, I_0, S, S^+ are the direct sums of the corresponding groups for each simple component.

In [9, proof of Lemma 2.6] Jacobinski proved that $I(\Lambda) \supset S^+(C, \mathfrak{f})$ if A satisfies the Eichler condition, so we have a chain of ideal groups

$$I_0(C, \mathfrak{f}) \supset I(\Lambda') \supset I(\Lambda) \supset S^+(C, \mathfrak{f}).$$

Now take $\Lambda = ZG$, G an odd p-group of order p^n. In this case $S(C, \mathfrak{f}) = S^+(C, \mathfrak{f})$. We first show that $I_0(R_i, \mathfrak{f})/S(R_i, \mathfrak{f})$ is a (finite) p-group for $i > 1$. By (7.12) we must show that $u(R_i/p^n R_i)/u'(R_i)$ is a p-group. Now the numerator has order equal to $(p - 1)$ times a p-power. Since $u'(R_i)$ contains the subgroup of order $p - 1$ generated by the cyclotomic units $(1 - \omega^r)/(1 - \omega)$, r prime to p, it follows that $u(R_i/p^n R_i)/u'(R_i)$ is a p-group. (We have used (5.1) and (5.2).)

Now $A_1 = K_1 = Q$ and

$$I_0(Z, p^n Z)/S(Z, p^n Z) \cong u(Z/p^n Z)/\{\pm 1\}.$$

(4) This notation is explained in §7 just before (7.12).

The second paragraph of the proof of (5.4) shows that the A_1-component of any element of $I(\Lambda')$ modulo $I(\Lambda)$ has a generator which is integral and $\equiv 1 \pmod p$. It follows that $I(\Lambda')/I(\Lambda)$ is a p-group, as claimed.

6. **The case** $p = 2$. We now turn to the more difficult case where G is a finite 2-group, again setting $A = QG$, $\Lambda = ZG$, and keeping the notation of §3. For certain choices of G, the algebra A may fail to satisfy the Eichler condition, and thus we must use Theorem 3.11 rather than (3.7) in order to calculate $D(\Lambda)$. We have $D(\Lambda) \cong J(\Lambda')/J(\Lambda)$, where

$$J(\Lambda') = \{C \cdot N^*(x) \colon x = \text{unit in } E(\Lambda'_2 \dotplus \Lambda'_2)\},$$

$$J(\Lambda) = \{C \cdot N^*(x) \colon x = \text{unit in } E(\Lambda_2 \dotplus \Lambda_2)\}.$$

We are going to prove the analogue of Theorem 5.3 for the present case, namely:

(6.1) **Theorem.** *For any 2-group G, the order of $D(ZG)$ is a power of 2 (possibly equal to 1).*

To start with, we need a pair of preliminary results.

(6.2) **Theorem.** *Let G be a 2-group. For each i, the center K_i of the ith simple component A_i of A is a subfield of a cyclotomic field $Q(\omega)$, where ω is some 2^nth root of unity. There is a unique prime ideal P_i of R_i containing 2, and $R_i/P_i \cong Z/2Z$. Each $\alpha \in R_i$ which is prime to P_i satisfies the congruence $\alpha \equiv 1 \pmod{P_i}$.*

Proof. The field K_i is obtained from Q by adjoining values of irreducible complex characters of G (see [3] or [6]). Therefore, $K_i \subset Q(\omega)$ for some 2^nth root of unity ω. The remaining assertions are obvious, since 2 ramifies completely in $Q(\omega)$, hence also in K_i.

(6.3) **Theorem (Eichler [5]).** *Let $S = \operatorname{alg\,int}\{L\}$, and let Γ be a maximal S-order in the simple algebra B with center L. Assume that B satisfies the Eichler condition. Let $N \colon B \to L$ be the reduced norm map, and let $\mathfrak{a}$ be any two-sided ideal of Γ. Let $x \in \Gamma$ be such that*

$$N(x) \equiv \textit{unit of } S \pmod{S \cap \mathfrak{a}}.$$

Then there exists a unit u of Γ such that $x \equiv u \pmod{\mathfrak{a}}$.

Taking this result for granted, we continue with the proof of Theorem 6.1. We have seen in §3 that $E(A \dotplus A)$ is a semisimple algebra with center F, and that A satisfies the Eichler condition. Furthermore, $E(\Lambda' \dotplus \Lambda')$ is a maximal R-order in $E(A \dotplus A)$. To simplify the notation, let us write

$$E' = E(\Lambda' \dotplus \Lambda') = \sum_{i=1}^{m}{}^{\cdot}\, E^i, \quad \text{where } E^i = E(\Lambda_i \dotplus \Lambda_i).$$

Let $N_i^*\colon E(A_i \dotplus A_i) \to K_i$ be the reduced norm map, and define $N^*\colon E(A \dotplus A) \to F$ by setting $N^* = \dot{\Sigma} N_i^*$.

(6.4) **Lemma.** *There is an epimorphism of multiplicative groups*

$$\varphi\colon 1 + \text{rad } E'_2 \to J(\Lambda'),$$

given by $\varphi(y) = C \cdot N^*(y)$, $y \in 1 + \text{rad } E'_2$. *Here,* E'_2 *denotes the localization of* E' *at the rational prime* 2.

Proof. As in the proof of (5.4), it is clear that φ gives a multiplicative homomorphism of the group $1 + \text{rad } E'_2$ into the group $J(\Lambda')$. This latter group is generated by the set of elements

$$\{C \cdot N^*(x)\colon x \in E',\ x \text{ prime to } 2\}.$$

Hence, to show that φ is epic, it suffices to show that for each such x we can find an element $y \in 1 + \text{rad } E'_2$ such that $C \cdot N^*(y) = C \cdot N^*(x)$.

We observe that

$$E'_2 = \dot{\sum} (E^i)_2 = \dot{\sum} (E^i)_{P_i}, \qquad \text{rad } E'_2 = \dot{\sum} \text{rad } (E^i)_{P_i},$$

with the $\{P_i\}$ defined as in (6.2). The entire computation can be performed componentwise, so our problem reduces to the following:

Given $x_i \in E^i$ prime to P_i, show that there exists an element $y_i \in 1 +$ rad $(E^i)_{P_i}$ such that

$$R_i \cdot N_i^*(y_i) = R_i \cdot N_i^*(x_i). \tag{6.5}$$

Now $N_i^*(x_i) \in R_i$ is prime to P_i (since x_i is prime to P_i), and therefore (by (6.2)) $N_i^*(x_i) \equiv 1 \pmod{P_i}$. We shall now use (6.3) for the case when $B = E(A_i \dotplus A_i)$, $\Gamma = E^i = E(\Lambda_i \dotplus \Lambda_i)$, $L = K_i$, $S = R_i$, $\mathfrak{a} = P_i E^i$, $\mathfrak{a} \cap S = P_i$. It follows that there exists a unit $u_i \in E^i$ such that

$$x_i \equiv u_i \pmod{P_i E^i}.$$

Hence $u_i^{-1} x_i \equiv 1 \pmod{P_i E^i}$, and thus

$$u_i^{-1} x_i - 1 \in P_i E^i \subset \text{rad } (E^i)_{P_i}.$$

Let us now set $y_i = u_i^{-1} x_i$, so $y_i \in 1 + \text{rad } (E^i)_{P_i}$. Since $N_i^*(u_i)$ is a unit in R_i, it is clear that (6.5) is valid. Thus y_i has the desired properties, and if we set $y = \Sigma\, y_i$, then

$$y \in 1 + \text{rad } E'_2, \quad C \cdot N^*(y) = C \cdot N^*(x).$$

This completes the proof of the lemma.

Remark. If we had used Eichler's Theorem 6.3 for the case where p is odd, we would not have needed to use the result from (5.1) that A_i is a full matrix algebra over K_i.

It is now an easy matter to complete the proof that $|D(ZG)|$ is a power of 2. The argument given at the end of §5 carries over unchanged to the present case, and shows that φ induces an epimorphism

$$(1+Y')/(1+Y)^* \twoheadrightarrow J(\Lambda')/J(\Lambda) \cong D(\Lambda),$$

where

$$Y' = \text{rad } E'_2, \qquad Y = Y' \cap \{E(\Lambda \dotplus \Lambda)\}_2,$$

and $(1+Y)^*$ = normal closure of $1+Y$ in $1+Y'$. As in §5, the index $[1+Y': 1+Y]$ is a power of 2. Thus the order of $D(ZG)$ is also a power of 2, as claimed.

Next we shall sketch an alternate proof of (6.1), which parallels that given at the end of §5. Set $E' = E(\Lambda' \dotplus \Lambda')$ and $E = E(\Lambda \dotplus \Lambda)$. The centers of Λ' and E' are isomorphic to C, and we have $J(\Lambda') \cong I(E')$, $J(\Lambda) \cong I(E)$. Since KE' satisfies the Eichler condition, we have a chain

$$I_0(C, \mathfrak{f}) \supset J(\Lambda') \supset J(\Lambda) \supset S^+(C, \mathfrak{f}).$$

Let $|G| = 2^n$, $\mathfrak{f} = 2^n Z$. Using (7.12) and the fact that $|u(R_i/2^n R_i)|$ is a power of 2, we see that $I_0(R_i, 2^n R_i)/S(R_i, 2^n R_i)$ is a 2-group for all i. But $S(C, \mathfrak{f})/S^+(C, \mathfrak{f})$ is also a 2-group, whence so is $J(\Lambda')/J(\Lambda)$, as desired.

7. **Cyclic groups of order** $2p$. Throughout this section, let p be an odd prime, and let G be cyclic of order $2p$. We shall calculate $C(ZG)$ and $D(ZG)$ explicitly, and we begin with a few elementary results on units in cyclotomic fields. To fix the notation, let ω be a primitive pth root of unity, and set

$$\begin{aligned} &K = Q(\omega), \quad R = \text{alg int}\{K\} = Z[\omega], \quad P = (1-\omega)R, \\ &L = Q(\omega + \omega^{-1}), \qquad S = \text{alg int}\{L\} = Z[\omega + \omega^{-1}]. \end{aligned} \tag{7.1}$$

Then P is the unique prime ideal of R containing p, and we have

$$R/P \cong S/(P \cap S) \cong Z/pZ = \bar{Z} \quad (\text{say}). \tag{7.2}$$

For each $r \in Z$, define

$$\xi_r = (\omega^r - \omega^{-r})/(\omega - \omega^{-1}) \in S. \tag{7.3}$$

If $p \nmid r$, then ω is a power of ω^r, and hence $\xi_r^{-1} \in S$. Thus ξ_r is a unit in S whenever $p \nmid r$.

Furthermore, for each $r \in Z$ we obtain

$$\xi_r \equiv r \pmod{P}. \tag{7.4}$$

This follows at once from the equations

$$\xi_r = (\omega^{-r}/\omega^{-1}) \cdot (1-\omega^{2r})/(1-\omega^2) = \omega^{1-r}(1+\omega^2+\omega^4+\cdots+\omega^{2r-2})$$
$$\equiv r \pmod{P}, \quad \text{since } P = (1-\omega)R.$$

Next, an elementary calculation yields

$$\xi_r^2 - \xi_s^2 = \omega^{2-2r} \cdot \frac{\omega^{2(r-s)}-1}{\omega^2-1} \cdot \frac{\omega^{2(r+s)}-1}{\omega^2-1}.$$

If $r \equiv \pm s \pmod{p}$, this shows that $\xi_r^2 - \xi_s^2 = 0$. If $r \not\equiv \pm s \pmod{p}$, each of the three factors on the right-hand side of the above displayed equations is a unit in R, and hence $\xi_r^2 - \xi_s^2$ is a unit in S. We have thus proved the following result, which however will not be needed until §8:

(7.5) **Lemma.** *For $r, s \in Z$, define ξ_r, ξ_s as in* (7.3). *Then*

$$\xi_r^2 - \xi_s^2 = \begin{cases} 0, & r \equiv \pm s \pmod{p}, \\ \text{unit in } S, & \text{otherwise.} \end{cases}$$

We may describe the cyclic group G of order $2p$ in terms of generators and relations; thus

$$G = \langle \sigma, \tau \colon \sigma^p = 1,\ \tau^2 = 1,\ \sigma\tau = \tau\sigma \rangle.$$

Let $H = \langle \tau \rangle$ be the subgroup of G of order 2. Our main result here is as follows:

(7.6) **Theorem.** *Let $u'(R)$ be the image of $u(R)$ in $u(R/2R)$. Then*

$$D(ZG) \cong u(R/2R)/u'(R),$$

and consequently

$$|C(ZG)| = |C(R)|^2[u(R/2R) : u'(R)],$$

where $C(R)$ is the ideal class group of R.

In proving Theorem 7.6, as well as in later calculations, we shall need some remarks on pullback diagrams (= fibre products). Let $\mathfrak{a}$, $\mathfrak{b}$ be two-sided ideals of an arbitrary ring Λ with 1, not necessarily commutative. Then it is easily verified that there is a pullback diagram

$$\begin{array}{ccc} \dfrac{\Lambda}{\mathfrak{a} \cap \mathfrak{b}} & \longrightarrow & \dfrac{\Lambda}{\mathfrak{a}} \\ \downarrow & & \downarrow \\ \dfrac{\Lambda}{\mathfrak{b}} & \longrightarrow & \dfrac{\Lambda}{\mathfrak{a}+\mathfrak{b}} \end{array} \tag{7.7}$$

that is,

$$\Lambda/(\mathfrak{a} \cap \mathfrak{b}) \cong \{(x, y) \in (\Lambda/\mathfrak{a}) \dotplus (\Lambda/\mathfrak{b}) : x \equiv y \bmod (\mathfrak{a} + \mathfrak{b})\}.$$

We apply this to the case where $\Lambda = ZG$, with G cyclic of order $2p$ as above. We pick

$$(7.8) \qquad \mathfrak{a} = \Phi(\sigma) \cdot \Lambda, \qquad \mathfrak{b} = (\sigma - 1) \cdot \Lambda,$$

where $\Phi(X) = 1 + X + \cdots + X^{p-1}$ is the cyclotomic polynomial of order p. Since QG is free as $Q[\sigma]$-module, it is clear that $\mathfrak{a} \cap \mathfrak{b} = 0$. Furthermore, there are ring isomorphisms

$$\Lambda/\mathfrak{a} \cong RH, \qquad \Lambda/\mathfrak{b} \cong ZH, \qquad \Lambda/(\mathfrak{a} + \mathfrak{b}) \cong \overline{Z}H,$$

with $\overline{Z} = Z/pZ \cong R/P$. The first of these isomorphisms is obtained by letting $\sigma \to \omega$, the second by $\sigma \to 1$, the third by $\sigma \to 1$, $Z \to \overline{Z}$. The pullback diagram (7.7) becomes

$$(7.9) \qquad \begin{array}{ccc} \Lambda & \longrightarrow & RH \\ \downarrow & & \downarrow \\ ZH & \longrightarrow & \overline{Z}H \end{array}$$

We shall use bars to indicate images of elements of ZH or RH in $\overline{Z}H$. Tensoring with Q, we see that QG can be identified with $QH \dotplus KH$, and that the maximal order Λ' of QG may be identified with $Z^{(2)} \dotplus R^{(2)}$. Further,

$$(7.10) \qquad \Lambda \cong \{(\xi, \eta) \in ZH \dotplus RH : \overline{\xi} = \overline{\eta} \text{ in } \overline{Z}H\}.$$

Once Λ' has been identified with $Z^{(2)} \dotplus R^{(2)}$, it remains for us to describe Λ as a subring of Λ'. Now $QH \cong Q^{(2)}$, $KH \cong K^{(2)}$, and

$$ZH \cong \{(a_0, a_1) \in Z^{(2)} : a_0 \equiv a_1 \pmod{2Z}\},$$

$$RH \cong \{(b_0, b_1) \in R^{(2)} : b_0 \equiv b_1 \pmod{2R}\}.$$

It follows from these isomorphisms, together with (7.10), that we may identify Λ with the subring of Λ' consisting of all 4-tuples $(a_0, a_1, b_0, b_1) \in Z^{(2)} \dotplus R^{(2)}$ which satisfy the congruence conditions

$$(7.11) \qquad a_0 \equiv a_1 \pmod{2Z}, \quad b_0 \equiv b_1 \pmod{2R}, \quad a_0 \equiv b_0, \; a_1 \equiv b_1 \pmod{P}.$$

Now $D(\Lambda) \cong I(\Lambda')/I(\Lambda)$ by Theorem 3.7. In our particular case, $C = \Lambda'$ and the reduced norm map N is the identity map. Thus

$$I(\Lambda') = \{\Lambda' \cdot x : x \in u(\Lambda'_{2p})\}, \qquad I(\Lambda) = \{\Lambda' \cdot y : y \in u(\Lambda_{2p})\}.$$

Denote by $I_0(R, \mathfrak{c})$ the group of all principal R-ideals in K prime to the ideal $\mathfrak{c}$ of R, and define $I_0(Z, m)$ analogously as the group of Z-ideals in Q prime to the integer m. Since $\Lambda' = Z^{(2)} \dotplus R^{(2)}$, we have

$$I(\Lambda') = \{I_0(Z, 2p)\}^{(2)} \dotplus \{I_0(R, 2P)\}^{(2)}.$$

Before proceeding further, we require two lemmas, the first of which is a general result true for any algebraic number field Ω. Let $\mathfrak{O} = \text{alg int}\{\Omega\}$, let $\mathfrak{q}$ be a nonzero ideal of $\mathfrak{O}$, and write $\mathfrak{q} = \prod \mathfrak{p}_i^{e_i}$, $e_i > 0$, with the $\{\mathfrak{p}_i\}$ distinct prime ideals of $\mathfrak{O}$. Let v_i be the normalized $\mathfrak{p}_i$-adic valuation on Ω. For $x \in \Omega$, we shall write $x \equiv 1 \pmod{^* \mathfrak{q}}$, if and only if $v_i(x-1) \geq e_i$ for each $\mathfrak{p}_i$ dividing $\mathfrak{q}$. Let $I_0(\mathfrak{O}, \mathfrak{q})$ be the group of principal $\mathfrak{O}$-ideals in Ω prime to $\mathfrak{q}$, and let $S(\mathfrak{O}, \mathfrak{q})$ be the subgroup of $I_0(\mathfrak{O}, \mathfrak{q})$ defined by

$$S(\mathfrak{O}, \mathfrak{q}) = \{x\mathfrak{O} : x \in \Omega,\ x \neq 0,\ x \equiv 1 \bmod^* \mathfrak{q}\}.$$

(7.12) **Lemma.** *Let $\mathfrak{q}$ and $\mathfrak{r}$ be relatively prime ideals of $\mathfrak{O}$, and let $u'(\mathfrak{O})$ be the image of $u(\mathfrak{O})$ in $u(\mathfrak{O}/\mathfrak{q})$. Then*

$$\frac{I_0(\mathfrak{O}, \mathfrak{q}\mathfrak{r})}{I_0(\mathfrak{O}, \mathfrak{q}\mathfrak{r}) \cap S(\mathfrak{O}, \mathfrak{q})} \cong \frac{u(\mathfrak{O}/\mathfrak{q})}{u'(\mathfrak{O})}.$$

Hence $I_0(\mathfrak{O}, \mathfrak{q}\mathfrak{r}) \subset S(\mathfrak{O}, \mathfrak{q})$ whenever $u'(\mathfrak{O}) = u(\mathfrak{O}/\mathfrak{q})$.

Proof. Define a mapping $\kappa: I_0(\mathfrak{O}, \mathfrak{q}\mathfrak{r}) \to u(\mathfrak{O}/\mathfrak{q})/u'(\mathfrak{O})$ by setting $\kappa(x\mathfrak{O}) =$ coset containing $\bar{x}$, where the element $x \in \Omega$ prime to $\mathfrak{q}\mathfrak{r}$ maps onto $\bar{x} \in u(\mathfrak{O}/\mathfrak{q})$. Clearly κ is a well-defined homomorphism. Further, κ is epic, since given any $\alpha \in u(\mathfrak{O}/\mathfrak{q})$, we can find an element $x \in \Omega$ such that $\bar{x} = \alpha$, $x \equiv 1 \pmod{^* \mathfrak{r}}$, and then $\kappa(x\mathfrak{O})$ equals the coset containing α.

It remains to determine $\ker \kappa$. Let $\kappa(x\mathfrak{O}) = 1$, where $x \in \Omega$ is prime to $\mathfrak{q}\mathfrak{r}$. Then $\bar{x} \in u'(\mathfrak{O})$; that is, $x \equiv u \pmod{^* \mathfrak{q}}$ for some $u \in u(\mathfrak{O})$. But then $x\mathfrak{O} = xu^{-1} \cdot \mathfrak{O} \in S(\mathfrak{O}, \mathfrak{q})$, which proves that

$$\ker \kappa \subset I_0(\mathfrak{O}, \mathfrak{q}\mathfrak{r}) \cap S(\mathfrak{O}, \mathfrak{q}).$$

The reverse inclusion is obvious, so the lemma is established.

Now observe that $u'(R) = u(R/P)$ by (7.4), so by the second statement in the lemma we obtain

$$I_0(R, 2P) \subset S(R, P). \tag{7.13}$$

We shall use this in a moment.

(7.14) **Lemma.** *Let $u''(R)$ be the image in $u(R/2R)$ of the group $\{u \in u(R): u \equiv 1 \pmod{P}\}$. Then $u''(R) = u'(R)$.*

Proof.(5) Obviously $u''(R) \subset u'(R)$, and we must prove the reverse inclusion. Let $u \in u(R)$ be given; we need to find a unit v in R such that $v \equiv u \pmod{2R}$, $v \equiv 1 \pmod{P}$. Let θ be the automorphism of K defined by letting $\omega \to \omega^2$. Then

$$u^\theta \equiv u^2 \pmod{2R}, \qquad u^\theta \equiv u \pmod{P},$$

(5) The authors wish to thank Professor L. McCulloh for providing this simple proof.

and we need only choose $v = u^\theta/u$.

We shall now complete the proof of Theorem 7.6 by constructing an epimorphism

$$\psi : I(\Lambda') \to u(R/2R)/u'(R)$$

with kernel $I(\Lambda)$. Let bars denote images in $u(R/2R)$, and define

$$\psi(Za_0, Za_1, Rb_0, Rb_1) = \text{coset containing } \bar{b}_0/\bar{b}_1,$$

where $a_i \in u(Z_{2p})$, $b_i \in u(R_{2P})$. Clearly, ψ is a homomorphism. It is epic because the map $I_0(R, 2P) \to u(R/2R)/u'(R)$ is already epic, as follows from the proof of (7.12).

We see from (7.11) that $u(\Lambda_{2p})$ consists of all 4-tuples

$$(a_0, a_1, b_0, b_1), \qquad a_i \in u(Z_{2p}), \ b_i \in u(R_{2P}),$$

which satisfy the conditions

$$(7.15) \quad a_0 \equiv a_1 \ (\text{mod}^*\, 2Z), \quad b_0 \equiv b_1 \ (\text{mod}^*\, 2R), \quad a_0 \equiv b_0,\ a_1 \equiv b_1 \ (\text{mod}^*\, P).$$

For each such 4-tuple we have $\bar{b}_0 = \bar{b}_1$, whence $I(\Lambda) \subset \ker \psi$. To prove the reverse inclusion, let $(Za_0, Za_1, Rb_0, Rb_1) \in I(\Lambda')$ lie in the kernel of ψ. By (7.13), there exist elements $c_0, c_1 \in K$ such that

$$Rb_i a_i^{-1} = Rc_i, \quad c_i \equiv 1 \ (\text{mod}^*\, P), \qquad i = 1, 2.$$

Hence $Rb_i = Ra_i c_i$, so replacing b_i by $a_i c_i$, we may henceforth assume that $a_0 \equiv b_0$, $a_1 \equiv b_1$ (mod* P). But $\psi(Za_0, Za_1, Rb_0, Rb_1) = 1$ implies that $\bar{b}_0/\bar{b}_1 \in u'(R)$, so by (7.14) there exists a unit $u \in R$ such that $b_0 \equiv b_1 u$ (mod* $2R$), $u \equiv 1$ (mod P). Replacing b_1 by $b_1 u$ does not affect the congruence $a_1 \equiv b_1$ (mod* P), and permits us to assume that $b_0 \equiv b_1$ (mod* $2R$). Since trivially $a_0 \equiv a_1$ (mod* $2Z$), we may conclude that the 4-tuple (a_0, a_1, b_0, b_1) satisfies all of the conditions in (7.15), and hence that $(Za_0, Za_1, Rb_0, Rb_1) \in I(\Lambda)$. This completes the proof of the assertion about $D(\Lambda)$. The second assertion in (7.6) is then obvious, since $C(\Lambda') \cong C(R) \times C(R)$.

As a matter of fact, the epimorphism $\Lambda \to RH$ induces epimorphisms $D(ZG) \to D(RH)$, $C(ZG) \to C(RH)$. Ullom [16] has previously shown that

$$|D(RH)| = [u(R/2R) : u'(R)],$$

from which we may conclude that $D(ZG) \cong D(RH)$ and $C(ZG) \cong C(RH)$. These latter conclusions can also be obtained directly from (7.9) by use of Milnor's Mayer-Vietoris sequence (see [18]), but one is then left with the problem of calculating $D(RH)$.

8. Dihedral groups of order $2p$. Let G be the dihedral group of order $2p$, where p is an odd prime. We shall again use the notation introduced in (7.1), so that S is the ring of algebraic integers in the subfield L of the cyclotomic field K. The class group $C(S)$ is then the ideal class group of S. We shall prove

$$C(ZG) \cong C(S), \qquad D(ZG) = 1. \tag{8.1}$$

These results are immediate consequences of Lee's classification [10] of ZG-lattices. Our proof is more direct, but of course provides no information about the nonprojective ZG-lattices.

Throughout this section, we put

$$G = \langle \sigma, \tau\colon \sigma^p = 1, \tau^2 = 1, \tau\sigma\tau^{-1} = \sigma^{-1} \rangle, \qquad H = \langle \tau\colon \tau^2 = 1 \rangle,$$

and set $A = QG$, $\Lambda = ZG$. Let $x \to x'$ be the L-automorphism of K induced by the map $\omega \to \omega^{-1}$ (see (7.1)). We introduce the twisted group algebra

$$K\langle \tau \rangle = K \oplus K\tau, \qquad \tau x = x' \cdot \tau, \qquad x \in K.$$

Then $K\langle \tau \rangle$ is a simple algebra with center L, and is split by K:

$$K\langle \tau \rangle \cong \mathrm{Hom}_L(K, K).$$

The isomorphism is obtained by mapping $x + y\tau \in K\langle \tau \rangle$ onto $\phi_{x+y\tau}$, where $\phi_{x+y\tau}(u) = xu + yu'$, $u \in K$. On the other hand, $K\langle \tau \rangle$ is a cyclic algebra, and as such has a matrix representation over K given by

$$x + y\tau \to \begin{pmatrix} x & y \\ y' & x' \end{pmatrix}, \qquad x, y \in K.$$

If $N\colon K\langle \tau \rangle \to L$ is the reduced norm map, the above implies that

$$N(x + y\tau) = \begin{vmatrix} x & y \\ y' & x' \end{vmatrix} = xx' - yy'. \tag{8.2}$$

We may make $K\langle \tau \rangle$ into a QG-module by letting σ act as multiplication by ω on the coefficients from K. The decomposition of A into simple components is then given by $A = Q \dotplus Q \dotplus K\langle \tau \rangle$. The first summand affords the trivial representation $\sigma \to 1, \tau \to 1$, while the second gives $\sigma \to 1, \tau \to -1$.

Let us use a pullback diagram to describe the elements of Λ in terms of their projections into the simple components of A. Since the cyclic subgroup $[\sigma]$ generated by σ is normal in G, the ideals $\mathfrak{a}$, $\mathfrak{b}$ of Λ defined by (7.8) are two-sided ideals of Λ. Furthermore, $\mathfrak{a} \cap \mathfrak{b} = 0$ since QG is $Q[\sigma]$-free. Let

$$R\langle \tau \rangle = R \oplus R\tau \subset K\langle \tau \rangle.$$

Then there are ring isomorphisms

$$\Lambda/\mathfrak{b} \cong ZH, \qquad \Lambda/\mathfrak{a} \cong R\langle \tau \rangle, \qquad \Lambda/(\mathfrak{a} + \mathfrak{b}) \cong \bar{Z}H,$$

with $\bar{Z}$ as in (7.2). The pullback diagram (7.7) then becomes

$$(8.3)\qquad \begin{array}{ccc} \Lambda & \xrightarrow{\alpha} & R\langle\tau\rangle \\ \beta\downarrow & & \downarrow\mu \\ ZH & \longrightarrow & \overline{Z}H \end{array}$$

The map μ is given by $\mu(x + y\tau) = \overline{x} + \overline{y}\tau$, where $x \in R$ maps onto $\overline{x} \in R/P \cong \overline{Z}$. Since $x \equiv x'$ (mod P) for $x \in R$, we see that (mod P) τ commutes with the elements of R, and hence $\overline{Z}H$ is an ordinary untwisted group algebra. Further, α is just the projection of Λ into the simple component $K\langle\tau\rangle$, while β is the projection into $Q \dotplus Q$, once we identify QH with $Q \dotplus Q$.

In the isomorphism $QH \cong Q \dotplus Q$, the ring ZH is identified with

$$\{(a_0, a_1) \in Z \dotplus Z: a_0 \equiv a_1 \pmod 2\}.$$

It then follows from (8.3) that we may identify Λ with the ring of all triples $\{(a_0, a_1, x + y\tau) \in Z \dotplus Z \dotplus R\langle\tau\rangle\}$ satisfying the congruences

$$(8.4)\quad a_0 \equiv a_1 \pmod 2,\ x \equiv (a_0 + a_1)/2 \pmod P,\quad y \equiv (a_0 - a_1)/2 \pmod P.$$

Now let Λ' be a maximal Z-order in A containing Λ. Since each simple component of A is a full matrix algebra over its center, surely A satisfies the Eichler condition. Furthermore, the discussion in §3(i) shows that

$$\begin{aligned} I(\Lambda') &= \{C \cdot \alpha: \alpha \in F,\ \alpha \text{ prime to } 2p\} \\ &= \{(Zb_0, Zb_1, S\beta): b_0, b_1 \in Q,\ \beta \in L, \text{ with } b_0, b_1, \beta \text{ prime to } 2p\}. \end{aligned}$$

On the other hand, $I(C, 2p)$ consists of all C-ideals in F which are prime to $2p$. Therefore (by (3.7))

$$C(\Lambda') \cong I(C, 2p)/I(\Lambda') \cong C(Z) \times C(Z) \times C(S) \cong C(S),$$

where $C(Z)$ = ideal class group of Z, and so on. By (3.7) we have $D(\Lambda) \cong I(\Lambda')/I(\Lambda)$, where

$$I(\Lambda) = \{C \cdot N(x): x \in \Lambda,\ x \text{ prime to } 2p\}.$$

We shall show below that $I(\Lambda) = I(\Lambda')$. This will imply that $D(\Lambda) = 1$ and $C(\Lambda) \cong C(\Lambda') \cong C(S)$, as claimed in (8.1).

(8.5) **Lemma.** (i) *The map $\mu: R\langle\tau\rangle \to \overline{Z}H$ induces an epimorphism of unit groups*: $u(R\langle\tau\rangle) \twoheadrightarrow u(\overline{Z}H)$.

(ii) *Let $\beta \in S$ be prime to p. Then there exists an element $x + y\tau \in R\langle\tau\rangle$ such that $S \cdot N(x + y\tau) = S\beta$.*

Proof. (i) Each element of $\overline{Z}H$ is of the form $\overline{r} + \overline{s}\tau$, where $r, s \in Z$ have images $\overline{r}, \overline{s} \in \overline{Z}$. Since $(\overline{r} + \overline{s}\tau)(\overline{r} - \overline{s}\tau) = \overline{r}^2 - \overline{s}^2$, we see that $\overline{r} + \overline{s}\tau \in u(\overline{Z}H)$ if and only if $\overline{r}^2 \neq \overline{s}^2$. Given such a unit $\overline{r} + \overline{s}\tau$, we may form $\xi_r + \xi_s\tau \in R\langle\tau\rangle$. Now τ commutes with the ξ's, since they lie in S. It follows from (7.5) that $\xi_r + \xi_s\tau$ is a unit in $R\langle\tau\rangle$, and clearly $\mu(\xi_r + \xi_s\tau) = \bar{r} + \bar{s}\tau$, as desired.

(ii) We have $S/(P \cap S) \cong \bar{Z}$, so $\beta \equiv r \pmod{P}$ for some $r \in Z$ with $p \nmid r$. The element ξ_r given in (7.3) is a unit in S such that $\xi_r \equiv r \pmod{P}$. Setting $\alpha = \beta \cdot \xi_r^{-1}$, we obtain

$$S \cdot \beta = S \cdot \alpha, \qquad \alpha \equiv 1 \pmod{P}.$$

It then suffices to find an element $x + y\tau \in R\langle\tau\rangle$ such that $N(x + y\tau) = \alpha$. We may set

$$\alpha = 1 - y(\omega - \omega^{-1})^2/(\omega^2 + \omega^{-2})$$

for some $y \in S$. Then define $x = x_0 + x_1\omega$, $y = y_0 + y_1\omega$, where $x_0 = 1 - y$, $y_0 = \alpha - x_0$, $x_1 = -y_1 = (\omega + \omega^{-1})y_0/2$. It is then easily checked that $x + y\tau \in R\langle\tau\rangle$, and $N(x + y\tau) = xx' - yy' = \alpha$.

Let $\lambda \in \Lambda$ correspond to the triple $(a_0, a_1, x + y\tau)$; then

$$C \cdot N(\lambda) = Za_0 \dotplus Za_1 \dotplus S \cdot N(x + y\tau).$$

Further, λ is prime to $2p$ if and only if a_0, a_1, and $x + y\tau$ are prime to $2p$. Now $I(\Lambda')$ is generated by ideals of the form

$$Zb_0 \dotplus Zb_1 \dotplus S\beta, \qquad b_0, b_1, \beta \text{ prime to } 2p. \tag{8.6}$$

In order to show that $I(\Lambda') = I(\Lambda)$, we must show that given the ideal in (8.6), we can choose a_0, a_1, $x + y\tau$ prime to $2p$, and satisfying (8.4), such that

$$Za_0 = Zb_0, \quad Za_1 = Zb_1, \quad S \cdot N(x + y\tau) = S\beta. \tag{8.7}$$

Indeed, we pick $a_0 = b_0$, $a_1 = b_1$, and we must find $x + y\tau \in R\langle\tau\rangle$ such that

$$x \equiv (b_0 + b_1)/2, \quad y \equiv (b_0 - b_1)/2 \pmod{P}, \quad S \cdot N(x + y\tau) = S \cdot \beta.$$

Let $c_1 = (b_0 + b_1)/2$, $c_2 = (b_0 - b_1)/2$, so $c_1^2 - c_2^2 = b_0 b_1$, whence in Z we have $\bar{c}_1^2 - \bar{c}_2^2 \neq 0$. Thus $\bar{c}_1 + \bar{c}_2\tau \in u(\bar{Z}H)$. By (8.5)(ii) we may choose $x_0 + y_0\tau \in R\langle\tau\rangle$ such that $S \cdot N(x_0 + y_0\tau) = S \cdot \beta$. Hence $x_0x_0' - y_0y_0'$ is prime to p, whence $\bar{x}_0 + \bar{y}_0\tau \in u(\bar{Z}H)$. By (8.5)(i), we may choose $u \in u(R\langle\tau\rangle)$ such that $(\bar{x}_0 + \bar{y}_0\tau) \cdot \mu(u) = \bar{c}_1 + \bar{c}_2\tau$. Set $x + y\tau = (x_0 + y_0\tau)u$; since $N(u) \in u(S)$, this gives $S \cdot N(x + y\tau) = S \cdot N(x_0 + y_0\tau) = S \cdot \beta$, and furthermore, $\bar{x} + \bar{y}\tau = \bar{c}_1 + \bar{c}_2\tau$, that is, $x \equiv c_1 \pmod{P}$, $y \equiv c_2 \pmod{P}$, as desired. This completes the proof of (8.1).

9. The quaternion group of order 8. Our main result here is as follows:

(9.1) **Theorem.** *Let G be the quaternion group of order 8. Then $C(ZG)$ has order 2, and equals $D(ZG)$.*

We may write

$$G = \langle \sigma, \tau: \sigma^4 = 1, \ \sigma^2 = \tau^2, \ \tau\sigma\tau^{-1} = \sigma^{-1} \rangle.$$

The group algebra $A = QG$ splits into a direct sum of five simple components:

(9.2) $$A \cong Q^{(4)} \dotplus S, \qquad S = Q \oplus Qi \oplus Qj \oplus Qk.$$

The first four components correspond to the four distinct one-dimensional representations of G, namely $\sigma \to \pm 1$, $\tau \to \pm 1$. The fifth component S is the skewfield of rational quaternions, corresponding to the representation $\sigma \to i$, $\tau \to j$.

It is more convenient for us to use the isomorphism

(9.3) $$f\colon QG \cong QV \dotplus S,$$

where $V = \langle s, t\colon s^2 = t^2 = 1,\ st = ts\rangle$ is an abelian (2, 2)-group, and where the isomorphism f is given by $\sigma \to (s, i)$, $\tau \to (t, j)$. We will henceforth identify QG with $QV \dotplus S$, and view ordered pairs $(v, \alpha) \in QV \dotplus S$ as elements of QG.

Now let e_0, e_1, e_2, e_3 be the primitive idempotents of QV, given by $e_i = (1 \pm s)(1 \pm t)/4$. There is an isomorphism $g\colon QV \cong Q^{(4)}$, which we may specify by setting

(9.4) $$g^{-1}(x_0, x_1, x_2, x_3) = \sum_{i=0}^{3} x_i e_i, \qquad x_i \in Q.$$

Our original isomorphism (9.2) may be obtained thus:

$$QG \cong QV \dotplus S \cong Q^{(4)} \dotplus S.$$

In order to determine the image of Λ in $QV \dotplus S$, where $\Lambda = ZG$, we use a pullback diagram. Let $Q[\sigma^2]$ be the group algebra of the cyclic group $[\sigma^2]$ generated by the element σ^2 in the center of G. Set

$$\mathfrak{a} = \Lambda(\sigma^2 + 1), \qquad \mathfrak{b} = \Lambda(\sigma^2 - 1),$$

a pair of two-sided ideals in Λ. Since QG is $Q[\sigma^2]$-free, it follows at once that $\mathfrak{a} \cap \mathfrak{b} = 0$.

The projection $QG \to QV$ induces the mappings

$$\Lambda \to \Lambda/\mathfrak{b} \cong ZV, \quad \sigma \to s, \quad \tau \to t.$$

On the other hand, the projection $QG \to S$ gives

$$\Lambda \to \Lambda/\mathfrak{a} \cong H = Z \oplus Zi \oplus Zj \oplus Zk \subset S,$$

where $\sigma \to i$, $\tau \to j$. Then

$$\Lambda/(\mathfrak{a} + \mathfrak{b}) \cong \overline{ZV} \cong \overline{H},$$

where $\overline{Z} = Z/2Z$, $\overline{H} = H/2H$. Let $\gamma\colon \overline{ZV} \cong \overline{H}$ be the above isomorphism, chosen so that $\gamma(\bar{s}) = \bar{i}$, $\gamma(\bar{t}) = \bar{j}$, where bars denote images in $\overline{ZV}$ or $\overline{H}$, depending on the context.

The pullback diagram (7.7) now becomes

$$\begin{array}{ccccc} \Lambda & \longrightarrow & & & H \\ \downarrow & & & & \downarrow \\ ZV & \longrightarrow & \overline{Z}V & \underset{\gamma}{\longrightarrow} & \overline{H} \end{array} \tag{9.5}$$

so under the isomorphism f, we may identify Λ with the ring

(9.6) $$\{(v, \alpha): v \in ZV, \ \alpha \in H, \ \gamma(\overline{v}) = \overline{\alpha} \text{ in } \overline{H}\}.$$

Furthermore, in terms of the isomorphism $g: QV \cong Q^{(4)}$, it follows from (9.4) that

(9.7) $$g(ZV) = \{(x_0, x_1, x_2, x_3) \in Z^{(4)}: x_0 \pm x_1 \pm x_2 \pm x_3 \equiv 0 \pmod 4\},$$

where the $\pm$ signs are chosen so that the number of minus signs is even.

Now let $H' = H + Z \cdot \zeta$, $\zeta = (1 + i + j + k)/2$, so H' is a maximal order in S containing H. As is well known, H' is a noncommutative PID and so its class group $C(H')$ is trivial. Set $\Lambda' = g^{-1}(Z^{(4)}) \dotplus H'$, a maximal Z-order in $QV \dotplus S$ containing Λ; we have

$$C(\Lambda') \cong C(Z)^{(4)} \dotplus C(H') = 0.$$

This shows that $D(\Lambda) = C(\Lambda)$, and it remains for us to prove that $D(\Lambda)$ is of order 2. Since S is a totally definite quaternion algebra, A does not satisfy the Eichler condition, and so we must use the formula $D(\Lambda) \cong J(\Lambda')/J(\Lambda)$ in Theorem 3.11 to calculate $D(\Lambda)$.

From (9.2) or (9.3), we obtain

$$F \cong Q^{(4)} \dotplus Q, \qquad C \cong Z^{(4)} \dotplus Z,$$

where F is the center of $QV \dotplus S$, C the maximal Z-order in F. Since $C(\Lambda') = 0$, it follows from (3.11) that $J(\Lambda')$ coincides with the group $I(C, 2)$ of all C-ideals in F prime to the rational prime 2. For brevity, let I denote the group of all Z-ideals in Q prime to 2. Then $J(\Lambda') = I(C, 2) \cong I^{(5)}$.

Now let $(v, \alpha) \in QV \dotplus S$, viewed as element of QG. It is readily seen that the reduced norm $N(v, \alpha) \in F$ is given by

(9.8) $$N(v, \alpha) = (g(v), N\alpha) \in Q^{(4)} \dotplus Q, \qquad v \in QV, \ \alpha \in S,$$

where $N\alpha$ is the reduced norm of the quaternion α. (If $\alpha = x_0 + x_1 i + x_2 j + x_3 k$, then $N\alpha = \Sigma x_i^2$.)

Let $U_0 = \{1, i, j, k\}$, and let $\overline{U}_0$ be the image of U_0 in $u(\overline{H})$.

(9.9) **Lemma.** *There is an epimorphism*

$$\theta: u(Z_2V) \times u(H_2) \longrightarrow I^{(5)},$$

given by

$$\theta(v, \alpha) = Z^{(5)} \cdot N(v, \alpha) = (Z^{(4)} \cdot g(v), Z \cdot N\alpha), \qquad v \in u(Z_2V), \ \alpha \in u(H_2).$$

Furthermore, $\theta(v, \alpha) \in J(\Lambda)$ *whenever* $\gamma(\bar{v}) \in \bar{\alpha}\bar{U}_0$.

Proof. The group $I^{(5)}$ is generated by elements of the form

$$w = (Za_0, Za_1, Za_2, Za_3, Za_4), \qquad a_i \in Z,\ a_i \text{ odd}.$$

Given such an element, we may as well adjust the a's so that

$$a_i \equiv 1 \pmod 4, \qquad 0 \leq i \leq 3,\ a_4 > 0.$$

By (9.7), there exists an element $v \in ZV$ such that $g(v) = (a_0, a_1, a_2, a_3)$. On the other hand, $a_4 = N\alpha$ for some $\alpha \in H$, since every positive integer is a sum of four squares. Therefore we have $\theta(v, \alpha) = w$. Since the image of θ is a group containing the generators $\{w\}$ of $I^{(5)}$, we may conclude that θ is an epimorphism, as desired.

Next, suppose that $\gamma(\bar{v}) = \bar{\alpha}\bar{u}_0$ for some $u_0 \in U_0$. Then $(v, \alpha u_0) \in \Lambda$ by (9.6), and $\theta(v, \alpha u_0) = \theta(v, \alpha)$. Since $\theta(v, \alpha u_0) \in I(\Lambda) \subset J(\Lambda)$, the lemma is established.

Remark. Since θ is epic, and $\Lambda' \supset ZV \dotplus H$, we conclude that $I(\Lambda') = I^{(5)}$. But $I(\Lambda') \subset J(\Lambda')$ obviously holds true, and therefore $J(\Lambda') = I^{(5)} \cong I(C, 2)$, as previously stated.

We may next deduce from Lemma 9.9 the important fact that $|D(\Lambda)| \leq 2$. We shall define an epimorphism

$$\nu\colon I^{(5)} \longrightarrow u(\bar{H})/\bar{U}_0$$

by setting

$$\nu(w) = \gamma(\bar{v}) \cdot \bar{\alpha}^{-1} \cdot \bar{U}_0, \quad \text{where } w = \theta(v, \alpha),\ v \in u(Z_2V),\ \alpha \in u(H_2).$$

Let us show that $\nu(w)$ is well defined. The ideal $Z \cdot N(\alpha)$ is determined by w, so α is determined up to a factor from $u(H)$. But $u(H) = \{\pm 1, \pm i, \pm j, \pm k\}$, and $u(H)$ maps onto $\bar{U}_0 \subset u(\bar{H})$, so that $\bar{\alpha} \pmod{\bar{U}_0}$ is uniquely determined by w. Secondly, $g(v) \in Z^{(4)}$ is determined up to a factor from $u(Z^{(4)})$. The eight units $\{(\pm 1, \pm 1, \pm 1, \pm 1)\}$ in $Z^{(4)}$ are the images (under g) of the eight units $\{\pm 1, \pm s, \pm t, \pm st\}$ of ZV. Hence $\gamma(\bar{v})$ is determined up to a factor $\gamma(\bar{v}_0)$, with $v_0 \in u(ZV)$. But $\gamma(\bar{s}) = \bar{i}$, $\gamma(\bar{t}) = \bar{j}$, so $\gamma(\bar{v}_0) \in \bar{U}_0$ for each $v_0 \in u(ZV)$. Therefore $\gamma(\bar{v})$ is also uniquely determined in $u(\bar{H})/\bar{U}_0$, and so ν is well defined. Obviously, ν is an epimorphism of multiplicative groups.

The second statement in Lemma 9.9 tells us that $\ker \nu \subset J(\Lambda)$. Hence we have

$$[u(\bar{H}) : U_0] = [I^{(5)} : \ker \nu] = [J(\Lambda') : \ker \nu] \geq [J(\Lambda') : J(\Lambda)] = |D(\Lambda)|.$$

However, since $\bar{H} \cong \bar{Z}V$, we obtain $|u(\bar{H})| = |u(\bar{Z}V)| = 8$. On the other hand, $|\bar{U}_0| = 4$. Therefore, $|D(\Lambda)| \leq 2$, as claimed.

We are now ready to use the explicit definition:

$$J(\Lambda) = \{C \cdot N^*(x): x = \text{unit in } E(\Lambda_2 \dotplus \Lambda_2)\},$$

and begin by describing $E(\Lambda_2 \dotplus \Lambda_2)$ as a pullback. Indeed, from (9.5) we obtain a pullback diagram

$$\begin{array}{ccccc} E(\Lambda \dotplus \Lambda) & & \longrightarrow & & E(H \dotplus H) \\ \downarrow & & & & \downarrow \\ E(ZV \dotplus ZV) & \longrightarrow & E(\overline{ZV} \dotplus \overline{ZV}) & \longrightarrow & E(\overline{H} \dotplus \overline{H}) \end{array} \tag{9.10}$$

(It is obvious that this is a pullback, once we identify $E(\Lambda \dotplus \Lambda)$ with the matrix ring $(\Lambda)^{2\times 2}$, and so on.) Localizing (9.10) at the rational prime 2, we again obtain a pullback diagram

$$\begin{array}{ccccc} E(\Lambda_2 \dotplus \Lambda_2) & & \longrightarrow & & E(H_2 \dotplus H_2) \\ \downarrow & & & & \downarrow \\ E(Z_2V \dotplus Z_2V) & \longrightarrow & E(\overline{ZV} \dotplus \overline{ZV}) & \longrightarrow & E(\overline{H} \dotplus \overline{H}) \end{array} \tag{9.11}$$

To complete the proof of Theorem 9.1, it suffices to show that the element $w = (Z, Z, Z, Z, 3Z) \in J(\Lambda')$ does not lie in $J(\Lambda)$. Suppose to the contrary that $w \in J(\Lambda)$, so $w = Z^{(5)} \cdot N(x)$ for some unit $x \in E(\Lambda_2 \dotplus \Lambda_2)$. By (9.11), x corresponds to a pair of 2×2 matrices X, Y such that

$$\begin{cases} X = \text{unit in } (Z_2V)^{2\times 2}, \qquad Y = \text{unit in } (H_2)^{2\times 2}, \\ \gamma(\overline{X}) = \overline{Y} \quad \text{as matrices over } \overline{H}, \\ Z^{(4)} \cdot N(X) = (Z, Z, Z, Z), \qquad Z \cdot N(Y) = 3Z. \end{cases} \tag{9.12}$$

Since Z_2V is commutative, the reduced norm $N(X)$ is precisely $g(\det X)$. But $N(X)$ is a unit in $Z^{(4)}$, so $g(\det X) \in u(Z^{(4)})$; therefore (as shown in the proof of (9.9)), $\det X \in u(ZV)$. Hence $\gamma(\det \overline{X}) \in \gamma\{\overline{u(ZV)}\} = \overline{U}_0$. Therefore, we have

$$\det \overline{Y} \in \overline{U}_0. \tag{9.13}$$

Let us write $Y = \begin{pmatrix} \alpha & \beta \\ \gamma & \delta \end{pmatrix} \in (H_2)^{2\times 2}$. Since Y is invertible over H_2, it is also invertible over H'_2; but H'_2 is a local ring, so either α or γ must be a unit in H'_2. It follows from Lemma 3.3 that either α or γ is a unit in H_2. Suppose (without loss of generality) that $\alpha \in u(H_2)$. Then

$$N(Y) = N\begin{pmatrix} \alpha & \beta \\ \gamma & \delta \end{pmatrix} = N\begin{pmatrix} \alpha & \beta \\ 0 & \delta - \gamma\alpha^{-1}\beta \end{pmatrix} = N(\alpha\delta - \alpha\gamma\alpha^{-1}\beta).$$

The image of $\alpha\delta - \alpha\gamma\alpha^{-1}\beta$ in $\overline{H}$ is just $\det \overline{Y}$ (since $\overline{H}$ is commutative), and thus this image lies in $\overline{U}_0$ by (9.13). Therefore,

$$\alpha\delta - \alpha\gamma\alpha^{-1}\beta = u_0(1 + 2t) \quad \text{for some } t \in H_2.$$

But then

$$N(\alpha\delta - \alpha\gamma\alpha^{-1}\beta) = N(u_0) \cdot N(1 + 2t) = 1 + 2(t + t') + 4tt',$$

where t' is the conjugate of t in H. This shows that $N(Y)$ is a positive rational number, congruent to 1 (mod 4). The equation $Z \cdot N(Y) = 3Z$ is then impossible. This completes the proof that $J(\Lambda') \neq J(\Lambda)$, and that $D(\Lambda)$ has order 2.

REFERENCES

1. H. Bass, *Algebraic K-theory*, Math. Lecture Note Series, Benjamin, New York, 1968. MR 40 #2736.

2. N. Bourbaki, *Éléments de mathématique*. XXIII. Part I. *Les structures fondamentales de l'analyse*. Livre II: *Algèbre*. Chap. 8, Actualités Sci. Indust., no. 1261, Hermann, Paris, 1958. MR 20 #4576.

3. C. W. Curtis and I. Reiner, *Representation theory of finite groups and associative algebras*, Pure and Appl. Math., vol. XI, Interscience, New York, 1962; 2nd ed., 1966. MR 26 #2519.

4. M. Deuring, *Algebren*, Springer-Verlag, Berlin, 1935; Zweite korrigierte Auflage, Ergebnisse der Mathematik und ihrer Grenzgebiete, Band 41, 1968. MR 37 #4106.

5. M. Eichler, *Allgemeine Kongruenzklasseneinteilungen der Ideale einfachen Algebren über algebraischen Zahlkörpern und ihre L-Reihen*, J. Reine Angew. Math. 179 (1938), 227–251.

6. W. Feit, *Characters of finite groups*, Math. Lecture Note Series, Benjamin, New York, 1967. MR 36 #2715.

7. A. Fröhlich, *Ideals in an extension field as modules over the algebraic integers in a finite number field*, Math. Z. 74 (1960), 29–38. MR 22 #4708.

8. ———, *On the classgroup of integral group rings of finite Abelian groups*, Mathematika 16 (1969), 143–152. MR 41 #5512.

9. H. Jacobinski, *Genera and decompositions of lattices over orders*, Acta. Math. 121 (1968), 1–29. MR 40 #4294.

10. M. P. Lee, *Integral representations of dihedral groups of order 2p*, Trans. Amer. Math. Soc. 110 (1964), 213–231. MR 28 #139.

11. I. Reiner, *A survey of integral representation theory*, Bull. Amer. Math. Soc. 76 (1970), 159–227. MR 40 #7302.

12. K. W. Roggenkamp and V. H. Dyson, *Lattices over orders*. I, II, Lecture Notes in Math., nos. 115, 142, Springer-Verlag, Berlin and New York, 1970.

13. R. G. Swan, *Induced representations and projective modules*, Ann. of Math. (2) 71 (1960), 552–578. MR 25 #2131.

14. ———, *The Grothendieck ring of a finite group*, Topology 2 (1963), 85–110. MR 27 #3683.

15. R. G. Swan and E. G. Evans, *K-theory of finite groups and orders*, Lecture Notes in Math., no. 149, Springer-Verlag, Berlin and New York, 1970.

16. S. Ullom, *A note on the classgroup of integral group rings of some cyclic groups*, Mathematika 17 (1970), 79–81. MR 42 #4650.

17. A. Fröhlich, *The Picard group* (to appear).

18. I. Reiner and S. Ullom, *A Mayer-Vietoris sequence for class groups*, J. Algebra (to appear).

19. A. Fröhlich, I. Reiner and S. Ullom, *Picard groups and class groups* (to appear).

20. J. Martinet, *Modules sur l'algèbre du groupe quaternionien*, Ann. Sci. École Norm. Sup. 4 (1971), 399–408.

DEPARTMENT OF MATHEMATICS, UNIVERSITY OF ILLINOIS AT URBANA-CHAMPAIGN, URBANA, ILLINOIS 61801

A Mayer–Vietoris Sequence for Class Groups*

I. Reiner and S. Ullom

Department of Mathematics, University of Illinois, Urbana, Illinois 61801

Communicated by A. Fröhlich

Received December 14, 1971

1. Introduction

Let K be an algebraic number field, R a Dedekind domain with quotient field K, and Λ an R-order in a semisimple K-algebra A with 1. Let P range over all maximal ideals of R, and let R_P denote the localization of R at P, that is, the ring of P-integral elements of K. A *Λ-lattice* is a finitely generated left Λ-module which is R-torsionfree. The Λ-lattice X is called *locally free* if for each P, X_P is free as left Λ_P-module. Such a lattice X is automatically projective as left Λ-module. Let $\mathbf{P}(\Lambda)$ be the free abelian group generated by symbols $[X]$, one for each isomorphism class of locally free Λ-lattices X, modulo relations $[X] = [X'] + [X'']$ whenever $X \cong X' \dotplus X''$. Let $C(\Lambda)$ be the subgroup of $\mathbf{P}(\Lambda)$ consisting of all expressions $[X] - [Y]$, with X, Y locally free and $KX \cong KY$. We shall call $C(\Lambda)$ the *class group* of Λ, or more precisely, the "locally free class group" of Λ.

For example, when R is the ring of all algebraic integers in K, and $\Lambda = RG$ is the group ring of a finite group G, Swan [17] showed that *every* projective Λ-lattice is locally free. Hence, in this case, $C(\Lambda)$ is precisely the "reduced projective class group" of RG (see [15, 17]). In particular, $C(R)$ is the customary ideal class group of R.

Returning to the general case, an arbitrary R-order Λ may always be embedded in a maximal R-order Λ' in A. Since the locally free class group $C(\Lambda')$ is usually easy to calculate, it is of interest to compare $C(\Lambda)$ with $C(\Lambda')$. As shown by Swan [18] (see also Reiner–Ullom [16]), the map defined by $[X] \to [\Lambda' \otimes_\Lambda X]$ gives an epimorphism $C(\Lambda) \to C(\Lambda')$. Denote by $D(\Lambda)$ the kernel of this epimorphism, so that there is an exact sequence of finite abelian groups

$$0 \to D(\Lambda) \to C(\Lambda) \to C(\Lambda') \to 0.$$

* This research was partially supported by a contract with the National Science Foundation.

Jacobinski [9] proved that (up to isomorphism) $D(\Lambda)$ is independent of the choice of Λ'. We shall show here that the calculation of $D(\Lambda)$ can often be carried out by using an analog of Milnor's Mayer–Vietoris sequence (see Bass [1] and Milnor [12]). Milnor's sequence arises in the following context: Let us start with a pullback diagram of rings

$$\begin{array}{ccc} \Lambda & \xrightarrow{\psi_1} & \Lambda_1 \\ {\scriptstyle \psi_2}\downarrow & & \downarrow{\scriptstyle \varphi_1} \\ \Lambda_2 & \xrightarrow[\varphi_2]{} & \bar{\Lambda} \end{array}, \tag{1.1}$$

where each map is a ring homomorphism preserving the identity element. This means that we may identify Λ with a subring of $\Lambda_1 \dotplus \Lambda_2$, namely,

$$\Lambda = \{(\lambda_1, \lambda_2) : \lambda_1 \in \Lambda_1, \lambda_2 \in \Lambda_2, \varphi_1\lambda_1 = \varphi_2\lambda_2\}. \tag{1.2}$$

We call Λ the *pullback* or *fibre product* of Λ_1 and Λ_2, relative to the maps φ_1, φ_2. We shall assume throughout this paper that at least one of the maps φ_1 and φ_2 is surjective. Under this hypothesis, Milnor showed that the following Mayer–Vietoris sequence is exact

$$\begin{aligned} K_1(\Lambda) &\to K_1(\Lambda_1) \dotplus K_1(\Lambda_2) \to K_1(\bar{\Lambda}) \\ &\xrightarrow{\delta} K_0(\Lambda) \to K_0(\Lambda_1) \dotplus K_0(\Lambda_2) \to K_0(\bar{\Lambda}). \end{aligned} \tag{1.3}$$

In the special case where Λ is commutative, the above implies the exactness of the sequence

$$1 \to u(\Lambda) \to u(\Lambda_1) \times u(\Lambda_2) \to u(\bar{\Lambda}) \to \operatorname{Pic} \Lambda \to \operatorname{Pic}(\Lambda_1) \dotplus \operatorname{Pic}(\Lambda_2) \to \operatorname{Pic} \bar{\Lambda}, \tag{1.4}$$

where $u(\Lambda)$ denotes the group of units of the ring Λ, and where $\operatorname{Pic} \Lambda$ is the class group of the set of invertible Λ-modules, with multiplication given by $\otimes_\Lambda$.

Remark 1.5. Fibre products often arise as follows: Let I, J be two-sided ideals of the ring Λ. Then there is a pullback diagram

$$\begin{array}{ccc} \dfrac{\Lambda}{I \cap J} & \longrightarrow & \dfrac{\Lambda}{I} \\ \downarrow & & \downarrow{\scriptstyle \varphi_1} \\ \dfrac{\Lambda}{J} & \xrightarrow[\varphi_2]{} & \dfrac{\Lambda}{I + J} \end{array}, \tag{1.6}$$

where all of the maps are canonical. In this case, both φ_1 and φ_2 are surjective. In particular, when $I \cap J = 0$, we obtain the pullback diagram

$$\begin{array}{ccc} \Lambda & \longrightarrow & \dfrac{\Lambda}{I} \\ \downarrow & & \downarrow \varphi_1 \\ \dfrac{\Lambda}{J} & \xrightarrow{\varphi_2} & \dfrac{\Lambda}{I+J} \end{array} \quad . \tag{1.7}$$

Given a pullback diagram (1.1), we shall denote $\varphi_i\{u(\Lambda_i)\}$ by $u^*(\Lambda_i)$, $i = 1, 2$, so $u^*(\Lambda_i)$ is a subgroup of $u(\bar{\Lambda})$. Likewise, $GL_2{}^*(\Lambda_i)$ will denote the image of $GL_2(\Lambda_i)$ in $GL_2(\bar{\Lambda})$.

The following hypotheses will remain in force throughout this article.

Hypotheses 1.8. *Let R be a Dedekind ring whose quotient field K is an algebraic number field, and let A be a semisimple finite dimensional K-algebra. Let Λ be an R-order in A such that $K \cdot \Lambda = A$, and suppose that Λ is given by a pullback diagram* (1.1) *in which each Λ_i is an R-order in a K-algebra A_i, $i = 1, 2$. Assume that $\bar{\Lambda}$ is an R-torsion R-algebra, and that at least one of the maps φ_1 and φ_2 is surjective.*

We shall say that the algebra A satisfies the *Eichler condition* if no simple component of A is a totally definite quaternion algebra (see [9, 15]). In this case, we shall show that there is an exact sequence

$$1 \to u^*(\Lambda_1) \cdot u^*(\Lambda_2) \to u(\bar{\Lambda}) \to D(\Lambda) \to D(\Lambda_1) \dotplus D(\Lambda_2) \to 0. \tag{1.9}$$

When Λ is commutative, the algebra A automatically satisfies the Eichler condition, and it will follow from (1.9) that the sequence

$$1 \to u(\Lambda) \to u(\Lambda_1) \times u(\Lambda_2) \to u(\bar{\Lambda}) \to D(\Lambda) \to D(\Lambda_1) \dotplus D(\Lambda_2) \to 0 \tag{1.10}$$

is exact. Indeed, if $\bar{\Lambda}$ is commutative, and if either Λ_1 or Λ_2 is commutative, and if furthermore A satisfies the Eichler condition, then we shall show that (1.10) is exact.

Whether or not A satisfies the Eichler condition, we shall show the exactness of the sequence

$$1 \to GL_2{}^*(\Lambda_1) \cdot GL_2{}^*(\Lambda_2) \to GL_2(\bar{\Lambda}) \to D(\Lambda) \to D(\Lambda_1) \dotplus D(\Lambda_2) \to 0. \tag{1.11}$$

In particular, when $\bar{\Lambda}$ is commutative, we shall obtain an exact sequence

$$GL_2(\Lambda_1) \times GL_2(\Lambda_2) \xrightarrow{\overline{\det}} u(\bar{\Lambda}) \to D(\Lambda) \to D(\Lambda_1) \dotplus D(\Lambda_2) \to 0. \tag{1.12}$$

Section 7 will be devoted to several computational examples illustrating the usefulness of the above sequences. Finally, in Section 8, we shall prove the following result: As G ranges over any collection of abelian groups of composite order such that $|G| \to \infty$, also $|D(ZG)| \to \infty$. This was conjectured by Fröhlich [5] for G ranging over abelian p-groups, and was recently proved by him independently for that case.

2. Locally Free Lattices

In the products below, let P range over all maximal ideals of R, and let R_P be the localization of R at P, and $\hat{R}_P$ its P-adic completion. For the purposes of this article, it does not matter whether we use R_P or $\hat{R}_P$, the point being that if X is any left Λ-lattice then $R_P \otimes_R X$ is free if and only if $\hat{R}_P \otimes_R X$ is free (see [3]). With a slight abuse of terminology, therefore, we shall define the *idèle group* of Λ to be

$$\mathbf{J}(\Lambda) = \left\{ (a_P) \in \prod_P u(A) : a_P \in u(\Lambda_P) \text{ a.e.} \right\}.$$

Here, $u(A)$ denotes the group of units of A, and "a.e." means "almost everywhere," that is, for all but a finite number of maximal ideals P.

We embed $u(A)$ in $\mathbf{J}(\Lambda)$ diagonally, obtaining the subgroup A^* of principal idèles. Finally, let $U(\Lambda)$ be the group of unit idèles, defined by

$$U(\Lambda) = \prod u(\Lambda_P).$$

Now let X be any locally free left Λ-lattice in A. Then for each P, we may write $X_P = \Lambda_P x_P$ for some $x_P \in u(A)$, and of course x_P is determined only up to a left factor from $u(\Lambda_P)$. Hence X uniquely determines a coset $U(\Lambda) \cdot (x_P)$ in $\mathbf{J}(\Lambda)$. Further, if we are interested only in the Λ-isomorphism class of X, then we may replace X by Xa, where $a \in u(A)$. Hence the class of X uniquely determines the double coset $U(\Lambda) \cdot (x_P) \cdot A^*$ in $\mathbf{J}(\Lambda)$.

Conversely, given any such double coset, let (x_P) be any element thereof. Then $\Lambda_P x_P = \Lambda_P$ a.e., and hence if we set

$$X = \bigcap_P \Lambda_P x_P ,$$

then X is a Λ-lattice in A for which $X_P = \Lambda_P x_P$ (see [2]). This proves the following

Theorem 2.1. *There is a one-to-one correspondence between the set of*

Λ-isomorphism classes of locally free Λ-lattices in A, and the set of $(U(\Lambda), A^*)$-double cosets in $\mathbf{J}(\Lambda)$.

This result is well known, of course; see, for example, Faddeev [4] or Takahashi [20].

3. Fibre Products

As usual, we shall embed Λ in $K \otimes_R \Lambda$, and shall denote the latter ring by $K\Lambda$; thus $A = K\Lambda$. Likewise, we set $A_i = K\Lambda_i$, $i = 1, 2$. Now let P range over the maximal ideals of the Dedekind ring R. Keeping the hypotheses 1.8, we may "localize" diagram (1.1) by applying $R_P \otimes_R \cdot$ to each of its terms, so as to obtain a new pullback diagram. Thus, we have the identification

$$\Lambda_P = \{(\lambda_1, \lambda_2): \lambda_i \in (\Lambda_i)_P, \varphi_1\lambda_1 = \varphi_2\lambda_2\},$$

where now $\varphi_i : (\Lambda_i)_P \to \bar{\Lambda}_P$. On the other hand, $K \otimes_R \bar{\Lambda} = 0$, since $\bar{\Lambda}$ is R-torsion. Hence when we apply $K \otimes_R \cdot$ to (1.1), we get a pullback diagram

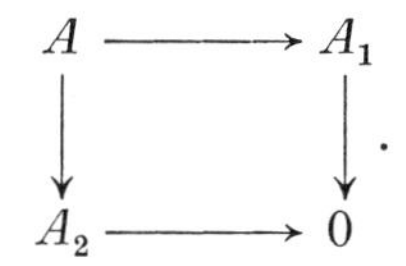

This shows that $A \cong A_1 \dotplus A_2$. Hereafter we shall identify A with $A_1 \dotplus A_2$; this is consistent with identifying Λ with a subring of $\Lambda_1 \dotplus \Lambda_2$, and there exist inclusions

$$\Lambda \subset \Lambda_1 \dotplus \Lambda_2 \subset A_1 \dotplus A_2 = A. \tag{3.1}$$

The maps $\psi_i : \Lambda \to \Lambda_i$ are then given by $\psi_i(\lambda_1, \lambda_2) = \lambda_i$, $i = 1, 2$. Note that A_1 and A_2 are algebra direct summands of A, so there exists a central idempotent $e \in A$ such that

$$A_1 = Ae, \qquad A_2 = A(1 - e).$$

Therefore, for $(\lambda_1, \lambda_2) \in \Lambda$, we have

$$\psi_1(\lambda_1, \lambda_2) = \lambda_1 e, \qquad \psi_2(\lambda_1, \lambda_2) = \lambda_2(1 - e).$$

We choose once and for all a maximal R-order Λ_i' in A_i containing Λ_i,

$i = 1, 2$, and set $\Lambda' = \Lambda_1' \dotplus \Lambda_2'$. Then Λ' is a maximal R-order in A containing Λ, and we have

$$\Lambda \subset \Lambda_1 \dotplus \Lambda_2 \subset \Lambda_1' \dotplus \Lambda_2' = \Lambda' \subset A.$$

Suppose now that X is a free Λ-lattice with free basis $\{b_k\}$. Clearly (for $i = 1, 2$), $\Lambda_i \otimes_\Lambda X$ is Λ_i-free with basis $\{1 \otimes b_k\}$, and $\bar{\Lambda} \otimes_\Lambda X$ is $\bar{\Lambda}$-free with basis $\{1 \otimes b_k\}$. There is a natural embedding of X in the A-module KX. Under this embedding, we may identify $\Lambda_1 \otimes_\Lambda X$ with $\Lambda_1 X$, the Λ_1-submodule of KX generated by $\{x_1 : (x_1, x_2) \in X\}$. Likewise, $\Lambda_2 \otimes_\Lambda X = \Lambda_2 X$, where $\Lambda_2 X$ is the Λ_2-submodule of KX generated by second components of elements of X. In particular, then,

$$\Lambda_1 X = \sum_k{}^{\cdot}\, \Lambda_1 \cdot e b_k, \qquad \Lambda_2 X = \sum_k{}^{\cdot}\, \Lambda_2 \cdot (1 - e)\, b_k,$$

which exhibits each $\Lambda_i X$ as a free Λ_i-submodule of KX.

The preceding discussion implies at once that if X is any projective Λ-lattice, not necessarily free, then each $\Lambda_i \otimes_\Lambda X$ is Λ_i-projective, and $\bar{\Lambda} \otimes_\Lambda X$ is $\bar{\Lambda}$-projective. Just as above, we may identify $\Lambda_i \otimes_\Lambda X$ with $\Lambda_i X$ computed within KX. Moreover, if X is locally free, then each $\Lambda_i \otimes_\Lambda X$ is a locally free Λ_i-lattice, and $\bar{\Lambda} \otimes_\Lambda X$ is a locally free $\bar{\Lambda}$-lattice. (Indeed, $\bar{\Lambda} \otimes_\Lambda X$ is $\bar{\Lambda}$-free, since $\bar{\Lambda}$ is an R-torsion R-algebra, and is therefore semilocal).

We have denoted by $C(\Lambda)$ the locally free class group of Λ. It follows from the above remarks that there is a well defined additive homomorphism

$$f' : C(\Lambda) \to C(\Lambda_1) \dotplus C(\Lambda_2),$$

given by

$$f'\{[X] - [Y]\} = ([\Lambda_1 \otimes_\Lambda X] - [\Lambda_1 \otimes_\Lambda Y], [\Lambda_2 \otimes_\Lambda X] - [\Lambda_2 \otimes_\Lambda Y]),$$

where X, Y are locally free Λ-lattices such that $KX \cong KY$. It is clear from the inclusions (3.1) and the identifications $\Lambda_i \otimes_\Lambda X \cong \Lambda_i \cdot X$, that the following diagram is commutative and has exact rows:

$$\begin{array}{ccccccccc}
0 & \longrightarrow & D(\Lambda) & \longrightarrow & C(\Lambda) & \longrightarrow & C(\Lambda') & \longrightarrow & 0 \\
 & & \downarrow f & & \downarrow f' & & \downarrow f'' & & \\
0 & \longrightarrow & D(\Lambda_1) \dotplus D(\Lambda_2) & \longrightarrow & C(\Lambda_1) \dotplus C(\Lambda_2) & \longrightarrow & C(\Lambda_1') \dotplus C(\Lambda_2') & \longrightarrow & 0.
\end{array} \tag{3.2}$$

Note that f'' is an isomorphism, and that f is induced by f'.

LEMMA 3.3. *The map* $f : D(\Lambda) \to D(\Lambda_1) \dotplus D(\Lambda_2)$ *defined above is an epimorphism.*

Proof. It suffices to show that f' is epic. Let Y be any locally free Λ-lattice; then Y is in the same genus as a free Λ-lattice. The arguments in Swan [17] (see also Swan–Evans [19]) then imply that

$$Y \cong \Lambda^{(m)} \dotplus X,$$

for some locally free Λ-lattice X in A. Furthermore, if X and X' are any pair of locally free Λ-lattices in A, there exists another such lattice X'' for which $X \dotplus X' \cong \Lambda \dotplus X''$. This implies at once that every element of $C(\Lambda)$ is expressible in the form $[X] - [\Lambda]$, with X some locally free Λ-lattice in A. (We may remark that $[X] - [\Lambda] = 0$ in $C(\Lambda)$ if and only if X is *stably isomorphic* to Λ, that is, $X \dotplus \Lambda^{(m)} \cong \Lambda \dotplus \Lambda^{(m)}$ for some m.)

Let us apply this discussion to the class groups $C(\Lambda_i)$, $i = 1, 2$. Each element of $C(\Lambda_1) \dotplus C(\Lambda_2)$ is then expressible as

$$([X_1] - [\Lambda_1], [X_2] - [\Lambda_2]),$$

with X_i a locally free Λ_i-lattice in A_i. In order to prove that f' is epic, we need only determine a locally free Λ-lattice X in A such that

$$\Lambda_1 \otimes_\Lambda X \cong X_1, \qquad \Lambda_2 \otimes_\Lambda X \cong X_2. \tag{3.4}$$

For each P, we may write

$$(X_1)_P = (\Lambda_1)_P \cdot x_P, \quad (X_2)_P = (\Lambda_2)_P \cdot y_P, \quad x_P \in u(A_1), \quad y_P \in u(A_2). \tag{3.5}$$

Then for almost all P, we have

$$x_P \in u((\Lambda_1)_P), \; y_P \in u((\Lambda_2)_P). \tag{3.6}$$

Now x_P and y_P are determined only up to unit factors; hence, at each maximal ideal P for which (3.6) holds, we may in fact choose $x_P = 1 \in \Lambda_1$, $y_P = 1 \in \Lambda_2$, and set $v_P = (x_P, y_P) = 1$ in Λ. On the other hand, at the finite number of P's where (3.6) fails to hold true, we use the fact that $A = A_1 \dotplus A_2$ in order to choose an element $v_P \in u(A)$ such that $v_P = (x_P, y_P)$. Thus $v_P \in u(A)$ for each P, and $v_P = (x_P, y_P) \in u(\Lambda_P)$ almost everywhere. Setting

$$X = \bigcap_P \Lambda_P v_P,$$

it follows from Section 2 that X is a Λ-lattice in A such that $X_P = \Lambda_P v_P$. Thus, X is locally free, and it remains for us to check (3.4). We already know that $\Lambda_1 \otimes_\Lambda X \cong \Lambda_1 X$, and we see readily that $\Lambda_1 X = X_1$ inside KX, since at each P we have the identifications

$$(\Lambda_1 X)_P = (\Lambda_1)_P X_P = (\Lambda_1)_P v_P = (\Lambda_1)_P x_P = (X_1)_P .$$

Thus $\Lambda_1 X = X_1$, and similarly $\Lambda_2 X = X_2$. Thus (3.4) is established, and the lemma is proved.

4. Action of Automorphisms on Fibre Products

Some of the material in the next two sections may be found in Bass [1] and Milnor [12]. For the benefit of the reader, however, we have included here most of the relevant proofs and details, especially since they are somewhat simpler in our special case. We shall continue to assume that hypotheses (1.8) hold, and we shall automatically make the identifications described in (3.1) and the paragraphs following (3.1).

Let X be any projective Λ-lattice, and apply the exact functor $\cdot \otimes_\Lambda X$ to the pullback diagram (1.1). This gives a new pullback diagram

$$\begin{array}{ccc} X & \longrightarrow & \Lambda_1 \otimes_\Lambda X \\ \downarrow & & \downarrow \\ \Lambda_2 \otimes_\Lambda X & \longrightarrow & \bar{\Lambda} \otimes_\Lambda X \end{array}, \tag{4.1}$$

in which $\Lambda_1 \otimes_\Lambda X$ is Λ_1-projective, $\Lambda_2 \otimes_\Lambda X$ is Λ_2-projective, and

$$\bar{\Lambda} \otimes_{\Lambda_2} (\Lambda_2 \otimes_\Lambda X) \cong \bar{\Lambda} \otimes_{\Lambda_1} (\Lambda_1 \otimes_\Lambda X). \tag{4.2}$$

We shall see below that the process may be reversed: starting with projective Λ_i-lattices X_i, $i = 1, 2$, such that

$$\theta\colon \bar{\Lambda} \otimes_{\Lambda_1} X_1 \cong \bar{\Lambda} \otimes_{\Lambda_2} X_2 , \tag{4.3}$$

we can construct a projective Λ-lattice X as the fibre product of X_1 and X_2 relative to θ. For the most part, we shall restrict our attention to the case of primary interest to us, namely that in which the modules involved are locally free.

Lemma 4.4. *For $i = 1, 2$, let X_i be a locally free Λ_i-lattice, and set $\bar{X}_i =$*

$\bar{\Lambda} \otimes_{\Lambda_i} X_i$. *Suppose that* $\bar{X}_1$ *and* $\bar{X}_2$ *are* $\bar{\Lambda}$*-isomorphic, and let* $\theta\colon \bar{X}_1 \cong \bar{X}_2$ *be a* $\bar{\Lambda}$*-isomorphism. Then define a* Λ*-lattice* X *by the pullback diagram*

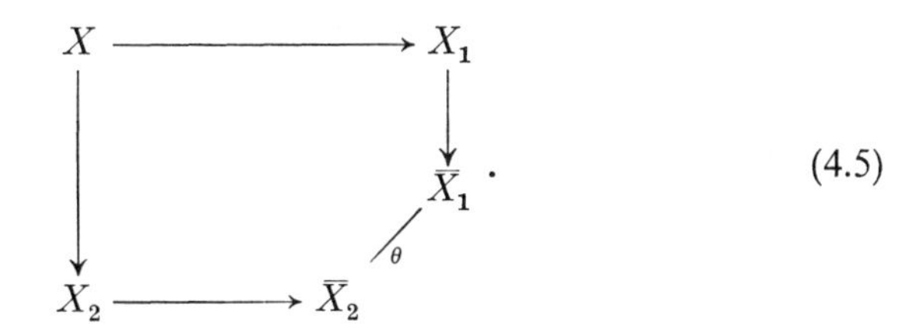

(4.5)

In other words, set

$$X = \{(x_1, x_2)\colon x_i \in X_i,\ \theta\bar{x}_1 = \bar{x}_2\}, \tag{4.6}$$

where $\bar{x}_i$ *is the image of* x_i *in* $\bar{X}_i$. *Then* X *is a locally free* Λ*-lattice such that* $\Lambda_i \otimes_\Lambda X \cong X_i$, $i = 1, 2$. *Indeed, when we identify* X *with a submodule of* $X_1 \dotplus X_2$, *and identify* $\Lambda_i \otimes_\Lambda X$ *with* $\Lambda_i X$ *in* KX, *we have equalities*

$$\Lambda_i \otimes_\Lambda X = \Lambda_i X = X_i, \qquad i = 1, 2.$$

Proof. The maps ψ_i, φ_i in (1.1) permit us to view each X_i and $\bar{X}_i$ as Λ-module. The isomorphism θ, and the maps $X_i \to \bar{X}_i$ $(i = 1, 2)$ in (4.5), are then Λ-homomorphisms. Hence, the pullback module X defined above is automatically a Λ-module. The action of Λ on X is given by the formula

$$(\lambda_1, \lambda_2)(x_1, x_2) = (\lambda_1 x_1, \lambda_2 x_2), \qquad (\lambda_1, \lambda_2) \in \Lambda, \qquad (x_1, x_2) \in X.$$

Since $\bar{X}_1$ and $\bar{X}_2$ are R-torsion, when we tensor (4.5) with K we obtain a pullback diagram

$$\begin{array}{ccc} KX & \longrightarrow & KX_1 \\ \downarrow & & \downarrow \\ KX_2 & \longrightarrow & 0 \end{array}.$$

Identifying KX with $KX_1 \dotplus KX_2$, we have embeddings

$$X \subset X_1 \dotplus X_2 \subset KX_1 \dotplus KX_2 = KX.$$

Furthermore, we shall identify $\Lambda_i \otimes_\Lambda X$ with $\Lambda_i X$, computed inside KX. Thus $\Lambda_1 X$ is the Λ_1-submodule of KX generated by

$$\{x_1 : (x_1, x_2) \in X \quad \text{for some} \quad x_2 \in X_2\}.$$

Clearly $\Lambda_1 X \subset X_1$. Likewise, we write $\Lambda_2 \otimes_\Lambda X = \Lambda_2 \cdot X \subset X_2$.

Now localize the diagram (4.5) at an arbitrary maximal ideal P of R. Rather than write the subscript P repeatedly, let us change notation and assume for the time being that R is a discrete valuation ring. Since each X_i was assumed locally free, we may write

$$X_1 = \sum_{i=1}^{n}{}^{\boldsymbol{\cdot}} \Lambda_1 u_i\,, \qquad X_2 = \sum_{j=1}^{m}{}^{\boldsymbol{\cdot}} \Lambda_2 v_j\,,$$

where the $\{u_i\}$ and $\{v_j\}$ are free bases. Therefore,

$$\bar{X}_1 = \sum_{i=1}^{n}{}^{\boldsymbol{\cdot}} \bar{\Lambda}\bar{u}_i\,, \qquad \bar{X}_2 = \sum_{j=1}^{m}{}^{\boldsymbol{\cdot}} \bar{\Lambda}\bar{v}_j\,,$$

where the $\{\bar{u}_i\}$ are the images in $\bar{X}_1$ of the elements $\{u_i\}$, and hence are a free $\bar{\Lambda}$-basis of $\bar{X}_1$. Likewise, the $\{\bar{v}_j\}$ are a free $\bar{\Lambda}$-basis of $\bar{X}_2$. By hypothesis, there exists a $\bar{\Lambda}$-isomorphism $\theta\colon \bar{X}_1 \cong \bar{X}_2$. Since the number of elements in a free $\bar{\Lambda}$-basis of $\bar{X}_1$ is an invariant of $\bar{X}_1$, this shows that $m = n$. Furthermore, the isomorphism θ can be specified by its action on the elements $\{\bar{u}_i\}$. We may write

$$\theta(\bar{u}_i) = \sum_{j=1}^{n} \xi_{ij}\bar{v}_j\,, \qquad 1 \leqslant i \leqslant n,$$

where (ξ_{ij}) is an invertible $n \times n$ matrix over $\bar{\Lambda}$. In other words, (ξ_{ij}) is an element of $GL_n(\bar{\Lambda})$.

Keeping the temporary hypothesis that R is a local ring, we now make use of the underlying hypothesis that at least one of the maps φ_1 and φ_2 is epic. Let us suppose that φ_2 is epic, noting that if instead φ_1 is epic, then we can give a corresponding proof using θ^{-1} in place of θ. Since Λ_2 is a semilocal ring, and $\varphi_2 : \Lambda_2 \to \bar{\Lambda}$ is epic, it follows easily (see Bass [1, p. 87]) that φ_2 induces an epimorphism $GL_n(\Lambda_2) \to GL_n(\bar{\Lambda})$. Hence there exists a Λ_2-automorphism σ of X_2 such that

$$\sigma(v_i) = \sum_{j=1}^{n} \sigma_{ij}v_j\,, \quad 1 \leqslant i \leqslant n, \qquad \text{where} \qquad \bar{\sigma}_{ij} = \xi_{ij}\,, \quad 1 \leqslant i, j \leqslant n.$$

Setting $w_i = \sigma(v_i)$, $1 \leqslant i \leqslant n$, we have

$$X_2 = \sum_{i=1}^{n}{}^{\boldsymbol{\cdot}} \Lambda_2 w_i\,, \qquad \theta(\bar{u}_i) = \bar{w}_i\,, \quad 1 \leqslant i \leqslant n.$$

It follows from (4.6) that the elements $\{(u_i\,, w_i)\colon 1 \leqslant i \leqslant n\}$ form a free Λ-basis for X. Furthermore, $\Lambda_1 X$ is the Λ_1-submodule of KX generated by the elements $\{u_i\}$, so $\Lambda_1 X = X_1$. Likewise $\Lambda_2 X = X_2$ in KX.

Returning to the global case, let R be any Dedekind ring. The above discussion shows that if X_1 and X_2 are locally free, and $\bar{X}_1 \cong \bar{X}_2$, then the Λ-module X defined by (4.6) is also locally free. Furthermore, $(\Lambda_i X)_P = (X_i)_P$ in KX, for $i = 1, 2$, and for all P. Therefore $\Lambda_i X = X_i$, $i = 1, 2$, and the lemma is established.

Remark 4.7. It is of interest to sketch a proof of Milnor's [12] slightly stronger version of the above lemma. Let X_1, X_2 be any projective lattices, let $\theta\colon \bar{X}_1 \cong \bar{X}_2$, and define X by (4.6). Milnor proves then that X is a projective Λ-lattice such that

$$\Lambda_i \otimes_\Lambda X \cong X_i, \qquad i = 1, 2. \tag{4.8}$$

To begin the proof, choose projective lattices Y_1, Y_2, such that $X_i \dotplus Y_i$ is Λ_i-free, $i = 1, 2$, and such that there is a $\bar{\Lambda}$-isomorphism $\eta\colon \bar{Y}_1 \cong \bar{Y}_2$. Relative to free $\bar{\Lambda}$-bases of $\bar{X}_1 \dotplus \bar{Y}_1$ and $\bar{X}_2 \dotplus \bar{Y}_2$, the $\bar{\Lambda}$-isomorphism (θ, η) determines an invertible matrix ξ over $\bar{\Lambda}$. Now choose free Λ_i-lattices $Z_i \cong X_i \dotplus Y_i$, $i = 1, 2$, and define a $\bar{\Lambda}$-isomorphism $\bar{Z}_1 \cong \bar{Z}_2$ by means of the matrix ξ^{-1}. Then $X_i \dotplus Y_i \dotplus Z_i$ is a free Λ_i-module, $i = 1, 2$, and there is a $\bar{\Lambda}$-isomorphism

$$\theta'\colon \bar{X}_1 \dotplus \bar{Y}_1 \dotplus \bar{Z}_1 \cong \bar{X}_2 \dotplus \bar{Y}_2 \dotplus \bar{Z}_2,$$

described by the matrix $\left(\begin{smallmatrix} \xi & 0 \\ 0 & \xi^{-1} \end{smallmatrix}\right)$. This latter matrix is easily expressed as a product of elementary matrices over $\bar{\Lambda}$. If we assume that $\varphi_2 : \Lambda_2 \to \bar{\Lambda}$ is epic, it is then clear that there exists an invertible matrix σ over Λ_2 which maps onto the matrix $\left(\begin{smallmatrix} \xi & 0 \\ 0 & \xi^{-1} \end{smallmatrix}\right)$. The reasoning given in the proof of Lemma 4.4 (for the local case) then shows that the fibre product F of $X_1 \dotplus Y_1 \dotplus Z_1$ and $X_2 \dotplus Y_2 \dotplus Z_2$, relative to θ', is a free Λ-lattice such that

$$\Lambda_i \otimes_\Lambda F \cong X_i \dotplus Y_i \dotplus Z_i, \qquad i = 1, 2.$$

On the other hand, $F \cong X \dotplus X'$, where X' is the fibre product of the pair $Y_1 \dotplus Z_1$ and $Y_2 \dotplus Z_2$, relative to the isomorphism (η, ξ^{-1}). Hence X is Λ-projective, and (4.8) holds true.

Remark 4.9. In proving Lemma 3.3, the crucial step was to show that given locally free Λ_i-lattices X_i in A_i, $i = 1, 2$, there exists a locally free Λ-lattice X in A such that $\Lambda_i \otimes_\Lambda X \cong X_i$, $i = 1, 2$. This result could have been obtained from Lemma 4.3 as follows: Since X_1 and X_2 are locally free of rank 1, the $\bar{\Lambda}$-modules $\bar{X}_1$ and $\bar{X}_2$ are locally isomorphic at each P. However, since $\bar{\Lambda}$ is a semilocal ring, this implies that there is a $\bar{\Lambda}$-isomorphism $\theta\colon \bar{X}_1 \cong \bar{X}_2$. The Λ-lattice X defined by (4.6) then has the desired properties. It should be noted, however, that in the proof of

Lemma 4.4 we used the hypothesis that either φ_1 or φ_2 is epic, whereas the proof of (3.3) did not require this hypothesis.

Let us return to the situation given in Lemma 4.4, where X is the fibre product of a pair of locally free lattices X_1 and X_2 relative to an isomorphism $\theta: \bar{X}_1 \cong \bar{X}_2$. Let Aut $\bar{X}_2$ denote the group of $\bar{\Lambda}$-automorphisms of $\bar{X}_2$. Then for each $\beta \in \operatorname{Aut} \bar{X}_2$, we may define a new Λ-lattice βX by the pullback diagram[1]

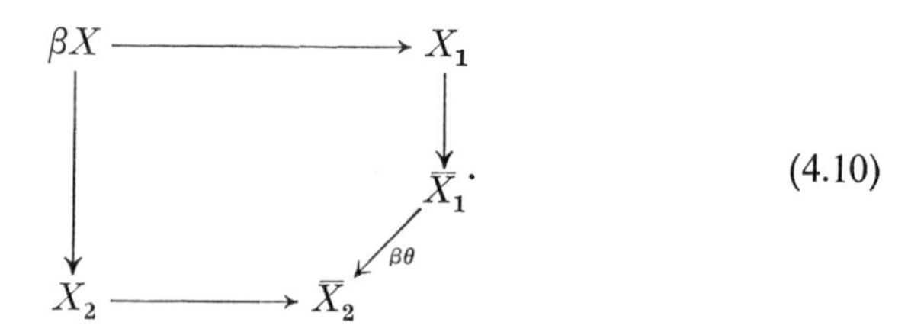

(4.10)

Thus (by definition)

$$\beta X = \{(x_1, x_2): x_i \in X_i, \beta\theta\bar{x}_1 = \bar{x}_2\}. \tag{4.11}$$

It follows immediately from Lemma 4.4 that βX is also a locally free Λ-lattice, and that

$$\Lambda_i \otimes_\Lambda \beta X \cong \Lambda_i \cdot \beta X = X_i, \qquad i = 1, 2.$$

Given another automorphism $\beta' \in \operatorname{Aut} \bar{X}_2$, we may apply β' to (4.10) so as to define a new fibre product $\beta'(\beta X)$. It is then clear that

$$\beta'(\beta X) \cong (\beta'\beta)X, \qquad \beta, \beta' \in \operatorname{Aut} \bar{X}_2,$$

and indeed $\beta'(\beta X) = (\beta'\beta)X$ as submodules of KX, after the canonical identifications.

We shall apply the above construction to the following case: let X be a given locally free Λ-lattice, and set

$$X_i = \Lambda_i \otimes_\Lambda X, \qquad \bar{X}_i = \bar{\Lambda} \otimes_{\Lambda_i} X_i, \qquad i = 1, 2. \tag{4.12}$$

There is then a canonical isomorphism $\theta: \bar{X}_1 \cong \bar{X}_2$, given by composition of maps

$$\begin{aligned} \bar{X}_1 &= \bar{\Lambda} \otimes_{\Lambda_1} (\Lambda_1 \otimes_\Lambda X) \cong \bar{\Lambda} \otimes_\Lambda X \\ &\cong \bar{\Lambda} \otimes_{\Lambda_2} (\Lambda_2 \otimes_\Lambda X) = \bar{X}_2. \end{aligned}$$

[1] We caution the reader that we have defined an action of Aut $\bar{X}_2$ on fibre products of the form (4.5), rather than on Λ-lattices X themselves.

The pullback diagram (4.1) may be rewritten in the form

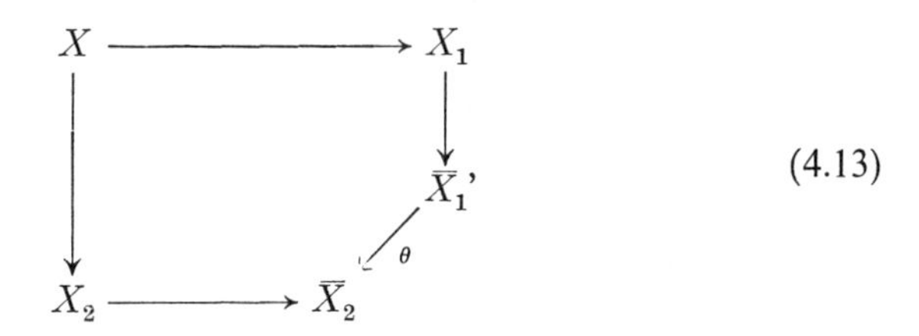

so in fact X is represented as a pullback of a pair of locally free lattices X_1 and X_2. For each $\beta \in \operatorname{Aut} \bar{X}_2$, define a new Λ-lattice βX by means of (4.10) and (4.11). It follows from Lemma 4.4 that each such βX is also a locally free Λ-lattice, and that there are equalities

$$\Lambda_i \cdot \beta X = \Lambda_i \cdot X = X_i, \qquad i = 1, 2,$$

holding true in KX.

LEMMA 4.14. *Let X and X' be locally free Λ-lattices, and define X_i', $\bar{X}_i'$ by formulas* (4.12), *with X' in place of X. Given Λ_i-isomorphisms $f_i : X_i \cong X_i'$, $i = 1, 2$, there is a Λ-isomorphism $X' \cong \beta X$ for some $\beta \in$* Aut $\bar{X}_2$.

Further, let B_2 denote the image of Aut X_2 *in* Aut $\bar{X}_2$, *and let B_1 denote the image of* Aut X_1 *in* Aut $\bar{X}_2$ *under the composite mapping*

$$\operatorname{Aut} X_1 \longrightarrow \operatorname{Aut} \bar{X}_1 \xrightarrow{\theta^*} \operatorname{Aut} \bar{X}_2,$$

where θ^* is induced by the canonical isomorphism $\theta: \bar{X}_1 \cong \bar{X}_2$. Let β, $\beta' \in$ Aut $\bar{X}_2$. *Then*

$$\beta X \cong \beta' X \text{ if and only if } \beta' \in B_2 \cdot \beta \cdot B_1. \tag{4.15}$$

Proof. Let X and X' be a given pair of locally free lattices, such that there exist Λ_i-isomorphisms $f_i : X_i \cong X_i'$, $i = 1, 2$. By (4.11) we may write

$$X = \{(x_1, x_2): x_i \in X_i, \theta\bar{x}_1 = \bar{x}_2\}, \qquad X' = \{(y_1, y_2): y_i \in X_i', \theta'\bar{y}_1 = \bar{y}_2\}.$$

Now each f_i induces a $\bar{\Lambda}$-isomorphism $\bar{f}_i : \bar{X}_i \cong \bar{X}_i'$, $i = 1, 2$. Hence we have

$$\begin{aligned} X' &= \{(f_1x_1, f_2x_2) : x_i \in X_i, \theta'(\overline{f_1x_1}) = \overline{f_2x_2}\} \\ &= \{(f_1x_1, f_2x_2) : x_i \in X_i, \theta'\bar{f}_1\bar{x}_1 = \bar{f}_2\bar{x}_2\}. \end{aligned}$$

Define $\bar{g}: \bar{X}_2 \cong \bar{X}_2'$ by setting $\bar{g}\theta = \theta'\bar{f}_1$, and let $\beta = \bar{f}_2^{-1} \cdot \bar{g} \in \operatorname{Aut} \bar{X}_2$. Then we may write

$$\begin{aligned} X' &= \{(f_1x_1, f_2x_2): x_i \in X_i, \bar{g}\theta\bar{x}_1 = \bar{f}_2\bar{x}_2\} \\ &= \{(f_1x_1, f_2x_2): x_i \in X_i, \beta\theta\bar{x}_1 = \bar{x}_2\}. \end{aligned}$$

It is then clear that the map defined by $(x_1, x_2) \to (f_1 x_1, f_2 x_2)$ gives the desired Λ-isomorphism $\beta X \cong X'$.

Turning to the second assertion in the lemma, let $\beta, \beta' \in \operatorname{Aut} \bar{X}_2$ and let there be given a Λ-isomorphism $\rho: \beta X \cong \beta' X$. Using the identifications

$$\Lambda_i \cdot \beta X = \Lambda_i \cdot \beta X' = X_i \text{ in } KX, \qquad i = 1, 2,$$

it follows at once that ρ induces automorphisms $f_i \in \operatorname{Aut} X_i$, $i = 1, 2$. But then

$$\rho(x_1, x_2) = (f_1 x_1, f_2 x_2), \qquad (x_1, x_2) \in \beta X.$$

Since $\rho(x_1, x_2) \in \beta' X$, we conclude that

$$\beta' \theta(\bar{f}_1 \bar{x}_1) = \bar{f}_2 \bar{x}_2 \qquad \text{for all} \quad (x_1, x_2) \in \beta X.$$

However, $\beta\theta\bar{x}_1 = \bar{x}_2$ whenever $(x_1, x_2) \in \beta X$, and consequently

$$\beta' \theta(\bar{f}_1 \bar{x}_1) = \bar{f}_2(\beta\theta\bar{x}_1) \qquad \text{whenever} \quad (x_1, x_2) \in \beta X.$$

Thus the $\bar{\Lambda}$-isomorphisms $\beta'\theta\bar{f}_1$ and $\bar{f}_2\beta\theta$ agree on all elements $\bar{x}_1$, where x_1 ranges over all first components of elements of βX. Since $\Lambda_1 \cdot \beta X = X_1$, these elements x_1 generate X_1 as Λ_1-module, and hence the elements $\bar{x}_1$ generate $\bar{X}_1$ as $\bar{\Lambda}$-module. Therefore

$$\beta'\theta\bar{f}_1 = \bar{f}_2\beta\theta \text{ on } \bar{X}_1,$$

that is,

$$\beta' = \bar{f}_2 \cdot \beta \cdot \theta\bar{f}_1^{-1}\theta^{-1} = \bar{f}_2 \cdot \beta \cdot (\theta^*\bar{f}_1^{-1}),$$

where the last equality follows from the definition of θ^*. We have thus shown that if $\beta X \cong \beta' X$, then $\beta' \in B_2 \cdot \beta \cdot B_1$. All of the above steps are reversible, so the proof of the lemma is completed.

Remarks 4.16. (i) Let X be any projective Λ-lattice, not necessarily locally free. Then X is given by a fibre product (4.1), with each $\Lambda_i \otimes X$ a projective Λ_i-lattice, $i = 1, 2$. If both $\Lambda_1 \otimes X$ and $\Lambda_2 \otimes X$ are locally free, it follows from (4.4) that also X is locally free. We have thus proved:

If in addition to the hypotheses (1.8) *we assume that every projective Λ_i-lattice is locally free, $i = 1, 2$, then also every projective Λ-lattice is locally free.*

(ii) The proof of Lemma 4.14 carries over unchanged to the more general situation in which X and X' are merely assumed to be projective Λ-lattices, not necessarily locally free. One need only use (4.7) in place of (4.4) in the relevant parts of the proof.

LEMMA 4.17. *Let Λ' be the maximal R-order in A containing Λ defined in* (3.1), *and let X be any locally free Λ-lattice such that $\Lambda' \otimes_\Lambda X$ is Λ'-free. Define $\bar{X}_2$ as in* (4.12). *Then for each $\beta \in \operatorname{Aut} \bar{X}_2$, the lattice $\Lambda' \otimes_\Lambda \beta\Lambda$ is also Λ'-free.*

Proof. We have the identifications in KX:

$$\Lambda' \otimes_\Lambda \beta X = \Lambda' \cdot \beta X, \qquad \Lambda' \otimes_\Lambda X = \Lambda' X.$$

Then

$$\begin{aligned}\Lambda' \cdot \beta X &= (\Lambda_1' \dotplus \Lambda_2') \cdot \beta X = \Lambda_1' \cdot \beta X \dotplus \Lambda_2' \cdot \beta X \\ &= \Lambda_1' \cdot \Lambda_1 \cdot \beta X \dotplus \Lambda_2' \cdot \Lambda_2 \cdot \beta X = \Lambda_1' \cdot \Lambda_1 X \dotplus \Lambda_2' \cdot \Lambda_2 X \\ &= \Lambda' \cdot X.\end{aligned}$$

(We have used the previously established result that $\Lambda_i \cdot \beta X = \Lambda_i \cdot X$ in KX). This completes the proof.[2]

Finally, we show

LEMMA 4.18. *Let X be any locally free* Λ-lattice, and define $\bar{X}_2$ by* (4.12). *Then for any $\beta, \gamma \in \operatorname{Aut} \bar{X}_2$, there are Λ-isomorphisms*

$$\beta X \dotplus \gamma X \cong X \dotplus \beta\gamma X \cong X \dotplus \gamma\beta X.$$

Proof. Let $Y = X \dotplus X$, and define Y_2, $\bar{Y}_2$ as in (4.12), with Y in place of X. Set

$$\sigma = \begin{pmatrix} \beta & 0 \\ 0 & \gamma \end{pmatrix}, \qquad \tau = \begin{pmatrix} 1 & 0 \\ 0 & \beta\gamma \end{pmatrix} \in \operatorname{Aut} \bar{Y}_2 .$$

Let End $\bar{X}_2$ be the ring of $\bar{\Lambda}$-endomorphisms of $\bar{X}_2$. Without loss of generality assume that $\varphi_2 : \Lambda_2 \to \bar{\Lambda}$ is epic; then the induced map $X_2 \to \bar{X}_2$ is also epic. But then every $\bar{\Lambda}$-endomorphism of $\bar{X}_2$ lifts to a Λ_2-endomorphism of X_2, since X_2 is Λ_2-projective. This shows that End X_2 maps onto End $\bar{X}_2$.

It is easily seen that there exists a product π of 2×2 elementary matrices over End $\bar{X}_2$, such that $\sigma = \pi\tau$. Each of these elementary matrices is the image of a corresponding elementary matrix over End X_2. Hence there exists an element $\rho \in \operatorname{Aut} Y_2$ which induces the automorphism π. It follows from (4.15) that $\sigma Y \cong \tau Y$ as Λ-lattices. However,

$$\sigma Y \cong \beta X \dotplus \gamma X, \qquad \tau Y \cong X \dotplus \beta\gamma X,$$

and so the proof is complete.

[2] The same proof works if we assume X to be Λ-projective rather than locally free.

We shall apply all of the above discussion to the case where $X = \Lambda$, identifying both $\bar{X}_1$ and $\bar{X}_2$ with $\bar{\Lambda}$. Let $\operatorname{Aut} \bar{\Lambda}$ denote the group of automorphisms of the left $\bar{\Lambda}$-module $\bar{\Lambda}$. For each unit $u \in u(\bar{\Lambda})$, let r_u be the right multiplication by u on $\bar{\Lambda}$. There is then an anti-isomorphism $u(\bar{\Lambda}) \to \operatorname{Aut} \bar{\Lambda}$, given by the map $u \to r_u$. The Λ-lattice $r_u\Lambda$ defined as in (4.11) will therefore be denoted instead by the symbol Λu. Explicitly, we have

$$\Lambda u = \{(\lambda_1, \lambda_2) : \lambda_i \in \Lambda_i, (\varphi_1\lambda_1)u = \varphi_2\lambda_2\}, \tag{4.19}$$

where φ_1, φ_2 are the maps occurring in (1.1). Let us collect all of the information so far obtained about these lattices $\{\Lambda u\}$ into a single lemma.

LEMMA 4.20. *For each $u \in u(\bar{\Lambda})$, define Λu by formula* (4.19). *Then*

(i) *Each Λu is a locally free Λ-lattice such that*

$$\Lambda_i \otimes_\Lambda \Lambda u \cong \Lambda_i \cdot \Lambda u = \Lambda_i, \qquad i = 1, 2,$$

the last equality holding true in the K-algebra A.

(ii) *For $u, u' \in u(\bar{\Lambda})$, we have*

$$\Lambda u \dotplus \Lambda u' \cong \Lambda \dotplus \Lambda uu' \cong \Lambda \dotplus \Lambda u'u.$$

(iii) *Let Λ' be a maximal R-order in A containing Λ, defined as in* (3.1). *Then*

$$\Lambda' \otimes_\Lambda \Lambda u \cong \Lambda' \cdot \Lambda u = \Lambda',$$

the equality holding true inside A.

(iv) *Let X be any projective Λ-lattice such that $\Lambda_i \otimes_\Lambda X \cong \Lambda_i$, $i = 1, 2$. Then $X \cong \Lambda u$ for some $u \in u(\bar{\Lambda})$.*

(v) *Let $u^*(\Lambda_i)$ denote the image of $u(\Lambda_i)$ in $u(\bar{\Lambda})$, and let $u, u' \in u(\bar{\Lambda})$. Then $\Lambda u \cong \Lambda u'$ if and only if $u' \in u^*(\Lambda_1) \cdot u \cdot u^*(\Lambda_2)$.*

Proof. Part (i) is a special case of (4.4), part (ii) a special case of (4.18), and part (iii) a special case of (4.17). Part (iv) is a special case of the first assertion in Lemma 4.14, once we take the remarks in (4.16) into account. Finally, part (v) is a special case of (4.15), except that we must reverse the order of the factors on the right hand side of (4.15) because of the anti-isomorphism between $\operatorname{Aut} \bar{\Lambda}$ and $u(\bar{\Lambda})$.

We are now ready to define a connecting homomorphism

$$\partial : u(\bar{\Lambda}) \to D(\Lambda),$$

by setting

$$\partial u = [\Lambda u] - [\Lambda]. \tag{4.21}$$

Here, $D(\Lambda)$ is the kernel of the surjection $C(\Lambda) \to C(\Lambda')$, where Λ' is a maximal R-order containing Λ, defined as in (3.1). It is clear from (4.20, iii) that $\partial u \in D(\Lambda)$. From (4.20, ii) we obtain

$$\partial(uu') = \partial u + \partial u', \qquad u, u' \in u(\bar{\Lambda}),$$

so ∂ is indeed a homomorphism. This homomorphism ∂ will play a vital role in the next section.

5. Analog of the Mayer–Vietoris Sequence

Keeping the hypotheses (1.8), let us assume to begin with that A satisfies the Eichler condition. Then so do A_1 and A_2, since $A = A_1 \oplus A_2$. In this case, Jacobinski [9] showed that cancellation is possible, that is, if X and Y are projective Λ-lattices such that $X \dotplus \Lambda^{(r)} \cong Y \dotplus \Lambda^{(r)}$, then $X \cong Y$. Each element of $C(\Lambda)$ is then of the form $[X] - [\Lambda]$, with X a locally free Λ-lattice in A; further, $[X] - [\Lambda] = 0$ if and only if $X \cong \Lambda$. Thus the order of $C(\Lambda)$ equals the number of Λ-isomorphism classes of locally free lattices in A.

Let $u^*(\Lambda_i)$ denote the image of $u(\Lambda_i)$ in $u(\bar{\Lambda})$, $i = 1, 2$. We shall prove the exactness of the following sequence:

$$u^*(\Lambda_1) \cdot u^*(\Lambda_2) \to u(\bar{\Lambda}) \xrightarrow{\partial} D(\Lambda) \xrightarrow{f} D(\Lambda_1) \dotplus D(\Lambda_2) \to 0, \tag{5.1}$$

where ∂ is defined by formula (4.21). We have shown in (3.3) that f is epic. Further, we proved in (4.20, i) that $\Lambda_i \otimes_\Lambda \Lambda u \cong \Lambda_i$, $i = 1, 2$, so $f \cdot \partial = 0$. Next, let $[X] - [\Lambda] \in \ker f$, where X is a locally free Λ-lattice in A. Then $\Lambda_i \otimes_\Lambda X \cong \Lambda_i$, $i = 1, 2$, so by (4.20, iv) we have $X \cong \Lambda u$ for some $u \in u(\bar{\Lambda})$. Therefore $[X] - [\Lambda] = \partial u$, which completes the proof that $\ker f = \operatorname{im} \partial$.

Let us now determine the kernel of ∂. For $u \in u(\bar{\Lambda})$, we have $\partial u = 0$ if and only if $\Lambda u \cong \Lambda$. But by (4.20, v), $\Lambda u \cong \Lambda$ if and only if $u \in u^*(\Lambda_1)\, u^*(\Lambda_2)$. Thus the sequence (5.1) is exact.

From the exactness of (5.1), we deduce at once that $u^*(\Lambda_1)\, u^*(\Lambda_2)$ is a normal subgroup of $u(\bar{\Lambda})$, a fact which was not *a priori* obvious. Since

$$C(\Lambda') \cong C(\Lambda_1) \dotplus C(\Lambda_2),$$

we may also conclude that

$$|C(\Lambda)|/|C(\Lambda_1)|\,|C(\Lambda_2)| = [u(\bar{\Lambda}) : u^*(\Lambda_1) \cdot u^*(\Lambda_2)].$$

Keeping the hypothesis that A satisfies the Eichler condition, let us temporarily make the simplifying assumption that $\bar{\Lambda}$ is commutative, and assume in addition that either Λ_1 or Λ_2 is commutative, say Λ_1.

If ψ_1 and ψ_2 are the maps given in (1.1), then we may define a group homomorphism $\psi : u(\Lambda) \to u(\Lambda_1) \times u(\Lambda_2)$ by setting

$$\psi(u) = (\psi_1 u^{-1}, \psi_2 u), \qquad u \in u(\Lambda).$$

Likewise, there is a homomorphism $\varphi : u(\Lambda_1) \times u(\Lambda_2) \to u(\bar{\Lambda})$, given by $\varphi(u_1, u_2) = \bar{u}_1 \bar{u}_2$. We claim that the sequence

$$1 \to u(\Lambda) \xrightarrow{\psi} u(\Lambda_1) \times u(\Lambda_2) \xrightarrow{\varphi} u(\bar{\Lambda}) \xrightarrow{\partial} D(\Lambda) \to D(\Lambda_1) \dotplus D(\Lambda_2) \to 0 \tag{5.2}$$

is exact. Since the image of φ equals $u^*(\Lambda_1)\, u^*(\Lambda_2)$, it follows from (5.1) that the sequence is exact at $u(\bar{\Lambda})$. Further, if $\varphi(u_1, u_2) = 1$, then $\bar{u}_1^{-1} = \bar{u}_2$, so by (1.2) there exists an element $u \in \Lambda$ such that $\psi_1(u) = u_1^{-1}$, $\psi_2(u) = u_2$. But then u is a unit in Λ, and $\psi(u) = (u_1, u_2)$. The argument is reversible, and proves that $\ker \varphi = \operatorname{im} \psi$. Finally, it is clear that ψ is monic.

Passing now to the general case where Λ need not satisfy the Eichler condition, let us set $E = \operatorname{End}_\Lambda(\Lambda \dotplus \Lambda)$, $E_i = \operatorname{End}_{\Lambda_i}(\Lambda_i \dotplus \Lambda_i)$, $\bar{E} = \operatorname{End}_{\bar{\Lambda}}(\bar{\Lambda} \dotplus \bar{\Lambda})$. Since we may view E as the ring of all 2×2 matrices over Λ, it is clear that E is expressible as a fibre product

$$\begin{array}{ccc} E & \longrightarrow & E_1 \\ \downarrow & & \downarrow \\ E_2 & \longrightarrow & \bar{E} \end{array} .$$

Furthermore, E always satisfies the Eichler condition, and as shown by Jacobinski [9], $C(E) \cong C(\Lambda)$, $D(E) \cong D(\Lambda)$, $D(E_i) \cong D(\Lambda_i)$, $i = 1, 2$. Let us write $u(\bar{E})$ in more conventional notation as $GL_2(\bar{\Lambda})$, and let $GL_2{}^*(\Lambda_i)$ denote the image of $GL_2(\Lambda_i)$ in $GL_2(\bar{\Lambda})$. Then by (5.1) there is an exact sequence

$$1 \to GL_2{}^*(\Lambda_1) \cdot GL_2{}^*(\Lambda_2) \to GL_2(\bar{\Lambda}) \xrightarrow{\partial'} D(\Lambda) \to D(\Lambda_1) \dotplus D(\Lambda_2) \to 0. \tag{5.3}$$

If we make the simplifying assumption that $\bar{\Lambda}$ is commutative, we can improve on this as follows. Let us denote the composite map

$$GL_2(\Lambda_i) \longrightarrow GL_2(\bar{\Lambda}) \xrightarrow{\det} u(\bar{\Lambda})$$

by $\overline{\det}$. Let us assume that the map φ_1 is surjective, and now suppose $M \in GL_2(\bar{\Lambda})$. Since $\bar{\Lambda}$ is a semilocal ring, we can find[3] a product M_1 of elementary matrices over $\bar{\Lambda}$, and a unit $u \in u(\bar{\Lambda})$, such that

$$M = \begin{pmatrix} 1 & 0 \\ 0 & u \end{pmatrix} M_1 .$$

[3] See [19, Proposition 8.5].

But M_1 clearly lies in $GL_2{}^*(\Lambda_1)$, so

$$\begin{aligned}\partial' M = \partial' \begin{pmatrix}1 & 0\\ 0 & u\end{pmatrix} + \partial' M_1 &= \left[(\Lambda \dotplus \Lambda)\begin{pmatrix}1 & 0\\ 0 & u\end{pmatrix}\right] - [\Lambda \dotplus \Lambda]\\ &= [\Lambda \dotplus \Lambda u] - [\Lambda \dotplus \Lambda] = [\Lambda u] - [\Lambda]\\ &= \partial u.\end{aligned}$$

Here, $\partial : u(\bar{\Lambda}) \to D(\Lambda)$ is our previously defined map and we have just shown that im $\partial' \subset$ im ∂. The reverse inclusion is obvious, and so the sequence

$$u(\bar{\Lambda}) \xrightarrow{\partial} D(\Lambda) \to D(\Lambda_1) \dotplus D(\Lambda_2) \to 0$$

is exact in this case.

Still assuming that $\bar{\Lambda}$ is commutative, let us show that ker $\partial =$ im $\{\overline{\det}\}$, where

$$\overline{\det} : GL_2(\Lambda_1) \times GL_2(\Lambda_2) \to u(\bar{\Lambda}).$$

If $u \in u(\bar{\Lambda})$ and $\partial u = 0$, then $\partial'(\begin{smallmatrix}1 & 0\\ 0 & u\end{smallmatrix}) = 0$. It follows from (5.3) that there exist matrices $N_i \in GL_2(\Lambda_i)$, $i = 1, 2$, such that $(\begin{smallmatrix}1 & 0\\ 0 & u\end{smallmatrix}) = N_1 \cdot N_2$. But then $u = (\det N_1)(\overline{\det N_2})$, which proves that ker $\partial \subset$ im$\{\overline{\det}\}$. For the reverse inclusion, suppose that $u = (\det N_1)(\det N_2)$, with $N_i \in GL_2(\Lambda_i)$. Then $M = (\begin{smallmatrix}1 & 0\\ 0 & u\end{smallmatrix})^{-1} \bar{N}_1 \bar{N}_2 \in GL_2(\bar{\Lambda})$ is a matrix of determinant 1, so by the argument of the preceding paragraph M is a product of elementary matrices over $\bar{\Lambda}$. But as above this implies that $\partial' M = 0$, whence also $\partial u = \partial'(\begin{smallmatrix}1 & 0\\ 0 & u\end{smallmatrix}) = 0$. We have, therefore, shown that the sequence

$$GL_2(\Lambda_1) \times GL_2(\Lambda_2) \xrightarrow{\overline{\det}} u(\bar{\Lambda}) \to D(\Lambda) \to D(\Lambda_1) \dotplus D(\Lambda_2) \to 0 \tag{5.4}$$

is exact whenever $\bar{\Lambda}$ is commutative.

6. Milnor's Mayer–Vietoris Sequence

Before proceeding to specific computations, let us briefly discuss the relation between Milnor's sequence and those given in Section 5. Let Λ be a fibre product of Λ_1 and Λ_2, as in (1.8), so by Milnor's Theorem (see [1, p. 481] or [12]), there is an exact sequence

$$\begin{aligned}K_1(\Lambda) \to K_1(\Lambda_1) \dotplus K_1(\Lambda_2) \to K_1(\bar{\Lambda}) \xrightarrow{\delta} K_0(\Lambda)\\ \to K_0(\Lambda_1) \dotplus K_0(\Lambda_2) \to K_0(\bar{\Lambda}).\end{aligned} \tag{6.1}$$

The connecting homomorphism δ is given as follows: The general element of $K_1(\bar{\Lambda})$ is expressible as a pair $[\bar{\Lambda}^{(n)}, \beta]$, with $\beta \in \text{Aut}\,\bar{\Lambda}^{(n)}$. Then

$$\delta[\bar{\Lambda}^{(n)}, \beta] = [\beta\Lambda^{(n)}] - [\Lambda^{(n)}],$$

using the notation of Section 4. The results in (4.4) and (4.17) show that $\beta\Lambda^{(n)}$ is locally free, and $\Lambda' \otimes_\Lambda \beta\Lambda^{(n)} \cong \Lambda'^{(n)}$. Hence the image of δ lies in the subgroup $D(\Lambda)$ of $K_0(\Lambda)$. Here, $K_0(\Lambda)$ is the Grothendieck group of the category of all projective Λ-lattices.

We claim that the sequence

$$K_1(\bar{\Lambda}) \xrightarrow{\delta} D(\Lambda) \xrightarrow{f} D(\Lambda_1) \dotplus D(\Lambda_2) \to 0$$

is exact. We have already seen in (3.3) that f is epic. It is clear that $f \cdot \delta = 0$. Further, if $x \in D(\Lambda)$ is such that $f(x) = 0$, then also $f(x) = 0$ in $K_0(\Lambda_1) \dotplus K_0(\Lambda_2)$,
whence by (6.1) we may conclude that x lies in the image of δ.

Combining this sequence with (6.1), it follows that

$$K_1(\Lambda) \to K_1(\Lambda_1) \dotplus K_1(\Lambda_2) \to K_1(\bar{\Lambda}) \xrightarrow{\delta} D(\Lambda) \xrightarrow{f} D(\Lambda_1) \dotplus D(\Lambda_2) \to 0 \tag{6.2}$$

is exact. (We could have avoided using (3.3), by observing that $K_0(\bar{\Lambda})$ is Z-free because of the Krull–Schmidt theorem.)

Furthermore, since $\bar{\Lambda}$ is semilocal, there is an epimorphism $u(\bar{\Lambda}) \to K_1(\bar{\Lambda})$, given by $u \to [\bar{\Lambda}, r_u]$, with r_u the right multiplication by u on $\bar{\Lambda}$. Hence we obtain a commutative diagram

$$\begin{array}{ccc} K_1(\bar{\Lambda}) & \xrightarrow{\delta} & D(\Lambda) \\ \uparrow & \nearrow_{\partial} & \\ u(\bar{\Lambda}) & & \end{array} \tag{6.3}$$

and $\ker f = \text{im}\,\delta = \text{im}\,\partial$. Even if we use (6.2), it is somewhat awkward to determine $\ker \partial$ explicitly. Indeed, let $u \in u(\bar{\Lambda})$. Then

$$\delta \begin{pmatrix} 1 & 0 \\ 0 & u \end{pmatrix} = \partial u,$$

whence $u \in \ker \partial$ if and only if $\left(\begin{smallmatrix} 1 & 0 \\ 0 & u \end{smallmatrix}\right)$ lies in the image of $K_1(\Lambda_1) \dotplus K_1(\Lambda_2)$ in $K_1(\bar{\Lambda})$. Now each Λ_i is an R-order, so (see [1, 19]) the maps $GL_2(\Lambda_i) \to K_1(\Lambda_i)$,

$i = 1, 2$, are epic. Hence, $u \in \ker \partial$ if and only if there exist matrices $N_i \in GL_2(\Lambda_i)$, $i = 1, 2$, such that

$$\begin{pmatrix} 1 & 0 \\ 0 & u \end{pmatrix} \cdot \bar{N}_1 \bar{N}_2 \in E(\bar{\Lambda}),$$

where $E(\bar{\Lambda})$ is the normal subgroup of $GL(\bar{\Lambda})$ generated by all elementary matrices. This criterion is not a useful one in practice, unless $\bar{\Lambda}$ is commutative or local. It is not clear to us how to deduce (5.3) directly from these considerations, for example.

If $\bar{\Lambda}$ is commutative, it is easy to obtain from (6.2) the exact sequence given in (5.4). When $\bar{\Lambda}$ is not commutative, we could obtain (5.1) from (6.2) and (6.3), by examining the proof of the exactness of (6.2). Of course, the relevant part of the proof will be just a more general version of the arguments given in Section 4.

7. Applications

We shall now give several illustrations of the usefulness of the exact sequences (1.9) and (1.12).

a. *Cyclic Group of Order 2p*

Let p be an odd prime, and let $\Lambda = ZG$, where

$$G = \langle x, y \colon x^p = 1,\ y^2 = 1,\ xy = yx \rangle,$$

a cyclic group of order $2p$. Let $H = \langle y \rangle$, and set

$$I = (x^{p-1} + x^{p-2} + \cdots + x + 1)\Lambda, \qquad J = (x - 1)\Lambda. \tag{7.1}$$

Since Λ is $Z[x]$-free, it is clear that $I \cap J = 0$, and that $\bar{\Lambda}$ is Z-torsion. Let $R = Z[\omega]$, where ω is a primitive pth root of 1 over Q, and let $\bar{Z} = Z/pZ \cong R/(1 - \omega)R$. The pullback diagram (1.7) becomes

$$\begin{array}{ccc} \Lambda & \longrightarrow & RH \\ \downarrow & & \downarrow \\ ZH & \longrightarrow & \bar{Z}H \end{array} .$$

Since $A = QG$ satisfies the Eichler condition (and indeed is commutative), we may apply (1.10) to obtain the exact sequence

$$u(ZH) \times u(RH) \to u(\bar{Z}H) \to D(\Lambda) \to D(ZH) \dotplus D(RH) \to 0. \tag{7.2}$$

Now $D(ZH) = 0$, since every projective ZH-module is free (see [14]). Alternatively, there is a pullback diagram

$$\begin{array}{ccc} ZH & \longrightarrow & Z \\ \downarrow & & \downarrow \\ Z & \longrightarrow & Z/2Z \end{array}$$

obtained from (1.7) by using the ideals $(y+1)ZH$ and $(y-1)ZH$ of ZH. Hence by (1.10),

$$u(Z) \times u(Z) \to u(Z/2Z) \to D(ZH) \to D(Z) \dotplus D(Z) \to 0$$

is exact, which implies at once that $D(ZH) = 0$.

Likewise, there is a pullback diagram

$$\begin{array}{ccc} RH & \longrightarrow & R \\ \downarrow & & \downarrow \\ R & \longrightarrow & R/2R \end{array},$$

obtained from (1.7) by using the ideals $(y+1)RH$, $(y-1)RH$. Hence, the sequence

$$u(R) \times u(R) \to u(R/2R) \to D(RH) \to D(R) \dotplus D(R) \to 0$$

is exact. But $D(R) = 0$ since R is a maximal order. If we denote by $u^*(R)$ the image of $u(R)$ in $u(R/2R)$, we then obtain

$$D(RH) \cong u(R/2R)/u^*(R).$$

This agrees with an earlier result of Ullom [21], who showed that the above groups have the same order.

Since $u(ZH) \subset u(RH)$, it remains for us to calculate the image $u^*(RH)$ of $u(RH)$ in $u(\bar{Z}H)$. As a matter of fact, it was shown in [16] that $u^*(RH) = u(\bar{Z}H)$. Indeed, a typical element of $u(\bar{Z}H)$ is of the form $\bar{r} + \bar{s}y$, where r, $s \in Z$ and $r \not\equiv \pm s \pmod{p}$. For such a pair r, s, set

$$\xi_r = (\omega^r - \omega^{-r})/(\omega - \omega^{-1}),$$

and define ξ_s likewise. In [16] we showed that $\xi_r + \xi_s y \in u(RH)$ and that $\xi_r + \xi_s y \to \bar{r} + \bar{s}y$. Hence $u^*(RH) = u(\bar{Z}H)$, and consequently (7.2) yields the formula

$$D(\Lambda) \cong D(RH) \cong u(R/2R)/u^*(R).$$

This was obtained in a more computational manner in [16].

b. *Quaternion Group of Order* 8

Let $\Lambda = ZG$, where

$$G = \langle x, y: x^4 = 1, x^2 = y^2, yxy^{-1} = x^{-1}\rangle,$$

the quaternion group of order 8. Let $H = Z \oplus Zi \oplus Zj \oplus Zk$ be the ring of integral quaternions. Choose $I = (x^2 + 1)\Lambda$, $J = (x^2 - 1)\Lambda$. Then (1.7) becomes

$$\begin{array}{ccc} \Lambda & \xrightarrow{\psi_1} & H \\ {\scriptstyle \psi_2}\downarrow & & \downarrow \\ ZV & \longrightarrow & \bar{Z}V \end{array},$$

where $V = \langle s\rangle \times \langle t\rangle$ is a (2,2)-group, $\bar{Z} = Z/2Z$, and the maps are given thus

$$\psi_1(x) = i,\ \psi_1(y) = j; \qquad \psi_2(x) = s,\ \psi_2(y) = t.$$

In this case $A = QG$ fails to satisfy the Eichler condition. However, $\bar{\Lambda}$ is commutative, so by (1.12) there is an exact sequence

$$GL_2(ZV) \times GL_2(H) \xrightarrow{\overline{\det}} u(\bar{Z}V) \to D(ZG) \longrightarrow D(H) \dotplus D(ZV) \to 0.$$

There is an obvious pullback diagram

$$\begin{array}{ccc} ZV & \longrightarrow & Z[s] \\ \downarrow & & \downarrow \\ Z[s] & \longrightarrow & Z/2Z \end{array},$$

from which we easily deduce that $D(ZV) = 0$. It is folklore that $D(H) = 0$ (in fact, every locally free H-lattice is free). Thus

$$D(ZG) \cong u(\bar{Z}V)/\overline{\det}\{GL_2(ZV) \times GL_2(H)\}.$$

Now ZV is commutative, which readily implies that $\overline{\det}\, GL_2(ZV)$ coincides with $u^*(ZV)$, the image of $u(ZV)$ in $u(\bar{Z}V)$. Since

$$u(ZV) = \{\pm 1, \pm s, \pm t, \pm st\},$$

it follows that $u^*(ZV) = \{1, s, t, st\}$ is a subgroup of index 2 in $u(\bar{Z}V)$. But by using Dieudonné determinants (see [16]), it is not hard to verify that the unit $1 + s + t$ of $\bar{Z}V$ does not lie in $\overline{\det}\, GL_2(H)$. We may then conclude that $|D(ZG)| = 2$, a result proved in [16] in a more complicated manner. Martinet

[11a] showed there are exactly two nonisomorphic projective rank one ZG-lattices; he showed furthermore that the nonprincipal one is not stably free.

c. *Dihedral Group of Order* $2p$

Let $\Lambda = ZG$, where

$$G = \langle x, y \colon x^p = 1, y^2 = 1, yxy^{-1} = x^{-1} \rangle$$

is the dihedral group of order $2p$, with p an odd prime. Let $K = Q(\omega)$, $R = Z[\omega]$, where ω is a primitive pth root of unity over Q, and let σ be the automorphism of R defined by the map $\omega \to \omega^{-1}$. Finally, let L be the subfield of K fixed by σ, and S the ring of all algebraic integers in L.

Now let $H = \langle y \rangle$, and set $R \circ H = R \oplus Ry$, the twisted group ring of H over R, with $y^2 = 1$, $y\alpha = \alpha^\sigma y$, $\alpha \in R$. Choosing I and J as in (7.1), the pullback diagram (1.7) becomes

$$\begin{array}{ccc} ZG & \longrightarrow & R \circ H \\ \downarrow & & \downarrow \\ ZH & \longrightarrow & \bar{Z}H \end{array}, \qquad (7.3)$$

where $\bar{Z} = Z/pZ$, and $\bar{Z}H$ is the ordinary (untwisted) group algebra.

Since QG satisfies the Eichler condition, and $\bar{\Lambda}$ is commutative, there is an exact sequence

$$u(ZH) \times u(R \circ H) \to u(\bar{Z}H) \to D(ZG) \to D(ZH) \dotplus D(R \circ H) \to 0.$$

The first factor $u(ZH)$ may be omitted, since its image in $u(\bar{Z}H)$ lies in the image $u^*(R \circ H)$ of $u(RH)$. From [14] we have $C(ZH) = D(ZH) = 0$. On the other hand, it was shown in [11] that the ring $R \circ H$ is hereditary, and $C(R \circ H) \cong C(S)$. Thus $D(R \circ H) = 0$, and so we obtain the isomorphism

$$D(ZG) \cong u(\bar{Z}H)/u^*(R \circ H).$$

As in our first application (a), let $\bar{a} + \bar{b}y \in u(\bar{Z}H)$, and put $\xi_a = (\omega^a - \omega^{-a})/(\omega - \omega^{-1})$. Then $(\xi_a)^\sigma = \xi_a$, so $\xi_a + \xi_b y$ is a unit in $R \circ H$ whose image in $u(\bar{Z}H)$ is $\bar{a} + \bar{b}y$. Thus, $u^*(R \circ H) = u(\bar{Z}H)$, which proves that $D(ZG) = 0$, and consequently that

$$C(ZG) \cong C(S).$$

Remarks. (i) This result can be obtained directly from Lee's classification [11] of ZG-lattices. It was also derived in [16] by another approach, which nevertheless made use of the pullback diagram (7.3).

(ii) It is possible to handle the more general case when G is a nonabelian group of order pq, say

$$G = \langle x, y \colon x^p = 1, y^q = 1, yxy^{-1} = x^r \rangle.$$

Here, p is any odd prime, q is any divisor of $p - 1$, and r is a primitive qth root of 1 mod p. A partial approach to the calculation of $D(ZG)$ is given in [13], where ZG-lattices are classified. However, even without using this classification, it is possible to calculate $D(ZG)$ explicitly, by using an analog of (7.3) and the exact sequence associated with it. It turns out that there is an epimorphism of $D(ZG)$ onto $D(ZH)$, where $H = \langle y \rangle$, and that the kernel of this epimorphism is cyclic of order $q/(q, 2)$. Details may be found in [7].

d. *Dihedral Group of Order* 8

Let $\Lambda = ZG$, where

$$G = \langle x, y \colon x^4 = 1, y^2 = 1, yxy^{-1} = x^{-1} \rangle$$

is the dihedral group of order 8. We are going to prove that every projective Λ-lattice is free. Since $\underset{\sim}{Q}G$ satisfies the Eichler condition, the cancellation law holds for projective Λ-lattices. Hence it suffices to show that $C(\Lambda) = 0$.

Let $R = Z[i]$, $i^2 = -1$, and let $\bar{\alpha}$ denote the complex conjugate of the element $\alpha \in R$. Set $H = \langle y \rangle$, $R \circ H = R \oplus Ry$ = twisted group ring of H over R, with

$$y^2 = 1, \qquad y\alpha = \bar{\alpha} y, \qquad \alpha \in R.$$

Choose I, J, V and $\bar{Z}$ as in example (b) above. The pullback diagram (1.7) becomes

$$\begin{array}{ccc} ZG & \longrightarrow & R \circ H \\ \downarrow & & \downarrow \\ ZV & \longrightarrow & \bar{Z}V \end{array},$$

and there is an exact sequence

$$u(ZV) \times u(R \circ H) \xrightarrow{\varphi} u(\bar{Z}V) \to D(\Lambda) \to D(ZV) \dotplus D(R \circ H) \to 0.$$

We have seen in (b) that $D(ZV) = 0$. Let us show that φ is surjective. The group $u(\bar{Z}V)$ is of order 8, while the subgroup $u^*(ZV)$ has order 4. Furthermore, $(1 + i) + y$ is a unit in $R \circ H$, and its image in $u(\bar{Z}V)$ does not lie in $u^*(ZV)$. This proves that φ is surjective, and, therefore, $D(\Lambda) \cong D(R \circ H)$.

We shall use Jacobinski's formula (see [9, 16]) to compute $D(R \circ H)$.

Set $K = Q(i)$; then $K \circ H$ is isomorphic to the algebra of 2×2 matrices over Q. The isomorphism may be given by

$$f\colon K \circ H \cong \operatorname{Hom}_Q(K, K),$$

where for $\alpha \in K$, $f(\alpha)$ is multiplication by α on K, and where $f(y)$ is complex conjugation on K. Treat f as an identification, and set $\Lambda_1' = \operatorname{Hom}_Z(R, R)$, a maximal Z-order in $K \circ H$ containing $R \circ H$.

Let $N : K \circ H \to Q$ be the reduced norm map, and let the subscript 2 denote localization at the rational prime 2. Then Jacobinski's formula expresses $D(R \circ H)$ as a quotient of two ideal groups

$$D(R \circ H) = \frac{\{Z \cdot N(\xi) : \xi \in u((\Lambda_1')_2)\}}{\{Z \cdot N(\eta) : \eta \in u((R \circ H)_2)\}}.$$

If $\xi = \left(\begin{smallmatrix} a & 0 \\ 0 & 1 \end{smallmatrix}\right)$, where $a \in u(Z_2)$, then $N(\xi) = a$. Hence the numerator in the above formula is precisely the ideal group generated by $\{Za\colon a \in Z,\ a \text{ odd}\}$.

We shall prove that $D(R \circ H) = 0$ by showing that every odd rational integer $2m + 1$ is expressible as $N(\eta)$ for some $\eta \in u((R \circ H)_2)$. Indeed, if $\eta \in R \circ H$ is such that $N(\eta) = 2m + 1$, then η is automatically a unit in $(R \circ H)_2$. Now it is easily checked that

$$N(\alpha + \beta y) = \alpha\bar{\alpha} - \beta\bar{\beta}, \qquad \alpha + \beta y \in R \circ H.$$

Hence choosing $\eta = (m + 1) + my$, we have

$$N(\eta) = (m + 1)^2 - m^2 = 2m + 1,$$

as desired. This completes the proof that $D(R \circ H) = 0$, and shows therefore that $D(\Lambda) = 0$.

Finally, if Λ' is any maximal Z-order in QG, then

$$\Lambda' \cong Z^{(4)} \dotplus \operatorname{Hom}_Z(R, R),$$

which easily implies that $C(\Lambda') = 0$. Therefore $C(\Lambda) \cong D(\Lambda) = 0$, as claimed above.

e. *Center of ZG, where G is Dihedral of Order 2p*

Keep the notation of example (c). We shall calculate the order of $D(C)$, where C is the center of ZG. Given any fibre product (1.1) satisfying hypotheses (1.8), we may easily determine $c(\Lambda)$, the center of Λ. Indeed,

$K \cdot c(\Lambda) = c(K\Lambda) = c(K\Lambda_1) \dotplus c(K\Lambda_2)$, whence $c(\Lambda)$ is given by a pullback diagram

$$\begin{array}{ccc} c(\Lambda) & \longrightarrow & c(\Lambda_1) \\ \downarrow & & \downarrow \\ c(\Lambda_2) & \longrightarrow & c(\bar{\Lambda}) \end{array} .$$

In particular, we obtain from (7.3) the pullback diagram

$$\begin{array}{ccc} C & \longrightarrow & S \\ \downarrow & & \downarrow \varphi_1 \\ ZH & \xrightarrow[\varphi_2]{} & \bar{Z}H \end{array} .$$

Here we have an example of a fibre product in which the map φ_2 is surjective, but φ_1 is not surjective.

We thus have an exact sequence

$$u(S) \times u(ZH) \to u(\bar{Z}H) \to D(C) \to D(S) \dotplus D(ZH) \to 0.$$

Now $D(S) = 0$, since S is a maximal order in L. We have already noted that $D(ZH) = 0$. Thus we have the isomorphism

$$D(C) \cong u(\bar{Z}H)/u^*(S) \cdot u^*(ZH).$$

Let $P_0 = S \cap (1 - \omega)R$, so $S/P_0 \cong \bar{Z}$. The mapping φ_1 is then given by the composite

$$S \to S/P_0 \cong \bar{Z} \subset \bar{Z}H.$$

Let $u \in S$ have image $\bar{u}$ in $\bar{Z}$; we know from [16] that the map $u \to \bar{u}$ gives an epimorphism of $u(S)$ onto $u(\bar{Z})$.

Let ρ: $\bar{Z}H \cong \bar{Z} \dotplus \bar{Z}$ be the ring isomorphism defined via

$$\bar{Z}H \cong \frac{\bar{Z}[y]}{(y^2 - 1)} \cong \frac{\bar{Z}[y]}{(y - 1)} \dotplus \frac{\bar{Z}[y]}{(y + 1)} \cong \bar{Z} \dotplus \bar{Z}.$$

Then for $u \in u(S)$, we have

$$\rho(\bar{u}) = (\bar{u}, \bar{u}) \in u(\bar{Z}) \times u(\bar{Z}).$$

On the other hand, it is easily seen that $u(ZH) = \{\pm 1, \pm y\}$, so $\rho\{u^*(ZH)\}$ consists of the 4 elements $\pm(1, \pm 1)$. Therefore

$$u(\bar{Z}H)/u^*(S)\, u^*(ZH) \cong u(\bar{Z})/\{\pm 1\},$$

a cyclic group of order $(p - 1)/2$. This proves that $D(C)$ is cyclic of order $(p - 1)/2$, a result obtained independently by Fröhlich in [6].

8. Abelian Groups

We have already remarked that for a cyclic group G of prime order p, it follows from [14] that $D(ZG) = 0$. In contradistinction to this fact, Kervaire and Murthy [10] recently showed that

$$|D(ZG)| \to \infty \text{ as } |G| \to \infty, \tag{8.1}$$

where G ranges over cyclic p-groups, p a fixed prime. Fröhlich [5] proved that (8.1) remains valid for the cases $p = 2, 3$, when G ranges over all elementary p-groups. We shall now generalize these results.

Theorem 8.2. *Let $\{G_i\}$ be any sequence of abelian groups of composite order, such that $|G_i| \to \infty$. Then also $|D(ZG_i)| \to \infty$.*

An independent proof of this theorem, for the case where each G_i is a p-group, has been communicated to the authors by Fröhlich (see [6a]). His calculations are somewhat more explicit than ours, but in essence both his proof and ours depend on a basic idea used by Kervaire and Murthy [10] and by Ullom [21]. The proof of Theorem 8.2 splits into two cases, depending on whether $|G_i|$ is a prime power or not. In either case, we shall make use of the following earlier result (see [5, 16]):

$$|D(ZH)| \text{ divides } |D(ZG)| \quad \text{if } H \text{ is a factor group of } G. \tag{8.3}$$

We begin the proof of Theorem 8.2 by handling the case of p-groups. Let us first dispose of the case where $p = 2$. By [5], (8.1) is valid when G ranges over elementary 2-groups; on the other hand, by [10] (8.1) holds when G ranges over cyclic 2-groups. But any abelian 2-group of large order has either a large elementary factor group or a large cyclic factor group. It follows at once from (8.3) that (8.1) is valid with G ranging over arbitrary abelian 2-groups. Thus it suffices for us to concentrate on the case where p is odd. We shall prove the following theorem, which of course implies the validity of (8.1) for abelian p-groups G.

Theorem 8.4. *Let p be an odd prime, and let G_0 be an abelian group of order p^{n+1}, where $n \geqslant 0$. Then G_0 contains a subgroup G of order p^n such that*

$$|D(ZG_0)| \textit{ is a multiple of } |D(ZG)| \cdot p^r,$$

with

$$r = \tfrac{1}{2}(p^n - 1) - n.$$

Proof. *Step* 1. Let $\langle x \rangle$ be any cyclic direct factor of G_0, and write

$$G_0 = \langle x \rangle \times H, \qquad G = \langle x^p \rangle \times H,$$

where x is of order p^m, $m \geqslant 1$. Let ω be a primitive p^mth root of 1 over Q, and set $R = Z[\omega]$, $\bar{Z} = Z/pZ \cong R/(1-\omega)R$. Let c be the automorphism of R induced by letting $\omega \to \omega^{-1}$, so for each $\alpha \in R$, α^c equals the complex conjugate $\bar{\alpha}$ of α. We shall also denote by c the automorphism of ZG_0 defined by

$$c\left(\sum_{G_0} a_g \cdot g\right) = \sum_{G_0} a_g \cdot g^{-1}, \qquad a_g \in Z. \tag{8.5}$$

Now let $\Lambda = ZG_0$, and define the ideals

$$I = \Phi_{p^m}(x) \cdot \Lambda, \qquad J = (x^{p^{m-1}} - 1) \cdot \Lambda,$$

where $\Phi_k(x)$ denotes the cyclotomic polynomial of order k and degree $\varphi(k)$. Our usual pullback diagram (1.7) is then

$$\begin{array}{ccc} ZG_0 & \longrightarrow & RH \\ \downarrow & & \downarrow \\ ZG & \longrightarrow & \bar{Z}G \end{array} \quad . \tag{8.6}$$

Since $c(I) = I$ and $c(J) = J$, the involution c of ZG_0 induces involutions (also denoted by c) of the rings RH, ZG, and $\bar{Z}G$. Indeed,

$$c\left(\sum_{H} \alpha_h \cdot h\right) = \sum_{H} \bar{\alpha}_h \cdot h^{-1}, \qquad \alpha_h \in R, \tag{8.7}$$

whereas the action of c on ZG and $\bar{Z}G$ is given as in (8.5). It is then easily seen that all of the maps in (8.6) are c-linear.

Step 2. We now set

$$N = \sum_{g \in G-1}^{\cdot} \bar{Z}(g-1), \qquad U = 1 + N, \qquad U^+ = \{u \in U : cu = u\}.$$

Then N is the augmentation ideal of $\bar{Z}G$, and $u(\bar{Z}G)$ is the inverse image of $u(\bar{Z})$ under the augmentation map $\bar{Z}G \to \bar{Z}G/N \cong \bar{Z}$. Therefore $u(\bar{Z}G) \cong u(\bar{Z}) \times U$.

For the sake of clarity, we shall denote the image of G in $u(\bar{Z}G)$ by $\tilde{G}$. Obviously

$$|\tilde{G}| = p^n, \qquad \tilde{G} \cap U^+ = 1, \qquad |U| = |N| = p^{p^n - 1}.$$

Furthermore,

$$U^+ = \left\{1 + w : w = \sum a_g(g - 1 + g^{-1} - 1),\ a_g \in \bar{Z}\right\},$$

where there is one summand in w for each pair (g, g^{-1}) from G, $g \neq 1$. Hence $|U^+| = p^{(p^n-1)/2}$. But then

$$[U: \tilde{G}U^+] = |U|/|\tilde{G}||U^+| = p^r, \qquad r = \tfrac{1}{2}(p^n - 1) - n. \tag{8.8}$$

We shall use this formula later in the proof.

Step 3. Let us recall two basic results.

LEMMA 8.9. *Let ζ be a primitive p^k*th *root of* 1, $k \geqslant 0$. *Then every unit of $Z[\zeta]$ is a power of ζ times a real unit u_0.*

LEMMA 8.10. *Let the algebraic number field L be a splitting field for the abelian group G, and let S be any subring of the ring of all algebraic integers in L. Then every unit of SG of finite order is the product of a root of unity in S and an element of G.*

The first result is due to Kummer (see [22]), the second to G. Higman (see [3]).

Suppose in our particular case that our group G is of exponent p^k, where $k \geqslant 1$, and let $S = Z[\zeta]$ where ζ is a primitive p^kth root of 1. Extend the involution c from ZG to SG as in (8.7). We claim that for each $u \in u(SG)$, there exist $\zeta^i \in S$ and $g \in G$ such that

$$c(u \cdot g\zeta^i) = u \cdot g\zeta^i. \tag{8.11}$$

To prove this, set $L = Q(\zeta)$, a splitting field for G. Then G has p^n distinct one-dimensional representations χ_i in S, and the corresponding central idempotents $\{e_i\}$ in LG are given by

$$e_i = |G|^{-1} \sum_G \overline{\chi_i(g)} \cdot g.$$

It is then clear that $c(e_i) = e_i$, and that $\sum^{\cdot} Se_i$ is the (unique) maximal order in LG containing SG.

Now write $u = \sum u_i e_i$, $u_i \in S$. Then each u_i is a unit in S, and $cu = \sum c(u_i) \cdot e_i$, so

$$u/cu = \sum (u_i/\bar{u}_i)\, e_i .$$

But by (8.9), each u_i is a root of unity times a real unit, so each $u_i/\bar{u}_i$ is a root of unity and hence has finite order. Thus u/cu is a unit of SG of finite order, and so by (8.10) we have

$$u/cu = \pm\zeta^{-2i}g^{-2} \tag{8.12}$$

for some i and some $g \in G$. Once we show that the plus sign must occur, we immediately deduce (8.11). Now there are ring homomorphisms

$$SG \to \bar{Z}G \to \bar{Z},$$

the first being induced by $S \to S/(1-\zeta)S \cong \bar{Z}$, the second being the augmentation map. If $\rho : SG \to \bar{Z}$ is the composite map, then $\rho(cu) = \rho u$. On the other hand, $\rho(\pm\zeta^{-2i}g^{-2}) = \pm 1$. Hence the minus sign cannot occur in (8.12), and the validity of (8.11) for some i and g is established.

Step 4. Starting with the pullback diagram (8.6), our exact sequence (1.10) becomes

$$u(ZG) \times u(RH) \xrightarrow{\varphi} u(\bar{Z}G) \to D(ZG_0) \to D(ZG) \dotplus D(RH) \to 0.$$

Therefore

$$|D(ZG_0)| = |D(ZG)| \cdot |D(RH)| \cdot [u(\bar{Z}G)\text{: image of } \varphi],$$

so to prove Theorem 8.4, it suffices to prove that the last factor on the right is a multiple of p^r. By virtue of (8.8), it is enough to prove that

$$\text{image of } \varphi \subset u(\bar{Z}) \times (\tilde{G}U^+), \tag{8.13}$$

when we identify $u(\bar{Z}G)$ with $u(\bar{Z}) \times U$.

First let $u \in u(RH)$; by the previous step, we have

$$c(u \cdot \omega^i h) = u \cdot \omega^i h$$

for some i and some $h \in H$. Since $\varphi : u(RH) \to u(\bar{Z}G)$ is c-linear, and

$$u(\bar{Z}G)^+ = u(\bar{Z}) \times U^+,$$

we have $\varphi(u \cdot \omega^i h) \in u(\bar{Z}) \times U^+$. Noting that $\varphi(\omega^i h) \in \tilde{G}$, it follows that $\varphi(u) \in u(\bar{Z}) \times \tilde{G}U^+$. A similar argument works for $\varphi(u(ZG))$, which establishes (8.13) and completes the proof of Theorem 8.4.

The next step in the proof of Theorem 8.2 deals with the case where $|G_i|$ has at least two distinct prime divisors, and by (8.3) it will be enough to consider the situation in which G is cyclic of order pq, where p and q are distinct primes. Let us write

$$p^f \equiv 1 \pmod{q}, \quad fg = q-1; \qquad q^{f'} \equiv 1 \pmod{p}, \quad f'g' = p-1,$$

where f is the order of p mod q, and f' the order of q mod p. We define

$$N(p,q) = \begin{cases} (p^{f/2}+1)^g/q, & f \text{ even}, \\ (p^f-1)^{g/2}/q, & f \text{ odd}. \end{cases} \tag{8.14}$$

Obviously $N(p, q)$ is always an integer. Likewise define $N(q, p)$ by corresponding formulae, by interchanging p and q and using f', g' in place of f, g. We shall now establish

THEOREM 8.15. (i) *Let G be cyclic of order pq, where p and q are distinct odd primes. Then*

$$N(p, q) \cdot N(q, p) \textit{ divides } 4 \cdot |D(ZG)|.$$

(ii) *Let G be cyclic of order $2p$, where p is an odd prime. Then $N(2, p)$ divides $|D(ZG)|$.*

Proof. *Step* 1. Let p and q be distinct odd primes, and set

$$\begin{cases} R = Z[\omega], & \omega = \text{primitive } p\text{th root of } 1, \quad K = Q(\omega), \\ S = R[\theta], & \theta = \text{primitive } q\text{th root of } 1, \\ \mathbf{F}_n = \text{finite field with } n \text{ elements.} \end{cases}$$

Form the polynomial ring $Z[x, y]$ in two variables. The ideals

$$(x - 1, y^q - 1) \quad \text{and} \quad (\Phi_p(x), y^q - 1)$$

of this ring have intersection $(x^p - 1, y^q - 1)$ and sum $(p, x - 1, y^q - 1)$. Thus we have the pullback diagram (see (1.6))

$$\begin{array}{ccc} \dfrac{Z[x, y]}{(x^p - 1, y^q - 1)} & \longrightarrow & \dfrac{Z[x, y]}{(\Phi_p(x), y^q - 1)} \\ \big\downarrow & & \big\downarrow \varphi \\ \dfrac{Z[x, y]}{(x - 1, y^q - 1)} & \longrightarrow & \dfrac{Z[x, y]}{(p, x - 1, y^q - 1)} \end{array}. \tag{8.16}$$

Define an involution c on all the above quotient rings by setting $cx = x^{-1}$, $cy = y^{-1}$. Denote by H the cyclic group of order q generated by the class of y. Then diagram (8.16) may be written as

$$\begin{array}{ccc} ZG & \longrightarrow & RH \\ \big\downarrow & & \big\downarrow \varphi \\ ZH & \longrightarrow & \mathbf{F}_pH \end{array}.$$

The following sequence is therefore exact:

$$u(ZH) \times u(RH) \to u(\mathbf{F}_pH) \to D(ZG) \to D(ZH) \dotplus D(RH) \to 0.$$

Therefore, since $D(ZH) = 0$ and $u(ZH) \subset u(RH)$, we obtain

$$|D(ZG)| = |D(RH)| \cdot [u(\mathbf{F}_p H): u^*(RH)]. \tag{8.17}$$

In order to investigate $u^*(RH)$ in the above formula, we begin by observing that

$$RH = \frac{Z[x, y]}{(\Phi_p(x), y^q - 1)} \subset \frac{Z[x, y]}{(\Phi_p(x), y - 1)} \dotplus \frac{Z[x, y]}{(\Phi_p(x), \Phi_q(y))}$$
$$\cong R \dotplus S \cong \text{maximal } Z\text{-order of } KH.$$

(The last step uses the isomorphism $K \dotplus K(\theta) \cong KH$.) Of course, all of the indicated summands are mapped onto themselves by the involution c.

In a similar fashion, we obtain

$$\mathbf{F}_p H \cong \frac{\mathbf{F}_p[y]}{(y^q - 1)} \cong \frac{\mathbf{F}_p[y]}{(y - 1)} \dotplus \frac{\mathbf{F}_p[y]}{(\Phi_q(y))}.$$

The map φ in (8.16) extends to a map $\tilde{\varphi} : R \dotplus S \to \mathbf{F}_p H$, and

$$\tilde{\varphi}(R) \subset \mathbf{F}_p[y]/(y - 1) \cong \mathbf{F}_p, \qquad \tilde{\varphi}(S) \subset \mathbf{F}_p[y]/(\Phi_q(y)) \underset{\text{def}}{=} \bar{\Lambda}.$$

Now $\tilde{\varphi}$ maps the cyclotomic units $\{(1 - \omega^i)/(1 - \omega),\ 1 \leqslant i \leqslant p - 1\}$ of R onto $u(\mathbf{F}_p)$. Since these are also units of RH, it follows that $u(\mathbf{F}_p) \subset u^*(RH)$. Therefore,

$$[u(\mathbf{F}_p H): u^*(RH)] = n[u(\bar{\Lambda}): u^*(S)], \tag{8.18}$$

where n is some positive integer, and $u^*(S)$ denotes the image of $u(S)$ in $u(\bar{\Lambda})$. In Steps 2 and 3 we shall investigate $[u(\bar{\Lambda}): u^*(S)]$.

Step 2. From the fact that f is the order of p mod q, it follows (see [22]) that

$$(y^q - 1) = (y - 1)\, \varphi_1(y) \cdots \varphi_g(y) \text{ in } \mathbf{F}_p[y],$$

where $\varphi_1(y), \ldots, \varphi_g(y)$ are irreducible polynomials of degree f. Therefore,

$$\frac{\mathbf{F}_p[y]}{(\Phi_q(y))} \cong \dot{\sum_{i=1}^{g}} \frac{\mathbf{F}_p[y]}{(\varphi_i(y))},$$

and $\mathbf{F}_p[y]/(\varphi_i(y))$ is a field F_i, with $(F_i : \mathbf{F}_p) = f$. The action of c on $\mathbf{F}_p H$ induces a permutation on the fields $F_1, \ldots, F_g$; namely, $c : F_i \to F_j$ if the roots of φ_j are the inverses of those of φ_i. Hence if f is even, $c(F_i) = F_i$, $1 \leqslant i \leqslant g$, and the subfield of F_i fixed by c is of degree $f/2$ over $\mathbf{F}_p$. If f is odd, then g is even, and the fields $\{F_1, \ldots, F_g\}$ may be partitioned into $g/2$ conjugate pairs $\{F_i, c(F_i)\}$.

Set $u^+(\bar{\Lambda}) = \{u \in u(\bar{\Lambda}): cu = u\}$. The preceding discussion then yields the formula

$$|u^+(\bar{\Lambda})| = \begin{cases}(p^{f/2} - 1)^g, & f \text{ even},\\ (p^f - 1)^{g/2}, & f \text{ odd}.\end{cases} \tag{8.19}$$

Step 3. For any abelian multiplicative group M with an involution c, let

$$M^+ = \{m \in M: cm = m\},\ M^- = \{m \in M: cm = m^{-1}\},\ (M^-)^2 = \{m^2: m \in M^-\}.$$

It is easily checked that the map $m \to m/cm$ induces a monomorphism

$$\frac{M}{M^+M^-} \to \frac{M^-}{(M^-)^2}. \tag{8.20}$$

We apply this to the case where $M = S$, and begin by determining M^- explicitly. We claim that M^- = group of roots of unity of S. Clearly M^- contains all roots of unity of S. Let $u \in M^-$, so $cu = u^{-1}$. Since $cu = \bar{u}$, it follows that $u\bar{u} = 1$, whence $|u|^2 = 1$, so $|u| = 1$. But likewise $|u^\sigma| = 1$ for each algebraic conjugate u^σ of u, so (see [22]) u is a root of unity.

In the case where $M = S$, the above discussion shows that M^- is a cyclic group of even order, so $[M^-: (M^-)^2] = 2$. It follows from (8.20) that[4]

$$[u(S): u^+(S)\, u^-(S)] = 1 \text{ or } 2.$$

Therefore the images of $u^+(S)$ and $u^-(S)$ in $u(\bar{\Lambda})$ generate a subgroup of $u^*(S)$ of index 1 or 2. These images lie in $u^+(\bar{\Lambda})$ and $u^-(\bar{\Lambda})$, respectively, and the image of $u^-(S)$ is in fact $\{\pm\theta^j: 0 \leqslant j \leqslant q-1\}$. Therefore, $2[u(\bar{\Lambda}): u^*(S)]$ is a multiple of

$$[u(\bar{\Lambda}): u^+(\bar{\Lambda}) \cdot \text{image of } u^-(S)]$$

hence also a multiple of

$$|u(\bar{\Lambda})|/|u^+(\bar{\Lambda})| \cdot q.$$

Using (8.19) and the fact that $|u(\bar{\Lambda})| = (p^f - 1)^g$, we see that this last expression is precisely $N(p, q)$. Taking (8.18) into account, we have now proved that

$$2[u(\mathbf{F}_pH): u^*(RH)] \text{ is a multiple of } N(p, q).$$

[4] Hasse [8, Satz 27] proved the deeper result that the index is exactly two when p and q are both odd primes.

The second factor in formula (8.17) has now been handled, so in order to complete the proof of Theorem 8.15 (i), we need only verify that

$$2 \cdot |D(RH)| \text{ is a multiple of } N(q, p). \tag{8.21}$$

Step 4. In order to calculate $|D(RH)|$, we use a new pullback diagram (see (1.6))

$$\begin{array}{ccc} \dfrac{Z[x, y]}{(\Phi_p(x), y^q - 1)} & \longrightarrow & \dfrac{Z[x, y]}{(\Phi_p(x), \Phi_q(y))} \\ \downarrow & & \downarrow \\ \dfrac{Z[x, y]}{(\Phi_p(x), y - 1)} & \longrightarrow & \dfrac{Z[x, y]}{(q, y - 1, \Phi_p(x))} \end{array}.$$

With the obvious isomorphisms, we may rewrite this as

$$\begin{array}{ccc} RH & \longrightarrow & S \\ \downarrow & & \downarrow \\ R & \longrightarrow & \bar{\Lambda}' \end{array}, \tag{8.22}$$

where $\bar{\Lambda}' \cong \mathbf{F}_q[x]/(\Phi_p(x))$. Note that $\bar{\Lambda}'$ has the same relation to the pair $\{q, p\}$ as $\bar{\Lambda}$ does to $\{p, q\}$. From (8.22) we obtain the exact sequence

$$u(R) \times u(S) \to u(\bar{\Lambda}') \to D(RH) \to 0,$$

and so

$$|D(RH)| = [u(\bar{\Lambda}') : u^*(S)], \tag{8.23}$$

where $u^*(S)$ is the image of $u(S)$ in $u(\bar{\Lambda}')$. Just as in Steps 2 and 3, we conclude that $2[u(\bar{\Lambda}') : u^*(S)]$ is a multiple of $N(q, p)$. Hence (8.21) is proved, and we are through with the first part of Theorem 8.15.

Step 5. Finally, let p be an odd prime, and let G be cyclic of order $2p$, H cyclic of order 2. In Section 7(a) we showed that

$$D(RH) \cong u(R/2R)/u^*(R).$$

If we take $q = 2$ in Step 4, then $\bar{\Lambda}' \cong R/2R$, and $S = R[-1] = R$, so formula (8.23) is still valid in this case. By (8.9) we have $u(S) = u(S^+)\,u(S^-)$. It follows, therefore, that $|D(RH)|$ is a multiple of $N(2, p)$. Since (8.17) also holds in the present case, we may conclude that $|D(ZG)|$ is a multiple of $N(2, p)$. This completes the proof of part (ii) of Theorem 8.15.

In order to complete the proof of Theorem 8.2, we derive some crude lower bounds on $|D(ZG)|$.

LEMMA 8.24. *Let p be any odd prime. Then*

(i) *If p^2 divides $|G|$, where $p > 3$, then $|D(ZG)| \geqslant p$.*

(ii) *If p^N divides $|G|$, where $N \geqslant 4$, then $|D(ZG)| > N$.*

(iii) *If 2^N divides $|G|$, where $N \geqslant 25$, then $|D(ZG)| > N$.*

(iv) *Let $p \geqslant 47$, $2 \leqslant q < p$, q a prime. If qp divides $|G|$, then*

$$|D(ZG)| > p.$$

Proof. (i). Suppose that p^2 divides $|G|$, where $p > 3$. By (8.3) and (8.4), we conclude that $|D(ZG)|$ is a multiple of p^r, where

$$r = \tfrac{1}{2}(p-1) - 1 \geqslant 1.$$

(ii) Let $N \geqslant 4$, and suppose that p^N divides $|G|$. By (8.3) and repeated use of (8.4), we find that $|D(ZG)|$ is a multiple of p^s, where

$$s = \sum_{j=1}^{N-1} \{\tfrac{1}{2}(p^{N-j} - 1) - (N-j)\}.$$

The minimum value of s (as function of p) occurs when $p = 3$; in this case it is easily checked that $s > 3^{N-2}$ (when $N \geqslant 4$). Therefore,

$$|D(ZG)| > p^{3^{N-2}} > N.$$

(iii) Let $N \geqslant 25$, and let 2^N divide G. Then for some $m \geqslant N^{1/2}$, either G has a subgroup C_{2^m} which is cyclic of order 2^m, or else G has a subgroup E_{2^m} which is an elementary abelian 2-group on m generators. In the first case, by (8.3) it is clear that $|D(ZG)| \geqslant |D(ZC_{2^m})|$. However, according to [10],

$$|D(ZC_{2^m})| = 2^r, \qquad r = 2^{m-1} - \tfrac{1}{2}m(m-1) - 1.$$

It is easily checked that $2^r > m^2$ if $m \geqslant 5$, whence $2^r > N$ if $N \geqslant 25$.

On the other hand, by [5] we have

$$|D(ZE_{2^m})| = 2^s, \qquad s = (m-4)\cdot 2^{m-1} + m + 2,$$

and this is clearly greater than $|D(ZC_{2^m})|$ for $m \geqslant 5$.

(iv) Let $p \geqslant 47$, and suppose first that $2p$ divides $|G|$. Then

$$|D(ZG)| \geqslant N(2,p) = \begin{cases} (2^{f'/2}+1)^{g'}/p, & f' \text{ even},\\ (2^{f'}-1)^{g'/2}/p, & f' \text{ odd}, \end{cases}$$

where f' is the order of 2 mod p, and $f'g' = p - 1$. For f' even, we have

$$(2^{f'/2} + 1)^{g'}/p > 2^{(p-1)/2}/p > p, \qquad \text{if} \quad p \geqslant 19.$$

For f' odd, we have $f' \geqslant 3$, and hence

$$(2^{f'} - 1)^{g'/2}/p > 2^{(f'-1)g'/2}/p > 2^{(p-1)/4}/p > p \qquad \text{if} \quad p \geqslant 47.$$

These same lower bounds can be used when qp divides $|G|$, when q is an odd prime $< p$. This completes the proof of the lemma.

Now we prove Theorem 8.2. Let $\{G_i\}$ be any sequence of abelian groups of composite order, such that $n_i = |G_i| \to \infty$. Let $N \geqslant 50$ be any positive integer, and choose i_0 so that

$$n_i > \{N!\}^N \qquad \text{for} \quad i \geqslant i_0 . \tag{8.25}$$

Then we claim that

$$|D(ZG)| \geqslant N, \qquad i \geqslant i_0 . \tag{8.26}$$

Indeed, let $i \geqslant i_0$, so that (8.25) is valid, and let p_i denote the largest prime factor of n_i.

Case 1. $p_i \geqslant N$. If $p_i^2 \mid n_i$, then (8.24 i) implies that

$$|D(ZG)| \geqslant p_i \geqslant N. \tag{8.27}$$

If $p_i^2 \nmid n_i$, then *since n_i is composite*, there is a prime q such that $2 \leqslant q < p_i$, and $qp_i \mid n_i$. Then $p_i \geqslant N \geqslant 50$, so by (8.24 iv) we again find (8.27).

Case 2. $p_i < N$. Since the largest prime factor of n_i is $< N$, it follows from (8.25) that $q^N \mid n_i$ for some prime q. Now use (8.24 ii, iii) to show that $|D(ZG_i)| > N$, as claimed.

We have thus completed the proof of Theorem 8.2. Actually, we may easily obtain better lower bounds for $|D(ZG)|$. Even when G is cyclic of order $2p$, p an odd prime, it is not hard to see that $|D(ZG)| > p^{(\lambda p/\ln p)-1}$ for some positive constant λ. Similar refinements may be made for the other cases considered in Lemma 8.24.

References

1. H. Bass, "Algebraic K-Theory," Mathematics Lecture Note Series, Benjamin, New York, 1968.
2. N. Bourbaki, "Algèbre Commutative," Chapitre 7, Hermann, Paris, 1965.
3. C. W. Curtis and I. Reiner, "Representation Theory of Finite Groups and Associative Algebras," Pure and Applied Mathematics, 2nd ed., Vol. XI, Interscience, New York, 1962; 1966; MR **26** #2519.

4. D. K. FADDEEV, On the semigroup of genera in the theory of integer representations, *Izv. Akad. Nauk SSSR Ser. Mat.* **28** (1964), 475–478; *Amer. Math. Soc. Transl.* **64** (1967), 97–101; MR **28** #5089.
5. A. FRÖHLICH, On the classgroup of integral group rings of finite abelian groups, *Mathematika* **16** (1969), 143–152; MR **41** #5512.
6. A. FRÖHLICH, The Picard group of noncommutative rings, in particular of orders, *Trans. Amer. Math. Soc.* **180** (1973), 1–46.
6a. A. FRÖHLICH, On the classgroup of integral grouprings of finite abelian groups II, *Mathematika* **19** (1972), 51–56.
7. S. GALOVICH, I. REINER, AND S. ULLOM, Class groups for integral representations of metacyclic groups, *Mathematika* **19** (1972), 105–111.
8. H. HASSE, "Über die Klassenzahl abelscher Zahlkörper," Akademie-Verlag, Berlin, 1952.
9. H. JACOBINSKI, Genera and decomposition of lattices over orders, *Acta Math.* **121** (1968), 1–29.
10. M. A. KERVAIRE AND M. P. MURTHY, On the projective class group of cyclic groups of prime power order, to appear.
11. M. P. LEE, Integral representations of dihedral groups of order $2p$, *Trans. Amer. Math. Soc.* **110** (1964), 213–231; MR **28** #139.
11a. J. MARTINET, Modules sur l'algèbre du groupe quaternionien, *Ann. Sci. École Norm. Sup.* **4** (1971), 399–408.
12. J. MILNOR, "Introduction to Algebraic K-Theory," Annals of Mathematical Studies No. 72, Princeton University Press, Princeton, NJ, 1971.
13. L. C. PU, Integral representations of non-abelian groups of order pq, *Michigan Math. J.* **12** (1965), 231–246; MR **31** #2321.
14. I. REINER, Integral representations of cyclic groups of prime order, *Proc. Amer. Math. Soc.* **8** (1957), 142–146; MR **18** #717.
15. I. REINER, A survey of integral representation theory, *Bull. Amer. Math. Soc.* **76** (1970), 159–227; MR **40** #7302.
16. I. REINER AND S. ULLOM, Class groups of integral group rings, *Trans. Amer. Math. Soc.* **170** (1972), 1–30; MR **46** #3605.
17. R. G. SWAN, Induced representations and projective modules, *Ann. of Math.* **71** (1960), 552–578; MR **25** #2131.
18. R. G. SWAN, The Grothendieck ring of a finite group, *Topology* **2** (1963), 85–110; MR **27** #3683.
19. R. G. SWAN AND E. G. EVANS, "K-Theory of Finite Groups and Orders," Lecture Notes No. 149, Springer-Verlag, Berlin/New York, 1970.
20. S. TAKAHASHI, Arithmetic of group representations, *Tôhoku Math. J.* **11** (1959), 216–246; MR **22** #733.
21. S. ULLOM, A note on the classgroup of integral group rings of some cyclic groups, *Mathematika* **17** (1970), 79–81; MR **42** #4650.
22. E. WEISS, "Algebraic Number Theory," McGraw-Hill, New York, 1963.

HEREDITARY ORDERS

Irving Reiner

ABSTRACT

This paper giver an alternate approach to some theorems of Jacobinski [1] on hereditary orders. Assume R is a Dedekind domain with quotient field K, and let Λ be an R–order in a separable K–algebra A. The ***locally free class group*** $Cl(\Lambda)$ is the abelian group generated by the classes $[M]$ with M a left Λ module in the same genus as Λ and addition defined by

$$[M] + [M'] = [M"] \quad \text{whenever} \quad M \oplus M' \simeq \Lambda \oplus M".$$

If Λ' is an order containing Λ then there is a homomorphism

$$\beta : Cl(\Lambda) \to Cl(\Lambda'), \quad \beta[X] = [\Lambda' \otimes_\Lambda X] \quad \text{for} \quad X \in Cl(\Lambda).$$

(2) Theorem. *Let Λ be a hereditary order and Λ' an order containing Λ. Then the map $\beta : Cl(\Lambda) \to Cl(\Lambda')$ is an isomorphism.*

The proof uses now standard local–global approximation methods for lattices, Roiter's lemma and Schanuel's lemma.

(9) Theorem. *Let Λ be a hereditary order and M any non–zero Λ–lattice. Then $End_\Lambda(M)$ is a hereditary R–order in $End_A(KM)$.*

Two Λ–lattices M, N are in the same ***restricted genus*** if they are in the same genus and in addition, there is some maximal order Λ' containing Λ such that $\Lambda' M \simeq \Lambda' N$.

Theorem (Jacobinski[1]). *Let Λ be a hereditary R–order in the K–separable algebra A, where K is a global field and where $End_A(KM)$ satisfies the Eichler condition relative to R. If M and N are Λ lattices in the same restricted genus then $M \simeq N$.*

Reference

[1] H. Jacobinski, *Two remarks about hereditary orders*, Proc. Amer. Math. Soc., **28** (1971)1–8.

CLASS GROUPS AND PICARD GROUPS OF ORDERS

By A. FRÖHLICH, I. REINER, *and* S. ULLOM†

[Received 16 July 1973]

Introduction

Let A be an R-order in a separable K-algebra KA, where R is a Dedekind domain with quotient field K. Fröhlich ([**3**]) has introduced a 'locally free Picard group', denoted below by LFP(A), whose elements are classes of invertible A-A-bimodules in KA which are locally free as one-sided modules. On the other hand, Reiner and Ullom ([**6**]) have investigated a 'locally free class group' Cl(A), whose elements are stable isomorphism classes of locally free rank-one left A-modules.

When A is a commutative order, it is well known (see [**1**]) that $\mathrm{LFP}(A) \cong \mathrm{Cl}(A)$. However, we show below (see (1.25)) that even when A need not be commutative there is a homomorphism $\theta\colon \mathrm{LFP}(A) \to \mathrm{Cl}(A)$. We shall see that $\ker\theta$ and $\operatorname{cok}\theta$ involve important properties of the order A and its completions. Specifically, $\ker\theta$ gives information about the automorphism group of A, under the hypotheses that stable isomorphism coincides with isomorphism, for locally free A-modules. Further, we can compute $\operatorname{cok}\theta$ explicitly when K is an algebraic number field, by using properties of reduced norms.

We also investigate the behaviour of Cl(A) and the homomorphism θ, when A is replaced by a matrix ring $M_n(A)$, or more generally by $E \otimes_R A$, where E is an Azumaya R-algebra. Section 2 is devoted to this topic, and among other results we obtain a generalization of a theorem of Rosenberg and Zelinsky ([**7**]) concerning the automorphism group of $M_n(A)$ (see (2.17)).

In § 3, we deal with the case where A is a maximal order and K is an algebraic number field. The main result here (Theorem 3.4) gives information about $\ker\theta$ and $\operatorname{cok}\theta$ in terms of ideal class groups of R.

As an application of our theorems, in § 4 we treat the case where A is the integral group ring $\mathbf{Z}G$, with G a dihedral group of order $2p$. We determine the group of outer automorphisms of A which fix its centre, and also find the set of $\mathbf{Z}$-orders in $\mathbf{Q}G$ which are locally conjugate to A.

† The last two authors wish to thank the Science Research Council for support during part of the time when this research was performed.

Finally, in §5 we compute $\operatorname{cok}\theta$ explicitly in the case where K is an algebraic number field.

1. Basic concepts

We shall use the following notation throughout this paper.

R = Dedekind domain with quotient field K.
p = maximal ideal of R (or possibly a prime of K).
R_p = p-adic completion of R, K_p = p-adic completion of K.
R-lattice = finitely generated torsion-free R-module.
A = R-order (= R-lattice which is a ring whose centre contains R).
KA = $K \otimes_R A$, regarding A as embedded in KA.
C = centre of A, KC = centre of KA.
A_p = $R_p \otimes_R A$ = R_p-order.
Left A-lattice = left A-module which is an R-lattice.
M_p = completion of A-lattice $M = R_p \otimes_R M \cong A_p \otimes_A M$.
$M^{(n)}$ = external direct sum of n copies of M.
${}_AX_B$ = left A-, right B-bimodule.
(X) = bimodule isomorphism class of bimodule X.
$[M]$ = left isomorphism class of left A-module ${}_AM$.
B^* = group of units of a ring B.

Let us recall some facts about bimodules (Bass, [**1**]; Fröhlich, [**3**]). A bimodule ${}_AX_B$ is called *invertible* if there exist a bimodule ${}_BY_A$ and bimodule isomorphisms

$$X \otimes_B Y \cong A, \quad Y \otimes_A X \cong B \tag{1.1}$$

satisfying some obvious compatibility relations. The class (X) of an invertible bimodule ${}_AX_B$ uniquely determines its inverse class (Y), and indeed $(Y) = (X^{-1})$, where

$$X^{-1} = \operatorname{Hom}_A(X, A) \cong \operatorname{Hom}_B(X, B). \tag{1.2}$$

Each invertible ${}_AX_B$ is a projective A-lattice, and

$$B \cong \operatorname{Hom}_A(X, X) \cong \operatorname{Hom}_A(Y, Y). \tag{1.3}$$

Thus, every left A-endomorphism of X is given by a right multiplication by some element of B.

Conversely, starting with a progenerator ${}_AX$ for the category of left A-modules, we may define B as $\operatorname{Hom}_A(X, X)$, and then ${}_AX_B$ is invertible, with inverse given by (1.2).

(1.4) Definition. A left A-lattice M is *locally free of rank n* (notation: $M \in \mathrm{LF}_n(A)$) if for each maximal ideal p of R, there is a left

A_p-isomorphism $M_p \cong A_p^{(n)}$. More precisely, $\mathrm{LF}_n(A)$ will denote the set of isomorphism classes $\{M\}$ of such lattices.

We may remark that if $M_p \cong A_p^{(n)}$ for even one choice of p, then

$$(KM)_p \cong K_p.M_p \cong K_p.A_p^{(n)} \cong ((KA)^{(n)})_p.$$

By the Noether–Deuring theorem (Curtis and Reiner, [**2**]), this implies that $KM \cong K.A^{(n)}$. Thus M is embeddable in $A^{(n)}$ so that its quotient is an R-torsion A-module. In particular, every class in $\mathrm{LF}_1(A)$ contains an element of $S(A)$, where $S(A)$ hereafter denotes the set of locally free left A-lattices in KA. Furthermore, for each $M \in \mathrm{LF}_n(A)$ we have $K'.M \cong K'.A^{(n)}$ whenever K' is an extension field of K.

Turning our attention back to bimodules ${}_AX_B$, we shall be concerned with whether ${}_AX \in \mathrm{LF}_n(A)$, that is, whether X is locally free as a left A-module.

(1.5) LEMMA. *Let A and B be R-orders, and let ${}_AX_B$ be invertible. Then ${}_AX \in \mathrm{LF}_1(A)$ if and only if X_B is a locally free right B-module of rank* 1.

Proof. Since ${}_AX_B$ is invertible, also X_p is invertible as A_p-B_p-bimodule, for each p. Changing notation temporarily, let R denote a local ring. We need only show that ${}_AX \cong A$ if and only if $X_B \cong B$. But this is obvious: if ${}_AX \cong A$, then $B = \mathrm{Hom}_A(X, X) \cong A_r$, where A_r denotes the ring of right multiplications by elements of A acting on X; therefore $X = 1.A_r \cong B$, as desired. The reverse argument is equally simple.

(1.6) REMARK. Let ${}_AX_B$ be invertible, with ${}_AX \in \mathrm{LF}_n(A)$. For the moment, let R denote a local ring, so now $X \cong A^{(n)}$, the left A-module consisting of all n-component row vectors over A. Then $B = \mathrm{Hom}_A(X, X)$, the ring of all $n \times n$ matrices over A, acting on the right on X. Thus X cannot be a free right B-module except when $n = 1$. Returning to the case of general R, we have thus proved that if ${}_AX_B$ is invertible and lies in $\mathrm{LF}_n(A)$, then X_B is locally free if and only if $n = 1$.

By virtue of (1.5), whenever we speak of an invertible bimodule ${}_AX_B$ as being locally free of rank 1, we need not specify whether we are viewing X as left A-module or as right B-module. We record some obvious facts.

(1.7) LEMMA. *Let A, B, E be R-orders, and let ${}_AX_B, {}_BY_E$ be invertible bimodules which are locally free of rank* 1. *Then X^{-1} (as defined in* (1.2)) *and $X \otimes_B Y$ are also invertible locally free rank-*1 *bimodules. Further, if ${}_BM \in \mathrm{LF}_1(B)$, then $X \otimes_B M \in \mathrm{LF}_1(A)$, and the class $\{X \otimes M\}$ depends only on the classes (X) and $\{M\}$.*

Proof. Localize and call R a local ring; then the proof reduces to the case where ${}_AX = A$, $B = A_r$, and so on, in the notation used in the proof of (1.5). The conclusions are then obvious in this case.

Fröhlich ([**3**]) introduced the *locally free Picard group* of A, denoting it by LF Picent(A). Here we shall use the shorter notation LFP(A). This is the group of all bimodule classes (X) such that

(1.8) $$ {}_AX_A \text{ is invertible}, \quad {}_AX \in \mathrm{LF}_1(A), $$

and

(1.9) $$ cx = xc \quad \text{for all } x \in X,\ c \in C. $$

The group operation is given by the formula $(X)(Y) = (X \otimes_A Y)$, and it follows at once from (1.7) that LFP(A) is indeed a group.

It is also convenient to introduce the *locally free ideal group*

(1.10) $$ \mathrm{LFI}(A) = \{{}_AX_A \colon X \subseteq KA,\ X \text{ invertible},\ {}_AX \in \mathrm{LF}_1(A)\}, $$

with multiplication given by computing $X.Y$ inside the algebra KA, for any $X, Y \in \mathrm{LFI}(A)$. This concept enables us to exhibit LFP(A) as a kind of 'ideal class group' by virtue of

(1.11) LEMMA. *Let KA be a separable K-algebra. Then there is an exact sequence of groups*

$$ (KC)^* \xrightarrow{\alpha} \mathrm{LFI}(A) \xrightarrow{\beta} \mathrm{LFP}(A) \longrightarrow 1, $$

where

$$ \alpha(x) = Ax, \quad x \in (KC)^*, \quad \beta(X) = (X), \quad X \in \mathrm{LFI}(A). $$

Proof. Clearly α is a homomorphism and $\beta\alpha = 1$. Since KA is a separable K-algebra, it follows (Fröhlich, [**3**], Theorem 4, Corollary 2) that each invertible ${}_AX_A$ satisfying (1.9) is isomorphic to a bimodule contained in KA. This implies at once that β is surjective. The fact that β is a homomorphism and the exactness of the sequence are both direct consequences of the above-quoted result of Fröhlich.

We now define the *normalizer* of A in KA by

(1.12) $$ N(A) = \{x \in (KA)^* \colon xAx^{-1} = A\}. $$

(1.13) LEMMA. *Let $x \in (KA)^*$, where KA is semisimple. Then $x \in N(A)$ if and only if $xAx^{-1} \subseteq A$.*

Proof. The result is clear in one direction. For the other, let $x \in (KA)^*$ be such that $xAx^{-1} \subseteq A$. The R-order ideal $[A : Ax]$ equals R.Norm(x), where Norm(x) is the ordinary norm of x, computed by letting x act on a K-basis of KA. This norm is the same whether x acts on the left or on the right on such a K-basis (since KA is semisimple), and thus $[A : Ax] = [A : xA]$. But if $xAx^{-1} \subseteq A$, then $xA \subseteq Ax$. Therefore $xA = Ax$, so $x \in N(A)$ as claimed.

This result enables us to simplify our definition of LFI(A).

(1.14) LEMMA. *Let ${}_AX_A$ be a bimodule contained in the semisimple algebra KA, and suppose that ${}_AX \in \mathrm{LF}_n(A)$. Then $n = 1$ and X is an invertible bimodule. Consequently*

$$\mathrm{LFI}(A) = \{{}_AX_A : X \subseteq KA,\ {}_AX \in \mathrm{LF}_1(A)\}.$$

Proof. Suppose that ${}_AX \in \mathrm{LF}_n(A)$; the discussion following (1.4) shows that $KX \cong K.A^{(n)}$. Hence $n = 1$ since $KX \subseteq KA$ by hypothesis. In order to prove that X is invertible, it suffices to verify that X_p is invertible for each p. Changing notation, assume that R is local and that $X = Ax$ for some $x \in (KA)^*$. Since X is a right A-module, we obtain $Ax.A \subseteq Ax$, whence $xAx^{-1} \subseteq A$. Therefore $x \in N(A)$ by (1.13). But then $x^{-1}A$ is also an A–A-bimodule and is the desired inverse of xA. This completes the proof.

Our ultimate aim is to relate the one-sided and two-sided structures of A-modules, and we give next a criterion for the left A-isomorphism of a pair of A–A-bimodules. Let $\mathrm{Aut}_R(A)$ be the group of all R-automorphisms of A, and $\mathrm{In}(A)$ the subgroup of all inner automorphisms of A. Then put

$$\text{(1.15)} \qquad \begin{cases} \mathrm{Autcent}(A) = \{f \in \mathrm{Aut}_R(A) : f(c) = c,\ c \in C\}, \\ \mathrm{Outcent}(A) = \mathrm{Autcent}(A)/\mathrm{In}(A), \end{cases}$$

where C is the centre of A. For $f, g \in \mathrm{Aut}_R(A)$, let ${}_fA_g$ be the A–A-bimodule having the additive structure of A, but with the action of A 'twisted' by f on the left and by g on the right. For $f \in \mathrm{Autcent}(A)$, put $\omega(f) = {}_1A_f$. Then ω induces a monomorphism of groups

$$\omega : \mathrm{Outcent}(A) \to \mathrm{LFP}(A),$$

and we have (Fröhlich, [**3**], for example)

(1.16) LEMMA. *Let $(X), (Y) \in$ LFP(A). Then ${}_AX \cong {}_AY$ if and only if $(Y) \in (X).\mathrm{im}\,\omega$. In particular, ${}_AX \cong {}_AA$ if and only if X is bimodule-isomorphic to ${}_1A_f$ for some $f \in$ Autcent(A).*

Finally, Fröhlich ([**3**]) proved that if KA is a separable K-algebra, then Outcent(A) is closely related to the normalizer $N(A)$ defined by (1.12). Specifically, there is an isomorphism

$$\text{(1.17)} \qquad \frac{N(A)}{A^*.(KC)^*} \cong \mathrm{Outcent}(A),$$

induced by letting $x \in N(A)$ map onto the class of the automorphism $a \to xax^{-1}$, $a \in A$. Thus there is a monomorphism

$$\frac{N(A)}{A^*.(KC)^*} \xrightarrow{\omega'} \mathrm{LFP}(A), \tag{1.18}$$

induced by mapping $x \in N(A)$ onto the class $(Ax) \in \mathrm{LFP}(A)$.

The discussion following (1.4) shows that every class in $\mathrm{LF}_1(A)$ contains an A-lattice in $S(A)$, the set of locally free rank-1 left A-lattices in KA. For $M \in S(A)$, the *right order* of M is defined as

$$O_r(M) = \{a \in KM : M.a \subseteq M\},$$

an R-order in KA. Clearly

$$\{O_r(M)\}_p = O_r(M_p) \quad \text{for all } p \quad \text{and} \quad O_r(Ma) = a^{-1}.O_r(M).a,\ a \in (KA)^*.$$

Since M is locally free, for each p we have $M_p = A_p m_p$ for some $m_p \in (K_pA)^*$, and then $(O_r(M))_p = m_p{}^{-1}A_p m_p$. Furthermore, if M is replaced by the isomorphic module Ma, where $a \in (KA)^*$, then $O_r(M)$ is replaced by $a^{-1}.O_r(M).a$, a conjugate order. This suggests the following definition.

(1.19) DEFINITION. Let A and B be R-orders in KA. We call them *locally conjugate*, and write $B \in L\text{-Conj}(A)$, if for each p there exists an element $t_p \in (K_pA)^*$ such that $B_p = t_p.A_p t_p{}^{-1}$. If $a \in (KA)^*$, then for each $B \in L\text{-Conj}(A)$ also $aBa^{-1} \in \text{L-Conj}(A)$, so $(KA)^*$ acts on L-Conj(A) by conjugation. Let L-Conj$(A)/(KA)^*$ denote the set of orbits of L-Conj(A) under this action.

We have shown above that there is a well-defined map

$$\tau' : \mathrm{LF}_1(A) \to \text{L-Conj}(A)/(KA)^*,$$

given by

$$\tau'\{M\} = \text{orbit containing } O_r(M), \quad \text{for } M \in S(A).$$

Now let $(X) \in \mathrm{LFP}(A)$, $\{M\} \in \mathrm{LF}_1(A)$. By virtue of (1.7), the class $\{X \otimes_A M\}$ is well defined in $\mathrm{LF}_1(A)$. If we define $(X).\{M\} = \{X \otimes M\}$, then the group $\mathrm{LFP}(A)$ acts on the left on the set $\mathrm{LF}_1(A)$, and clearly this action is associative.

(1.20) THEOREM. *Let KA be a separable K-algebra. There is an 'exact' sequence*

$$1 \longrightarrow \mathrm{Outcent}(A) \xrightarrow{\omega} \mathrm{LFP}(A) \xrightarrow{\theta'} \mathrm{LF}_1(A)$$

$$\xrightarrow{\tau'} \text{L-Conj}(A)/(KA)^* \longrightarrow 1,$$

where ω and τ' are as defined previously, and where $\theta'(X) = \{X\}$ for $(X) \in \mathrm{LFP}(A)$.

'Exactness' means that the following four conditions hold:

(i) ω is a monomorphism of groups.

(ii) For $(X), (Y) \in \mathrm{LFP}(A)$, we have $\theta'(X) = \theta'(Y)$ if and only if $(Y) \in (X).\mathrm{im}\,\omega$.

(iii) For $\{M\}, \{N\} \in \mathrm{LF}_1(A)$, we have $\tau'\{M\} = \tau'\{N\}$ if and only if $\{M\} \in \mathrm{LFP}(A).\{N\}$.

(iv) Each R-order $B \in \text{L-Conj}(A)$ is of the form $\tau'\{M\}$ for some $\{M\} \in \mathrm{LF}_1(A)$.

Proof. Note first that θ' is well defined, since the bimodule class (X) uniquely determines the module class $\{X\}$. Parts (i) and (ii) are obvious from (1.16) and the discussion preceding it. To prove (iii), we show first that if $N \in S(A)$ and $(X) \in \mathrm{LFP}(A)$, then $\tau'\{X \otimes_A N\} = \tau'\{N\}$. By (1.11) we may assume that X is an invertible A–A-bimodule inside KA. There is then a left A-isomorphism $X \otimes_A N \cong X.N$, where $X.N$ is computed inside KA. Consequently

$$\tau'\{X \otimes N\} \text{ is the orbit of } O_r(X.N) \text{ in L-Conj}(A)/(KA)^*.$$

If we put $M = X.N$, then clearly $O_r(M) \supseteq O_r(N)$. On the other hand, we have $N = X^{-1}.M$, so the reverse inclusion holds. Thus $O_r(M) = O_r(N)$, which proves that $\tau'\{M\} = \tau'\{N\}$, as desired.

Conversely, let $M, N \in S(A)$ be such that $O_r(M)$ is conjugate to $O_r(N)$. Replacing M by a module isomorphic to it, we may assume that $O_r(M) = O_r(N)$. For each p, we may write

$$M_p = A_p m_p, \quad N_p = A_p n_p, \quad m_p, n_p \in (K_p A)^*,$$

with $m_p = n_p = 1$ a.e. (that is, for all but finitely many p). Since $O_r(M_p) = O_r(N_p)$, we have

$$m_p^{-1} A_p m_p = n_p^{-1} A_p n_p \quad \text{for all } p,$$

whence $m_p n_p^{-1} \in N(A_p)$ for all p. Let us define

$$X = \bigcap_p A_p m_p n_p^{-1},$$

where such an intersection is to be interpreted as being formed within KA. Then $(X) \in \mathrm{LFP}(A)$ since for each p, $X_p = A_p m_p n_p^{-1}$ is an A_p–A_p-bimodule. Further $M = X.N$, since this holds true locally at each p. Thus $M \cong X \otimes_A N$, and the proof of (iii) is completed.

Lastly, let $B \in \text{L-Conj}(A)$. Then for each p we may write

$$B_p = t_p^{-1} A_p t_p \quad \text{for some } t_p \in (K_p A)^*,$$

with $t_p = 1$ a.e. Put $M = \bigcap A_p t_p$, an R-lattice in KA such that $M_p = A_p t_p$ for each p. Then M is a left A-lattice, since for each p, M_p

is a left A_p-lattice. Further,

$$\{O_r(M)\}_p = t_p^{-1} A_p t_p = B_p \quad \text{for all } p,$$

whence $B = O_r(M)$, with $\{M\} \in \mathrm{LF}_1(A)$, as desired. This completes the proof of the theorem.

We are now ready to define various class groups of A based on left A-modules, rather than on A–A-bimodules. To begin with, let $K_0(A)$ be the Grothendieck group of the category of projective left A-lattices, that is $K_0(A)$ has one generator $[M]$ for each isomorphism class $\{M\}$ of projective left A-lattices, with relations given by direct sum. Likewise, let $K_0^{\mathrm{LF}}(A)$ be the Grothendieck group of the category of locally free left A-lattices.

(1.21) Lemma. *Every locally free left A-lattice is projective. Furthermore, if M and N are A-lattices such that both M and $M \dotplus N$ are locally free, then so is N.*

Proof. Let $M \in \mathrm{LF}_m(A)$, and let L be any left A-module. Then for each p,

$$(\mathrm{Ext}_A^1(M, L))_p \cong \mathrm{Ext}_{A_p}^1(M_p, L_p) = 0$$

since M_p is left A_p-free. Therefore $\mathrm{Ext}_A^1(M, L) = 0$ for all L, and so M is left A-projective.

Now let $M \in \mathrm{LF}_m(A)$, $M \dotplus N \in \mathrm{LF}_{m+n}(A)$. For each p, we have

$$A_p{}^{(m)} \dotplus N_p \cong A_p{}^{(m+n)}.$$

Since the Krull–Schmidt theorem holds for finitely generated A_p-modules, it follows from the above isomorphism that $N_p \cong A_p^{(n)}$. Therefore $N \in \mathrm{LF}_n(A)$, and the lemma is proved.

Let M, M' be projective A-lattices; then $[M] = [M']$ in $K_0(A)$ if and only if M and M' are *stably isomorphic*, that is, $M \dotplus A^{(r)} \cong M' \dotplus A^{(r)}$ for some r. It follows at once from this fact, together with (1.21), that there is a monomorphism $K_0^{\mathrm{LF}}(A) \to K_0(A)$, given by $[M] \to [M]$, M locally free. We shall view $K_0^{\mathrm{LF}}(A)$ as a subgroup of $K_0(A)$ hereafter.

There is a homomorphism $K_0^{\mathrm{LF}}(A) \to \mathbf{Z}$, defined by mapping $[M]$ onto m whenever $M \in \mathrm{LF}_m(A)$. The kernel of this homomorphism will be called the *locally free class group* of A (notation: $\mathrm{Cl}(A)$). Thus $\mathrm{Cl}(A)$ is the subgroup of $K_0^{\mathrm{LF}}(A)$ consisting of all expressions $[M]$–$[M']$, with M, M' locally free of the same rank. We have $[M]$–$[M'] = 0$ in $\mathrm{Cl}(A)$ if and only if M and M' are stably isomorphic.

The following lemma shows that the study of locally free A-lattices can be reduced to the study of rank-1 lattices.

(1.22) LEMMA (Swan, [**8**]). *Let KA be a separable K-algebra. Then given any $M \in \mathrm{LF}_m(A)$, $N \in \mathrm{LF}_n(A)$, there exists $L \in \mathrm{LF}_1(A)$ such that*

$$M \dotplus N \cong A^{(m+n-1)} \dotplus L.$$

(1.23) COROLLARY. *Let KA be a separable K-algebra. For $L \in \mathrm{LF}_1(A)$ put*

$$\mu\{L\} = [A] - [L] \in \mathrm{Cl}(A).$$

Then every element of $\mathrm{Cl}(A)$ is of the form $\mu\{L\}$ for some $L \in \mathrm{LF}_1(A)$, and $\mu\{L\} = 0$ if and only if L is stably isomorphic to A.

Proof. Let $[M]-[M'] \in \mathrm{Cl}(A)$, where $M, M' \in \mathrm{LF}_m(A)$. Since M is A-projective, there exists a (projective) A-lattice N such that $M \dotplus N \cong A^{(r)}$ for some r. Then N is locally free by (1.21). Therefore

$$[M]-[M'] = [A^{(r)}] - [M' \dotplus N] \in \mathrm{Cl}(A).$$

However, $M' \dotplus N \in \mathrm{LF}_r(A)$, so by (1.22) there exists $L \in \mathrm{LF}_1(A)$ such that $M' \dotplus N \cong A^{(r-1)} \dotplus L$. Therefore

$$[M]-[M'] = [A]-[L] = \mu\{L\},$$

as desired. The last statement in the corollary is obvious.

Still assuming that KA is a separable K-algebra, we may describe the operation of addition in $\mathrm{Cl}(A)$ as follows: given any $L, L' \in \mathrm{LF}_1(A)$, by (1.22) there exists $L'' \in \mathrm{LF}_1(A)$ such that $L \dotplus L' \cong A \dotplus L''$. Then

$$[A]-[L]+[A]-[L'] = [A]-[L''],$$

so we have

$$\mu\{L\} + \mu\{L'\} = \mu\{L''\}, \quad \text{where } L \dotplus L' \cong A \dotplus L''. \tag{1.24}$$

We shall say that the *cancellation law* holds for locally free left A-lattices if stable isomorphism of such lattices implies their isomorphism. An important result due to Jacobinski ([**5**]; see also Fröhlich, [**4**], and Swan–Evans, [**10**]) tells us that when KA is a semisimple algebra over an algebraic number field K, then the cancellation law holds for locally free A-lattices whenever KA satisfies the *Eichler condition* (that is, no simple component of KA is a totally definite quaternion algebra). Indeed, the cancellation law holds for lattices in $\mathrm{LF}_n(A)$ when $n > 1$, regardless of whether the Eichler condition is satisfied. Let $\mu\colon \mathrm{LF}_1(A) \to \mathrm{Cl}(A)$ be the map defined in (1.23). We saw there that μ is always surjective. It is now clear that μ is injective if and only if the cancellation law holds for lattices in $\mathrm{LF}_1(A)$.

Whether or not the cancellation law holds for locally free A-lattices, we can define a map $\theta\colon \mathrm{LFP}(A) \to \mathrm{Cl}(A)$ by requiring that the following diagram commute:

$$\begin{array}{ccc} & \overset{\theta'}{\nearrow} & \mathrm{LF}_1(A) \\ \mathrm{LFP}(A) & & \downarrow \mu \\ & \underset{\theta}{\searrow} & \mathrm{Cl}(A) \end{array}$$

Thus

$$\theta(X) = [A]-[X] = \mu\{X\}, \quad (X) \in \mathrm{LFP}(A).$$

(1.25) THEOREM. *Let KA be a separable K-algebra. Then θ is a homomorphism of groups, that is,*

$$\theta((X)(X')) = \theta(X)+\theta(X'), \quad \textit{for all } (X), (X') \in \mathrm{LFP}(A).$$

Proof. By (1.11) we may assume that X and X' are in the group $\mathrm{LFI}(A)$ defined in (1.10) or (1.14). We wish to prove that

$$\mu\{X \otimes_A X'\} = \mu\{X\}+\mu\{X'\}. \tag{1.26}$$

By (1.24), this is equivalent to showing that

$$X+X' \cong A+X \otimes_A X' \tag{1.27}$$

with both sides as left A-modules. Replacing X by aX if need be, with $a \in R$, we may assume hereafter that $X \subseteq A$. Thus there is an exact sequence of bimodules

$$0 \to X \to A \to T \to 0, \tag{1.28}$$

with T an R-torsion A–A-bimodule. For each p, there is a left A_p-isomorphism $X'_p \cong A_p$, and hence also a left A_p-isomorphism

$$(T \otimes_A X')_p \cong T_p \otimes_{A_p} X'_p \cong T_p.$$

But $T \otimes X'$ and T are R-torsion A-modules, and consequently it follows that $T \otimes X'$, T are isomorphic as left A-modules.

On the other hand, $\cdot \otimes_A X'$ is an exact functor since X' is left A-projective. Applying this functor to the exact sequence (1.28) and using the left A-isomorphism $T \otimes X' \cong T$, we obtain another exact sequence of left A-modules

$$0 \to X \otimes_A X' \to X' \to T \to 0. \tag{1.29}$$

We may now use Schanuel's lemma (see [**1**]) for the pair of sequences (1.28) and (1.29), and this immediately yields (1.27). This completes the proof of the theorem.

(1.30) REMARK. Let KA be a separable K-algebra. The preceding proof (of formula (1.26)) does not use the bimodule structure of X', so we obtain at once

$$\mu\{X \otimes_A M\} = \mu\{X\}+\mu\{M\}, \quad (X) \in \mathrm{LFP}(A),\ M \in \mathrm{LF}_1(A). \tag{1.31}$$

This formula has a natural interpretation, as follows. By means of the homomorphism θ, we may view the additive group $\mathrm{Cl}(A)$ as a left module relative to the multiplicative group $\mathrm{LFP}(A)$. Then

$$\mu\{X\}+\mu\{M\} = \theta(X)+\mu\{M\} = (X).\mu\{M\}.$$

On the other hand, as in (1.20), $\mathrm{LFP}(A)$ acts on the set $\mathrm{LF}_1(A)$, according to the formula $(X).\{M\} = \{X \otimes_A M\}$. Thus (1.31) may be rewritten as

$$\mu((X).\{M\}) = (X).\mu\{M\},$$

which asserts that the actions of $\mathrm{LFP}(A)$ on $\mathrm{LF}_1(A)$ and $\mathrm{Cl}(A)$ are consistent with the map μ.

We come at last to the main result of this section.

(1.32) THEOREM. *Let KA be a separable K-algebra such that the cancellation law holds for lattices in* $\mathrm{LF}_1(A)$. *Then there is an exact sequence of groups*

$$1 \longrightarrow \mathrm{Outcent}\, A \xrightarrow{\ \omega\ } \mathrm{LFP}(A) \xrightarrow{\ \theta\ } \mathrm{Cl}(A) \longrightarrow \mathrm{cok}\,\theta \longrightarrow 1,$$

and there is a bijection

$$\mathrm{cok}\,\theta \leftrightarrow \text{L-Conj}(A)/(KA)^*.$$

Proof. In this case, the map μ in (1.23) is a bijection. The result is then merely a restatement of Theorem 1.20, since we have just shown in (1.25) that θ is a homomorphism.

(1.33) COROLLARY. *Keeping the hypotheses of* (1.32), *assume further that A is commutative. Then*

$$\mathrm{Picent}(A) \cong \mathrm{LFP}(A) \cong \mathrm{Cl}(A),$$

where $\mathrm{Picent}(A)$ *is the group of classes of invertible bimodules X satisfying* (1.9), *with multiplication defined via* $\otimes_A$.

Proof. Since A is commutative, every $(X) \in \mathrm{Picent}(A)$ is such that $X \in \mathrm{LF}_1(A)$ (Bass, [**1**]), and thus $\mathrm{Picent}(A) = \mathrm{LFP}(A)$ in this case. Further, L-Conj(A) consists of the single order A, since if $B \in$ L-Conj(A), then $B_p = A_p$ for all p, whence $B = A$. Thus $\mathrm{cok}\,\theta = 1$. But Outcent(A) is clearly trivial, so it follows from (1.32) that θ maps $\mathrm{LFP}(A)$ isomorphically on $\mathrm{Cl}(A)$, as desired.

2. Matrix rings over orders

Throughout this section let A be an R-order in a separable K-algebra KA and let E be an Azumaya R-algebra. (This means that E is a central separable R-algebra (see [**1a**]). We shall eventually restrict our attention to the case where $E = M_n(R)$, the ring of all $n \times n$ matrices over R.) Our aim here is to compare the basic exact sequences (1.20) and (1.32) for A with those for $E \otimes_R A$. For brevity, we shall usually write $\otimes$ instead of $\otimes_R$.

(2.1) THEOREM. *Let E be any Azumaya R-algebra, and let KA be a separable K-algebra. The functor $E \otimes_R \cdot$ induces maps f_1, f_2, f'_3, f'_4 making the following diagram commute:*

$$\begin{array}{ccccccc} 1 & \longrightarrow & \mathrm{Outcent}(A) & \xrightarrow{\omega} & \mathrm{LFP}(A) & \xrightarrow{\theta'} & \\ & & \downarrow f_1 & & \downarrow f_2 & & \\ 1 & \longrightarrow & \mathrm{Outcent}(E \otimes A) & \xrightarrow{\omega_1} & \mathrm{LFP}(E \otimes A) & \xrightarrow{\theta'_1} & \end{array}$$

$$\begin{array}{ccccc} \mathrm{LF}_1(A) & \xrightarrow{\tau'} & \text{L-Conj}(A)/(KA)^* & \longrightarrow & 1 \\ \downarrow f'_3 & & \downarrow f'_4 & & \\ \mathrm{LF}_1(E \otimes A) & \xrightarrow{\tau'_1} & \text{L-Conj}(E \otimes A)/(K(E \otimes A))^* & \longrightarrow & 1 \end{array}$$

where the rows are 'exact' as in (1.20). The map f_2 is an isomorphism of groups and f_1 is a monomorphism.

Proof. (i) Each $h \in \mathrm{Aut}_R(A)$ maps onto an element $1 \otimes h \in \mathrm{Aut}_R(E \otimes A)$. If $h \in \mathrm{Autcent}(A)$, then $1 \otimes h \in \mathrm{Autcent}(E \otimes A)$, since $E \otimes A$ has centre $1 \otimes C$ (Fröhlich, [**3**]). The homomorphism f_1 is induced by the mapping $h \to 1 \otimes h$, $h \in \mathrm{Autcent}(A)$.

(ii) Let Picent(A) be the multiplicative group of classes (X) of invertible A–A-bimodules satisfying (1.9), with multiplication given via $\otimes_A$. Fröhlich ([**3**]) proved that there is an isomorphism

$$f\colon \mathrm{Picent}(A) \cong \mathrm{Picent}(E \otimes A), \tag{2.2}$$

given by $(X) \to (E \otimes X)$. Let f_2 be the restriction of f to the subgroup LFP(A) of Picent(A). Clearly f_2 carries LFP(A) into LFP($E \otimes A$), and we must show that this map is surjective. Every element of LFP($E \otimes A$) is of the form $E \otimes X$ for some $(X) \in \mathrm{Picent}(A)$ by (2.2). For each prime p, there is a left $(E \otimes A)_p$-isomorphism $(E \otimes X)_p \cong (E \otimes A)_p$. Restricting the operator domain to $1 \otimes A_p$, and letting r denote the R-rank of E, we obtain a left A_p-isomorphism $X_p^{(r)} \cong A_p^{(r)}$. Therefore $X_p \cong A_p$, since the Krull–Schmidt theorem holds for left A_p-lattices. Thus $(X) \in \mathrm{LFP}(A)$, and we have shown that

$$f_2\colon \mathrm{LFP}(A) \cong \mathrm{LFP}(E \otimes A).$$

(iii) For any $h \in \text{Autcent}(A)$, we have

$$(2.3) \qquad (f_2\omega)h = f_2({}_1A_h) = E \otimes {}_1A_h \cong {}_1(E \otimes A)_{1\otimes h} = (\omega_1 f_1)h.$$

This proves that the first square in (2.1) is commutative and therefore f_1 is a monomorphism.

(iv) We now define $f_3'\{X\} = \{E \otimes X\}$, for $\{X\} \in \text{LF}_1(A)$, and we let $f_4'(B)$ be the orbit of $E \otimes B$, for $B \in \text{L-Conj}(A)$. It is then clear that the middle square in (2.1) commutes. Now let $X \subseteq KA$ be such that $X \in \text{LF}_1(A)$, and let $B = O_r(X)$. In order to show that the right-hand square in (2.1) is commutative, we need only verify that

$$(2.4) \qquad O_r(E \otimes X) = E \otimes B.$$

It suffices to establish this when R is a local ring. If $\{e_i\}$ is an R-basis of E, we have

$$E \otimes X = \textstyle\sum^{\cdot} e_i \otimes X, \quad K(E \otimes A) = \sum^{\cdot} e_i \otimes KA.$$

Let $\xi = \sum e_i \otimes w_i$, $w_i \in KA$. Then $\xi \in O_r(E \otimes X)$ if and only if

$$(1 \otimes\ )\xi \subseteq E \otimes X.$$

This occurs if and only if $Xw_i \subseteq X$ for each i, that is, if and only if $\xi \in E \otimes B$. This establishes (2.4), and completes the proof of the theorem.

(2.5) Corollary. *Let f_3: $\text{Cl}(A) \to \text{Cl}(E \otimes A)$ be the homomorphism induced by the functor $E \otimes_R \cdot$. Then there is a commutative diagram of groups*

$$\begin{array}{ccc} \text{LFP}(A) & \xrightarrow{\theta} & \text{Cl}(A) \\ {\scriptstyle f_2}\downarrow & & \downarrow{\scriptstyle f_3} \\ \text{LFP}(E \otimes A) & \xrightarrow{\theta_1} & \text{Cl}(E \otimes A). \end{array}$$

Proof. This follows at once from (2.1) by using the fact that the square

$$\begin{array}{ccc} \text{LF}_1(A) & \xrightarrow{\mu} & \text{Cl}(A) \\ {\scriptstyle f_3'}\downarrow & & \downarrow{\scriptstyle f_3} \\ \text{LF}_1(E \otimes A) & \xrightarrow{\mu_1} & \text{Cl}(E \otimes A) \end{array}$$

is commutative, together with the formulas $\theta = \mu\theta'$, $\theta_1 = \mu_1\theta_1'$.

(2.6) Theorem. *Keeping the hypotheses of* (2.1), *assume further that the cancellation law holds for both locally free A-lattices and locally free $E \otimes A$-lattices. Then there is a commutative diagram of groups, with exact* *rows:*

$$\begin{array}{ccccccc}
1 & \longrightarrow & \mathrm{Outcent}(A) & \xrightarrow{\omega} & \mathrm{LFP}(A) & \xrightarrow{\theta} & \\
 & & \downarrow{\scriptstyle f_1} & & \downarrow{\scriptstyle f_2} & & \\
1 & \longrightarrow & \mathrm{Outcent}(E\otimes A) & \xrightarrow{\omega_1} & \mathrm{LFP}(E\otimes A) & \xrightarrow{\theta_1} &
\end{array}$$

$$\begin{array}{ccccc}
\mathrm{Cl}(\mathrm{A}) & \longrightarrow & \mathrm{cok}\,\theta & \longrightarrow & 1 \\
\downarrow{\scriptstyle f_3} & & \downarrow{\scriptstyle f_4} & & \\
\mathrm{Cl}(E\otimes A) & \longrightarrow & \mathrm{cok}\,\theta_1 & \longrightarrow & 1,
\end{array}$$

where f_3 is as in (2.5) and f_4 is induced by f_3. The map f_2 is an isomorphism and f_1 a monomorphism.

Proof. This follows at once from (1.32), (2.1), and (2.5).

REMARK. In the special case where $E = M_n(R)$, we shall show below in (2.8) that if the cancellation law holds for $\mathrm{LF}_1(A)$, then it also holds for $\mathrm{LF}_1(E\otimes A)$. On the other hand, $K(E\otimes A) = M_n(KA)$, and if K is an algebraic number field then $M_n(KA)$ satisfies the Eichler condition whenever $n > 1$ (whether or not KA satisfies the Eichler condition). Thus the cancellation law necessarily holds for $\mathrm{LF}_1(E\otimes A)$ whenever $n > 1$ and K is an algebraic number field.

(2.7) THEOREM. *Under the hypotheses of (2.6), there is an exact sequence of groups*

$$1 \longrightarrow \mathrm{Outcent}(A) \xrightarrow{f_1} \mathrm{Outcent}(E\otimes A) \longrightarrow \ker f_3 \longrightarrow$$

$$\mathrm{cok}\,\theta \xrightarrow{f_4} \mathrm{cok}\,\theta_1 \longrightarrow \mathrm{cok} f_3 \longrightarrow 0.$$

Proof. From (2.6) we obtain a commutative diagram with exact rows:

$$\begin{array}{ccccccccc}
 & & 0 & \longrightarrow & \mathrm{LFP}(A) & \longrightarrow & \mathrm{LFP}(E\otimes A) & \longrightarrow & 0 \\
 & & \downarrow & & \downarrow{\scriptstyle \theta} & & \downarrow{\scriptstyle \theta_1} & & \\
0 & \longrightarrow & \ker f_3 & \longrightarrow & \mathrm{Cl}(A) & \longrightarrow & \mathrm{Cl}(E\otimes A). & &
\end{array}$$

The groups in the bottom row are abelian, so the 'snake lemma' ([**1**]) can be applied to yield the exact sequence in (2.7). The last term $\mathrm{cok} f_3$ occurs, since $\mathrm{cok} f_3 \cong \mathrm{cok} f_4$ by a 'snake lemma' argument applied to the diagram in (2.6).

From here on, we shall restrict our attention to the case where E is a matrix ring $M_n(R)$, making use of the fact that the ring $E \otimes A$ is Morita-equivalent to A ([**1**]). Let $V = R^{(n)}$, viewed as left E-module. Then there is a bijection λ between the set of isomorphism classes of left A-modules and the corresponding set for $E \otimes A$, given by

$$\lambda\{M\} = \{V \otimes_R M\}, \quad \text{for } M \text{ a left } A\text{-module}.$$

(2.8) LEMMA. *Let KA be a separable K-algebra, let $E = M_n(R)$ and let λ be as above. Then*

(i) *M is projective if and only if $\lambda\{M\}$ is projective;*

(ii) *$M \in \mathrm{LF}_{nk}(A)$ if and only if $\lambda\{M\} \in LF_k(E \otimes A)$;*

(iii) *if the cancellation law holds for locally free left A-lattices, then it also holds for locally free left $E \otimes A$ lattices;*

(iv) *λ induces isomorphisms*

$$K_0(A) \cong K_0(E \otimes A), \quad \mathrm{Cl}(A) \cong \mathrm{Cl}(E \otimes A).$$

Proof. We use repeatedly the following facts:

$V \otimes M$, $M^{(n)}$ are isomorphic as A-modules,

$(V \otimes M)^{(n)}$, $E \otimes M$ are isomorphic as $E \otimes A$-modules.

Hence if M is a direct summand of $A^{(r)}$, then $(V \otimes M)^{(n)}$ is a direct summand of $(E \otimes M)^{(r)}$, and so $V \otimes M$ is projective. Conversely, if $V \otimes M$ is a direct summand of $(E \otimes M)^{(r)}$, then restricting the operator domain to A, we see that $M^{(n)}$ is a direct summand of $A^{(rn^2)}$, so M is projective. This proves (i), and an analogous argument (see (1.21), for example) proves (ii).

Next, let M and N be A-projective modules such that $\lambda\{M\}$ and $\lambda\{N\}$ are stably isomorphic. Then for some r,

$$V \otimes M \dotplus E \otimes A^{(r)} \cong V \otimes N \dotplus E \otimes A^{(r)},$$

that is,

$$V \otimes (M \dotplus A^{(nr)}) \cong V \otimes (N \dotplus A^{(nr)}).$$

Since λ is bijective, this gives $M \dotplus A^{(nr)} \cong N \dotplus A^{(nr)}$, so M is stably isomorphic to N. This proves (iii), and also shows that $K_0(A) \cong K_0(E \otimes A)$.

Finally, we define $g\colon \mathrm{Cl}(A) \to \mathrm{Cl}(E \otimes A)$ by putting

$$g([A]-[M]) = [V \otimes A]-[V \otimes M], \quad \text{for } \{M\} \in \mathrm{LF}_1(A). \tag{2.9}$$

Since

$$[V \otimes A]-[V \otimes M] = [E \otimes A]-[V^{(n-1)} \otimes A + V \otimes M],$$

it is clear that g carries $\mathrm{Cl}(A)$ into $\mathrm{Cl}(E \otimes A)$. Then g is a well-defined monomorphism, being the restriction to $\mathrm{Cl}(A)$ of the isomorphism $K_0(A) \cong K_0(E \otimes A)$.

It remains for us to show that g is surjective. Each $\xi \in \mathrm{Cl}(E \otimes A)$ is of the form

$$\xi = [E \otimes A] - [V \otimes M],$$

where M is a left A-lattice such that $V \otimes M \in \mathrm{LF}_1(E \otimes A)$. By (ii) we may conclude that $M \in \mathrm{LF}_n(A)$, and so by (1.22) there exists an $L \in \mathrm{LF}_1(A)$ such that $M \cong A^{(n-1)} \dotplus L$. Therefore

$$\begin{aligned}\xi &= [V^{(n)} \otimes A] - [V \otimes A^{(n-1)} \dotplus V \otimes L] \\ &= [V \otimes A] - [V \otimes L],\end{aligned}$$

which lies in the image of g. This completes the proof of the lemma.

The preceding lemma enables us to draw several interesting consequences from Theorems 2.6 and 2.7. Let us introduce the notation

(2.10) $\quad (\mathrm{Cl}(A))_n = \{x \in \mathrm{Cl}(A) : nx = 0\}, \quad (\mathrm{Cl}(A))^n = \{x^n : x \in \mathrm{Cl}(A)\}.$

Let $E = M_n(R)$, and identify $E \otimes A$ with $M_n(A)$ and $(K(E \otimes A))^*$ with $\mathrm{GL}_n(KA)$.

(2.11) THEOREM. *Let $E = M_n(R)$, and assume that KA is a separable K-algebra, and that the cancellation law holds for locally free left A-*lattices. *Let $f_3 \colon \mathrm{Cl}(A) \to \mathrm{Cl}(E \otimes A)$ be the homomorphism induced by the functor $E \otimes_R \cdot$, and let ω, ω_1 be as in* (2.1). *Then*†

(2.12) $\quad \ker f_3 \cong (\mathrm{Cl}(A))_n, \quad \operatorname{cok} f_3 \cong \mathrm{Cl}(A)/(\mathrm{Cl}(A))^n$

and†

(2.13) $\quad \operatorname{im} \omega_1 = \{f_2(x) : x \in \mathrm{LFP}(A),\ x^n \in \operatorname{im} \omega\}.$

Furthermore, there is an exact sequence of groups

(2.14) $\quad 1 \longrightarrow \mathrm{Outcent}(A) \xrightarrow{f_1} \mathrm{Outcent}(M_n(A)) \longrightarrow (\mathrm{Cl}(A))_n \longrightarrow$

$$\operatorname{cok} \theta \longrightarrow \operatorname{cok} \theta_1 \longrightarrow \mathrm{Cl}(A)/(\mathrm{Cl}(A))^n \longrightarrow 0,$$

with bijections

(2.15) $\quad \operatorname{cok} \theta \leftrightarrow \text{L-Conj}(A/(KA)^*), \quad \operatorname{cok} \theta_1 \leftrightarrow \text{L-Conj}(M_n(A))/\mathrm{GL}_n(KA).$

Proof. Let $g \colon \mathrm{Cl}(A) \cong \mathrm{Cl}(E \otimes A)$ be the isomorphism given by (2.9). We claim that $f_3 = n.g$. Indeed, for any $M \in \mathrm{LF}_1(A)$ we have

$$\begin{aligned}f_3([A] - [M]) &= [E \otimes A] - [E \otimes M] \\ &= [V^{(n)} \otimes A] - [V^{(n)} \otimes M] = n.g([A] - [M]).\end{aligned}$$

This shows that $f_3 = n.g$, and (2.12) is then clear.

† See (2.18) and (2.19).

Next let $x \in \mathrm{LFP}(A)$ be such that $x^n \in \operatorname{im} \omega$. Then $\theta(x^n) = 0$ by (2.6), and therefore $n.\theta(x) = 0$. Thus $\theta(x) \in \ker f_3$, and so $(\theta_1 f_2)x = (f_3\theta)x = 0$. This shows that $f_2(x) \in \ker \theta_1$, so by (2.6) we obtain $f_2(x) \in \operatorname{im} \omega_1$. We have now proved that the left-hand expression in (2.13) contains the right-hand one. For the reverse inclusion, let $y \in \operatorname{im} \omega_1$ and put $y = f_2(x)$, where $x \in \mathrm{LFP}(A)$. Then

$$(f_3\theta)x = (\theta_1 f_2)x = \theta_1(y) = 0.$$

and so $\theta(x) \in \ker f_3$. Therefore $n.\theta(x) = 0$, whence $\theta(x^n) = 0$, and thus $x^n \in \operatorname{im} \omega$. This completes the proof of (2.13).

As for the remaining statements in the theorem, (2.14) is now a consequence of (2.7) and (2.12), while (2.15) is a restatement of the last part of (1.32).

(2.16) Corollary. *Keep the hypotheses of* (2.11). *If* LFP(A) *is a finite group*† *of exponent* n, *then there is an isomorphism*

$$\omega_1\colon \mathrm{Outcent}(M_n(A)) \cong \mathrm{LFP}(M_n(A)).$$

Proof. The hypothesis implies that for each $x \in \mathrm{LFP}(A)$ we have $x^n = 0$, so by (2.13) we may conclude that $\operatorname{im} \omega_1 = \operatorname{im} f_2$. But f_2 is surjective, whence so is ω_1. This gives the result, since ω_1 is already known to be injective.

(2.17) Corollary. *Let* A *be a commutative order, and keep the hypotheses of* (2.11). *Then for each* n

$$\mathrm{Outcent}(M_n(A)) \cong (\mathrm{Cl}(A))_n,$$

and there is a bijection

$$\text{L-Conj}(M_n(A))/\mathrm{GL}_n(KA) \leftrightarrow \mathrm{Cl}(A)/(\mathrm{Cl}(A))^n.$$

Proof. In (2.14) we have $\mathrm{Outcent}(A) = 0$, $\operatorname{cok} \theta = 0$, which gives the result. Rosenberg and Zelinsky ([**7**]) proved this corollary for the special case where $A = R$.

(2.18) Remark. Supposing that KA is a separable K-algebra, and that $E = M_n(R)$, let us investigate briefly what happens when we drop the hypothesis that the cancellation law holds for locally free A-lattices. Of course Theorem 2.1 and Lemma 2.8 remain valid, as does the first paragraph in the proof of (2.11). Consequently (2.12) is still true.

We now show

$$\operatorname{im} \omega_1 \supseteq \{f_2(x)\colon x \in \mathrm{LFP}(A),\ x^n \in \operatorname{im} \omega\}. \tag{2.19}$$

† If K is an algebraic number field, then LFP(A) is finite (Fröhlich, [**3**]).

For let $(X) \in \mathrm{LFP}(A)$ be such that $(X)^n \in \operatorname{im} \omega$. Then the n-fold tensor product $X \otimes_A \dots \otimes_A X$ is isomorphic to A as left A-module. Hence by (1.27) we have

$$X^{(n)} \cong A^{(n-1)} \dot{+} X \otimes \dots \otimes X \cong A^{(n)}$$

as left A-modules. Therefore

$$E \otimes X \cong V \otimes X^{(n)} \cong V \otimes A^{(n)} \cong E \otimes A$$

as left $E \otimes A$-modules. By (1.16) we may conclude that

$$f_2(X) = (E \otimes X) \in \operatorname{im} \omega_1,$$

which establishes (2.19).

The proof of (2.16) remains valid in this case, provided we use the inclusion (2.19) in place of the equality (2.13). Thus (2.16) is still true in the present situation.

3. Maximal orders

To begin with, let A be a maximal R-order in a separable K-algebra KA, where K is an algebraic number field. The order A and all of the groups $\mathrm{Outcent}(A)$, $\mathrm{LFP}(A)$, $\mathrm{Cl}(A)$, et cetera, split according to the decomposition of KA into simple components. It thus suffices to restrict our attention to simple algebras. From here on in this section, let KA be a simple algebra with centre K, and then R is the centre of A.

The prime p of K is said to *ramify* in KA if $(KA)_p$ is not a full matrix algebra over K_p. At each unramified prime p, $A_p \cong M_n(R_p)$, where $n^2 = (KA : K)$. Let $I(R)$ denote the group of R-ideals in K, and $I_A(R)$ its subgroup

$$\{R\alpha \colon \alpha \in K,\ \alpha_p > 0 \text{ at each real infinite prime } p \text{ of } K \text{ ramified in } KA\}$$

Then define

$$\mathrm{Cl}_A(R) = I(R)/I_A(R), \tag{3.1}$$

a modification of the usual class group $\mathrm{Cl}(R)$.

Let $\nu \colon KA \to K$ be the reduced norm map. For each $a \in (KA)^*$, it is well known that $\nu(a)_p > 0$ at each real prime p ramified in KA. For any left A-lattice X spanning KA, define the *reduced norm* of X as

$$\nu(X) = \sum_{x \in X} R.\nu(x) \in I(R).$$

Each such X is necessarily a locally free A-lattice, and ν gives a surjection $\mathrm{LF}_1(A) \to I(R)$, inducing a surjection

$$\bar{\nu} \colon \mathrm{LF}_1(A) \to \mathrm{Cl}_A(R).$$

Further, $\bar{\nu}(X)$ depends only on the stable isomorphism class of X, and

$$\bar{\nu}(X'') = \bar{\nu}(X).\bar{\nu}(X') \quad \text{whenever } X \dotplus X' \cong A \dotplus X''.$$

If we define ρ: $(\mathrm{Cl}(A) \to Cl_A(R)$ by

$$\rho\{[A]-[X]\} = \bar{\nu}(X), \quad \{X\} \in \mathrm{LF}_1(A),$$

it follows from the preceding remarks that ρ is a well-defined homomorphism. By a theorem of Eichler (see also Swan, [**9**]) we have

$$\rho\colon \mathrm{Cl}(A) \cong \mathrm{Cl}_A(R), \tag{3.2}$$

whether or not KA satisfies the Eichler condition.

Let $(KA : K) = n^2$ and define a homomorphism

$$t\colon \mathrm{Cl}(R) \to \mathrm{Cl}_A(R) \tag{3.3}$$

by letting the class of the R-ideal J in K map onto the class of J^n in $\mathrm{Cl}_A(R)$. Note that if $\mathrm{Cl}(R) \neq \mathrm{Cl}_A(R)$, then some real prime of K *must* ramify in KA, so n is even. This shows that t is always well defined. In terms of t, we may describe the cokernel of the homomorphism θ: $\mathrm{LFP}(A) \to \mathrm{Cl}(A)$. Our main result is

(3.4) THEOREM. *Let A be a maximal R-order in the central simple K-algebra KA, and let $\mathrm{Cl}_A(R)$ and t be as above. Let S be the set of finite primes of K ramified in KA. Then there is an exact sequence of groups*

$$1 \to \ker t \to \ker\theta \to \prod_{p\in S} \mathrm{LFP}(A_p) \to \operatorname{cok} t \to \operatorname{cok}\theta \to 0. \tag{3.5}$$

Further, we have:

(i) $\mathrm{LFP}(A_p) \cong \mathrm{Outcent}(A_p) \cong N(A_p)/(K_p)^*.(A_p)^*$;

(ii) *if KA satisfies the Eichler condition, then* $\ker\theta \cong \mathrm{Outcent}(A)$;

(iii) *if* $\mathrm{Cl}(R) = \mathrm{Cl}_A(R)$, *then using the notation in* (2.10), *we have* $\ker t = (\mathrm{Cl}(R))_n$, $\operatorname{cok} t = \mathrm{Cl}(R)/(\mathrm{Cl}(R))^n$;

(iv) *if A is a separable R-algebra, then* $\ker t \cong \ker\theta$, $\operatorname{cok} t \cong \operatorname{cok}\theta$.

As a first step in the proof, we establish

(3.6) PROPOSITION. *Let KA be any semisimple algebra over an algebraic number field K, and let A be any R-order in KA, not necessarily maximal. Let S_0 be the set of all finite primes p of R at which A_p is not central-separable as C_p-algebra, where C is the centre of A. Then there is an exact sequence of groups*

$$1 \longrightarrow \mathrm{Picent}(C) \xrightarrow{\ \gamma\ } \mathrm{LFP}(A) \xrightarrow{\ \delta\ } \prod_{p\in S_0} \mathrm{LFP}(A_p) \longrightarrow 1.$$

Proof. The elements of Picent(C) are of the form (J), where J is an invertible C-lattice spanning KC. The map γ is defined by $\gamma(J) = JA$. Fröhlich ([**3**]) showed exactness of the above sequence, with Picent in place of each LFP. However, each such J is locally free, whence $\operatorname{im}\gamma \subseteq \mathrm{LFP}(A)$. To prove δ surjective, let $(X_p) \in \mathrm{LFP}(A_p), p \in S_0$. By [**3**], there exists $(Y) \in \mathrm{Picent}(A)$ such that $(Y_p) = (X_p), p \in S_0$. Then Y_p, X_p, A_p are isomorphic as left A-modules, for $p \in S_0$. But for $p \notin S_0$, A_p is central-separable over C_p, and in this case $Y_p \cong A_p$ because $KY = KA$. Thus $(Y) \in \mathrm{LFP}(A_p)$ and $\delta(Y) = \prod_{p \in S_0} (X_p)$, as desired.

We turn now to the proof of Theorem 3.4, and let A be a maximal order in a central simple K-algebra. The set S_0 in (3.6) may be taken to be the set S in (3.4). Using (3.2), we then obtain a commutative diagram with exact rows

$$\begin{array}{ccccccccc} 1 & \longrightarrow & \mathrm{Cl}(R) & \xrightarrow{\gamma} & \mathrm{LFP}(A) & \xrightarrow{\delta} & \prod_{p \in S} \mathrm{LFP}(A_p) & \longrightarrow & 1 \\ & & \downarrow t & & \downarrow \theta & & \downarrow & & \\ 1 & \longrightarrow & \mathrm{Cl}_A(R) & \xrightarrow{\rho^{-1}} & \mathrm{Cl}(A) & \longrightarrow & 1. & & \end{array}$$

We should verify that $t = \rho\theta\gamma$; for any R-ideal J in K, we have

$$(\rho\theta\gamma)\,(\text{class of } J) = \rho\{[A]-[JA]\} = \bar{\nu}(JA)$$

$$= J^n.\bar{\nu}(A) = \text{class of } J^n \text{ in } \mathrm{Cl}_A(R) = t(\text{class of } J).$$

We now apply the snake lemma to this diagram, to get the exact sequence (3.5).

Since R_p is already a local ring, any $(X) \in \mathrm{LFP}(A_p)$ is such that X_p, A_p are isomorphic as left A_p-modules. Then the map

$$\mathrm{Outcent}(A_p) \to \mathrm{LFP}(A_p)$$

is an isomorphism by (1.20). The rest of (i) follows from (1.17). We have proved (ii) in Theorem 1.32. Part (iii) is obvious from the definition of the homomorphism t. Finally, if A is central-separable over R, then the set S is empty, and so (iv) is an immediate consequence of the exactness of (3.5). This completes the proof of Theorem 3.4.

We have the following corollary to Theorem 3.4.

(3.7) Corollary. *Let*

$$\prod_{p \in S} \mathrm{LFP}(A_p) \xrightarrow{\psi} \operatorname{cok} t \longrightarrow \operatorname{cok}\theta \longrightarrow 0,$$

as in (3.5). *Then*

$$\operatorname{cok}\theta \cong \operatorname{cok} t/\operatorname{im}\psi \cong \frac{\mathrm{Cl}_A(R)}{(\mathrm{Cl}(R))^n . T},$$

where T *is the subgroup of* $\mathrm{Cl}_A(R)$ *generated by the classes of ideals* $\{[p]^{n_p}: p \in S\}$, *where for* $p \in S$, K_pA *is an* $n_p \times n_p$ *matrix algebra over a skew field.*

Proof. Let $K_pA = M_{n_p}(D_p)$, where D_p is a skewfield. Then we may assume that $A_p = M_{n_p}(\Delta)$, where Δ is a maximal R_p-order in D_p.

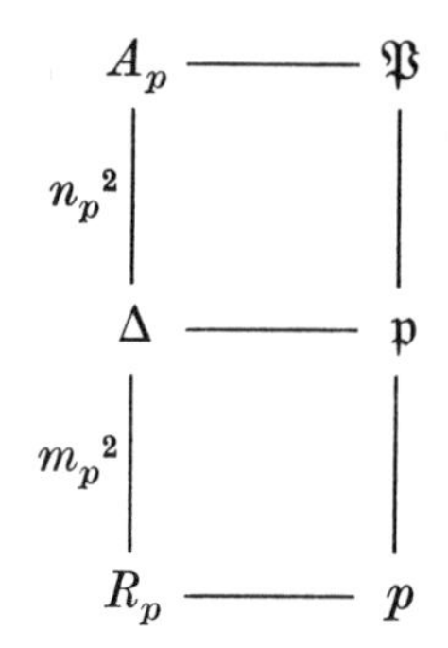

We have $\mathfrak{P} = \mathfrak{p}.A_p$ and LFP(A_p) has generator ($\mathfrak{P}$) of order m_p. Then $\nu(\mathfrak{P}) = \nu(\mathfrak{p}A_p) = \nu_{\Delta/R_p}(\mathfrak{p})^{n_p} = p^{n_p}$. Thus $\psi(\mathfrak{P}) = [p]^{n_p}$, and the result follows.

4. Further notation. Dihedral group example

Hereafter assume that K is an algebraic number field. Let $A \subseteq B \subseteq KA$, where B is a maximal R-order in KA (so $KB = KA$). Then $B = \sum^{\cdot} B_i$, with B_i a maximal R-order in the ith simple component KB_i of KA. We let F_i be the centre of KB_i, R_i be the integral closure of R in F_i, $(KB_i : F_i) = m_i^2$, ν_i: $KB_i \to F_i$ be the reduced norm map.

Define $I(R_i)$, $I_{B_i}(R_i)$ as in § 3, and let ρ_i: $\mathrm{Cl}(B_i) \cong \mathrm{Cl}_{B_i}(R_i)$ be the isomorphism induced by ν_i, as in (3.2). Let t_i: $\mathrm{Cl}(R_i) \to \mathrm{Cl}_{B_i}(R_i)$ be given as in (3.3).

In the proof of Theorem 3.4, we showed that the diagram

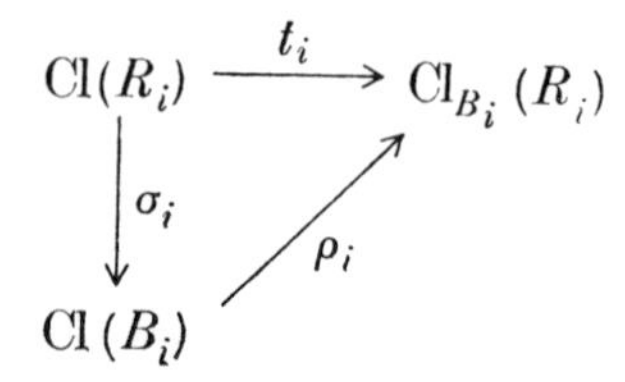

commutes, where $\sigma_i(J_i) = [J_i B_i]$, J_i an R_i-ideal in F_i. Forming direct sums over the index i, we obtain a commutative diagram

(4.1)
$$\begin{array}{ccc} \mathrm{Cl}(C') & \xrightarrow{t} & \mathrm{Cl}_B(C') \\ \downarrow{\scriptstyle\sigma} & \nearrow{\scriptstyle\rho} & \\ \mathrm{Cl}(B) & & \end{array}$$

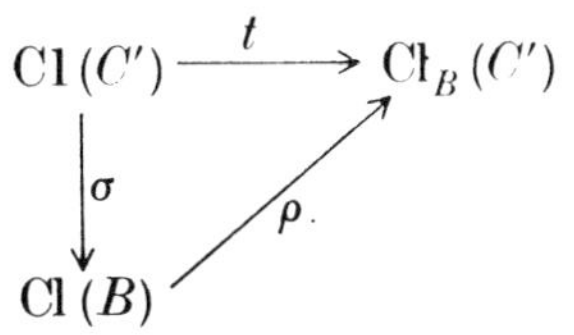

where $C' = \sum^{\cdot} R_i$ is the centre of B.

Next we observe that there are 'change of rings' homomorphisms

$$\varphi\colon \mathrm{Picent}(C) \to \mathrm{Picent}(C') \cong \mathrm{Cl}(C'), \quad \eta\colon \mathrm{Cl}(A) \to \mathrm{Cl}(B),$$

with φ induced by the functor $C' \otimes_C \cdot$, and η by $B \otimes_A \cdot$. Both φ and η are surjective (Reiner–Ullom, [**6**], or Theorem 5.3), and it follows at once from the associativity of tensor products that the diagram below is commutative:

(4.2)
$$\begin{array}{ccccccc} & & \mathrm{Picent}(C) & \xrightarrow{\varphi} & \mathrm{Cl}(C') & \xrightarrow{t} & \mathrm{Cl}_B(C') \\ & \swarrow{\scriptstyle\gamma} & & & \downarrow{\scriptstyle\sigma} & & \downarrow{\scriptstyle 1} \\ \mathrm{LFP}(A) & \xrightarrow{\theta} & \mathrm{Cl}(A) & \xrightarrow{\eta} & \mathrm{Cl}(B) & \xrightarrow{\rho} & \mathrm{Cl}_B(C') \end{array}$$

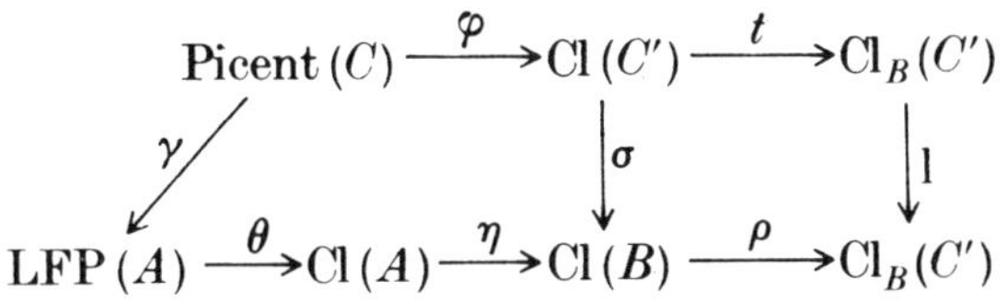

Here γ is defined by $\gamma(J) = (JA)$ for $(J) \in \mathrm{Picent}(C)$.

To illustrate the techniques so far developed, we investigate the relationship between $\mathrm{LFP}(A)$ and $\mathrm{Cl}(A)$ for the case where A is the integral group ring $\mathbf{Z}G$ of a dihedral group G. Among other results, we obtain an explicit formula for $\mathrm{Outcent}(A)$.

(4.3) Theorem. *Let G be the dihedral group of order $2p$, where p is an odd prime, and let $A = \mathbf{Z}G$. Let $\theta\colon \mathrm{LFP}(A) \to \mathrm{Cl}(A)$ be the homomorphism defined in* (1.25). *Denote by S the ring of all algebraic integers in the maximal real subfield L of $\mathbf{Q}(\sqrt[p]{1})$, and define $(\mathrm{Cl}(S))_2$ and $(\mathrm{Cl}(S))^2$ as in* (2.10) *with $n = 2$. Then $\mathrm{Outcent}(A) \cong \ker\theta$, and $\ker\theta$ is a split extension of $(\mathrm{Cl}(S))_2$ by a cyclic group of order $(p-1)/2$. Furthermore,* $\operatorname{cok}\theta = \mathrm{Cl}(S)/(\mathrm{Cl}(S))^2$.

Proof. For $A = \mathbf{Z}G$ we have

$$\mathbf{Q}A = \mathbf{Q}G \cong \mathbf{Q} + \mathbf{Q} + M_2(L),$$

so $\mathbf{Q}A$ satisfies the Eichler condition. The discussion following (1.24) then shows that the cancellation law holds for locally free A-lattices.

It follows from Theorem 1.32 that $\text{Outcent}(A) \cong \ker\theta$, and that there is a bijection $\text{cok}\,\theta \leftrightarrow \text{L-Conj}(A/(\mathbf{Q}A)^*)$.

Now let B be a maximal $\mathbf{Z}$-order in $\mathbf{Q}A$ containing A, and let C' be the centre of B. Since C' is the integral closure of $\mathbf{Z}$ in the centre of $\mathbf{Q}A$, we have $C' \cong \mathbf{Z}\dotplus\mathbf{Z}\dotplus S$. In terms of the notation introduced at the beginning of this section, we obtain

$$\text{Cl}(C') \cong \text{Cl}(S) \cong \text{Cl}_B(C'),$$

since $\text{Cl}(\mathbf{Z}) = 0$ and since no real prime of L ramifies in $M_2(L)$.

We may now consider the commutative diagram (4.2). As observed in (3.6), the map γ is always injective. For the special case where $A = \mathbf{Z}G$, however, the following additional information is available:

(i) γ is surjective,

(ii) $\ker\varphi$ is cyclic of order $(p-1)/2$ and is a direct factor of $\text{Picent}(C)$,

(iii) η is an isomorphism.

(The first two are due to Fröhlich ([**3**]), the last to Reiner and Ullom ([**6**]), who also give a proof of (ii).) We have already observed that ρ is an isomorphism in this case. The diagram (4.2) then becomes

$$\begin{array}{ccccc} \text{Picent}(C) & \xrightarrow{\quad\varphi\quad} & \text{Cl}(S) & \xrightarrow{\;t\;} & \text{Cl}(S)\\ \big\downarrow{\scriptstyle\gamma} & & \big\downarrow{\scriptstyle\sigma} & & \big\downarrow{\scriptstyle 1}\\ \text{LFP}(A) & \xrightarrow{\;\theta'\;} & \text{Cl}(A)\cong\text{Cl}(B) & \cong & \text{Cl}(S)\end{array} \tag{4.4}$$

where γ is an isomorphism, and where $t[J] = [J]^2$ for $[J] \in \text{Cl}(S)$. It is then clear that $\ker\theta \cong \ker t\varphi$. However, there is an exact sequence of groups

$$1 \to \ker\varphi \to \ker t\psi \to \ker t \to \text{cok}\,\varphi \to \text{cok}\,t\varphi \to \text{cok}\,t \to 1. \tag{4.5}$$

However, $\text{cok}\,\varphi = 1$ since (in general) φ is surjective, and $\ker t = (\text{Cl}(S))_2$. Thus there is an exact sequence

$$1 \to \ker\varphi \to \ker\theta \to (\text{Cl}(S))_2 \to 1.$$

It follows from (ii) above that the sequence is split, and therefore $\ker\theta$ is a split extension of $(\text{Cl}(S))_2$ by a cyclic group of order $(p-1)/2$, as claimed.

Finally, from (4.4) and (4.5) we obtain

$$\text{cok}\,\theta \cong \text{cok}\,t\varphi = \text{cok}\,t = \text{Cl}(S)/(\text{Cl}(S))^2.$$

This completes the proof of the theorem.

5. Idele groups and a formula for cok θ

In this section K is an algebraic number field and R its ring of algebraic integers. Throughout KA is a finite-dimensional semisimple K-algebra with order A. We could thus without loss of generality take $R = \mathbf{Z}$ and $K = \mathbf{Q}$. We shall translate our basic concepts into idele theoretic language. This will in particular yield an alternative definition of our basic homomorphism θ. As an application we shall derive a formula for $\operatorname{cok}\theta$.

Throughout, the symbol p ranges over the finite and infinite primes of K, subscripts denoting completion. If X is a lattice and p an infinite prime then X_p is defined to be $(KX)_p$. The *idele group* $J(KA)$ of KA is the subgroup of the product $\prod (KA)_p^*$ of elements α with $\alpha_p \in A_p^*$ for almost all p, that is, with at most finitely many exceptions. The idele group has as subgroup the group

$$U(A) = \prod A_p^* \tag{5.1}$$

of unit ideles and we define the topology on $J(KA)$ by postulating that $U(A)$ be an open subgroup, and the subgroup topology on $U(A)$ to coincide with the product topology, as given by (5.1). Note that the topological group $J(KA)$ is independent of the particular choice of the order spanning KA. In the usual manner we embed $(KA)^*$ diagonally in $J(KA)$.

Analogously we define the idele group $J(KC)$, where $C = \operatorname{cent}(A)$, and view this as embedded in $J(KA)$, via the inclusion map $KC \subseteq KA$. On the other hand, if

$$KC = \prod_i F_i \tag{5.2}$$

is the product of algebraic number fields F_i, then

$$J(KC) = \prod_i J(F_i),$$

where the $J(F_i)$ are the usual idele groups.

Recall the definition of the normalizer of A_p, namely

$$N(A_p) = \{x \in (KA)_p^* : x^{-1}A_p x = A_p\},$$

which extends formally also to infinite primes. The *idele normalizer* of A (cf. [**3**]) is then given by

$$\bar{N}(A) = J(KA) \cap \prod_p N(A_p),$$

that is, $\bar{N}(A)$ is the subgroup of $\prod_p N(A_p)$ of elements α with $\alpha_p \in A_p^*$ almost everywhere.

If now M is an R-lattice spanning KA and $\alpha \in J(KA)$, then there is a unique R-lattice, denoted by $M\alpha$, so that

$$(M\alpha)_p = M_p\alpha_p \quad \text{for all } p.$$

(5.3) THEOREM. *The map*

$$\alpha \mapsto A\alpha \quad (\alpha \in J(KA))$$

yields bijections of sets

(i) $U(A)\backslash J(KA) \xrightarrow{\sim} S(A)$ (*set of locally free rank*-1 *left* A-*lattices in* KA),

(ii) $U(A)\backslash J(KA)/A^* \xrightarrow{\sim} \mathrm{LF}_1(A)$,

and isomorphisms of groups

(iii) $\bar{N}(A)/U(A) \xrightarrow{\sim} \mathrm{LFI}(A)$,

(iv) $\bar{N}(A)/U(A)(KC)^* \xrightarrow{\sim} \mathrm{LFP}(A)$.

Moreover the composition of (ii) *with* $\mu\colon \mathrm{LF}_1(A) \to \mathrm{Cl}(A)$ (cf. (1.23)) *yields an isomorphism of groups*

(v) $J(KA)/J'(KA)U(A)(KA)^* \xrightarrow{\sim} \mathrm{Cl}(A)$,

where $J'(KA)$ *is the closure of the commutator group of* $J(KA)$. *Also the diagram*

$$\text{(vi)} \qquad \begin{array}{ccc} \bar{N}(A)/U(A)(KC)^* & \xrightarrow{\theta^*} & J(KA)/J'(KA)U(A)(KA)^* \\ \downarrow & & \downarrow \\ \mathrm{LFP}(A) & \xrightarrow{\theta} & \mathrm{Cl}(A) \end{array}$$

commutes, where

$$\theta^*(\alpha U(A)(KC)^*) = \alpha J'(KA)U(A)(KA)^*.$$

(5.4) REMARK. One can in fact establish the commutativity of (vi) without knowing that the map θ is a homomorphism. This will then follow from the fact that θ^* clearly is a homomorphism and that (iv) and (v) are isomorphisms. This yields a new proof of (1.24).

Proof of Theorem 5.3. (i) and (ii) are, of course, known but we shall outline a proof.

Clearly $A(\alpha\beta) = (A\alpha)\beta$, $A1 = A$. Hence $A\alpha = A\alpha'$ precisely when $A\alpha'\alpha^{-1} = A$, that is, $\alpha'\alpha^{-1} \in U(A)$. Thus the fibres of the map

$$\alpha \mapsto A\alpha$$

are the right cosets of $U(A)$ in $J(KA)$. We thus get an injection $U(A)\backslash J(KA) \to S(A)$, observing that $A\alpha$ is a left A-lattice spanning KA and that, for all p, we have $(A\alpha)_p = A_p\alpha_p \cong A_p$; hence $A\alpha \in S(A)$.

Conversely, if $M \in S(A)$, then, for all p, $M_p = A_p\alpha_p$, $\alpha_p \in (KA)_p^*$ and, for almost all p, even $M_p = A_p$, so that we may then take $\alpha_p = 1$. Hence $M = A\alpha$, $\alpha \in J(KA)$. Thus (i) is a bijection.

Clearly, for $\alpha \in J(KA)$, $a \in (KA)^*$, right multiplication by a yields an isomorphism $A\alpha \cong A\alpha a$. Thus we get a surjection (ii). Moreover if we are given an isomorphism $A\alpha \cong A\beta$, then this extends to an automorphism of the left KA-module KA, that is, is given by right multiplication by an element a of $(KA)^*$. Thus $A\beta = A\alpha a$; hence, as we already know, $\beta = \mu\alpha a$, $\mu \in U(A)$. This shows that (ii) is a bijection.

Next observe that $A\alpha$ is a A–A-bimodule precisely when $\alpha \in \bar{N}(A)$. Thus (i) yields a bijection (iii). This is in fact an isomorphism of groups as, for $\alpha \in \bar{N}(A)$,

$$(A\alpha)(A\beta) = A(\alpha A\beta) = A(A\alpha\beta) = A\alpha\beta.$$

In the same way as we went from (i) to (ii), we also get from (iii) to (iv), noting that automorphisms of the KA–KA-bimodule KA are given by multiplication by elements $c \in (KC)^*$.

Finally (v) is established by Fröhlich in [**4**]. The commutativity of (vi) is obvious from the definitions.

(5.5) Corollary. $\operatorname{Ker}\theta \cong \bar{N}(A)/[\bar{N}(A) \cap (J'(KA)(KA)^*)]U(A)$.

(5.6) Corollary. $\operatorname{Cok}\theta \cong J(KA)/J'(KA)\bar{N}(A)(KA)^*$.

We shall now consider the reduced norm

$$\nu\colon (KA)^* \to (KC)^*,$$

which yields a continuous homomorphism

$$\nu\colon J(KA) \to J(KC),$$

again called the reduced norm. We shall need to use

(5.7) Theorem ([**4**]). *The reduced norm yields an isomorphism*

$$J(KA)/J'(KA)U(A)(KA)^* \cong J(KC)/\nu(U(A))(KC)^*$$

and $\nu(U(A))$ is an open subgroup of $J(KC)$.

We shall now derive

(5.8) Theorem. *The reduced norm yields an isomorphism*

$$\operatorname{cok}\theta \cong J(KC)/\nu(\bar{N}(A))(KC)^*$$

and $\nu(\bar{N}(A))$ is an open subgroup of $J(KC)$.

Proof. As $\bar{N}(A) \supseteq U(A)$, also $\nu(\bar{N}(A)) \supseteq \nu(U(A))$; and this, by (5.7), implies that $\nu(\bar{N}(A))$ is open.

For the required isomorphism we use (5.6) and the commutative diagram with exact rows, and columns induced by ν,

$$\begin{array}{ccccccccc}
1 & \longrightarrow & \dfrac{\bar{N}(A)}{[\bar{N}(A)\cap(J'(KA)(KA)^*)]U(A)} & \longrightarrow & \dfrac{J(KA)}{J'(KA)U(A)(KA)^*} & \longrightarrow & \dfrac{J(KA)}{J'(KA)\bar{N}(A)(KA)^*} & \longrightarrow & 1 \\
 & & \downarrow{\scriptstyle \nu_1} & & \downarrow{\scriptstyle \nu_2} & & \downarrow{\scriptstyle \nu_3} & & \\
1 & \longrightarrow & \dfrac{\nu\bar{N}(A))}{[\nu(\bar{N}(A))\cap(KC)^*]\nu(U(A))} & \longrightarrow & \dfrac{J(KC)}{(KC)^*\nu(U(A))} & \longrightarrow & \dfrac{J(KC)}{(KC)^*\nu(\bar{N}(A))} & \longrightarrow & 1
\end{array}$$

We have to show that ν_3 is an isomorphism. But this follows immediately from the fact that ν_1 is obviously surjective and that, by (5.7), ν_2 is an isomorphism.

We can now, in the standard manner of translation from idele language to ideal language, reformulate the isomorphism of (5.8) in terms of ideal groups.

Let C' be the integral closure of C in KC. Then the product (5.2) yields a product

$$C' = \prod_i R_i, \tag{5.9}$$

where R_i is the ring of algebraic integers of F_i. Now we choose once and for all a finite set S containing all infinite primes, all finite primes p which are ramified in any of the fields F_i and all finite primes p for which A_p is not a product of full matrix rings over the $R_{i,p}$. In other words, in the notation given at the start of § 4, let

$$KA = \prod_i KB_i, \tag{5.10}$$

with simple algebras KB_i, and with $F_i = \mathrm{cent}(KB_i)$. Then for $p \notin S$, the prime p is finite, $F_{i,p}$ is non-ramified over K_p, and

$$A_p = \prod_i A_{i,p},$$

where $A_{i,p}$ is a full matrix ring over $R_{i,p}$ spanning $(KB_i)_p$.

With the given choice of S, let $I(C', S)$ be the group of invertible fractional ideals X of C' with $X_p = C'_p$ for all $p \in S$. The reduced norm, or rather more precisely the composition

$$(KC)^* \longrightarrow (KA)^* \xrightarrow{\nu} (KC)^*,$$

defines an endomorphism

$$\nu' \colon I(C', S) \to I(C', S).$$

Alternatively we may, in view of (5.9), put

$$I(C', S) = \prod_i I(R_i, S),$$

where $I(R_i, S)$ is the group of fractional ideals of R_i, 'prime to S'. On the component $I(R_i, S)$, the map ν' is given by $X \to X^{m_i}$, where m_i^2 is the rank of the simple algebra KB_i (cf. (5.10)) over F_i. Thus, in particular,

(5.11) $$\nu'(I(C', S)) = \prod_i I(R_i, S)^{m_i}.$$

Next let

$$J_S(C') = \{\alpha \in J(KC) \colon \alpha_p = 1 \text{ if } p \notin S\} = \prod_S (KC)_p^*,$$

$$J^S = \{\alpha \in J(KC) \colon \alpha_p = 1 \text{ if } p \in S\}.$$

Then every idele $\alpha \in J(KC)$ can be written uniquely in the form

$$\alpha = \alpha_S \alpha^S, \quad \alpha_S \in J_S(C'),\ \alpha^S \in J^S.$$

For $p \notin S$ note that $C_p = C'_p$. We write

$$U^S = \prod_{p \notin S} C_p^*.$$

Then the map

$$\alpha \to C'\alpha, \quad \alpha \in J^S,$$

yields an isomorphism

(5.12) $$J^S/U^S \simeq I(C', S).$$

Finally let

$$W = \{C'.a^S \colon a \in KC,\ a_p \in \nu(N(A_p)) \text{ for all } p \in S\}.$$

Then we get

(5.13) Theorem.

$$\operatorname{cok} \theta \cong I(C', S)/W.\nu'(I(C', S)).$$

In particular, every element of $\operatorname{cok}\theta$ *has order dividing the least common multiple of the* m_i.

Proof. We shall use (5.8) and prove that the idele class group given there is isomorphic to the class group of $I(C', S)$ given in (5.13).

First recall that, by (5.8), $\nu(\bar{N}(A))$ is open. Hence, and by the weak approximation theorem,

$$J_S(C') \subseteq J^S\nu(\bar{N}(A))(KC)^* = J(KC).$$

Thus the composition

$$\lambda\colon J^S \to J(KC) \to J(KC)/\nu(\bar{N}(A))(KC)^*$$

is a surjective homomorphism, that is, we get an isomorphism

$$(5.14) \qquad J^S/J^S \cap [\nu(\bar{N}(A))(KC)^*] \cong J(KC)/\nu(\bar{N}(A))(KC)^*.$$

Let $G = \prod_{p \in S} N(A_p)$. Thus $G \subseteq \bar{N}(A)$. Also, for $p \notin S$, we know that A_p is a product of matrix rings, hence $N(A_p) = A_p^*(KC)_p^*$. Always $(KC)_p^* \subseteq N(A_p)$. Therefore

$$\bar{N}(A) = \prod_{p \notin S} A_p^*.G.J(KC)$$

and hence

$$\nu(\bar{N}(A))(KC)^* = \nu\left(\prod_{p \notin S} A_p^*\right).\nu(G).\nu(J(KC))(KC)^*.$$

However, for $p \notin S$, the reduced norm is a surjection $A_p^* \to C_p^*$, whence

$$\nu\left(\prod_{p \notin S} A_p^*\right) = U^S.$$

Also $J_S(C') \subseteq G$, hence $\nu(G).\nu(J(KC)) = \nu(G)\nu(J^S)$. Therefore we now have

$$\nu(\bar{N}(A))(KC)^* = U^S.\nu(J^S).\nu(G)(KC)^*.$$

Hence,

$$(5.15) \qquad J^S \cap [\nu(\bar{N}(A))(KC)^*] = U^S.\nu(J^S).[\nu(G)(KC)^* \cap J^S].$$

We shall now show that

$$(5.16) \quad [\nu(G)(KC)^*] \cap J^S = \{a^S \colon a \in (KC)^*,\ a_p \in \nu(N(A_p^*)) \text{ for all } p \in S\}.$$

In fact, an element a^S, as on the right-hand side, can be written as

$$a^S = a a_S^{-1},$$

where $a \in (KC)^*$ and $a_S^{-1} \in \nu(G)$. Conversely, if $\alpha \in J^S$ and $\alpha = a\nu(g)$, $g \in G$, $a \in (KC)^*$, then $\nu(g) \in J_S(C')$; hence $\alpha = a^S$, $\nu(g)^{-1} = a_S$.

From (5.8), (5.14), and (5.15) we have an isomorphism

$$J^S/U^S.\nu(J^S).[\nu(G)(KC)^* \cap J^S] \cong \operatorname{cok}\theta.$$

By (5.12), we get an isomorphism

$$(5.17) \qquad I(C', S)/\pi(\nu(J^S)).\pi[\nu(G)(KC)^* \cap J^S] \cong \operatorname{cok}\theta,$$

where π is the surjection $J^S \to I(C', S)'$

However, trivially, $\pi(\nu(J^S)) = \nu'(I(C', S))$, and, by (5.16),

$$\pi[\nu(G)(KC)^* \cap J^S] = W.$$

Thus the theorem is now seen to follow from (5.17).

REFERENCES

1. H. Bass, *Algebraic K-theory*, Math. Lecture Note Series (Benjamin, New York, 1968).

1a. *Lectures on topics in algebraic K-theory* (Tata Institute, Bombay, 1967).

2. C. W. Curtis and I. Reiner, *Representation theory of finite groups and associative algebras*, Pure and Appl. Math., Vol. XI (Interscience, New York, 1962; 2nd edn, 1966).

3. A. Fröhlich, 'The Picard group of non-commutative rings, in particular of orders', *Trans. Amer. Math. Soc.* 180 (1973) 1–45.

4. 'Locally free modules over arithmetic orders', *J. Math.* (*Crelle*), to appear.

5. H. Jacobinski, 'Genera and decompositions of lattices over orders', *Acta Math.* 121(1968) 1–29.

6. I. Reiner and S. Ullom, 'Class groups of integral group rings', *Trans. Amer. Math. Soc.* 170 (1972) 1–30.

7. A. Rosenberg and D. Zelinsky, 'Automorphisms of separable algebras', *Pacific J. Math.* 11 (1961) 1109–17.

8. R. G. Swan, 'Induced representations and projective modules', *Ann. of Math.* (2) 71 (1960) 552–78.

9. —— 'Projective modules over group rings and maximal orders', ibid. (2) 76 (1962) 55–61.

10. —— and E. G. Evans, *K-theory of finite groups and orders*, Lecture Notes No. 149 (Springer-Verlag, Berlin, 1970).

A. Fröhlich
King's College, London

I. Reiner and S. Ullom
University of Illinois
Urbana–Champaign

INVARIANTS OF INTEGRAL REPRESENTATIONS

IRVING REINER

Let ZG be the integral group ring of a finite group G. A ZG-lattice is a left G-module with a finite free Z-basis. In order to classify ZG-lattices, one seeks a full set of isomorphism invariants of a ZG-lattice M. Such invariants are obtained here for the special case where G is cyclic of order p^2, where p is prime. This yields a complete classification of the integral representations of G. There are also several results on extensions of lattices, which are of independent interest and apply to more general situations.

Two ZG-lattices M and N are placed in the same genus if their p-adic completions M_p and N_p are Z_pG-isomorphic. One first gives a full set of genus invariants of a ZG-lattice. There is then the remaining problem, considerably more difficult in this case, of finding additional invariants which distinguish the isomorphism classes within a genus. Generally speaking, such additional invariants are some sort of ideal classes. In the present case, these invariants will be a pair of ideal classes in rings of cyclotomic integers, together with two new types of invariants: an element in some factor group of the group of units of some finite ring, and a quadratic residue character (mod p).

For arbitrary finite groups G, the classification of ZG-lattices has been carried out in relatively few cases. The problem has been solved for G of prime order p or dihedral of order $2p$. It was also solved for the case of an elementary abelian (2, 2)-group, and for the alternating group A_4 (see [10a] for references).

The main results of the present article deal with the case where G is cyclic of order p^2, where p is prime. In Theorem 7.3 below, there is a full list of all indecomposable ZG-lattices, up to isomorphism. Theorem 7.8 then gives a full set of invariants for the isomorphism class of a finite direct sum of indecomposable lattices.

Sections 1 and 2 contain preliminary remarks about extensions of lattices over orders. Sections 3 and 5 consider the following problem: given two lattices M and N over some order, find a full set of isomorphism invariants for a direct sum of extensions of lattices in the genus of N by lattices in the genus of M. The results of these sections are applied in §§ 4 and 6 to the special case of ZG-lattices, where G is any cyclic p-group, p prime. Finally, § 7 is devoted to detailed calculations for the case where G is cyclic of order p^2.

Throughout the article, R will denote a Dedekind ring whose quotient field K is an algebraic number field, and Λ will be an R-order in a finite dimensional semisimple K-algebra A. For P a maximal ideal of R, the subscript P in R_P, K_P, Λ_P, etc., denotes P-adic completion. Let $S(\Lambda)$ be a finite nonempty set of P's, such that Λ_P is a maximal R_P-order in A_P for each $P \notin S(\Lambda)$; such a set can always be chosen. (In the special case where $\Lambda = RG$, $S(\Lambda)$ need only be picked so as to include all prime ideal divisors of the order of G.) A *Λ-lattice* is a left Λ-module, finitely generated and torsionfree (hence projective) over R. Two Λ-lattices M, N are in the same *genus* if $M_P \cong N_P$ as Λ_P-modules for all P (or equivalently, for all $P \in S(\Lambda)$). For M a Λ-lattice, $\mathrm{End}_\Lambda(M)$ denotes its endomorphism ring, and $M^{(n)}$ the external direct sum of n copies of M. Let $\coprod M_i$ denote the external direct sum of a collection of modules $\{M_i\}$.

1. Generalities about extensions of modules. We briefly review some known facts about extensions (see, for example, [3] and [16]). Let Λ be an arbitrary ring, and let M, N be left Λ-modules. We shall write $\mathrm{Ext}(N, M)$ instead of $\mathrm{Ext}^1_\Lambda(N, M)$ for brevity, when there is no danger of confusion. Let

$$\Gamma = \mathrm{End}_\Lambda(M),\ \Delta = \mathrm{End}_\Lambda(N)\ ,$$

and view $\mathrm{Ext}(N, M)$ as a (Γ, Δ)-bimodule. For later use, we need to know explicitly how Γ and Δ act on $\mathrm{Ext}(N, M)$.

Consider a Λ-exact sequence

$$\xi\colon 0 \longrightarrow M \xrightarrow{\mu} X \xrightarrow{\nu} N \longrightarrow 0\ , \qquad \xi \in \mathrm{Ext}(N, M)\ .$$

For each $\gamma \in \Gamma$, we may form the pushout ${}_\gamma X$ of the pair of maps $\gamma\colon M \to M$, $\mu\colon M \to X$, so

$${}_\gamma X = (X \oplus M)/\{(\mu m, -\gamma m)\colon m \in M\}\ .$$

Then we obtain a commutative diagram with exact rows:

$$\begin{array}{lccccccccc} \xi\colon & 0 & \longrightarrow & M & \xrightarrow{\mu} & X & \xrightarrow{\nu} & N & \longrightarrow & 0 \\ & & & \gamma\downarrow & & \varphi\downarrow & & 1\downarrow & & \\ \gamma\xi\colon & 0 & \longrightarrow & M & \longrightarrow & {}_\gamma X & \longrightarrow & N & \longrightarrow & 0\ , \end{array} \tag{1.1}$$

and the bottom row corresponds to the extension class $\gamma\xi \in \mathrm{Ext}(N, M)$. Applying the Snake Lemma to the above (see [11, Exercise 2.8]), we obtain

$$\ker\gamma \cong \ker\varphi\ , \quad \mathrm{cok}\,\gamma \cong \mathrm{cok}\,\varphi\ . \tag{1.2}$$

Analogously, given any $\delta \in \Delta$, let

$$X_\delta = \{(x, n) \in X \oplus N : \nu x = \delta n\} ,$$

the pullback of the pair of maps $\nu : X \to N$, $\delta : N \to N$. Then we obtain a commutative diagram with exact rows:

$$\begin{array}{ccccccccc} \xi: 0 & \longrightarrow & M & \xrightarrow{\mu} & X & \xrightarrow{\nu} & N & \longrightarrow & 0 \\ & & {\scriptstyle 1}\uparrow & & {\scriptstyle \psi}\uparrow & & {\scriptstyle \delta}\uparrow & & \\ \xi\delta: 0 & \longrightarrow & M & \longrightarrow & X_\delta & \longrightarrow & N & \longrightarrow & 0 , \end{array}$$

and the bottom now gives the extension class $\xi\delta$. By the Snake Lemma,

$$\ker \psi \cong \ker \delta , \quad \operatorname{cok} \psi \cong \operatorname{cok} \delta .$$

Formules such as $(\gamma\gamma')\xi = \gamma(\gamma'\xi)$ are easily verified, and yield Λ-isomorphisms

$${}_\gamma X \cong X \text{ if } \gamma \in \operatorname{Aut}(M) , \quad X_\delta \cong X \text{ if } \delta \in \operatorname{Aut}(N) ,$$

where Aut means $\operatorname{Aut}_\Lambda$.

For later use, an alternative description of the action Δ on $\operatorname{Ext}(N, M)$ is important. Consider a Λ-exact sequence

$$0 \longrightarrow L \xrightarrow{i} P \longrightarrow N \longrightarrow 0$$

in which P is Λ-projective. Applying $\operatorname{Hom}_\Lambda(\cdot, M)$, we obtain an exact sequence of additive groups

$$0 \longrightarrow \operatorname{Hom}(N, M) \longrightarrow \operatorname{Hom}(P, M) \xrightarrow{i^*} \operatorname{Hom}(L, M) \longrightarrow \operatorname{Ext}^1_\Lambda(N, M) \longrightarrow 0 ,$$

and thus

$$\operatorname{Ext}(N, M) \cong \operatorname{Hom}(L, M)/\operatorname{im} i^* .$$

Each $\xi \in \operatorname{Ext}(N, M)$ is thus of the form $\bar{f}$, where $f \in \operatorname{Hom}(L, M)$ and where $\bar{f}$ denotes its image in $\operatorname{cok} i^*$. Now let $\delta \in \Delta$; we can lift δ to a map $\delta_1 \in \operatorname{End}(P)$, and δ_1 then induces a map $\delta_2 \in \operatorname{End}(L)$ for which the following diagram commutes:

$$\begin{array}{ccccccccc} 0 & \longrightarrow & L & \longrightarrow & P & \longrightarrow & N & \longrightarrow & 0 \\ & & {\scriptstyle \delta_2}\downarrow & & {\scriptstyle \delta_1}\downarrow & & {\scriptstyle \delta}\downarrow & & \\ 0 & \longrightarrow & L & \longrightarrow & P & \longrightarrow & N & \longrightarrow & 0 . \end{array}$$

Of course End L acts from the right on $\operatorname{Hom}(L, M)$, and for $\xi = \bar{f}$ as above, we have $\xi\delta = \overline{f\delta_2}$ in $\operatorname{Ext}(N, M)$.

PROPOSITION 1.3. *For* $i = 1, 2$, *let* M_i *and* N_i *be* Λ-*modules, and let* $\xi_i \in \operatorname{Ext}^1_\Lambda(N_i, M_i)$ *determine a* Λ*-module* X_i. *Assume that* $\operatorname{Hom}_\Lambda(M_1, N_2) = 0$. *Then* $X_1 \cong X_2$ *if and only if*

$$(1.4) \qquad \gamma\xi_1 = \xi_2\delta \text{ for some } \Lambda\text{-isomorphisms } \gamma\colon M_1 \cong M_2,\ \delta\colon N_1 \cong N_2\,.$$

Proof. Let $\varphi \in \operatorname{Hom}(X_1, X_2)$, and consider the diagram

$$\begin{array}{lccccccccc}
\xi_1\colon & 0 & \longrightarrow & M_1 & \xrightarrow{\mu_1} & X_1 & \xrightarrow{\nu_1} & N_1 & \longrightarrow & 0 \\
 & & & & & \Big\downarrow{\scriptstyle\varphi} & & & & \\
\xi_2\colon & 0 & \longrightarrow & M_2 & \xrightarrow{\mu_2} & X_2 & \xrightarrow{\nu_2} & N_2 & \longrightarrow & 0\,.
\end{array}$$

Since $\operatorname{Hom}(M_1, N_2) = 0$ by hypothesis, we have $\nu_2\varphi\mu_1 = 0$. Therefore $\varphi\mu_1(M_1) \subset \operatorname{im} \mu_2$, so φ induces maps γ, δ making the following diagram commute:

$$\begin{array}{ccccccccc}
0 & \longrightarrow & M_1 & \longrightarrow & X_1 & \longrightarrow & N_1 & \longrightarrow & 0 \\
 & & \Big\downarrow{\scriptstyle\gamma} & & \Big\downarrow{\scriptstyle\varphi} & & \Big\downarrow{\scriptstyle\delta} & & \\
0 & \longrightarrow & M_2 & \longrightarrow & X_2 & \longrightarrow & N_2 & \longrightarrow & 0\,.
\end{array}$$

But this means that $\gamma\xi_1 = \xi_2\delta$ in $\operatorname{Ext}(N_1, M_2)$. Furthermore, by the Snake Lemma, φ is an isomorphism if and only if both γ and δ are isomorphisms. Hence (1.4) holds if $X_1 \cong X_2$.

Conversely, assume that (1.4) is true; since $\gamma\xi_1 = \xi_2\delta$, there exists a commutative diagram

$$\begin{array}{ccccccccc}
0 & \longrightarrow & M_1 & \longrightarrow & X_1 & \longrightarrow & N_1 & \longrightarrow & 0 \\
 & & \Big\downarrow{\scriptstyle\gamma} & & \Big\downarrow{\scriptstyle\psi_1} & & \Big\downarrow{\scriptstyle 1} & & \\
0 & \longrightarrow & M_2 & \longrightarrow & Y_1 & \longrightarrow & N_1 & \longrightarrow & 0 \\
 & & \Big\downarrow{\scriptstyle 1} & & \Big\downarrow{\scriptstyle\psi_2} & & \Big\downarrow{\scriptstyle 1} & & \\
0 & \longrightarrow & M_2 & \longrightarrow & Y_2 & \longrightarrow & N_1 & \longrightarrow & 0 \\
 & & \Big\downarrow{\scriptstyle 1} & & \Big\downarrow{\scriptstyle\psi_3} & & \Big\downarrow{\scriptstyle\delta} & & \\
0 & \longrightarrow & M_2 & \longrightarrow & X_2 & \longrightarrow & N_2 & \longrightarrow & 0\,.
\end{array}$$

But γ and δ are isomorphisms, whence so is each ψ_i. Thus $X_1 \cong X_2$, as desired. (This part of the argument does not require the hypothesis that $\operatorname{Hom}(M_1, N_2) = 0$.)

COROLLARY 1.5. *Let* M, N *be* Λ*-modules such that* $\operatorname{Hom}(M, N) = 0$. *Let* $\xi_i \in \operatorname{Ext}(N, M)$ *determine a* Λ*-module* X_i, $i = 1, 2$. *Then* $X_1 \cong X_2$ *if and only if*

$$(1.6) \qquad \gamma\xi_1 = \xi_2\delta \text{ for some } \gamma \in \operatorname{Aut}(M),\ \delta \in \operatorname{Aut}(N)\,.$$

We shall call ξ_1 and ξ_2 *strongly equivalent* (notation: $\xi_1 \approx \xi_2$) whenever condition (1.6) is satisfied.

2. **Extensions of lattices.** Keeping the notation used in the introduction, let Λ be an R-order in the semisimple K-algebra A. Choose a nonempty set $S(\Lambda)$ of maximal ideals P of R, such that for each $P \in S(\Lambda)$, the P-adic completion Λ_P is a maximal R_P-order in A_P. Now let M and N be Λ-lattices, so M_P and N_P are Λ_P-lattices. For $P \in S(\Lambda)$, the maximal order Λ_P is hereditary, and so the Λ_P-lattice N_P is Λ_P-projective (see [11, (21.5)]); thus $\operatorname{Ext}_{\Lambda_P}(N_P, M_P) = 0$ for each $P \in S(\Lambda)$.

Now consider $\operatorname{Ext}^1_\Lambda(N, M)$, which we will denote for brevity by $\operatorname{Ext}(N, M)$ when there is no danger of confusion. Then $\operatorname{Ext}(N, M)$ is a finitely generated torsion R-module, with no torsion at the maximal ideals $P \in S(\Lambda)$. As in [4, (75.22)], we have

$$\operatorname{Ext}^1_\Lambda(N, M) \cong \prod_{P \notin S(\Lambda)} \operatorname{Ext}^1_{\Lambda_P}(N_P, M_P) . \tag{2.1}$$

The following analogue of Schanuel's Lemma will be useful:

LEMMA 2.2. *Let X, X', Y, Y' be Λ-lattices, and let T be an R-torsion Λ-module such that $T_P = 0$ for each $P \in S(\Lambda)$. Suppose that there exist a pair of Λ-exact sequences*

$$0 \longrightarrow X' \longrightarrow X \xrightarrow{f} T \longrightarrow 0 , \quad 0 \longrightarrow Y' \longrightarrow Y \xrightarrow{g} T \longrightarrow 0 .$$

Then there is a Λ-isomorphism

$$X \oplus Y' \cong X' \oplus Y .$$

Proof. Let W be the pullback of the pair of maps f, g. Then we obtain a commutative diagram of Λ-modules, with exact rows and columns:

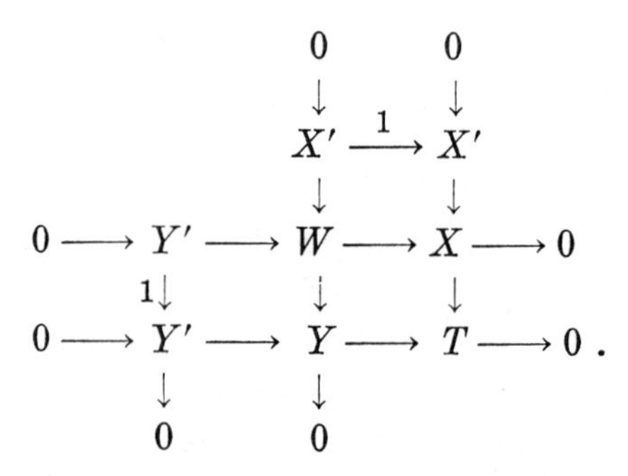

At each $P \in S(\Lambda)$, we have $T_P = 0$ by hypothesis. However, the process of forming P-adic completions preserves commutativity and exactness, since R_P is R-flat. Hence both of the Λ-exact sequences

$$(2.3) \quad 0 \longrightarrow X' \longrightarrow W \longrightarrow Y \longrightarrow 0, \ 0 \longrightarrow Y' \longrightarrow W \longrightarrow X \longrightarrow 0 ,$$

are split at each $P \in S(\Lambda)$. On the other hand, for $P \notin S(\Lambda)$ we know that Λ_P is a maximal order, so the Λ_P-lattices X_P, Y_P are Λ_P-projective. Hence the sequences (2.3) are also split at each $P \notin S(\Lambda)$. Therefore they split at every P, and hence split globally (see [11, (3.20)]). This gives

$$W \cong X' \oplus Y , \quad W \cong X \oplus Y' ,$$

and proves the result. This result is due to Roiter [15].

We shall apply this lemma to the following situation. Each $\xi \in \mathrm{Ext}(N, M)$ determines a Λ-exact sequence

$$\xi\colon \ 0 \longrightarrow M \longrightarrow X \longrightarrow N \longrightarrow 0 ,$$

with X unique up to isomorphism. The sequence is R-split since N is R-projective, and so $X \cong M \oplus N$ as R-modules. Thus X is itself a Λ-lattice, called an *extension* of N by M. There is an embedding $M \to K \otimes_R M$, given by $m \to 1 \otimes m$ for $m \in M$; we shall always identify M with its image $1 \otimes M$, so that $K \otimes_R M$ may be written as KM. We shall set

$$\Gamma = \mathrm{End}_\Lambda(M) , \quad \Delta = \mathrm{End}_\Lambda(N) .$$

Then $K\Gamma = \mathrm{End}_A(KM)$, and Γ is an R-order in the semisimple K-algebra $K\Gamma$. Likewise $K\Delta = \mathrm{End}_A(KN)$, and Δ is an R-order in the semisimple K-algebra $K\Delta$. For each $\gamma \in \Gamma$, $\delta \in \Delta$, we may form the Λ-lattices ${}_\gamma X$ and X_δ as in § 1. We now prove

(2.4) Exchange Formula. *Let X and Y be a pair of extensions of N by M, and let $\gamma \in \mathrm{End}(M)$ satisfy the condition*

$$(2.5) \qquad \gamma_P \in \mathrm{Aut}(M_P) \ \textit{for each} \ P \in S(\Lambda) .$$

Then there is a Λ-isomorphism

$$(2.6) \qquad X \oplus {}_\gamma Y \cong {}_\gamma X \oplus Y .$$

Proof. For each $P \in S(\Lambda)$, we have

$$(\ker \gamma)_P \cong \ker(\gamma_P) = 0 , \quad (\mathrm{cok}\, \gamma)_P \cong \mathrm{cok}(\gamma_P) = 0 .$$

Now $\ker \gamma$ is an R-submodule of the Λ-lattice M, and thus $\ker \gamma$ is itself an R-lattice. Since $(\ker \gamma)_P = 0$ for at least one P (namely, for any $P \in S(\Lambda)$), it follows that $\ker \gamma = 0$.

From (1.1) and (1.2) we obtain Λ-exact sequences

$$0 \longrightarrow X \longrightarrow {}_{\gamma}X \longrightarrow \operatorname{cok}\gamma \longrightarrow 0\,,\quad 0 \longrightarrow Y \longrightarrow {}_{\gamma}Y \longrightarrow \operatorname{cok}\gamma \longrightarrow 0\,,$$

where $\operatorname{cok}\gamma = M/\gamma(M)$. But $(\operatorname{cok}\gamma)_P = 0$ for each $P \in S(\Lambda)$, so we may apply Lemma 2.2 to the above sequences. This gives the isomorphism in (2.6), and completes the proof.

In the same manner, we obtain

(2.7) Absorption Formula. *Let X be an extension of N by M, and let $\gamma \in \operatorname{End}(M)$ satisfy condition* (2.5). *Then*

$$X \oplus M \cong {}_{\gamma}X \oplus M\,.$$

Proof. Apply Lemma 2.2 to the pair of exact sequences

$$0 \longrightarrow X \longrightarrow {}_{\gamma}X \longrightarrow \operatorname{cok}\gamma \longrightarrow 0\,,\quad 0 \longrightarrow M \xrightarrow{\gamma} M \longrightarrow \operatorname{cok}\gamma \longrightarrow 0\,.$$

Remark 2.8. There are obvious analogues of (2.6) and (2.7), in which we start with an element $\delta \in \operatorname{End}(N)$ such that $\delta_P \in \operatorname{Aut}(N_P)$ for each $P \in S(\Lambda)$.

Now let M, N be Λ-lattices, and let $M' \vee M$, $N' \vee N$. It is clear from (2.1) that $\operatorname{Ext}(N', M') \cong \operatorname{Ext}(N, M)$. In fact, by Roiter's Lemma (see [11, (27.1)]), we can find Λ-exact sequences

$$(2.9)\quad 0 \longrightarrow M \xrightarrow{\varphi} M' \longrightarrow T \longrightarrow 0\,,\quad 0 \longrightarrow N' \xrightarrow{\psi'} N \longrightarrow U \longrightarrow 0\,,$$

in which $T_P = 0$ and $U_P = 0$ for all $P \in S(\Lambda)$. The pair (ϕ, ψ') induces an isomorphism

$$(2.10)\qquad t\colon \operatorname{Ext}(N, M) \cong \operatorname{Ext}(N', M')\,,$$

(hereafter called a *standard* isomorphism), which may be described explicitly as follows: if $\xi \in \operatorname{Ext}(N, M)$, then $t(\xi) = \phi\xi\psi'$ (in the notation of § 1). Thus, if ξ determines the Λ-lattice X (up to isomorphism), then $t(\xi)$ determines the Λ-lattice ${}_{\phi}(X)_{\psi'}$, which lies in the same genus as X.

Lemma 2.11. *The inverse of a standard isomorphism is also a standard isomorphism.*

Proof. We may choose a nonzero proper ideal $\mathfrak{a}$ of R, all of whose prime ideal factors lie in $S(\Lambda)$, such that $\mathfrak{a}\cdot\operatorname{Ext}(N, M) = 0$. If $\mu \in \operatorname{End}(M)$ is such that $\mu - 1 \in \mathfrak{a}\cdot\operatorname{End}(M)$, it then follows that μ acts as the identity map on $\operatorname{Ext}(N, M)$.

Let t be a standard isomorphism as in (2.10), induced from the pair of maps (ϕ, ψ') as in (2.9). Since ϕ_P is an isomorphism for each $P \in S(\Lambda)$, we can find a map $\phi' \in \operatorname{Hom}(M', M)$ such that ϕ'_P approximates ϕ_P^{-1} at each $P \in S(\Lambda)$; indeed, we can choose ϕ' so that

$$\phi' \cdot \phi \equiv 1 \bmod \mathfrak{a} \cdot \operatorname{End}(M) .$$

Then ϕ' is an inclusion, and $\phi'\phi$ acts as 1 on $\operatorname{Ext}(N, M)$. Likewise, we may choose an inclusion $\psi: N \to N'$ such that $\psi'\psi$ acts as 1 on $\operatorname{Ext}(N, M)$. The pair (ϕ', ψ) then induces a standard isomorphism $t': \operatorname{Ext}(N', M') \cong \operatorname{Ext}(N, M)$ such that $t't = 1$. This completes the proof.

We wish to determine all isomorphism classes of Λ-lattices X which are extensions of a given lattice N by another given lattice M. Let us show that under suitable hypotheses on M and N, this determination depends only upon the genera of M and N. A Λ-lattice M is called an *Eichler lattice* if $\operatorname{End}_A(KM)$ satisfies the Eichler condition over R (see [11, (38.1)]). This condition depends only on the A-module KM and on the underlying ring of integers R. (In the special case where $R = \text{alg. int. }\{K\}$, M is an Eichler lattice if and only if no simple component of $\operatorname{End}_A(KM)$ is a totally definite quaternion algebra.) Of course, M is an Eichler lattice wherever $\operatorname{End}_A(KM)$ is a direct sum of matrix algebras over fields.

We now establish

THEOREM 2.12. *Let M and N be Λ-lattices such that $M \oplus N$ is an Eichler lattice, and let $M' \vee M$, $N' \vee N$. Let*

$$t: \operatorname{Ext}(N, M) \cong \operatorname{Ext}(N', M')$$

be a standard isomorphism as in (2.10). *Then t induces a one-to-one correspondence between the set of isomorphism classes of extensions of N by M, and that of extensions of N' by M'.*

Proof. Each Λ-lattice X, which is an extension of N by M, determines an extension class $\xi \in \operatorname{Ext}(N, M)$. Two X's which yield the same ξ must be isomorphic to one another, but the converse of this statement need not be true. (Herein lies the difficulty in the proof.) In any case, given the extension X, let ξ be its extension class; set $\xi' = t(\xi) \in \operatorname{Ext}(N', M')$, and let ξ' determine the Λ-lattice X' (up to isomorphism). Then X' is an extension of N' by M', and $X' \vee X$. Now let Y be another extension of N by M, and let Y' be the corresponding extension of N' by M'. We must prove that $X \cong Y$ if and only if $X' \cong Y'$. (Note that every extension of N' by M' comes from some X, by virtue of Lemma 2.11.)

It suffices to prove the implication in one direction, since by (2.11) the inverse of a standard isomorphism is again standard. Furthermore, every standard isomorphism can be expressed as a product of two standard isomorphisms, each of which involves a change of only one of the "variables" M and N. It therefore suffices to prove the desired result for the case in which there is a change in only one variable, say M. Thus, let us start with an inclusion $\phi: M \longrightarrow M'$ as in (2.9), such that $(\operatorname{cok} \phi)_P = 0$ for all $P \in S(\Lambda)$. Given an exact sequence

$$0 \longrightarrow M \xrightarrow{\mu} X \longrightarrow N \longrightarrow 0,$$

define a Λ-module X' as the pushout of the pair of maps (μ, ϕ). We then obtain a commutative diagram of Λ-modules, with exact rows:

$$\begin{array}{ccccccccc} 0 & \longrightarrow & M & \xrightarrow{\mu} & X & \longrightarrow & N & \longrightarrow & 0 \\ & & \phi\downarrow & & \downarrow & & 1\downarrow & & \\ 0 & \longrightarrow & M' & \longrightarrow & X' & \longrightarrow & N & \longrightarrow & 0 . \end{array} \tag{2.13}$$

Then X' is precisely the Λ-lattice determined by X as above, by means of the standard isomorphism $t: \operatorname{Ext}(N, M) \cong \operatorname{Ext}(N, M')$ induced by ϕ. Let Y be another extension of N by M, and let Y' denote the extension of N by M' corresponding to Y. It then suffices for us to prove that $X' \cong Y'$ whenever $X \cong Y$.

Applying the Snake Lemma to (2.13), we obtain an exact sequence of Λ-modules

$$0 \longrightarrow X \longrightarrow X' \longrightarrow \operatorname{cok} \phi \longrightarrow 0,$$

with $(\operatorname{cok} \phi)_P = 0$ for all $P \in S(\Lambda)$. Likewise, there is an exact sequence

$$0 \longrightarrow Y \longrightarrow Y' \longrightarrow \operatorname{cok} \phi \longrightarrow 0 .$$

Therefore we obtain

$$X \oplus Y' \cong X' \oplus Y \tag{2.14}$$

by Lemma 2.2.

Suppose now that $X \cong Y$; since $X' \vee X$ and $Y' \vee Y$, the lattices X, X', Y, Y' are in the same genus, and we may rewrite (2.14) as

$$X \oplus Y' \cong X \oplus X' . \tag{2.15}$$

Clearly $KX \cong K(M \oplus N)$, and thus X is an Eichler lattice (since $M \oplus N$ is an Eichler lattice by hypothesis). By Jacobinski's Cancellation Theorem [8], we may then conclude from (2.15) that $X' \cong Y'$. This completes the proof of the theorem.

REMARKS. (i) It seems likely that the conclusion of the theorem holds true whether or not $M \oplus N$ is an Eichler lattice.

(ii) Suppose that $\operatorname{Hom}(M, N) = 0$. By (1.5), there is a one-to-one correspondence between the set of all isomorphism classes of Λ-lattices X which are extensions of N by M, and the set of orbits of the bimodule $\operatorname{Ext}(N, M)$ under the left action of $\operatorname{Aut}(N)$ and the right action of $\operatorname{Aut}(M)$. By definition, two elements of $\operatorname{Ext}(N, M)$ are *strongly equivalent* if they lie in the same orbit. The preceding theorem then shows, in this case where $\operatorname{Hom}(M, N) = 0$ and where $M \oplus N$ is an Eichler lattice, that the orbits depend only upon the genera of M and N. Indeed, we have shown above that under these hypotheses, standard isomorphisms preserve strong equivalence.

(iii) In the special cases of interest in §§ 4-7, one can prove (2.12) directly without using Jacobinski's Cancellation Theorem (see [13], for example).

The author wishes to thank Professor Jacobinski for some helpful conversations, which led to a considerable simplification of the original proof of Theorem 2.12.

3. Direct sums of extensions. As in § 2, let Λ be a R-order in a semisimple K-algebra A, where K is an algebraic number field. Given Λ-lattices M, N with $\operatorname{Hom}_\Lambda(M, N) = 0$, we wish to classify up to isomorphism all extensions of a direct sum of copies of N by a direct sum of copies of M. Let $\xi_1, \xi_2 \in \operatorname{Ext}^1(N^{(s)}, M^{(r)})$, and let ξ_i determine the extension Y_i of $N^{(s)}$ by $M^{(r)}$. Since $\operatorname{Hom}_\Lambda(M^{(r)}, N^{(s)}) = 0$, we may apply (1.5) to obtain

PROPOSITION 3.1. *The Λ-lattices Y_1, Y_2 are isomorphic if and only if*

$$\alpha\xi_1 = \xi_2\beta \text{ for some } \alpha \in \operatorname{Aut} M^{(r)},\ \beta \in \operatorname{Aut} N^{(s)}\,. \tag{3.2}$$

As before, call ξ_1 *strongly equivalent* to ξ_2 (notation: $\xi_1 \approx \xi_2$) whenever condition (3.2) is satisfied. We may rewrite this condition in a more convenient form, as follows: there is an isomorphism

$$\operatorname{Ext}(N^{(s)}, M^{(r)}) \cong (\operatorname{Ext}(N, M))^{r\times s}\,,$$

where the right hand expression denotes the set of all $r \times s$ matrices with entries in $\operatorname{Ext}(N, M)$. If we put

$$\Gamma = \operatorname{End}_\Lambda(M)\,, \quad \Delta = \operatorname{End}_\Lambda(N)\,,$$

acting from the left on M and N, respectively, then we may identify $\operatorname{Aut} M^{(r)}$ with $GL(r, \Gamma)$, and $\operatorname{Aut} N^{(s)}$ with $GL(s, \Delta)$. Then $(\operatorname{Ext}(N, M))^{r\times s}$ is a left $GL(r, \Gamma)$-, right $GL(s, \Delta)$-bimodule, and $\xi_1 \approx \xi_2$ if and only if $\alpha\xi_1 = \xi_2\beta$ for some $\alpha \in GL(r, \Gamma)$, $\beta \in GL(s, \Delta)$.

As a matter of fact, we may choose a nonzero ideal $\mathfrak{a}$ of R, involving only prime ideals P from the set $S(\Lambda)$, such that $\mathfrak{a}\cdot \operatorname{Ext}(N, M) = 0$. Then Γ acts on $\operatorname{Ext}(N, M)$ via the map $\Gamma \to \bar{\Gamma}$, where $\bar{\Gamma} = \Gamma/\mathfrak{a}\Gamma$. Hence $GL(r, \Gamma)$ acts on $(\operatorname{Ext}(N, M))^{r\times s}$ via the map $GL(r, \Gamma) \to GL(r, \bar{\Gamma})$. A corresponding result holds for Δ.

We are thus faced with the question of determining the orbits of $(\operatorname{Ext}(N, M))^{r\times s}$ under the actions of $GL(r, \Gamma)$ and $GL(s, \Delta)$. We cannot hope to specify these orbits in general, but we shall see that they can be determined in some interesting special cases which arise in practice. Before proceeding with this determination, however, it is desirable to adopt a slightly more general point of view.

Let M and N be as above, and let $M_i \vee M$, $N_j \vee N$ for $1 \leqq i \leqq r$, $1 \leqq j \leqq s$. By hypothesis $\operatorname{Hom}(M, N) = 0$, so also $\operatorname{Hom}(M_i, N_j) = 0$ for all i, j. Now let $\xi \in \operatorname{Ext}(\coprod N_j, \coprod M_i)$ determine an extension X. It follows from § 1 that a full set of isomorphism invariants of X are the isomorphism classes of $\coprod M_i$ and $\coprod N_j$, and the strong equivalence class of ξ. Further, since $\coprod M_i \vee M^{(r)}$ and $\coprod N_j \vee N^{(s)}$, there is a standard isomorphism

$$t\colon \operatorname{Ext}(\coprod N_j, \coprod M_i) \cong \operatorname{Ext}(N^{(s)}, M^{(r)})$$

as in (2.10). If we assume that both $M^{(r)}$ and $N^{(s)}$ are Eichler lattices, then by (2.11) t gives a one-to-one correspondence between strong equivalence classes in these two Ext's. We remark in passing that $M^{(r)}$ is necessarily an Eichler lattice if $r > 1$.

As a consequence, we deduce

PROPOSITION 3.3. *Let M, N be Eichler lattices such that $\operatorname{Hom}(M, N) = 0$, and let $M_i \vee M$, $N_i \vee N$, $1 \leqq i \leqq r$. For each i, let $\xi_i \in \operatorname{Ext}(N_i, M_i)$ determine an extension X_i of N_i by M_i, and let $t_i\colon \operatorname{Ext}(N_i, M_i) \cong \operatorname{Ext}(N, M)$ be a standard isomorphism. Then a full set of isomorphism invariants of $\coprod X_i$ are the isomorphism classes of $\coprod M_i$ and $\coprod N_i$, and the strong equivalence class of*

$$\operatorname{diag}(t_1(\xi_1), \cdots, t_r(\xi_r))$$

in $\operatorname{Ext}(N^{(r)}, M^{(r)})$.

Proof. The element $\operatorname{diag}(\xi_1, \cdots, \xi_r) \in \operatorname{Ext}(\coprod N_i, \coprod M_i)$ determines the extension $\coprod X_i$ of $\coprod N_i$ by $\coprod M_i$. There is a standard isomorphism

$$\operatorname{Ext}(\coprod N_i, \coprod M_i) \cong \operatorname{Ext}(N^{(r)}, M^{(r)})$$

which carries $\operatorname{diag}(\xi_1, \cdots, \xi_r)$ onto $\operatorname{diag}(t_1(\xi_1), \cdots, t_r(\xi_r))$. The proposition then follows at once from the above discussion.

4. **Cyclic p-groups.** We consider here the special case where G is cyclic of order p^κ, where p is prime and $\kappa \geqq 1$. We shall identify ZG with the ring $\Lambda_\kappa = Z[x]/(x^{p^\kappa} - 1)$, which we denote by Λ for brevity when there is no danger of confusion. Let $\Phi_i(x)$ be the cyclotomic polynomial of order p^i and degree $\phi(p^i)$, $0 \leqq i \leqq \kappa$. Let ω_i denote a primitive p^i-th root of 1, and set

$$K_i = Q(\omega_i),\ R_i = \text{alg. int.}\{K_i\} = Z[\omega_i],\ P_i = (1 - \omega_i)R_i .$$

Then $R_i \cong Z[x]/(\Phi_i(x))$, a factor ring of Λ, so every R_i-module may be viewed as Λ-module.

Given a Λ-lattice M, let

$$L = \{m \in M: (x^{p^{\kappa-1}} - 1)m = 0\} .$$

Thus L is a $\Lambda_{\kappa-1}$-lattice, and it is easily verified that M/L is an R_κ-lattice. Assuming that we can classify all L's, the problem of finding all Λ-lattices M becomes one of determining the extensions of R_κ-lattices by such L's. This procedure works well for $\kappa = 1, 2$ (see [1], [7]), but gives only partial results for $\kappa > 2$.

Let us first establish a basic result due to Diederichsen [5]:

PROPOSITION 4.1. *Let $1 \leqq j \leqq \kappa$, and let L be a Λ-lattice such that $(x^{p^{j-1}} - 1)L = 0$. Then*

$$\operatorname{Ext}^1_\Lambda(R_j, L) \cong L/pL .$$

Proof. From the exact sequence $0 \to \Phi_j(x)\Lambda \to \Lambda \to R_j \to 0$ we obtain

$$\operatorname{Ext}(R_j, L) \cong \operatorname{Hom}(\Phi_j(x)\Lambda, L)/\text{image of } \operatorname{Hom}(\Lambda, L) .$$

Each Λ-homomorphism $f: \Phi_j(x)\Lambda \to L$ is completely determined by the image $f(\Phi_j(x))$ in L; this image may be any element of L which is annihilated by the Λ-annihilator of the ideal $\Phi_j(x)\Lambda$. This Λ-annihilator is $\{\prod_{n=j+1}^{\kappa} \Phi_n(x)\}\cdot(x^{p^{j-1}}-1)\Lambda$, which annihilates L by hypothesis. Thus *every* element of L may serve as the image $f(\Phi_j(x))$, and so $\operatorname{Hom}(\Phi_j(x)\Lambda, L) \cong L$. In this isomorphism, the image of $\operatorname{Hom}(\Lambda, L)$ is precisely $\Phi_j(x)L$. But

$$\Phi_j(x) = \sum_{i=0}^{p-1} x^{p^{j-1}\cdot i} ,$$

which acts on L as multiplication by p. Therefore $\operatorname{Ext}(R_j, L) \cong L/pL$, as claimed.

We shall consider the problem of classifying extensions of R_j-lattices by R_i-lattices, where $0 \leqq i < j \leqq \kappa$. However, a slightly more general situation can be handled by the same methods, and this extra generality will be needed later. Let E be any Z-torsionfree factor ring of Λ_{j-1}, so E is a Z-order in a Q-algebra which is a subsum of $\coprod_{i=0}^{pj-1} K_i$. Let J be the kernel of the surjection $Z[x] \to E$. If $a \cdot f(x) \in J$, where $a \in Z$ is nonzero and $f(x) \in Z[x]$, then also $f(x) \in J$ since E is Z-torsionfree. This implies readily that J is a principal ideal $(h(x))$, generated by a primitive polynomial $h(x) \in J$ of least degree. Since $x^{p^{j-1}} - 1 \in J$, we find that $h(x)$ divides $x^{p^{j-1}} - 1$, so $h(x)$ is monic. This shows that E is of the form $Z[x]/(h(x))$, for some monic divisor $h(x)$ of $x^{p^{j-1}} - 1$ in $Z[x]$.

Let E be as above; an E-lattice L is called *locally free* of rank r if $L \vee E^{(r)}$. (Note that all R_j-lattices are necessarily locally free.) We intend to classify extensions of R_j-lattices by locally free E-lattices. From (4.1) we have

$$\operatorname{Ext}^1_\Lambda(R_j, E) \cong E/pE = \bar{E}\,(\text{say}) .$$

Let $\bar{Z} = Z/pZ$; the surjection $\Lambda_{j-1} \to E$ induces a surjection $\bar{\Lambda}_{j-1} \to \bar{E}$. Here

$$\bar{\Lambda}_{j-1} = \bar{Z}[x]/(x^{p^{j-1}} - 1) \cong \bar{Z}[\lambda]/(\lambda^{p^{j-1}}), \text{ where } \lambda = 1 - x .$$

Thus $\bar{E}$ is a factor of a local ring $\bar{\Lambda}_{j-1}$, and hence is itself a local ring of the form

$$\bar{E} \cong \bar{Z}[\lambda]/(\lambda^e), \text{ where } e = \deg h(x) .$$

The action of E on $\operatorname{Ext}(R_j, E)$ is given via the surjection $E \to \bar{E}$. On the other hand, $\Phi_j(x) = p$ in Λ_{j-1}, hence also in E, so there is a ring surjection $R_j \to \bar{E}$. Then R_j acts on $\operatorname{Ext}(R_j, E)$ via this surjection. Now let N be any R_j-lattice. By Steinitz's Theorem, we may write $N \cong \coprod_{k=1}^{s} \mathfrak{c}_k$ where each $\mathfrak{c}_k$ is an R_j-ideal in K_j. The isomorphism class of N is determined by its rank s and its *Steinitz class* (namely, $\Pi\mathfrak{c}_k$ computed inside K_j). Analogously (see [11, Exercise 27.7]), a locally free E-lattice L may be written as $L \cong \coprod_{k=1}^{r} \mathfrak{b}_k$, where each $\mathfrak{b}_k$ is an E-lattice in $Q \otimes_Z E$ $(=QE)$ such that $\mathfrak{b}_k \vee E$. The isomorphism class of L is determined by its rank r and its Steinitz class (that is, the isomorphism class of $\Pi\mathfrak{b}_k$ computed inside QE).

Suppose that L and N are given, and let $\xi \in \operatorname{Ext}^1(N, L)$ determine a Λ-lattice X. We wish to classify all such X's up to isomorphism. We have

$$\operatorname{Ext}(N, L) \cong \operatorname{Ext}(R_j^{(s)}, E^{(r)}) \cong \{\operatorname{Ext}(R_j, E)\}^{r \times s} \cong \bar{E}^{r \times s} ,$$

where $\bar{E}^{r\times s}$ denotes the set of all $r \times s$ matrices over $\bar{E}$. Note that $\operatorname{Hom}_\Lambda(E, R_j) = 0$ since $\Phi_j(x)$ annihilates R_j, but acts as multiplication by p on the Z-torsionfree Λ-lattice E. Furthermore, both

R_j and E have commutative endomorphism rings, hence are Eichler lattices. If $t\colon \mathrm{Ext}\,(N, L) \cong \bar{E}^{r\times s}$ is the isomorphism given above, it follows from § 3 that a full set of invariants of the isomorphism class of X are

(i) The rank s and Steinitz class of N,

(ii) The rank r and Steinitz class of L, and

(iii) The strong equivalence class of $t(\xi)$ in $\bar{E}^{r\times s}$.

We shall assume that the problem of classifying all lattices N and L can be solved somehow. To classify all R_j-lattices, we must determine all R_j-ideal classes in K_j, and we assume that this has been done by standard methods of algebraic number theory. To classify all L's, we need to determine all classes of locally free E-ideals in QE. This is a difficult problem when $j \geqq 3$, and can be handled to some extent by the recent methods due to Galovich [6], Kervaire-Murthy [9], and Ullom [18], [19].

Supposing then that N and L are known, we shall concentrate on the problem of determining all strong equivalence classes in $\bar{E}^{r\times s}$. There are homomorphisms

$$GL(r, E) \longrightarrow GL(r, \bar{E}),\ GL(s, R_j) \longrightarrow GL(s, \bar{E}),$$

induced by the ring surjections $E \longrightarrow \bar{E}$, $R_j \longrightarrow \bar{E}$. The strong equivalence classes in $\bar{E}^{r\times s}$ are then the orbits in $\bar{E}^{r\times s}$ under the actions of $GL(r, E)$ on the left, and $GL(s, R_j)$ on the right. In the next section, we shall treat a somewhat more general version of the question of finding all strong equivalence classes.

5. Strong equivalence classes. Throughout this section, let Γ and Δ be a pair of commutative rings, and let

$$\varphi\colon \Gamma \longrightarrow \bar{\Gamma},\ \psi\colon \Delta \longrightarrow \bar{\Gamma},$$

be a pair of ring surjections. We assume that $\bar{\Gamma}$ is a local principal ideal ring, whose distinct ideals are given by $\{\lambda^k\bar{\Gamma}\colon 0 \leqq k \leqq e\}$, with $\lambda^e\bar{\Gamma} = 0$. Here, e is assumed finite and nonzero. Let $\bar{\Gamma}^{m\times n}$ consist of all $m \times n$ matrices with entries in $\bar{\Gamma}$. The maps φ, ψ induce homomorphisms

$$\varphi_*\colon GL(m, \Gamma) \longrightarrow GL(m, \bar{\Gamma}),\ \psi_*\colon GL(n, \Delta) \longrightarrow GL(n, \bar{\Gamma}), \tag{5.1}$$

which permit us to view $\bar{\Gamma}^{m\times n}$ as a left $GL(m, \Gamma)$-, right $GL(n, \Delta)$-bimodule. As suggested by our earlier considerations, we call two elements $\xi, \xi' \in \bar{\Gamma}^{m\times n}$ *strongly equivalent* (notation: $\xi \approx \xi'$) if $\xi' = \alpha\xi\beta$ for some $\alpha \in GL(m, \Gamma)$, $\beta \in GL(n, \Delta)$; here, α acts as $\varphi_*(\alpha)$, and β as $\psi_*(\beta)$. We wish to determine the strong equivalence classes in $\bar{\Gamma}^{m\times n}$.

(We have already encountered this problem in § 4, where we had a pair of rings R_j and E, with ring surjections $R_j \to \bar{E}$, $E \to \bar{E}$, and where $\bar{E}$ was a local principal ideal ring. In order to classify all extensions of an R_j-lattice of rank s by a locally free E-lattice of rank r, we needed to determine the strong equivalence classes of $\bar{E}^{r \times s}$ under the actions of $GL(r, E)$ and $GL(s, R_j)$.)

Returning to the more general case, we note that if $\xi \approx \xi'$ in $\bar{\Gamma}^{m \times n}$, then ξ is equivalent to ξ' in the usual (weaker) sense, that is, $\xi' = \mu\xi\nu$ for some $\mu, \nu \in GL(\bar{\Gamma})$. We can use the machinery of elementary divisors over the commutative principal ideal ring $\bar{\Gamma}$; these elementary divisors may be chosen to be powers of the prime element λ. Letting el. div. (ξ) denote the set of elementary divisors of ξ, we have at once

PROPOSITION 5.2. *If* $\xi \approx \xi'$, *then* el. div. $(\xi) =$ el. div. (ξ').

As before, let $u(\bar{\Gamma})$ denote the group of units of $\bar{\Gamma}$. The next two lemmas are simple but basic:

LEMMA 5.3. *For* $u \in u(\bar{\Gamma})$, *let* D_u *denote a diagonal matrix in* $GL(m, \bar{\Gamma})$ *with diagonal entries* $u, u^{-1}, 1, \cdots, 1$, *arranged in any order. Let* D'_u *denote an analogous matrix in* $GL(n, \bar{\Gamma})$. *Then for any* $\xi \in \bar{\Gamma}^{m \times n}$,

$$\xi \approx D_u\xi , \quad \xi \approx \xi D'_u .$$

Proof. There is an identity

$$(5.4) \qquad \begin{pmatrix} u & 0 \\ 0 & u^{-1} \end{pmatrix} = \begin{pmatrix} 1 & u^{-1} \\ 0 & 1 \end{pmatrix} \begin{pmatrix} 1 & 0 \\ 1 & 1 \end{pmatrix} \begin{pmatrix} 1 & u^{-1} - 1 \\ 0 & 1 \end{pmatrix} \begin{pmatrix} 1 & 0 \\ -u & 1 \end{pmatrix}$$

in $GL(2, \bar{\Gamma})$. This implies that D_u is expressible as a product of elementary matrices in $GL(m, \bar{\Gamma})$. Each factor is the image of an elementary matrix in $GL(m, \Gamma)$, so $\xi \approx D_u\xi$. An analogous argument proves that $\xi \approx \xi D'_u$.

LEMMA 5.5. *If* $m \leqq n$, *then each* $\xi \in \bar{\Gamma}^{m \times n}$ *is strongly equivalent to a matrix* $[D\ 0]$, *where*

$$(5.6) \quad D = \operatorname{diag}(\lambda^{k_1}u_1, \cdots, \lambda^{k_m}u_m) , \ 0 \leqq k_1 \leqq \cdots \leqq k_m \leqq e,\ u_i \in u(\bar{\Gamma}) .$$

If $m \geqq n$, *then* $\xi \approx \begin{bmatrix} D' \\ 0 \end{bmatrix}$, *where* D' *is a diagonal* $n \times n$ *matrix of the above type.*

Proof. Let $\xi \in \bar{\Gamma}^{m \times n}$, where $m \leqq n$. Since $\bar{\Gamma}$ is a local principal ideal ring, we can bring ξ into the form $[D\ 0]$, with D as above, by a sequence of left and right multiplications by elementary matrices in $GL(\bar{\Gamma})$. Each such elementary matrix lies in either $\mathrm{im}(\varphi_*)$ or $\mathrm{im}(\psi_*)$, and thus $\xi \approx [D\ 0]$ as claimed. An analogous proof is valid for the case where $m \geqq n$.

Suppose now that $\xi \in \bar{\Gamma}^{m\times n}$; for convenience of notation let us assume that $m \leqq n$, and let $\xi \approx [D\ 0]$ with D as in (5.6). Then obviously

$$\text{el. div.}\,(\xi) = \{\lambda^{k_1}, \cdots, \lambda^{k_m}\}\,.$$

It follows at once from (5.2) that the set $\{\lambda^{k_1}, \cdots, \lambda^{k_m}\}$ is an invariant of the strong equivalence class of ξ. Let us show at once that this is the *only* invariant when $m \neq n$.

PROPOSITION 5.7. *Let* $\xi, \xi' \in \bar{\Gamma}^{m\times n}$, *where* $m \neq n$. *Then* $\xi \approx \xi'$ *if and only if* el. div. $(\xi) =$ el. div. (ξ').

Proof. By (5.2) it suffices to show that if $m \neq n$, then ξ is determined up to strong equivalence by its set of elementary divisors. For convenience of notation, assume that $m \leqq n$, and write $\xi \approx [D\ 0]$, with D as in (5.6). By (5.3) we have

$$[D\ 0] \approx [D\ 0]\cdot\mathrm{diag}\,(u_1^{-1}, u_2^{-1}, \cdots, u_m^{-1}, \underbrace{u_1\cdots u_m, 1, \cdots, 1}_{n-m})\,.$$

This gives

$$\xi \approx [D_1\ 0] \text{ where } D_1 = \mathrm{diag}\,(\lambda^{k_1}, \cdots, \lambda^{k_m})\,,$$

and so the strong equivalence class of ξ is determined by el. div. (ξ). This completes the proof.

We are now ready to turn to the question as to when two elements ξ and ξ' in $\bar{\Gamma}^{m\times m}$ are strongly equivalent. By (5.2), it suffices to treat the case where ξ and ξ' have the same elementary divisors. We shall see that there is exactly one additional invariant needed for this case. To begin with, we introduce the following notation: let $\xi \in \bar{\Gamma}^{m\times m}$, and suppose that $\xi \approx D$, where D is given by (5.6). We set

$$\Gamma' = \bar{\Gamma}/\lambda^{e-k_m}\bar{\Gamma},\ U = u(\Gamma')/u^*(\Gamma)u^*(\Delta)\,, \tag{5.8}$$

where $u^*(\Gamma)$ denotes the image of $u(\Gamma)$ in $u(\Gamma')$, and $u^*(\Delta)$ the image of $u(\Delta)$. Define

$$u(\xi) = \text{image of } u_1\cdots u_m \text{ in } U\,. \tag{5.9}$$

The main result of this section is as follows:

THEOREM 5.10. *Let* $\xi, \xi' \in \bar{\Gamma}^{m\times m}$. *Then* $\xi \approx \xi'$ *if and only if*
(i) el. div. $(\xi) =$ el. div. (ξ'), *and*
(ii) $u(\xi) = u(\xi')$ *in* U.

Proof. Supposing that conditions (i) and (ii) are satisfied, let

$$\xi \approx D, \ \xi \approx \operatorname{diag}(\lambda^{k_1}u'_1, \cdots, \lambda^{k_m}u'_m), \ u'_i \in u(\bar{\Gamma}) ,$$

where D is given by (5.6). Setting $u = \Pi u_i$, $u' = \Pi u'_i$, it follows from the proof of (5.7) that

$$(5.11) \quad \xi \approx \operatorname{diag}(\lambda^{k_1}, \cdots, \lambda^{k_{m-1}}, \lambda^{k_m}u), \ \xi' \approx \operatorname{diag}(\lambda^{k_1}, \cdots, \lambda^{k_{m-1}}, \lambda^{k_m}u') .$$

By virtue of (ii), there exist elements $\gamma \in u(\Gamma)$, $\delta \in u(\Delta)$, such that $u' = \gamma u \delta$ in Γ'. But then

$$\lambda^{k_m}u' = \gamma \cdot \lambda^{k_m}u \cdot \delta \quad \text{in } \bar{\Gamma} ,$$

so

$$\operatorname{diag}(1, \cdots, 1, \gamma) \cdot \operatorname{diag}(\lambda^{k_1}, \cdots, \lambda^{k_{m-1}}, \lambda^{k_m}u) \cdot \operatorname{diag}(1, \cdots, 1, \delta)$$
$$= \operatorname{diag}(\lambda^{k_1}, \cdots, \lambda^{k_{m-1}}, \lambda^{k_m}u') .$$

Therefore $\xi' \approx \xi$, as desired.

Conversely, assume that $\xi \approx \xi'$, so (i) holds by (5.2). In proving (ii), we may assume without loss of generality that ξ and ξ' are equal (respectively) to the diagonal matrices listed in (5.11). Since $\xi \approx \xi'$, we have $\mu\xi' = \xi\nu$ for some $\mu \in GL(m, \Gamma)$, $\nu \in GL(m, \Delta)$. It is tempting to take determinants of both sides, but this procedure fails because $\lambda^e = 0$ in $\bar{\Gamma}$. Instead, we proceed as follows: let $D_0 = \operatorname{diag}(\lambda^{k_1}, \cdots, \lambda^{k_m})$, and put

$$\mu_1 = \varphi_*(\mu) \cdot \operatorname{diag}(1, \cdots, 1, u'), \ \nu_1 = \operatorname{diag}(1, \cdots, 1, u) \cdot \psi_*(\nu) .$$

The equation $\mu\xi' = \xi\nu$ then becomes $\mu_1 \cdot D_0 = D_0 \cdot \nu_1$. By (5.12) below, this implies that $(\det \mu_1)\lambda^{k_m} = (\det \nu_1)\lambda^{k_m}$. But $\det \mu_1 = u' \cdot \varphi(\det \mu)$, and $\det \nu_1 = u \cdot \psi(\det \nu)$. Therefore the images of u and u' in $u(\Gamma')$ differ by a factor from $u^*(\Gamma)u^*(\Delta)$, which shows that $u(\xi) = u(\xi')$ in U, and completes the proof.

It remains for us to establish the following amusing result on determinants:

PROPOSITION 5.12. *Let* R *be an arbitrary commutative ring, and let* $D = \operatorname{diag}(\xi_1, \cdots, \xi_m)$ *be a matrix over* R *such that*

$$r_1\xi_1 = \cdots = r_{m-1}\xi_{m-1} = \xi_m$$

for some elements $r_i \in R$. *Let* $X, Y \in R^{m\times m}$ *be matrices for which* $XD = DY$. *Then*

$$(\det X)\cdot\xi_m = (\det Y)\cdot\xi_m$$

in R.

Proof. Let $X = (x_{ij})$, $Y = (y_{ij})$. The equation $XD = DY$ gives

$$x_{ij}\xi_j = \xi_i y_{ij},\ 1 \leqq i,\ j \leqq m\ .$$

Let $\pi\colon 1 \to i_1, \cdots, m \to i_m$, be a permutation of the symbols $\{1, \cdots, m\}$. A typical term in the expansion of $\det X$ is of the form $\pm x_{1i_1}\cdots x_{mi_m}$, and we need only show that for each π we have

$$x_{1i_1}\cdots x_{mi_m}\xi_m = y_{1i_1}\cdots y_{mi_m}\xi_m\ . \tag{5.13}$$

Write π as a product of cycles, and suppose by way of illustration that (a, b, c) is a 3-cycle occuring as a factor of π. Then

$$\begin{aligned} x_{ab}x_{bc}x_{ca}\xi_m &= r_a\cdot x_{ab}x_{bc}x_{ca}\xi_a = r_a x_{ab}x_{bc}\xi_c y_{ca} = r_a x_{ab}\xi_b y_{bc}y_{ca} \\ &= r_a\xi_a y_{ab}y_{bc}y_{ca} = y_{ab}y_{bc}y_{ca}\xi_m\ . \end{aligned}$$

The same procedure applies to each cycle occuring in π, which establishes (5.13), and completes the proof of the proposition.

The special case where $\Gamma = \Delta = Z$, $\bar{\Gamma} = Z/(p^e)$, p prime, is of interest. For a matrix $X \in Z^{m\times n}$, let p-el. div. (X) be the powers of p occurring in the ordinary elementary divisors of X (over Z). If X is square, write $\det X =$ (power of p)$\cdot u_X$, where $p \nmid u_X$. (Take $u_X = 1$ if $\det X = 0$.) For $X, Y, \in Z^{m\times n}$, we write $X \approx Y$ if

$$Y \equiv PXQ \pmod{p^e}$$

for some $P \in GL(m, Z)$, $Q \in GL(n, Z)$. From (5.10) we obtain

COROLLARY 5.14. *Let* $X, Y \in Z^{m\times n}$. *Then* $X \approx Y$ *if and only if*
(i) p-el. div. $(X) = p$-el. div. (Y), *and*
(ii) *when* $m = n$,

$$u_Y \equiv \pm u_X \pmod{p^{e-k}}\ ,$$

where k *is the maximum of the exponents of the* p*-elementary divisors of* X. *(If* $k \geqq e$, *condition* (ii) *is automatically satisfied.)*

For the particular cases needed in §§ 6-7, one can easily deduce (5.7) and (5.10) as special cases of the results of Jacobinski [8]. However, it seemed desirable to give here a self-contained proof of (5.7) and (5.10).

6. **Invariants of direct sums of extensions.** We now return to the study of integral representations of a cyclic group G of

order p^κ, keeping the notation of § 4. Let N be an R_j-lattice of rank s, and L a locally free E-lattice of rank r. We have seen that

$$\mathrm{Ext}^1_{ZG}(N, L) \cong \mathrm{Ext}^1_{ZG}(R_j^{(s)}, E^{(r)}) \cong \bar{E}^{r\times s},$$

where $\bar{E} \cong \bar{Z}[\lambda]/(\lambda^e)$ is a local principal ideal ring. Each extension X of N by L determines a class $\xi_X \in \bar{E}^{r\times s}$, and an element $u(\xi_X)$ in a factor group of the group of units of some quotient ring of $\bar{E}$ (see (5.9)). It follows from the results of §§ 4, 5 that a full set of isomorphism invariants of X are as follows:

(i) The rank s of N, and its Steinitz class,
(ii) The rank r of L, and its Steinitz class,
(iii) The elementary divisors of the matrix ξ_X,
(iv) For the case $r = s$ only, the element $u(\xi_X)$.

Since G is a p-group, the genus of X is completely determined by the p-adic completion X_p. In the local case, however, the ideal classes occurring above are trivial, as is the group in which $u(\xi_X)$ lies. Therefore the genus invariants of X are just r, s, and el. div. (ξ_X). Furthermore, by (5.5) the extension X must decompose into a direct sum of ideals $\mathfrak{b}$ of R_j, locally free ideals $\mathfrak{c}$ of E, and nonsplit extensions of $\mathfrak{c}$ by $\mathfrak{b}$. Let us denote by $(\mathfrak{b}, \mathfrak{c}; \lambda^k u)$ an extension of $\mathfrak{c}$ by $\mathfrak{b}$ corresponding to the extension class $\lambda^k u \in \bar{E}$, where $0 \leqq k < e$, $u \in u(\bar{E})$, and we have chosen some standard isomorphism $\mathrm{Ext}(\mathfrak{c}, \mathfrak{b}) \cong \bar{E}$. By (1.5), the lattice $(\mathfrak{b}, \mathfrak{c}; \lambda^k u)$ is indecomposable since $\lambda^k u \neq 0$ in $\bar{E}$.

Some further notation will be useful below. Let us set $E' = \bar{E}/\lambda^m\bar{E} \cong \bar{Z}[\lambda]/(\lambda^m)$, where $1 \leqq m \leqq e$. There are ring surjections $E \to E'$, $R_j \to E'$; let $u^*(E)$ denote the image of $u(E)$ in $u(E')$, and define $u^*(R_j)$ analogously. We now set

$$U_m = u(E')/u^*(E)u^*(R_j) . \tag{6.1}$$

It follows from the above discussion that a full set of isomorphism invariants of $(\mathfrak{b}, \mathfrak{c}; \lambda^k u)$ are the isomorphism classes of $\mathfrak{b}$ and $\mathfrak{c}$, the integer k, and the image of u in U_{e-k}. The genus of X depends only on k.

We may remark that the group $u(E')$ is easily described, namely,

$$u(E') \cong u(\bar{Z}) \times \prod_{i=1}^{m-1}{}' \langle 1 + \lambda^i \rangle ,$$

where in the product i ranges over the integers between 1 and $m - 1$ which are prime to p. On the other hand, the calculation of $u^*(E)$ and $u^*(R_j)$ is considerably more difficult, and the results so far known are given in [6], [9], [18], and [19]. It is easily verified that $u^*(R_j)$ contains the factor $u(\bar{Z})$, and further that $1+\lambda \in u^*(R_j)$ since $1+\lambda = x$. It follows at once that U_1 and U_2 are trivial for all p.

In the special case where $E = R_i$ with $i < j$, we claim that $u^*(R_j) \subset u^*(R_i)$, and hence that

$$U_m = u(E')/u^*(R_i) .$$

Indeed, as pointed out in [6], there is a commutative diagram

$$\begin{array}{ccc} u(R_j) & \xrightarrow{N} & u(R_i) \\ & \underset{\theta_j}{\searrow} & \downarrow \theta_i \\ & & u(\bar{R}_i) , \end{array}$$

where $\bar{R}_i = R_i/pR_i$ and N is the relative norm map. Hence $\operatorname{im}\theta_j \subset \operatorname{im}\theta_i$, which implies that $u^*(R_j) \subset u^*(R_i)$ in $u(E')$.

The structure of U_m has been studied in detail by Galovich [6] and Kervaire-Murthy [9], especially for the case of regular primes. An odd prime p is *regular* if the ideal class number of R_1 is relatively prime to p (see [2]). For regular p, we have

$$u^*(R_j) \cong u(\bar{Z}) \times \langle 1+\lambda \rangle \times \prod \langle 1+\lambda^{2i} + \alpha_i \lambda^{2i+1} \rangle , \tag{6.2}$$

where each $\alpha_i \in E'$, and where i ranges over all integers from 1 to $[(m-1)/2]$ which are prime to p. Furthermore, $u^*(E) \subset u^*(R_j)$ in this case, so U_m is of order $p^{f(m)}$, where $f(m)$ is the number of odd integers among $3, 5, \cdots, m-1$ which are prime to p.

Some additional information is available for the special case where $j = 2$; here, $e \leqq p$ and U_m is an elementary abelian p-group. Let $\delta(k)$ be the number of Bernoulli numbers among $B_1, B_2, \cdots, B_k$ whose numerators are divisible by p. Then (see [2]) the prime p is regular if and only if $\delta((p\text{-}3)/2) = 0$. Call p *properly irregular* if p divides the class number of R_1 but not that of $Z[\omega_1 + \omega_1^{-1}]$. For such p, one must omit from the formula (6.2) all those factors $1 + \lambda^{2i} + \alpha_i\lambda^{2i+1}$ for which $2i \leqq p-3$ and the numerator of B_i is a multiple of p. Thus for properly irregular primes p, U_m is elementary abelian of order $p^{g(m)}$, where

$$g(m) = \begin{cases} [(m-2)/2] + \delta[(m-1)/2] , & 0 \leqq m \leqq p-2 , \\ (p-3)/2 + \delta((p-3)/2) , & m = p-1 , \ p . \end{cases} \tag{6.3}$$

Here, we must interpret the greatest integer function $[(m-2)/2]$ as 0 when $m < 2$. Further, for $j = 2$, U_m is trivial when $p = 2$.

For the case where $E = Z[x]/(x^p - 1)$ and $j = 2$, it is known (see [6], [9], [19]) that $u^*(E) = u^*(R_2)$ for all m and all regular or properly irregular primes p. It seems likely that a corresponding result holds for $j > 2$ for arbitrary E, for all primes p (in this connection, see [19]).

From the results stated earlier in this section, we obtain

THEOREM 6.3a. *Consider the direct sum*

$$Y = \coprod_{k=1}^{b} \mathfrak{b}'_k \oplus \coprod_{n=1}^{c} \mathfrak{c}'_n \oplus \coprod_{i=1}^{d} (\mathfrak{b}_i, \mathfrak{c}_i; \lambda^{k_i} u_i) , \tag{6.4}$$

where each $\mathfrak{b}$ *is a locally free E-ideal, each* $\mathfrak{c}$ *an* R_j*-ideal, and* $0 \leqq k_i < e$, $u_i \in u(\bar{E})$ *for each i. We may view Y as an extension* $0 \to Y_0 \to Y \to Y_1 \to 0$, *where*

$$Y_0 = \coprod \mathfrak{b}'_k \oplus \coprod \mathfrak{b}_i , \quad Y_1 = \coprod \mathfrak{c}'_n \oplus \coprod \mathfrak{c}_i ,$$

corresponding to the $(b + d) \times (c + d)$ *matrix* $\begin{pmatrix} D & 0 \\ 0 & 0 \end{pmatrix}$ *over* $\bar{E}$, *with* $D = \operatorname{diag}(\lambda^{k_1} u_1, \cdots, \lambda^{k_d} u_d)$. *Define* U_m *as in* (6.1), *with*

$$m = \operatorname{Min}\{e - k_i : 1 \leqq i \leqq d\} .$$

Then we have

(i) *The genus of Y is determined by the integer* $b + d$ (*=E-rank of* Y_0), *the integer* $c + d$ (*=*R_j*-rank of* Y_1), *and the set of exponents* $\{k_i\}$.

(ii) *The additional invariants of the isomorphism class of Y, needed to determine this class, are the isomorphism classes of*

$$\Pi \mathfrak{b}'_k \cdot \Pi \mathfrak{b}_i \quad \text{and} \quad \Pi \mathfrak{c}'_n \cdot \Pi \mathfrak{c}_i ,$$

and one further invariant which occurs only when $b = c = 0$, *namely the image of* $u_1 \cdots u_d$ *in* U_m.

Several remarks are in order concerning the above result. First of all, we note that

$$\mathfrak{b}' \oplus (\mathfrak{b}, \mathfrak{c}; \lambda^k u) \cong \mathfrak{b}' \oplus (\mathfrak{b}, \mathfrak{c}; \lambda^k) \tag{6.5}$$

as a consequence of the Absorption Formula (2.7). Namely, choose $w \in E$ with image $u \in \bar{E}$, so then

$$(\mathfrak{b}, \mathfrak{c}; \lambda^k u) = {}_w(\mathfrak{b}, \mathfrak{c}; \lambda^k)$$

in the notation of § 2. Since $\mathfrak{b}'/w\mathfrak{b}' \cong \mathfrak{b}/w\mathfrak{b}$ because $\mathfrak{b} \vee \mathfrak{b}'$, formula (6.5) follows from the proof of (2.7). Likewise, we have

$$\mathfrak{c}' \oplus (\mathfrak{b}, \mathfrak{c}; \lambda^k u) \cong \mathfrak{c}' \oplus (\mathfrak{b}, \mathfrak{c}; \lambda^k)$$

always, by using the fact that R_j maps onto E. Thus, if either b

or c is nonzero, we may replace each u_i in (6.4) by 1 without affecting the isomorphism class of Y. This agrees with our previous result.

Next, suppose that $b = c = 0$, and suppose the summands of Y numbered so that $k_d = \text{Max}\,\{k_i\}$. The Exchange Formula (2.4) gives $Y \cong W \oplus X$, where

$$W = \coprod_1^{d-1} (\mathfrak{b}_i, \mathfrak{c}_i; \lambda^{k_i}) , \quad X = (\mathfrak{b}_d, \mathfrak{c}_d; \lambda^{k_d} u) ,$$

and $u = u_1 \cdots u_d$. Let $X' = (\mathfrak{b}_d, \mathfrak{c}_d; \lambda^{k_d} u')$. Our previous result then takes the form of a Cancellation Theorem, namely,

$$(6.6) \qquad W \oplus X \cong W \oplus X' \text{ if and only if } X \cong X' .$$

This is of special interest in that it applies to a situation in which the summands lie in different genera. We may also deduce (6.6) from Jacobinski's Cancellation Theorem [8, § 4] if desired.

To conclude these remarks, we may point out that the results of § 5 yield a slightly more general cancellation theorem, as follows: let Λ be any R-lattice in a semisimple K-algebra A, and let M, N be Λ-lattices with commutative Λ-endomorphism rings Γ, Δ, respectively. For $i = 1, \cdots, d$, let X_i be an extension of N by M corresponding to the class $\xi_i \in \text{Ext}(N, M)$. Suppose that for each i, we may write $\xi_d = \gamma_i \xi_i \delta_i$ for some $\gamma_i \in \Gamma$, $\delta_i \in \Delta$, and let X' be any Λ-lattice. Then

$$\coprod_1^{d-1} X_i \oplus X_d \cong \coprod_1^{d-1} X_i \oplus X' \text{ if and only if } X_d \cong X' .$$

Further, the same result holds if each X_i is replaced by a lattice in its genus.

7. **Cyclic groups of order p^2.** We shall now determine a full set of isomorphism invariants of ZG-lattices, where G is cyclic of order p^2. To simplify the notation, we set

$$R = Z[\omega_1],\ S = Z[\omega_2],\ E = Z[x]/(x^p - 1),\ \bar{E} = E/pE \cong \bar{Z}[\lambda]/(\lambda^p) ,$$

where $\bar{Z} = Z/pZ$ and $\lambda = 1 - x$. By § 4, every E-lattice is an extension of an R-lattice by a Z-lattice. In this case, we have $\text{Ext}\,(R, Z) \cong \bar{Z}$, and $u(R)$ maps onto $u(\bar{Z})$. Thus by § 6, the only indecomposable E-lattices are Z, $\mathfrak{b}$, and $E(\mathfrak{b}) = (Z, \mathfrak{b}; 1)$, where $\mathfrak{b}$ ranges over a full set of representatives of the h_R ideal classes of R. Here, $(Z, \mathfrak{b}; 1)$ denotes an extension of $\mathfrak{b}$ by Z corresponding to the extension class $\bar{1} \in \bar{Z}$, using a standard isomorphism $\text{Ext}\,(\mathfrak{b}, Z) \cong \bar{Z}$. We note that $E(\mathfrak{b}) \vee E$, so $E(\mathfrak{b})$ is a locally free E-lattice of rank 1; conversely, every such lattice is isomorphic to some $E(\mathfrak{b})$.

By § 6, every E-lattice L is of the form

$$(7.1) \qquad L \cong Z^{(a)} \oplus \coprod_{i=1}^{b} \mathfrak{b}_i' \oplus \coprod_{j=1}^{c} E(\mathfrak{b}_j),$$

and a full set of isomorphism invariants of L are the integers a, b, c (the genus invariants), and the ideal class of $\Pi\mathfrak{b}_i' \cdot \Pi\mathfrak{b}_j$.

Now let M be any ZG-lattice. By § 4, M is an extension of an S-lattice N by an E-lattice L. A full set of isomorphism invariants of M are the isomorphism class of L (just determined above), the isomorphism class of N, and the strong equivalence class in $\operatorname{Ext}(N, L)$ containing the extension class of M. Of course, N is determined up to isomorphism by its S-rank and Steinitz class. Furthermore, in calculating strong equivalence classes in $\operatorname{Ext}(N, L)$, we may replace N by any lattice in its genus, and likewise for L. Thus it suffices to treat the case where L is a sum of copies of Z, R, and E, and where N is S-free. However, our conclusions can be stated more neatly as an answer to the following equivalent question: what are the isomorphism invariants of a direct sum of indecomposable ZG-lattices?

As shown in [7], a ZG-lattice M is indecomposable if and only if M_p is indecomposable. The indecomposable Z_pG-lattices can be determined explicitly by considering strong equivalence in the local case (see [1] or [7]; the case $p = 2$ is treated in [14] and [17]). Rather than repeat the local argument here, we just state the conclusion: every indecomposable ZG-lattice is in the same genus as one (and only one) of the following $4p + 1$ indecomposable ZG-lattices:

$$(7.2) \qquad \begin{cases} Z, R, E, S, (Z, S; 1), \\ (E, S; \lambda^r),\ 0 \leqq r \leqq p-1, \\ (Z \oplus E, S; 1 \oplus \lambda^r),\ 1 \leqq r \leqq p-2, \\ (R, S; \lambda^r),\ 0 \leqq r \leqq p-2, \\ (Z \oplus R, S; 1 \oplus \lambda^r),\ 0 \leqq r \leqq p-2. \end{cases}$$

Here $(Z, S; 1)$ represents an extension of S by Z with class $\bar{1} \in \bar{Z}$, using the isomorphism $\operatorname{Ext}(S, Z) \cong \bar{Z}$. Further, $(Z \oplus E, S; 1 \oplus \lambda^r)$ denotes an extension of S by $Z \oplus E$ with class $(1, \lambda^r) \in \bar{Z} \oplus \bar{E}$, using the isomorphism $\operatorname{Ext}(S, Z \oplus E) \cong \bar{Z} \oplus \bar{E}$. Analogous definitions hold for the other cases.

A full set of nonisomorphic indecomposable ZG-lattices may now be obtained from (7.2), by finding all isomorphism classes in each of the genera occurring in (7.2). This was done in [12], but we take this opportunity to correct a misstatement in that article. Let us denote by $\tilde{U}_m$ a full set of representatives u in $u(\bar{R})$ or $u(\bar{E})$

of the elements of the factor group U_m, where the u's are chosen so that $u \equiv 1 \pmod{\lambda}$. Recall that for $1 \leqq m \leqq p-1$, U_m denotes the group of units of $\bar{Z}[\lambda]/(\lambda^m)$ modulo the image of $u(R)$, while U_p denotes $u(\bar{E})$ modulo the images of $u(S)$ and $u(E)$. The notations U_m, U_p are then consistent with those introduced in § 6. Finally, let n_0 be some fixed quadratic nonresidue (mod p).

THEOREM 7.3. *Let* $\mathfrak{b}$ *range over a full set of representatives of the* h_R *ideal classes of* R, *and* $\mathfrak{c}$ *likewise for the* h_S *ideal classes of* S. *A full list of nonisomorphic indecomposable* ZG*-lattices is as follows*:

(a) $Z, \mathfrak{b}, E(\mathfrak{b}), \mathfrak{c}, (Z, \mathfrak{c}; 1)$.
(b) $(E(\mathfrak{b}), \mathfrak{c}; \lambda^r u)$, $u \in \tilde{U}_{p-r}$, $0 \leqq r \leqq p-1$.
(c) $(Z \oplus E(\mathfrak{b}), \mathfrak{c}; 1 \oplus \lambda^r u)$, $u \in \tilde{U}_{p-1-r}$, $1 \leqq r \leqq p-2$.
(d) *If* $p \equiv 1 \pmod 4$, $(Z \oplus E(\mathfrak{b}), \mathfrak{c}; 1 \oplus \lambda^r u n_0)$, $u \in \tilde{U}_{p-1-r}$, $1 \leqq r \leqq p-2$.
(e) $(\mathfrak{b}, \mathfrak{c}; \lambda^r u)$, $u \in \tilde{U}_{p-1-r}$, $0 \leqq r \leqq p-2$.
(f) $(Z \oplus \mathfrak{b}, \mathfrak{c}; 1 \oplus \lambda^r u)$, $u \in \tilde{U}_{p-1-r}$, $0 \leqq r \leqq p-2$.

Proof. Observe first that $\mathfrak{b}$ gives all isomorphism classes in the genus of R, and $\mathfrak{c}$ these in the genus of S. Further $E(\mathfrak{b})$ gives all isomorphism classes in the genus of E. It remains for us to check strong equivalence classes in each of the remaining cases, and for this it suffices to treat the cases where $\mathfrak{b} = R$, $\mathfrak{c} = S$ and $E(\mathfrak{b}) = E$.

Next, we have $\operatorname{Ext}(S, Z) \cong \bar{Z}$, and $u(S)$ maps onto $u(\bar{Z})$. Hence there is only one nonzero strong equivalence class in $\operatorname{Ext}(S, Z)$, so all nonsplit extensions of S by Z are mutually isomorphic. Also, $\operatorname{Ext}(S, E) \cong \bar{E} \cong \bar{Z}[\lambda]/(\lambda^p)$, and by § 6 the nonzero strong equivalence classes in $\operatorname{Ext}(S, E)$ are represented by $\{\lambda^r u : u \in \tilde{U}_{p-r}, 0 \leqq r \leqq p-1\}$. This gives the lattices described in (b). A similar argument yields those in (e).

Consider next the classification of lattices in the genus of $(Z \oplus R, S; 1 \oplus \lambda^r)$, where $0 \leqq r \leqq p-2$. The following observation is needed both here and later: each E-lattice L is expressible as an extension $0 \to L_0 \to L \xrightarrow{\theta} L_1 \to 0$, with L_0 a Z-lattice uniquely determined inside L, and L_1 an R-lattice. The map θ induces surjections $\bar{L} \to \bar{L}_1$, $\operatorname{Ext}(S, L) \to \operatorname{Ext}(S, L_1)$, where bars denote reduction mod p, and where the surjections are consistent with the isomorphisms $\operatorname{Ext}(S, L) \cong \bar{L}$, $\operatorname{Ext}(S, L_1) \cong \bar{L}_1$. Now let M be an extension of an S-lattice N by L, so L (and hence also L_0) are uniquely determined inside M. Then there is a commutative diagram

$$\begin{array}{ccccccccc} 0 & \longrightarrow & L & \longrightarrow & M & \longrightarrow & N & \longrightarrow & 0 \\ & & \downarrow{\scriptstyle\theta} & & \downarrow & & \downarrow{\scriptstyle 1} & & \\ 0 & \longrightarrow & L_1 & \longrightarrow & M^* & \longrightarrow & N & \longrightarrow & 0, \end{array}$$

giving rise to a ZG-exact sequence

$$0 \longrightarrow L_0 \longrightarrow M \longrightarrow M^* \longrightarrow 0 .$$

The isomorphism class of M uniquely determines that of M^*, and the extension class of M^* in $\mathrm{Ext}(N, L_1)$ is the image of the class of M in $\mathrm{Ext}(N, L)$, under the map induced by θ.

Suppose in particular that $M = (Z \oplus R, S; 1 \oplus \lambda^r u)$; then $M^* \cong (R, S; \lambda^r u)$, and thus the image of u in U_{p-1-r} is an isomorphism invariant of M. Conversely, any M' in the genus of M may be written as $(Z \oplus \mathfrak{b}, \mathfrak{c}; q \oplus \lambda^r u')$, where $q \in u(\bar{Z})$. Then $M' \cong (Z \oplus \mathfrak{b}, \mathfrak{c}; q\beta \oplus \lambda^r \alpha u' \beta)$ for any $\alpha \in u(R)$, $\beta \in u(S)$. Choose β so that $q\beta = \bar{1}$ in $\bar{Z}$, and then choose α so that $\alpha u' \beta$ lies in $\tilde{U}_{p-1-r}$. This proves that M' is isomorphic to one of the lattices in (f), and therefore (f) gives a full list of nonisomorphic indecomposable ZG-lattices in the genus of $(Z \oplus R, S; 1 \oplus \lambda^r)$.

Turning to the most difficult case, we have $\mathrm{Ext}(S, Z \oplus E) \cong \bar{Z} \oplus \bar{E}$, and we must determine strong equivalence classes in $\bar{Z} \oplus \bar{E}$ under the actions of $\mathrm{Aut}(Z \oplus E)$ and $\mathrm{Aut}(S)$. We may represent the elements of $\bar{Z} \oplus \bar{E}$ as vectors $\binom{\bar{z}}{\bar{e}}$ on which $\mathrm{Aut}(Z \oplus E)$ acts from the left, and $\mathrm{Aut}(S)$ from the right. Let $\Phi(x)$ denote the cyclotomic polynomial of order p. Since $\mathrm{Hom}(E, Z) \cong Z$ and $\mathrm{Hom}(Z, E) \cong \Phi(x)E$, we obtain

$$\text{(7.4)} \qquad \mathrm{End}(Z \oplus E) = \left\{ f : f = \begin{pmatrix} a & b \\ \Phi(x)c & d \end{pmatrix}, \ a, b \in Z, \ c, d \in E \right\} .$$

There is a fiber product diagram

$$\text{(7.5)} \qquad \begin{array}{ccc} E & \xrightarrow{\varphi_1} & Z \\ {\scriptstyle \varphi_2}\downarrow & & \downarrow \\ R & \longrightarrow & \bar{Z} , \end{array}$$

and an E-exact sequence

$$0 \longrightarrow Z \oplus Z \xrightarrow{1 \oplus \Phi(x)} Z \oplus E \longrightarrow R \longrightarrow 0 .$$

Each $f \in \mathrm{End}(Z \oplus E)$, given as in (7.4), induces a map f_1 on $Z \oplus Z$ and a map f_2 on R, where

$$f_1 = \begin{pmatrix} a & b \\ p\varphi_1(c) & \varphi_1(d) \end{pmatrix}, \quad f_2 = \text{multiplication by } \varphi_2(d) .$$

Clearly, f is an automorphism if and only if $f_1 \in GL(2, Z)$ and $\varphi_2(d) \in u(R)$. Furthermore, for each $\alpha \in u(R)$ there exists an automorphism

f such that $\varphi_2(d) = \alpha$. Also, for each matrix $\mu \in GL(2, Z)$ whose (2, 1) entry is divisible by p, we can find an automorphism f such that $f_1 = \mu$.

Now let M be a lattice in the genus of $(Z \oplus E, S; 1 \oplus \lambda^r)$ with extension class $\begin{pmatrix}\bar{z}\\ \bar{e}\end{pmatrix}$. Since $\begin{pmatrix}\bar{z}\\ \bar{e}\end{pmatrix} \approx \begin{pmatrix}\bar{1}\\ \bar{e}_1\end{pmatrix}$ for some $\bar{e}_1$, we may hereafter assume that $\bar{z} = \bar{1}$. Factoring out the submodule $Z \oplus Z$ of M as before, we obtain an extension M^* of S by R, with extension class $\bar{\varphi}_2(\bar{e})$, where $\bar{\varphi}_2 \colon \bar{E} \to \bar{R}$ is induced from φ_2. The isomorphism class of M^* is determined from that of M. In particular, if the extension class of M is $\begin{pmatrix}\bar{1}\\ \lambda^r u\end{pmatrix}$, where $1 \leqq r \leqq p-2$ and $u \in u(\bar{E})$, then the extension class of M^* is $\lambda^r u$, viewed as element of $\bar{R}$. Therefore the image of u in U_{p-1-r} is an isomorphism invariant of M. We shall see that when $p \equiv 3 \bmod 4$, this image of u and the integer r are a full set of isomorphism invariants of M. On the other hand, when $p \equiv 1 \bmod 4$, an additional invariant will be needed, namely the quadratic character of $u \pmod{\lambda}$ viewed as element of $u(\bar{Z})$.

Let $f \in \operatorname{Aut}(Z \oplus E)$, $s \in u(S)$, and let $1 \leqq r \leqq p-2$. The equation

$$\begin{pmatrix} a & b \\ \Phi(x)c & d\end{pmatrix}\begin{pmatrix}\bar{1}\\ \lambda^r u\end{pmatrix}\cdot s = \begin{pmatrix}\bar{1}\\ \lambda^r u'\end{pmatrix} \tag{7.5'}$$

becomes (since $\lambda^r b = 0$ in $\bar{Z}$)

$$a s \bar{1} = \bar{1} \text{ in } \bar{Z},\ \lambda^r u' = s d \lambda^r u + \Phi(x)\bar{c} s \text{ in } \bar{E}\,.$$

Since $\det f_1 = \pm 1$, we have $ad \equiv \pm 1 \pmod{\lambda}$ in $\bar{E}$. Thus we obtain

$$u' \equiv sdu \equiv \pm a^{-2} u \pmod{\lambda}\,. \tag{7.6}$$

If $p \equiv 3 \pmod 4$, then as a ranges over all integers prime to p so does $\pm a^{-2}$, and (7.6) imposes no condition on $u \pmod{\lambda}$. However, if $p \equiv 1 \pmod 4$, then $\pm a^{-2}$ is always a quadratic residue $(\operatorname{mod} p)$. It follows from (7.6) that the quadratic character of $u \pmod{\lambda}$ is an invariant of the strong equivalence class of $\begin{pmatrix}\bar{1}\\ \lambda^r u\end{pmatrix}$, and is therefore an isomorphism invariant of M. This argument, together with the discussion in the preceding paragraph, shows that no two of the lattices listed in (c) and (d) can be isomorphic.

To complete the proof of the theorem, we must show that a given lattice M with extension class $\begin{pmatrix}\bar{1}\\ \lambda^r u\end{pmatrix}$, where $1 \leqq r \leqq p-2$ and $u \in u(\bar{E})$, is isomorphic to one of the lattices in (c) and (d). Choosing $a = 1$, $b = 0$, $d = 1$ in (7.5'), we see that we can change

$\lambda^r u$ modulo λ^{p-1} without affecting the strong equivalence class of $\begin{pmatrix}\bar{1}\\ \lambda^r u\end{pmatrix}$. Now suppose that $u \equiv \pm q^2 \pmod{\lambda}$, where $q \in Z$; we may choose $\rho \in u(R)$, $s \in u(S)$, such that $\rho \equiv s \equiv q^{-1} \pmod{\lambda}$. There exists an $f \in \operatorname{Aut}(Z \oplus E)$, given as in (7.4), with $\varphi_2(d) = \rho$ and $\det f_1 = \pm 1$. Therefore $ad \equiv \pm 1 \pmod{\lambda}$, and so $a \equiv \pm q \pmod{p}$. Thus (7.6) yields

$$u' \equiv (\pm a^{-2})(\pm q^2) \equiv 1 \pmod{\lambda}\ .$$

Now choose $\tilde{u} \in \tilde{U}_{p-1-r}$ so that $\tilde{u}$ and u' have the same image in U_{p-1-r}, so $\tilde{u} \equiv \alpha u' \pmod{\lambda^{p-1}}$ for some $\alpha \in u(R)$. Then $\alpha \equiv 1 \pmod{\lambda}$, since $u' \equiv \tilde{u} \equiv 1 \pmod{\lambda}$. It follows from (7.5) that $\alpha = \varphi_2(d)$ for some $d \in u(E)$. Then $d \cdot \lambda^r u' \equiv \lambda^r \tilde{u} \pmod{\lambda^{p-1}}$, which shows that

$$\begin{pmatrix}\bar{1}\\ \lambda^r u\end{pmatrix} \approx \begin{pmatrix}\bar{1}\\ \lambda^r u'\end{pmatrix} \approx \begin{pmatrix}\bar{1}\\ \lambda^r \tilde{u}\end{pmatrix},$$

as desired. On the other hand, when $p \equiv 1 \pmod{4}$ and $u \pmod{\lambda}$ is not a square in $u(\bar{Z})$, then $u \equiv n_0 q^2 \pmod{\lambda}$ for some $q \in Z$. The above reasoning shows that

$$\begin{pmatrix}\bar{1}\\ \lambda^r u\end{pmatrix} \approx \begin{pmatrix}\bar{1}\\ \lambda^r \tilde{u} n_0\end{pmatrix},$$

so M is isomorphic to a lattice of type (d). This completes the proof of the theorem.

COROLLARY 7.7. *The number of isomorphism classes of indecomposable ZG-lattices equals*

$$1 + 2h_R + 2h_S + h_R h_S(3N_1 + |U_p| + \varepsilon_p(N_1 - |U_{p-1}|)),$$

where

$$N_1 = \sum_{r=0}^{p-2} |U_{p-1-r}|,$$

and $\varepsilon_p = 2$ if $p \equiv 1 \pmod{4}$, $\varepsilon_p = 1$ otherwise. If p is a regular odd prime (or if $p = 2$), then $|U_m| = p^{[(m-2)/2]}$ for $0 \leqq m \leqq p-1$, where the greatest integer function is interpreted as 0 if $m < 2$. Further, $|U_p| = |U_{p-1}|$ if p is regular or properly irregular; in the latter case, $|U_m| = p^{g(m)}$ where g is given by (6.3).

Proof. In (7.3) there are $1 + 2h_R + 2h_S$ lattices of type (a), and $h_R h_S N_1$ lattices for each of types (e) and (f). Further, there are $h_R h_S(N_1 + |U_p|)$ lattices of type (b), and $\varepsilon_p h_R h_S(N_1 - |U_{p-1}|)$ of types (c) and (d). This gives the desired result.

We note that for $p = 2, 3, 5$, the number of indecomposable ZG-lattices equals 9, 13, 40, respectively.

We are now ready to give a full set of isomorphism invariants for a direct sum M of indecomposable lattices chosen from the list in (7.3). Since the Krull-Schmidt-Azumaya Theorem holds for Z_pG-lattices, it is clear that the number of summands in the genus of each of the $4p + 1$ types in (7.2) must be an invariant. This gives us a set of $4p + 1$ nonnegative integers, which are precisely the genus invariants of M. Furthermore, the ideal class of the product of all R-ideals $\mathfrak{b}$ occurring in the various summands must be an isomorphism invariant of M. Likewise, the ideal class of the product of all S-ideals $\mathfrak{c}$ which occur is another invariant.

Now let M be a direct sum of indecomposable ZG-lattices chosen from the list (a)-(f) in (7.3). For each summand of type (b)-(f), the symbol u or un_0 occurring therein may be viewed as an element of $u(\bar{E})$. We may then form the product $u_0(M)$ of all u's and un_0's which occur in the summands of M of types (b)-(f); if there are no such summands, we set $u_0(M) = 1$. Let $r_1(M)$ be the largest exponent r which occurs in any type (b) summand, and let $r_2(M)$ be the largest exponent r among all summands of types (c), (d), (e), and (f). (Choose $r_1(M) = p$ if M has no summand of type (b), and choose $r_2(M) = p - 1$ if M has no summands of types (c)-(f).) The main result of this article is as follows:

THEOREM 7.8. *Let M be a direct sum of indecomposable ZG-lattices, which we may assume are of the types listed in* (7.3). *In terms of the above notation, a full set of isomorphism invariants of M consists of:*

(i) *The $4p + 1$ genus invariants of M,*

(ii) *The R- and S-ideal classes associated with M,*

(iii) *If M has no summand of types $\mathfrak{b}$, $E(\mathfrak{b})$, $\mathfrak{c}$, $(Z, \mathfrak{c}; 1)$, and if $r_1(M) \leqq r_2(M)$, the isomorphism invariant given by the image of $u_0(M)$ in U_{p-1-r_2}, whereas if $r_1(M) > r_2(M)$, the invariant given by the image of $u_0(M)$ in U_{p-r_1}, and*

(iv) *If $p \equiv 1 \pmod 4$, and if M has no summand of types Z, $E(\mathfrak{b})$, $(Z, \mathfrak{c}; 1)$, $(E(\mathfrak{b}), \mathfrak{c}; \lambda^r u)$ or $(Z \oplus \mathfrak{b}, \mathfrak{c}; 1 \oplus \lambda^r u)$, the isomorphism invariant given by the quadratic character of the image of $u_0(M)$ in $u(\bar{Z})$.*

Proof. Step 1. We have already remarked that the isomorphism class of M determines the invariants listed in (i) and (ii), and that the only remaining invariants needed to determine M up to isomorphism are those which characterize the strong equivalence class of M. In this step (the hardest of all), we suppose that M is as in (iii), and proceed to show that the proposed invariant is indeed an isomorphism invariant of M. Define $M^* = M/L_0$ as in the proof of (7.3); then M^* must be a direct sum of lattices in the genus of $(R, S; \lambda^r)$ for various r, because of the hypotheses on M. It follows from § 6 that the image of $u_0(M)$ in U_m is an isomorphism invariant of M^* (and hence also of M), where

$$m = p - 1 - \operatorname{Max}\{r\} = p - 1 - \operatorname{Max}\{r_1, r_2\} .$$

Thus we see that if $r_1 \leqq r_2$, then the image of $u_0(M)$ in U_{p-1-r_2} is an isomorphism invariant of M, as claimed.

Now let $r_1 > r_2$, and suppose M is as in (iii), so M is a direct sum of lattices in the genera of

$$(7.9)\quad Z,\ (Z \oplus R, S; \lambda^r),\ (R, S; \lambda^r),\ (Z \oplus E, S; 1 \oplus \lambda^r),\ (E, S; \lambda^r)$$

for various r's. Viewing M as an extension of a free S-lattice by a direct sum of copies of Z, R, and E, the extension class ξ_M of M has the form

$$\left(\begin{array}{cccc} 0 & 0 & 0 & 0 \\ I & 0 & 0 & 0 \\ 0 & 0 & I & 0 \\ \hline D_1 & 0 & 0 & 0 \\ 0 & D_2 & 0 & 0 \\ \hline 0 & 0 & D_3 & 0 \\ 0 & 0 & 0 & D_4 \end{array}\right) .$$

The top row corresponds to summands of type Z; each D_i is a diagonal matrix with diagonal entries of the form $\lambda^r u$ or $\lambda^r u n_0$; the four columns correspond (respectively) to the last four types of summands listed in (7.9). Changing notation slightly, we may then write

$$\xi_M = \begin{bmatrix} H & 0 & 0 \\ 0 & I & 0 \\ D_{12} & 0 & 0 \\ 0 & D_3 & 0 \\ 0 & 0 & D_4 \end{bmatrix}, \quad H = \begin{bmatrix} 0 & 0 \\ I & 0 \end{bmatrix}, \quad D_{12} = \begin{bmatrix} D_1 & 0 \\ 0 & D_2 \end{bmatrix} .$$

We must show that the image of $u_0(M)$ in U_{p-r_2} is an invariant of the strong equivalence class of ξ_M.

The endomorphism ring of $Z^{(a)} \oplus R^{(b)} \oplus E^{(c)}$ consists of all matrices

$$f = \begin{bmatrix} A_{11} & 0 & A_{13} \\ 0 & A_{22} & A_{23} \\ \Phi(x)A_{31} & \lambda A_{32} & A_{33} \end{bmatrix},$$

where the rows have entries in Z, R, and E, respectively. As in the proof of (7.3), f is an automorphism if and only if

$$(7.10) \quad \begin{bmatrix} A_{11} & A_{13} \\ p\varphi_1(A_{31}) & \varphi_1(A_{33}) \end{bmatrix} \in GL(Z)\,, \quad \begin{bmatrix} A_{22} & A_{23} \\ \lambda\varphi_2(A_{32}) & \varphi_2(A_{33}) \end{bmatrix} \in GL(R)\,,$$

where the φ_i are induced from those in (7.5).

Now suppose that $\xi_M \approx \xi_{M'}$, where $\xi_{M'}$ has the same form as ξ_M, but with diagonal entries $\lambda^r u'$ or $\lambda^r u' n_0$. Then we obtain

$$\begin{bmatrix} B_{11} & B_{12} & 0 & B_{14} & B_{15} \\ B_{21} & B_{22} & 0 & B_{24} & B_{25} \\ 0 & 0 & B_{33} & B_{34} & B_{35} \\ \Phi(x)B_{41} & \Phi(x)B_{42} & \lambda B_{43} & B_{44} & B_{45} \\ \Phi(x)B_{51} & \Phi(x)B_{52} & \lambda B_{53} & B_{54} & B_{55} \end{bmatrix} \begin{bmatrix} H & 0 & 0 \\ 0 & I & 0 \\ D_{12} & 0 & 0 \\ 0 & D_3 & 0 \\ 0 & 0 & D_4 \end{bmatrix}$$

$$= \begin{bmatrix} H & 0 & 0 \\ 0 & I & 0 \\ D'_{12} & 0 & 0 \\ 0 & D'_3 & 0 \\ 0 & 0 & D'_4 \end{bmatrix} \begin{bmatrix} S_{11} & S_{12} & S_{13} \\ S_{21} & S_{22} & S_{23} \\ S_{31} & S_{32} & S_{33} \end{bmatrix},$$

where $[S_{ij}]^{3\times 3} \in GL(S)$. The (4, 1) block in the left hand product equals $\Phi(x)B_{41}H + \lambda B_{43}D_{12}$. However, $\Phi(x)$ is a multiple of λ^{p-1} in $\bar{E}$, and each diagonal entry of D_{12} is of the form $\lambda^r u$ for some $r \leqq r_2$. We may therefore write this (4, 1) block as

$$(\lambda^{p-1-r_2}C_{41} + \lambda B_{43})D_{12}$$

for some C_{41}. The same procedure can be carried out for the blocks in positions (5, 1), (4, 2), and (4, 3). Setting $k = p - 1 - r_2$ for brevity, we obtain

$$(7.11) \quad \begin{bmatrix} B_{33} & B_{34} & B_{35} \\ \lambda^k C_{41} + \lambda B_{43} & \lambda^k C_{42} + B_{44} & B_{45} \\ \lambda^k C_{51} + \lambda B_{53} & \lambda^k C_{52} + B_{54} & B_{55} \end{bmatrix} \cdot \operatorname{diag}(D_{12}, D_3, D_4)$$
$$= \operatorname{diag}(D'_{12}, D'_3, D'_4) \cdot [S_{ij}]^{3\times 3}\,.$$

Now $r_1 > r_2 \geqq 0$ gives $r_1 \geqq 1$; thus for each $\rho \in \bar{R}$, the product $\lambda^{r_1}\rho$ is unambiguously defined inside $\bar{E}$. The method of proof of (5.10) then shows that

$$\lambda^{r_1}\beta u_0(M) = \lambda^{r_1}\sigma u_0(M') \text{ in } \bar{E}\ ,$$

where β is the determinant of the first matrix appearing in (7.11), and $\sigma = \det[S_{ij}] \in u(S)$. However, $r_1 + k = r_1 + p - 1 - r_2 \geqq p$ so $\lambda^{r_1+k} = 0$ in $\bar{E}$. Therefore $\lambda^{r_1}\beta = \lambda^{r_1}\beta^*$, where

$$\beta^* = \det\begin{bmatrix} B_{33} & B_{34} & B_{35} \\ \varphi_2(\lambda B_{43}) & \varphi_2(B_{44}) & \varphi_2(B_{45}) \\ \varphi_2(\lambda B_{53}) & \varphi_2(B_{54}) & \varphi_2(B_{55}) \end{bmatrix} \in u(R)\ .$$

This shows that $\beta^* u_0(M) = \sigma u_0(M')$ in $\bar{Z}[\lambda]/(\lambda^{p-r_1})$, so therefore $u_0(M)$ and $u_0(M')$ have the same image in U_{p-r_1}, as desired. Thus when $r_1 > r_2$, the image of $u_0(M)$ in U_{p-r_1} is an isomorphism invariant of M.

Step 2. Suppose next that the hypotheses of (iv) are satisfied. Then M is a direct sum of lattices in the genera of

$$R,\ S,\ (R, S; \lambda^r),\ (Z \oplus E, S; 1 \oplus \lambda^r)\ , \tag{7.12}$$

and we may write the extension class ξ_M of M in the form

$$\xi_M = \begin{bmatrix} 0 & I \\ H & 0 \\ 0 & D \end{bmatrix},$$

with D a diagonal matrix with entries $\lambda^r u$. The first column corresponds to summands of the first three types in (7.12), and the second column to the last type. If $\xi_{M'}$ has the same form as ξ_M, then the strong equivalence $\xi_M \approx \xi_{M'}$ yields an equation

$$\begin{bmatrix} A_{11} & 0 & A_{13} \\ 0 & A_{22} & A_{23} \\ \Phi(x)A_{31} & \lambda A_{32} & A_{33} \end{bmatrix} \begin{bmatrix} 0 & I \\ H & 0 \\ 0 & D \end{bmatrix} = \begin{bmatrix} 0 & I \\ H' & 0 \\ 0 & D' \end{bmatrix} \begin{bmatrix} S_{11} & S_{12} \\ S_{21} & S_{22} \end{bmatrix}.$$

But $A_{13}D = 0$ over $\bar{Z}$, since every diagonal entry of D is a multiple of λ, and $\lambda\bar{Z} = 0$. Thus we obtain

$$A_{11} = S_{22} \text{ over } \bar{Z},\quad \Phi(x)A_{31} + A_{33}D = D'S_{22}\ .$$

Consequently $|A_{11}| = |S_{22}|$ in $\bar{Z}$, and

$$\lambda^{p-2}|A_{33}|u_0(M) \equiv \lambda^{p-2}|S_{22}|u_0(M') \pmod{\lambda^{p-1}}\ .$$

On the other hand, $|A_{11}|\,|\varphi_1(A_{33})| \equiv \pm 1 \pmod p$ by (7.10), so we have

$$u_0(M) \equiv \pm\,|A_{11}|^2 u_0(M') \pmod \lambda\,.$$

This proves that in case (iv), the quadratic character of the image of $u_0(M)$ in $u(\bar{Z})$ is an isomorphism invariant of M. (This argument is an obvious extension of that given in the proof of (7.3).)

Step 3. To complete the proof, we must show that the set of invariants (i)-(iv) do indeed determine M up to isomorphism. We shall accomplish this by repeated use of the Absorption and Exchange Formulas of § 2, and for this purpose we need a collection of short exact sequences. For brevity of notation we omit the 0's at either end of such sequences, agreeing that the first arrow is assumed monic, the second arrow epic.

We have already pointed out that every element in a factor group U_k can be represented by an element u in E or R, such that $u \equiv 1 \pmod \lambda$. For such u, we have $uZ = Z$, and thus

$$R/uR \cong (E/Z)/u(E/Z) = (E/Z)/(uE/Z) \cong E/uE\,. \tag{7.13}$$

Likewise, $\mathfrak{b}/u\mathfrak{b} \cong E(\mathfrak{b}')/uE(\mathfrak{b}')$ always. Further, for $u \equiv 1 \pmod \lambda$ there are exact sequences

$$R \longrightarrow R \longrightarrow R/uR,\ (R,\,S;\,\lambda^r) \longrightarrow (R,\,S;\,\lambda^r u) \longrightarrow R/uR\,,$$

$$(Z \oplus E,\,S;\,1 \oplus \lambda^r) \longrightarrow (Z \oplus E,\,S;\,1 \oplus \lambda^r u \longrightarrow E/uE\,,$$

and so on. If M is a direct sum of indecomposable lattices of the types listed in (7.3), it thus follows from the existence of such exact sequences that we may concentrate all of the u's in any preassigned summand of M, without affecting the isomorphism class of M. This means that we can set all but one of the u's equal to 1, and replace the remaining u by the product of all of the original u's. (Caution: this does *not* enable us to move the n_0's occuring in type (d) summands!) Furthermore, if either $\mathfrak{b}$ or $E(\mathfrak{b})$ occurs as summand, then the Absorption Formula permits us to make *every* u equal to 1, without affecting the isomorphism class of M.

Next, there is a surjection $S \to \bar{E}$, so for each $u \in u(\bar{E})$ we can find an element $v \in S$ such that $\bar{v} = u^{-1}$ in $\bar{E}$; then v acts on $\bar{E}$ as multiplication by u^{-1}. From the commutative diagram

$$\begin{array}{ccccc} R & \longrightarrow & (R,\,S;\,\lambda^r u) & \longrightarrow & S \\ {\scriptstyle 1}\uparrow & & \uparrow & & \uparrow{\scriptstyle v} \\ R & \longrightarrow & (R,\,S;\,\lambda^r) & \longrightarrow & S \end{array}$$

we obtain an E-exact sequence

$$(R, S; \lambda^r) \longrightarrow (R, S; \lambda^r u) \longrightarrow S/vS .$$

Likewise, there are exact sequences

$$S \longrightarrow S \longrightarrow S/vS, \ (Z, S; 1) \longrightarrow (Z, S; 1) \longrightarrow S/vS ,$$

$$(Z \oplus E, S; 1 \oplus \lambda^r) \longrightarrow (Z \oplus E, S; u \oplus \lambda^r u) \longrightarrow S/vS ,$$

and so on. Note also that

$$(Z \oplus E, S; v \oplus \lambda^r v) \cong (Z \oplus E, S; 1 \oplus \lambda^r v)$$

whenever $v \equiv 1 \pmod{\lambda}$. It thus follows (by the Absorption Formula) that if either $\mathfrak{c}$ or $(Z, \mathfrak{c}; 1)$ is a summand of M, then we can replace u by 1 in every summand of M in which u's occur, without affecting the isomorphism class of M. This completes the proof that if M has any summand of the types in (iii), then we can eliminate all of the u's. On the other hand, if M has no such summand, then by Step 1 the image of $u_0(M)$ in either U_{p-1-r_2} or U_{p-r_1} is an isomorphism invariant of M.

Step 4. Suppose finally that $p \equiv 1 \pmod 4$. There are exact sequences

$$Z \longrightarrow Z \longrightarrow Z/n_0 Z, \ (Z, S; 1) \longrightarrow (Z, S; 1) \longrightarrow Z/n_0 Z ,$$

$$(7.14) \quad (Z \oplus R, S; 1 \oplus \lambda^r u) \longrightarrow (Z \oplus R, S; n_0 \oplus \lambda^r u) \longrightarrow Z/n_0 Z .$$

Choose $v_0 \in u(S)$ with $v_0 = n_0$ in $u(\bar{Z})$; then u and uv_0^{-1} have the same image in U_{p-1-r} for each r, and therefore

$$(Z \oplus R, S; 1 \oplus \lambda^r u) \cong (Z \oplus R, S; 1 \oplus \lambda^r uv_0^{-1}) \cong (Z \oplus R, S; v_0 \oplus \lambda^r u)$$
$$\cong (Z \oplus R, S; n_0 \oplus \lambda^r u) .$$

Thus (7.14) yields an exact sequence

$$(Z \oplus R, S; 1 \oplus \lambda^r u) \longrightarrow (Z \oplus R, S; 1 \oplus \lambda^r u) \longrightarrow Z/n_0 Z .$$

Now let $u \equiv 1 \pmod{\lambda}$, and let us denote by $[un_0]$ an element $u_1 \in u(\bar{E})$ such that $u_1 \equiv 1 \pmod{\lambda}$ and $u_1 = un_0$ in U_{p-1-r} (for some given r). Then we have

$$(Z \oplus E, S; n_0^{-1} \oplus \lambda^r u_1) \cong (Z \oplus E, S; 1 \oplus \lambda^r u_1 v_0)$$
$$\cong (Z \oplus E, S; 1 \oplus \lambda^r un_0) ;$$

the second isomorphism is valid because $u_1 v_0$ and un_0 have the same image in $u(\bar{Z})$, as well as the same image in U_{p-1-r}. Thus the exact sequence

$$(Z \oplus E, S; n_0^{-1} \oplus \lambda^r u_1) \longrightarrow (Z \oplus E, S; 1 \oplus \lambda^r u_1) \longrightarrow Z/n_0 Z$$

may be rewritten as

$$(7.15) \quad (Z \oplus E, S; 1 \oplus \lambda^r u n_0) \longrightarrow (Z \oplus E, S; 1 \oplus \lambda^r [u n_0]) \longrightarrow Z/n_0 Z .$$

Finally, there are exact sequences

$$(E, S; \lambda^r u) \longrightarrow (E, S; \lambda^r u n_0) \longrightarrow E/n_0 E ,$$
$$(Z \oplus E, S; 1 \oplus \lambda^r u) \longrightarrow (Z \oplus E, S; 1 \oplus \lambda^r u n_0) \longrightarrow E/n_0 E .$$

It now follows from (7.15), and the other sequences listed above, that if M contains any summand of the types listed in (iv), then the isomorphism class of M is unchanged if we replace un_0 by $[un_0]$ in every type (d) summand of M. In any case, if both u and u' are congruent to $1 \pmod \lambda$, then (7.15) gives

$$(Z \oplus E, S; 1 \oplus \lambda^r u n_0) \oplus (Z \oplus E, S; 1 \oplus \lambda^s u' n_0)$$
$$\cong (Z \oplus E, S; 1 \oplus \lambda^r [u n_0]) \oplus (Z \oplus E, S; 1 \oplus \lambda^s [u' n_0]) .$$

Hence, we can always eliminate any even number of type (d) summands of M. Further, if M contains no summands of the types listed in (iv), then we have shown in Step 2 that the quadratic character of the image of $u_0(M)$ in $u(\bar{Z})$ is an isomorphism invariant of M.

In view of the various changes which we have described in Steps 3 and 4, it is now clear that the invariants listed in (i)-(iv) completely determine the isomorphism class of M. This completes the proof of the theorem.

To conclude, we remark that many of the above results can be generalized to extensions of R_j-lattices by a direct sum of locally free lattices over several orders which are factor rings of $Z[x]/(x^{p^{j-1}}-1)$. In particular, we can classify all Λ_κ-lattices M for which QM is a direct sum of copies of Z, R_i, and R_j, where $1 \leqq i < j \leqq \kappa$. It is known (see [1]) that there are only finitely many isomorphism classes of indecomposable lattices of this type. However, this gives only a partial classification of the integral representations of a cyclic group of order p^3, since for G cyclic of order p^2, there exist ZG-lattices which are not direct sums of locally free lattices of the types just mentioned.

Even for G cyclic of order p^2, a further question remains: given a ZG-lattice M, how can one calculate the isomorphism invariants of M intrinsically, without first expressing M as a direct sum of indecomposable lattices? Such a calculation would undoubtedly help to clarify the structure of M.

1. S. D. Berman and P. M. Gudivok, *Indecomposable representations of finite groups over the ring of p-adic integers*, Izv. Akad. Nauk, SSSR Ser. Mat., **28** (1964), 875–910; English transl., Amer. Math. Soc. Transl. (2) **50** (1966), 77–113.
2. Z. I. Borevich and I. R. Shafarevich, *Number Theory*, Academic Press, New York, 1966.
3. H. Cartan and S. Eilenberg, *Homological Algebra*, Princeton University Press, Princeton, NJ, 1956.
4. C. W. Curtis and I. Reiner, *Representation theory of finite groups and associative algebras*, Pure and Appl. Math., vol. XI, Interscience, New York, 1962; 2nd ed., 1966.
5. F. E. Diederichsen, *Über die Ausreduktion ganzzahliger Gruppendarstellungen bei arithmetischer Äquivalenz*, Abh. Math. Sem. Univ. Hamburg, **13** (1940), 357–412.
6. S. Galovich, *The class group of a cyclic p-group*, J. Algebra, **30** (1974), 368–387.
7. A. Heller and I. Reiner, *Representations of cyclic groups in rings of integers. I. II*, Ann. of Math., (2) **76** (1962), 73–92; (2) **77** (1963), 318–328.
8. H. Jacobinski, *Genera and decompositions of lattices over orders*, Acta Math., **121** (1968), 1–29.
9. M. A. Kervaire and M. P. Murthy, *On the projective class group of cyclic groups of prime power order*, Comment. Math. Helvetici, **52** (1977), 415–452.
10. I. Reiner, *Integral representations of cyclic groups of prime order*, Proc. Amer. Math. Soc., **8** (1957), 142–146.
10a, ———, *A survey of integral representation theory*, Bull. Amer. Math. Soc., **76** (1970), 159–227.
11. I. Reiner, *Maximal Orders*, Academic Press, London, 1975.
12. ———, *Integral representations of cyclic groups of order p^2*, Proc. Amer. Math. Soc., **58** (1976), 8–12.
13. ———, *Indecomposable integral representations of cyclic p-groups*, Proc. Philidelphia Conference 1976, Dekker Lecture Notes **37** (1977), 425–445.
14. A. V. Roiter, *On representations of the cyclic group of fourth order by integral matrices*, Vestnik Leningrad. Univ., **15** (1960), no. 19, 65–74. (Russian)
15. ———, *On integral representations belonging to a genus*, Izv. Akad. Nauk, SSSR, Ser. Mat., **30** (1966), 1315–1324; English transl. Amer. Math. Soc. Transl. (2) **71** (1968), 49–59.
16. J. J. Rotman, *Notes on homological algebra*, van Nostrand Reinhold, New York, 1970.
17. A. Troy, *Integral representations of cyclic groups of order p^2*, Ph. D. Thesis, University of Illinois, Urbana, IL., 1961.
18. S. Ullom, *Fine structure of class groups of cyclic p-groups*, J. Algebra, **48** (1977).
19. ———, *Class groups of cyclotomic fields and group rings*, J. London Math. Soc., (to appear).

Received January 4, 1977 and in revised form November 10, 1977. This research was partially supported by the National Science Foundation.

University of Illinois
Urbana, IL 61801

INTEGRAL REPRESENTATIONS AND DIAGRAMS

E. L. Green and I. Reiner

ABSTRACT

This paper considers questions concerning representations of "graphs" and the relation of this to the problem of determining rings of "finite–type"; that is rings having only a finite number of indecomposable representations. Methods of this paper extend those used previously by Gabriel, Dlab–Ringel, and Bernstein–Gelfand–Ponomarev.

Consider a finite directed graph $\mathcal{M}$ with verticies $\{\alpha\}$ and directed edges (arrows) $\alpha \to \beta$. To each vertex α is attached a ring k_α and to each arrow from α to β is attached a (k_β, k_α)–bimodule ${}_\beta M_\alpha$. A representation of $\mathcal{M}$ assigns to each vertex α a left k_α–space V_α, and to each arrow from α to β a k_β homomorphism ${}_\beta M_\alpha \otimes_{k_\alpha} V_\alpha \to V_\beta$. One defines homomorphisms, equivalence, and direct sums of representations in the natural way. The main problem in this subject is the determination of the graphs having only a finite number of indecomposable representations. This has been treated previously in the case that each k_α is a skewfield. In this paper, k_α is possibly a local ring.

Let R be a Dedekind ring with quotient field K, A a finite dimensional, separable K–algebra, and Λ an R-order in A. Denote by $\mathcal{L}(\Lambda)$ the category of left Λ–lattices; that is Λ–modules which are finitely generated and torsion free over R.

Theorem (Drozd–Roiter). *Let A be a commutative separable K–algebra, where K is a global field. Let Λ be an R–order in A, and Λ' a maximal R–order in A containing Λ. Then $\mathcal{L}(\Lambda)$ is of finite type if and only if the Λ–module Λ'/Λ can be generated by two elements and its radical can be generated by one element.*

The problem of deciding whether $\mathcal{L}(\Lambda)$ is of finite type reduces to the case that R is a complete discrete valuation ring; in case Λ is commutative, this can be reduced further to the case that Λ is a local ring; that is Λ modulo its radical is a field. Since Λ'/Λ is an R–torsion module, it is artinian. The proof of the theorem is carried out by examining a certain category $\mathcal{A}$ constructed using artinian rings.

Let I be an ideal of Λ' contained in Λ (for example the conductor of Λ' into Λ) and let

$$\Delta = \Lambda/I, \qquad \Gamma = \Lambda'/I,$$

so there is an inclusion $\phi : \Delta \to \Gamma$. The category $\mathcal{A} = \mathcal{A}(\Delta, \Gamma, \phi)$ has objects (X, Y, f) where

i. X is a finitely generated projective Δ–module and Y is a finitely generated projective Γ–module;

ii. $f : X \to Y$ is a Δ–homomorphism for which $Y = \Gamma\mathcal{I}m(f)$ and $\ker f \subset (\mathrm{rad}\Delta)X$.

A morphism in $\mathcal{A}$ is a pair $(\alpha, \beta) : (X_1, Y_1, f_1) \to (X_2, Y_2, f_2)$ where α is a Δ homomorphism $X_1 \to X_2$ and β is a Γ–homomorphism $Y_1 \to Y_2$ such that $f_2\alpha = \beta f_1$.

A correspondence F is given from $\mathcal{L}(\Lambda)$ to $\mathcal{A}$ which is not a functor but which preserves isomorphisms, direct sums and some other properties. The authors reduce the proof of the Drozd–Roiter theorem to verifying the following.

Theorem. *Let Δ, Γ be as above, and let $\mathbf{r} = rad(\Delta)$ and $k = \Delta/\mathbf{r}$. Then $\mathcal{A}(\Delta, \Gamma, \phi)$ has only a finite number of non–isomorphic indecomposable objects if and only if*

$$dim_k \Gamma/\mathbf{r}\Gamma \leq 3 \quad \text{and} \quad dim_k(\mathbf{r}\Gamma + \Delta)/(\mathbf{r}^2\Gamma + \Delta) \leq 2.$$

The proof of this occupies most of the paper; it is carried out by considering several cases, making reductions and dealing directly with them by detailed matrix computations.

LIFTING ISOMORPHISMS OF MODULES

IRVING REINER

Throughout this note, let R be a discrete valuation ring with prime element π, residue class field $\bar{R}$, and quotient field K. Let Λ be an R-order in a finite dimensional K-algebra A. A Λ-*lattice* is an R-free finitely generated left Λ-module. For $k > 0$, we set

$$R_k = R/\pi^k R, \quad \Lambda_k = \Lambda/\pi^k\Lambda, \quad M_k = M/\pi^k M,$$

where M is any Λ-lattice. Obviously, for Λ-lattices M and N,

$$M \cong N \text{ as } \Lambda\text{-lattices} \Rightarrow M_k \cong N_k \text{ as } \Lambda_k\text{-modules for each } k > 0.$$

Maranda [**1**] and D. G. Higman [**3**] considered the reverse implication, and proved

THEOREM. *Let Λ be an R-order in a separable K-algebra A. Then there exists a positive integer k (which depends on Λ) with the following property: for each pair of Λ-lattices M and N,*

$$M_k \cong N_k \text{ as } \Lambda_k\text{-modules} \Rightarrow M \cong N \text{ as } \Lambda\text{-lattices}.$$

Indeed, it suffices to choose k so that

$$\pi^{k-1} \cdot H^1(\Lambda, \operatorname{Hom}_R(M, N)) = 0.$$

Maranda proved this result for the special case where Λ is the integral group ring RG of a finite group G. The general case was treated by D. G. Higman, who showed the existence of a nonzero ideal $i(\Lambda)$ of R such that

$$i(\Lambda) \cdot H^1(\Lambda, \mathrm{T}) = 0$$

for all Λ-bimodules T. This readily implies that the integer k (occurring above) can be chosen independently of M and N. Higman's construction depends on the fact that if A is any separable K-algebra, then $H^1(A, Y) = 0$ for all A-bimodules Y. For details of the proof, see Curtis-Reiner [**2**, §§ 75, 76].

The aim of this note is to prove an analogue of the Maranda-Higman Theorem, in which the K-algebra A need not be separable, nor even semisimple. (As a matter of fact, the proof below applies equally well to the situation in which Λ is an R-algebra, finitely generated as an R-module, and does not require that Λ have a unity element. However, this seemingly more general case readily reduces to the case where Λ is an R-order in a K-algebra A.)

Received January 4, 1978 and in revised form March 7, 1978. This work was partially supported by a research grant from the National Science Foundation.

THEOREM. *Let Λ be an R-order in an arbitrary finite dimensional K-algebra A, and let M, N be Λ-lattices. If $M_k \cong N_k$ as Λ_k-modules for each $k > 0$, then $M \cong N$ as Λ-modules.*

Proof. We set $\mathrm{T} = \mathrm{Hom}_R(M, N)$, a Λ-bimodule. A *derivation* $D : \Lambda \to \mathrm{T}$ is an R-homomorphism such that

$$D(xy) = xD(y) + D(x)y \quad \text{for all } x, y \in \Lambda.$$

Let $\mathrm{Der}(\Lambda, \mathrm{T})$ be the R-module consisting of all derivations from Λ to T; it is clearly finitely generated over R.

For each $k > 0$, we are given a Λ_k-isomorphism $\varphi_k : M_k \cong N_k$. Since M is R-projective, we can find a map $\theta_k \in \mathrm{Hom}_R(M, N)$ making the following diagram commute:

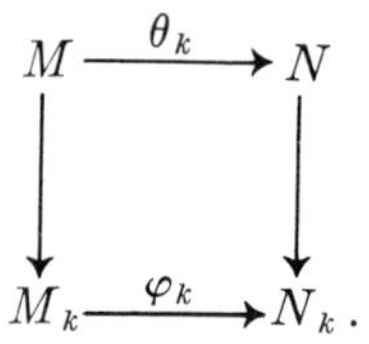

$$\begin{array}{ccc} M & \xrightarrow{\theta_k} & N \\ \downarrow & & \downarrow \\ M_k & \xrightarrow{\varphi_k} & N_k . \end{array}$$

Then $N = \theta_k(M) + \pi^k N$, so θ_k is surjective by Nakayama's Lemma. But M and N have the same R-rank, namely the R_k-rank of M_k; thus θ_k must be an R-isomorphism. The commutativity of the diagram implies that for each $x \in \Lambda$,

$$\theta_k(xm) - x\theta_k(m) \in \pi^k N \quad \text{for all } m \in M.$$

Hence for each $m \in M$, we may write

$$\theta_k(xm) - x\theta_k(m) = \pi^k n \quad \text{for some } n \in N,$$

and m uniquely determines n because N is R-free. We may therefore define an R-homomorphism $D_k : \Lambda \to \mathrm{T}$ by setting

$$D_k(x) = \pi^{-k}(\theta_k x - x\theta_k), \quad x \in \Lambda.$$

It is easily checked that D_k is a derivation.

Now consider the R-submodule $\mathscr{D}$ of $\mathrm{Der}(\Lambda, \mathrm{T})$ generated by $D_1, D_2, \ldots$. Since $\mathrm{Der}(\Lambda, \mathrm{T})$ is finitely generated as R-module, and R is noetherian, it follows that $\mathscr{D}$ is also finitely generated over R. Hence there exists a positive integer k such that D_k is an R-linear combination of $D_1, \ldots, D_{k-1}$, say

$$D_k = \alpha_1 D_1 + \ldots + \alpha_{k-1} D_{k-1}, \quad \text{with each } \alpha_i \in R.$$

Using the definition of the D's, this gives

$$\pi^{-k}(\theta_k x - x\theta_k) = \textstyle\sum_1^{k-1} \alpha_i \pi^{-i}(\theta_i x - x\theta_i) \quad \text{for each } x \in \Lambda.$$

Multiplying by π^k and setting

$$\theta' = \theta_k - \textstyle\sum_1^{k-1} \alpha_i \pi^{k-i} \theta_i,$$

we obtain $\theta' x = x\theta'$ for each $x \in \Lambda$. Thus $\theta' \in \mathrm{Hom}_\Lambda(M, N)$, and $\theta' \equiv \theta_k (\mathrm{mod}\ \pi)$. But then θ' is an R-isomorphism of M onto N, since θ_k is such an isomorphism. This shows that $M \cong N$ as Λ-modules, and completes the proof.

The preceding result shows that two Λ-lattices M, N are isomorphic if and only if $M_k \cong N_k$ for each k. There does not seem to be any obvious way to find a single choice for k, depending on Λ, M and N, such that $M_k \cong N_k$ implies that $M \cong N$.

We may also show that decomposability of lattices can be handled in the same manner, provided that R is complete. Let us prove

THEOREM. *Let R be a complete discrete valuation ring, and let Λ be an R-order in an arbitrary finite dimensional K-algebra A. Let M be a Λ-lattice such that M_k is Λ_k-decomposable for each $k > 0$. Then M is decomposable as a Λ-lattice.*

Proof. For each $k > 0$, let $\varphi_k : M_k \to M_k$ be a nontrivial idempotent in the endomorphism ring $\mathrm{End}_{\Lambda_k}(M_k)$. As above, we may choose $\theta_k \in \mathrm{End}_R(M)$ so that the diagram

$$\begin{array}{ccc} M & \xrightarrow{\theta_k} & M \\ \downarrow & & \downarrow \\ M_k & \xrightarrow{\varphi_k} & M_k \end{array}$$

is commutative. The proof of the preceding theorem shows that for some $n > 0$, there exists a map $\theta' \in \mathrm{End}_\Lambda(M)$ such that $\theta' \equiv \theta_n$ (mod π). Hence both θ' and θ_n are liftings of φ_n (mod π), and therefore

$$\{(\theta')^2 - \theta'\} M \subseteq \pi M.$$

We now set $E = \mathrm{End}_\Lambda(M)$, $\bar{E} = E/\pi E$, and let ψ be the image of θ' in $\bar{E}$. The preceding inclusion shows that $(\theta')^2 - \theta'$ lies in πE, whence $\psi^2 = \psi$ in $\bar{E}$. But then ψ is a nontrivial idempotent, since ψ coincides with φ_n (mod π). On the other hand, the method of "lifting idempotents" then implies that there exists a nontrivial idempotent $\mu \in E$ whose image in $\bar{E}$ is ψ. Therefore $M = \mu M \oplus (1 - \mu) M$ gives a nontrivial decomposition of M, and the result follows.

For the case of orders in separable algebras, the above technique is due to Heller, and improves the method originally used by Maranda. For references, see Curtis-Reiner [**2**, § 76].

REFERENCES

1. J. M. Maranda, *On p-adic integral representations of finite groups*, Can. J. Math. *5* (1953), 344–355.

2. C. W. Curtis and I. Reiner, *Representation theory of finite groups and associative algebras* (Wiley and Sons, New York, 1962).
3. D. G. Higman, *On representations of orders over Dedekind domains*, Can. J. Math. *12* (1960), 107–125.

University of Illinois,
Urbana, Illinois

On Diederichsen's Formula for Extensions of Lattices

IRVING REINER*

Department of Mathematics, University of Illinois at Urbana–Champaign, Urbana, Illinois 61801

Communicated by Walter Feit

Received September 12, 1978

1. INTRODUCTION

Let G be a finite cyclic group of order g, and let ZG be its integral group ring. A *ZG-lattice* is a left ZG-module which is finitely generated and torsionfree as Z-module (and hence has a finite free Z-basis). Let ω_n denote a primitive n-th root of unity over Q, and let $\Phi_n(x)$ be the cyclotomic polynomial of order n and degree $\varphi(n)$, where φ is the Euler φ-function. Let

$$K_n = Q(\omega_n), \qquad R_n = Z[\omega_n] = \text{alg. int.}\{K_n\},$$

so R_n is the ring of all algebraic integers in the cyclotomic field K_n. We shall always identify ZG with the ring $\Lambda = Z[x]/(x^g - 1)$ by mapping a generator of G onto x. Then for each integer n dividing g, we have

$$R_n \cong Z[x]/(\Phi_n(x)) \cong \Lambda/\Phi_n(x)\Lambda,$$

and thus R_n may be viewed as ZG-lattice. Indeed, every R_n-module becomes a ZG-module by virtue of the homomorphism $ZG \to R_n$.

In Diederichsen's fundamental article [2] on integral representations of cyclic groups, one of the vital steps was the computation of $\text{Ext}^1_{ZG}(R_n, R_m)$ when m and n are divisors of g. Diederichsen proved the following remarkable results:

$\text{Ext}^1_{ZG}(R_n, R_m) = 0$ unless $m/n = p^t$ for some prime p and some nonzero $t \in Z$.

If $m/n = p^t$ for some prime p and some nonzero integer t (possibly negative), then $\text{Ext}^1_{ZG}(R_n, R_m)$ is isomorphic to the direct sum of $\varphi(l)$ copies of $\bar{Z}$, where $\bar{Z} = Z/pZ$ and $l = \min(m, n)$.

Diederichsen's proof, though elementary in nature, depended on complicated manipulations with roots of unity. In this article we shall give an extremely

* This work was partially supported by a research grant from the National Science Foundation.

simple proof of Diederichsen's results. Indeed, in case (2) we shall establish the stronger result that

$$\mathrm{Ext}^1_{ZG}(R_n, R_m) \cong R_l/pR_l, \qquad l = \min(m, n). \tag{3}$$

Since R_l is Z-free on $\varphi(l)$ generators, (2) follows at once from (3). We shall prove (1) and (3) in the next section, and then in Section 3 we shall apply these facts to the question of classifying all ZG-lattices which are extensions of R_n-lattices by R_m-lattices.

As references for the results from algebraic number theory used below, we cite van der Waerden [5] and Weiss [6]. For the homological algebra needed, see Curtis-Reiner [1] and Rotman [4]. The techniques used in Section 3 may be found in Reiner [3].

2. Calculation of Ext

For the remainder of this article, Ext^1 will be denoted by Ext. We need an easy result:

(4) Lemma. *Let $\alpha \in \Lambda$, where Λ is any ring, and let $\bar{\Lambda} = \Lambda/\Lambda\alpha$. Put $J = \{\beta \in \Lambda : \beta\alpha = 0\}$. Then for each left Λ-module M we have*

$$\mathrm{Ext}(\bar{\Lambda}, M) \cong M'/\alpha M, \qquad \textit{where} \quad M' = \{m \in M : Jm = 0\}.$$

When Λ is commutative, the above is a Λ-isomorphism.

Proof. The exact sequence of left Λ-modules

$$0 \to \Lambda\alpha \xrightarrow{i} \Lambda \to \bar{\Lambda} \to 0$$

yields a long exact sequence

$$0 \to \mathrm{Hom}(\bar{\Lambda}, M) \to \mathrm{Hom}(\Lambda, M) \xrightarrow{i^*} \mathrm{Hom}(\Lambda\alpha, M)$$
$$\to \mathrm{Ext}(\bar{\Lambda}, M) \to \mathrm{Ext}(\Lambda, M) \to \cdots.$$

(We have omitted the subscript Λ on Hom and Ext, for convenience.) But $\mathrm{Ext}(\Lambda, M) = 0$, so

$$\mathrm{Ext}(\bar{\Lambda}, M) \cong \mathrm{Hom}(\Lambda\alpha, M)/i^*\{\mathrm{Hom}(\Lambda, M)\}.$$

Now each $f \in \mathrm{Hom}(\Lambda\alpha, M)$ is determined by the image $f(\alpha) \in M$. This image may be chosen arbitrarily in M, subject to the condition that it be annihilated by J. Thus there is an additive isomorphism $\mathrm{Hom}(\Lambda\alpha, M) \cong M'$, given by $f \to f(\alpha)$, $f \in \mathrm{Hom}(\Lambda\alpha, M)$. Now $\mathrm{Hom}(\Lambda, M) \cong M$, so the image of i^* is precisely αM. This completes the proof that $\mathrm{Ext}(\bar{\Lambda}, M) \cong M'/\alpha M$. The final assertion of the lemma is then obvious.

Now let G be cyclic of order g, and let m and n be divisors of g. We show at once that

$$\mathrm{Ext}_{ZG}(R_n, R_n) = 0. \tag{5}$$

Indeed, given any ZG-exact sequence

$$0 \to R_n \to M \to R_n \to 0$$

in which M is a ZG-module, we have $\Phi_n \cdot M \subseteq R_n$, so $\Phi_n{}^2 M = 0$. (We write Φ_n instead of $\Phi_n(x)$ for convenience.) Thus M is annihilated by the G.C.D. of $\Phi_n{}^2$ and $x^g - 1$ in $Z[x]$. This G.C.D. is $\Phi_n Z[x]$, so M may be viewed as R_n-module. But then the above sequence is R_n-split, hence also ZG-split, and so (5) is proved.

Let us now compute $\mathrm{Ext}(R_n, R_m)$ when $m \neq n$, where Ext means Ext^1_{ZG}. Let $\Psi_n(x) \in Z[x]$ be defined by the equation $\Phi_n \Psi_n = x^g - 1$; then $\Psi_n ZG$ is the annihilator of Φ_n in ZG. Since Φ_m is a factor of Ψ_n, it follows that $\Psi_n R_m = 0$. But then

$$\mathrm{Ext}(R_n, R_m) \cong R_m/\Phi_n(x) R_m$$

by (4). This gives

$$\mathrm{Ext}(R_n, R_m) \cong Z[x]/(\Phi_n(x), \Phi_m(x)), \tag{6}$$

whence $\mathrm{Ext}(R_n, R_m) \cong \mathrm{Ext}(R_m, R_n)$. We could have predicted this last isomorphism in advance, by a general argument about contragredients of ZG-lattices.

It is rather complicated to compute $\mathrm{Ext}(R_n, R_m)$ directly from (6), so instead we pass to localizations. As shown in [1, Theorem 75.25], $\mathrm{Ext}(R_n, R_m)$ is a finitely generated Z-module annihilated by g. It is therefore equal to the direct sum of its p-primary components, with p ranging over all prime divisors of g. This gives

$$\mathrm{Ext}_{ZG}(R_n, R_m) \cong \coprod_{p \mid g} \mathrm{Ext}_{Z_pG}((R_n)_p, (R_m)_p), \tag{7}$$

where the subscript p denotes localization at p. From (6) we obtain

$$\mathrm{Ext}_{Z_pG}((R_n)_p, (R_m)_p) \cong Z_p[x]/(\Phi_n, \Phi_m). \tag{8}$$

Keep p fixed, and write $m = p^a r$, $n = p^b s$, where $p \nmid rs$. Let $\bar{Z} = Z/pZ$, and let bars denote images mod p.

Then we have

$$\bar{\Phi}_m = (\bar{\Phi}_r)^{\varphi(p^a)}, \qquad \bar{\Phi}_n = (\bar{\Phi}_s)^{\varphi(p^b)}. \tag{9}$$

Indeed, the zeros of Φ_m are the $\varphi(m)$ primitive roots of 1 over Q, while the zeros of $\bar{\Phi}_m$ are the $\varphi(r)$ primitive roots of 1 over $\bar{Z}$, each occurring with multiplicity $\varphi(p^a)$. This proves the first equality in (9), and the second comes from the first by renaming the variables.

Using the above, let us verify that

$$\operatorname{Ext}_{Z_pG}((R_n)_p, (R_m)_p) = 0 \text{ except when } m/n = p^t \text{ for some nonzero } t \in Z. \tag{10}$$

(We are still assuming that $m \neq n$.) If m/n is not a power of p, then $r \neq s$. Hence $(\bar{\Phi}_r, \bar{\Phi}_s) = 1$ in $\bar{Z}[x]$, since every zero of $\bar{\Phi}_r$ is a primitive r-th root of 1 over $\bar{Z}$, while every zero of $\bar{\Phi}_s$ is a primitive s-th root of 1. From (9) we conclude that $(\bar{\Phi}_m, \bar{\Phi}_n) = 1$ in $\bar{Z}[x]$. Let Ω be the resultant of Φ_m and Φ_n in $Z[x]$. Then

$$\Omega = \pm \prod_{k,j} (\alpha_k - \beta_j) \in Z,$$

where α_k ranges over the zeros of Φ_m, and β_j over the zeros of Φ_n. Therefore

$$\bar{\Omega} = \pm \prod_{k,j} (\bar{\alpha}_k - \bar{\beta}_j) \in \bar{Z}.$$

Since $(\bar{\Phi}_m, \bar{\Phi}_n) = 1$, each factor $\alpha_k - \beta_j$ is nonzero, so $\bar{\Omega} \neq 0$. Hence Ω is a unit in Z_p. But $\Omega \in (\Phi_m, \Phi_n)$, which shows that $Z_p[x] = (\Phi_m, \Phi_n)$ whenever $r \neq s$. Thus we have established (10).

We now treat the case where $r = s$ and $m \neq n$, and begin by showing that $p \in (\Phi_n, \Phi_m)$. There is no loss of generality in assuming that $m < n$, so $a < b$. Put $y = x^{s \cdot p^{b-1}}$. Since x acts as ω_m on R_m, it follows that y acts as 1 on R_m. On the other hand, x acts as ω_n on R_n, so y acts on R_n as multiplication by a primitive p-th root of 1. Hence $1 + y + \cdots + y^{p-1}$ annihilates R_n. But $\operatorname{Ext}(R_n, R_m)$ is a ZG-bimodule since ZG is commutative. The right action of ZG on Ext arises from the left action of ZG on R_n, while the left action of ZG on Ext comes from from its left action on R_m. Further, the two actions of ZG on Ext are identical: $\gamma\xi = \xi\gamma$ for all $\gamma \in ZG$, $\xi \in \operatorname{Ext}$. Thus $1 + y + \cdots + y^{p-1}$ acts on Ext as the zero map (from the right) and as multiplication by p (from the left). Therefore $p \cdot \operatorname{Ext}(R_n, R_m) = 0$, and so p lies in the ideal (Φ_n, Φ_m). We thus have

$$\operatorname{Ext}_{ZG}(R_n, R_m) \cong Z[x]/(\Phi_n, \Phi_m, p) \cong \bar{Z}[x]/(\bar{\Phi}_n, \bar{\Phi}_m).$$

But when $m < n$, $\bar{\Phi}_n$ is a power of $\bar{\Phi}_m$ by (9). Thus we obtain

$$\operatorname{Ext}_{ZG}(R_n, R_m) \cong \bar{Z}[x]/(\bar{\Phi}_m) = Z[x]/(p, \Phi_m) \cong R_m/pR_m. \tag{11}$$

This establishes formula (3).

We have also completed the proof of (1) and (2). Indeed, we have already shown that $\mathrm{Ext}(R_n, R_n) = 0$, so suppose now that $m \neq n$. We examine the quotients m/n and n/m; if neither one is a prime power, then $\mathrm{Ext}(R_n, R_m) = 0$ by (7) and (10). On the other hand, if $m/n = p^t$ with $t \in Z$, $t \neq 0$, then we have just established (3), and then (2) follows by use of (7).

3. Indecomposable Lattices

We keep the notation of the earlier sections. An *R_n-lattice* N is a finitely generated R_n-torsionfree R_n-module, and as such N is automatically a ZG-lattice. The isomorphism class of N is completely determined by two invariants: the R_n-rank ν of N, and the Steinitz class of N. (The *Steinitz class* of N is the ideal class of the product $\prod \mathfrak{b}_i$ when N is expressed as an external direct sum $\coprod_{i=1}^{\nu} \mathfrak{b}_i$ of fractional R_n-ideals $\mathfrak{b}_i$.)

Now let M be an R_m-lattice of R_m-rank μ. We wish to classify all ZG-lattices L for which there is an exact sequence

$$\xi: 0 \to M \to L \to N \to 0, \tag{12}$$

where $\xi \in \mathrm{Ext}(N, M)$. In short, we wish to classify all extensions of R_n-lattices by R_m-lattices. Now M is in the same genus as the free R_m-module $(R_m)^{\mu}$, and N as $(R_n)^{\nu}$. It follows at once from (7) that

$$\mathrm{Ext}(N, M) \cong \mathrm{Ext}((R_n)^{\nu}, (R_m)^{\mu}) \cong E^{\mu \times \nu}, \tag{13}$$

where $E = \mathrm{Ext}(R_n, R_m)$, and $E^{\mu\times\nu}$ devotes the additive group of all $\mu \times \nu$ matrices with entries in E.

Consider first the trivial case where $m = n$, so $E = 0$, and L is just an R_n-lattice. The isomorphism invariants of L are its rank and its Steinitz class (the product of the Steinitz classes of M and N). In this particular case, the isomorphism class of L does not determine those of M and N.

For the remainder of this section, we assume that $m \neq n$. We shall use the results of [3] freely, usually without explicit citation. As shown in [3], a full set of isomorphism invariants of L are

(a) The R_m-rank μ and the Steinitz class of M,
(b) The R_n-rank ν and the Steinitz class of N, and (14)
(c) The orbit of ξ in $\mathrm{Ext}_{ZG}(N, M)$ under the actions of $\mathrm{Aut}_{R_n}(N)$ and $\mathrm{Aut}_{R_m}(M)$ on $\mathrm{Ext}_{ZG}(N, M)$.

(This depends essentially on the fact that $\mathrm{Hom}_{ZG}(M, N) = 0$.)

We may easily handle the case where neither m/n nor n/m is a prime power (assuming always that $m \neq n$). We have shown that $\mathrm{Ext}(R_n, R_m) = 0$ in this case, so the orbit invariant of L is trivial. Indeed, we have $L \cong M \oplus N$, and (14 a, b) give a full set of isomorphism invariants of L.

Let us turn now to the only case where any difficulty arises, that where $m \neq n$ and $m/n = p^t$, $t \neq 0$. There is no real loss of generality in assuming that $m < n$, by the following argument. Given an exact sequence (12), we have $QL \cong QM \oplus QN$ as QG-modules, and thus QL contains a QG-submodule $V \cong QN$. There is then an exact sequence

$$0 \to L \cap V \to L \to L/(L \cap V) \to 0$$

in which $L \cap V$ is an R_n-lattice and $L/(L \cap V)$ an R_m-lattice. Thus we obtain a new extension problem in which the roles of m and n are reversed. We caution that $L \cap V$ need not be isomorphic to N, however, nor $L/(L \cap V)$ to M.

So now let

$$m = p^a s, \qquad n = p^b s, \quad p \nmid s, \quad a < b,$$

so $\mathrm{Ext}(R_n, R_m) \cong R_m/pR_m = \bar{R}_m$ (say). Suppose that the invariants (14, a, b) are given. We are then left with the problem of finding all orbits in $\mathrm{Ext}(N, M)$, as described in (14 c). As shown in [3], there is a bijection between this set of orbits, and the corresponding set when we replace N by the module $(R_n)^\nu$ in its genus, and likewise when M is replaced by $(R_m)^\mu$. In view of (13), we must therefore determine the orbits of $\bar{R}_m^{\mu\times\nu}$ under the actions of $GL(\mu, R_m)$ from the left and $GL(\nu, R_n)$ from the right. It follows from (11) that $\bar{R}_m$ is also a factor ring of R_n, so both of the groups $GL(\mu, R_m)$ and $GL(\nu, R_n)$ act on $\bar{R}_m^{\mu\times\nu}$ by means of the ring surjections $R_m \to \bar{R}_m$, $R_n \to \bar{R}_m$.

Let

$$\bar{\Phi}_s(x) = \prod_{i=1}^{\sigma} f_i(x), \qquad f_i(x) \text{ irreducible in } \bar{Z}[x],$$

where as usual $\bar{Z} = Z/pZ$. Each $f_i(x)$ has degree f, where f is the order of p (mod s), and we have $f \cdot \sigma = \varphi(s)$. By (9) and (11) we obtain

$$\bar{R}_m \cong \bar{Z}[x]/(\bar{\Phi}_m) \cong \coprod_{i=1}^{\sigma} \bar{Z}[x]/(f_i^{\varphi(p^a)}),$$

so $\bar{R}_m$ is a semilocal ring which is a direct sum of σ local rings. For convenience of notation, let

$$S_i = \bar{Z}[x]/(f_i^e), \qquad e = \varphi(p^a), \qquad 1 \leqslant i \leqslant \sigma.$$

Then each S_i is a local principal ideal ring, with maximal ideal $f_i S_i$.

Each $\xi \in \bar{R}_m^{\mu\times\nu}$ can be written as $\xi = \xi_1 + \cdots + \xi_\sigma$, with $\xi_i \in S_i^{\mu\times\nu}$. For each i, the matrix ξ_i has entries in the local principal ideal ring S_i. By an *elementary* operation on a matrix we shall mean adding a multiple of one row to another, or of one column to another. Now each elementary operation over S_i lifts to one over R_m; indeed, if $\{\psi_1, \dots, \psi_\sigma\}$ are elementary matrices, with $\psi_i \in S_i^{\mu\times\mu}$, then we can find an elementary matrix $\psi \in R_m^{\mu\times\mu}$, such that ψ maps onto $\psi_1 + \cdots + \psi_\sigma$ under the map induced by the surjection $R_m \to \bar{R}_m$. A corresponding result holds when μ is replaced by ν, and m by n.

We may now apply elementary row and column operations (over S_i) to the matrix ξ_i, so as to bring ξ_i into the following canonical form (we illustrate the case where $\mu \leqslant \nu$):

$$\xi_i = [D\ 0], \qquad D = \operatorname{diag}(f_i^{e_{i1}}, f_i^{e_{i2}}, \dots, f_i^{e_{i\mu}} u_i), \qquad 1 \leqslant i \leqslant \sigma, \tag{15}$$

where

$$0 \leqslant e_{i1} \leqslant e_{i2} \leqslant \cdots \leqslant e_{i\mu} \leqslant e = \varphi(p^a), \qquad \text{and} \qquad u_i \in u(S_i). \tag{16}$$

Here, $u(S_i)$ denotes the group of units of S_i. Furthermore, since for $u \in u(S_i)$ we can express $\operatorname{diag}(u, u^{-1})$ as a product of elementary matrices, it follows that we can choose $u_i = 1$ whenever $\mu \neq \nu$.

The preceding discussion shows that, without changing the orbit of ξ in $\bar{R}_m^{\mu\times\nu}$, we may assume that each ξ_i has the canonical form given in (15) and (16), with $u_i = 1$ if $\mu \neq \nu$. For each i, the exponents $e_{i1}, \dots, e_{i\mu}$ are precisely the exponents which occur in the elementary divisors (in the usual sense) of the matrix ξ_i over the principal ideal ring S_i. It is then clear that the set of exponents

$$\{e_{ij} \colon 1 \leqslant i \leqslant \sigma,\ 1 \leqslant j \leqslant \min(\mu, \nu)\} \tag{17}$$

are invariants of the orbit of ξ; we shall call them the *elementary divisor exponents* of the orbit of ξ.

From the fact that we can diagonalize ξ, it follows at once that the only indecomposable L's are those where the ordered pair $\{\mu, \nu\}$ is one of $\{1, 0\}$, $\{0, 1\}$ or $\{1, 1\}$. In the first two cases, L is an ideal of R_m or R_n, respectively. In the case where $\mu = \nu = 1$, L is a nonsplit extension

$$\xi \colon 0 \to \mathfrak{a} \to L \to \mathfrak{b} \to 0,$$

with $\mathfrak{a}$ an ideal of R_m, $\mathfrak{b}$ an ideal of R_n, and $\xi \in \operatorname{Ext}(\mathfrak{b}, \mathfrak{a}) \cong \bar{R}_m$. We denote this extension L by the symbol $(\mathfrak{a}, \mathfrak{b}; \xi)$. Here, identifying ξ with its image in $\bar{R}_m$, we may write

$$\xi = \sum_{i=1}^{\sigma} f_i^{e_{i1}} u_i, \qquad \text{where } 0 \leqslant e_{i1} \leqslant e,\ u_i \in u(S_i), \qquad 1 \leqslant i \leqslant \sigma, \tag{18}$$

and where $e_{i1} < e$ for at least one value of i. From the preceding discussion (see also [3, Section 5]) we obtain

(19) THEOREM. *A full set of isomorphism invariants for the nonsplit extension* $(\mathfrak{a}, \mathfrak{b}; \xi)$, *with* ξ *as in* (18), *is given by*

(a) *the ideal classes of* $\mathfrak{a}$ *and* $\mathfrak{b}$,

(b) *the elementary divisor exponents* $e_{11}, e_{21}, \ldots, e_{\sigma 1}$, *and*

(c) *the image of the* σ*-tuple* $(u_1, \ldots, u_\sigma)$ *in the group*

$$\left\{\prod_{i=1}^{\sigma} u(S_i f_i^{e-e_{i1}} S_i)\right\} \Big/ \{\text{images of } u(R_m) \text{ and } u(R_n)\}.$$

In fact, we may even specify the deviation from the Krull-Schmidt-Azumaya Theorem in this case; namely, we may decide precisely when two direct sums of indecomposables are isomorphic. Let L be any extension of an R_n-lattice N by an R_m-lattice M, as in (12). We have shown above that (up to isomorphism)

$$L = \coprod_{j=1}^{\alpha} (\mathfrak{a}_j, \mathfrak{b}_j; \xi^{(j)}) \oplus \coprod_{j=\alpha+1}^{\beta} \mathfrak{a}_j \oplus \coprod_{j=\alpha+1}^{\gamma} \mathfrak{b}_j, \tag{20}$$

where each $\mathfrak{a}_j$ is an ideal of R_m, each $\mathfrak{b}_j$ is an ideal of R_n, and where

$$\xi^{(j)} = (f_1^{e_{1j}} u_{1j}, \ldots, f_\sigma^{e_{\sigma j}} u_{\sigma j}) \in \coprod_{i=1}^{\sigma} S_i,$$

with

$$0 \leqslant \min(e_{1j}, \ldots, e_{\sigma j}) < e = \varphi(p^a), \qquad u_{ij} \in u(S_i), \quad 1 \leqslant j \leqslant \alpha.$$

From [3, Theorem 5.10] we obtain

(21) THEOREM. *Let* L *be any extension of an* R_n*-lattice* N *by an* R_m*-lattice* M, *and let* L *be expressed as a direct sum of indecomposables as in* (20). *The following are invariants of the isomorphism class of* L:

(a) *the integer* β *and the ideal class of* $\prod_{j=1}^{\beta} \mathfrak{a}_j$ (*these give the invariants of* M),

(b) *the integer* γ *and the ideal class of* $\prod_{j=1}^{\gamma} \mathfrak{b}_j$ (*these give the invariants of* N),

(c) *the set of elementary divisor exponents* $\{e_{ij}: 1 \leqslant i \leqslant \sigma, 1 \leqslant j \leqslant \alpha\}$.

If either $\beta \geqslant \alpha + 1$ *or* $\gamma \geqslant \alpha + 1$ (*in other words, either some* $\mathfrak{a}_j$ *or some* $\mathfrak{b}_j$ *is a summand of* L), *then* (a), (b) *and* (c) *are a full set of invariants, and determine* L *up to isomorphism.*

On the other hand, when $L = \coprod_{j=1}^{\alpha} (\mathfrak{a}_j, \mathfrak{b}_j; \xi^{(j)})$, *then a full set of isomorphism invariants consists of those in* (a)–(c), *together with the following additional invariant*:

(d) *the image of the* σ*-tuple*

$$\left(\prod_{j=1}^{\alpha} u_{1j}, \prod_{j=1}^{\alpha} u_{2j}, \ldots, \prod_{j=1}^{\alpha} u_{\sigma j}\right)$$

in the factor group

$$U = \left\{\prod_{i=1}^{\sigma} u(S_i/f_i^{m_i} S_i)\right\} \Big/ \{\textit{images of } u(R_m) \textit{ and } u(R_n)\},$$

where

$$m_i = \min(e - e_{i1}, e - e_{i2}, ..., e - e_{i\alpha}), \qquad 1 \leqslant i \leqslant \sigma.$$

In practice, it may be extremely difficult to determine the group U explicitly. For example, let $g = p^2$, $m = p$, $n = p^2$. Then $\bar{\Phi}_m = (1 - x)^{p-1}$ in $\bar{Z}[x]$, so $\bar{R}_m = \bar{Z}[x]/(1 - x)^{p-1}$. Thus when $\sigma = 1$,

$$U = u(\bar{Z}[x]/(1 - x)^{m_1})/\text{image of } u(R_m),$$

where $m_1 = \min(e - e_1, e - e_2, ..., e - e_\alpha)$. (It turns out that the image of $u(R_m)$ contains the image of $u(R_n)$ is this case.) Then U is an elementary p-group whose rank depends on whether p is a regular prime or not (see [3]).

Note added in proof. A somewhat different proof of Diederichsen's results may be found in the article by S. D. Berman, "Integral representations of a cyclic group containing two irreducible rational components," which appeared in the Čebotarev Memorial Volume, Izdat. Kazan Univ., Kazan (1964), pp. 18–29. Also, Hyman Bass recently sent me a preprint of his article "The Grothendieck group of the category of abelian group automorphisms of finite order," in which he independently proves Diederichsen's results.

References

1. C. W. Curtis and I. Reiner, "Representation Theory of Finite Groups and Associative Algebras," 2nd ed., Interscience, New York, 1966.
2. F. E. Diederichsen, Über die Ausreduktion ganzzahliger Gruppendarstellungen bei arithmetischer Aquivalenz, *Abh. Math. Sem. Univ. Hamburg* **13** (1940), 357–412.
3. I. Reiner, Invariants of integral representations, *Pacific J. Math.* **78** (1978), 467–501.
4. J. Rotman, "Notes on Homological Algebra," Van Nostrand and Reinhold, Princeton, N.J. and New York, 1970.
5. B. L. van der Waerden, "Modern Algebra," Vol. 1, Ungar, New York, 1948.
6. E. Weiss, "Algebraic Number Theory," McGraw–Hill, New York, 1963.

Zeta Functions of Arithmetic Orders and Solomon's Conjectures

Colin J. Bushnell[1] and Irving Reiner[2]*

[1] Department of Mathematics, University of London King's College, Strand, London WC2R 2LS, England

[2] Department of Mathematics, University of Illinois, Urbana, Illinois 61801, U.S.A.

Introduction

In a recent article [6], Solomon introduced a zeta function which counts sublattices of a given lattice over an order Λ. Let us recall the definition of this zeta function. Throughout this paper, let A be a finite dimensional semisimple algebra over the rational field $\mathbb{Q}$ or over a p-adic field $\mathbb{Q}_p$, and let Λ be an order in A. When A is a $\mathbb{Q}$-algebra, Λ is a $\mathbb{Z}$-order in A; when A is a $\mathbb{Q}_p$-algebra, Λ is a $\mathbb{Z}_p$-order in A, where $\mathbb{Z}_p$ is the ring of p-adic integers. The subscript p always indicates p-adic *completion*.

Now let V be a finitely generated left A-module, and let L be a full Λ-lattice in V, that is, L is a finitely generated Λ-submodule of V such that $\mathbb{Q}\cdot L = V$. Solomon's zeta function is defined as

$$\zeta_\Lambda(L; s) = \sum_{N \subseteq L} (L: N)^{-s}, \tag{1}$$

where the sum extends over all full Λ-lattices N in L. Here, $(L: N)$ denotes the index of N in L, and s is a complex variable. We shall usually write $\zeta(L; s)$ instead of $\zeta_\Lambda(L; s)$, unless there is danger of confusion.

Solomon's zeta function is an obvious generalization of the classical Dedekind zeta function $\zeta_K(s)$ of an algebraic number field K, and the analogy carries over to the Euler product formula as well. If Λ is a $\mathbb{Z}$-order in the $\mathbb{Q}$-algebra A, and L is a full Λ-lattice in the A-module V, then for each rational prime p, L_p is a full Λ_p-lattice in the A_p-module V_p. We may therefore form the "local" zeta function $\zeta_{\Lambda_p}(L_p; s)$ defined as in (1). Solomon proved that there is an Euler product identity

$$\zeta_\Lambda(L; s) = \prod_p \zeta_{\Lambda_p}(L_p; s), \tag{2}$$

and made a conjecture as to the nature of the factors $\zeta(L_p; s)$. He gave an *ad hoc* definition, in terms of Dedekind zeta functions, of a zeta function $\zeta_A(V; s)$ which

* The research of the second author was supported by the National Science Foundation

depends only on the A-module V, and not on the order Λ or the Λ-lattice L. Solomon proved that the quotient

$$\zeta(L_p; s)/p\text{-part of } \zeta(V; s)$$

is 1 for almost all primes p. He conjectured that for each p, the quotient is a polynomial in p^{-s} with coefficients in $\mathbb{Z}$. One object of this paper is to prove a stronger version of this conjecture.

Tate's thesis [8] shows how to deal with Dedekind zeta functions by means of certain "zeta integrals" on the idele group. Formula (2) suggests that Tate's approach can be extended to cover the present case. Our first task is to give a brief development of Solomon's theory from this point of view; this is relatively straightforward, necessitating only a mild generalization of classical material (see [9], for example). This new approach leads to a machinery for computing with Solomon's zeta function which is, in many circumstances, much more efficient than the elementary algebraic and combinatorial methods employed by Solomon.

Further, Tate's thesis has inspired far-reaching developments in harmonic analysis on adele groups, and in particular has led to a greater understanding of the concept of zeta functions in very general contexts. Specializing this theory to the present case, we obtain a pleasing local functional equation for Solomon's zeta function of the integral group ring RG of a finite group G over a p-adic ring of integers R.

For the information on integral representations and maximal orders used below, we refer to [4, 1] and [9]. For the machinery of ideles and zeta integrals, see [8, 9] and [3].

§1. Solomon's Zeta Function

As above, let A be a finite dimensional semisimple $\mathbb{Q}$-algebra, and Λ an order in A. Given a full Λ-lattice L in a finitely generated left A-module V, we define $\zeta_\Lambda(L; s)$ by (1). Next, we fix some full Λ-lattice M in V. The *genus* of M consists of all full Λ-lattices N in V such that for each rational prime p, there is a Λ_p-isomorphism $N_p \cong M_p$. We write $N \vee M$ to indicate that N lies in the genus of M. Now define

$$Z_\Lambda(L, M; s) = \sum_{\substack{N \subseteq L \\ N \vee M}} (L; N)^{-s}, \tag{3}$$

the sum extending over all full Λ-lattices N in L such that N lies in the genus of M.

There are only finitely many genera of full Λ-lattices in V, so we can pick a finite set $\mathscr{S}$ of representatives of these genera, and then clearly

$$\zeta L; s) = \sum_{M \in \mathscr{S}} Z(L, M; s). \tag{4}$$

We ought to say something about the convergence of these series. We follow Solomon in this, but phrase the argument in a way suitable for later developments. Suppose first that $A=V=\mathbb{Q}$, $\Lambda=L=\mathbb{Z}$. Then $\zeta_{\mathbb{Z}}(\mathbb{Z}; s)$, denoted briefly by $\zeta_{\mathbb{Z}}(s)$, is just the ordinary Riemann zeta function. Next, let $A=\mathbb{Q}$, $V=\mathbb{Q}^n$, $\Lambda=\mathbb{Z}$, $L=\mathbb{Z}^n$. Each full Λ-sublattice N of L is then of the form $N=Lx$, where x is an integral $n\times n$ matrix of nonzero determinant (that is, $x\in M_n(\mathbb{Z})\cap GL_n(\mathbb{Q})$). Further, $Lx=Ly$ if and only if $xy^{-1}\in GL_n(\mathbb{Z})$. Since the index $(L\colon Lx)$ equals $|\det x|$, we obtain

$$\zeta_{\mathbb{Z}}(\mathbb{Z}^n; s)=\sum_x |\det x|^{-s},$$

where x ranges over a full set of representatives of the right coset space $GL_n(\mathbb{Z})\backslash(M_n(\mathbb{Z})\cap GL_n(\mathbb{Q}))$. A set of coset representatives is given by the integral matrices $x=(a_{ij})$ in "Hermite normal form":

$$a_{ii}\geqq 1, 1\leqq i\leqq n; \quad a_{ij}=0 \text{ if } i>j; \quad 0\leqq a_{ij}<a_{jj} \text{ if } i<j.$$

We may rearrange $\zeta(\mathbb{Z}^n; s)$ as a Dirichlet series

$$\zeta(\mathbb{Z}^n; s)=\sum_{m=1}^{\infty} c(m)\, m^{-s}.$$

The general coefficient $c(m)$ is then given by

$$c(m)=\sum_{(d_1,\ldots,d_n)} \prod_{i=2}^{n} d_i^{(i-1)},$$

the sum extending over all n-tuples $(d_1,\ldots,d_n)$ of positive integers whose product is m. From this it is easy to see that

$$\zeta(\mathbb{Z}^n; s)=\zeta_{\mathbb{Z}}(s)\,\zeta_{\mathbb{Z}}(s-1)\ldots\zeta_{\mathbb{Z}}(s-(n-1)). \tag{5}$$

Returning to the case where Λ and A are arbitrary, the comparison test for convergence then gives

Proposition 1. *Let Λ be any order in the $\mathbb{Q}$-algebra A, and let L, M be full Λ-lattices in an A-module V. Then both series $\zeta_\Lambda(L; s)$ and $\zeta_\Lambda(L, M; s)$ converge absolutely in the half-plane* $\operatorname{Re} s>\dim_{\mathbb{Q}} V$, *and define holomorphic functions there.*

On the other hand, suppose that A is a $\mathbb{Q}_p$-algebra. We have already defined $\zeta_\Lambda(L; s)$, where Λ is a $\mathbb{Z}_p$-order in A and L is a full Λ-lattice in an A-module V. Given a full Λ-lattice M in V, we modify the Definition (3) in this local case, and set

$$Z_\Lambda(L, M; s)=\sum_{\substack{N\subseteq L\\ N\cong M}} (L:N)^{-s}, \tag{6}$$

the sum extending over all full sublattices N of L such that $N\cong M$. (This is the correct analogue of (3), since in the local case there is only one prime p involved.) We can again express $\zeta(L; s)$ as a finite sum $\sum_M Z(L, M; s)$, where now M ranges over a full set of representatives of the isomorphism classes of full Λ-lattices in V.

§2. The Transition to Ideles

§2.1. The Local Case

Suppose that A is a $\mathbb{Q}_p$-algebra, let V be a left A-module, and let L, M and N denote full A-lattices in V. We shall treat the local zeta functions $\zeta(L;s)$ and $Z(L,M;s)$ in a manner similar to the example in §1 where $A=\mathbb{Q}$, $V=\mathbb{Q}^n$. We set

$$B=\operatorname{End}_A V, \quad B^\times=\operatorname{Aut}_A V=\text{units of } B, \tag{7}$$

and view V as an (A,B)-bimodule. Now define

$$\{M:N\}=\{x\in B: Mx\subseteq N\}. \tag{8}$$

This is a full $\mathbb{Z}_p$-lattice in B, and furthermore

$$\{M:M\}=\{x\in B: Mx\subseteq M\}=\text{right order of } M \text{ in } B. \tag{9}$$

Then $\{M:M\}$ is a $\mathbb{Z}_p$-order in B, and its group of units is precisely $\operatorname{Aut}_A M$, denoted by $\operatorname{Aut} M$ for brevity.

If $N\subseteq L$ and $N\cong M$, then $N=Mx$ for some $x\in\{M:L\}\cap B^\times$, with x unique mod Aut M. This gives at once

$$Z(L,M;s)=\sum_x (L:Mx)^{-s},$$

the sum extending over all $x\in(\operatorname{Aut} M)\backslash(\{M:L\}\cap B^\times)$. In order to compute the index $(L:Mx)$, it is convenient to extend the notation $(X:Y)$ so that it is defined for *every* pair of full $\mathbb{Z}_p$-lattices X, Y in V. Specifically, we put

$$(X:Y)=(X:X\cap Y)/(Y:X\cap Y).$$

This symbol is then skew-symmetric, and satisfies

$$(X:Y)=(X:W)(W:Y)$$

if W is also a full $\mathbb{Z}_p$-lattice in V.

Now let us put

$$\|x\|_V=(Nx:N), \qquad x\in B^\times,$$

where N is any full $\mathbb{Z}_p$-lattice in V. This norm $\|x\|$ is independent of the choice of N, and is multiplicative. The norm map gives a homomorphism from $B^\times$ into the multiplicative group of positive rationals, and indeed, into the cyclic subgroup generated by p. Further, $\|x\|=1$ whenever x is a unit in some $\mathbb{Z}_p$-order in B. We now have

$$(L:Mx)=(L:M)(M:Mx)=(L:M)\,\|x\|_V^{-1},$$

and therefore

$$Z(L,M;s)=(L:M)^{-s}\sum_x \|x\|_V^s, \tag{10}$$

where x ranges over the coset space $(\operatorname{Aut} M)\backslash(\{M:L\}\cap B^\times)$ as before.

This formula for $Z(L, M; s)$ can be given another form when we take the topology of the situation into account. The $\mathbb{Q}_p$-algebra B inherits a topology from its structure as a finite dimensional $\mathbb{Q}_p$-space, making B a locally compact topological algebra. Any full $\mathbb{Z}_p$-lattice in B is a compact open neighborhood of 0. The units $B^\times$ form an open subset of B, and with the subset topology, they form a locally compact topological group. If Γ is any $\mathbb{Z}_p$-order in B, its unit group $\Gamma^\times$ is a compact open subgroup of $B^\times$. Now choose a Haar measure $d^\times x$ on $B^\times$; this measure is automatically bi-invariant, being of the form $\|x\|_B^{-1}\, dx$ for a Haar measure dx on B.

For each full $\mathbb{Z}_p$-lattice M in V, the measure $\mu(\mathrm{Aut}_\Lambda M)$ is finite and nonzero. Now consider

$$\int_{B^\times} \Phi(x)\, \|x\|_V^s\, d^\times x,$$

where $\Phi = \Phi_{\{M:L\}}$ is the characteristic function in B of the lattice $\{M:L\}$. By Fubini's Theorem we have, in the domain of absolute convergence of the integral,

$$\int_{B^\times} \Phi(x)\, \|x\|_V^s\, d^\times x = \sum_y \int_{\mathrm{Aut}\, M} \Phi(x\, y)\, \|x\, y\|_V^s\, d^\times x,$$

the sum extending over all $y \in (\mathrm{Aut}\, M)\backslash B^\times$. But the inner integral equals $\mu(\mathrm{Aut}\, M) \cdot \|y\|_V^s$ if $y \in \{M:L\}$, and is zero if $y \notin \{M:L\}$. Therefore

$$\int_{B^\times} \Phi(x)\, \|x\|_V^s d^\times x = \mu(\mathrm{Aut}\, M) \cdot \sum_y \|y\|_V^s,$$

where y ranges over $(\mathrm{Aut}\, M)\backslash(\{M:L\} \cap B^\times)$. Comparison with (10) yields

$$Z(L, M; s) = \mu(\mathrm{Aut}\, M)^{-1} (L:M)^{-s} \int_{B^\times} \Phi_{\{M:L\}}(x)\, \|x\|_V^s\, d^\times x. \tag{11}$$

§ 2.2 The Global Case

Suppose now that A is a $\mathbb{Q}$-algebra. In order to extend the above treatment to global zeta functions, we must introduce ideles. The constructions we shall give are closely related to the standard ones (see [9], for example), except that we omit the archimedean prime, which plays no role in our considerations.

To begin with, let B be any finite dimensional semisimple $\mathbb{Q}$-algebra, and let Γ be a $\mathbb{Z}$-order in B. The ring $\mathrm{Ad}\, B$ of *finite adeles* of B is the topological restricted direct product of the algebras B_p, p ranging over all prime numbers, with respect to the p-adic completions Γ_p. This definition is independent of the choice of the order Γ, and makes $\mathrm{Ad}\, B$ into a locally compact topological ring. We set

$$\mathrm{Ad}\, \Gamma = \prod_p \Gamma_p,$$

a compact open subring of $\mathrm{Ad}\,B$. Similarly, for any full $\mathbb{Z}$-lattice X in B, we define

$$\mathrm{Ad}\,X=\prod_p X_p,$$

and this is a compact open additive subgroup of $\mathrm{Ad}\,B$.

Likewise, we can form the group $J(B)$ of *finite ideles* of B as the topological restricted direct product of the groups $B_p^\times$ with respect to the subgroups $\Gamma_p^\times$. (Here, $\Gamma_p^\times$ denotes the group of units of the ring Γ_p.) Again, $J(B)$ is independent of Γ, and is a locally compact topological group. We write

$$U(\Gamma)=\prod_p \Gamma_p^\times,$$

a compact open subgroup of $J(B)$.

There is a natural continuous embedding $J(B)\to \mathrm{Ad}\,B$, but $J(B)$ does not have the subset topology.

We suppose now that A is a $\mathbb{Q}$-algebra as before, and V an A-module. Set $B=\mathrm{End}_A V$; then the group $J(B)$ acts on the set of full Λ-lattices in V. Indeed, given such a lattice M, and an idele $x=(x_p)\in J(B)$, we define a lattice Mx by the requirement that

$$(Mx)_p=M_p x_p \qquad \text{for all } p.$$

(To do this, we of course identify B_p with $\mathrm{End}_{A_p} V_p$. If $\Gamma=\mathrm{End}_\Lambda M$, then Γ is an order in B, and $\Gamma_p=\mathrm{End}_{\Lambda_p} M_p$. We have $x_p\in\Gamma_p^\times$ for almost all p, and then $M_p x_p=M_p$ for each such p. It follows that Mx is a well defined full Γ-lattice in V.)

The action of the idele group $J(B)$ on the set of full Λ-lattices in V is transitive on the lattices in each genus. The lattices in the genus of M are precisely the lattices Mx, with $x\in J(B)$. Further, $Mx=My$ if and only if $xy^{-1}\in U(\Gamma)$.

Now define

$$\|x\|_V=(Mx:\mathrm{M}), \qquad x\in J(B),$$

where M is any full $\mathbb{Z}$-lattice in V. Then $\|x\|_V$ is independent of the choice of M, and just as in the local case, we obtain the formula

$$Z(L,M;s)=\mu(U(\Gamma))^{-1}(L:M)^{-s}\int_{J(B)}\Phi_{\{M:L\}}(x)\,\|x\|_V^s\,d^\times x. \tag{12}$$

Here, $\Gamma=\mathrm{End}_\Lambda M$, $\Phi_{\{M:L\}}$ is the characteristic function of $\mathrm{Ad}\{M:L\}$ in $\mathrm{Ad}\,B$, and μ denotes the measure of a set with respect to the Haar measure $d^\times x$ on $J(B)$.

§2.3 Euler Product

Let us now sketch how one may relate formulas (11) and (12) into an Euler product. Returning to our general notation, let B be a $\mathbb{Q}$-algebra, and Γ an order in B. A basis for the open sets in $J(B)$ is provided by sets of the form $E = \prod_p E_p$, where E_p is open in $B_p^\times$ for all p, and $E_p = \Gamma_p^\times$ for almost all p. We can construct a Haar measure $d^\times x$ on $J(B)$ by choosing a Haar measure $d^\times x_p$ on $B_p^\times$ for each p, in such a manner that $\int_{\Gamma_p^\times} d^\times x_p = 1$ for almost all p. We then put

$$\int_E d^\times x = \prod_p \int_{E_p} d^\times x_p.$$

This product has only finitely many factors different from 1. The measure $d^\times x$ is thereby uniquely determined, and any Haar measure on $J(B)$ can be put in this form.

(Alternatively, for any finite set S of primes, we can put

$$J_S(B) = \prod_{p \in S} B_p^\times \times \prod_{p \notin S} \Gamma_p^\times.$$

This is an open subgroup of $J(B)$, and its subset topology is the product topology. On each $J_S(B)$ we can take a product measure, in a coherent fashion (relative to increasing the size of S). Then $J(B)$ is the topological direct limit of the $J_S(B)$, and the limit of the product measures is a Haar measure on $J(B)$.)

To apply this in §2.2, let $B = \operatorname{End}_A V$, and let $\Gamma = \operatorname{End}_A M$, where M is a full A-lattice in the A-module V. We choose a measure $d^\times x_p$ on $B_p^\times$ as above. Now consider finite sets S of primes which include all those primes p for which either $\{M_p : L_p\} \neq \Gamma_p$ or $\int_{\Gamma_p^\times} d^\times x_p \neq 1$. In formula (12), the integrand and the factors outside the integral can all be expressed as products of the corresponding local objects. Specifically, for $x = (x_p) \in J(B)$, we have

$$\Phi_{\{M:L\}}(x) = \prod_p \Phi_{\{M_p:L_p\}}(x_p), \qquad \|x\|_V = \prod_p \|x_p\|_{V_p},$$

and furthermore

$$\mu(U(\Gamma)) = \prod_p \mu_p(\Gamma_p^\times), \qquad (L:M) = \prod_p (L_p : M_p).$$

Therefore

$$\int_{J_S(B)} \Phi_{\{M:L\}}(x)\, \|x\|_V^s d^\times x = \prod_{p \in S} \mu_p(\Gamma_p^\times)(L_p : M_p)^s Z_{A_p}(L_p, M_p; s).$$

Taking the limit over increasing sequences of such sets S, we obtain (in any domain of absolute and uniform convergence)

$$\int_{J(B)} \Phi_{\{M:L\}}(x)\, \|x\|_V^s\, d^\times x = \mu(U(\Gamma))\,(L:M)^s \prod_p Z_{A_p}(L_p, M_p; s),$$

that is,

$$Z_A(L, M; s) = \prod_p Z_{A_p}(L_p, M_p; s). \tag{13}$$

The computations in §3.3 will give the values of almost all the factors in the right hand product, and from this it will be easy to see that formula (13) is valid in some half-plane $\operatorname{Re} s > \alpha$.

(It should be remarked that formula (13) can be obtained directly from the Definitions (3) and (6), without use of ideles. Indeed, any full Λ-lattice N in L in the genus of M determines a collection $\{N_p\}$ of lattices, such that for each p, N_p is a full Λ_p-lattice in L_p satisfying $N_p \cong M_p$. Since $N_p = M_p$ for almost all p, formula (13) is an easy consequence of this remark; see [6, Lemma 6].)

One can assemble these product formulas (13) for various M's, to obtain an Euler product for $\zeta_\Lambda(L; s)$. First of all, Λ_p is a maximal order in A_p for almost all p. For each such p, any two full Λ_p-lattices in V_p are isomorphic, and hence

$$Z_{\Lambda_p}(L_p, M_p; s) = \zeta_{\Lambda_p}(L_p; s)$$

for any L and M. Using the action of $J(B)$ on Λ-lattices, it is easy to see that each genus of full Λ-lattices in V can be represented by a lattice M with the property that

$$M_p = L_p \text{ whenever } \Lambda_p \text{ is a maximal order.}$$

Let $\mathcal{R}$ be a complete set of genus representatives of this type. For each prime p, we may choose a subset $\mathcal{R}_p$ of $\{M_p : M \in \mathcal{R}\}$, such that $\mathcal{R}_p$ is a complete set of representatives of the isomorphism classes of full Λ_p-lattices in V_p. If S denotes the (finite) set of primes p for which Λ_p is *not* a maximal order, then we have a natural bijection between $\mathcal{R}$ and $\prod_{p \in S} \mathcal{R}_p$. Therefore we obtain

$$\begin{aligned}
\prod_p \zeta(L_p; s) &= \prod_p \sum_{M_p \in \mathcal{R}_p} Z(L_p, M_p; s) \\
&= \prod_{p \notin S} \zeta(L_p; s) \cdot \prod_{p \in S} \sum_{M_p \in \mathcal{R}_p} Z(L_p, M_p; s) \\
&= \prod_{p \notin S} \zeta(L_p; s) \cdot \sum_{M \in \mathcal{R}} \prod_{p \in S} Z(L_p, M_p; s) \\
&= \sum_{M \in \mathcal{R}} \prod_p Z(L_p, M_p; s) = \sum_{M \in \mathcal{R}} Z(L, M; s) = \zeta(L; s).
\end{aligned}$$

Thus we have proved the Euler product formula (2) given in §1. Again, we should point out that this formula can be obtained directly from the definition of the zeta function, without use of integrals; see [6, Lemma 6]. Nevertheless, the power of this analytic approach will become apparent in the later sections.

§3. Examples

§3.1. Decomposable Orders

If the order Λ is decomposable into a direct sum of orders: $\Lambda = \Lambda_1 \oplus \Lambda_2$, then every Λ-lattice L has a corresponding decomposition: $L = L_1 \oplus L_2$, where L_i

$=\Lambda_i L$. Then clearly

$$\zeta_\Lambda(L;s)=\zeta_{\Lambda_1}(L_1;s)\cdot\zeta_{\Lambda_2}(L_2;s),$$

and for any M,

$$Z(L,M;s)=Z(L_1,M_1;s)\cdot Z(L_2,M_2;s).$$

These formulas hold both locally and globally.

§3.2. Genus

It is clear that, both locally and globally, $Z(L,M;s)$ depends only upon the isomorphism classes of the Λ-lattices L and M. The Euler product formula (13) therefore shows that the global $Z(L,M;s)$ depends only upon the *genera* of L and M.

§3.3. Maximal Orders

Let Λ be a maximal order in the $\mathbb{Q}$-algebra A, and let L, M be full Λ-lattices in the A-module V. Since any two full Λ-lattices in V are necessarily in the same genus, we have

$$Z(L,M;s)=Z(L,L;s)=\zeta(L;s)=\zeta(M;s).$$

This shows that $\zeta(L;s)$ depends only on V, and not on L. The same applies locally. Moreover, any two maximal orders in A_p are isomorphic (indeed, conjugate), and hence have the same zeta functions relative to V_p. Therefore, both locally and globally, $\zeta_\Lambda(L;s)$ depends on the A-module V, and not upon the choice of L and Λ.

The above remarks, together with formula (2) and §3.1, show that we can compute the zeta function of any maximal order once we can handle the case of a maximal order Λ in a simple $\mathbb{Q}_p$-algebra A. For the remainder of this subsection, we may therefore take

$$A=M_m(D),\qquad D=\text{division algebra with center } F,\quad \dim_F D=e^2.$$

Then F is a finite extension of $\mathbb{Q}_p$; let R be its valuation ring, and $P=\pi R$ the maximal ideal of R. We set $\mathcal{N}P=(R:P)$, the number of elements in R/P. Let $\| \ \|_F$ be the valuation on F, normalized so that

$$\|\alpha\|_F=(R\alpha:R)\quad\text{for nonzero}\quad \alpha\in F,\ \|\pi\|_F=(\mathcal{N}P)^{-1}.$$

Suppose now that $V=W^k$, where W is the simple left A-module D^m. We may then assume that

$$\Lambda=M_m(\Delta),\quad \Delta=\text{unique maximal } R\text{-order in } D,\quad L=(\Delta^m)^k$$

Let ξ be a prime element of Δ, so $\Delta\xi$ is the unique maximal ideal of Δ, and $(\Delta:\Delta\xi)=(\mathcal{N}P)^e$. To complete our list of notation, we note that

$$B=\operatorname{End}_\Delta V\cong M_k(D),\qquad \Gamma=\operatorname{End}_\Delta L\cong M_k(\Delta),$$

and we treat these isomorphisms as identifications. The "index function" $\|x\|_V$, for $x\in B^\times$, is given by

$$\|x\|_V=\|nrx\|_F^{me}=(R\cdot nrx:R)^{me},$$

where nr is the reduced norm from B to its center F.

Now choose the Haar measure $d^\times x$ on $B^\times$ so that $\Gamma^\times$ has measure 1. Then

$$\zeta_\Delta(L;s)=Z_\Delta(L,L;s)=\int_{B^\times}\Phi(x)\,\|x\|_V^s\,d^\times x,$$

where Φ is the characteristic function of Γ in B. Reversing our earlier procedure, we obtain

$$\int_{B^\times}\Phi(x)\,\|x\|_V^s\,d^\times x=\sum_x\|nrx\|_F^{mes},\tag{15}$$

where the summation extends over $x\in GL_k(\Delta)\backslash(M_k(\Delta)\cap GL_k(D))$. (This coset space is precisely $\Gamma^\times\backslash(\Gamma\cap B^\times)$.) We again use the Hermite normal form to obtain a set of representatives of this coset space. Since Δ is a non-commutative discrete valuation ring with prime element ξ, it is easily seen that the coset representatives may be taken of the form $x=(a_{ij})\in GL_k(D)$, where the $\{a_{ij}\}$ satisfy the following restrictions:

$a_{ii}=\xi^{r_i}$, $r_i\geq 0$, $1\leq i\leq k$; $a_{ij}=0$ if $j<i$,

a_{ij} ranges over a full set of representatives of Δ mod $\xi^{r_j}\Delta$, if $i<j$.

Note that for $i<j$, there are $(\Delta:\xi^{r_j}\Delta)$ choices for a_{ij}, that is, $(\mathcal{N}P)^{er_j}$ choices. If $x=(a_{ij})$, then

$$\|nrx\|_F=(\mathcal{N}P)^{-(r_1+\dots+r_k)}.$$

It now follows that

$$\sum_x\|nrx\|_F^{mes}=\sum_{n=0}^{\infty}c(n)(\mathcal{N}P)^{-mesn},$$

where for $n\geq 0$, the coefficient $c(n)$ is given by

$$c(n)=\sum_{(r_1,\dots,r_k)}\prod_{i=2}^{k}(\mathcal{N}P)^{r_i(i-1)}.$$

The sum extends over all k-tuples $(r_1,\dots,r_k)$ of nonnegative integers whose sum is n.

In the special case where $A=D=F=V$, we have $e=m=k=l$, and then

$$\zeta_\Lambda(\Lambda;s)=\zeta_R(s)=\sum_{n=0}^{\infty}(\mathcal{N}P)^{-sn}=(1-\mathcal{N}P^{-s})^{-1},$$

the Dedekind zeta function of the local field F.

In the general case, we easily obtain Hey's formula

$$\zeta_\Lambda(L;s)=\prod_{j=0}^{k-1}\zeta_R(mes-ej). \tag{16}$$

§3.4. Group Rings

Suppose to start with that Λ is a $\mathbb{Z}$-order in a $\mathbb{Q}$-algebra A, and L, M are full Λ-lattices in an A-module V. The Euler product formula (13) reduces the calculation of $Z_\Lambda(L, M; s)$ to the corresponding calculation in the local case. Now let Λ' be a maximal order in A containing Λ. We can form $\Lambda'L$ computed inside V; then $\Lambda'L$ is a full Λ'-lattice in V, and we have

$$Z_{\Lambda'}(\Lambda'L, \Lambda'M; s)=\zeta_{\Lambda'}(\Lambda'L; s),$$

by virtue of the discussion in §3.3. Thus

$$Z_\Lambda(L, M; s)/\zeta_{\Lambda'}(\Lambda'L; s)=\prod_p\{Z_{\Lambda_p}(L_p, M_p; s)/\zeta_{\Lambda'_p}(\Lambda'_pL_p; s)\}.$$

Now $\zeta(\Lambda'L; s)$ can be computed readily from §3.3, so in order to calculate $Z_\Lambda(L, M; s)$ and $\zeta_\Lambda(L; s)$, we need to determine

$$Z_{\Lambda_p}(L_p, M_p; s)/\zeta_{\Lambda'_p}(\Lambda'_pL_p; s) \tag{17}$$

for each p. This quotient is identically 1 for almost all p, namely, for those p such that $\Lambda_p=\Lambda'_p$. The quotient depends only on L, M and Λ, and not on the choice of Λ'.

Let us give an explicit example in which we calculate various zeta functions. We take

$$A=\mathbb{Q}G,\ \Lambda=\mathbb{Z}G,\ V=A,\ L=\Lambda,$$

where G is a cyclic group of prime order p. Let ω be a primitive p-th root of 1. Then $A\cong\mathbb{Q}\oplus\mathbb{Q}(\omega)$ as $\mathbb{Q}$-algebras, and such an isomorphism identifies the unique maximal order Λ' in A with $\mathbb{Z}\oplus\mathbb{Z}[\omega]$. We have $\Lambda_q=\Lambda'_q$ for all primes q different from p, and so

$$Z_\Lambda(\Lambda, M; s)/\zeta_{\Lambda'}(\Lambda'; s)=Z_{\Lambda_p}(\Lambda_p, M_p; s)/\zeta_{\Lambda'_p}(\Lambda'_p; s).$$

Further, there are precisely two genera of full Λ-lattices in A, and these can be represented by Λ and Λ'. Therefore

$$\zeta_\Lambda(\Lambda; s)=Z_\Lambda(\Lambda, \Lambda; s)+Z_\Lambda(\Lambda, \Lambda'; s),$$

so we need to compute

$$Z_{\Lambda_p}(\Lambda_p, \Lambda_p; s) \quad \text{and} \quad Z_{\Lambda_p}(\Lambda_p, \Lambda'_p; s).$$

We intend to use the methods of §2 for this purpose. Note that

$$\zeta_{\Lambda'}(\Lambda'; s) = \zeta_{\mathbb{Z}}(s) \cdot \zeta_{\mathbb{Z}[\omega]}(s),$$

the product of the Dedekind zeta functions of the fields $\mathbb{Q}$ and $\mathbb{Q}(\omega)$, respectively.

Since the remainder of our computation is carried out for the local case, it is better to change notation rather than repeat the subscript p incessantly. For the rest of this subsection, let

$$A = \mathbb{Q}_p G, \Lambda = \mathbb{Z}_p G, F = \mathbb{Q}_p(\omega), R = \mathbb{Z}_p[\omega].$$

We have identifications

$$A = \mathbb{Q}_p \oplus F, \Lambda' = \mathbb{Z}_p \oplus R.$$

The standard fibre product diagram for Λ then yields an identification

$$\Lambda = \{(\alpha, \beta) : \alpha \in \mathbb{Z}_p, \beta \in R, \alpha (\mathrm{mod}\, p) = \beta (\mathrm{mod}\, (1-\omega))\}.$$

Case 1. Calculation of $Z_\Lambda(\Lambda, \Lambda; s)$.

In this case we have $L = M = \Lambda$, $V = A$, and clearly

$$\mathrm{End}_A V = A, \mathrm{End}_\Lambda M = \Lambda, A^\times = \mathbb{Q}_p^\times \times F^\times.$$

The Haar measure $d^\times x$ on $A^\times$ may be taken as $(d^\times \alpha)(d^\times \beta)$, with $d^\times \alpha$ the Haar measure on $\mathbb{Q}_p^\times$, $d^\times \beta$ that on $F^\times$, normalized so that

$$\int_{\mathbb{Z}_p^\times} d^\times \alpha = 1, \int_{R^\times} d^\times \beta = 1.$$

Formula (11) gives

$$Z(\Lambda, \Lambda; s) = (p-1) \int_{\Lambda \cap A^\times} \|x\|_A^s \, d^\times x, \tag{18}$$

the factor $(p-1)$ arising from the measure of $\Lambda^\times$, $\mu(\Lambda^\times) = (\Lambda'^\times : \Lambda^\times)^{-1} = (p-1)^{-1}$.

The unique maximal ideal of the local ring Λ is given by

$$\mathrm{rad}\, \Lambda = p\mathbb{Z}_p \oplus (1-\omega) R,$$

and $\Lambda/\mathrm{rad}\, \Lambda \cong \mathbb{Z}/p\mathbb{Z}$. Thus, the units of Λ are precisely the elements of Λ not contained in $\mathrm{rad}\, \Lambda$. Hence we may write the range of integration in (18) as a disjoint union

$$\Lambda \cap A^\times = \Lambda^\times \cup (A^\times \cap \mathrm{rad}\, \Lambda).$$

This gives

$$Z(\Lambda, \Lambda; s) = 1 + (p-1) \int_{(\operatorname{rad} \Lambda) \cap A^\times} \|\alpha\|_{\mathbb{Q}_p}^s \|\beta\|_F^s d^\times \alpha d^\times \beta.$$

The latter integral is easily seen to be $p^{-2s}(1-p^{-s})^{-2}$, so we obtain

$$Z(\Lambda, \Lambda; s) = 1 + (p-1) p^{-2s}(1-p^{-s})^{-2} = (1-2p^{-s}+p^{1-2s})(1-p^{-s})^{-2}.$$

The last factor $(1-p^{-s})^{-2}$ is just $\zeta_{\Lambda'}(\Lambda'; s)$.

Case 2. Calculation $Z_\Lambda(\Lambda, \Lambda'; s)$.

Here we choose $V = A$, $L = \Lambda$, $M = \Lambda'$. Then

$$\operatorname{End}_\Lambda M = \Lambda', \{M:L\} = \{\Lambda' : \Lambda\} = \operatorname{rad} \Lambda.$$

Formula (11) gives

$$Z(\Lambda, \Lambda'; s) = p^s \int_{(\operatorname{rad} \Lambda) \cap A^\times} \|\alpha\|_{\mathbb{Q}_p}^s \|\beta\|_F^s d^\times \alpha d^\times \beta,$$

since $(\Lambda')^\times$ has measure 1 and $(\Lambda' : \Lambda) = p$. As above, we obtain

$$Z(\Lambda, \Lambda'; s) = p^s \cdot p^{-2s}(1-p^{-s})^{-2} = p^{-s}(1-p^{-s})^{-2}.$$

Now let $\mathcal{M} = \mathbb{Z} \oplus \mathbb{Z}[\omega]$ denote the maximal order in $\mathbb{Q}G$, so

$$\zeta_{\mathcal{M}}(\mathcal{M}; s) = \zeta_{\mathbb{Z}}(s) \cdot \zeta_{\mathbb{Z}[\omega]}(s),$$

as remarked earlier. Then our calculations yield

$$\zeta_\Lambda(\Lambda; s)/\zeta_{\mathcal{M}}(\mathcal{M}; s) = (1-2p^{-s}+p^{1-2s}) + p^{-s},$$

so

$$\zeta_\Lambda(\Lambda; s) = (1-p^{-s}+p^{1-2s})\, \zeta_{\mathbb{Z}}(s)\, \zeta_{\mathbb{Z}[\omega]}(s).$$

This example was worked out in [6] and [5] by other methods.

§4. Solomon's Conjectures

§4.1. Main Theorem

In this subsection we work purely locally, and assume throughout that Λ is a $\mathbb{Z}_p$-order in a finite dimensional semisimple $\mathbb{Q}_p$-algebra A. Let L, M be full Λ-lattices in a finitely generated left A-module V. Let Λ' be a maximal $\mathbb{Z}_p$-order in A. Our first main result is as follows.

Theorem 1. *With the above notation, the function*

$$Z_\Lambda(L, M; s)/\zeta_{\Lambda'}(\Lambda' L; s)$$

is a polynomial in the argument p^{-s}, *with rational integer coefficients. Here,*

$$Z_\Lambda(L, M; s) = \sum_{\substack{N \subseteq L \\ N \cong M}} (L:N)^{-s}, \qquad N = \Lambda\text{-}lattice,$$

and

$$\zeta_{\Lambda'}(\Lambda' L; s) = \sum_{N \subseteq \Lambda' L} (\Lambda' L:N)^{-s}, \qquad N = \Lambda'\text{-}lattice.$$

Remark. The above theorem reflects the fact that (in the terminology of [3]) the "Euler factor" attached to the trivial representation of $\mathrm{Aut}_A V$ is $\zeta_{\Lambda'}(\Lambda' L; s)$.

To begin the proof of Theorem 1, recall that a Schwartz-Bruhat (or "test") function on a finite dimensional $\mathbb{Q}_p$-space W is a locally constant complex-valued function on W with compact support. (These functions are called "standard" in [9]). We intend to establish

Lemma. *Let* Φ *be a Schwartz-Bruhat function on* $\mathrm{End}_A V$. *Then the function*

$$\{\zeta_{\Lambda'}(\Lambda' L; s)\}^{-1} \int_{\mathrm{Aut}_A V} \Phi(x) \|x\|_V^s \, d^\times x$$

is a polynomial in p^s *and* p^{-s} *with complex coefficients.*

Before proving this Lemma, let us show how to deduce Theorem 1 from it. Using formula (11), we see that the Lemma implies that

$$\{\zeta_{\Lambda'}(\Lambda' L; s)\}^{-1} Z_\Lambda(L, M; s) \in \mathbb{C}[p^s, p^{-s}].$$

But $\{\zeta_{\Lambda'}(\Lambda' L, s)\}^{-1} \in \mathbb{Z}[p^{-s}]$ by §3.3. Further, we know that[1] $Z_\Lambda(L, M; s) \in \mathbb{Z}[[p^{-s}]]$, by virtue of its original definition as an infinite series $\sum_N (L:N)^{-s}$. Therefore

$$Z_\Lambda(L, M; s)/\zeta_{\Lambda'}(\Lambda' L; s) \in \mathbb{C}[p^s, p^{-s}] \cap \mathbb{Z}[[p^{-s}]].$$

This intersection is precisely $\mathbb{Z}[p^{-s}]$, and the Theorem is proved.

To prove the Lemma, we must first recall something about the nature of Schwartz-Bruhat functions on a finite-dimensional $\mathbb{Q}_p$-vector space W. As in [9], it is easy enough to see that, given such a function Φ, there are full $\mathbb{Z}_p$-lattices X, Y in W, with $X \subset Y$, such that Φ is constant on cosets $\mathrm{mod}\, X$ and is identically zero outside Y. Therefore the vector space $\mathscr{S}(W)$ of Schwartz-Bruhat functions on W is spanned by the characteristic functions of "spheres", that is, sets of the form $w + X$, for $w \in W$ and X a full $\mathbb{Z}_p$-lattice in W. One can refine this somewhat. For, given two full $\mathbb{Z}_p$-lattices X_0, X in W, there is an integral power p^f of p so that $p^f X_0 \subset X$. So $\mathscr{S}(W)$ is spanned by the characteristic functions of spheres of the form $w + p^f X_0$, for varying $w \in W$, $f \in \mathbb{Z}$, $f \geq 0$, and a fixed lattice X_0.

Now return to the notation of the Lemma, and set

$$B = \mathrm{End}_A V, \; \Gamma' = \mathrm{End}_{\Lambda'}(\Lambda' L),$$

[1] $\mathbb{Z}[[p^{-s}]]$ denotes the ring of formal power series in p^{-s} with coefficients in $\mathbb{Z}$

so Γ' is a maximal order in B. We observe next that for $f \in \mathbb{Z}$,

$$\int_{B^\times} \Phi(x) \|x\|_V^s d^\times x = \int_{B^\times} \Phi(p^f x) \|p^f x\|_V^s d^\times x$$
$$= \|p^f\|_V^s \int_{B^\times} \Phi(p^f x) \|x\|_V^s d^\times x.$$

The support of the function $x \mapsto \Phi(p^f x)$, $x \in B$, is p^{-f} supp Φ. Further, $\|p^f\|_V$ is an integral power p^l for some $l \in \mathbb{Z}$. Hence, at the cost of introducing a factor p^{ls} for some $l \in \mathbb{Z}$, we can assume that the support of Φ is contained in the lattice Γ'. By linearity, we can therefore assume that Φ is the characteristic function of a sphere $\alpha + p^f \Gamma'$ for some $\alpha \in \Gamma'$ and some $f \geqq 0$. Since Γ' is a maximal order in B, such a function Φ decomposes as a product of corresponding functions over the simple components of B. Thus the integral likewise decomposes as a product of integrals of the same type over the simple components of B. There is a corresponding decomposition of $\zeta_{\Lambda'}(\Lambda' L; s)$, so for the rest of the proof of the Lemma, we may as well assume that B is *simple*. This reduces the problem to the situation of §3.3, so we adopt the following notation:

$$A = M_m(D), \quad D = \text{division algebra with center } F, \dim_F D = e^2,$$
$$\Lambda' = M_m(\Delta), \quad \Delta = \text{unique maximal order in } D,$$
$$V = D^{mk}, \Lambda' L = \Delta^{mk}.$$

Then

$$B = \operatorname{End}_A V \cong M_k(D), \Gamma' = \operatorname{End}_{\Lambda'}(\Lambda' L) \cong M_k(\Delta),$$

so

$$B^\times = GL_k(D), (\Gamma')^\times = GL_k(\Delta).$$

As in §3.3, let R be the valuation ring of F, $P = \pi R$ the maximal ideal of R, and $\mathcal{N}P = (R:P)$. Further, let

$$\|x\|_V = \|nr\, x\|_F^{me}, \quad x \in B^\times,$$

and let ξ denote a prime element of Δ. Finally, let Φ be the characteristic function in B of a sphere $\alpha + p^f \Gamma'$, for some $\alpha \in \Gamma'$ and some $f \geqq 0$.

In order to prove the Lemma, it suffices to show

$$\{\zeta_{\Lambda'}(\Lambda' L; s)\}^{-1} \int_{(\alpha + p^f \Gamma') \cap B^\times} \|nr\, x\|_F^{mes} d^\times x \in \mathbb{C}[p^{-s}]. \tag{19}$$

We can translate the range of integration on either side by a factor from $(\Gamma')^\times$ without changing the value of the integral. This allows us to reduce α to diagonal form

$$\alpha = \operatorname{diag}(\xi^{j_1}, \xi^{j_2}, \ldots, \xi^{j_r}, 0, \ldots, 0),$$

where $j_1 \leqq j_2 \leqq \ldots \leqq j_r$, and no ξ^{j_i} is divisible by p^f. Let $\beta = \operatorname{diag}(\beta_1, \ldots, \beta_k)$ where $\beta_i = \xi^{-j_i}$ for $1 \leqq i \leqq r$, and $\beta_j = 1$ for $j > r$. Then our integral is a power of p^{-s}

multiplied by the same integral taken over the set $(\beta\alpha+\beta p^f\Gamma')\cap B^\times$. Notice that $\beta p^f\Gamma'\subseteq\Gamma'$. We may now decompose $\beta\alpha+\beta p^f\Gamma'$ as a disjoint union of spheres $\gamma+p^f\Gamma'$. From the way we have constructed them, these elements γ, when reduced to diagonal form as for α, have the shape $\operatorname{diag}(1,1,\dots,1,0,\dots,0)$, with r ones on the diagonal. We just have to check that

$$\{\zeta_{\Lambda'}(\Lambda' L;s)\}^{-1}\cdot\int_{(\gamma+p^f\Gamma')\cap B^\times}\|x\|_V^s\,d^\times x$$

is a polynomial in p^{-s}. Now, the integral is unchanged in value when we translate the range of integration on the left by an element of $(\Gamma')^\times$. The range of integration itself is invariant under left-translation by elements of $(\Gamma')^\times$ which are $\equiv 1\,(\operatorname{mod} p^f)$. Such elements form a subgroup of $(\Gamma')^\times$ of finite index. Therefore our integral is a finite non-zero constant multiple of

$$\int_{(\Gamma')^\times\cdot\{(\gamma+p^f\Gamma')\cap B^\times}\|x\|_V^s\,d^\times x.$$

This new range of integration is a disjoint union of cosets $(\Gamma')^\times\delta$, where δ runs over those matrices in Γ' satisfying the Hermite normal form conditions of §3.3, and looking like

$$\begin{bmatrix} 1 & 0\dots0 & & & \\ 0 & 1 \ddots \vdots & & & \\ \vdots & \ddots\ddots 0 & & & \\ 0\dots & 0\;1 & & & \\ \vdots & & 0 & \xi^{j_{r+1}} & \\ \vdots & & & \ddots & \ddots \\ 0\dots & \dots & & 0\dots & \xi^{j_k} \end{bmatrix},$$

with all entries, apart from the r diagonal ones, divisible by p^f.

We recall that $\|x\|_V$ is just $\|nrx\|_F^{me}$. Suppose that $p^f\Delta=\xi^a\Delta$. Then our integral can be arranged as a power series

$$(\mathcal{N}P)^{-ames(k-r)}\sum_{n=0}^{\infty}c(n)(\mathcal{N}P)^{-mesn}, \tag{20}$$

where for $n\geq 0$, the coefficient $c(n)$ is given by

$$c(n)=\sum_{(b_{r+1},\dots,b_k)}\prod_{i=r+1}^{k}(\mathcal{N}P)^{eb_i(i-1)}.$$

Here, the summation extends over all $(k-r)$-tuples $(b_{r+1},\dots,b_k)$ of nonnegative integers whose sum is n. Apart from the first factor, the expression in (20) is precisely the product of the last $k-r$ factors in the expression for $\zeta_{\Lambda'}(\Lambda' L;s)$ obtained from formula (16) in §3.3. This completes the proof of the Lemma, and of Theorem 1 as well.

§4.2 Proof of Solomon's First Conjecture

In order to prove this conjecture, we begin by reviewing Solomon's definition of $\zeta_V(s)$. Suppose for the moment that A is a finite dimensional *simple* $\mathbb{Q}$-algebra, W a simple left A-module, and $V=W^k$. We have $A=M_m(D)$ for some division algebra D whose center F is a finite extension of $\mathbb{Q}$. Let $\dim_F D=e^2$, and let R be the ring of integers in F, so that $\zeta_R(s)=\zeta_R(R;s)$ is the Dedekind zeta function of F. Then by definition

$$\zeta_V(s)=\prod_{j=0}^{ke-1}\zeta_R(mes-j).$$

If A is a semisimple $\mathbb{Q}$-algebra, V a finitely generated left A-module, we write $A=\prod_i A_i$, with the A_i simple, $V_i=A_i V$, and then define

$$\zeta_V(s)=\prod_i \zeta_{V_i}(s).$$

Exactly the same definition may be used when A is a $\mathbb{Q}_p$-algebra. So, in our original global situation, we can define $\zeta_{V_p}(s)$ by passing to the completion.

On the other hand, the Euler product for $\zeta_R(s)$ can be arranged as a product over the rational primes p:

$$\zeta_R(s)=\prod_p\prod_{P|p}(1-\mathcal{N}P^{-s})^{-1},$$

where the inner product is taken over those primes P of F above p. We can now assemble these p-factors to give us an Euler product

$$\zeta_V(s)=\prod_p \zeta_{V,p}(s),$$

$\zeta_{V,p}(s)$ being the p-part of $\zeta_V(s)$.

Lemma. $\zeta_{V_p}(s)=\zeta_{V,p}(s)$.

Proof. We need only prove this under the assumption that A is simple, as above. Let P range over the primes of F above p. Then

$$\zeta_{V,p}(s)=\prod_P\prod_{j=0}^{ke-1}\zeta_{R_P}(mes-j),$$

where, of course $\zeta_{R_P}(s)=(1-\mathcal{N}P^{-s})^{-1}$.

We have $A_p=\prod_P A_P$, and each A_P is a simple $\mathbb{Q}_p$-algebra. Indeed, $A_P=M_{m_P}(D(P))$, for some central F_P-division algebra $D(P)$. If we put $e_P^2=\dim_{F_P} D(P)$, we have $m_P e_P=m e$. The module $V_P=A_P\cdot V_p$ is $D(P)^{m_P k_P}$, identifying the simple A_P-module with $D(P)^{m_P}$. However, V_P is also the completion of V at P, and comparison of dimensions shows that the integer k_P satisfies $k_P e_P=k e$.

Therefore

$$\zeta_{V_P}(s)=\prod_{j=0}^{k_P e_P-1}\zeta_{R_P}(m_P e_P s-j)=\prod_{j=0}^{ke-1}\zeta_{R_P}(mes-j).$$

Since, by definition, $\zeta_{V_p}(s)=\prod_P \zeta_{V_P}(s)$, the Lemma follows.

Now take a maximal order Λ' in the simple $\mathbb{Q}$-algebra A, and choose a full Λ'-lattice L in V. We can now show that $\zeta_{\Lambda'_p}(L_p; s)/\zeta_{V_p}(s)\in\mathbb{Z}[p^{-s}]$. For this, it is enough to show $\zeta_{\Lambda'_P}(L_P; s)/\zeta_{V_P}(s)\in\mathbb{Z}[p^{-s}]$ for each P. But formula (16) gives an explicit expression for $\zeta_{\Lambda'_P}(L_P; s)$. Using the notation above, it is

$$\zeta_{\Lambda'_P}(L_P; s)=\prod_{j=0}^{k_P-1}\zeta_{R_P}(m_P e_P s-e_P j).$$

We know that $m_P e_P=me$, so the quotient $\zeta_{\Lambda'_P}/\zeta_{V_P}$ is a product of functions $(1-\mathcal{N}P^{n-mes})$, for various positive integers n, and certainly lies in $\mathbb{Z}[p^{-s}]$.

Now let A be a semisimple $\mathbb{Q}$-algebra, Λ' a maximal order in A. By using the above result at each simple component of A, we conclude that

$$\zeta_{\Lambda'_p}(L'_p;\ s)/\zeta_{V_p}(s)\in\mathbb{Z}[p^{-s}] \qquad \text{for each } p,$$

for each full Λ'-lattice L' in V. Combining this fact with Theorem 1 and formula (4), we obtain the following result:

Corollary (Solomon's First Conjecture). *Let Λ be a $\mathbb{Z}$-order in a finite dimensional semisimple $\mathbb{Q}$-algebra A, and let L, M be full Λ-lattices in a finitely generated left A-module V. Then for each prime p, both*

$$Z_{\Lambda_p}(L_p, M_p; s)/\zeta_{V,p}(s) \quad \text{and} \quad \zeta_{\Lambda_p}(L_p; s)/\zeta_{V,p}(s)$$

lie in $\mathbb{Z}[p^{-s}]$.

§4.3 Solomon's Second Conjecture

This second conjecture concerns the functions $Z_\Lambda(L, M; s)$ in the local case. Let Λ be an order in the $\mathbb{Q}_p$-algebra A, and let $L_1,\ldots,L_n$ be a complete set of non-isomorphic full Λ-lattices in a given A-module V. Consider the $n\times n$ matrix

$$\mathbf{B}=[Z_\Lambda(L_i, L_j; s)]_{1\leqq i,\, j\leqq n}.$$

Solomon constructed an $n\times n$ matrix $\mathbf{A}$, with entries in $\mathbb{Z}[p^{-s}]$, such that $\mathbf{AB}=\mathbf{I}$. This implies at once that $(\det\mathbf{B})^{-1}\in\mathbb{Z}[p^{-s}]$. Solomon's second conjecture is that $(\det\mathbf{B})^{-1}$ has the form

$$(\det\mathbf{B})^{-1}=\prod_i(1-p^{a_i-b_i s}), \tag{21}$$

a finite product in which each a_i is a non-negative integer, and each b_i a positive integer ([7]).

We support this conjecture by an easy application of Theorem 1. Let Λ' be a maximal order in A; by §3.3, we may write (for each i)

$$1/\zeta_{\Lambda'}(\Lambda' L_i; s) = \prod_n (1 - p^{c_n - d_n s}) = G(s), \qquad \text{say,}$$

where $G(s)$ is a finite product in which the integers c_n, d_n satisfy $c_n \geqq 0$, $d_n > 0$ for each n. Note that $G(s)$ is independent of i, by §3.3. On the other hand, Theorem 1 tells us that for each i and j,

$$Z_\Lambda(L_i, L_j; s) = F_{ij}(s)/G(s),$$

where each $F_{ij}(s) \in \mathbb{Z}[p^{-s}]$. Hence

$$\det \mathbf{B} = \{\det[F_{ij}(s)]\}\, G(s)^{-n},$$

so $(\det \mathbf{B})^{-1} = G(s)^n / F(s)$, where $F(s) = \det[F_{ij}(s)] \in \mathbb{Z}[p^{-s}]$. Since we already know that $(\det \mathbf{B})^{-1} \in \mathbb{Z}[p^{-s}]$, it follows that $(\det \mathbf{B})^{-1}$ is a divisor of a product of the type occurring in (21).

§5. Local Functional Equations

§5.1 Simple Algebras

For this section, let A be a finite-dimensional *simple* $\mathbb{Q}_p$-algebra, and F a subfield of the center of A containing $\mathbb{Q}_p$, so that we may think of A as an F-algebra. We write $\|x\|$ for what we would previously have called $\|x\|_A$. Let t denote the trace $A \to F$, and let $\lambda \in F^\times$. Then the pairing

$$(x, y) \mapsto \lambda\, t(x y), \qquad x, y \in A,$$

is nondegenerate, and identifies A with its linear dual $\mathrm{Hom}_F(A, F)$.

Let χ be a non-trivial continuous character of the additive group of F. Then the pairing

$$(x, y) \mapsto \chi(\lambda\, t(x y))$$

is also nondegenerate, and identifies A with its *Pontrjagin dual* $\hat{A}$. With this identification, we can define the *Fourier transform* $\hat{\Phi}$ of a Schwartz-Bruhat function Φ on A by

$$\hat{\Phi}(y) = \int_A \Phi(x)\, \chi(\lambda\, t(x y))\, dx.$$

Here, dx is a certain Haar measure on A. It is easy to verify that $\hat{\Phi}$ is again a Schwartz-Bruhat function on A. The Haar measure dx is chosen so that the Fourier inversion formula $\hat{\hat{\Phi}}(x) = \Phi(-x)$ holds. This measure is thereby uniquely determined, relative to the given identification of A with $\hat{A}$, and is called the *self-dual measure* for this identification.

Local Functional Equation. With the notation as above, let Φ, Ψ be Schwartz-Bruhat functions on A. Then, for a Haar measure $d^\times x$ on $A^\times$, the zeta integral $Z(\Phi; s)$ defined by

$$Z(\Phi; s) = \int_{A^\times} \Phi(x) \, \|x\|^s \, d^\times x$$

admits analytic continuation to a meromorphic function on the whole s-plane, and

$$\frac{Z(\Phi; s)}{Z(\hat{\Phi}; 1-s)} = \frac{Z(\Psi; s)}{Z(\hat{\Psi}; 1-s)}.$$

This theorem is in Tate's thesis [8] in the case A a field. Tate's proof works equally well when A is a division ring. When A is an arbitrary simple algebra, the result can be obtained as a special case of [3, p. 125, 9.6], by taking π to be the trivial representation of $A^\times$, and observing that our s is ns in [3], where n^2 is the dimension of A over its center. The results of [3], however, are much more general (and complicated) than we need here, so we include a simple *ad hoc* proof of the above result as an Appendix.

It is still useful to remark, at this stage, that the existence of analytic continuation for $Z(\Phi; s)$ follows from our explicit computations in §3.3 and the Lemma of §4.1.

§5.2 *Semisimple Algebras*

Now let F be a finite field extension of $\mathbb{Q}_p$, and let A be a finite dimensional semisimple F-algebra. As before, if Φ is a Schwartz-Bruhat function on A and $d^\times x$ is a Haar measure on $A^\times$, write

$$Z(\Phi; s) = \int_{A^\times} \Phi(x) \, \|x\|_A^s \, d^\times x.$$

Let $A = \prod_{i=1}^{r} A_i$, where each A_i is a simple algebra. Let t_i denote the trace $A_i \to F$, $\lambda_i \in F^\times$, and put $t = \sum_{i=1}^{r} \lambda_i t_i \colon A \to F$. The pairing $(x, y) \mapsto t(xy)$ identifies A with its linear dual, and composing it with a non-trivial additive character χ of F induces a topological group isomorphism $A \cong \hat{A}$. This isomorphism is the direct product of isomorphisms $A_i \cong \hat{A}_i$. Using this identification, we define Fourier transforms as before, namely.

$$\hat{\Phi}(y) = \int_A \Phi(x) \, \chi(t(xy)) \, dx.$$

Again, dx is the self-dual Haar measure on A for the given identification of $\hat{A}$ with A. We have $dx = dx_1 \, dx_2 \ldots dx_r$, where dx_i is the self-dual Haar measure on A_i relative to the pairing $(x_i, y_i) \to \chi(\lambda_i t_i(x_i y_i))$, $x_i, y_i \in A_i$.

Call a Schwartz-Bruhat function Φ on A *decomposable* if it can be written in the form

$$\Phi=\Phi_1\otimes\Phi_2\otimes\ldots\otimes\Phi_r\colon(x_1,\ldots,x_r)\mapsto\prod_{i=1}^{r}\Phi_i(x_i),$$

for Schwartz-Bruhat functions Φ_i on A_i. With the measures as above, we have $\hat{\Phi}=\hat{\Phi}_1\otimes\ldots\otimes\hat{\Phi}_r$ for such a Φ. We choose Haar measures $d^\times x_i$ on $A_i^\times$ so that $d^\times x=(d^\times x_1)(d^\times x_2)\ldots(d^\times x_r)$. Then

$$Z(\Phi;s)=\prod_{i=1}^{r}Z(\Phi_i;s).$$

Further, if Ψ is another decomposable Schwartz-Bruhat function on A, the result of §5.1 implies that

$$\frac{Z(\Phi;s)}{Z(\hat{\Phi};1-s)}=\frac{Z(\Psi;s)}{Z(\hat{\Psi};1-s)}.$$

The existence of analytic continuation for $Z(\Phi;s)$ follows as before, whether or not Φ is decomposable. We also have, by linearity,

$$\frac{Z(\Phi;s)}{Z(\hat{\Phi};1-s)}=\frac{Z(\Psi;s)}{Z(\hat{\Psi};1-s)}=\frac{Z(\Phi+\Psi;s)}{Z(\hat{\Phi}+\hat{\Psi};1-s)}.$$

We assert that any Schwartz-Bruhat function on A is a finite sum of decomposable functions. For, choose a full $\mathbb{Z}_p$-lattice L_i in each A_i, and put $L=\bigoplus_{i=1}^{r}L_i$. Any Schwartz-Bruhat function is a linear combination of characteristic functions of spheres $x+p^fL$, $f\geqq 0$, $x\in A$. Such a characteristic function is certainly decomposable. We thus conclude:

Proposition 2. *Given any Schwartz-Bruhat functions Ψ, Φ on A, the zeta-integrals $Z(\Phi;s)$, $Z(\Psi;s)$ admit analytic continuation to meromorphic functions of s, and satisfy:*

$$\frac{Z(\Phi;s)}{Z(\hat{\Phi};1-s)}=\frac{Z(\Psi;s)}{Z(\hat{\Psi};1-s)}.$$

§5.3 Group Rings

Let G be a finite group, F a finite extension field of $\mathbb{Q}_p$, R the ring of integers of F, and P the maximal ideal of R. We set $A=FG$, $\Lambda=RG$, and let Λ' be a maximal order in A containing Λ. *Changing our earlier notation somewhat*, we define

$$\zeta_\Lambda(s)=\sum_{N\subseteq\Lambda}(\Lambda:N)^{-s},\qquad Z_\Lambda(M;s)=\sum_{\substack{N\subseteq\Lambda\\ N\cong M}}(\Lambda:N)^{-s},\tag{22}$$

where M is some fixed full left Λ-lattice in A. In both summations, N ranges over full sublattices of Λ (that is, full left ideals of Λ). In terms of our earlier notation, $\zeta_\Lambda(s)$ would have been written as $\zeta_\Lambda(\Lambda; s)$, and $Z_\Lambda(M; s)$ as $Z_\Lambda(\Lambda, M; s)$.

Now let

$$\{M:\Lambda\}=\{x\in A: Mx\subseteq \Lambda\}\cong \mathrm{Hom}_\Lambda(M,\Lambda).$$

Let $x\mapsto x^*$ be the involution on A which fixes F, and maps each element of G onto its inverse. We set

$$\tilde{M}=\{M:\Lambda\}^*,$$

so (after obvious identifications) $\tilde{M}$ is a full left Λ-lattice in A, called the *contragredient* of M.

Our main result is as follows:

Theorem 2. *With the notation as above, we have*

$$Z_\Lambda(M; s)/Z_\Lambda(\tilde{M}; 1-s)=(\Lambda':\Lambda)^{1-2s}\zeta_{\Lambda'}(s)/\zeta_{\Lambda'}(1-s),$$
$$\zeta_\Lambda(s)/\zeta_\Lambda(1-s)=(\Lambda':\Lambda)^{1-2s}\zeta_{\Lambda'}(s)/\zeta_{\Lambda'}(1-s).$$

To begin the proof, let $t: A\to F$ be the F-linear map sending 1_G to 1, and all other elements of G to 0. Then t is a linear combination of traces, as in §5.2. We identify A with its linear dual via the pairing $(x, y)\mapsto t(xy)$. On the other hand, choose a continuous additive character χ on F which is trivial on R but not on P^{-1}. We then identify A with its Pontrjagin dual $\hat{A}$ via the pairing $(x, y)\mapsto \chi(t(xy))$.

These pairings have some particularly useful properties. If L is a full R-lattice in A, we have

$$\{x\in A: Lx\subseteq \Lambda\}=\{x\in A: t\ (Lx)\subseteq R\}\cong \mathrm{Hom}_R(L, R);$$

we treat the isomorphism as an identification. In particular, Λ is self-dual, and the linear dual of M is $\{M:\Lambda\}$. Moreover $\{M:\Lambda\}$ is the "orthogonal complement" of M under the pairing $(x, y)\mapsto \chi(t(xy))$.

Now we apply the functional equation of Proposition 2 to particular choices of the functions Φ, Ψ. Fourier transforms are to be taken with respect to the self-dual measure on A. We choose a Haar measure $d^\times x$ on $A^\times$. For measurable sets $E\subset A$, $E'\subset A^\times$, it will be convenient to write

$$\mu(E)=\int_E dx, \qquad \mu^\times(E')=\int_{E'} d^\times x.$$

Let Ψ be the characteristic function of the maximal order Λ'. Then, using formula (11),

$$Z(\Psi; s)=\mu^\times(\Lambda'^\times)\,\zeta_{\Lambda'}(s).$$

A simple computation shows that $\hat{\Psi}$ is the characteristic function of $\{\Lambda':\Lambda\}$, multiplied by the constant factor $\mu(\Lambda')$. Further, the measure $\mu(\Lambda')$ is just $(\Lambda':\Lambda)\,\mu(\Lambda)$. Fourier inversion applied to the characteristic function of Λ shows that $\mu(\Lambda)=1$, and therefore $\mu(\Lambda')=(\Lambda':\Lambda)$. On the other hand, $\{\Lambda':\Lambda\}$ is a full left Λ'-lattice in A, so $\{\Lambda':\Lambda\}=\Lambda'\alpha$ for some $\alpha\in A^\times$. Altogether, we obtain

$$\hat{\Psi}(x)=(\Lambda':\Lambda)\,\Psi(x\alpha^{-1}),$$

$$Z(\hat{\Psi};s)=(\Lambda':\Lambda)\int_{A^\times}\Psi(x\alpha^{-1})\,\|x\|^s d^\times x=(\Lambda':\Lambda)\int_{A^\times}\Psi(x)\,\|x\alpha\|^s d^\times x$$

$$=(\Lambda':\Lambda)\,\|\alpha\|^s\int_{A^\times}\Psi(x)\,\|x\|^s d^\times x=(\Lambda':\Lambda)\,\mu^\times(\Lambda'^\times)\,\|\alpha\|^s\,\zeta_{\Lambda'}(s).$$

However,

$$\|\alpha\|=(\Lambda'\alpha:\Lambda')=(\Lambda'\alpha:\Lambda)(\Lambda:\Lambda')$$

There is a natural duality (with respect to F/R) between the finite R-modules Λ'/Λ and $\Lambda/\Lambda'\alpha$, since $\Lambda'\alpha=\{\Lambda':\Lambda\}$. Therefore these two modules have the same cardinality, and it follows that $\|\alpha\|=(\Lambda':\Lambda)^{-2}$. Consequently we obtain

$$Z(\hat{\Psi};s)=(\Lambda':\Lambda)^{1-2s}\,\mu^\times(\Lambda'^\times)\,\zeta_{\Lambda'}(s),$$

and therefore

$$Z(\Psi;s)/Z(\hat{\Psi};1-s)=(\Lambda':\Lambda)^{1-2s}\,\zeta_{\Lambda'}(s)/\zeta_{\Lambda'}(1-s).$$

Now let Φ be the characteristic function of $\tilde{M}$, where M is a full left Λ-lattice in A. Then

$$Z(\Phi;s)=\int_{A^\times}\Phi(x)\,\|x\|^s d^\times x=\int_{A^\times}\Phi(x^*)\,\|x^*\|^s d^\times x,$$

since the map $x\mapsto x^*$ leaves the measure invariant. But $\|x^*\|=\|x\|$, so

$$Z(\Phi;s)=\int_{A^\times}\Phi(x^*)\,\|x\|^s d^\times x.$$

Since $x\mapsto\Phi(x^*)$ is the characteristic function of $\{M:\Lambda\}$, we obtain

$$Z(\Phi;s)=\mu^\times(\mathrm{Aut}_\Lambda M)\cdot(\Lambda:M)^s\cdot Z_\Lambda(M;s).$$

The Fourier transform of Φ is the characteristic function of $\{\tilde{M}:\Lambda\}$, multiplied by the constant factor $\mu(\tilde{M})$, and this factor is equal to the index $(\tilde{M}:\Lambda)$. Therefore

$$Z(\hat{\Phi};s)=(\tilde{M}:\Lambda)(\Lambda:\tilde{M})^s\,\mu^\times(\mathrm{Aut}_\Lambda\tilde{M})\,Z_\Lambda(\tilde{M};s).$$

Now $\mathrm{Aut}_\Lambda\tilde{M}=\mathrm{Aut}_\Lambda M$ since $\tilde{M}$ is the contragredient of M, and also $(\tilde{M}:\Lambda)=(\Lambda:M)$ since $\tilde{M}$ is the dual of M and $\tilde{\Lambda}=\Lambda$. Therefore we obtain

$$Z(\Phi;s)/Z(\hat{\Phi};1-s)=Z_\Lambda(M;s)/Z_\Lambda(\tilde{M};1-s),$$

which establishes the first assertion in Theorem 2.

Next, as M ranges over a complete set of non-isomorphic full left Λ-lattices in A, so does $\tilde{M}$ (since $\tilde{\tilde{M}}=M$). Hence when we sum the above equation over M, we obtain the second assertion of Theorem 2.

Remarks. 1) One can obviously generalize this argument to any order Λ which is self-dual under an appropriate trace form. This yields a functional equation of the form

$$(\Lambda':\Lambda)^{1-2s}\zeta_{\Lambda'}(s)/\zeta_{\Lambda'}(1-s)=\zeta_{\Lambda}(s)/\zeta^{r}_{\Lambda}(s),$$

where $\zeta^{r}_{\Lambda}(s)$ denotes the zeta function formed with *right* Λ-lattices. We do not know whether $\zeta_{\Lambda}=\zeta^{r}_{\Lambda}$ always.

2) The quotient $\zeta_{\Lambda'}(s)/\zeta_{\Lambda'}(1-s)$ is easily computed using §3.3.

3) One can rearrange the functional equation of Theorem 2 by setting

$$\phi(s)=\zeta_{\Lambda}(s)/\zeta_{\Lambda'}(s).$$

The Theorem then asserts that

$$\phi(s)=(\Lambda':\Lambda)^{1-2s}\phi(1-s).$$

We known that $\phi(s)$ is a polynomial in p^{-s}, with integer coefficients and constant term 1. The functional equation expresses a symmetry among the coefficients of this polynomial, and shows in particular that its highest term is $p^{n(1-2s)}$, where $(\Lambda':\Lambda)=p^{n}$. This is consistent with a "Riemann hypothesis" for $\zeta_{\Lambda}(s)$. Note that $n=1$ in the Example in §3.4.

4) In [1], a functional equation is given for $\zeta_{\Lambda'}(s)$, when Λ' is a maximal order in a division algebra. From this, it is now easy to deduce a functional equation for $\zeta_{RG}(RG;s)$ when R is the ring of integers of an algebraic number field.

5) For other functional equations, see [2].

Appendix

Proof of the Local Functional Equation

We give here a straightforward proof of the "Local Functional Equation" of §5.1. We use the notation of that subsection, so that in particular, A is a *simple* $\mathbb{Q}_p$-algebra. The theorem states that, for any Schwartz-Bruhat function Φ on A, the zeta-integral $Z(\Phi;s)$ admits analytic continuation to a meromorphic function of s, and, given any other Schwartz-Bruhat function Ψ on A, we have

$$\frac{Z(\Phi;s)}{Z(\hat{\Phi};1-s)}=\frac{Z(\Psi;s)}{Z(\hat{\Psi};1-s)}.$$

As we have already observed, the existence of analytic continuation follows from §3.3 and the Lemma of §4.1. Further, by linearity, it is enough to check the functional equation when Φ, Ψ are taken from a set which spans the vector space of Schwartz-Bruhat functions on A. A convenient spanning set is obtained by choosing a maximal order Λ in A, and taking the characteristic functions of all sets of the form $\alpha+M$, where $\alpha\in A$, and M is a full left Λ-lattice in A. Any such lattice M is Λ-free of rank 1, so we only have to check the functional equation for pairs Φ, Ψ of the following types:

(i) Φ arbitrary, Ψ given by $\Psi(x)=\Phi(x\alpha)$, for some $\alpha\in A^{\times}$;

(this case implies the functional equation when Φ and Ψ are both characteristic functions of Λ-lattices)

(ii) Φ is the characteristic function of a full left Λ-lattice M, and Ψ is the characteristic function of $\alpha+M$, for some $\alpha\in A$.

Case (i) is easy. For,

$$Z(\Psi;s)=\int_{A^{\times}}\Phi(x\alpha)\,\|x\|^{s}d^{\times}x=\int\Phi(x)\,\|x\alpha^{-1}\|^{s}d^{\times}x$$
$$=\|\alpha\|^{-s}Z(\Phi;s).$$

On the other hand,

$$\hat{\Psi}(y)=\int_{A}\Phi(x\alpha)\,\chi(\lambda\,t(x\,y))\,dx$$
$$=\|\alpha\|^{-1}\int_{A}\Phi(x)\,\chi(\lambda\,t(x\alpha^{-1}y))\,dx,$$

since $d(x\alpha)=\|\alpha\|\cdot dx$, and this last expression is just $\|\alpha\|^{-1}\hat{\Phi}(\alpha^{-1}y)$. Therefore we obtain $Z(\hat{\Psi};s)=\|\alpha\|^{s-1}Z(\hat{\Phi};s)$, and the result.

Remark. The symmetry of the trace form enables the same proof to be applied to the case

(i)′ $\Psi(x)=\Phi(\alpha x)$, for some $\alpha\in A^{\times}$.

In Case (ii), we put $N=\Lambda\alpha+M$, and work by induction on the Jordan-Hölder length of the finite Λ-module N/M. When this length is zero, we have $\Phi=\Psi$, and there is nothing to prove. So we assume that $N\neq M$, and choose a set of coset representatives of N/M containing α. It will be convenient to write Φ_{β} for the characteristic function of $\beta+M$, so that $\Psi=\Phi_{\alpha}$, $\Phi=\Phi_{0}$. We divide our set of coset representatives into two subsets E_1, E_2 by:

$$\beta\in E_1 \quad \text{if } \Lambda\beta+M=N,$$
$$\beta\in E_2 \quad \text{otherwise.}$$

Let Ξ be the characteristic function of N so that

$$\Xi=\sum_{\beta\in N/M}\Phi_{\beta}=\sum_{\beta\in E_1}\Phi_{\beta}+\sum_{\gamma\in E_2}\Phi_{\gamma},$$

and likewise for the Fourier transforms. By case (i), we have $Z(\Xi;s)/Z(\hat{\Xi};1-s)=Z(\Phi;s)/Z(\hat{\Phi};1-s)$, and by inductive hypothesis,

$$Z(\Phi_{\gamma};s)/Z(\hat{\Phi}_{\gamma};1-s)=Z(\Phi;s)/Z(\hat{\Phi};1-s) \qquad \text{for all } \gamma\in E_2.$$

Lemma. *Let $\beta\in E_1$. Then there exists $u\in\Lambda^{\times}$ such that $\beta+M=u\alpha+M$.*

The proof will be given later. However, since $u\alpha+M=u(\alpha+M)$, this lemma and Case (i)′ together imply that $Z(\Phi_{\beta};s)/Z(\hat{\Phi}_{\beta};1-s)=Z(\Phi_{\alpha};s)/Z(\hat{\Phi}_{\alpha};1-s)$ for all $\beta \in E_1$. So, we take the equations

$$\sum_{\beta \in E_1} Z(\Phi_\beta; s) = Z(\Xi; s) - \sum_{\gamma \in E_2} Z(\Phi_\gamma; s),$$

$$\sum_{\beta \in E_1} Z(\Phi_\beta; 1-s) = Z(\hat{\Xi}; 1-s) - \sum_{\gamma \in E_2} Z(\hat{\Phi}_\gamma; 1-s)$$

and divide them. The resulting equation reduces to

$$Z(\Phi_\alpha; s)/Z(\hat{\Phi}_\alpha; 1-s) = Z(\Phi; s)/Z(\hat{\Phi}; 1-s),$$

as required.

It remains only to prove the lemma. Since $N \simeq \Lambda$ as left Λ-module, the lemma is equivalent to

Lemma′. *Let I be a full left ideal of Λ, and let $x, y \in \Lambda$ satisfy $\Lambda x + I = \Lambda y + I = \Lambda$. Then there exists $u \in \Lambda^\times$ such that $ux + I = y + I$.*

Proof. It is clearly enough to treat the case $x = 1$, where we have to show that y is congruent, modulo I, to a unit of Λ. Let J denote the Jacobson radical of Λ, so that $\bar{\Lambda} = \Lambda/J$ can be identified with a full matrix algebra $M_n(k)$, for some field k and some $n \geqq 1$.

Assume to start with that $I \supseteq J$. Then we can reduce $\bmod J$, put $\bar{I} = I/J$, $\bar{y} = y$ $(\bmod J)$, when we have $\bar{\Lambda}\bar{y} + \bar{I} = \bar{\Lambda}$. Applying an inner automorphism of Λ, we can assume that $\bar{I} = \bar{\Lambda} e_r$, where e_r is the matrix whose first r diagonal entries are 1, with all other entries zero. The equation $\bar{\Lambda}\bar{y} + \bar{I} = \bar{\Lambda}$ now implies that the last $n-r$ columns of $\bar{y}$ span a k-vector space of dimension $n-r$, from which it follows that $\bar{y}$ is congruent, modulo $\bar{I}$, to a nonsingular matrix. Since Λ is a complete semilocal ring, we can lift units mod the radical to obtain $u \in \Lambda^\times$ such that $u \equiv y \ (\bmod I)$.

In the general case, we can always find $u \in \Lambda^\times$ such that $y \equiv u (\bmod I + J)$. This gives us $z \in I$ such that $(y - z) \equiv u (\bmod J)$. It follows that $(y - z) \in \Lambda^\times$, and of course $y \equiv (y - z) (\bmod I)$.

This completes the proof of the lemma, and of the functional equation.

References

1. Deuring, M.: Algebren, 2nd ed., Berlin-Heidelberg-New York: Springer 1968
2. Galkin, V.M.: ζ-function of some one-dimensional rings. Izv. Akad. Nauk SSSR Ser. Mat. **37**, 3–19 (1973), [Russ.]. Engl. Transl.: Math. USSR-Izv. **7**, 1–17 (1973)
3. Godement, R., Jacquet, H.: Zeta-functions of simple algebras. Lecture Notes in Mathematics **260**. Berlin-Heidelberg-New York: Springer 1972
4. Reiner, I.: Maximal Orders. London-New York: Academic Press 1975
5. Reiner, I.: Zeta functions of integral representations. Comm. Algebra (to appear)
6. Solomon, L.: Zeta-functions and integral representation theory. Advances in Math. **26**, 306–326 (1977)

7. Solomon, L.: Partially ordered sets with colors. In: Relations between combinatories and other parts of mathematics. Proceedings of a Symposium (Columbus, 1978), pp. 309–329. Proceedings of Symposia in Pure Mathematics XXXIV. Providence, Rhode Island: American Mathematical Society 1979
8. Tate, J.T.: Fourier analysis in local fields and Hecke's zeta-functions. Thesis, Princeton University 1950 (=Cassels, J.W.S. & Fröhlich, A. (eds.): Algebraic Number Theory, pp. 305–347. London: Academic Press 1967
9. Weil, A.: Basic Number Theory. Berlin-Heidelberg-New York: Springer 1974

Received October 13, 1979

Zeta Functions of Integral Representations

Irving Reiner

University of Illinois
Urbana, Illinois

1. Introduction

Let R be the ring of algebraic integers in an algebraic number field, or a localization or $\mathcal{P}$-adic completion of such a ring. Let RG be the integral group ring of a finite group G; an $RG-$ *lattice* is a left RG-module L which is finitely generated and R-torsionfree. A *full* sublattice of L is an RG-submodule M such that M and L span the same vector space over the field of quotients of R; for each such M, let $(L:M)$ denote the number of elements of the finite RG-module L/M.

L. Solomon [1] recently introduced a zeta function $\zeta_L(s)$ associated with a given RG-lattice L, namely,

$$\zeta_L(s) = \sum_M (L:M)^{-s}, \tag{1}$$

the sum extending over all full sublattices M in L. The series converges for $\Re(s)$ sufficiently large.

Let $\mathcal{P}$ range over all maximal ideals of R, and let the subscript $\mathcal{P}$ denote localization at $\mathcal{P}$. Put

$$\zeta_{L_\mathcal{P}}(s) = \sum_{M_\mathcal{P}} (L_\mathcal{P}:M_\mathcal{P})^{-s},$$

the "$\mathcal{P}$-part" of $\zeta_L(s)$, where $M_\mathcal{P}$ ranges over all the full $R_\mathcal{P}G$-sublattices of $L_\mathcal{P}$. As shown by Solomon, one has

$$\zeta_L(s) = \prod_\mathcal{P} \zeta_{L_\mathcal{P}}(s), \tag{2}$$

and furthermore $\zeta_{L_\mathcal{P}}(s)$ is unchanged when the local ring $R_\mathcal{P}$ is replaced by its $\mathcal{P}$-adic completion. This reduces the calculation of $\zeta_L(s)$ to the local (complete) case. Whenever $|G|$ is a unit in $R_\mathcal{P}$, the order $R_\mathcal{P}G$ is maximal, and $\zeta_{L_\mathcal{P}}(s)$ is easily expressed in terms of Dedekind zeta functions of fields. The difficulty in the evaluation of $\zeta_L(s)$ is thus concentrated at those $\mathcal{P}$'s which divide $|G|$. Solomon proved that $\zeta_{L_\mathcal{P}}(s)$ is always a rational function of p^{-s}, where p is the unique rational prime contained in $\mathcal{P}$. As an illustration of his methods, he calculated $\zeta_L(s)$ for the case $L = ZG$, with G a cyclic group of prime order p. We present here a simpler procedure for calculating this zeta function, and apply it to the more difficult case where G is cyclic of order p^2. The same method will work for any cyclic p-group, but the computational details become rather complicated.

Suppose now that G is a cyclic group with generator x of order p^n, where p is a prime and $n \geq 1$; we identify the integral group ring ZG with $Z[x]/(x^{p^n} - 1)$ from now on. For q a rational prime, let

$$Z_q = \{a/b : a, b \in Z, \quad (b, q) = 1\},$$

the *localization* of Z at q. If $q \neq p$, then $Z_q G$ is a maximal Z_q-order in the group algebra QG. Now

$$QG = K_0 \oplus K_1 \oplus \cdots \oplus K_n, \quad K_i = Q(\omega_i),$$

where ω_i is a primitive p^i-th root of 1. Put $S_i = Z[\omega_i]$, the ring of all algebraic integers in K_i. Then

$$Z_q G = (S_0)_q \oplus \cdots \oplus (S_n)_q,$$

and each full sublattice of $Z_q G$ is of the form $\oplus \mathcal{A}_i$, where $\mathcal{A}_i$ is an arbitrary ideal of $(S_i)_q$. It follows at once that

$$\zeta_{Z_q G}(s) = \prod_{i=0}^{n} \{\sum_i N(\mathcal{A}_i)^{-s}\},$$

where $N(\mathcal{A}_i) = ((S_i)_q : \mathcal{A}_i) =$ norm of $\mathcal{A}_i$. Thus $\zeta_{Z_q G}(s)$ is precisely the q-part of the product $\prod_{i=0}^{n} \zeta_{K_i}(s)$, where $\zeta_{K_i}(s)$ is the usual Dedekind zeta function of the field K_i. In view of formula (2), it remains for us to calculate $\zeta_{Z_p G}(s)$. We begin this calculation in the next section.

2. Fiber products.

Let G be a cyclic group of order p^n, H cyclic of order p^{n-1}, where $n \geq 1$ and p is prime. Put

$$R = Z_p, \quad \bar{R} = R/pR \cong Z/pZ, \quad S = R[\omega],$$

where ω is a primitive p^n-th root of 1. Then S is a local ring with maximal ideal $(1 - \omega)S$, and $S/(1 - \omega)S \cong \bar{R}$. We identify S with $R[x]/(\Phi(x))$, where $\Phi(x)$ is the cyclotomic polynomial of order p^n and degree $\varphi(p^n)$. There is a fiber product diagram

$$\text{(3)} \qquad \begin{array}{ccccc} & RG & \xrightarrow{f_1} & S \\ & \downarrow{\scriptstyle f_2} & & \downarrow{\scriptstyle g_1} \\ RH \cong & R[x]/(x^{p^{n-1}} - 1) & \xrightarrow{g_2} & S/(1-\omega)^{p^{n-1}} S \cong \bar{R}H \end{array}$$

in which all maps are canonical ring surjections. We wish to find all full sublattices of RG, that is, all left ideals I of RG for which $Q \otimes_R I = QG$. It will be helpful to consider a somewhat more general situation.

Let us start with a fiber product diagram of rings

$$\begin{array}{ccc} A & \xrightarrow{f_1} & A_1 \\ \downarrow{\scriptstyle f_2} & & \downarrow{\scriptstyle g_1} \\ A_2 & \xrightarrow{g_2} & \bar{A} \end{array}$$

in which all maps are ring surjections. By definition,

$$A = \{(a_1, a_2) \,:\, a_i \in A_i, \quad g_1(a_1) = g_2(a_2)\}.$$

If I is a left ideal of A, put $I_k = f_k(I)$, a left ideal of A_k, $k = 1, 2$. Let $\bar{I} = g_1 f_1(I) = g_1(I_1) = g_2(I_2)$, a left ideal of $\bar{A}$. Clearly the ideal I uniquely determines I_1 and I_2, but not conversely.

Suppose now that $I_1 = A_1\alpha$, a principal ideal; then $(\alpha, \beta) \in I$ for some $\beta \in I_2$. For any $(x, y) \in I$, we may write

$$(x, y) = a(\alpha, \beta) + (0, y')$$

for some $a \in A$ and $y' \in I_2$. We put

$$J_2 = \{c \in A_2 \,:\, (0, c) \in I\}.$$

Given any $a_2 \in A_2$, there exists an $a \in A$ with $f_2(a) = a_2$. Thence for $c \in J_2$,

$$a(0, c) \in I \;\Rightarrow\; (0, a_2 c) \in I \;\Rightarrow\; a_2 c \in J_2,$$

which shows that J_2 is a left ideal of A_2. We have then

$$I = A(\alpha, \beta) + (0, J_2), \text{ where } f_1(I) = A_1\alpha. \tag{4}$$

We note that $g_2(J_2) = 0$ since $(0, J_2) \subseteq I$. Further, $g_2(\beta) = g_1(\alpha)$ in $\bar{A}$ since $(\alpha, \beta) \in I$. The ideal is thus determined by the following data:

i)an ideal J_2 of A_2 such that $g_2(J_2) = 0$,

ii)a generator α of a principal ideal $A_1\alpha$ of A_1,

iii)an element $\beta \in A_2$ such that $g_2(\beta) = g_1(\alpha)$.

Clearly, β is uniquely determined mod J_2.

There is a further condition implicit in the above construction. Let

$$D = \{a \in A \,:\, f_1(a)\alpha = 0\},$$

a left ideal of A. Then $D \cdot (\alpha, \beta) \subseteq I$; but

$$D \cdot (\alpha, \beta) = \{(f_1(a)\alpha, f_2(a)\beta) \,:\, a \in D\} = (0, f_2(D)\beta),$$

which gives

$$f_2(D)\beta \in J_2. \tag{5}$$

We apply these results to the diagram (3). Let I be a full ideal of RG; then $I_1 = f_1(I)$ is an ideal of the principal ideal domain S, so $I_1 = \alpha S$ where $\alpha = (1-\omega)^r$ for some $r \geq 0$. The annihilator of α in RG is the principal ideal $\Phi(x)RG$, that is $D = \Phi(x)RG$. Since $x^{p^{n-1}}$ acts as 1 on RH, it follows that $\Phi(x)$ acts as multiplication by p on RH. Condition (5) then gives $p\beta \in J_2$. Furthermore, $g_2(J_2) = 0$ implies that $J_2 \subseteq p \cdot RH$, so we may set $J_2 = pJ$ for some ideal J of RH. Therefore

$$I = RG(\alpha, \beta) + (0, pJ),$$

where J is a full ideal in RH. The element $\beta \in RH$ is unique mod pJ, and staisfies the conditions

$$\beta \in J, \quad \bar{\beta} = \bar{\alpha} \text{ in } \bar{R}H,$$

where $\bar{\alpha}$ denotes the image of α in $\bar{R}H$ (by abuse of notation).

Our next step is to determine how many ideals correspond to a given choice

$$\alpha = (1-\omega)^r, \quad r \geq 0, \quad J = \text{ full ideal in } RH.$$

The number of such ideals I is precisely the number of solutions of

$$\beta \equiv \bar{\alpha} \bmod p \cdot RH, \quad \beta \equiv 0 \bmod J,$$

which are distinct mod pJ. If there exists a solution for β, then necessarily $\bar{\alpha} \in \bar{J}$, where

$$\bar{J} = (J + p \cdot RH)/p \cdot RH \ = \ \text{image of } J \text{ in } \bar{R}H.$$

If $\bar{\alpha} \in \bar{J}$, there is at least one solution for β_0, and all other solutions are obtained by increasing β_0 by any element of $J \cap p \cdot RH$. The number of β's distinct mod pJ is therefore equal to $(J \cap p \cdot RH \ : \ pJ)$. This number equals

$$\frac{(J \ : \ pJ)}{(J \ : \ J \cap p \cdot RH)} = \frac{p^{\text{rank } J}}{(J + p \cdot RH \ : \ p \cdot RH)} = \frac{p^{p^{n-1}}}{|\bar{J}|},$$

since rank $J = |H| = p^{n-1}$ because J is a full ideal in RH. Now

$$\bar{R}H \cong \bar{R}[\lambda]/(\lambda^{p^{n-1}}), \quad \lambda = \text{ image of } 1-x,$$

so we may write

$$\bar{J} = \lambda^k \bar{R}H, \text{ where } 0 \leq k \leq p^{n-1}, \tag{8}$$

and then

$$|\bar{J}| = p^{p^{n-1}-k}.$$

Thus the number of distinct ideals I of RG corresponding to a given choice of J and r is precisely p^k. Note that $r \geq k$ because $\alpha \in \bar{J}$.

Finally we observe that

$$RG = RG(1,1) + (0, p \cdot RH),$$

which readily implies that

$$(RG : I) = (S : S\alpha)(RH : J) = p^r(RH : J). \tag{9}$$

We may now give an inductive procedure for the calculation of $\zeta_{RG}(s)$. Suppose that we know all the ideals J of RH, and for each J we calculate the integer k occurring in (8). Then there are p^k ideals I given by (6), where $\alpha = (1-\omega)^r$ and $r \geq k$. Thus

$$\zeta_{RG}(s) = \sum_J \sum_{r=k}^{\infty} p^k \cdot (p^r(RH : J))^{-s}. \tag{10}$$

As we shall see, this formula is easy to use when $n = 1$, and is somewhat harder to apply when $n = 2$.

3. The cyclic group of order p.

We apply the results of §2 to the case where G is cyclic of order p, so (3) becomes

$$\begin{array}{ccccc} RG & \longrightarrow & S & = & R[\omega] \\ \downarrow & & \downarrow & & \\ R & \longrightarrow & \bar{R} & \cong & S/(1-\omega)S, \end{array}$$

with ω a primitive p-th root of 1, and $R = Z_p$, the localization of Z at p. Let J be an ideal in R, $\bar{J}$ its image in $\bar{R}$. The full ideals I of RG have the form

$$I = RG(\alpha, \beta) + (0, pJ), \quad \alpha = (1-\omega)^r, \tag{11}$$

where $\bar{\alpha} \in \bar{J}$. There are p such ideals when $\bar{J} = 0$ (and $k = 1$, using the notation of (8)); on the other hand, there is a unique such ideal when $\bar{J} = \bar{R}$ and $k = 0$. Put $J = p^t R$, $t \geq 0$, so $\bar{J} = 0$ except when $t = 0$. Formula (10) then yields

$$\zeta_{RG}(s) = \sum_{r=0}^{\infty} (p^r)^{-s} + \sum_{t=1}^{\infty} \sum_{r=1}^{\infty} p \cdot (p^{r+t})^{-s},$$

the first summation corresponding to the case $t = 0$, $\bar{J} = \bar{R}$, the second to the case where $t > 0$ and $\bar{J} = \bar{R}$. Let $y = p^{-s}$; then

$$\begin{aligned}\zeta_{RG}(s) &= \sum_{r=0}^{\infty} y^r + p \cdot \sum_{r,t=1}^{\infty} y^{r+t} \\ &= \frac{1}{(1-y)} + py^2(1 + 2y + 3y^2 + \cdots) \\ &= \frac{1}{(1-y)} + \frac{py^2}{(1-y)^2} = \frac{(1-y+py^2)}{(1-y)^2}.\end{aligned}$$

Thus

$$\zeta_{RG}(s) = \frac{(1 - p^{-s} + p^{1-2s})}{(1-p^{-s})^2},$$

which agrees with Solomon's result.

Before turning to the case of cyclic groups of order p^2, let us calculate $\bar{I}$, the image of I in $\bar{R}G$. This will be needed in order to apply the inductive procedure described at the end of §2. Put

$$\bar{R}G = \bar{R}[\lambda]/(\lambda^p), \quad \lambda = 1 - x, \quad \bar{I} = \lambda^k \bar{R}G, \quad 0 \le k \le p.$$

We have

$$\Phi(x) = x^{p-1} + \cdots + x + 1 \equiv \lambda^{p-1} \bmod p,$$

and

$$\Phi(x)(1,1) = (0,p), \quad (1-x)^r(1,1) = ((1-\omega)^r, 0), \quad r > 0. \tag{12}$$

Case 1. $J = R$, $r \ge 0$.

When $J = R$ and $r = 0$, then $I = RG$, $\bar{I} = \bar{R}G$, $k = 0$. When $J = R$ and $r > 0$, then

$$I = RG((1-\omega)^r, \beta) + (0, pR).$$

Since $p|\beta$, we may take $\beta = 0$. From (12) we find at once that $k = Min(r, p-1)$. This formula also holds when $r = 0$. The ideal I is uniquely determined by r.

Case 2. $J = pR$, $r \ge 1$.

Here

$$I = RG((1-\omega)^r, \beta) + (0, p^2R),$$

and $(0, p^2R) \subseteq p \cdot RG$, so $\bar{I}$ is unchanged if we omit the summand $(0, p^2R)$. Since β is determined mod p^2, we may take $\beta = mp$, $m = 0, 1, \ldots, p-1$. Now (mp, mp) is in $p \cdot RG$, so we may replace I by $RG((1-\omega)^r - mp, 0)$ without affecting $\bar{I}$. Let m_0 be the unique integer such that

$$(1-\omega)^{p-1} \equiv m_0 p \bmod (1-\omega)^p, \quad 0 \le m_0 \le p-1.$$

Then $\bar{I} = \lambda^k \bar{R}G$, where

$$k = \begin{cases} r & \text{if } r < p-1 \\ p-1 & \text{if } r = p-1 \text{ and } m \neq m_0 \\ p & \text{if } r = p-1 \text{ and } m = m_0 \\ p-1 & \text{if } r > p-1 \text{ and } m \neq 0 \\ p & \text{if } r > p-1 \text{ and } m = 0. \end{cases}$$

Thus we have $k = Min(r, p-1)$, except for the cases where

$$I = RG((1-\omega)^{p-1}, m_0 p) + (0, p^2 R),$$
$$I = RG((1-\omega)^r, 0) + (0, p^2 R), \quad r \geq p,$$

and in these cases $k = p$.

Case 3. $J = p^t R$, $t > 1$, $r \geq 1$.

Here

$$I = RG((1-\omega)^r, \beta) + (0, p^{t+1} R),$$

and $\beta \in J$. Modulo $p \cdot RG$, we may take $\beta = 0$ and omit the second summand. This gives $k = Min(r, p)$, and there are p ideals corresponding to the pair r, t.

4. The cyclic group of order p^2.

Let us use the results of the preceding two sections to calculate $\zeta_{RG}(s)$ for G cyclic of order p^2. Let H be cyclic of order p, and let ω be a primitive p^2-th root of 1. Diagram (3) becomes

$$\begin{array}{ccccccc} RG & \longrightarrow & S & = & R[\omega] & & \\ \downarrow & & \downarrow & & & & \\ RH & \longrightarrow & \bar{R}H & \cong & S/(1-\omega)^p S & \cong & \bar{R}[\lambda]/(\lambda^p), \end{array}$$

where $\lambda = 1-x$. Let I range over all full ideals of RH, and for each I put $\bar{I} = \lambda^k \bar{R}H$, where $0 \leq k \leq p$. The full ideals of RG are then given by

$$M = RG((1-\omega)^d, \gamma) + (0, pI), \quad d \geq k,$$

and there are p^k choices of γ which yield distinct ideals M. If we put

$$I = RH((1-\omega_1)^r, \beta) + (0, p^{t+1} R),$$

where ω_1 is a primitive p-th root of 1, then we have

$$(RG : M) = p^d (RH : I) = p^{d+r+t}.$$

For a given pair r, t, there are p distinct choices of I, except that there is only one choice when $t = 0$.

In §3 we have calculated k as a function of the parameters r, t which are associated with the ideal I. Let us now use formula (10) for $\zeta_{RG}(s)$, dividing the sum into three parts S_1, S_2, S_3 corresponding to the cases 1–3 in §3. Thus S_1 consists of the terms where $t = 0$, S_2 where $t = 1$, and S_3 where $t > 0$. Then we have

$$S_1 = \sum_{r=0}^{\infty} \sum_{d \geq k} p^k (p^{d+r})^{-s}, \quad k = Min(r, p-1).$$

Let $y = p^{-s}$. Then

$$\begin{aligned} S_1 &= \sum_{r=0}^{\infty} p^k y^r y^k (1-y)^{-1} \\ &= \sum_{r=0}^{p-1} p^r y^r y^r (y-1)^{-1} + \sum_{r=p}^{\infty} p^{p-1} y^r y^{p-1} (1-y)^{-1} \end{aligned}$$

so we obtain

$$S_1 = (1-y)^{-1} \left\{ \frac{1-(py^2)^p}{1-py^2} + \frac{p^{p-1}y^{2p-1}}{1-y} \right\}. \tag{13}$$

Likewise

$$S_3 = p \cdot \sum_{r=1}^{\infty} \sum_{t=2}^{\infty} \sum_{d \geq k} p^k (p^{d+r+t})^{-s}, \quad k = Min(r, p),$$

the factor p arising from the choice of p ideals I associated with the parameters r and t. A calculation like that above yields

$$S_3 = \frac{py^2}{(1-y)^2} \left\{ \frac{py^2(1-(py^2)^p)}{1-py^2} + \frac{p^p y^{2p+1}}{1-y} \right\}. \tag{14}$$

Finally

$$\begin{aligned} S_2 = p &\sum_{r=1}^{p-2} \sum_{d \geq r} p^r (p^{d+r+1})^{-s} \\ &+ (p-1) \sum_{r=p-1}^{\infty} \sum_{d \geq p-1} p^{p-1} (p^{d+r+1})^{-s} \\ &+ \sum_{r=p-1}^{\infty} \sum_{d \geq p} p^p (p^{d+r+1})^{-s}. \end{aligned}$$

The factors p and $p-1$ before the summations signs arise from the number of choices of the ideal I; for the range $1 \leq r \leq p-2$ we have $k = r$, while for the range

$r \geq p-1$ there are $p-1$ choices of I for which $k = p-1$ and one choice of I for which $k = p$. We readily obtain

$$S_2 = (1-y)^{-1}\left\{\frac{p^2y^3(1-(py^2)^{p-2})}{1-py^2} + \frac{(p-1)p^{p-1}y^{2p-1}}{1-y} + \frac{p^py^{2p}}{1-y}\right\}. \tag{15}$$

As remarked above, we have

$$\zeta_{RG}(s) = S_1 + S_2 + S_3, \quad \text{with } y = p^{-s},$$

where the S_i are given by (13)–(15). As predicted by Solomon's results, $\zeta_{RG}(s)$ is a rational function of y. Furthermore, the discussion in §1 yields

$$\zeta_{ZG}(s) = \frac{\zeta_{RG}(s)}{(1-p^{-s})^{-3}} \cdot \zeta(s)\zeta_{K_1}(s)\zeta_{K_2}(s),$$

where K_i is the cyclotomic field of p^i-th roots of 1, $i = 1, 2$. The first factor $\zeta_{RG}(s)(1-p^{-s})^3$ is then $(1-y)^3(S_1+S_2+S_3)$, which is easily seen to be a *polynomial* in y, as conjectured by Solomon. This polynomial is of degree $2p+2$ in y, and can be computed explicitly from (13)–(15), though there does not seem to be any neat expression for it.

Acknowledgement. The above research was supported in part by the National Science Foundation.

Reference

1. L. Solomon, Zeta functions and integral representation theory, Advances in Math. **26**(1977), 306–326.

Received: March 1979

L-functions of arithmetic orders and asymptotic distribution of ideals

By *Colin J. Bushnell* at London and *Irving Reiner* at Urbana*)

Introduction

Let Λ denote a $\mathbb{Z}$-order in a finite-dimensional semisimple $\mathbb{Q}$-algebra A. In our earlier paper [1] (but see also [3]), we developed the theory of Solomon's zeta function:

$$\zeta_\Lambda(s) = \sum_X (\Lambda : X)^{-s}, \qquad \mathrm{Re}(s) > 1,$$

where s is a complex variable, and X ranges over all left ideals of Λ of finite index $(\Lambda : X)$ in Λ. One can also define "partial zeta functions" by means of the same sort of series, but restricting X to certain sets of ideals of Λ. We shall mainly be concerned with the case in which X ranges over a fixed stable isomorphism class. As in the classical case, where Λ is the ring of integers in an algebraic number field, the best approach to these partial zeta functions is via L-series. We define these in § 1. Following the methods of [1], we are able to recognise our L-functions as certain idelic integrals, this fact being reflected in the existence of an Euler product expansion, as discussed in § 2.

For the present paper, the main task is to show that our L-series admit analytic continuation, and then to discover something about their singularities. We follow the strategy of Godement and Jacquet [7] here. There is a "universal" or "standard" L-function, depending only on the algebra A, which can be described completely in terms of classical L-functions. The construction of this object occupies §§ 3 and 4. The L-function attached to an order then differs from the appropriate standard L-function only by an elementary function, which can be described in purely local terms. (The whole process is, in fact, rather similar to the proof of Solomon's conjecture in [1].) This gives us enough information (§ 5, § 6) about our L-series, and hence about partial zeta functions, to prove that *the left ideals of Λ in a given genus are asymptotically uniformly distributed among the stable isomorphism classes in that genus.*

---

*) The research for this paper was performed while the second author was a Senior Visiting Research Fellow at King's College London, partially supported by the Science Research Council. He wishes to acknowledge this support, as well as the hospitality of King's College, during his sabbatical leave from the University of Illinois. He also thanks the N[illegible]al Science Foundation for its support at other times.

Of course, for many algebras A, there is a distinction between isomorphism and stable isomorphism for Λ-lattices. When this is the case, our methods give no information about distribution in isomorphism classes. In § 9, we discuss briefly the simplest case in which this can arise. This section is based largely on work of Eichler, and shows that distribution in isomorphism classes is not necessarily uniform, but the discrepancy is easily described.

In principle, the whole procedure applies equally well to sublattices of a fixed Λ-lattice, rather than just to ideals of Λ. The obstructions to this generalisation are mainly notational, but in § 8 we give a sketch of the theory for the case of lattices in a free Λ-module.

Finally, we note that our L-functions also satisfy functional equations. We will discuss this topic in a future article, continuing the present one. *)

Throughout this paper, methods and terminology are basically those of [1], but we make some minor changes of notation. We also take this opportunity to correct some misprints in [1]:

p. 143, line 11: $L=(\Delta^m)^k$, p. 144, line 19: $a_{ij}=0$ if $j<i$,

p. 148, lines 9, 10: are called "standard" in [9],

p. 157, line 7: $\|\alpha\|=(\Lambda'\alpha:\Lambda')=(\Lambda'\alpha:\Lambda)(\Lambda:\Lambda')$.

§ 1. Basic definitions

Throughout this paper, A denotes a finite-dimensional semisimple algebra over either the field $\mathbb{Q}$ of rational numbers or a p-adic rational field $\mathbb{Q}_p$, for some prime number p. We denote by Λ an order in A. Thus Λ is a $\mathbb{Z}$-order when A is a $\mathbb{Q}$-algebra, a $\mathbb{Z}_p$-order when A is a $\mathbb{Q}_p$-algebra. Here and throughout, the subscript p denotes p-adic *completion*, not localisation. We use a similar convention for primes of an algebraic number field.

When considering ideals M of Λ, we always restrict our attention to those for which the group index $(\Lambda:M)$ is *finite*. We use this index symbol in the generalised sense: if, for example, X and Y are $\mathbb{Z}$-lattices spanning the same $\mathbb{Q}$-vector space, we put

$$(X:Y)=\frac{(X:X\cap Y)}{(Y:X\cap Y)}.$$

The symbol $(X:Y)$ is therefore unambiguously defined whether or not X contains Y.

We assume for the rest of this section that Λ is a $\mathbb{Z}$-order in the finite-dimensional semisimple $\mathbb{Q}$-algebra A. Solomon [12] introduced the zeta-function $\zeta_\Lambda(s)$, defined by

$$(1.1)\qquad \zeta_\Lambda(s)=\sum_X(\Lambda:X)^{-s},\qquad \mathrm{Re}(s)>1,$$

where s is a complex variable, and where X ranges over all left ideals of Λ of finite index $(\Lambda:X)$. In order to analyse $\zeta_\Lambda(s)$, it is convenient to introduce various "partial zeta-functions", in which X is restricted to lie in certain sets of left ideals of Λ, as we now explain.

*) Functional equations of L-functions of arithmetic orders, to appear in this journal.

Given a left ideal M in a $\mathbb{Z}$-order Λ, the *genus* of M, denoted by $g(M)$, consists of all left Λ-lattices X such that $X_p \cong M_p$ as Λ_p-modules, for all prime numbers p. As is well-known (Jacobinski [8], Fröhlich [6]), the genus $g(M)$ can be partitioned into a finite number of stable isomorphism classes $[M_1], \ldots, [M_h]$. Here, two lattices $X, Y \in g(M)$ are called *stably isomorphic* (notation: $[X]=[Y]$) if $X \oplus M \cong Y \oplus M$. (In most cases, stable isomorphism implies ordinary isomorphism. Indeed, this holds if the algebra A satisfies the "Eichler condition", namely, no simple component of A is a totally definite quaternion algebra.)

We now define *partial zeta-functions*

$$Z_\Lambda(g(M), s), \quad Z_\Lambda([M], s), \quad Z_\Lambda(M, s)$$

by series such as (1. 1), but where X is restricted to the genus of M, the stable isomorphism class $[M]$ of M in $g(M)$, and the isomorphism class of M, respectively.

Caution. This represents a change in notation from [1], where $Z_\Lambda(M, s)$ was used to denote what we now call $Z_\Lambda(g(M), s)$.

As in [1], it is easily shown that these series all converge well for $\mathrm{Re}(s) > 1$, and we have

$$\text{(1. 2)} \qquad \zeta_\Lambda(s) = \sum_M Z_\Lambda(g(M), s), \quad Z_\Lambda(g(M), s) = \sum_{i=1}^{h} Z_\Lambda([M_i], s),$$

where M ranges over a (finite) set of representatives of the genera of left ideals of Λ, and where for fixed M, the stable isomorphism classes in $g(M)$ are $[M_1], \ldots, [M_h]$.

In the special case $M = \Lambda$, the stable isomorphism classes in $g(\Lambda)$ form a finite abelian group $\mathrm{Cl}(\Lambda)$, the *locally free class group* of Λ. Addition in $\mathrm{Cl}(\Lambda)$ is given by $[X]+[X'] = [X'']$, for $X, X' \in g(\Lambda)$, where $[X'']$ is defined by the condition $X \oplus X' \cong X'' \oplus \Lambda$. For properties of $\mathrm{Cl}(\Lambda)$, see [6].

In the general case where M is an arbitrary left ideal of Λ, we let $\Gamma = \mathrm{End}_\Lambda(M)$. Then Γ is another order in A, and we may view M as a right Γ-module. As shown by Jacobinski [8], there is a bijection between $\mathrm{Cl}(\Gamma)$ and the set of stable isomorphism classes in $g(M)$, given by $[X] \mapsto [MX]$ for each $[X] \in \mathrm{Cl}(\Gamma)$. This bijection enables us to impose a structure of finite abelian group on the set of stable isomorphism classes in $g(M)$. The addition is therefore given by the formula $[M_1]+[M_2]=[M_3]$, where $M_3 \in g(M)$ is any Λ-lattice such that $M_1 \oplus M_2 \cong M_3 \oplus M$. This group, denoted by $\mathrm{Cl}(M)$, will be called the *genus class group* of M. Of course, this group structure on $\mathrm{Cl}(M)$ given by the isomorphism $\mathrm{Cl}(\Gamma) \cong \mathrm{Cl}(M)$ depends on the choice of M in its genus, but this will not trouble us.

Now let ψ be a character of $\mathrm{Cl}(M)$, that is, a homomorphism from $\mathrm{Cl}(M)$ to the unit circle in the complex numbers $\mathbb{C}$. We now define L-functions by the formula

$$\text{(1. 3)} \qquad L_\Lambda(M, s, \psi) = \sum_{\substack{X \subseteq \Lambda \\ X \in g(M)}} \psi[X] (\Lambda : X)^{-s}, \qquad \mathrm{Re}(s) > 1,$$

where s is a complex variable. This series converges absolutely for $\mathrm{Re}(s) > 1$, uniformly on compact subsets, by comparison with the series (1. 1). Thus (1. 3) defines a holomorphic function of s for $\mathrm{Re}(s) > 1$. This L-function is clearly a generalisation of the classical L-functions occurring in algebraic number theory. For properties of classical L-functions, especially those associated with ray-class characters (or equivalently, idele-class characters of finite order), we refer the reader to [4], [9], [10] and [13].

We may rewrite (1. 3) in the form

$$L_A(M, s, \psi) = \sum_{[N] \in \mathrm{Cl}(M)} \psi[N]\, Z_A([N], s), \tag{1. 4}$$

where $Z_A([N], s)$ is the partial zeta-function defined earlier. The standard orthogonality relations for characters of finite groups allow us to invert (1. 4), and we obtain

$$Z_A([N], s) = h^{-1} \sum_{\psi} \psi^{-1}[N]\, L_A(M, s, \psi) \tag{1. 5}$$

for each stable isomorphism class $[N]$ in $g(M)$. Here, ψ ranges over all characters of $\mathrm{Cl}(M)$, and $h = |\mathrm{Cl}(M)|$.

This last identity suggests that, just as in the classical case, we can use information about L-functions to study the partial zeta-function $Z_A([N], s)$. Unfortunately, our methods do not seem to yield any information about $Z_A(N, s)$, except when stable isomorphism is identical with ordinary isomorphism.

To conclude this section, we show how each character ψ of $\mathrm{Cl}(M)$ may be viewed as an idele-class character of A. As pointed out earlier, there is an isomorphism $\mathrm{Cl}(M) \cong \mathrm{Cl}(\Gamma)$, where $\Gamma = \mathrm{End}_A(M)$, so ψ may be viewed as a character of $\mathrm{Cl}(\Gamma)$. We shall now recall Fröhlich's formula for the locally free class group $\mathrm{Cl}(\Gamma)$ of an order Γ in a $\mathbb{Q}$-algebra A. Let $J(A)$ denote the full idele group of A, formed by using both the finite and infinite primes of $\mathbb{Q}$. (This represents a change of notation from [1], where $J(A)$ was used for the group of "finite ideles".) Let $J'(A)$ denote the closure of the commutator subgroup of $J(A)$. The group $A^\times$ of invertible elements of A is always identified with the group of principal ideles in $J(A)$. Fröhlich [6] showed that

$$\mathrm{Cl}(\Gamma) \cong J(A)/J'(A) \cdot A^\times \cdot U(\Gamma), \tag{1. 6}$$

where $U(\Gamma)$ is the group of unit ideles of the order Γ, defined by

$$U(\Gamma) = (\mathbb{R} \otimes_{\mathbb{Q}} A)^\times \times \prod_p \Gamma_p^\times, \tag{1. 7}$$

with p ranging over all prime numbers. The isomorphism (1. 6) is induced by the map $x \mapsto [\Gamma x]$, where, for $x = (x_p) \in J(A)$, Γx is the unique Γ-lattice in A such that $(\Gamma x)_p = \Gamma_p x_p$, for all prime numbers p.

Following Fröhlich, we give an alternative description of $\mathrm{Cl}(\Gamma)$ by using the reduced norm map $\mathrm{nr}_{A/C}\colon A^\times \to C^\times$, where C is the centre of A. For convenience, we write nr instead of $\mathrm{nr}_{A/C}$ if there is no danger of confusion. The reduced norm map extends to a continuous homomorphism $\mathrm{nr}\colon J(A) \to J(C)$, with kernel $J'(A)$. It follows from the Hasse-Schilling-Maass norm theorem that

$$C^\times \cdot \mathrm{nr}\,(J(A)) = J(C), \quad \mathrm{nr}\,(A^\times) = C^\times \cap \mathrm{nr}\,(J(A)),$$

so

$$J(C)/C^\times = C^\times \cdot \mathrm{nr}\,(J(A))/C^\times = \mathrm{nr}\,(J(A))/\mathrm{nr}\,(A^\times).$$

Therefore nr induces isomorphisms

$$(1.8) \qquad J(A)/J'(A)\cdot A^{\times} \cong J(C)/C^{\times},$$

$$(1.9) \qquad J(A)/J'(A)\cdot A^{\times}\cdot U(\Gamma) \cong J(C)/C^{\times}\cdot \mathrm{nr}\,(U(\Gamma)).$$

Suppose now that M is a left ideal of an order Λ in a $\mathbb{Q}$-algebra A, and let $\Gamma = \mathrm{End}_{\Lambda}(M)$ as above. We shall be concerned with characters ψ of the genus class group $\mathrm{Cl}(M)$. By virtue of the isomorphism $\mathrm{Cl}(M) \cong \mathrm{Cl}(\Gamma)$ and formula (1.6), each such ψ can be viewed as a continuous character (also denoted by ψ) of the idele group $J(A)$. This idele character ψ is then trivial on $J'(A)\cdot A^{\times}\cdot U(\Gamma)$. Furthermore, it follows from (1.9) that we may write

$$(1.10) \qquad \psi(x) = \tilde{\psi}(\mathrm{nr}\,x), \qquad x \in J(A),$$

for some uniquely determined continuous character $\tilde{\psi}$ of the idele group $J(C)$, such that $\tilde{\psi}$ is trivial on $C^{\times}$ and on the open subgroup $\mathrm{nr}\,(U(\Gamma))$ of $J(C)$. We shall use this interpretation of characters of the genus class group $\mathrm{Cl}(M)$ in all of our later calculations.

§ 2. Euler products

In formula (13) of [1], we obtained an Euler product expansion for the partial zeta-function $Z_{\Lambda}(g(M), s)$, where M is a left ideal in a $\mathbb{Z}$-order Λ. This formula generalises immediately to L-functions, with virtually the same proof, once we have defined local L-functions.

To give this definition, let us temporarily change notation, and assume that Λ is a $\mathbb{Z}_p$-order in a $\mathbb{Q}_p$-algebra A. Let M be a left ideal of Λ, of finite index in Λ as always, and set $\Gamma = \mathrm{End}_{\Lambda}(M)$. Let ψ be a continuous character of $A^{\times}$ which is trivial on $\Gamma^{\times}$. We define the local L-function

$$(2.1) \qquad L_{\Lambda}(M, s, \psi) = \sum_{X} \psi(X)\,(\Lambda : X)^{-s}, \qquad \mathrm{Re}(s) > 1,$$

where the sum extends over all left ideals X of Λ such that $X \cong M$. Here, $\psi(X)$ is defined by the formula

$$\psi(X) = \psi(x), \quad \text{for any} \quad x \in A^{\times} \quad \text{such that} \quad X = Mx.$$

Since ψ is trivial on $\Gamma^{\times} = \mathrm{Aut}_{\Lambda}(M)$, $\psi(X)$ is independent of the choice of x in the above formula. Then $L_{\Lambda}(M, s, \psi)$ is a power series in p^{-s}, with coefficients in the ring $\mathbb{Z}[\psi]$ generated by adjoining to $\mathbb{Z}$ the values of ψ (which are of course roots of unity).

We now return to the global situation of § 1, where A is a $\mathbb{Q}$-algebra, and let Λ, M, Γ, ψ be as in § 1. The Archimedean prime of $\mathbb{Q}$ plays no role here, and it will be more convenient for us to work with the group $J_{\mathrm{fin}}(A)$ of *finite ideles* of A. This group is defined as the restricted direct product of the groups $A_p^{\times}$, with respect to the subgroups $\Lambda_p^{\times}$, with p ranging over all prime numbers. This definition is independent of the choice of order Λ. We may view $J_{\mathrm{fin}}(A)$ as a subgroup of $J(A)$.

Similarly, we may form the adele ring $\mathrm{Ad}(A)$ of A, and the ring $\mathrm{Ad}_{\mathrm{fin}}(A)$ of finite adeles of A. If X is any $\mathbb{Z}$-lattice spanning A over $\mathbb{Q}$, we put

$$\mathrm{Ad}_{\mathrm{fin}}(X)=\prod_p X_p,$$

where p ranges over all prime numbers. Then $\mathrm{Ad}_{\mathrm{fin}}(X)$ can be viewed as an additive subgroup of $\mathrm{Ad}(A)$. Likewise we set

$$U_{\mathrm{fin}}(\Gamma)=\prod_p \Gamma_p^{\times},$$

viewed as a subgroup of $J_{\mathrm{fin}}(A)$. Clearly

$$U_{\mathrm{fin}}(\Gamma)=J_{\mathrm{fin}}(A)\cap U(\Gamma).$$

We now set

$$\{M:\Lambda\}=\{x\in A:\ Mx\subseteqq\Lambda\}, \tag{2. 2}$$

and this is a (Γ, Λ)-bimodule in A, which spans A over $\mathbb{Q}$. Then as in [1] we have

$$\mathrm{Ad}_{\mathrm{fin}}\{M:\Lambda\}=\prod_p \{M_p:\Lambda_p\}.$$

Finally, let $\Phi_{\{M:\Lambda\}}$ denote the characteristic function of $\mathrm{Ad}_{\mathrm{fin}}\{M:\Lambda\}$ in $\mathrm{Ad}_{\mathrm{fin}}(A)$, and likewise $\Phi_{\{M_p:\Lambda_p\}}$ the characteristic function of $\{M_p:\Lambda_p\}$ in A_p.

Let ψ be a character of the genus class group $\mathrm{Cl}(M)$. As shown at the end of § 1, ψ defines a continuous character of $J(A)$, also written as ψ. We now put

$$\psi_{\mathrm{fin}}=\psi|J_{\mathrm{fin}}(A),\qquad \psi_p=\psi|A_p^{\times},$$

for each prime number p. We then obtain

$$\psi_{\mathrm{fin}}=\prod_p \psi_p, \tag{2. 3}$$

with the customary abuse of notation. The character ψ of $J(A)$ is trivial on the subgroup $(\mathbb{R}\otimes_{\mathbb{Q}}A)^{\times}$ of $J(A)$, by the definition of $U(\Gamma)$.

Just as in the proof of formula (12) of [1], we obtain

$$L_{\Lambda}(M,s,\psi)=\mu^{\times}(U_{\mathrm{fin}}(\Gamma))^{-1}(\Lambda:M)^{-s}\int_{J_{\mathrm{fin}}(A)}\Phi_{\{M:\Lambda\}}(x)\,\psi_{\mathrm{fin}}(x)\,\|x\|^s d^{\times}x, \tag{2. 4}$$

where $d^{\times}x$ is a Haar measure on $J_{\mathrm{fin}}(A)$, $\mu^{\times}$ denotes the measure of a set with respect to $d^{\times}x$, and $\|x\|$ is defined by the formula

$$\|x\|=(N:Nx)^{-1},$$

for any $\mathbb{Z}$-lattice N spanning A over $\mathbb{Q}$.

The character ψ is trivial on $U(\Gamma)$, so ψ_p is trivial on $\Gamma_p^{\times}$, for all p. Therefore our formula (2. 1) provides a definition of $L_{\Lambda_p}(M_p,s,\psi_p)$, for each prime number p. Just as in [1], formula (11), we obtain

$$L_{\Lambda_p}(M_p,s,\psi_p)=\mu_p^{\times}(\Gamma_p^{\times})^{-1}(\Lambda_p:M_p)^{-s}\int_{A_p^{\times}}\Phi_{\{M_p:\Lambda_p\}}(x)\,\psi_p(x)\,\|x\|^s d^{\times}x, \tag{2. 5}$$

where $d^{\times}x$ is now a Haar measure on $A_p^{\times}$, and $\|x\|=(N:Nx)^{-1}$ for any $\mathbb{Z}_p$-lattice N spanning A_p over $\mathbb{Q}_p$. Further, as in [1], formula (13), we have the following *Euler product formula* for L-functions

$$L_{\Lambda}(M,s,\psi)=\prod_p L_{\Lambda_p}(M_p,s,\psi_p),\qquad \mathrm{Re}(s)>1. \tag{2. 6}$$

(2. 7) Remark. Suppose that A is an F-algebra, where F is an algebraic number field with ring of integers R, and let Λ be an R-order in A. Then for each non-Archimedean prime P of R, we may form the P-adic completions A_P, Λ_P, and so on. If M is a left ideal of Λ, and ψ a character of $\mathrm{Cl}(M)$, we may proceed as above, using P in place of p at each step. In particular, ψ_{fin} has a decomposition $\psi_{\mathrm{fin}} = \prod_P \psi_P$, where $\psi_P = \psi | A_P^\times$. The function $L_{\Lambda_P}(M_P, s, \psi_P)$ is defined as in (2. 1), and this L-function satisfies a formula analogous to (2. 5). We also have an Euler product as in (2. 6), with p replaced by P, where now P ranges over all maximal ideals of R. Furthermore, we obtain

$$(2.\ 8) \qquad L_{\Lambda_p}(M_p, s, \psi_p) = \prod_{P|p} L_{\Lambda_P}(M_P, s, \psi_P),$$

where the product extends over all primes P of F lying above the rational prime p.

§ 3. Standard L-functions (local case)

Suppose first that A is a Q-algebra, and that ψ is any continuous character of finite order[1]) of the idele group $J(A)$ which is trivial on $A^\times$. We wish to define a "standard" L-function $L_A(s, \psi)$ which depends only on A and ψ. Our ultimate aim is to compare the L-function $L_A(M, s, \psi)$, defined in § 1, with the standard L-function $L_A(s, \psi)$. This comparison is one of the key steps in the present paper, and generalises our earlier results on zeta-functions (see [1], Theorem 1).

For the moment, we keep the assumption that A is a Q-algebra, and that ψ is a continuous character of $J(A)$ of finite order, trivial on $A^\times$. We intend to define $L_A(s, \psi)$ as an Euler product

$$(3.\ 1) \qquad L_A(s, \psi) = \prod_p L_{A_p}(s, \psi_p),$$

where p ranges over all rational prime numbers, and $\psi_p = \psi | A_p^\times$. In this section, we construct the local standard L-functions $L_{A_p}(s, \psi_p)$. We follow [7], § 3, in this, by specifying $L_{A_p}(s, \psi_p)$ via a set of axioms, and then proving that such a function exists.

Changing notation for the rest of this section, we now assume that A is a Q_p-algebra, and that ψ is a continuous character of finite order of the group $A^\times$. Let $\mathcal{S}(A)$ be the space of Schwartz-Bruhat functions on A, that is, the space of complex-valued locally constant functions on A with compact support. For $\Phi \in \mathcal{S}(A)$, we define

$$(3.\ 2) \qquad Z(\Phi, s, \psi) = \int_{A^\times} \Phi(x)\, \psi(x)\, \|x\|^s d^\times x,$$

where $d^\times x$ is a Haar measure on $A^\times$. This integral converges to a holomorphic function of s, at least for $\mathrm{Re}(s) > 1$. The axiomatic conditions which characterise our proposed standard L-function $L_A(s, \psi)$ may now be stated in terms of the integrals $Z(\Phi, s, \psi)$, and are as follows:

---

[1]) Throughout we consider only characters of finite order of $J(A)$ (in the global case) or of $A^\times$ (in the local case), since these arise from the module theory. Our treatment extends to general characters with virtually no change, and even, with only minor modifications, to quasicharacters.

C1) *There is a polynomial* $f(X) \in \mathbf{C}[X]$ *with* $f(0)=1$, *such that* $L_A(s, \psi)=f(p^{-s})^{-1}$.

C2) *There exists* $\Phi \in \mathscr{S}(A)$ *such that* $L_A(s, \psi)=Z(\Phi, s, \psi)$.

C3) *For any* $\Phi \in \mathscr{S}(A)$, *the function* $L_A(s, \psi)^{-1} \cdot Z(\Phi, s, \psi)$ *is a polynomial in* p^s *and* p^{-s} *with constant (complex) coefficients.*

We shall show below that such a function $L_A(s, \psi)$ exists by giving an explicit formula for it. In any case, it is clear that if such a function does exist, it must be unique.

We begin by reducing the construction of $L_A(s, \psi)$ to the case of simple algebras, by means of

(3. 3) Proposition. *Let* $A=\prod_{i=1}^{r} A_i$ *be a finite-dimensional semisimple* $\mathbf{Q}_p$*-algebra with simple components* A_i, *and let* $\psi_i=\psi|A_i^{\times}$, $1 \leqq i \leqq r$. *Suppose that, for each i, there exists a standard L-function* $L_{A_i}(s, \psi_i)$ *satisfying* C1—C3. *Then the standard L-function* $L_A(s, \psi)$ *also exists, and is given by*

$$L_A(s, \psi)=\prod_{i=1}^{r} L_{A_i}(s, \psi_i).$$

Proof. We need only check that the above product satisfies C1—C3. Here, C1 and C2 are obviously satisfied, while C3 follows readily from the fact that

$$\mathscr{S}(A)=\mathscr{S}(A_1) \otimes_{\mathbf{C}} \cdots \otimes_{\mathbf{C}} \mathscr{S}(A_r).$$

It therefore suffices for us to define $L_A(s, \psi)$ when A is a *simple* $\mathbf{Q}_p$-algebra. As a first step in this direction, we prove the following local version of (1. 10).

(3. 4) Lemma. *Let* A *be a finite-dimensional simple* $\mathbf{Q}_p$*-algebra with centre* C. *Let* R *be the valuation ring of the field* C, *and let* Λ' *be a maximal order in* A. *Then for each continuous character* ψ *of* $A^{\times}$ *of finite order, there is a unique character* $\tilde{\psi}$ *of* $C^{\times}$ *such that* $\psi=\tilde{\psi} \circ \mathrm{nr}$, *where* nr *is the reduced norm map from* $A^{\times}$ *to* $C^{\times}$. *Further*

$$\mathrm{nr}(A^{\times})=C^{\times}, \quad \mathrm{nr}(\Lambda'^{\times})=R^{\times}. \tag{3. 5}$$

Thus ψ *is trivial on* $(\Lambda')^{\times}$ *if and only if* $\tilde{\psi}$ *is trivial on* $R^{\times}$.

Proof. The assertions in (3. 5) are well-known, and follow readily from the description of maximal orders given in [11]. Further, the Nakayama-Matsushima theorem states that the commutator subgroup of $A^{\times}$ coincides with the kernel of the reduced norm. Thus there is an isomorphism

$$A^{\times}/[A^{\times}, A^{\times}] \cong C^{\times},$$

induced by the reduced norm map. Since ψ is trivial on $[A^{\times}, A^{\times}]$, it now follows that $\psi=\tilde{\psi} \circ \mathrm{nr}$ for some uniquely determined character $\tilde{\psi}$ of $C^{\times}$. The remaining statements of the lemma are now obvious.

The character $\tilde{\psi}$ of $C^{\times}$ is called *unramified* if it is trivial on $R^{\times}$, and *ramified* otherwise. We extend this terminology in the obvious way by saying ψ is unramified if and only if $\tilde{\psi}$ is unramified, that is, if and only if ψ is trivial on $(\Lambda')^{\times}$, for some (or equivalently, any) maximal order Λ'. We now prove our first main result.

(3. 6) Theorem. *Let A be a simple $\mathbb{Q}_p$-algebra with centre C. Let R be the valuation ring of C, and $\mathfrak{p}=\pi R$ the maximal ideal of R. Put:*

$$\mathcal{N}\mathfrak{p}=(R:\mathfrak{p}),$$

so that $\mathcal{N}$ is the usual counting norm. Write $A=M_k(D)$, where D is a division algebra with centre C, and let

$$\dim_C D=e^2, \quad \dim_C A=n^2, \quad n=ke.$$

Given a continuous character ψ of $A^\times$ of finite order, let $\tilde{\psi}$ be the character of $C^\times$ such that $\psi=\tilde{\psi}\circ \mathrm{nr}$, as in (3. 4). Define

$$L_C(s,\tilde{\psi})=\begin{cases}(1-\tilde{\psi}(\pi)\mathcal{N}\mathfrak{p}^{-s})^{-1}, & \text{if } \tilde{\psi} \text{ is unramified,}\\ 1, & \text{otherwise,}\end{cases} \tag{3.7}$$

and

$$L_A(s,\psi)=\prod_{\substack{j=0\\ j\equiv 0 \,(\mathrm{mod}\, e)}}^{n-1} L_C(ns-j,\tilde{\psi}). \tag{3.8}$$

Then $L_A(s,\psi)$ is the standard L-function satisfying the conditions C1—C3.

Remark. The notations in (3. 7), (3. 8) are consistent, in that $L_C(s,\tilde{\psi})$ is the standard L-function for the pair C, $\tilde{\psi}$.

Proof. The proof splits into two distinct cases, depending on whether ψ is unramified or not. We start with the harder case in which ψ is unramified. When ψ is the trivial character of $A^\times$, the expression in (3. 8) is precisely the zeta-function $\zeta_{\Lambda'}(s)$, where Λ' is a maximal order in A. Indeed, this is just a reformulation of Hey's formula for $\zeta_{\Lambda'}(s)$. See [1], formula (16). Further, the fact that $\zeta_{\Lambda'}(s)$ satisfies C1—C3 in this case where $\psi=1$ is just a restatement of [1], Theorem 1. See also [3], § 5. 2.

We shall now generalise the preceding remarks to the case of an arbitrary unramified character ψ.

(3. 9) Lemma. *Let Λ' be a maximal order in A, and let ψ be any unramified character of $A^\times$. Then*

$$L_{\Lambda'}(\Lambda',s,\psi)=L_A(s,\psi),$$

where $L_{\Lambda'}(\Lambda',s,\psi)$ is the L-function of Λ' defined in (2. 1), and $L_A(s,\psi)$ is as in (3. 8).

Proof. We shall imitate the proof of Hey's formula for $\zeta_{\Lambda'}(s)$ given in [1], § 3. 3. Keeping the notation of (3. 4) and (3. 6), let Δ denote the unique maximal R-order in D. Then Δ is a noncommutative discrete valuation ring, and we may choose its prime element ξ so that $\mathrm{nr}_{D/C}(\xi)=\pi$, where π is a prime element of R. We have $\xi\Delta=\Delta\xi$, and $(\Delta:\Delta\xi)=\mathcal{N}\mathfrak{p}^e$. Any two choices of the maximal order Λ' are conjugate in A, and it follows that $L_{\Lambda'}(\Lambda',s,\psi)$ is independent of the choice of Λ'. Thus we may assume that $\Lambda'=M_k(\Delta)$. Since every left ideal of Λ' is principal, and since ψ is trivial on $(\Lambda')^\times$, the L-function $L_{\Lambda'}(\Lambda',s,\psi)$ can be written as

$$L_{\Lambda'}(\Lambda',s,\psi)=\sum_x \psi(x)\,(\Lambda':\Lambda'x)^{-s}, \qquad \mathrm{Re}(s)>1.$$

Here, x ranges over a full set of representatives of the orbits of $\Lambda' \cap A^{\times}$ under the left action of $(\Lambda')^{\times}$. Since $(\Lambda')^{\times} = GL_k(\Delta)$, we may take these orbit representatives in Hermite canonical form $x = (a_{ij}) \in M_k(\Delta)$, where the a_{ij} satisfy the restrictions

$$\begin{cases} a_{ii} = \xi^{r_i}, \quad r_i \geqq 0, \\ a_{ij} = 0 \ \text{if} \ j < i, \\ a_{ij} \ \text{ranges over a full set of representatives of} \ \Delta \bmod \xi^{r_j}\Delta, \ \text{if} \ j > i. \end{cases}$$

Now we have

$$\begin{aligned} \mathrm{nr}_{A/C}(x) &= (\mathrm{nr}_{D/C}\,\xi)^{r_1 + \cdots + r_k} = \pi^{\Sigma r_i}, \\ (\Lambda' : \Lambda' x) &= (\Delta : \Delta\xi)^{ke(r_1 + \cdots + r_k)} = \mathcal{N}\mathfrak{p}^{n\Sigma r_i}. \end{aligned}$$

For a given choice of the parameters $r_1, \ldots, r_k$, there are

$$\mathcal{N}\mathfrak{p}^{e(r_2 + 2r_3 + \cdots + (k-1)r_k)}$$

choices for the coset representatives x. Then $L_{\Lambda'}(\Lambda', s, \psi)$ is equal to

$$\sum_{r_1, \ldots, r_k \geqq 0} \tilde{\psi}(\pi)^{r_1 + \cdots + r_k} \mathcal{N}\mathfrak{p}^{e(r_2 + \cdots + (k-1)r_k)} \mathcal{N}\mathfrak{p}^{-ns(r_1 + \cdots + r_k)}.$$

This multiple series is easily summed, and gives precisely the desired formula

$$(3.\,10) \qquad L_{\Lambda'}(\Lambda', s, \psi) = \prod_{\substack{j=0 \\ j \equiv 0 \ (\mathrm{mod}\ e)}}^{n-1} (1 - \tilde{\psi}(\pi)\mathcal{N}\mathfrak{p}^{-(ns-j)})^{-1}$$

for $\mathrm{Re}(s) > 1$. This completes the proof of the lemma.

(3. 11) Corollary. *Keep the above notation. For some* $\varepsilon > 0$, *the function* $L_A(s, \psi)$ *is holomorphic and non-zero for* $\mathrm{Re}(s) > 1 - \varepsilon$.

To complete the proof of (3. 6) for the case where ψ is unramified, we must show that the function $L_A(s, \psi)$ in (3. 8) satisfies the conditions C1—C3. Of these, C1 is obviously true. Condition C2 also holds, because by (3. 9) and (2. 4),

$$L_A(s, \psi) = L_{\Lambda'}(\Lambda', s, \psi) = \mu^{\times}(\Lambda'^{\times})^{-1} Z(\Phi_{\{\Lambda' : \Lambda'\}}, s, \psi).$$

It remains to verify C3, and here we follow the proof in [1] of the lemma on p. 148. Let $\Phi \in \mathcal{S}(A)$; we may express Φ as a $\mathbb{C}$-linear combination of a finite number of characteristic functions of "spheres" $w + p^f \Lambda'$, where f is some fixed positive integer (depending on Φ), and w ranges over some finite set of elements of A. It then suffices to show that

$$L_A(s, \psi)^{-1} \int_{(w + p^f \Lambda') \cap A^{\times}} \psi(x)\, \|x\|^s d^{\times} x$$

lies in $\mathbb{C}[p^s, p^{-s}]$. Replacing x by $p^m x$, for some m, changes the integral by a factor of the form $c \cdot (p^{-s})^l$, for some non-zero c and some $l \in \mathbb{Z}$. This does not affect the validity of the desired result. Likewise, replacing x by ux, with $u \in (\Lambda')^{\times}$, has no effect on the integral, since ψ is trivial on $(\Lambda')^{\times}$ by hypothesis. As in [1], the problem can then be reduced to proving that

$$L_A(s, \psi)^{-1} \int_{(\Lambda')^{\times} \cdot \{(\gamma + p^f \Lambda') \cap A^{\times}\}} \psi(x)\, \|x\|^s d^{\times} x$$

lies in $\mathbb{C}[p^s, p^{-s}]$, where now $\Lambda' = M_k(\Delta)$ as in the proof of (3. 9), and where γ is a diagonal matrix with 1's and 0's along the main diagonal. The above integral is then converted into a series by reversing the argument used in establishing formula (2. 5). This series can be summed explicitly, as in the proof of (3. 9), and the sum can be compared with the expression (3. 8) for $L_A(s, \psi)$ to yield the desired result. This completes the proof of (3. 6) for the case of ψ unramified.

We turn finally to the case where ψ is ramified, that is, ψ is not trivial on $(\Lambda')^\times$ for any maximal order Λ' in A. In this case, the assertion is that $L_A(s, \psi)$ is identically 1, so C1 surely holds. To prove C2, we note that $\ker(\psi|(\Lambda')^\times)$ is an open subgroup of $(\Lambda')^\times$ of finite index. Therefore $\ker(\psi) \supseteq 1 + \mathfrak{f}\Lambda'$ for some proper ideal $\mathfrak{f}$ of R. Let Φ be the characteristic function of the set $1 + \mathfrak{f}\Lambda'$ in A. Then $\Phi \in \mathscr{S}(A)$, and

$$Z(\Phi, s, \psi) = \mu^\times(1 + \mathfrak{f}\Lambda').$$

This measure is non-zero and finite, which shows that the constant function 1 is of the form $Z(\Phi', s, \psi)$ for some $\Phi' \in \mathscr{S}(A)$, as desired.

We now establish C3, that is, $Z(\Phi, s, \psi) \in C[p^s, p^{-s}]$ for every $\Phi \in \mathscr{S}(A)$. As in the above proof for unramified characters (see also [3], § 5. 2), it suffices to do this when Φ is the characteristic function of a set $\gamma + \mathfrak{f}'\Lambda'$, where $\mathfrak{f}'$ is an ideal of R contained in the ideal $\mathfrak{f}$ above, and where γ is a diagonal matrix of the form

$$\gamma = \operatorname{diag}(1, 1, \ldots, 1, 0, \ldots, 0) \in M_k(\Delta) = \Lambda'.$$

Suppose that the first r diagonal entries are 1, the rest 0. If $r = k$, then $\gamma = 1_A$, and we have just seen that $Z(\Phi, s, \psi)$ is constant in this case. On the other hand, we show that $Z(\Phi, s, \psi) = 0$ if $r < k$. Indeed, if $r < k$ we choose $y \in (\Lambda')^\times$ of the form $y = \operatorname{diag}(1, \ldots, 1, u)$ with $u \in \Delta^\times$ chosen so that

$$\psi(y) = \tilde{\psi}(\operatorname{nr}_{A/C} y) = \tilde{\psi}(\operatorname{nr}_{D/C} u) \neq 1.$$

Then

$$Z(\Phi, s, \psi) = \int_{A^\times} \Phi(x)\,\psi(x)\,\|x\|^s d^\times x = \int_{A^\times} \Phi(yx)\,\psi(yx)\,\|yx\|^s d^\times x = \psi(y)\, Z(\Phi, s, \psi),$$

since $\Phi(yx) = \Phi(x)$, $x \in A$, and since $\|y\| = 1$. But $\psi(y) \neq 1$, which forces $Z(\Phi, s, \psi)$ to be zero, as desired. This completes the proof of (3. 6).

Let us apply this theorem to prove our next major result, which compares the local L-function $L_\Lambda(M, s, \psi)$ defined in (2. 1) with the standard L-function $L_A(s, \psi)$. This result, which generalises [1], Theorem 1, is as follows:

(3. 12) Theorem. *Let Λ be an order in a finite-dimensional semisimple Q_p-algebra A, M a left ideal of Λ, and ψ a character of $A^\times$ which is trivial on $\operatorname{Aut}_\Lambda(M)$. Then there exists a polynomial*

$$f(s) \in \mathbb{Z}[\psi][p^{-s}],$$

where $\mathbb{Z}[\psi]$ is the ring obtained by adjoining to $\mathbb{Z}$ all the values of ψ, such that

$$L_\Lambda(M, s, \psi) = f(s)\, L_A(s, \psi).$$

Further, $f(s)$ is identically 1 if Λ is a maximal order.

Proof. For $\mathrm{Re}(s)>1$, let us define $f(s)=L_A(M, s, \psi)/L_A(s, \psi)$. Both of these L-functions lie in the ring $\mathbb{Z}[\psi]\,[[p^{-s}]]$ of formal power series in p^{-s} with coefficients in the ring $\mathbb{Z}[\psi]$. Further, $L_A(s, \psi)^{-1}\in\mathbb{Z}[\psi]\,[p^{-s}]$, and it follows that

$$f(s)\in\mathbb{Z}[\psi]\,[[p^{-s}]].$$

On the other hand, we may use (2. 5) to express $L_A(M, s, \psi)$ in terms of a zeta-integral $Z(\Phi, s, \psi)$, and it then follows from C3 that $f(s)\in\mathbb{C}[p^s, p^{-s}]$. Therefore

$$f(s)\in\mathbb{C}[p^s, p^{-s}]\cap\mathbb{Z}[\psi]\,[[p^{-s}]]=\mathbb{Z}[\psi]\,[p^{-s}],$$

as required.

If Λ is a maximal order, then $M\cong\Lambda$ as Λ-modules, and so $L_\Lambda(M, s, \psi)$ is equal to $L_\Lambda(\Lambda, s, \psi)$, which is the same as $L_A(s, \psi)$, by (3. 9). Observe that when Λ is a maximal order, the restriction of ψ to any simple component of A is necessarily unramified.

(3. 13) Corollary. *Each local L-function admits analytic continuation to a meromorphic function on the whole s-plane. Further, there exists* $\varepsilon>0$ *such that* $L_\Lambda(M, s, \psi)$ *is holomorphic for* $\mathrm{Re}(s)>1-\varepsilon$.

Proof. Immediate from (3. 11) and (3. 12).

From C3 and (3. 6) we also deduce

(3. 14) Corollary. *Let* A *be a finite-dimensional semisimple* $\mathbb{Q}_p$*-algebra, and* ψ *a continuous character of* $A^\times$ *of finite order. Then the function* $Z(\Phi, s, \psi)$, *defined by* (3. 2), *admits analytic continuation to a meromorphic function of s. Moreover, there exists* $\varepsilon>0$ *such that, for all* $\Phi\in\mathcal{S}(A)$, *the integral* (3. 2) *converges absolutely to a holomorphic function of s in the region* $\mathrm{Re}(s)>1-\varepsilon$.

§ 4. Standard *L*-functions (global case)

Throughout this section, A is a finite-dimensional semisimple $\mathbb{Q}$-algebra, and ψ is a continuous character of $J(A)$ of finite order which is trivial on $A^\times$. For each prime number p, the local standard L-function $L_{A_p}(s, \psi_p)$ has been defined in § 3. Now we have to show that the Euler product (3. 1) converges well enough to give a definition of the desired global standard L-function $L_A(s, \psi)$. We must also establish certain analytic properties of $L_A(s, \psi)$, with a view to applying these to the study of our global L-functions $L_\Lambda(M, s, \psi)$ in the next section.

(4. 1) Theorem. *Let* $A=\prod_{i=1}^{r} A_i$, *where the* A_i *are simple algebras, and let* $\psi_i=\psi|J(A_i)$, $1\leqq i\leqq r$.

(i) *The Euler product* (3. 1) *converges absolutely, uniformly on compact sets, in the region* $\mathrm{Re}(s)>1$, *to a holomorphic function* $L_A(s, \psi)$.

Moreover

$$L_A(s, \psi)=\prod_{i=1}^{r} L_{A_i}(s, \psi_i). \tag{4. 2}$$

(ii) *The function* $L_A(s, \psi)$ *admits analytic continuation to a meromorphic function of* s *on the whole* s*-plane.*

(iii) *Let* $t = t(\psi)$ *be the number of indices* i *for which* ψ_i *is the trivial character. Then* $L_A(s, \psi)$ *has a pole of order* t *at* $s=1$. *If* $t(\psi)=0$, *then* $L_A(s, \psi)$ *is an entire function of* s, *and* $L_A(1, \psi) \neq 0$.

Proof. We start with the case in which the algebra A is *simple*. Let C denote the centre of A, and let R be the ring of algebraic integers in C. Let $\dim_C(A) = n^2$, and let $\tilde{\psi}$ be the idele-class character of C such that $\psi = \tilde{\psi} \circ \mathrm{nr}_{A/C}$, as in (1. 10). For each prime number p, we have

$$A_p = \prod_{\mathfrak{p}|p} A_{\mathfrak{p}},$$

where the product is taken over all maximal ideals $\mathfrak{p}$ of R lying over p. Each $A_{\mathfrak{p}}$ is a central simple $C_{\mathfrak{p}}$-algebra, of dimension n^2, and is isomorphic to a full ring of matrices over some central $C_{\mathfrak{p}}$-division algebra of dimension $e_{\mathfrak{p}}^2$, say. It is well-known that there are only finitely many $\mathfrak{p}$ for which $e_{\mathfrak{p}} > 1$. By (3. 3) we have

$$L_{A_p}(s, \psi_p) = \prod_{\mathfrak{p}|p} L_{A_{\mathfrak{p}}}(s, \psi_{\mathfrak{p}}), \tag{4. 3}$$

where, as before, $\psi_{\mathfrak{p}} = \psi | A_{\mathfrak{p}}^{\times}$. The Euler product (3. 1) therefore amounts to

$$\prod_{\mathfrak{p}} L_{A_{\mathfrak{p}}}(s, \psi_{\mathfrak{p}}), \tag{4. 4}$$

where $\mathfrak{p}$ ranges over all maximal ideals of R. We have shown in (3. 6) that

$$L_{A_{\mathfrak{p}}}(s, \psi_{\mathfrak{p}}) = \prod_{\substack{j=0 \\ j \equiv 0 \pmod{e_{\mathfrak{p}}}}}^{n-1} L_C(ns-j, \tilde{\psi}_{\mathfrak{p}}), \tag{4. 5}$$

with $L_{C_{\mathfrak{p}}}(s, \tilde{\psi}_{\mathfrak{p}})$ defined as in (3. 7). We may form

$$L_C(s, \tilde{\psi}) = \prod_{\mathfrak{p}} L_{C_{\mathfrak{p}}}(s, \tilde{\psi}_{\mathfrak{p}}), \tag{4. 6}$$

(product over all maximal ideals of R), which is the classical L-function attached to the idele-class character $\tilde{\psi}$ of C. We need the following basic properties of $L_C(s, \tilde{\psi})$, the proofs of which may be found in [9], [10] and [13]:

Li) *The product* (4. 6) *converges absolutely, uniformly on compact sets, in the region* $\mathrm{Re}(s) > 1$.

Lii) *The function* $L_C(s, \tilde{\psi})$ *defined by* (4. 6) *admits analytic continuation to a meromorphic function of* s, *with no zeros in the region* $\mathrm{Re}(s) > 1$.

Liii) *If* $\tilde{\psi}$ *is the trivial character, then* $L_C(s, \tilde{\psi})$ *has a simple pole at* $s=1$, *while otherwise* $L_C(s, \tilde{\psi})$ *is entire and* $L_C(1, \tilde{\psi}) \neq 0$.

We now define an auxiliary function:

$$K_A(s, \psi) = \prod_{j=0}^{n-1} L_C(ns-j, \tilde{\psi}), \tag{4. 7}$$

and its local analogue:

$$K_{A_{\mathfrak{p}}}(s, \psi_{\mathfrak{p}}) = \prod_{j=0}^{n-1} L_{C_{\mathfrak{p}}}(ns-j, \tilde{\psi}_{\mathfrak{p}}). \tag{4. 8}$$

The properties Li)—Liii) imply immediately that

Ki) *The Euler product* $\prod_{\mathfrak{p}} K_{A_{\mathfrak{p}}}(s, \psi_{\mathfrak{p}})$ *converges absolutely, uniformly on compact sets, to* $K_A(s, \psi)$ *in the region* $\mathrm{Re}(s) > 1$.

Kii) $K_A(s, \psi)$ *admits analytic continuation to a meromorphic function of* s.

Kiii) *If* ψ, *or equivalently* $\tilde{\psi}$, *is the trivial character, then* $K_A(s, \psi)$ *has a simple pole at* $s=1$.

Kiv) *If* ψ *is non-trivial, then* $K_A(s, \psi)$ *is an entire function of* s, *and* $K_A(1, \psi) \neq 0$.

Comparing (4. 7), (4. 8) with (4. 5), we find that

$$L_A(s, \psi) = K_A(s, \psi) \prod_{\mathfrak{p}} f_{\mathfrak{p}}(s), \tag{4. 9}$$

where, for each $\mathfrak{p}$, $f_{\mathfrak{p}}(s)$ is a polynomial in p^{-s}, where p is the prime number lying below $\mathfrak{p}$. Explicitly:

$$f_{\mathfrak{p}}(s) = \prod_{\substack{j=0 \\ j \not\equiv 0 \pmod{e_{\mathfrak{p}}}}}^{n-1} \left(1 - \tilde{\psi}_{\mathfrak{p}}(\pi_{\mathfrak{p}}) \mathcal{N}\mathfrak{p}^{-(ns-j)}\right) \quad \text{if } \tilde{\psi}_{\mathfrak{p}} \text{ is unramified,} \tag{4. 10}$$

$$f_{\mathfrak{p}}(s) = 1 \quad \text{if } \tilde{\psi}_{\mathfrak{p}} \text{ is ramified.} \tag{4. 11}$$

(Here, $\pi_{\mathfrak{p}}$ denotes some prime element of the discrete valuation ring $R_{\mathfrak{p}}$.) For each $\mathfrak{p}$, therefore, we have $f_{\mathfrak{p}}(1) \neq 0$. Also, $f_{\mathfrak{p}}(s)$ is identically 1 unless $e_{\mathfrak{p}} > 1$ and $\tilde{\psi}_{\mathfrak{p}}$ is unramified. It follows that $L_A(s, \psi)/K_A(s, \psi)$ is a *finite* product of exponential polynomials $f_{\mathfrak{p}}(s)$, and is therefore continuable to an entire function of s which does not vanish at $s=1$. The required properties of $L_A(s, \psi)$ now follow from Ki)—Kiv), and we have finished the proof of (4. 1) for the case of A simple.

Now we treat the general case, where A is semisimple. By (3. 3) we have

$$L_{A_p}(s, \psi_p) = \prod_{i=1}^{r} L_{A_{i,p}}(s, \psi_{i,p}).$$

We know that the Euler products for the $L_{A_i}(s, \psi_i)$ converge absolutely for $\mathrm{Re}(s) > 1$, so we may rearrange them to give

$$\prod_{i=1}^{r} \prod_{p} L_{A_{i,p}}(s, \psi_{i,p}) = \prod_{p} \prod_{i=1}^{r} L_{A_{i,p}}(s, \psi_{i,p}) = \prod_{p} L_{A_p}(s, \psi_p),$$

which establishes the convergence of this last product, and also the truth of (4. 2). All the statements concerning $L_A(s, \psi)$ now follow from the simple case.

§ 5. *L*-functions of orders

We now return to the global situation of § 1, where Λ is an order in the finite-dimensional semisimple Q-algebra A, M is a left ideal of Λ, and ψ is a character of the genus class group $\mathrm{Cl}(M)$ of M. Let $\Gamma = \mathrm{End}_{\Lambda}(M)$, so that Γ is also an order in A. As shown in § 1, we can view ψ as a continuous character of $J(A)$ which is trivial on $J'(A) \cdot A^{\times} \cdot U(\Gamma)$. Let $L_A(s, \psi)$ be the standard L-function defined in § 4. We are now ready to prove one of our main results.

(5. 1) Theorem. i) *The global L-function* $L_A(M, s, \psi)$ *admits analytic continuation to a meromorphic function on the whole s-plane.*

ii) *Let* $A = \prod_{i=1}^{r} A_i$, *where the* A_i *are simple algebras, and let* ψ_i *be the restriction of* ψ *to* $J(A_i)$, *for* $1 \leqq i \leqq r$. *If exactly* $t = t(\psi)$ *of the characters* ψ_i *are trivial, then* $L_A(M, s, \psi)$ *has a pole of order at most t at* $s=1$.

iii) *If the character* ψ *is non-trivial, then* $L_A(M, s, \psi)$ *has a pole of order at most* $r-1$ *at* $s=1$. *If* $t(\psi)=0$, *then* $L_A(M, s, \psi)$ *is an entire function.*

iv) *If* Λ *is a maximal order (so that* ψ_p *is unramified for all p), then*

$$L_A(M, s, \psi) = L_A(s, \psi).$$

Proof. From the Euler products (2. 6) and (3. 1), we have, for $\mathrm{Re}(s) > 1$:

$$L_A(M, s, \psi)/L_A(s, \psi) = \prod_p f_p(s),$$

where for each prime number p,

$$f_p(s) = L_{A_p}(M_p, s, \psi_p)/L_{A_p}(s, \psi_p).$$

By (3. 12), each $f_p(s)$ is a polynomial in p^{-s} with coefficients in $\mathbb{Z}[\psi_p]$. Further, $f_p(s)$ is identically 1 for almost all primes p; indeed, $f_p(s)=1$ by (3. 12) whenever Λ_p is a maximal order. Thus we obtain

$$L_A(M, s, \psi) = L_A(s, \psi) \prod_p f_p(s). \tag{5. 2}$$

In view of the properties of $L_A(s, \psi)$ given in (4. 1), this identity immediately implies all the assertions of our theorem.

It seems difficult to determine the exact order of the pole $s=1$ for the function $L_A(M, s, \psi)$, since we do not know (in general) whether $f_p(1) \neq 0$.

(5. 3) Corollary. *The partial zeta-function* $Z_A([M], s)$ *defined in* § 1 *admits analytic continuation to a meromorphic function of s, with a pole of order at most r at* $s=1$.

Proof. This is clear from (5. 1) and formula (1. 5).

In (6. 1), we shall show that the order of this pole is precisely r.

Our methods yield no direct information about the partial zeta-function $Z_A(M, s)$, except in those cases where stable isomorphism implies ordinary isomorphism. It seems highly probable that the following conjecture holds:

(5. 4) Conjecture. *Each partial zeta-function* $Z_A(M, s)$ *admits analytic continuation to a meromorphic function of s.*

§ 6. Behaviour at $s=1$

Throughout this section, let Λ be an order in the finite-dimensional semisimple $\mathbb{Q}$-algebra A, and let M be a left ideal of Λ.

(6. 1) Proposition. *Each of the functions* $\zeta_A(s)$, $Z_A(g(M), s)$ *and* $Z_A([M], s)$ *has a pole of order precisely r at* $s=1$, *where r is the number of simple components of the algebra A. The same is true for* $Z_A(M, s)$ *if Conjecture* (5. 4) *holds.*

Proof. Let Λ' denote a maximal order in A. Since every left ideal of Λ' lies in the genus $g(\Lambda')$, we have

$$\zeta_{\Lambda'}(s) = Z_{\Lambda'}(g(\Lambda'), s) = L_{\Lambda'}(\Lambda', s, \psi_0),$$

where ψ_0 is the trivial character of $\mathrm{Cl}(\Lambda')$ (or of $J(A)$). But

$$L_{\Lambda'}(\Lambda', s, \psi_0) = L_A(s, \psi_0) \tag{6. 2}$$

by (5. 1) (iv). It now follows from (4. 1) that $\zeta_{\Lambda'}(s)$ has a pole of order r at $s=1$.

Now let Λ, Δ be orders in A, and let M be a left ideal of Λ, N a left ideal of Δ. Then for $\mathrm{Re}(s) > 1$,

$$Z_\Lambda(M, s) = \sum_{\substack{X \subseteq \Lambda \\ X \cong M}} (\Lambda : X)^{-s} = (\Lambda : M)^{-s} \sum_x \|x\|^s,$$

where x ranges over the orbits of $\{M : \Lambda\} \cap A^\times$ under the left action of the group $\mathrm{Aut}_\Lambda(M)$ (see (2. 2) and [1], § 2). Similarly,

$$Z_\Delta(N, s) = (\Delta : N)^{-s} \sum_y \|y\|^s,$$

the sum extending over orbit representatives y of $\{N : \Delta\} \cap A^\times$ modulo $\mathrm{Aut}_\Delta(N)$. Now consider the group

$$W = \mathrm{Aut}_\Lambda(M) \cap \mathrm{Aut}_\Delta(N).$$

Lemma. *Let $\Gamma_1 \supset \Gamma_2$ be orders in a finite-dimensional semisimple $\mathbb{Q}$-algebra B. Then the unit index $(\Gamma_1^\times : \Gamma_2^\times)$ is finite.*

Proof. There is a positive integer a so that Γ_2 contains $a\Gamma_1$, and therefore also the order $\mathbb{Z} + a\Gamma_1$. The index $(\Gamma_1^\times : \Gamma_2^\times)$ is then at most the order of the unit group of the finite ring $\Gamma_1/a\Gamma_1$.

Since W is the group of units of the order $\mathrm{End}_\Lambda(M) \cap \mathrm{End}_\Delta(N)$, it follows from the lemma that W is of finite index in both $\mathrm{Aut}_\Lambda(M)$ and $\mathrm{Aut}_\Delta(N)$. So we get

$$Z_\Lambda(M, s) = c_1 (\Lambda : M)^{-s} \sum_{x \in W \backslash (\{M : \Lambda\} \cap A^\times)} \|x\|^s,$$

for some positive constant c_1. There is a similar expression for $Z_\Delta(N, s)$ involving a constant $c_2 > 0$. We may now choose a positive integer q such that $q \cdot \{N : \Delta\} \subseteq \{M : \Lambda\}$. If $\dim_{\mathbb{Q}}(A) = n$, and if s is real, $s > 1$, then

$$\begin{aligned} Z_\Delta(N, s) &= c_2 q^{ns} (\Delta : N)^{-s} \sum_{y \in W \backslash \{N : \Delta\} \cap A^\times} \|qy\|^s \\ &= c_2 q^{ns} (\Delta : N)^{-s} \sum_{y \in W \backslash q\{N : \Delta\} \cap A^\times} \|y\|^s \\ &\leqq c_2 q^{ns} (\Delta : N)^{-s} c_1^{-1} (\Lambda : M)^s Z_\Lambda(M, s). \end{aligned}$$

Therefore, given $\varepsilon > 0$, there exists a constant $c > 0$ such that

$$Z_\Delta(N, s) \leqq c Z_\Lambda(M, s), \qquad 1 < s < 1 + \varepsilon. \tag{6. 3}$$

Now let $\{M_i\}$ be any finite collection of left ideals of Λ, and let $\{N_j\}$ be a finite collection of left ideals of Δ. Summing the inequalities (6. 3) for M_i and N_j, we find a constant $c'>0$ such that

$$\sum_j Z_\Delta(N_j, s) \leqq c' \sum_i Z_\Lambda(M_i, s), \qquad 1<s<1+\varepsilon.$$

The reverse inequality, with a different constant factor, holds by symmetry. Therefore any two such sums $\sum_i Z_\Lambda(M_i, s)$, $\sum_j Z_\Delta(N_j, s)$, which are meromorphic at $s=1$, must have poles of the same order at $s=1$. Since we already know that $\zeta_{\Lambda'}(s)$ has a pole of order r at $s=1$, the proposition follows at once.

For M a left ideal in an order Λ, we wish to compare the behaviour of $Z_\Lambda(g(M), s)$ and $\zeta_{\Lambda'}(s)$ at $s=1$, where Λ' is a maximal order in A. We define

$$(6.\ 4) \qquad d_M = \lim_{s\to 1} Z_\Lambda(g(M), s)/\zeta_{\Lambda'}(s),$$

a positive constant by virtue of (6. 1). This number is independent of the choice of Λ', by virtue of (6. 2), and is surprisingly easy to calculate.

(6. 5) Theorem. *Let Λ be an order in the finite-dimensional semisimple $\mathbb{Q}$-algebra A, Λ' a maximal order in A, and M a left ideal of Λ. The constant d_M defined in* (6. 4) *is given by*

$$(6.\ 6) \qquad d_M = (\Lambda : M)^{-1} (\{M : \Lambda\} : \Lambda') \prod_p (\Lambda_p'^{\times} : \mathrm{Aut}_{\Lambda_p}(M_p)).$$

Remark. In the above formula, $(\Lambda_p'^{\times} : \mathrm{Aut}_{\Lambda_p}(M_p))=1$ whenever $\mathrm{End}_{\Lambda_p}(M_p)$ is a maximal order in A_p. This is the case for almost all primes p, so the product extends over only finitely many p. Further, examples show that this product need not equal the global unit index $(\Lambda'^{\times} : \mathrm{Aut}_\Lambda(M))$.

Proof. For each prime number p, we have by (3. 12) or [1], Th. 1,

$$Z_{\Lambda_p}(M_p, s) = \zeta_{\Lambda'}(s)\, f_p(s),$$

for some polynomial $f_p(s) \in \mathbb{Z}[p^{-s}]$. Further, $f_p(s)$ is identically 1 if Λ_p is a maximal order. For $\mathrm{Re}(s)>1$, the Euler products for $Z_\Lambda(g(M), s)$ and $\zeta_{\Lambda'}(s)$ then imply that

$$Z_\Lambda(g(M), s)/\zeta_{\Lambda'}(s) = \prod_p f_p(s).$$

By analytic continuation, the same formula holds for all s, and we have

$$d_M = \prod_p f_p(1).$$

Therefore it suffices to prove that

$$(6.\ 7) \qquad f_p(1) = (\Lambda_p : M_p)^{-1} (\{M_p : \Lambda_p\} : \Lambda_p') (\Lambda_p'^{\times} : \mathrm{Aut}_{\Lambda_p}(M_p))$$

for each p. It follows readily from (3. 14) that the formula

$$Z_{\Lambda_p}(M_p, s) = \mu^\times (\mathrm{Aut}_{\Lambda_p}(M_p))^{-1} (\Lambda_p : M_p)^{-s} \int_{A_p^\times} \Phi_{\{M_p : \Lambda_p\}}(x)\, \|x\|^s d^\times x$$

(compare with (2. 5)) is valid in the region $\mathrm{Re}(s)>1-\varepsilon$, for some $\varepsilon>0$.

We may rewrite the integral as

$$(6.8) \qquad \int_{\{M_p:\Lambda_p\}\cap\Lambda_p^\times} \|x\|^{s-1}\,dx,$$

where dx is the additive Haar measure on A_p for which $d^\times x = \|x\|^{-1}\cdot dx$. To evaluate this integral at $s=1$, we need the following bit of folklore:

(6. 9) Lemma. *Let B be a finite-dimensional semisimple $\mathbf{Q}_p$-algebra. Then the set $B - B^\times = \{x \in B\colon x \notin B^\times\}$ has measure zero, with respect to any Haar measure on B.*

We postpone the proof of this to the end of the current section. To continue with the present argument, the integral (6. 8) is just

$$\int_{\{M_p:\Lambda_p\}} \|x\|^{s-1}\,dx,$$

and therefore

$$Z_{\Lambda_p}(M_p, 1) = \mu^\times(\mathrm{Aut}_{\Lambda_p}(M_p))^{-1}\,(\Lambda_p\colon M_p)^{-1}\,\mu(\{M_p\colon \Lambda_p\}),$$

where μ denotes the measure of a set with respect to dx. Likewise

$$\zeta_{\Lambda'_p}(1) = \mu^\times(\Lambda_p'^\times)^{-1}\,\mu(\Lambda'_p),$$

and therefore

$$f_p(1) = Z_{\Lambda_p}(M_p, 1)/\zeta_{\Lambda'_p}(1) = \mu^\times(\Lambda_p'^\times)\,\mu^\times(\mathrm{Aut}_{\Lambda_p}(M_p))^{-1}\,\mu(\{M_p\colon\Lambda_p\})\,\mu(\Lambda'_p)^{-1}\,(\Lambda_p\colon M_p)^{-1}.$$

The quotients of measures are group indices, and (6. 7) is proved.

Now let us write

$$\zeta_{\Lambda'}(s) = c_A(s-1)^{-r} + c'_A\,(s-1)^{1-r} + \cdots$$

for the Laurent expansion of $\zeta_{\Lambda'}(s)$ about $s=1$. We know that r is the number of simple components of the algebra A. By virtue of formula (6. 2) (see also [1], § 3), we may compute the positive constant c_A explicitly, in terms of values of Dedekind zeta-functions and various constants associated with the algebraic number fields occurring as the centres of the simple components of A. For convenience, we reproduce this formula below in (6. 12). By Proposition (6. 1), the Laurent expansion of $Z_\Lambda(g(M), s)$ has leading term $d_M c_A(s-1)^{-r}$, where d_M is given by (6. 6). We are now ready to give our main result:

(6. 10) Theorem. *Let Λ be an order in a finite-dimensional semisimple $\mathbf{Q}$-algebra A, and suppose that A has r simple components. Let Λ' be a maximal order in A, and let*

$$c_A = \lim_{s\to 1}(s-1)^r\,\zeta_{\Lambda'}(s).$$

Let M be a left ideal of Λ, and let h_M be the number of stable isomorphism classes in the genus $g(M)$. Then

(i) *For each stable isomorphism class $[N]$ in $g(M)$, the partial zeta function $Z_\Lambda([N], s)$ has a pole of order r at $s=1$, and its Laurent expansion about $s=1$ has leading term*

$$c_A d_M h_M^{-1}(s-1)^{-r},$$

with d_M given by (6. 6)

(ii) *For positive T tending to ∞, the number of left ideals X of Λ such that*

$$(\Lambda : X) \leqq T, \quad X \in g(M), \quad [X] = [N],$$

is asymptotically equal to

$$\frac{c_A d_M}{h_M} \frac{1}{(r-1)!} T(\log T)^{r-1}.$$

In particular, the left ideals of Λ in a given genus $g(M)$ are asymptotically uniformly distributed among the stable isomorphism classes in that genus.

Proof. Formulas (1. 5), (6. 4) and (5. 1, iii) show that $Z_\Lambda([N], s)$ has a pole of order r at $s=1$, and that the leading term of its Laurent expansion is $c_A d_M h_M^{-1}(s-1)^{-r}$. Further, it follows from (4. 7), (4. 9) and standard results on classical L-functions that $L_A(s, \psi)$ is holomorphic for $\mathrm{Re}(s)=1$ except for the pole at $s=1$. Since each $L_\Lambda(M, s, \psi)$ is the product of $L_A(s, \psi)$ and a holomorphic function, the same is true for $L_\Lambda(M, s, \psi)$, and hence for $Z_\Lambda([N], s)$. Assertion (ii) now follows from (i), using the Tauberian theorem of Delange and Ikehara, in the form quoted in [10], p. 464, Th. 1.

(6. 11) Corollary. *We have* $\lim_{s \to 1} (s-1)^r \zeta_\Lambda(s) = c_A \sum_M d_M$, *where M ranges over a (finite) set of representatives of the genera of left ideals of Λ.*

Now let us evaluate the constant c_A appearing in (6. 10) and (6. 11). Let Λ' be a maximal order in the Q-algebra A, and let A_i, $1 \leqq i \leqq r$, be the simple components of A. We may express Λ' as a direct product $\Lambda_1 \times \cdots \times \Lambda_r$, where Λ_i is a maximal order in A_i. Then

$$\zeta_{\Lambda'}(s) = \prod_{i=1}^{r} \zeta_{\Lambda_i}(s),$$

and each $\zeta_{\Lambda_i}(s)$ has a simple pole at $s=1$. If c_i denotes the residue of $\zeta_{\Lambda_i}(s)$ at $s=1$, we have

$$c_A = \prod_{i=1}^{r} c_i,$$

and so it suffices to calculate each c_i.

Changing notation, suppose now that Λ' is a maximal order in a simple Q-algebra A with centre F. Let R be the ring of integers of F, and let $\dim_F(A) = n^2$. Denote by $\zeta_R(s)$ the Dedekind zeta-function associated with R, and let c_F be the residue of $\zeta_R(s)$ at $s=1$. As is well-known ([9], p. 161), we have

$$c_F = \frac{2^{r_1} (2\pi)^{r_2} \operatorname{Reg}(F)}{w_F |d_F|^{\frac{1}{2}}} h_R,$$

where F has r_1 real primes and r_2 imaginary primes, w_F is the number of roots of unity in F, $\operatorname{Reg}(F)$ is the regulator of F, d_F is the absolute discriminant of F, and h_R is the ideal class number of R.

For each maximal ideal $\mathfrak{p}$ of R, we can form $\mathfrak{p}$-adic completions $A_\mathfrak{p}$, $F_\mathfrak{p}$, etc. Then $A_\mathfrak{p}$ is a central simple $F_\mathfrak{p}$-algebra of dimension n^2, and we may write

$$A_\mathfrak{p} \cong M_{\kappa_\mathfrak{p}}(D_\mathfrak{p}),$$

where $D_\mathfrak{p}$ is a division algebra with centre $F_\mathfrak{p}$. Let $\dim_{F_\mathfrak{p}}(D_\mathfrak{p}) = e_\mathfrak{p}{}^2$, so that $n = \kappa_\mathfrak{p} e_\mathfrak{p}$. Then $e_\mathfrak{p} = 1$ for almost all $\mathfrak{p}$. By virtue of formulas (4. 3)—(4. 5), we have

$$\zeta_{\Lambda'}(s) = \prod_{j=0}^{n-1} \zeta_R(ns-j) \cdot \prod_{\substack{\mathfrak{p} \\ e_\mathfrak{p} > 1}} \prod_{\substack{j=0 \\ j \not\equiv 0 \, (\mathrm{mod}\, e_\mathfrak{p})}}^{n-1} (1 - \mathcal{N}\mathfrak{p}^{-(ns-j)}).$$

The factor $\zeta_R(ns-j)$ corresponding to the value $j = n-1$ has a simple pole at $s=1$ with residue c_F/n. The other factors are holomorphic at $s=1$, so we obtain

$$\lim_{s \to 1} (s-1)\, \zeta_{\Lambda'}(s) = \frac{c_F}{n} \zeta_R(2) \cdots \zeta_R(n) \prod_{\substack{\mathfrak{p} \\ e_\mathfrak{p} > 1}} \prod_{\substack{j=1 \\ j \not\equiv 0 \, (\mathrm{mod}\, e_\mathfrak{p})}}^{n-1} (1 - \mathcal{N}\mathfrak{p}^{-j}). \tag{6. 12}$$

(To simplify this expression, we have used the fact that $e_\mathfrak{p}$ divides n, whence $e_\mathfrak{p}$ divides $(n-j)$ if and only if $e_\mathfrak{p}$ divides j.)

(6. 13) *Proof of* (6. 9). We prove the lemma under the assumption that the algebra B is *simple*, the general case then being immediate. We let $B = M_n(D)$, where D is a finite-dimensional Q_p-division algebra, and we may as well assume that $n \geqq 2$, since the case $n=1$ is trivial. Let Δ be the maximal order in D, and let ξ be a prime element of Δ. Let $(\Delta : \Delta\xi) = q$, and let $|\cdot|$ be the absolute value on D for which $|\xi| = q^{-1}$. In other words, $|\cdot| = \|\cdot\|_D$. If we are given a left vector space V over D, and a D-basis $v_1, \ldots, v_r$ of V, we extend $|\cdot|$ to a function on V by defining

$$\left| \sum_{i=1}^{r} x_i v_i \right| = \max_{1 \leqq i \leqq r} |x_i|,$$

where, of course, the coefficients x_i lie in D. Let μ_D be the (additive) Haar measure on D for which $\mu_D(\Delta) = 1$. Identifying V with D^r via the basis $v_1, \ldots, v_r$, we get the product measure μ_V on V. Then, for fixed $v \in V$ and $N \geqq 0$, we have

$$\mu_V(\{u \in V : |u - v| \leqq q^{-N}\}) = q^{-rN}. \tag{6. 14}$$

We have to show that the set $M_n(D) - GL_n(D)$ of singular matrices in B has measure zero. This set is the union of the sets S_i, for $1 \leqq i \leqq n$, where S_i is the set of matrices with the property that the i-th row is a left D-linear combination of the remaining rows. Therefore it is enough to show that $\mu_B(S_i) = 0$, for all i. We may as well take $i=n$, for notational convenience. Let V be the left D-space of "vectors" (x_{ij}, λ_k), for $1 \leqq i \leqq n-1$, $1 \leqq j \leqq n$, $1 \leqq k \leqq n-1$. Then $\dim_D(V) = n^2 - 1$. We have a surjection

$$F : V \to S_n$$

which maps the vector (x_{ij}, λ_k) to the matrix whose i, j entry is x_{ij}, for $1 \leqq i \leqq n-1$, and whose (n, j) entry is

$$\sum_{i=1}^{n-1} \lambda_i x_{ij}.$$

Let L be the Δ-lattice in V consisting of the vectors with all their entries x_{ij}, λ_k, lying in Δ. Since V is the union of the $\xi^{-m}L$, for $m \geqq 0$, it is enough to show that

$$\mu_B(F(\xi^{-m}L))=0$$

for all $m \geqq 0$. Here, μ_B is the product measure on the D-space B, defined via the obvious D-basis of B. Let N be a large positive integer, and view m as fixed. Let

$$v=(x_{ij}, \lambda_k) \in \xi^{-m}L, \quad v'=(x'_{ij}, \lambda'_k) \in V,$$

and suppose that $|v-v'| \leqq q^{-N}$. Then

$$|F(v)-F(v')| \leqq \max \{q^{-N}, |\sum_i (\lambda_i x_{ij} - \lambda'_i x'_{ij})|: 1 \leqq j \leqq n\}.$$

But

$$\left|\sum_{i=1}^{n-1} \lambda_i x_{ij} - \lambda'_i x'_{ij}\right| \leqq \max \{|\lambda_i - \lambda'_i|\,|x_{ij}|, |\lambda'_i|\,|x_{ij}-x'_{ij}|\} \leqq q^{m-N}.$$

Thus

$$|F(v)-F(v')| \leqq q^{m-N}.$$

Therefore, for fixed $v \in \xi^{-m}L$, the set

$$E_v=\{F(v'): |v-v'| \leqq q^{-N}\}$$

has measure $\leqq q^{n^2(m-N)}$, by (6. 14). Now let v range over a set T of coset representatives of $\xi^{-m}L$ modulo $\xi^N L$. Then T has $q^{(m+N)(n^2-1)}$ elements. Moreover, $\xi^{-m}L$ is the union, over $v \in T$, of the sets $\{v' \in V: |v-v'| \leqq q^{-N}\}$. It follows that

$$\mu_B(F(\xi^{-m}L)) \leqq q^{(m+N)(n^2-1)} \cdot q^{n^2(m-N)} = q^{(2n^2-1)m-N}.$$

The integers m, n are fixed. This inequality holds for all large positive integers N, and therefore $\mu_B(F(\xi^{-m}L))=0$, as required. This completes the proof of (6. 9).

§ 7. Examples

We give here some calculations of L-functions, in order to present a few concrete cases.

(7. 1) Example. Let G be a cyclic group of prime order p, and put $A=\mathbb{Q}G$, $\Lambda=\mathbb{Z}G$. Let $K=\mathbb{Q}(\omega)$, $R=\mathbb{Z}[\omega]$, where ω is a primitive p-th root of unity. We can identify A with $\mathbb{Q} \times K$, and then $\Lambda'=\mathbb{Z} \times R$ is the unique maximal order in A. There are exactly two genera of left ideals of Λ, namely $g(\Lambda)$ and $g(\Lambda')$, where Λ' is viewed as a Λ-module. We wish to calculate $L_\Lambda(\Lambda, s, \psi)$ and $L_\Lambda(\Lambda', s, \psi)$.

As is well-known, the obvious maps $\Lambda \to \Lambda' \to R$ induce isomorphisms

$$\mathrm{Cl}(\Lambda) \cong \mathrm{Cl}(\Lambda') \cong \mathrm{Cl}(R),$$

so the above character ψ is necessarily unramified. We have

$$L_{\Lambda'}(\Lambda', s, \psi)=\zeta_{\mathbb{Z}}(s)\, L_K(s, \psi), \tag{7. 2}$$

where $\zeta_{\mathbb{Z}}(s)$ is the Riemann zeta-function, and $L_K(s, \psi)$ is the classical L-function

$$L_K(s, \psi)=\sum_X \psi(X)\,(R:X)^{-s},$$

X ranging over all ideals of R. This coincides with the definition given for $L_K(s, \psi)$ in § 4.

Next, $\Lambda_q=\Lambda'_q$ for all primes $q\neq p$, so from the Euler product formulas we have

$$L_A(\Lambda, s, \psi) / L_{A'}(\Lambda', s, \psi) = L_{A_p}(\Lambda_p, s, \psi_p) / L_{A'_p}(\Lambda'_p, s, \psi_p). \tag{7. 3}$$

There is a unique maximal ideal $\mathfrak{p}$ of R lying above p, and indeed $\mathfrak{p}$ is principal, and it follows that ψ_p is identically 1. Therefore the right hand side of (7. 3) is

$$Z_{A_p}(\Lambda_p, s)/\zeta_{A'_p}(s).$$

This has been evaluated in [1], § 3. 4, and equals $(1-2p^{-s}+p^{1-2s})$. Thus

$$L_A(\Lambda, s, \psi)=(1-2p^{-s}+p^{1-2s})\,\zeta_Z(s)\,L_K(s, \psi). \tag{7. 4}$$

The factor $\zeta_Z(s)\cdot L_K(s, \psi)$ is precisely the standard L-function $L_A(s, \psi)$.

A similar argument yields

$$L_A(\Lambda', s, \psi)=p^{-s}\,\zeta_Z(s)\,L_K(s, \psi). \tag{7. 5}$$

Now let us calculate the partial zeta-function $Z_\Lambda(M, s)$, where $M\in g(\Lambda)$. Notice that stable isomorphism is the same as ordinary isomorphism here, since A is commutative. Let $h=|\mathrm{Cl}(\Lambda)|=|\mathrm{Cl}(R)|$. Then

$$Z_\Lambda(M, s)=h^{-1}\,\zeta_Z(s)\,(1-2p^{-s}+p^{1-2s})\sum_{\psi}\psi^{-1}[M]\,L_K(s, \psi),$$

where ψ ranges over all characters of $\mathrm{Cl}(\Lambda)$, that is, of $\mathrm{Cl}(R)$. Now, each $M\in g(\Lambda)$ corresponds to an ideal $\mathfrak{a}$ of R. Indeed, M is a non-split extension of $\mathfrak{a}$ by $\mathbb{Z}$, and the identification $\mathrm{Cl}(\Lambda)=\mathrm{Cl}(R)$ gives $\psi[M]=\psi(\mathfrak{a})$. However

$$h^{-1}\sum_{\psi}\psi^{-1}(\mathfrak{a})\,L_K(s, \psi)=Z_R(\mathfrak{a}, s)=\sum_{\substack{X\subseteq R\\ X\cong\mathfrak{a}}}(R:X)^{-s}.$$

Thus we obtain

$$Z_\Lambda(M, s)=(1-2p^{-s}+p^{1-2s})\,\zeta_Z(s)\,Z_R(\mathfrak{a}, s). \tag{7. 6}$$

Furthermore we have

$$Z_\Lambda(g(\Lambda), s)=(1-2p^{-s}+p^{1-2s})\,\zeta_{\Lambda'}(s).$$

Let $c_A=\lim_{s\to 1}(s-1)^2\,\zeta_{\Lambda'}(s)=\lim_{s\to 1}(s-1)\,\zeta_R(s)$. Then

$$\lim_{s\to 1}(s-1)^2\,Z_\Lambda(g(\Lambda), s)=d_\Lambda c_A,$$

where $d_\Lambda=(\Lambda:\Lambda')\,(\Lambda'^{\times}_p:\Lambda^{\times}_p)=(p-1)/p$. Observe that

$$1-2p^{-s}+p^{1-2s}|_{s=1}=(p-1)/p,$$

as expected.

(7. 7) Example. Now let A be the rational quaternion division algebra with $\mathbb{Q}$-basis $1, i, j, k$, where $k=ij=-ji$, and $i^2=j^2=-1$. Let Λ be the order

$$\mathbb{Z}\oplus\mathbb{Z}i\oplus\mathbb{Z}j\oplus\mathbb{Z}k,$$

and put

$$\Lambda'=\Lambda+\mathbb{Z}u, \quad\text{where}\quad u=\tfrac{1}{2}(1+i+j+k). \tag{7. 8}$$

Then Λ' is a maximal order in A containing Λ. From (6. 2) and formulas (4. 3)—(4. 5) we have

$$\zeta_{\Lambda'}(s)=\zeta_Z(2s)\,\zeta_Z(2s-1)\,(1-2^{1-2s}). \tag{7. 9}$$

On the other hand, $\Lambda_q=\Lambda'_q$ for all primes $q\neq 2$, while Λ'_2 is a maximal order in the quaternion skewfield A_2. Then

$$\zeta_{\Lambda'_2}(s)=\zeta_{Z_2}(2s)=(1-2^{-2s})^{-1}. \tag{7. 10}$$

Further we obtain

$$\frac{Z_\Lambda(g(\Lambda),s)}{\zeta_{\Lambda'}(s)}=\frac{Z_{\Lambda_2}(\Lambda_2,s)}{\zeta_{\Lambda'_2}(s)}=1+2^{1-2s}. \tag{7. 11}$$

Let us sketch the proof of this formula (7. 11), which depends on calculating $Z_{\Lambda_2}(\Lambda_2,s)$. By (2. 5),

$$Z_{\Lambda_2}(\Lambda_2,s)=\mu^\times(\Lambda_2^\times)^{-1}\int_{\Lambda_2\cap A_2^\times}\|x\|^s d^\times x.$$

We normalise the multiplicative Haar measure so that $\mu^\times(\Lambda_2'^\times)=1$, and then

$$\mu^\times(\Lambda_2^\times)=(\Lambda_2'^\times:\Lambda_2^\times)^{-1}=1/3.$$

Next we find easily that the Jacobson radical of Λ_2 is $\Lambda'_2\cdot(1+i)$, so

$$\int_{\Lambda_2}\|x\|^s d^\times x=\int_{\Lambda_2^\times}\|x\|^s d^\times x+\int_{\Lambda'_2(1+i)}\|x\|^s d^\times x,$$

omitting the intersection with $A_2^\times$ for the sake of brevity. The first integral is just the measure of $\Lambda_2^\times$, namely 1/3, while for the second integral we use the fact that $\|1+i\|=\frac{1}{4}$ to obtain

$$\int_{\Lambda'_2(1+i)}\|x\|^s d^\times x=4^{-s}\int_{\Lambda'_2}\|x\|^s d^\times x=4^{-s}\mu^\times(\Lambda_2'^\times)\,\zeta_{\Lambda'_2}(s)=4^{-s}(1-2^{-2s})^{-1}.$$

Together, these formulas give

$$Z_{\Lambda_2}(\Lambda_2,s)=(1+2^{1-2s})\,(1-2^{-2s})^{-1},$$

and this yields the second equality in (7. 11).

In this case we have

$$d_\Lambda=(\Lambda_2:\Lambda'_2)\,(\Lambda_2'^\times:\Lambda_2^\times)=3/2.$$

Similar arguments give

$$\frac{Z_\Lambda(g(\Lambda'),s)}{\zeta_{\Lambda'}(s)}=\frac{Z_{\Lambda_2}(\Lambda'_2,s)}{\zeta_{\Lambda'_2}(s)}=2^{-s}.$$

(7. 12) Example. Let $A=M_2(Q)$, and for a fixed prime number p, let Λ be the order in A consisting of all matrices $\begin{pmatrix} a & b \\ c & d\end{pmatrix}\in M_2(Z)$ for which $b\equiv 0\ (\mathrm{mod}\, p)$. Then $\Lambda'=M_2(Z)$ is a maximal order, and

$$\zeta_{\Lambda'}(s)=\zeta_Z(2s)\,\zeta_Z(2s-1).$$

Then $Z_\Lambda(g(\Lambda), s)/\zeta_{\Lambda'}(s) = Z_{\Lambda_p}(\Lambda_p, s)/\zeta_{\Lambda'_p}(s)$. This last quotient has been calculated in [2], and is equal to $(1+p^{1-2s})$. Therefore

$$Z_\Lambda(g(\Lambda), s) = (1+p^{1-2s})\, \zeta_{\Lambda'}(s).$$

Further, we find here that

$$d_\Lambda = (\Lambda : \Lambda')\, (\Lambda_p'^{\times} : \Lambda_p^{\times}) = (p+1)/p.$$

§ 8. Lattices in free A-modules

Let A be a finite-dimensional semisimple Q-algebra, Λ an order in A, and let $V = A^{(v)}$ be a free left A-module on v generators. We shall consider left Λ-lattices M in V, always restricting our attention to the case where $Q \cdot M = V$. The concepts of genus $g(M)$, stable isomorphism classes in $g(M)$, and so on, carry over readily to this case. Specifically, for X a Λ-lattice in V, write $X \in g(M)$ if $X_p \cong M_p$ as Λ_p-modules for each prime p. For X, $Y \in g(M)$, write $[X] = [Y]$, and call X, Y *stably isomorphic*, if $X \oplus M \cong Y \oplus M$. However, if $v \geqq 2$, it is known that $[X] = [Y]$ implies $X \cong Y$, see [6]. Let $\Gamma = \mathrm{End}_\Lambda(M)$, so now Γ is an order in the algebra $B = \mathrm{End}_A(V)$.

Given Λ-lattices M, N, in V, we define

$$Z_\Lambda(N, g(M), s) = \sum_X (N : X)^{-s}, \tag{8. 1}$$

the sum extending over all Λ-sublattices X of N such that $X \in g(M)$. Likewise, if Λ' is a maximal order in A, and N' is a Λ'-lattice in V, we put

$$\zeta_{\Lambda'}(N', s) = \sum_X (N' : X)^{-s}, \tag{8. 2}$$

where this sum extends over all Λ'-sublattices X of N'. Observe that all Λ'-lattices in V lie in the same genus of Λ'-modules, namely $g_{\Lambda'}(\Lambda'^{(v)})$.

We introduce the function $K_V(s, \psi_0)$, where ψ_0 is the trivial character (of $J(B)$), just as in (4. 3). (This function already occurs in [1] and [12], where it was denoted by $\zeta_V(s)$.) Explicitly, let A_i, $1 \leqq i \leqq r$, be the simple components of A. Let F_i be the centre of A_i, R_i the ring of algebraic integers in F_i, and let $n_i^2 = \dim_{F_i}(A_i)$. Writing $\zeta_{R_i}(s)$ for the Dedekind zeta-function of R_i, we define[2])

$$K_V(s, \psi_0) = \prod_{i=1}^{r} \prod_{j=0}^{n_i v - 1} \zeta_{R_i}(n_i s - j). \tag{8. 3}$$

As shown in [1], formula (16), the analogues of (4. 4) and (6. 2) hold in this situation, and we have

$$\zeta_{\Lambda'}(\Lambda'^{(v)}, s) = K_V(s, \psi_0) \prod_p f_p(s), \tag{8. 4}$$

where p ranges over the finitely many prime numbers for which A_p is not a split semisimple Q_p-algebra, and where $f_p(s) \in Z[p^{-s}]$. One may give $f_p(s)$ quite explicitly. Further, there is $\varepsilon > 0$, depending only on A and v, such that $f_p(s)$ is non-vanishing in the region $\mathrm{Re}(s) > v - \varepsilon$.

---

[2]) Observe that $K_V(s, \psi_0) = K_A(s, \psi_0)$ when $v = 1$.

It follows readily that $\zeta_{\Lambda'}(\Lambda'^{(v)}, s)$ is holomorphic for $\mathrm{Re}(s) > v$, and has a pole of order r at $s = v$. The series (8. 1), (8. 2) converge for $\mathrm{Re}(s) > v$, and define holomorphic functions for s in that range. Further, as shown in [1], Th. 1, the quotient

$$Z_\Lambda(N, g(M), s)/\zeta_{\Lambda'}(\Lambda'^{(v)}, s)$$

is expressible as a finite product $\prod_p g_p(s)$, with $g_p(s) \in \mathbb{Z}[p^{-s}]$, for each p. We find that each partial zeta-function $Z_\Lambda(N, g(M), s)$ and $Z_\Lambda(N, [M], s)$ (defined in the obvious way) admits analytic continuation and has a pole of order r at $s = v$.

We now put

$$(8.\ 5) \qquad d_{N,M} = \lim_{s \to v} Z_\Lambda(N, g(M), s)/\zeta_{\Lambda'}(\Lambda'^{(v)}, s).$$

As in the proof of (6. 5), we obtain

$$(8.\ 6) \quad d_{N,M} = (N : M)^{-1} (\{M : N\} : M_v(\Lambda')) \prod_p (GL_v(\Lambda'_p) : \mathrm{Aut}_{\Lambda_p}(M_p)),$$

the product taken over the finitely many primes p for which $\mathrm{End}_{\Lambda_p}(M_p)$ is not a maximal order. In the above, we identify $\mathrm{End}_A(V)$ with the matrix algebra $M_v(A)$, and then $\mathrm{End}_{\Lambda'}(\Lambda'^{(v)})$ is identified with $M_v(\Lambda')$. Further,

$$\{M : N\} = \{x \in B : \ Mx \subseteqq N\},$$

where $B = \mathrm{End}_A(V)$, as above. Now let

$$(8.\ 7) \qquad c_V = \lim_{s \to v} (s - v)^r \, \zeta_{\Lambda'}(\Lambda'^{(v)}, s),$$

a positive constant which can be computed explicitly. Then the Laurent expansion of $Z_\Lambda(N, g(M), s)$ about $s = v$ has leading term

$$c_V \, d_{N,M} (s - v)^{-r}.$$

In order to use the Delange-Ikehara Tauberian theorem, we make a change of variable. Put

$$Y(s) = Z_\Lambda(N, g(M), vs),$$

so that $Y(s)$ has a pole of order r at $s = 1$, and the leading term of its Laurent expansion about $s = 1$ is $c_V \cdot d_{N,M} \cdot v^{-r} \cdot (s - 1)^{-r}$. This yields the following analogue of Theorem (6. 10):

(8. 8) Theorem. *Let Λ be an order in a finite-dimensional semisimple $\mathbb{Q}$-algebra A, and suppose that A has r simple components. Let M and N be left Λ-lattices spanning a free A-module V on v generators. Let c_V be as in (8. 7). Let $[M_i]$, $1 \leqq i \leqq h_M$, be the stable isomorphism classes in $g(M)$, and let*

$$Z_\Lambda(N, [M_i], s) = \sum_X (N : X)^{-s},$$

where X ranges over all Λ-sublattices of N such that $X \in g(M)$, $[X] = [M_i]$. Then

(i) *$Z_\Lambda(N, [M_i], s)$ has a pole of order r at $s = v$, and its Laurent expansion about $s = v$ has leading term*

$$c_V \, d_{N,M} h_M^{-1} (s - v)^{-r},$$

where $d_{N,M}$ is given by (8. 6).

(ii) *Let* $[M_i]$ *be a fixed stable isomorphism class in* $g(M)$. *Then for positive* T *tending to infinity, the number of* A*-sublattices* X *of* N *such that*

$$(N:X) \leqq T, \quad X \in g(M), \quad [X]=[M_i],$$

is asymptotically equal to

$$\frac{c_V\, d_{N,M}}{h_M\, v^r} \cdot \frac{T(\log T)^{r-1}}{(r-1)!}.$$

Remark. A somewhat similar formalism applies to lattices in an arbitrary finitely generated left A-module W. However, the distribution properties are most strongly influenced by that simple A-module which occurs in W with the greatest multiplicity.

§ 9. Totally definite quaternion algebras

Now we return to our study of ideals of an order Λ. We consider here the simplest case in which stable isomorphism does not imply isomorphism for Λ-lattices. For this, we let F be a totally real algebraic number field with ring of integers R, and we take A to be a totally definite F-quaternion algebra. That is, A is a 4-dimensional central F-division algebra, and for each real place v of F, $A_v = A \otimes_F F_v$ is isomorphic to the skewfield of Hamiltonian quaternions.

To start with, we let Λ' be a maximal order in A. This case has been considered by Eichler in [5], and we now summarise his results. The zeta function $\zeta_{\Lambda'}(s)$ has a simple pole at $s=1$ with residue

$$h_R\, \zeta_R(2)\, 2^{n-2}\, \mathrm{Reg}(F)\, |d_F|^{-\frac{1}{2}} \prod_{\mathfrak{p} \in S} (1 - \mathcal{N}\mathfrak{p}^{-1}),$$

as in (6. 12), and with the same notation, except that we have written $r_1 = n = [F:\mathbb{Q}]$, and S denotes the finite set of maximal ideals $\mathfrak{p}$ of R for which $A_\mathfrak{p}$ is a skewfield.

On the other hand, let $M_1, \ldots, M_{h'}$ be a full set of representatives of the isomorphism classes of left ideals of Λ'. (Notice that there is only one genus of Λ'-ideals here.) Then we have

$$\zeta_{\Lambda'}(s) = \sum_{i=1}^{h'} Z_{\Lambda'}(M_i, s).$$

As usual, we may rewrite $Z_{\Lambda'}(M_i, s)$ in the form

$$(10.\ 1) \qquad Z_{\Lambda'}(M_i, s) = (\Lambda' : M_i)^{-s} \sum_{x \in \Gamma_i^\times \setminus (M_i : \Lambda')} \|x\|^s, \quad \mathrm{Re}(s) > 1,$$

where Γ_i denotes the order $\mathrm{End}_{\Lambda'}(M_i)$, and $\|x\| = (\Lambda' x : \Lambda')$. One can estimate the partial sums of the coefficients of the Dirichlet series (9. 1) directly. The situation is indeed formally very similar to the standard case in which Λ' is replaced by the ring of integers of a number field, as discussed in [9], VI and VIII. In our case, for each prime number p, we have our "p-index" map $\|\cdot\|_p : A_p^\times \to \mathbb{R}^\times$, given by $\|x\|_p = (\Lambda'_p x : \Lambda'_p)$, $x \in A_p^\times$. We may regard this as the composite of the absolute algebra norm from A_p to $\mathbb{Q}_p$, and the canonical absolute value on $\mathbb{Q}_p$. This latter point of view also applies to the Archimedean prime, and gives us a homomorphism $\|\cdot\|_\infty : (A \otimes_\mathbb{Q} \mathbb{R})^\times \to \mathbb{R}^\times$. By the product formula for valuations of $\mathbb{Q}$, we have

$$\|x\| = \prod_p \|x\|_p = \|x\|_\infty^{-1}, \qquad x \in A^\times.$$

Now let us view A as embedded in $A \otimes_Q \mathbb{R}$. Let G denote the kernel of $\|\cdot\|_\infty$ in $(A \otimes_Q \mathbb{R})^\times$. Then G contains $R^\times$, and one sees easily that the quotient $G/R^\times$ is compact[3]). For $T>0$, the number of terms $\|x\|^s$ in the sum (9. 1) with $\|x\| \geqq T^{-1}$ is the number of points of the lattice $\{M_i : \Lambda'\}$ lying in a fundamental domain for the left action of $\Gamma_i^\times$ on the space

$$\{y \in (A \otimes_Q \mathbb{R})^\times : \|y\|_\infty \leqq T\}.$$

One verifies that this number is, up to a comparatively insignificant error, proportional to the volume of such a domain. For fixed i, it is therefore proportional to T, and for fixed T, inversely proportional to the *finite* group index

$$w_i = (\Gamma_i^\times : R^\times).$$

The asymptotic behaviour, as $T \to \infty$, of this partial sum of coefficients, determines the behaviour of $Z_{\Lambda'}(M_i, s)$ at $s=1$, and one finds

(9. 2) Theorem (Eichler, [5]). (i) *The function $Z_{\Lambda'}(M_i, s)$ admits analytic continuation to a meromorphic function of s. It has a simple pole at $s=1$, with residue*

$$\frac{1}{w_i}(2\pi)^{2n}\, 2^{n-3}\, \mathrm{Reg}(F)\, |d_F|^{-2} \prod_{\mathfrak{p} \in S} \mathcal{N}\mathfrak{p}^{-1}.$$

(ii) $(2\pi)^{2n} \sum_{i=1}^{h'} \frac{1}{w_i} = 2\zeta_R(2)\, h_R\, |d_F|^{3/2} \prod_{\mathfrak{p}\in S} (\mathcal{N}\mathfrak{p} - 1).$

Part (ii) follows from comparing the two expressions for the residue of $\zeta_{\Lambda'}(s)$ at $s=1$.

In terms of distributions, the asymptotic density of left ideals of Λ' which are isomorphic to M_i is therefore c/w_i, for some constant c independent of i. Examples show that the index w_i can vary with i, so one need not have uniform distribution in isomorphism classes. Of course, the distribution in stable isomorphism classes is still uniform.

It is not hard to extend Eichler's calculations to cover the case of a general order in A:

(9. 3) Theorem. *Let Λ be an order in A, and M a left ideal of Λ. Let $L_1, \ldots, L_h$ be a full set of representatives of the isomorphism classes in $g(M)$. Then $Z_\Lambda(L_j, s)$ admits analytic continuation to a meromorphic function of s. It has a simple pole at $s=1$, with residue*

$$\frac{1}{q_j} \frac{(M^{-1} : \Lambda)}{(\Lambda : M)(\Lambda' : \Lambda)} (2\pi)^{2n}\, 2^{n-3}\, \mathrm{Reg}(F)\, |d_F|^{-2} \prod_{\mathfrak{p}\in S} \mathcal{N}\mathfrak{p}^{-1},$$

where $q_j = (\mathrm{Aut}(L_j) : R^\times)$ (a generalised group index). Moreover

$$\sum_{j=1}^{h} \frac{1}{q_j} = (U(\Lambda') : U(\mathrm{End}(L_j)))\, d_M \sum_{i=1}^{h'} \frac{1}{w_i},$$

where d_M is as in (6. 6).

The second formula again follows by comparing two expressions for the residue of $\zeta_\Lambda(s)$ at $s=1$. The asymptotic density of left ideals of Λ which are isomorphic to L_j is c'/q_j, where c' is independent of j. Again, q_j may vary with j.

---

[3]) This is essentially the reason for the failure of the stability theorem for Λ'-lattices. See [6] and M. Kneser, Starke Approximation in algebraischen Gruppen, J. reine angew. Math. **218** (1965), 190—203.

References

[1] *C. J. Bushnell, I. Reiner*, Zeta functions of arithmetic orders and Solomon's conjectures, Math. Z. **173** (1980), 135—161.
[2] *C. J. Bushnell, I. Reiner*, Zeta functions of hereditary orders and integral group rings, Texas Tech University Math. Series, to appear.
[3] *C. J. Bushnell, I. Reiner*, Zeta-functions of orders, Orders and their applications (K. Roggenkamp, ed.), Berlin-Heidelberg-New York, Visiting Scholar's Lectures 1980, **14**, (1981).
[4] *J. W. S. Cassels, A. Fröhlich* (ed.), Algebraic Number Theory, London 1967.
[5] *M. Eichler*, Idealklassenzahl total definierter Quaternionenalgebren, Math. Z. **43** (1938), 102—109.
[6] *A. Fröhlich*, Locally free modules over arithmetic orders, J. reine angew. Math. **274/275** (1975), 112—138.
[7] *R. Godement, H. Jacquet*, Zeta-functions of simple algebras, Lecture Notes in Mathematics **260**, Berlin-Heidelberg-New York 1972.
[8] *H. Jacobinski*, Genera and decomposition of lattices, Acta Mathematica **121** (1968), 1—29.
[9] *S. Lang*, Algebraic Number Theory, Reading, Mass. 1970.
[10] *W. Narkiewicz*, Elementary and Analytic Theory of Algebraic Numbers, Warsaw 1974.
[11] *I. Reiner*, Maximal Orders, London-New York 1975.
[12] *L. Solomon*, Zeta functions and integral representation theory, Advances in Mathematics **26** (1977), 306—326.
[13] *A. Weil*, Basic Number Theory, Berlin-Heidelberg-New York 1974.

University of London King's College, Department of Mathematics, Strand, London WC2R 2LS, GB

Department of Mathematics, University of Illinois, Urbana IL 61801, USA

Eingegangen 23. Januar 1981

Functional equations for L-functions of arithmetic orders

By *Colin J. Bushnell* at London and *Irving Reiner**) at Urbana

---

This paper is written as a continuation of [2], with consecutively numbered sections. Thus a reference to (n.m.), with $n \leqq 9$, means [2], (n.m.). Any unexplained notations are as in [2], but we shall avoid these wherever practicable. The bibliography is independent of that of [2].

As before, A denotes a finite-dimensional semisimple algebra over either the rational number field $\mathbb{Q}$ *(global case)* or some p-adic rational field $\mathbb{Q}_p$ *(local case)*, and Λ is an order in A. Thus Λ is a $\mathbb{Z}$-order in the global case, a $\mathbb{Z}_p$-order in the local case.

The first part of the paper is concerned with the local case. Here we let M be a left ideal of Λ such that the index $(\Lambda : M)$ is finite. Let ψ be a character (i.e., a continuous homomorphism to the unit circle in $\mathbb{C}$) of $A^\times$ which is of finite order and is trivial on the subgroup $\mathrm{Aut}_\Lambda(M)$. Then the local L-function $L_\Lambda(M, s, \psi)$ is defined, as in (2. 1). More generally, for any character ψ of $A^\times$ of finite order, and any function $\Phi \in \mathscr{S}(A)$, the space of Schwartz-Bruhat functions on A, we have the zeta-integral

$$Z(\Phi, s, \psi) = \int_{A^\times} \Phi(x)\, \psi(x)\, \|x\|^s\, d^\times x, \qquad \mathrm{Re}(s) > 1,$$

as in (3. 2). Here, $\|x\| = (\Lambda : \Lambda x)^{-1}$, and $d^\times x$ is a Haar measure on $A^\times$. The functions $Z(\Phi, s, \psi)$ admit analytic continuation, and satisfy a functional equation. This is basically well-known. We have already established the existence of analytic continuation in § 3, and we give a brief summary of the functional equation in § 10. This functional equation gives rise to a "local constant" or symmetry factor $\varepsilon_A(s, \psi)$, which we here evaluate completely. This is done in two stages. We first give a formula for it (in § 10) in terms of a certain "non-abelian congruence Gauss sum" $\tau(\psi)$. This is a straightforward calculation based on the functional equation and the construction of the standard L-function $L_A(s, \psi)$ in § 3. Later, in § 13, we evaluate $\tau(\psi)$ in terms of classical Gauss sums. This is partly taken from unpublished joint work of the first author and A. Fröhlich.

---

*) The research for this paper was performed while the second author was a Senior Visiting Research Fellow at King's College London, partially supported by the Science Research Council. He wishes to acknowledge this support, as well as the hospitality of King's College, during his sabbatical leave from the University of Illinois. He also thanks the National Science Foundation for its support at other times.

More immediately, the functional equation for $Z(\Phi, s, \psi)$ implies a functional equation for $L_\Lambda(M, s, \psi)$, which is particularly interesting when Λ is the integral group ring RG of a finite group G over the valuation ring R in a p-adic local field. We work this out in § 11 and indicate some of the global implications.

Staying in the local case, we recall from § 3 that $L_\Lambda(M, s, \psi)$ is the product of the standard L-function $L_A(s, \psi)$ and a polynomial function $f(s) \in \mathbb{C}[p^{-s}]$. In some cases, the functional equation yields some information about the apparently inaccessible function $f(s)$. This is the subject of § 12.

In § 14, we go to the global case. Our primary interest is the functional equation for $L_\Lambda(M, s, \psi)$ when Λ is the group ring RG of a finite group G over the ring R of integers in some algebraic number field. To get this, we need the functional equation for the global standard L-function $L_A(s, \psi)$. This can be extracted from the literature. However, the local calculations of §§ 10, 13 enable us to give a very straightforward proof, for a general semisimple algebra A, based on the classical case of A a field. We also indicate another formulation of the functional equation, and outline some of the history of the result. Finally, in § 15, we obtain the desired functional equation for $L_\Lambda(M, s, \psi)$, when $\Lambda = RG$. This is pleasingly simple in form, and implies a similar functional equation for the partial zeta-function $Z_{RG}([M], s)$.

§ 10. Local functional equations and Gauss sums

Let A be a finite-dimensional semisimple $\mathbb{Q}_p$-algebra, and let ψ be a continuous character of $A^\times$. For $x \in A$, we define $\|x\| = (X : Xx)^{-1}$, where X is any $\mathbb{Z}_p$-lattice spanning A over $\mathbb{Q}_p$. Let $d^\times x$ be a Haar measure on $A^\times$. Let $\mathcal{S}(A)$ be the space of Schwartz-Bruhat functions on A (that is, complex-valued locally constant functions of compact support). Then, for $\Phi \in \mathcal{S}(A)$, we have the zeta-integral

$$Z(\Phi, s, \psi) = \int_{A^\times} \Phi(x)\, \psi(x)\, \|x\|^s\, d^\times x, \tag{10.1}$$

as in (3. 2). We proved in (3. 14) that $Z(\Phi, s, \psi)$ admits analytic continuation to a meromorphic function on the whole complex s-plane. We proceed to deduce a functional equation for $Z(\Phi, s, \psi)$, and later apply this to our local L-function $L_\Lambda(M, s, \psi)$ attached to a left ideal M of an order Λ in A.

It is sufficient here to treat only the case of A a simple algebra in any detail. The transition to the semisimple case is then quite straightforward, and we outline the process at the end of the present section. So, until further notice, A denotes a finite-dimensional *simple* $\mathbb{Q}_p$-algebra. Let C be the centre of A, so that C is a field. Let ϑ_A, or just ϑ, denote the canonical continuous character of the additive group of A defined by

$$\vartheta_A(x) = \exp(2\pi i\, \mathrm{tr}_{A/\mathbb{Q}_p}(x)), \quad x \in A. \tag{10.2}$$

Here, $\mathrm{tr}_{A/\mathbb{Q}_p}$ denotes the absolute reduced trace defined by

$$\mathrm{tr}_{A/\mathbb{Q}_p} = \mathrm{Tr}_{C/\mathbb{Q}_p} \circ \mathrm{tr}_{A/C}, \tag{10.3}$$

where $\mathrm{tr}_{A/C} = \mathrm{tr}$ is the ordinary reduced trace from A to its centre C, and $\mathrm{Tr}_{C/\mathbb{Q}_p}$ is the field trace. Further, for $\alpha \in \mathbb{Q}_p$, we interpret $\exp(2\pi i \cdot \alpha)$ as $\exp(2\pi i \cdot \alpha_0)$, where $\alpha_0 \in \mathbb{Q}$ is the principal part of α.

The pairing $A \times A \to C^{\times}$ given by $(x, y) \mapsto \theta(xy)$, $x, y \in A$, is nondegenerate and symmetric. It allows us to identify the locally compact abelian group A with its Pontrjagin dual $\hat{A}$. Let dx be the self-dual Haar measure on A for this identification. This measure is determined as follows. For $\Phi \in \mathscr{S}(A)$, we define the *Fourier transform* $\hat{\Phi}$ of Φ by

$$\hat{\Phi}(y) = \int_A \Phi(x)\, \theta(xy)\, dx, \quad y \in A. \tag{10.4}$$

Then $\hat{\Phi} \in \mathscr{S}(A)$ also, and the defining property of the self-dual measure dx is the validity of the *Fourier inversion formula*

$$\hat{\hat{\Phi}}(x) = \Phi(-x), \quad \Phi \in \mathscr{S}(A),\ x \in A. \tag{10.5}$$

This certainly determines the measure uniquely. In later sections, we shall need to indicate the dependence of the Fourier transform on the character θ, when we shall write θ_A in place of θ, and use the notation

$$\hat{\Phi} = \mathscr{F}_A(\Phi). \tag{10.6}$$

Keeping the notation (10. 1)—(10. 5), we can now state the functional equation for zeta-integrals.

(10. 7) Theorem. *Let A be a finite-dimensional simple Q_p-algebra, and ψ a continuous character of $A^{\times}$. Then:*

(i) *For each $\Phi \in \mathscr{S}(A)$, the function $Z(\Phi, s, \psi)$ admits analytic continuation to a meromorphic function on the whole complex s-plane.*

(ii) *Let $\Phi, \Psi \in \mathscr{S}(A)$, and let $\bar{\psi}$ be the complex conjugate of ψ. Then*

$$Z(\Phi, s, \psi)\, Z(\hat{\Psi}, 1-s, \bar{\psi}) = Z(\hat{\Phi}, 1-s, \bar{\psi})\, Z(\Psi, s, \psi)$$

for all s.

Proof. Statement (i) has already been proved in (3. 14). To prove (ii), we may appeal to [5], 9. 6. Alternatively, the elementary proof given in [1], Appendix, can be easily extended to cover the case where the character ψ is unramified. That proof fails for ψ ramified, because then some of the functions $Z(\Phi, s, \psi)$ occurring in the proof will be identically zero. We may nevertheless give an elementary proof of (ii) for the case of ψ ramified. In this case, the integral (10. 1) converges uniformly for all s, as follows readily from the proof in § 3 that $L_A(s, \psi) = 1$. Therefore the two sides of the functional equation (ii) can be compared directly. This comparison uses the techniques of Tate's thesis, and is written out in [5], § 4, when A is a division algebra. For convenience, we summarise the argument briefly. We take the multiplicative Haar measure to be $d^{\times}x = \|x\|^{-1} \cdot dx$, where dx is the self-dual Haar measure on A. We write out the left-hand side of the functional equation:

$$\begin{aligned} Z(\Phi, s, \psi)\, Z(\hat{\Psi}, 1-s, \bar{\psi}) &= \int_{A^{\times}} \Phi(x)\, \psi(x)\, \|x\|^s\, d^{\times}x \int_{A^{\times}} \hat{\Psi}(y)\, \bar{\psi}(y)\, \|y\|^{1-s}\, d^{\times}y \\ &= \int_{A^{\times}} \{\Phi(xy)\, \psi(xy)\, \|xy\|^s \int_{A^{\times}} \hat{\Psi}(y)\, \bar{\psi}(y)\, \|y\|^{1-s}\, d^{\times}y\}\, d^{\times}x \\ &= \iint_{(A^{\times}) \times A} \Phi(xy)\, \hat{\Psi}(y)\, \psi(x)\, \|x\|^s\, d^{\times}x\, dy \quad \text{(see (6. 9))} \\ &= \int_{A^{\times}} \{\psi(x)\, \|x\|^s \int_A \Phi(xy)\, \hat{\Psi}(y)\, dy\}\, d^{\times}x. \end{aligned} \tag{10.8}$$

Similarly, the right-hand side of the functional equation comes down to

$$\int_{A^\times} \{\psi(x)\, \|x\|^{s-1} \int_A \hat{\Phi}(yx^{-1})\, \Psi(y)\, dy\}\, d^\times x, \tag{10. 9}$$

replacing x by x^{-1} after the first step. Using the definition of the Fourier transform, one finds readily that

$$\int_A \Phi(xy)\, \hat{\Psi}(y)\, dy = \int_A \Psi(z) \int_A \Phi(xy)\, \theta(yz)\, dy\, dz.$$

The inner integral here is just $\|x\|^{-1} \cdot \hat{\Phi}(zx^{-1})$. Substituting for the inner integral in (10. 8), we obtain (10. 9), thereby proving (ii).

Now let us define

$$\Xi(\Phi, s, \psi) = L_A(s, \psi)^{-1}\, Z(\Phi, s, \psi), \quad \Phi \in \mathscr{S}(A). \tag{10. 10}$$

From the defining property C3 of $L_A(s, \psi)$ in § 3, we see that $\Xi(\Phi, s, \psi) \in \mathbb{C}[p^s, p^{-s}]$. The functional equation (10. 7) (ii) shows that there is a meromorphic function $\varepsilon_A(s, \psi)$, independent of Φ, such that

$$\Xi(\hat{\Phi}, 1-s, \bar{\psi}) = \varepsilon_A(s, \psi)\, \Xi(\Phi, s, \psi), \quad \Phi \in \mathscr{S}(A). \tag{10. 11}$$

Observe that the definition of $\varepsilon_A(s, \psi)$ is also independent of the choice of multiplicative Haar measure $d^\times x$. The main object of this section is to calculate $\varepsilon_A(s, \psi)$ explicitly in terms of a certain "non-abelian Gauss sum" $\tau(\psi)$, to be defined below.

We begin by introducing additional notation. We let $A = M_k(D)$, where D is a skew-field with centre C, and $\dim_C(D) = e^2$. Let Λ be a maximal order in A. It will be easy to see that all of our calculations are independent of the choice of Λ, so we take $\Lambda = M_k(\Delta)$, where Δ is the unique maximal order in D. All the ideals of Δ are necessarily two-sided, and they are the powers of the unique maximal ideal $\mathfrak{P}$ of Δ. The two-sided fractional ideals of Λ in A form a free abelian group generated by $\mathfrak{P}\Lambda$. We now define the *absolute inverse different* $\mathfrak{D}_A^{-1}$ of A by

$$\mathfrak{D}_A^{-1} = \{x \in A : \mathrm{tr}_{A/\mathbb{Q}_p}(xy) \in \mathbb{Z}_p, \quad \text{for all} \quad y \in \Lambda\}. \tag{10. 12}$$

This is a fractional Λ-ideal in A, and its inverse $\mathfrak{D}_A$, or just $\mathfrak{D}$, is the *absolute different* of A. As is well-known (see [10], § 20 and Ex. 25. 1), we have

$$\mathfrak{D}_A = \mathfrak{P}^{e-1}\, \mathfrak{D}_C \Lambda, \tag{10. 13}$$

where $\mathfrak{D}_C$ coincides with the usual absolute different of the field C. We note also that, with θ_A defined as in (10. 2), we have

$$\mathfrak{D}_A^{-1} = \{x \in A : \theta_A(xy) = 1, \quad \text{for all} \quad y \in \Lambda\}. \tag{10. 14}$$

It follows easily from (10. 14) that the self-dual Haar measure on A is the one for which

$$\int_\Lambda dx = \mathscr{N}\mathfrak{D}^{-\frac{1}{2}}, \tag{10. 15}$$

where, if $\mathfrak{b}$ is a fractional Λ-ideal in A, we put

$$\mathscr{N}_A \mathfrak{b} = \mathscr{N}\mathfrak{b} = (\Lambda : \mathfrak{b}). \tag{10. 16}$$

Now let ψ be a character of $A^\times$, as above, and let us define the *conductor* $\mathfrak{f}(\psi)$ of ψ in the usual manner. If ψ is *unramified* (that is, ψ is trivial on $\Lambda^\times$), we set $\mathfrak{f}(\psi) = \Lambda$. If ψ is *ramified*, we let $\mathfrak{f}(\psi)$ be the largest two-sided ideal $\mathfrak{A}$ of Λ such that ψ is trivial on the subgroup $1+\mathfrak{A}=\{1+a: a\in\mathfrak{A}\}$ of $\Lambda^\times$. In this case where ψ is ramified, $\mathfrak{f}(\psi)$ is a proper ideal of Λ. To simplify later notation, we agree to interpret $1+\mathfrak{f}(\psi)$ as $\Lambda^\times$ when ψ is unramified. Further, when ψ is unramified, we may view ψ as a function on the group of two-sided fractional ideals of Λ in A, and also as a function on the set of one-sided Λ-lattices in A.

We now define the *Gauss sum* $\tau(\psi)$ associated with the character ψ by the formula

$$(10.17)\qquad \tau(\psi)=\sum_{x\in\Lambda^\times/1+\mathfrak{f}(\psi)}\psi(c^{-1}x)\,\theta_A(c^{-1}x),$$

where c is any element of Λ such that

$$(10.18)\qquad c\Lambda=\mathfrak{D}_A\mathfrak{f}(\psi).$$

The sum in (10. 17) extends over a full set of representatives x of the cosets of $\Lambda^\times$ modulo the normal subgroup $1+\mathfrak{f}(\psi)$. When ψ is unramified, we have agreed to interpret $1+\mathfrak{f}(\psi)$ as $\Lambda^\times$, and in this case we obtain

$$\tau(\psi)=\psi(c^{-1})=\psi(\mathfrak{D}_A^{-1}),$$

since θ_A is trivial on $\mathfrak{D}^{-1}$.

We ought to verify that the above definition of $\tau(\psi)$ is independent of the choices x and c. While this can easily be done directly, it will follow readily from the proof of the next theorem.

We also note that if Λ is replaced by another maximal order Λ', then Λ is conjugate to Λ', and the ideals $\mathfrak{f}(\psi)$ and $\mathfrak{D}_A$ are replaced by their corresponding conjugates. Since both ψ and θ_A are invariant under conjugation by elements of $A^\times$, it follows at once that the complex number $\tau(\psi)$ is independent of the choice of maximal order.

The main result of this section is:

(10. 19) Theorem. *Keeping the above notation, the function $\varepsilon_A(s,\psi)$ defined by* (10. 11) *is given by*

$$\varepsilon_A(s,\psi)=\mathcal{N}(\mathfrak{D}_A\,\mathfrak{f}(\psi))^{\frac{1}{2}-s}\frac{\tau(\bar{\psi})}{\mathcal{N}\mathfrak{f}(\psi)^{\frac{1}{2}}}.$$

Proof. We begin with the easier case in which ψ is unramified, and we have to prove that

$$(10.20)\qquad \varepsilon_A(s,\psi)=\psi(\mathfrak{D}_A)\,\mathcal{N}\mathfrak{D}_A^{\frac{1}{2}-s}.$$

We saw in (3. 9) that $L_A(s,\psi)=\mu^\times(\Lambda^\times)^{-1}Z(\Phi,s,\psi)$, where $\mu^\times$ is the measure of a set with respect to $d^\times x$, and Φ is the characteristic function of Λ in A. For this Φ, we therefore have $\Xi(\Phi,s,\psi)=\mu^\times(\Lambda^\times)$. On the other hand, the Fourier transform $\hat{\Phi}$ of Φ is given by

$$\hat{\Phi}=(\text{measure of }\Lambda)\cdot(\text{characteristic function of }\mathfrak{D}^{-1}).$$

If $c\in\Lambda$ is chosen so that $c\Lambda=\mathfrak{D}$, then by (10. 15) we obtain

$$\hat{\Phi}(x)=\mathcal{N}\mathfrak{D}^{-\frac{1}{2}}\,\Phi(cx),\qquad x\in A.$$

Therefore

$$Z(\hat{\Phi}, s, \bar{\psi}) = \mathcal{N}\mathfrak{D}^{-\frac{1}{2}} \int_{A^\times} \Phi(cx)\, \bar{\psi}(x)\, \|x\|^s\, d^\times x$$

$$= \mathcal{N}\mathfrak{D}^{-\frac{1}{2}}\, \psi(c)\, \|c\|^{-s}\, Z(\Phi, s, \bar{\psi})$$

$$= \mathcal{N}\mathfrak{D}^{-\frac{1}{2}}\, \psi(c)\, \|c\|^{-s}\, \mu^\times(\Lambda^\times)\, L_A(s, \bar{\psi}).$$

But $\|c\| = (\Lambda : \Lambda c)^{-1} = \mathcal{N}\mathfrak{D}^{-1}$, so from (10. 10) we obtain

$$\Xi(\hat{\Phi}, s, \bar{\psi}) = \mathcal{N}\mathfrak{D}^{s-\frac{1}{2}}\, \psi(c)\, \mu^\times(\Lambda^\times).$$

Replacing s by $1-s$, and using (10. 11), we obtain (10. 20) as desired.

We turn now to the more difficult case in which ψ is *ramified*: $\mathfrak{f}(\psi) \neq \Lambda$. We write $\mathfrak{f} = \mathfrak{f}(\psi)$, for short. It follows immediately from the definition (10. 1) that

$$Z(\Psi, s, \psi) = \mu^\times(1+\mathfrak{f}),$$

where Ψ is the characteristic function of $1+\mathfrak{f}$. Therefore by (3. 6)

$$\Xi(\Psi, s, \psi) = \mu^\times(1+\mathfrak{f}), \tag{10. 21}$$

and we proceed to calculate $\Xi(\hat{\Psi}, s, \bar{\psi})$. Notice that $\mathfrak{f}(\bar{\psi}) = \mathfrak{f}$ also. Let Φ be the characteristic function of Λ, as above, and let Ψ_0 be the characteristic function of $\mathfrak{f}$. Then

$$\hat{\Psi}(y) = \int_A \Psi(x)\, \theta(xy)\, dx = \int_A \Psi_0(x-1)\, \theta(xy)\, dx$$

$$= \int_A \Psi_0(x)\, \theta((x+1)\, y)\, dx = \theta(y)\, \hat{\Psi}_0(y).$$

But $\hat{\Psi}_0$ is the characteristic function of $\mathfrak{D}^{-1}\mathfrak{f}^{-1}$, multiplied by the (additive) measure of $\mathfrak{f}$, and so

$$\hat{\Psi}_0(y) = \mathcal{N}\mathfrak{D}^{-\frac{1}{2}}\, \mathcal{N}\mathfrak{f}^{-1}\, \Phi(cy), \qquad y \in A,$$

where c is *any* element of Λ such that $c\Lambda = \mathfrak{D}\mathfrak{f}$. Therefore

$$Z(\hat{\Psi}, s, \bar{\psi}) = \mathcal{N}\mathfrak{D}^{-\frac{1}{2}}\, \mathcal{N}\mathfrak{f}^{-1} \int_{A^\times} \Phi(cy)\, \theta(y)\, \bar{\psi}(y)\, \|y\|^s\, d^\times y$$

$$= \mathcal{N}(\mathfrak{D}\mathfrak{f})^{s-\frac{1}{2}}\, \mathcal{N}\mathfrak{f}^{-\frac{1}{2}} \int_{A^\times} \Phi(y)\, \theta(c^{-1}y)\, \bar{\psi}(c^{-1}y)\, \|y\|^s\, d^\times y,$$

since $\|c\| = \mathcal{N}(\mathfrak{D}\mathfrak{f})^{-1}$. We shall now evaluate this integral, denoted by I for brevity. We have $I = I_1 + I_2$, where

$$I_1 = \int_{\Lambda^\times} \bar{\psi}(c^{-1}y)\, \theta(c^{-1}y)\, d^\times y,$$

$$I_2 = \int_{(\Lambda \setminus \Lambda) \cap A^\times} \bar{\psi}(c^{-1}y)\, \theta(c^{-1}y)\, \|y\|^s\, d^\times y.$$

We intend to prove that $I_2 = 0$, and that $I_1 = \mu^\times(1+\mathfrak{f}) \cdot \tau(\bar{\psi})$.

The ranges of integration in both I_1 and I_2 are invariant under translation by elements of $\Lambda^\times$, and it follows easily that both integrals are independent of the choice of the element c, subject to the condition (10. 18). We assert that this element c may be chosen to lie in the centre C of A. We will show in (13. 3) that there is an ideal $\mathfrak{a}$ of the valuation ring of C such that $\mathfrak{D}\mathfrak{f}=\mathfrak{a}\Lambda$, and this implies the existence of such an element c. To prove $I_2=0$, we assume that $c\in C$. The symmetry of the reduced trace then implies

$$\text{(10. 22)}\qquad \theta(c^{-1}yz)=\theta(c^{-1}zy)\quad y, z\in A.$$

We may write $(\Lambda\setminus\Lambda^\times)\cap A^\times$ as a countable disjoint union of sets $x\Lambda^\times$, for certain $x\in(\Lambda\setminus\Lambda^\times)\cap A^\times$. It suffices therefore to show that

$$I(x)=\int_{x\Lambda^\times}\bar{\psi}(c^{-1}y)\,\theta(c^{-1}y)\,\|y\|^s\,d^\times y=0$$

whenever $x\in\Lambda\cap A^\times$, $x\notin\Lambda^\times$. Using (10. 22), the integral $I(x)$ is unchanged when the range of integration is replaced by $\Lambda^\times x$. We may thus replace x by uxu', for any $u, u'\in\Lambda^\times$, and this allows us to assume that $x=\text{diag}(\xi^{r_1}, \xi^{r_2},\ldots, \xi^{r_k})$, where $\xi\in D$, $\xi\Delta=\mathfrak{P}$, and the r_i are non-negative integers, with $r_k>0$. Suppose first that $\mathfrak{f}=\mathfrak{P}^f\Lambda$, with $f>1$. Set

$$z=\text{diag}(0, 0,\ldots, \lambda\xi^{f-1}),$$

where $\lambda\in\Delta^\times$ is chosen so that $\psi(1+z)\neq 1$. This is possible by the definition of $\mathfrak{f}$, and because $\text{nr}_{A/C}(1+z)=\text{nr}_{D/C}(1+\lambda\xi^{f-1})$. Certainly $1+z\in\Lambda^\times$, and we have

$$\begin{aligned}I(x)&=\int_{x\Lambda^\times}\bar{\psi}(c^{-1}y(1+z))\,\theta(c^{-1}y(1+z))\,\|y\|^s\,d^\times y\\&=\bar{\psi}(1+z)\int_{x\Lambda^\times}\bar{\psi}(c^{-1}y)\,\theta(c^{-1}y)\,\theta(c^{-1}yz)\,\|y\|^s\,d^\times y.\end{aligned}$$

Since $\|1+z\|=1$, this follows from the translation invariance of the measure. Now write $y=xu$, with $u\in\Lambda^\times$. Then $\theta(c^{-1}yz)=\theta(c^{-1}xuz)=\theta(c^{-1}zxu)=1$, since $zxu\in\mathfrak{f}$, $c^{-1}zxu\in\mathfrak{D}^{-1}$, and θ is trivial on $\mathfrak{D}^{-1}$. Therefore $I(x)=\bar{\psi}(1+z)\,I(x)$, and so $I(x)=0$, as required.

Now suppose that $\mathfrak{f}=\mathfrak{P}\Lambda$. We choose $z\in\Lambda$ of the form

$$z=\text{diag}(0, 0,\ldots, 0, \lambda),$$

where $\lambda\in\Delta^\times$ is chosen so that $1+\lambda\in\Delta^\times$ and $\psi(1+z)\neq 1$. Such a choice is possible, except when $(\Delta:\mathfrak{P})=2$, and then the case $\mathfrak{f}=\mathfrak{P}\Lambda$ cannot arise. The above argument now carries over, word for word, with this choice of z.

Overall, we now know that

$$Z(\hat{\Psi}, s, \bar{\psi})=\mathcal{N}(\mathfrak{D}\mathfrak{f})^{s-\frac{1}{2}}\,\mathcal{N}\mathfrak{f}^{-\frac{1}{2}}\int_{\Lambda^\times}\bar{\psi}(c^{-1}y)\,\theta(c^{-1}y)\,d^\times y,$$

where (*changing notation*) c is any element of A such that $c\Lambda=\mathfrak{D}\mathfrak{f}$. We may write

$$\Lambda^\times=\bigcup_x x(1+\mathfrak{f})\quad\text{(disjoint union)}$$

so that x ranges over a set of coset representatives of $\Lambda^\times$ mod $(1+\mathfrak{f})$. Then, since $\bar{\psi}$ is trivial on $1+\mathfrak{f}$, and θ is trivial on $c^{-1}x\mathfrak{f}$, for $x\in\Lambda^\times$, we obtain

$$Z(\hat{\Psi}, s, \bar{\psi})=\mathcal{N}(\mathfrak{D}\mathfrak{f})^{s-\frac{1}{2}}\,\mathcal{N}\mathfrak{f}^{-\frac{1}{2}}\,\mu^\times(1+\mathfrak{f})\sum_x\bar{\psi}(c^{-1}x)\,\theta(c^{-1}x)=\Xi(\hat{\Psi}, s, \bar{\psi}).$$

Using (10. 17), (10. 11) and (10. 21), we now have

$$\varepsilon_A(s, \psi) = \mathcal{N}(\mathfrak{D}\mathfrak{f})^{\frac{1}{2}-s} \cdot \frac{\tau(\bar{\psi})}{\mathcal{N}\mathfrak{f}^{\frac{1}{2}}},$$

which completes the proof of (10. 19).

Now choose a function $\Phi \in \mathcal{S}(A)$ such that $\Xi(\Phi, s, \psi)$ is not identically zero. We have already produced several examples of such functions. From (10. 11) we have

$$\Xi(\hat{\Phi}, 1-s, \bar{\psi}) = \varepsilon_A(s, \psi)\, \Xi(\Phi, s, \psi),$$

$$\Xi(\hat{\hat{\Phi}}, 1-s, \psi) = \varepsilon_A(s, \bar{\psi})\, \Xi(\hat{\Phi}, s, \bar{\psi}).$$

Using (10. 5), we find that $\Xi(\hat{\hat{\Phi}}, s, \psi) = \psi(-1) \cdot \Xi(\Phi, s, \psi)$, and we obtain the identity

$$\tau(\bar{\psi})\, \tau(\psi) = \psi(-1)\, \mathcal{N}\mathfrak{f}(\psi).$$

It is immediate from (10. 17) that $\overline{\tau(\psi)} = \psi(-1)\, \tau(\bar{\psi})$, so we have proved

(10. 23) Corollary. *The Gauss sum $\tau(\psi)$ is a complex number of absolute value $\mathcal{N}\mathfrak{f}(\psi)^{\frac{1}{2}}$. In particular, $\tau(\psi) \neq 0$.*

We conclude this section by discussing the case where the $\mathbb{Q}_p$-algebra A is *semisimple*, say $A = \prod_{i=1}^{r} A_i$, where each A_i is a simple algebra. Let θ_{A_i} be the canonical character of A_i, $1 \leqq i \leqq r$, as in (10. 2). We define the canonical character θ_A of A by

$$\theta_A = \prod_{i=1}^{r} \theta_{A_i}. \tag{10. 24}$$

The self-dual Haar measure dx on A is then the product of the self-dual measures on the A_i. Defining the Fourier transform exactly as in (10. 4), we see at once that Theorem (10. 7) remains valid, using the fact that

$$\mathcal{S}(A) = \mathcal{S}(A_1) \otimes \cdots \otimes \mathcal{S}(A_r).$$

That is, $\mathcal{S}(A)$ is spanned by products of Schwartz-Bruhat functions on the A_i. (The argument is exactly as in [1], § 5. 2.) Further, if we define $\Xi(\Phi, s, \psi)$ as in (10. 10), for a function $\Phi \in \mathcal{S}(A)$ and a character ψ of $A^\times$, then (10. 11) remains valid. By choosing $\Phi = \Pi\, \Phi_i$, with $\Phi_i \in \mathcal{S}(A_i)$, we obtain

$$\varepsilon_A(s, \psi) = \prod_{i=1}^{r} \varepsilon_{A_i}(s, \psi_i), \tag{10. 25}$$

where ψ_i is the restriction of ψ to $A_i^\times$.

One can define differents and conductors as before, relative to the choice of a maximal order Λ in A, and these are naturally expressed as products of the corresponding objects for the A_i. Formula (10. 17) then defines the Gauss sum $\tau(\psi)$, and clearly

$$\tau(\psi) = \prod_{i=1}^{r} \tau(\psi_i). \tag{10. 26}$$

Further, Theorem (10. 19) and Corollary (10. 23) hold without change.

For later use, we restate the generalised version of (10. 19) as follows:

(10. 27) Theorem. *Let A be a finite-dimensional semisimple $\mathbb{Q}_p$-algebra, and ψ a continuous character of $A^\times$. Let Λ be a maximal order in A, and let $\mathfrak{D}_A$ be the absolute different of A, and $\mathfrak{f}(\psi)$ the conductor of ψ, defined relative to Λ. Then for each $\Phi \in \mathcal{S}(A)$ we have*

$$L_A(1-s, \bar{\psi})^{-1} Z(\hat{\Phi}, 1-s, \bar{\psi}) = \mathcal{N}_A(\mathfrak{D}_A \mathfrak{f}(\psi))^{\frac{1}{2}-s} \frac{\tau(\bar{\psi})}{\mathcal{N}_A \mathfrak{f}(\psi)^{\frac{1}{2}}} L_A(s, \psi)^{-1} Z(\Phi, s, \psi),$$

where, for a two-sided fractional ideal $\mathfrak{b}$ of Λ in A, we put $\mathcal{N}_A \mathfrak{b} = (\Lambda : \mathfrak{b})$.

§ 11. Local functional equations for L-functions of integral group rings

To start with, let A be any finite-dimensional semisimple $\mathbb{Q}_p$-algebra, and let Λ be a $\mathbb{Z}_p$-order in A. Let M be a left ideal of Λ, of finite index $(\Lambda : M)$ in Λ. Let θ_A be the canonical additive character of A defined in (10. 2) and (10. 24), and set

$$M^\perp = \{x \in A : \theta_A(yx) = 1, \ y \in M\}. \tag{11. 1}$$

Since $\theta_A(yx) = \theta_A(xy)$, for $x, y \in A$, it follows that $M^\perp$ is a right Λ-module. Moreover, the identification of A with $\hat{A}$ induced by θ_A also identifies $A/M^\perp$ with the Pontrjagin dual $\hat{M}$ of M, and $M^\perp$ with $(A/M)^\wedge$. Thus $M^\perp$ is a compact open subgroup of A, and is therefore a Λ-lattice. In particular, $M^\perp$ is isomorphic to a right ideal of Λ.

Now let $\Gamma = \operatorname{End}_\Lambda(M)$, and we view M as a (Λ, Γ)-bimodule. Then $\Gamma \subseteq \operatorname{End}_\Lambda(M^\perp)$, and so $M^\perp$ is a (Γ, Λ)-bimodule. Iterating the process, we get $\operatorname{End}_\Lambda(M^\perp) \subseteq \operatorname{End}_\Lambda(M^{\perp\perp})$, and since $M^{\perp\perp} = M$, we have $\Gamma = \operatorname{End}_\Lambda(M^\perp)$. Suppose now that ψ is a character of $A^\times$ which is trivial on $\Gamma^\times = \operatorname{Aut}_\Lambda(M)$. Then the L-function $L_\Lambda(M, s, \psi)$ is defined, as in (2. 1). We may also define another L-function based on *right* ideals of Λ as follows:

$$L_\Lambda^{(r)}(M^\perp, s, \psi) = \sum \psi(Y)\,(\Lambda : Y)^{-s}, \quad \operatorname{Re}(s) > 1, \tag{11. 2}$$

where the sum is taken over all right ideals Y of Λ which are isomorphic to $M^\perp$. The symbol $\psi(Y)$ denotes $\psi(y)$, for any $y \in A^\times$ such that $Y = yM^\perp$. Since ψ is trivial on $\operatorname{Aut}_\Lambda(M^\perp)$, this is well-defined. The functional equation (10. 27) then yields a relation between

$$\frac{L_\Lambda(M, s, \psi)}{L_A(s, \psi)} \quad \text{and} \quad \frac{L_\Lambda^{(r)}(M^\perp, 1-s, \bar{\psi})}{L_A(1-s, \bar{\psi})}.$$

This relation is particularly interesting when Λ is a group ring, because then we may use the standard anti-automorphism of Λ to identify $L_\Lambda^{(r)}(M^\perp, s, \psi)$ with an L-function defined via left ideals of Λ. This procedure was carried out in [1], § 5. 3 for zeta-functions, and we now extend this calculation to L-functions.

Suppose now that F is a finite field extension of $\mathbb{Q}_p$, with valuation ring R. Let G be a finite group, and put $A = FG$, $\Lambda = RG$. Denote by $*$ the standard anti-automorphism of A defined by

$$\Big(\sum_{g \in G} \alpha_g g\Big)^* = \sum_{g \in G} \alpha_g g^{-1}, \quad \alpha_g \in F.$$

For a function f on A, let f^* denote the function given by

(11. 3) $$f^*(x)=f(x^*), \quad x \in A.$$

We use a similar notation for functions on G and $A^\times$.

Now let M be a left ideal of Λ, with $(\Lambda: M)$ finite, as always, and write

(11. 4) $$M^{-1}=\{M:\Lambda\}=\{x \in A: Mx \subseteq \Lambda\}.$$

Then

(11. 5) $$M^{-1}=\mathrm{Hom}_\Lambda(M, \Lambda) \cong \mathrm{Hom}_R(M, R),$$

with the isomorphism arising from the Frobenius Reciprocity Theorem. Note that M^{-1} is a right Λ-lattice, and that $(M^{-1})^{-1}=M$, so the notation M^{-1} is justified, to some extent. We now put

(11. 6) $$\check{M}=(M^{-1})^*,$$

the *contragredient* of M, which is a left Λ-lattice. If $\Gamma=\mathrm{End}_\Lambda(M)$, the equality $(M^{-1})^{-1}=M$ implies that $\Gamma=\mathrm{End}_\Lambda(M^{-1})$ also, and hence that $\Gamma^*=\mathrm{End}_\Lambda(\check{M})$. Therefore, if ψ is a character of $A^\times$ which is trivial on $\Gamma^\times$, then ψ^*, defined by (11. 3), is also a character of $A^\times$ and it is trivial on $\Gamma^{*\times}=\mathrm{Aut}_\Lambda(\check{M})$. Both of the L-functions $L_\Lambda(M, s, \psi)$ and $L_\Lambda(\check{M}, s, \bar{\psi}^*)$ are therefore defined, as in (2. 1).

Let us summarise the notation to be used in stating and proving the main result of this section.

(11. 7) Notation. G = a finite group of order n;
F = a finite field extension of $\mathbb{Q}_p$, with valuation ring R;
$\mathfrak{D}_F$ = the absolute different of F; c_F is some element of R with $c_F R=\mathfrak{D}_F$;
Λ = RG, $A=FG$, Λ' = some maximal order in A containing Λ;
M = a left ideal of Λ, with contragredient $\check{M}$ (see (11. 6));
ψ = a character of $A^\times$ which is trivial on $\mathrm{Aut}_\Lambda(M)$; ψ^* is defined by (11. 3);
$\mathfrak{f}$ = $\mathfrak{f}(\psi)$ is the conductor of ψ, a 2-sided ideal of Λ';
tr = the reduced trace $A \to C$, where C is the centre of A;
b = $\mathrm{tr}(1) \in C$;
$\tau(\psi)$ = the Gauss sum attached to ψ, as in (10. 17) and (10. 26);
$\mathcal{N}\mathfrak{b}=(\Lambda':\mathfrak{b})$, for any ideal $\mathfrak{b}$ of Λ'.

(11. 8) Theorem. *With the notation listed in* (11. 7), *we have*

$$L_\Lambda(1-s, \bar{\psi}^*)^{-1}\, L_\Lambda(\check{M}, 1-s, \bar{\psi}^*)$$
$$=\{(\Lambda':\Lambda)\, \mathcal{N}\mathfrak{f}(\psi)^{-\frac{1}{2}}\}^{2s-1}\, \bar{\psi}(nc_F b^{-1}) \frac{\tau(\bar{\psi})}{\mathcal{N}\mathfrak{f}(\psi)^{\frac{1}{2}}}\, L_\Lambda(s, \psi)^{-1}\, L_\Lambda(M, s, \psi).$$

Moreover, $L_A(s, \psi^*)=L_A(s, \psi)$, *for any continuous character* ψ *of* $A^\times$.

(11. 9) Remarks. (i) Since $R^\times \subseteq \mathrm{Aut}_\Lambda(M)$, the character ψ is trivial on $R^\times$, and so the value $\bar{\psi}(nc_F b^{-1})$ is independent of the choice of element c_F generating $\mathfrak{D}_F$.

(ii) If ψ is unramified, then $\mathfrak{f}=\Lambda'$, and $\tau(\bar{\psi})=\psi(\mathfrak{D}_A)$, where $\mathfrak{D}_A$ is the absolute different of A, as in (10. 12). Since $\mathfrak{D}_A$ has a factorisation $\mathfrak{D}_A=\mathfrak{D}_{A/F}\cdot\mathfrak{D}_F$, where the relative different $\mathfrak{D}_{A/F}$ is defined, in the obvious way, with respect to $\mathrm{tr}_{A/F}=\mathrm{Tr}_{C/F}\cdot\mathrm{tr}_{A/C}$, we obtain as a special case of Theorem (11. 8) the formula

$$L_A(1-s,\bar{\psi}^*)^{-1}\,L_A(\check{M},1-s,\bar{\psi}^*)=(\Lambda':\Lambda)^{2s-1}\,\bar{\psi}(nb^{-1}\,\mathfrak{D}_{A/F}^{-1})\,L_A(s,\psi)^{-1}\,L_A(M,s,\psi). \tag{11. 10}$$

Further, in this case, we have $L_A(s,\psi)=L_{\Lambda'}(\Lambda',s,\psi)$, by (3. 9). Specialising further to the case of ψ trivial, we get

$$\zeta_{\Lambda'}(1-s)^{-1}\,Z_\Lambda(\check{M},1-s)=(\Lambda':\Lambda)^{2s-1}\,\zeta_{\Lambda'}(s)^{-1}\,Z_\Lambda(M,s),$$

which is Theorem 2 of [1].

(iii) Suppose that $n\in R^\times$. Then $\Lambda=\Lambda'$, and if we take $M=\Lambda$ in (11. 10), we find that $\psi(\mathfrak{D}_{A/F})=1$ for all unramified characters ψ of $A^\times$. This implies that $\mathfrak{D}_{A/F}=\Lambda'(=\Lambda)$, or equivalently that A is a direct product of full matrix algebras over *fields*, each of which is *unramified* over F. (This result is well-known, and can be proved in a number of different ways. For example, it follows from the observation that the discriminant of RG with respect to R is a power of the ideal nR.)

We now prove Theorem (11. 8), and begin by introducing a new additive character θ_G of A. This character has the advantage that its properties closely reflect the algebraic situation. We let $t: FG\longrightarrow F$ be the F-linear map defined by

$$t\Big(\sum_{g\in G}\alpha_g g\Big)=\alpha_1,\quad \alpha_g\in F.$$

Now let χ be a continuous character of F which is trivial on R, but not on $\mathfrak{p}^{-1}$, where $\mathfrak{p}$ denotes the prime ideal of R. Any such χ is of the form

$$\chi(\alpha)=\theta_F(c_F^{-1}\alpha),\quad \alpha\in F,$$

for some $c_F\in R$ such that $c_F R=\mathfrak{D}_F$, and where θ_F is defined by (10. 2). Taking such a character χ, we define

$$\theta_G=\chi\cdot t.$$

It is then easily seen that for any left ideal M of Λ, we have

$$M^{-1}=\{x\in A:\theta_G(yx)=1,\ \text{for all}\ y\in M\}, \tag{11. 11}$$

and a corresponding formula holds for right ideals of Λ. Note also that $t(yx)=t(xy)$, and hence $\theta_G(yx)=\theta_G(xy)$, for all $x, y\in A$.

The pairing $(x,y)\mapsto\theta_G(xy)$, $x,y\in A$, is again nondegenerate, and identifies A with $\hat{A}$. Therefore there is a Haar measure $d_G x$ on A which is self-dual for this identification. Since $\Lambda^{-1}=\Lambda$, this measure is determined by the condition

$$\int_\Lambda d_G x=1.$$

Now let $d_A x$ denote the self-dual Haar measure for the canonical character θ_A of A defined in (10. 2) (generalised to the semisimple case, as after (10. 24)). Then we have

$$d_G x=\mathcal{N}_A\mathfrak{D}_A^{\frac{1}{2}}(\Lambda':\Lambda)\,d_A x.$$

Let $\mathscr{F}_G$ denote the Fourier transform taken with respect to θ_G; that is,

$$(\mathscr{F}_G \Phi)(y) = \int_A \Phi(x)\, \theta_G(xy)\, d_G x, \qquad y \in A,\ \Phi \in \mathscr{S}(A).$$

This transform has some useful properties. If Φ is the characteristic function of a left ideal M of Λ, then by (11. 11) we have

$$\text{(11. 12)} \qquad \begin{aligned} \mathscr{F}_G \Phi &= (\Lambda : M)^{-1} \cdot \text{characteristic function of } M^{-1}; \\ (\mathscr{F}_G \Phi)^* &= (\Lambda : M)^{-1} \cdot \text{characteristic function of } \check{M}. \end{aligned}$$

We need to know the relation between $\mathscr{F}_G \Phi$ and $\mathscr{F}_A \Phi$ (see (10. 6)). Suppose that $A = \prod_{i=1}^{r} A_i$, where each A_i is a simple algebra with centre C_i, and let $n_i^2 = \dim_{C_i}(A_i)$. Then $t(x) = n^{-1} \cdot \mathrm{Tr}_{A/F}(x) = \mathrm{Tr}_{A/F}(n^{-1} x)$, $x \in A$, where $\mathrm{Tr}_{A/F}$ is the trace of the regular representation of A over F. The element $b = \mathrm{tr}_{A/C}(1) \in C$ has component n_i in C_i, for each i, and therefore $\mathrm{Tr}_{A/F}(x) = \mathrm{tr}_{A/F}(bx)$, $x \in A$. Combining this with the definitions of θ_G, θ_A, we obtain

$$\text{(11. 13)} \qquad \theta_G(x) = \theta_A(c_F^{-1} \cdot n^{-1} \cdot b \cdot x), \qquad x \in A.$$

Since the element $c_F n b^{-1}$ is central in A, we obtain, for $\Phi \in \mathscr{S}(A)$,

$$\begin{aligned} (\mathscr{F}_G \Phi)(y) &= \mathscr{N}\mathfrak{D}^{\frac{1}{2}} (\Lambda' : \Lambda) \int_A \Phi(x)\, \theta_A(c_F^{-1} n^{-1} bxy)\, d_A x \\ &= \mathscr{N}\mathfrak{D}^{\frac{1}{2}} (\Lambda' : \Lambda)\, \hat{\Phi}(c_F^{-1} n^{-1} by), \qquad y \in A, \end{aligned}$$

where $\hat{\Phi} = \mathscr{F}_A \Phi$. Consequently we have

$$\begin{aligned} Z(\mathscr{F}_G \Phi, s, \bar{\psi}) &= \mathscr{N}\mathfrak{D}^{\frac{1}{2}} (\Lambda' : \Lambda) \int_{A^\times} \hat{\Phi}(c_F^{-1} n^{-1} bx)\, \bar{\psi}(x)\, \|x\|^s\, d^\times x \\ &= \mathscr{N}\mathfrak{D}^{\frac{1}{2}} (\Lambda' : \Lambda)\, \bar{\psi}(c_F n b^{-1})\, \|c_F n b^{-1}\|^s\, Z(\hat{\Phi}, s, \bar{\psi}). \end{aligned}$$

The functional equation (10. 27) now reads

$$\text{(11. 14)} \qquad \begin{aligned} &L_A(1-s, \bar{\psi})^{-1}\, Z(\mathscr{F}_G \Phi, 1-s, \bar{\psi}) \\ &= \mathscr{N}\mathfrak{D}^{\frac{1}{2}} (\Lambda' : \Lambda)\, \bar{\psi}(c_F n b^{-1})\, \|c_F n b^{-1}\|^{1-s}\, \varepsilon_A(s, \psi)\, L_A(s, \psi)^{-1}\, Z(\Phi, s, \psi), \end{aligned}$$

where

$$\varepsilon_A(s, \psi) = \mathscr{N}(\mathfrak{D}\mathfrak{f}(\psi))^{\frac{1}{2}-s} \frac{\tau(\bar{\psi})}{\mathscr{N}\mathfrak{f}(\psi)^{\frac{1}{2}}},$$

as in (10. 19) and (10. 25). Observe next that the measure $d^\times x$ is invariant under $x \mapsto x^*$, since $x^{**} = x$, for all $x \in A$. Therefore

$$Z(\Phi^*, s, \psi^*) = Z(\Phi, s, \psi), \qquad \Phi \in \mathscr{S}(A).$$

It follows that the function $L_A(s, \psi^*)$ also satisfies the conditions C1–C3 of § 3 which characterise $L_A(s, \psi)$, and we have

$$L_A(s, \psi^*) = L_A(s, \psi).$$

Now let M be as in (11. 7), and let $\Phi=\Phi_{M^{-1}}$ be the characteristic function of M^{-1}. Thus

$$Z(\Phi, s, \psi)=\mu^{\times}(\mathrm{Aut}_{\Lambda}(M))\,(\Lambda:M)^{s}\,L_{\Lambda}(M, s, \psi),$$

by (2. 5). Since $(\Lambda:M^{-1})=(M:\Lambda)$, and $(M^{-1})^{-1}=M$, we conclude from (11. 12) that

$$\mathscr{F}_G(\Phi)=(\Lambda:M)\,\Phi_M, \quad \text{and} \quad (\mathscr{F}_G\Phi)^*=(\Lambda:M)\,\Phi_{M^*}.$$

But $M^*=(\check{M})^{-1}$, so

$$\{\mathscr{F}_G\Phi\}^*=(\Lambda:M)\,\Phi_{(\check{M})^{-1}}.$$

Consequently we obtain

$$\begin{aligned} Z(\mathscr{F}_G\Phi, s, \bar{\psi}) &= Z(\{\mathscr{F}_G\Phi\}^*, s, \bar{\psi}^*) \\ &= (\Lambda:M)\,\mu^{\times}(\mathrm{Aut}_{\Lambda}(\check{M}))\,(\Lambda:\check{M})^{s}\,L_{\Lambda}(\check{M}, s, \bar{\psi}^*). \end{aligned}$$

Now, $\mathrm{Aut}_{\Lambda}(\check{M})=\{\mathrm{Aut}_{\Lambda}(M)\}^*$, and so $\mu^{\times}(\mathrm{Aut}_{\Lambda}(\check{M}))=\mu^{\times}(\mathrm{Aut}_{\Lambda}(M))$. Also

$$(\Lambda:\check{M})=(\Lambda:M^{-1})=(\Lambda:M)^{-1},$$

so substituting into (11. 14) we obtain

$$\begin{aligned} &L_{\Lambda}(1-s, \bar{\psi}^*)^{-1}\,L_{\Lambda}(\check{M}, 1-s, \bar{\psi}^*) \\ &\quad = L_{\Lambda}(s,\psi)^{-1}\,L_{\Lambda}(M,s,\psi)\,\mathscr{N}\mathfrak{D}^{\frac{1}{2}}(\Lambda':\Lambda)\,\bar{\psi}(c_F nb^{-1})\,\|c_F nb^{-1}\|^{1-s}\,\varepsilon_A(s,\psi) \\ &\quad = L_{\Lambda}(s,\psi)^{-1}\,L_{\Lambda}(M,s,\psi)\,\mathscr{N}\mathfrak{D}^{1-s}(\Lambda':\Lambda)\,\bar{\psi}(c_F nb^{-1})\,\|c_F nb^{-1}\|^{1-s}\,\mathscr{N}\mathfrak{f}(\psi)^{\frac{1}{2}-s}\,\frac{\tau(\bar{\psi})}{\mathscr{N}\mathfrak{f}(\psi)^{\frac{1}{2}}}. \end{aligned}$$

But $\mathscr{N}\mathfrak{D}^{1-s}\cdot\|c_F nb^{-1}\|^{1-s}=\mathscr{N}(\mathfrak{D}^{-1}c_F nb^{-1})^{s-1}$, and we have $\mathfrak{D}^{-1}c_F nb^{-1}=\mathfrak{D}_{A/F}^{-1}\cdot nb^{-1}$. Since $t=n^{-1}b\cdot\mathrm{tr}_{A/F}$, the dual $\tilde{\Lambda}'$ of Λ' with respect to t is precisely $\mathfrak{D}_{A/F}^{-1}\cdot nb^{-1}$. Thus

$$\mathscr{N}(\mathfrak{D}^{-1}c_F nb^{-1})=(\Lambda':\tilde{\Lambda}')=(\Lambda':\Lambda)\,(\Lambda:\tilde{\Lambda}')=(\Lambda':\Lambda)^2,$$

since Λ is self-dual with respect to t. This gives

$$\begin{aligned} &L_{\Lambda}(1-s, \bar{\psi}^*)^{-1}\,L_{\Lambda}(\check{M}, 1-s, \bar{\psi}^*) \\ &\quad = \{(\Lambda':\Lambda)\,\mathscr{N}\mathfrak{f}(\psi)^{-\frac{1}{2}}\}^{2s-1}\,\bar{\psi}(c_F nb^{-1})\,\frac{\tau(\bar{\psi})}{\mathscr{N}\mathfrak{f}(\psi)^{\frac{1}{2}}}\,L_{\Lambda}(s, \psi)^{-1}\,L_{\Lambda}(M, s, \psi), \end{aligned}$$

and thereby proves the theorem.

We can now combine these local functional equations into a preliminary form of a global functional equation. For this, we let F be an algebraic number field with ring of integers R. We set $\Lambda=RG$, $A=FG$, where G is a finite group, and we let Λ' be a maximal order in A. We take a left ideal M of Λ, and a character ψ of the genus class group $\mathrm{Cl}(M)$. We have a commutative diagram

$$\begin{array}{ccc} \mathrm{Cl}(M) & \xrightarrow{\approx} & \mathrm{Cl}(\check{M}) \\ \uparrow & & \uparrow \\ J(A) & \xrightarrow{\approx} & J(A) \end{array}$$

in which $\check{M}$ denotes the contragredient of M, defined exactly as in the local case: $\check{M}=\{M:\Lambda\}^*$. The isomorphism $\mathrm{Cl}(M)\to\mathrm{Cl}(\check{M})$ is given by $N\mapsto\check{N}$. The map

$J(A) \to \mathrm{Cl}(M)$ is the canonical surjection induced by $x \mapsto Mx$, $x \in J(A)$, as described in § 1. The map $J(A) \to \mathrm{Cl}(\check{M})$ is induced by $y \mapsto \check{M}y$, and the isomorphism $J(A) \to J(A)$ which makes the diagram commute is $x \mapsto (x^{-1})^*$. Thus, when we view ψ as an idele class character of A, this diagram gives us a natural interpretation of $\bar{\psi}^*$ as a character of $\mathrm{Cl}(\check{M})$. Hence the L-functions $L_A(M, s, \psi)$ and $L_A(\check{M}, s, \bar{\psi}^*)$ are defined as in (1.3).

Let $\mathfrak{p}$ denote a maximal ideal of R, and write $\psi_{\mathfrak{p}} = \psi | A_{\mathfrak{p}}^{\times}$, where we view ψ as a character of $J(A)$. Theorem (11.8) gives us a functional equation relating

$$L_{A_{\mathfrak{p}}}(1-s, \bar{\psi}_{\mathfrak{p}}^*)^{-1} L_{A_{\mathfrak{p}}}(\check{M}_{\mathfrak{p}}, 1-s, \bar{\psi}_{\mathfrak{p}}^*) \quad \text{and} \quad L_{A_{\mathfrak{p}}}(s, \psi_{\mathfrak{p}})^{-1} L_{A_{\mathfrak{p}}}(M_{\mathfrak{p}}, s, \psi_{\mathfrak{p}}).$$

These quotients are identically one for almost all $\mathfrak{p}$, namely those which are relatively prime to the order n of the group G. Therefore we may take the product of the functional equations (11.8) over all $\mathfrak{p}$, to get an equation relating $L_A(1-s, \bar{\psi}^*)^{-1} L_A(\check{M}, 1-s, \bar{\psi}^*)$ and $L_A(s, \psi)^{-1} L_A(M, s, \psi)$, where these functions are defined by analytic continuation, as in §§ 4 and 5. This equation involves among other things the factors

$$\prod_{\mathfrak{p}} \bar{\psi}_{\mathfrak{p}}(n), \quad \prod_{\mathfrak{p}} \bar{\psi}_{\mathfrak{p}}(b_{\mathfrak{p}}^{-1}),$$

where the products are taken over all maximal ideals $\mathfrak{p}$ of R. We use the obvious notation here: $b_{\mathfrak{p}} = \mathrm{tr}_{A_{\mathfrak{p}}}(1)$. We assert that these two products are both one. In the first place, n is a principal idele of A, and any character of $A_{\infty}^{\times} = (A \otimes_{\mathbb{Q}} \mathbb{R})^{\times}$ of finite order vanishes on n, since n is real and positive in each of the fields making up the centre of the algebra A_{∞}. Thus

$$1 = \psi(n) = \prod_{\mathfrak{p}} \psi_{\mathfrak{p}}(n).$$

For the second product, we define $b = \mathrm{tr}_{A/C}(1)$, an element of the centre C of A. Then, viewing b as an element of $A_{\mathfrak{p}}$, we still have $b = \mathrm{tr}_{A_{\mathfrak{p}}/C_{\mathfrak{p}}}(1) = b_{\mathfrak{p}}$. The same argument now applies to give

$$1 = \psi(b) = \prod_{\mathfrak{p}} \psi_{\mathfrak{p}}(b_{\mathfrak{p}}).$$

We use the maximal order $\Lambda'_{\mathfrak{p}}$ to define the local objects $\mathfrak{f}(\psi_{\mathfrak{p}})$, $\tau(\psi_{\mathfrak{p}})$, $\mathfrak{D}_{\mathfrak{p}}$, for each $\mathfrak{p}$. The continuity of the character ψ implies that $\psi_{\mathfrak{p}}$ is an unramified character of $A_{\mathfrak{p}}^{\times}$, for almost all $\mathfrak{p}$. Also, $\mathfrak{D}_{\mathfrak{p}} = \Lambda'_{\mathfrak{p}}$ for almost all $\mathfrak{p}$. Therefore $\tau(\psi_{\mathfrak{p}}) = 1$, and $\mathfrak{f}(\psi_{\mathfrak{p}}) = \Lambda'_{\mathfrak{p}}$, for almost all $\mathfrak{p}$. Hence we may define the *global Gauss sum* $\tau(\psi)$ by

$$\tau(\psi) = \prod_{\mathfrak{p}} \tau(\psi_{\mathfrak{p}})$$

(product over all $\mathfrak{p}$). Also, we can define the *global conductor* $\mathfrak{f}(\psi)$ to be the unique ideal of Λ' such that $\mathfrak{f}(\psi)\Lambda'_{\mathfrak{p}} = \mathfrak{f}(\psi_{\mathfrak{p}})$ for all $\mathfrak{p}$. Then the numbers $\tau(\psi)$ and $\mathcal{N}\mathfrak{f}(\psi) = (\Lambda' : \mathfrak{f}(\psi))$ are independent of the choice of Λ'. In all, we obtain

$$(11.15) \quad L_A(1-s, \bar{\psi}^*)^{-1} L_A(\check{M}, 1-s, \bar{\psi}^*)$$
$$= \{(\Lambda' : \Lambda)\, \mathcal{N}\mathfrak{f}(\psi)^{-\frac{1}{2}}\}^{2s-1}\, \bar{\psi}(\mathfrak{D}_F) \frac{\tau(\bar{\psi})}{\mathcal{N}\mathfrak{f}(\psi)^{\frac{1}{2}}} L_A(s, \psi)^{-1} L_A(M, s, \psi).$$

Here, $\mathfrak{D}_F$ is the absolute different of the field F. The local analogue implies via (3. 1) that $L_A(s, \psi^*) = L_A(s, \psi)$, and we may rearrange (11. 15) to give

$$(11.16)\quad \frac{L_A(\check{M}, 1-s, \bar{\psi}^*)}{L_A(M, s, \psi)} = \frac{L_A(1-s, \bar{\psi})}{L_A(s, \psi)} \{(\Lambda' : \Lambda)\, \mathcal{N}\mathfrak{f}(\psi)^{-\frac{1}{2}}\}^{2s-1}\, \bar{\psi}(\mathfrak{D}_F) \frac{\tau(\bar{\psi})}{\mathcal{N}\mathfrak{f}(\psi)^{\frac{1}{2}}}.$$

The quotient $L_A(1-s, \bar{\psi})/L_A(s, \psi)$ can be computed explicitly to give a self-contained form for the functional equation of $L_A(M, s, \psi)$ in this global case. We return to this matter in § 14.

§ 12. Quotients of zeta-functions

Throughout this section, A denotes a finite-dimensional semisimple Q_p-algebra, Λ an order in A, and Λ' a maximal order in A. Let M be a left ideal of Λ. Then we have the zeta-functions

$$Z_\Lambda(M, s) = \sum_{\substack{X \subseteq \Lambda \\ X \simeq M}} (\Lambda : X)^{-s}, \quad \zeta_{\Lambda'}(s) = \sum_{X \subseteq \Lambda'} (\Lambda' : X)^{-s},$$

or in term of our local L-series,

$$Z_\Lambda(M, s) = L_\Lambda(M, s, \psi_0), \quad \zeta_{\Lambda'}(s) = L_{\Lambda'}(\Lambda', s, \psi_0),$$

where ψ_0 is the trivial character of $A^\times$. By (3. 12), the function $f(s)$ defined by

$$f(s) = Z_\Lambda(M, s)/\zeta_{\Lambda'}(s)$$

lies in $Z[p^{-s}]$. The object of this section is to find the highest power of p^{-s} occurring in $f(s)$, along with its coefficient. We shall then specialise to the case of Λ a group ring, and thereby strengthen the result of [1], p. 158, Remark 3.

For the moment, let M be an arbitrary Z_p-lattice which spans A over Q_p. We define an integer $a(M)$ by

(12. 1) $\qquad p^{a(M)} = \text{Min}\,\{(\Lambda : \Lambda x) : x \in M \cap A^\times\}$, or equivalently,

(12. 2) $\qquad p^{-a(M)} = \text{Max}\,\{\|x\| : x \in M \cap A^\times\}$.

For each $k \in Z$, set

(12. 3) $\qquad S_k = S_k(M) = \{x \in M \cap A^\times : \|x\| = p^k\}$.

Then S_k is empty if $k > -a(M)$, while $S_{-a(M)}$ is not empty. The set S_k is invariant under translation by $\text{Aut}_\Gamma(M)$, for any order Γ which stabilises M. Therefore S_k is open in both A and $A^\times$. Moreover, it is a finite union of orbits of $\text{Aut}_\Gamma(M)$, and it is therefore compact. We choose a Haar measure $d^\times x$ on $A^\times$, and write $\mu^\times$ for the measure of a set with respect to $d^\times x$. We define

(12. 4) $\qquad \lambda(M) = \mu^\times(S_{-a(M)}(M))$.

Then $\lambda(M)$ is finite and positive. Using this notation, we now establish

(12. 5) Lemma. *Let M be a Z_p-lattice which spans A over Q_p. Then*

$$\int_{M \cap A^\times} \|x\|^s\, d^\times x = \lambda(M)\, p^{-a(M)s}\, \{1 + c_1 p^{-s} + c_2 p^{-2s} + \cdots\},$$

where the c_j, $j \geqq 1$, are constants. This expression is valid in some region $\text{Re}(s) > 1 - \varepsilon$, $\varepsilon > 0$.

Proof. We have

$$\int_{M \cap A^\times} \|x\|^s d^\times x = \sum_{k=-\infty}^{-a(M)} \int_{S_k} \|x\|^s d^\times x = \sum_{k=-\infty}^{-a(M)} \mu^\times (S_k) p^{ks},$$

and this gives the result. The convergence property comes from (3. 14).

Now let θ be a continuous character of the additive group of A with the following properties:

(i) the pairing $(x, y) \mapsto \theta(xy)$, $x, y \in A$, is nondegenerate, and therefore identifies A with its Pontrjagin dual $\hat{A}$;

(ii) $\theta(yx) = \theta(xy)$, for all $x, y \in A$.

(For example, the character θ_A defined in (10. 2) has these properties.) Let dx be the self-dual Haar measure on A for the identification of A with $\hat{A}$ induced by θ. Write μ^+ for the measure of a set with respect to dx. For $\Phi \in \mathcal{S}(A)$, we have the Fourier transform

$$\hat{\Phi}(y) = \int_A \Phi(x)\, \theta(xy)\, dx, \quad y \in A,$$

and the zeta-integral

$$Z(\Phi, s) = \int_{A^\times} \Phi(x) \|x\|^s d^\times x$$

in the usual manner. By (3. 14) or [1], § 5. 2, $Z(\Phi, s)$ is continuable to a meromorphic function of s. The functional equation

$$Z(\Phi, s)\, Z(\hat{\Psi}, 1-s) = Z(\hat{\Phi}, 1-s)\, Z(\Psi, s), \quad \Phi, \Psi \in \mathcal{S}(A),$$

still holds in this generality. Indeed, the proof in [1], Appendix, applies without change. Moreover, an easy exercise shows that there is no real gain in generality compared to (10. 7).

We now let M be a left ideal of the order Λ, and define $\{M:\Lambda\}$ as in (2. 2), namely by $\{M:\Lambda\} = \{x \in A \colon Mx \subseteq \Lambda\}$. Let

(12. 6) $$N = \{y \in A \colon \theta(xy) = 1, \text{ for all } x \in \{M:\Lambda\}\}.$$

For brevity, we may reasonably write $N = \{M:\Lambda\}^\perp$. This N is a left Λ-lattice, and we may identify A/N with the Pontrjagin dual $\{M:\Lambda\}^\wedge$ of $\{M:\Lambda\}$.

On the other hand, $\Lambda'^\perp$ is a left Λ'-lattice, and therefore

(12. 7) $$\Lambda'^\perp = \Lambda'\alpha, \quad \text{for some} \quad \alpha \in A^\times.$$

The main result of this section is:

(12. 8) Theorem. *Let Λ be an order in A, Λ' a maximal order in A, and M a left ideal of Λ. Set*

$$f(s) = Z_\Lambda(M, s)/\zeta_{\Lambda'}(s) \in \mathbb{Z}[p^{-s}].$$

Then

$$f(s)=c_M\,\|\alpha\|^{s-1}(\Lambda:M)^{-s}\,p^{-a(N)(1-s)}+\textit{lower powers of }p^{-s},$$

where c_M is the positive constant

$$c_M=\mu^\times(\mathrm{Aut}_\Lambda(M))^{-1}\,\lambda(N)\,(\{M:\Lambda\}:\Lambda'),$$

and where $a(N)$, $\lambda(N)$ and α are defined by (12. 1), (12. 4), (12. 6) *and* (12. 7).

Proof. Let Φ be the characteristic function of $\{M:\Lambda\}$, and Ψ that of Λ'. By (2. 5) we have

$$Z_\Lambda(M,s)=\mu^\times(\mathrm{Aut}_\Lambda(M))^{-1}\,(\Lambda:M)^{-s}\,Z(\Phi,s),$$
$$\zeta_{\Lambda'}(s)=\mu^\times(\Lambda'^\times)^{-1}\,Z(\Psi,s).$$

Since

$$\hat{\Phi}=\mu^+(\{M:\Lambda\})\cdot\text{characteristic function of }N,$$

it follows from (12. 5) that

$$Z(\hat{\Phi},1-s)=\mu^+(\{M:\Lambda\})\,\lambda(N)\,p^{-a(N)(1-s)}\,\{1+b_1p^s+b_2p^{2s}+\cdots\}.$$

On the other hand,

$$\hat{\Psi}(x)=\mu^+(\Lambda')\,\Psi(x\alpha^{-1}),$$

and therefore

$$Z(\hat{\Psi},s)=\mu^+(\Lambda')\,\|\alpha\|^s\,\mu^\times(\Lambda'^\times)\,\zeta_{\Lambda'}(s)$$
$$=\mu^+(\Lambda')\,\mu^\times(\Lambda'^\times)\,\|\alpha\|^s\,(1+e_1p^{-s}+\cdots),$$

by [1], § 3. 3, or by (3. 6) and (3. 9). This gives

$$Z(\hat{\Psi},1-s)=\mu^+(\Lambda')\,\mu^\times(\Lambda'^\times)\,\|\alpha\|^{1-s}(1+e_1'p^s+e_2'p^{2s}+\cdots).$$

Combining these formulas we obtain

$$\frac{Z(\hat{\Phi},1-s)}{Z(\hat{\Psi},1-s)}=(\{M:\Lambda\}:\Lambda')\,\lambda(N)\,\mu^\times(\Lambda'^\times)^{-1}\,\|\alpha\|^{s-1}\,p^{-a(N)(1-s)}\,\{1+e_1''p^s+\cdots\}.$$

The functional equation now yields

$$f(s)=\frac{Z_\Lambda(M,s)}{\zeta_{\Lambda'}(s)}=\mu^\times(\Lambda'^\times)\,\mu^\times(\mathrm{Aut}_\Lambda(M))^{-1}\,(\Lambda:M)^{-s}\,\frac{Z(\Phi,s)}{Z(\Psi,s)}$$
$$=\mu^\times(\Lambda'^\times)\,\mu^\times(\mathrm{Aut}_\Lambda(M))^{-1}\,(\Lambda:M)^{-s}\,\frac{Z(\hat{\Phi},1-s)}{Z(\hat{\Psi},1-s)}$$
$$=c_M(\Lambda:M)^{-s}\,\|\alpha\|^{s-1}\,p^{-a(N)(1-s)}\,\{1+e_1''p^s+e_2''p^{2s}+\cdots\}.$$

This implies the theorem.

We now turn to the special case of $\Lambda=RG$, and use the notation of (11. 7). For θ, we take a character θ_G as defined in the proof of (11. 8). In this situation, we have Λ self-dual, and $N\cong M$. Moreover, as in the proof of (11. 8), we have

$$\|\alpha\|=(\Lambda':\Lambda)^{-2}.$$

(See also [1], p. 157.) Using this, we prove

(12. 9) Corollary. (i) *Let $\Lambda = RG$ as in* (11. 7), *and let M be a left ideal of Λ. Then*

$$\frac{Z_\Lambda(M, s)}{\zeta_{\Lambda'}(s)} = c_M(\Lambda' : \Lambda)^{2-2s}(\Lambda : M)^{-s}\, p^{-a(M)(1-s)} + \textit{lower powers of } p^{-s},$$

where now

$$c_M = \mu^\times(\mathrm{Aut}_\Lambda(M))^{-1}\, \lambda(M)\, (\{M : \Lambda\} : \Lambda').$$

(ii) *We have*

$$\frac{Z_\Lambda(\Lambda, s)}{\zeta_{\Lambda'}(s)} = (\Lambda' : \Lambda)^{1-2s} + \textit{lower powers of } p^{-s}.$$

(iii) *Suppose that M is not a principal left ideal of Λ. Let $(\Lambda' : \Lambda) = p^m$, and write $Z_\Lambda(M, s)/\zeta_{\Lambda'}(s) = cp^{-ls} +$ lower powers of p^{-s}, where $c \neq 0$. Then we have $l < 2m$.*

Proof. (i) Immediate from the above remarks.

(ii) When $M = \Lambda$, we have $\mu^\times(\mathrm{Aut}_\Lambda(M)) = \mu^\times(\Lambda^\times) = \lambda(M)$.

(iii) Suppose that M is non-principal, and put $(\Lambda : M) = p^b$, $b > 0$. For $x \in M \cap A^\times$, we have $\Lambda x \subseteqq M \subseteqq \Lambda$, whence p^b divides $(\Lambda : \Lambda x)$. Moreover, we have $M \neq \Lambda x$, and so p^b is a *proper* divisor of $(\Lambda : \Lambda x)$. Therefore $b < a(M)$. The largest power of p^{-s} occurring in $Z_\Lambda(M, s)/\zeta_{\Lambda'}(s)$ is therefore $(\Lambda' : \Lambda)^{-2s} \cdot p^{(a(M)-b)s}$. Since $a(M) - b > 0$, this gives the result.

Remark. One can also prove (iii) by taking the functional equation in the form of [1], Theorem 2. Then, in our present notation, the highest term of $Z_\Lambda(M, s)/\zeta_{\Lambda'}(s)$ is

$$(\Lambda' : \Lambda)^{1-2s}$$

multiplied by the lowest term in p^s occurring in $Z_\Lambda(\check{M}, 1-s)/\zeta_{\Lambda'}(1-s)$. This lowest term is a constant if and only if the power series $Z_\Lambda(\check{M}, s)$ in p^{-s} has non-zero constant term. This is the case if and only if $\check{M}$ (or equivalently M) is principal.

§ 13. Evaluation of the Gauss sum

Now we let A be a finite-dimensional *simple* $\mathbb{Q}_p$-algebra. The object of this section is to evaluate the Gauss sum $\tau(\psi)$ of (10. 17) in terms of Gauss sums attached to characters of the centre of A. The first main result is (13. 4), which shows how to reduce the problem to the case of A a division algebra. The division algebra case is dealt with in (13. 7) and (13. 8). These last two results are taken from unpublished joint work of Bushnell and A. Fröhlich.

It is convenient to introduce the following notation. We let $A = M_k(D)$, where D is a skewfield. We let F be the centre of A, and we put $\dim_F(D) = e^2$. We let R denote the valuation ring in F, and Δ the unique maximal order in D. We write $\Lambda = M_k(\Delta)$. This is a convenient choice of maximal order in A but, as we remarked in § 10, everything we do is independent of this. We let $\mathfrak{p}$ be the maximal ideal of R, and $\mathfrak{P}$ that of Λ. Then the group of two-sided fractional ideals of Λ in A is free abelian and is generated by $\mathfrak{P}\Lambda$. We let θ_A be the canonical additive character of A defined in (10. 2), and the symbols θ_F, θ_D are defined likewise.

Now we let ψ_F be a continuous character of $F^\times$ of finite order, and we define

$$\psi_A = \psi_F \circ \mathrm{nr}_{A/F}. \tag{13. 1}$$

This also tells us how to define ψ_D. As before, $\mathfrak{D}_A$ is the absolute different of A, and $\mathfrak{f}(\psi_A)$ is the conductor of ψ_A, defined relative to the maximal order Λ in A. We also define the *Swan conductor* $\mathrm{sw}(\psi_A)$ of ψ_A by

$$\mathrm{sw}(\psi_A) = \begin{cases} \Lambda & \text{if } \mathfrak{f}(\psi_A) = \Lambda, \\ \mathfrak{P}^{-1}\mathfrak{f}(\psi_A) & \text{otherwise.} \end{cases} \tag{13. 2}$$

(13. 3) Proposition. *With the above notation, we have*

(i) $\mathrm{sw}(\psi_A) = \mathrm{sw}(\psi_F) \cdot \Lambda$.

(ii) *If* $\mathfrak{f}(\psi_F) \neq R$, *or equivalently if* $\mathfrak{f}(\psi_A) \neq \Lambda$, *then* $\mathfrak{D}_A\, \mathfrak{f}(\psi_A) = \mathfrak{D}_F\, \mathfrak{f}(\psi_F) \cdot \Lambda$.

Proof. We showed in (3. 4) that $\mathfrak{f}(\psi_A) = \Lambda$ if and only if $\mathfrak{f}(\psi_F) = R$, so (i) is obvious when ψ_A is unramified. Next, the reduced norm $\mathrm{nr}_{A/F}$ induces a surjection $\Lambda^\times/(1+\mathfrak{P}\Lambda) \to R^\times/(1+\mathfrak{p})$, so if $\mathfrak{f}(\psi_F) = \mathfrak{p}$, then we have $\mathfrak{f}(\psi_A) = \mathfrak{P}\Lambda$. This implies (i) when $\mathrm{sw}(\psi_F) = R$, so let us henceforward assume that $\mathrm{sw}(\psi_F) \subseteq \mathfrak{p}$.

First we have to examine the effect of the reduced norm on congruence unit groups. Let $\mathfrak{A}$ be a proper 2-sided ideal of Λ. Then $1+\mathfrak{A}$ is a proper subgroup of $\Lambda^\times$, and one sees easily that $\mathrm{nr}_{A/F}(1+\mathfrak{A}) \subseteq 1 + (\mathfrak{A} \cap R)$, which is a proper subgroup of $R^\times$. Now let $\mathfrak{a}$ be a proper ideal of R. The reduced norm therefore satisfies

$$\mathrm{nr}(1+\mathfrak{a}\Lambda) \subseteq (1+\mathfrak{a}), \quad \mathrm{nr}(1+\mathfrak{P}\mathfrak{a}\Lambda) \subseteq (1+\mathfrak{p}\mathfrak{a}),$$

and it induces a homomorphism

$$(1+\mathfrak{a}\Lambda)/(1+\mathfrak{P}\mathfrak{a}\Lambda) \to (1+\mathfrak{a})/(1+\mathfrak{p}\mathfrak{a}).$$

We assert that this homomorphism is surjective. For, let E be a maximal subfield of A which is unramified over F. Replacing E by a conjugate if necessary, we may assume that Λ contains the valuation ring S of E. The restriction of $\mathrm{nr}_{A/F}$ to E is the field norm $\mathcal{N}_{E/F}$, and we have a commutative diagram:

$$\begin{array}{ccc} (1+\mathfrak{a}\Lambda)/(1+\mathfrak{P}\mathfrak{a}\Lambda) & \xrightarrow{\mathrm{nr}} & (1+\mathfrak{a})/(1+\mathfrak{p}\mathfrak{a}) \\ \uparrow\cup & \nearrow_{\mathcal{N}_{E/F}} & \\ (1+\mathfrak{a}S)/(1+\mathfrak{p}\mathfrak{a}S) & & \end{array}$$

The map $\mathcal{N}_{E/F}$ here is surjective, as one readily verifies, and therefore $\mathrm{nr}_{A/F}$ maps $(1+\mathfrak{a}\Lambda)/(1+\mathfrak{P}\mathfrak{a}\Lambda)$ onto $(1+\mathfrak{a})/(1+\mathfrak{p}\mathfrak{a})$, as claimed. Now we apply this with $\mathfrak{a} = \mathrm{sw}(\psi_F) \subseteq p$. Then $\mathfrak{p}\mathfrak{a} = \mathfrak{f}(\psi_F)$. We see that ψ_A is trivial on $1+\mathfrak{P}\mathfrak{a}\Lambda$, but is not trivial on $1+\mathfrak{a}\Lambda$. This means that $\mathrm{sw}(\psi_A) = \mathfrak{a}\Lambda = \mathrm{sw}(\psi_F)\,\Lambda$, as required.

To prove (ii), we use the fact that $\mathfrak{D}_A = \mathfrak{D}_F \cdot \mathfrak{P}^{e-1} \cdot \Lambda$ (see (10. 13)). Then

$$\mathfrak{D}_F \cdot \mathfrak{f}(\psi_F) \cdot \Lambda = \mathfrak{D}_F \cdot \mathfrak{p} \cdot \mathrm{sw}(\psi_F) \cdot \Lambda = \mathfrak{D}_F \cdot \mathfrak{P}^{e-1} \cdot \mathfrak{P} \cdot \mathrm{sw}(\psi_A) = \mathfrak{D}_A \cdot \mathfrak{f}(\psi_A).$$

If $\mathfrak{A}$ is a two-sided ideal of Λ, we write $\mathcal{N}_A\mathfrak{A}=(\Lambda:\mathfrak{A})$, and use similar conventions for ideals of Δ and of R.

(13. 4) Theorem. *With the above notation, we have*

$$\frac{\tau(\psi_A)}{\mathcal{N}_A\,\mathfrak{f}(\psi_A)^{\frac{1}{2}}}=\left(\frac{\tau(\psi_D)}{\mathcal{N}_D\,\mathfrak{f}(\psi_D)^{\frac{1}{2}}}\right)^k.$$

Proof. In the case of ψ_F (or equivalently ψ_A) unramified, we have

$$\tau(\psi_D)=\bar{\psi}_D(\mathfrak{D}_D),\ \tau(\psi_A)=\bar{\psi}_A(\mathfrak{D}_A)=\bar{\psi}_A(\mathfrak{D}_D\Lambda)=\bar{\psi}_D(\mathfrak{D}_D)^k,$$

and hence the result. So we assume that $\mathfrak{f}(\psi_A)\neq\Lambda$. Write $\mathfrak{f}(\psi_D)=\mathfrak{f}=\mathfrak{P}^f$, so that $f\geqq 1$, and $\mathfrak{f}(\psi_A)=\mathfrak{f}\Lambda$, by (13. 3) (ii). In the definition of the Gauss sum

$$\tau(\psi_A)=\sum_{X\in\Lambda^\times/(1+\mathfrak{f}(\psi_A))}\psi_A(c^{-1}X)\,\theta_A(c^{-1}X),$$

the element c merely has to satisfy $c\Lambda=\mathfrak{D}_A\,\mathfrak{f}(\psi_A)$, and so by (13. 3) (ii) we may choose $c\in F$. We also write

$$A'=M_{k-1}(D),\ \Lambda'=M_{k-1}(\Delta),$$

and view A' as embedded in A in the "bottom right hand corner". We partition the matrix X:

$$X=\begin{pmatrix}x_{11} & \mathbf{a}\\ \mathbf{b} & X'\end{pmatrix}=X(x_{11},\mathbf{a},\mathbf{b},X'),$$

and rewrite $\tau(\psi_A)$ as a sum over x_{11}, $\mathbf{a}$, $\mathbf{b}$, and X'. Then x_{11} ranges over $\Delta\,(\text{mod }\mathfrak{f})$, $\mathbf{a}$ and $\mathbf{b}$ over Δ^{k-1} (mod $\mathfrak{f}$), and X' over Λ' (mod $\mathfrak{f}\Lambda'$), subject to the condition $X(x_{11},\mathbf{a},\mathbf{b},X')\in\Lambda^\times$. We have $\theta_A(c^{-1}X(x_{11},\mathbf{a},\mathbf{b},X'))=\theta_D(c^{-1}x_{11})\,\theta_{A'}(c^{-1}X')$, so we may rearrange the sum as

$$\tau(\psi_A)=\sum_{x_{11},\mathbf{b},X'}\sum_{\mathbf{a}}\psi_A\left(c^{-1}\begin{pmatrix}x_{11} & \mathbf{a}\\ \mathbf{b} & X'\end{pmatrix}\right)\theta_A\left(c^{-1}\begin{pmatrix}x_{11} & \mathbf{a}\\ \mathbf{b} & X'\end{pmatrix}\right)$$

$$(13.\ 5)\qquad =\sum_{x_{11},\mathbf{b},X'}\theta_{A'}(c^{-1}X')\,\theta_D(c^{-1}x_{11})\,\psi_A(c^{-1})\sum_{\mathbf{a}}\psi_A\begin{pmatrix}x_{11} & \mathbf{a}\\ \mathbf{b} & X'\end{pmatrix}.$$

Let us write $S=S(x_{11},\mathbf{b},X')$ for the inner sum, and we show

$$(13.\ 6)\qquad S=\begin{cases}\mathcal{N}_D\mathfrak{f}^{k-1}\,\psi_D(x_{11})\,\psi_{A'}(X') & \text{if}\quad x_{11}\in\Delta^\times,\ \mathbf{b}=\mathbf{0},\ X'\in\Lambda'^\times,\\ 0 & \textit{otherwise}.\end{cases}$$

Granting this, (13. 5) becomes

$$\tau(\psi_A)=\mathcal{N}_D\mathfrak{f}^{k-1}\sum_{\substack{x_{11}\in\Delta^\times\ (\text{mod }1+\mathfrak{f})\\ X'\in\Lambda'^\times\ (\text{mod }1+\mathfrak{f}\Lambda')}}\psi_D(c^{-1}x_{11})\,\theta_D(c^{-1}x_{11})\,\psi_{A'}(c^{-1}X')\,\theta_{A'}(c^{-1}X')$$

$$=\kappa\,\tau(\psi_D)\,\tau(\psi_{A'}),\quad\text{for some }\kappa>0.$$

By induction, $\tau(\psi_A)=\kappa\tau(\psi_D)^k$, for some $\kappa>0$. Since $\tau(\psi_A)/\mathcal{N}_A\,\mathfrak{f}(\psi_A)^{\frac{1}{2}}$ has absolute value 1, the theorem follows.

Now we only have to prove (13. 6). Replacing X' by $X'V$, $V \in \Lambda'^{\times}$ affects neither the range of summation in the definition of S, nor the validity of (13. 6). Likewise, we may replace X' by UX', $U \in \Lambda'^{\times}$. This means we can assume that

$$X' = \operatorname{diag}(\xi^{j_2}, \dots, \xi^{j_k}), \quad 0 \leqq j_2 \leqq \cdots \leqq j_k,$$

where $\xi \in \Delta$, $\xi\Delta = \mathfrak{P}$. Let us assume to start with that $f > 1$. Then suppose $j_k > 0$. We choose $\lambda \in \Delta^{\times}$ such that $\psi_D(1 + \lambda\xi^{f-1}) \neq 1$. Then

$$(*) \qquad \psi_D(1+\lambda\xi^{f-1})\, S = \sum_{\mathbf{a}} \psi_A\left(\begin{pmatrix} x_{11} & \mathbf{a} \\ \mathbf{b} & X' \end{pmatrix} \cdot \begin{pmatrix} 1 & 0 \\ 0 & V \end{pmatrix}\right)$$

where $V = \operatorname{diag}(1, 1, \dots, 1, 1 + \lambda\xi^{f-1}) \in \Lambda'$. Multiplying out the matrix, we get

$$\begin{pmatrix} x_{11} & \mathbf{a}' \\ \mathbf{b} & X'V \end{pmatrix},$$

where $\mathbf{a}'$ differs from $\mathbf{a}$ only in its last entry: $a_k' = a_k(1 + \lambda\xi^{f-1})$. So the process $\mathbf{a} \mapsto \mathbf{a}'$ is a bijection of $\Delta^{k-1} \pmod{\mathfrak{f}}$ with itself. Moreover, $X'V \equiv X' \pmod{\mathfrak{f}}$. Therefore this sum $(*)$ is just S. By the choice of λ, this gives $S = 0$.

A similar argument gives the same result when $f = 1$ and $j_k > 0$. One has to replace $1 + \lambda\xi^{f-1}$ by $\mu \in \Delta^{\times}$ such that $\mu - 1 \in \Delta^{\times}$ and $\psi_D(\mu) \neq 1$.

This leaves the case $j_2 = \cdots = j_k = 0$, or equivalently, $X' \in \Lambda'^{\times}$. We may translate on the right by the matrix $\begin{pmatrix} 1 & 0 \\ 0 & X'^{-1} \end{pmatrix}$ to reduce to the case $X' = 1$. We can also replace $\mathbf{b}$ by $U\mathbf{b}$, $U \in \Lambda'^{\times}$, without changing anything. Thus we may take $\mathbf{b} = (\xi^j, 0, \dots, 0)$, and then

$$S = \sum_{\mathbf{a}} \psi_D(x_{11} - x_{12}\xi^j) = \mathcal{N}_D\mathfrak{f}^{k-2} \sum_{x_{12}} \psi_D(x_{11} - x_{12}\xi^j),$$

where $\mathbf{a} = (x_{12}, \dots, x_{1k})$. If $j = 0$, this sum extends over all $x_{12} \in \Delta \pmod{\mathfrak{f}}$ such that $x_{11} - x_{12} \in \Delta^{\times}$. Thus $x_{11} - x_{12}$ ranges over the whole of $\Delta^{\times} \pmod{1 + \mathfrak{f}}$, and the sum is zero. Now take $1 \leqq j \leqq f - 1$ (whence $f > 1$). Then, in the last sum, x_{12} ranges over the whole of $\Delta \pmod{\mathfrak{f}}$. We must have $x_{11} \in \Delta^{\times}$, since otherwise the sum is empty, and hence zero. Thus $x_{11} - x_{12}\xi^j$ ranges over all units of Δ congruent to $x_{11} \bmod \mathfrak{P}^j$. Each value $\pmod{1 + \mathfrak{f}}$ is taken an equal number of times. By the hypothesis on j, ψ_D is a non-trivial character of $(1 + \mathfrak{P}^j)/(1 + \mathfrak{f})$, and the sum vanishes again. This leaves the case $j = f$, or rather $\mathbf{b} = 0$. The sum is now $\mathcal{N}_D\mathfrak{f}^{k-1}\, \psi_D(x_{11})$ if $x_{11} \in \Delta^{\times}$, and zero otherwise (being empty). Translating back to arbitrary $X' \in \Lambda'^{\times}$, we get (13. 6), and complete the proof of (13. 4).

Remark. The proof of (13. 4) shows that the Gauss sum $\tau(\psi_A)$ reduces to

$$\tau(\psi_A) = \sum_{x \in A^{\times} \cap P/(1 + \mathfrak{f}\Lambda) \cap P} \psi_A(c^{-1}x)\, \theta_A(c^{-1}x),$$

where P is the group of upper triangular matrices in $A^{\times} = GL_k(D)$, provided we take c to be a diagonal matrix. Let T be the group of diagonal matrices in P with entries in F. Thus T is a maximal split torus. The centraliser Z of T in $GL_k(D)$ is the group of all diagonal matrices. If N denotes the normaliser of T in $GL_k(D)$, the *relative Weyl group* $W = N/Z$

is isomorphic to Σ_k, the symmetric group on k letters. Indeed, we view W as the group of permutation matrices in A, whence $W \subseteq A^\times$. The relative Bruhat decomposition (see [6], p. 220 for a convenient summary) gives

$$A^\times = \dot{\bigcup_{w \in W}} PwP \quad \text{(disjoint union).}$$

The proof of (13. 4) shows that the matrices

$$\bigcup_{\substack{w \in W \\ w \neq 1}} PwP,$$

taken together, contribute nothing to the Gauss sum. One may show that each of the "cells" PwP individually contributes zero to the sum, if $w \neq 1$.

The problem of evaluating the Gauss sum is now effectively reduced to the skewfield case $A = D$. When the characters are unramified, we find immediately (see (10. 13)) that

(13. 7) Proposition. *If ψ_F is unramified, we have*

$$\bar{\psi}_D(\mathfrak{P})\, \tau(\psi_D) = (\bar{\psi}_F(\mathfrak{p})\, \tau(\psi_F))^e.$$

Otherwise, the result is

(13. 8) Theorem. *Suppose that the character ψ_F is ramified:* $\mathfrak{f}(\psi_F) \neq R$. *Then*

$$\frac{\tau(\psi_D)}{\mathcal{N}_D\, \mathfrak{f}(\psi_D)^{\frac{1}{2}}} = (-1)^{e-1} \left(\frac{\tau(\psi_F)}{\mathcal{N}_F\, \mathfrak{f}(\psi_F)^{\frac{1}{2}}} \right)^e.$$

Remark. This is purely a statement on the *arguments* of the Gauss sums, since both sides of this equation have absolute value 1 (see (10. 23)).

Proof. This requires some preliminary results and background. Above all, we need some relations between abelian Gauss sums, and these are most conveniently derived as consequences of the formalism of Galois Gauss sums, which we now recall. For a complete treatment of this topic, see [9]. We start by choosing an algebraic closure $\bar{\mathbb{Q}}_p$ of $\mathbb{Q}_p$, and we restrict attention to finite extensions $F/\mathbb{Q}_p$ with $F \subset \bar{\mathbb{Q}}_p$. For such F, we write $\Omega_F = \mathrm{Gal}(\bar{\mathbb{Q}}_p/F)$, and think of this as a topological group, with its Krull topology. For any finite-dimensional continuous linear representation ρ of Ω_F (over $\mathbb{C}$), the *Galois Gauss sum* $\tau(\rho)$ is defined. This is a complex number, determined uniquely by the following three conditions. First, we let ρ, ρ' be representations of Ω_F. Then the direct sum $\rho + \rho'$ is defined, and we have

$$\tau(\rho + \rho') = \tau(\rho)\, \tau(\rho').$$

This relation enables one to speak of $\tau(\rho)$, even when ρ is a *virtual* representation (i.e. formal difference of representations) of Ω_F. Second, we let E/F be a finite extension, and ρ a virtual representation of Ω_E. Then the induced representation of Ω_F:

$$\mathrm{Ind}_{E/F}(\rho) = \mathrm{Ind}_{\Omega_E}^{\Omega_F}(\rho)$$

is defined, and we have

$$\tau(\mathrm{Ind}_{E/F}(\rho)) = \tau(\rho),$$

provided ρ is of dimension zero. Brauer induction theory now shows that $\tau(\rho)$ is uniquely determined once one specifies the values $\tau(\chi)$, where χ ranges over all continuous homomorphisms $\Omega_E \to \mathbb{C}^\times$, with E ranging over all finite extensions of $\mathbb{Q}_p$ in $\bar{\mathbb{Q}}_p$. For such E, the Artin Reciprocity map a_E gives a continuous embedding of $E^\times$ in the maximal abelian quotient Ω_E^{ab} of Ω_E. Taking duals, $\chi \mapsto \chi \circ a_E$ identifies $\mathrm{Hom}(\Omega_E, \mathbb{C}^\times)$ with the group of continuous characters of $E^\times$ of finite order. The third condition is now

$$\tau(\chi) = \tau(\chi \circ a_E),$$

for any continuous homomorphism $\chi : \Omega_E \to \mathbb{C}^\times$, where the Gauss sum $\tau(\chi \circ a_E)$ is defined by (10. 17).

As a matter of notational convenience, we shall not distinguish between χ and $\chi \circ a_E$. We also write ι_E for the trivial homomorphism $\Omega_E \to \mathbb{C}^\times$, so that $\tau(\iota_E) = 1$. We set

$$\rho_{E/F} = \mathrm{Ind}_{E/F}(\iota_E). \tag{13. 9}$$

With this apparatus, we can now treat (13. 8) in the case $\mathfrak{f}(\psi_F) = \mathfrak{p}$. Here, $\mathfrak{f}(\psi_D) = \mathfrak{P}$, and

$$\tau(\psi_D) = \sum_{x \in \Delta^\times/1+\mathfrak{P}} \psi_D(c^{-1}x)\, \theta_D(c^{-1}x).$$

As before, we may choose $c \in F$ such that $cR = \mathfrak{p}\mathfrak{D}_F = \mathfrak{f}(\psi_F)\,\mathfrak{D}_F$. Let E be a maximal subfield of D such that E/F is unramified. If S denotes the valuation ring of E, then the prime of S is $\mathfrak{p}S$, and the canonical injection $S/\mathfrak{p}S \to \Delta/\mathfrak{P}$ is a bijection. We may therefore take our representatives x of $\Delta^\times$ (mod $1+\mathfrak{P}$) from S, and then

$$\tau(\psi_D) = \sum_{x \in S^\times/1+\mathfrak{p}S} \psi_D(c^{-1}x)\, \theta_D(c^{-1}x).$$

However, the restriction of the reduced norm $\mathrm{nr}_{D/F}$ to E is the field norm $\mathcal{N}_{E/F}$, and the restriction of the reduced trace $\mathrm{tr}_{D/F}$ is the field trace $\mathrm{Tr}_{E/F}$. So, defining

$$\psi_E = \psi_F \circ \mathcal{N}_{E/F},\ \theta_E = \theta_F \circ \mathrm{Tr}_{E/F}$$

(which is consistent with our standard usage), we have $cS = \mathfrak{D}_E\,\mathfrak{f}(\psi_E)$, and hence $\tau(\psi_D) = \tau(\psi_E)$. Now we may use the dual Artin isomorphism to view ψ_F as a character of Ω_F, and ψ_E as a character of Ω_E, when they are related by

$$\psi_E = \psi_F | \Omega_E \quad \text{(restriction of representations)}.$$

We now have $\tau(\psi_E) = \tau(\psi_E - \iota_E) = \tau(\mathrm{Ind}_{E/F}(\psi_E - \iota_E))$. But $\mathrm{Ind}_{E/F}(\iota_E) = \rho_{E/F}$, and

$$\mathrm{Ind}_{E/F}(\psi_F | \Omega_E) = \mathrm{Ind}_{E/F}(\psi_E) = \psi_F \otimes \rho_{E/F}$$

(tensor product of representations). Of course, the representation $\rho_{E/F}$ is the sum of all characters χ of the group $G = \mathrm{Gal}(E/F)$, which is cyclic, since E/F is unramified. Thus

$$\tau(\psi_E) = \prod_{\chi \in \hat{G}} \tau(\psi_F \cdot \chi)/\tau(\chi).$$

Now we think of ψ_F and χ as characters of $F^\times$. The χ's are unramified characters, so $\tau(\chi) = \bar{\chi}(\mathfrak{D}_F)$, and one sees easily that $\tau(\psi_F \cdot \chi) = \chi(c^{-1})\,\tau(\psi_F) = \bar{\chi}(\mathfrak{D}_F)\,\bar{\chi}(\mathfrak{p})\,\tau(\psi_F)$. Therefore

$$\tau(\psi_E) = \tau(\psi_F)^{[E:F]} \prod_{\chi \in \hat{G}} \bar{\chi}(\mathfrak{p}).$$

The character $\prod_{\chi} \chi$ of G is trivial if G is of odd order, and is of order two otherwise, since G is cyclic. Therefore $\prod_{\chi} \chi(\mathfrak{p}) = (-1)^{[E:F]+1} = (-1)^{e+1}$, and we have proved

$$\tau(\psi_D) = \tau(\psi_E) = (-1)^{e+1}\, \tau(\psi_F)^e,$$

exactly as required for (13. 8).

Remark. The relation between $\tau(\psi_E)$ and $\tau(\psi_F)$ in this case is a reformulation of a classical result due to Davenport and Hasse.

For the rest of the proof, therefore, we assume that $\mathfrak{f}(\psi_F) \subseteq \mathfrak{p}^2$. Hence, by (13. 3), we have $\mathfrak{f}(\psi_D) \subseteq \mathfrak{P}^2$. The basic tool is the following:

(13. 10) Proposition. *Let* $\mathfrak{A}$ *be an ideal of* Δ *such that* $\mathfrak{f} = \mathfrak{f}(\psi_D)$ *contains* $\mathfrak{A}^2$. *Let* $\mathfrak{B} = \mathfrak{A}^{-1}\mathfrak{f}$. *Then there exists* $c \in \Delta$ *such that*

(i) $c\Delta = \mathfrak{f}\mathfrak{D}_D$, *and*

(ii) *if* $x \in \mathfrak{A}$, *then* $\psi_D(1+x) = \bar{\theta}_D(c^{-1}x)$.

(iii) *For any* c *satisfying* (i) *and* (ii), *we have*

$$\tau(\psi_D) = \kappa \sum_{x \in (1+\mathfrak{B})/(1+\mathfrak{A})} \psi_D(c^{-1}x)\, \theta_D(c^{-1}x),$$

for some $\kappa > 0$.

(iv) *Moreover, there is an element* $c \in F$ *which satisfies* (i) *and* (ii).

Proof. For $x, x' \in \Delta$ we have

$$(1+x)(1+x') \equiv 1 + x + x' \pmod{\mathfrak{A}^2},$$

$$\operatorname{nr}(1+x) \equiv 1 + \operatorname{tr}(x) \pmod{\mathfrak{A}^2}.$$

Therefore the map $y \mapsto \psi_D(1+y)$ defines a continuous character of the group $\mathfrak{A}$, which is trivial on $\mathfrak{f}$ but not on $\mathfrak{P}^{-1}\mathfrak{f}$. Moreover, the value of $\psi_D(1+y)$ depends only on $\operatorname{tr}(y)$, $y \in \mathfrak{A}$. Consequently, there is a character ϕ of the compact open subgroup $\operatorname{tr}(\mathfrak{A})$ of F such that $\psi_D(1+y) = \phi(\operatorname{tr}(y))$, for $y \in \mathfrak{A}$. This ϕ can be extended to a character of F, and it is therefore of the form $\phi(z) = \theta_F(-c^{-1}z) = \bar{\theta}_F(c^{-1}z)$, for some $c \in F^\times$. Then, for $y \in \mathfrak{A}$, we have $\psi_D(1+y) = \bar{\theta}_F(c^{-1}\operatorname{tr}(y)) = \bar{\theta}_D(c^{-1}y)$. Since θ_D is trivial on $\mathfrak{D}_D^{-1}$ but not on $\mathfrak{P}^{-1}\mathfrak{D}_D^{-1}$, we see that $c\Delta = \mathfrak{D}_D\mathfrak{f}$. This proves (i), (ii), (iv).

Now we just take $c \in \Delta$ as in (i) and (ii). By (i), we can use this element c to form the Gauss sum. We write our coset representatives x of $\Delta^\times$ mod $1+\mathfrak{f}$ in the form $x = yz$, where $y \in \Delta^\times/(1+\mathfrak{A})$, $z \in (1+\mathfrak{A})/(1+\mathfrak{f})$. Then

$$\tau(\psi_D) = \sum_y \sum_z \psi_D(c^{-1}yz)\, \theta_D(c^{-1}yz)$$

$$= \sum_y \psi_D(c^{-1}y)\, \theta_D(c^{-1}y) \sum_z \psi_D(z)\, \theta_D(c^{-1}y(z-1)).$$

The term in the inner sum is $\theta_D(c^{-1}y(z-1) - c^{-1}(z-1)) = \theta_D(c^{-1}(y-1)(z-1))$, by (ii). Now let $u = z-1$. Then the inner sum becomes

$$\sum_{u \in \mathfrak{A}/\mathfrak{f}} \theta_D(c^{-1}(y-1)u).$$

The character $u \mapsto \theta_D(c^{-1}(y-1)u)$ of $\mathfrak{A}/\mathfrak{f}$ is trivial if and only if $c^{-1}(y-1)\mathfrak{A}$ is contained in $\mathfrak{D}_D^{-1}$, or equivalently $(y-1) \in \mathfrak{B}$. So this sum is 0 if $y-1 \notin \mathfrak{B}$, and equals $(\mathfrak{A}:\mathfrak{f})$ otherwise. Taking $\kappa = (\mathfrak{A}:\mathfrak{f})$, we obtain (iii).

We now return to the proof of (13. 8) with the assumption that $\mathfrak{f}(\psi_F) \subseteqq \mathfrak{p}^2$. We have to examine various cases which are distinguished by the parities of the Swan conductors mod squares. Let us assume first that $\mathrm{sw}(\psi_D)$ is not the square of an ideal of Δ. Therefore e is odd, and $\mathrm{sw}(\psi_F)$ is not a square, by (13. 3), while $\mathfrak{f}(\psi_D)$, $\mathfrak{f}(\psi_F)$ are squares. Let $\mathfrak{a}$ be the ideal of R such that $\mathfrak{a}^2 = \mathfrak{f}(\psi_F)$. By (13. 10) applied to the case $D = F$, we have an element $c \in F$ such that $cR = \mathfrak{D}_F \, \mathfrak{f}(\psi_F)$ and $\psi_F(1+x) = \bar{\theta}_F(c^{-1}x)$, $x \in \mathfrak{a}$. Then

$$\tau(\psi_F) = \kappa \psi_F(c^{-1}) \, \theta_F(c^{-1}), \tag{13. 11}$$

for some $\kappa > 0$. Now let $\mathfrak{A}$ be the ideal of Δ such that $\mathfrak{A}^2 = \mathfrak{f}(\psi_D)$. One sees that $\mathfrak{A} \cap R \subseteqq \mathfrak{a}$, whence, for $x \in \mathfrak{A}$, we have

$$\mathrm{nr}(1+x) \equiv 1 + \mathrm{tr}(x) \pmod{\mathfrak{f}(\psi_D)},$$

$$\psi_D(1+x) = \psi_F(\mathrm{nr}(1+x)) = \bar{\theta}_F(c^{-1}\,\mathrm{tr}(x)) = \bar{\theta}_D(c^{-1}x).$$

Of course, $c\Delta = \mathfrak{D}_D \, \mathfrak{f}(\psi_D)$, and hence

$$\tau(\psi_D) = \kappa \psi_D(c^{-1}) \, \theta_D(c^{-1}),$$

for some $\kappa > 0$. But $c \in F$, so $\theta_D(c^{-1}) = \theta_F(c^{-1})^e$, $\psi_D(c^{-1}) = \psi_F(c^{-1})^e$, and comparing with (13. 11) we find

$$\tau(\psi_D) = \kappa \tau(\psi_F)^e,$$

for some $\kappa > 0$. Since e is odd, this proves (13. 8) in this case.

For the next case, we assume that $\mathrm{sw}(\psi_F)$ is the square of an ideal of R. Therefore, by (13. 3), $\mathrm{sw}(\psi_D)$ is also a square. Indeed, let $\mathfrak{b}^2 = \mathrm{sw}(\psi_F)$, $\mathfrak{B} = \mathfrak{b}\Delta$, so that $\mathfrak{B}^2 = \mathrm{sw}(\psi_D)$. Write $\mathfrak{A} = \mathfrak{B}^{-1} \, \mathfrak{f}(\psi_D)$, $\mathfrak{a} = \mathfrak{b}^{-1} \, \mathfrak{f}(\psi_F)$. We apply (13. 10) with this choice of $\mathfrak{a}$ to get $c \in F$ such that $\psi_F(1+x) = \bar{\theta}_F(c^{-1}x)$ for $x \in \mathfrak{a}$. Observe that $\mathfrak{A} = \mathfrak{P}\mathfrak{B}$, $\mathfrak{a} = \mathfrak{p}\mathfrak{b}$, which implies $\mathfrak{A} \cap R = \mathfrak{a}$. It follows that

$$\psi_D(1+x) = \bar{\theta}_D(c^{-1}x), \quad x \in \mathfrak{A},$$

and that

$$\tau(\psi_D) = \kappa \sum_{x \in (1+\mathfrak{B})/(1+\mathfrak{A})} \psi_D(c^{-1}x) \, \theta_D(c^{-1}x),$$

for some $\kappa > 0$. Now let E be a maximal subfield of D which is unramified over F. Let S be the valuation ring in E. We obtain

$$\tau(\psi_E) = \kappa \sum_{x \in (1+\mathfrak{b}S)/(1+\mathfrak{a}S)} \psi_E(c^{-1}) \, \theta_E(c^{-1}x)$$

for some $\kappa > 0$. However, as x ranges over a set of representatives of $(1+\mathfrak{b}S)/(1+\mathfrak{a}S)$, it also ranges over $(1+\mathfrak{B})/(1+\mathfrak{A})$. This gives $\tau(\psi_D) = \kappa \tau(\psi_E)$, for $\kappa > 0$. We can compute $\tau(\psi_E)$ exactly as we did in the case $\mathfrak{f}(\psi_F) = \mathfrak{p}$, using the formalism of Galois Gauss sums, and we find $\tau(\psi_E) = (-1)^{e+1} \cdot \tau(\psi_F)^e$, exactly as before. This proves the theorem in the present case.

This leaves only the case in which $\mathrm{sw}(\psi_D)$ is a square, but $\mathrm{sw}(\psi_F)$ is not a square. Therefore, by (13. 3), the index e of D is *even*. We have only to show that

$$\tau(\psi_D) = -\kappa \tau(\psi_F)^e, \tag{13. 12}$$

for some $\kappa > 0$.

By the usual sort of argument, we find $c \in F$ such that

$$\tau(\psi_F) = \kappa \psi_F(c^{-1})\, \theta_F(c^{-1}),$$
$$\tau(\psi_D) = \kappa' \sum_{x \in (1+\mathfrak{A})/(1+\mathfrak{P}\mathfrak{A})} \psi_D(c^{-1}x)\, \theta_D(c^{-1}x),$$

for $\kappa, \kappa' > 0$, and where $\mathfrak{A}^2 = \mathrm{sw}(\psi_D)$. We evaluate this sum by making a special choice of representatives of $(1+\mathfrak{A})/(1+\mathfrak{P}\mathfrak{A})$.

(13. 13) Lemma. *Let E be a maximal subfield of D which is unramified over F. Then there exists $\xi \in D$ such that*

(i) $\xi \Delta = \mathfrak{P}$;

(ii) $\xi E \xi^{-1} = E$;

(iii) *the map* $x \mapsto \xi x \xi^{-1}$, $x \in E$, *generates* $\mathrm{Gal}(E/F)$;

(iv) $\xi^e \in F$.

Proof. Since E is a maximal subfield of D, the centraliser of $E^\times$ in $D^\times$ is $E^\times$ itself. The Skolem-Noether theorem gives an isomorphism $N/E^\times \cong \mathrm{Gal}(E/F)$, where N is the normaliser of $E^\times$ in $D^\times$. If conjugation by $\gamma \in D^\times$ induces the Frobenius of E/F, one knows from [11], p. 138, that $\gamma\Delta = \mathfrak{P}^r$, where r/e (mod $\mathbb{Z}$) is the Hasse invariant of D. Thus we can find ξ of the form $\gamma^s\delta$, $\delta \in F$, to satisfy (i)—(iii). The element ξ^e then centralises E, and is fixed by conjugation by ξ. Therefore $\xi^e \in F$ as required. (For a more detailed proof, see [10], Theorem 14. 5.)

Now take $\mathfrak{A}$ as above, and write $\mathfrak{A} = \mathfrak{P}^t$, and put $d = \xi^t$, with ξ as in (13. 12). By hypothesis, we have $t \equiv \frac{1}{2}e$ (mod e), whence $d \notin F$, but $d^2 \in F$. Therefore $x \mapsto dxd^{-1}$, $x \in E$, is an F-automorphism of E with fixed field L, where $[E: L] = 2$. Let μ_E denote the group of roots of unity in E of order prime to p, the residual characteristic of F. Then the elements $1+ud$, for $u \in \mu_E \cup \{0\}$, form a set of representatives of $1+\mathfrak{A} \bmod (1+\mathfrak{P}\mathfrak{A})$, and we have

$$\tau(\psi_D) = \kappa \sum_{u \in \mu_E \cup \{0\}} \psi_D(c^{-1}(1+ud))\, \theta_D(c^{-1}(1+ud)).$$

We observe that, for $u \in \mu_E$, we have $(ud)^2 = \mathcal{N}_{E/L}(u)\, d^2$. Further, if S is the valuation ring in L, then $d^2 S = \mathfrak{p} S$, whence $E_u = L(ud)$ is a ramified quadratic extension of L, and indeed a maximal subfield of D. Now

$$\begin{aligned} \mathrm{nr}(1+ud) &= \mathcal{N}_{L/F}(\mathcal{N}_{E_u/L}(1+ud)) = \mathcal{N}_{L/F}(1 - \mathcal{N}_{E/L}(u)\, d^2) \\ &\equiv 1 - \mathrm{Tr}_{L/F}(\mathcal{N}_{E/L}(u))\, d^2 \pmod{\mathfrak{f}(\psi_F)}. \end{aligned}$$

Therefore

$$\psi_D(1+ud) = \psi_F(1 - \mathrm{Tr}_{L/F}(\mathcal{N}_{E/L}(u))\, d^2) = \theta_F(c^{-1}\, \mathrm{Tr}_{L/F}(\mathcal{N}_{E/L}(u))\, d^2).$$

On the other hand, $\mathrm{tr}(ud) = \mathrm{Tr}_{L/F}(\mathrm{Tr}_{E_u/L}(ud)) = 0$, and therefore

$$\theta_D(c^{-1}(1+ud)) = \theta_D(c^{-1}).$$

The Gauss sum reduces to

$$\tau(\psi_D) = \kappa \psi_D(c^{-1})\, \theta_D(c^{-1}) \sum_{u \in \mu_E \cup \{0\}} \theta_F(c^{-1} d^2\, \mathrm{Tr}_{L/F}(\mathcal{N}_{E/L}(u))).$$

The first factor we recognise as $\tau(\psi_F)^e$, and we need only show that the sum is real and negative. We can reduce this to a problem in finite fields. Write $k(E)$, $k(L)$, $k(F)$ for the residue class fields of E, L, F respectively. Computing the ideal $c^{-1}d^2R$, we see that, for $x \in R$, the quantity $\theta_F(c^{-1}d^2x)$ depends only on $x \bmod \mathfrak{p}$. Let $\tilde{y}$ denote the reduction of $y \bmod \mathfrak{p}$ for y lying in the valuation ring of E. We define

$$\phi(\tilde{x}) = \theta_F(c^{-1}d^2x), \quad x \in R.$$

Then ϕ is a non-trivial additive character of $k(F)$, and we have

$$\theta_F(c^{-1}d^2\operatorname{Tr}_{L/F}(\mathcal{N}_{E/L}(u))) = \phi(\operatorname{Tr}_{k(L)/k(F)}(\mathcal{N}_{k(E)/k(L)}(\tilde{u}))),$$

for $u \in \mu_E \cup \{0\}$. Therefore we are to evaluate the sum

$$(13.\ 14) \qquad \sum_{x \in k(E)} \phi(\operatorname{Tr}_{k(L)/k(F)}(\mathcal{N}_{k(E)/k(L)}(x))).$$

Write $q = |k(F)|$, so that $|k(E)| = q^e$, $|k(L)| = q^{\frac{1}{2}e}$. Then any $a \in k(F)^\times$ is the trace of $q^{\frac{1}{2}e-1}$ distinct elements of $k(L)^\times$, each of which is the norm of $q^{\frac{1}{2}e}+1$ elements of $k(E)$. On the other hand, $\operatorname{Ker}(\operatorname{Tr}_{k(L)/k(F)})$ has $q^{\frac{1}{2}e-1}-1$ nonzero elements. The sum (13. 4) is therefore

$$(q^{\frac{1}{2}e}+1)\,q^{\frac{1}{2}e-1} \sum_{a \in k(F)^\times} \phi(a) + (q^{\frac{1}{2}e}+1)\,(q^{\frac{1}{2}e-1}-1)+1$$

$$= -q^{\frac{1}{2}e-1}(q^{\frac{1}{2}e}+1) + (q^{\frac{1}{2}e}+1)\,(q^{\frac{1}{2}e-1}-1)+1 = -q^{\frac{1}{2}e},$$

since $\sum_{a \in k(F)} \phi(a) = -1$. This proves (13. 12) and completes the proof of (13. 8).

We now summarise the results of this section:

(13. 15) Theorem. *Let $F/\mathbf{Q}_p$ be a finite field extension, and let A be a finite dimensional central simple F-algebra. Write $A = M_k(D)$, where D is a central F-division algebra of dimension e^2. Let ψ_F be a continuous character of $F^\times$ of finite order, and let $\psi_A = \psi_F \circ \operatorname{nr}_{A/F}$. Then, if ψ_F is unramified, we have*

$$\tau(\psi_A) = \bar{\psi}_F(\mathfrak{p})^{ke-k}\,\tau(\psi_F)^{ke},$$

where $\mathfrak{p}$ is the prime of the valuation ring of F. Otherwise,

$$\frac{\tau(\psi_A)}{\mathcal{N}_A \mathfrak{f}(\psi_A)^{\frac{1}{2}}} = (-1)^{ke-k}\left(\frac{\tau(\psi_F)}{\mathcal{N}_F \mathfrak{f}(\psi_F)^{\frac{1}{2}}}\right)^{ke}.$$

Of course, one may deduce the relationship between the norms of the conductors from (13. 3), to obtain a relationship the Gauss sums themselves, rather than just their arguments. This also gives a relation between the ε's, which we exploit in the proof of (14. 7) below.

§ 14. Global functional equation

Our main remaining task is to give a self-contained version of the functional equation (11. 16) for the global L-function $L_A(M, s, \psi)$ attached to an ideal M of an integral group ring A. This requires us to work out the functional equation of the associated standard L-function $L_A(s, \psi)$. It is this matter which concerns us here. It will then be very simple to deduce the desired functional equation in § 15.

We now let A be any finite-dimensional semisimple $\mathbb{Q}$-algebra, and ψ a continuous homomorphism from the idele group $J(A)$ to $\mathbb{C}^\times$, with finite image, which is trivial on $A^\times$. (The finiteness hypothesis is not really necessary, but it does make the exposition slightly shorter.) The global standard L-function $L_A(s, \psi)$ is then defined as in §§ 3, 4, and we view it as a meromorphic function of s via analytic continuation, as in (4. 1). We need first to introduce some "Euler factors" corresponding to the Archimedean primes.

Therefore, we take a finite-dimensional semisimple $\mathbb{R}$-algebra B, and a continuous character ϕ of $B^\times$ of finite order. We let $B = \prod_{i=1}^{r} B_i$, where each B_i is a simple algebra, and $\phi_i = \phi | B_i^\times$. We initially decide that

$$(14.\ 1) \qquad \begin{cases} L_B(s, \phi) = \prod_{i=1}^{r} L_{B_i}(s, \phi_i), \\ \varepsilon_B(s, \phi) = \prod_{i=1}^{r} \varepsilon_{B_i}(s, \phi_i). \end{cases}$$

We proceed to define L_B and ε_B under the assumption that B is *simple*. Thus $B \cong M_k(K)$, where $K = \mathbb{R}$, $\mathbb{C}$ or $\mathbb{H}$ (the skewfield of Hamiltonian quaternions). When $K = \mathbb{C}$ or $\mathbb{H}$, the character ϕ is trivial. Otherwise, we have either $\phi = \phi_0 \circ \det$, or $\phi = \phi_1 \circ \det$, where det denotes the determinant, and

$$\phi_0(y) = 1, \qquad y \in \mathbb{R}^\times,$$
$$\phi_1(y) = \operatorname{sign}(y) = y/|y|, \quad y \in \mathbb{R}^\times.$$

Starting with the case $k = 1$, we define

$$(14.\ 2) \qquad \begin{cases} L_{\mathbb{R}}(s, \phi_0) = \pi^{-s/2}\, \Gamma\left(\dfrac{s}{2}\right), \\ L_{\mathbb{R}}(s, \phi_1) = \pi^{-(1+s)/2}\, \Gamma\left(\dfrac{1+s}{2}\right), \\ L_{\mathbb{C}}(s, \phi) \ = (2\pi)^{1-s}\, \Gamma(s), \\ L_{\mathbb{H}}(s, \phi) \ = (2\pi)^{-2s}\, \Gamma(2s), \end{cases}$$

where Γ denotes Euler's gamma-function:

$$\Gamma(s) = \int_0^\infty x^s e^{-x} \frac{dx}{x}, \qquad \operatorname{Re}(s) > 0,$$

continued to a meromorphic function of s, and π is what it always was. Now we extend to the general case:

$$(14.\ 3) \qquad \begin{cases} L_{M_k(\mathbb{R})}(s, \phi_i \circ \det) = \prod_{j=0}^{k-1} L_{\mathbb{R}}(ks - j, \phi_i), \quad i = 0, 1, \\ L_{M_k(K)}(s, \phi) = \prod_{j=0}^{k-1} L_K(ks - j, \phi') \qquad \text{for } K = \mathbb{C}, \mathbb{H}. \end{cases}$$

Here, ϕ, ϕ' denote trivial characters when $K=\mathbb{C}$ or $\mathbb{H}$. *Still assuming B is simple*, we define

$$\varepsilon_B(s,\phi)=\begin{cases}(-i)^k & \text{if } B\cong M_k(\mathbb{R}) \text{ and } \phi=\phi_1\circ\det,\\ 1 & \text{otherwise.}\end{cases} \tag{14. 4}$$

For our $\mathbb{Q}$-algebra A, we write $A_\infty=A\otimes_{\mathbb{Q}}\mathbb{R}$, and set $\psi_\infty=\psi|A_\infty^\times$. The local factors $L_{A_\infty}(s,\psi_\infty)$, $\varepsilon_{A_\infty}(s,\psi_\infty)$ are defined by (14. 1)—(14. 4), and we put

$$\tilde{L}_A(s,\psi)=L_{A_\infty}(s,\psi_\infty)\,L_A(s,\psi), \tag{14. 5}$$

$$\varepsilon_A(s,\psi)=\varepsilon_{A_\infty}(s,\psi_\infty)\prod_p \varepsilon_{A_p}(s,\psi_p), \tag{14. 6}$$

where the product extends over all prime numbers p, and ε_{A_p} is defined by (10. 11) extended to the semisimple case, as in (10. 25).

(14. 7) Theorem. *Let A be a finite-dimensional semisimple $\mathbb{Q}$-algebra, and ψ a continuous character of $J(A)$ of finite order and which is trivial on $A^\times$. Then, with the notations* (14. 5), (14. 6), *we have*

$$\tilde{L}_A(s,\psi)=\varepsilon_A(s,\psi)\,\tilde{L}_A(1-s,\bar{\psi}).$$

Proof. In view of our definitions, we can assume that A is *simple*. Let F be the centre of A, and $n^2=\dim_F(A)$. We give a short proof, based on our extensive local computations and on the well-known case $A=F$. We mention some other approaches below.

Let R be the ring of integers of F, and let $\mathfrak{p}$ range over all places, Archimedean and non-Archimedean, of F. Then we have

$$\tilde{L}_A(s,\psi)=\prod_{\mathfrak{p}} L_{A_\mathfrak{p}}(s,\psi_\mathfrak{p}),\quad \psi_\mathfrak{p}=\psi|A_\mathfrak{p}^\times,\quad \varepsilon_A(s,\psi)=\prod_{\mathfrak{p}}\varepsilon_{A_\mathfrak{p}}(s,\psi_\mathfrak{p}).$$

As in the proof of (4. 1), it is convenient to introduce an auxiliary function

$$\tilde{K}_A(s,\psi)=\prod_{j=0}^{n-1}\tilde{L}_F(ns-j,\tilde{\psi}),$$

where $\tilde{\psi}$ is the unique idele-class character of F such that $\psi=\tilde{\psi}\circ\mathrm{nr}$, as in (1. 10). It is well known that (see [8], [14] or [15])

$$\tilde{L}_F(s,\tilde{\psi})=\varepsilon_F(s,\tilde{\psi})\,\tilde{L}_F(1-s,\bar{\tilde{\psi}}),$$

and this implies immediately that

$$\tilde{K}_A(s,\psi)=\tilde{K}_A(1-s,\bar{\psi})\prod_{j=0}^{n-1}\varepsilon_F(ns-j,\tilde{\psi}). \tag{14. 8}$$

For each place $\mathfrak{p}$ of F, we have $A_\mathfrak{p}\cong M_{k(\mathfrak{p})}(D(\mathfrak{p}))$, where $D(\mathfrak{p})$ is a central $F_\mathfrak{p}$-division algebra of dimension $e(\mathfrak{p})^2$, say, and we have $n=k(\mathfrak{p})\,e(\mathfrak{p})$. We write $\tilde{\psi}_\mathfrak{p}=\tilde{\psi}|F_\mathfrak{p}^\times$. Let S denote the set of Archimedean places of F, and T the set of non-Archimedean places of F. By (3. 6) and the definitions of the Archimedean Euler factors, we have

$$\tilde{K}_A(s,\psi)=\tilde{L}_A(s,\psi)\,\Omega(s,\psi), \tag{14. 9}$$

where

$$\Omega(s,\psi)=\prod_{\substack{\mathfrak{p}\in T\\ \mathfrak{p}\nmid\mathfrak{f}}}\left\{\prod_{\substack{j=0\\ j\not\equiv 0 \pmod{e(\mathfrak{p})}}}^{n-1}(1-\tilde{\psi}(\mathfrak{p})\,\mathcal{N}\mathfrak{p}^{j-ns})^{-1}\right\}\prod_{\mathfrak{p}\in S}\left\{L_{A_\mathfrak{p}}(s,\psi_\mathfrak{p})^{-1}\prod_{j=0}^{n-1}L_{F_\mathfrak{p}}(ns-j,\tilde{\psi}_\mathfrak{p})\right\}.$$

Here we have written $\mathfrak{f}=\mathfrak{f}(\tilde{\psi})$ for the conductor of $\tilde{\psi}$, and $\mathcal{N}$ for the counting norm on ideals of R. Substituting in (14. 8) we obtain

$$\tilde{L}_A(s,\psi)=\tilde{L}_A(1-s,\bar{\psi})\left\{\prod_{j=0}^{n-1}\varepsilon_F(ns-j,\tilde{\psi})\right\}\frac{\Omega(1-s,\bar{\psi})}{\Omega(s,\psi)},$$

and we are reduced to proving

$$\varepsilon_A(s,\psi)=\frac{\Omega(1-s,\bar{\psi})}{\Omega(s,\psi)}\prod_{j=0}^{n-1}\varepsilon_F(ns-j,\tilde{\psi}), \tag{14. 10}$$

where Ω is defined by (14. 9). We do this by comparing the contributions to either side of (14. 10) from each place $\mathfrak{p}$ of F. Let us start with the case $\mathfrak{p}\in T$, $\mathfrak{p}\nmid\mathfrak{f}$. The Ω-factors in (14. 10) contribute

$$\prod_{\substack{j=0\\ j\not\equiv 0\,(\mathrm{mod}\; e(\mathfrak{p}))}}^{n-1}(1-\tilde{\psi}(\mathfrak{p})\,\mathcal{N}\mathfrak{p}^{j-ns})\,(1-\bar{\tilde{\psi}}(\mathfrak{p})\,\mathcal{N}\mathfrak{p}^{ns-j})^{-1},$$

after an obvious simplification of the denominator. This product reduces to

$$\{-\tilde{\psi}(\mathfrak{p})\,\mathcal{N}\mathfrak{p}^{n(\frac{1}{2}-s)}\}^{k(\mathfrak{p})(e(\mathfrak{p})-1)}$$

For these $\mathfrak{p}$, we have $\varepsilon_{F_\mathfrak{p}}(s,\tilde{\psi}_\mathfrak{p})=\tilde{\psi}(\mathfrak{D}_\mathfrak{p})\,\mathcal{N}\mathfrak{D}_\mathfrak{p}^{\frac{1}{2}-s}$, by (10. 19), where $\mathfrak{D}_\mathfrak{p}=\mathfrak{D}_{F_\mathfrak{p}}$ is the absolute different of $F_\mathfrak{p}$. Therefore

$$\prod_{j=0}^{n-1}\varepsilon_{F_\mathfrak{p}}(ns-j,\tilde{\psi}_\mathfrak{p})=\tilde{\psi}(\mathfrak{D}_\mathfrak{p})^n\,\mathcal{N}\mathfrak{D}_\mathfrak{p}^{n^2(\frac{1}{2}-s)},$$

and the $\mathfrak{p}$-part of the right hand side of (14. 10) is

$$\{-\tilde{\psi}(\mathfrak{p})\,\mathcal{N}\mathfrak{p}^{n(\frac{1}{2}-s)}\}^{k(\mathfrak{p})(e(\mathfrak{p})-1)}\,\mathcal{N}\mathfrak{D}_\mathfrak{p}^{n^2(\frac{1}{2}-s)}\,\tilde{\psi}(\mathfrak{D}_\mathfrak{p})^n. \tag{14. 11}$$

On the other hand

$$\varepsilon_{A_\mathfrak{p}}(s,\psi_\mathfrak{p})=\mathcal{N}_{A_\mathfrak{p}}(\mathfrak{D}_{A_\mathfrak{p}})^{\frac{1}{2}-s}\,\psi_\mathfrak{p}(\mathfrak{D}_{A_\mathfrak{p}}),$$

where $\mathfrak{D}_{A_\mathfrak{p}}$ is the absolute different of $A_\mathfrak{p}$ relative to some choice of maximal order in $A_\mathfrak{p}$, as in (10. 12). The explicit determination (10. 13) of $\mathfrak{D}_{A_\mathfrak{p}}$ shows immediately that

$$\varepsilon_{A_\mathfrak{p}}(s,\psi_\mathfrak{p})=\{\tilde{\psi}(\mathfrak{p})\,\mathcal{N}\mathfrak{p}^{n(\frac{1}{2}-s)}\}^{k(\mathfrak{p})(e(\mathfrak{p})-1)}\,\mathcal{N}\mathfrak{D}_\mathfrak{p}^{n^2(\frac{1}{2}-s)}\,\tilde{\psi}(\mathfrak{D}_\mathfrak{p})^n.$$

In other words, the $\mathfrak{p}$-parts of the two sides of (14. 10) differ only by a factor $(-1)^{k(\mathfrak{p})(e(\mathfrak{p})-1)}$.

Now let us treat the case of $\mathfrak{p}$ dividing $\mathfrak{f}$. Then the characters $\psi_\mathfrak{p}$, $\tilde{\psi}_\mathfrak{p}$ are ramified, and we just have to appeal to (10. 19), (13. 3) (ii) and (13. 15) to show that the $\mathfrak{p}$-parts of the two sides of (14. 10) again only differ by a factor of $(-1)^{k(\mathfrak{p})(e(\mathfrak{p})-1)}$.

This leaves the case of $\mathfrak{p}$ Archimedean. We have to compare $\varepsilon_{A_\mathfrak{p}}(s,\psi_\mathfrak{p})$ and

(14. 12)

$$\left\{\prod_{j=0}^{n-1}\varepsilon_{F_\mathfrak{p}}(ns-j,\tilde{\psi}_\mathfrak{p})\right\}L_{A_\mathfrak{p}}(s,\psi_\mathfrak{p})\,L_{A_\mathfrak{p}}(1-s,\bar{\psi}_\mathfrak{p})^{-1}\prod_{j=0}^{n-1}\{L_{F_\mathfrak{p}}(n(1-s)-j,\bar{\tilde{\psi}}_\mathfrak{p})\,L_{F_\mathfrak{p}}(ns-j,\tilde{\psi}_\mathfrak{p})^{-1}\}.$$

The expression in L-functions is identically one in the case $e(\mathfrak{p})=1$ (i.e. $A_{\mathfrak{p}} \cong M_n(\mathbb{R})$ or $M_n(\mathbb{C})$), and also the product of ε's is $\varepsilon_{A_{\mathfrak{p}}}(s, \psi_{\mathfrak{p}})$. The $\mathfrak{p}$-parts of the two sides of (14. 10) are thus the same in this case, and hence differ by the factor $1=(-1)^{k(\mathfrak{p})(e(\mathfrak{p})-1)}$.

We are left with the case of $\mathfrak{p}$ Archimedean, $e(\mathfrak{p})=2$, $A_{\mathfrak{p}} \cong M_{\frac{n}{2}}(\mathbb{H})$. There are two sub-cases, depending on whether $\tilde{\psi}_{\mathfrak{p}}$ is trivial or not. We take the case of $\tilde{\psi}_{\mathfrak{p}}$ trivial. The other case is exactly similar, so we omit it. The argument depends on the following standard properties of the gamma-function:

$$\Gamma(s+1)=s\Gamma(s). \tag{14. 13}$$

$$\Gamma(s)=\pi^{-\frac{1}{2}}2^{s-1}\,\Gamma\left(\frac{s}{2}\right)\Gamma\left(\frac{1+s}{2}\right). \tag{14. 14}$$

All the ε's are 1 here. For the L-functions, we have

$$L_{A_{\mathfrak{p}}}(s, \psi_{\mathfrak{p}})=\prod_{j=0}^{k-1}(2\pi)^{-2(ks-j)}\,\Gamma(2(ks-j))$$

where $2k=n$. The first factor gives a product $(2\pi)^{-2k^2s+k(k-1)}$. We apply (14. 14) to get

$$\begin{aligned}\Gamma(2(ks-j)) &= \pi^{-\frac{1}{2}}2^{2ks-2j-1}\,\Gamma\left(\frac{2ks-2j}{2}\right)\Gamma\left(\frac{2ks-2j+1}{2}\right)\\ &= \pi^{-\frac{1}{2}}2^{2ks-2j-1}\left(\frac{2ks-2j-1}{2}\right)\Gamma\left(\frac{2ks-2j}{2}\right)\Gamma\left(\frac{2ks-2j-1}{2}\right).\end{aligned}$$

Taking the product over all j, and incorporating the power of 2π above, we find

$$L_{A_{\mathfrak{p}}}(s, \psi_{\mathfrak{p}})=\pi^{-n/4}\,\pi^{-2k^2s+k(k-1)}\,2^{-k}\prod_{j=0}^{k-1}\left(\frac{2ks-2j-1}{2}\right)\prod_{j=0}^{n-1}\Gamma\left(\frac{ns-j}{2}\right).$$

On the other hand

$$\prod_{j=0}^{n-1}L_{F_{\mathfrak{p}}}(ns-j, \tilde{\psi}_{\mathfrak{p}})=\pi^{-\frac{1}{2}n^2s+n(n-1)/4}\prod_{j=0}^{n-1}\Gamma\left(\frac{ns-j}{2}\right),$$

and so

$$L_{A_{\mathfrak{p}}}(s, \psi_{\mathfrak{p}})=(2\pi)^{-k}\prod_{j=0}^{k-1}(ks-j-\tfrac{1}{2})\prod_{j=0}^{n-1}L_{F_{\mathfrak{p}}}(ns-j, \tilde{\psi}_{\mathfrak{p}}).$$

Therefore (14. 12) reduces to

$$\prod_{j=0}^{k-1}\left(\frac{ks-j-\frac{1}{2}}{k-j-\frac{1}{2}-ks}\right)=(-1)^k=(-1)^{k(\mathfrak{p})(e(\mathfrak{p})-1)}.$$

Putting all these computations together, we have proved

$$\varepsilon_A(s, \psi)=\frac{\Omega(1-s, \tilde{\psi})}{\Omega(s, \psi)}\prod_{j=0}^{n-1}\varepsilon_F(ns-j, \tilde{\psi})\prod_{\mathfrak{p}}(-1)^{k(\mathfrak{p})(e(\mathfrak{p})-1)},$$

where the product extends over all places $\mathfrak{p}$ of F. We just have to show that this product is 1, or that

$$\sum_{\mathfrak{p}}k(\mathfrak{p})\,(e(\mathfrak{p})-1)\equiv 0 \pmod 2.$$

It is enough to prove this when A is a division algebra. Also, it is trivial when n is odd. We assume therefore that n is even. The only odd terms in the sum come from those $\mathfrak{p}$ for which $k(\mathfrak{p})$ is odd. Such places are characterised by the fact that the Hasse invariant $h_{\mathfrak{p}}(A)$ of A at $\mathfrak{p}$ is of the form $ab^{-1}2^{-t}$ (mod $\mathbb{Z}$), where a and b are odd integers, and 2^t divides n exactly. The Hasse reciprocity law in the form

$$\sum_{\mathfrak{p}} h_{\mathfrak{p}}(A) \equiv 0 \pmod{\mathbb{Z}}$$

implies that there are an even number of such places. This completes the proof of (14. 7).

We now outline a slightly more general formulation of (14. 7) along the lines of Tate's thesis. First we need the local theory at the Archimedean places. For this, we let B be a finite-dimensional *simple* $\mathbb{R}$-algebra, and ϕ a continuous character of $B^\times$ of finite order. (Again, this hypothesis is not at all serious.) We can still form zeta integrals

$$Z(\Phi, s, \phi) = \int_{B^\times} \Phi(x)\,\phi(x)\,\|x\|^s\,d^\times x, \qquad \mathrm{Re}(s) > 1,$$

for suitable functions Φ, where $\|\cdot\|$ is the topological module of the action of $B^\times$ on B. To explain the term "suitable", we let $K = \mathbb{R}$, $\mathbb{C}$ or $\mathbb{H}$, and $B = M_k(K)$. Then K has a unique involution "bar" such that $x\bar{x}$ is real and positive for all $x \in K^\times$. We extend this to an involution on B by setting $\overline{(x_{ij})} = {}^t(\bar{x}_{ij})$, where t denotes transposition of matrices. Suitable functions are those of the form

$$\Phi(x) = P(x)\exp\left(-\pi\,\mathrm{tr}_{B/\mathbb{R}}(\bar{x}x)\right),$$

where $P(x)$ is a polynomial function on B, $\delta = 2$ if $K = \mathbb{H}$, $\delta = 1$ otherwise, and tr_B is the reduced trace from B to its centre. These functions form a vector space $\mathcal{S}(B)$. By taking the $\Phi_0 \in \mathcal{S}(B)$ for which $P = 1$ when ϕ is the trivial character, $P(x) = \det(x)$ otherwise, we get

$$Z(\Phi_0, s, \phi) = c\,L_B(s, \phi),$$

where c is a positive constant depending on the choice of Haar measure $d^\times x$. See [15], X, § 3. One may show that $L_B(s, \phi)^{-1} \cdot Z(\Phi, s, \phi)$ is a polynomial in s, for any $\Phi \in \mathcal{S}(B)$. Further, $\mathbb{R}$ has a canonical additive character $\theta_{\mathbb{R}}$ defined by

$$\theta_{\mathbb{R}}: x \mapsto \exp(-2\pi i x), \qquad x \in \mathbb{R},$$

which we lift to B in the usual way:

$$\theta_B = \theta_{\mathbb{R}} \circ \mathrm{tr}_{B/\mathbb{R}},$$

exactly as in the non-Archimedean case. This can be used to identify B with $\hat{B}$. We define Fourier transforms as before:

$$\hat{\Phi}(y) = \int_B \Phi(x)\,\theta_B(xy)\,dx, \qquad y \in B,\ \Phi \in \mathcal{S}(B).$$

Then $\hat{\Phi} \in \mathcal{S}(B)$, and there is a unique self-dual Haar measure dx on B defined as before. There is a local functional equation analogous to (10. 7). One defines $\varepsilon_B(s, \phi)$ as in (10. 11), and its value is given by (14. 4).

Now let A be a central simple F-algebra, and ψ an idele-class character of A, exactly as in (14. 7) above. We choose a maximal order Λ in A, and define a space $\mathscr{S}(A)$ of functions on the adele ring $\mathrm{Ad}(A)$ of A as follows. It is to be the space spanned by functions of the form

$$\Phi=\prod_{\mathfrak{p}} \Phi_{\mathfrak{p}}: x \mapsto \prod_{\mathfrak{p}} \Phi_{\mathfrak{p}}(x_{\mathfrak{p}}) \quad x=(x_{\mathfrak{p}}) \in \mathrm{Ad}(A),$$

where $\Phi_{\mathfrak{p}} \in \mathscr{S}(A_{\mathfrak{p}})$ for all places $\mathfrak{p}$ of F, and $\Phi_{\mathfrak{p}}$ is the characteristic function of $\Lambda_{\mathfrak{p}}$ for almost all $\mathfrak{p}$. This definition is independent of the choice of Λ. We can now form

$$Z(\Phi, s, \psi)=\int_{J(A)} \Phi(x)\, \psi(x)\, \|x\|^s\, d^{\times} x, \quad \mathrm{Re}(s)>1,$$

for $\Phi \in \mathscr{S}(A)$. Here, $\|\cdot\|$ is the module of the action of $J(A)$ on $\mathrm{Ad}(A)$. If the function Φ is of the form $\prod_{\mathfrak{p}} \Phi_{\mathfrak{p}}$, then

$$Z(\Phi, s, \psi)=\prod_{\mathfrak{p}} Z(\Phi_{\mathfrak{p}}, s, \psi_{\mathfrak{p}}), \quad \mathrm{Re}(s)>1.$$

We also have a canonical character θ_A of $\mathrm{Ad}(A)$ given by

$$\theta_A: x \mapsto \prod_{\mathfrak{p}} \theta_{A_{\mathfrak{p}}}(x_{\mathfrak{p}}), \quad x=(x_{\mathfrak{p}}) \in \mathrm{Ad}(A).$$

This identifies $\mathrm{Ad}(A)$ with $\mathrm{Ad}(A)\hat{}$, and moreover identifies A with $(\mathrm{Ad}(A)/A)\hat{}$. There is a self-dual Haar measure dx on $\mathrm{Ad}(A)$ for this identification, and it is the product of the local self-dual measures. Thus, if $\Phi=\prod_{\mathfrak{p}} \Phi_{\mathfrak{p}} \in \mathscr{S}(A)$, then the Fourier transform $\hat{\Phi}$ of Φ is given by

$$\hat{\Phi}=\prod_{\mathfrak{p}} \hat{\Phi}_{\mathfrak{p}}.$$

One may choose $\Phi \in \mathscr{S}(A)$ so that

$$Z(\Phi, s, \psi)=\tilde{L}_A(s, \psi),$$

and the local functional equations show that

$$Z(\hat{\Phi}, s, \bar{\psi})=\varepsilon_A(1-s, \psi)\, \tilde{L}_A(s, \bar{\psi}),$$

with ε_A as in (14. 6). Now one easily combines (14. 7) and (10. 7) to obtain

(14. 15) Corollary. $Z(\hat{\Phi}, 1-s, \bar{\psi})=Z(\Phi, s, \psi)$, *for all* $\Phi \in \mathscr{S}(A)$.

There is an extensive literature on the subject of the global functional equation. The first proof of (14. 7) is due to Eichler [3]. Eichler's proof is global, and extends Hecke's treatment of the standard case of A a field via a detailed analysis of the structure of maximal orders in A. He gives a formula for $\varepsilon_A(s, \psi)$ in terms of a global non-abelian Gauss sum. He also has the expression for $L_A(s, \psi)$ in terms of $L_F(s, \tilde{\psi})$, so he obtains a formula for ε_A in terms of ε_F, by comparing functional equations.

It would be more pleasing to give a direct proof of (14. 15), and then deduce (14. 7) from the local theory. (This would not, of course, be sufficient in itself to prove (13. 15).) This is quite easy when A is a division algebra. Tate's proof for the case of A a field (especially as written out in [15]) extends unchanged. When A has divisors of zero however, there are severe difficulties associated with the crucial application of the Poisson summation formula. This problem is overcome (partially) in [12], in the context of more general representations of $J(A)$, when A is a full matrix algebra. Fujisaki [4] gives a different treatment of the general case, involving a reduction to the division algebra case. He works in terms of zeta-integrals, but only deals with L_A, and leaves ε_A in terms of a non-abelian Gauss sum. Among many other papers on the subject, we mention [7] and [13]. Our standard reference [5] does not treat one-dimensional representations in the global case.

Finally we remark that all the papers just cited agree on the use of the symbol s, which is n^{-1} times our s, where $n^2 = \dim_F(A)$. We defend our non-standard choice on the grounds that it arises naturally from the module theory, which is our primary interest.

§ 15. Global functional equation for integral group rings

Now we return to our global functional equation (11. 16). Our notations are the obvious global analogues of (11. 7). To summarise:

(15. 1) Notation. F = an algebraic number field with ring of integers R;
$\mathfrak{D}_F$ = the absolute different of F; $\mathcal{N}$ = the counting norm on ideals of R;
G = a finite group of order n;
Λ = RG; $A = FG$; Λ' = a maximal order in A; M = a left ideal of Λ;
$\check{M}$ = the contragredient of M: $\check{M} = \mathrm{Hom}_\Lambda(M, \Lambda)^*$, where $*$ is the standard anti-automorphism of A;
ψ = a character of $\mathrm{Cl}(M)$, viewed as an idele-class character of A;
$\bar{\psi}^*$ is defined as an idele-class character, and may be viewed as a character of $\mathrm{Cl}(\check{M})$ via $\bar{\psi}^*(\check{N}) = \psi(N)$, for $N \in g(M)$;
$\mathfrak{f}(\psi)$ = the conductor of ψ, an ideal of Λ';
$\mathfrak{D}_A$ = the absolute different of A, an ideal of Λ';
$\mathcal{N}_A$ = the counting norm on ideals of Λ';
tr = the reduced trace of A; $b = \mathrm{tr}(1)$, an element of the centre of A.

From (11. 16) we have

$$\frac{L_\Lambda(\check{M}, 1-s, \bar{\psi}^*)}{L_\Lambda(M, s, \psi)} = \frac{L_A(1-s, \bar{\psi})}{L_A(s, \psi)} \{(\Lambda' : \Lambda)\, \mathcal{N}_A \mathfrak{f}(\psi)^{-\frac{1}{2}}\}^{2s-1} \frac{\tau(\bar{\psi})}{\mathcal{N}_A \mathfrak{f}(\psi)^{\frac{1}{2}}} \bar{\psi}(\mathfrak{D}_F).$$

By (14. 7) we have

$$\frac{L_A(1-s, \bar{\psi})}{L_A(s, \psi)} = \frac{L_{A_\infty}(s, \psi_\infty)}{L_{A_\infty}(1-s, \bar{\psi}_\infty)} \varepsilon_A(s, \psi)^{-1},$$

where $A_\infty = A \otimes_{\mathbb{Q}} \mathbb{R}$, $\psi_\infty = \psi | A_\infty^\times$, and L_{A_∞} is defined by (14. 1)—(14. 3), ε_A by (14. 4), (14. 6), (10. 11) and (10. 25). Let us consider the contribution from each non-Archimedean place $\mathfrak{p}$ of F to the expression

$$\text{(15. 2)} \qquad \varepsilon_A(s, \psi)^{-1} \{(\Lambda' : \Lambda)\, \mathcal{N}_A \mathfrak{f}(\psi)^{-\frac{1}{2}}\}^{2s-1} \bar{\psi}(\mathfrak{D}_F) \frac{\tau(\bar{\psi})}{\mathcal{N}_A \mathfrak{f}(\psi)^{\frac{1}{2}}}.$$

By (10. 19), this contribution is

$$(\Lambda'_\mathfrak{p} : \Lambda_\mathfrak{p})^{2s-1} \cdot \mathcal{N}_{A_\mathfrak{p}} \mathfrak{D}_{A_\mathfrak{p}}^{s-\frac{1}{2}} \cdot \bar{\psi}_\mathfrak{p}(\mathfrak{D}_{F_\mathfrak{p}}).$$

We factorise $\mathfrak{D}_{A_\mathfrak{p}} = \mathfrak{D}_{A_\mathfrak{p}/F_\mathfrak{p}} \cdot \mathfrak{D}_{F_\mathfrak{p}}$. Then $\mathcal{N}_{A_\mathfrak{p}}(\mathfrak{D}_{F_\mathfrak{p}}) = \mathcal{N}\mathfrak{D}^n_{F_\mathfrak{p}}$. We have essentially calculated $\mathcal{N}_{A_\mathfrak{p}}(\mathfrak{D}_{A_\mathfrak{p}/F_\mathfrak{p}})$ in the proof of (11. 8). It is given by

$$\mathcal{N}_{A_\mathfrak{p}}(\mathfrak{D}_{A_\mathfrak{p}/F_\mathfrak{p}}) = (\Lambda'_\mathfrak{p} : \Lambda_\mathfrak{p})^{-2} \, \|nb^{-1}\|^{-1}_{A_\mathfrak{p}}.$$

Taking account of this, (15. 2) has $\mathfrak{p}$-part

$$\mathcal{N}\mathfrak{D}^{n(s-\frac{1}{2})}_{F_\mathfrak{p}} \cdot \|nb^{-1}\|^{\frac{1}{2}-s}_{A_\mathfrak{p}} \cdot \bar{\psi}_\mathfrak{p}(\mathfrak{D}_{F_\mathfrak{p}}).$$

Taking the product over all $\mathfrak{p}$, we find that (15. 2) is equal to

$$|\mathfrak{d}_F|^{n(s-\frac{1}{2})} \, \bar{\psi}(\mathfrak{D}_F) \, c^{\frac{1}{2}-s} \, \varepsilon_{A_\infty}(s, \psi_\infty),$$

where $\mathfrak{d}_F$ is the absolute discriminant of F, and the constant c is

$$c = \prod_\mathfrak{p} \|nb^{-1}\|_{A_\mathfrak{p}}$$

(product over all non-Archimedean places $\mathfrak{p}$). Thus c^{-1} is the determinant of the element nb^{-1}, viewed as an endomorphism of the $\mathbb{Q}$-vector space FG. We write $c^{-1} = \det_{FG}(nb^{-1})$. Clearly $\det_{FG}(n) = n^{n[F:\mathbb{Q}]}$. Next, the element $b = \operatorname{tr}(1)$ in fact lies in $\mathbb{Q}G$, as is evident from the fact that the reduced trace is unchanged by extension of the ground field. Thus $\det_{FG}(b) = \det_{\mathbb{Q}G}(b)^{[F:\mathbb{Q}]}$. But we may also think of b as a $\mathbb{C}$-endomorphism of $\mathbb{C}G$, when we write it $b_\mathbb{C}$. Then clearly

$$\det_{\mathbb{Q}G}(b) = \det_{\mathbb{C}G}(b) = \prod_\chi d(\chi)^{d(\chi)^2},$$

where χ ranges over all absolutely irreducible characters of the finite group G, and $d(\chi)$ is the degree of χ. Therefore

$$c = \{n^{-n} \prod_\chi d(\chi)^{d(\chi)^2}\}^{[F:\mathbb{Q}]}.$$

(It will be clear from (15. 6) below that the quantity c^{-1} is an *integer*.)

Now we have to deal with the Archimedean contribution $\varepsilon_{A_\infty}(s, \psi_\infty)$. The idele-class character ψ has to be trivial on the group $U(\operatorname{End}_\Lambda(M))$ of unit ideles of the order $\operatorname{End}_\Lambda(M)$, and this contains $A^\times_\infty$ as a subgroup (see (1. 7)). It follows that the character ψ_∞ is trivial, and $\varepsilon_{A_\infty}(s, \psi_\infty) = 1$. To sum up, we have proved:

(15. 3) Theorem. *In addition to the notation* (15. 1), *we let* $\mathfrak{d}_F$ *be the absolute discriminant of* F, $A_\infty = A \otimes_\mathbb{Q} \mathbb{R}$, $\psi_\infty = \psi | A^\times_\infty$ *(when* ψ *is viewed as an idele-class character), and we let* χ *range over the absolutely irreducible characters of the finite group* G. *We write* $d(\chi)$ *for the degree of* χ. *Then*

(i) *the character* ψ_∞ *is trivial*;

(ii) $L_\Lambda(M, s, \psi) \, L_{A_\infty}(s, \psi_\infty)$

$$= L_\Lambda(\check{M}, 1-s, \bar{\psi}^*) \, L_{A_\infty}(1-s, \bar{\psi}_\infty) \, \psi(\mathfrak{D}_F) \, \{|\mathfrak{d}_F|^{n/[F:\mathbb{Q}]} \, n^n \prod_\chi d(\chi)^{-d(\chi)^2}\}^{(\frac{1}{2}-s)[F:\mathbb{Q}]}.$$

We recall that $L_{A_\infty}(s, \psi_\infty)$ is a product of gamma-functions and exponentials given explicitly by (14. 1)—(14. 3). Observe that only the functions $L_A(M, s, \psi)$, $L_A(\check{M}, s, \bar{\psi}^*)$ themselves and the constant factor $\psi(\mathfrak{D}_F)$ actually depend on either M or ψ. The factor $\psi(\mathfrak{D}_F)$ can be absorbed as follows. The modules $\check{M}$ and $\mathfrak{D}_F^{-1}\check{M}$ lie in the same genus of Λ-modules, and we have an isomorphism $\mathrm{Cl}(\check{M}) \cong \mathrm{Cl}(\mathfrak{D}_F^{-1}\check{M})$ given by $N \mapsto \mathfrak{D}_F^{-1}N$, $N \in g(\check{M})$. We may view $\bar{\psi}^*$ as a character of $\mathrm{Cl}(\mathfrak{D}_F^{-1}\check{M})$ via this isomorphism, and this does not change $\bar{\psi}^*$ when viewed as a character of $J(A)$. We then have

$$L_\Lambda(\check{M}, s, \bar{\psi}^*)\, \psi(\mathfrak{D}_F) = L_\Lambda(\mathfrak{D}_F^{-1}\check{M}, s, \bar{\psi}^*),$$

and the functional equation (15. 3) (ii) reads

$$\text{(15. 4)} \qquad L_\Lambda(M, s, \psi)\, L_{A_\infty}(s, \psi_\infty)$$
$$= L_\Lambda(\mathfrak{D}_F^{-1}\check{M}, 1-s, \bar{\psi}^*)\, L_{A_\infty}(1-s, \bar{\psi}_\infty^*)\, \{|\mathfrak{d}_F|^{n/[F:\mathbb{Q}]}\, n^n \prod_\chi d(\chi)^{-d(\chi)^2}\}^{(\frac{1}{2}-s)[F:\mathbb{Q}]}$$

Taking account of formula (1. 5), we have proved the first statement of

(15. 5) Corollary. *Define*

$$E(s) = \{|\mathfrak{d}_F|^{n/[F:\mathbb{Q}]}\, n^n \prod_\chi d(\chi)^{-d(\chi)^2}\}^{(\frac{1}{2}-s)[F:\mathbb{Q}]},$$

and let ϕ_0 denote the trivial character of $A_\infty^\times$, where $A_\infty = FG \otimes_{\mathbb{Q}} \mathbb{R}$. Then, for each left ideal N of $\Lambda = RG$, we have

$$\text{(i)} \quad Z_\Lambda([N], s)\, L_{A_\infty}(s, \phi_0) = Z_\Lambda([\mathfrak{D}_F^{-1}\check{N}], 1-s)\, L_{A_\infty}(1-s, \phi_0)\, E(s),$$

where $Z_\Lambda([N], s)$ is defined by (1. 2). *Moreover,*

$$\text{(ii)} \quad Z_\Lambda(g(M), s)\, L_{A_\infty}(s, \phi_0) = Z_\Lambda(g(\check{M}), 1-s)\, L_{A_\infty}(1-s, \phi_0)\, E(s),$$

and

$$\text{(iii)} \qquad \zeta_\Lambda(s)\, L_{A_\infty}(s, \phi_0) = \zeta_\Lambda(1-s)\, L_{A_\infty}(1-s, \phi_0)\, E(s).$$

Proof. The Λ-modules $\mathfrak{D}_F^{-1}\check{M}$ and $\check{M}$ lie in the same genus. Therefore, as N ranges over a set of representatives of the stable isomorphism classes in the genus $g(M)$, $\mathfrak{D}_F^{-1}\check{N}$ ranges over a set of representatives of the stable isomorphism classes in $g(\check{M})$. Therefore, summing the equations (i) over such a set of representatives, we obtain (ii). (Alternatively, we may take ψ trivial in (15. 3).) Statement (iii) follows from summing (ii) over a set of representatives M of the genera of left ideals of Λ.

If Λ' is a maximal order in $A = FG$, we know that $\zeta_{\Lambda'}(s) = L_A(s, \psi_0)$, where ψ_0 is the trivial character of $J(A)$. Therefore, when we compare the global functional equation (15. 5) (iii) with the local functional equation (11. 10) (or [1], Theorem 2), we obtain confirmation of the well known formula (which we had to prove earlier)

$$\text{(15. 6)} \qquad \{n^n \prod_\chi d(\chi)^{-d(\chi)^2}\}^{[F:\mathbb{Q}]} = \mathcal{N}_A \mathfrak{D}_{A/F} \cdot (\Lambda' : \Lambda)^2.$$

This concludes our discussion of functional equations.

References

[1] *C. J. Bushnell, I. Reiner*, Zeta-functions of arithmetic orders and Solomon's conjectures, Math. Z. **173** (1980), 135—161.

[2] *C. J. Bushnell, I. Reiner*, *L*-functions of arithmetic orders and asymptotic distribution of ideals, J. reine angew. Math. **327** (1981), 156—183.

[3] *M. Eichler*, Allgemeine Kongruenzklasseneinteilungen der Ideale einfacher Algebren über algebraischen Zahlkörpern und ihre *L*-Reihen, J. reine angew. Math. **178** (1938), 227—251.

[4] *G. Fujisaki*, On *L*-functions of simple algebras over the field of rational numbers, J. Fac. Sci. Univ. Tokyo, Sect. I, **9** (1962), 293—311.

[5] *R. Godement, H. Jacquet*, Zeta-functions of simple algebras, Lecture Notes in Mathematics **260**, Berlin-Heidelberg-New York 1972.

[6] *J. E. Humphreys*, Linear Algebraic Groups, Graduate Texts in Mathematics **21**, Berlin-Heidelberg-New York 1975.

[7] *M. Kinoshita*, On the ζ-functions of a total matrix algebra over the field of rational numbers, J. Math. Soc. Japan **17** (1965), 374—408.

[8] *S. Lang*, Algebraic Number Theory, Reading, Massachusetts 1970.

[9] *J. Martinet*, Character theory and Artin *L*-functions, Algebraic Number Fields (A. Fröhlich, ed.), London-New York 1977, 1—88.

[10] *I. Reiner*, Maximal Orders, London-New York 1975.

[11] *J-P. Serre*, Local class field theory, Algebraic Number Theory (Cassels, J.W.S. & Fröhlich, A., edd.), London 1967, 129—162.

[12] *H. Strassberg*, *L*-functions for $GL(n)$, Math. Ann. **245** (1979), 23—36.

[13] *T. Tamagawa*, On the ζ-functions of a division algebra, Ann. Math. **77** (1963), 387—405.

[14] *J. Tate*, Fourier analysis in number fields and Hecke's zeta-functions, Thesis Princeton 1950, also: Algebraic Number Theory (Cassels, J.W.S. & Fröhlich, A., edd.), London 1967, 305—347.

[15] *A. Weil*, Basic Number Theory, Berlin-Heidelberg-New York 1974.

University of London King's College, Department of Mathematics, Strand, London WC2R 2LS, GB

University of Illinois, Department of Mathematics, Urbana, IL 61801, USA

Eingegangen 23. März 1981

The Prime Ideal Theorem in Non-Commutative Arithmetic

Colin J. Bushnell[1]* and Irving Reiner[2]*

[1] Department of Mathematics, King's College, Strand, London WC2R2LS, England
[2] Department of Mathematics, University of Illinois, 1409 W. Green Street, Urbana, Illinois 61801, U.S.A.

We begin by recalling the prime ideal theorem from algebraic number theory (see Lang [8] and Narkiewicz [9] for proofs). This theorem describes the asymptotic distribution of prime ideals among ideal classes. Let R be the ring of all algebraic integers in an algebraic number field F, and let $Cl(R)$ be the ideal class group of R. The elements of $Cl(R)$ are the isomorphism classes $[X]$ of non-zero ideals X of R, with $[X]=[Y]$ if and only if $X=Y\alpha$ for some $\alpha\in F$. Let $h(R)=|Cl(R)|$ be the *ideal class number* of R.

If X is a non-zero ideal of R, we put $NX=(R:X)$, the index of X in R, and this is necessarily finite. Let s be a complex variable, and write $\sigma=\mathrm{Re}(s)$ for the real part of s. As usual, the notation NX^{-s} means $(NX)^{-s}$. We have:

Prime Ideal Theorem. *Given an ideal class $c\in Cl(R)$, let P range over all prime ideals of R whose ideal class is c. Then:*

(i) *We have*

$$\sum_P NP^{-s}=\frac{1}{h(R)}\log\frac{1}{s-1}+G_0(s),$$

where $G_0(s)$ is holomorphic in an open region containing the closed half-plane $\sigma\geqq 1$.

(ii) *For $T>0$, the number of prime ideals P of R with $NP\leqq T$ and $[P]=c$ is asymptotically*

$$h(R)^{-1}\,T/\log T+o(T/\log T)$$

as $T\to\infty$.

The principal term in (i) is independent of the ideal class c, so one may say that the prime ideals of R are uniformly distributed among the ideal classes. Further, (ii) follows from (i) by applying a Tauberian theorem.

This result implies at once that every ideal class c of R contains infinitely many prime ideals P. A slight refinement shows that any c contains infinitely many P for which NP is a rational prime.

* Research partially supported by the National Science Foundation.

The aim of the present paper is to generalize the prime ideal theorem to the non-commutative case. Let A be a finite dimensional semisimple $\mathbf{Q}$-algebra, and let Λ be a $\mathbf{Z}$-order in A, so that Λ will play the role of the ring of integers R. We always restrict our attention to left ideals X of Λ such that the index $NX=(\Lambda:X)$ is *finite*. The ideal theory of Λ is more involved than that of R, in that the set of ideals must first be divided into *genera*, and then each genus can be divided into ideal classes. We shall concentrate on ideals in the genus $g(\Lambda)$ of Λ itself (see §1 for the definitions), and sketch the general case in §6. It may be noted in passing that ideals outside $g(\Lambda)$ make a comparatively insignificant contribution (see §7). The obvious analogues of the prime ideals of R are here the maximal left ideals of Λ, and in the first instance we consider the distribution of these. Let $Cl(\Lambda)$ be the locally free class group of Λ, and let $h(\Lambda)=|Cl(\Lambda)|$. Each left ideal $X\in g(\Lambda)$ determines a class $[X]\in Cl(\Lambda)$.

Our first main result is the following generalization of the prime ideal theorem.

Theorem A. *Let Λ be an order in a simple $\mathbf{Q}$-algebra A, and let $c\in Cl(\Lambda)$. Let X range over all maximal left Λ-ideals in $g(\Lambda)$ with $[X]=c$. Then*

(i) $\sum_X NX^{-s}=h(\Lambda)^{-1}\log\frac{1}{s-1}+G_0(s),$

where $G_0(s)$ is holomorphic for $\sigma\geqq 1$.

(ii) *Let $T>0$. The number of such ideals X for which $NX\leqq T$ is asymptotically*

$$h(\Lambda)^{-1}T/\log T+o(T/\log T)$$

as $T\to\infty$.

We shall establish Theorem A in the course of proving the more general Theorem B, stated below, which applies to semisimple, rather than just simple, algebras. Theorem A implies that for an order Λ in a *simple* $\mathbf{Q}$-algebra, every ideal class $c\in Cl(\Lambda)$ contains infinitely many maximal left ideals of Λ. We now give an easy example showing that the corresponding result need not be valid for semisimple algebras. Let F, R be as above, and put $A=F\oplus F$, $\Lambda=R\oplus R$. Then $Cl(\Lambda)=Cl(R)\times Cl(R)$, and we take $c=(c_1,c_2)\in Cl(\Lambda)$, with c_1, $c_2\in Cl(R)$ both non-trivial. The maximal ideals of Λ are precisely the ideals $P\oplus R$, $R\oplus P$, as P ranges over maximal ideals of R. Clearly no such ideal can have class c, so c contains *no* maximal ideal of Λ at all.

Our second main result, Theorem B below, includes Theorem A as a special case. Theorem B, and its strengthened versions Theorem C and D, give a precise quantitative formula which completely explains this example, and provides the correct extension of Theorem A to the semisimple case. Instead of restricting our attention to maximal left ideals of Λ, we fix a positive integer k and consider those left ideals X of Λ for which the composition length $l(\Lambda/X)$ of the finite Λ-module Λ/X is k.

Theorem B. *Let Λ be an order in a semisimple $\mathbf{Q}$-algebra A. Given an integer $k\geqq 1$ and a class $c\in Cl(\Lambda)$, let X range over all left Λ-ideals such that $X\in g(\Lambda)$, $l(\Lambda/X)=k$, and $[X]=c$. Then:*

(i) *There is an integer* $\mathsf{N}(k,c)\geqq 0$ *such that for* $\sigma>1$,

$$\sum_X NX^{-s}=\frac{\mathsf{N}(k,c)}{k!\,h(\Lambda)}\left(\log\frac{1}{s-1}\right)^k+O\left(\log\frac{1}{s-1}\right)^{k-1}$$

as $s\to 1$.

(ii) *The result is best possible in the following sense:*

(a) *If* $\mathsf{N}(k,c)>0$, *there are infinitely many left* Λ*-ideals* $X\in g(\Lambda)$ *with* $l(\Lambda/X)=k$ *and* $[X]=c$.

(b) *If* $\mathsf{N}(k,c)=0$, *there are no such ideals* X.

We give sufficient information about the error term in (i) to be able to apply a Tauberian theorem and thereby obtain an analogue of Theorem A(ii). We prove Theorem B using a "modified L-function" attached to Λ, which generalizes a function used by Delange, Selberg and others, to study integers with a given number of prime factors.

The fact that Theorem A may fail for semisimple algebras was pointed out by Drozd [5] and Jacobinski [7], in proving a conjecture of Roiter. They showed essentially that when the semisimple algebra A has r simple components, then every class $c\in Cl(\Lambda)$ contains infinitely many left ideals X of Λ for which $X\in g(\Lambda)$, $l(\Lambda/X)=r$, and Λ/X is a semisimple Λ-module. Both Drozd and Jacobinski assumed that the algebra A satisfies the Eichler condition, in which case the stable isomorphism classes in $g(\Lambda)$ are ordinary isomorphism classes.

We shall see in Theorem 5.19 below that the "distribution coefficient" $\mathsf{N}(k,c)$ occurring in Theorem B can be given explicitly in terms of class numbers, combinatorial expressions, and information about whether the class c and its homomorphic images are trivial or not. Theorem B, and the stronger Theorems C and D, improve on the Drozd-Jacobinski result, in that they provide quantitative information (which incidentally implies a qualitative improvement also).

In §6, we show how to generalize our results to arbitrary Λ-lattices, and make some further general remarks in §7.

This paper was written while the first author was on sabbatical leave from King's College London, and visiting the University of Illinois as a participant in a Special Year in Algebra and Algebraic Number, partially supported by the NSF. He wishes to acknowledge this support, as well as the hospitality of the University of Illinois.

§1. Background

We shall need various concepts and results from the theory of orders and their L-functions. We now review these, and at the same time establish some permanent notation. Everything we need is in [2], though for a more detailed version of the idelic approach to class groups, we refer the reader to [6].

Let A be a finite-dimensional semisimple algebra over the field $\mathbf{Q}$ of rational numbers, and let Λ be a $\mathbf{Z}$-order in A. (We could use any algebraic number field as base field here, but there would be no gain in generality and only a minor gain in precision.) For a prime number p, let A_p be the p-adic completion of A, Λ_p the p-adic completion of Λ, and so on.

Recall that the *genus* $g(\Lambda)$ of Λ consists of all left ideals X of Λ such that $X_p \cong \Lambda_p$ as Λ_p-modules for all prime numbers p. (Strictly speaking, $g(\Lambda)$ consists of all left Λ-*lattices* with this property, but any such lattice is isomorphic to a left Λ-ideal, so there is no loss in generality here.) Two ideals $X, X' \in g(\Lambda)$ are *stably isomorphic* if $X \oplus \Lambda \cong X' \oplus \Lambda$. Let $[X]$ denote the stable isomorphism class of X. We remark that if A satisfies the Eichler condition, then $[X]=[X']$ if and only if $X \cong X'$.

The set of stable isomorphism classes $\{[X]: X \in g(\Lambda)\}$ forms a finite abelian group $Cl(\Lambda)$, called the *locally free class group* of Λ. Throughout we put

$$h(\Lambda) = |Cl(\Lambda)|.$$

Now let ψ be a character of $Cl(\Lambda)$, that is, a homomorphism $Cl(\Lambda) \to \mathbf{C}^\times$. We recall from [2] the definition of the L-function $L_A(\Lambda, s, \psi)$, which we here abbreviate as $L_A(s, \psi)$. Let s be a complex variable, $\sigma = \mathrm{Re}(s)$. For a left ideal X of Λ, we put $NX = (\Lambda : X) = |\Lambda/X|$, and consider only ideals for which this is finite. For $X \in g(\Lambda)$, we write $\psi(X)$ rather than $\psi([X])$, for brevity. We define

$$L_A(s, \psi) = \sum_{X \in g(\Lambda)} \psi(X) NX^{-s}, \quad \sigma > 1. \tag{1.1}$$

By [2] (5.1), we know that $L_A(s, \psi)$ admits analytic continuation to a meromorphic function on the whole s-plane.

Let $J(A)$ denote the idele group of A. There is a canonical surjection τ: $J(A) \to Cl(\Lambda)$, which carries an idele $x = (x_p) \in J(A)$ to the class $[X] \in Cl(\Lambda)$, where X is the unique Λ-submodule of A such that $X_p = \Lambda_p x_p$ for all p. Then

$$\mathrm{Ker}(\tau) = J'(A) \cdot A^\times \cdot U(\Lambda),$$

where $J'(A)$ is the closure of the commutator subgroup of $J(A)$, $A^\times$ is embedded in $J(A)$ as the group of principal ideles, and $U(\Lambda)$ is the group of *unit ideles* of Λ defined by

$$U(\Lambda) = \prod_p \Lambda_p^\times.$$

Just this once, we here allow p to range over all *places* of $\mathbf{Q}$, including ∞, with the convention that $\Lambda_\infty = A_\infty$. Thus we have

$$J(A)/J'(A) \cdot A^\times \cdot U(\Lambda) \cong Cl(\Lambda). \tag{1.2}$$

Given a character ψ of $Cl(\Lambda)$, the map $\psi \circ \tau$ is a continuous character from $J(A)$ to $\mathbf{C}^\times$, with finite image. This character $\psi \circ \tau$ of $J(A)$ will be denoted simply by ψ. For each prime number p, there is a canonical inclusion $A_p^\times \subset J(A)$, and we may therefore define ψ_p to be the restriction of ψ to $A_p^\times$. Then $\psi_p(\Lambda_p^\times) = \{1\}$, so for any principal left ideal $Y = \Lambda_p y$ of Λ_p, with $y \in A_p^\times$, we may define $\psi_p(Y)$ unambiguously by the formula $\psi_p(Y) = \psi_p(y)$. It follows at once that

$$\psi(X) = \prod_p \psi_p(X_p) \quad \text{for all } X \in g(\Lambda). \tag{1.3}$$

Next we recall the definition of the local L-function:

$$L_{\Lambda_p}(s, \psi_p) = \sum_Y \psi_p(Y) NY^{-s}, \tag{1.4}$$

where Y ranges over all principal left ideals of Λ_p (with $NY = (\Lambda_p : Y)$ finite, as always), and $\psi_p(Y)$ is defined as above. This series converges to a holomorphic function of s in the region $\sigma > 1 - (\dim_{\mathbf{Q}} A)^{-1}$. From [2], §2 there is an Euler product formula

$$L_\Lambda(s, \psi) = \prod_p L_{\Lambda_p}(s, \psi_p), \quad \sigma > 1, \tag{1.5}$$

where p ranges over all prime numbers. This product converges absolutely and uniformly on compact subsets of the region $\sigma > 1$.

We now choose a *finite* set S of "bad" primes p, such that for all $p \notin S$ we have

(1.6) (i) A_p is a split semisimple $\mathbf{Q}_p$-algebra (i.e., a direct product of full matrix algebras over fields), and

(ii) Λ_p is a maximal $\mathbf{Z}_p$-order in A_p.

Such a set S can always be found.

A left ideal X of Λ will be called "prime to S" if NX is relatively prime to every $p \in S$. Notice that for such X, we have $X_p = \Lambda_p$ if $p \in S$. On the other hand $X_p \cong \Lambda_p$ for $p \notin S$, since then Λ_p is a maximal order. Thus $X \in g(\Lambda)$. Given a character ψ of $Cl(\Lambda)$, we define a new L-function

$$L_\Lambda^{(S)}(s, \psi) = \sum_X \psi(X) NX^{-s}, \quad \sigma > 1, \tag{1.7}$$

where X now ranges over all left Λ-ideals prime to S. The analogue of (1.5) is

$$L_\Lambda^{(S)}(s, \psi) = \prod_{p \notin S} L_{\Lambda_p}(s, \psi_p), \quad \sigma > 1. \tag{1.8}$$

From [2], §§3, 4, we know that $L_\Lambda^{(S)}(s, \psi)$ also admits analytic continuation to a meromorphic function on the whole s-plane.

We now write A as a direct sum of its simple components:

$$A = \bigoplus_{i=1}^r A_i, \tag{1.9}$$

which gives us a decomposition

$$J(A) = \prod_{i=1}^r J(A_i).$$

Viewing ψ as a character of $J(A)$, we let ψ_i be its restriction to $J(A_i)$. Thus $\psi = \prod \psi_i$, with an obvious interpretation of the product. Let us define

(1.10) t_ψ = number of indices i, $1 \leqq i \leqq r$, such that ψ_i is trivial. Then $0 \leqq t_\psi \leqq r$, and indeed $t_\psi = r$ if and only if ψ is the trivial character of $J(A)$ (or of $Cl(\Lambda)$). By [2], (3.11), (4.1), we have

(1.11) **Theorem.** *The function* $L_\Lambda^{(S)}(s,\psi)$ *is holomorphic in the region* $\sigma>1-\varepsilon$ *for some* $\varepsilon>0$, *except for a pole of order* t_ψ *at* $s=1$. *It has no zero in the closed region* $\sigma \geqq 1$.

Following the terminology in [9], we denote by $G_0(s)$ any function of s which is holomorphic in an open region containing the closed half-plane $\sigma \geqq 1$ (briefly, we say that $G_0(s)$ is holomorphic for $\sigma \geqq 1$). If $f(s)$, $g(s)$ are functions holomorphic for $\sigma>1$, we write

(1.12) $$f(s)=g(s)+G_0(s) \quad \text{if } f(s)-g(s) \text{ is holomorphic for } \sigma \geqq 1.$$

More generally, we introduce the notation $G_{k-1}(s)$, where k is any positive integer. By definition, $G_{k-1}(s)$ denotes any function of the form

(1.13) $$G_{k-1}(s)=\sum_{j=0}^{k-1} f_j(s)\left(\log\frac{1}{s-1}\right)^j,$$

for functions $f_j(s)$ which are $G_0(s)$.

From (1.11), we have

(1.14) $$\log L_\Lambda^{(S)}(s,\psi)=t_\psi \log\frac{1}{s-1}+G_0(s),$$

in the sense of (1.12).

§2. Modified *L*-functions (local case)

We shall now introduce a modified *L*-function $L_\Lambda(s,\psi;z)$ depending on two complex variables s and z. This new function will enable us to keep track of the composition length $l(\Lambda/X)$ for a Λ-ideal X. In this section, we develop the local theory of these modified *L*-functions only so far as is needed for our later applications.

Let B denote a finite-dimensional semisimple $\mathbf{Q}_p$-algebra, and let Δ be a $\mathbf{Z}_p$-order in B. Let $\theta\colon B^\times \to \mathbf{C}^\times$ be a continuous character of finite order, such that $\theta(\Delta^\times)=\{1\}$. For any principal left ideal $X=\Delta x$ of Δ, $x \in B^\times$, the expression $\theta(X)$ is unambiguously defined by the formula $\theta(X)=\theta(x)$.

For a left Δ-ideal X (always of finite index NX in Δ), let $l(\Delta/X)$ be the Δ-composition length of the finite Δ-module Δ/X. Then $l(\Delta/X)$ is finite, and $l(\Delta/X)=0$ if and only if $\Delta=X$.

Now let z be a complex variable. We agree once and for all that we consider only values of z for which $|z|\leqq 1$. Now define a *modified L-function* by the formula

(2.1) $$L_\Delta(s,\theta;z)=\sum_X \theta(X)\cdot NX^{-s}\cdot z^{l(\Delta/X)},$$

where X ranges over all principal left ideals of Δ. By comparison with the ordinary *L*-series $L_\Delta(s,\theta)$, the series (2.1) converges absolutely and uniformly on compact subsets of the region $\sigma>1-\varepsilon$, $|z|\leqq 1$, for some $\varepsilon>0$. Thus $L_\Delta(s,\theta;z)$ is holomorphic in s and z in that region. We may rearrange the series (2.1) to obtain

(2.2) $$L_\Delta(s, \theta; z) = 1 + \sum_{m=1}^{\infty} f_m(s, \theta) z^m.$$

Each coefficient $f_m(s, \theta)$ is a polynomial in p^{-s} with coefficients in the ring $\mathbf{Z}[\theta]$ generated by adjoining to $\mathbf{Z}$ the finite set $\theta(B^\times)$ of roots of unity.

If Δ' is an order in another $\mathbf{Q}_p$-algebra B', and θ' is a character of $B'^\times$ trivial on $\Delta'^\times$, then clearly

(2.3) $$L_{\Delta \oplus \Delta'}(s, \theta \times \theta'; z) = L_\Delta(s, \theta; z) L_{\Delta'}(s, \theta'; z).$$

We now treat in detail the special case where B is a *split simple* $\mathbf{Q}_p$-algebra, and where Δ is a *maximal* order in B. We may then take

(2.4) $$B = M_n(F), \qquad \Delta = M_n(R),$$

for some $n \geqq 1$, where F is a finite extension field of $\mathbf{Q}_p$ with valuation ring R. Let P be the prime ideal of R, and set

$$q = NP = (R : P),$$

which is a power of p. Alse we put

$$d = \dim_{\mathbf{Q}_p}(B) \geqq n.$$

As in [2] (3.4), we may write

$$\theta(x) = \tilde{\theta}(\det x), \qquad \text{for } x \in B^\times = GL_n(F),$$

where $\tilde{\theta}$ is some uniquely determined character of $F^\times$. Further, $\tilde{\theta}$ is *unramified* (that is, $\tilde{\theta}(R^\times) = \{1\}$), since $\theta(\Delta^\times) = \{1\}$ by hypothesis.

With this notation, it follows from [2] (3.6), (3.9) that we have

(2.5) $$L_\Delta(s, \theta) = \prod_{j=0}^{n-1} (1 - \tilde{\theta}(P) q^{j-ns})^{-1}, \qquad \sigma > 1 - \frac{1}{n}.$$

We now show that analogously

(2.6) $$L_\Delta(s, \theta; z) = \prod_{j=0}^{n-1} (1 - \tilde{\theta}(P) q^{j-ns} z)^{-1}, \qquad \sigma > 1 - \frac{1}{n}.$$

This may be proved directly in exactly the same manner as (2.5). However, (2.6) may be deduced quickly from (2.5) as follows. Let X be a principal left ideal of Δ, with $l(\Delta/X) = m$. Since the ring Δ has (up to isomorphism) a unique simple left module, namely $(R/P)^{(n)}$, we obtain

$$NX = q^{nm} = q^{nl(\Delta/X)}.$$

Now write $z = q^w$, where $w = \log z / \log q$. Then

$$\theta(X) NX^{-s} z^{l(\Delta/X)} = \theta(X) q^{-mns} q^{mw} = \theta(X) NX^{-s'},$$

where $s' = s - n^{-1} w$. This gives $L_\Delta(s, \theta; z) = L_\Delta(s', \theta)$, and evaluating the latter by (2.5), we obtain (2.6).

Continuing with this example, it is clear from (2.6) that $L_\Delta(s, \theta; z)$ has no poles or zeros for $\sigma > 1 - \frac{1}{n}$, $|z| \leqq 1$. We take logarithms, and expand these as power series, to obtain

$$\log L_\Delta(s, \theta; z) = \sum_{m=1}^{\infty} g_m(s, \theta) z^m, \tag{2.7}$$

where

$$g_m(s, \theta) = \frac{1}{m} \tilde{\theta}(P)^m \sum_{j=0}^{n-1} q^{m(j-ns)}, \qquad m \geqq 1. \tag{2.8}$$

We shall now estimate the coefficients $g_m(s, \theta)$, for $\sigma > 1 - (4d)^{-1}$, noting that $1 - (4d)^{-1} \geqq 1 - (4n)^{-1}$. For such s, $\operatorname{Re}(j - ns) \leqq -\frac{3}{4}$, for $0 \leqq j \leqq n-1$, so we obtain

$$|g_m(s, \theta)| < \frac{n}{m} q^{-3m/4} \leqq d p^{-3m/4}, \qquad \sigma > 1 - (4d)^{-1}, \; m \geqq 1. \tag{2.9}$$

On the other hand, we have

$$|g_1(s, \theta)| \leqq d p^{-\sigma}, \qquad \sigma > 1. \tag{2.10}$$

Combining these estimates with (2.3), we obtain the following information about the semisimple case:

(2.11) **Proposition.** *Let B be a split semisimple $\mathbf{Q}_p$-algebra of dimension d, and let Δ be a maximal order in B. Let θ be a character of $B^\times$ of finite order such that $\theta(\Delta^\times) = \{1\}$. Then the modified L-function $L_\Delta(s, \theta; z)$ defined in* (2.1) *is holomorphic in the region*

$$\sigma > 1 - (4d)^{-1}, \qquad |z| \leqq 1,$$

and has no zero there. Further, in this region we may write

$$\log L_\Delta(s, \theta; z) = \sum_{m=1}^{\infty} g_m(s, \theta) z^m,$$

where the coefficients $g_m(s, \theta)$ are holomorphic for $\sigma > 1 - (4d)^{-1}$. They satisfy the bounds given in (2.9), (2.10) *above.*

§ 3. The Global Case

We shall now consider modified L-functions in the global case. They will play a key role in the proof of Theorem B stated in the introduction. As in § 1, let Λ be an order in a $\mathbf{Q}$-algebra A. If X is a left ideal of Λ (always of finite index NX in Λ), let $l(\Lambda/X)$ denote the Λ-composition length of the finite Λ-module Λ/X. Let z be a complex variable, always subject to the restriction $|z| \leqq 1$. For each character ψ of $Cl(\Lambda)$, the *global modified L-function* is defined by

$$L_\Lambda(s, \psi; z) = \sum_X \psi(X) \cdot NX^{-s} \cdot z^{l(\Lambda/X)}, \qquad \sigma > 1, \tag{3.1}$$

where X ranges over all left Λ-ideals in the genus $g(\Lambda)$. Comparing this series with $L_\Lambda(s, \psi_0)$, where ψ_0 is the trivial character of $Cl(\Lambda)$, we see that (3.1) converges absolutely and uniformly on compact subsets of the region $\sigma > 1$, $|z| \leqq 1$, and defines a function which is holomorphic in s and z on the interior of that region.

Because of this absolute convergence, any subsum of (3.1) is holomorphic (for $\sigma > 1$, $|z| \leqq 1$), so when we arrange (3.1) as a power series in z,

$$L_\Lambda(s, \psi; z) = 1 + \sum_{m=1}^{\infty} f_m(s, \psi) z^m, \tag{3.2}$$

the coefficients $f_m(s, \psi)$ converge to holomorphic functions for $\sigma > 1$. For each $m \geqq 1$, let us put

$$I(m, \Lambda) = \{X: X = \text{left } \Lambda\text{-ideal}, X \in g(\Lambda), l(\Lambda/X) = m\}. \tag{3.3}$$

Since the coefficient of z^m in (3.1) comes exactly from those ideals X in $I(m, \Lambda)$, we obtain

$$f_m(s, \psi) = \sum_{X \in I(m, \Lambda)} \psi(X) NX^{-s}, \quad \sigma > 1. \tag{3.4}$$

We intend to describe the analytic behaviour of $f_m(s, \psi)$ near $s = 1$, and begin with an easy result:

(3.5) **Proposition.** *There is an Euler product expansion*

$$L_\Lambda(s, \psi; z) = \prod_p L_{\Lambda_p}(s, \psi_p; z), \quad \sigma > 1, |z| \leqq 1,$$

where the local modified L-functions are defined as in (2.1). *This product is absolutely and uniformly convergent on compact subsets of the region* $\sigma > 1$, $|z| \leqq 1$.

Proof. The absolute convergence of the product is determined by that of the series $\sum_p |L_{\Lambda_p}(s, \psi_p; z) - 1|$. Taking absolute values term by term here, we obtain a subsum of the series $L_\Lambda(\sigma, \psi_0, |z|)$, where ψ_0 is the trivial character of $Cl(\Lambda)$. This last series certainly converges well for $\sigma > 1$, $|z| \leqq 1$.

This leaves only the task of proving (3.5) as an identity between formal Dirichlet series. Each left ideal X of Λ in $g(\Lambda)$ determines a collection $\{X_p: p \text{ prime}\}$ of principal left ideals X_p of Λ_p, with $X_p = \Lambda_p$ for all but a finite number of p. Each such collection determines an X. We have $NX = \prod_p NX_p$. Also, Λ/X is a **Z**-torsion module, whose p-primary component $(\Lambda/X)_p$ coincides with Λ_p/X_p. It follows that

$$\Lambda/X = \oplus \Lambda_p/X_p, \quad l(\Lambda/X) = \sum_p l(\Lambda_p/X_p). \tag{3.6}$$

These facts, taken in conjunction with (1.3), prove (3.5). Indeed, this argument is an obvious generalization of the proof of (1.5).

Now let S be chosen as in (1.6). Define

$$L_A^{(S)}(s, \psi; z) = \sum_X \psi(X) NX^{-s} z^{l(A/X)}, \quad \sigma > 1, \tag{3.7}$$

where X ranges over all left Λ-ideals prime to S, as in (1.7). We have already remarked that any X prime to S necessarily lies in $g(\Lambda)$. The convergence properties of (3.7) are clearly the same (or at least as good) as those of (3.1). As in (3.5), we have

$$L_A^{(S)}(s, \psi; z) = \prod_{p \notin S} L_{A_p}(s, \psi_p; z), \quad \sigma > 1, \ |z| \leqq 1. \tag{3.8}$$

Taking principal value logarithms, this yields

$$\log L_A^{(S)}(s, \psi; z) = \sum_{p \notin S} \log L_{A_p}(s, \psi_p; z). \tag{3.9}$$

By (2.11), with $d = \dim_{\mathbf{Q}}(A)$, we may write

$$\log L_{A_p}(s, \psi_p; z) = \sum_{m=1}^{\infty} g_{m,p}(s, \psi_p) z^m, \quad p \notin S,$$

where

$$|g_{m,p}(s, \psi_p)| < dp^{-3m/4}, \quad m \geqq 1, \ \sigma > 1 - (4d)^{-1},$$
$$|g_{1,p}(s, \psi_p)| < dp^{-\sigma}, \quad \sigma > 1.$$

It follows that the double series

$$\sum_{p \notin S} \sum_{m=2}^{\infty} g_{m,p}(s, \psi_p) \tag{3.10}$$

is dominated by the convergent double series

$$\sum_{p \notin S} \sum_{m=2}^{\infty} dp^{-3m/4}$$

in the range $\sigma > 1 - (4d)^{-1}$. Therefore, in this range, all subsums of (3.10) converges absolutely to holomorphic functions of s. This applies in particular to the series

$$h_m(s, \psi) = \sum_{p \notin S} g_{m,p}(s, \psi_p), \quad m \geqq 2,$$

and also to $\sum_{m=2}^{\infty} h_m(s, \psi)$.

On the other hand, we set

$$h_1(s, \psi) = \sum_{p \notin S} g_{1,p}(s, \psi_p),$$

and this converges to a holomorphic function of s for $\sigma > 1$. We can now write (3.8) in the form

$$\log L_A^{(S)}(s, \psi; z) = h_1(s, \psi) z + \sum_{m=2}^{\infty} h_m(s, \psi) z^m, \quad \sigma > 1, \ |z| \leqq 1. \tag{3.12}$$

We now set $z=1$ in (3.12). The left hand side is $\log L_\Lambda^{(S)}(s, \psi)$, and (1.14) gives us

$$h_1(s, \psi)=t_\psi \log \frac{1}{s-1}+G_0(s) \tag{3.13}$$

in the sense of (1.12).

It is now an easy task to estimate the functions $f_m(s, \psi)$ of (3.2), (3.4):

(3.14) **Theorem.** *For each $k \geqq 1$, let $f_k(s, \psi)$ be the coefficient of z^k in $L_\Lambda(s, \psi; z)$. Then we have, for each $k \geqq 1$,*

$$f_k(s, \psi)=\left(t_\psi \log \frac{1}{s-1}\right)^k \Big/ k!+G_{k-1}(s),$$

for some function $G_{k-1}(s)$ as in (1.13).

Proof. By (3.5) and (3.8), we have

$$L_\Lambda(s, \psi; z)=L_{\Lambda}^{(S)}(s, \psi; z) \prod_{p \in S} L_{\Lambda_p}(s, \psi_p; z), \qquad \sigma>1. \tag{3.15}$$

For each $p \in S$, (2.2) allows us to express $L_{\Lambda_p}(s, \psi_p; z)$ as a power series in z with coefficients which are holomorphic for $\sigma>0$. It follows that

$$\prod_{p \in S} L_{\Lambda_p}(s, \psi_p; z)=1+\sum_{m=1}^{\infty} u_m(s, \psi) z^m, \tag{3.16}$$

at least for $\sigma>1$, $|z| \leqq 1$, and each coefficient is holomorphic for $\sigma>0$.

Next, (3.12) and (3.13) give

$$\log L_\Lambda^{(S)}(s, \psi; z)=z t_\psi \log \frac{1}{s-1}+\sum_{m=1}^{\infty} v_m(s, \psi) z^m, \qquad \sigma>1,$$

where each v_m is holomorphic for $\sigma \geqq 1$. Exponentiating, we obtain

$$L_\Lambda^{(S)}(s, \psi; z)=\exp \left(z t_\psi \log \frac{1}{s-1}\right) \exp \left(\sum v_m z^m\right).$$

We may expand both exponentials in powers of z and rearrange terms at will, since all the series involved converge absolutely for $\sigma>1$, $|z| \leqq 1$. This gives

$$L_\Lambda^{(s)}(s, \psi; z)=\left\{\sum_{k=0}^{\infty}\left(z t_\psi \log \frac{1}{s-1}\right)^k \Big/ k!\right\}\left\{1+\sum_{m=1}^{\infty} w_m(s, \psi) z^m\right\} \tag{3.17}$$

for $\sigma>1$, $|z| \leqq 1$, with each $w_m(s, \psi)$ holomorphic for $\sigma \geqq 1$. (Indeed, each w_m is a finite sum of finite products of v_m's, so it is certainly holomorphic in an open set containing $\sigma \geqq 1$.) Now we substitute (3.16), (3.17) into (3.15), and extract the coefficient of z^k. This immediately gives the required estimate for $f_k(s, \psi)$, and proves Theorem 3.14.

We can now state and prove the main result of this section, which is the key to the proof of Theorem B.

(3.18) **Theorem.** *For an integer $k \geqq 1$, and a class $c \in Cl(\Lambda)$,* define

$$\mathsf{N}(k, c) = \sum_{\psi} \psi(c)\, t_{\psi}^{k}, \tag{3.19}$$

where ψ ranges over all characters of the finite abelian group $Cl(\Lambda)$. Then $\mathsf{N}(k, c)$ is real, and

$$\sum_{\substack{X \in I(k, \Lambda) \\ [X] = c}} NX^{-s} = \frac{\mathsf{N}(k, c)}{k!\, h(\Lambda)} \left(\log \frac{1}{s-1}\right)^{k} + G_{k-1}(s), \tag{3.20}$$

where $I(k, \Lambda)$ is defined in (3.3), *and $G_{k-1}(s)$ is of the type* (1.13).

Proof. For a character ψ of $Cl(\Lambda)$, it is clear that $t_{\psi^{-1}} = t_{\psi}$, so certainly $\overline{\mathsf{N}(k, c)} = \mathsf{N}(k, c)$, and $\mathsf{N}(k, c)$ is therefore real. By (3.4), (3.14), we may write

$$\sum_{X \in I(k, \Lambda)} \psi(X)\, NX^{-s} = \frac{t_{\psi}^{k} \left(\log \dfrac{1}{s-1}\right)^{k}}{k!} + G_{k-1}(s). \tag{3.21}$$

Multiply both sides by $\overline{\psi(c)}$ and sum over ψ. Since

$$\sum_{\psi} \overline{\psi(c)}\, \psi(X) = \begin{cases} h(\Lambda) & \text{if } [X] = c, \\ 0 & \text{otherwise,} \end{cases}$$

we obtain the desired formula.

Now we can apply the Delange-Ikehara Tauberian theorem (see Appendix II of [9]) to obtain:

(3.22) **Corollary.** *Let $T > 0$. For an integer $k \geqq 1$, and a class $c \in Cl(\Lambda)$, the number of left Λ-ideals X with*

$$NX \leqq T, \quad X \in g(\Lambda), \quad [X] = c, \quad l(\Lambda/X) = k$$

is asymptotically

$$\left(\frac{\mathsf{N}(k, c)}{(k-1)!\, h(\Lambda)} + o(1)\right) \frac{T(\log \log T)^{k-1}}{\log T}$$

as $T \to \infty$.

The quantity $\mathsf{N}(k, c)$ is real, and clearly it is non-negative because of formula (3.20), but the "explicit" formula (3.19) does not give a clear idea of its dependence on k and c. Most important, it is hard to decide when it is zero and when not. However, a certain amount of information can be extracted at this stage.

First, $\mathsf{N}(k, c)$ is a non-negative *integer.* For, let m be the exponent of the finite abelian group $Cl(\Lambda)$, and let $E = \mathbf{Q}(e^{2\pi i/m})$ be the field of m-th roots of unity. Then every character ψ of $Cl(\Lambda)$ takes values in E. Further, $\mathsf{N}(k, c)$ is an algebraic integer in E. If $\omega \in \mathrm{Gal}(E/\mathbf{Q})$, we have $t_{\psi^{\omega}} = t_{\psi}$ for any ψ, so $\mathsf{N}(k, c)^{\omega} = \mathsf{N}(k, c)$. Thus $\mathsf{N}(k, c)$ is indeed a rational integer.

At this point we may prove Theorem A (given in the introduction). Let A now be a *simple* **Q**-algebra, and ψ a character of $Cl(\Lambda)$. Then $0 \leqq t_\psi \leqq 1$, and $t_\psi = 1$ if and only if ψ is the trivial character. It follows that $\mathsf{N}(k, c) = 1$ for all $k \geqq 1$ and all $c \in Cl(\Lambda)$. Therefore (3.18) and (3.22) give:

(3.23) **Theorem.** *Let A be a simple* **Q**-*algebra, $k \geqq 1$ an integer, and $c \in Cl(\Lambda)$. Let $I(k, \Lambda)$ be the set of left ideals X of Λ with $X \in g(\Lambda)$ and such that Λ/X has composition length k. Let $h(\Lambda) = |Cl(\Lambda)|$. Then*

(i) $$\sum_{\substack{X \in I(k, \Lambda) \\ [X] = c}} NX^{-s} = \left(\log \frac{1}{s-1}\right)^k \Big/ k!\, h(\Lambda) + G_{k-1}(s),$$

where $G_{k-1}(s)$ is a function of type (1.13).

(ii) *Let $T > 0$. The number of $X \in I(k, \Lambda)$ with $NX \leqq T$ and $[X] = c$ is asymptotically equal to*

$$\left(\frac{1}{(k-1)!\, h(\Lambda)} + o(1)\right) \frac{T(\log\log T)^{k-1}}{\log T}$$

as $T \to \infty$.

The special case $k = 1$ of (3.23) gives Theorem A at once.

§ 4. Roiter Ideals

As in § 3, Λ is an order in the semisimple **Q**-algebra A. A *Roiter ideal* of Λ is a left Λ-ideal X such that Λ/X is a direct sum of simple Λ-modules $V_1, \ldots, V_t$, say, such that

$$\text{(4.1)} \qquad \operatorname{ann}_{\mathbf{Z}}(V_i) + \operatorname{ann}_{\mathbf{Z}}(V_j) = \mathbf{Z} \quad \text{if } i \neq j.$$

Here, the annihilator $\operatorname{ann}_{\mathbf{Z}}(V_i)$ is defined to be $\{a \in \mathbf{Z} : aV_i = 0\}$, and this is a prime ideal of **Z**. Equivalently, X is a Roiter ideal if, for each p dividing NX, Λ_p/X_p is a simple Λ-module (=simple Λ_p-module). The terminology is suggested by the proof of Roiter's Theorem (see [4], pp. 660–663).

Our aim here is to show that the Roiter ideals X in $g(\Lambda)$ with $X \in I(k, \Lambda)$, $[X] = c$, already contribute the principal term on the right hand side of formula (3.20). We let S again denote a finite set of "bad" primes as in (1.6), and for each $k \geqq 0$ we define

(4.2) $\qquad R(k) = R(k, \Lambda, S) =$ the set of all Roiter ideals X of Λ, with X prime to S and $l(\Lambda/X) = k$.

We have already remarked that any X prime to S lies in $g(\Lambda)$. We intend to show that

$$\text{(4.3)} \qquad \sum_{\substack{x \in R(k) \\ [X] = c}} NX^{-s} = \frac{\mathsf{N}(k, c)}{k!\, h(\Lambda)} \left(\log \frac{1}{s-1}\right)^k + G_{k-1}(s),$$

for each $k \geqq 1$ and each class $c \in Cl(\Lambda)$. As is clear from the proof of (3.18), it is sufficient to show, for any character ψ of $Cl(\Lambda)$, that we have

(4.4)
$$\sum_{X\in R(k)} \psi(X)NX^{-s} = \left(t_\psi \log \frac{1}{s-1}\right)^k \Big/ k! + G_{k-1}(s)$$

for some function $G_{k-1}(s)$ of type (1.13).

To begin with, we define a new modified L-function. Let

(4.5)
$$R_\infty = \bigcup_{k=0}^{\infty} R(K),$$

so that R_∞ is the set of all Roiter ideals of Λ prime to S. Then we define

(4.6)
$$L^*_\Lambda(s, \psi; z) = \sum_{X\in R_\infty}' \psi(X)NX^{-s}z^{l(\Lambda/X)}, \qquad \sigma>1,\ |z|\leqq 1.$$

Convergence is guaranteed by comparison with $L_\Lambda(s, \psi; z)$. The coefficient of z^k in $L^*_\Lambda(s, \psi; z)$ is precisely the sum in the left hand side of (4.4).

Just as before, there is an Euler product

(4.7)
$$L^*_\Lambda(s, \psi; z) = \prod_{p\notin S} L^*_{\Lambda_p}(s, \psi_p; z),$$

where now

(4.8)
$$L^*_{\Lambda_p}(s, \psi_p; z) = 1 + z\sum_{Y_1} \psi_p(Y_1)NY^{-s}, \qquad p\notin S,$$

where Y_1 ranges over all left Λ_p-ideals with $l(\Lambda_p/Y_1)=1$. Any such Y_1 is principal, since $p\notin S$. In fact, if as in § 3 we write

$$\log L_{\Lambda_p}(s, \psi_p; z) = \sum_{m=1}^{\infty} g_{m,p}(s, \psi_p)\, z^m,$$

then we have the identity

$$L^*_{\Lambda_p}(s, \psi_p; z) = 1 + g_{1,p}(s, \psi_p)z, \qquad p\notin S.$$

However, we have seen that $|g_{1,p}(s, \psi_p)|\leqq dp^{-\sigma}$ for $\sigma>1$, where $d=\dim_{\mathbf{Q}}(A)$, and this establishes the convergence of (4.7). (Incidentally, we can have equality in this estimate of $g_{1,p}$.) We may write

$$\begin{aligned}\log L^*_\Lambda(s, \psi; z) &= \sum_{p\notin S} \log L^*_{\Lambda_p}(s, \psi_p; z)\\ &= zh_1(s, \psi) + \sum_{p\notin S}\sum_{m=2}^{\infty} \frac{(-1)^{m-1}}{m} g_{1,p}(s, \psi_p)^m z^m,\end{aligned}$$

where $h_1(s, \psi) = \sum_{p\notin S} g_{1,p}(s, \psi_p)$, as in § 3. The double series here is absolutely convergent for $\sigma>1-(4d)^{-1}$, $|z|<d^{-1}$, because of the estimate $|g_{1,p}(s, \psi_p)|\leqq dp^{-3/4}$ for $\sigma>1-(4d)^{-1}$. This gives us

(4.9)
$$\log L^*_\Lambda(s, \psi; z) = zh_1(s, \psi) + \sum_{m=2}^{\infty} h^*_m(s, \psi)z^m, \qquad \sigma>1,\ |z|<d^{-1},$$

$$h_m^*(s, \psi) = \frac{1}{m}(-1)^{m-1} \sum_{p \notin S} g_{1,p}(s, \psi)^m.$$

This defines h_m^* as a holomorphic function of s for $\sigma > 1-(4d)^{-1}$. Formula (3.13) gives us an estimate for $h_1(s, \psi)$. We now exponentiate (4.9), as in the proof of (3.14), and rearrange as a power series in z. There are no convergence difficulties in the region $\sigma > 1$, $|z| < d^{-1}$. This proves (4.4), and with it (4.3). To summarize:

(4.10) **Theorem.** *Let $k \geqq 1$, and $c \in Cl(\Lambda)$. Let X range over all left Roiter ideals of Λ which are prime to S and satisfy $l(\Lambda/X) = k$, $[X] = c$. Then*

(i) $\sum_X NX^{-s} = \frac{\mathsf{N}(k, c)}{k!\, h(\Lambda)} \left(\log \frac{1}{s-1}\right)^k + G_{k-1}(s)$, *and*

(ii) *for $T > 0$, the number of such ideals X with $NX \leqq T$ is asymptotically equal to*

$$\left(\frac{\mathsf{N}(k, c)}{(k-1)!\, h(\Lambda)} + o(1)\right) \frac{T(\log \log T)^{k-1}}{\log T}$$

as $T \to \infty$.

Here, (ii) follows from (i) on applying the Delange-Ikehara Tauberian theorem as before.

(4.11) *Remark.* It is an easy exercise to see that the same result holds if the hypothesis "X prime to S" is replaced by "$X \in g(\Lambda)$".

§5. The Main Theorems

We have now proved the version of Theorem B, part (i), stated in the introduction, and even improved it by giving a more precise version of the error term. The first aim of this section is to strengthen it further by giving a more useful formula for the distribution coefficient $\mathsf{N}(k, c)$. We state the final version of the result as Theorem C below. The second aim is to use this new formula for $\mathsf{N}(k, c)$ to prove Theorem D, which is a strengthened version of Theorem B, part (ii).

Let $Cl^*(\Lambda)$ denote the group of characters of $Cl(\Lambda)$. For $k \geqq 1$ and for a class $c \in Cl(\Lambda)$, we have defined

$$\mathsf{N}(k, c) = \sum_{\psi \in Cl^*(\Lambda)} \psi(c)\, t_\psi^k, \tag{5.1}$$

where t_ψ is given by (1.10). Let

$$A = A_1 \oplus \ldots \oplus A_r$$

be the decomposition of A into simple components. For convenience of notation, let $R = \{1, \ldots, r\}$ be the index set for the simple components. Given any non-empty (unordered) subset E of R, we define

$$A_E = \bigoplus_{i \in E} A_i. \tag{5.2}$$

There is a canonical projection $\pi_E\colon A \to A_E$; let $\Lambda_E = \pi_E(\Lambda)$, an order in A_E. Thus there are surjections

$$\pi_E\colon A \to A_E, \quad \Lambda \to \Lambda_E, \quad Cl(\Lambda) \to Cl(\Lambda_E). \tag{5.3}$$

The last of these surjections induces an injection on the character groups

$$\pi_E^*\colon Cl^*(\Lambda_E) \to Cl^*(\Lambda). \tag{5.4}$$

For each $\psi \in Cl^*(\Lambda)$, we define

$$K(\psi) = \{i \in R\colon \psi_i = \text{trivial character of } J(A_i)\}, \qquad \text{so } t_\psi = |K(\psi)|. \tag{5.5}$$

We use a similar definition for $K(\theta)$, when $\theta \in Cl^*(\Lambda_E)$. For each $E \subset R$, we put $E' = R - E$, its complement in R.

Remark. We shall need to consider expressions such as $Cl^*(\Lambda_{E'})$, and it may happen that E' is the empty set $\varnothing$. We now adopt the following conventions:

(i) $Cl(\Lambda_\phi) = \{1\}$, $Cl^*(\Lambda_\phi) = \{1\}$, $\pi_\phi\colon Cl(\Lambda) \to Cl(\Lambda_\phi)$ is the trivial map,

(ii) For $x \in Cl^*(\Lambda_\phi)$, $t_\psi = 0$ and $K(\psi) = \varnothing$.

We are now ready to describe the image of the injection π_E^* defined in (5.4).

(5.6) **Proposition.** *Let E be a subset of R. Then π_E^* induces a canonical bijective homomorphism*

$$Cl^*(\Lambda_E) \leftrightarrow \{\psi \in Cl^*(\Lambda)\colon K(\psi) \supset E'\}.$$

Further,

$$K(\pi_E^* \theta) = K(\theta) \cup E' \quad \textit{for each } \theta \in Cl^*(\Lambda_E). \tag{5.7}$$

Proof. The result is trivially true if either E or E' is empty, by virtue of the preceding remark and the fact that a character ψ of $Cl(\Lambda)$ is trivial if and only if $K(\psi) = R$. For the rest of the proof, assume both E, E' non-empty. It is clear from the definition of π_E^* that (5.7) holds, and that $K(\psi) \supset E'$ for each $\psi \in \operatorname{im} \pi_E^*$. It suffices to prove the converse, so let $\psi \in Cl^*(\Lambda)$ be such that $K(\psi) \supset E'$. We must show that $\psi = \pi_E^* \theta$ for some $\theta \in Cl^*(\Lambda_E)$, or equivalently, that ψ is trivial on the kernel of the surjection $\pi_E\colon Cl(\Lambda) \to Cl(\Lambda_E)$.

We shall use formula (1.2) for both $Cl(\Lambda)$ and $Cl(\Lambda_E)$, in order to determine this kernel. We treat the isomorphism $J(A) \cong J(A_{E'}) \times J(A_E)$ as an identification. Let $c \in Cl(\Lambda)$ be such that $\pi_E(c) = 1$; we must prove that $\psi(c) = 1$. Since $\pi_E(c) = 1$, formula (1.2) shows that c is represented by an idele (x', x) with

$$x' \in J(A_{E'}), \quad x = yzu \in J(A_E) \qquad \text{where } y \in J'(A_E),\ z \in A_E^\times,\ u \in U(\Lambda_E).$$

We may certainly choose elements $y_0 \in J'(A)$, $z_0 \in A^\times$, such that $y_0 = (y', y)$, $z_0 = (z', z)$ in the identification $J(A) = J(A_{E'}) \times J(A_E)$. We also wish to choose $u_0 \in U(\Lambda)$ such that $u_0 = (u', u)$. This will be possible provided we know that the map $U(\Lambda) \to U(\Lambda_E)$ is surjective, and by virtue of the definition of $U(\Lambda)$, it suffices to check this locally. But for each p, the map

$$(\Lambda_p)^\times \to (\Lambda_{p,E})^\times$$

is indeed surjective (see [4], Exercise 5.12). Thus, the required u_0 exists.

But now we use the hypothesis that $K(\psi) \supset E'$, so the restriction of ψ to $J(A_{E'})$ is trivial. Then we obtain

$$\psi(c) = \psi(x', x) = \psi(1, x) = \psi(y' z' u', yzu) = \psi(y_0 z_0 u_0) = 1,$$

as desired. This completes the proof of the proposition.

We single out a special case:

(5.8) **Corollary.** *The map π_E^* gives a canonical bijection*

$$\{\theta \in Cl^*(\Lambda_E) \colon t_\theta = 0\} \leftrightarrow \{\psi \in Cl^*(\Lambda);\ K(\psi) = E'\}.$$

Further, for each $c \in Cl(\Lambda)$, let $c_E = \pi_E(c) \in Cl(\Lambda_E)$. Then for this bijection we have

$$\psi(c) = \theta(c_E).$$

Let us introduce additional notation: for each non-empty subset E of R, put

$$(5.9) \qquad \begin{cases} c_E = \pi_E(c), & c \in Cl(\Lambda);\ h_E = |Cl(\Lambda_E)|, \\ \delta_E = \delta_E(c) = \begin{cases} 1 & \text{if } c_E \text{ is trivial,} \\ 0 & \text{otherwise.} \end{cases} \end{cases}$$

Further, for the case where $E = \varnothing$, define

$$(5.10) \qquad \delta_\phi = 1, \quad h_\phi = 1, \quad c_\phi = 1 \in Cl(\Lambda_\phi) = \{1\}.$$

We wish to calculate $\mathsf{N}(k, c)$, and we remark that when $r = 1$ (and A is simple), we have already seen at the end of § 3 that $\mathsf{N}(k, c) = 1$ for all $k \geqq 1$ and all $c \in Cl(\Lambda)$. To handle the semisimple case, let us keep the class c fixed throughout, and let us set

$$(5.11) \qquad F(c, R) = \sum_{\substack{\psi \in Cl^*(\Lambda) \\ t_\psi = 0}} \psi(c).$$

More generally, for $E \subset R$, define

$$(5.12) \qquad F(c, E) = F(c_E, E) = \sum_{\substack{\theta \in Cl^*(\Lambda_E) \\ t_\theta = 0}} \theta(c_E).$$

Note that $F(c, \varnothing) = 1$. We now prove:

(5.13) **Proposition.** *We have*

$$F(c, R) = \sum_{E \subset R} (-1)^{|R - E|} \delta_E h_E,$$

where E ranges over all subsets of R, and where δ_E, h_E are defined by (5.9) *and* (5.10).

Proof. The result is trivially true for $R=\varnothing$. Next, when $R=\{1\}$, there are just two choices for E, namely $\varnothing$ and $\{1\}$, so

$$\sum_{E\subset R}(-1)^{|R-E|}\,\delta_E h_E=\delta_R h_R-1.$$

On the other hand, for $\psi\in Cl^*(\Lambda)$ we have $t_\psi=1$ if ψ is the trivial character, and $t_\psi=0$ otherwise. This gives

$$F(c,R)=\sum_{\psi\text{ nontrivial}}\psi(c)=-1+\sum_{\text{all }\psi}\psi(c)=-1+\delta_R h_R,$$

so the proposition holds in this case.

Now let $R=\{1,\dots,r\}$, where $r>1$. Then

$$F(c,R)+\sum_{\substack{\psi\in Cl^*(\Lambda)\\ t_\psi>0}}\psi(c)=\sum_{\text{all }\psi}\psi(c)=\delta_R h_R.$$

Further

$$\sum_{t_\psi>0}\psi(c)=\sum_{E\subsetneqq R}\Big\{\sum_{\substack{\psi\in Cl^*(\Lambda)\\ K(\psi)=E'}}\psi(c)\Big\},$$

where $E'=R-E$ as usual. But by (5.8),

$$\sum_{\substack{\psi\in Cl^*(\Lambda)\\ K(\psi)=E'}}\psi(c)=\sum_{\substack{\theta\in Cl^*(\Lambda_E)\\ t_\theta=0}}\theta(c_E)=F(c_E,E).$$

Using the induction hypothesis, we obtain

(5.14)
$$F(c,R)=\delta_R h_R-\sum_{E\subsetneqq R}\Big\{\sum_{F\subset E}(-1)^{|E-F|}\,\delta_F h_F\Big\}.$$

We now show that for each $F\subset R$, the coefficient of $\delta_F h_F$, on the right above, is precisely $(-1)^{|R-F|}$. This will establish the desired formula, and prove the proposition.

For $F=R$, $\delta_R h_R$ appears with coefficient 1 on the right in (5.14), which agrees with $(-1)^{|R-R|}$. Now let $F\subsetneqq R$; then the coefficient of $\delta_F h_F$ on the right in (5.14) is

(5.15)
$$-\sum_E(-1)^{|E-F|},$$

where E ranges over all proper subsets of R containing F. If $|F|=f$, then E is obtaining by adjoining to F any proper subset of the $r-f$ elements of $R-F$. The sum in (5.15) is thus equal to

(5.16)
$$\sum_{n=0}^{r-f-1}(-1)^n\binom{r-f}{n}.$$

Comparing the above with the binomial expansion of $(1-1)^{r-f}$, we see that the sum (5.16) equals $(-1)^{r-f}$. Thus, the coefficient of $\delta_F h_F$ on the right in (5.14) is $(-1)^{|R-F|}$, as claimed, and the proposition is established.

Before stating our explicit formula for $\mathsf{N}(k, c)$, we need some notation from elementary combinatorics. We define the combinatorial expressions $s(k, n)$ for integers $k \geqq 1$, $n \geqq 0$, by the formulas

$$(5.17)\qquad \begin{cases} s(k,0)=0,\ s(k,1)=1,\ s(k,2)=2^k-2.1^k,\ \ldots, \\ s(k,n)=\sum_{m=1}^{n}(-1)^{n-m}\binom{n}{m}m^k, \qquad \text{for } n\geqq 1. \end{cases}$$

The integers $\{s(k, n)/n!\}$ are the *Stirling numbers of the second kind* (see [3] or [10]), and $s(k, n)$ is the number of surjective maps from a set with k elements onto a set with n elements. If D denotes the differentiation operator $\frac{d}{dx}$, it is easily seen that

$$s(k,n)=(xD)^k(x-1)^n|_{x=1}.$$

This implies at once:

(5.18) **Proposition.** *$s(k, n)=0$ if $k<n$, while $s(k, n)$ is a positive multiple of $n!$ if $k \geqq n \geqq 1$.*

We may now state one our main results, which gives an explicit formula for $\mathsf{N}(k, c)$ in terms of these integers $s(k, n)$. As we see at once, the formula shows that $\mathsf{N}(k, c)$ is always a non-negative integer, and that it depends on c only to the extent of knowing which of the homomorphic images $\{c_E\colon E \subset R\}$ are trivial. In terms of the notation (5.9), we have:

(5.19) **Theorem.** *Let $k \geqq 1$ and let $c \in Cl(\Lambda)$. Then the distribution coefficient $\mathsf{N}(k, c)$ defined in (5.1) is given by*

$$(5.20)\qquad \mathsf{N}(k,c)=\sum_{E\subset R} s(k, r-|E|)\,\delta_E h_E,$$

where $R=\{1, \ldots, r\}$, and where E ranges over all subsets of R, including the empty set $\varnothing$.

Proof. The summand corresponding to $E=R$ is $s(k,0)\delta_R h_R$, which is zero; thus, we could as well let E range over all proper subsets of R, if desired. We remark also that the formula holds automatically when $r=1$, since then both sides of (5.20) have the value 1.

We now treat the general case. Let $E'=R-E$ as usual. The characters $\psi \in Cl^*(\Lambda)$ for which $t_\psi=0$ contribute nothing to the sum (5.1) for $\mathsf{N}(k, c)$. Therefore

$$\begin{aligned} \mathsf{N}(k,c) &= \sum_{E\subsetneqq R}\Big\{\sum_{\substack{\psi\in Cl^*(\Lambda)\\ K(\psi)=E'}}\psi(c)\,|E'|^k\Big\} \\ &= \sum_{E\subsetneqq R}\Big\{\sum_{\substack{\theta\in Cl^*(\Lambda_E)\\ t_\theta=0}}\theta(c_E)\Big\}\,|E'|^k \qquad \text{by (5.8)} \\ &= \sum_{E\subsetneqq R}\Big\{\sum_{F\subset E}(-1)^{|E|-|F|}\delta_F h_F\Big\}(r-|E|)^k \qquad \text{by (5.13).} \end{aligned}$$

For fixed $F \subsetneqq R$, the coefficient of $\delta_F h_F$ in the above is (compare with the proof of (5.16)!):

$$\sum_{\substack{E \supset F \\ E \subsetneqq R}} (r-|E|)^k (-1)^{|E|-|F|} = \sum_{n=0}^{r-f-1} (-1)^n \binom{r-f}{n} (r-n-f)^k,$$

where $f=|F|$. The right hand expression is precisely $s(k, r-|F|)$, and the theorem is established.

Taking into account Theorems 3.18, 5.19, and the remarks at the end of § 3, we have now established the following strengthened version of Theorem B, part (i):

Theorem C. *Let Λ be an order in a semisimple* **Q**-*algebra A with r simple components $A_1, \ldots, A_r$. Given an integer $k \geqq 1$ and a class $c \in Cl(\Lambda)$, let X range over all left ideals of Λ such that*

$$X \in g(\Lambda), \quad [X]=c, \quad l(\Lambda/X)=k.$$

For each subset E of $\{1, \ldots, r\}$, let Λ_E be the projection of Λ into $\bigoplus_{i \in E} A_i$, and let $c_E \in Cl(\Lambda_E)$ be the image of c, with the convention that $Cl(\Lambda_\phi)=\{1\}$. Let

$$\delta_E = \begin{cases} 1, & c_E=1 \\ 0, & \text{otherwise} \end{cases}, \quad h_E=|Cl(\Lambda_E)|,$$

and define the integers $s(k, n)$ by (5.17). Set

$$\mathsf{N}(k, c)=\sum_E s(k, r-|E|)\, h_E \delta_E,$$

a non-negative integer. Then we have

(i) *For $\sigma > 1$,*

$$\sum_X NX^{-s} = \frac{\mathsf{N}(k, c)}{h(\Lambda) k!} \left(\log \frac{1}{s-1}\right)^k + \sum_{j=0}^{k-1} f_j(s) \left(\log \frac{1}{s-1}\right)^j,$$

where each $f_j(s)$ is holomorphic in an open region containing the closed half-plane $\sigma \geqq 1$.

(ii) *Let $T>0$. Then the number of such ideals X, for which $NX \leqq T$, is asymptotically equal to*

$$\left(\frac{\mathsf{N}(k, c)}{(k-1)!\, h(\Lambda)} + o(1)\right) \frac{T(\log \log T)^{k-1}}{\log T}$$

as $T \to \infty$.

As remarked earlier, (ii) follows from (i) by use of the Delange-Ikehara Tauberian theorem of [9], Appendix II.

We must still establish part (ii) of Theorem B, showing that the result is best possible. Before doing so, however, we list some obvious consequences of formula (5.20) for $\mathsf{N}(k, c)$.

(5.21) **Corollary.** *Let* $k \geqq 1$ *and let* $c \in Cl(\Lambda)$.

(i) $\mathsf{N}(k,c)$ *is always a non-negative integer.*

(ii) $\mathsf{N}(k,c)>0$ *if and only if there is a subset E of R such that*

$$c_E=1 \text{ in } Cl(\Lambda_E), \quad \text{and} \quad k+|E|\geqq r.$$

(iii) *If* $k\geqq r$, *then* $\mathsf{N}(k,c)>0$ *for all* $c\in Cl(\Lambda)$.

(iv) *If* c_0 *is the trivial class in* $Cl(\Lambda)$, *then* $\mathsf{N}(c_0,k)>0$ *for all* $k\geqq 1$.

Proof. All of the terms occurring on the right hand side of (5.20) are non-negative integers. Further, $s(k,r-|E|)>0$ if and only if $k\geqq r-|E|>0$. Also, $\delta_E h_E>0$ if and only if c_E is trivial in $Cl(\Lambda_E)$. This proves (i) and (ii). For (iii), note that $E=\varnothing$ satisfies the conditions in (ii) whenever $k\geqq r$. Finally, (iv) is clear from (ii).

(5.22) **Corollary.** *Let* $k\geqq 1$, *and let* c_0 *be the trivial class in* $Cl(\Lambda)$. *Then*

$$\mathsf{N}(k,c_0)\geqq \mathsf{N}(k,c) \quad \text{for all } c\in Cl(\Lambda).$$

Furthermore, for $c\in Cl(\Lambda)$, *equality holds if and only if* c_E *is trivial for every* $(r-1)$-*tuple E chosen from* $\{1,\dots,r\}$.

Proof. The first assertion is obvious from (5.20) and (5.21). For the second assertion, merely use the fact that if c_E is trivial, then so is c_F for each $F\subset E$.

Finally, we prove a sharper version of part (ii), Theorem B, showing that Theorem C is best possible.

Theorem D. *Let* Λ *be an order in a semisimple* **Q**-*algebra A, as in Theorem C. Let* $k\geqq 1$ *and* $c\in Cl(\Lambda)$ *be given, and let* $\mathsf{N}(k,c)$ *be the "distribution coefficient" given by* (5.1) *and* (5.20). *Then*

a) *If* $\mathsf{N}(k,c)>0$, *there are infinitely many Roiter ideals*[1] *X of* Λ *such that*

$$X\in g(\Lambda), \quad [X]=c, \quad l(\Lambda/X)=k.$$

Indeed, for $T>0$, *the number of such ideals X, for which* $NX\leqq T$, *is asymptotically equal to*

$$\left(\frac{\mathsf{N}(k,c)}{(k-1)!\,h(\Lambda)}+o(1)\right)\cdot\frac{T(\log\log T)^{k-1}}{\log T}$$

as $T\to\infty$. *Moreover, the same result holds if we restrict further to ideals which are prime to any preassigned finite set* S_1 *of prime numbers.*

b) *If* $\mathsf{N}(k,c)=0$, *there are no left ideals* $Y\in g(\Lambda)$ *such that* $[Y]=c$ *and* $l(\Lambda/Y)=k$.

Proof. Part (a) has already been proved in Theorem 4.12(ii). To prove (b), suppose that $\mathsf{N}(k,c)=0$. Let X be any left ideal of Λ, not necessarily a Roiter ideal, and suppose that $X\in g(\Lambda)$ and $[X]=c$. We shall show that $l(\Lambda/X)>k$, which will establish part (b) of the Theorem.

---

[1] As in §4, a left ideal X of Λ is a *Roiter ideal* if Λ/X is a direct sum of simple Λ-modules with relatively prime **Z**-annihilators.

Let us define

$$m=\mathrm{Max}\{|E|: E\subset R, \delta_E(c)=1\},$$

where $R=\{1,\ldots,r\}$. Since $\mathsf{N}(k,c)=0$, it follows from (5.21) that

$$m\leqq r-1 \quad \text{and} \quad k<r-m.$$

We intend to prove:

$$X\in g(\Lambda), \qquad [X]=c\Rightarrow l(\Lambda/X)\geqq r-m,$$

which gives the desired result.

For each subset E of R, let $X_E=\pi_E(X)$, the projection of X into the algebra A_E. Then $X_E\in g(\Lambda_E)$, and we have $[X_E]=c_E$ in $Cl(\Lambda_E)$. In particular, if E is a singleton set $\{i\}$, we write Λ_i, c_i, X_i instead of Λ_E, c_E, X_E. By Proposition 5.24 below, we have

(5.23) $$l(\Lambda/X)=\sum_{i=1}^{r} l(\Lambda_i/X_i).$$

Now let $E\subset R$, $|E|\geqq m+1$. Then c_E is non-trivial, so $X_E\neq\Lambda_E$. Therefore

$$1\leqq l(\Lambda_E/X_E)=\sum_{i\in E} l(\Lambda_i/X_i)$$

by (5.23). Thus $l(\Lambda_i/X_i)\geqq 1$ for some $i\in E$. It follows that $l(\Lambda_i/X_i)=0$ for at most m values of i in the set R. But then $l(\Lambda/X)\geqq r-m$ by (5.23), which completes the proof of the Theorem.

It remains for us to establish

(5.24) **Proposition.** *Let $A=A_1\oplus\ldots\oplus A_r$, and let Λ_i be the projection of Λ into A_i. Let $X\in g(\Lambda)$ be a left ideal of Λ, and let X_i be the projection of X in Λ_i, so $X_i\in g(\Lambda_i)$. Then*

$$l(\Lambda/X)=\sum_{i=1}^{r} l(\Lambda_i/X_i).$$

Proof. It suffices to prove the result for the case $r=2$, if we assume the summands are semisimple algebras instead of simple algebras. Further, by (3.6) we can compute composition lengths by first localizing, and reducing to the situation where X is a principal left ideal.

Changing notation, let $B=B_1\oplus B_2$, with B_i semisimple $\mathbf{Q}_p$-algebras, and let Δ be a $\mathbf{Z}_p$-order in B. Let π_i: $B\to B_i$ be the projection map $(i=1,2)$, and let $\Delta_i=\pi_i(\Delta)$, so there is an inclusion $\Delta\subset\Delta_1\oplus\Delta_2=\Delta'$, with Δ' another order in B. Let X be a left ideal of Δ of the form $X=\Delta x$, with $x\in B^\times$. If we write $x=(x_1,x_2)$, with $x_i\in\Delta_i\cap B_i$, then $\pi_i(X)=X_i=\Delta_i x_i$ for $i=1,2$, and we are now trying to prove that

(5.25) $$l(\Delta/\Delta x)=l(\Delta_1/\Delta_1 x_1)+l(\Delta_2/\Delta_2 x_2).$$

Note that $\Delta_i/\Delta_i x_i$ is a left Δ-module by means of the surjection $\Delta\to\Delta_i$, and that the Δ-composition length of $\Delta_i/\Delta_i x_i$ coincides with its Δ_i-composition length. Now

$$\Delta'/\Delta' x \cong \Delta_1/\Delta_1 x_1 \oplus \Delta_2/\Delta_2 x_2$$

as Δ-modules, so to establish (5.25) and the proposition, it will suffice to prove that

$$l(\Delta/\Delta x) = l(\Delta'/\Delta' x). \tag{5.26}$$

However, there is a left Δ-isomorphism $\Delta' \cong \Delta' x$, given by right multiplication by x. This isomorphism carries Δ onto Δx, and induces a Δ-isomorphism $\Delta'/\Delta \cong \Delta' x/\Delta x$. Consequently

$$l(\Delta'/\Delta) = l(\Delta' x/\Delta x). \tag{5.27}$$

On the other hand,

$$l(\Delta'/\Delta' x) + l(\Delta' x/\Delta x) = l(\Delta'/\Delta x) = l(\Delta'/\Delta) + l(\Delta/\Delta x).$$

But then using (5.27), we obtain the desired equality (5.26). This completes the proof of the proposition.

§6. Arbitrary Lattices

We now investigate the behavior of sublattices of an arbitrary left Λ-lattice M (not necessarily an ideal of Λ). Throughout this section, let

$$\Gamma = \mathrm{End}_\Lambda(M), \qquad B = \mathrm{End}_A(\mathbf{Q} \otimes_{\mathbf{Z}} M), \tag{6.1}$$

so Γ is a $\mathbf{Z}$-order in the semisimple finite dimensional $\mathbf{Q}$-algebra B.

Let $J(B)$ be the idele group of B. Each Λ-lattice X in the genus $g(M)$ is isomorphic to a lattice of the type $M\alpha$, where $\alpha = (\alpha_p) \in J(B)$, and where $M\alpha$ is the Λ-module in $\mathbf{Q} \otimes_{\mathbf{Z}} M$ defined by the conditions

$$(M\alpha)_p = M_p \alpha_p$$

for each prime p. Then $\Gamma\alpha$ is a left Γ-lattice in B, and $\Gamma\alpha \in g(\Gamma)$. The correspondence

$$M\alpha \in g(M) \leftrightarrow \Gamma\alpha \in g(\Gamma)$$

preserves isomorphisms and inclusions. It also preserves stable isomorphism, and induces an isomorphism

$$\mu\colon\ Cl(M) \cong Cl(\Gamma), \tag{6.2}$$

where $\mu[M\alpha] = [\Gamma\alpha]$, $\alpha \in J(B)$.

For our later purposes, it would be convenient to know that the Λ-composition length $l_\Lambda(M/M\alpha)$ coincides with the Γ-composition length $l_\Gamma(\Gamma/\Gamma\alpha)$, assuming α is an integral idele. This need not be true, however. The difficulty arises from the fact that Λ-modules between M and $M\alpha$ need not lie in $g(M)$, and the bijection described above no longer applies to such modules.

To overcome this difficulty, let S be a finite set of primes chosen as in §1. We now have

(6.3) **Proposition.** *Let X range over all Λ-submodules of M such that $(M:X)$ is prime to S, and let X' range over all left ideals of Γ such that $(\Gamma:X')$ is prime to S. Then $X \in g(M)$, $X' \in g(\Gamma)$, and the bijection $X \leftrightarrow X'$ gives an equality*

$$l_\Lambda(M/X) = l_\Gamma(\Gamma/X').$$

Furthermore, if X' is a Roiter ideal of Γ, then X is a Roiter submodule of M (that is, M/X is a direct sum of simple left Λ-modules, whose **Z***-annihilators are pairwise coprime.)*

Proof. If $(M:X)$ is prime to S, then for each $p \in S$ we have $X_p \cong M_p$, so $X \in g(M)$. If $X \subset Y \subset M$, where Y is also a Λ-lattice, then $(M:Y)$ is prime to S, and $Y \in g(M)$. The assertions of the proposition are now obvious from the bijection $g(M) \leftrightarrow g(\Gamma)$ described above.

From this result, we obtain

(6.4) **Drozd-Jacobinski Theorem.** *Let M be an arbitrary left Λ-lattice, and suppose that the algebra B defined in* (6.1) *satisfies the Eichler condition. Let q be the number of simple components of B, and let N be a given Λ-lattice in $g(M)$. Then M contains infinitely many Λ-sublattices X such that $X \cong N$ and M/X is a direct sum of at most q simple left Λ-modules, whose* **Z***-annihilators are pairwise coprime.*

Proof. Since B satisfies the Eichler condition (by hypothesis), the stable isomorphism classes in $g(\Gamma)$ coincide with ordinary isomorphism classes. Choose $c' \in Cl(\Gamma)$ to be the image of the class $[N] \in g(M)$, where N is given. By (5.21) and Theorem D, there exist infinitely many Roiter ideals X' of Γ such that $(\Gamma:X')$ is prime to S, $l_\Gamma(\Gamma/X) \leqq q$, and $[X]=c'$. The desired conclusion then follows from (6.3).

The above result may be rephrased thus: under the hypotheses of the theorem, for each $N \in g(M)$ there are infinitely many embeddings θ: $N \to M$ for which $M/\theta(N)$ is a direct sum of at most q simple modules, with pairwise coprime annihilators. Roiter's conjecture was originally stated in this form.

We now use our results from §5 to sharpen the preceding. To begin with, we need the analogue of the non-negative integers $\mathsf{N}(k, c)$. Let

$$B = B_1 \oplus \ldots \oplus B_q$$

be the decomposition of B into simple components, and let $\{e_i\}$ be the corresponding primitive central idempotents. Given $c' \in Cl(\Gamma)$ and a subset E of the index set $\{1, \ldots, q\}$, let Γ_E be the projection of Γ into B, where $B_E = \bigoplus_{i \in E} B_i$, and let $c'_E \in Cl(\Gamma_E)$ be the image of c' under the surjection $Cl(\Gamma) \to Cl(\Gamma_E)$. We may then define δ_E, h_E, etc., as in §5, and can then define $\mathsf{N}(k, c')$ as in (5.1) and (5.20). We now give a sharper version of (6.4):

(6.5) **Theorem.** *Let M be an arbitrary left Λ-lattice, and let $\Gamma = \mathrm{End}_\Lambda(M)$. Given a class $c \in Cl(M)$, let c' be its image in $Cl(\Gamma)$ under the isomorphism μ in* (6.2). *Define $\mathsf{N}(k, c')$ as in the discussion above. Then*

a) *If* $\mathsf{N}(k, c')>0$, *there are infinitely many* Λ*-lattices* X *contained in* M *such that* $X \in g(M)$, $[X]=c$, *and* M/X *is a direct sum of* k *simple left* Λ*-modules with pairwise coprime annihilators.*

b) *The result is best possible, in the sense that if* $\mathsf{N}(k, c')=0$, *there are no* Λ*-sublattices* X *of* M *such that*

$$X \in g(M), \quad [X]=c, \quad l_\Lambda(M/X) \leqq k.$$

Proof. Part (a) follows at once from Theorem D and the bijection given in (6.3). It remains to prove (b) above. For each subset E of the index set $\{1, \ldots, q\}$, where q is the number of simple components of B, define

$$M_E = \{\sum_{i \in I} x e_i : x \in M\}.$$

Then $\Gamma_E = \operatorname{End}_\Lambda(M_E)$, and the surjection $M \to M_E$ induces a surjection $Cl(M) \to Cl(M_E)$. There is a commutative diagram

$$\begin{array}{rccc} \mu: & Cl(M) & \longrightarrow & Cl(\Gamma) \\ & \downarrow & & \downarrow \\ \mu_E: & Cl(M_E) & \longrightarrow & Cl(\Gamma_E), \end{array}$$

where the μ's are isomorphisms. If $c \in Cl(M)$ maps onto $c_E \in Cl(M_E)$, it follows that we can compute $\mathsf{N}(k, c')$ by using the classes $\{c_E\}$ in place of the classes $\{c'_E\}$.

Next, we observe that for $X \in g(M)$, the analogue of (5.23) remains valid, namely,

$$l_\Lambda(M/X) = \sum_{i=1}^{q} l_\Lambda(M_i/X_i).$$

In fact, the proof of (5.24) carries over readily to the present situation. The rest of the argument is almost identical with that used to prove Theorem D, part (b), and we omit it. This finishes the proof of Theorem 6.5.

We may hope to extend our earlier results so as to estimate $\sum (M:X)^{-s}$, where X ranges over various collections of submodules of the Λ-lattice M. The difficulties here are largely notational, and arise from the fact that in the bijection given in (6.3), there is no simple relation connecting $(M:M\alpha)$ and $(\Gamma:\Gamma\alpha)$. One way of avoiding this difficulty is to restrict one's attention to the case where A is a simple $\mathbf{Q}$-algebra, in which case $(M:M\alpha)$ is a power $(\Gamma:\Gamma\alpha)^t$, with an exponent t independent of α. On the other hand, we may also choose A to be any semisimple algebra, and restrict M to be a Λ-sublattice of a free A-module. Explicit formulas can then be obtained, in the same manner as in [2], §8, but we shall not carry out this routine calculation here.

§7. Further Remarks

One can also generalize the main results of the paper in a slightly different direction from §6. This leads to some interesting phenomena, which we now sketch. We take a left ideal M of Λ, of finite index $N(M)$ in Λ, as always, and consider the distribution of *all* Λ-ideals in $g(M)$, rather than just those contained in M itself.

Indeed, we let X range over all left Λ-ideals such that

$$X \in g(M), \quad l(\Lambda/X)=k, \quad [X]=c, \tag{7.1}$$

for some fixed integer $k \geqq 1$, and some fixed stable isomorphism class c in the genus class group $Cl(M)$. We try to estimate the behavior of the function

$$\sum_X NX^{-s}, \quad \text{where} \quad NX=(\Lambda : X), \tag{7.2}$$

near $s=1$. As in the standard case, we may introduce a modified L-function

$$L_\Lambda(M,s,\psi;z)=\sum_{X\in g(M)} \psi(X)\cdot NX^{-s}\cdot z^{l(\Lambda/X)}, \qquad \sigma>1, \tag{7.3}$$

for any character ψ of $Cl(M)$. We let

$$m=\inf\{n: \text{there exists } X\in g(M) \text{ with } l(\Lambda/X)=n\}. \tag{7.4}$$

Then we may rearrange (7.3) to give

$$L_\Lambda(M,s,\psi;z)=\sum_{k=m}^{\infty} f_k(M,s,\psi)z^k, \tag{7.5}$$

where

$$f_k(M,s,\psi)=\sum_{\substack{X\in g(M)\\ l(\Lambda/X)=k}} \psi(X)NX^{-s}, \qquad \sigma>1. \tag{7.6}$$

As before, we can deduce the behavior of the function (7.2) from that of $f_k(M,s,\psi)$.

We may again view ψ as a character of $J(A)$, and form local L-functions

$$L_{\Lambda_p}(M_p,s,\psi_p;z)=\sum_{Y\cong M_p} \psi_p(Y)\cdot NY^{-s}\cdot z^{l(\Lambda_p/Y)}, \tag{7.7}$$

where $\psi_p(Y)=\psi_p(y)$, for any $y\in A_p^\times$ with $M_p y=Y$. We put

$$m(p)=\inf\{n: \text{there exists } Y\cong M_p \text{ with } l(\Lambda_p/Y)=n\}, \tag{7.8}$$

so that

$$m=\sum_p m(p). \tag{7.9}$$

Of course, $m(p)=0$ for all but a finite number of p. We may rearrange (7.7) to give

$$L_{\Lambda_p}(M_p,s,\psi_p;z)=\sum_{k=m(p)}^{\infty} u_k(M_p,s,\psi_p)z^k, \tag{7.10}$$

and each u_k is a polynomial in p^{-s}. Again we have an Euler product

(7.11) $$L_{\Lambda}(M,s,\psi;z)=\prod_{p} L_{\Lambda_p}(M_p,s,\psi_p;z).$$

We take a finite set S of bad primes satisfying (1.6), and such that also

(7.12) $$M_p=\Lambda_p \quad \text{for all } p\notin S.$$

Thus $m(p)=0$ for $p\notin S$. We define $L_{\Lambda}^{(S)}(M,s,\psi;z)$ in the obvious way, and it still has the crucial property

(7.13) $$\log L_{\Lambda}^{(S)}(M,s,\psi;z)=t_{\psi}\log\frac{1}{s-1}+G_0(s).$$

We now proceed exactly as in the proof of (3.14) to obtain

(7.14) $$f_k(M,s,\psi)=\frac{1}{(k-m)!}\left(t_{\psi}\log\frac{1}{s-1}\right)^{k-m}\prod_{p\in S}u_{m(p)}(M_p,s,\psi_p)+G_{k-m-1}(s).$$

The functions $u_{m(p)}(M_p,s,\psi_p)$ are not identically zero, but it is of some importance here to know whether they vanish at $s=1$. We can say nothing about this. Nevertheless, we now get an estimate for the function (7.2), namely,

(7.15) $$\sum_{\substack{X\in g(M)\\ [X]=c,\, l(\Lambda/X)=k}} NX^{-s}=\frac{\mathsf{N}(M,k,c,s)}{(k-m)!\,|Cl(M)|}\left(\log\left(\frac{1}{s-1}\right)\right)^{k-m}+G_{k-m-1}(s),$$

where

(7.16) $$\mathsf{N}(M,k,c,s)=\sum_{\psi\in Cl^*(M)}\{\bar{\psi}(c)t_{\psi}^{k-m}\prod_{p\in S}u_{m(p)}(M_p,s,\psi_p)\}.$$

If we knew that $\mathsf{N}(M,k,1,c)\neq 0$, we could apply the Tauberian Theorem as before.

Despite all this uncertainty, there is one useful consequence, already mentioned in the introduction.

(7.17) **Proposition.** *Let X range over all left Λ-ideals such that $X\notin g(\Lambda)$, $l(\Lambda/X)=k$, for some fixed integer k. Then*

$$\sum_{X} NX^{-s}=G_{k-1}(s).$$

Remark. By way of contrast, let ψ_0 be the trivial character of $Cl(\Lambda)$, and take $\psi=\psi_0$ in (3.14). This gives

$$\sum_{\substack{X\in g(\Lambda)\\ l(\Lambda/X)=k}} NX^{-s}=r^k\left(\log\left(\frac{1}{s-1}\right)\right)^k\Big/k!+G_{k-1}(s),$$

where we recall that r is the number of simple components of the algebra A.

Proof of (7.17). There are only finitely many genera of left ideals of Λ (of finite index in Λ). We take (7.14) with ψ the trivial character of $Cl(M)$, and sum over a set or representatives of the genera, excluding $g(\Lambda)$. If $g(M)\neq g(\Lambda)$, the integer m defined by (7.4) is $\geqq 1$. The result follows.

References

1. Bushnell, C.J., Reiner, I.: Zeta functions of arithmetic orders and Solomon's conjectures. Math. Z. **173**, 135–161 (1980)
2. Bushnell, C.J., Reiner, I.: *L*-functions of arithmetic orders and asymptotic distribution of ideals. J. Reine Angew. Math. **327**, 156–183 (1981)
3. Comtet, L.: Advanced Combinatorics. Dordrecht: Reidel 1974
4. Curtis, C.W., Reiner, I.: Methods of Representation Theory, Vol. I. New York: Wiley 1981
5. Drozd, Yu.A.: The distribution of maximal sublattices. Mat. Zametki **6**, 19–24 (1969) [Russian]. Engl. Transl.: Math. Notes **6**, 469–471 (1969)
6. Fröhlich, A.: Locally free modules over arithmetic orders, J. Reine Angew. Math. **274/275**, 112–138 (1975)
7. Jacobinski, H.: On embedding of lattices belonging to the same genus. Proc. Amer. Math. Soc. **24**, 134–136 (1970)
8. Lang, S.: Algebraic Number Theory. Reading, Massachusetts: Addison-Wesley 1970
9. Narkiewicz, W.: Elementary and Analytic Theory of Algebraic Numbers. Warsaw: Polish Scientific Publishers 1974
10. Tomescu, I.: Introduction to Combinatorics. London: Collet 1975

Received March 3, 1982

LEFT-VS-RIGHT ZETA-FUNCTIONS

By COLIN J. BUSHNELL *and* IRVING REINER*

[Received 18th April 1983]

WE suppose to begin with that Λ is a $\mathbb{Z}$-order in a finite-dimensional semisimple $\mathbb{Q}$-algebra A. In earlier articles [**1**]–[**3**], we developed the theory of Solomon's zeta-function

$$\zeta_\Lambda(s) = \sum_X (\Lambda\colon\ X)^{-s}, \qquad \mathrm{Re}\,(s) > 1,$$

where s is a complex variable and X ranges over all left ideals of Λ such that the index $(\Lambda\colon\ X)$ of X in Λ is finite. As originally shown in [**5**], there is an Euler product expansion

$$\zeta_\Lambda(s) = \prod_p \zeta_{\Lambda_p}(s), \qquad \mathrm{Re}\,(a) > 1,$$

where p ranges over all prime numbers and Λ_p is the p-adic completion $\Lambda \otimes_{\mathbb{Z}} \mathbb{Z}_p$ of Λ at p. The local zeta-functions $\zeta_{\Lambda_p}(s)$ are defined in the same way as the global one $\zeta_\Lambda(s)$. If Λ' is a maximal order in A, we know from [**1**] that the quotient $\phi_p(s) = \zeta_{\Lambda_p}(s)/\zeta_{\Lambda'_p}(s)$ is a polynomial in p^{-s} with rational integer coefficients. Moreover, $\phi_p(s) = 1$ whenever Λ_p is a maximal order, and this holds for all but a finite number of primes p.

The local zeta-functions $\zeta_{\Lambda'_p}(s)$ are known explicitly (see [**1**], § 3.3), so that in general the calculation of $\zeta_\Lambda(s)$ reduces to a finite number of local problems. Indeed, the fact that $\phi_p(s) \in \mathbb{Z}[p^{-s}]$ reduces the local problem to finding a finite number of coefficients.

We therefore turn to the local case and let Λ be a $\mathbb{Z}_p$-order spanning a finite-dimensional semisimple $\mathbb{Q}_p$-algebra A. In [**1**], we gave examples of calculations of $\zeta_\Lambda(s)$. These were based on decomposing $\zeta_\Lambda(s)$ as a finite sum of "partial" zeta-functions, each of which only counts ideals in a fixed isomorphism class. Various (basically analytic) methods can then be used to calculate the partial zeta-functions. See [**2**] and [**6**] for other calculations.

In this paper, we develop an elementary method for calculating the first few terms of the total zeta-function $\zeta_\Lambda(s)$. The method rapidly becomes unmanageable as one increases the desired number of terms, but it is very efficient initially. In particular, it gives $\zeta_\Lambda(s)$, when Λ is the group ring

* The research for this paper was performed while the second author was visiting King's College London. He wishes to thank King's College for its hospitality, and also the SERC and the Center for Advanced Study of the University of Illinois for their support.

$\mathbb{Z}_pG$ of a dihedral group G of order $2p$, much more quickly than our earlier calculation in [**2**]. (The functional equation described in [**1**] is of assistance here.)

Our main application, however, concerns questions of left-right symmetry. *Changing notation*, we can attach *two* total zeta-functions (a left one and a right one) to our order Λ (in either the local or the global case):

$$\zeta_\Lambda^l(s) = \sum_{\Lambda^X} (\Lambda\colon X)^{-s}, \tag{0.1}$$

$$\zeta_\Lambda^r(s) = \sum_{Y_\Lambda} (\Lambda\colon Y)^{-s}, \tag{0.2}$$

where X runs over the left ideals of Λ, and Y over the right ideals of Λ (always of finite index in Λ). One has $\zeta_\Lambda^l(s) = \zeta_\Lambda^r(s)$ in all examples previously computed. In particular, it holds if Λ is commutative, or maximal (even hereditary), or a group ring. The last follows from the existence of an anti-automorphism for a group ring. (We shall omit the superscripts l, r in such cases.) We produce in this paper a $\mathbb{Z}_p$-order Λ for which $\zeta_\Lambda^l(s) \neq \zeta_\Lambda^r(s)$. It is then a simple matter to construct a $\mathbb{Z}$-order with the same property.

We note before starting that, in this context, we need only show $\zeta_\Lambda^l(s) \neq \zeta_\Lambda^r(s)$ as *formal* Dirichlet series, since the two series have a common region of convergence and consequently cannot represent the same function unless they are identical (see [**7**], § 9.6). Indeed, in the local case, we need only invoke the corresponding property of power series, because then the zeta-functions are power series in p^{-s}.

Notation. For the remainder of this article, we use the following notation:

K = a finite extension of the field $\mathbb{Q}_p$ of p-adic numbers;
R = the valuation ring in K;
P = the maximal ideal of R;
$\bar{R} = R/P$; $q = |\bar{R}|$;
A = a finite-dimensional semisimple K-algebra;
Λ = an R-order in A;
Λ' = a maximal order in A containing Λ;
J = the Jacobson radical of Λ;
$\bar{\Lambda} = \Lambda/J$;
$\bar{X} = X/JX$, for any left Λ-lattice X.

§ 1. Maximal ideals

Using this local notation, we calculate the contribution to $\zeta_\Lambda^l(s)$ arising from the *maximal* left ideals M of Λ, i.e.

$$\sum_M (\Lambda\colon M)^{-s}.$$

Let $\{V_1, \ldots, V_n\}$ be a basic set of simple left Λ-modules, that is, a full set of representatives of the n distinct isomorphism classes of simple left Λ-modules. Each V_i is a finite $\bar{\Lambda}$-module, and

$$E_i = \mathrm{End}_{\Lambda}(V_i) \tag{1.1}$$

is a finite field containing $\bar{R}$. We put

$$q_i = |E_i|. \tag{1.2}$$

Let m_i be the multiplicity of V_i in $\bar{\Lambda}$, so that

$$\bar{\Lambda} \simeq \coprod_{i=1}^{n} V_i^{(m_i)} \qquad \text{(as left } \Lambda\text{-module)}, \tag{1.3}$$

and

$$\bar{\Lambda} \simeq \prod_{i=1}^{n} M_{m_i}(E_i) \quad \text{(as ring)}. \tag{1.4}$$

If M is a maximal left ideal of Λ, then Λ/M is isomorphic to some V_i. For each i, we now count the set of maximal ideals M with $\Lambda/M \simeq V_i$.

(1.5) LEMMA. *Let L be a left Λ-lattice, and let*

$$\bar{L} = L/JL \simeq \coprod_{i=1}^{n} V_i^{(k_i)}.$$

Then, for fixed i, the number of distinct maximal sublattices X of L such that $L/X \simeq V_i$ is given by

$$(q_i^{k_i} - 1)/(q_i - 1),$$

where q_i is defined by (1.2).

Proof. We show first that the set of X with $L/X \simeq V_i$ is in bijection with the set of maximal E_i-subspaces of $E_i^{(k_i)}$.

The radical J of Λ annihilates each V_j, so if L/X is simple, then X contains JL. Thus the lattices X with $L/X \simeq V_i$ correspond bijectively with the submodules Y of $\bar{L}$ such that $\bar{L}/Y \simeq V_i$. Each such Y contains $\coprod_{j \neq i} V_j^{(k_j)}$, so these modules Y correspond bijectively with the maximal submodules of $V_i^{(k_i)}$. However, Λ acts on V_i via a surjection $\Lambda \to \mathrm{End}_{E_i}(V_i)$, and tensoring with V_i induces a Morita equivalence between the category of left E_i-vector spaces and the category of left $\mathrm{End}_{E_i}(V_i)$-modules. Such an equivalence preserves lattices of submodules, and the module $V_i^{(k_i)}$ corresponds to $E_i^{(k_i)}$. It follows that the maximal Λ-submodules of $V_i^{(k_i)}$ are in bijection with the maximal E_i-subspaces of $E_i^{(k_i)}$, as required for the first step.

The lemma itself now follows, since $E_i^{(k_i)}$ has precisely $(q_i^{k_i}-1)(q_i-1)$ subspaces of codimension one.

Applying the lemma to the case $L=\Lambda$, and using the notation (1.1)–(1.3), we obtain

$$\sum_M (\Lambda\colon M)^{-s} = \sum_{i=1}^n \frac{q_i^{m_i}-1}{q_i-1}|V_i|^{-s} = \sum_{i=1}^n \frac{q_i^{m_i}-1}{q_i-1} q_i^{-m_is}. \tag{1.6}$$

Here, M ranges over the maximal left ideals of Λ, and we note that $|V_i|=q_i^{m_i}$.

We can do exactly the same for right ideals. If we put

$$W_i = \mathrm{Hom}_R\,(V_i, \bar{R}),$$

then $\{W_1, \ldots, W_n\}$ is a basic set of simple right Λ-modules. As right Λ-module we have

$$\bar{\Lambda} \simeq \coprod_{i=1}^n W_i^{(m_i)},$$

with the same m_i as in (1.3) and (1.4), giving

$$\sum_N (\Lambda\colon N)^{-s} = \sum_{i=1}^n \frac{q_i^{m_i}-1}{q_i-1} q_i^{-m_is}. \tag{1.7}$$

Here, N runs through the maximal right ideals of Λ. The expression (1.7) is identical with the corresponding one (1.6) for left ideals.

EXAMPLE. The formula (1.6) is particularly interesting when Λ is the group ring RG of a finite group G. Here, $\zeta_\Lambda^l(s)=\zeta_\Lambda^r(s)$, and

$$\zeta_\Lambda(s) = 1 - aq^{-s} + \text{higher powers of } q^{-s},$$

where a is the number of maximal left ideals M of Λ with $\Lambda/M \simeq \bar{R}$ as $\bar{R}$-module. Any simple Λ-module V of $\bar{R}$-dimension one occurs just once in $\bar{\Lambda}$, and its G-structure is given by a homomorphism $G \to \bar{R}$. It follows that

$$a = |\mathrm{Hom}\,(G, \bar{R}^\times)|.$$

On the other hand, the calculation of $\zeta_{\Lambda'}(s)$ (in **[1]** or **[5]**) gives

$$\zeta_{\Lambda'}(s) = (1-q^{-s})^{-b}(1+cq^{-2s}+\cdots),$$

for some constant c, where b is the number of Wedderburn components of the group algebra KG which are totally ramified field extensions of K. Thus

$$\phi(s) = \zeta_\Lambda(s)/\zeta_{\Lambda'}(s) = 1 + (a-b)q^{-s} + \text{higher powers of } q^{-s}.$$

The functional equation ([**1**], p. 158) says that

$$\phi(s)=(\Lambda'\colon\ \Lambda)^{1-2s}\phi(1-s), \tag{1.8}$$

so we know *four* terms in the polynomial $\phi(s)$:

$$\phi(s)=1+(a-b)q^{-s}+\cdots+(\Lambda'\colon \Lambda)^{1-2s}((a-b)q^{s-1}+1).$$

Of course, if it happens that $(\Lambda'\colon\ \Lambda)=q$, then there are only three terms in $\phi(s)$ anyway, and

$$\phi(s)=1+(a-b)q^{-s}+q^{1-2s}.$$

In particular, if $|G|=p$ and $R=\mathbb{Z}_p$, we have $(\Lambda'\colon\ \Lambda)=p$. In this case, $a=1$ and $b=2$, since

$$\mathbb{Q}_pG\simeq\mathbb{Q}_p\times\mathbb{Q}_p(\omega),$$

for a primitive p-th root of unity ω. Therefore in this case

$$\phi(s)=1-p^{-s}+p^{1-2s},$$

and

$$\zeta_{\mathbb{Z}_pG}(s)=(1-p^{-s}+p^{1-2s})(1-p^{-s})^{-2},$$

in accordance with the calculations in [**1**] or [**5**].

§ 2. Ideals of level two

For a left or right Λ-ideal X of finite index in Λ, we write $l(\Lambda/X)$ for the composition length of the finite Λ-module Λ/X, and refer to this integer as the *level* of X. Thus maximal ideals are those of level one. We intend to calculate the contribution to $\zeta^l_\Lambda(s)$ from ideals of level 2. This information, together with some partial results on level 3, will be adequate for our applications.

A left Λ-ideal of level 2 is a maximal submodule of a maximal left ideal. For a given X of level 2, let $h(X)$ denote the number of distinct maximal left ideals containing X. Given a maximal left ideal M, define

$$f(M)=\sum_{X\subset M}(\Lambda\colon X)^{-s}, \tag{2.1}$$

where X ranges over the maximal submodules of M. We have

$$\sum_X(\Lambda\colon\ X)^{-s}=\sum_M f(M)-\sum_X(h(X)-1)(\Lambda\colon\ X)^{-s}, \tag{2.2}$$

where X (in both cases) ranges over the left ideals of Λ of level 2, and M ranges over the maximal left ideals of Λ. Of course, in the last series, only those X lying in at least two distinct maximal ideals actually occur.

We begin by evaluating $f(M)$, retaining the notation (1.2)–(1.4). We have

$$f(M)=(\Lambda:\ M)^{-s}\sum_{X\subset M}(M:\ X)^{-s}.$$

Set

$$\bar{M}=M/JM\simeq\coprod_{j=1}^{n}V_j^{(t_j(M))}. \tag{2.3}$$

By (1.5), there are $(q_j^{t_j(M)}-1)/(q_j-1)$ distinct modules $X\subset M$ with $M/X\simeq V_j$. This gives

$$f(M)=(\Lambda:\ M)^{-s}\sum_{j=1}^{n}\{(q_j^{t_j(M)}-1)/(q_j-1)\}\,|V_j|^{-s}. \tag{2.4}$$

The multiplicities $t_j(M)$ in (2.3) depend only on the isomorphism class of the simple module Λ/M. Indeed, if $\Lambda/M\simeq\Lambda/M'$, Schanuel's Lemma gives an isomorphism $M\simeq M'$. It follows that $\bar{M}\simeq\bar{M}'$, and that $t_j(M)=t_j(M')$ for all j. We proceed to calculate these multiplicities.

We choose a set $\{e_1,\ldots,e_n\}$ of mutually orthogonal idempotents in Λ, such that $\bar{\Lambda}e_i\simeq V_i$, for each i. We put

$$M=Je_i\oplus\Lambda(1-e_i),$$

so that $\Lambda/M\simeq V_i$, and M is thus a maximal left ideal of $\Lambda=\Lambda e_i\oplus\Lambda(1-e_i)$. It suffices to calculate $\bar{M}$ for this M. Now

$$\bar{M}=Je_i/J^2e_i\oplus\bar{\Lambda}(1-e_i),$$

and of course

$$\bar{\Lambda}(1-e_i)\simeq\coprod_{j=1}^{n}V_j^{(m_j-\delta_{ji})},$$

where m_j is as in (1.3) and δ_{ji} is the Kronecker delta. Setting

$$Je_i/J^2e_i\simeq\coprod_{j=1}^{n}V_j^{(r_{ji})}, \tag{2.5}$$

we obtain, for this M,

$$t_j(M)=r_{ji}+m_j-\delta_{ji}.$$

The same therefore holds for *any* M such that $\Lambda/M\simeq V_i$. There are $(q_i^{m_i}-1)/(q_i-1)$ ideals M with $\Lambda/M\simeq V_i$, and for each of these we have

$$f(M)=|V_i|^{-s}\sum_{j=1}^{n}\frac{q_j^{r_{ji}+m_j-\delta_{ji}}-1}{q_j-1}\,|V_j|^{-s}.$$

Thus

$$\sum_{M}^{n} f(M) = \sum_{i,j=1}^{n} \frac{(q_i^{m_i}-1)(q_j^{r_{ji}+m_j-\delta_{ji}}-1)}{(q_i-1)(q_j-1)} |V_i|^{-s}\,|V_j|^{-s}, \tag{2.6}$$

where M runs over the maximal left ideals of Λ.

We turn to the evaluation of the second term on the right hand side of (2.2), namely

$$\sum_{l(\Lambda/X)=2} (h(X)-1)(\Lambda:\ X)^{-s}. \tag{2.7}$$

(2.8) LEMMA. *Let M, M' be distinct maximal left ideals of Λ, and let $X = M \cap M'$. Then:*

(i) *X is of level 2;*

(ii) *if $\Lambda/M \neq \Lambda/M'$, then M, M' are the only maximal left ideals of Λ which contain X;*

(iii) *if $\Lambda/M \simeq \Lambda/M'$, then any maximal left ideal M'' of Λ containing X also satisfies $\Lambda/M'' \simeq \Lambda/M$, and further, $X = M \cap M' = M \cap M''$.*

Proof. For (i), we have $M/X = M/(M \cap M') \simeq (M+M')/M' = \Lambda/M'$, which is simple. Therefore $l(\Lambda/X) = l(\Lambda/M) + l(M/X) = 2$.

Suppose now that $\Lambda/M \simeq \Lambda/M'$, and that $M'' \supset X$. A set of composition factors for Λ/X is Λ/M, $M/X \simeq \Lambda/M'$. However, Λ/M'' is also a composition factor of Λ/X, so $\Lambda/M'' \simeq \Lambda/M$. Further, $X \subset M \cap M''$, which also has level 2, and it follows that $X = M \cap M''$. This proves (iii).

Finally suppose that Λ/M is not isomorphic to Λ/M'. Let $X \subset M''$, with $M'' \neq M$ or M'. Comparing composition factors, Λ/M'' must be isomorphic to either Λ/M or Λ/M', say $\Lambda/M'' \simeq \Lambda/M$. Then $X = M \cap M''$, by (i), and (iii) implies that $\Lambda/M' \simeq \Lambda/M$, which is not the case. Thus $M'' = M$ or M', as required for (ii).

We evaluate the sum (2.7) by breaking it into two parts. For those ideals X of the form

$$X = M \cap M', \qquad \Lambda/M \simeq V_i, \qquad \Lambda/M' \simeq V_j, \qquad i \neq j,$$

we have $h(X) = 2$ by (2.8) (ii). Since there are $(q_i^{m_i}-1)/(q_i-1)$ choices for the ideal M with $\Lambda/M \simeq V_i$, the contribution to the sum from these ideals X is

$$\sum_{1 \le i < j \le n} \frac{(q_i^{m_i}-1)(q_j^{m_j}-1)}{(q_i-1)(q_j-1)} |V_i|^{-s}\,|V_j|^{-s}. \tag{2.9}$$

We may of course rewrite the summation as

$$\tfrac{1}{2} \sum_{\substack{i,j=1 \\ i \neq j}}^{n} .$$

Now consider the ideals X of the form

$$X = M \cap M', \qquad \Lambda/M \simeq \Lambda/M' \simeq V_i. \tag{2.10}$$

Since J is the intersection of all the maximal left ideals, we have $X \supset J$, and Λ/X is semisimple. Indeed, $\Lambda/X \simeq V_i^{(2)}$. Arguing as in the proof of (1.5), the set of such ideals X is in bijection with the set of subspaces of $E_i^{(m_i)}$ of E_i-codimension 2. For fixed i, there are

$$\begin{bmatrix} m_i \\ 2 \end{bmatrix}_{q_i} = \frac{(q_i^{m_i}-1)(q_i^{m_i}-q_i)}{(q_i^2-1)(q_i^2-q_i)}$$

such subspaces. If X corresponds to the subspace S of $E_i^{(m_i)}$, the maximal left ideals containing X are in bijection with the maximal subspaces of $E_i^{(m_i)}$ containing S. There are q_i+1 such subspaces, so $h(X) = q_i + 1$, for X as in (2.10). Letting i vary, the contribution to the sum (2.7) from all ideals X of the type (2.10) is given by

$$\sum_{i=1}^{n} q_i \begin{bmatrix} m_i \\ 2 \end{bmatrix}_{q_i} |V_i|^{-2s}. \tag{2.11}$$

We can now combine this with (2.6) and (2.9) to obtain the final result:

(2.12) THEOREM. *The contribution to* $\zeta_\Lambda^l(s)$ *from left ideals* X *with* $l(\Lambda/X) = 2$ *is given by*

$$\sum_{i,j=1}^{n} \frac{(q_i^{m_i}-1)(q_j^{r_{ji}+m_j-\delta_{ji}}-1)}{(q_i-1)(q_j-1)} |V_i|^{-s}\,|V_j|^{-s}$$
$$-\tfrac{1}{2}\sum_{\substack{i,j=1\\ i\neq j}}^{n} \frac{(q_i^{m_i}-1)(q_j^{m_j}-1)}{(q_i-1)(q_j-1)} |V_i|^{-s}\,|V_j|^{-s}$$
$$-\sum_{i=1}^{n} q_i \frac{(q_i^{m_i}-1)(q_i^{m_i}-q_i)}{(q_i^2-1)(q_i^2-q_i)} |V_i|^{-2s},$$

with notation as in (1.1)–(1.4), (2.5).

Remarks. (1) The terms in $\zeta_\Lambda^l(s)$ coming from maximal ideals depend only on the Wedderburn structure of the finite semisimple ring $\bar{\Lambda}$. Those coming from ideals of level 2 depend only on this ring structure and the $\bar{\Lambda}$-bimodule structure of J/J^2, as in (2.5). Observe in this connection that the choice of the idempotent e_i, subject to $\bar{\Lambda}e_i \simeq V_i$, is immaterial: the left Λ-module Je_i/J^2e_i depends (up to isomorphism) only on the isomorphism class of $\bar{\Lambda}e_i$.

(2) We can perform the same analysis for right ideals of Λ. If N is a maximal right ideal, we set

$$g(N) = \sum_{\substack{Y \subset N\\ l(N/Y)=1}} (\Lambda:\ Y)^{-s}, \tag{2.13}$$

and if Y is a right ideal of level 2,

$$h(Y) = \textit{the number of maximal right ideals containing } Y.$$

As before

$$\sum_{\substack{Y_\Lambda \\ l(\Lambda/Y)=2}} (\Lambda\colon Y)^{-s} = \sum_{\substack{N_\Lambda \\ l(\Lambda/N)=1}} g(N) - \sum_{Y} (h(Y)-1)(\Lambda\colon Y)^{-s}.$$

The explicit calculation of the left hand "correction" term (2.7) given in (2.9) and (2.11) shows that it depends only on the Wedderburn structure (1.4) of $\bar{\Lambda}$. This is left-right symmetric, and so

$$\sum_{\substack{\Lambda^X \\ l(\Lambda/X)=2}} (h(X)-1)(\Lambda\colon X)^{-s} = \sum_{\substack{Y_\Lambda \\ l(\Lambda/Y)=2}} (h(Y)-1)(\Lambda\colon Y)^{-s}. \quad (2.14)$$

In other words, at level $\leqslant 2$, any difference between $\zeta^l_\Lambda(s)$ and $\zeta^r_\Lambda(s)$ is to be sought in the "principal" terms (2.6), (2.13). More precisely,

$$\zeta^l_\Lambda(s) - \zeta^r_\Lambda(s) = \sum_M f(M) - \sum_N g(N) + \textit{terms from level} \geqslant 3. \quad (2.15)$$

§ 3. The dihedral group case

We let p be an odd prime number, and put

$$G = \langle x, y \mid x^p = y^2 = 1, y^{-1}xy = x^{-1} \rangle,$$

the dihedral group of order $2p$. We take $R = \mathbb{Z}_p$, $K = \mathbb{Q}_p$, $\Lambda = RG$, $A = KG$. We shall evaluate $\zeta^l_\Lambda(s) = \zeta^r_\Lambda(s) = \zeta_\Lambda(s)$ explicitly by using the machinery of § 2, together with the functional equation (1.8).

In this case, a maximal order Λ' is isomorphic to $R \times R \times M_2(S)$, where S is the valuation ring in the field $\mathbb{Q}_p(\omega + \omega^{-1})$, for a p-th root of unity ω. Therefore ([1], § 3.3),

$$\zeta_{\Lambda'}(s)^{-1} = (1-p^{-s})^2(1-p^{-2s})(1-p^{1-2s}). \quad (3.1)$$

Further, $(\Lambda'\colon \Lambda) = p^3$, and (1.8) shows that $\phi(s) = \zeta_\Lambda(s)/\zeta_{\Lambda'}(s)$ is a polynomial of degree 6 in p^{-s}. Indeed

$$\phi(s) = p^{3-6s}\phi(1-s). \quad (3.2)$$

Thus, to compute $\phi(s)$ and $\zeta_\Lambda(s)$ completely, we only need to find the coefficients of p^{-ns} in $\zeta_\Lambda(s)$ for $n \leqslant 3$.

We find readily that

$$J = p\Lambda + (x-1)\Lambda, \quad \bar{\Lambda} = \bar{R}\bar{e}_1 \oplus \bar{R}\bar{e}_2,$$

where the idempotents $e_1, e_2 \in \Lambda$ are

$$e_1 = (1+y)/2, \qquad e_2 = (1-y)/2, \qquad \bar{e}_i = e_i \ (\mathrm{mod}\ J).$$

There are just two simple left modules V_1, V_2, and $V_i \simeq \bar{R}\bar{e}_i$. The group element x acts trivially on both, while y acts trivially on V_1, but as multiplication by -1 (in $\bar{R}$) on V_2. From (1.6) we obtain

$$\zeta_\Lambda(s) = 1 + 2p^{-s} + \sum_{l(\Lambda/X)=2} (\Lambda:\ X)^{-s} + \sum_{l(\Lambda/W)=3} (\Lambda:\ W)^{-s} + \cdots$$

Let M_i be the unique maximal ideal of Λ with $\Lambda/M_i \simeq V_i$, $i = 1, 2$. We shall show that, up to level 2, the lattice of left ideals of Λ looks like

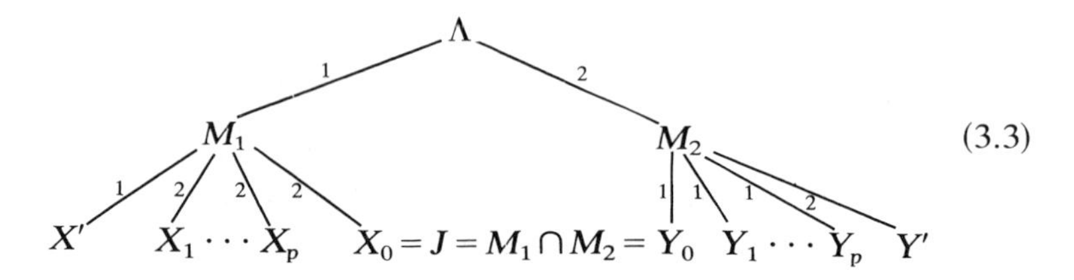

(3.3)

The numbers 1, 2 attached to a containment line in this diagram indicate the isomorphism class V_1 or V_2 of the quotient. We will eventually describe the next level in the diagram.

We begin by determining the structure of J/J^2. We have

$$J = p\Lambda + (x-1)\Lambda, \qquad J^2 = p^2\Lambda + p(x-1)\Lambda + (x-1)^2\Lambda,$$

so $\bar{J} = J/J^2$ can be identified with the $\bar{R}$-vector space with basis $\{p\bar{e}_1, p\bar{e}_2, (x-1)\bar{e}_1, (x-1)\bar{e}_2\}$. The action of the group elements on $\bar{J}$ is

$$y(x-1)\bar{e}_1 = (x^{-1}-1)\bar{e}_1 = -(x-1)\bar{e}_1, \qquad yp\bar{e}_1 = p\bar{e}_1,$$

$$y(x-1)\bar{e}_2 = (x^{-1}-1)y\bar{e}_2 = (x-1)\bar{e}_2, \qquad yp\bar{e}_2 = -p\bar{e}_2.$$

The element x acts trivially on all these basis elements. Therefore

$$\bar{J}e_1 \simeq V_1 \oplus V_2 \simeq \bar{J}e_2. \tag{3.4}$$

This gives us

$$M_1 = Je_1 \oplus \Lambda e_2, \quad \bar{M}_1 = \bar{J}e_1 \oplus \bar{\Lambda}e_2 \simeq V_1 \oplus V_2^{(2)},$$

which, via (1.5), verifies the left hand half of (3.3). The right hand half is treated similarly. The only ideal of level two contained in both M_1 and M_2 is $J = M_1 \cap M_2$, as shown.

We now determine the ideals of level 3 by examining the maximal submodules of the ideals of level 2. Taking first the ideal $J = X_0 = Y_0$, we have $\bar{J} \simeq V_1^{(2)} \oplus V_2^{(2)}$, and so J contains just $2(p+1)$ maximal submodules.

Now let us deal with X'. Since $JM_1 \subset X' \subset M_1$, and $M_1/X' \simeq V_1$, we have

$$M_1 = Je_1 \oplus \Lambda e_2, \qquad X' = (R(x-1)+J^2)e_1 \oplus \Lambda e_2,$$
$$JX' = (J(x-1)+J^3)e_1 \oplus Je_2.$$

However, one sees easily that

$$(R(x-1)+J^2)+(J(x-1)+J^3) = (R(x-1)+Rp^2)+(J(x-1)+J^3),$$

and since y acts as -1 on the image of $(x-1)e_1$, we get $\bar{X}' \simeq V_1 \oplus V_2^{(2)}$. Therefore X' has $p+2$ maximal submodules W, with $X'/W \simeq V_1$ for one of them, while $X'/W \simeq V_2$ for the remaining $p+1$ of them.

The procedure for $X_1, \ldots, X_p$ is simplified by the fact that $X_1 \simeq X_2 \simeq \cdots \simeq X_p$. From the exact sequence

$$0 \to M_1/X_i \to \Lambda/X_i \to \Lambda/M_1 \to 0, \qquad 1 \leq i \leq p,$$

we see that Λ/X_i is a *nonsplit* extension of V_1 by V_2. (If it were split, we would have $X_i \subset M_2$, which is not the case.) Using the exact sequence $0 \to M_1 \to \Lambda \to V_1 \to 0$, we have

$$\mathrm{Ext}^1_\Lambda (V_1, V_2) \simeq \mathrm{Hom}_\Lambda (M_1, V_2)/\mathrm{im}\,(\mathrm{Hom}_\Lambda (\Lambda, V_2)),$$

where $\mathrm{Hom}_\Lambda(\Lambda, V_2)$ maps to $\mathrm{Hom}_\Lambda (M_1, V_2)$ by restriction. This gives, by the simplicity of V_2,

$$\mathrm{Ext}^1_\Lambda(V_1, V_2) \simeq \mathrm{Hom}_\Lambda (\bar{M}_1, V_2)/\mathrm{im}\,(\mathrm{Hom}_\Lambda (\bar{\Lambda}, V_2)).$$

As $\bar{M}_1 \simeq V_1 \oplus V_2^{(2)}$, $\bar{\Lambda} \simeq V_1 \oplus V_2$, $\Lambda/M_1 \simeq V_1$, we obtain

$$\mathrm{Ext}^1_\Lambda (V_1, V_2) \simeq \bar{R}^{(2)}/\bar{R} \simeq \bar{R}.$$

The split extension $V_1 \oplus V_2$ corresponds to $0 \in \bar{R}$. On the other hand, $\mathrm{Aut}_\Lambda (V_2) \simeq \bar{R}^\times$ acts transitively on $\bar{R} - \{0\}$, so all nonsplit extensions of V_1 by V_2 are mutually isomorphic as modules (see [**4**], (34.5)). Finally, the isomorphism class of Λ/X_i determines that of X_i by Schanuel's Lemma, and this proves the assertion that $X_1 \simeq \cdots \simeq X_p$.

We can now choose a convenient $X_1 \subset M_1$ with $M_1/X_1 \simeq V_2$, namely

$$X_1 = (pR+J^2)e_1 \oplus \Lambda e_2,$$

giving

$$JX_1 = (pJ+J^3)e_1 \oplus Je_2.$$

One finds that $(pR+J^2)+(pJ+J^3) = (pR+(x-1)^2R)+(pJ+J^3)$, and y acts trivially on the image of $(x-1)^2e_1$ in $\bar{X}_1$. Consequently $\bar{X}_1 \simeq V_1^{(2)} \oplus V_2$, and X_1 has exactly $p+2$ distinct maximal submodules. The same therefore holds for $X_2, \ldots, X_p$.

Let us now consider intersections of the ideals $X', X_0, X_1, \ldots, X_p$. The intersection of any pair is an ideal of level 3. Since $\bar{M}_1 \simeq V_1 \oplus V_2^{(2)}$, each

$X_i \cap X_j$, $0 \leq i \neq j \leq p$, lies in precisely $p+1$ maximal submodules of M_1, arguing as in (1.5) and § 2. In other words, the intersections $X_i \cap X_j$, $0 \leq i \neq j \leq p$, all coincide. This common intersection cannot be X', by consideration of composition factors. It follows also that the $p+1$ intersections $X' \cap X_i$ are all distinct, and each is different from the common intersection $X_i \cap X_j$.

A corresponding calculation can be carried out for the maximal submodules $Y', Y_0, \ldots, Y_p$ of M_2. We omit this. Further, if $X = X'$ or X_i, $Y = Y'$ or Y_j, $0 \leq i, j \leq p$, we have $X/X \cap Y \simeq (X+Y)/Y$. If $X \neq X_0$ and $Y \neq Y_0$, we have $X+Y = \Lambda$, and it follows that $l(\Lambda/X \cap Y) = 4$. We may therefore neglect these intersections.

Altogether, the total number of ideals $W \subset M_1$ of level 3 is

$$\begin{aligned} a &= (p+2) + p(p+2) + 2(p+1) - (p+1) - p \\ &= p^2 + 3p + 3; \end{aligned}$$

because X' contains $p+2$ ideals W, each $X_1, \ldots, X_p$ contains $p+2$ ideals W, X_0 contains $2(p+1)$ ideals W, the $(p+1)$ intersections $X' \cap X_0, X' \cap X_1, \ldots, X' \cap X_p$ have each been counted twice, and the common intersection $X_i \cap X_j$ has been counted $p+1$ times. The omitted calculation shows that M_2 also contains just a ideals of level 3. However, the $2(p+1)$ maximal submodules of $J = X_0 = Y_0$ occur in both M_1 and M_2, and these are the only level 3 ideals in both M_1 and M_2. Consequently

$$\begin{aligned} \sum_{l(\Lambda/W)=3} (\Lambda : W)^{-s} &= (2a - 2(p+1))p^{-3s} \\ &= 2(p^2+p+2)p^{-3s}. \end{aligned}$$

Therefore

$$\zeta_\Lambda(s) = 1 + 2p^{-s} + (2p+3)p^{-2s} + 2(p^2+2p+2)p^{-3s} + \cdots,$$

and

$$\begin{aligned} \phi(s) = \zeta_\Lambda(s)/\zeta_{\Lambda'}(s) &= \zeta_\Lambda(s)(1-p^{-s})^2(1-p^{-2s})(1-p^{1-2s}) \\ &= 1 + (p-1)p^{-2s} + 2p^2p^{-3s} + \cdots, \end{aligned}$$

using (3.1). The functional equation (3.2) now yields

$$\phi(s) = 1 + (p-1)p^{-2s} + 2p^2p^{-3s} + p(p-1)p^{-4s} + p^3p^{-6s},$$

in agreement with the formula in [**3**].

§ 4. An asymmetric example

We return to the general local notation of the end of the introduction. It will also be useful to set

$$\bar{P} = P/P^2.$$

Of course, $\bar{P} \simeq \bar{R}$ as $\bar{R}$-module.

We now produce an example of an R-order Λ for which $\zeta_\Lambda^l(s) \neq \zeta_\Lambda^r(s)$. As we remarked in the introduction, we need only show that ζ_Λ^l, ζ_Λ^r have different coefficients of q^{-ms}, for some m. Here, $m=3$ will do. The calculation is based on the formula (2.15), which we now recall:

$$\zeta_\Lambda^l(s) - \zeta_\Lambda^r(s) = \sum_M f(M) - \sum_N g(N) + \textit{terms from level} \geqslant 3.$$

The notation is as in § 2. In particular, M runs over the maximal left ideals of Λ, and N runs over the maximal right ideals.

Our order Λ is the R-order in $M_4(K)$ defined symbolically by

$$\Lambda = \left[\begin{array}{cc|cc} R & R & R - & R \\ R & R & R - & R \\ \hline P & P & R \diagdown & P \\ P & P & P & R \end{array}\right] \tag{4.1}$$

where a line connecting two entries indicates that they are congruent modulo P. Explicitly, Λ consists of all 4×4 matrices (a_{ij}) over R which satisfy

$$\left.\begin{array}{lll} a_{ij} \in P \quad \text{if} \quad i \geqslant 3 \quad \text{and} \quad i \neq j, & & \\ a_{13} \equiv a_{14}, & a_{23} \equiv a_{24}, & a_{33} \equiv a_{44}, \end{array}\right\} \tag{4.2}$$

all congruences being modulo P.

This Λ is indeed an R-order spanning $M_4(K)$, and one finds readily that its Jacobson radical is

$$J = \left[\begin{array}{cc|cc} P & P & R - & R \\ P & P & R - & R \\ \hline P & P & P & P \\ P & P & P & P \end{array}\right]. \tag{4.3}$$

Therefore, as ring

$$\bar{\Lambda} = \Lambda/J \simeq M_2(\bar{R}) \times \bar{R}.$$

Thus Λ has just two non-isomorphic simple left modules V_1, V_3, and two non-isomorphic simple right modules W_1, W_3, numbered so that

$$\bar{\Lambda} \simeq V_1^{(2)} \oplus V_3 \quad \text{(as left } \Lambda\text{-module)}, \tag{4.4}$$

$$\bar{\Lambda} \simeq W_1^{(2)} \oplus W_3 \quad \text{(as right } \Lambda\text{-module)}. \tag{4.5}$$

Therefore $|V_1| = |W_1| = q^2$, $|V_3| = |W_3| = q$.

As idempotents in Λ, we take the diagonal matrices

$$e_1 = \text{diag}\,(1, 0, 0, 0), \qquad e_2 = \text{diag}\,(0, 1, 0, 0), \qquad e_3 = \text{diag}\,(0, 0, 1, 1), \tag{4.6}$$

so that $\bar{\Lambda}e_1 \simeq \bar{\Lambda}e_2 \simeq V_1$, $\bar{\Lambda}e_3 \simeq V_3$, $e_1\bar{\Lambda} \simeq e_2\bar{\Lambda} \simeq W_1$, $e_3\bar{\Lambda} \simeq W_3$.

Computing directly from (4.3), we find

$$J^2=\left[\begin{array}{cc|cc} P & P & P & P \\ P & P & P & P \\ \hline P^2 & P^2 & P-P \\ P^2 & P^2 & P-P \end{array}\right]$$

where a connecting line now indicates congruence mod P^2. We may therefore identify $\bar{J}=J/J^2$ with the additive group of matrices

$$\bar{J}=\left[\begin{array}{cc|cc} 0 & 0 & \bar{R}-\bar{R} \\ 0 & 0 & \bar{R}-\bar{R} \\ \hline \bar{P} & \bar{P} & 0 & \bar{P} \\ \bar{P} & \bar{P} & 0 & \bar{P} \end{array}\right], \tag{4.7}$$

with the connecting lines indicating equality of entries. The obvious left and right actions of Λ on this group coincide with the natural actions of Λ on $\bar{J}$. Therefore

$$\bar{J}\simeq V_1\oplus V_3^{(6)} \qquad \text{(as left } \Lambda\text{-module)}, \tag{4.8}$$

$$\bar{J}\simeq W_1^{(2)}\oplus W_3^{(4)} \qquad \text{(as right } \Lambda\text{-module)}. \tag{4.9}$$

As we shall see, this asymmetry in the structure of $\bar{J}$ induces an asymmetry in the zeta-functions of Λ.

Applying (1.5) to (4.4), we see that Λ has exactly $q+1$ maximal left ideals $M_1, \ldots, M_{q+1}$ satisfying $\Lambda/M_i\simeq V_1$, and a unique maximal left ideal M' such that $\Lambda/M'\simeq V_3$. The M_i are all mutually isomorphic by Schanuel's Lemma. The lattices of maximal submodules of these maximal ideals look like

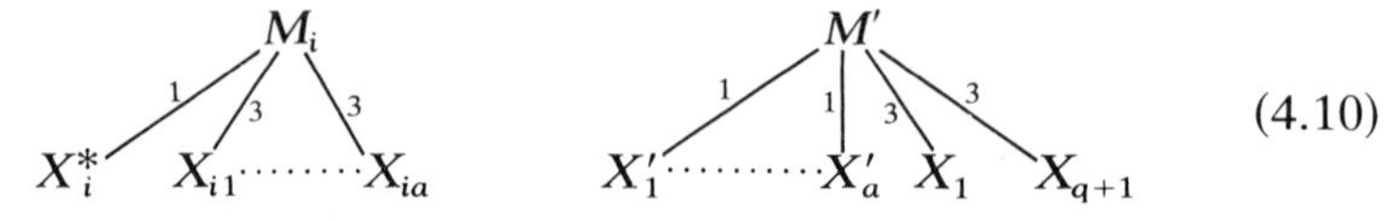

(4.10)

where a figure 1, 3 on a containment line indicates the isomorphism class V_1, V_3 of the quotient. (Notice that M_i and M_j may (and indeed do) have some maximal submodules in common, but that this will not affect our calculations, because of (2.15).) The integer a which appears in (4.10) is

$$a=q^2+q+1.$$

We have $(\Lambda\colon M')=q$, $(\Lambda\colon M_i)=(\Lambda\colon X_j)=q^2$, $(\Lambda\colon X_{ik})=(\Lambda\colon X_m')=q^3$, $(\Lambda\colon X_t^*)=q^4$, for all values of i, j, k, m, t.

To verify the assertions made in (4.10), we choose (to start with)

$$M_1=Je_1\oplus\Lambda e_2\oplus\Lambda e_3,$$

$$\bar{M}_1 = \bar{J}e_1 \oplus \bar{\Lambda}e_2 \oplus \bar{\Lambda}e_3 \simeq V_1 \oplus V_3^{(3)}.$$

Thus M_1 has the indicated pattern of maximal submodules, by (1.5), and the same result follows for each M_i, as $M_i \simeq M_1$.

On the other hand

$$M' = \Lambda e_1 \oplus \Lambda e_2 \oplus J e_3,$$
$$\bar{M}' = \bar{\Lambda}e_1 \oplus \bar{\Lambda}e_2 \oplus \bar{J}e_3 \simeq V_1^{(3)} \oplus V_3^{(2)}.$$

This verifies the right hand half of (4.10). Altogether, we now know that

$$\sum_{\substack{\Lambda^M \\ l(\Lambda/M)=1}} f(M) = (q+1)\{q^{-4s} + aq^{-3s}\} + (q+1)q^{-2s} + aq^{-3s}$$
$$= (q+1)q^{-2s} + (q^3+3q^2+3q+2)q^{-3s} + (q+1)q^{-4s}. \quad (4.11)$$

Now we turn our attention to right ideals. From (4.5), there are exactly $q+1$ maximal right ideals $N_1, \ldots, N_{q+1}$ satisfying $\Lambda/N_i \simeq W_1$, and any two of these are isomorphic. In addition, there is a unique right ideal N' with $\Lambda/N' \simeq W_3$. The right-handed version of (4.10) looks like

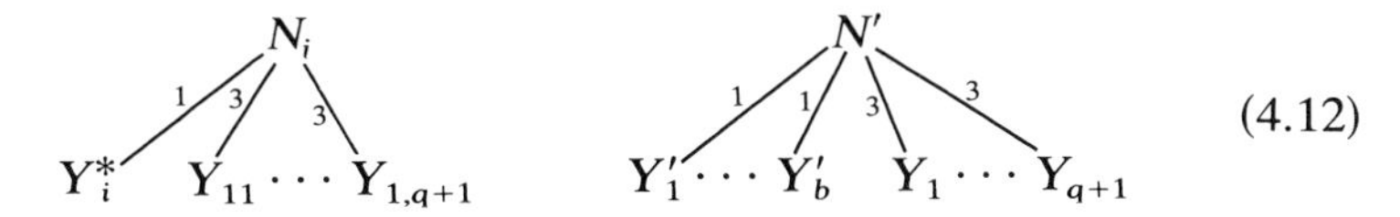

(4.12)

with the same conventions as before, and where

$$b = q^3 + q^2 + q + 1.$$

The verification is similar. We take

$$N_1 = e_1 J \oplus e_2\Lambda \oplus e_3\Lambda, \quad \text{giving} \quad N_1/N_1J \simeq W_1 \oplus W_3^{(2)}.$$

Also

$$N' = e_1\Lambda \oplus e_2\Lambda \oplus e_3 J \quad \text{giving} \quad N'/N'J \simeq W_1^{(4)} \oplus W_3^{(2)}.$$

The assertions contained in (4.12) now follow from the right hand analogue of (1.5). Therefore

$$\sum_{\substack{N_\Lambda \\ l(\Lambda/N)=1}} g(N) = (q+1)q^{-4s} + (q+1)^2 q^{-3s} + bq^{-3s} + (q+1)q^{-2s}$$
$$= (q+1)q^{-2s} + (q^3+2q^2+3q+2)q^{-3s} + (q+1)q^{-4s}.$$

Applying (2.15), we get

$$\zeta_\Lambda^l(s) - \zeta_\Lambda^r(s) = q^2 q^{-3s} + \text{terms from level} \geqslant 3. \quad (4.13)$$

We next prove that

$$\sum_{\substack{\Lambda^X \\ l(\Lambda/X)\geq 3}} (\Lambda\colon X)^{-s} - \sum_{\substack{Y_\Lambda \\ l(\Lambda/Y)\geq 3}} (\Lambda\colon Y)^{-s} = cq^{-4s} + \text{higher powers of } q^{-s}, \tag{4.14}$$

for some constant c. The two series on the left hand side of (4.14) both start with a q^{-3s} term. The coefficient of q^{-3s} in $\zeta^l_\Lambda(s)$ is the number of left ideals X of Λ such that Λ/X has composition factors $V_3^{(3)}$. A similar remark applies to $\zeta^r_\Lambda(s)$. The equation (4.14) follows from

(4.15) LEMMA. *The order Λ possesses a unique left ideal X such that Λ/X has composition type $V_3^{(3)}$, and a unique right ideal Y such that Λ/Y has composition type $W_3^{(3)}$.*

Proof. We start with left ideals. Any left ideal X of the required type is a maximal submodule of some X_i (in the notation of (4.10)). We have $M' \supset X_i \supset JM'$, for all i, and $M'/JM' \simeq V_1^{(3)} \oplus V_3^{(2)}$. Thus, if $i \neq j$, $X_i \cap X_j$ is the unique ideal $Z \subset M'$ such that $M'/Z \simeq V_3^{(2)}$. In other words, all the intersections $X_i \cap X_j$, $i \neq j$, coincide. The composition type of Λ/Z is $V_3^{(3)}$.

We now prove that each X_i has a unique submodule Z_i such that $X_i/Z_i \simeq V_3$. This Z_i will then be the common intersection $Z = X_i \cap X_j$, and the lemma (for left ideals) will follow. We recall that

$$M' = \Lambda e_1 \oplus \Lambda e_2 \oplus Je_3, \qquad X_i \supset JM', \qquad M'/X_i \simeq V_3.$$

In terms of matrices, we have

$$M' = \left[\begin{array}{cc|cc} R & R & R & -R \\ R & R & R & -R \\ \hline P & P & P & P \\ P & P & P & P \end{array}\right], \qquad JM' = \left[\begin{array}{cc|cc} P & P & P & P \\ P & P & P & P \\ \hline P & P & P & -P \\ P & P & P & -P \end{array}\right].$$

It follows that X_i looks like

$$\left[\begin{array}{cc|c} R & R & R-R \\ R & R & R-R \\ \hline P & P & \multirow{2}{*}{A_i} \\ P & P & \end{array}\right],$$

where A_i is some R-sublattice of $M_2(P)$, of index q, and containing all 2×2 matrices of the form

$$\begin{bmatrix} P-P \\ P-P \end{bmatrix}.$$

Remark. Although we shall not need to know the precise nature of the lattice A_i, it is not difficult to determine this. Indeed, A_i consists of all 2×2 matrices $(a_{ij})\in M_2(P)$ satisfying a congruence relation of the sort $\lambda_i a_{11}+\mu_i a_{21}\equiv\lambda_i a_{12}+\mu_i a_{22} \pmod{P^2}$, for certain λ_i, $\mu_i\in R$ which vary with i.

Returning to the proof, we find

$$JX_i=\left[\begin{array}{cc|cc} P & P & R & -R \\ P & P & R & -R \\ \hline P & P & P & P \\ P & P & P & P \end{array}\right]\left[\begin{array}{cc|cc} R & R & R & -R \\ R & R & R & -R \\ \hline P & P & \multicolumn{2}{c}{\multirow{2}{*}{A_i}} \\ P & P & & \end{array}\right]$$

$$=\left[\begin{array}{cc|cc} \multicolumn{2}{c|}{*} & \multicolumn{2}{c}{*} \\ \hline P & P & P & -P \\ P & P & P & -P \end{array}\right],$$

where we use $*$ to denote a 2×2 block which we do not need to know. The lower right hand block has the asserted value since all matrices in $M_2(P)\cdot A_i$ have entries in P^2. Therefore $X_i/JX_i\simeq V_1^{(k)}\oplus V_3$, for some integer k. Thus X_i does indeed have a unique submodule Z_i such that $X_i/Z_i\simeq V_3$.

The situation for right ideals is very similar. Here,

$$N'/N'J\simeq W_1^{(4)}\oplus W_3^{(2)},$$

so the intersections $Y_i\cap Y_j$, $i\neq j$, all coincide (in the notation of (4.12)). The composition type of $\Lambda/(Y_i\cap Y_j)$ is $W_3^{(3)}$. Again, it is enough to show that each Y_i possesses a unique submodule T_i such that $Y_i/T_i\simeq W_3$. Here,

$$Y_i=\left[\begin{array}{cc|cc} R & R & R & -R \\ R & R & R & -R \\ \hline P & P & \multicolumn{2}{c}{\multirow{2}{*}{B_i}} \\ P & P & & \end{array}\right],$$

for an R-sublattice B_i of $M_2(P)$ of index q which contains all matrices of the form

$$\begin{bmatrix} P & -P \\ P & -P \end{bmatrix}.$$

(With suitable numbering, it happens that $B_i=A_i$.) This time,

$$Y_1J=\left[\begin{array}{cc|cc} \multicolumn{2}{c|}{\multirow{2}{*}{$*$}} & R & -R \\ & & R & -R \\ \hline \multicolumn{2}{c|}{\multirow{2}{*}{$*$}} & P & P \\ & & P & P \end{array}\right],$$

giving $Y_i/Y_iJ \simeq W_1^{(m)} \oplus W_3$, for some integer m. This proves (4.15) for right ideals, and completes the proof of (4.14).

Altogether, we now have proved:

(4.16) THEOREM. *Let* Λ *be the* R*-order specified by* (4.1), (4.2). *Then* $\zeta_\Lambda^l(s) \neq \zeta_\Lambda^r(s)$. *More precisely,*

$$\zeta_\Lambda^l(s) - \zeta_\Lambda^r(s) = q^2 q^{-3s} + \textit{higher powers of } q^{-s}.$$

This result holds, in particular, when $R = \mathbb{Z}_p$. We can now define a $\mathbb{Z}$-order $\mathcal{A}$ in $M_4(\mathbb{Q})$ by

$$\mathcal{A}_p = \Lambda, \qquad \mathcal{A}_{p'} = M_4(\mathbb{Z}_{p'}), \quad \textit{for } p' \textit{ prime}, \quad p' \neq p.$$

Then

$$\zeta_{\mathcal{A}}^l(s) = \zeta_\Lambda^l(s) \prod_{p' \neq p} \prod_{j=0}^{3} (1 - p'^{j-4s})^{-1}$$

$$\zeta_{\mathcal{A}}^r(s) = \zeta_\Lambda^r(s) \prod_{p' \neq p} \prod_{j=0}^{3} (1 - p'^{j-4s})^{-1}.$$

Therefore $\zeta_{\mathcal{A}}^l \neq \zeta_{\mathcal{A}}^r$.

Asymmetry of ideal theory in algebras over fields has been studied. The authors wish to thank G. Janusz for pointing out the following example of an $\bar{R}$-algebra

$$B = \begin{pmatrix} \bar{R} & 0 & 0 \\ \bar{R} & \bar{R} & 0 \\ \bar{R} & 0 & \bar{R} \end{pmatrix},$$

where the connecting line denotes equality, which is left but not right uniserial. The R-order Λ in $M_3(K)$ given by

$$\Lambda = \begin{pmatrix} R & P & P \\ R & R & P \\ R & P & R \end{pmatrix},$$

where the connecting line is now congruence mod P, has a two-sided ideal $I = P \cdot M_3(R)$ such that $\Lambda/I \simeq B$. One might hope that the asymmetry of B would lift to an asymmetry among the zeta-functions of Λ. However, Λ is conjugate to its own transpose (in $M_3(K)$) via the diagonal matrix diag $(\pi, 1, 1)$, where $\pi R = P$. Therefore Λ has an anti-automorphism, and consequenctly $\zeta_\Lambda^l(s) = \zeta_\Lambda^r(s)$.

REFERENCES

1. C. J. Bushnell and I. Reiner, 'Zeta-functions of arithmetic orders and Solomon's conjectures', *Mathematische Zeitschrift* 173 (1980), 135–161.

2. ——, *Zeta-functions of hereditary orders and integral group rings*, Texas Tech University Mathematics Series 14 (1981), 71–94.

3. ——, 'L-functions of arithmetic orders and asymptotic distribution of ideals', *Journal für die reine und angewandte Mathematik* 327 (1981), 156–183.

4. C. W. Curtis and I. Reiner, *Methods of Representation Theory I.* Wiley, 1981. New York.

5. L. Solomon, 'Zeta-functions and integral representation theory', *Advances in Mathematics* 26 (1977), 306–326.

6. I. Reiner, 'Zeta-functions of integral representations', *Communications in Algebra* 8 (1980), 911–925.

7. E. C. Titchmarsh, *The theory of functions.* Oxford University Press, 2nd ed, (1939), London.

Department of Mathematics,
King's College, Strand,
London WC2R 2LS, UK.

University of Illinois,
Department of Mathematics,
1409 West Green Street,
Urbana, IL 61801, USA.

Analytic continuation of partial zeta functions of arithmetic orders

By *Colin J. Bushnell* at London and *Irving Reiner**) at Urbana

---

We return here to a question left unsettled in our earlier paper [2]. To fix the notation, let A be a finite-dimensional semisimple $\mathbb{Q}$-algebra, with r simple direct factors, and let Λ be a $\mathbb{Z}$-order in A. The (total) *zeta-function* of Λ is defined by

$$\zeta_\Lambda(s) = \sum_{X \subset \Lambda} (\Lambda : X)^{-s}, \quad \operatorname{Re}(s) > 1,$$

where X ranges over all left ideals of Λ of finite index $(\Lambda : X)$ in Λ. There are also various partial zeta-functions attached to Λ, of which the "finest" is given as follows. Let M be some left ideal of Λ (we always assume that M spans A over $\mathbb{Q}$), and set

$$Z_\Lambda(M, s) = \sum_{X \cong M} (\Lambda : X)^{-s}, \quad \operatorname{Re}(s) > 1,$$

where X now ranges over those left ideals of Λ which are isomorphic to M. If M is allowed to range over a set of representatives of the isomorphism classes of left Λ-ideals (there are only finitely many such, by the Jordan-Zassenhaus theorem), we have

$$\zeta_\Lambda(s) = \sum_M Z_\Lambda(M, s).$$

We conjectured in [2], (5. 4), that $Z_\Lambda(M, s)$ admits analytic continuation to a meromorphic function on the whole s-plane. Once this is known, it is easy to show that it has a pole of order r at $s = 1$. In [2], we proved this conjecture when A satisfies the Eichler condition, and also for the opposite extreme when A is a totally definite quaternion algebra. Indeed, when A is a totally definite quaternion algebra, this is an old result of Eichler if Λ is a maximal order, while the same approach extends easily to the case of arbitrary Λ. The primary aim of this paper is to prove the conjecture for all algebras A.

---

*) The research for this paper was performed while the second author was visiting King's College London. He wishes to thank King's College for its hospitality, and also the SERC and the Center for Advanced Study of the University of Illinois for their support.

To this end, it is convenient (and interesting) to generalise the problem slightly. To motivate this, we rearrange the above expression for $Z_\Lambda(M, s)$. First, let G be any group, and H_1, H_2 subgroups of G. We say that H_1 is *commensurable* with H_2 if $H_1 \cap H_2$ is of finite index in both H_1 and H_2. If this is so, the *generalised index*

$$(H_1 : H_2) = (H_1 : H_1 \cap H_2)/(H_2 : H_1 \cap H_2)$$

is defined, and is a positive rational number. In our situation above, we put $\Gamma_M = \mathrm{End}_\Lambda(M)$, and view M as a right Γ_M-module. Identifying $\mathrm{End}_A(A)$ with A in the usual way, Γ_M becomes an order in A. We also put

$$M^{-1} = \mathrm{Hom}_\Lambda(M, \Lambda) = \{x \in A : Mx \subset \Lambda\}.$$

Then M^{-1} is a left Γ_M-module, spanning A over $\mathbb{Q}$. Any ideal X of Λ with $X \cong M$ can be written $X = Mx$, for some $x \in M^{-1} \cap A^\times$. This element x is determined by X up to multiplication on the left by an element of $\Gamma_M^\times = \mathrm{Aut}_\Lambda(M)$. The quantity $\|x\| = (M : Mx)^{-1}$ depends only on x, and we have

$$Z_\Lambda(M, s) = (\Lambda : M)^{-s} \sum_{x \in \Gamma_M^\times \setminus M^{-1} \cap A^\times} \|x\|^s, \quad \mathrm{Re}(s) > 1.$$

This prompts us to consider the following more general situation. We take a pair (W, L) consisting of a $\mathbb{Z}$-lattice L which spans A over $\mathbb{Q}$, and an *arithmetic subgroup* W of $A^\times$. (This means that W is a subgroup of $A^\times$ which is commensurable with $\Lambda^\times$ for some (or, equivalently, any) order Λ in A.) We further impose the condition that

$$wy \in L, \quad \textit{for all} \quad w \in W, \; y \in L.$$

Thus W is a group of automorphisms of the lattice L, and we may form

$$f(s) = f_{(W,L)}(s) = \sum_{x \in W \setminus L \cap A^\times} \|x\|^s.$$

Comparison with $Z_\Lambda(M, s)$, for suitable Λ, M, shows that the Dirichlet series $f(s)$ converges for $\mathrm{Re}(s) > 1$. In these terms, we have $Z_\Lambda(M, s) = (\Lambda : M)^{-s} f_{(\Gamma_M^\times, M^{-1})}(s)$, using our earlier notation. Our conjecture on $Z_\Lambda(M, s)$ now follows from

Theorem A. *Let A be a finite-dimensional semisimple $\mathbb{Q}$-algebra with r simple direct factors. Let L be a $\mathbb{Z}$-lattice spanning A over $\mathbb{Q}$, and W an arithmetic subgroup of $A^\times$ such that $wL = L$ for all $w \in W$. Then the function $f_{(W,L)}(s)$ admits analytic continuation to a meromorphic function on the whole s-plane, and it has a pole of order r at $s = 1$. Further, if Λ' is any maximal order in A, we have*

$$\lim_{s \to 1} (s-1)^r f_{(W,L)}(s) = (\Lambda' : L)^{-1} (\Lambda'^\times : W)\, b_{\Lambda'},$$

where the non-zero constant $b_{\Lambda'}$ depends only on the maximal order Λ'. The quantity $(\Lambda'^\times : W)\, b_{\Lambda'}$ depends only on W, and not on the choice of Λ'.

It would be cumbersome to say much more about the constant $b_{\Lambda'}$ at this stage. We shall later determine it explicitly for the case of a simple algebra (see (2.5), (3.9) below). The general case then follows from

Theorem B. *In the situation of Theorem* A, *let* $A_1, \ldots, A_r$ *be the simple direct factors of* A. *Let* Λ'_i *be a maximal order in* A_i, *for each* i, *such that* $\Lambda' = \prod_{i=1}^{r} \Lambda'_i$. *Then we have* $b_{\Lambda'} = \prod_{i=1}^{r} b_{\Lambda'_i}$.

The proof of Theorem A is based on reducing to the case of a simple algebra. This reduction, however, forces us to consider a yet more general class of functions, as in the following:

Theorem C. *Let* A *be a finite-dimensional semisimple* $\mathbb{Q}$*-algebra, and* L *a* $\mathbb{Z}$*-lattice spanning* A *over* $\mathbb{Q}$. *Let* W *be an arithmetic subgroup of* $A^\times$ *such that* $WL = L$. *Let* $m \in A$ *be such that* $wm - m \in L$ *for all* $w \in W$, *so that* W *acts on the coset* $m + L$. *Define*

$$F(s) = F_{(W, L, m)}(s) = \sum_{x \in W \backslash ((m+L) \cap A^\times)} \|x\|^s, \quad \operatorname{Re}(s) > 1.$$

Then $F(s)$ *admits analytic continuation to a meromorphic function on the whole s-plane. It has a pole of order* r *at* $s = 1$, *where* r *is the number of simple direct factors of* A. *Let* Λ' *be a maximal order in* A. *We have*

$$\lim_{s \to 1} (s-1)^r F(s) = (\Lambda' : L)^{-1} (\Lambda'^\times : W)\, b_{\Lambda'},$$

where the non-zero constant $b_{\Lambda'}$ *depends only on* Λ' (*and not on the triple* (W, L, m)). *The quantity* $(\Lambda'^\times : W)\, b_{\Lambda'}$ *depends only on* W, *and not on the choice of* Λ'.

Theorem A is then a special case of Theorem C. Notice that the functions $F(s)$ in Theorem C are partial generalisations of the Hurwitz zeta function:

$$\zeta(s; x) = \sum_{n=1}^{\infty} (n+x)^{-s}, \quad \operatorname{Re}(s) > 1, \quad 0 \leqq x < 1.$$

The proof of Theorem C occupies the first three sections of the paper. In §4, we derive some consequences of Theorem A. That result enables us first to give a complete account of the asymptotic distribution of ideals in an order among the isomorphism classes, thereby refining Theorem 6.10 of [2]. In a different direction, if M is a left Λ-ideal, $\Gamma_M = \operatorname{End}_\Lambda(M)$ as above, we can attach to the isomorphism class of M the *weight* $(\Lambda^\times : \Gamma_M^\times)$. The principal results of §4 describe the behaviour of the sum of these weights over a stable isomorphism class. The depth of this is indicated by the fact that, when A satisfies the Eichler condition, it reduces to Fröhlich's formula for the cardinality of the "kernel group" of an order.

The results contained in this paper also have implications for the more general zeta-functions of the type considered in [1]. For example, let V be a finitely generated left A-module, spanned by Λ-lattices L, L'. We set $B = \operatorname{End}_A(V)$, and view V as a right B-module. Define

$$Z_\Lambda(L, L'; s) = \sum_{\substack{X \cong L' \\ X \subset L}} (L : X)^{-s},$$

where X ranges over all Λ-submodules of L which are isomorphic to L'. We rearrange this series by setting

$$\Gamma = \mathrm{End}_\Lambda(L) \quad \text{(an order in } B\text{)},$$

$$M = \{x \in B : L'x \subset L\},$$

$$\|x\|_V = (L : Lx)^{-1}, \quad x \in B^\times.$$

The quantity $\|x\|_V$ depends only on x and V, and, moreover, if A is simple, $\|x\|_V = \|x\|_B^a$, for some positive $a \in \mathbb{Q}$. We have

$$Z_\Lambda(L, L'; s) = (L : L')^{-s} \sum_{x \in \Gamma^\times \setminus M \cap B^\times} \|x\|_V^s .$$

This series is therefore a finite sum of functions $F_{(W, N, m)}(as)$, for various $a \in \mathbb{Q}$, $m \in B$, lattices $N \subset B$, and arithmetic subgroups W of $B^\times$. The analytic properties of the functions $Z_\Lambda(L, L'; s)$ can therefore be deduced from those of the functions F of Theorem C.

Finally, we take this opportunity of correcting some misprints in [2] and [3]:

[2] p. 174 line 2: *left ideals* X *of* Λ.

p. 182 (9. 3) Theorem: the displayed formulas should be

$$\frac{1}{q_j} \frac{(M^{-1} : \Lambda)}{(\Lambda : M)(\Lambda' : \Lambda)} (2\pi)^{2n} 2^{n-3} \operatorname{Reg}(F) |d_F|^{-2} \prod_{P \in S} NP^{-1}, \quad \text{and}$$

$$\sum_{j=1}^{h} \frac{1}{q_j} = (U(\Lambda') : U(\mathrm{End}(L_j))) \sum_{i=1}^{h'} \frac{1}{w_i}$$

respectively.

[3] p. 119 line 18 et seq: remove δ, and tr_B should be $\mathrm{tr}_{B/\mathbb{R}}$, the reduced trace from B to $\mathbb{R}$.

§ 1. Reduction to the simple case

We use the notation of Theorem C. If W_0 is a subgroup of W of finite index, then immediately

$$(1.1) \qquad F_{(W_0, L, m)}(s) = (W : W_0)\, F_{(W, L, m)}(s).$$

Therefore, in proving Theorem C, we can always replace W by a convenient subgroup of finite index.

Now let

$$A = \prod_{i=1}^{r} A_i$$

be the Wedderburn decomposition of A into a product of simple algebras A_i, and use the subscript i to denote projection into A_i. We may choose a $\mathbb{Z}$-sublattice N of L

so that $(L:N)$ is finite and $N=\prod_{i=1}^{r} N_i$. For example, we can take $N=a\Lambda'$, for some positive integer a. Replacing W by a subgroup of finite index we may further assume that

$$\text{(1.2)} \quad \text{(i)} \quad W=\prod_{i=1}^{r} W_i; \qquad \text{(ii)} \quad W_iN_i=N_i, \quad \textit{for all } i;$$

$$\text{(iii)} \quad w_im_i-m_i\in N_i \quad \textit{for all } i \textit{ and all } w_i\in W_i;$$

$$\text{(iv)} \quad W \textit{ acts trivially on } L/N.$$

Each W_i is automatically an arithmetic subgroup of $A_i^\times$. Letting $l\in L$ range over a full set of coset representatives of L/N, we have

$$m+L=\bigcup(m+l+N) \quad \text{(disjoint union).}$$

Further, $(w_i-1)(m_i+l_i)\in N_i$ for all w, i and l. We now have

$$F_{(W,L,m)}(s)=\sum_{l\in L/N} F_{(W,N,m+l)}(s),$$

and

$$F_{(W,N,m+l)}(s)=\prod_{i=1}^{r} F_{(W_i,N_i,m_i+l_i)}(s).$$

If we prove Theorem C for the case of a simple algebra, we then obtain the analytic continuation of $F_{(W,L,m)}(s)$, and furthermore

$$\begin{aligned}\lim_{s\to 1}(s-1)^r F_{(W,L,m)}(s) &= \sum_{l\in L/N}\prod_{i=1}^{r}\lim_{s\to 1}(s-1)F_{(W_i,N_i,m_i+l_i)}(s)\\ &=(L:N)\prod_{i=1}^{r}(\Lambda_i':N_i)^{-1}(\Lambda_i'^\times:W_i)\,b_{\Lambda_i'}\\ &=(\Lambda':L)^{-1}(\Lambda'^\times:W)\,b_{\Lambda'},\end{aligned}$$

thereby proving Theorem C in general, and with it Theorems A and B.

This discussion also shows that in proving Theorem C for a simple algebra, we can always replace W by a subgroup of finite index, and L by a sublattice. Further, if α is an invertible element of the centre of A, then

$$F_{(W,\alpha L,\alpha m)}(s)=\|\alpha\|^s F_{(W,L,m)}(s).$$

Since $\|\alpha\|(\Lambda':L)^{-1}=(\Lambda':\alpha L)^{-1}$, the result for the triple (W,L,m) will follow from that for $(W,\alpha L,\alpha m)$. In all, we are reduced to proving:

Theorem C′. *Let A be a finite-dimensional central simple K-algebra, where K is an algebraic number field with ring of integers R. Let Λ' be a maximal R-order in A, and $\mathfrak{c}$ a non-zero ideal of R. Let $m\in\Lambda'$, and let*

$$W=\{u\in\Lambda'^\times : u\equiv 1 \pmod{\mathfrak{c}\Lambda'}\},$$

an arithmetic subgroup of $A^\times$. Then $F(s)=F_{(W,\mathfrak{c}\Lambda',m)}(s)$ admits analytic continuation to a meromorphic function on the whole s-plane. It has a simple pole at $s=1$ with residue $(\Lambda':\mathfrak{c}\Lambda')^{-1}(\Lambda'^\times:W)\,b_{\Lambda'}$, where the constant $b_{\Lambda'}$ is independent of $\mathfrak{c}$ and m. If Λ'' is another maximal order in A, then $b_{\Lambda''}=(\Lambda'^\times:\Lambda''^\times)\,b_{\Lambda'}$.

§ 2. The Eichler case

We now prove Theorem C′ under the additional hypothesis that the simple algebra *A satisfies the Eichler condition*. This amounts to saying that A is not a totally definite quaternion algebra. *Changing notation*, we let Λ denote a maximal order in A for the remainder of this section.

The basic plan of the proof is to compare the series $F(s)=F_{(W,\mathfrak{c}\Lambda,m)}(s)$ with an idelic integral, of the type considered in [2], whose analytic properties are more accessible.

We need some preliminaries and notation. Let P range over the maximal ideals of R, and let A_P, Λ_P etc. denote P-adic *completions*. Let $J(A)$ denote the group of *finite ideles of A*: this is the topological restricted direct product of the groups $A_P^\times$ with respect to the subgroups $\Lambda_P^\times$. Its natural restricted product topology makes it into a locally compact group. We view $A^\times$ as a subgroup of $J(A)$ in the obvious way. We also have

$$U(\Lambda)=\prod_P \Lambda_P^\times,$$

which is a compact open subgroup of $J(A)$. This has a "congruence subgroup" which is also open in $J(A)$:

$$U_\mathfrak{c}(\Lambda)=\{(x_P)\in U(\Lambda): x_P\equiv 1 \,(\mathrm{mod}\ \mathfrak{c}\Lambda_P),\quad \textit{for all } P\}.$$

Notice that the condition on x_P here is vacuous if P does not divide $\mathfrak{c}$. The index $(U(\Lambda):U_\mathfrak{c}(\Lambda))$ is finite, and clearly

$$W=U_\mathfrak{c}(\Lambda)\cap A^\times.$$

Let nr denote the reduced norm map $A^\times\to K^\times$, and also its extension to a continuous homomorphism $J(A)\to J(K)$. Set

$$A'=\{x\in A^\times : nr(x)=1\},$$

$$J'(A)=\{x\in J(A): nr(x)=1\}.$$

Since A satisfies the Eichler condition (and we have omitted the Archimedean completions from our definition of $J(A)$), A' is dense in $J'(A)$. Thus, since $U_\mathfrak{c}(\Lambda)$ is open in $J(A)$, we have $J'(A)\subset U_\mathfrak{c}(\Lambda)\cdot A^\times$. Since $J'(A)$ contains the commutator subgroup of $J(A)$, the set $A^\times\cdot U_\mathfrak{c}(\Lambda)=U_\mathfrak{c}(\Lambda)\cdot A^\times$ is a group, and indeed an open normal subgroup of $J(A)$. The quotient

$$H_\mathfrak{c}=J(A)/A^\times U_\mathfrak{c}(\Lambda) \tag{2.1}$$

is a discrete abelian group.

The class group $Cl(\Lambda)$ of Λ is given by

$$Cl(\Lambda)=J(A)/A^\times U(\Lambda)$$

(see [5]), and this is a finite abelian group. We have a natural map $H_\mathfrak{c}\to Cl(\Lambda)$, which is surjective and whose kernel has order $(A^\times U(\Lambda): A^\times U_\mathfrak{c}(\Lambda))$. We have a canonical surjection $U(\Lambda)/U_\mathfrak{c}(\Lambda)\to A^\times U(\Lambda)/A^\times U_\mathfrak{c}(\Lambda)=U(\Lambda)\,A^\times/A^\times U_\mathfrak{c}(\Lambda)$. Each fibre here is in bijection with $\Lambda^\times/W$, and therefore we have

$$(A^\times U(\Lambda): A^\times U_\mathfrak{c}(\Lambda))=(U(\Lambda): U_\mathfrak{c}(\Lambda))\,(\Lambda^\times : W)^{-1}.$$

This is finite, and so therefore is H_c. For brevity, we put

$$|H_c| = h_c. \tag{2.2}$$

Now we recall some facts about idelic zeta-integrals. See also [1] and [2], § 2. Let $\mathrm{Ad}(A)$ denote the ring of *finite adeles* of A: this is the topological restricted direct product of the rings A_P with respect to the subrings Λ_P, with P ranging over all maximal ideals of R. We consider functions $\Phi : \mathrm{Ad}(A) \to \mathbb{C}$ of the following form. For each P, we take a locally constant compactly supported function $\Phi_P : A_P \to \mathbb{C}$ such that, for all but a finite number of primes P, the function Φ_P is the characteristic function of Λ_P. Then

$$x \mapsto \Phi(x) = \prod_P \Phi_P(x_P), \quad x = (x_P) \in \mathrm{Ad}(A),$$

defines a continuous function on $\mathrm{Ad}(A)$. Notice that, by the definitions, this infinite product is effectively finite. We may view $J(A)$ as a subset of $\mathrm{Ad}(A)$, when Φ defines a function $J(A) \to \mathbb{C}$, again denoted by Φ, which is continuous in the idele topology.

Let $\psi : J(A) \to \mathbb{C}^\times$ be a continuous homomorphism with finite image. We write ψ_P for the restriction of ψ to $A_P^\times \subset J(A)$. Then, by continuity, ψ_P is trivial on $\Lambda_P^\times$ for almost all P. Next, for $x = (x_P) \in J(A)$, we set

$$\|x\| = \prod_P \|x_P\|_P, \quad \text{where} \quad \|x_P\|_P = (\Lambda_P : \Lambda_P x)^{-1}.$$

This product is again effectively finite, and coincides with our earlier definition of $\|x\|$ when $x \in A^\times$.

Finally, we take a Haar measure $d^\times x$ on $J(A)$. For definiteness, we choose it as follows. For each P, let $d_P^\times x$ be the Haar measure on $A_P^\times$ for which

$$\int_{\Lambda_P^\times} d_P^\times x = 1.$$

We then take for $d^\times x$ the unique Haar measure on $J(A)$ whose restriction to $U(\Lambda)$ is the product of the measures $d_P^\times x$. This gives

$$\int_{U(\Lambda)} d^\times x = 1.$$

Now we can define

$$Z(\Phi, \psi, s) = \int_{J(A)} \Phi(x)\, \psi(x)\, \|x\|^s\, d^\times x, \quad \mathrm{Re}(s) > 1.$$

The proof outlined in [2] (see [6], Ch. IX for more details) shows

$$Z(\Phi, \psi, s) = \prod_P Z(\Phi_P, \psi_P, s), \quad \mathrm{Re}(s) > 1,$$

where

$$Z(\Phi_P, \psi_P, s) = \int_{A_P^\times} \Phi_P(x)\, \psi_P(x)\, \|x\|_P^s\, d_P^\times x.$$

For the definitions and properties of the standard L-functions $L_A(s, \psi)$ and $L_{A_P}(s, \psi_P)$, see [2]. We summarise the relevant properties of the zeta-integrals:

(2. 3) Lemma. (i) *For each P, the quotient $\theta_P(s)=Z(\Phi_P, \psi_P, s)\, L_{\Lambda_P}(s, \psi_P)^{-1}$ is a polynomial in p^s and p^{-s}, where p is the rational prime below P. Further, θ_P is identically 1 if ψ_P is trivial on $\Lambda_P^\times$ and Φ_P is the characteristic function of Λ_P.*

(ii) *We have $\theta_P(s)=1$ for almost all P, and*

$$Z(\Phi, \psi, s)=\theta(s)\, L_A(s, \psi), \quad \text{where} \quad \theta(s)=\prod_P \theta_P(s).$$

(iii) *$Z(\Phi, \psi, s)$ admits analytic continuation to a meromorphic function on the whole s-plane. It is holomorphic if ψ is non-trivial. If $\psi=1$, $Z(\Phi, \psi, s)$ has a simple pole at $s=1$ with residue*

$$\theta(1) \lim_{s \to 1} (s-1)\, \zeta_A(s).$$

Remark. When ψ is trivial, $Z(\Phi, \psi, s)$ will, in general, have other singularities in the region $\mathrm{Re}(s) \leqq 1$ other than the pole at $s=1$.

Proof. (i) The first statement comes from the definition of the local standard L-function in [2], § 3. The second follows from [2], (3. 9) and (3. 6) (in particular the formula on p. 165, line 22).

(ii) is immediate from (i) and the Euler products for $Z(\Phi, \psi, s)$ and $L_A(s, \psi)$ (see [2], §4).

(iii) The analytic continuation and the behaviour at $s=1$ follow from [2], (4. 1). The residue formula is now immediate, given that if ψ is trivial, then $L_A(s, \psi)=\zeta_A(s)$. This follows from [2], (3. 9), (1. 3) and (2. 6).

We now take a special function $\Phi=\prod_P \Phi_P$ on $\mathrm{Ad}(A)$ given by

(2. 4) $\qquad \Phi_P=$ *the characteristic function of* $m+\mathfrak{c}\Lambda_P$.

We have $m \in \mathfrak{c}\Lambda_P=\Lambda_P$ if P does not divide $\mathfrak{c}$, so that Φ_P is the characteristic function of Λ_P for almost all P. Now let ψ range over the character group $\mathrm{Hom}(H_\mathfrak{c}, \mathbb{C}^\times)$ of the finite abelian group $H_\mathfrak{c}$. We may view any such ψ as a continuous character of $J(A)$ which is trivial on $A^\times U_\mathfrak{c}(\Lambda)$. We have, for $x \in J(A)$,

$$\sum_\psi \psi(x)=\begin{cases} h_\mathfrak{c} & \text{if } x \in A^\times U_\mathfrak{c}(\Lambda), \\ 0 & \text{otherwise.} \end{cases}$$

Therefore, with this Φ and the same range of ψ,

$$\sum_\psi Z(\Phi, \psi, s)=h_\mathfrak{c} \int_{A^\times U_\mathfrak{c}(\Lambda)} \Phi(x)\, \|x\|^s\, d^\times x.$$

The left-hand side here is continuable to a meromorphic function on the whole s-plane. It has a simple pole at $s=1$ with residue equal to that of $Z(\Phi, \psi_0, s)$, where ψ_0 is the trivial character. In the right-hand side, both the integrand and the range of integration are invariant under left translation by the compact open subgroup $U_\mathfrak{c}(\Lambda)$ of $J(A)$. Thus

$$\int_{A^\times U_\mathfrak{c}(\Lambda)} \Phi(x)\, \|x\|^s\, d^\times x=\mu^\times(U_\mathfrak{c}(\Lambda)) \sum_{x \in U_\mathfrak{c}(\Lambda) \setminus A^\times U_\mathfrak{c}(\Lambda)} \Phi(x)\, \|x\|^s.$$

Here, $\mu^\times(U_\mathfrak{c}(\Lambda))$ is the measure of $U_\mathfrak{c}(\Lambda)$ with respect to the measure $d^\times x$. By our choice of $d^\times x$, we have

$$\mu^\times(U_\mathfrak{c}(\Lambda)) = (U(\Lambda) : U_\mathfrak{c}(\Lambda))^{-1}\,\mu^\times(U(\Lambda)) = (U(\Lambda) : U_\mathfrak{c}(\Lambda))^{-1}.$$

We have $A^\times U_\mathfrak{c}(\Lambda) = U_\mathfrak{c}(\Lambda)\,A^\times$, and a canonical identification $U_\mathfrak{c}(\Lambda)\backslash U_\mathfrak{c}(\Lambda)A^\times = W\backslash A^\times$. On $A^\times$, Φ is just the characteristic function of $(m+\mathfrak{c}\Lambda)\cap A^\times$. Thus this last series is precisely $F(s)$, and therefore

$$F(s) = h_\mathfrak{c}^{-1}(U(\Lambda) : U_\mathfrak{c}(\Lambda)) \sum_\psi Z(\Phi, \psi, s),$$

where ψ ranges over the characters of $H_\mathfrak{c}$. This gives the analytic continuation of $F(s)$. To complete the proof of Theorem C′ in this case, we just have to compute the residue of $F(s)$ at $s=1$.

Let S be the set of primes P which divide $\mathfrak{c}$. If ψ_0 denotes the trivial character of $J(A)$, we write $Z(\Phi, s) = Z(\Phi, \psi_0, s)$, and $Z(\Phi_P, s) = Z(\Phi_P, \psi_{0,P}, s)$, for brevity. If $P\notin S$, we have

$$Z(\Phi_P, s) = \zeta_{\Lambda_P}(s),$$

by the definitions of S, Φ, and (2. 3). It also follows from (2. 3) that

$$\operatorname*{res}_{s=1} Z(\Phi, s) = \prod_{P\in S} Z(\Phi_P, 1)\operatorname*{res}_{s=1}\prod_{P\notin S}\zeta_{\Lambda_P}(s).$$

Of course, we view this last infinite product as a meromorphic function of s via analytic continuation.

To calculate $Z(\Phi_P, 1)$ for $P\in S$, we let $d_P x$ be that Haar measure on A_P for which $d_P^\times x = \|x\|_P^{-1} d_P x$. Recall that $d_P^\times x$ gives $\Lambda_P^\times$ total measure 1. Therefore

$$Z(\Phi_P, 1) = \int_{A_P^\times}\Phi_P(x)\,\|x\|_P\,d_P^\times x = \int_{A_P}\Phi_P(x)\,d_P x$$

(by [2], (6. 9)), and this is just the measure, with respect to $d_P x$, of the support of the characteristic function Φ_P. Thus, using μ to denote measure with respect to $d_P x$, we have

$$Z(\Phi_P, 1) = \mu(m+\mathfrak{c}\Lambda_P) = \mu(\mathfrak{c}\Lambda_P) = \mu(\Lambda_P)\,(\Lambda_P : \mathfrak{c}\Lambda_P)^{-1}.$$

However, if Ψ_P denotes the characteristic function of Λ_P, (2. 3) gives $Z(\Psi_P, s) = \zeta_{\Lambda_P}(s)$, and we obtain, in the same way, $Z(\Psi_P, 1) = \zeta_{\Lambda_P}(1) = \mu(\Lambda_P)$. Altogether, we now have

$$\begin{aligned}\lim_{s\to 1}(s-1)F(s) &= h_\mathfrak{c}^{-1}(U(\Lambda) : U_\mathfrak{c}(\Lambda))\prod_{P\in S}(\Lambda_P : \mathfrak{c}\Lambda_P)^{-1}\lim_{s\to 1}(s-1)\,\zeta_\Lambda(s)\\ &= h_\mathfrak{c}^{-1}(U(\Lambda) : U_\mathfrak{c}(\Lambda))\,(\Lambda : \mathfrak{c}\Lambda)^{-1}\lim_{s\to 1}\zeta_\Lambda(s).\end{aligned}$$

Finally, let $h_\Lambda = |Cl(\Lambda)|$, the class number of Λ. We have already noted that

$$h_\mathfrak{c} = h_\Lambda(U(\Lambda) : U_\mathfrak{c}(\Lambda))\,(A^\times : W)^{-1}.$$

This completes the proof of Theorem C′ in this case, and also yields

(2. 5) Corollary. *Suppose that A satisfies the Eichler condition, and let Λ' be a maximal order in A. Then the constant $b_{\Lambda'}$ in Theorem C′ is*

$$h_{\Lambda'}^{-1}\lim_{s\to 1}(s-1)\,\zeta_{\Lambda'}(s), \quad \textit{where}\quad h_{\Lambda'} = |Cl(\Lambda')|.$$

Both the residue of $\zeta_{\Lambda'}(s)$ and the class number $h_{\Lambda'}$ depend only on A, and not on the choice of Λ'.

The second statement here is well-known. The residue of $\zeta_{\Lambda'}(s)$ can be found from the explicit formula for $\zeta_{\Lambda'}(s)$. Indeed, we wrote it down in [2], (6. 12).

§ 3. Totally definite quaternion algebras

We are left with the task of proving Theorem C′ when our simple $\mathbb{Q}$-algebra A does not satisfy the Eichler condition, in other words when A is a totally definite quaternion algebra. In fact, the reduction of Theorem C to Theorem C′ is of no assistance here, so we prove Theorem C directly. The methods here are necessarily very different from the ones of § 2. Those idelic methods derived from Tate's thesis, involving separation of $A^{\times}$-cosets using linear characters, can give no information concerning the functions $Z_{\Lambda}(M, s)$ in the absence of the Eichler condition. We therefore employ the older method of Hecke. Indeed, this section follows very closely Hecke's proof of the functional equation of the Dedekind zeta-function, and has very few novel features.

We need first to summarise, in a suitable form, some results from elementary harmonic analysis. Let V be a finite-dimensional real vector space equipped with a positive definite symmetric bilinear form

$$\langle\, ,\rangle : V \times V \to \mathbb{R}.$$

Choose a Haar measure dx on V. If L is a lattice in V (i.e. a discrete subgroup of V which spans V over R), we write $\mu(V/L)$ for the volume (with respect to dx) of a fundamental domain for V modulo L. Then, for lattices L_1, L_2 in V, we put

$$(L_1 : L_2) = \mu(V/L_2)\, \mu(V/L_1)^{-1}.$$

This is independent of the choice of dx, and coincides with the usual generalised index when L_1 and L_2 are commensurable.

The pairing $(x, y) \mapsto e^{-2\pi i \langle x, y\rangle}$, $x, y \in V$, identifies V with its Pontrjagin dual, and we now take for dx the corresponding self-dual Haar measure on V. Thus, if h is a suitable complex-valued function on V (i.e. h is smooth and $\langle x, x\rangle^N |h(x)| \to 0$ as $\langle x, x\rangle \to \infty$, for all N), and we define its Fourier transform $\hat{h}$ by

$$\hat{h}(y) = \int_V h(x)\, e^{-2\pi i \langle x, y\rangle}\, dx, \quad y \in V,$$

then the Fourier inversion formula $\hat{\hat{h}}(x) = h(-x)$ holds for all $x \in V$.

In particular, if ϕ is the function

$$\phi(x) = e^{-\pi \langle x, x\rangle}, \quad x \in V, \tag{3. 1}$$

we have $\hat{\phi}(x) = \phi(x)$. (This reduces to a standard calculation on choosing an orthonormal basis for V.)

For a lattice L in V, we define

$$L^* = \{y \in V: \langle x, y\rangle \in \mathbb{Z} \quad \textit{for all} \quad x \in L\}. \tag{3.2}$$

This is another lattice in V, and the Poisson summation formula says

$$\sum_{x \in L} h(x) = (L^* : L)^{-\frac{1}{2}} \sum_{y \in L^*} \hat{h}(y), \tag{3.3}$$

for any suitable function h as above. More generally, if $b \in GL(V)$ is an automorphism of V, we can define the module $\|b\|$ of b by the equation

$$\int_V f(bx)\,dx = \|b\|^{-1} \int_V f(x)\,dx,$$

for any measurable function f on V. (In fact, $\|b\| = |\det(b)|$.) The Fourier transform of the function $x \mapsto h(bx)$, $x \in V$, is

$$y \mapsto \|b\|^{-1}\,\hat{h}(b^{-t}y),$$

where the "transpose" $b^t \in GL(V)$ is given by

$$\langle bx, y\rangle = \langle x, b^t y\rangle, \quad x,\ y \in V.$$

Applying (3. 3) to this function, we get

$$\sum_{x \in L} h(bx) = \|b\|^{-1}\,(L^* : L)^{-\frac{1}{2}} \sum_{y \in L^*} \hat{h}(b^{-t}y). \tag{3.4}$$

(When $V = \mathbb{R}^n$ with its standard bilinear form, and $L = \mathbb{Z}^n$, (3. 3) is standard. The general case of (3. 3) then follows easily from the corresponding special case of (3. 4).)

We shall need the following superficially more general version of (3. 3).

(3. 5) Lemma. *Let L be a lattice in V, and $m \in V$. Then, for any suitable function h on V, we have*

$$\sum_{x \in m+L} h(x) = (L^* : L)^{-\frac{1}{2}} \sum_{y \in L^*} e^{2\pi i \langle y, m\rangle}\,\hat{h}(y).$$

Further, if $b \in GL(V)$, we have

$$\sum_{x \in m+L} h(bx) = \|b\|^{-1}(L^* : L)^{-\frac{1}{2}} \sum_{y \in L^*} e^{2\pi i \langle y, m\rangle}\,\hat{h}(b^{-t}y).$$

Proof. Let h' be the function $h'(x) = h(m+x)$, so that

$$\sum_{x \in L} h'(x) = \sum_{x \in m+L} h(x).$$

The Fourier transform of h' is $y \mapsto e^{2\pi i \langle y, m\rangle}\,\hat{h}(y)$. The first statement now follows from (3. 3) applied to the function h', and the second is immediate.

Now we return to the proof of Theorem C when A is a totally definite quaternion algebra. Let K be the centre of A, so that K is a totally real algebraic number field, A is a 4-dimensional K-division algebra, and for each real place v of K, A_v is isomorphic to the $\mathbb{R}$-algebra $\mathbb{H}$ of Hamiltonian quaternions. Thus

$$A_\infty = A \otimes_{\mathbb{Q}} \mathbb{R} = \prod_v A_v \cong \mathbb{H}^n,$$

where $n = [K : \mathbb{Q}]$.

Let "bar" denote the standard involution on A: this is the unique involution on A whose space of symmetric elements is precisely the centre of A. This extends by linearity to an involution (also denoted "bar") on A_∞, which stabilises each factor A_v, and on A_v coincides with the standard involution on $\mathbb{H}$.

We define a nondegenerate symmetric $\mathbb{Q}$-bilinear form

$$\langle\ ,\ \rangle : A \times A \rightarrow \mathbb{Q},\ \langle x, y\rangle = \mathrm{tr}_{A/\mathbb{Q}}(\bar{x}y),$$

where $\mathrm{tr}_{A/\mathbb{Q}}$ is the composite of the reduced trace $A \rightarrow K$ and the field trace $\mathrm{Tr}_{K/\mathbb{Q}}: K \rightarrow \mathbb{Q}$. This extends by linearity to a positive definite symmetric $\mathbb{R}$-bilinear form

$$\langle\ ,\ \rangle : A_\infty \times A_\infty \rightarrow \mathbb{R}.$$

We apply the above theory to the pair $(A_\infty, \langle\ ,\ \rangle)$. We set

$$\phi(x) = e^{-\pi\langle x, x\rangle},\quad x \in A_\infty.$$

We take Fourier transforms with respect to the self-dual Haar measure on A_∞, and this gives $\hat{\phi} = \phi$, as above.

Further, there is a uniquely determined Haar measure $d^\times x$ on $A_\infty^\times$ such that

$$\int_{A_\infty^\times} \phi(x)\, \|x\|^s\, d^\times x = \{(2\pi)^{-2s}\,\Gamma(2s)\}^n,\quad \mathrm{Re}(s) > 0,$$

where $\|x\|$ is the module of $x \in A_\infty^\times$ viewed as a linear automorphism of A_∞, and Γ is Euler's Γ-function. (See the proof of (3. 7) below for more details.) This module $\|x\|$ of $x = (x_v) \in A_\infty^\times$ is given explicitly by

$$\|x\| = \prod_v nr_v(x_v)^2,$$

where nr_v is the reduced norm map from $A_v = \mathbb{H}$ to $\mathbb{R}$.

We now take a triple (W, L, m) as in Theorem C. Via the diagonal embedding $A \rightarrow A_\infty$, L becomes a lattice in A_∞. Notice that, for $x \in A^\times$, we have $\|x\| = (L : Lx)$, which is *inverse* to our previous use of this symbol. We put

$$\tilde{F}_{(W, L, m)}(s) = \tilde{F}(s) = F_{(W, L, m)}(s)\ \{(2\pi)^{-2s}\,\Gamma(2s)\}^n,$$

and examine this function. Immediately, for $y \in A_\infty^\times$, we have

$$\|y\|^{-s}\, \{(2\pi)^{-2s}\,\Gamma(2s)\}^n = \int_{A_\infty^\times} \phi(xy)\, \|x\|^s\, d^\times x,$$

so

$$\tilde{F}_{(W, L, m)}(s) = \sum_{y \in W \setminus (m+L) \cap A^\times} \int_{A_\infty^\times} \phi(xy)\, \|x\|^s\, d^\times x,\quad \mathrm{Re}(s) > 1.$$

Absolute uniform convergence on compact subsets of the half-plane $\mathrm{Re}(s) > 1$ ensures that we may rewrite this as

$$\tilde{F}(s) = \int_{A_\infty^\times} \sum_{y \in W \setminus (m+L) \cap A^\times} \phi(xy)\, \|x\|^s\, d^\times x.$$

Now we apply a variant of Fubini's theorem. There is a uniquely determined left $A_\infty^\times$-invariant measure $d\dot{x}$ on the homogeneous space $A_\infty^\times/W$ such that

$$\int_{A_\infty^\times} k(x)\, d^\times x = \int_{A_\infty^\times/W} \sum_{w\in W} k(xw)\, d\dot{x}$$

for any measurable function k on $A_\infty^\times$. Thus

$$\tilde{F}(s) = \int_{A_\infty^\times/W} \|x\|^s \sum_{y\in(m+L)\cap A^\times} \phi(xy)\, d\dot{x}, \quad \mathrm{Re}(s) > 1.$$

(Here we use the fact that $\|w\| = 1$ for all $w \in W$.)

The map $x \mapsto \|x\|$ is a surjective homomorphism from $A_\infty^\times$ to the group $\mathbb{R}_+^\times$ of positive real numbers, and it is split by $t \mapsto t^{1/4n}$, $t \in \mathbb{R}_+^\times$. We therefore have

$$G \times \mathbb{R}_+^\times \cong A_\infty^\times, \ (g, t) \mapsto t^{1/4n} g,$$

where G is the kernel of $\|\cdot\|$. If dt is the Lebesgue measure on $\mathbb{R}$, this gives us $d^\times x = dg \cdot \frac{dt}{t}$, for some Haar measure dg on G. We have $W \subset G$, and so

$$d\dot{x} = d\dot{g} \cdot \frac{dt}{t},$$

for some left G-invariant measure $d\dot{g}$ on G/W.

We transform our expression for $\tilde{F}(s)$ in accordance with all this to obtain

$$\tilde{F}(s) = \int_0^\infty t^s \int_{G/W} \sum_{y\in(m+L)\cap A^\times} \phi(t^{1/4n} gy)\, d\dot{g} \frac{dt}{t}, \quad \mathrm{Re}(s) > 1.$$

The next step is to split this t-integral into two parts. By comparison with an earlier expression for $\tilde{F}(s)$, the tail end

$$\int_1^\infty t^s \int_{G/W} \sum_y \phi(t^{1/4n} gy)\, d\dot{g} \frac{dt}{t} \tag{3. 6}$$

converges absolutely and uniformly on compact subsets of the s-plane, and so defines an everywhere holomorphic function of s. This leaves us with

$$E(s) = \int_0^1 t^s \int_{G/W} \sum_{y\in(m+L)\cap A^\times} \phi(t^{1/4n} gy)\, d\dot{g} \frac{dt}{t}.$$

The set $(m+L)\cap A^\times$ is just the set of non-zero points of the coset $m+L$. Indeed, it is the whole of $m+L$ unless $m \in L$. We consequently define

$$\delta_m = \begin{cases} 1 & \text{if } m \in L, \\ 0 & \text{otherwise.} \end{cases}$$

We now wish to apply the Poisson summation formula (3. 5) which here says that

$$\sum_{y\in m+L} \phi(t^{1/4n} gy) = t^{-1} (L^* : L)^{-\frac{1}{2}} \sum_{z\in L^*} e^{2\pi i \langle z, m\rangle} \phi(t^{-1/4n} \bar{g}^{-1} z).$$

(We use here the fact that $\hat{\phi}=\phi$, and also that the transpose operation on $A_\infty^\times$ defined by the pairing $\langle\ ,\ \rangle$ is simply $x \mapsto \bar{x}$.) We have to describe the set $L^* \cap A_\infty^\times$. Since the pairing $\langle\ ,\ \rangle : A \times A \to Q$ is nonsingular, any $x \in A_\infty$ satisfying $\langle x, A\rangle \subset Q$ must actually lie in A. It follows that $L^* \subset A$, and therefore $L^* \cap A_\infty^\times = L^* \cap A^\times = L^* - \{0\}$. Finally, the map $g \mapsto \bar{g}^{-1}$ defines a homeomorphism $G/W \cong G/\bar{W}$, and transforms the measure $d\dot{g}$ into a left G-invariant measure d^*g, say, on $G/\bar{W}$. In all, we have

$$E(s)=\int_0^1 t^s \int_{G/W} \{-\delta_m + t^{-1}(L^*:L)^{-\frac{1}{2}}\}\, d\dot{g}\,\frac{dt}{t}$$

$$+\int_0^1 \int_{G/\bar{W}} t^{s-1}(L^*:L)^{-\frac{1}{2}} \sum_{z\in L^*\cap A^\times} e^{2\pi i\langle z,m\rangle}\,\phi(t^{-1/4n}gz)\, d^*g\,\frac{dt}{t}.$$

In the last integral, we replace t by t^{-1} to obtain

$$\int_1^\infty \int_{G/\bar{W}} t^{1-s}(L^*:L)^{-\frac{1}{2}} \sum_{z\in L^*\cap A^\times} e^{2\pi i\langle z,m\rangle}\,\phi(t^{1/4n}gz)\, d^*g\,\frac{dt}{t}.$$

This is majorised by an integral of the same type as (3. 6), and therefore defines an everywhere holomorphic function of s. In all, we have

$$\tilde{F}(s)=\mu^\times(G/W)\left\{\frac{-\delta_m}{s-1}+\frac{(L^*:L)^{-\frac{1}{2}}}{s-1}\right\}+I(s),$$

where $I(s)$ is everywhere holomorphic and

$$\mu^\times(G/W)=\int_{G/W} d\dot{g}.$$

This gives the analytic continuation of $\tilde{F}(s)$, with a simple pole at $s=1$, where the residue is

$$\lim_{s\to 1}(s-1)\,\tilde{F}(s)=\mu^\times(G/W)\,(L^*:L)^{-\frac{1}{2}}.$$

(3. 7) Lemma. *Let $d^\times x$ be the Haar measure on $A_\infty^\times$ for which*

$$\int_{A_\infty^\times} \phi(x)\,\|x\|^s\, d^\times x=\{(2\pi)^{-2s}\,\Gamma(2s)\}^n.$$

Let G be the kernel of the module $\|\cdot\| : A_\infty^\times \to \mathbb{R}_+^\times$, and dg the Haar measure on G such that $d^\times x = dg\cdot\frac{dt}{t}$, for the Lebesgue measure dt on $\mathbb{R}$, under the isomorphism $G\times\phi_+^\times \cong A_\infty^\times$ given by $(g,t)\mapsto t^{1/4n}g$. Let $d\dot{g}$ be the left G-invariant measure on G/W such that

$$\int_G k(g)\,dg=\int_{G/W}\sum_{w\in W} k(gw)\,d\dot{g},$$

for any measurable function k on G. Then

$$\mu^\times(G/W)=\int_{G/W} d\dot{g}=2^{n-3}\,\mathrm{Reg}(K)\,(W:R^\times)^{-1},$$

where $\mathrm{Reg}(K)$ *is the regulator of the field K, and R is the ring of integers of K.*

(3. 8) Lemma. *Let L be a lattice in A, and S the set of maximal ideals of R which ramify in A. Then*

$$(L^* : L)^{-\frac{1}{2}} = (\Lambda' : L)^{-1}\, d_K^{-2} \prod_{P \in S} \mathbb{N}P^{-1},$$

where Λ' is any maximal order in A, d_K is the discriminant of the field K (which is necessarily positive), and $\mathbb{N}P = (R : P)$.

Postponing the proofs of these lemmas for the moment, we now write down the residue of the function $F(s) = F_{(W, L, m)}(s)$. Let Λ' be a maximal order in A. Then, from (3. 7), (3. 8),

$$\lim_{s \to 1} (s-1)\, F_{(W, L, m)}(s) = (\Lambda' : L)^{-1} (\Lambda'^{\times} : W)\, b_{\Lambda'}, \quad \text{where}$$

$$b_{\Lambda'} = (2\pi)^{2n}\, 2^{n-3}\, \mathrm{Reg}(K)\, (\Lambda'^{\times} : R^{\times})^{-1}\, d_K^{-2} \prod_{P \in S} \mathbb{N}P^{-1}. \tag{3. 9}$$

This constant certainly has the property demanded of it by Theorem C. Notice that $\Lambda'^{\times}$ is a discrete subgroup of G, and it is well-known (and not hard to prove: see the proof of (3. 7) below) that $G/R^{\times}$ is compact. Thus the index $(\Lambda'^{\times} : R^{\times})$ is finite.

Proof of (3. 7). (This calculation, in an essentially equivalent form, is contained in [4]. However, for the convenience of the reader, and because it contains some useful ideas, we include this account.) For $x, y \in \mathbb{H}$, we put $\langle x, y\rangle = \mathrm{tr}(\bar{x}y)$, where tr is the reduced trace map from $\mathbb{H}$ to $\mathbb{R}$. Let C denote the kernel of the module $\|\cdot\|_{\mathbb{H}}$ on $\mathbb{H}^{\times}$. We identify $\mathbb{H}^{\times}$ with $C \times \mathbb{R}_+^{\times}$ via the map $(c, t) \mapsto t^{1/4}c$. We let $d^{\times}y = dc \cdot \frac{dt}{t}$, where dt is the Lebesgue measure on $\mathbb{R}$ and dc is the Haar measure on the compact group C for which

$$\int_C dc = \frac{1}{2}.$$

One finds readily that

$$\int_{\mathbb{H}^{\times}} e^{-\pi\langle y, y\rangle} \|y\|^s\, d^{\times}y = \frac{1}{2} \int_0^{\infty} e^{-2\pi\sqrt{t}}\, t^s \frac{dt}{t} = (2\pi)^{-2s}\,\Gamma(2s).$$

For each real place v of K, we choose an isomorphism $\mathbb{H} \cong A_v$ which thus transforms $d^{\times}y$ into a Haar measure $d^{\times}x_v$ on $A_v^{\times}$, and this measure is independent of the choice of isomorphism. We let $d^{\times}x$ be the product of the measures $d^{\times}x_v$, and then

$$\int_{A_\infty^{\times}} \phi(x)\, \|x\|^s\, d^{\times}x = \{(2\pi)^{-2s}\,\Gamma(2s)\}^n,$$

so that $d^{\times}x$ is the required measure on $A_\infty^{\times}$.

We calculate $\mu^{\times}(G/R^{\times})$, where R is the ring of integers in K. The general case of the lemma is then immediate. For each real place v of K, let $|\cdot|_v$ be the associated normalised absolute value on $K^{\times}$ (or $K_v^{\times}$). Also, let $K_{v,+}$ be the group of positive elements in $K_v^{\times}$. If C_v denotes the kernel of the module $\|\cdot\|_v$ on $A_v^{\times}$, we have an exact sequence

$$1 \to \prod_v C_v \to A_\infty^{\times} \to \prod_v K_{v,+} \to 1,$$

in which the surjection is the product of the modules $\|\cdot\|_v$. The already specified measures on $A_\infty^\times$ and the C_v induce the natural measure on $\prod_v K_{v,+}$. This sequence restricts to an exact sequence

$$1 \to \prod_v C_v \to G \xrightarrow{j} G_0 \to 1,$$

where G_0 is the kernel of the map $(x_v) \mapsto \prod |x_v|_v$ in $\prod K_{v,+}$. The natural embedding of R in A induces an embedding $R^\times \to G$. The intersection $R^\times \cap \prod_v C_v$ is the torsion subgroup $\{\pm 1\}$ of $R^\times$. The group $\prod C_v$ has measure 2^{-n}, so

$$\mu^\times(G/R^\times) = 2^{-(n+1)} \mu^\times(G_0/j(R^\times)),$$

where j is the surjection $G \to G_0$ above. However, the composite map $R^\times \to G \to G_0$ is *not* the natural map $u \mapsto (|u|_v)$ of $R^\times$ in G_0. Indeed, we have a commutative diagram

$$\begin{array}{ccc} K^\times & \longrightarrow & A_\infty^\times \\ {\scriptstyle \prod|\cdot|_v}\downarrow & & \downarrow{\scriptstyle \prod\|\cdot\|_v} \\ \prod_v K_{v,+} & \xrightarrow{\ 4\ } & \prod_v K_{v,+} \end{array}$$

where the bottom map raises elements of $K_{v,+}$ to the 4-th power. The volume of G_0 modulo the natural image of $R^\times$ is, by definition, the regulator $\operatorname{Reg}(K)$ of K. Therefore $\mu^\times(G_0/j(R^\times)) = 4^{n-1} \operatorname{Reg}(K)$. The lemma follows.

Proof of (3. 8). Since "bar" is an involution, we have $(L^* : L) = (\bar{L}^* : L)$. Also, $\bar{L}^*$ is the ordinary reduced trace dual of L in A:

$$\bar{L}^* = \{x \in A : \operatorname{tr}_{A/\mathbb{Q}}(xL) \subset \mathbb{Z}\}.$$

Choose a maximal order Λ' in A. Then

$$(\bar{L}^* : L) = (\Lambda' : L)^2 (\bar{\Lambda}'^* : \Lambda').$$

The reduced trace dual $\bar{\Lambda}'^*$ of Λ' is the *inverse different* $\mathfrak{D}_A^{-1}$ of A (relative to the maximal order Λ'). It is a two-sided Λ'-lattice, and one may calculate it locally. If P is a maximal ideal of R which is unramified in A, then $\mathfrak{D}_{A,P}^{-1} = \mathfrak{D}_{K_P}^{-1} \Lambda'_P$, where $\mathfrak{D}_{K_P}^{-1}$ is the usual absolute inverse different of the local field K_P. If P ramifies in A, then Λ'_P is a non-commutative discrete valuation ring with maximal ideal $\mathfrak{P}$, and

$$\mathfrak{D}_{A,P}^{-1} = \mathfrak{P}^{-1} \mathfrak{D}_{K_P}^{-1} \Lambda'_P = P^{-1} \mathfrak{D}_{K_P}^{-1} \mathfrak{P} \Lambda'_P.$$

The index $(\Lambda'_P : \mathfrak{P})$ is $(R : P)^2$, and the lemma follows.

§ 4. Some consequences

We now list some corollaries of Theorems A and B. At this stage, we have a complete account of the constant $b_{\Lambda'}$, from (2. 5), (3. 9) and Theorem B. There is no new information in this, however. The most interesting applications rely only on the existence and formal properties of $b_{\Lambda'}$.

First, we give the complete version of the asymptotic distribution result [2], (6. 10).

(4. 1) Theorem. *Let A be a finite-dimensional semisimple $\mathbf{Q}$-algebra with r simple components. Let Λ be an order in A, Λ' a maximal order in A, and M a left ideal of Λ (such that $\mathbf{Q}M = A$). Let $M^{-1} = \{x \in A : Mx \subset \Lambda\}$, and $\Gamma_M = \mathrm{End}_\Lambda(M)$ (viewed as an order in A). For $t > 0$, let $N(M, T)$ denote the number of left ideals X of Λ which satisfy*

$$\text{(i) } X \cong M, \quad \textit{and} \quad \text{(ii) } (\Lambda : X) \leqq T.$$

Then

$$N(M, T) = \left\{ \frac{1}{(r-1)!} \frac{(M^{-1} : \Lambda)}{(\Lambda : M)(\Lambda' : \Lambda)} (\Lambda'^{\times} : \Gamma_M^{\times})\, b_{\Lambda'} + o(1) \right\} T(\log T)^{r-1}$$

as $T \to \infty$, where $b_{\Lambda'}$ is the constant in Theorem A.

The singular behaviour at $s = 1$ of the function $Z_\Lambda(M, s)$ is given by Theorem A and the expression $Z_\Lambda(M, s) = (\Lambda : M)^{-s} f_{(\Gamma_M^{\times}, M^{-1})}(s)$. (4. 1) now follows from the Delange-Ikehara Tauberian theorem, just as in [2]. (Here is a situation in which it is useful to know $b_{\Lambda'}$ explicitly.)

Again take A and Λ' as in (4. 1). Let $L_1, L_2, \ldots, L_{h'}$ be a set of representatives of the isomorphism classes of left ideals of Λ'. Thus

$$\zeta_{\Lambda'}(s) = \sum_{i=1}^{h'} Z_{\Lambda'}(L_i, s).$$

Put $\Lambda_i = \mathrm{End}_{\Lambda'}(L_i)$, $L_i^{-1} = \{x \in A : L_i x \subset \Lambda'\}$. Since L_i is a projective Λ'-module, we have $(L_i^{-1} : \Lambda') = (\Lambda' : L_i)$. Theorem A gives the relation

$$\lim_{s \to 1} (s-1)^r \zeta_{\Lambda'}(s) = b_{\Lambda'} \sum_{i=1}^{h'} (\Lambda'^{\times} : \Lambda_i^{\times}). \tag{4. 2}$$

The left hand side of (4. 2) can be written down explicitly in terms of the structure of A. See the discussion in [2], § 6, where this quantity is called c_A.

(4. 3) Remark. When A is a totally definite quaternion algebra, and we substitute in (4. 2) the known value of the residue of $\zeta_{\Lambda'}(s)$ and the value of $b_{\Lambda'}$ given by (3. 9), we obtain the "mass formula" of Eichler [4]. (This formula is quoted in [2], (9. 2).) On the other hand, when A satisfies the Eichler condition, (4. 2) is essentially tautological.

We can derive analogues of (4. 2) for ideals in an arbitrary order Λ in A. Indeed, we can make much finer statements. Let M, Γ_M and M^{-1} be as in (4. 1). The quantity $(M^{-1} : \Lambda)(\Lambda : M)^{-1}$ depends only on the *genus* $g(M)$ of M. (Recall that X is said to lie in the genus of M if $X_P \cong M_P$ as Λ_P-modules for all P.) Let $M_1, M_2, \ldots, M_{e_M}$ be a set of representatives of the isomorphism classes in the *stable isomorphism class* $[M]$ of M. (Recall that X is said to be stably isomorphic to M if X lies in the genus of M and, for some (or equivalently any) Y in the genus of M, we have $X \oplus Y \cong M \oplus Y$.) By definition

$$Z_\Lambda([M], s) = \sum_{j=1}^{e_M} Z_\Lambda(M_j, s),$$

and therefore

$$(4.4)\qquad \lim_{s\to 1}(s-1)^r Z_\Lambda([M],s)=b_{\Lambda'}\frac{(M^{-1}:\Lambda)}{(\Lambda:M)(\Lambda':\Lambda)}\sum_{j=1}^{e_M}(\Lambda'^\times:\Gamma_j^\times),$$

where, for brevity, $\Gamma_j=\Gamma_{M_j}$. However, from [2], (6. 10), we know that

$$(4.5)\qquad \lim_{s\to 1}(s-1)^r Z_\Lambda([M],s)=h_M^{-1}\frac{(M^{-1}:\Lambda)}{(\Lambda:M)(\Lambda':\Lambda)}\big(U(\Lambda'):U(\Gamma_M)\big)\lim_{s\to 1}(s-1)^r\zeta_{\Lambda'}(s).$$

Here, h_M is the number of stable isomorphism classes in the genus of M. (These stable classes form a finite abelian group $Cl(M)$, which is isomorphic to $Cl(\Gamma_M)$.) The unit idele index $\big(U(\Lambda'):U(\Gamma_M)\big)$ clearly depends only on the genus of M. Combining (4. 5) with (4. 2), we obtain

$$(4.6)\qquad h_M\sum_{j=1}^{e_M}(\Lambda'^\times:\Gamma_j^\times)=\big(U(\Lambda'):U(\Gamma_M)\big)\sum_{i=1}^{h'}(\Lambda'^\times:\Lambda_i^\times).$$

We can compare this for two stable classes, eliminate the contribution from the maximal order Λ', and thereby obtain the following completely symmetric result.

(4. 7) Theorem (Weight formula for stable isomorphism classes). *Let A be as in* (4. 1). *Let Λ_i be an order in A, and M_i a left ideal if Λ_i, for $i=1, 2$. Let $M_{i1}, M_{i2},\ldots, M_{ie_i}$ be a set of representatives of the isomorphism classes in the stable isomorphism class $[M_i]$ of M_i. Put $\Gamma_i=\Gamma_{M_i}$, $\Gamma_{ij}=\Gamma_{M_{ij}}$. Then*

$$\frac{\big(U(\Gamma_1):U(\Gamma_2)\big)}{(\Lambda_1^\times:\Lambda_2^\times)}h_{M_1}\sum_{j=1}^{e_1}(\Lambda_1^\times:\Gamma_{1j}^\times)=h_{M_2}\sum_{j=1}^{e_2}(\Lambda_2^\times:\Gamma_{2j}^\times),$$

where $h_{M_i}=|Cl(\Gamma_i)|$ is the number of stable classes in the genus of M_i.

Certain special cases of this are of interest. For example, take $\Lambda_1=\Lambda_2=\Lambda$, to obtain

(4. 8) Corollary. $\big(U(\Gamma_{M_1}):U(\Gamma_{M_2})\big)h_{M_1}\sum_{j=1}^{e_1}(\Lambda^\times:\Gamma_{1j}^\times)=h_{M_2}\sum_{j=1}^{e_2}(\Lambda^\times:\Gamma_{2j}^\times),$

If we specialise further and take $M_2\in g(M_1)$, we obtain the very appealing result:

(4. 9) Corollary. *Let M, M' be left ideals of Λ lying in the same genus. Let $M_1, M_2,\ldots, M_e$* (resp. *$M_1', M_2',\ldots, M_{e'}'$*) *be a set of representatives of the isomorphism classes in the stable isomorphism class of M* (resp. *M'*). *Put $\Gamma_j=\Gamma_{M_j}$, $\Gamma_j'=\Gamma_{M_j'}$. Then*

$$\sum_{j=1}^{e}(\Lambda^\times:\Gamma_j^\times)=\sum_{j=1}^{e'}(\Lambda^\times:\Gamma_j'^\times).$$

If we think of the index $(\Lambda^\times:\Gamma_M^\times)$ as a "weight" attached to the isomorphism class of M, then (4. 9) says that weighted isomorphism classes are uniformly distributed among the stable classes in a fixed genus.

(4. 10) Remark. When A satisfies the Eichler condition, the unit index $(\Lambda^\times:\Gamma_M^\times)$ depends only on the genus of M (exercise!). In (4. 6) we get $e_1=e_2=1$. If we take $\Lambda_2=M_2=\Lambda'$, (a maximal order), then (4. 6) reduces to Fröhlich's formula (see [5]):

$$h_M/h_{\Lambda'}=|Cl(\Gamma_M)|/|Cl(\Lambda')|=\big(\Lambda^\times U(\Lambda'):\Lambda^\times U(\Gamma_M)\big).$$

References

[1] *C. J. Bushnell and I. Reiner*, Zeta functions of arithmetic orders and Solomon's conjectures, Math. Z. **173** (1980), 135—161.

[2] *C. J. Bushnell and I. Reiner*, L functions of arithmetic orders and asymptotic distribution of ideals, J. reine angew. Math. **327** (1981), 156—183.

[3] *C. J. Bushnell and I. Reiner*, Functional equations for L-functions of arithmetic orders, J. reine angew. Math. **329** (1981), 88—123.

[4] *M. Eichler*, Idealklassenzahl total definiter Quaternionenalgebren, Math. Z. **43** (1938), 102—109.

[5] *A. Fröhlich*, Locally free modules over arithmetic orders, J. reine angew. Math. **274/275** (1975), 112—138.

[6] *A. Weil*, Basic Number Theory, Berlin-Heidelberg-New York 1974.

Department of Mathematics, King's College, Strand, London WC2R 2LS

University of Illinois, Department of Mathematics, 1409 West Green Street, Urbana, Illinois 61801

Eingegangen 11. Juli 1983

Functional equations for Hurwitz series and partial zeta functions of orders

By *Colin J. Bushnell**) at London and *Irving Reiner**) at Urbana

§ 1. Introduction

Let A be a finite-dimensional semisimple $\mathbf{Q}$-algebra, Λ a $\mathbf{Z}$-order in A, and M a left ideal of Λ such that $\mathbf{Q}M = A$. Recall the definition of the *partial zeta function*

$$Z_\Lambda(M, s) = \sum_{X \cong M} (\Lambda : X)^{-s}, \tag{1.1}$$

where s is a complex variable, and the summation extends over all left ideals X of Λ for which $X \cong M$ as Λ-modules. This is a Dirichlet series with abscissa of convergence $\mathrm{Re}(s) = 1$. If we write

$$Z_\Lambda(M, s) = \sum_{n=1}^{\infty} c_n n^{-s},$$

$$N(M, x) = \sum_{n \leqq x} c_n, \quad x \geqq 1,$$

the function $N(M, x)$ counts the left ideals of Λ which are isomorphic to M:

$$N(M, x) = |\{X \subset \Lambda : X \cong M \quad \textit{and} \quad (\Lambda : X) \leqq x\}|. \tag{1.2}$$

The behaviour of the counting function $N(M, x)$ and the analytic properties of $Z_\Lambda(M, s)$ are, of course, intimately related.

In our earlier papers [2], [4], we showed that $Z_\Lambda(M, s)$ can be continued analytically to a meromorphic function on the whole s-plane. It has a pole of order r at $s = 1$, where r is the number of simple components of the algebra A. Using the Delange-Ikehara Tauberian theorem, we deduced that

$$N(M, x) = c\,x(\log x)^{r-1} + o(x(\log x)^{r-1}) \quad \textit{as} \quad x \to \infty, \tag{1.3}$$

for a certain constant c which we computed explicitly in terms of M.

*) The authors wish to thank the National Science Foundation for its support of this research, which was performed while the first-named author was visiting the University of Illinois at Urbana-Champaign. The final version of this paper was prepared while the second-named author was visiting King's College London, and partially supported by the Science and Engineering Research Council.

The object of this paper is to find a *functional equation* for $Z_A(M, s)$. This will be both more general and more precise than our earlier results of this type in [3]. However, the results and methods of [3] continue to play an essential role in our proofs. This functional equation is of great interest in itself, but it also contains information about the counting function $N(M, x)$. In the sequel [5] to this paper, we will use it, together with a Tauberian theorem of Landau [8], [9], to get an estimate

$$(1.4) \qquad N(M, x) = \sum_{j=0}^{r-1} b_j x (\log x)^j + O(x^{1-\delta}(\log x)^{r-1}) \quad \textit{as} \quad x \to \infty .$$

The coefficients $\{b_j\}$ are given in terms of the Laurent expansion of $Z_A(M, s)$ at $s=1$, and $\delta > 0$ is given explicitly in terms of A.

Here we develop further the methods of [4] and study $Z_A(M, s)$ by introducing a more general class of functions attached to A. These are the *Hurwitz series* $F(s; W, L, m)$ whose definition is recalled below in §2. The advantage of this approach is that a Hurwitz series for A can be expressed in terms of other Hurwitz series which are attached to the simple components of A. The partial zeta function $Z_A(M, s)$ usually cannot be expressed in terms of partial zeta functions for simple components.

It is for Hurwitz series that we produce a functional equation. The proof proceeds by reducing to the case of a simple algebra A, and then it has to be divided into two sub-cases. In the first, we assume that A satisfies the Eichler condition, i.e. A is not a totally definite quaternion algebra. In this case, we can express any Hurwitz series for A in terms of idelic zeta integrals and use the machinery of [1]—[4]. In the other case, this is not possible because of the lack of a suitable strong approximation theorem. We use instead a global method based on Hecke's treatment of the Dedekind zeta function. This division into cases seems to be what forces one to reduce to the simple case in the first place.

It would be aesthetically preferable to have a unified proof, and this could also lead to new results. There are some indications that the Hecke method can be extended to cover the general case. The problem is basically well-known: one runs into severe technical difficulties associated with the application of the Poisson summation formula when the algebra A has zero-divisors. Cogdell [6] overcame these for the case $A = M_2(\mathbf{Q})$. Unpublished calculations of Rosaline Turner extend this to a few more cases. However, this evidence seems to show that, lacking a new idea, the construction of a general proof along these lines would be a burdensome task indeed.

This paper is arranged as follows. In §2, we give the preliminary definitions and state the main theorem. In §3, we perform the reduction to simple algebras. The Eichler case is handled in §5, after results and notation concerning zeta integrals are assembled in §4. The non-Eichler case is treated in §6. This completes the proof of the functional equation, and with it the main business of the paper. However, when one specialises the functional equation so that one side is a partial zeta function, it is not clear that the other side is expressible as a partial zeta function. In many interesting special cases it is, but this is false in general. We discuss this matter briefly in §7. All consideration of (1.4) is postponed to [5].

For background, we refer the reader to [7] and [10], as well as our earlier papers [1]—[4]. Throughout, we use the following abbreviations, notation and terminology:

a.e. = almost everywhere = for all but a finite number of exceptions;

f.g. = finitely generated;

f.d. = finite-dimensional;

$\Lambda^{\times}$ = the group of units of a ring Λ.

A *full Z-lattice* in a f.d. **Q**-vector space V is a f.g. submodule L of V which spans V over **Q**. We use the same terminology when **Q** is replaced by $\mathbf{Q}_p$ and **Z** by $\mathbf{Z}_p$. When **Q** is replaced by **R**, we also demand that L be a *discrete* subgroup of V. A **Z**-*order* in a f.d. **Q**-algebra A is a subring of A which is also a full **Z**-lattice in A. Likewise for $\mathbf{Z}_p$-orders.

§ 2. Hurwitz series and the functional equation

Let A be a f.d. semisimple **Q**-algebra. In this section, we define the notion of a Hurwitz series attached to A, and recall from [4] a few of the properties of these functions. We then state our main result on them, to the effect that they satisfy a certain functional equation. This implies, as a special case, a functional equation for the partial zeta function $Z_{\Lambda}(M, s)$ of (1. 1).

We need a few preliminaries. Let G be a group, and H_1, H_2 subgroups of G. Recall that H_1 is said to be *commensurate* with H_2 if $H_1 \cap H_2$ is of finite index in both H_1 and H_2. The *generalised group index*

$$(H_1 : H_2) = (H_1 : H_1 \cap H_2)/(H_2 : H_1 \cap H_2)$$

is then well-defined. For example, two full **Z**-lattices in the same f.d. **Q**-vector space V are commensurate subgroups of V.

Recall that a subgroup W of $A^{\times}$ is called an *arithmetic subgroup* if it is commensurate with $\Lambda^{\times}$, for some (equivalently any) **Z**-order Λ in A. A (*left*) *admissible triple* (W, L, m) in A consists of an arithmetic subgroup W of $A^{\times}$, a full **Z**-lattice L in A and an element $m \in A$, such that $WL = L$ and $W(m+L) = m+L$. These conditions mean that

$$(2.\ 1) \qquad wL = L \quad \textit{and} \quad wm - m \in L \quad \textit{for all} \quad w \in W.$$

Define a *right* admissible triple (m, L, W) in the same way, except that W acts on L and $m+L$ from the right.

For $x \in A^{\times}$, define

$$(2.\ 2) \qquad \|x\| = (Mx : M),$$

for any full **Z**-lattice M in A. This depends only on x, not on M. Since A is semisimple, we also have $(xM : M) = \|x\|$, for $x \in A^{\times}$. Moreover, $\|w\| = 1$ if w is an element of an arithmetic subgroup of $A^{\times}$. We write $\|x\|_A$ in place of $\|x\|$ when we need to specify the algebra A.

Given an admissible triple (W, L, m) in A, the *Hurwitz series* $F(s; W, L, m)$ is defined by

$$F(s; W, L, m) = \sum_{x \in W \backslash (m+L) \cap A^{\times}} \|x\|^{s}, \tag{2. 3}$$

where s is a complex variable and the sum extends over a set of representatives x of the W-orbits in $(m+L) \cap A^{\times}$. Likewise, for a right admissible triple (m, L, W), we define $F(s; m, L, W)$ in the obvious way.

From [4], we know

(2. 4) Theorem. (i) *The series* (2. 3) *converges absolutely, uniformly on compact sets, to a holomorphic function of s in the region* $\mathrm{Re}(s) > 1$.

(ii) *The Hurwitz series* (2. 3) *admits analytic continuation to a meromorphic function on the whole s-plane. If* $A = \prod_{i=1}^{r} A_i$, *where the* A_i *are simple algebras, then* $F(s; W, L, m)$ *has a pole of order r at* $s=1$.

For example, let Λ be a $\mathbf{Z}$-order in A, and M a full left ideal of Λ. We can form the partial zeta function $Z_{\Lambda}(M, s)$ as in (1. 1). Define

$$M^{-1} = \{x \in A : Mx \subset \Lambda\}, \tag{2. 5}$$

$$\mathrm{Aut}_{\Lambda} M = \{x \in A^{\times} : Mx = M\}. \tag{2. 6}$$

Then M^{-1} is a full $\mathbf{Z}$-lattice in A and $\mathrm{Aut}_{\Lambda} M$ is an arithmetic subgroup of $A^{\times}$ which acts on M^{-1} on the left. Thus $(\mathrm{Aut}_{\Lambda} M, M^{-1}, 0)$ is a left admissible triple in A, and by [1], p. 138, we have

$$Z_{\Lambda}(M, s) = (\Lambda : M)^{-s} F(s; \mathrm{Aut}_{\Lambda} M, M^{-1}, 0). \tag{2. 7}$$

The functional equation for a Hurwitz series $F(s)$ will involve "gamma factors", which should be regarded as a contribution from the Archimedean prime ∞ of $\mathbf{Q}$. They are defined as follows. Let Γ denote Euler's gamma function, and set

$$\gamma_{\mathbf{R}}(s) = \pi^{-\frac{s}{2}} \Gamma\left(\frac{s}{2}\right), \quad \gamma_{\mathbf{C}}(s) = (2\pi)^{1-s} \Gamma(s), \quad \gamma_{\mathbf{H}}(s) = (2\pi)^{-2s} \Gamma(2s), \tag{2. 8}$$

where $\mathbf{H}$ is the algebra of Hamiltonian quaternions. If $\mathbf{B}$ is a f.d. simple $\mathbf{R}$-algebra, we have $\mathbf{B} \cong M_n(\mathbf{K})$ for some $n \geqq 1$, where $\mathbf{K} = \mathbf{R}$, $\mathbf{C}$ or $\mathbf{H}$. We define

$$\gamma_{\mathbf{B}}(s) = \prod_{j=0}^{n-1} \gamma_{\mathbf{K}}(ns - j). \tag{2. 9}$$

Finally, for a f.d. semisimple $\mathbf{Q}$-algebra A, we put

$$A_{\infty} = A \otimes_{\mathbf{Q}} \mathbf{R} = \prod_i \mathbf{B}_i, \tag{2. 10}$$

where the $\mathbf{B}_i$ are simple $\mathbf{R}$-algebras, and define

$$\gamma_A(s) = \prod_i \gamma_{\mathbf{B}_i}(s). \tag{2. 11}$$

For an admissible triple (W, L, m) in A, we define

(2. 12) $$\tilde{F}(s; W, L, m) = \gamma_A(s)\, F(s; W, L, m),$$

and refer to such a function as an *extended Hurwitz series*. The functional equation below is much neater when expressed in terms of the functions $\tilde{F}$ rather than F. The situation is precisely parallel to the classical functional equation of zeta functions and L-functions of algebraic number fields.

Now let $\operatorname{tr}_{A/\mathbf{Q}}: A \to \mathbf{Q}$ be the reduced trace map (see [10], §9). Define $\theta_A: A \to \mathbf{C}^\times$ by

(2. 13) $$\theta_A(x) = \exp(2\pi i \cdot \operatorname{tr}_{A/\mathbf{Q}}(x)), \quad x \in A.$$

We have $\theta_A(x+y) = \theta_A(x)\,\theta_A(y)$ and $\theta_A(yx) = \theta_A(xy)$, for $x, y \in A$. If L is a full $\mathbf{Z}$-lattice in A, define

(2. 14) $$L^\perp = \{x \in A : \theta_A(xL) = \{1\}\} = \{x \in A : \operatorname{tr}_{A/\mathbf{Q}}(xL) \subset \mathbf{Z}\}.$$

The reduced trace pairing is nondegenerate, so this is another full $\mathbf{Z}$-lattice in A, and $L^{\perp\perp} = L$. Moreover, for any other full $\mathbf{Z}$-lattice M in A, we have

(2. 15) $$(M : L) = (L^\perp : M^\perp).$$

Further, if W is an arithmetic subgroup of $A^\times$ such that $WL = L$, then $L^\perp W = L^\perp$.

We now have all we need to state our main result.

(2. 16) Theorem (Functional equation for Hurwitz series). *Let A be a f.d. semisimple $\mathbf{Q}$-algebra, and (W, L, m) a left admissible triple in A. There is a $\mathbf{Z}$-lattice M in A satisfying*

(2. 17) $\quad m \in M, \; L \subset M, \; WM = M,$ *and W acts trivially on M/L.*

For any such M, and any $b \in L^\perp$, $(b, M^\perp, W)$ is a right admissible triple in A, and

(2. 18) $$\tilde{F}(s; W, L, m) = (L^\perp : L)^{-\frac{1}{2}} \sum_{b \in L^\perp/M^\perp} \theta_A(mb)\, \tilde{F}(1-s; b, M^\perp, W).$$

(2. 19) Remarks. (i) The existence of a lattice M satisfying (2. 17) is straightforward: just take $M = \mathbf{Z}m + L$. The admissibility of the triples $(b, M^\perp, W)$, $b \in L^\perp$, is immediate.

(ii) The left hand side of (2. 18) is certainly independent of the choice of M. We shall give a direct proof below (in (3. 10)) that the same applies to the right hand side. In fact, this is a useful step in the proof of (2. 18).

(iii) It will be convenient to have a slightly different version of the constant $(L^\perp : L)^{-\frac{1}{2}}$. Let Λ be a maximal order in A, and define

(2. 20) $$d_A = (\Lambda^\perp : \Lambda).$$

This is a positive integer, called the *absolute discriminant* of A. It is independent of the choice of Λ, and can be computed explicitly in terms of A: seee [10], §25. We have $(L^\perp : L) = (L^\perp : \Lambda^\perp)(\Lambda^\perp : \Lambda)(\Lambda : L)$, so by (2. 15),

(2. 21) $$(L^\perp : L) = (\Lambda : L)^2 d_A, \quad \text{and} \quad (L^\perp : L)^{-\frac{1}{2}} = (L : \Lambda)\, d_A^{-\frac{1}{2}}.$$

(iv) The fact that the right hand side of (2. 18) involves a linear combination of Hurwitz series, rather than a single one of them, is not really surprising. Consider the classical *Hurwitz zeta function*

$$\zeta(s;a)=\sum_{n=0}^{\infty}(n+a)^{-s},\quad \mathrm{Re}(s)>1,$$

where $0<a\leqq 1$. By [11], p. 269, $\zeta(s;a)$ satisfies a functional equation (for $\mathrm{Re}(s)<0$)

$$\zeta(s;a)=2\,\Gamma(1-s)\,(2\pi)^{s-1}\sum_{n=1}^{\infty}\left(\sin\frac{\pi s}{2}\cos 2\pi a n+\cos\frac{\pi s}{2}\sin 2\pi a n\right)n^{s-1}.$$

If we take $a=\frac{1}{k}$, $k\geqq 2$ an integer, this expresses $\zeta(s;a)$ as a linear combination of the $\zeta\left(1-s;\frac{j}{k}\right)$, $0<j\leqq k$, much as in (2. 18). Note that

$$\zeta(s;k^{-1})+\zeta(s;1-k^{-1})=F(s;\{1\},\mathbf{Z},k^{-1}).$$

If, in (2. 16), we have $m=0$, then we may take $M=L$ to get the special case

(2. 22) $$\tilde{F}(s;W,L,0)=(L^{\perp}:L)^{-\frac{1}{2}}\,\tilde{F}(1-s;0,L^{\perp},W).$$

Then, by (2. 5)—(2. 7) we have

(2. 23) Corollary (Functional equation for partial zeta functions). *Let Λ be a $\mathbf{Z}$-order in a f.d. semisimple $\mathbf{Q}$-algebra A, and let M be a full left Λ-lattice in A. Let Λ' be a maximal $\mathbf{Z}$-order in A. Then*

$$\gamma_A(s)\,Z_\Lambda(M,s)=(\Lambda:M)^{-s}\,(M^{-1}:\Lambda')\,d_A^{-\frac{1}{2}}\,\gamma_A(1-s)\,F(1-s;0,(M^{-1})^{\perp},\mathrm{Aut}_\Lambda M).$$

In certain cases, it is possible to express the Hurwitz series

$$F(1-s;0,(M^{-1})^{\perp},\mathrm{Aut}_\Lambda M)$$

in terms of a partial zeta function $Z_\Lambda(M',1-s)$, for some right Λ-lattice M'. The crucial point seems to be whether or not $(M^{-1})^{\perp}=(M')^{-1}$, for some M'. In general, there are algebraic obstructions to such an equation. We return to this matter in §7 below. In the case where $\Lambda=RG$, the group ring of a finite group G over the ring R of integers in an algebraic number field, we have $(M^{-1})^{\perp}=(M^{\perp})^{-1}$, and we obtain a functional equation for $Z_\Lambda(M,s)$ in which the right hand side involves the contragredient of M. Using this, we may recover the functional equations given in [3], §15. However, it should be pointed out that the machinery of [3], especially that developed in §14, plays an essential role in our proof of (2. 18).

§3. Reduction to the simple case

Our first task is to reduce the proof of (2. 16) to the case of a rather special sort of triple in a simple algebra. We start with some easy formal properties of the Hurwitz series, all of which are immediate consequences of the definitions. Throughout, A denotes a f.d. semisimple $\mathbf{Q}$-algebra, unless the contrary is specified.

(3. 1) Lemma. *Let (W, L, m) be an admissible triple in A, and W' an arithmetic subgroup of $A^\times$ such that (W', L, m) is also admissible. Then*

$$F(s; W', L, m) = (W : W')\, F(s; W, L, m).$$

(3. 2) Lemma. *Let (W, L, m) be an admissible triple in A, and let M, N be full W-stable lattices in A, such that*

$$M \supset L \supset N, \quad m \in M, \quad \text{and } W \text{ acts trivially on } M/N.$$

Then, for each $l \in L$, the triple $(W, N, l+m)$ is admissible and

$$F(s; W, L, m) = \sum_{l \in L/N} F(s; W, N, l+m).$$

(3. 3) Lemma. *For $i = 1, 2$, let (W_i, L_i, m_i) be an admissible triple in the algebra A_i. Then $(W_1 \times W_2, L_1 \times L_2, (m_1, m_2))$ is an admissible triple in $A_1 \times A_2$, and*

$$F(s; W_1 \times W_2, L_1 \times L_2, (m_1, m_2)) = F(s; W_1, L_1, m_1)\, F(s; W_2, L_2, m_2).$$

Remark. (3. 1)—(3. 3) and (3. 8) below all remain true when F is replaced by $\tilde{F}$.

(3. 4) Notation. Let $A = \prod_{i=1}^{r} A_i$, where the A_i are *simple* algebras, and write $x \mapsto x_i$ for the canonical projection $A \to A_i$. If E is a subset of A, write E_i for its image in A_i. In particular, if Λ is a maximal order in A, we have $\Lambda = \prod \Lambda_i$, and Λ_i is a maximal order in A_i.

Now let (W, L, m) be an admissible triple in A. We say it is *decomposable* if

$$W = \prod_{i=1}^{r} W_i, \quad L = \prod_{i=1}^{r} L_i. \tag{3. 5}$$

If this is satisfied, then (W_i, L_i, m_i) is an admissible triple in A_i, for each i, and (3. 3) gives

$$F(s; W, L, m) = \prod_{i=1}^{r} F(s; W_i, L_i, m_i). \tag{3. 6}$$

It is often more convenient to write $(W, L, m)_i$ in place of (W_i, L_i, m_i).

The standard example of a decomposable triple is a congruence triple. To define this, we need a maximal order Λ in A, and a positive integer c. An admissible triple (W, L, m) in A is a *congruence triple* (*relative to* Λ, *with conductor* c), if

$$L \text{ is a left } \Lambda\text{-lattice}, \; cm \in L, \text{ and } W = W_c = (1 + c\Lambda) \cap \Lambda^\times. \tag{3. 7}$$

Of course, W_c is an arithmetic subgroup of $A^\times$ for any positive integer c. Any such congruence triple is decomposable, and each factor $(W, L, m)_i$ is a congruence triple in A_i relative to the maximal order Λ_i and with conductor c.

(3. 8) Proposition. *Let (W, L, m) be an admissible triple in A, and Λ a maximal order in A. There is a left Λ-lattice M such that $M \supset L$, $m \in M$, and there is a positive integer c such that $cM \subset L$. If $W_c = (1 + c\Lambda) \cap \Lambda^\times$, we have*

$$\begin{aligned} F(s; W, L, m) &= (W_c : W) \sum_{l \in L/cM} F(s; W_c, cM, m+l) \\ &= (W_c : W) \sum_{l \in L/cM} \prod_{i=1}^{r} F(s; (W_c, cM, m+l)_i). \end{aligned}$$

The existence of M and c are immediate, and the expressions for $F(s;W,L,m)$ follow from (3. 1)—(3. 3). Note that each $(W_c, cM, m+l)_i$ is a congruence triple relative to Λ_i with conductor c.

We can restate this more informally:

(3. 9) Corollary. *Any Hurwitz series attached to A can be written as a finite linear combination of products of Hurwitz series of congruence triples in the simple components of A.*

Now we come to the preliminary reduction step in the proof of (2. 16).

(3. 10) Proposition. *In the situation of* (2. 16), *let M_1 be another* **Z**-*lattice satisfying* (2. 17). *Then*

$$\sum_{b\in L^\perp/M^\perp} \theta_A(mb)\, F(s; b, M^\perp, W) = \sum_{b_1\in L^\perp/M_1^\perp} \theta_A(mb_1)\, F(s; b_1, M_1^\perp, W).$$

Therefore the right hand side of (2. 18) *is independent of the choice of the lattice M satisfying* (2. 17).

Proof. If M and M_1 both satisfy (2. 17), then so does $M_2 = M + M_1$. Therefore we need only the prove the proposition under the additional hypothesis that $M_1 \supset M$. Otherwise, we just compare M with M_2, and then M_1 with M_2.

So, assuming $M_1 \supset M$, we choose coset decompositions

$$L^\perp = \dot{\bigcup_b}\, (b + M^\perp), \quad M^\perp = \dot{\bigcup_d}\, (d + M_1^\perp),$$

and we can take the coset representatives b_1 of $L^\perp/M_1^\perp$ in the form $b+d$. Then

$$\sum_{b_1\in L^\perp/M_1^\perp} \theta_A(mb_1)\, F(s; b_1, M_1^\perp, W) = \sum_{b\in L^\perp/M^\perp} \theta_A(mb) \sum_{d\in M^\perp/M_1^\perp} \theta_A(md)\, F(s; b+d, M_1^\perp, W).$$

However, since $m \in M$ and $d \in M^\perp$, we have $\theta_A(md) = 1$, and the result follows from (3. 2).

(3. 11) Theorem. *Suppose that* (2. 18) *holds under the following additional hypotheses*:

(i) *A is a simple algebra*;

(ii) *(W, L, m) is a congruence triple relative to some maximal order Λ and conductor c*;

(iii) *M is a left Λ-lattice and $cM \subset L$.*

Then (2. 18) *holds in general.*

Remark. Given (i) and (ii), one can always find M satisfying (iii) and (2. 17), e.g. $M = \Lambda m + L$.

Proof. In the general situation, we take an admissible triple (W, L, m) and a $\mathbf{Z}$-lattice M satisfying (2. 17). We choose a maximal order Λ, a left Λ-lattice P containing M, and a positive integer c such that $cP \subset L$. The group $W_c = (1 + c\Lambda) \cap \Lambda^{\times}$ acts trivially on P/cP, so (W_c, L, m) is an admissible triple relative to which M still satisfies (2. 17). By (3. 1), the functional equation holds for (W, L, m) and M if and only if it holds for (W_c, L, m) and M. Therefore, we may as well take $W = W_c$. By (3.10), we can also take $M = P$. Now we put $N = cP (= cM)$, and use the notation (3. 4). By (3. 8), we have

$$\tilde{F}(s;\, W, L, m) = \sum_{l \in L/N} \left\{ \prod_{i=1}^{r} \tilde{F}(s;\, (W, N, m+l)_i) \right\}$$

and each $(W, N, m+l)_i$ is a congruence triple. Indeed, the additional hypotheses (i)—(iii) now apply to $(W, N, m+l)_i$ and M_i, so

$$\tilde{F}(s;\, W, L, m) = \sum_{l \in L/N} \prod_{i=1}^{r} \{(N_i^{\perp} : N_i)^{-\frac{1}{2}}$$
$$\times \sum_{b_i \in N_i^{\perp}/M_i^{\perp}} \theta_{A_i}((m_i + l_i)\, b_i)\, \tilde{F}(1-s;\, b_i, M_i^{\perp}, W_i) \Big\}.$$

Here, $N_i^{\perp}$ denotes the orthogonal complement of N_i with respect to θ_{A_i}. Since θ_A is the product of the θ_{A_i}, we have

$$N^{\perp} = \prod_{i=1}^{r} N_i^{\perp}, \quad N^{\perp}/M^{\perp} = \prod_{i=1}^{r} N_i^{\perp}/M_i^{\perp}.$$

We can therefore interchange the product and the inner sum above to get

$$\tilde{F}(s;\, W, L, m) = (N^{\perp} : N)^{-\frac{1}{2}} \sum_{l \in L/N} \sum_{b \in N^{\perp}/M^{\perp}} \theta_A((m+l)\, b)\, \tilde{F}(1-s;\, b, M^{\perp}, W)$$
$$= (N^{\perp} : N)^{-\frac{1}{2}} \sum_{b \in N^{\perp}/M^{\perp}} \theta_A(mb)\, \tilde{F}(1-s;\, b, M^{\perp}, W) \sum_{l \in L/N} \theta_A(lb).$$

Now, for $b \in N^{\perp}$,

$$\sum_{l \in L/N} \theta_A(lb) = \begin{cases} (L:N) & \text{if } b \in L, \\ 0 & \text{if } b \notin L. \end{cases}$$

By (2. 21), we have $(L:N)(N^{\perp}:N)^{-\frac{1}{2}} = (L^{\perp}:L)^{-\frac{1}{2}}$, so we have proved (3. 11).

§ 4. Zeta integrals

We collect here the notation and results concerning idelic zeta integrals which we shall need for our proof. In this section, A is a f.d. *simple* $\mathbf{Q}$-algebra with centre K, and $\dim_K A = n^2$, $n \geqq 1$. Let Λ be some maximal order in A.

We write $\operatorname{Ad} A$ and $J(A)$ respectively for the groups of *finite* adeles and ideles of A, as in [4], § 2. With Λ as above, if L is a full Λ-lattice in A, we set

(4. 1) $$U(\Lambda) = \prod_P \Lambda_P^{\times}, \quad \operatorname{Ad} L = \prod_P L_P.$$

Here, P ranges over the non-Archimedean places of K, and the subscript P denotes P-adic completion. Thus $U(\Lambda)$ and $\mathrm{Ad}\, L$ are compact open subgroups of $J(A)$ and $\mathrm{Ad}\, A$ respectively.

If $x \in J(A)$, we write $\|x\| = ((\mathrm{Ad}\, L)x : \mathrm{Ad}\, L)$, for any L as above. If we view $A^{\times}$ as embedded in $J(A)$ "on the diagonal", this definition coincides with (2. 2) for $x \in A^{\times}$. Let $\mathscr{S}(A)$ denote the space of locally constant compactly supported functions $\Phi : \mathrm{Ad}\, A \to \mathbf{C}$. This is spanned, over $\mathbf{C}$, by functions of the form

$$\Phi(x) = \prod_P \Phi_P(x_P), \quad x = (x_P) \in \mathrm{Ad}\, A, \tag{4. 2}$$

where, for each P, the function Φ_P is an element of the space $\mathscr{S}(A_P)$ of locally constant compactly supported functions $A_P \to \mathbf{C}$, and Φ_P is the characteristic function of Λ_P a.e.

We next let $\psi : J(A) \to \mathbf{C}^{\times}$ be a continuous character with finite image, which is trivial on $A^{\times}$. (This last hypothesis was inadvertently omitted in the discussion in [4].) We put

$$Z(\Phi, \psi, s) = \int_{J(A)} \Phi(x)\, \psi(x)\, \|x\|^s\, d^{\times} x, \quad \Phi \in \mathscr{S}(A), \ \mathrm{Re}(s) > 1,$$

where $d^{\times} x$ is a Haar measure on $J(A)$. The precise choice of $d^{\times} x$ is immaterial, but, for convenience, we take the one for which

$$\int_{U(\Lambda)} d^{\times} x = 1. \tag{4. 3}$$

From [4] (2. 3) and [2] (3. 8), we have

(4. 4) Theorem. *Let χ be the unique character of $J(K)$ for which $\psi = \chi \cdot nr$, where nr denotes the reduced norm map $J(A) \to J(K)$. Then χ is a Hecke character of K, and let $L_K(s, \chi)$ denote its L-function. Then, for any $\Phi \in \mathscr{S}(A)$, we have*

$$Z(\Phi, \psi, s) = \prod_{j=0}^{n-1} L_K(ns - j, \chi) \cdot \prod_P g_P(s), \tag{4. 5}$$

where $g_P(s) \in \mathbf{C}[p^s, p^{-s}]$ for all P (p being the rational prime below P). Moreover, $g_P = 1$ a.e.

In particular,

(i) *$Z(\Phi, \psi, s)$ admits analytic continuation to a meromorphic function on the whole s-plane;*

(ii) *if ψ is nontrivial, then $Z(\Phi, \psi, s)$ is an entire function of s;*

(iii) *if ψ is trivial, then $Z(\Phi, \psi, s)$ is holomorphic except possibly for simple poles at $s = \dfrac{j}{n}$, $1 \leqq j \leqq n$.*

We now need to extract some information from [3] concerning functional equations. Let dx be the Haar measure on $\mathrm{Ad}\,A$ for which

$$(4.6)\qquad \int_{\mathrm{Ad}\,A} dx = d_A^{-\frac{1}{2}},$$

where d_A is as in (2. 20). Let $\theta_A : \mathrm{Ad}\,A \to \mathbf{C}^\times$ be the character given by

$$\theta_A(x) = \prod_P \theta_P(x_P); \quad x = (x_P) \in \mathrm{Ad}\,A,$$

where $\theta_P(y) = \exp(2\pi i \operatorname{tr}_{A_P/\mathbf{Q}_p}(y))$, $y \in A_P$. Here, tr denotes the reduced trace map, and, for $\alpha \in \mathbf{Q}_p$, $\exp(2\pi i \alpha)$ means $\exp(2\pi i \alpha_0)$, $\alpha_0 \in \mathbf{Q}$ being the principal part of α. This definition coincides with (2. 13) when $x \in A \subset \mathrm{Ad}\,A$. The pairing $(x, y) \mapsto \theta_A(xy)$, $x, y \in \mathrm{Ad}\,A$, is nondegenerate and identifies $\mathrm{Ad}\,A$ with its Pontrjagin dual. If L is a full Λ-lattice in A, we can form $\mathrm{Ad}\,L$ and take its orthogonal complement $(\mathrm{Ad}\,L)^\perp$ with respect to this pairing. If $L^\perp$ is defined by (2. 14), one verifies easily that

$$(4.7)\qquad (\mathrm{Ad}\,L)^\perp = \mathrm{Ad}(L^\perp),$$

so we use the symbol $\mathrm{Ad}\,L^\perp$ for this group.

For $\Phi \in \mathscr{S}(A)$, we put

$$(4.8)\qquad \hat{\Phi}(y) = \int_{\mathrm{Ad}\,A} \Phi(x)\, \theta_A(xy)\, dx, \quad y \in \mathrm{Ad}\,A.$$

Then $\hat{\Phi} \in \mathscr{S}(A)$, and we have

(4. 9) Theorem. *Let A, ψ, be as above, and define $\gamma_A(s)$ as in §2. Then, for all $\Phi \in \mathscr{S}(A)$, we have*

$$\gamma_A(s)\, Z(\Phi, \psi, s) = \gamma_A(1-s)\, Z(\hat{\Phi}, \psi^{-1}, 1-s).$$

Proof. Let $A_\infty = A \otimes_{\mathbf{Q}} \mathbf{R}$, and put $\tilde{J}(A) = J(A) \times A_\infty^\times$, the full idele group of A. Then ψ is effectively a continuous character of $\tilde{J}(A)$ which is trivial on both $A_\infty^\times$ and the diagonal image of $A^\times$ in $\tilde{J}(A)$. The assertion now follows immediately from [3], (14. 15).

(4. 10) Remark. The measure dx defined in (4. 6) is the self-dual Haar measure on $\mathrm{Ad}\,A$ for the pairing $(x, y) \mapsto \theta_A(xy)$. It is therefore the correct measure for defining the Fourier transform $\hat{\Phi}$ as in (4. 8).

§5. Proof of the functional equation in the Eichler case

In order to prove the functional equation (2. 18) for the Hurwitz series $F(s; W, L, m)$, it suffices by (3. 11) to treat the following special case:

(5. 1) (i) *A is a simple algebra,*

(ii) *(W, L, m) is a congruence triple relative to Λ with conductor c,*

(iii) *the lattice M is a left Λ-lattice and $cM \subset L$.*

Here, Λ is some maximal order in A, and c is a positive integer.

We shall also assume in this section that *A satisfies the Eichler condition*, i.e. A is not a totally definite quaternion algebra. (The non-Eichler case will be handled in §6.) All the notation of §4 remains in force.

The equation we have to prove is

$$(5.2)\qquad \tilde{F}(s; W, L, m) = (L: \Lambda)\, d_A^{-\frac{1}{2}} \sum_{b \in L^\perp/M^\perp} \theta_A(mb)\, \tilde{F}(1-s; b, M^\perp, W).$$

From now on, we think of A as embedded in Ad A and $A^\times$ as embedded in $J(A)$. Now define

$$(5.3)\qquad U_c(\Lambda) = \{(x_P) \in U(\Lambda): x_P \equiv 1 \ (\mathrm{mod}\ c\,\Lambda_P) \quad \textit{for all } P\}.$$

This is an open subgroup of $J(A)$, of finite index in $U(\Lambda)$. We have $\Lambda^\times = A^\times \cap U(\Lambda)$, and $W = W_c$ by definition, so

$$(5.4)\qquad W = A^\times \cap U_c(\Lambda).$$

Since A satisfies the Eichler condition, and since $U_c(\Lambda)$ is open, it follows that $A^\times \cdot U_c(\Lambda)$ contains the commutator subgroup of $J(A)$ (see [4] p. 165). Therefore $A^\times \cdot U_c(\Lambda)$ is an open normal subgroup of $J(A)$, and the quotient

$$H_c = J(A)/A^\times \cdot U_c(\Lambda)$$

is a finite abelian group whose order we denote by h_c. Let $H_c^* = \mathrm{Hom}(H_c, \mathbf{C}^\times)$ be the character group of H_c. If $\psi \in H_c^*$, we may view ψ as a continuous character of $J(A)$ which is trivial on $A^\times$. It certainly also has finite image. This enables us to form zeta integrals with ψ, and use the machinery of §4.

If E is a subset of Ad A, we write

$$\mathbf{cf}_E = \textit{the characteristic function of } E.$$

If E happens to be compact and open, then $\mathbf{cf}_E \in \mathscr{S}(A)$. In particular, $m + \mathrm{Ad}\, L$ is compact and open, so $\mathbf{cf}_{m+\mathrm{Ad}\,L} \in \mathscr{S}(A)$, and we may form the zeta integral

$$Z(\mathbf{cf}_{m+\mathrm{Ad}\,L}, \psi, s) = \int_{J(A)} \mathbf{cf}_{m+\mathrm{Ad}\,L}(x)\, \psi(x)\, \|x\|^s\, d^\times x$$

for each $\psi \in H_c^*$, viewed as a character of $J(A)$.

We recall from [4] p. 168 that

$$(5.5)\qquad F(s; W, L, m) = h_c^{-1} (U(\Lambda): U_c(\Lambda)) \sum_{\psi \in H_c^*} Z(\mathbf{cf}_{m+\mathrm{Ad}\,L}, \psi, s).$$

To abbreviate the notation, let us put

$$\tilde{Z}(\Phi, \psi, s) = \gamma_A(s)\, Z(\Phi, \psi, s), \qquad \Phi \in \mathscr{S}(A).$$

Multiplying (5.5) by $\gamma_A(s)$, we obtain

$$(5.6)\qquad \tilde{F}(s; W, L, m) = h_c^{-1} (U(\Lambda): U_c(\Lambda)) \sum_{\psi \in H_c^*} \tilde{Z}(\mathbf{cf}_{m+\mathrm{Ad}\,L}, \psi, s).$$

Via analytic continuation, this equation is valid on the whole s-plane.

We shall also need the right-handed version of this identity. For $b \in L^\perp$, the triple $(b, M^\perp, W)$ is a (right) congruence triple, relative to Λ and with conductor c. Therefore

$$(5.7)\qquad \tilde{F}(s; b, M^\perp, W) = h_c^{-1} (U(\Lambda): U_c(\Lambda)) \sum_{\psi \in H_c^*} \tilde{Z}(\mathbf{cf}_{b+\mathrm{Ad}\,M}, \psi, s).$$

Returning to the identity (5. 6), and using the functional equation (4. 9), we obtain

$$(5.8)\qquad \tilde{F}(s; W, L, m) = h_c^{-1}(U(\Lambda):U_c(\Lambda)) \sum_{\psi\in H_c^*} \tilde{Z}(\widehat{\mathbf{cf}}_{m+\mathrm{Ad}\,L}, \psi, 1-s),$$

noting that as ψ ranges over H_c^*, so does ψ^{-1}.

We now must calculate the Fourier transform $\widehat{\mathbf{cf}}_{m+\mathrm{Ad}\,L}$, as in (4. 8), in order to recognise the right hand side of (5. 8) in terms of extended Hurwitz series. Since

$$\mathbf{cf}_{m+\mathrm{Ad}\,L}(x) = \mathbf{cf}_{\mathrm{Ad}\,L}(x-m) \quad \textit{for} \quad x\in \mathrm{Ad}\,A,$$

we obtain for $y\in \mathrm{Ad}\,A$:

$$\widehat{\mathbf{cf}}_{m+\mathrm{Ad}\,L}(y) = \int_{\mathrm{Ad}\,A} \mathbf{cf}_{\mathrm{Ad}\,L}(x-m)\,\theta(xy)\,dx = \int_{\mathrm{Ad}\,A} \mathbf{cf}_{\mathrm{Ad}\,L}(x)\,\theta((x+m)\,y)\,dx$$
$$= \theta(my)\,\widehat{\mathbf{cf}}_{\mathrm{Ad}\,L}(y).$$

(We write θ in place of θ_A, for convenience.) We show next that

$$(5.9)\qquad \widehat{\mathbf{cf}}_{\mathrm{Ad}\,L} = \mu(\mathrm{Ad}\,L)\,\mathbf{cf}_{\mathrm{Ad}\,L^\perp},$$

where μ denotes measure with respect do dx. By definition,

$$\widehat{\mathbf{cf}}_{\mathrm{Ad}\,L}(y) = \int_{\mathrm{Ad}\,L} \theta(xy)\,dx, \quad \textit{for} \quad y\in \mathrm{Ad}\,A.$$

If $y\in \mathrm{Ad}\,L^\perp$, then $\theta(xy)=1$ for all $x\in \mathrm{Ad}\,L$, so the above integral takes the value $\mu(\mathrm{Ad}\,L)$. On the other hand, if $y\notin \mathrm{Ad}\,L^\perp$, then there is an $x_0\in \mathrm{Ad}\,L$ such that $\theta(x_0 y)\neq 1$, since θ induces a perfect self-duality on $\mathrm{Ad}\,A$. Therefore

$$\int_{\mathrm{Ad}\,L} \theta(xy)\,dx = \int_{\mathrm{Ad}\,L} \theta((x+x_0)\,y)\,dx = \theta(x_0 y) \int_{\mathrm{Ad}\,L} \theta(xy)\,dx.$$

It follows that $\int_{\mathrm{Ad}\,L} \theta(xy)\,dx = 0$, which completes the proof of (5. 9).

We claim next that if L_1 and L_2 are full Λ-lattices in A, then

$$(5.10)\qquad (\mathrm{Ad}\,L_1 : \mathrm{Ad}\,L_2) = (L_1 : L_2).$$

It suffices to verify this when $L_1 \supset L_2$; but then the canonical map $L_1/L_2 \to \mathrm{Ad}\,L_1/\mathrm{Ad}\,L_2$ is an isomorphism, so (5. 10) is proved. Therefore

$$(5.11)\qquad \mu(\mathrm{Ad}\,L) = (\mathrm{Ad}\,\Lambda : \mathrm{Ad}\,L)^{-1}\,\mu(\mathrm{Ad}\,\Lambda) = (\Lambda : L)^{-1}\,d_A^{-\frac{1}{2}},$$

by the way we chose the measure dx (see (4. 6)).

Putting all this together, we obtain

$$\widehat{\mathbf{cf}}_{m+\mathrm{Ad}\,L}(y) = \mu(\mathrm{Ad}\,L)\,\theta(my)\,\mathbf{cf}_{\mathrm{Ad}\,L^\perp}(y), \quad y\in \mathrm{Ad}\,A.$$

Since $m\in M\subset \mathrm{Ad}\,M$, the quantity $\theta(my)$ depends only on the coset $y+\mathrm{Ad}\,M^\perp$, for $y\in \mathrm{Ad}\,A$, or, if $y\in A$, only on the coset $y+M^\perp$. We identify $\mathrm{Ad}\,L^\perp/\mathrm{Ad}\,M^\perp$ with $L^\perp/M^\perp$, as in the proof of (5. 10), and we now have

$$\widehat{\mathbf{cf}}_{m+\mathrm{Ad}\,L} = \mu(\mathrm{Ad}\,L) \sum_{b\in L^\perp/M^\perp} \theta(mb)\,\mathbf{cf}_{b+\mathrm{Ad}\,M^\perp}.$$

Substituting into (5. 8), the right hand side becomes

$$h_c^{-1}(U(\Lambda):U_c(\Lambda)) \sum_{\psi\in H_c^*} \sum_{b\in L^\perp/M^\perp} \theta(mb)\,\tilde{Z}(\mathbf{cf}_{b+\mathrm{Ad}\,M^\perp}, \psi, 1-s).$$

Now we reverse the order of summation, and use (5. 7) and (5. 11), to obtain the expression occurring on the right hand side of (5. 2). This establishes the functional equation for the special case described in (5. 1), when A satisfies the Eichler condition.

§ 6. Totally definite quaternion algebras

In this section, we shall complete the proof of the functional equation (2. 18) for Hurwitz series. It only remains to treat the case in which the algebra A is a totally definite quaternion algebra. The extra restrictions placed on the admissible triple (W, L, m) in (3. 11) play no further role, and can be ignored.

Throughout this section, we assume that A is a *totally definite quaternion algebra.* Thus A is simple, and

$$A_\infty = A \otimes_{\mathbf{Q}} \mathbf{R} \cong \mathbf{H}^{(n)}, \tag{6. 1}$$

the direct product of n copies of the algebra $\mathbf{H}$ of Hamiltonian quaternions. Note that $\dim_{\mathbf{Q}} A = 4n$. The techniques needed in this section are rather different from those used in § 5. There we embedded A in its ring $\operatorname{Ad} A$ of finite adeles, and all of the analysis took place in $\operatorname{Ad} A$ (and in the associated idele group $J(A)$). In the present case this is ineffectual. The strong approximation theorem (for the kernel of the reduced norm in A) no longer holds, and this means that we cannot express Hurwitz series for the totally definite quaternion algebra A in terms of zeta integrals. Instead, we shall embed A in the $\mathbf{R}$-algebra A_∞, and work inside $A_\infty^\times$. This requires a new list of notation, which we now give.

Let tr denote the *reduced trace* from A_∞ to $\mathbf{R}$, and put

$$\theta_\infty(x) = \exp(-2\pi i \operatorname{tr}(x)), \quad x \in A_\infty.$$

Note that, for $x \in A$, we have $\theta_\infty(x) = \theta_A(-x) = \theta_A(x)^{-1}$, where θ_A is given by (2. 13). The pairing $(x, y) \mapsto \theta_\infty(xy)$ again induces a perfect self-duality of A_∞. In particular, if L is a full $\mathbf{Z}$-lattice in A, and $L^\perp$ is defined by (2. 14), then we also have

$$L^\perp = \{x \in A_\infty : \theta_\infty(xL) = \{1\}\}.$$

For $x \in A_\infty^\times$, we define a positive real number $\|x\|_\infty$, abbreviated to $\|x\|$ when there is no danger of confusion, by the formula

$$\int_{A_\infty} f(xy)\, dy = \int_{A_\infty} f(yx)\, dy = \|x\|_\infty^{-1} \int_{A_\infty} f(y)\, dy$$

for any Haar measure dy and any measurable function f on A_∞. It is easy to see that if L is a full $\mathbf{Z}$-lattice in $A \subset A_\infty$, and $x \in A^\times$, then $\|x\|_\infty = (L : Lx)$. Thus, for $x \in A^\times$, we have $\|x\|_\infty = \|x\|_A^{-1}$.

Now let "bar" be the involution on A_∞ which is the product of the standard involutions on the factors $\mathbf{H}$ in (6. 1). Define

$$\phi(x) = \exp(-\pi \cdot \operatorname{tr}(x\bar{x})), \quad x \in A_\infty.$$

There is a unique Haar measure $d^\times x$ on $A_\infty^\times$ such that

$$(6.2) \qquad \int_{A_\infty^\times} \phi(x)\, \|x\|_\infty^s\, d^\times x = \gamma_A(s) = \{(2\pi)^{-2s}\,\Gamma(2s)\}^n,$$

where n is given by (6. 1).

Let $\mathbf{R}_+^\times$ be the multiplicative group of positive real numbers. There is a homomorphism

$$A_\infty^\times \to \mathbf{R}_+^\times, \text{ given by } x \mapsto \|x\|_\infty, \; x \in A_\infty^\times,$$

whose kernel we shall denote by G. There is a decomposition

$$A_\infty^\times = G \times \mathbf{R}_+^\times, \text{ given by } t^{\frac{1}{4n}} g \leftrightarrow (g, t).$$

Hence there is a unique bi-invariant Haar measure dg on G such that

$$d^\times x = dg \cdot \frac{dt}{t},$$

where dt is the usual Lebesgue measure on $\mathbf{R}$.

Now let (W, L, m) be any admissible triple in the totally definite quaternion algebra A, and let M be any lattice in A satisfying (2. 17), that is,

$$m \in M, \; L \subset M, \; WM = M, \text{ and } W \text{ acts trivially on } M/L.$$

We have to prove the functional equation (2. 18), which we now write as

$$(6.3) \quad \tilde{F}(s; W, L, m) = (L^\perp : L)^{-\frac{1}{2}} \sum_{b \in L^\perp/M^\perp} \theta_\infty(-mb)\, \tilde{F}(1-s; b, M^\perp, W).$$

The arithmetic subgroup W is contained in G, and there is a unique left G-invariant measure $d\dot{g}$ on G/W such that

$$\int_G F(g)\, dg = \int_{G/W} \sum_{w \in W} F(gw)\, d\dot{g}$$

for any measurable function F on G. Likewise, there is a unique right G invariant measure $d\ddot{g}$ on $W \backslash G$ such that

$$\int_G F(g)\, dg = \int_{W\backslash G} \sum_{w \in W} F(wg)\, d\ddot{g}$$

These measures are connected by the homeomorphism on G induced by $g \mapsto g^{-1}$, so the measures

$$\int_{G/W} d\dot{g} \text{ and } \int_{W\backslash G} d\ddot{g}$$

are equal. We denote their common value by $\mu^\times(G/W)$; this is finite and non-zero, and is given explicitly in [4], Lemma 3. 7.

By [4] p. 173, we have

$$\text{(6. 4)}\quad \tilde{F}(s;\, W,\, L,\, m)=\mu^{\times}(G/W)\left\{\frac{-\delta(m,\, L)}{s}+\frac{(L^{\perp}:L)^{-\frac{1}{2}}}{s-1}\right\}+I_1(s)+I_2(s),$$

where we define

$$\delta(m,\, L)=\begin{cases}1 & \textit{if} \quad m\in L,\\ 0 & \textit{if} \quad m\notin L,\end{cases}$$

and where $I_1(s)$ and $I_2(s)$ are given as follows:

$$\text{(6. 5)}\quad I_1(s)=\int_1^{\infty} t^s \int_{G/W} \sum_{x\in(m+L)\cap A^{\times}} \phi(t^{\frac{1}{4n}}\, g x)\, d\dot{g}\, \frac{dt}{t},$$

$$I_2(s)=\int_1^{\infty} t^{1-s} \int_{W\backslash G} (L:L)^{-\frac{1}{2}} \sum_{y\in L^{\perp}\cap A^{\times}} \theta_{\infty}(-my)\, \phi(t^{\frac{1}{4n}}\, y g)\, d\ddot{g}\, \frac{dt}{t}.$$

The integrals I_1, I_2 converge absolutely on the whole s-plane, uniformly on compact sets, to holomorphic functions of s. We can rewrite $I_2(s)$ in the more convenient form

$$\text{(6. 6)}\quad I_2(s)=\int_1^{\infty} t^{1-s} \int_{W\backslash G} (L^{\perp}:L)^{-\frac{1}{2}} \sum_{b\in L^{\perp}/M^{\perp}} \theta_{\infty}(-mb) \times \sum_{y\in(b+M^{\perp})\cap A^{\times}} \phi(t^{\frac{1}{4n}}\, y g)\, d\ddot{g}\, \frac{dt}{t}.$$

(6. 7) Remark. There are certain differences between (6. 4) and the version in [4]. First, the term $-\dfrac{\delta_a}{s-1}$ in p. 173 line 13 of [4] should read $-\dfrac{\delta_a}{s}$. Second, in [4], we identified A_{∞} with its dual by using the pairing $(x,\, y)\mapsto \theta_{\infty}(x\bar{y})$. Since $\phi(x)=\phi(\bar{x})$, it is not difficult to check that (6. 4) is equivalent to that given in [4].

For $b\in L^{\perp}$, there is a "right-handed" version of (6. 4):

(6. 8)

$$\tilde{F}(s;\, b,\, M^{\perp},\, W)=\mu^{\times}(G/W)\left\{\frac{-\delta(b,\, M^{\perp})}{s}+\frac{(M:M^{\perp})^{-\frac{1}{2}}}{s-1}\right\}+I_3(s)+I_4(s),$$

(noting that $M^{\perp\perp}=M$), where

$$\text{(6. 9)}\quad I_3(s)=\int_1^{\infty} t^s \int_{W\backslash G} \sum_{x\in(b+M^{\perp})\cap A^{\times}} \phi(t^{\frac{1}{4n}}\, x g)\, d\ddot{g}\, \frac{dt}{t},$$

$$\text{(6. 10)}\quad I_4(s)=\int_1^{\infty} t^{1-s} \int_{G/W} (M:M^{\perp})^{-\frac{1}{2}} \sum_{b'\in M/L} \theta_{\infty}(-bb') \times \sum_{y\in(b'+L)\cap A^{\times}} \phi(t^{\frac{1}{4n}}\, g y)\, d\dot{g}\, \frac{dt}{t}.$$

We use the above to evaluate the sum appearing on the right hand side of (6. 3). Writing s in place of $1-s$, this sum is

$$\sum_{b \in L^\perp/M^\perp} \theta_\infty(-mb)\, \tilde{F}(s; b, M^\perp, W). \tag{6. 11}$$

By virtue of formula (6. 8), the above sum splits into three parts. The contribution from the polar terms in (6. 8) is

$$\sum_{b \in L^\perp/M^\perp} \mu^\times(G/W)\, \theta_\infty(-mb) \left\{ \frac{-\delta(b, M^\perp)}{s} + \frac{(M : M^\perp)^{-\frac{1}{2}}}{s-1} \right\}. \tag{6. 12}$$

Now we use

$$\sum_{b \in L^\perp/M^\perp} \theta_\infty(-mb)\, \delta(b, M^\perp) = 1,$$

$$\sum_{b \in L^\perp/M^\perp} \theta_\infty(-mb) = (L^\perp : M^\perp)\, \delta(m, L).$$

By (2. 21), we have $(L^\perp : M^\perp)(M : M^\perp)^{-\frac{1}{2}} = (L : L^\perp)^{-\frac{1}{2}}$, so (6. 12) becomes

$$\mu^\times(G/W) \left\{ \frac{-1}{s} - \frac{(L : L^\perp)^{-\frac{1}{2}}}{s-1} \right\}. \tag{6. 13}$$

Next, the contribution to (6. 11) from the terms $I_3(s)$ in formula (6. 8) is equal to

$$\int_1^\infty t^s \int_{W \backslash G} \sum_{b \in L^\perp/M^\perp} \theta_\infty(-mb) \sum_{x \in (b+M^\perp) \cap A^\times} \phi(t^{\frac{1}{4n}} xg)\, d\ddot{g} \frac{dt}{t}. \tag{6. 14}$$

Finally, the $I_4(s)$ terms contribute

$$\int_1^\infty t^{1-s} \int_{G/W} (M : M^\perp)^{-\frac{1}{2}} \sum_{\substack{b \in L^\perp/M^\perp \\ b' \in M/L}} \theta_\infty(-mb)\, \theta_\infty(-bb') \times \sum_{y \in (b'+L) \cap A^\times} \phi(t^{\frac{1}{4n}} gy)\, d\dot{g} \frac{dt}{t}. \tag{6. 15}$$

However, $\theta_\infty(-mb) = \theta_\infty(-bm)$, and

$$\sum_{b \in L^\perp/M^\perp} \theta_\infty(-b(m+b')) = \begin{cases} (L^\perp : M^\perp) & \text{if } m+b' \in L, \\ 0 & \text{if } m+b' \notin L. \end{cases}$$

Since $\phi(-y) = \phi(y)$, (6. 15) now becomes

$$(L : L^\perp)^{-\frac{1}{2}} \int_1^\infty t^{1-s} \int_{G/W} \sum_{y \in (m+L) \cap A^\times} \phi(t^{\frac{1}{4n}} gy)\, d\dot{g} \frac{dt}{t}. \tag{6. 16}$$

The expression (6. 11) is the sum of the terms (6. 12), (6. 13) and (6. 16). Multiplying (6. 11) by $(L^\perp : L)^{-\frac{1}{2}}$ and replacing s by $1-s$, we thus obtain the right hand side of (6. 4). But this establishes the desired identity (6. 3), and completes the proof of Theorem (2. 16).

§ 7. Zeta functions of orders

We have also completed the proof of Corollary (2. 23), which gives the functional equation for partial zeta functions, namely

$$(7.1)\quad \gamma_A(s)\,Z_\Lambda(M,s)=(\Lambda:M)^{-s}(M^{-1}:\Lambda')\,d_A^{-\frac{1}{2}}\,\gamma_A(1-s)\,F(1-s;0,(M^{-1})^{\perp},\mathrm{Aut}_\Lambda M).$$

Here, Λ is a **Z**-order in the f.d. semisimple **Q**-algebra A, M is a full left Λ-lattice in A, and Λ' is a maximal **Z**-order in A. Further, M^{-1} and $(M^{-1})^{\perp}$ are defined as in (2. 5) and (2. 14), that is,

$$M^{-1}=\{x\in A: Mx\subset\Lambda\},\quad (M^{-1})^{\perp}=\{x\in A:\mathrm{tr}_{A/\mathbf{Q}}(M^{-1}x)\subset\mathbf{Z}\}.$$

By summing (7. 1) over various finite sets of Λ-lattices M, we obtain functional equations for other types of zeta functions, such as $\zeta_\Lambda(s)$, $Z_\Lambda(g(M),s)$, $Z_\Lambda([M],s)$ attached to M and Λ (see [2], § 1 for definitions). We shall not carry out this computation here.

It would certainly be more pleasing to have (7. 1) in a "closed form", involving only zeta functions and elementary functions like the gamma factors. The difficulty arises in trying to express the Hurwitz series $F(s;0,(M^{-1})^{\perp},\mathrm{Aut}_\Lambda M)$ in terms of zeta functions. For this purpose, we would need to find a full *right* Λ-lattice P in A such that $(M^{-1})^{\perp}=P^{-1}$. Here we define

$$(7.2)\qquad P^{-1}=\{x\in A: xP\subset\Lambda\},$$

the right-handed analogue of (2. 5). If such a P could be found, we would have

$$(7.3)\qquad F(s;0,(M^{-1})^{\perp},\mathrm{Aut}_\Lambda M)=(\Lambda:P)^s\,(\mathrm{Aut}_\Lambda P:\mathrm{Aut}_\Lambda M)\,Z_\Lambda(P,s),$$

where now

$$Z_\Lambda(P,s)=\sum_Y(\Lambda:Y)^{-s},\quad \mathrm{Re}(s)>1,$$

with Y ranging over all right ideals of Λ such that $Y\cong P$.

Such a lattice P can always be found in the special cases where Λ is either a maximal order or the group ring RG of a finite group G over the ring R of integers in an algebraic number field. Indeed, in both these cases, any full left Λ-lattice N in A can be expressed in the form $N=Q^{-1}$, for some full right Λ-lattice Q in A. For arbitrary orders Λ, however, this need not be so, as the following example demonstrates.

We need some preliminaries. Let Λ be a **Z**-order in some f.d. semisimple **Q**-algebra A, and let M be a full left Λ-lattice in A. Then $(M^{-1})^{-1}$ is another full left Λ-lattice, and we have a canonical inclusion $M\subset(M^{-1})^{-1}$. We say that M is *Λ-reflexive* if $M=(M^{-1})^{-1}$.

(7. 4) Lemma. *Let M be a full left Λ-lattice in A. Then $M=P^{-1}$, for some full right Λ-lattice P in A, if and only if M is Λ-reflexive.*

Proof. If M is Λ-reflexive, then $M=(M^{-1})^{-1}$, and we may take $P=M^{-1}$. Conversely, suppose that $M=P^{-1}$. Then $M^{-1}=(P^{-1})^{-1}$, and hence $P\subset(P^{-1})^{-1}=M^{-1}$. Taking inverses again, we get

$$P^{-1}\supset((P^{-1})^{-1})^{-1}=(M^{-1})^{-1}\supset M=P^{-1}.$$

Thus $M=(M^{-1})^{-1}$, and M is Λ-reflexive, as desired.

(7. 5) Example. We now give an example of an order Λ in the matrix algebra $A = M_3(\mathbf{Q})$, and a full left Λ-lattice M in A, such that $(M^{-1})^{\perp}$ is not Λ-reflexive. Via (7. 4), this will imply the impossibility of an equation of the form (7. 3) in this case.

Fix a prime number p, and write $P = p\mathbf{Z}$. Take Λ to be the order

$$\Lambda = \begin{pmatrix} \mathbf{Z} & P & P \\ \mathbf{Z} & \mathbf{Z} & P \\ \mathbf{Z} & P & \mathbf{Z} \end{pmatrix}.$$

Here, the slant bar indicates congruence $(\mathrm{mod}\, p)$. In other words, Λ is the set of all matrices $(a_{ij}) \in M_3(\mathbf{Z})$ whose entries satisfy

$$a_{11} \equiv a_{22} \,(\mathrm{mod}\, p), \quad \textit{and} \quad a_{12} \equiv a_{13} \equiv a_{23} \equiv a_{32} \equiv 0 \,(\mathrm{mod}\, p).$$

Taking $M = \Lambda$ as left Λ-module, we see that $M^{-1} = \Lambda_{\Lambda}$, and

$$(M^{-1})^{\perp} = \Lambda^{\perp} = \{x \in A : \mathrm{Tr}(\Lambda x) \subset \mathbf{Z}\}.$$

A routine calculation shows that $\Lambda^{\perp} \neq ((\Lambda^{\perp})^{-1})^{-1}$, so $\Lambda^{\perp}$ is *not* of the form N^{-1}, for any full right Λ-lattice N in A.

References

[1] *C. J. Bushnell* and *I. Reiner*, Zeta functions of arithmetic orders and Solomon's conjectures, Math. Z. **173** (1980), 135—161.
[2] *C. J. Bushnell* and *I. Reiner*, *L*-functions of arithmetic orders and asymptotic distribution of ideals, J. reine angew. Math. **327** (1981), 156—183.
[3] *C. J. Bushnell* and *I. Reiner*, Functional equations for *L*-functions of arithmetic orders, J. reine angew. Math. **329** (1981), 88—123.
[4] *C. J. Bushnell* and *I. Reiner*, Analytic continuation of partial zeta functions of arithmetic orders, J. reine angew. Math. **349** (1984), 160—178.
[5] *C. J. Bushnell* and *I. Reiner*, New asymptotic formulas for the distribution of left ideals of orders, J. reine angew. Math. **364** (1986), 149—170.
[6] *R. Cogdell*, Congruence zeta functions for $M_2(\mathbf{Q})$ and their associated modular forms, Math. Ann. **266** (1983), 141—198.
[7] *C. W. Curtis* and *I. Reiner*, Methods of representation theory. I, New York 1981.
[8] *E. Landau*, Über die Anzahl der Gitterpunkte in gewissen Bereichen. II, Gött. Nachr. 1915, 209—243.
[9] *E. Landau*, Ausgewählte Abhandlungen zur Gitterpunktlehre (A. Walfisz, ed.), Berlin 1962.
[10] *I. Reiner*, Maximal orders, London-New York 1975.
[11] *E. T. Whittaker* and *G. N. Watson*, A course of modern analysis, 4th ed., Cambridge 1927.

Department of Mathematics, King's College, Strand, London WC2R 2LS, England

University of Illinois, Department of Mathematics, 1409 West Green Street, Urbana, IL 61801, U.S.A.

Eingegangen 4. September 1984, in revidierter Form 24. Juni 1985

New asymptotic formulas for the distribution of left ideals of orders

By *Colin J. Bushnell**) at London and *Irving Reiner**) at Urbana

---

Let Λ be a $\mathbf{Z}$-order in a finite-dimensional semisimple $\mathbf{Q}$-algebra A, and let M be a full left ideal of Λ. For $x>1$, let $N(M,x)$ denote the number of left ideals X of Λ such that $X\cong M$ as Λ-modules, and $(\Lambda:X)\leqq x$. The principal object of this paper is to prove the result announced in [6], namely that $N(M,x)$ satisfies an asymptotic formula

$$(*)\qquad N(M,x)=\sum_{j=0}^{r-1} b_i x(\log x)^j + O(x^{1-\delta}(\log x)^{r-1}) \quad as \quad x\to\infty.$$

Here, the $\{b_j\}$ are certain constants, r is the number of simple components of the algebra A, and $\delta=\dfrac{2}{1+\dim_{\mathbf{Q}} A}$.

The technique of proof is to apply a Tauberian theorem of E. Landau [9], [10] to the partial zeta function

$$Z_\Lambda(M,s)=\sum_{X\cong M}(\Lambda:X)^{-s},\quad \mathrm{Re}(s)>1,$$

where X ranges over all left ideals of Λ which are isomorphic to M as Λ-modules. Landau's theorem gives an asymptotic formula of the required type for the partial sums of the coefficients of a (generalised) Dirichlet series, provided that the series satisfies a certain long list of hypotheses. The main group of these hypotheses demands that the series be continuable to a meromorphic function on the whole s-plane, and that it satisfy a certain functional equation. Such results for the series $Z_\Lambda(M,s)$ are available from [6].

Thus the main labour of this paper lies in recalling Landau's theorem and checking that all its hypotheses are satisfied by partial zeta functions. We do this, for a wider class of functions, in § 9 and § 10.

---

*) The authors wish to thank the National Science Foundation for its support of this research, which was performed while the first-named author was visiting the University of Illinois at Urbana-Champaign. The final version of this paper was prepared while the second-named author was visiting King's College London, and partially supported by the Science and Engineering Research Council.

The "principal term" from (*), namely

$$R(M, x) = \sum_{j=0}^{r-1} b_j x (\log x)^j,$$

is an interesting numerical invariant of the ideal M. In terms of the partial zeta function, $R(M, x)$ is the residue at $s=1$ of the function $\frac{x^s Z_\Lambda(M, s)}{s}$. The coefficients $\{b_j\}$ can therefore be computed in terms of the Laurent coefficients of $Z_\Lambda(M, s)$ at $s=1$. The leading coefficient, b_{r-1}, is equal to $\frac{a_{-r}}{(r-1)!}$, where a_{-r} is the leading Laurent coefficient of $Z_\Lambda(M, s)$ (this function has a pole of order r at $s=1$). One can give a fairly good account of the number a_{-r}: see [5]. In many situations, it depends only on the genus of M, rather than its actual isomorphism class.

This is as far as one can go in the direction of explicit computation. The other coefficients b_j, for $0 \leqq j \leqq r-2$, do vary with the isomorphism class of M, even in very simple cases. One can sometimes "calculate" them, via the Laurent expansion and various explicit formulas, but the expressions obtained involve such quantities as higher Laurent coefficients and special values of Dedekind zeta functions and Hecke L-series. This feature persists even when one simplifies the problem by passing from $Z_\Lambda(M, s)$ to the more tractable functions $\zeta_\Lambda(s)$, $Z_\Lambda(g(M), s)$, $Z_\Lambda([M], s)$ defined in [3] § 1. However, one can say something of the behaviour of $R(M, x)$ as a function of M. Actually, it is easier to treat the stable zeta function

$$Z_\Lambda([M], s) = \sum_X (\Lambda : X)^{-s},$$

where X now ranges over all left ideals of Λ which are *stably isomorphic* to M, i.e. $X \oplus M \cong M \oplus M$. (Of course, stable isomorphism implies ordinary isomorphism, in many cases.) We get an asymptotic estimate, just like (*), for the number of ideals in a given stable isomorphism class, involving the quantity

$$R([M], x) = \textit{the residue of } \frac{x^s Z_\Lambda([M], s)}{s} \textit{ at } s=1.$$

This gives another numerical invariant of the ideal M, which behaves well under passage to homomorphic images. We discuss this matter, and all others relating just to partial zeta functions, in § 11.

At various points in the argument, we have to consider the distribution of the poles of partial zeta functions, and more generally of Hurwitz series. With the exception of the r-fold pole at $s=1$, these make no contribution to our results, but this seems a good opportunity to set down some of the facts pertaining to them. We do this briefly in § 12.

This paper is written as a continuation of [6], and its sections are numbered in the same sequence. Results and definitions from [6] are referred to simply by number. However, the two papers have independent lists of references. Notation and terminology are carried over from [6], and we also use:

$$s = \sigma + it \textit{ is a complex variable, with } \sigma, t \in \mathbf{R};$$

$$\log^j x = (\log x)^j, \ j \in \mathbf{Z}.$$

§ 8. Preliminaries

We recall one fundamental definition from § 2. Let A be a f.d. semisimple **Q**-algebra, and (W, L, m) a left admissible triple in A. Thus $m \in A$, L is a full **Z**-lattice in A, and W is an arithmetic subgroup of $A^{\times}$ such that $WL = L$ and $W(m+L) = m+L$. The *Hurwitz series* attached to (W, L, m) is

$$F(s; W, L, m) = \sum_{x \in W \backslash (m+L) \cap A^{\times}} \|x\|^{s}, \quad \sigma > 1,$$

where $\|x\| = (Lx : L)$. From (2. 4), we know that $F(s) = F(s; W, L, m)$ admits analytic continuation to a meromorphic function on the whole s-plane. Let $\gamma_A(s)$ be the product of gamma functions etc. defined in § 2. Then, by (2. 18), we have the *functional equation*

$$\gamma_A(s)\, F(s) = \sum_{i=1}^{n} \lambda_i \gamma_A(1-s)\, F_i(1-s), \tag{8. 1}$$

where $\lambda_i \in \mathbf{C}$ are constants, and each $F_i(s)$ is the Hurwitz series of some (right) admissible triple in A.

We now recall some elementary facts concerning Dirichlet series. By a *Dirichlet series*, we mean a series of the form

$$\sum_{n=1}^{\infty} c_n l_n^{-s}, \tag{8. 2}$$

with complex coefficients $\{c_n\}$, in which

$$0 < l_1 < l_2 < \cdots, \quad \textit{and} \quad \lim_{n \to \infty} l_n = \infty.$$

If each $l_n = n$, we call this an *ordinary Dirichlet series*. A series (8. 2) either diverges everywhere or else it converges in a half-plane $\sigma > \sigma_0$. and its sum function

$$f(s) = \sum_{n=1}^{\infty} c_n l_n^{-s}, \quad \sigma > \sigma_0,$$

is holomorphic there. If σ_0 is finite, the sum function $f(s)$ determines the terms $c_n l_n^{-s}$ of the Dirichlet series uniquely, apart from inserting or deleting terms with $c_n = 0$, and then renumbering. The proof of the corresponding fact for ordinary Dirichlet series (see [14], § 9. 6) can easily be generalised to the present case. We deduce:

(8. 3) Lemma. *Let $f(s)$ be a function represented by a Dirichlet series*

$$f(s) = \sum_{n=1}^{\infty} c_n l_n^{-s},$$

in some non-empty half-plane $\sigma > \sigma_0$. Then the function

$$N_f(x) = \sum_{l_n \leqq x} c_n, \quad x > 0,$$

depends only on f, and not on the choice of Dirichlet series representation.

Now we return to Hurwitz series and prove

(8. 4) Lemma. *Let (W, L, m) be an admissible triple in A. The associated Hurwitz series has a Dirichlet series representation*

$$F(s; W, L, m) = \sum_{n=1}^{\infty} c_n l_n^{-s}$$

which is absolutely convergent for $\sigma > 1$, and in which the coefficients $\{c_n\}$ are non-negative integers.

Proof. We find an order Λ in A and a positive integer b such that $m + L \subset b^{-1}\Lambda$. For $y \in \Lambda \cap A^{\times}$, we have $\|y\|^{-1} = (\Lambda : \Lambda y)$, which is a positive integer. Hence, for $x \in (m+L) \cap A^{\times}$, we have

$$\|x\| = \|b\|^{-1}\,\|bx\| = b^d n^{-1},$$

for some $n \in \mathbf{Z}$, $n > 0$, and where $d = \dim_{\mathbf{Q}} A$. The definition now gives us

$$F(s; W, L, m) = \sum_{n=1}^{\infty} c_n (b^{-d} n)^{-s},$$

for certain non-negative integers $\{c_n\}$. The convergence is given by (2. 4).

§ 9. The main theorem

This section is devoted to proving an asymptotic formula for partial sums of coefficients of a Dirichlet series $f(s)$, under the hypothesis that $f(s)$ is expressible in terms of Hurwitz series attached to a f.d. semisimple $\mathbf{Q}$-algebra A. The case $A = \mathbf{Q}$ is comparatively simple, but must be treated separately. The more interesting situation, where $d = \dim_{\mathbf{Q}} A > 1$, will follow from Landau's Theorem, once we verify that its numerous hypotheses hold in our case. We postpone until § 11 the application of the results of this section to the case of partial zeta functions.

The key result of this article is:

(9. 1) Main Theorem. *Let A be a f.d. semisimple $\mathbf{Q}$-algebra with r simple components, and let*

$$d = \dim_{\mathbf{Q}} A \geqq 1, \quad \delta = \frac{2}{d+1}. \tag{9. 2}$$

Let $f(s)$ be a meromorphic function of the form

$$f(s) = \sum_{i=1}^{m} v_i k_i^s F_i(s), \tag{9. 3}$$

where the $\{v_i\}$ are complex constants, the $\{k_i\}$ positive constants, and where each $F_i(s)$ is the Hurwitz series of some (left or right) admissible triple in A.

The function $f(s)$ has a Dirichlet series representation

$$f(s)=\sum_{n=1}^{\infty} c_n l_n^{-s}, \quad \sigma>1, \tag{9.4}$$

by (8. 4). *There is an asymptotic formula*

$$\sum_{l_n \leqq x} c_n = R(x)+O(x^{1-\delta}\log^{r-1} x) \quad as \quad x\to\infty, \tag{9.5}$$

where

$$R(x)=\textit{the residue of } \frac{x^s f(s)}{s} \quad at \quad s=1. \tag{9.6}$$

(9. 7) Remark. It follows from (2. 4) that $f(s)$ can be continued analytically to a meromorphic function on the whole s-plane, and that it has a pole of order $\leqq r$ at $s=1$. The residue $R(x)$ in (9. 6) is easily expressible in terms of the Laurent expansion

$$f(s)=\sum_{i=-r}^{\infty} a_i(s-1)^i \tag{9.8}$$

about $s=1$. Then we have

$$R(x)=\sum_{j=0}^{r-1} b_j x \log^j x, \tag{9.9}$$

where for $0\leqq j\leqq r-1$,

$$b_j=\frac{1}{j!}\{a_{-(j+1)}-a_{-(j+2)}+a_{-(j+3)}-\cdots+(-1)^{r-1-j}a_{-r}\}. \tag{9.10}$$

In particular, suppose that $f(s)$ has a pole of order k at $s=1$, where $1\leqq k\leqq r$. Then $a_i=0$ for $-r\leqq i\leqq -(k+1)$, and $a_{-k}\neq 0$, so in this case therefore

$$R(x)=x\left(\frac{a_{-k}\log^{k-1}x}{(k-1)!}+\textit{lower powers of } \log x\right). \tag{9.11}$$

Note that $R(x)=0$ if $k=0$, that is, if $f(s)$ is holomorphic at $s=1$.

As a first step in the proof of our main theorem we show:

(9. 12) Proposition. *Let $f(s)$ be defined as in* (9. 3), *and suppose that $f(s)$ has a pole at $s=s_0$, where $s_0\neq 1$. Then there is a simple component A_i of A, of dimension n_i^2 over its centre, such that $s_0=\frac{j}{n_i}$, for some integer j, $0<j<n_i$. Consequently, $f(s)$ has no poles in the region $1-\delta\leqq\sigma\leqq 1$, except (possibly) at $s=1$, where δ is as in* (9. 2).

Proof. By (3. 8) we may express $f(s)$ in terms of Hurwitz series of congruence triples in the simple components A_i of A. It thus suffices to prove the result when A is a simple algebra, and $f(s)=F(s; W, L, m)$, with (W, L, m) a congruence triple in A (see (3. 7)). When A satisfies the Eichler condition, the assertion follows from (5. 5) and (4. 4). On the other hand, when A is a totally definite quaternion algebra, we conclude from formula (6. 4) that $\gamma_A(s)\, f(s)$ has a simple pole at $s=1$, and possibly another at $s=0$, but no others. Since $\gamma_A(s)$ contains $\Gamma(2s)$ as a factor (see (6. 2)), it has a pole at $s=0$. It follows that $f(s)$ must be holomorphic at $s=0$, so the only pole of $f(s)$ is at $s=1$.

Now let A be any f.d. semisimple $\mathbf{Q}$-algebra, and let $d=\dim_{\mathbf{Q}} A \geqq 1$. If s_0 is a pole of $f(s)$ and $s_0 \neq 1$, then by the above s_0 is real and $s_0 \leqq 1-\frac{1}{n_i}$, for some positive integer n_i such that $n_i^2 \leqq d$. It follows that $s_0 < 1-\delta$ except when $n_i = d = 1$, that is, when $A=\mathbf{Q}$. But in this case, $f(s)$ is holomorphic everywhere except at $s=1$, by the preceding remarks. This completes the proof of the proposition.

Continuing with our proof, we note that the hypotheses and conclusions of (9. 1) are linear in $f(s)$, that is, if the theorem holds for $f_1(s)$ and $f_2(s)$, it also holds for $v_1 f_1(s) + v_2 f_2(s)$ with $v_1, v_2 \in \mathbf{C}$. It thus suffices to prove (9. 1) for functions of the form $f(s) = k^s F(s)$, where $k>0$ and $F(s)$ is a Hurwitz series. We first treat the special case where $A=\mathbf{Q}$, to which the general method does not apply.

(9. 13) Proposition. *Theorem* (9. 1) *holds when* $A=\mathbf{Q}$ (i.e. *when* $d=1$).

Proof. In this case $d=r=\delta=1$. We may take $f(s)=k^s F(s; W, L, m)$, and there are only two choices for the arithmetic subgroup W, namely $\{1\}$ and $\{\pm 1\}$. By (3. 1) it suffices to treat the case $W=\{1\}$. Since L is a full $\mathbf{Z}$-lattice in $\mathbf{Q}$, we have $L=c\mathbf{Z}$ for some positive $c \in \mathbf{Q}$. Therefore

$$f(s) = k^s \sum_{n \in \mathbf{Z}}{}' |m+cn|^{-s},$$

the prime indicating that we must omit the integer n (if any) for which $m+cn=0$. Incorporating the factor k^s inside the summation, we obtain

$$f(s) = \sum_{n \in \mathbf{Z}}{}' |a+bn|^{-s}$$

where now $a, b \in \mathbf{R}$ and $b=\frac{c}{k}>0$.

Equation (9. 5), which we are trying to prove, now reads as follows:

$$\sum_{|a+bn| \leqq x}{}' 1 = R(x) + O(1) \quad as \quad x \to \infty, \tag{9. 14}$$

where $R(x)$ is the residue of $\frac{x^s f(s)}{s}$ at the simple pole $s=1$. The left hand side of (9. 14) is clearly of the form $\frac{2x}{b} + O(1)$ as $x \to \infty$. On the other hand, we have

$$\begin{aligned} R(x) &= x \; (\textit{residue of } k^s F(s; W, c\mathbf{Z}, m) \textit{ at } s=1) \\ &= \frac{2xk}{c} \end{aligned}$$

by [5], Theorem C. Since $b=\frac{c}{k}$, we obtain $R(x)=\frac{2x}{b}$, which proves (9. 14) and the proposition.

We now quote Landau's Theorem, on which the proof of the general case of (9. 1) is based. The theorem originally appeared in [9], and has also been reprinted in [10], pages 30—64. We start with a Dirichlet series

(9. 15) $$f(s) = \sum_{n=1}^{\infty} c_n l_n^{-s}$$

as in (8. 2). Landau's Theorem, which is in two parts (9. 26) and (9. 33), gives asymptotic formulas for the partial sum

(9. 16) $$N_f(x) = \sum_{l_n \leqq x} c_n, \quad x > 0.$$

The first part gives, after a long list of notation and hypotheses, an asymptotic formula for $N_f(x)$ as $x \to \infty$. In this formula, the error term is slightly weaker than desired. The second part of the theorem gives a better estimate for the error term, once some additional hypotheses are imposed. For the functions (9. 3), in which we are interested, it will turn out that via the functional equation (8. 1), the first part of Landau's Theorem will imply the validity of these additional hypotheses. We will thus be able to use the second part of the theorem to obtain the desired term in (9. 5).

Starting with the series (9. 15), we assume that the following hypotheses are satisfied:

(9. 17) Hypotheses. (i) *For some* $\beta > 0$, *the series* (9. 15) *converges absolutely for* $\sigma > \beta$. *The sum* $f(s)$ *is thus holomorphic in the half-plane* $\sigma > \beta$.

(ii) $f(s)$ *admits analytic continuation to a meromorphic function on the whole s-plane. Moreover, for any pair* $\sigma_1 < \sigma_2$, $f(s)$ *has only finitely many poles in the vertical strip* $\sigma_1 < \sigma < \sigma_2$.

(iii) *For any* $\sigma_1 < \sigma_2$, *there is a constant* $c = c(\sigma_1, \sigma_2)$ *such that*

(9. 18) $f(s) = O(e^{c|t|})$ *as* $|t| \to \infty$, *uniformly in the strip* $\sigma_1 \leqq \sigma \leqq \sigma_2$.

(Equivalently, $e^{-c|t|} f(s)$ *is bounded in the region* $\sigma_1 \leqq \sigma \leqq \sigma_2$, *once we exclude a neighbourhood of each pole of* $f(s)$.)

(iv) *For* $\sigma < 0$, *the function* $f(s)$ *satisfies a functional equation*

(9. 19) $$f(s) \cdot \prod_{i=1}^{h} \Gamma(\alpha_i + \beta_i s) = \left\{ \sum_{n=1}^{\infty} e_n \lambda_n^s \right\} \cdot \prod_{j=1}^{l} \Gamma(\gamma_j - \delta_j s),$$

where the parameters $\{h, l, \alpha_i, \beta_i, \text{etc.}\}$ *satisfy the conditions*

(9. 20) $$\begin{cases} h \geqq 1,\ l \geqq 1;\ \beta_1, \ldots, \beta_h > 0;\ \delta_1, \ldots, \delta_l > 0; \\ \alpha_1, \ldots, \alpha_h,\ \gamma_1, \ldots, \gamma_l \ \textit{real}; \\ \sum_{i=1}^{h} \beta_i = \sum_{j=1}^{l} \delta_j; \end{cases}$$

(9. 21) $$\begin{cases} 0 < \lambda_1 < \lambda_2 < \cdots;\ \lim_{n \to \infty} \lambda_n = \infty;\ e_1, e_2, \ldots \ \textit{complex}; \\ \sum_{n=1}^{\infty} e_n \lambda_n^s \ \textit{is absolutely convergent for}\ \sigma < 0. \end{cases}$$

(v) *We now define*

(9. 22)
$$q=\sum_{j=1}^{l}\gamma_j-\sum_{i=1}^{h}\alpha_i+\frac{h-l}{2},$$

and assume further that

(9. 23)
$$q>\frac{1}{2}.$$

Define the positive constant μ *by the formula*

(9. 24)
$$\mu=\frac{2q-1}{2q+1}\beta.$$

(vi) *Finally, we assume that there is another Dirichlet series*

(9. 25)
$$g(s)=\sum_{n=1}^{\infty}d_n l_n^{-s}, \quad \textit{where each} \quad d_n\geqq 0,$$

with the same $\{l_n\}$ *as in* (9. 15), *such that* $|c_n|\leqq d_n$ *for each n. Assume that* $g(s)$ *satisfies the same hypotheses as* $f(s)$, *namely* (i)—(v) *above, with the same values of* β *and* q, *but possibly different values of the parameters* $\{h, l, \alpha_i$, etc.$\}$.

Remark. If each $c_n\geqq 0$ in (9. 15), we can choose $g(s)=f(s)$ in (vi).

We may now state the first part of Landau's Theorem:

(9. 26) Theorem. *Given* $f(s)$ *as in* (9. 15), *keep all of the notation and hypotheses in* (9. 17). *Then, for any* $\varepsilon>0$, *we have an asymptotic formula*

(9. 27)
$$\sum_{l_n\leqq x}c_n=R(x)+O(x^{\mu+\varepsilon}) \quad \textit{as} \quad x\to\infty.$$

Here, $R(x)$ *is defined as the sum of the residues of the function* $\frac{x^s f(s)}{s}$ *at the poles of* $f(s)$ *in the region*

(9. 28)
$$\mu\leqq\sigma\leqq\beta.$$

Landau also proved that the error term $O(x^{\mu+\varepsilon})$ in (9. 27) can be improved to $O(x^{\mu}\log^{\tau}x)$ for some τ, provided that the following extra hypotheses are satisfied:

(9. 29) Additional Hypotheses. *Keeping the notation and hypotheses in* (9. 17), *suppose that the functional equation, which the function* $g(s)$ *in* (9. 17) (vi) *is assumed to satisfy, has the form*

(9. 30)
$$g(s)\cdot\prod_{i=1}^{h'}\Gamma(\alpha_i'+\beta_i's)=\left\{\sum_{n=1}^{\infty}e_n'(\lambda_n')^s\right\}\cdot\prod_{j=1}^{l'}\Gamma(\gamma_j'-\delta_j's).$$

We now assume that for some constant $\varrho\geqq 0$, *we have*

(9. 31)
$$\sum_{\lambda_n\leqq x}|e_n|\lambda_n^{\beta}=O(x^{\beta}\log^{\varrho}x) \quad \textit{as} \quad x\to\infty,$$

(9. 32)
$$\sum_{\lambda_n'\leqq x}|e_n'|(\lambda_n')^{\beta}=O(x^{\beta}\log^{\varrho}x) \quad \textit{as} \quad x\to\infty.$$

With these additional hypotheses in force, we may now state the second part of Landau's Theorem:

(9. 33) Theorem. *Keeping all of the above notation, and hypotheses* (9. 17) *and* (9. 29), *there is an asymptotic formula*

$$\sum_{l_n \leqq x} c_n = R(x) + O(x^{\mu} \log^{\tau} x) \quad as \quad x \to \infty. \tag{9. 34}$$

Here, μ *is given by* (9. 24), $R(x)$ *is as in* (9. 26), *and* $\tau = \mathrm{Max}(\varrho, k-1)$, *where the integer* k *is defined as follows:* $k=0$ *if* $g(s)$ *is holomorphic at* $s=\beta$, *while otherwise* k *is the order of the pole of* $g(s)$ *at* $s=\beta$.

Remark. As Landau points out, (9. 34) remains true if we replace $R(x)$ by $R_1(x)$, which is defined to be the sum of the residues of $\dfrac{x^s f(s)}{s}$ at the poles of $f(s)$ in the larger region $0<\sigma \leqq \beta$. For, $f(s)$ has only finitely many poles in the strip $0<\sigma<\mu$. If s_0 is one of these, it is elementary to check that the residue of $\dfrac{x^s f(s)}{s}$ at $s=s_0$ is $O(x^{\mu-\varepsilon})$ as $x \to \infty$, for some $\varepsilon>0$.

We now apply all of this to the proof of (9. 1) for the case $d>1$, where $d=\dim_{\mathbf{Q}} A$, noting that (9. 13) settles the case $d=1$. Since the hypotheses and conclusions of (9. 1) are linear in $f(s)$, it is sufficient to restrict our attention to the case where $f(s)=k^s F(s)$, for some $k>0$ and some Hurwitz series $F(s)$ attached to A. We shall prove

(9. 35) Theorem. *Let* A *be a f.d. semisimple* $\mathbf{Q}$*-algebra with* r *simple components, and let* $d=\dim_{\mathbf{Q}} A>1$, $\delta=\dfrac{2}{1+d}$. *Let* $f(s)=k^s F(s)$, *where* $k>0$ *and* $F(s)$ *is the Hurwitz series of some (left or right) admissible triple in* A. *Then* $f(s)$ *satisfies all of the hypotheses of Landau's Theorem* (9. 33), *with the following values of the parameters*:

$$\beta=1, \quad q=\frac{1}{2}d, \quad \mu=1-\delta, \quad \varrho=\tau=r-1. \tag{9. 36}$$

(9. 37) Remarks. (i) We shall postpone to § 10 the verification that the growth condition (9. 18) is satisfied by our function $f(s)$, since this involves rather different techniques.

(ii) In the notation of (9. 35), the vertical strip (9. 28) now becomes $1-\delta \leqq \sigma \leqq 1$. We have already shown in (9. 12) that the only pole of $f(s)$ in this strip occurs at $s=1$. Thus the function $R(x)$ defined in (9. 26) is precisely the residue of $\dfrac{x^s f(s)}{s}$ at $s=1$, which agrees with its definition in (9. 6). Thus, once we establish (9. 35), our Main Theorem (9. 1) for $d>1$ is a consequence of Landau's Theorem (9. 33).

(iii) By symmetry, it suffices to prove (9. 35) for the case where $F(s)$ is the Hurwitz series of a *left* admissible triple in A.

We are now ready to prove (9. 35).

Step 1. Let $f(s)=k^s F(s)$ as in (9. 35), with $F(s)$ the Hurwitz series of a left admissible triple in A. By (8. 4), $f(s)$ is represented by a Dirichlet series (9. 4) with non-negative coefficients $\{c_n\}$, and the series is absolutely convergent for $\sigma>1$. We may thus take $\beta=1$ in (9. 17 i), and we may choose $g(s)=f(s)$ in (9. 17 vi). Note that hypothesis (9. 17 ii) is satisfied, by virtue of (2. 4) and (9. 12). As remarked in (9. 37 ii) above, the growth condition (9. 18) will be proved in § 10.

By (8. 1), we have a functional equation for $f(s)$ of the form

$$\gamma_A(s)\, f(s)=\gamma_A(1-s) \sum_{m=1}^{t} \lambda_m k^s F_m(1-s),$$

with complex constants $\{\lambda_m\}$, and where each $F_m(s)$ is a Hurwitz series of a right admissible triple in A. While the precise version (2. 18) of (8. 1) is not required here, the explicit definition of $\gamma_A(s)$ via formulas (2. 8)—(2. 11) will be vital. We may write

$$\gamma_A(s)=d_1 \cdot d_2^s \cdot \prod_{i=1}^{h} \Gamma(\alpha_i+\beta_i s),$$

where d_1, d_2 and each β_i are positive, the $\{\alpha_i\}$ are real, and $h\geqq 1$. Absorbing the factors d_1, d_2^s and d_2^{1-s} into the summation $\sum \lambda_m k^s F_m(1-s)$ *and then changing notation*, we obtain a functional equation

$$f(s)\cdot \prod_{i=1}^{h} \Gamma(\alpha_i+\beta_i s)=\left\{\sum_{m=1}^{t} \lambda_m k^s F_m(1-s)\right\}\cdot \prod_{i=1}^{h} \Gamma(\alpha_i+\beta_i-\beta_i s), \tag{9. 38}$$

and the series $\sum \lambda_m k^s F_m(1-s)$, viewed as a Dirichlet series with s replaced by $1-s$, is absolutely convergent for $\sigma<0$. Comparing (9. 38) with (9. 19), we see at once that condition (9. 17 iv) is satisfied with $h=l$.

We must next compute the constant

$$q=q_A=\sum_{i=1}^{h} (\alpha_i+\beta_i-\alpha_i)=\sum_{i=1}^{h} \beta_i$$

defined by (9. 22). We may write $q_A=\sum_{\mathbf{B}} q_{\mathbf{B}}$, with $\mathbf{B}$ ranging over the simple components of the $\mathbf{R}$-algebra $A_\infty=A\otimes_{\mathbf{Q}}\mathbf{R}$, and where $q_{\mathbf{B}}$ is the contribution to $\sum \beta_i$ coming from the gamma-factors in $\gamma_{\mathbf{B}}$. We may write $\mathbf{B}\cong M_n(\mathbf{K})$, where $\mathbf{K}=\mathbf{R}$, $\mathbf{C}$ or $\mathbf{H}$; set $d_{\mathbf{K}}=\frac{1}{2}\dim_{\mathbf{R}}\mathbf{K}$. Then, by definition,

$$\gamma_{\mathbf{B}}(s)=\prod_{j=0}^{n-1} \gamma_{\mathbf{K}}(ns-j),\quad \gamma_{\mathbf{B}}(1-s)=\prod_{j=0}^{n-1} \gamma_{\mathbf{K}}(n-j-ns),$$

and $\gamma_{\mathbf{K}}(s)=c_1 c_2^s \Gamma(d_{\mathbf{K}} s)$, for some positive c_1, c_2. Therefore

$$q_{\mathbf{B}}=d_{\mathbf{K}} \sum_{j=0}^{n-1} (n-j+j)=d_{\mathbf{K}} n^2=\frac{1}{2}\dim_{\mathbf{R}}\mathbf{B}.$$

Since $\gamma_A(s)$ is the product of the $\gamma_{\mathbf{B}}(s)$, it follows that

$$q_A = \sum_{\mathbf{B}}' q_{\mathbf{B}} = \frac{1}{2} \dim_{\mathbf{R}} A_\infty = \frac{1}{2} d,$$

where $d = \dim_{\mathbf{Q}} A$. Thus $q > \frac{1}{2}$, since $d > 1$ by hypotheses. Further, since $\beta = 1$, we obtain

$$\mu = \frac{d-1}{d+1} = 1 - \frac{2}{d+1} = 1 - \delta,$$

as claimed.

We have now verified that $f(s)$ satisfies *all* of the hypotheses in (9. 17), except for (9. 17 iii) which we treat in the next section. These same hypotheses also hold for the function $g(s)$ in (9. 17 vi), since we have chosen $g(s) = f(s)$.

Step 2. Let $f(s) = k^s F(s)$, as above, and let $f(s)$ have the Dirichlet series expansion (9. 15). By the first part (9. 27) of Landau's Theorem, we obtain

$$\sum_{l_n \leqq x} c_n = R(x) + O(x^{1-\delta+\varepsilon}) \quad \text{as} \quad x \to \infty,$$

for each $\varepsilon > 0$. As pointed out in Remark (9. 37 ii), $R(x)$ is the residue of $\frac{x^s f(s)}{s}$ at $s = 1$. Since $f(s)$ has a pole of order r at $s = 1$, it follows from (9. 11) that

$$R(x) = O(x \log^{r-1} x) \quad \text{as} \quad x \to \infty.$$

Since $\delta > 0$, we thus obtain

$$\sum_{l_n \leqq x} c_n = O(x \log^{r-1} x) \quad \text{as} \quad x \to \infty. \tag{9. 39}$$

By symmetry, the same estimate (9. 39) holds when $F(s)$ is the Hurwitz series of a right admissible triple in A.

Step 3. In order to apply Landau's Theorem (9. 33), it remains for us to verify condition (9. 31) with the choice $\varrho = r - 1$. Consider any one of the terms $k^s F_m(1-s)$ occurring on the right hand side of (9. 38), and set

$$G(s) = k^{1-s} F_m(s) = \sum_{n=1}^{\infty} b_n \mu_n^{-s}, \quad \sigma > 1.$$

Since $F_m(s)$ is the Hurwitz series of a right admissible triple, we deduce from Step 2 that

$$\sum_{\mu_n \leqq x} b_n = O(x \log^{r-1} x) \quad \text{as} \quad x \to \infty. \tag{9. 40}$$

On the other hand,

$$k^s F_m(1-s) = G(1-s) = \sum_{n=1} (b_n \mu_n^{-1}) \mu_n^s,$$

and since each $b_n \mu_n^{-1} > 0$, the left hand side of (9. 31) becomes

$$\sum_{\mu_n \leqq x} (b_n \mu_n^{-1}) \mu_n = \sum_{\mu_n \leqq x} b_n.$$

Thus (9. 31) holds with $\varrho = r - 1$, by (9. 40).

Finally, we note that $f(s)$ has a pole of order r at $s=1$, so the constant τ in (9. 33) is precisely $r-1$. This completes the proof of Theorem (9. 35), and of our Main Theorem (9. 1), except of course for the verification of condition (9. 17 iii).

§ 10. Growth in vertical strips

We shall now complete the proof of Theorem (9. 35), and hence also the proof of our Main Theorem (9. 1), by verifying the growth condition (9. 18) for the function $f(s)=k^s F(s)$, where $k>0$ and $F(s)$ is the Hurwitz series of some admissible triple in A. The factor k^s is totally irrelevant here, since both it and its reciprocal are bounded in any vertical strip. Thus we need only prove

(10. 1) Proposition. *Let $F(s)$ be the Hurwitz series of some left or right admissible triple in the f.d. semisimple $\mathbf{Q}$-algebra A. Let $\sigma_1<\sigma_2$. There exists $c=c(\sigma_1, \sigma_2)$ such that $F(s)=O(e^{c|t|})$ as $|t|\to\infty$, uniformly in the strip $\sigma_1\leqq\sigma\leqq\sigma_2$.*

This is implied by the stronger result:

(10. 2) Theorem. *With the hypotheses of* (10. 1), *there exists $b=b(\sigma_1, \sigma_2)$ such that*

$$F(s)=O(|t|^b) \quad as \quad |t|\to\infty, \tag{10. 3}$$

uniformly in the strip $\sigma_1\leqq\sigma\leqq\sigma_2$. (In other words, $|t|^{-b}F(\sigma+it)$ is bounded for $\sigma_1\leqq\sigma\leqq\sigma_2$, outside of some compact set.)

Remark. Theorem (10. 2) says that the function $F(s)$ is of "finite order", in the sense of the theory of functions representable by a Dirichlet series (see [14], § 9. 4). The constant $b(\sigma_1, \sigma_2)$ in (10. 2) will depend on the algebra A, as well as on σ_1 and σ_2. It can, however, be chosen independently of the Hurwitz series $F(s)$. This will be clear from the proof, but we will make no use of the fact.

By symmetry, we need only prove (10. 2) for the Hurwitz series of a *left* admissible triple. We can, moreover, reduce to the case in which A is *simple* and the triple is a *congruence triple* in A, by (3. 9). The first step is to prove:

(10. 4) Theorem. *Theorem* (10. 2) *is valid when A is a totally definite quaternion algebra.*

When A is not a totally definite quaternion algebra, (5. 5) shows that the Hurwitz series of a congruence triple in A is a finite linear combination of zeta integrals $Z(\Phi, \psi, s)$, as in § 4. Theorem (10. 2) therefore follows from (10. 4) and:

(10. 5) Theorem. *Let A be a f.d. simple $\mathbf{Q}$-algebra, and let $\sigma_1<\sigma_2$. There exists a constant $b=b(\sigma_1, \sigma_2)$ such that, for any $\Phi\in\mathcal{S}(A)$ and any continuous character ψ of $J(A)$ which is trivial on $A^\times$ and has finite image, we have*

$$Z(\Phi, \psi, s)=O(|t|^b) \quad as \quad |t|\to\infty,$$

uniformly in the strip $\sigma_1\leqq\sigma\leqq\sigma_2$. (Here, we view $Z(\Phi, \psi, s)$ as a meromorphic function on the whole s-plane via analytic continuation: *see* (4. 4 i).)

Remark. (10. 5) holds, with the same proof, when ψ is any quasicharacter of the full idele group $J(A)\times A_\infty^\times$ which is trivial on the image of $A^\times$. This extra generality merely requires more notation, and is not useful here.

We first recall a simple property of the gamma function:

(10. 6) Lemma. *Given $\sigma_1 < \sigma_2$ and $\delta > 0$, let $\mathscr{D}$ be the region in the s-plane defined by $\sigma_1 \leqq \sigma \leqq \sigma_2$, $|t| > \delta$. There exist constants $c_2 > c_1 > 0$, which depend on $\mathscr{D}$, such that*

$$c_1 |t|^{\sigma - \frac{1}{2}} e^{-\pi \frac{|t|}{2}} \leqq |\Gamma(s)| \leqq c_2 |t|^{\sigma - \frac{1}{2}} e^{-\pi \frac{|t|}{2}}$$

for all $s \in \mathscr{D}$.

Proof. Assume first that $\sigma_1 > 0$. By Stirling's Formula (see [1], p. 166 or [15], (12. 33)) we have

$$\log \Gamma(s) = \frac{1}{2} \log 2\pi + \left(s - \frac{1}{2}\right) \log s - s + J(s),$$

where the error term $J(s)$ tends to zero as $|s| \to \infty$, uniformly in the region $\sigma \geqq \sigma_1$. (Here we take the principal value of the logarithm, which is real on the positive real axis.) Therefore

$$\log |\Gamma(s)| = \mathrm{Re}(\log \Gamma(s)) = \frac{1}{2} \log 2\pi + \left(\sigma - \frac{1}{2}\right) \log |s| - t \arg s - \sigma + \mathrm{Re}(J(s)).$$

In the region $\mathscr{D}$, both terms $-\sigma$ and $\mathrm{Re}(J(s))$ are uniformly bounded, while as $|t| \to \infty$,

$$\log |s| - \log |t| = O(t^{-2}), \quad \text{and} \quad t \arg s - \pi \frac{|t|}{2} = O(1),$$

uniformly in $\mathscr{D}$. This gives the desired inequalities on $|\Gamma(s)|$ for the case $\sigma_1 > 0$. To move σ_1 to the left, we just use the functional equation $s\Gamma(s) = \Gamma(s+1)$ repeatedly. This completes the proof.

We now begin the proof of Theorem (10. 4), so until further notice, we take A to be a totally definite quaternion algebra. We use the notation of § 6, so in particular

$$(10.\ 7) \qquad \begin{cases} \dim_{\mathbf{Q}} A = 4n, & \gamma_A(s) = \{(2\pi)^{-2s} \Gamma(2s)\}^n, \\ \tilde{F}(s) = \gamma_A(s) F(s), & \text{where} \quad F(s) = F(s; W, L, m). \end{cases}$$

The proof will proceed in three steps:

(i) Use formula (6. 4) to estimate $\tilde{F}(s)$ in the region $\mathscr{D}$ of (10. 6). Then use (10. 6) to get a rough estimate for $F(s)$.

(ii) From the definition of the Hurwitz series $F(s)$, it is clear that $F(s)$ is bounded on any vertical line $\sigma = \sigma_0 > 1$. By using the functional equation (8. 1), we deduce that (10. 3) holds for $s = \sigma_1 + it$, with $\sigma_1 < 0$.

(iii) The rough estimates for $F(s)$ in (i), together with the finer estimates in (ii), allow us to use the Phragmen-Lindelöf Theorem to obtain the desired estimate (10. 3) in an arbitrary vertical strip.

Embarking on this programme, we note first that the integrals $I_1(s)$ and $I_2(s)$ occurring in formula (6. 4) for $\tilde{F}(s)$, converge absolutely on the whole s-plane. An obvious argument shows that both $I_1(s)$ and $I_2(s)$ are bounded in any vertical strip. In view of the formula for $\gamma_A(s)$ in (10. 7), we obtain the following result by using (10. 6):

(10. 8) Proposition. *For any $\sigma_1 < \sigma_2$, there is a constant $b = b(\sigma_1, \sigma_2)$ such that $F(s) = O(e^{b|t|})$ as $|t| \to \infty$, uniformly in the strip $\sigma_1 \leqq \sigma \leqq \sigma_2$.*

We next prove:

(10. 9) Proposition. *Let $F(s)=F(s;\,W,\,L,\,m)$. Given any $\sigma_1<0$, there is a constant $c=c(\sigma_1)$ such that $F(\sigma_1+it)=O(|t|^c)$ as $|t|\to\infty$.*

Proof. The definition gives us

(10. 10) $\qquad F(s)$ *is bounded on any line* $\sigma=\sigma_0>1$.

The functional equation (8. 1) can be rewritten as

$$F(s)=\left\{\frac{\gamma_A(1-s)}{\gamma_A(s)}\right\}\sum_{m=1}^{k}\lambda_m F_m(1-s),$$

for certain constants $\{\lambda_m\}$ and certain Hurwitz series $\{F_m\}$. For $\sigma=\sigma_1<0$, we have $\mathrm{Re}(1-s)=1-\sigma_1>1$, so the terms $F_m(1-s)$ are uniformly bounded. Furthermore,

$$\frac{\gamma_A(1-s)}{\gamma_A(s)}=\left\{(2\pi)^{4s-2}\,\frac{\Gamma(2-2s)}{\Gamma(2s)}\right\}^n,$$

and on the line $\sigma=\sigma_1$, the right hand expression is bounded by a power $|t|^c$ for some c, by (10. 6). This completes the proof.

For the final step, we recall

(10. 11) Phragmen-Lindelöf Theorem. *Let $f(s)$ be holomorphic in the region $\mathscr{R}$ defined by $\sigma_1\leqq\sigma\leqq\sigma_2$, $t\geqq\delta$, where $\delta>0$. Suppose that $f(s)$ is bounded on the two half-lines*

(10. 12) $\qquad \sigma=\sigma_1,\quad t\geqq\delta\quad$ *and* $\quad\sigma=\sigma_2,\quad t\geqq\delta.$

Assume further that there is a constant d such that

(10. 13) $\qquad f(s)=O(e^{dt})\quad$ *as* $\quad t\to\infty,$

uniformly in the region $\mathscr{R}$. Then $f(s)$ is bounded in $\mathscr{R}$.

This is proved in [8], p. 703, and a slightly more powerful version is given in [11], p. 234.

We are now ready to finish the proof of Theorem (10. 4), so we start with an arbitrary vertical strip $\sigma_1\leqq\sigma\leqq\sigma_2$. Decreasing σ_1 if need be, we may assume for the proof that $\sigma_1<0$, and similarly that $\sigma_2>1$. Let us choose $\delta=1$, noting that all the poles of $F(s)$ lie on the real axis. We may choose the constant c in (10. 9) to be a positive integer, and we apply (10. 11) to the function

$$f(s)=(s-2\sigma_2)^{-c}\,F(s).$$

Then $f(s)$ is bounded on the lines (10. 12), by (10. 9) and (10. 10). Further, condition (10. 13) holds by virtue of (10. 8). It thus follows from (10. 11) that $f(s)$ is bounded in the region $\sigma_1\leqq\sigma\leqq\sigma_2$, $t\geqq1$. Therefore $F(s)$ is $O(|t|^c)$ in this region. To deal with the bottom half of the strip, we appeal to symmetry. This completes the proof of Theorem (10. 4).

It remains for us to prove Theorem (10. 5), so now let A be any f.d. *simple* **Q**-algebra; it is irrelevant whether or not A is a totally definite quaternion algebra. The proof of (10. 5) is based on the following standard result from algebraic number theory, which is indeed a special case of the result we are trying to prove.

(10. 14) Theorem. *Let K be an algebraic number field and χ a Hecke character of K. Let $L_K(\chi, s)$ be the Hecke L-function of χ, and let $\sigma_1 < \sigma_2$. There exists $c = c(\sigma_1, \sigma_2)$ such that*

$$L_K(\chi, s) = O(|t|^c) \quad as \quad |t| \to \infty,$$

uniformly in the strip $\sigma_1 \leqq \sigma \leqq \sigma_2$.

Proof. See [12], p. 317. (This can be proved in just the same way as (10. 4). Indeed, our proof of (10. 4) is modelled on this classical result.)

Now return to our simple algebra A. Let K be the centre of A, and put $n^2 = \dim_K A$, $n \geqq 1$. Let ψ be a character of $J(A)$ as in (10. 5). As in (4. 4), we find a Hecke character χ of K such that $\psi = \chi \cdot nr$, where $nr: J(A) \to J(K)$ is the reduced norm map. (Strictly speaking, χ is a character of the full idele group $J(K) \times K_\infty^\times$, trivial on $K^\times$. Here, we restrict it $J(K)$.) For any $\Phi \in \mathcal{S}(A)$, (4. 4) gives us

$$Z(\Phi, \psi, s) = \prod_{j=0}^{n-1} L_K(\chi, ns - j) \cdot \prod_P g_P(s).$$

Here, P ranges over the non-Archimedean primes of K, and $g_P(s)$ lies in $\mathbf{C}[p^s, p^{-s}]$, where p is the rational prime below P. Moreover, $g_P(s)$ is identically one for all but finitely many P. The product of the functions $g_P(s)$ above is therefore uniformly $O(1)$ in any vertical strip, and (10. 5) now follows immediately from (10. 14).

This finally completes the proof of our Main Theorem (9. 1).

§ 11. Asymptotic distribution of ideals in orders

We can now read off from (9. 1) our principal result concerning ideals of orders.

(11. 1) Theorem. *Let A be a f.d. semisimple $\mathbf{Q}$-algebra with r simple components. Let Λ be a $\mathbf{Z}$-order in A and M a full left ideal of Λ. For $x \geqq 1$, let $N(M, x)$ be the number of left ideals X of Λ with $(\Lambda : X) \leqq x$ and $X \cong M$ as Λ-modules. Let*

(11. 2)
$$Z_\Lambda(M, s) = \sum_{i=-r}^{\infty} a_i (s-1)^i$$

be the Laurent expansion of $Z_\Lambda(M, s)$ about $s = 1$. Then

(11. 3)
$$N(M, x) = R(M, x) + O(x^{1-\delta} \log^{r-1} x) \quad as \quad x \to \infty,$$

where

$$\delta = \frac{2}{1 + \dim_{\mathbf{Q}} A},$$

$$R(M, x) = \text{the residue of } \frac{x^s Z_\Lambda(M, s)}{s} \quad at \quad s = 1$$

$$= \sum_{j=0}^{r-1} b_j x \log^j x,$$

and

$$b_j = \frac{1}{j!} \{a_{-(j+1)} - a_{-(j+2)} + \cdots + (-1)^{r-j-1} a_{-r}\}, \quad 0 \leqq j \leqq r-1.$$

Proof. We have

$$Z_\Lambda(M, s) = \sum_{n=1}^{\infty} c_n n^{-s}, \quad \mathrm{Re}(s) > 1,$$

$$N(M, x) = \sum_{1 \leqq n \leqq x} c_n,$$

for certain non-negative integers $\{c_n\}$. By (2. 7), we have

$$Z_\Lambda(M, s) = (\Lambda : M)^{-s} F(s; \mathrm{Aut}_\Lambda M, M^{-1}, 0),$$

and everything follows from (9. 1) and (9. 10).

(11. 4) Corollary. *If* $r=1$, i.e. *if* A *is a simple algebra, we have*

(11. 5) $$N(M, x) = a_{-1} x + O(x^{1-\delta}) \quad \text{as} \quad x \to \infty.$$

On the other hand, if $r \geqq 2$,

(11. 6) $$N(M, x) = \frac{a_{-r}}{(r-1)!} x \log^{r-1} x + O(x \log^{r-2} x) \quad \text{as} \quad x \to \infty.$$

Notice that (11. 6) coincides with [5] Th. 4. 1 except for the greatly improved error term. This new error term, at least in the case $r \geqq 2$, is usually best possible.

Likewise, we can apply (9. 1) to the various other zeta functions $\zeta_\Lambda(s)$, $Z_\Lambda(g(M), s)$, $Z_\Lambda([M], s)$ (see [3] § 1) attached to Λ and M. Each of these is a finite sum of partial zeta functions $Z_\Lambda(M', s)$, so we get asymptotic estimates for the numbers of ideals in various other classes. For example, we have

$$\zeta_\Lambda(s) = \sum_X (\Lambda : X)^{-s}, \quad \mathrm{Re}(s) > 1,$$

where X ranges over all full left ideals of Λ. Let $N_\Lambda(x)$ be the number of such ideals X with $(\Lambda : X) \leqq x$. Then

(11. 7) $$N_\Lambda(x) = R_\Lambda(x) + O(x^{1-\delta} \log^{r-1} x) \quad \text{as} \quad x \to \infty,$$

where $R_\Lambda(x)$ = the residue of $\dfrac{x^s \zeta_\Lambda(s)}{s}$ at $s=1$. One can give a fairly explicit account of $\zeta_\Lambda(s)$ in terms of classical zeta functions — see [2], [3] — but this is not really helpful when it comes to calculating the coefficients of $R_\Lambda(x)$ beyond the leading one.

We now recall some module theory. Let M, M' be full left ideals of Λ. Then M, M' are said to lie in the same *genus* (notation: $M' \in g(M)$ or $g(M') = g(M)$) if $M'_p \cong M_p$ as Λ_p-modules for all prime numbers p. We divide the set of left Λ-ideals in $g(M)$ into *stable isomorphism classes* $[M']$ by $[M'] = [M'']$ if and only if $M' \oplus M \cong M'' \oplus M$. Put

$$Z_\Lambda([M'], s) = \sum_{[X] = [M']} (\Lambda : X)^{-s},$$

X ranging over the left Λ-ideals in the genus of M with X stably isomorphic to M'. Likewise, let $N([M'], x)$ be the number of such ideals X with $(\Lambda : X) \leqq x$. From (11. 1) we have

(11. 8) $$N([M'], x) = R([M'], x) + O(x^{1-\delta} \log^{r-1} x),$$

where $R([M'], x)$ is the residue of $\dfrac{x^s Z_A([M'], s)}{s}$ at $s=1$. We know from [3], Th. 6. 10, that the leading Laurent coefficient of $Z_A([M], s)$ depends only on $g(M)$, so

(11. 9) Proposition. *Let M_1, M_2 be full left ideals of Λ lying in the same genus. Then*

$$N([M_1], x) - N([M_2], x) = \begin{cases} O(x \log^{r-2} x) & \text{if } r \geqq 2, \\ O(x^{1-\delta}) & \text{if } r = 1. \end{cases}$$

In the case $r \geqq 2$ of (11. 9), the error term is, in general, best possible. In other words, if $r \geqq 2$, the coefficient of $x \log^j x$, for $j \leqq r-2$, in $R([M], x)$ is liable to vary with the stable isomorphism class of M, rather than just its genus. This is demonstrated by the following simple example.

(11. 10) Example. Let K be an algebraic number field with ring of integers R. Suppose that the ideal class group of R has order 2. Let $A = \mathbf{Q} \times K$, $\Lambda = \mathbf{Z} \times R$. The full left ideals of Λ all lie in $g(\Lambda)$, and there are just two isomorphism classes, represented by Λ and $\mathbf{Z} \times I$ respectively, where I is any non-principal ideal of R. Isomorphism and stable isomorphism of Λ-ideals coincide here, so we consider $N(\Lambda, x)$ and $N(\mathbf{Z} \times I, x)$. Let χ be the nontrivial character of the ideal class group $\mathrm{Cl}\, R$ of R, and let $L_K(s, \chi)$ be its usual L-function. It is easy to see that

$$Z_\Lambda(\Lambda, s) = \frac{1}{2} \zeta_{\mathbf{Q}}(s) (\zeta_K(s) + L_K(s, \chi)),$$

$$Z_\Lambda(\mathbf{Z} \times I, s) = \frac{1}{2} \zeta_{\mathbf{Q}}(s) (\zeta_K(s) - L_K(s, \chi)),$$

where $\zeta_{\mathbf{Q}}$, ζ_K are respectively the Riemann zeta function and the Dedekind zeta function of K. The residue of $\dfrac{x^s(Z_\Lambda(\Lambda, s) - Z_\Lambda(\mathbf{Z} \times I, s))}{s}$ at $s=1$ is $L_K(1, \chi)$, and class field theory gives $L_K(1, \chi) > 0$. Therefore we have a nontrivial deviation from uniformity of distribution of ideals in isomorphism (= stable isomorphism) classes:

$$N(\Lambda, x) - N(\mathbf{Z} \times I, x) = x L_K(1, \chi) + O(x^{1-\delta} \log x),$$

with $\delta = \dfrac{2}{1 + [K : \mathbf{Q}]}$. Indeed here, *the residue $R(M, x)$ actually determines the ideal M*, up to isomorphism.

Of course, $R([M], x)$ does not always determine $[M]$. This follows from (11. 9) when the algebra A is simple. There are plenty of other examples.

We now turn to another aspect of the behaviour of $R([M], x)$ as an invariant of the stable isomorphism class $[M]$. For this, we need to recall the notion of the *genus class group* $\mathrm{Cl}\, M$ *of* M. Let $J(A)$ be the finite idele group of A, as in § 4. It is more convenient here to regard $J(A)$ as a subgroup of $\prod A_p^\times$, with p ranging over the prime numbers. For $x = (x_p) \in J(A)$, define a left Λ-lattice Mx in A by $(Mx)_p = M_p x_p$ for all p. Then a full left Λ-lattice M' in A lies in $g(M)$ if and only if $M' = Mx$, for some $x \in J(A)$.

Now write $\mathrm{Cl}\,M$ for the (finite) set of stable isomorphism classes $[M']$ of left Λ-ideals M' in $g(M)$. Then $x \mapsto Mx$ gives a surjective map $J(A) \to \mathrm{Cl}\,M$. The fibres of this map are all cosets of a certain open normal subgroup of $J(A)$ containing $A^{\times}$ and the commutator subgroup of $J(A)$. Thus $\mathrm{Cl}\,M$ has a unique abelian group structure which makes the map $J(A) \to \mathrm{Cl}\,M$ into a group homomorphism. Explicitly, this group structure amounts to

$$[M_1] \cdot [M_2] = [M_3], \quad M_1, M_2 \in g(M), \quad \textit{where} \quad M_1 \oplus M_2 \cong M_3 \oplus M.$$

(The case $M = \Lambda$ is discussed fully in [7]. The general case is little different.)

Now let $A = \prod_{i=1}^{r} A_i$, where the A_i are simple algebras. Put $\mathscr{R} = \{1, 2, \ldots, r\}$, and for a non-empty subset E of $\mathscr{R}$, write

$$A_E = \prod_{i \in E} A_i.$$

Write π_E for the canonical surjection $A \to A_E$, and put $\Lambda_E = \pi_E(\Lambda)$, $M_E = \pi_E(M)$. Then Λ_E is an order in A_E, and M_E is a full left ideal of Λ_E. Moreover, π_E induces a surjective homomorphism $\mathrm{Cl}\,M \to \mathrm{Cl}\,M_E$, which we also denote by π_E. Indeed, if $E' = R - E$, we get a surjection $(\pi_E \times \pi_{E'}): \mathrm{Cl}\,M \to \mathrm{Cl}\,M_E \times \mathrm{Cl}\,M_{E'}$, but this is not necessarily injective. (These projection maps are discussed more fully in [4], § 5.)

Now let M_1, M_2 be full left ideals of Λ lying in the same genus $g(M)$, and let k be an integer, $1 \leqq k \leqq r$. We say that M_1 is *k-equivalent* to M_2 if $\pi_E[M_1] = \pi_E[M_2]$ for all subsets E of $\mathscr{R}$ with $|E| \leqq k$. (For $k = 0$, we say that any two left Λ-ideals in the same genus are 0-equivalent.) Note that M_1 and M_2 are r-equivalent if and only if they are stably isomorphic.

(11. 11) Theorem. *Let A be a f.d. semisimple $\mathbf{Q}$-algebra with r simple components, as above, and let $0 \leqq k \leqq r$. Let M_1, M_2 be full left ideals of Λ lying in the same genus, and suppose that they are k-equivalent. Then*

$$(11.12) \qquad N([M_1], x) - N([M_2], x) = \begin{cases} O(x \log^{r-k-2} x) & \textit{if} \ \ 0 \leqq k \leqq r-2, \\ O(x^{1-\delta} \log^{r-1} x) & \textit{if} \ \ k = r-1, \\ 0 & \textit{if} \ \ k = r, \end{cases}$$

where $\delta = \dfrac{2}{1 + \dim_{\mathbf{Q}} A}$.

(11. 13) Remark. One can restate this in terms of residues. It simply says that

$$\frac{R([M_1], x) - R([M_2], x)}{x}$$

is a polynomial in $\log x$ of degree $\leqq r - k - 2$.

Proof. By (9. 1), we need only show that $Z_\Lambda([M_1], s) - Z_\Lambda([M_2], s)$ has a pole of order $\leqq r - k - 1$ at $s = 1$.

Let $(\mathrm{Cl}\, M)^*$ denote the character group of the finite abelian group $\mathrm{Cl}\, M$. A character $\psi \in (\mathrm{Cl}\, M)^*$ may be viewed as a continuous character of $J(\Lambda)$ via the surjection $J(\Lambda) \to \mathrm{Cl}\, M$ described above. As an idele character, ψ then has finite image and is trivial on $A^\times$.

The surjection $\pi_E : \mathrm{Cl}\, M \to \mathrm{Cl}\, M_E$, $E \subset \mathscr{R}$, induces an injection

$$\pi_E^* : (\mathrm{Cl}\, M_E)^* \to (\mathrm{Cl}\, M)^*.$$

In terms of idele characters, this map π_E^* amounts to composition with the canonical projection $J(\Lambda) \to J(\Lambda_E)$. One can calculate the image of π_E^*, as in [4]. If ψ is a character of $J(\Lambda)$, we write ψ_i for the restriction of ψ to $J(\Lambda_i)$. Let $K(\psi)$ denote the set of $i \in \mathscr{R}$ for which ψ_i is trivial. Then

$$\mathrm{Image}(\pi_E^*) = \{\psi \in (\mathrm{Cl}\, M)^* : K(\psi) \supset \mathscr{R} - E\}.$$

Writing $t(\psi) = |K(\psi)|$, we therefore have

(11. 14) Lemma. *Let M_1, M_2 be left Λ-ideals in the same genus $g(M)$. Then M_1, M_2 are k-equivalent if and only if $\psi[M_1] = \psi[M_2]$ for all $\psi \in (\mathrm{Cl}\, M)^*$ such that $t(\psi) \geqq r - k$.*

We now recall from [3] the definition of the L-function $L_\Lambda(M, \psi, s)$, where $\psi \in (\mathrm{Cl}\, M)^*$:

$$L_\Lambda(M, \psi, s) = \sum_{X \in g(M)} \psi[X]\, (\Lambda : X)^{-s}, \quad \mathrm{Re}(s) > 1.$$

Here, X ranges over all left ideals of Λ in $g(M)$. This L-function is a finite linear combination of partial zeta functions $Z_\Lambda(M', s)$. Moreover (see [3], (1. 5))

$$Z_\Lambda([M_1], s) = h_M^{-1} \sum_{\psi \in (\mathrm{Cl}\, M)^*} \psi[M_1]^{-1}\, L_\Lambda(M, \psi, s),$$

for any $M_1 \in g(M)$. Here, $h_M = |\mathrm{Cl}\, M|$. If M_1, M_2 are k-equivalent, we therefore have by (11. 14):

$$Z_\Lambda([M_1], s) - Z_\Lambda([M_2], s) = h_M^{-1} \sum_{\substack{\psi \in (\mathrm{Cl}\, M)^* \\ t(\psi) \leqq r-k-1}} \{\psi[M_1]^{-1} - \psi[M_2]^{-1}\}\, L_\Lambda(M, \psi, s).$$

However, by [3], (5. 1), $L_\Lambda(M, \psi, s)$ has a pole of order $\leqq t(\psi)$ at $s = 1$. Hence $Z_\Lambda([M_1], s) - Z_\Lambda([M_2], s)$ has a pole of order $\leqq r - k - 1$ at $s = 1$, as required.

It is worth pointing out one special case of (11. 11). Two ideals M_1, M_2 in the same genus are $(r-1)$-equivalent if and only if they have stably isomorphic images in all *proper* homomorphic images of Λ. In this case, we have

$$(11.\ 15) \qquad N([M_1], x) - N([M_2], x) = O(x^\kappa \log^{r-1} x),$$

for a certain $\kappa < 1$. It is natural to ask whether the converse holds: given two left Λ-ideals M_1, M_2 in the same genus satisfying (11. 15), are they necessarily $(r-1)$-equivalent? More generally, one could ask for a converse of (11. 11). There is one fairly trivial obstruction to this: if M_1 and M_2 are related by an *automorphism* of Λ, then (11. 15) will hold irrespective of any k-equivalence. Indeed, if K is a quadratic numberfield with class number 3, and Λ is a maximal order in $A = \mathbf{Q} \times K$, it is easy to find full Λ-ideals M_1, M_2 satisfying (11. 15) but which are not 1-equivalent. The question as to whether this is the only obstruction seems rather hard, even at the level of maximal orders in commutative algebras.

§ 12. Extra poles of Hurwitz series

We return here to the question, already considered in Proposition (9. 12), of the location of the poles of a function $f(s) = \sum v_i k_i^s F_i(s)$ defined as in (9. 3), with each $F_i(s)$ a Hurwitz series attached to the algebra A. By an *extra pole* of $f(s)$ we mean a pole at $s = \varrho$, where $\varrho \neq 1$. By (9. 12), these extra poles (if any) must lie on the real axis, in the interval $0 < \sigma < 1 - \delta$, where $\delta = \dfrac{2}{1 + \dim_{\mathbf{Q}} A}$. Because of the remark following (9. 33), they make no contribution (except to the error term) in the asymptotic formulas with which we have been mainly concerned here. However, this seems a convenient place to set down a few of their basic properties.

If $f(s)$ (as above) does have a pole at $s = \varrho \neq 1$, then some Hurwitz series $F_i(s)$ must also have a pole at $s = \varrho$. By (3. 9), $F_i(s)$ is expressible as a linear combination of products of Hurwitz series attached to simple components of A, so one of these series also has a pole at $s = \varrho$. We shall now determine, *for a simple algebra A*, necessary and sufficient conditions for the existence of a Hurwitz series having an extra pole. The corresponding problem for semisimple algebras seems more difficult, because of our lack of knowledge about zeros of Hurwitz series.

The main result of the section is as follows:

(12. 1) Theorem. *Let A be a f.d. simple* **Q***-algebra, let* Λ *be a maximal* **Z***-order in A, and denote by* $\zeta_\Lambda(s)$ *the (total) zeta function of* Λ*. Let* ϱ *be real,* $0 < \varrho < 1$*. Then:*

(a) *There exists a Hurwitz series F(s) attached to A, with a pole at* $s = \varrho$*, if and only if* $\zeta_\Lambda(s)$ *has a pole at* $s = \varrho$*.*

(b) $\zeta_\Lambda(s)$ *has a pole at* $s = \varrho$ *in precisely the following cases:*

1. $A \cong M_2(\mathbf{Q})$, $\varrho = \dfrac{1}{2}$;

2. $A \cong M_n(\mathbf{Q})$, $n \geqq 3$, *and* $\varrho = 1 - \dfrac{1}{n}$ *or* $1 - \dfrac{2}{n}$;

3. $A \cong M_n(K)$, $n \geqq 2$, *where K is an imaginary quadratic field and* $\varrho = 1 - \dfrac{1}{n}$;

4. $A \cong M_m(\mathbf{H})$, $m \geqq 2$, *where* **H** *is a quaternion division algebra with centre* **Q**, *and* $\varrho = 1 - \dfrac{1}{m}$.

Proof. Part (a). Since $\zeta_\Lambda(s)=\sum(\Lambda:X)^{-s}$, where X ranges over all full left ideals of Λ, we can express $\zeta_\Lambda(s)$ as a finite sum of Hurwitz series $F(s)$. Thus, if $\zeta_\Lambda(s)$ has a pole at $s=\varrho$, so does some $F(s)$.

Conversely, let $F(s)$ be a Hurwitz series attached to A with a pole at $s=\varrho$. The proof of (9. 12) shows that A cannot be a totally definite quaternion algebra, so we may now assume that A satisfies the Eichler condition. We use (3. 8) to express $F(s)$ in terms of Hurwitz series of congruence triples, one of which must then also have a pole at $s=\varrho$. It then follows from (5. 5) that $Z(\Phi, \psi, s)$ has a pole at $s=\varrho$, for some $\Phi\in\mathscr{S}(A)$ and some character ψ of $J(A)$ of the type considered in § 4. Since A is a simple algebra, $Z(\Phi, \psi, s)$ is holomorphic everywhere by (4. 4 ii), unless ψ is the trivial character ψ_0 of $J(A)$. Thus $Z(\Phi, \psi_0, s)$ must have a pole at $s=\varrho$ for some $\Phi\in\mathscr{S}(A)$.

On the other hand, for each $\Phi\in\mathscr{S}(A)$ we have, by (4. 5) and [2] p. 148,

$$Z(\Phi, \psi_0, s)=\zeta_\Lambda(s)\cdot\prod_p \phi_p(s), \tag{12. 2}$$

where each $\phi_p(s)\in\mathbf{C}[p^s, p^{-s}]$, and where $\phi_p(s)$ is identically 1 for almost all p. Thus $\zeta_\Lambda(s)$ must have a pole at $s=\varrho$, as claimed.

Part (b). Suppose that $\zeta_\Lambda(s)$ has a pole at $s=\varrho+1$, and let us use the explicit formula for $\zeta_\Lambda(s)$ from [2]. Let $\dim_K A=n^2$, where K is the centre of A. Let R be the ring of all algebraic integers in K, and put $\mathbf{N}P=(R:P)$, where P ranges over the maximal ideals of R. For each P, the *local index* e_P of A at P is defined by

$$A_P\cong M_{n(P)}(D(P)),\quad \dim_{K_P} D(P)=e_P^2,$$

where $D(P)$ is a division algebra with centre K_P. Let $\zeta_K(s)$ be the Dedekind zeta function of the field K. From [2], § 3. 3 we have

$$\zeta_\Lambda(s)=\prod_{j=0}^{n-1}\zeta_K(ns-j)\cdot\prod_P\prod_{\substack{j=0\\ j\equiv 0\,(\mathrm{mod}\,e_P)}}^{n-1}(1-\mathbf{N}P^{j-ns}). \tag{12. 3}$$

Using this formula, we can find the poles of $\zeta_\Lambda(s)$ fairly easily. The only pole of $\zeta_K(s)$ is a simple one at $s=1$, but $\zeta_K(s)$ has a k-fold zero at $s=0$, where k is the torsion-free rank of the unit group $R^\times$ (so $k\geqq 1$, except that $k=0$ if $K=\mathbf{Q}$ or if K is an imaginary quadratic field). Further, $\zeta_K(s)$ vanishes at the even negative integers for all K, and also at the odd negative integers if K is not totally real.

This tells us about the poles of the product of the zeta functions in (12. 3). The rest of the argument, which we omit, just depends on the ramification theory of simple algebras over the number field K, that is, on the standard theory of the Brauer group of K (see [13]).

It seems likely that in the cases listed in (12. 1 b), *every* Hurwitz series attached to Λ has a pole at $s=\varrho$. However, we shall not pursue this matter here.

References

[1] *L. V. Ahlfors*, Complex Analysis, New York-Toronto-London 1953.
[2] *C. J. Bushnell* and *I. Reiner*, Zeta functions of arithmetic orders and Solomon's conjectures, Math. Z. **173** (1980), 135—161.
[3] *C. J. Bushnell* and *I. Reiner*, *L*-functions of arithmetic orders and asymptotic distribution of ideals, J. reine angew. Math. **327** (1981), 156—183.
[4] *C. J. Bushnell* and *I. Reiner*, The prime ideal theorem in noncommutative arithmetic, Math. Z. **181** (1982), 143—170.
[5] *C. J. Bushnell* and *I. Reiner*, Analytic continuation of partial zeta functions of arithmetic orders, J. reine angew. Math. **349** (1984), 160—178.
[6] *C. J. Bushnell* and *I. Reiner*, Functional equations for Hurwitz series and partial zeta functions of orders, J. reine angew. Math. **364** (1986), 130—148.
[7] *A. Fröhlich*, Locally free modules over arithmetic orders, J. reine angew. Math. **274/275** (1975), 112—138.
[8] *E. Landau*, Über die Anzahl der Gitterpunkte in gewissen Bereichen, Gött. Nachr. (Math.-Phys. Klasse) (1912), 685—772.
[9] *E. Landau*, Über die Anzahl der Gitterpunkte in gewissen Bereichen. II, Gött. Nachr. (Math.-Phys. Klasse) (1915), 209—263.
[10] *E. Landau*, Ausgewählte Abhandlungen zur Gitterpunktlehre (A. Walfisz, ed.), Berlin 1962.
[11] *S. Lang*, Complex Analysis, Reading, Mass. 1977.
[12] *W. Narkiewicz*, Elementary and analytic theory of algebraic numbers, Warsaw 1974.
[13] *I. Reiner*, Maximal Orders, London-New York 1975.
[14] *E. C. Titchmarsh*, Theory of Functions, Oxford 1939.
[15] *E. T. Whittaker* and *G. N. Watson*, A Course of Modern Analysis, 4th ed., Cambridge 1927.

Department of Mathematics, King's College, Strand, London WC2R 2LS, England

University of Illinois, Department of Mathematics, 1409 West Green Street, Urbana, Illinois 61801, U.S.A.

Eingegangen 4. Oktober 1984, in revidierter Form 24. Juni 1985

Zeta Functions and Composition Factors for Arithmetic Orders

Colin J. Bushnell[1] * and Irving Reiner[2] **
[1] Department of Mathematics, King's College, Strand, London WC2R 2LS
[2] University of Illinois, Department of Mathematics, 1409 West Green Street, Urbana, IL 61801, USA

§ 1.

In a series of earlier papers, summarised in [5], we have developed the theory of zeta functions which count certain left ideals X of an arithmetic order Λ according to the index $(\Lambda:X)$ of X in Λ. In this paper, we present the basic theory of a more precise type of zeta function which counts according to the *Λ-composition factors* of the finite Λ-module Λ/X. Functions somewhat of this kind were used for an entirely different purpose in [3], but it was not necessary to develop any general theory there.

We consider only the local case here, so let A be a finite-dimensional semisimple Q_p-algebra, and let Λ be a Z_p-order in A. Let $F_1, \ldots, F_k$ be a complete set of representatives of the isomorphism classes of simple left Λ-modules. If F is any finite left Λ-module, we define, for $1 \leqq i \leqq k$,

(1.1) $l_i(F)$ = *the multiplicity of F_i as a composition factor of F.*

Now let $z_1, \ldots, z_k$ be complex variables. If M, M' are full left Λ-lattices in A with $M \supset M'$, we set

$$z^{l(M/M')} = \prod_{i=1}^{k} z_i^{l_i(M/M')}. \tag{1.2}$$

Fixing such a lattice M, we define

$$Z_\Lambda^L(M; z) = \sum_{X \simeq M} z^{l(\Lambda/X)}, \tag{1.3}$$

where X ranges over those left ideals of Λ which are Λ-isomorphic to M. Similarly, let

---

* Research of both authors partially supported by NSF
** Part of the research for this paper was performed while the second-named author was visiting King's College London and partially supported by SERC

(1.4) $$\zeta_\Lambda^L(z)=\sum_X z^{l(\Lambda/X)}$$

with X ranging over all full left ideals of Λ. (The superscript "L" indicates that we consider *left* ideals and composition structures. We omit it when there is no fear of confusion.) The series (1.3), (1.4) converge provided that all $|z_i|$ are sufficiently small. If we replace $z^{l(\Lambda/X)}$ by $(\Lambda:X)^{-s}$, where s is a single complex variable, then (1.3) and (1.4) define respectively the "ordinary" zeta functions $Z_\Lambda(M,s)$ and $\zeta_\Lambda(s)$ as in [5]. These ordinary zeta functions are indeed special cases of (1.3) and (1.4): one may choose $z_1, \ldots, z_k$ to obtain $Z_\Lambda^L(M;z)=Z_\Lambda(M,s)$ and $\zeta_\Lambda^L(z)=\zeta_\Lambda(s)$.

We shall not give here an exhaustive treatment of these new zeta functions. We confine ourselves to generalising the main results of [1]. In particular, we show that $Z_\Lambda(M;z)$ is a rational function in $z_1, \ldots, z_k$ whose denominator depends in an explicit and straightforward way on Λ, but is independent of M. Moreover, when Λ is a group ring, we obtain functional equations which sharpen those of [1].

As before, the technique is to express the zeta functions in terms of certain *zeta integrals*. These integrals are only marginally more general than the ones we have previously used. However, the key step involves describing the composition structures attached to Λ in terms of an *unramified quasicharacter* of the multiplicative group $A^\times$ of the semisimple algebra A. This process generalises the classical theorem of Brauer and Nesbitt on the existence of the decomposition map for group characters (and gives a very short proof of that result). Moreover, it reveals a surprising wealth of detail in the relation between the analysis on $A^\times$ and the ideal theory of Λ, at the level of classical algebraic K-theory. This is exemplified by the purely algebraic Corollary (3.16).

§ 2. Generalised Zeta Integrals

We need to establish notation and recall some results from [1] and [2]. To start with, suppose that A is a finite-dimensional *simple* Q_p-algebra with centre K. Thus $A \simeq M_m(D)$, where D is a central K-division algebra with dimension e^2, $e \geqq 1$. Put $n=me$. Let $\theta_K: K \to C^\times$ be the "Iwasawa-Tate character" of K. This is the composition of the field trace Tr_{K/Q_p} and the canonical maps

$$Q_p \to Q_p/Z_p \to Q/Z \to C^\times.$$

Thus θ_K is a continuous character of the additive group of K, and the largest fractional ideal (of the discrete valuation ring in K) contained in $\mathrm{Ker}(\theta_K)$ is the *inverse absolute different* D_K^{-1} of K.

Put $\theta_A=\theta_K \cdot \mathrm{tr}$, where $\mathrm{tr}: A \to K$ is the reduced trace map. Let Λ' be some maximal order in A, and set

(2.1) $$\hat{\Lambda}'=\{x \in A: \theta_A(xy)=1, \text{ for all } x \in \Lambda'\}.$$

This is the inverse different of Λ' (denoted by $\underline{\underline{D}}_{\Lambda'}^{-1}$ in [2]). There exists $\delta_{\Lambda'} \in \Lambda'$ such that

(2.2) $$\hat{\Lambda}'=\delta_{\Lambda'}^{-1}\Lambda'=\Lambda'\delta_{\Lambda'}^{-1}.$$

This determines $\delta_{\Lambda'}$ up to multiplication (on left or right) by a unit of Λ'. The index

$$d_A = (\hat{\Lambda}' : \Lambda') = (\Lambda' : \delta_{\Lambda'} \Lambda') \tag{2.3}$$

depends only on A, and not on the choice of Λ'. See [7], §20 for more discussion of this.

Let $\zeta_{\Lambda'}(s)$ be the ordinary zeta function of Λ', as in [1]. Hey's formula ([1], §3.3) says that

$$\zeta_{\Lambda'}(s) = \prod_{h=0}^{m-1} (1 - q^{e(h-ms)})^{-1}, \tag{2.4}$$

where q is the cardinality of the residue class field of K.

For $x \in A^\times$, we put $\|x\|_A = (P : Px)^{-1}$, where P is any full Z_p-lattice in A. This is an integral power of q^n. Finally, let dx be the Haar measure on A for which

$$\int_{\Lambda'} dx = d_A^{-\frac{1}{2}}, \tag{2.5}$$

where Λ' is any maximal order in A.

Now we change notation, and let A be a finite-dimensional *semisimple* algebra over Q_p, say

$$A = \prod_{j=1}^{r} A_j, \tag{2.6}$$

where the A_j are simple algebras. We use the following notation throughout:

(2.7) *Notation.* K_j = the center of A_j, $A_j \simeq M_{m_j}(D_j)$, where D_j is a central K_j-division algebra of dimension e_j^2, and $n_j = m_j e_j$. Let q_j be the cardinality of the residue class field of K_j. For $x \in A$, denote by x_j the component of x in A_j. Write $\|x\|_j = \|x_j\|_{A_j}$, $x \in A^\times$. If Λ' is a maximal order in A, put $\Lambda'_j = \Lambda' \cap A_j$, so that Λ'_j is a maximal order in A_j, and $\Lambda' = \prod_j \Lambda'_j$.

We let θ_A be the product of the θ_{A_j}, and dx the product of the dx_j, as defined by (2.5) in the simple case. We can define $\hat{\Lambda}'$ and d_A as before, and these are the products of the corresponding objects for the A_j. The relation (2.5) still holds. We can likewise find elements $\delta_{\Lambda'}$ to satisfy (2.2).

Now let $\chi: A^\times \to C^\times$ be a continuous homomorphism, i.e., a *quasicharacter* of $A^\times$. We say that χ is *unramified* if it is trivial on $\Lambda'^\times$ for some (equivalently, any) maximal order Λ' in A.

(2.8) **Proposition.** *Let χ be an unramified quasicharacter of $A^\times$. There exist $s_1, \ldots, s_r \in C$ such that*

$$\chi(x) = \prod_{j=1}^{r} \|x\|_j^{s_j}, \quad \textit{for all } x \in A^\times. \tag{2.9}$$

Moreover, χ determines s_j uniquely modulo $2\pi\sqrt{-1}\, n_j \log q_j \cdot Z$.

Proof. The quasicharacter χ is the product of its restrictions to A_j, $1 \leqq j \leqq r$, and each of these restrictions is an unramified quasicharacter. We may therefore assume that A is *simple*, and use the notation (2.7) without the subscripts j. The reduced norm map $nr: A^\times \to K^\times$ is surjective, and its kernel is the commutator subgroup of $A^\times$ ([6], Th. 7.49). Moreover, we have $nr(\Lambda'^\times) = R^\times$, where R is the discrete valuation ring in K and Λ' is any maximal order in A. Thus $\chi = \psi \cdot nr$, where ψ is a quasicharacter of $K^\times$ which is trivial on $R^\times$. The valuation $\| \ \|_K$ gives an isomorphism of $K^\times/R^\times$ with q^Z (the cyclic group generated by q), so $\psi(x) = \|x\|_K^s$, for some $s \in C$ uniquely determined modulo $2\pi\sqrt{-1}\log q \cdot Z$. Moreover, $\chi(x) = \|nr(x)\|_K^s = \|x\|_A^{s/n}$, and the result follows.

If χ is defined by (2.9), we write

$$\chi = \chi^A_{s_1, \ldots, s_r}. \tag{2.10}$$

In the special case $s_1 = \ldots = s_r = s$, we abbreviate this to χ^A_s, so that

$$\chi^A_s(x) = \|x\|^s_A, \qquad x \in A^\times. \tag{2.11}$$

Next, let $S(A)$ be the space of complex-valued, locally constant, compactly supported functions on A. We have $S(A) = S(A_1) \otimes_C \ldots \otimes_C S(A_r)$, or equivalently, $S(A)$ is spanned by functions Φ of the form

$$\Phi(x) = \prod_{j=1}^{r} \Phi_j(x_j), \qquad x \in A, \ \Phi_j \in S(A_j), \ 1 \leqq j \leqq r. \tag{2.12}$$

We take some Haar measure $d^\times x$ on $A^\times$ and define

$$Z(\Phi, \chi) = \int_{A^\times} \Phi(x)\,\chi(x)\,d^\times x, \qquad \Phi \in S(A), \tag{2.13}$$

where χ is an unramified quasicharacter of $A^\times$. If $\chi = \chi^A_{s_1, \ldots, s_r}$, the integral (2.13) converges provided all $\operatorname{Re}(s_j) \geqq 1$, but we shall see later that (2.13) extends, via analytic continuation, to all χ.

(2.14) *Remarks.* 1) One can, of course, define $Z(\Phi, \chi)$ for any quasicharacter χ of $A^\times$, in the same manner. Using the techniques of §3 below, one can describe the "modified L-functions" of [3] in terms of these more general zeta integrals. We do not pursue this aspect here.

2) If we take $\chi = \chi^A_s$ in (2.13), we obtain $Z(\Phi, \chi^A_s) = Z(\Phi, s)$ in the notation of [1].

(2.15) **Proposition.** *Let $\Phi \in S(A)$ and let χ be an unramified quasicharacter of $A^\times$. Then*

$$Z(\Phi, \chi) = f(\Phi, \chi) \prod_{j=1}^{r} \zeta_{A_j}(s_j), \tag{2.16}$$

where $\chi = \chi^A_{s_1, \ldots, s_r}$, and $f(\Phi, \chi) \in C[q_j^{\pm n_j s_j}]_{1 \leqq j \leqq r}$.

Proof. For fixed Φ, the right hand side of (2.16) admits analytic continuation, so it is enough to prove the result under the additional hypothesis that all

$\operatorname{Re}(s_j)>1$. By linearity, we can also take Φ to be of the form (2.12). We then get $Z(\Phi,\chi)=\prod_j Z(\Phi_j,\chi^A_{s_j})$. However, $Z(\Phi_j,\chi^A_{s_j})=Z(\Phi_j,s_j)$, in the notation of [1]. The result therefore follows from [1], Lemma p. 148.

We now define

$$\hat{\Phi}(y)=\int_A \Phi(x)\,\theta_A(xy)\,dx,\qquad \Phi\in S(A),\ y\in A. \tag{2.17}$$

The function $\hat{\Phi}$ (the Fourier transform of Φ) again lies in $S(A)$ and $\hat{\hat{\Phi}}(x)=\Phi(-x)$ for all $x\in A$ and all $\Phi\in S(A)$, by the choice of dx.

Reducing to the one variable case, just as in the proof of (2.15) above, and applying the results of [2], §10, we obtain the *functional equation*:

$$f(\hat{\Phi},\chi^A_1\chi^{-1})=f(\Phi,\chi)(\chi^A_{-\frac{1}{2}}\chi)(\delta_{A'}) \tag{2.18}$$

for all unramified quasicharacters χ of $A^\times$ and all $\Phi\in S(A)$. Here, f is defined by (2.16), and $\delta_{A'}=(\delta_{A'_j})$ where $\delta_{A'_j}$ is given by (2.2). Since the quasicharacter $\chi^A_{-\frac{1}{2}}\chi$ is unramified, the quantity $(\chi^A_{-\frac{1}{2}}\chi)(\delta_{A'})$ is independent of the choice of $\delta_{A'}$, and indeed of the maximal order Λ'.

§3. Zeta Functions and Composition Factors

Let A be a finite-dimensional semisimple Q_p-algebra as in (2.6) and (2.7), and let Λ be a Z_p-order in A. As in §1, let $F_1,\dots,F_k$ be a complete set of representatives of the isomorphism classes of simple left Λ-modules. We extend the notation of §1 by setting

$$l_i(M/M')=l_i(M/M\cap M')-l_i(M'/M\cap M'), \tag{3.1}$$

for any pair M, M' of full left Λ-lattices in A. We correspondingly define

$$z^{l(M/M')}=\prod_{i=1}^{k} z_i^{l_i(M/M')}. \tag{3.2}$$

Thus $z^{l(M/M')}\cdot z^{l(M'/M)}=1$, and

$$z^{l(M/M'')}=z^{l(M/M')}\cdot z^{l(M'/M'')}$$

for any triple M, M', M'' of full left Λ-lattices in A.

We fix a full left Λ-lattice M in A, and consider the function $Z^L_\Lambda(M;z)$ defined by (1.3). The series (1.3), (1.4) converge provided that all $|z_i|$ are sufficiently small, but we shall show below that they admit analytic continuation. In the domain of converge, we can rearrange (1.3) as a power series

$$Z_\Lambda(M;z)=\sum_u a_\Lambda(M;u)\,z^u, \tag{3.3}$$

where $u=(u_1,\dots,u_k)$ ranges over all k-tuples of non-negative integers, $z^u=\prod z_i^{u_i}$, and $a_\Lambda(M;u)$ is the number of left ideals X of Λ such that $X \simeq M$ as Λ-modules and Λ/X has composition factors $\{F_1^{u_1}, \ldots, F_k^{u_k}\}$.

As an example, and for future use, we consider the special case of a *maximal order* Λ' in A. Any full left ideal of Λ' is principal, so we only have to consider the function $Z_{\Lambda'}(\Lambda'; z) = \zeta_{\Lambda'}(z)$. We have $k = r$, the number of simple components of A, and we enumerate the simple Λ'-modules $F'_1, \ldots, F'_r$ so that F'_j is a simple Λ'_j-module, viewed as a Λ'-module via the canonical projection $\Lambda' \to \Lambda'_j$. If Y is a full left Λ'-ideal, then $Y = \prod_j Y_j$, and Y_j is full left Λ'_j-ideal. Moreover, $l_j(\Lambda'/Y) = l_j(\Lambda'_j/Y_j)$, so

$$\zeta_{\Lambda'}(z) = \prod_{j=1}^{r} \zeta_{\Lambda'_j}(z_j). \tag{3.4}$$

Now, since Λ'_j has a unique simple left module (namely F'_j), the composition factors of Λ'_j/Y_j are just $F_j'^{\,l_j j}$, where $\# F_j'^{\,l_j} = (\Lambda'_j : Y_j)$. However, $\# F'_j = q_j^{n_j}$, so

$$z_j^{l_j(\Lambda'_j/Y_j)} = (\Lambda'_j : Y_j)^{\log z_j / n_j \log q_j}.$$

Hey's formula (2.4) now gives us

$$\zeta_{\Lambda'_j}(z_j) = \prod_{h=0}^{m_j - 1} (1 - q_j^{h e_j} z_j)^{-1}. \tag{3.5}$$

Returning to Λ and M as above, we define

$$M^{-1} = \{x \in A : Mx \subset \Lambda\}. \tag{3.6}$$

This is a full right Λ-lattice in A, and it is invariant under multiplication (on the left) by the group

$$\operatorname{Aut}_\Lambda M = \{x \in A^\times : Mx = M\}. \tag{3.7}$$

Arguing as in [1] §2 for ordinary zeta functions, we get the identity

$$Z_\Lambda^L(M; z) = \sum_{x \in \operatorname{Aut}_\Lambda M \backslash M^{-1} \cap A^\times} z^{l(M/Mx)}. \tag{3.8}$$

The next step is to describe more fully the function $x \mapsto z^{l(m/Mx)}$, $x \in A^\times$.

(3.9) **Proposition.** *Let M, M' be full left Λ-lattices in A. Then*

$$z^{l(M/Mx)} = z^{l(M'/M'x)} \quad \textit{for all } x \in A^\times.$$

Moreover, the map $\chi_\Lambda : A^\times \to C^\times$ given by

$$\chi_\Lambda(x) = z^{l(M/Mx)}, \qquad x \in A^\times, \tag{3.10}$$

is an unramified quasicharacter of $A^\times$.

Proof. It is enough to treat the case where $M \supset M'$. Right multiplication by $x \in A^\times$ then yields a Λ-isomorphism $M/M' \simeq Mx/M'x$. Therefore

$$z^{l(M/M')} = z^{l(Mx/M'x)}.$$

Further,

$$z^{l(M/Mx)} = z^{l(M/M')}\, z^{l(M'/M'x)}\, z^{l(M'x/Mx)} = z^{l(M'/M'x)},$$

and this proves the first assertion.

It follows that the definition (3.10) of χ_Λ is independent of the choice of Λ-lattice M so, for $x, y \in A^\times$, we have

$$\chi_\Lambda(xy) = z^{l(M/Mxy)} = z^{l(M/Mx)}\, z^{l(Mx/Mxy)} = \chi_\Lambda(x)\,\chi_\Lambda(y).$$

Thus $\chi_\Lambda\colon A^\times \to C^\times$ is at least a homomorphism. Moreover, if Λ' is a maximal order in A and $x \in \Lambda'^\times$, we have $\Lambda' x = \Lambda'$ and so $\chi_\Lambda(x) = z^{l(\Lambda'/\Lambda' x)} = 1$. This proves that χ_Λ is locally constant (hence continuous) and also that it is unramified.

(3.11) *Remark.* The first part of (3.9) says simply that the Λ-composition factors of M/Mx depend only on $x \in A^\times$ and not on the lattice M. Exactly the same proof gives the following more general result: *let M_1, M_2 be left Λ-lattices spanning the same A-module V, and let $x \in \operatorname{Aut}_A(V)$ be such that $M_1 x \subset M_1$ and $M_2 x \subset M_2$. Then the Λ-modules $M_1/M_1 x$ and $M_2/M_2 x$ have the same Λ-composition factors.* As a special case, let K be a p-adic local field with discrete valuation ring R, and let G be a finite group. We take $A = KG$, $\Lambda = RG$, and we let V be a simple KG-module. Let π be a prime element of R, and put $k = R/\pi R$. Then the kG-composition factors (which are the same as the RG-composition factors) of $M/\pi M$ are independent of the choice of RG-lattice M spanning V.

To return to our main theme, we can combine (3.9) with (3.8) to obtain.

(3.12) **Corollary.** *Let $\mu^\times$ denote measure with respect to $d^\times x$. Then*

$$Z_\Lambda^L(M; z) = \mu^\times(\operatorname{Aut}_\Lambda M)^{-1}\, z^{l(\Lambda/M)}\, Z(\Phi_M, \chi_\Lambda),$$

where $\Phi_M \in S(A)$ is the characteristic function of M^{-1} and χ_Λ is given by (3.10).

It is now necessary to express χ_Λ in the form (2.10). We choose a maximal order $\Lambda' \supset \Lambda$. For each j, $1 \leqq j \leqq r$, we let F_j' be the simple left Λ'-module obtained by viewing the unique (up to isomorphism) simple left Λ_j'-module as a Λ'-module via the canonical projection $\Lambda' \to \Lambda_j'$. Thus $F_1', \ldots, F_r'$ is a complete set of representatives of the isomorphism classes of simple left Λ'-modules. Put

$$d_{ij} = l_i(F_j'), \qquad 1 \leqq i \leqq k,\ 1 \leqq j \leqq r. \tag{3.13}$$

(3.14) **Proposition.** *We have $\chi_\Lambda = \chi^A_{s_1, \ldots, s_r}$, where*

$$s_j = \frac{-1}{n_j \log q_j} \sum_{i=1}^{k} d_{ij} \log z_i, \qquad 1 \leqq j \leqq r.$$

(*By* (2.8), *the choice of* $\log z_i$ *here is immaterial.*)

Proof. The quasicharacter χ_Λ is completely determined by its values on $\Lambda' \cap A^\times$. For $x \in \Lambda' \cap A^\times$, we have $\Lambda'/\Lambda' x = \prod_j \Lambda_j'/\Lambda_j' x_j$. The Λ'-composition factors of $\Lambda_j'/\Lambda_j' x_j$ are $F_j'^{\,\mu_j}$, where

$$(\# F_j')^{\mu_j} = (\Lambda_j' : \Lambda_j' x_j) = \|x\|_j^{-1}.$$

However, $\# F_j' = q_j^{n_j}$, so $\mu_j = -\log \|x\|_j / n_j \log q_j$. Therefore

$$l_i(\Lambda'/\Lambda' x) = \sum_{j=1}^{r} d_{ij}\mu_j = -\sum_{j=1}^{r} d_{ij} \log \|x\|_j / n_j \log q_j.$$

Hence

$$z^{l(\Lambda'/\Lambda' x)} = \chi_\Lambda(x) = \exp\left(-\sum_{i=1}^{k} \log z_i \sum_{j=1}^{r} d_{ij} \log \|x\|_j / n_j \log q_j\right),$$

and the result now follows.

Remark. The character χ_Λ, and hence also its parameters $s_1, \ldots, s_r$, depend only on Λ and not on the choice of maximal order Λ' containing Λ. The same therefore applies to the quantities

$$t_j = q_j^{-n_j s_j} = \prod_{i=1}^{k} z_i^{d_{ij}}, \quad 1 \leqq j \leqq r. \tag{3.15}$$

We can assume that the z_i are algebraically independent, so the multiplicities d_{ij} are also independent of Λ'. We have therefore proved the following purely algebraic result:

(3.16) **Corollary.** *Let A be a finite-dimensional semisimple Q_p-algebra, $A = \prod_{j=1}^{r} A_j$, where the A_j are simple algebras. Let Λ be a Z_p-order in A, and let Λ', Λ'' be maximal orders in A containing Λ. For each j, $1 \leqq j \leqq r$, let F_j' (resp. F_j'') be a simple left module over $\Lambda' \cap A_j$ (resp. $\Lambda'' \cap A_j$). When viewed as Λ-modules in the canonical way, the modules F_j' and F_j'' have the same Λ-composition factors.*

In the situation of (3.16), it is not generally true that $F_j' \simeq F_j''$ as Λ-modules. For an example of this, take $A = M_2(Q_p)$, and let Λ be the order of matrices over Z_p which are upper triangular modulo pZ_p. This order has two simple modules (up to isomorphism): call them F_1 and F_2. There are just two maximal orders Λ', Λ'' containing Λ, with simple modules F', F'' respectively. As Λ-modules, these are given by *nonsplit* exact sequences:

$$0 \to F_1 \to F' \to F_2 \to 0, \quad 0 \to F_2 \to F'' \to F_1 \to 0.$$

Since the F_i are simple over Λ, this implies that $F' \not\simeq F''$.

Let us now return to zeta functions, and prove our first main result on them. This is the generalisation of "Solomon's Conjecture" to this situation.

(3.17) **Theorem.** *Let A be a finite-dimensional semisimple Q_p-algebra, let Λ be a Z_p-order in A, and let M be a full left Λ-lattice in A. Let Λ' be a maximal order in A containing Λ. Then*

$$Z_\Lambda^L(M; z) = z^{l(\Lambda/M)} f_\Lambda^L(M; z)\, \zeta_{\Lambda'}(t),$$

for some $f_\Lambda^L(M; z) \in Z[t_j^{\pm 1}]_{1 \leqq j \leqq r}$, such that $z^{l(\Lambda/M)} f_\Lambda^L(M; z) \in Z[z_1, \ldots, z_k]$, and where $t = (t_1, \ldots, t_r)$, t_j being given by (3.15).

Proof. We use (3.12) to express $Z^L_\Lambda(M; z)$ in terms of the zeta integral $Z(\Phi_M, \chi_A)$, where $\chi_A = \chi^A_{s_1, \ldots, s_r}$ as in (3.14). Using (3.15), (2.15), (3.4) and (3.5), we get $Z(\Phi_M, \chi_A) = g(s_1, \ldots, s_r)\, \zeta_{\Lambda'}(t)$, where $g(s_1, \ldots, s_r) \in C[t_j^{\pm 1}]$. We view this function g as a function of z, via (3.14), and we denote it by $f^L_\Lambda(M; z)$. This lies in $C[t_j^{\pm 1}]$. However, by definition, $Z_\Lambda(M; z) \in Z[\![z_1, \ldots, z_k]\!]$, and we know that $\zeta_{\Lambda'}(t)^{-1} \in Z[t_1, \ldots, t_r] \subset Z[z_1, \ldots, z_k]$. Hence

$$\zeta_{\Lambda'}(t)^{-1}\, Z_\Lambda(M; z) = z^{l(\Lambda/M)} f^L_\Lambda(M; z) \in Z[\![z_1, \ldots, z_k]\!] \cap C[t_j^{\pm 1}] = Z[z_1, \ldots, z_k].$$

It now follows that $f^L_\Lambda(M; z) \in Z[t_j^{\pm 1}]$, as required.

(3.18) **Corollary.** *We have*

$$\zeta^L_\Lambda(z) = \phi^L_\Lambda(z)\, \zeta^L_{\Lambda'}(t),$$

for some $\phi^L_\Lambda(z) \in z[z_1, \ldots, z_k]$.

Proof. We have $\zeta^L_\Lambda(z) = \sum_M Z^L_\Lambda(M; z)$, where M runs over a complete set of representatives of the isomorphism classes of full left Λ-lattices in A. There are only finitely many such classes, and the result is immediate.

Now we look back at the power series expansion (3.3)

$$Z^L_\Lambda(M; z) = \sum_u a_\Lambda(M; u)\, z^u,$$

where $u = (u_1, \ldots, u_k)$ runs over all k-tuples of non-negative integers. By (3.5) and (3.17), we also have

$$\zeta^L_{\Lambda'}(t)^{-1} = \sum_{v \in V} b(v)\, z^v,$$

$$z^{l(\Lambda/M)} f^L_\Lambda(M; z) = \sum_{w \in W} c(w)\, z^w$$

for certain *finite* sets V, W of k-tuples v, w of non-negative integers. For any k-tuple w, therefore,

$$\sum_{\substack{u+v=w \\ v \in V}} a_\Lambda(M, u)\, b(v) = c(w).$$

In particular,

$$\sum_{\substack{u+v=w \\ v \in V}} a_\Lambda(M, u)\, b(v) = 0 \quad \text{if } w \notin W. \tag{3.19}$$

Thus (3.19) holds for all but finitely many w, in particular it holds if $\sum_i w_i$ is sufficiently large. Hence, beyond a certain point, the coefficients $a_\Lambda(M, u)$ satisfy a linear recurrence relation. The coefficients $b(v)$ of this relation depend only on A and on the multiplicities d_{ij}. In particular, they are independent of M.

The power series coefficients of $\zeta_\Lambda(z)$ also satisfy (3.19), for an appropriate finite set W.

§ 4. Group Rings and Functional Equation

In this section, we let K be a p-adic local field with discrete valuation ring R. We let G be a finite group of order g, and we consider only the algebra $A=KG$ and the order $\Lambda=RG$. Otherwise, we use the notation of the preceding sections, particularly (2.6) and (2.7).

We first have to consider some questions of duality, and the right-handed analogues of our zeta functions. Let M be a full left Λ-lattice in A, and define

(4.1) $$\hat{M}=\{x\in A:\theta_A(xy)=1,\ \text{for all}\ y\in M\}.$$

This is a full right Λ-lattice in A. Moreover, if we define M^{-1} by (3.6), there exists $\alpha\in A$ (independent of M) such that

(4.2) $$M^{-1}=\alpha\hat{M}.$$

Indeed, we may take $\alpha=(\alpha_j)$ of the following form:

(4.3) $$\alpha_j=\delta_K g\, n_j^{-1}\ \sharp A_j, \qquad \text{where}\ \delta_K\in R,\ \delta_K R=D_K.$$

See [2], § 10 for more discussion of this. Let $\Lambda'\supset\Lambda$ be a maximal order. Since $\Lambda^{-1}=\Lambda$, and $\hat{\Lambda}\supset\hat{\Lambda}'\supset\Lambda'$, we conclude that $\alpha\in\Lambda$. Also, α lies in the *centre* of A. Everything we do will be independent of the choice of δ_K in (4.3).

We can perform similar operations on full right Λ-lattices N in A, and we get the identity $N^{-1}=\alpha\hat{N}$. Moreover,

$$(M^{-1})^{-1}=M, \qquad \hat{\hat{M}}=M, \qquad \mathrm{Aut}_\Lambda(M^{-1})=\mathrm{Aut}_\Lambda M.$$

If $M_1\supset M_2$ are full left Λ-lattices in A, we have $M_2^{-1}/M_1^{-1}\simeq\hat{M}_2/\hat{M}_1$ as right Λ-modules, and these are isomorphic to the Pontrjagin dual of M_1/M_2, with its canonical right Λ-module structure.

Again take a maximal order $\Lambda'\supset\Lambda$ as above, and choose $\delta_{\Lambda'}\in\Lambda'$ as in (2.2). We have $\Lambda=\Lambda^{-1}\supset\Lambda'^{-1}=\alpha\delta_{\Lambda'}^{-1}\Lambda'$. Since Λ' is the smallest two-sided Λ'-lattice containing Λ, duality implies that $\Gamma=(\Lambda')^{-1}$ is the largest two-sided Λ'-lattice contained in Λ. The formula

(4.4) $$(\Lambda')^{-1}=\Gamma=\alpha\delta_{\Lambda'}^{-1}\Lambda'$$

is the *Jacobinski conductor formula* (see [6], p. 548).

Now let F be any finite left or right Λ-module, and denote its Pontrjagin dual by $\hat{F}$. This is a right or left Λ-module. If $F_1,\dots,F_k$ are as before, then $\hat{F}_1,\dots,\hat{F}_k$ is a complete set of representatives of the isomorphism classes of simple right Λ-modules. If F is a finite right Λ-module, we put

(4.5) $r_i(F)=$ *the multiplicity of* $\hat{F}_i$ *as a composition factor of* F.

Using the same variables $z_1, \dots, z_k$ as before, we define

$$z^{r(N/N')} = \prod_{i=1}^{k} z_i^{r_i(N/N')},$$

where N, N' are full right Λ-lattices in A. We define zeta functions $Z_\Lambda^R(N;z)$ and $\zeta_\Lambda^R(z)$ just as before. Notice however that if we apply the same procedure to a maximal order Λ', we get the identity

$$\zeta_{\Lambda'}^L(t) = \zeta_{\Lambda'}^R(t).$$

The definition of r_i gives

$$z^{l(M/M')} = z^{r(\hat{M}'/\hat{M})} = z^{r(M'^{-1}/M^{-1})} \tag{4.6}$$

for left Λ-lattices M, M'. In particular, with Γ as in (4.4), we have $\Lambda'/\Lambda = (\Lambda/\Gamma)\hat{}$, so

$$z^{l(\Lambda/\Gamma)} = z^{r(\Lambda'/\Lambda)}, \quad \text{and} \quad (\Lambda':\Gamma) = (\Lambda':\Lambda)^2. \tag{4.7}$$

Further, if χ_Λ is defined by (3.10), we have

$$\chi_\Lambda(x) = z^{l(M/Mx)} = z^{r(x^{-1}M^{-1}/M^{-1})} = z^{r(M^{-1}/xM^{-1})}.$$

In other words,

$$\chi_\Lambda(x) = z^{r(N/xN)}, \qquad x \in A^\times, \tag{4.8}$$

for any full right Λ-lattice N in A.

We now come to the main result of this section. If M (resp. N) is a full left (resp. right) Λ-lattice in A, and if Λ' is any maximal order in A, define

$$\begin{aligned} \phi_\Lambda^L(M;z) &= \zeta_{\Lambda'}(t)^{-1} Z_\Lambda^L(M;z), \\ \phi_\Lambda^R(N;z) &= \zeta_{\Lambda'}(t)^{-1} Z_\Lambda^R(N;z), \end{aligned} \tag{4.9}$$

where $t = (t_j)$ is given by (3.15). These functions lie in $Z[z_1, \dots, z_k]$, by (3.17).

(4.10) **Theorem.** *Let $A = KG$, $\Lambda = RG$ as above. Let Λ' be a maximal order in A containing Λ. Write $f_i = \# F_i$, $fz = (f_1 z_1, \dots, f_k z_k)$,*

$$(fz)^{-1} = (f_1^{-1} z_1^{-1}, \dots, f_k^{-1} z_k^{-1}),$$

Then

$$(\Lambda':\Lambda)\, z^{l(\Lambda'/\Lambda)}\, z^{r(\Lambda'/\Lambda)}\, \phi_\Lambda^R(M^{-1}; (fz)^{-1}) = \phi_\Lambda^L(M;z), \tag{4.11}$$

for any full left Λ-lattice M in A.

Remarks. 1) The factor $z^{l(\Lambda'/\Lambda)} z^{r(\Lambda'/\Lambda)}$ is just $z^{l(\Lambda'/\Gamma)} = \chi_\Lambda(\alpha\delta_{\Lambda'}^{-1})$, by (4.4) and (4.7). This is certainly independent of the choice of Λ'. However, we will later see by example that the quantity $z^{l(\Lambda'/\Lambda)}$ does depend on Λ'.

2) When we specialise z to get $Z_\Lambda^L(M;z)=Z_\Lambda(M,s)$, the functional equation (4.11) becomes $(\Lambda':\Lambda)^{1-2s}\,\phi_\Lambda(M^{-1},1-s)=\phi_\Lambda(M,s)$, as in [1].

3) The equation (4.11) gives a bound on the *degree* in each z_i of the polynomials ϕ. Indeed,

$$\deg_{z_i}(\phi_\Lambda^L(M;z))\leqq l_i(\Lambda'/\Lambda)+r_i(\Lambda'/\Lambda).$$

As M ranges over a complete set of representatives for the isomorphism classes of full left Λ-lattices in A, M^{-1} does the same for full right Λ-lattices. Summing (4.11) over such a set of lattices M, we obtain

(4.12) **Corollary.** *Let $\phi_\Lambda^L(z)=\zeta_{L'}(t)^{-1}\,\zeta_\Lambda^L(z)$, and define $\phi_\Lambda^R(z)$ similarly. Then*

$$(\Lambda':\Lambda)\,z^{l(\Lambda'/\Lambda)}\,z^{r(\Lambda'/\Lambda)}\,\phi_\Lambda^R((fz)^{-1})=\phi_\Lambda^L(z).$$

Proof of (4.10). We start with the identity (3.12), and calculate the Fourier transform $\hat{\Phi}_M$, using the definition (2.17). This transform is easily seen to be

$$\hat{\Phi}_M=(\Lambda':M^{-1})^{-1}\,d_A^{-\frac{1}{2}}\cdot\textit{characteristic function of }(M^{-1})^\wedge.$$

Since $(M^{-1})^\wedge=\alpha^{-1}M$, we have

$$\begin{aligned}Z(\hat{\Phi}_M,\chi)&=(\Lambda':M^{-1})^{-1}\,d_A^{-\frac{1}{2}}\int_{\alpha^{-1}M}\chi(x)\,d^\times x\\&=\chi(\alpha^{-1})(\Lambda':M^{-1})^{-1}\,d_A^{-\frac{1}{2}}\int_M\chi(x)\,d^\times x\end{aligned}$$

where χ is any unramified quasicharacter of $A^\times$ for which the integral converges. Using analytic continuation (if necessary) and the right-handed analogue of (3.12), we deduce

$$\begin{aligned}Z(\hat{\Phi}_M,\chi_A)=\chi_A(\alpha)^{-1}(\Lambda':M^{-1})^{-1}\,d_A^{-\frac{1}{2}}\,z^{-r(\Lambda/M^{-1})}\\\cdot\mu^\times(\operatorname{Aut}_\Lambda M^{-1})\,Z_\Lambda^R(M^{-1};z).\end{aligned}\tag{4.13}$$

Replacing z_i by $(f_iz_i)^{-1}$ in (3.14) replaces s_j by $1-s_j$, so we have

$$Z(\hat{\Phi}_M,\chi_1^A\chi_A^{-1})=\chi_1^A(\alpha^{-1})\,\chi_A(\alpha)(\Lambda':M^{-1})^{-1}\,d_A^{-\frac{1}{2}}(fz)^{r(\Lambda/M^{-1})}\,\mu^\times\,Z_\Lambda^R(M^{-1};(fz)^{-1}),$$

where $\mu^\times=\mu^\times(\operatorname{Aut}_\Lambda M^{-1})=\mu^\times(\operatorname{Aut}_\Lambda M)$. We now substitute in the functional equation (2.18) to get

$$\begin{aligned}&\chi_1^A(\alpha^{-1})\,\chi_A(\alpha)(\Lambda':M^{-1})^{-1}\,d_A^{-\frac{1}{2}}(fz)^{r(\Lambda/M^{-1})}\,\phi_\Lambda^R(M^{-1};(fz)^{-1})\\&\quad=z^{-l(\Lambda/M)}\,\chi_{-\frac{1}{2}}^A(\delta_{\Lambda'})\,\chi_A(\delta_{\Lambda'})\,\phi_\Lambda^L(M;z).\end{aligned}$$

Now, $z^{r(\Lambda/M^{-1})}=z^{l(M/\Lambda)}$, and $f^{r(\Lambda/M^{-1})}=(\Lambda:M^{-1})$. We also have $d_A^{-\frac{1}{2}}=\chi_{\frac{1}{2}}^A(\delta_{\Lambda'})$. Taking all this into account, we get

$$\chi_1^A(\alpha^{-1}\delta_{\Lambda'})\,\chi_A(\alpha\delta_{\Lambda'}^{-1})(\Lambda':\Lambda)^{-1}\,\phi_\Lambda^R(M^{-1};(fz)^{-1})=\phi_\Lambda^L(M;z).$$

Next, $\chi_1^A(\alpha^{-1}\delta_{\Lambda'})=(\Lambda':\alpha^{-1}\delta_{\Lambda'}\Lambda')^{-1}=(\Lambda':\Gamma)=(\Lambda':\Lambda)^2$, by (4.7). Moreover, $\chi_A(\alpha\delta_{\Lambda'}^{-1})=z^{l(\Lambda'/\Gamma)}=z^{l(\Lambda'/\Lambda)}\,z^{r(\Lambda'/\Lambda)}$, and the result is proved.

We can express the functional equation (4.11) purely in terms of zeta functions of left Λ-lattices, by using the canonical involution $x \mapsto x^*$ on A:

$$x^* = \sum_{\gamma \in G} x_\gamma \gamma^{-1}, \qquad \text{where } x = \sum_{\gamma \in G} x_\gamma \gamma \in A.$$

Given a right Λ-module Y, we can form the left Λ-module Y^* by setting $x y = y x^*$, $x \in \Lambda$, $y \in Y$. We define z^* by

$$z_i^* = z_j, \qquad \text{where } (F_i)^* \simeq F_j.$$

This gives $z^{r(N/N')} = (z^*)^{l(N^*/N'^*)}$, and so

$$Z_\Lambda^R(N; z) = Z_\Lambda^L(N^*; z^*), \qquad \zeta_\Lambda^R(z) = \zeta_\Lambda^L(z^*),$$

for full right Λ-lattices N, N' in A. The process $z \mapsto z^*$ does not change $\zeta_{\Lambda'}(t)$, so we get

(4.14) **Corollary.** *Let M be a full left Λ-lattice in A, and define $\check{M} = (M^{-1})^*$. Then, for any maximal order Λ' in A containing Λ, we have*

$$(\Lambda' : \Lambda)\, z^{l(\Lambda'/\Lambda)} (z^*)^{l(\Lambda'^*/\Lambda)}\, \phi_\Lambda^L(\check{M}; (f^* z^*)^{-1}) = \phi_\Lambda^L(M; z),$$
$$(\Lambda' : \Lambda)\, z^{l(\Lambda'/\Lambda)} (z^*)^{l(\Lambda'^*/\Lambda)}\, \phi_\Lambda^L((f^* z^*)^{-1}) = \phi_\Lambda^L(z).$$

Example. We take $\mathrm{K} = \mathrm{Q}_p$ and G = the dihedral group of order $2p$, where $p \geqq 3$, and work out $\phi_\Lambda^{\mathrm{L}}(z)$. We write

$$G = \langle x, y \mid x^p = y^2 = 1,\ y x y^{-1} = x^{-1} \rangle.$$

Thus $\Lambda = Z_p G$ has two simple modules F_1, F_2. These are of dimension one over F_p, and y acts as $+1$ on F_1, and as -1 on F_2. Let L denote the field $Q_p(\omega + \omega^{-1})$, where ω is a primitive p-th root of unity. Thus $A = Q_p \times Q_p \times M_2(L)$, and we think of the factor $Q_p \times Q_p$ as $Q_p C$, where $C = G/\langle x \rangle$. The group ring Λ is given by a fibre diagram:

$$\begin{array}{ccc} \Lambda & \longrightarrow & H \\ \downarrow & & \downarrow \\ Z_p C & \longrightarrow & F_p C \end{array}$$

for a certain non-maximal hereditary order H in $M_2(L)$. Thus Λ is contained in the order $H' = Z_p C \times H$, and the fibre diagram shows that

$$z^{l(H'/\Lambda)} = z_1 z_2.$$

The order H is contained in just two maximal orders in $M_2(L)$, so there are exactly two maximal orders Λ', Λ'' in A which contain Λ. One finds, for suitable numbering, that

$$z^{l(\Lambda'/\Lambda)} = z_1^2 z_2, \qquad z^{l(\Lambda''/\Lambda)} = z_1 z_2^2.$$

Moreover, $z^* = z$, and $\Lambda''^* = \Lambda'$. Thus

$$z^{l(\Lambda'/\Lambda)} = z^{*\,l(\Lambda'^*/\Lambda)} = z_1^3 z_2^3, \qquad (\Lambda' : \Lambda) = p^3.$$

It is easy to see that

$$\zeta_{\Lambda'}(t)^{-1} = (1-z_1)(1-z_2)(1-z_1 z_2)(1-p z_1 z_2).$$

We follow the calculation in [4], and find without effort that

$$\zeta_\Lambda^L(z) = 1 + z_1 + z_2 + z_1^2 + z_2^2 + (2p+1)\, z_1 z_2 + z_1^3 + z_2^3 + \alpha z_1^2 z_2 + \beta z_1 z_2^2 + \ldots,$$

where the omitted terms all have degree $\geqq 4$, and α, β are non-negative integers satisfying

$$\alpha + \beta = 2(p^2 + p + 1).$$

Omitting all terms of degree $\geqq 4$, we get

$$\phi_\Lambda^L(z) = 1 + p z_1 z_2 + z_1^2 z_2(\alpha - 2p - 1) + z_1 z_2^2(\beta - 2p - 1) + \ldots.$$

Therefore

$$\begin{aligned}(\Lambda' : \Lambda)\, z^{l(\Lambda'/\Lambda)}\, z^{*\,l(\Lambda'^*/\Lambda)}\, \phi_\Lambda^L((f^* z^*)^{-1} \\ = p^3 z_1^3 z_2^3 + p^2 z_1^2 z_2^2 + z_1 z_2^2(\alpha - 2p - 1) + z_1^2 z_2(\beta - 2p - 1) + \ldots,\end{aligned}$$

this time omitting terms of degree $\leqq 2$. It follows from the functional equation (4.14) that $\alpha = \beta$, hence that $\alpha = \beta = p^2 + p + 1$. Therefore

$$\begin{aligned}\zeta_\Lambda^L(z) &= \phi_\Lambda^L(z)\, \zeta_{\Lambda'}(t) \\ &= \{1 + p z_1 z_2 + (p^2 - p)\, z_1 z_2(z_1 + z_2) + p^2 z_1^2 z_2^2 + p^3 z_1^3 z_2^3\} \\ &\quad \cdot \{(1-z_1)(1-z_2)(1-z_1 z_2)(1-p z_1 z_2)\}^{-1}.\end{aligned}$$

References

1. Bushnell, C.J., Reiner, I.: Zeta functions of arithmetic orders and Solomon's conjectures. Math. Z. **173**, 135–161 (1980)
2. Bushnell, C.J., Reiner, I.: Functional equations for *L*-functions of arithmetic orders. J. Reine Angew. Math. **329**, 88–124 (1981)
3. Bushnell, C.J., Reiner, I.: Prime ideal theorem in noncommutative arithmetic. Math. Z. **181**, 143–170 (1982)
4. Bushnell, C.J., Reiner, I.: Left-vs-right zeta functions. Q. J. Math., Oxf. (2) **35**, 1–19 (1984)
5. Bushnell, C.J., Reiner, I.: A survey of analytic methods in noncommutative number theory. Orders and their applications – Proceedings Oberwolfach 1984 (Roggenkamp, K., Reiner, I., ed.). Lecture Notes in Math. **1142**, pp. 50–87. Berlin Heidelberg New York: Springer 1985
6. Curtis, C.W., Reiner, I.: Methods of representation theory. Vol. I. New York: Wiley-Interscience 1981
7. Reiner, I.: Maximal orders. London-New York: Academic Press 1975

Received February 6, 1986

A Note on the Editor

After completing his Ph.D. at the University of Oregon in 1965, Gerald J. Janusz spent one year as a member of the Institute for Advanced Study and two years as an instructor at the University of Chicago before joining the mathematics department of the University of Illinois at Urbana–Champaign in 1968. He has published over thirty research papers in representation theory of finite groups and algebras, Brauer groups, and other areas of algebra. He is the author of *Algebraic Number Fields* (1973) and co-author with Neal McCoy of the 1987 edition of *Introduction to Modern Algebra.* For eight years he was on the editorial board of *Communications in Algebra*, and is presently a member of the editorial board of *Contemporary Mathematics* and managing editor of the *Illinois Journal of Mathematics.*